# NFPA 70

# *National Electrical Code*®

## 2002 Edition

This edition of NFPA 70, *National Electrical Code*, was prepared by the National Electrical Code Committee and acted on by the National Fire Protection Association, Inc., at its May Association Technical Meeting held May 13–17, 2001, in Anaheim, CA. It was issued by the Standards Council on July 13, 2001, with an effective date of August 2, 2001, and supersedes all previous editions.

This edition of NFPA 70 was approved as an American National Standard on August 2, 2001.

### History and Development of the *National Electrical Code*

The National Fire Protection Association has acted as sponsor of the *National Electrical Code* since 1911. The original *Code* document was developed in 1897 as a result of the united efforts of various insurance, electrical, architectural, and allied interests.

In accordance with the provisions of the NFPA Regulations Governing Committee Projects, a National Electrical Code Committee Report on Proposals containing proposed amendments to the 1999 *National Electrical Code* was published by the NFPA in July 2000. This report recorded the actions of the various Code-Making Panels and the Correlating Committee of the National Electrical Code Committee on each proposal that had been made to revise the 1999 *Code*. The report was circulated to all members of the National Electrical Code Committee and was made available to other interested NFPA members and to the public for review and comment. Following the close of the public comment period, the Code-Making Panels met, acted on each comment, and reported their action to the Correlating Committee. The NFPA published the National Electrical Code Committee Report on Comments in April 2001, which recorded the actions of the Code-Making Panels and the Correlating Committee on each public comment to the National Electrical Code Committee Report on Proposals. The National Electrical Code Committee Report on Proposals and the National Electrical Code Committee Report on Comments were presented to the 2001 May Association Technical Meeting for adoption.

NFPA has an Electrical Section that provides particular opportunity for NFPA members interested in electrical safety to become better informed and to contribute to the development of the *National Electrical Code* and other NFPA electrical standards. Each of the Code-Making Panels and the Chairman of the Correlating Committee reported their recommendations to meetings of the Electrical Section at the 2001 NFPA World Fire Safety Congress and Exposition. The Electrical Section thus had opportunity to discuss and review the report of the National Electrical Code Committee prior to the adoption of this edition of the *Code* by the Association at its 2001 May Technical Session.

This 2002 edition supersedes all other previous editions, supplements, and printings dated 1897, 1899, 1901, 1903, 1904, 1905, 1907, 1909, 1911, 1913, 1915, 1918, 1920, 1923, 1925, 1926, 1928, 1930, 1931, 1933, 1935, 1937, 1940, 1942, 1943, 1947, 1949, 1951, 1953, 1954, 1955, 1956, 1957, 1958, 1959, 1962, 1965, 1968, 1971, 1975, 1978, 1981, 1984, 1987, 1990, 1993, 1996, and 1999.

New or revised technical content in this edition is indicated by a vertical rule next to the paragraph, table, or figure in which the change occurred.

This *Code* is purely advisory as far as NFPA and ANSI are concerned but is offered for use in law and for regulatory purposes in the interest of life and property protection. Anyone noticing any errors should notify the Secretary of the National Electrical Code Committee at the NFPA Executive Office.

# Contents

ARTICLE

 80   Administration and Enforcement ............... **70**– 23

 90   Introduction ....................................... **70**– 29

## Chapter 1 General

100   Definitions ......................................... **70**– 33

   I. General ......................................... **70**– 33
  II. Over 600 Volts, Nominal ..................... **70**– 39

110   Requirements for Electrical Installations ....... **70**– 40

    I. General ......................................... **70**– 40
   II. 600 Volts, Nominal, or Less ................. **70**– 43
  III. Over 600 Volts, Nominal ..................... **70**– 45
  IV. Tunnel Installations over 600 Volts,
     Nominal ....................................... **70**– 47

## Chapter 2 Wiring and Protection

200   Use and Identification of Grounded
     Conductors ....................................... **70**– 49

210   Branch Circuits .................................... **70**– 51

    I. General Provisions ............................ **70**– 51
   II. Branch-Circuit Ratings ........................ **70**– 54
  III. Required Outlets .............................. **70**– 57

215   Feeders ............................................ **70**– 60

220   Branch-Circuit, Feeder, and Service
     Calculations ...................................... **70**– 61

    I. General ......................................... **70**– 61
   II. Feeders and Services .......................... **70**– 63
  III. Optional Calculations for Computing
      Feeder and Service Loads .................... **70**– 66
  IV. Method for Computing Farm Loads ......... **70**– 69

225   Outside Branch Circuits and Feeders .......... **70**– 69

    I. General ......................................... **70**– 70
   II. More Than One Building or Other
      Structure ..................................... **70**– 72
  III. Over 600 Volts ............................... **70**– 74

230   Services ........................................... **70**– 75

    I. General ......................................... **70**– 75
   II. Overhead Service-Drop Conductors ......... **70**– 77
  III. Underground Service-Lateral
      Conductors ................................... **70**– 78
  IV. Service-Entrance Conductors ................ **70**– 78
   V. Service Equipment — General .............. **70**– 81
  VI. Service Equipment — Disconnecting
      Means ........................................ **70**– 81
  VII. Service Equipment — Overcurrent
      Protection .................................... **70**– 82
  VIII. Services Exceeding 600 Volts, Nominal .... **70**– 84

240   Overcurrent Protection ........................... **70**– 85

    I. General ......................................... **70**– 85
   II. Location ........................................ **70**– 88

ARTICLE

  III. Enclosures ..................................... **70**– 91
  IV. Disconnecting and Guarding ................. **70**– 92
   V. Plug Fuses, Fuseholders, and Adapters ..... **70**– 92
  VI. Cartridge Fuses and Fuseholders ............ **70**– 93
  VII. Circuit Breakers .............................. **70**– 93
  VIII. Supervised Industrial Installations ........... **70**– 94
  IX. Overcurrent Protection Over 600 Volts,
     Nominal ...................................... **70**– 95

250   Grounding ......................................... **70**– 95

    I. General ......................................... **70**– 95
   II. Circuit and System Grounding .............. **70**– 98
  III. Grounding Electrode System and
      Grounding Electrode Conductor ............ **70**–104
  IV. Enclosure, Raceway, and Service Cable
      Grounding ................................... **70**–108
   V. Bonding ...................................... **70**–108
  VI. Equipment Grounding and Equipment
      Grounding Conductors ...................... **70**–111
  VII. Methods of Equipment Grounding .......... **70**–115
  VIII. Direct-Current Systems ...................... **70**–117
  IX. Instruments, Meters, and Relays ............ **70**–118
   X. Grounding of Systems and Circuits of
      1 kV and Over (High Voltage) ............... **70**–119

280   Surge Arresters ................................... **70**–120

    I. General ......................................... **70**–120
   II. Installation ..................................... **70**–121
  III. Connecting Surge Arresters ................... **70**–121

285   Transient Voltage Surge Suppressors:
     TVSSs ............................................ **70**–122

    I. General ......................................... **70**–122
   II. Installation ..................................... **70**–122
  III. Connecting Transient Voltage Surge
      Suppressors ................................... **70**–122

## Chapter 3 Wiring Methods and Materials

300   Wiring Methods ................................... **70**–123

    I. General Requirements ......................... **70**–123
   II. Requirements for Over 600 Volts,
      Nominal ...................................... **70**–132

310   Conductors for General Wiring ................. **70**–133

312   Cabinets, Cutout Boxes, and Meter Socket
     Enclosures ........................................ **70**–161

    I. Installation ..................................... **70**–161
   II. Construction Specifications ................... **70**–162

314   Outlet, Device, Pull, and Junction Boxes;
     Conduit Bodies; Fittings; and Manholes ........ **70**–164

    I. Scope and General ............................ **70**–164
   II. Installation ..................................... **70**–165
  III. Construction Specifications ................... **70**–170
  IV. Manholes and Other Electric
      Enclosures Intended for Personnel
      Entry ......................................... **70**–171

ARTICLE

    V. Pull and Junction Boxes for Use on
    Systems Over 600 Volts, Nominal ............ **70**–172

320   Armored Cable: Type AC ...................... **70**–173

    I. General ......................................... **70**–173
    II. Installation ................................... **70**–173
    III. Construction Specifications .................. **70**–174

322   Flat Cable Assemblies: Type FC ............... **70**–174

    I. General ......................................... **70**–174
    II. Installation ................................... **70**–174
    III. Construction ................................. **70**–175

324   Flat Conductor Cable: Type FCC .............. **70**–175

    I. General ......................................... **70**–175
    II. Installation ................................... **70**–176
    III. Construction ................................. **70**–177

326   Integrated Gas Spacer Cable: Type IGS ....... **70**–178

    I. General ......................................... **70**–178
    II. Installation ................................... **70**–178
    III. Construction Specifications .................. **70**–178

328   Medium Voltage Cable: Type MV ............. **70**–179

    I. General ......................................... **70**–179
    II. Installation ................................... **70**–179
    III. Construction Specifications .................. **70**–179

330   Metal-Clad Cable: Type MC .................... **70**–179

    I. General ......................................... **70**–179
    II. Installation ................................... **70**–179
    III. Construction Specifications .................. **70**–181

332   Mineral-Insulated, Metal-Sheathed Cable:
    Type MI ......................................... **70**–181

    I. General ......................................... **70**–181
    II. Installation ................................... **70**–181
    III. Construction Specifications .................. **70**–182

334   Nonmetallic-Sheathed Cable: Types NM,
    NMC, and NMS ................................. **70**–182

    I. General ......................................... **70**–182
    II. Installation ................................... **70**–182
    III. Construction Specifications .................. **70**–184

336   Power and Control Tray Cable: Type TC ...... **70**–184

    I. General ......................................... **70**–184
    II. Installation ................................... **70**–185
    III. Construction Specifications .................. **70**–185

338   Service-Entrance Cable: Types SE and
    USE .............................................. **70**–186

    I. General ......................................... **70**–186
    II. Installation ................................... **70**–186
    III. Construction ................................. **70**–186

340   Underground Feeder and Branch-Circuit
    Cable: Type UF ................................. **70**–187

    I. General ......................................... **70**–187
    II. Installation ................................... **70**–187
    III. Construction Specifications .................. **70**–187

ARTICLE

342   Intermediate Metal Conduit: Type IMC ........ **70**–187

    I. General ......................................... **70**–187
    II. Installation ................................... **70**–188
    III. Construction Specifications .................. **70**–189

344   Rigid Metal Conduit: Type RMC .............. **70**–189

    I. General ......................................... **70**–189
    II. Installation ................................... **70**–189
    III. Construction Specifications .................. **70**–191

348   Flexible Metal Conduit: Type FMC ........... **70**–191

    I. General ......................................... **70**–191
    II. Installation ................................... **70**–191

350   Liquidtight Flexible Metal Conduit: Type
    LFMC ............................................ **70**–192

    I. General ......................................... **70**–192
    II. Installation ................................... **70**–193
    III. Construction Specifications .................. **70**–193

352   Rigid Nonmetallic Conduit: Type RNC ........ **70**–194

    I. General ......................................... **70**–194
    II. Installation ................................... **70**–194
    III. Construction Specifications .................. **70**–195

354   Nonmetallic Underground Conduit with
    Conductors: Type NUCC ....................... **70**–197

    I. General ......................................... **70**–197
    II. Installation ................................... **70**–197
    III. Construction Specifications .................. **70**–198

356   Liquidtight Flexible Nonmetallic Conduit:
    Type LFNC ...................................... **70**–198

    I. General ......................................... **70**–198
    II. Installation ................................... **70**–198
    III. Construction Specifications .................. **70**–199

358   Electrical Metallic Tubing: Type EMT ......... **70**–200

    I. General ......................................... **70**–200
    II. Installation ................................... **70**–200
    III. Construction Specifications .................. **70**–201

360   Flexible Metallic Tubing: Type FMT .......... **70**–201

    I. General ......................................... **70**–201
    II. Installation ................................... **70**–201
    III. Construction Specifications .................. **70**–202

362   Electrical Nonmetallic Tubing: Type ENT .... **70**–202

    I. General ......................................... **70**–202
    II. Installation ................................... **70**–202
    III. Construction Specifications .................. **70**–204

366   Auxiliary Gutters ............................... **70**–204

368   Busways ......................................... **70**–206

    I. General Requirements ......................... **70**–206
    II. Requirements for Over 600 Volts,
    Nominal ......................................... **70**–207

370   Cablebus ......................................... **70**–208

372   Cellular Concrete Floor Raceways ............. **70**–209

ARTICLE

374    Cellular Metal Floor Raceways ................ **70**–210

     I. Installation .......................... **70**–210
     II. Construction Specifications ................ **70**–211

376    Metal Wireways ........................... **70**–211

     I. General ............................. **70**–211
     II. Installation ......................... **70**–211
     III. Construction Specifications ............. **70**–212

378    Nonmetallic Wireways ..................... **70**–212

     I. General ............................. **70**–212
     II. Installation ......................... **70**–212
     III. Construction Specifications ............. **70**–213

380    Multioutlet Assembly ...................... **70**–213

382    Nonmetallic Extensions .................... **70**–214

     I. General ............................. **70**–214
     II. Installation ......................... **70**–214

384    Strut-Type Channel Raceway ............... **70**–215

     I. General ............................. **70**–215
     II. Installation ......................... **70**–215
     III. Construction Specifications ............. **70**–216

386    Surface Metal Raceways ................... **70**–216

     I. General ............................. **70**–216
     II. Installation ......................... **70**–216
     III. Construction Specifications ............. **70**–217

388    Surface Nonmetallic Raceways ............. **70**–217

     I. General ............................. **70**–217
     II. Installation ......................... **70**–217
     III. Construction Specifications ............. **70**–218

390    Underfloor Raceways ...................... **70**–218

392    Cable Trays .............................. **70**–219

394    Concealed Knob-and-Tube Wiring .......... **70**–225

     I. General ............................. **70**–225
     II. Installation ......................... **70**–226
     III. Construction Specifications ............. **70**–226

396    Messenger Supported Wiring ............... **70**–227

     I. General ............................. **70**–227
     II. Installation ......................... **70**–227

398    Open Wiring on Insulators ................. **70**–227

     I. General ............................. **70**–227
     II. Installation ......................... **70**–227
     III. Construction Specifications ............. **70**–229

## Chapter 4 Equipment for General Use

400    Flexible Cords and Cables ................. **70**–231

     I. General ............................. **70**–231
     II. Construction Specifications ............. **70**–239
     III. Portable Cables Over 600 Volts,
        Nominal ........................... **70**–240

402    Fixture Wires ............................ **70**–241

ARTICLE

404    Switches ................................. **70**–244

     I. Installation ......................... **70**–244
     II. Construction Specifications ............. **70**–247

406    Receptacles, Cord Connectors, and
      Attachment Plugs (Caps) ................. **70**–247

408    Switchboards and Panelboards ............. **70**–250

     I. General ............................. **70**–250
     II. Switchboards ........................ **70**–251
     III. Panelboards ........................ **70**–252
     IV. Construction Specifications ............ **70**–253

410    Luminaires (Lighting Fixtures),
      Lampholders, and Lamps ................. **70**–254

     I. General ............................. **70**–254
     II. Luminaire (Fixture) Locations ........... **70**–254
     III. Provisions at Luminaire (Fixture)
         Outlet Boxes, Canopies, and Pans .......... **70**–255
     IV. Luminaire (Fixture) Supports ........... **70**–256
     V. Grounding .......................... **70**–257
     VI. Wiring of Luminaires (Fixtures) ......... **70**–257
     VII. Construction of Luminaires (Fixtures) ...... **70**–259
     VIII. Installation of Lampholders ............. **70**–260
     IX. Construction of Lampholders ........... **70**–260
     X. Lamps and Auxiliary Equipment ........... **70**–260
     XI. Special Provisions for Flush and
        Recessed Luminaires (Fixtures) ............ **70**–260
     XII. Construction of Flush and Recessed
        Luminaires (Fixtures) .................. **70**–261
     XIII. Special Provisions for
        Electric-Discharge Lighting Systems of
        1000 Volts or Less ..................... **70**–261
     XIV. Special Provisions for
        Electric-Discharge Lighting Systems of
        More Than 1000 Volts .................. **70**–262
     XV. Lighting Track ....................... **70**–263

411    Lighting Systems Operating at 30 Volts or
      Less ..................................... **70**–264

422    Appliances ............................... **70**–264

     I. General ............................. **70**–264
     II. Installation ......................... **70**–264
     III. Disconnecting Means .................. **70**–267
     IV. Construction ........................ **70**–267
     V. Marking ............................ **70**–268

424    Fixed Electric Space-Heating Equipment ...... **70**–269

     I. General ............................. **70**–269
     II. Installation ......................... **70**–269
     III. Control and Protection of Fixed
        Electric Space-Heating Equipment .......... **70**–270
     IV. Marking of Heating Equipment .......... **70**–272
     V. Electric Space-Heating Cables ............ **70**–272
     VI. Duct Heaters ....................... **70**–274
     VII. Resistance-Type Boilers ............... **70**–274
     VIII. Electrode-Type Boilers ............... **70**–275
     IX. Electric Radiant Heating Panels and
        Heating Panel Sets .................... **70**–276

426    Fixed Outdoor Electric Deicing and
      Snow-Melting Equipment ................. **70**–278

     I. General ............................. **70**–278

ARTICLE

   II. Installation ........................................ 70–279
  III. Resistance Heating Elements ................ 70–279
   IV. Impedance Heating .......................... 70–280
    V. Skin-Effect Heating ........................ 70–280
   VI. Control and Protection ...................... 70–281

427 Fixed Electric Heating Equipment for
     Pipelines and Vessels .......................... 70–281

    I. General ........................................ 70–281
   II. Installation .................................... 70–282
  III. Resistance Heating Elements ................ 70–282
   IV. Impedance Heating .......................... 70–283
    V. Induction Heating ............................ 70–283
   VI. Skin-Effect Heating .......................... 70–283
  VII. Control and Protection ...................... 70–284

430 Motors, Motor Circuits, and Controllers ....... 70–284

     I. General ........................................ 70–284
    II. Motor Circuit Conductors .................... 70–290
   III. Motor and Branch-Circuit Overload
        Protection ...................................... 70–292
   IV. Motor Branch-Circuit Short-Circuit and
        Ground-Fault Protection ...................... 70–295
    V. Motor Feeder Short-Circuit and
        Ground-Fault Protection .................... 70–298
   VI. Motor Control Circuits ........................ 70–299
  VII. Motor Controllers ............................ 70–301
 VIII. Motor Control Centers ........................ 70–302
   IX. Disconnecting Means ........................ 70–304
    X. Over 600 Volts, Nominal .................... 70–307
   XI. Protection of Live Parts — All
        Voltages ........................................ 70–308
  XII. Grounding — All Voltages .................. 70–308
 XIII. Tables ............................................ 70–309

440 Air-Conditioning and Refrigerating
     Equipment ........................................ 70–312

     I. General ........................................ 70–312
    II. Disconnecting Means ........................ 70–314
   III. Branch-Circuit Short-Circuit and
        Ground-Fault Protection ...................... 70–315
   IV. Branch-Circuit Conductors .................... 70–316
    V. Controllers for Motor-Compressors .......... 70–316
   VI. Motor-Compressor and Branch-Circuit
        Overload Protection .......................... 70–317
  VII. Provisions for Room Air Conditioners ...... 70–318

445 Generators ........................................ 70–319

450 Transformers and Transformer Vaults
     (Including Secondary Ties) .................... 70–320

     I. General Provisions ............................ 70–320
    II. Specific Provisions Applicable to
        Different Types of Transformers .............. 70–324
   III. Transformer Vaults .......................... 70–326

455 Phase Converters ................................ 70–327

     I. General ........................................ 70–327
    II. Specific Provisions Applicable to
        Different Types of Phase Converters ........ 70–328

460 Capacitors ........................................ 70–329

     I. 600 Volts, Nominal, and Under ............ 70–329

ARTICLE

   II. Over 600 Volts, Nominal .................... 70–329

470 Resistors and Reactors (For Rheostats,
     See 430.82.) .................................... 70–330

     I. 600 Volts, Nominal, and Under ............ 70–330
    II. Over 600 Volts, Nominal .................... 70–331

480 Storage Batteries ................................ 70–331

490 Equipment, Over 600 Volts, Nominal .......... 70–332

     I. General ........................................ 70–332
    II. Equipment — Specific Provisions .......... 70–332
   III. Equipment — Metal-Enclosed Power
        Switchgear and Industrial Control
        Assemblies .................................... 70–335
   IV. Mobile and Portable Equipment ............ 70–337
    V. Electrode-Type Boilers ...................... 70–337

## Chapter 5 Special Occupancies

500 Hazardous (Classified) Locations, Classes
     I, II, and III, Divisions 1 and 2 ................ 70–339

501 Class I Locations ................................ 70–347

502 Class II Locations ................................ 70–356

503 Class III Locations .............................. 70–362

504 Intrinsically Safe Systems ...................... 70–364

505 Class I, Zone 0, 1, and 2 Locations ............ 70–367

510 Hazardous (Classified) Locations —
     Specific ............................................ 70–380

511 Commercial Garages, Repair and Storage ..... 70–380

513 Aircraft Hangars ................................ 70–382

514 Motor Fuel Dispensing Facilities .............. 70–385

515 Bulk Storage Plants ............................ 70–389

516 Spray Application, Dipping, and Coating
     Processes .......................................... 70–393

517 Health Care Facilities .......................... 70–399

     I. General ........................................ 70–399
    II. Wiring and Protection ........................ 70–402
   III. Essential Electrical System .................. 70–405
   IV. Inhalation Anesthetizing Locations .......... 70–411
    V. X-Ray Installations .......................... 70–414
   VI. Communications, Signaling Systems,
        Data Systems, Fire Alarm Systems, and
        Systems Less Than 120 Volts, Nominal ..... 70–415
  VII. Isolated Power Systems ...................... 70–416

518 Places of Assembly .............................. 70–417

520 Theaters, Audience Areas of Motion
     Picture and Television Studios,
     Performance Areas, and Similar Locations ..... 70–418

     I. General ........................................ 70–418
    II. Fixed Stage Switchboards .................... 70–419

ARTICLE

   III. Fixed Stage Equipment Other Than
       Switchboards ................................. **70**–421
   IV. Portable Switchboards on Stage ............. **70**–422
    V. Portable Stage Equipment Other Than
       Switchboards ................................. **70**–425
   VI. Dressing Rooms ............................. **70**–426
  VII. Grounding ................................. **70**–426

525   Carnivals, Circuses, Fairs, and Similar
      Events ............................................. **70**–427

    I. General Requirements ........................ **70**–427
   II. Power Sources .............................. **70**–427
  III. Wiring Methods ............................ **70**–427
  IV. Grounding and Bonding ...................... **70**–428

527   Temporary Installations ........................... **70**–429

530   Motion Picture and Television Studios
      and Similar Locations ........................... **70**–430

    I. General .................................... **70**–430
   II. Stage or Set ............................... **70**–431
  III. Dressing Rooms ............................ **70**–433
  IV. Viewing, Cutting, and Patching Tables ..... **70**–434
   V. Cellulose Nitrate Film Storage Vaults ....... **70**–434
  VI. Substations ................................ **70**–434

540   Motion Picture Projection Rooms .............. **70**–434

    I. General .................................... **70**–434
   II. Equipment and Projectors of the
      Professional Type ........................... **70**–434
  III. Nonprofessional Projectors .................. **70**–435
  IV. Audio Signal Processing, Amplification,
      and Reproduction Equipment ................. **70**–435

545   Manufactured Buildings ......................... **70**–436

547   Agricultural Buildings ............................ **70**–437

550   Mobile Homes, Manufactured Homes, and
      Mobile Home Parks .............................. **70**–439

    I. General .................................... **70**–439
   II. Mobile and Manufactured Homes .......... **70**–440
  III. Services and Feeders ........................ **70**–447

551   Recreational Vehicles and Recreational
      Vehicle Parks ................................... **70**–449

    I. General .................................... **70**–449
   II. Low-Voltage Systems ....................... **70**–450
  III. Combination Electrical Systems ............. **70**–452
  IV. Other Power Sources ........................ **70**–452
   V. Nominal 120-Volt or 120/240-Volt
      Systems ...................................... **70**–453
  VI. Factory Tests .............................. **70**–460
  VII. Recreational Vehicle Parks ................... **70**–460

552   Park Trailers .................................... **70**–463

    I. General .................................... **70**–463
   II. Low-Voltage Systems ....................... **70**–463
  III. Combination Electrical Systems ............. **70**–464
  IV. Nominal 120-Volt or 120/240-Volt
      Systems ...................................... **70**–465
   V. Factory Tests .............................. **70**–472

ARTICLE

553   Floating Buildings .............................. **70**–472

    I. General .................................... **70**–472
   II. Services and Feeders ........................ **70**–472
  III. Grounding ................................. **70**–472

555   Marinas and Boatyards .......................... **70**–473

## Chapter 6 Special Equipment

600   Electric Signs and Outline Lighting ............ **70**–477

    I. General .................................... **70**–477
   II. Field-Installed Skeleton Tubing ............. **70**–480

604   Manufactured Wiring Systems .................. **70**–481

605   Office Furnishings (Consisting of Lighting
      Accessories and Wired Partitions) .............. **70**–482

610   Cranes and Hoists .............................. **70**–483

    I. General .................................... **70**–483
   II. Wiring .................................... **70**–483
  III. Contact Conductors .......................... **70**–486
  IV. Disconnecting Means ......................... **70**–487
   V. Overcurrent Protection ....................... **70**–487
  VI. Control .................................... **70**–488
  VII. Grounding ................................. **70**–488

620   Elevators, Dumbwaiters, Escalators,
      Moving Walks, Wheelchair Lifts, and
      Stairway Chair Lifts ............................ **70**–488

    I. General .................................... **70**–488
   II. Conductors ................................ **70**–490
  III. Wiring .................................... **70**–491
  IV. Installation of Conductors .................... **70**–494
   V. Traveling Cables ............................ **70**–494
  VI. Disconnecting Means and Control .......... **70**–495
  VII. Overcurrent Protection ...................... **70**–496
 VIII. Machine Rooms, Control Rooms,
      Machinery Spaces, and Control Spaces ...... **70**–497
  IX. Grounding ................................. **70**–497
   X. Emergency and Standby Power
      Systems ...................................... **70**–497

625   Electric Vehicle Charging System .............. **70**–498

    I. General .................................... **70**–498
   II. Wiring Methods ............................. **70**–498
  III. Equipment Construction ...................... **70**–499
  IV. Control and Protection ....................... **70**–499
   V. Electric Vehicle Supply Equipment
      Locations .................................... **70**–500

630   Electric Welders ................................ **70**–502

    I. General .................................... **70**–502
   II. Arc Welders ............................... **70**–502
  III. Resistance Welders .......................... **70**–503
  IV. Welding Cable .............................. **70**–504

640   Audio Signal Processing, Amplification,
      and Reproduction Equipment ................... **70**–504

    I. General .................................... **70**–504
   II. Permanent Audio System Installations ...... **70**–507

ARTICLE

    III. Portable and Temporary Audio System
        Installations ..................................... **70**–508

645    Information Technology Equipment ............ **70**–509

647    Sensitive Electronic Equipment ................. **70**–511

650    Pipe Organs ........................................ **70**–512

660    X-Ray Equipment ................................. **70**–513

    I. General ........................................ **70**–513
    II. Control ........................................ **70**–514
    III. Transformers and Capacitors ................. **70**–514
    IV. Guarding and Grounding ...................... **70**–514

665    Induction and Dielectric Heating
    Equipment ......................................... **70**–515

    I. General ......................................... **70**–515
    II. Guarding, Grounding, and Labeling ......... **70**–516

668    Electrolytic Cells .................................. **70**–516

669    Electroplating ...................................... **70**–519

670    Industrial Machinery ............................. **70**–520

675    Electrically Driven or Controlled
    Irrigation Machines .............................. **70**–521

    I. General ......................................... **70**–521
    II. Center Pivot Irrigation Machines ............ **70**–523

680    Swimming Pools, Fountains, and Similar
    Installations ....................................... **70**–523

    I. General ......................................... **70**–523
    II. Permanently Installed Pools ................. **70**–526
    III. Storable Pools ................................ **70**–532
    IV. Spas and Hot Tubs .......................... **70**–533
    V. Fountains .................................... **70**–535
    VI. Pools and Tubs for Therapeutic Use ........ **70**–536
    VII. Hydromassage Bathtubs ...................... **70**–537

685    Integrated Electrical Systems ................... **70**–537

    I. General ......................................... **70**–537
    II. Orderly Shutdown ............................ **70**–538

690    Solar Photovoltaic Systems ..................... **70**–538

    I. General ......................................... **70**–538
    II. Circuit Requirements .......................... **70**–540
    III. Disconnecting Means ......................... **70**–542
    IV. Wiring Methods .............................. **70**–543
    V. Grounding .................................... **70**–544
    VI. Marking ..................................... **70**–545
    VII. Connection to Other Sources ................. **70**–545
    VIII. Storage Batteries ............................. **70**–546
    IX. Systems Over 600 Volts ..................... **70**–547

692    Fuel Cell Systems ............................... **70**–547

    I. General ......................................... **70**–547
    II. Circuit Requirements .......................... **70**–548
    III. Disconnecting Means ......................... **70**–549
    IV. Wiring Methods .............................. **70**–549
    V. Grounding .................................... **70**–549
    VI. Marking ..................................... **70**–549
    VII. Connection to Other Circuits ................. **70**–549

ARTICLE

    VIII. Outputs Over 600 Volts ........................ **70**–550

695    Fire Pumps ........................................ **70**–550

## Chapter 7 Special Conditions

700    Emergency Systems .............................. **70**–555

    I. General ......................................... **70**–555
    II. Circuit Wiring ................................. **70**–556
    III. Sources of Power .............................. **70**–557
    IV. Emergency System Circuits for
        Lighting and Power ......................... **70**–558
    V. Control — Emergency Lighting
        Circuits ..................................... **70**–559
    VI. Overcurrent Protection ........................ **70**–559

701    Legally Required Standby Systems ............ **70**–559

    I. General ......................................... **70**–559
    II. Circuit Wiring ................................. **70**–560
    III. Sources of Power .............................. **70**–561
    IV. Overcurrent Protection ........................ **70**–562

702    Optional Standby Systems ...................... **70**–562

    I. General ......................................... **70**–562
    II. Circuit Wiring ................................. **70**–563
    III. Grounding .................................... **70**–563

705    Interconnected Electric Power Production
    Sources ............................................ **70**–563

720    Circuits and Equipment Operating at Less
    Than 50 Volts .................................... **70**–565

725    Class 1, Class 2, and Class 3
    Remote-Control, Signaling, and
    Power-Limited Circuits .......................... **70**–565

    I. General ......................................... **70**–565
    II. Class 1 Circuits ............................... **70**–566
    III. Class 2 and Class 3 Circuits ................. **70**–568

727    Instrumentation Tray Cable: Type ITC ......... **70**–574

760    Fire Alarm Systems ............................. **70**–575

    I General ......................................... **70**–575
    II. Non–Power-Limited Fire Alarm
        (NPLFA) Circuits ........................... **70**–576
    III. Power-Limited Fire Alarm (PLFA)
        Circuits ..................................... **70**–579

770    Optical Fiber Cables and Raceways ........... **70**–583

    I. General ......................................... **70**–583
    II. Protection ..................................... **70**–584
    III. Cables Within Buildings ...................... **70**–584

780    Closed-Loop and Programmed Power
    Distribution ....................................... **70**–587

## Chapter 8 Communications Systems

800    Communications Circuits ........................ **70**–589

    I. General ......................................... **70**–589
    II. Conductors Outside and Entering
        Buildings .................................... **70**–590

ARTICLE

    III. Protection ........................................ **70**–591
    IV. Grounding Methods ........................... **70**–592
    V. Communications Wires and Cables
        Within Buildings ............................. **70**–593

810    Radio and Television Equipment ............... **70**–597

    I. General ......................................... **70**–597
    II. Receiving Equipment — Antenna
       Systems ........................................ **70**–598
    III. Amateur Transmitting and Receiving
        Stations — Antenna Systems ................ **70**–600
    IV. Interior Installation — Transmitting
       Stations ....................................... **70**–601

820    Community Antenna Television and Radio
    Distribution Systems ............................. **70**–601

    I. General ......................................... **70**–601
    II. Cables Outside and Entering Buildings ..... **70**–602
    III. Protection ..................................... **70**–603
    IV. Grounding Methods ........................... **70**–603
    V. Cables Within Buildings ...................... **70**–604

830    Network-Powered Broadband
    Communications Systems ........................ **70**–607

    I. General ......................................... **70**–607
    II. Cables Outside and Entering Buildings ..... **70**–609
    III. Protection ..................................... **70**–612
    IV. Grounding Methods ........................... **70**–613
    V. Wiring Methods Within Buildings ........... **70**–614

TABLES

## Chapter 9 Tables

1    Percent of Cross Section of Conduit and
    Tubing for Conductors ........................... **70**–617

TABLES

4    Dimensions and Percent Area of Conduit
    and Tubing (Areas of Conduit or Tubing
    for the Combinations of Wires Permitted
    in Table 1, Chapter 9) ........................... **70**–617

5    Dimensions of Insulated Conductors and
    Fixture Wires ..................................... **70**–622

5A    Compact Aluminum Building Wire
    Nominal Dimensions* and Areas .............. **70**–624

8    Conductor Properties ............................. **70**–625

9    Alternating-Current Resistance and
    Reactance for 600-Volt Cables, 3-Phase,
    60 Hz, 75°C (167°F) — Three Single
    Conductors in Conduit ........................... **70**–626

11(A)    Class 2 and Class 3 Alternating-Current
    Power Source Limitations ........................ **70**–628

11(B)    Class 2 and Class 3 Direct-Current Power
    Source Limitations ............................... **70**–628

12(A)    PLFA Alternating-Current Power Source
    Limitations ....................................... **70**–629

12(B)    PLFA Direct-Current Power Source
    Limitations ....................................... **70**–629

    Annex A ............................................. **70**–631

    Annex B ............................................. **70**–635

    Annex C ............................................. **70**–645

    Annex D ............................................. **70**–667

    Annex E ............................................. **70**–675

    Annex F ............................................. **70**–677

    Index ................................................ **70**–681

# NATIONAL ELECTRICAL CODE COMMITTEE

*These lists represent the membership at the time each Committee was balloted on the text of this edition. Since that time, changes in the membership may have occurred. A key to classifications is found at the back of this book.*

## Technical Correlating Committee

**D. Harold Ware,** *Chair* [IM]
Libra Electric Co., OK
Rep. National Electrical Contractors Association

**Mark W. Earley**, *Secretary*
National Fire Protection Association, MA
(nonvoting)

**Jean A. O'Connor,** *Recording Secretary*
National Fire Protection Association, MA
(nonvoting)

**Richard Berman,** Underwriters Laboratories Inc., IL [RT]
**James E. Brunssen,** Telcordia Technologies, Inc., NJ [UT]
Rep. Alliance for Telecommunications Industry Solutions
**Michael I. Callanan,** Nat'l. Joint Apprentice & Training Committee, PA [L]
Rep. International Brotherhood of Electrical Workers
**Philip H. Cox,** Int'l. Association of Electrical Inspectors, TX [E]
Rep. International Association of Electrical Inspectors
**William R. Drake,** Marinco, CA [M]
**Antonio Macias,** AMERIC DF, Mexico [U]
**James T. Pauley,** Square D Co., KY [M]
Rep. National Electrical Manufacturers Association
**Joseph E. Pipkin,** U.S. Dept. of Labor OSHA, DC [E]
**John W. Troglia,** Edison Electric Institute, WI [UT]
Rep. Edison Electric Institute
**Craig M. Wellman,** DuPont Engineering, DE [U]
Rep. American Chemistry Council

### Alternates

**Jeffrey Boksiner,** Telcordia Technologies, Inc., NJ [UT]
Rep. Alliance for Telecommunications Industry Solutions
(Alt. to J. E. Brunssen)

**James M. Daly,** General Cable, NY [M]
Rep. National Electrical Manufacturers Association
(Alt. to J. T. Pauley)
**John R. Kovacik,** Underwriters Laboratories, IL [RT]
(Alt. to R. Berman)
**Edward C. Lawry,** WI [E]
Rep. International Association of Electrical Inspectors
(Alt. to P. H. Cox)
**William M. Lewis,** Eli Lilly & Co., IN [U]Rep. American Chemistry Council
(Alt. to C. M. Wellman)
**Neil F. LaBrake, Jr.,** Niagara Mohawk Power Corp., NY [UT]
Rep. Edison Electric Institute
(Alt. to J. W. Troglia)
**Michael D. Toman,** MEGA Power Electrical Services, Inc., MD [IM]
Rep. National Electrical Contractors Association
(Alt. To D. H. Ware)

### Nonvoting

**Richard G. Biermann,** Biermann Electric Co. Inc., IA
(Member Emeritus)

## CODE-MAKING PANEL NO. 1

### Articles 90, 100, 110

**John D. Minick,** *Chair*
Nat'l. Electrical Manufacturers Assn., TX [M]
Rep. National Electrical Manufacturers Association

**Michael A. Anthony,** University of Michigan, MI [U]
Rep. The Association of Higher Education Facilities Officers
**Philip H. Cox,** Int'l. Association of Electrical Inspectors, TX [E]
Rep. International Association of Electrical Inspectors
**David A. Dini,** Underwriters Laboratories Inc. IL [RT]
**William T. Fiske,** Intertek Testing Services N.A. Inc., NY [RT]
**H. Landis Floyd, II,** The DuPont Company, DE [U]
Rep. Institute of Electrical & Electronics Engineers, Inc.
**David L. Hittinger,** IEC of Greater Cincinnati, OH [IM]
Rep. Independent Electrical Contractors, Inc.
**Don B. Ivory,** Idaho Electrical JATC, ID [L]
Rep. International Brotherhood of Electrical Workers
**Antonio Macias,** AMERIC DF, Mexico [U]
**Randall R. McCarver,** Telcordia Technologies, Inc., NJ [UT]
Rep. Alliance for Telecommunications Industry Solutions
**Ralph C. Prichard,** Hercules, Inc., DE [U]
Rep. American Chemistry Council

**H. Brooke Stauffer,** Nat'l. Electrical Contractors Assn., MD [IM]
Rep. National Electrical Contractors Association
**John W. Troglia,** Edison Electric Institute, WI [UT]
Rep. Edison Electric Institute

### Alternates

**Louis A. Barrios, Jr.,** Equilon Enterprises, LLC, TX [U]
Rep. American Chemistry Council
(Alt. To R. C. Prichard)
**David P. Brown,** Baltimore Gas & Electric Co., MD [UT]
Rep. Edison Electric Institute
(Alt. to J. W. Troglia)
**Timothy Lee Emory,** Emory Electric, Inc., NC [IM]
Rep. Independent Electrical Contractors, Inc.
(Alt. to D. L. Hittinger)
**Russell J. Helmick, Jr.,** City of Irvine, CA [E]

Rep. International Association of Electrical Inspectors
(Alt. to P. H. Cox)

**Mahbub Hoque,** Telcordia Technologies, NJ [UT]
Rep. Alliance for Telecommunications Industry Solutions
(Alt. to R. R. McCarver)

**Donald H. McCullough, II,** Westinghouse Savannah River Co., SC [U]
Rep. Institute of Electrical & Electronics Engineers, Inc.
(Alt. to H. L. Floyd, II)

**Larry Miller,** Nat'l. Electrical Manufacturers Assn., VA [M]
Rep. National Electrical Manufacturers Association
(Alt. to J. D. Minick)

**Ricky L. Oakland,** IBEW, WY [L]
Rep. International Brotherhood of Electrical Workers
(Alt. to D. B. Ivory)

**William L. Schallhammer,** Underwriters Laboratories, Inc., IL [RT]
(Alt. to D. A. Dini)

### Nonvoting

**Ark Tsisserev,** City of Vancouver, BC
Rep. Canadian Standards Association International

**William Wusinich,** U.S. Dept of Labor, PA [E]

## CODE-MAKING PANEL NO. 2

### Articles 210, 215, 220, Chapter 9, Annex D
### Examples 1 through 6

**James W. Carpenter,** *Chair*
North Carolina Dept. of Insurance, NC [E]
Rep. International Association of Electrical Inspectors

**Richard W. Becker,** Engineered Electrical Systems, Inc., WA [U]
Rep. Institute of Electrical & Electronics Engineers, Inc.

**Thomas L. Harman,** University of Houston/Clear Lake, TX [SE]

**Bernard Mericle,** IBEW, Local Union 236, NY [L]
Rep. International Brotherhood of Electrical Workers

**Robert E. Moore,** TECO Energy, FL [UT]
Rep. Edison Electric Institute

**Donald A. Nissen,** Underwriters Laboratories, Inc., IL [RT]

**James T. Pauley,** Square D Co., KY [M]
Rep. National Electrical Manufacturers Association

**Joseph Patterson Roché,** Celanese Acetate, SC [U]
Rep. American Chemistry Council

**Albert F. Sidhom,** U.S. Army Corps of Engineers, CA [U]

**Michael D. Toman,** MEGA Power Electrical Services, Inc., MD [IM]
Rep. National Electrical Contractors Association

**Robert G. Wilkinson,** Independent Electrical Contractors of Houston, Inc., TX [IM]
Rep. Independent Electrical Contractors, Inc.

**Ernie Howell,** IEC, Rocky Mountain Chapter, CO [IM]
Rep. Independent Electrical Contractors
(Alt. to R. G. Wilkinson)

**James R. Jones,** University of Alabama at Birmingham, AL [U]
Rep. Institute of Electrical & Electronics Engineers, Inc.
(Alt. to R. W. Becker)

**Daniel J. Kissane,** Pass & Seymour/Legrand, NY [M]
Rep. National Electrical Manufacturers Association
(Alt. to J. T. Pauley)

**Charles D. Marshall, Jr.,** IBEW Local 948, MI [L]
Rep. International Brotherhood of Electrical Workers
(Alt. to B. Mericle)

**Susan W. Porter,** Underwriters Laboratories Inc., NY [RT]
(Alt. to D. A. Nissen)

**J. Morris Trimmer,** University of Florida, FL [SE]
(Alt. to T. L. Harman)

**Joseph E. Wiehagen,** Nat'l. Assn. of Home Builders, MD [IM]
Rep. National Association of Home Builders
(Voting Alt. to NAHB Rep.)

### Alternates

**Ernest S. Broome,** City of Knoxville, TN [E]
Rep. International Association of Electrical Inspectors
(Alt. to J. W. Carpenter)

**Charles G. Crawford,** TXU Electric, TX [UT]
Rep. Edison Electric Institute
(Alt. to R. E. Moore)

### Nonvoting

**Douglas A. Lee,** U.S. Consumer Product Safety Commission, MD

**Andrew M. Trotta,** U.S. Consumer Product Safety Commission, MD

## CODE-MAKING PANEL NO. 3

### Articles 300, 527, 690, 692

**Raymond W. Weber,** *Chair*
Dept. of Commerce, WI [E]
Rep. International Association of Electrical Inspectors

**Joseph J. Andrews,** Electrical Safety Resources, Inc., SC [U]
Rep. Institute of Electrical & Electronics Engineers, Inc.

**Charles W. Beile,** Allied Tube & Conduit/Tyco, IL [M]
Rep. National Electrical Manufacturers Association

**Ward I. Bower,** Sandia Nat'l. Laboratories, NM [U]
Rep. Solar Energy Industries Association
(VL 690)

**Paul Casparro,** Scranton Electricians JATC, PA [L]
Rep. International Brotherhood of Electrical Workers

**Charles W. Forsberg,** OH [M]
Rep. Society of the Plastics Industry Inc.

**Jack A. Gruber,** Wheatland Tube Co., PA [M]
Rep. American Iron and Steel Institute

**Dennis B. Horman,** PacifiCorp, UT [UT]
Rep. Edison Electric Institute

**Kenneth Krastins,** Plug Power, Inc., NY [M]
Rep. US Fuel Cell Council
(VL 691)

**George M. Kreiner,** Underwriters Laboratories Inc., IL [RT]
**Ronald E. Maassen,** Lemberg Electric Co., Inc., WI [IM]
   Rep. National Electrical Contractors Association
**Steven J. Owen,** AL [IM]
   Rep. Associated Builders and Contractors, Inc.
**David A. Pace,** Olin Corporation, AL [U]
   Rep. American Chemistry Council

### Alternates

**Martin J. Brett, Jr.,** Wheatland Tube Co., NJ [M]
   Rep. American Iron and Steel Institute
   (Alt. to J. A. Gruber)
**Les Easter**, Allied Tube and Conduit, IL [M]
   Rep. National Electrical Manufacturers Association
   (Alt. to C. W. Beile)
**James D. Erwin,** Celanese, Ltd., TX [U]
   Rep. American Chemistry Council
   (Alt. to D. A. Pace)
**Palmer L. Hickman,** IBEW Local 380, PA [L]
   Rep. International Brotherhood of Electrical Workers
   (Alt. to P. Casparro)
**David H. Kendall,** Carlon, Lamson & Sessions, OH [M]

Rep. Society of the Plastics Industry Inc.
   (Alt. to C. W. Forsberg)
**Mark C. Ode,** Underwriters Laboratories, Inc., NC [RT]
   (Alt. to G. M. Kreiner)
**Richard P. Owen,** City of St. Paul, MN [E]
   Rep. International Association of Electrical Inspectors
   (Alt. to R. W. Weber)
**Roger S. Passmore,** Davis Electrical Constructors, Inc., SC [IM]
   Rep. Associated Builders and Contractors, Inc.
   (Alt. to S. J. Owen)
**George E. Richey,** Consumers Energy, MI [UT]
   Rep. Edison Electric Institute
   (Alt. to D. B. Horman)
**Melvin K. Sanders,** Things Electrical Co., Inc. dba (TECo., Inc), IA [U]
   Rep. Institute of Electrical & Electronics Engineers, Inc.
   (Alt. to J. J. Andrews)
**Robert H. Wills,** Advanced Energy Systems Inc., NH [U]
   Rep. Solar Energy Industries Association
   (VL 690)
   (Alt. to W. I. Bower)

## CODE-MAKING PANEL NO. 4

### Articles 225, 230

**C. John Beck,** *Chair* [UT]
Pacific Gas and Electric Co., CA
Rep. Edison Electric Institute

**Malcolm Allison,** Ferraz Shawmut, MA [M]
**K. W. Carrick,** Electrical Engineering Professional Services, Inc., MS [U]
   Rep. Institute of Electrical & Electronics Engineers, Inc.
**Floyd C. Ferris,** New York Board of Fire Underwriters, NY [E]
**Howard D. Hughes,** Hughes Electric Co. Inc., AR [IM]
   Rep. National Electrical Contractors Association
**William M. Lewis,** Eli Lilly & Co., IN [U]
   Rep. American Chemistry Council
**Junior L. Owings,** State of Oregon, OR [E]
   Rep. International Association of Electrical Inspectors
**Robert J. Pollock,** Underwriters Laboratories Inc., IL, RT
**Mark H. Sumrall,** IBEW Local 527, TX [L]
   Rep. International Brotherhood of Electrical Workers
**John W. Young,** Siemens Energy & Automation, Inc., GA [M]
   Rep. National Electrical Manufacturers Association
**Vincent Zinnante,** Advantage Electric, Inc., TX [IM]
   Rep. Independent Electrical Contractors, Inc.

**Terry D. Cole,** Hamer Electric/Associated Electrical Consultants, Inc., WA [IM]
   Rep. Independent Electrical Contractors, Inc.
   (Alt. to V. Zinnante)
**Dennis Darling,** Ayres, Lewis, Norris, & May, Inc., MI [U]
   Rep. Institute of Electrical & Electronics Engineers, Inc.
   (Alt. to K. W. Carrick)
**Timothy Owens,** City of San Diego, CA [E]
   Rep. International Association of Electrical Inspectors
   (Alt. to J. L. Owings)
**Philip M. Piqueira,** General Electric Co., CT [M]
   Rep. National Electrical Manufacturers Association
   (Alt. to J. W. Young)
**John A. Sigmund,** PPG Industries, Inc., LA [U]
   Rep. American Chemistry Council
   (Alt. to W. M. Lewis)
**William J. Tipton,** IBEW, OH [L]
   Rep. International Brotherhood of Electrical Workers
   (Alt. to M. H. Sumrall)

### Alternates

**Thomas L. Adams,** Exelon Corporation, IL [UT]
   Rep. Edison Electric Institute
   (Alt. to C. J. Beck)

## CODE-MAKING PANEL NO. 5

### Articles 200, 250, 280, 285

**Ronald J. Toomer,** *Chair* [IM]
Toomer Electrical Co. Inc., LA
Rep. National Electrical Contractors Association

**Jeffrey Boksiner,** Telcordia Technologies, Inc., NJ [UT]
   Rep. Alliance for Telecommunications Industry Solutions
**David T. Brender,** Copper Development Association, Inc., NY [M]
   Rep. Copper Development Association Inc.

**Martin J. Brett, Jr.,** Wheatland Tube Co., NJ [M]
   Rep. American Iron and Steel Institute
**Elio L. Checca,** U.S. Dept. of Labor, VA [E]
**Paul Dobrowsky,** Eastman Kodak Co., NY [U]
   Rep. American Chemistry Council

**Gerald L. Hadeen,** Alflex Corp., CA [M]
    Rep. The Aluminum Association
**Dan Hammel,** Int'l. Brotherhood of Electrical Workers, IA [L]
    Rep. International Brotherhood of Electrical Workers
**Michael J. Johnston,** Int'l. Association of Electrical Inspectors, Inc., TX [E]
    Rep. International Association of Electrical Inspectors
**Charles F. Mello,** Electro-Test, Inc., OR [IM]
    Rep. International Electrical Testing Association Inc.
**Elliot Rappaport,** Electro Technology Consultants, Inc., MI [U]
    Rep. Institute of Electrical & Electronics Engineers, Inc.
**Ted G. Robertson,** Robertson Electric, Inc., TX, [IM]
    Rep. Independent Electrical Contractors, Inc.
**Walter Skuggevig,** Underwriters Laboratories, Inc., NY [RT]
**Gregory J. Steinman,** Thomas & Betts Corp., TN [M]
    Rep. National Electrical Manufacturers Association
**Robert G. Stoll,** Thomas Associates, Inc., OH [M]
    Rep. Power Tool Institute, Inc.
**C. Douglas White,** Reliant Energy, TX [UT]
    Rep. Edison Electric Institute

### Alternates

**Martin D. Adams,** Adams Electric, Inc., CO [IM]
    Rep. National Electrical Contractors Association
    (Alt. to R. J. Toomer)
**David A. Dini,** Underwriters Laboratories Inc, IL [RT]
    (Alt. to W. Skuggevig)
**Timothy Edwards,** Alcan Cable Co., GA [M]
    Rep. The Aluminum Association
    (Alt. to G. L. Hadeen)

**Garfield B. Gwyn,** Gwyn Electrical & Plumbing Co., NC [IM]
    Rep. Independent Electrical Contractors, Inc.
    (Alt. to T. G. Robertson)
**Ronald Lai,** FCI Electrical, NH [M]
    Rep. National Electrical Manufacturers Association
    (Alt. to G. J. Steinman)
**Dennis E. Lammert,** Ameren Services, MO [UT]
    Rep. Edison Electric Institute
    (Alt. to C. D. White)
**Richard E. Loyd,** R&N Associates, AR [M]
    Rep. American Iron and Steel Institute
    (Alt. to M. J. Brett, Jr.)
**Daleep C. Mohla,** Union Carbide Corp., TX [U]
    Rep. Institute of Electrical & Electronics Engineers, Inc.
    (Alt. to E. Rappaport)
**Thomas E. Moore,** City of Norton, OH [E]
    Rep. International Association of Electrical Inspectors
    (Alt. to M. J. Johnston)
**David Peot,** Ryobi, SC [M]
    Rep. Power Tool Institute, Inc.
    (Alt. to R. G. Stroll)
**Thomas J. Shea,** IBEW Local 405, IA [L]
    Rep. International Brotherhood of Electrical Workers
    (Alt. to D. Hammel)
**J. Philip Simmons,** Simmons Electrical Services, WA [M]
    (Alt. to D. T. Brender)
**Michael K. Toney,** Equistar Chemicals, LP, TX [U]
    Rep. American Chemistry Council
    (Alt. to P. Dobrowsky)

## CODE-MAKING PANEL NO. 6

### Articles 310, 400, 402, Chapter 9 Tables 5 through 9

**Stephen J. Thorwegen, Jr.,** *Chair* [IM]
Fisk Electric Co., TX
Rep. National Electrical Contractors Association

**Julian R. Burns,** Burns Electrical Contractors, Inc., NC [IM]
    Rep. Independent Electrical Contractors, Inc.
**William C. Ferrell,** IBEW, OH [L]
    Rep. International Brotherhood of Electrical Workers
**Samuel B. Friedman,** BICC General, RI [M]
    Rep. National Electrical Manufacturers Association
**Steven Galan,** Underwriters Laboratories Inc., NY [RT]
**Ravindra H. Ganatra,** Alcan Cable, GA [M]
    Rep. The Aluminum Association
**David G. Komassa,** Wisconsin Electric Power Co., WI [UT]
    Rep. Edison Electric Institute
**L. Bruce McClung,** Union Carbide Corp., WV [U]
    Rep. Institute of Electrical & Electronics Engineers, Inc.
**Dale W. Pettigrew,** Cognis Corp., OH [U]
    Rep. American Chemistry Council
**Oran P. Post,** City of Cuyahoga Falls, OH [E]
    Rep. International Association of Electrical Inspectors
**Joseph Zimnoch,** The Okonite Company, NJ [M]
    Rep. Copper Development Association Inc.

### Alternates

**Kenneth L. Brotherton,** IBEW Local 683, OH [L]
    Rep. International Brotherhood of Electrical Workers
    (Alt. to W. C. Ferrell)
**James M. Daly,** General Cable, NY [M]
    Rep. Copper Development Association, Inc.
    (Alt. to J. Zimnoch)

**G. W. "Jerry" Kent**, Kent Electric Systems, TX [IM]
    Rep. Independent Electrical Contractors
    (Alt. to J. R. Burns)
**Danny P. Liggett,** DuPont Engineering, DE [U]
    Rep. American Chemistry Council
    (Alt. to D. W. Pettigrew)
**Lowell S. Lisker,** American Insulated Wire Corp., RI [M]
    Rep. National Electrical Manufacturers Association
    (Alt. to S. B. Friedman)
**Harry J. Sassaman,** Forest Electric Corp., NJ [IM]
    Rep. National Electrical Contractors Association
    (Alt. to S. J. Thorwegen, Jr.)
**John Stacey,** City of St Louis, MO [E]
    Rep. International Association of Electrical Inspectors
    (Alt. to O. P. Post)
**Fred Truban,** American Electric Power, OH [UT]
    Rep. Edison Electric Institute
    (Alt. to D. G. Komassa)
**Donald A. Voltz,** Mustang Engineering, Inc., TX [U]
    Rep. Institute of Electrical & Electronics Engineers, Inc.
    (Alt. to L.B. McClung)
**Austin D. Wetherell,** Underwriters Laboratories Inc., NY [RT]
    (Alt. to S. Galan)

## CODE-MAKING PANEL NO. 7

### Articles 320, 322, 324, 326, 328, 330, 332, 334, 336, 338, 340, 382, 394, 396, 398

**Thomas E. Trainor,** *Chair*
City of San Diego, CA [E]
Rep. International Association of Electrical Inspectors

**James M. Daly,** General Cable, NY [M]
Rep. National Electrical Manufacturers Association
**Brian Ensign,** Intertek Testing Services, N.A. Inc., NY [RT]
**Chris Fahrenthold,** MH Technologies, TX [IM]
Rep. Independent Electrical Contractors, Inc.
**Robert L. Gotham,** Rose City Electric Co., Inc., OR [IM]
Rep. National Electrical Contractors Association
**Thomas J. Guida,** Underwriters Laboratories Inc., NY [RT]
**Ronald G. Nickson,** Nat'l. Multi Housing Council, DC [U]
Rep. National Multi Housing Council
**Bruce W. Nutt,** TXU Electric & Gas, TX [UT]
Rep. Edison Electric Institute
**John E. Propst,** Equilon Enterprises LLC, TX [U]
Rep. American Chemistry Concil
**Thomas G. Rodgers,** Dickens & Assoc. Inc., FL [L]
Rep. International Brotherhood of Electrical Workers
**David E. Schumacher,** All County Electric Co., IA [IM]
Rep. Associated Builders and Contractors, Inc.
**H. R. Stewart,** HRS Consulting, TX [U]
Rep. Institute of Electrical & Electronics Engineers, Inc.
**George A. Straniero,** AFC Cable Systems, Inc., NJ [M]
Rep. Copper Development Association Inc.
**Robert S. Strength,** Product Safety Management Inc., FL [M]
Rep. Society of the Plastics Industry Inc.
**Richard Temblador,** Alflex Corporation, CA [M]
Rep. The Aluminum Association

**Arthur Buxbaum,** San Diego Building Inspection Dept., CA [E]
Rep. International Association of Electrical Inspectors
(Alt. to T. E. Trainor)
**John J. Cangemi,** Underwriters Laboratories, Inc., NY [RT]
(Alt. to T. J. Guida)
**James V. Fitzgerald,** The Okonite Co., NJ [M]
Rep. National Electrical Manufacturers Association
(Alt. to J. M. Daly)
**Ravindra H. Ganatra,** Alcan Cable, GA [M]
Rep. The Aluminum Association
(Alt. to R. Temblador)
**Herman J. Hall,** TX [M]
Rep. Society of the Plastics Industry Inc.
(Alt. to R. S. Strength)
**Greg Hall,** Better-Way Electric, Inc., CO [IM]
Rep. Independent Electrical Contractors, Inc.
(Alt. to C. Fahrenthold)
**Dennis A. Nielsen,** Southdown, Inc., CA [U]
Rep. Institute of Electrical & Electronics Engineers, Inc.
(Alt. to H. R. Stewart)
**John Thomas Thompson,** A.B.C. Marathon Electrical Co., Inc., AL [IM]
Rep. Associated Builders and Contractors, Inc.
Alt. to D. E. Schumacher)
**David B. West,** Duke Energy Inc., NC [UT]
Rep. Edison Electric Institute
(Alt. to B. W. Nutt)
**Thomas H. Wood,** Cecil B. Wood Inc., IL [IM]
Rep. National Electrical Contractors Association
(Alt. to R. L. Gotham)

### Alternates

**Harry C. Brown,** IBEW Local 606, FL [L]
Rep. International Brotherhood of Electrical Workers
(Alt. to T. G. Rodgers)

## CODE-MAKING PANEL NO. 8

### Articles 342, 344, 348, 350, 352, 354, 356, 358, 360, 362, 366, 368, 370, 372, 374, 376, 378, 380, 384, 386, 388, 390, 392
### Chapter 9 Tables 1 through 4, and Appendix C

**Kenneth E. Jannot,** *Chair* [UT]
Detroit Edison, MI
Rep. Edison Electric Institute

**Richard Berman,** Underwriters Laboratories Inc., IL [RT]
**John S. Corry,** Corry Electric Inc., CA [IM]
Rep. Associated Builders and Contractors, Inc.
**Robert W. Cox,** Astra Zeneca, DE [U]
Rep. American Chemistry Council
**George R. Dauberger,** Thomas & Betts Corporation, TN [M]
Rep. National Electrical Manufacturers Association
**James C. Dollins,** AFC Cable Systems, MA [M]
Rep. The Aluminum Association
**M. Shan Griffith,** Brown & Root, Inc., TX [U]
Rep. Institute of Electrical & Electronics Engineers, Inc.
**David H. Kendall,** Carlon, Lamson & Sessions, OH [M]
Rep. Society of the Plastics Industry Inc.
**Wayne A. Lilly,** City of Harrisonburg, VA [E]
Rep. International Association of Electrical Inspectors
**Richard E. Loyd,** R&N Associates, AR [M]
Rep. American Iron and Steel Institute

**Stephen P. Poholski,** Newkirk Electric Associates, Inc., MI [IM]
Rep. National Electrical Contractors Association
**C. Ernest Reynolds,** Hatfield-Reynolds Electric Co., AZ [IM]
Rep. Independent Electrical Contractors, Inc.
**Dennis L. Rowe,** NY Board of Fire Underwriters, NY [E]
Rep. New York Board of Fire Underwriters
**Ray R. Simpson,** Int'l. Brotherhood of Electrical Workers, IN [L]
Rep. International Brotherhood of Electrical Workers

### Alternates

**Jimmy R. Bonds,** Oklahoma State Dept of Health, OK [E]
Rep. International Association of Electrical Inspectors
(Alt. to W. A. Lilly)
**Kenneth E. Christ,** Solutie, Inc., MO [U]
Rep. American Chemistry Council
(Alt. to R. W. Cox)

**Joseph G. Dabe,** City of St. Paul, MN [L]
Rep. International Brotherhood of Electrical Workers
(Alt. to R. R. Simpson)
**Ron Duren,** PacifiCorp, WA [UT]
Rep. Edison Electric Institute
(Alt. to K. E. Jannot)
**Charles W. Forsberg,** OH [M]
Rep. Society of the Plastics Industry Inc.
(Alt. to D. H. Kendall)
**Jack A. Gruber,** Wheatland Tube Co., PA [M]
Rep. American Iron and Steel Institute
(Alt. to R. E. Loyd)
**Alan Manche,** Schneider Electric/Square D Company, KY [M
Rep. National Electrical Manufacturers Association
(Alt. to G. R. Dauberger)

**Von Dewayne Stelljes, Jr.,** Wayne's Electric, Inc., CO [IM]
Rep. Independent Electrical Contractors
(Alt. to C. E. Reynolds)
**Richard Temblador,** Alflex Corporation, CA, [M]
Rep. The Aluminum Association
(Alt. to J. C. Dollins)
**Ronald J. Toomer,** Toomer Electrical Co. Inc., LA [IM]
Rep. National Electrical Contractors Association
(Alt. to S. P. Poholski)
**James Van Den Heuvel,** West Electric Inc., WI [IM]
Rep. Associated Builders and Contractors, Inc.
(Alt. to J. S. Corry)
**William C. Wagner,** Underwriters Laboratories, Inc., NY [RT]
(Alt. to R. Berman)

## CODE-MAKING PANEL NO. 9

### Articles 312, 314, 404, 408

**Timothy M. Croushore,** *Chair* [UT]
Allegheny Power Service Corp., PA
Rep. Edison Electric Institute

**Boyd H. Culp,** Phillips Petroleum Co., OK [U]
Rep. American Chemistry Council
**Dale R. Deming,** Am Electric Co. (T&B), MI [M]
**Frederic P. Hartwell,** Hartwell Electrical Services, MA [SE]
**Jeffrey H. Hidaka,** Underwriters Laboratories, IL [RT]
**Robert J. Kaemmerlen,** Kaemmerlen Electric Co., MO [IM]
Rep. National Electrical Contractors Association
**Thomas J. LeMay,** LeMay Electric, Inc., GA [IM]
Rep. Independent Electrical Contractors, Inc.
**Anthony Montuori,** The New York Board of Fire Underwriters, NY [E]
Rep. International Association of Electrical Inspectors
**Ronald H. Reed,** Square D Company, KY [M]
Rep. National Electrical Manufacturers Association
**Sukanta Sengupta,** FMC Corp., NJ [U]
Rep. Institute of Electrical & Electronics Engineers, Inc.
**Paul Welnak,** IBEW Local 494, WI [L]
Rep. International Brotherhood of Electrical Workers

**Alternates**

**Mark R. Berner,** PP&L, Inc., PA [UT]
Rep. Edision Electric Institute
(Alt. to T. M. Croushore)
**Jeff Bernson,** IBEW, IL [L]
Rep. International Brotherhood of Electrical Workers
(Alt. to P. Welnak)
**Donald Offerdahl,** ND State Electrical Board, ND [E]
Rep. International Association of Electrical Inspectors
(Alt. to A. Montuori)
**Bradford D. Rupp,** Allied Moulded Products, Inc., OH [M]
Rep. National Electrical Manufacturers Association
Alt. to R. H. Reed
**Jerome W. Seigel,** CT [U]
Rep. Institute of Electrical & Electronics Engineers, Inc.
(Alt. to S. Sengupta)
**Michael W. Wedel,** Phillips Petroleum Company, TX [U]
Rep. American Chemistry Council
(Alt. to B. H. Culp)

## CODE-MAKING PANEL NO. 10

### Articles 240, 780

**James T. Dollard, Jr.** *Chair* [L]
IBEW, PA
Rep. International Brotherhood of Electrical Workers

**Charles K. Blizard,** American Electrical Testing Co., Inc., MA [IM]
Rep. International Electrical Testing Association Inc.
**Madeline Borthick,** IEC of Houston, TX [IM]
Rep. Independent Electrical Contractors, Inc.
**John E. Brezan,** Lehigh Valley Electrical Inspection Service, PA [E]
Rep. International Association of Electrical Inspectors
**Robert J. Deaton,** Union Carbide Corp., TX [U]
Rep. Institute of Electrical & Electronics Engineers, Inc.

**Charles K. Eldridge,** Indianapolis Power & Light Co., IN [UT]
Rep. Edison Electric Institute
**Carl J. Fredericks,** The Dow Chemical Co., TX [U]
Rep. American Chemistry Council
**Don W. Jhonson,** Interior Electric, Inc., FL [IM]
Rep. National Electrical Contractors Association
**Clive W. Kimblin,** Cutler-Hammer, Inc., PA [M]
Rep. National Electrical Manufacturers Association
**Arden L. Munson,** Hussmann Corp., MO [M]
Rep. Air Conditioning and Refrigeration Institute

**George J. Ockuly,** MO [M]

**John A. Zaplatosch,** Underwriters Laboratories Inc., IL [RT]

### Alternates

**David E. Chartrand,** Middle Department Inspection Agency, Inc., NY [E]
Rep. International Association of Electrical Inspectors
(Alt. to J. E. Brezan)

**George D. Gregory,** Square D Co., IA [M]
Rep. National Electrical Manufacturers Association
(Alt. to C. W. Kimblin)

**Roderic L. Hageman,** Prit Service, Inc., IL [IM]
Rep. International Electrical Testing Association Inc.
(Alt. to C. K. Blizard)

**Charles D. Hughes,** Westinghouse Savannah River Co., SC [U]
Rep. Institute of Electrical & Electronics Engineers, Inc.
(Alt. to R. J. Deaton)

**Randy Jones,** IBEW Local 934, TN [L]
Rep. International Brotherhood of Electrical Workers
(Alt. to J. T. Dollard, Jr.)

**Kris Mantravadi,** La Roche Industries, LA [U]
Rep. American Chemistry Council
(Alt. to C. J. Fredericks)

**Paul J. Notarian,** Underwriters Laboratories Inc., NY [RT]
(Alt. to J. A. Zaplatosch)

**Vincent J. Saporita,** Cooper Bussmann, MO [M]
(Alt. to G. J. Ockuly)

**Steve A. Struble,** Freeman's Electric Service, Inc., SD [IM]
Rep. Independent Electrical Contractors
(Alt. to M. Borthick)

**John Tolbert,** Bristol Compressors, VA [M]
Rep. Air Conditioning and Refrigeration Institute
(Alt. to A. L. Munson)

**Leslie R. Zielke,** South Carolina Electric and Gas Co., SC [UT]
Rep. Edison Electric Institute
(Alt. to C. K. Eldridge)

### Nonvoting

**Rick C. Gilmour,** Canadian Standards Assn., ON

## CODE-MAKING PANEL NO. 11

### Articles 430, 440, 670, Appendix D Example D8

**Thomas H. Wood,** *Chair*
Cecil B Wood Inc., IL [IM]
Rep. National Electrical Contractors Association

**Rick L. Bunch,** Tecumseh Products Co., MI [M]
Rep. Air Conditioning and Refrigeration Institute

**Louis D. Closson,** Intertek Testing Services, N.A. Inc., NY [RT]

**Joe David Cox,** Eastman Chemical Co., TN [U]
Rep. American Chemistry Council

**Thomas J. Garvey,** State of Wisconsin, WI [E]
Rep. International Association of Electrical Inspectors

**Paul S. Hamer,** Chevron Research & Technology Co., CA [U]
Rep. American Petroleum Institute

**Michael D. Landolfi,** Landolfi Electric Co. Inc., NJ [IM]
Rep. Associated Builders and Contractors, Inc.

**James M. Naughton,** IBEW Local 103, MA [L]
Rep. International Brotherhood of Electrical Workers

**Richard A. Rasmussen,** Underwriters Laboratories, Inc., NC [RT]

**Vincent J. Saporita,** Cooper Bussmann, MO [M]

**Lynn F. Saunders,** GM Worldwide Facilities Group, MI [U]
Rep. Institute of Electrical & Electronics Engineers, Inc.

**Charles B. Schram,** Scottsdale, AZ [SE]

**M. Edward Thomas,** Alabama Power Co., AL [UT]
Rep. Edison Electric Institute

**Ron Widup,** Shermco Industries, Inc., TX [IM]
Rep. International Electrical Testing Association Inc.

**James R. Wright,** Siemens-Furnas Controls, IL [M]
Rep. National Electrical Manufacturers Association

### Alternates

**Frederick Bried,** Equilon Enterprises LLC, TX [U]
Rep. American Petroleum Institute
(Alt. to P. S. Hamer)

**Michael D'Amico,** IBEW Local 488, CT [L]
Rep. International Brotherhood of Electrical Workers
(Alt. to J. M. Naughton)

**Elwood J. Dodge,** Addison Products Co., FL [M]
Rep. Air Conditioning and Refrigeration Institute
(Alt. to R. L. Bunch)

**Stanley Folz,** Folz Electric, Inc., IL [IM]
Rep. National Electrical Contractors Association
(Alt. to T. H. Wood)

**William D. Glover,** PPG Industries, Inc., WV [U]
Rep. American Chemistry Council
(Alt. to J. D. Cox)

**Paul E. Guidry,** Fluor Daniel, Inc., TX [IM]
Rep. Associated Builders and Contractors, Inc.
(Alt. to M. D. Landolfi)

**Leo H. Haas, Jr.,** Reliant Energy HLP, TX [UT]
Rep. Edison Electric Institute
(Alt. to M. E. Thomas)

**Robert J. Keough,** U.S. Electrical Motors, MO [M]
Rep. National Electrical Manufacturers Association
(Alt. to J. R. Wright)

**George J. Ockuly,** MO [M]
(Alt. to V. J. Saporita)

**Carl Radcliffe,** Underwriters Laboratories Inc., NC [RT]
(Alt. to R. A. Rasmussen)

**John A. Schultz,** Minnesota Board of Electricity, MN [E]
Rep. International Association of Electrical Inspectors
(Alt. to T. J. Garvey)

**Arthur J. Smith, III,** Waldemar S. Nelson and Co., Inc., LA [U]
Rep. Institute of Electrical & Electronics Engineers, Inc.
(Alt. to L. F. Saunders)

### Nonvoting

**Nino Mancini,**
Rep. Canadian Standards Association International, ON

## CODE-MAKING PANEL NO. 12

**Articles 426, 427, 610, 620, 625, 630, 645, 660, 665, 668, 669, 685, and Appendix D Examples D9 and D10**

**Charles M. Trout,** *Chair* [IM]
Maron Electric Co., FL
Rep. National Electrical Contractors Association

**Thomas M. Burke,** Underwriters Laboratories Inc., CA  [RT]
**Andre R. Cartal,** Borough of Princeton, NJ  [E]
   Rep. International Association of Electrical Inspectors
**James F. Cook,** Eagle Electric Manufacturing, NY  [M]
   Rep. National Electrical Manufacturers Association
**Kent B. Givens,** Aluminum Co. of America, TX  [M]
   Rep. The Aluminum Association
   (VL 427, 610, 625, 630, 645, 646, 660, 665, 668, 669 and 685)
**Bill Hanthorn,** BICC General Pyrotenax Cables Ltd., ON  [M]
   Rep. Copper Development Association Inc.
**Robert A. Jones,** Independent Electrical Contractors, TX  [IM]
   Rep. Independent Electrical Contractors, Inc.
**William J. Kelly,** Eastman Kodak Co., NY  [U]
   Rep. Institute of Electrical & Electronics Engineers, Inc.
**Richard H. Laney,** Siecor Corp., NC  [U]
**Nick Marchitto,** Otis Elevator Co., CT  [M]
   Rep. National Elevator, Inc.
   (VL 610, 620 and 630)
**John H. Mortimer,** Inductotherm Corp., NJ  [M]
   (VL 665)
**Norbert Poch,** IBM Corp., MN  [U]
   Rep. Information Technology Industry Council
   (VL 645)
**Ronald L. Purvis,** Georgia Power Co., GA  [UT]
   Rep. Edison Electric Institute
**David R. Quave,** Int'l. Brotherhood of Electrical Workers, MS  [L]
   Rep. International Brotherhood of Electrical Workers
**Robert H. Reuss,** Morris Material Handling, LLC, WI  [M]
   (VL 610)
**T. Neil Thorla,** Inland Steel Co., IN  [U]
   Rep. Association of Iron & Steel Engineers
   (VL 610, 620, and 630)
**Craig B. Toepfer,** Ford Motor Co., MI  [U]
   Rep. Society of Automotive Engineers
   (VL 625)
**Kenneth P. White,** Olin Corp., NY  [U]
   Rep. American Chemistry Council

### Alternates

**Scott Cline,** McMurtrey Electric, Inc., CA  [IM]
   Rep. National Electrical Contractors Association
   (Alt. to C. M. Trout)

**Kenneth Hartwig,** Daimler Chrysler, MI  [U]
   Rep. Society of Automotive Engineers
   (Alt. to C. B. Toepfer)
   (VL 625)
**Jeffrey H. Hidaka,** Underwriters Laboratories, IL  [RT]
   (Alt. to T. M. Burke)
**Robert E. Johnson,** Motorola, MA  [U]
   Rep. Information Technology Industry Council
   (Alt. to N. Poch)
   (VL 645)
**Andy Juhasz,** Kone Inc., IL  [M]
   Rep. National Elevator Industry Inc.
   (Alt. to N. Marchitto)
   (VL 610, 620 and 630)
**Roger D. McDaniel,** Georgia Power Co., GA  [UT]
   Rep. Edision Electric Institute
   (Alt. to R. L. Purvis)
**Harold C. Ohde,** Int'l. Brotherhood of Electrical Workers, IL  [L]
   Rep. International Brotherhood of Electrical Workers
   (Alt. to D. R. Quave)
**Robert C. Oldham, Jr.,** Reynolds Metals Co., VA  [M]
   Rep. The Aluminum Association
   (Alt. to K. B. Givens)
   (VL 427, 610, 625, 630, 645, 646, 660, 665, 668, 669, and 685)
**Merritt D. Redick,** M. Redick & Associates, CA  [U]
   Rep. Institute of Electrical & Electronics Engineers, Inc.
   (Alt. to W. J. Kelly)
**James J. Rogers,** Massachusetts St. Board of Electrical Examiners, MA  [E]
   Rep. International Association of Electrical Inspectors
   (Alt. to A. R. Cartal)
**George S. Tidden,** IEC, TX  [IM]
   Rep. Independent Electrical Contractors, Inc.
   (Alt. to R. A. Jones)
**Robert C. Turner,** Inductotherm Corp., NJ  [M]
   (Alt. to J. H. Mortimer)
   (VL 665)
**James E. Winfrey,** Square D Co., NC  [M]
   Rep. National Electrical Manufactuers Association
   (Alt. to J. F. Cook)

## CODE-MAKING PANEL NO. 13

**Articles 450, 455, 460, 470, 490**

**William T. O'Grady,** *Chair* [RT]
Underwriters Laboratories Inc., NY

**Tarry L. Baker,** Broward County Board of Rules and Appeals, FL  [E]
   Rep. International Association of Electrical Inspectors
**William A. Brunner,** IBEW, ND  [L]
   Rep. International Brotherhood of Electrical Workers
**James C. Carroll,** Square D Co., TN  [M]
   Rep. National Electrical Manufacturers Association
**William B. Crist,** Houston Stafford Electric Company, TX  [IM]
   Rep. Independent Electrical Contractors, Inc.

**O. L. Davis,** Manzano Western, Inc., NM  [IM]
   Rep. National Electrical Contractors Association
**Richard P. Fogarty, Jr.,** Consolidated Edison Co. of N.Y., Inc., NY  [UT]
   Rep. Edison Electric Institute
**Milton D. Robinson,** Milt Robinson Engineering Co., IN  [U]
   Rep. Institute of Electrical & Electronics Engineers, Inc.
**LaVerne E. Stetson,** U.S. Dept. of Agriculture, NE  [U]
   Rep. American Society of Agricultural Engineers

**Lou G. Willoughby,** ALCOA Inc., OH [M]
Rep. The Aluminum Association
**Ralph H. Young,** Eastman Chemical Co., TN [U]
Rep. American Chemistry Council

### Alternates

**Douglas Elkins,** Exxon Chemical Co., TX [U]
Rep. American Chemistry Council
(Alt. to R. H. Young)
**Timothy D. Holleman,** AC Corp., NC [IM]
Rep. Independent Electrical Contractors, Inc.
(Alt. to W. B. Crist)
**Barry N. Hornberger,** PECO Energy Co., PA [UT]
Rep. Edison Electric Institute
(Alt. to R. P. Fogarty, Jr.)
**Richard Lofton,** IBEW Local 280, OR [L]
Rep. International Brotherhood of Electrical Workers
(Alt. to W. A. Brunner)
**Arthur Mastromarino,** Underwriters Laboratories Inc., NY [RT]
(Alt. to W. T. O'Grady)

**Gene Morehart,** Acme Electric Corp., NC [M]
Rep. National Electrical Manufacturers Association
(Alt. to J. C. Carroll)
**Robert L. Simpson,** Simpson Electrical Engineering Co., GA [U]
Rep. Institute of Electrical & Electronics Engineers, Inc.
(Alt. to M. D. Robinson)
**Monte Szendre,** Wilson construction Co., OR [IM]
Rep. National Electrical Contractors Association
(Alt. to O. L. Davis)
**Gerald W. Williams,** County of Ventura, CA [E]
Rep. International Association of Electrical Inspectors
(Alt. to T. L. Baker)
**Ivan L. Winsett,** Ronk Electrical Industries, GA [U]
Rep. American Society of Agricultural Engineers
(Alt. to L. E. Stetson)

## CODE-MAKING PANEL NO. 14
### Articles 500, 501, 502, 503, 504, 505, 510, 511, 513, 514, 515 and 516

**Donald R. Cook,** *Chair*
Shelby County Bldg. Inspections, AL [E]
Rep. International Association of Electrical Inspectors

**Robert B. Alexander,** Fluor Daniel, Inc., CA [IM]
Rep. Associated Builders and Contractors, Inc.
**Edward M. Briesch,** Underwriters Laboratories Inc., IL [RT]
**Al Engler,** EGS Electrical Group, IL [M]
Rep. International Society for Measurement and Control
**Mark Goodman,** BP (ARCO), CA [U]
Rep. American Petroleum Institute
**Doug Jagunich,** Intertek Testing Services, N.A., Inc., MN [RT]
**Joseph H. Kuczka,** Killark Electric Mfg. Co., MO [M]
Rep. National Electrical Manufacturers Association
**William G. Lawrence, Jr.,** FM Global, MA [I]
**Mike O'Meara,** Arizona Public Service Co., AZ [UT]
Rep. Edison Electric Institute
**Mark G. Saban,** Saban Electric, IL [IM]
Rep. National Electrical Contractors Association
**David Wechsler,** Union Carbide Corp., WV [U]
Rep. American Chemistry Council
**James A. Weldon,** IBEW Local 728, FL [L]
Rep. International Brotherhood of Electrical Workers
**Donald W. Zipse,** Zipse Electrical Engineering Inc., PA [U]
Rep. Institute of Electrical & Electronics Engineers, Inc.

### Alternates

**Alonza W. Ballard,** Crouse-Hinds, NY [M]
Rep. National Electrical Manufacturers Association
(Alt. to J. H. Kuczka)
**James D. Cospolich,** Waldemar S. Nelson & Co. Inc., LA [U]
Rep. Institute of Electrical & Electronics Engineers, Inc.
(Alt. to D. W. Zipse)
**William T. Fiske,** Intertek Testing Services N.A. Inc., NY [RT]
(Alt. to D. Jagunich)
**Larry E. Fuhrman,** City of Titusville, FL [E]
Rep. International Association of Electrical Inspectors
(Alt. to D. R. Cook)

**Paul T. Kelly,** Underwriters Laboratories Inc., IL [RT]
(Alt. to E. M. Briesch)
**Harold C. Kronz,** IBEW Local 308, FL [L]
Rep. International Brotherhood of Electrical Workers
(Alt. to J. A. Weldon)
**Michael E. McNeil,** FMC Corp/Bio Polymer, ME [U]
Rep. American Chemistry Council
(Alt. to D. Wechsler)
**Thomas F. Mueller,** Southern Company Generation, AL [UT]
Rep. Edison Electric Institute
(Alt. to M. O'Meara)
**Peter T. Schimmoeller,** FM Global, MA [I]
(Alt. to W. G. Lawrence, Jr.)
**Ted H. Schnaare,** Rosemount Inc., MN [M]
Rep. International Society for Measurement and Control
(Alt. to A. Engler)
**Francis M. Stone, Jr.,** Shell Exploration and Production Co., TX, [U]
Rep. American Petroleum Institute
(Alt. to M. Goodman)
**Mark C. Wirfs,** R & W Engineering Inc., OR [U]
Rep. Grain Elevator & Processing Society
(Voting Alt. to GEAPS Rep.)

### Nonvoting

**Eduardo N. Solano,** Estudio Ingeniero Solano S. A., Argentia [SE]
**Fred K. Walker,** U.S. Air Force, FL [U]
Rep. TC on Airport Facilities

## CODE-MAKING PANEL 15

### Articles 445, 480, 518, 520, 525, 530, 540, 647, 695, 700, 701, 702, 705

**Robert C. Duncan,** *Chair* [E]
Reedy Creek Improvement District, FL
Rep. International Association of Electrical Inspectors

**Peter W. Amos,** Consolidated Edison Co. of N.Y., Inc., NY [UT]
Rep. Edison Electric Institute
**James L. Boyer,** Firetrol, Inc., NC [M]
Rep. National Electrical Manufacturers Association
**Brian Burrows,** IBEW Local 98, PA [L]
Rep. International Brotherhood of Electrical Workers
**Tom Dunn,** Butler Amusements, CA [U]
Rep. Outdoor Amusement Business Association, Inc.
(VL 525)
**George W. Flach,** Flach Consultants, LA [SE]
**Michael V. Glenn,** Longview Fibre Co., WA [U]
Rep. Institute of Electrical & Electronics Engineers, Inc.
**Marcelo M. Hirschler,** GBH Int'l., CA [SE]
**George Thomas Howard,** George Thomas Howard &
Associates, NV [SE]
**Gordon S. Johnson,** Dundee, FL [M]
Rep. Electrical Generating Systems Association
**Robert J. Kakalec,** Telcordia Technologies, NJ [UT]
Rep. Alliance for Telecommunications Industry Solutions
(VL 445, 480, 700, 701, 702, and 705)
**Jack W. Kalbfeld,** Kalico Technology Inc., NY [SE]
(VL 518, 520, 525, 530, and 540)
**Michael B. Klein,** Consentini Associates DC, LLP, MD [IM]
Rep. Illuminating Engineering Society of North America
**John R. Kovacik,** Underwriters Laboratories Inc., IL [RT]
**Edwin S. Kramer,** Radio City Music Hall, NY [L]
Rep. Int'l. Association of Theatrical Stage Employees
**Michael A. Lanni,** Universal Studios, CA [U]
Rep. Motion Picture Association of America, Inc.
**Dennis W. Marshall,** TAG Electric Co., TX [IM]
Rep. Independent Electrical Contractors
**Steven H. Pasternack,** Intertek Testing Services, N.A. Inc., NY [RT]
**Richard Sobel,** Quantum Electric Corp., NY [IM]
Rep. National Electrical Contractors Association

**Dale A. Triffo,** Equilon Enterprises LLC, TX [U]
Rep. American Chemistry Council
**Kenneth E. Vannice,** NSI Corp., OR [M]
Rep. US Institute for Theatre Technology

#### Alternates

**Mike Grunwald,** IBEW Local 76, WA [L]
Rep. International Brotherhood of Electrical Workers
(Alt. to B. Burrows)
**Mitchell K. Hefter,** Rosco Entertainment Technology, OR [IM]
Rep. Illuminating Engineering Society of North America
(Alt. to M. B. Klein)
**Natalie J. McCord,** AEP, Public Service Co. of Oklahoma, OK [UT]
Rep. Edison Electric Institute
(Alt. to P. W. Amos)
Rep. National Electrical Manufacturers Association
(Alt. to J. L. Boyer)
**Michael D. Skinner,** CBS Studio Center, CA [U]
Rep. Motion Picture Association of America, Inc.
(Alt. to M. A. Lanni)
**Steven R. Terry,** Production Resource Group LLC, Fourth Phase, NJ [U]
Rep. US Institute for Theatre Technology
(Alt. to K. E. Vannice)
**Herbert V. Whittall,** Electrical Generating Systems Assn., FL [M]
Rep. Electrical Generating Systems Association
(Alt. to G. S. Johnson)
**Harold F. Willman,** City of Lakewood, CO [E]
Rep. International Association of Electrical Inspectors
(Alt. to R. C. Duncan)

## CODE-MAKING PANEL NO. 16

### Articles 640, 650, 720, 725, 727, 760, 770, 800, 810, 820, 830, and Chapter 9 Tables 11(a) and (b) and 12(a) and (b)

**Stanley D. Kahn,** *Chair* [IM]
Tri-City Electric Co., Inc., CA
Rep. National Electrical Contractors Association

**James E. Brunssen,** Telcordia Technologies, Inc., NJ [UT]
Rep. Alliance for Telecommunications Industry Solutions
**Orren E. Cameron, III,** United States Dept. of Agriculture, DC [UT]
**Loren M. Caudill,** The DuPont Company, DE [M]
Rep. Society of the Plastics Industry Inc.
**Gerald Lee Dorna,** Belden Wire & Cable, IN [M]
Rep. Insulated Cable Engineers Association Inc.
**Roland W. Gubisch,** Intertek Testing Services, N.A. Inc., MA [RT]
**Lee C. Hewitt,** Underwriters Laboratories Inc., IL [RT]
**William K. Hopple,** Simplex Time Recorder Co., CA [M]
Rep. National Electrical Manufacturers Association

**Robert L. Hughes,** DuPont, TN [U]
Rep. American Chemistry Council
**Steven C. Johnson,** Time Warner Cable, CO [UT]
Rep. National Cable Television Association
**Ronald G. Jones,** Ronald G. Jones, TX [U]
Rep. Institute of Electrical & Electronics Engineers, Inc.
**Stanley Kaufman,** Lucent Technologies, GA [M]
**Michael A. Lanni,** Universal Studios, CA [U]
Rep. Motion Picture Association of America, Inc.
**John Mangan,** Medford City Hall, MA [E]
Rep. International Association of Electrical Inspectors
**J. Jeffrey Moore,** Industrial Risk Insurers, OH [I]
Rep. Industrial Risk Insurers

**James W. Romlein,** MIS Labs, WI [M]
Rep. Building Industry Consulting Service International
**Arthur E. Schlueter, Jr.,** A. E. Schlueter Pipe Organ Co., GA [M]
(VL 640, 650, 720, and 725)
**Steven M. Speer,** IBEW Local 640, AZ [L]
Rep. International Brotherhood of Electrical Workers
**Kyle E. Todd,** Entergy – GSU, TX [UT]
Rep. Edison Electric Institute
**Inder L. Wadehra,** IBM Corp., NC [U]
**Melvin J. Wierenga,** Wierenga & Associates, MI [IM]

### Alternates

**Ronald P. Cantrell,** IBEW Local 72, TX [L]
Rep. International Brotherhood of Electrical Workers
(Alt. to S. M. Speer)
**Larry Chan,** City of New Orleans, LA [E]
Rep. International Association of Electrical Inspectors
(Alt. to J. Mangan)
**Chrysanthos Chrysanthou,** Telcordia Technologies, Inc., NJ [UT]
Rep. Alliance for Telecommunications Industry Solutions
(Alt. to J. E. Brunssen)
**A. William Coaker,** A. W. Coaker and Associates, Inc., OH [M]
Rep. Society of the Plastics Industry Inc.
(Alt. to L. M. Caudill)
**Gilbert J. Diaz,** Intertek Testing Services, N.A. Inc., TX [RT]
(Alt. to R. W. Gubisch)
**Richard S. Houghten,** MI [M]
(Alt. to A. E. Schlueter, Jr.)
(VL 640, 650, 720, and 725)

**Robert Jensen,** dbi - Telecommunication Infrastructure Design, TX [M]
Rep. Building Industry Consulting Service International
(Alt. to J. W. Romlein)
**William J. McCoy,** Verizon Wireless, TX [U]
Rep. Institute of Electrical & Electronics Engineers, Inc.
(Alt. to R. G. Jones)
**W. D. Pirkle,** Pirkle Electric Co., Inc., GA [IM]
Rep. National Electrical Contractors Association
(Alt. to S. D. Kahn)
**Bradley C. Rowe,** Underwriters Laboratories Inc., IL [RT]
(Alt. to L. C. Hewitt)
**Oleh Sniezko,** AT&T Broadband, CO [UT]
Rep. National Cable Television Association
(Alt. to S.C. Johnson)
**Sondra K. Todd,** Western Resources Co., KS [UT]
Rep. Edison Electric Institute
(Alt. to K. E. Todd)
**Lawrence J. Wenzel,** Industrial Risk Insurers, CT [I]
(Alt. to J. J. Moore)
**Kevin D. Wilhelm,** Eli Lilly and Co., IN [U]
Rep. American Chemistry Council
(Alt. to R. L. Hughes)
**Joe Rao,** RAO Electric Co., FL [IM]
Rep. Independent Electrical Contractors, Inc.
(Voting Alt. to IEC Rep.)

### Nonvoting

**Irving Mande,** Edwards Systems Technology, CT

## CODE-MAKING PANEL NO. 17

### Article 517

**Robert E. Bernd,** *Chair* [RT]
Underwriters Laboratories Inc., IL

**Steve Campolo,** Leviton Manufacturing Co., Inc., NY [M]
Rep. National Electrical Manufacturers Association
**Thomas C. Clark,** Clark Electrical Construction, Inc., TX [IM]
Rep. Associated Builders and Contractors, Inc.
**James R. Duncan,** Sparling, WA [U]
Rep. Institute of Electrical & Electronics Engineers, Inc.
**Douglas S. Erickson,** American Society for Healthcare Engineering, VI [U]
Rep. American Society for Healthcare Engineering
**James W. Hillebrand,** Byron Electric Co., KY [IM]
Rep. National Electrical Contractors Association
**James A. Meyer,** Pettis Memorial VA Hospital, CA [C]
Rep. American Society of Anesthesiologists
**Hugh O. Nash, Jr.,** Nash Lipsey Burch, LLC, TN [SE]
**Donald J. Sheratt,** Intertek Testing Services, N.A. Inc., MA [RT]
**Richard H. Smith,** OG&E Electric Services, OK [UT]
Rep. Edison Electric Institute
**Jeffrey L. Steplowski,** U.S. Dept. of Veterans Affairs, DC [U]
**Mike Velvikis,** High Voltage Maintenance Corp., WI [IM]
Rep. International Electrical Testing Association Inc.
**Walter N. Vernon, IV,** Mazzetti & Associates Inc., CA [U]
Rep. NFPA Health Care Section
**Andrew White,** WFJEATC Local Union 3 IBEW, NY [L]
Rep. International Brotherhood of Electrical Workers

### Alternates

**Matthew B. Dozier,** Smith Seckman Reid, Inc., TN [U]
Rep. Institute of Electrical & Electronics Engineers, Inc.
(Alt. to J. R. Duncan)
**Banks Hattaway,** Hattaway Brothers Inc., AL [IM]
Rep. Associated Builders and Contractors, Inc.
(Alt. to T. C. Clark)
**Stephen D. Hewson,** Underwriters Laboratories Inc., IL [RT]
(Alt. to R. E. Bernd)
**Stanley D. Kahn,** Tri-City Electric Co., Inc., CA [IM]
Rep. National Electrical Contractors Association
(Alt. to J. W. Hillebrand)
**Paul L. LeVasseur,** Bay City JEATC, MI [L]
Rep. International Brotherhood of Electrical Workers
(Alt. to A. White)
**David K. Norton,** U.S. Dept. of Veterans Affairs, DC [U]
(Alt. to J. L. Steplowski)
**Gaylen D. Rogers,** DFCM, UT [E]
Rep. International Association of Electrical Inspectors
(Voting Alt. to IAEI Rep.)

## CODE-MAKING PANEL NO. 18
### Articles 406, 410, 411, 600, 605
**Wayne Brinkmeyer,** *Chair* [IM]
Biddle Electric Corp., TX
Rep. National Electrical Contractors Association

**Michael N. Ber,** IEC, Houston, TX [IM]
Rep. Independent Electrical Contractors, Inc.
**Robert L. Cochran,** City of Costa Mesa, CA [E]
Rep. International Association of Electrical Inspectors
**Rudy T. Elam,** Systems Engineering Services, TN [U]
Rep. Institute of Electrical & Electronics Engineers, Inc.
**Kenneth F. Kempel,** Underwriters Laboratories Inc., NC [RT]
**Thomas J. Lynch,** IBEW Local 99, RI [L]
Rep. International Brotherhood of Electrical Workers
**Bernard J. Mezger,** American Lighting Assn., NY [M]
Rep. American Lighting Association
(VL 410 and 411)
**James F. Pierce,** Intertek Testing Services NA Inc., OR [RT]
**Saul Rosenbaum,** Leviton Mfg. Co. Inc., NY [M]
Rep. National Electrical Manufacturers Association
**Carl T. Wall,** Alabama Power Co., AL [UT]
Rep. Edison Electric Institute
**Jack Wells,** Pass & Seymour/Legrand, NY [M]

### Alternates

**Mark R. Berner,** PP&L, Inc., PA [UT]
Rep. Edison Electric Institute
(Alt. to C. T. Wall)
**Robert T. Carlock,** R. T. Carlock Co., TN [IM]
Rep. Independent Electrical Contractors
(Alt. to M. N. Ber)

**Howard D. Hughes,** Hughes Electric Co. Inc., AR [IM]
Rep. National Electrical Contractors Association
(Alt. to W. Brinkmeyer)
**Stephen G. Kieffer,** Kieffer & Co., Inc., WI [M]
Rep. International Sign Association
(Voting Alt. to ISA Rep.)
(VL 600)
**Steven A. Larson,** BWXT Y-12, TN [U]
Rep. Institute of Electrical & Electronics Engineers, Inc.
(Alt. to R. T. Elam)
**John J. Mahal,** Underwriters Laboratories Inc., IL [RT]
(Alt. to K. F. Kempel)
**Don Miletich,** Cooper Lighting, IL [M]
Rep. National Electrical Manufacturers Association
(Alt. to S. Rosenbaum)
**Michael S. O'Boyle,** Lightolier, Div. of Genlyte Thomas Group, MA [M]
Rep. American Lighting Association
(Alt. to B. J. Mezger)
(VL 410 and 411)
**Charles M. Trout,** Maron Electric Co., FL [IM]
Rep. National Electrical Contractors Association
(Alt. to W. Brinkmeyer)

## CODE-MAKING PANEL NO. 19
### Articles 545, 547, 550, 551, 552, 553, 555, 604, 675
**Robert A. McCullough,** *Chair* [E]
Ocean County Construction Inspection Dept., NJ
Rep. International Association of Electrical Inspectors

**Barry Bauman,** Alliant Energy, WI [U]
Rep. American Society of Agricultural Engineers
**James W. Finch,** Kampgrounds of America, Inc., MT [U]
(VL 550, 551, 552, and 555)
**Bruce A. Hopkins,** Recreation Vehicle Industry Assn., VA [M]
Rep. Recreation Vehicle Industry Association
(VL 550, 551, and 552)
**Steven Johnson,** IBEW, CA [L]
Rep. International Brotherhood of Electrical Workers
**Robert L. LaRocca,** Underwriters Laboratories Inc., NY [RT]
**Timothy P. McNeive,** Thomas & Betts Corp., TN [M]
Rep. National Electrical Manufacturers Association
**Leslie Sabin,** San Diego Gas & Electric Co., CA, UT
Rep. Edison Electric Institute
**Charles F. Shy,** AC Corp./Electrical Division, NC [IM]
Rep. Independent Electrical Contractors
**Dick Veenstra,** Fleetwood Enterprises, Inc., CA [M]
Rep. Manufactured Housing Institute
(VL 550, 551, and 552)
**Kenneth Weakley,** Mountain Electric, Inc., CA, [IM]
Rep. National Electrical Contractors Association
**Michael L. Zieman,** RADCO, CA [RT]
(VL 545, 550, 551, and 552)

### Alternates

**Glenn H. Ankenbrand,** Conectiv Power, MD [UT]
Rep. Edison Electric Institute
(Alt. to L. Sabin)
**Steven Blais,** EGS - Electrical Group, IL [M]
Rep. National Electrical Manufacturers Association
(Alt. to T. P. McNieve)
**James K. Hinrichs,** St. of Washington, WA [E]
Rep. International Association of Electrical Inspectors
(Alt. to R. A. McCullough)
**John Mikel,** Skyline Corporation, IN [M]
Rep. Manufactured Housing Institute
(Alt. to D. Veenstra)
(VL 550, 551, and 552)
**John Pabian,** Underwriters Laboratories Inc., IL [RT]
(Alt. to R. L. Larocca)
**Kent Perkins,** Recreation Vehicle Industry Assn., VA [M]
Rep. Recreation Vehicle Industry Association
(Alt. to B. A. Hopkins)
(VL 550, 551, and 552)
**Homer Staves,** Kampgrounds of America, Inc., MT [U]
(Alt. to J. W. Finch)
(VL 550, 551, 552, and 555)

**LaVerne E. Stetson,** U.S. Dept. of Agriculture, NE [U]
  Rep. American Society of Agricultural Engineers
  (Alt. to B. Bauman)
**David N. Tilmont,** IBEW Local 952, CA [L]
  Rep. International Brotherhood of Electrical Workers
  (Alt. to S. Johnson)
**Raymond F. Tucker,** RADCO, CA [RT]

(Alt. to M. L. Zieman)
(VL 545, 550, 551 and 552)
**David Gorin,** Nat'l. Assn. of RV Parks & Campgrounds, VA [U]
  Rep. Nat'l. Assn. of RV Parks & Campgrounds
  (Voting Alt. to ARVC rep.)
  (VL 550, 551, and 552)

## CODE-MAKING PANEL NO. 20

### Articles 422, 424, 680

**Robert M. Milatovich,** *Chair* [E]
Clark County, NV
Rep. International Association of Electrical Inspectors

**Scott Cline,** McMurtrey Electric, Inc., CA [M]
  Rep. National Electrical Contractors Association
**Robert J. Egan,** Int'l. Brotherhood of Electrical Workers, MO [L]
  Rep. International Brotherhood of Electrical Workers
**Christopher Gill,** New York Board of Fire Underwriters, NY [E]
**Walter Koessel,** Intertek Testing Services, N.A. Inc., MO [RT]
**Neil F. LaBrake, Jr.,** Niagara Mohawk Power Corp., NY [UT]
  Rep. Edison Electric Institute
**James N. Pearse,** Leviton Mfrg. Co. Inc., NY [M]
  Rep. National Electrical Manufacturers Association
**Marcos Ramirez,** Mr. Electric Service Co., Inc., NY [IM]
  Rep. Independent Electrical Contractors, Inc.
**Anthony Sardina,** Carrier Corp., NY [M]
  Rep. Air Conditioning and Refrigeration Institute
  (VL 422, 424, and 426)
**Donald J. Talka,** UL Intrnational Germany GmbH [RT]
**John T. Weizeorick,** WI [M]
  Rep. Association of Home Appliance Manufacturers
  (VL 422, 424, and 426)
**Lee L. West,** Balboa Instruments, CA [M]
  Rep. National Spa and Pool Institute
  (VL 680)
**Robert M. Yurkanin,** Electran Process Int'l. Inc., NJ [U]
  Rep. Institute of Electrical & Electronics Engineers, Inc.

Rep. International Association of Electrical Inspectors
  (Alt. to R. M. Milatovich)
**Dennis L. Baker,** Springs & Sons Electrical Cont. Inc., AZ [IM]
  Rep. Independent Electrical Contractors
  (Alt. to M. Ramirez)
**Bruce R. Hirsch,** Baltimore Gas & Electric Co., MD [UT]
  Rep. Edison Electric Institute
  (Alt. to N. F. LaBrake, Jr.)
**Mark Leimbeck,** Underwriters Laboratories, Inc., IL [RT]
  (Alt. to D. J. Talka)
**Tom McDonald,** Hubbell, Inc., CT [M]
  Rep. National Electrical Manufacturers Association
  (Alt. to J. N. Pearse)
**Stephen P. Schoemehl,** Int'l. Brotherhood of Electrical Workers, MO [L]
  Rep. International Brotherhood of Electrical Workers
  (Alt. to R. J. Egan)
**Robert E. Wisenburg,** Coates Heater Co., Inc., WA [M]
  Rep. National Spa and Pool Institute
  (Alt. to L. L. West)
  (VL 680)

### Nonvoting

**Rick C. Gilmour,** Canadian Standards Assn., ON
**William H. King, Jr.,** U.S. Consumer Product Safety Commission, MD
**Andrew M. Trotta,** U.S. Consumer Product Safety Commission, MD
  (Alt. to W. H. King, Jr.)

### Alternates

**Xen George Anchales,** San Bernardino County/Dept. of Bldg. & Safety, CA [E]

## NFPA Electrical Engineering Division Technical Staff

**Mark W. Earley,** Assistant Vice President/Chief Electrical Engineer
**John M. Caloggero,** Principal Electrical Specialist
**Kenneth G. Mastrullo,** Senior Electrical Specialist
**Jean A. O'Connor,** Electrical Project Specialist/Support Supervisor

**Lee F. Richardson,** Senior Electrical Engineer
**Jeffrey S. Sargent,** Senior Electrical Specialist
**Joseph V. Sheehan,** Principal Electrical Engineer

## NFPA Staff Editor

**Joyce G. Grandy,** Senior Project Editor

NOTE: Membership on a committee shall not in and of itself constitute an endorsement of the Association or any document developed by the committee on which the member serves.

**Committee Scope**: This Committee shall have primary responsibility for documents on minimizing the risk of electricity as a source of electric shock and as a potential ignition source of fires and explosions. It shall also be responsible for text to minimize the propagation of fire and explosions due to electrical installations.

## NFPA 70

## *National Electrical Code*®

### 2002 Edition

### ARTICLE 80
### Administration and Enforcement

*This article is informative unless specifically adopted by the local jurisdiction adopting the National Electrical Code*®. *(See 80.5.)*

**80.1 Scope.** The following functions are covered:

(1) The inspection of electrical installations as covered by 90.2
(2) The investigation of fires caused by electrical installations
(3) The review of construction plans, drawings, and specifications for electrical systems
(4) The design, alteration, modification, construction, maintenance, and testing of electrical systems and equipment
(5) The regulation and control of electrical installations at special events including but not limited to exhibits, trade shows, amusement parks, and other similar special occupancies

**80.2 Definitions.**

**Authority Having Jurisdiction.** The organization, office, or individual responsible for approving equipment, materials, an installation, or a procedure.

**Chief Electrical Inspector.** An electrical inspector who either is the authority having jurisdiction or is designated by the authority having jurisdiction and is responsible for administering the requirements of this *Code.*

**Electrical Inspector.** An individual meeting the requirements of 80.27 and authorized to perform electrical inspections.

**80.3 Purpose.** The purpose of this article shall be to provide requirements for administration and enforcement of the *National Electrical Code.*

**80.5 Adoption.** Article 80 shall not apply unless specifically adopted by the local jurisdiction adopting the *National Electrical Code.*

**80.7 Title.** The title of this *Code* shall be NFPA 70, *National Electrical Code*®, of the National Fire Protection Association. The short title of this *Code* shall be the *NEC*®.

**80.9 Application.**

**(A) New Installations.** This *Code* applies to new installations. Buildings with construction permits dated after adoption of this *Code* shall comply with its requirements.

**(B) Existing Installations.** Existing electrical installations that do not comply with the provisions of this *Code* shall be permitted to be continued in use unless the authority having jurisdiction determines that the lack of conformity with this *Code* presents an imminent danger to occupants. Where changes are required for correction of hazards, a reasonable amount of time shall be given for compliance, depending on the degree of the hazard.

**(C) Additions, Alterations, or Repairs.** Additions, alterations, or repairs to any building, structure, or premises shall conform to that required of a new building without requiring the existing building to comply with all the requirements of this *Code.* Additions, alterations, installations, or repairs shall not cause an existing building to become unsafe or to adversely affect the performance of the building as determined by the authority having jurisdiction. Electrical wiring added to an existing service, feeder, or branch circuit shall not result in an installation that violates the provisions of the *Code* in force at the time the additions are made.

**80.11 Occupancy of Building or Structure.**

**(A) New Construction.** No newly constructed building shall be occupied in whole or in part in violation of the provisions of this *Code.*

**(B) Existing Buildings.** Existing buildings that are occupied at the time of adoption of this *Code* shall be permitted to remain in use provided the following conditions apply:

(1) The occupancy classification remains unchanged
(2) There exists no condition deemed hazardous to life or property that would constitute an imminent danger

**80.13 Authority.** Where used in this article, the term *authority having jurisdiction* shall include the chief electrical inspector or other individuals designated by the governing body. This *Code* shall be administered and enforced by the authority having jurisdiction designated by the governing authority as follows.

(1) The authority having jurisdiction shall be permitted to render interpretations of this *Code* in order to provide clarification to its requirements, as permitted by 90.4.
(2) When the use of any electrical equipment or its installations is found to be dangerous to human life or property, the authority having jurisdiction shall be empowered to have the premises disconnected from its source of electric supply, as established by the Board.

When such equipment or installation has been so condemned or disconnected, a notice shall be placed thereon listing the causes for the condemnation, the disconnection, or both and the penalty under 80.23 for the unlawful use thereof. Written notice of such condemnation or disconnection and the causes therefor shall be given within 24 hours to the owners, the occupant, or both, of such building, structure, or premises. It shall be unlawful for any person to remove said notice, to reconnect the electric equipment to its source of electric supply, or to use or permit to be used electric power in any such electric equipment until such causes for the condemnation or disconnection have been remedied to the satisfaction of the inspection authorities.

(3) The authority having jurisdiction shall be permitted to delegate to other qualified individuals such powers as necessary for the proper administration and enforcement of this *Code*.

(4) Police, fire, and other enforcement agencies shall have authority to render necessary assistance in the enforcement of this *Code* when requested to do so by the authority having jurisdiction.

(5) The authority having jurisdiction shall be authorized to inspect, at all reasonable times, any building or premises for dangerous or hazardous conditions or equipment as set forth in this *Code*. The authority having jurisdiction shall be permitted to order any person(s) to remove or remedy such dangerous or hazardous condition or equipment. Any person(s) failing to comply with such order shall be in violation of this *Code*.

(6) Where the authority having jurisdiction deems that conditions hazardous to life and property exist, he or she shall be permitted to require that such hazardous conditions in violation of this *Code* be corrected.

(7) To the full extent permitted by law, any authority having jurisdiction engaged in inspection work shall be authorized at all reasonable times to enter and examine any building, structure, or premises for the purpose of making electrical inspections. Before entering a premises, the authority having jurisdiction shall obtain the consent of the occupant thereof or obtain a court warrant authorizing entry for the purpose of inspection except in those instances where an emergency exists. As used in this section, *emergency* means circumstances that the authority having jurisdiction knows, or has reason to believe, exist and that reasonably can constitute immediate danger to persons or property.

(8) Persons authorized to enter and inspect buildings, structures, and premises as herein set forth shall be identified by proper credentials issued by this governing authority.

(9) Persons shall not interfere with an authority having jurisdiction carrying out any duties or functions prescribed by this *Code*.

(10) Persons shall not use a badge, uniform, or other credentials to impersonate the authority having jurisdiction.

(11) The authority having jurisdiction shall be permitted to investigate the cause, origin, and circumstances of any fire, explosion, or other hazardous condition.

(12) The authority having jurisdiction shall be permitted to require plans and specifications to ensure compliance with this *Code*.

(13) Whenever any installation subject to inspection prior to use is covered or concealed without having first been inspected, the authority having jurisdiction shall be permitted to require that such work be exposed for inspection. The authority having jurisdiction shall be notified when the installation is ready for inspection and shall conduct the inspection within ___ days.

(14) The authority having jurisdiction shall be permitted to order the immediate evacuation of any occupied building deemed unsafe when such building has hazardous conditions that present imminent danger to building occupants.

(15) The authority having jurisdiction shall be permitted to waive specific requirements in this *Code* or permit alternative methods where it is assured that equivalent objectives can be achieved by establishing and maintaining effective safety. Technical documentation shall be submitted to the authority having jurisdiction to demonstrate equivalency and that the system, method, or device is approved for the intended purpose.

(16) Each application for a waiver of a specific electrical requirement shall be filed with the authority having jurisdiction and shall be accompanied by such evidence, letters, statements, results of tests, or other supporting information as required to justify the request. The authority having jurisdiction shall keep a record of actions on such applications, and a signed copy of the authority having jurisdiction's decision shall be provided for the applicant.

**80.15 Electrical Board.**

**(A) Creation of the Electrical Board.** There is hereby created the Electrical Board of the _____ of _____, hereinafter designated as the Board.

**(B) Appointments.** Board members shall be appointed by the Governor with the advice and consent of the Senate (or by the Mayor with the advice and consent of the Council, or the equivalent).

(1) Members of the Board shall be chosen in a manner to reflect a balanced representation of individuals or orga-

nizations. The Chair of the Board shall be elected by the Board membership.

(2) The Chief Electrical Inspector in the jurisdiction adopting this Article authorized in 80.15(B)(3)(a) shall be the nonvoting secretary of the Board. Where the Chief Electrical Inspector of a local municipality serves a Board at a state level, he or she shall be permitted to serve as a voting member of the Board.

(3) The board shall consist of not fewer than five voting members. Board members shall be selected from the following:

    a. Chief Electrical Inspector from a local government (for State Board only)

    b. An electrical contractor operating in the jurisdiction

    c. A licensed professional engineer engaged primarily in the design or maintenance of electrical installations

    d. A journeyman electrician

(4) Additional membership shall be selected from the following:

    a. A master (supervising) electrician

    b. The Fire Marshal (or Fire Chief)

    c. A representative of the property/casualty insurance industry

    d. A representative of an electric power utility operating in the jurisdiction

    e. A representative of electrical manufacturers primarily and actively engaged in producing materials, fittings, devices, appliances, luminaires (fixtures), or apparatus used as part of or in connection with electrical installations

    f. A member of the labor organization that represents the primary electrical workforce

    g. A member from the public who is not affiliated with any other designated group

    h. A representative of a telecommunications utility operating in the jurisdiction

**(C) Terms.** Of the members first appointed, _____ shall be appointed for a term of 1 year, _____ for a term of 2 years, _____ for a term of 3 years, and _____ for a term of 4 years, and thereafter each appointment shall be for a term of 4 years or until a successor is appointed. The Chair of the Board shall be appointed for a term not to exceed _____ years.

**(D) Compensation.** Each appointed member shall receive the sum of _____ dollars ($_____) for each day during which the member attends a meeting of the Board and, in addition thereto, shall be reimbursed for direct lodging, travel, and meal expenses as covered by policies and procedures established by the jurisdiction.

**(E) Quorum.** A quorum as established by the Board operating procedures shall be required to conduct Board business. The Board shall hold such meetings as necessary to carry out the purposes of Article 80. The Chair or a majority of the members of the Board shall have the authority to call meetings of the Board.

**(F) Duties.** It shall be the duty of the Board to:

(1) Adopt the necessary rules and regulations to administer and enforce Article 80.

(2) Establish qualifications of electrical inspectors.

(3) Revoke or suspend the recognition of any inspector's certificate for the jurisdiction.

(4) After advance notice of the public hearings and the execution of such hearings, as established by law, the Board is authorized to establish and update the provisions for the safety of electrical installations to conform with the current edition of the *National Electrical Code* (NFPA 70) and other nationally recognized safety standards for electrical installations.

(5) Establish procedures for recognition of electrical safety standards and acceptance of equipment conforming to these standards.

**(G) Appeals.**

(1) Review of Decisions. Any person, firm, or corporation may register an appeal with the Board for a review of any decision of the Chief Electrical Inspector or of any Electrical Inspector, provided that such appeal is made in writing within fifteen (15) days after such person, firm, or corporation shall have been notified. Upon receipt of such appeal, said Board shall, if requested by the person making the appeal, hold a public hearing and proceed to determine whether the action of the Board, or of the Chief Electrical Inspector, or of the Electrical Inspector complies with this law and, within fifteen (15) days after receipt of the appeal or after holding the hearing, shall make a decision in accordance with its findings.

(2) Conditions. Any person shall be permitted to appeal a decision of the authority having jurisdiction to the Board when it is claimed that any one or more of the following conditions exist:

    a. The true intent of the codes or ordinances described in this *Code* has been incorrectly interpreted.

    b. The provisions of the codes or ordinances do not fully apply.

    c. A decision is unreasonable or arbitrary as it applies to alternatives or new materials.

(3) Submission of Appeals. A written appeal, outlining the *Code* provision from which relief is sought and the remedy proposed, shall be submitted to the authority having jurisdiction within 15 calendar days of notification of violation.

**(H) Meetings and Records.** Meetings and records of the Board shall conform to the following:

(1) Meetings of the Board shall be open to the public as required by law.
(2) Records of meetings of the Board shall be available for review during normal business hours, as required by law.

**80.17 Records and Reports.** The authority having jurisdiction shall retain records in accordance with 80.17(A) and (B).

**(A) Retention.** The authority having jurisdiction shall keep a record of all electrical inspections, including the date of such inspections and a summary of any violations found to exist, the date of the services of notices, and a record of the final disposition of all violations. All required records shall be maintained until their usefulness has been served or as otherwise required by law.

**(B) Availability.** A record of examinations, approvals, and variances granted shall be maintained by the authority having jurisdiction and shall be available for public review as prescribed by law during normal business hours.

**80.19 Permits and Approvals.** Permits and approvals shall conform to 80.19(A) through (H).

**(A) Application.**

(1) Activity authorized by a permit issued under this *Code* shall be conducted by the permittee or the permittee's agents or employees in compliance with all requirements of this *Code* applicable thereto and in accordance with the approved plans and specifications. No permit issued under this *Code* shall be interpreted to justify a violation of any provision of this *Code* or any other applicable law or regulation. Any addition or alteration of approved plans or specifications shall be approved in advance by the authority having jurisdiction, as evidenced by the issuance of a new or amended permit.
(2) A copy of the permit shall be posted or otherwise readily accessible at each work site or carried by the permit holder as specified by the authority having jurisdiction.

**(B) Content.** Permits shall be issued by the authority having jurisdiction and shall bear the name and signature of the authority having jurisdiction or that of the authority having jurisdiction's designated representative. In addition, the permit shall indicate the following:

(1) Operation or activities for which the permit is issued
(2) Address or location where the operation or activity is to be conducted

(3) Name and address of the permittee
(4) Permit number and date of issuance
(5) Period of validity of the permit
(6) Inspection requirements

**(C) Issuance of Permits.** The authority having jurisdiction shall be authorized to establish and issue permits, certificates, notices, and approvals, or orders pertaining to electrical safety hazards pursuant to 80.23, except that no permit shall be required to execute any of the classes of electrical work specified in the following:

(1) Installation or replacement of equipment such as lamps and of electric utilization equipment approved for connection to suitable permanently installed receptacles. Replacement of flush or snap switches, fuses, lamp sockets, and receptacles, and other minor maintenance and repair work, such as replacing worn cords and tightening connections on a wiring device
(2) The process of manufacturing, testing, servicing, or repairing electric equipment or apparatus

**(D) Annual Permits.** In lieu of an individual permit for each installation or alteration, an annual permit shall, upon application, be issued to any person, firm, or corporation regularly employing one or more employees for the installation, alteration, and maintenance of electric equipment in or on buildings or premises owned or occupied by the applicant for the permit. Upon application, an electrical contractor as agent for the owner or tenant shall be issued an annual permit. The applicant shall keep records of all work done, and such records shall be transmitted periodically to the Electrical Inspector.

**(E) Fees.** Any political subdivision that has been provided for electrical inspection in accordance with the provisions of Article 80 may establish fees that shall be paid by the applicant for a permit before the permit is issued.

**(F) Inspection and Approvals.**

(1) Upon the completion of any installation of electrical equipment that has been made under a permit other than an annual permit, it shall be the duty of the person, firm, or corporation making the installation to notify the Electrical Inspector having jurisdiction, who shall inspect the work within a reasonable time.
(2) Where the Inspector finds the installation to be in conformity with the statutes of all applicable local ordinances and all rules and regulations, the Inspector shall issue to the person, firm, or corporation making the installation a certificate of approval, with duplicate copy for delivery to the owner, authorizing the connection to the supply of electricity and shall send written notice of such authorization to the supplier of electric service. When a certificate of temporary approval is

issued authorizing the connection of an installation, such certificates shall be issued to expire at a time to be stated therein and shall be revocable by the Electrical Inspector for cause.

(3) When any portion of the electrical installation within the jurisdiction of an Electrical Inspector is to be hidden from view by the permanent placement of parts of the building, the person, firm, or corporation installing the equipment shall notify the Electrical Inspector, and such equipment shall not be concealed until it has been approved by the Electrical Inspector or until _____ days have elapsed from the time of such notification, provided that on large installations, where the concealment of equipment proceeds continuously, the person, firm, or corporation installing the equipment shall give the Electrical Inspector due notice in advance, and inspections shall be made periodically during the progress of the work.

(4) At regular intervals, the Electrical Inspector having jurisdiction shall visit all buildings and premises where work may be done under annual permits and shall inspect all electric equipment installed under such permits since the date of the previous inspection. The Electrical Inspector shall issue a certificate of approval for such work as is found to be in conformity with the provisions of Article 80 and all applicable ordinances, orders, rules, and regulations, after payments of all required fees.

(5) If, upon inspection, any installation is found not to be fully in conformity with the provisions of Article 80, and all applicable ordinances, rules, and regulations, the Inspector making the inspection shall at once forward to the person, firm, or corporation making the installation a written notice stating the defects that have been found to exist.

**(G) Revocation of Permits.** Revocation of permits shall conform to the following:

(1) The authority having jurisdiction shall be permitted to revoke a permit or approval issued if any violation of this *Code* is found upon inspection or in case there have been any false statements or misrepresentations submitted in the application or plans on which the permit or approval was based.

(2) Any attempt to defraud or otherwise deliberately or knowingly design, install, service, maintain, operate, sell, represent for sale, falsify records, reports, or applications, or other related activity in violation of the requirements prescribed by this *Code* shall be a violation of this *Code*. Such violations shall be cause for immediate suspension or revocation of any related licenses, certificates, or permits issued by this jurisdiction. In addition, any such violation shall be subject to

any other criminal or civil penalties as available by the laws of this jurisdiction.

(3) Revocation shall be constituted when the permittee is duly notified by the authority having jurisdiction.

(4) Any person who engages in any business, operation, or occupation, or uses any premises, after the permit issued therefor has been suspended or revoked pursuant to the provisions of this *Code*, and before such suspended permit has been reinstated or a new permit issued, shall be in violation of this *Code*.

(5) A permit shall be predicated upon compliance with the requirements of this *Code* and shall constitute written authority issued by the authority having jurisdiction to install electrical equipment. Any permit issued under this *Code* shall not take the place of any other license or permit required by other regulations or laws of this jurisdiction.

(6) The authority having jurisdiction shall be permitted to require an inspection prior to the issuance of a permit.

(7) A permit issued under this *Code* shall continue until revoked or for the period of time designated on the permit. The permit shall be issued to one person or business only and for the location or purpose described in the permit. Any change that affects any of the conditions of the permit shall require a new or amended permit.

**(H) Applications and Extensions.** Applications and extensions of permits shall conform to the following:

(1) The authority having jurisdiction shall be permitted to grant an extension of the permit time period upon presentation by the permittee of a satisfactory reason for failure to start or complete the work or activity authorized by the permit.

(2) Applications for permits shall be made to the authority having jurisdiction on forms provided by the jurisdiction and shall include the applicant's answers in full to inquiries set forth on such forms. Applications for permits shall be accompanied by such data as required by the authority having jurisdiction, such as plans and specifications, location, and so forth. Fees shall be determined as required by local laws.

(3) The authority having jurisdiction shall review all applications submitted and issue permits as required. If an application for a permit is rejected by the authority having jurisdiction, the applicant shall be advised of the reasons for such rejection. Permits for activities requiring evidence of financial responsibility by the jurisdiction shall not be issued unless proof of required financial responsibility is furnished.

**80.21 Plans Review.** Review of plans and specifications shall conform to 80.21(A) through (C).

**(A) Authority.** For new construction, modification, or rehabilitation, the authority having jurisdiction shall be permitted to review construction documents and drawings.

**(B) Responsibility of the Applicant.** It shall be the responsibility of the applicant to ensure the following:

(1) The construction documents include all of the electrical requirements.
(2) The construction documents and drawings are correct and in compliance with the applicable codes and standards.

**(C) Responsibility of the Authority Having Jurisdiction.** It shall be the responsibility of the authority having jurisdiction to promulgate rules that cover the following:

(1) Review of construction documents and drawings within established time frames for the purpose of acceptance or to provide reasons for nonacceptance
(2) Review and approval by the authority having jurisdiction shall not relieve the applicant of the responsibility of compliance with this *Code*.
(3) Where field conditions necessitate any substantial change from the approved plan, the authority having jurisdiction shall be permitted to require that the corrected plans be submitted for approval.

**80.23 Notice of Violations, Penalties.** Notice of violations and penalties shall conform to 80.23(A) and (B).

**(A) Violations.**

(1) Whenever the authority having jurisdiction determines that there are violations of this *Code*, a written notice shall be issued to confirm such findings.
(2) Any order or notice issued pursuant to this *Code* shall be served upon the owner, operator, occupant, or other person responsible for the condition or violation, either by personal service or mail or by delivering the same to, and leaving it with, some person of responsibility upon the premises. For unattended or abandoned locations, a copy of such order or notice shall be posted on the premises in a conspicuous place at or near the entrance to such premises and the order or notice shall be mailed by registered or certified mail, with return receipt requested, to the last known address of the owner, occupant, or both.

**(B) Penalties.**

(1) Any person who fails to comply with the provisions of this *Code* or who fails to carry out an order made pursuant to this *Code* or violates any condition attached to a permit, approval, or certificate shall be subject to the penalties established by this jurisdiction.
(2) Failure to comply with the time limits of an abatement notice or other corrective notice issued by the authority

having jurisdiction shall result in each day that such violation continues being regarded as a new and separate offense.
(3) Any person, firm, or corporation who shall willfully violate any of the applicable provisions of this article shall be guilty of a misdemeanor and, upon conviction thereof, shall be punished by a fine of not less than _____dollars ($_____) or more than _____dollars ($_____) for each offense, together with the costs of prosecution, imprisonment, or both, for not less than _____(_____) days or more than _____ (_____) days.

**80.25 Connection to Electricity Supply.** Connections to the electric supply shall conform to 80.25(A) through (E).

**(A) Authorization.** Except where work is done under an annual permit and except as otherwise provided in 80.25, it shall be unlawful for any person, firm, or corporation to make connection to a supply of electricity or to supply electricity to any electric equipment installation for which a permit is required or that has been disconnected or ordered to be disconnected.

**(B) Special Consideration.** By special permission of the authority having jurisdiction, temporary power shall be permitted to be supplied to the premises for specific needs of the construction project. The Board shall determine what needs are permitted under this provision.

**(C) Notification.** If, within _____ business days after the Electrical Inspector is notified of the completion of an installation of electric equipment, other than a temporary approval installation, the Electrical Inspector has neither authorized connection nor disapproved the installation, the supplier of electricity is authorized to make connections and supply electricity to such installation.

**(D) Other Territories.** If an installation or electric equipment is located in any territory where an Electrical Inspector has not been authorized or is not required to make inspections, the supplier of electricity is authorized to make connections and supply electricity to such installations.

**(E) Disconnection.** Where a connection is made to an installation that has not been inspected, as outlined in the preceding paragraphs of this section, the supplier of electricity shall immediately report such connection to the Chief Electrical Inspector. If, upon subsequent inspection, it is found that the installation is not in conformity with the provisions of Article 80, the Chief Electrical Inspector shall notify the person, firm, or corporation making the installation to rectify the defects and, if such work is not completed within fifteen (15) business days or a longer period as may be specified by the Board, the Board shall have the authority to cause the disconnection of that portion of the installation that is not in conformity.

### 80.27 Inspector's Qualifications.

**(A) Certificate.** All electrical inspectors shall be certified by a nationally recognized inspector certification program accepted by the Board. The certification program shall specifically qualify the inspector in electrical inspections. No person shall be employed as an Electrical Inspector unless that person is the holder of an Electrical Inspector's certificate of qualification issued by the Board, except that any person who on the date on which this law went into effect was serving as a legally appointed Electrical Inspector of _____ shall, upon application and payment of the prescribed fee and without examination, be issued a special certificate permitting him or her to continue to serve as an Electrical Inspector in the same territory.

**(B) Experience.** Electrical inspector applicants shall demonstrate the following:

(1) Have a demonstrated knowledge of the standard materials and methods used in the installation of electric equipment
(2) Be well versed in the approved methods of construction for safety to persons and property
(3) Be well versed in the statutes of _____ relating to electrical work and the *National Electrical Code*, as approved by the American National Standards Institute
(4) Have had at least ____ years' experience as an Electrical Inspector or ____ years in the installation of electrical equipment. In lieu of such experience, the applicant shall be a graduate in electrical engineering or of a similar curriculum of a college or university considered by the Board as having suitable requirements for graduation and shall have had two years' practical electrical experience.

**(C) Recertification.** Electrical inspectors shall be recertified as established by provisions of the applicable certification program.

**(D) Revocation and Suspension of Authority.** The Board shall have the authority to revoke an inspector's authority to conduct inspections within a jurisdiction.

### 80.29 Liability for Damages.
Article 80 shall not be construed to affect the responsibility or liability of any party owning, designing, operating, controlling, or installing any electric equipment for damages to persons or property caused by a defect therein, nor shall the _____ or any of its employees be held as assuming any such liability by reason of the inspection, reinspection, or other examination authorized.

### 80.31 Validity.
If any section, subsection, sentence, clause, or phrase of Article 80 is for any reason held to be unconstitutional, such decision shall not affect the validity of the remaining portions of Article 80.

### 80.33 Repeal of Conflicting Acts.
All acts or parts of acts in conflict with the provisions of Article 80 are hereby repealed.

### 80.35 Effective Date.
Article 80 shall take effect _____ (_____) days after its passage and publication.

## ARTICLE 90
## Introduction

### 90.1 Purpose.

**(A) Practical Safeguarding.** The purpose of this *Code* is the practical safeguarding of persons and property from hazards arising from the use of electricity.

**(B) Adequacy.** This *Code* contains provisions that are considered necessary for safety. Compliance therewith and proper maintenance will result in an installation that is essentially free from hazard but not necessarily efficient, convenient, or adequate for good service or future expansion of electrical use.

> FPN: Hazards often occur because of overloading of wiring systems by methods or usage not in conformity with this *Code*. This occurs because initial wiring did not provide for increases in the use of electricity. An initial adequate installation and reasonable provisions for system changes will provide for future increases in the use of electricity.

**(C) Intention.** This *Code* is not intended as a design specification or an instruction manual for untrained persons.

**(D) Relation to International Standards.** The requirements in this *Code* address the fundamental principles of protection for safety contained in Section 131 of International Electrotechnical Commission Standard 60364-1, *Electrical Installations of Buildings*.

> FPN: IEC 60364-1, Section 131, contains fundamental principles of protection for safety that encompass protection against electric shock, protection against thermal effects, protection against overcurrent, protection against fault currents, and protection against overvoltage. All of these potential hazards are addressed by the requirements in this *Code*.

### 90.2 Scope.

**(A) Covered.** This *Code* covers the installation of electric conductors, electric equipment, signaling and communications conductors and equipment, and fiber optic cables and raceways for the following:

(1) Public and private premises, including buildings, structures, mobile homes, recreational vehicles, and floating buildings
(2) Yards, lots, parking lots, carnivals, and industrial substations

   FPN: For additional information concerning such installations in an industrial or multibuilding complex, see ANSI C2-1997, *National Electrical Safety Code.*

(3) Installations of conductors and equipment that connect to the supply of electricity
(4) Installations used by the electric utility, such as office buildings, warehouses, garages, machine shops, and recreational buildings, that are not an integral part of a generating plant, substation, or control center

**(B)  Not Covered.** This *Code* does not cover the following:

(1) Installations in ships, watercraft other than floating buildings, railway rolling stock, aircraft, or automotive vehicles other than mobile homes and recreational vehicles

   FPN: Although the scope of this *Code* indicates that the *Code* does not cover installations in ships, portions of this *Code* are incorporated by reference into Title 46, *Code of Federal Regulations*, Parts 110–113.

(2) Installations under ground in mines and self-propelled mobile surface mining machinery and its attendant electrical trailing cable
(3) Installations of railways for generation, transformation, transmission, or distribution of power used exclusively for operation of rolling stock or installations used exclusively for signaling and communications purposes
(4) Installations of communications equipment under the exclusive control of communications utilities located outdoors or in building spaces used exclusively for such installations
(5) Installations under the exclusive control of an electric utility where such installations

   a. Consist of service drops or service laterals, and associated metering, or
   b. Are located in legally established easements, rights-of-way, or by other agreements either designated by or recognized by public service commissions, utility commissions, or other regulatory agencies having jurisdiction for such installations, or
   c. Are on property owned or leased by the electric utility for the purpose of communications, metering, generation, control, transformation, transmission, or distribution of electric energy.

**(C)  Special Permission.** The authority having jurisdiction for enforcing this *Code* may grant exception for the instal-

lation of conductors and equipment that are not under the exclusive control of the electric utilities and are used to connect the electric utility supply system to the service-entrance conductors of the premises served, provided such installations are outside a building or terminate immediately inside a building wall.

**90.3  Code Arrangement.** This *Code* is divided into the introduction and nine chapters, as shown in Figure 90.3. Chapters 1, 2, 3, and 4 apply generally; Chapters 5, 6 and 7 apply to special occupancies, special equipment, or other special conditions. These latter chapters supplement or modify the general rules. Chapters 1 through 4 apply except as amended by Chapters 5, 6, and 7 for the particular conditions.

   Chapter 8 covers communications systems and is not subject to the requirements of Chapters 1 through 7 except where the requirements are specifically referenced in Chapter 8.

   Chapter 9 consists of tables.

   Annexes are not part of the requirements of this *Code* but are included for informational purposes only.

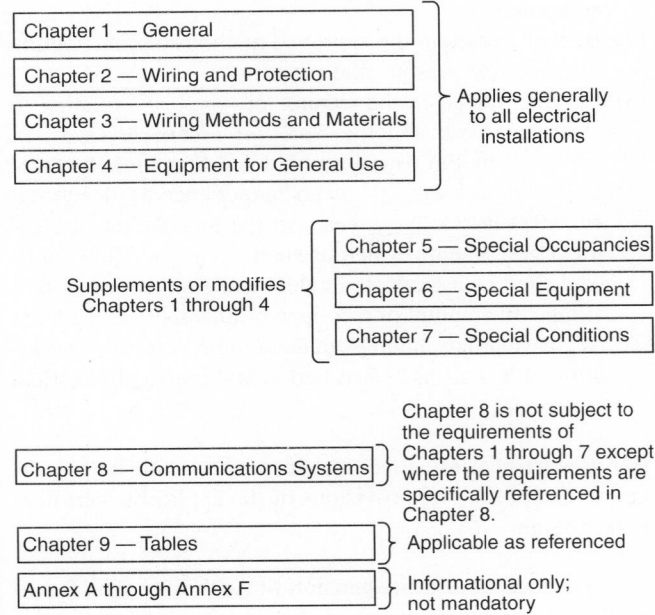

**Figure 90.3  Code arrangement.**

**90.4  Enforcement.** This *Code* is intended to be suitable for mandatory application by governmental bodies that exercise legal jurisdiction over electrical installations, including signaling and communications systems, and for use by insurance inspectors. The authority having jurisdiction for enforcement of the *Code* has the responsibility for making interpretations of the rules, for deciding on the approval of equipment and materials, and for granting the special permission contemplated in a number of the rules.

By special permission, the authority having jurisdiction may waive specific requirements in this *Code* or permit alternative methods where it is assured that equivalent objectives can be achieved by establishing and maintaining effective safety.

This *Code* may require new products, constructions, or materials that may not yet be available at the time the *Code* is adopted. In such event, the authority having jurisdiction may permit the use of the products, constructions, or materials that comply with the most recent previous edition of this *Code* adopted by the jurisdiction.

**90.5 Mandatory Rules, Permissive Rules, and Explanatory Material.**

**(A) Mandatory Rules.** Mandatory rules of this *Code* are those that identify actions that are specifically required or prohibited and are characterized by the use of the terms *shall* or *shall not*.

**(B) Permissive Rules.** Permissive rules of this *Code* are those that identify actions that are allowed but not required, are normally used to describe options or alternative methods, and are characterized by the use of the terms *shall be permitted* or *shall not be required*.

**(C) Explanatory Material.** Explanatory material, such as references to other standards, references to related sections of this *Code,* or information related to a *Code* rule, is included in this *Code* in the form of fine print notes (FPNs). Fine print notes are informational only and are not enforceable as requirements of this *Code*.

> FPN: The format and language used in this *Code* follows guidelines established by NFPA and published in the *NEC Style Manual*. Copies of this manual can be obtained from NFPA.

**90.6 Formal Interpretations.** To promote uniformity of interpretation and application of the provisions of this *Code,* formal interpretation procedures have been established and are found in the NFPA Regulations Governing Committee Projects.

**90.7 Examination of Equipment for Safety.** For specific items of equipment and materials referred to in this *Code*, examinations for safety made under standard conditions provide a basis for approval where the record is made generally available through promulgation by organizations properly equipped and qualified for experimental testing, inspections of the run of goods at factories, and service-value determination through field inspections. This avoids the necessity for repetition of examinations by different examiners, frequently with inadequate facilities for such

work, and the confusion that would result from conflicting reports on the suitability of devices and materials examined for a given purpose.

It is the intent of this *Code* that factory-installed internal wiring or the construction of equipment need not be inspected at the time of installation of the equipment, except to detect alterations or damage, if the equipment has been listed by a qualified electrical testing laboratory that is recognized as having the facilities described in the preceding paragraph and that requires suitability for installation in accordance with this *Code*.

> FPN No. 1: See requirements in 110.3.
>
> FPN No. 2: *Listed* is defined in Article 100.
>
> FPN No. 3: Annex A contains an informative list of product safety standards for electrical equipment.

**90.8 Wiring Planning.**

**(A) Future Expansion and Convenience.** Plans and specifications that provide ample space in raceways, spare raceways, and additional spaces allow for future increases in the use of electricity. Distribution centers located in readily accessible locations provide convenience and safety of operation.

**(B) Number of Circuits in Enclosures.** It is elsewhere provided in this *Code* that the number of wires and circuits confined in a single enclosure be varyingly restricted. Limiting the number of circuits in a single enclosure minimizes the effects from a short circuit or ground fault in one circuit.

**90.9 Units of Measurement.**

**(A) Measurement System of Preference.** For the purpose of this *Code*, metric units of measurement are in accordance with the modernized metric system known as the International System of Units (SI).

**(B) Dual System of Units.** The SI units shall appear first, and the inch-pound units shall immediately follow in parentheses. The conversion from the inch-pound units to SI units shall be based on hard conversion except as provided in 90.9(C).

**(C) Permitted Uses of Soft Conversion.** The cases given in 90.9(C)(1) through (4) shall not be required to use hard conversion and shall be permitted to use soft conversion.

**(1) Trade Sizes.** Where the actual measured size of a product is not the same as the nominal size, trade size designators shall be used rather than dimensions. Trade practices shall be followed in all cases.

**(2) Extracted Material.** Where material is extracted from another standard, the context of the original material shall not be compromised or violated. Any editing of the extracted text shall be confined to making the style consistent with that of the *NEC*.

**(3) Industry Practice.** Where industry practice is to express units in inch-pound units, the inclusion of SI units shall not be required.

**(4) Safety.** Where a negative impact on safety would result, hard conversion shall not be required.

**(D) Compliance.** The conversion from inch-pound units to SI units shall be permitted to be an approximate conversion. Compliance with the numbers shown in either the SI system or the inch-pound system shall constitute compliance with this *Code*.

FPN No. 1: Hard conversion is considered a change in dimensions or properties of an item into new sizes that might or might not be interchangeable with the sizes used in the original measurement. Soft conversion is considered a direct mathematical conversion and involves a change in the description of an existing measurement but not in the actual dimension.

FPN No. 2: SI conversions are based on IEEE/ASTM SI 10-1997, *Standard for the Use of the International System of Units (SI): The Modern Metric System.*

# Chapter 1 General

## ARTICLE 100
## Definitions

**Scope.** This article contains only those definitions essential to the proper application of this *Code*. It is not intended to include commonly defined general terms or commonly defined technical terms from related codes and standards. In general, only those terms that are used in two or more articles are defined in Article 100. Other definitions are included in the article in which they are used but may be referenced in Article 100.

Part I of this article contains definitions intended to apply wherever the terms are used throughout this *Code*. Part II contains definitions applicable only to the parts of articles specifically covering installations and equipment operating at over 600 volts, nominal.

## I. General

**Accessible (as applied to equipment).** Admitting close approach; not guarded by locked doors, elevation, or other effective means.

**Accessible (as applied to wiring methods).** Capable of being removed or exposed without damaging the building structure or finish or not permanently closed in by the structure or finish of the building.

**Accessible, Readily (Readily Accessible).** Capable of being reached quickly for operation, renewal, or inspections without requiring those to whom ready access is requisite to climb over or remove obstacles or to resort to portable ladders, and so forth.

**Ampacity.** The current, in amperes, that a conductor can carry continuously under the conditions of use without exceeding its temperature rating.

**Appliance.** Utilization equipment, generally other than industrial, that is normally built in standardized sizes or types and is installed or connected as a unit to perform one or more functions such as clothes washing, air conditioning, food mixing, deep frying, and so forth.

**Approved.** Acceptable to the authority having jurisdiction.

**Askarel.** A generic term for a group of nonflammable synthetic chlorinated hydrocarbons used as electrical insulating media. Askarels of various compositional types are used. Under arcing conditions, the gases produced, while consisting predominantly of noncombustible hydrogen chloride, can include varying amounts of combustible gases, depending on the askarel type.

**Attachment Plug (Plug Cap) (Plug).** A device that, by insertion in a receptacle, establishes a connection between the conductors of the attached flexible cord and the conductors connected permanently to the receptacle.

**Authority Having Jurisdiction.** The organization, office, or individual responsible for approving equipment, materials, an installation, or a procedure.

> FPN: The phrase "authority having jurisdiction" is used in NFPA documents in a broad manner, since jurisdictions and approval agencies vary, as do their responsibilities. Where public safety is primary, the authority having jurisdiction may be a federal, state, local, or other regional department or individual such as a fire chief; fire marshal; chief of a fire prevention bureau, labor department, or health department; building official; electrical inspector; or others having statutory authority. For insurance purposes, an insurance inspection department, rating bureau, or other insurance company representative may be the authority having jurisdiction. In many circumstances, the property owner or his or her designated agent assumes the role of the authority having jurisdiction; at government installations, the commanding officer or departmental official may be the authority having jurisdiction.

**Automatic.** Self-acting, operating by its own mechanism when actuated by some impersonal influence, as, for example, a change in current, pressure, temperature, or mechanical configuration.

**Bathroom.** An area including a basin with one or more of the following: a toilet, a tub, or a shower.

**Bonding (Bonded).** The permanent joining of metallic parts to form an electrically conductive path that ensures electrical continuity and the capacity to conduct safely any current likely to be imposed.

**Bonding Jumper.** A reliable conductor to ensure the required electrical conductivity between metal parts required to be electrically connected.

**Bonding Jumper, Equipment.** The connection between two or more portions of the equipment grounding conductor.

**Bonding Jumper, Main.** The connection between the grounded circuit conductor and the equipment grounding conductor at the service.

**Branch Circuit.** The circuit conductors between the final overcurrent device protecting the circuit and the outlet(s).

**Branch Circuit, Appliance.** A branch circuit that supplies energy to one or more outlets to which appliances are to be connected and that has no permanently connected luminaires (lighting fixtures) that are not a part of an appliance.

**Branch Circuit, General-Purpose.** A branch circuit that supplies two or more receptacles or outlets for lighting and appliances.

**Branch Circuit, Individual.** A branch circuit that supplies only one utilization equipment.

**Branch Circuit, Multiwire.** A branch circuit that consists of two or more ungrounded conductors that have a voltage between them, and a grounded conductor that has equal voltage between it and each ungrounded conductor of the circuit and that is connected to the neutral or grounded conductor of the system.

**Building.** A structure that stands alone or that is cut off from adjoining structures by fire walls with all openings therein protected by approved fire doors.

**Cabinet.** An enclosure that is designed for either surface mounting or flush mounting and is provided with a frame, mat, or trim in which a swinging door or doors are or can be hung.

**Circuit Breaker.** A device designed to open and close a circuit by nonautomatic means and to open the circuit automatically on a predetermined overcurrent without damage to itself when properly applied within its rating.

> FPN: The automatic opening means can be integral, direct acting with the circuit breaker, or remote from the circuit breaker.

*Adjustable (as applied to circuit breakers).* A qualifying term indicating that the circuit breaker can be set to trip at various values of current, time, or both, within a predetermined range.

*Instantaneous Trip (as applied to circuit breakers).* A qualifying term indicating that no delay is purposely introduced in the tripping action of the circuit breaker.

*Inverse Time (as applied to circuit breakers).* A qualifying term indicating that there is purposely introduced a delay in the tripping action of the circuit breaker, which delay decreases as the magnitude of the current increases.

*Nonadjustable (as applied to circuit breakers).* A qualifying term indicating that the circuit breaker does not have any adjustment to alter the value of current at which it will trip or the time required for its operation.

*Setting (of circuit breakers).* The value of current, time, or both, at which an adjustable circuit breaker is set to trip.

**Concealed.** Rendered inaccessible by the structure or finish of the building. Wires in concealed raceways are considered concealed, even though they may become accessible by withdrawing them.

**Conductor, Bare.** A conductor having no covering or electrical insulation whatsoever.

**Conductor, Covered.** A conductor encased within material of composition or thickness that is not recognized by this *Code* as electrical insulation.

**Conductor, Insulated.** A conductor encased within material of composition and thickness that is recognized by this *Code* as electrical insulation.

**Conduit Body.** A separate portion of a conduit or tubing system that provides access through a removable cover(s) to the interior of the system at a junction of two or more sections of the system or at a terminal point of the system.

Boxes such as FS and FD or larger cast or sheet metal boxes are not classified as conduit bodies.

**Connector, Pressure (Solderless).** A device that establishes a connection between two or more conductors or between one or more conductors and a terminal by means of mechanical pressure and without the use of solder.

**Continuous Load.** A load where the maximum current is expected to continue for 3 hours or more.

**Controller.** A device or group of devices that serves to govern, in some predetermined manner, the electric power delivered to the apparatus to which it is connected.

**Cooking Unit, Counter-Mounted.** A cooking appliance designed for mounting in or on a counter and consisting of one or more heating elements, internal wiring, and built-in or mountable controls.

**Copper-Clad Aluminum Conductors.** Conductors drawn from a copper-clad aluminum rod with the copper metallurgically bonded to an aluminum core. The copper forms a minimum of 10 percent of the cross-sectional area of a solid conductor or each strand of a stranded conductor.

**Cutout Box.** An enclosure designed for surface mounting that has swinging doors or covers secured directly to and telescoping with the walls of the box proper.

**Dead Front.** Without live parts exposed to a person on the operating side of the equipment.

**Demand Factor.** The ratio of the maximum demand of a system, or part of a system, to the total connected load of a system or the part of the system under consideration.

**Device.** A unit of an electrical system that is intended to carry but not utilize electric energy.

**Disconnecting Means.** A device, or group of devices, or other means by which the conductors of a circuit can be disconnected from their source of supply.

**Dusttight.** Constructed so that dust will not enter the enclosing case under specified test conditions.

**Duty, Continuous.** Operation at a substantially constant load for an indefinitely long time.

**Duty, Intermittent.** Operation for alternate intervals of (1) load and no load; or (2) load and rest; or (3) load, no load, and rest.

**Duty, Periodic.** Intermittent operation in which the load conditions are regularly recurrent.

**Duty, Short-Time.** Operation at a substantially constant load for a short and definite, specified time.

**Duty, Varying.** Operation at loads, and for intervals of time, both of which may be subject to wide variation.

**Dwelling Unit.** One or more rooms for the use of one or more persons as a housekeeping unit with space for eating, living, and sleeping, and permanent provisions for cooking and sanitation.

**Dwelling, One-Family.** A building that consists solely of one dwelling unit.

**Dwelling, Two-Family.** A building that consists solely of two dwelling units.

**Dwelling, Multifamily.** A building that contains three or more dwelling units.

**Electric Sign.** A fixed, stationary, or portable self-contained, electrically illuminated utilization equipment with words or symbols designed to convey information or attract attention.

**Enclosed.** Surrounded by a case, housing, fence, or wall(s) that prevents persons from accidentally contacting energized parts.

**Enclosure.** The case or housing of apparatus, or the fence or walls surrounding an installation to prevent personnel from accidentally contacting energized parts or to protect the equipment from physical damage.

FPN: See Table 430.91 for examples of enclosure types.

**Energized.** Electrically connected to a source of voltage.

**Equipment.** A general term including material, fittings, devices, appliances, luminaires (fixtures), apparatus, and the like used as a part of, or in connection with, an electrical installation.

**Explosionproof Apparatus.** Apparatus enclosed in a case that is capable of withstanding an explosion of a specified gas or vapor that may occur within it and of preventing the ignition of a specified gas or vapor surrounding the enclosure by sparks, flashes, or explosion of the gas or vapor within, and that operates at such an external temperature that a surrounding flammable atmosphere will not be ignited thereby.

FPN: For further information, see ANSI/UL 1203-1999, *Explosion-Proof and Dust-Ignition-Proof Electrical Equipment for Use in Hazardous (Classified) Locations.*

**Exposed (as applied to live parts).** Capable of being inadvertently touched or approached nearer than a safe distance by a person. It is applied to parts that are not suitably guarded, isolated, or insulated.

**Exposed (as applied to wiring methods).** On or attached to the surface or behind panels designed to allow access.

**Externally Operable.** Capable of being operated without exposing the operator to contact with live parts.

**Feeder.** All circuit conductors between the service equipment, the source of a separately derived system, or other power supply source and the final branch-circuit overcurrent device.

**Festoon Lighting.** A string of outdoor lights that is suspended between two points.

**Fitting.** An accessory such as a locknut, bushing, or other part of a wiring system that is intended primarily to perform a mechanical rather than an electrical function.

**Garage.** A building or portion of a building in which one or more self-propelled vehicles can be kept for use, sale, storage, rental, repair, exhibition, or demonstration purposes.

FPN: For commercial garages, repair and storage, see Article 511.

**Ground.** A conducting connection, whether intentional or accidental, between an electrical circuit or equipment and the earth or to some conducting body that serves in place of the earth.

**Grounded.** Connected to earth or to some conducting body that serves in place of the earth.

**Grounded, Effectively.** Intentionally connected to earth through a ground connection or connections of sufficiently low impedance and having sufficient current-carrying capacity to prevent the buildup of voltages that may result in undue hazards to connected equipment or to persons.

**Grounded Conductor.** A system or circuit conductor that is intentionally grounded.

**Grounding Conductor.** A conductor used to connect equipment or the grounded circuit of a wiring system to a grounding electrode or electrodes.

**Grounding Conductor, Equipment.** The conductor used to connect the non–current-carrying metal parts of equipment, raceways, and other enclosures to the system grounded conductor, the grounding electrode conductor, or both, at the service equipment or at the source of a separately derived system.

**Grounding Electrode Conductor.** The conductor used to connect the grounding electrode(s) to the equipment grounding conductor, to the grounded conductor, or to both, at the service, at each building or structure where supplied from a common service, or at the source of a separately derived system.

**Ground-Fault Circuit Interrupter.** A device intended for the protection of personnel that functions to de-energize a circuit or portion thereof within an established period of time when a current to ground exceeds the values established for a Class A device.

> FPN: Class A ground-fault circuit interrupters trip when the current to ground has a value in the range of 4 mA to 6 mA. For further information, see UL 943, *Standard for Ground-Fault Circuit Interrupters.*

**Ground-Fault Protection of Equipment.** A system intended to provide protection of equipment from damaging line-to-ground fault currents by operating to cause a disconnecting means to open all ungrounded conductors of the faulted circuit. This protection is provided at current levels less than those required to protect conductors from damage through the operation of a supply circuit overcurrent device.

**Guarded.** Covered, shielded, fenced, enclosed, or otherwise protected by means of suitable covers, casings, barriers, rails, screens, mats, or platforms to remove the likelihood of approach or contact by persons or objects to a point of danger.

**Hoistway.** Any shaftway, hatchway, well hole, or other vertical opening or space in which an elevator or dumbwaiter is designed to operate.

**Identified (as applied to equipment).** Recognizable as suitable for the specific purpose, function, use, environment, application, and so forth, where described in a particular *Code* requirement.

> FPN: Some examples of ways to determine suitability of equipment for a specific purpose, environment, or application include investigations by a qualified testing laboratory (listing and labeling), an inspection agency, or other organizations concerned with product evaluation.

**In Sight From (Within Sight From, Within Sight).** Where this *Code* specifies that one equipment shall be "in sight from," "within sight from," or "within sight," and so forth, of another equipment, the specified equipment is to be visible and not more than 15 m (50 ft) distant from the other.

**Interrupting Rating.** The highest current at rated voltage that a device is intended to interrupt under standard test conditions.

> FPN: Equipment intended to interrupt current at other than fault levels may have its interrupting rating implied in other ratings, such as horsepower or locked rotor current.

**Isolated (as applied to location).** Not readily accessible to persons unless special means for access are used.

**Labeled.** Equipment or materials to which has been attached a label, symbol, or other identifying mark of an organization that is acceptable to the authority having jurisdiction and concerned with product evaluation, that maintains periodic inspection of production of labeled equipment or materials, and by whose labeling the manufacturer indicates compliance with appropriate standards or performance in a specified manner.

**Lighting Outlet.** An outlet intended for the direct connection of a lampholder, a luminaire (lighting fixture), or a pendant cord terminating in a lampholder.

**Listed.** Equipment, materials, or services included in a list published by an organization that is acceptable to the authority having jurisdiction and concerned with evaluation of products or services, that maintains periodic inspection of production of listed equipment or materials or periodic evaluation of services, and whose listing states that the equipment, material, or services either meets appropriate designated standards or has been tested and found suitable for a specified purpose.

> FPN: The means for identifying listed equipment may vary for each organization concerned with product evaluation, some of which do not recognize equipment as listed unless it is also labeled. Use of the system employed by the listing organization allows the authority having jurisdiction to identify a listed product.

**Live Parts.** Energized conductive components.

**Location, Damp.** Locations protected from weather and not subject to saturation with water or other liquids but subject to moderate degrees of moisture. Examples of such locations include partially protected locations under canopies, marquees, roofed open porches, and like locations, and interior locations subject to moderate degrees of moisture, such as some basements, some barns, and some cold-storage warehouses.

**Location, Dry.** A location not normally subject to dampness or wetness. A location classified as dry may be temporarily subject to dampness or wetness, as in the case of a building under construction.

**Location, Wet.** Installations under ground or in concrete slabs or masonry in direct contact with the earth; in locations subject to saturation with water or other liquids, such as vehicle washing areas; and in unprotected locations exposed to weather.

**Luminaire.** A complete lighting unit consisting of a lamp or lamps together with the parts designed to distribute the light, to position and protect the lamps and ballast (where applicable), and to connect the lamps to the power supply.

**Metal-Enclosed Power Switchgear.** A switchgear assembly completely enclosed on all sides and top with sheet metal (except for ventilating openings and inspection windows) containing primary power circuit switching, interrupting devices, or both, with buses and connections. The assembly may include control and auxiliary devices. Access to the interior of the enclosure is provided by doors, removable covers, or both.

**Motor Control Center.** An assembly of one or more enclosed sections having a common power bus and principally containing motor control units.

**Multioutlet Assembly.** A type of surface, flush, or free-standing raceway designed to hold conductors and receptacles, assembled in the field or at the factory.

**Nonautomatic.** Action requiring personal intervention for its control. As applied to an electric controller, nonautomatic control does not necessarily imply a manual controller, but only that personal intervention is necessary.

**Nonincendive Circuit.** A circuit, other than field wiring, in which any arc or thermal effect produced under intended operating conditions of the equipment is not capable, under specified test conditions, of igniting the flammable gas–air, vapor–air, or dust–air mixture.

FPN: For test conditions, see ANSI/ISA-S12.12-1994, *Nonincendive Electrical Equipment for Use in Class I and II, Division 2 and Class III, Divisions 1 and 2 Hazardous (Classified) Locations.*

**Nonincendive Field Wiring.** Wiring that enters or leaves an equipment enclosure and, under normal operating conditions of the equipment, is not capable, due to arcing or thermal effects, of igniting the flammable gas–air, vapor–air, or dust–air mixture. Normal operation includes opening, shorting, or grounding the field wiring.

**Nonlinear Load.** A load where the wave shape of the steady-state current does not follow the wave shape of the applied voltage.

FPN: Electronic equipment, electronic/electric-discharge lighting, adjustable-speed drive systems, and similar equipment may be nonlinear loads.

**Outlet.** A point on the wiring system at which current is taken to supply utilization equipment.

**Outline Lighting.** An arrangement of incandescent lamps or electric-discharge lighting to outline or call attention to certain features such as the shape of a building or the decoration of a window.

**Overcurrent.** Any current in excess of the rated current of equipment or the ampacity of a conductor. It may result from overload, short circuit, or ground fault.

FPN: A current in excess of rating may be accommodated by certain equipment and conductors for a given set of conditions. Therefore the rules for overcurrent protection are specific for particular situations.

**Overload.** Operation of equipment in excess of normal, full-load rating, or of a conductor in excess of rated ampacity that, when it persists for a sufficient length of time, would cause damage or dangerous overheating. A fault, such as a short circuit or ground fault, is not an overload.

**Panelboard.** A single panel or group of panel units designed for assembly in the form of a single panel, including buses and automatic overcurrent devices, and equipped with or without switches for the control of light, heat, or power circuits; designed to be placed in a cabinet or cutout box placed in or against a wall, partition, or other support; and accessible only from the front.

**Plenum.** A compartment or chamber to which one or more air ducts are connected and that forms part of the air distribution system.

**Power Outlet.** An enclosed assembly that may include receptacles, circuit breakers, fuseholders, fused switches, buses, and watt-hour meter mounting means; intended to supply and control power to mobile homes, recreational vehicles, park trailers, or boats or to serve as a means for distributing power required to operate mobile or temporarily installed equipment.

**Premises Wiring (System).** That interior and exterior wiring, including power, lighting, control, and signal circuit wiring together with all their associated hardware, fittings, and wiring devices, both permanently and temporarily installed, that extends from the service point or source of power, such as a battery, a solar photovoltaic system, or a generator, transformer, or converter windings, to the outlet(s). Such wiring does not include wiring internal to appliances, luminaires (fixtures), motors, controllers, motor control centers, and similar equipment.

**Qualified Person.** One who has skills and knowledge related to the construction and operation of the electrical equipment and installations and has received safety training on the hazards involved.

**Raceway.** An enclosed channel of metal or nonmetallic materials designed expressly for holding wires, cables, or busbars, with additional functions as permitted in this *Code.* Raceways include, but are not limited to, rigid metal conduit, rigid nonmetallic conduit, intermediate metal conduit, liquidtight flexible conduit, flexible metallic tubing, flexible metal conduit, electrical nonmetallic tubing, electrical metallic tubing, underfloor raceways, cellular concrete floor raceways, cellular metal floor raceways, surface raceways, wireways, and busways.

**Rainproof.** Constructed, protected, or treated so as to prevent rain from interfering with the successful operation of the apparatus under specified test conditions.

**Raintight.** Constructed or protected so that exposure to a beating rain will not result in the entrance of water under specified test conditions.

**Receptacle.** A receptacle is a contact device installed at the outlet for the connection of an attachment plug. A single receptacle is a single contact device with no other contact device on the same yoke. A multiple receptacle is two or more contact devices on the same yoke.

**Receptacle Outlet.** An outlet where one or more receptacles are installed.

**Remote-Control Circuit.** Any electric circuit that controls any other circuit through a relay or an equivalent device.

**Sealable Equipment.** Equipment enclosed in a case or cabinet that is provided with a means of sealing or locking so that live parts cannot be made accessible without opening the enclosure. The equipment may or may not be operable without opening the enclosure.

**Separately Derived System.** A premises wiring system whose power is derived from a battery, from a solar photovoltaic system, or from a generator, transformer, or converter windings, and that has no direct electrical connection, including a solidly connected grounded circuit conductor, to supply conductors originating in another system.

**Service.** The conductors and equipment for delivering electric energy from the serving utility to the wiring system of the premises served.

**Service Cable.** Service conductors made up in the form of a cable.

**Service Conductors.** The conductors from the service point to the service disconnecting means.

**Service Drop.** The overhead service conductors from the last pole or other aerial support to and including the splices, if any, connecting to the service-entrance conductors at the building or other structure.

**Service-Entrance Conductors, Overhead System.** The service conductors between the terminals of the service equipment and a point usually outside the building, clear of building walls, where joined by tap or splice to the service drop.

**Service-Entrance Conductors, Underground System.** The service conductors between the terminals of the service equipment and the point of connection to the service lateral.

> FPN: Where service equipment is located outside the building walls, there may be no service-entrance conductors, or they may be entirely outside the building.

**Service Equipment.** The necessary equipment, usually consisting of a circuit breaker(s) or switch(es) and fuse(s) and their accessories, connected to the load end of service conductors to a building or other structure, or an otherwise designated area, and intended to constitute the main control and cutoff of the supply.

**Service Lateral.** The underground service conductors between the street main, including any risers at a pole or other structure or from transformers, and the first point of connection to the service-entrance conductors in a terminal box or meter or other enclosure, inside or outside the building wall. Where there is no terminal box, meter, or other enclosure, the point of connection is considered to be the point of entrance of the service conductors into the building.

**Service Point.** The point of connection between the facilities of the serving utility and the premises wiring.

**Show Window.** Any window used or designed to be used for the display of goods or advertising material, whether it is fully or partly enclosed or entirely open at the rear and whether or not it has a platform raised higher than the street floor level.

**Signaling Circuit.** Any electric circuit that energizes signaling equipment.

**Solar Photovoltaic System.** The total components and subsystems that, in combination, convert solar energy into electrical energy suitable for connection to a utilization load.

**Special Permission.** The written consent of the authority having jurisdiction.

**Structure.** That which is built or constructed.

**Switch, Bypass Isolation.** A manually operated device used in conjunction with a transfer switch to provide a means of directly connecting load conductors to a power source and of disconnecting the transfer switch.

**Switch, General-Use.** A switch intended for use in general distribution and branch circuits. It is rated in amperes, and it is capable of interrupting its rated current at its rated voltage.

**Switch, General-Use Snap.** A form of general-use switch constructed so that it can be installed in device boxes or on box covers, or otherwise used in conjunction with wiring systems recognized by this *Code*.

**Switch, Isolating.** A switch intended for isolating an electric circuit from the source of power. It has no interrupting rating, and it is intended to be operated only after the circuit has been opened by some other means.

**Switch, Motor-Circuit.** A switch rated in horsepower that is capable of interrupting the maximum operating overload current of a motor of the same horsepower rating as the switch at the rated voltage.

**Switch, Transfer.** An automatic or nonautomatic device for transferring one or more load conductor connections from one power source to another.

**Switchboard.** A large single panel, frame, or assembly of panels on which are mounted on the face, back, or both, switches, overcurrent and other protective devices, buses, and usually instruments. Switchboards are generally accessible from the rear as well as from the front and are not intended to be installed in cabinets.

**Thermal Protector (as applied to motors).** A protective device for assembly as an integral part of a motor or motor-compressor that, when properly applied, protects the motor against dangerous overheating due to overload and failure to start.

> FPN: The thermal protector may consist of one or more sensing elements integral with the motor or motor-compressor and an external control device.

**Thermally Protected (as applied to motors).** The words *Thermally Protected* appearing on the nameplate of a motor or motor-compressor indicate that the motor is provided with a thermal protector.

**Utilization Equipment.** Equipment that utilizes electric energy for electronic, electromechanical, chemical, heating, lighting, or similar purposes.

**Ventilated.** Provided with a means to permit circulation of air sufficient to remove an excess of heat, fumes, or vapors.

**Volatile Flammable Liquid.** A flammable liquid having a flash point below 38°C (100°F), or a flammable liquid whose temperature is above its flash point, or a Class II combustible liquid that has a vapor pressure not exceeding 276 kPa (40 psia) at 38°C (100°F) and whose temperature is above its flash point.

**Voltage (of a circuit).** The greatest root-mean-square (rms) (effective) difference of potential between any two conductors of the circuit concerned.

> FPN: Some systems, such as 3-phase 4-wire, single-phase 3-wire, and 3-wire direct current, may have various circuits of various voltages.

**Voltage, Nominal.** A nominal value assigned to a circuit or system for the purpose of conveniently designating its voltage class (e.g., 120/240 volts, 480Y/277 volts, 600 volts).

The actual voltage at which a circuit operates can vary from the nominal within a range that permits satisfactory operation of equipment.

> FPN: See ANSI C84.1-1995, *Voltage Ratings for Electric Power Systems and Equipment (60 Hz).*

**Voltage to Ground.** For grounded circuits, the voltage between the given conductor and that point or conductor of the circuit that is grounded; for ungrounded circuits, the greatest voltage between the given conductor and any other conductor of the circuit.

**Watertight.** Constructed so that moisture will not enter the enclosure under specified test conditions.

**Weatherproof.** Constructed or protected so that exposure to the weather will not interfere with successful operation.

> FPN: Rainproof, raintight, or watertight equipment can fulfill the requirements for weatherproof where varying weather conditions other than wetness, such as snow, ice, dust, or temperature extremes, are not a factor.

## II. Over 600 Volts, Nominal

Whereas the preceding definitions are intended to apply wherever the terms are used throughout this *Code*, the following definitions are applicable only to parts of the article specifically covering installations and equipment operating at over 600 volts, nominal.

**Electronically Actuated Fuse.** An overcurrent protective device that generally consists of a control module that provides current sensing, electronically derived time–current characteristics, energy to initiate tripping, and an interrupting module that interrupts current when an overcurrent occurs. Electronically actuated fuses may or may not operate in a current-limiting fashion, depending on the type of control selected.

**Fuse.** An overcurrent protective device with a circuit-opening fusible part that is heated and severed by the passage of overcurrent through it.

> FPN: A fuse comprises all the parts that form a unit capable of performing the prescribed functions. It may or may not be the complete device necessary to connect it into an electrical circuit.

*Controlled Vented Power Fuse.* A fuse with provision for controlling discharge circuit interruption such that no solid material may be exhausted into the surrounding atmosphere.

> FPN: The fuse is designed so that discharged gases will not ignite or damage insulation in the path of the discharge or propagate a flashover to or between grounded members or conduction members in the path of the discharge where the distance between the vent and such insulation or conduction members conforms to manufacturer's recommendations.

*Expulsion Fuse Unit (Expulsion Fuse).* A vented fuse unit in which the expulsion effect of gases produced by the arc and lining of the fuseholder, either alone or aided by a spring, extinguishes the arc.

*Nonvented Power Fuse.* A fuse without intentional provision for the escape of arc gases, liquids, or solid particles to the atmosphere during circuit interruption.

*Power Fuse Unit.* A vented, nonvented, or controlled vented fuse unit in which the arc is extinguished by being drawn through solid material, granular material, or liquid, either alone or aided by a spring.

*Vented Power Fuse.* A fuse with provision for the escape of arc gases, liquids, or solid particles to the surrounding atmosphere during circuit interruption.

**Multiple Fuse.** An assembly of two or more single-pole fuses.

**Switching Device.** A device designed to close, open, or both, one or more electric circuits.

*Circuit Breaker.* A switching device capable of making, carrying, and interrupting currents under normal circuit conditions, and also of making, carrying for a specified time, and interrupting currents under specified abnormal circuit conditions, such as those of short circuit.

*Cutout.* An assembly of a fuse support with either a fuseholder, fuse carrier, or disconnecting blade. The fuseholder or fuse carrier may include a conducting element (fuse link) or may act as the disconnecting blade by the inclusion of a nonfusible member.

*Disconnecting (or Isolating) Switch (Disconnector, Isolator).* A mechanical switching device used for isolating a circuit or equipment from a source of power.

*Disconnecting Means.* A device, group of devices, or other means whereby the conductors of a circuit can be disconnected from their source of supply.

*Interrupter Switch.* A switch capable of making, carrying, and interrupting specified currents.

*Oil Cutout (Oil-Filled Cutout).* A cutout in which all or part of the fuse support and its fuse link or disconnecting blade is mounted in oil with complete immersion of the contacts and the fusible portion of the conducting element (fuse link) so that arc interruption by severing of the fuse link or by opening of the contacts will occur under oil.

*Oil Switch.* A switch having contacts that operate under oil (or askarel or other suitable liquid).

*Regulator Bypass Switch.* A specific device or combination of devices designed to bypass a regulator.

## ARTICLE 110
## Requirements for Electrical Installations

### I. General

**110.1 Scope.** This article covers general requirements for the examination and approval, installation and use, access

to and spaces about electrical conductors and equipment, and tunnel installations.

**110.2 Approval.** The conductors and equipment required or permitted by this *Code* shall be acceptable only if approved.

> FPN: See 90.7, Examination of Equipment for Safety, and 110.3, Examination, Identification, Installation, and Use of Equipment. See definitions of *Approved, Identified, Labeled,* and *Listed.*

**110.3 Examination, Identification, Installation, and Use of Equipment.**

**(A) Examination.** In judging equipment, considerations such as the following shall be evaluated:

(1) Suitability for installation and use in conformity with the provisions of this *Code*

> FPN: Suitability of equipment use may be identified by a description marked on or provided with a product to identify the suitability of the product for a specific purpose, environment, or application. Suitability of equipment may be evidenced by listing or labeling.

(2) Mechanical strength and durability, including, for parts designed to enclose and protect other equipment, the adequacy of the protection thus provided
(3) Wire-bending and connection space
(4) Electrical insulation
(5) Heating effects under normal conditions of use and also under abnormal conditions likely to arise in service
(6) Arcing effects
(7) Classification by type, size, voltage, current capacity, and specific use
(8) Other factors that contribute to the practical safeguarding of persons using or likely to come in contact with the equipment

**(B) Installation and Use.** Listed or labeled equipment shall be installed and used in accordance with any instructions included in the listing or labeling.

**110.4 Voltages.** Throughout this *Code*, the voltage considered shall be that at which the circuit operates. The voltage rating of electrical equipment shall not be less than the nominal voltage of a circuit to which it is connected.

**110.5 Conductors.** Conductors normally used to carry current shall be of copper unless otherwise provided in this *Code.* Where the conductor material is not specified, the material and the sizes given in this *Code* shall apply to copper conductors. Where other materials are used, the size shall be changed accordingly.

> FPN: For aluminum and copper-clad aluminum conductors, see 310.15.

**110.6 Conductor Sizes.** Conductor sizes are expressed in American Wire Gage (AWG) or in circular mils.

**110.7 Insulation Integrity.** Completed wiring installations shall be free from short circuits and from grounds other than as required or permitted in Article 250.

**110.8 Wiring Methods.** Only wiring methods recognized as suitable are included in this *Code*. The recognized methods of wiring shall be permitted to be installed in any type of building or occupancy, except as otherwise provided in this *Code*.

**110.9 Interrupting Rating.** Equipment intended to interrupt current at fault levels shall have an interrupting rating sufficient for the nominal circuit voltage and the current that is available at the line terminals of the equipment.

Equipment intended to interrupt current at other than fault levels shall have an interrupting rating at nominal circuit voltage sufficient for the current that must be interrupted.

**110.10 Circuit Impedance and Other Characteristics.** The overcurrent protective devices, the total impedance, the component short-circuit current ratings, and other characteristics of the circuit to be protected shall be selected and coordinated to permit the circuit-protective devices used to clear a fault to do so without extensive damage to the electrical components of the circuit. This fault shall be assumed to be either between two or more of the circuit conductors or between any circuit conductor and the grounding conductor or enclosing metal raceway. Listed products applied in accordance with their listing shall be considered to meet the requirements of this section.

**110.11 Deteriorating Agents.** Unless identified for use in the operating environment, no conductors or equipment shall be located in damp or wet locations; where exposed to gases, fumes, vapors, liquids, or other agents that have a deteriorating effect on the conductors or equipment; or where exposed to excessive temperatures.

FPN No. 1: See 300.6 for protection against corrosion.

FPN No. 2: Some cleaning and lubricating compounds can cause severe deterioration of many plastic materials used for insulating and structural applications in equipment.

Equipment identified only as "dry locations," "Type 1," or "indoor use only" shall be protected against permanent damage from the weather during building construction.

**110.12 Mechanical Execution of Work.** Electrical equipment shall be installed in a neat and workmanlike manner.

**(A) Unused Openings.** Unused cable or raceway openings in boxes, raceways, auxiliary gutters, cabinets, cutout boxes, meter socket enclosures, equipment cases, or housings shall be effectively closed to afford protection substantially equivalent to the wall of the equipment. Where metallic plugs or plates are used with nonmetallic enclosures, they shall be recessed at least 6 mm (¼ in.) from the outer surface of the enclosure.

**(B) Subsurface Enclosures.** Conductors shall be racked to provide ready and safe access in underground and subsurface enclosures into which persons enter for installation and maintenance.

**(C) Integrity of Electrical Equipment and Connections.** Internal parts of electrical equipment, including busbars, wiring terminals, insulators, and other surfaces, shall not be damaged or contaminated by foreign materials such as paint, plaster, cleaners, abrasives, or corrosive residues. There shall be no damaged parts that may adversely affect safe operation or mechanical strength of the equipment such as parts that are broken; bent; cut; or deteriorated by corrosion, chemical action, or overheating.

**110.13 Mounting and Cooling of Equipment.**

**(A) Mounting.** Electrical equipment shall be firmly secured to the surface on which it is mounted. Wooden plugs driven into holes in masonry, concrete, plaster, or similar materials shall not be used.

**(B) Cooling.** Electrical equipment that depends on the natural circulation of air and convection principles for cooling of exposed surfaces shall be installed so that room airflow over such surfaces is not prevented by walls or by adjacent installed equipment. For equipment designed for floor mounting, clearance between top surfaces and adjacent surfaces shall be provided to dissipate rising warm air.

Electrical equipment provided with ventilating openings shall be installed so that walls or other obstructions do not prevent the free circulation of air through the equipment.

**110.14 Electrical Connections.** Because of different characteristics of dissimilar metals, devices such as pressure terminal or pressure splicing connectors and soldering lugs shall be identified for the material of the conductor and shall be properly installed and used. Conductors of dissimilar metals shall not be intermixed in a terminal or splicing connector where physical contact occurs between dissimilar conductors (such as copper and aluminum, copper and copper-clad aluminum, or aluminum and copper-clad aluminum), unless the device is identified for the purpose and conditions of use. Materials such as solder, fluxes, inhibitors, and compounds, where employed, shall be suitable for the use and shall be of a type that will not adversely affect the conductors, installation, or equipment.

> FPN: Many terminations and equipment are marked with a tightening torque.

**(A) Terminals.** Connection of conductors to terminal parts shall ensure a thoroughly good connection without damaging the conductors and shall be made by means of pressure connectors (including set-screw type), solder lugs, or splices to flexible leads. Connection by means of wire-binding screws or studs and nuts that have upturned lugs or the equivalent shall be permitted for 10 AWG or smaller conductors.

Terminals for more than one conductor and terminals used to connect aluminum shall be so identified.

**(B) Splices.** Conductors shall be spliced or joined with splicing devices identified for the use or by brazing, welding, or soldering with a fusible metal or alloy. Soldered splices shall first be spliced or joined so as to be mechanically and electrically secure without solder and then be soldered. All splices and joints and the free ends of conductors shall be covered with an insulation equivalent to that of the conductors or with an insulating device identified for the purpose.

Wire connectors or splicing means installed on conductors for direct burial shall be listed for such use.

**(C) Temperature Limitations.** The temperature rating associated with the ampacity of a conductor shall be selected and coordinated so as not to exceed the lowest temperature rating of any connected termination, conductor, or device. Conductors with temperature ratings higher than specified for terminations shall be permitted to be used for ampacity adjustment, correction, or both.

**(1) Equipment Provisions.** The determination of termination provisions of equipment shall be based on 110.14(C)(1)(a) or (C)(1)(b). Unless the equipment is listed and marked otherwise, conductor ampacities used in determining equipment termination provisions shall be based on Table 310.16 as appropriately modified by 310.15(B)(1) through (6).

(a) Termination provisions of equipment for circuits rated 100 amperes or less, or marked for 14 AWG through 1 AWG conductors, shall be used only for one of the following:

(1) Conductors rated 60°C (140°F)

(2) Conductors with higher temperature ratings, provided the ampacity of such conductors is determined based on the 60°C (140°F) ampacity of the conductor size used

(3) Conductors with higher temperature ratings if the equipment is listed and identified for use with such conductors

(4) For motors marked with design letters B, C, D, or E, conductors having an insulation rating of 75°C (167°F) or higher shall be permitted to be used provided the ampacity of such conductors does not exceed the 75°C (167°F) ampacity.

(b) Termination provisions of equipment for circuits rated over 100 amperes, or marked for conductors larger than 1 AWG, shall be used only for one of the following:

(1) Conductors rated 75°C (167°F)

(2) Conductors with higher temperature ratings, provided the ampacity of such conductors does not exceed the 75°C (167°F) ampacity of the conductor size used, or up to their ampacity if the equipment is listed and identified for use with such conductors

**(2) Separate Connector Provisions.** Separately installed pressure connectors shall be used with conductors at the ampacities not exceeding the ampacity at the listed and identified temperature rating of the connector.

> FPN: With respect to 110.14(C)(1) and (2), equipment markings or listing information may additionally restrict the sizing and temperature ratings of connected conductors.

**110.15 High-Leg Marking.** On a 4-wire, delta-connected system where the midpoint of one phase winding is grounded to supply lighting and similar loads, the conductor or busbar having the higher phase voltage to ground shall be durably and permanently marked by an outer finish that is orange in color or by other effective means. Such identification shall be placed at each point on the system where a connection is made if the grounded conductor is also present.

**110.16 Flash Protection.** Switchboards, panelboards, industrial control panels, and motor control centers that are in other than dwelling occupancies and are likely to require examination, adjustment, servicing, or maintenance while energized shall be field marked to warn qualified persons of potential electric arc flash hazards. The marking shall be located so as to be clearly visible to qualified persons before examination, adjustment, servicing, or maintenance of the equipment.

> FPN No. 1: NFPA 70E-2000, *Electrical Safety Requirements for Employee Workplaces*, provides assistance in determining severity of potential exposure, planning safe work practices, and selecting personal protective equipment.

> FPN No. 2: ANSI Z535.4-1998, *Product Safety Signs and Labels*, provides guidelines for the design of safety signs and labels for application to products.

**110.18 Arcing Parts.** Parts of electric equipment that in ordinary operation produce arcs, sparks, flames, or molten metal shall be enclosed or separated and isolated from all combustible material.

FPN: For hazardous (classified) locations, see Articles 500 through 517. For motors, see 430.14.

**110.19 Light and Power from Railway Conductors.** Circuits for lighting and power shall not be connected to any system that contains trolley wires with a ground return.

*Exception: Such circuit connections shall be permitted in car houses, power houses, or passenger and freight stations operated in connection with electric railways.*

**110.21 Marking.** The manufacturer's name, trademark, or other descriptive marking by which the organization responsible for the product can be identified shall be placed on all electric equipment. Other markings that indicate voltage, current, wattage, or other ratings shall be provided as specified elsewhere in this *Code*. The marking shall be of sufficient durability to withstand the environment involved.

**110.22 Identification of Disconnecting Means.** Each disconnecting means shall be legibly marked to indicate its purpose unless located and arranged so the purpose is evident. The marking shall be of sufficient durability to withstand the environment involved.

Where circuit breakers or fuses are applied in compliance with the series combination ratings marked on the equipment by the manufacturer, the equipment enclosure(s) shall be legibly marked in the field to indicate the equipment has been applied with a series combination rating. The marking shall be readily visible and state the following:

CAUTION — SERIES COMBINATION SYSTEM
RATED ___ AMPERES. IDENTIFIED
REPLACEMENT COMPONENTS REQUIRED.

FPN: See Section 240.86(A) for interrupting rating marking for end-use equipment.

**110.23 Current Transformers.** Unused current transformers associated with potentially energized circuits shall be short-circuited.

## II. 600 Volts, Nominal, or Less

**110.26 Spaces About Electrical Equipment.** Sufficient access and working space shall be provided and maintained about all electric equipment to permit ready and safe operation and maintenance of such equipment. Enclosures housing electrical apparatus that are controlled by lock and key shall be considered accessible to qualified persons.

**(A) Working Space.** Working space for equipment operating at 600 volts, nominal, or less to ground and likely to require examination, adjustment, servicing, or maintenance while energized shall comply with the dimensions of 110.26(A)(1), (2), and (3) or as required or permitted elsewhere in this *Code*.

**(1) Depth of Working Space.** The depth of the working space in the direction of live parts shall not be less than that specified in Table 110.26(A)(1) unless the requirements of 110.26(A)(1)(a), (b), or (c) are met. Distances shall be measured from the exposed live parts or from the enclosure or opening if the live parts are enclosed.

**Table 110.26(A)(1) Working Spaces**

| Nominal Voltage to Ground | Minimum Clear Distance | | |
|---|---|---|---|
| | Condition 1 | Condition 2 | Condition 3 |
| 0–150 | 900 mm (3 ft) | 900 mm (3 ft) | 900 mm (3 ft) |
| 151–600 | 900 mm (3 ft) | 1 m (3½ ft) | 1.2 m (4 ft) |

Note: Where the conditions are as follows:
**Condition 1** — Exposed live parts on one side and no live or grounded parts on the other side of the working space, or exposed live parts on both sides effectively guarded by suitable wood or other insulating materials. Insulated wire or insulated busbars operating at not over 300 volts to ground shall not be considered live parts.
**Condition 2** — Exposed live parts on one side and grounded parts on the other side. Concrete, brick, or tile walls shall be considered as grounded.
**Condition 3** — Exposed live parts on both sides of the work space (not guarded as provided in Condition 1) with the operator between.

(a) Dead-Front Assemblies. Working space shall not be required in the back or sides of assemblies, such as dead-front switchboards or motor control centers, where all connections and all renewable or adjustable parts, such as fuses or switches, are accessible from locations other than the back or sides. Where rear access is required to work on nonelectrical parts on the back of enclosed equipment, a minimum horizontal working space of 762 mm (30 in.) shall be provided.

(b) Low Voltage. By special permission, smaller working spaces shall be permitted where all uninsulated parts operate at not greater than 30 volts rms, 42 volts peak, or 60 volts dc.

(c) Existing Buildings. In existing buildings where electrical equipment is being replaced, Condition 2 working clearance shall be permitted between dead-front switchboards, panelboards, or motor control centers located across the aisle from each other where conditions of maintenance and supervision ensure that written procedures have been adopted to prohibit equipment on both sides of the aisle from being open at the same time and qualified persons who are authorized will service the installation.

**(2) Width of Working Space.** The width of the working space in front of the electric equipment shall be the width of the equipment or 750 mm (30 in.), whichever is greater.

In all cases, the work space shall permit at least a 90 degree opening of equipment doors or hinged panels.

**(3) Height of Working Space.** The work space shall be clear and extend from the grade, floor, or platform to the height required by 110.26(E). Within the height requirements of this section, other equipment that is associated with the electrical installation and is located above or below the electrical equipment shall be permitted to extend not more than 150 mm (6 in.) beyond the front of the electrical equipment.

**(B) Clear Spaces.** Working space required by this section shall not be used for storage. When normally enclosed live parts are exposed for inspection or servicing, the working space, if in a passageway or general open space, shall be suitably guarded.

**(C) Entrance to Working Space.**

**(1) Minimum Required.** At least one entrance of sufficient area shall be provided to give access to working space about electrical equipment.

**(2) Large Equipment.** For equipment rated 1200 amperes or more and over 1.8 m (6 ft) wide that contains overcurrent devices, switching devices, or control devices, there shall be one entrance to the required working space not less than 610 mm (24 in.) wide and 2.0 m (6½ ft) high at each end of the working space. Where the entrance has a personnel door(s), the door(s) shall open in the direction of egress and be equipped with panic bars, pressure plates, or other devices that are normally latched but open under simple pressure.

A single entrance to the required working space shall be permitted where either of the conditions in 110.26(C)(2)(a) or (b) is met.

(a) Unobstructed Exit. Where the location permits a continuous and unobstructed way of exit travel, a single entrance to the working space shall be permitted.

(b) Extra Working Space. Where the depth of the working space is twice that required by 110.26(A)(1), a single entrance shall be permitted. It shall be located so that the distance from the equipment to the nearest edge of the entrance is not less than the minimum clear distance specified in Table 110.26(A)(1) for equipment operating at that voltage and in that condition.

**(D) Illumination.** Illumination shall be provided for all working spaces about service equipment, switchboards, panelboards, or motor control centers installed indoors. Additional lighting outlets shall not be required where the work space is illuminated by an adjacent light source or as permitted by 210.70(A)(1), Exception No. 1, for switched receptacles. In electrical equipment rooms, the illumination shall not be controlled by automatic means only.

**(E) Headroom.** The minimum headroom of working spaces about service equipment, switchboards, panelboards, or motor control centers shall be 2.0 m (6½ ft). Where the electrical equipment exceeds 2.0 m (6½ ft) in height, the minimum headroom shall not be less than the height of the equipment.

*Exception: In existing dwelling units, service equipment or panelboards that do not exceed 200 amperes shall be permitted in spaces where the headroom is less than 2.0 m (6½ ft).*

**(F) Dedicated Equipment Space.** All switchboards, panelboards, distribution boards, and motor control centers shall be located in dedicated spaces and protected from damage.

*Exception: Control equipment that by its very nature or because of other rules of the Code must be adjacent to or within sight of its operating machinery shall be permitted in those locations.*

**(1) Indoor.** Indoor installations shall comply with 110.26(F)(1)(a) through (d).

(a) Dedicated Electrical Space. The space equal to the width and depth of the equipment and extending from the floor to a height of 1.8 m (6 ft) above the equipment or to the structural ceiling, whichever is lower, shall be dedicated to the electrical installation. No piping, ducts, leak protection apparatus, or other equipment foreign to the electrical installation shall be located in this zone.

*Exception: Suspended ceilings with removable panels shall be permitted within the 1.8-m (6-ft) zone.*

(b) Foreign Systems. The area above the dedicated space required by 110.26(F)(1)(a) shall be permitted to contain foreign systems, provided protection is installed to avoid damage to the electrical equipment from condensation, leaks, or breaks in such foreign systems.

(c) Sprinkler Protection. Sprinkler protection shall be permitted for the dedicated space where the piping complies with this section.

(d) Suspended Ceilings. A dropped, suspended, or similar ceiling that does not add strength to the building structure shall not be considered a structural ceiling.

**(2) Outdoor.** Outdoor electrical equipment shall be installed in suitable enclosures and shall be protected from accidental contact by unauthorized personnel, or by vehicular traffic, or by accidental spillage or leakage from piping systems. The working clearance space shall include the zone described in 110.26(A). No architectural appurtenance or other equipment shall be located in this zone.

**110.27 Guarding of Live Parts.**

**(A) Live Parts Guarded Against Accidental Contact.** Except as elsewhere required or permitted by this *Code,* live parts of electrical equipment operating at 50 volts or more shall be guarded against accidental contact by approved enclosures or by any of the following means:

(1) By location in a room, vault, or similar enclosure that is accessible only to qualified persons.
(2) By suitable permanent, substantial partitions or screens arranged so that only qualified persons have access to the space within reach of the live parts. Any openings in such partitions or screens shall be sized and located so that persons are not likely to come into accidental contact with the live parts or to bring conducting objects into contact with them.
(3) By location on a suitable balcony, gallery, or platform elevated and arranged so as to exclude unqualified persons.
(4) By elevation of 2.5 m (8 ft) or more above the floor or other working surface.

**(B) Prevent Physical Damage.** In locations where electric equipment is likely to be exposed to physical damage, enclosures or guards shall be so arranged and of such strength as to prevent such damage.

**(C) Warning Signs.** Entrances to rooms and other guarded locations that contain exposed live parts shall be marked with conspicuous warning signs forbidding unqualified persons to enter.

FPN: For motors, see 430.132 and 430.133. For over 600 volts, see 110.34.

### III. Over 600 Volts, Nominal

**110.30 General.** Conductors and equipment used on circuits over 600 volts, nominal, shall comply with Part I of this article and with the following sections, which supplement or modify Part I. In no case shall the provisions of this part apply to equipment on the supply side of the service point.

**110.31 Enclosure for Electrical Installations.** Electrical installations in a vault, room, or closet or in an area surrounded by a wall, screen, or fence, access to which is controlled by lock and key or other approved means, shall be considered to be accessible to qualified persons only. The type of enclosure used in a given case shall be designed and constructed according to the nature and degree of the hazard(s) associated with the installation.

For installations other than equipment as described in 110.31(D), a wall, screen, or fence shall be used to enclose an outdoor electrical installation to deter access by persons who are not qualified. A fence shall not be less than 2.1 m (7 ft) in height or a combination of 1.8 m (6 ft) or more of fence fabric and a 300-mm (1-ft) or more extension utilizing three or more strands of barbed wire or equivalent. The

distance from the fence to live parts shall be not less than given in Table 110.31.

**Table 110.31 Minimum Distance from Fence to Live Parts**

| Nominal Voltage | Minimum Distance to Live Parts | |
|---|---|---|
| | m | ft |
| 601–13,799 | 3.05 | 10 |
| 13,800–230,000 | 4.57 | 15 |
| Over 230,000 | 5.49 | 18 |

Note: For clearances of conductors for specific system voltages and typical BIL ratings, see ANSI C2-1997, *National Electrical Safety Code.*

FPN: See Article 450 for construction requirements for transformer vaults.

**(A) Fire Resistivity of Electrical Vaults.** The walls, roof, floors, and doorways of vaults containing conductors and equipment over 600 volts, nominal, shall be constructed of materials that have adequate structural strength for the conditions, with a minimum fire rating of 3 hours. The floors of vaults in contact with the earth shall be of concrete that is not less than 4 in. (102 mm) thick, but where the vault is constructed with a vacant space or other stories below it, the floor shall have adequate structural strength for the load imposed on it and a minimum fire resistance of 3 hours. For the purpose of this section, studs and wallboards shall not be considered acceptable.

**(B) Indoor Installations.**

**(1) In Places Accessible to Unqualified Persons.** Indoor electrical installations that are accessible to unqualified persons shall be made with metal-enclosed equipment. Metal-enclosed switchgear, unit substations, transformers, pull boxes, connection boxes, and other similar associated equipment shall be marked with appropriate caution signs. Openings in ventilated dry-type transformers or similar openings in other equipment shall be designed so that foreign objects inserted through these openings are deflected from energized parts.

**(2) In Places Accessible to Qualified Persons Only.** Indoor electrical installations considered accessible only to qualified persons in accordance with this section shall comply with 110.34, 110.36, and 490.24.

**(C) Outdoor Installations.**

**(1) In Places Accessible to Unqualified Persons.** Outdoor electrical installations that are open to unqualified persons shall comply with Article 225.

FPN: For clearances of conductors for system voltages over 600 volts, nominal, see ANSI C2-1997, *National Electrical Safety Code.*

**(2) In Places Accessible to Qualified Persons Only.** Outdoor electrical installations that have exposed live parts shall be accessible to qualified persons only in accordance with the first paragraph of this section and shall comply with 110.34, 110.36, and 490.24.

**(D) Enclosed Equipment Accessible to Unqualified Persons.** Ventilating or similar openings in equipment shall be designed so that foreign objects inserted through these openings are deflected from energized parts. Where exposed to physical damage from vehicular traffic, suitable guards shall be provided. Nonmetallic or metal-enclosed equipment located outdoors and accessible to the general public shall be designed so that exposed nuts or bolts cannot be readily removed, permitting access to live parts. Where nonmetallic or metal-enclosed equipment is accessible to the general public and the bottom of the enclosure is less than 2.5 m (8 ft) above the floor or grade level, the enclosure door or hinged cover shall be kept locked. Doors and covers of enclosures used solely as pull boxes, splice boxes, or junction boxes shall be locked, bolted, or screwed on. Underground box covers that weigh over 45.4 kg (100 lb) shall be considered as meeting this requirement.

**110.32 Work Space About Equipment.** Sufficient space shall be provided and maintained about electric equipment to permit ready and safe operation and maintenance of such equipment. Where energized parts are exposed, the minimum clear work space shall not be less than 2.0 m (6½ ft) high (measured vertically from the floor or platform) or less than 900 mm (3 ft) wide (measured parallel to the equipment). The depth shall be as required in 110.34(A). In all cases, the work space shall permit at least a 90 degree opening of doors or hinged panels.

**110.33 Entrance and Access to Work Space.**

**(A) Entrance.** At least one entrance not less than 610 mm (24 in.) wide and 2.0 m (6½ ft) high shall be provided to give access to the working space about electric equipment. Where the entrance has a personnel door(s), the door(s) shall open in the direction of egress and be equipped with panic bars, pressure plates, or other devices that are normally latched but open under simple pressure.

**(1) Large Equipment.** On switchboard and control panels exceeding 1.8 m (6 ft) in width, there shall be one entrance at each end of the equipment. A single entrance to the required working space shall be permitted where either of the conditions in 110.33(A)(1)(a) or (b) is met.

(a) Unobstructed Exit. Where the location permits a continuous and unobstructed way of exit travel, a single entrance to the working space shall be permitted.

(b) Extra Working Space. Where the depth of the working space is twice that required by 110.34(A), a single entrance shall be permitted. It shall be located so that the distance from the equipment to the nearest edge of the entrance is not less than the minimum clear distance specified in Table 110.34(A) for equipment operating at that voltage and in that condition.

**(2) Guarding.** Where bare energized parts at any voltage or insulated energized parts above 600 volts, nominal, to ground are located adjacent to such entrance, they shall be suitably guarded.

**(B) Access.** Permanent ladders or stairways shall be provided to give safe access to the working space around electric equipment installed on platforms, balconies, or mezzanine floors or in attic or roof rooms or spaces.

**110.34 Work Space and Guarding.**

**(A) Working Space.** Except as elsewhere required or permitted in this *Code,* the minimum clear working space in the direction of access to live parts of electrical equipment shall not be less than specified in Table 110.34(A). Distances shall be measured from the live parts, if such are exposed, or from the enclosure front or opening if such are enclosed.

*Exception: Working space shall not be required in back of equipment such as dead-front switchboards or control assemblies where there are no renewable or adjustable parts (such as fuses or switches) on the back and where all connections are accessible from locations other than the back. Where rear access is required to work on de-energized parts on the back of enclosed equipment, a minimum working space of 750 mm (30 in.) horizontally shall be provided.*

**(B) Separation from Low-Voltage Equipment.** Where switches, cutouts, or other equipment operating at 600 volts, nominal, or less are installed in a room or enclosure where there are exposed live parts or exposed wiring operating at over 600 volts, nominal, the high-voltage equipment shall be effectively separated from the space occupied by the low-voltage equipment by a suitable partition, fence, or screen.

*Exception: Switches or other equipment operating at 600 volts, nominal, or less and serving only equipment within the high-voltage vault, room, or enclosure shall be permitted to be installed in the high-voltage enclosure, room, or vault if accessible to qualified persons only.*

**(C) Locked Rooms or Enclosures.** The entrances to all buildings, rooms, or enclosures containing exposed live parts or exposed conductors operating at over 600 volts, nominal, shall be kept locked unless such entrances are under the observation of a qualified person at all times.

**Table 110.34(A) Minimum Depth of Clear Working Space at Electrical Equipment**

| Nominal Voltage to Ground | Minimum Clear Distance | | |
| --- | --- | --- | --- |
| | Condition 1 | Condition 2 | Condition 3 |
| 601–2500 V | 900 mm (3 ft) | 1.2 m (4 ft) | 1.5 m (5 ft) |
| 2501–9000 V | 1.2 m (4 ft) | 1.5 m (5 ft) | 1.8 m (6 ft) |
| 9001–25,000 V | 1.5 m (5 ft) | 1.8 m (6 ft) | 2.8 m (9 ft) |
| 25,001V–75 kV | 1.8 m (6 ft) | 2.5 m (8 ft) | 3.0 m (10 ft) |
| Above 75 kV | 2.5 m (8 ft) | 3.0 m (10 ft) | 3.7 m (12 ft) |

Note: Where the conditions are as follows:

**Condition 1**— Exposed live parts on one side and no live or grounded parts on the other side of the working space, or exposed live parts on both sides effectively guarded by suitable wood or other insulating materials. Insulated wire or insulated busbars operating at not over 300 volts shall not be considered live parts.

**Condition 2**— Exposed live parts on one side and grounded parts on the other side. Concrete, brick, or tile walls shall be considered as grounded surfaces.

**Condition 3**— Exposed live parts on both sides of the work space (not guarded as provided in Condition 1) with the operator between.

Where the voltage exceeds 600 volts, nominal, permanent and conspicuous warning signs shall be provided, reading as follows:

DANGER — HIGH VOLTAGE — KEEP OUT

**(D) Illumination.** Illumination shall be provided for all working spaces about electrical equipment. The lighting outlets shall be arranged so that persons changing lamps or making repairs on the lighting system are not endangered by live parts or other equipment.

The points of control shall be located so that persons are not likely to come in contact with any live part or moving part of the equipment while turning on the lights.

**(E) Elevation of Unguarded Live Parts.** Unguarded live parts above working space shall be maintained at elevations not less than required by Table 110.34(E).

**Table 110.34(E) Elevation of Unguarded Live Parts Above Working Space**

| Nominal Voltage Between Phases | Elevation | |
| --- | --- | --- |
| | m | ft |
| 601–7500 V | 2.8 | 9 |
| 7501–35,000 V | 2.9 | 9½ |
| Over 35 kV | 2.9 m + 9.5 mm/kV above 35 | 9½ ft + 0.37 in./kV above 35 |

**(F) Protection of Service Equipment, Metal-Enclosed Power Switchgear, and Industrial Control Assemblies.** Pipes or ducts foreign to the electrical installation and requiring periodic maintenance or whose malfunction would endanger the operation of the electrical system shall not be located in the vicinity of the service equipment, metal-enclosed power switchgear, or industrial control assemblies. Protection shall be provided where necessary to avoid damage from condensation leaks and breaks in such foreign systems. Piping and other facilities shall not be considered foreign if provided for fire protection of the electrical installation.

**110.36 Circuit Conductors.** Circuit conductors shall be permitted to be installed in raceways; in cable trays; as metal-clad cable, as bare wire, cable, and busbars; or as Type MV cables or conductors as provided in 300.37, 300.39, 300.40, and 300.50. Bare live conductors shall conform with 490.24.

Insulators, together with their mounting and conductor attachments, where used as supports for wires, single-conductor cables, or busbars, shall be capable of safely withstanding the maximum magnetic forces that would prevail when two or more conductors of a circuit were subjected to short-circuit current.

Open runs of insulated wires and cables that have a bare lead sheath or a braided outer covering shall be supported in a manner designed to prevent physical damage to the braid or sheath. Supports for lead-covered cables shall be designed to prevent electrolysis of the sheath.

**110.40 Temperature Limitations at Terminations.** Conductors shall be permitted to be terminated based on the 90°C (194°F) temperature rating and ampacity as given in Tables 310.67 through 310.86, unless otherwise identified.

## IV. Tunnel Installations Over 600 Volts, Nominal

**110.51 General.**

**(A) Covered.** The provisions of this part shall apply to the installation and use of high-voltage power distribution and utilization equipment that is portable, mobile, or both, such as substations, trailers, cars, mobile shovels, draglines, hoists, drills, dredges, compressors, pumps, conveyors, and underground excavators, and the like.

**(B) Other Articles.** The requirements of this part shall be additional to, or amendatory of, those prescribed in Articles 100 through 490 of this *Code*. Special attention shall be paid to Article 250.

**(C) Protection Against Physical Damage.** Conductors and cables in tunnels shall be located above the tunnel floor and so placed or guarded to protect them from physical damage.

**110.52 Overcurrent Protection.** Motor-operated equipment shall be protected from overcurrent in accordance with Article 430. Transformers shall be protected from overcurrent in accordance with Article 450.

**110.53 Conductors.** High-voltage conductors in tunnels shall be installed in metal conduit or other metal raceway, Type MC cable, or other approved multiconductor cable. Multiconductor portable cable shall be permitted to supply mobile equipment.

**110.54 Bonding and Equipment Grounding Conductors.**

**(A) Grounded and Bonded.** All non–current-carrying metal parts of electric equipment and all metal raceways and cable sheaths shall be effectively grounded and bonded to all metal pipes and rails at the portal and at intervals not exceeding 300 m (1000 ft) throughout the tunnel.

**(B) Equipment Grounding Conductors.** An equipment grounding conductor shall be run with circuit conductors inside the metal raceway or inside the multiconductor cable jacket. The equipment grounding conductor shall be permitted to be insulated or bare.

**110.55 Transformers, Switches, and Electrical Equipment.** All transformers, switches, motor controllers, motors, rec-

tifiers, and other equipment installed below ground shall be protected from physical damage by location or guarding.

**110.56 Energized Parts.** Bare terminals of transformers, switches, motor controllers, and other equipment shall be enclosed to prevent accidental contact with energized parts.

**110.57 Ventilation System Controls.** Electrical controls for the ventilation system shall be arranged so that the airflow can be reversed.

**110.58 Disconnecting Means.** A switch or circuit breaker that simultaneously opens all ungrounded conductors of the circuit shall be installed within sight of each transformer or motor location for disconnecting the transformer or motor. The switch or circuit breaker for a transformer shall have an ampere rating not less than the ampacity of the transformer supply conductors. The switch or circuit breaker for a motor shall comply with the applicable requirements of Article 430.

**110.59 Enclosures.** Enclosures for use in tunnels shall be dripproof, weatherproof, or submersible as required by the environmental conditions. Switch or contactor enclosures shall not be used as junction boxes or as raceways for conductors feeding through or tapping off to other switches, unless special designs are used to provide adequate space for this purpose.

# Chapter 2   Wiring and Protection

## ARTICLE 200
## Use and Identification of Grounded Conductors

**200.1 Scope.** This article provides requirements for the following:

(1)  Identification of terminals
(2)  Grounded conductors in premises wiring systems
(3)  Identification of grounded conductors

> FPN: See Article 100 for definitions of *Grounded Conductor* and *Grounding Conductor*.

**200.2 General.** All premises wiring systems, other than circuits and systems exempted or prohibited by 210.10, 215.7, 250.21, 250.22, 250.162, 503.13, 517.63, 668.11, 668.21, and 690.41, Exception, shall have a grounded conductor that is identified in accordance with 200.6.

The grounded conductor, where insulated, shall have insulation that is (1) suitable, other than color, for any ungrounded conductor of the same circuit on circuits of less than 1000 volts or impedance grounded neutral systems of 1 kV and over, or (2) rated not less than 600 volts for solidly grounded neutral systems of 1 kV and over as described in 250.184(A).

**200.3 Connection to Grounded System.** Premises wiring shall not be electrically connected to a supply system unless the latter contains, for any grounded conductor of the interior system, a corresponding conductor that is grounded. For the purpose of this section, *electrically connected* shall mean connected so as to be capable of carrying current, as distinguished from connection through electromagnetic induction.

**200.6 Means of Identifying Grounded Conductors.**

**(A) Sizes 6 AWG or Smaller.** An insulated grounded conductor of 6 AWG or smaller shall be identified by a continuous white or gray outer finish or by three continuous white stripes on other than green insulation along its entire length. Wires that have their outer covering finished to show a white or gray color but have colored tracer threads in the braid identifying the source of manufacture shall be considered as meeting the provisions of this section. Insulated grounded conductors shall also be permitted to be identified as follows:

(1)  The grounded conductor of a mineral-insulated, metal-sheathed cable shall be identified at the time of installation by distinctive marking at its terminations.

(2)  A single-conductor, sunlight-resistant, outdoor-rated cable used as a grounded conductor in photovoltaic power systems as permitted by 690.31 shall be identified at the time of installation by distinctive white marking at all terminations.
(3)  Fixture wire shall comply with the requirements for grounded conductor identification as specified in 402.8
(4)  For aerial cable, the identification shall be as above, or by means of a ridge located on the exterior of the cable so as to identify it.

**(B) Sizes Larger Than 6 AWG.** An insulated grounded conductor larger than 6 AWG shall be identified either by a continuous white or gray outer finish or by three continuous white stripes on other than green insulation along its entire length or at the time of installation by a distinctive white marking at its terminations. This marking shall encircle the conductor or insulation.

**(C) Flexible Cords.** An insulated conductor that is intended for use as a grounded conductor, where contained within a flexible cord, shall be identified by a white or gray outer finish or by three continuous white stripes on other than green insulation or by methods permitted by 400.22.

**(D) Grounded Conductors of Different Systems.** Where conductors of different systems are installed in the same raceway, cable, box, auxiliary gutter, or other type of enclosure, one system grounded conductor, if required, shall have an outer covering conforming to 200.6(A) or 200.6(B). Each other system grounded conductor shall have an outer covering of white with a readily distinguishable, different colored stripe other than green running along the insulation, or shall have other and different means of identification as allowed by 200.6(A) or (B) that will distinguish each system grounded conductor.

**(E) Grounded Conductors of Multiconductor Cables.** The insulated grounded conductors in a multiconductor cable shall be identified by a continuous white or gray outer finish or by three continuous white stripes on other than green insulation along its entire length. Multiconductor flat 4 AWG or larger shall be permitted to employ an external ridge on the grounded conductor.

*Exception No. 1: Where the conditions of maintenance and supervision ensure that only qualified persons service the installation, grounded conductors in multiconductor cables shall be permitted to be permanently identified at their terminations at the time of installation by a distinctive white marking or other equally effective means.*

*Exception No. 2: The grounded conductor of a multiconductor varnished-cloth-insulated cable shall be permitted to be identified at its terminations at the time of installation by a distinctive white marking or other equally effective means.*

> FPN: The color gray may have been used in the past as an ungrounded conductor. Care should be taken when working on existing systems.

### 200.7 Use of Insulation of a White or Gray Color or with Three Continuous White Stripes.

**(A) General.** The following shall be used only for the grounded circuit conductor, unless otherwise permitted in 200.7(B) and (C):

(1) A conductor with continuous white or gray covering
(2) A conductor with three continuous white stripes on other than green insulation
(3) A marking of white or gray color at the termination

**(B) Circuits of Less Than 50 Volts.** A conductor with white or gray color insulation or three continuous white stripes or having a marking of white or gray at the termination for circuits of less than 50 volts shall be required to be grounded only as required by 250.20(A).

**(C) Circuits of 50 Volts or More.** The use of insulation that is white or gray or that has three continuous white stripes for other than a grounded conductor for circuits of 50 volts or more shall be permitted only as in (1) through (3).

(1) If part of a cable assembly and where the insulation is permanently reidentified to indicate its use as an ungrounded conductor, by painting or other effective means at its termination, and at each location where the conductor is visible and accessible.
(2) Where a cable assembly contains an insulated conductor for single-pole, 3-way or 4-way switch loops and the conductor with white or gray insulation or a marking of three continuous white stripes is used for the supply to the switch but not as a return conductor from the switch to the switched outlet. In these applications, the conductor with white or gray insulation or with three continuous white stripes shall be permanently reidentified to indicate its use by painting or other effective means at its terminations and at each location where the conductor is visible and accessible.
(3) Where a flexible cord, having one conductor identified by a white or gray outer finish or three continuous white stripes or by any other means permitted by 400.22, is used for connecting an appliance or equipment permitted by 400.7. This shall apply to flexible cords connected to outlets whether or not the outlet is supplied by a circuit that has a grounded conductor.

> FPN: The color gray may have been used in the past as an ungrounded conductor. Care should be taken when working on existing systems.

### 200.9 Means of Identification of Terminals.
The identification of terminals to which a grounded conductor is to be connected shall be substantially white in color. The identification of other terminals shall be of a readily distinguishable different color.

*Exception: Where the conditions of maintenance and supervision ensure that only qualified persons service the installations, terminals for grounded conductors shall be permitted to be permanently identified at the time of installation by a distinctive white marking or other equally effective means.*

### 200.10 Identification of Terminals.

**(A) Device Terminals.** All devices, excluding panelboards, provided with terminals for the attachment of conductors and intended for connection to more than one side of the circuit shall have terminals properly marked for identification, unless the electrical connection of the terminal intended to be connected to the grounded conductor is clearly evident.

*Exception: Terminal identification shall not be required for devices that have a normal current rating of over 30 amperes, other than polarized attachment plugs and polarized receptacles for attachment plugs as required in 200.10(B).*

**(B) Receptacles, Plugs, and Connectors.** Receptacles, polarized attachment plugs, and cord connectors for plugs and polarized plugs shall have the terminal intended for connection to the grounded conductor identified as follows:

(1) Identification shall be by a metal or metal coating that is substantially white in color or by the word *white* or the letter *W* located adjacent to the identified terminal.
(2) If the terminal is not visible, the conductor entrance hole for the connection shall be colored white or marked with the word *white* or the letter *W*.

> FPN: See 250.126 for identification of wiring device equipment grounding conductor terminals.

**(C) Screw Shells.** For devices with screw shells, the terminal for the grounded conductor shall be the one connected to the screw shell.

**(D) Screw Shell Devices with Leads.** For screw shell devices with attached leads, the conductor attached to the screw shell shall have a white or gray finish. The outer finish of the other conductor shall be of a solid color that will not be confused with the white or gray finish used to identify the grounded conductor.

> FPN: The color gray may have been used in the past as an ungrounded conductor. Care should be taken when working on existing systems.

**(E) Appliances.** Appliances that have a single-pole switch or a single-pole overcurrent device in the line or any line-connected screw shell lampholders, and that are to be connected by (1) a permanent wiring method or (2) field-installed attachment plugs and cords with three or more wires (including the equipment grounding conductor), shall have means to identify the terminal for the grounded circuit conductor (if any).

**200.11 Polarity of Connections.** No grounded conductor shall be attached to any terminal or lead so as to reverse the designated polarity.

# ARTICLE 210
# Branch Circuits

## I. General Provisions

**210.1 Scope.** This article covers branch circuits except for branch circuits that supply only motor loads, which are covered in Article 430. Provisions of this article and Article 430 apply to branch circuits with combination loads.

**210.2 Other Articles for Specific-Purpose Branch Circuits.** Branch circuits shall comply with this article and also with the applicable provisions of other articles of this *Code.* The provisions for branch circuits supplying equipment in Table 210.2 amend or supplement the provisions in this article and shall apply to branch circuits referred to therein.

**210.3 Rating.** Branch circuits recognized by this article shall be rated in accordance with the maximum permitted ampere rating or setting of the overcurrent device. The rating for other than individual branch circuits shall be 15, 20, 30, 40, and 50 amperes. Where conductors of higher ampacity are used for any reason, the ampere rating or setting of the specified overcurrent device shall determine the circuit rating.

*Exception: Multioutlet branch circuits greater than 50 amperes shall be permitted to supply nonlighting outlet loads on industrial premises where conditions of maintenance and supervision ensure that only qualified persons service the equipment.*

**210.4 Multiwire Branch Circuits.**

**(A) General.** Branch circuits recognized by this article shall be permitted as multiwire circuits. A multiwire branch circuit shall be permitted to be considered as multiple circuits. All conductors shall originate from the same panelboard.

**Table 210.2 Specific-Purpose Branch Circuits**

| Equipment | Article | Section |
|---|---|---|
| Air-conditioning and refrigerating equipment | | 440.6, 440.31, 440.32 |
| Busways | | 368 |
| Circuits and equipment operating at less than 50 volts | 720 | |
| Central heating equipment other than fixed electric space-heating equipment | | 422.12 |
| Class 1, Class 2, and Class 3 remote-control, signaling, and power-limited circuits | 725 | |
| Closed-loop and programmed power distribution | 780 | |
| Cranes and hoists | | 610.42 |
| Electric signs and outline lighting | | 600.6 |
| Electric welders | 630 | |
| Elevators, dumbwaiters, escalators, moving walks, wheelchair lifts, and stairway chair lifts | | 620.61 |
| Fire alarm systems | 760 | |
| Fixed electric heating equipment for pipelines and vessels | | 427.4 |
| Fixed electric space-heating equipment | | 424.3 |
| Fixed outdoor electric deicing and snow-melting equipment | | 426.4 |
| Information technology equipment | | 645.5 |
| Infrared lamp industrial heating equipment | | 422.48, 424.3 |
| Induction and dielectric heating equipment | 665 | |
| Marinas and boatyards | | 555.19 |
| Mobile homes, manufactured homes, and mobile home parks | 550 | |
| Motion picture and television studios and similar locations | 530 | |
| Motors, motor circuits, and controllers | 430 | |
| Pipe organs | | 650.7 |
| Recreational vehicles and recreational vehicle parks | 551 | |
| Sound-recording and similar equipment | | 640.8 |
| Switchboards and panelboards | | 408.32 |
| Theaters, audience areas of motion picture and television studios, and similar locations | | 520.41, 520.52, 520.62 |
| X-ray equipment | | 660.2, 517.73 |

FPN: A 3-phase, 4-wire, wye-connected power system used to supply power to nonlinear loads may necessitate that the power system design allow for the possibility of high harmonic neutral currents.

**(B) Dwelling Units.** In dwelling units, a multiwire branch circuit supplying more than one device or equipment on the same yoke shall be provided with a means to disconnect simultaneously all ungrounded conductors at the panelboard where the branch circuit originated.

**(C) Line-to-Neutral Loads.** Multiwire branch circuits shall supply only line-to-neutral loads.

*Exception No. 1: A multiwire branch circuit that supplies only one utilization equipment.*

*Exception No. 2: Where all ungrounded conductors of the multiwire branch circuit are opened simultaneously by the branch-circuit overcurrent device.*

FPN: See 300.13(B) for continuity of grounded conductor on multiwire circuits.

**(D) Identification of Ungrounded Conductors.** Where more than one nominal voltage system exists in a building, each ungrounded conductor of a multiwire branch circuit, where accessible, shall be identified by phase and system. This means of identification shall be permitted to be by separate color coding, marking tape, tagging, or other approved means and shall be permanently posted at each branch-circuit panelboard.

**210.5 Identification for Branch Circuits.**

**(A) Grounded Conductor.** The grounded conductor of a branch circuit shall be identified in accordance with 200.6.

**(B) Equipment Grounding Conductor.** The equipment grounding conductor shall be identified in accordance with 250.119.

**210.6 Branch-Circuit Voltage Limitations.** The nominal voltage of branch circuits shall not exceed the values permitted by 210.6(A) through (E).

**(A) Occupancy Limitation.** In dwelling units and guest rooms of hotels, motels, and similar occupancies, the voltage shall not exceed 120 volts, nominal, between conductors that supply the terminals of the following:

(1) Luminaires (lighting fixtures)
(2) Cord-and-plug-connected loads 1440 volt-amperes, nominal, or less or less than ¼ hp

**(B) 120 Volts Between Conductors.** Circuits not exceeding 120 volts, nominal, between conductors shall be permitted to supply the following:

(1) The terminals of lampholders applied within their voltage ratings
(2) Auxiliary equipment of electric-discharge lamps
(3) Cord-and-plug-connected or permanently connected utilization equipment

**(C) 277 Volts to Ground.** Circuits exceeding 120 volts, nominal, between conductors and not exceeding 277 volts, nominal, to ground shall be permitted to supply the following:

(1) Listed electric-discharge luminaires (lighting fixtures)
(2) Listed incandescent luminaires (lighting fixtures), where supplied at 120 volts or less from the output of a stepdown autotransformer that is an integral component of the luminaire (fixture) and the outer shell terminal is electrically connected to a grounded conductor of the branch circuit
(3) Luminaires (lighting fixtures) equipped with mogul-base screw shell lampholders
(4) Lampholders, other than the screw shell type, applied within their voltage ratings
(5) Auxiliary equipment of electric-discharge lamps
(6) Cord-and-plug-connected or permanently connected utilization equipment

**(D) 600 Volts Between Conductors.** Circuits exceeding 277 volts, nominal, to ground and not exceeding 600 volts, nominal, between conductors shall be permitted to supply the following:

(1) The auxiliary equipment of electric-discharge lamps mounted in permanently installed luminaires (fixtures) where the luminaires (fixtures) are mounted in accordance with one of the following:
   a. Not less than a height of 6.7 m (22 ft) on poles or similar structures for the illumination of outdoor areas such as highways, roads, bridges, athletic fields, or parking lots
   b. Not less than a height of 5.5 m (18 ft) on other structures such as tunnels
(2) Cord-and-plug-connected or permanently connected utilization equipment

FPN: See 410.78 for auxiliary equipment limitations.

*Exception No. 1 to (B), (C), and (D): For lampholders of infrared industrial heating appliances as provided in 422.14.*

*Exception No. 2 to (B), (C), and (D): For railway properties as described in 110.19.*

**(E) Over 600 Volts Between Conductors.** Circuits exceeding 600 volts, nominal, between conductors shall be permitted to supply utilization equipment in installations where conditions of maintenance and supervision ensure that only qualified persons service the installation.

## 210.7 Branch Circuit Receptacle Requirements.

**(A) Receptacle Outlet Location.** Receptacle outlets shall be located in branch circuits in accordance with Part III of Article 210.

**(B) Receptacle Requirements.** Specific requirements for receptacles are covered in Article 406.

**(C) Multiple Branch Circuits.** Where more than one branch circuit supplies more than one receptacle on the same yoke, a means to simultaneously disconnect the ungrounded conductors supplying those receptacles shall be provided at the panelboard where the branch circuits originated.

## 210.8 Ground-Fault Circuit-Interrupter Protection for Personnel.

> FPN: See 215.9 for ground-fault circuit-interrupter protection for personnel on feeders.

**(A) Dwelling Units.** All 125-volt, single-phase, 15- and 20-ampere receptacles installed in the locations specified in (1) through (8) shall have ground-fault circuit-interrupter protection for personnel.

(1) Bathrooms

(2) Garages, and also accessory buildings that have a floor located at or below grade level not intended as habitable rooms and limited to storage areas, work areas, and areas of similar use

*Exception No. 1: Receptacles that are not readily accessible.*

*Exception No. 2: A single receptacle or a duplex receptacle for two appliances located within dedicated space for each appliance that, in normal use, is not easily moved from one place to another and that is cord-and-plug connected in accordance with 400.7(A)(6), (A)(7), or (A)(8).*

Receptacles installed under the exceptions to 210.8(A)(2) shall not be considered as meeting the requirements of 210.52(G).

(3) Outdoors

*Exception: Receptacles that are not readily accessible and are supplied by a dedicated branch circuit for electric snow-melting or deicing equipment shall be permitted to be installed in accordance with the applicable provisions of Article 426.*

(4) Crawl spaces — at or below grade level

(5) Unfinished basements — for purposes of this section, unfinished basements are defined as portions or areas of the basement not intended as habitable rooms and limited to storage areas, work areas, and the like

*Exception No. 1: Receptacles that are not readily accessible.*

*Exception No. 2: A single receptacle or a duplex receptacle for two appliances located within dedicated space for each appliance that, in normal use, is not easily moved from one place to another and that is cord-and-plug connected in accordance with 400.7(A)(6), (A)(7), or (A)(8).*

*Exception No. 3: A receptacle supplying only a permanently installed fire alarm or burglar alarm system shall not be required to have ground-fault circuit-interrupter protection.*

Receptacles installed under the exceptions to 210.8(A)(5) shall not be considered as meeting the requirements of 210.52(G).

(6) Kitchens — where the receptacles are installed to serve the countertop surfaces

(7) Wet bar sinks — where the receptacles are installed to serve the countertop surfaces and are located within 1.8 m (6 ft) of the outside edge of the wet bar sink.

(8) Boathouses

**(B) Other Than Dwelling Units.** All 125-volt, single-phase, 15- and 20-ampere receptacles installed in the locations specified in (1), (2), and (3) shall have ground-fault circuit-interrupter protection for personnel:

(1) Bathrooms

(2) Rooftops

*Exception: Receptacles that are not readily accessible and are supplied from a dedicated branch circuit for electric snow-melting or deicing equipment shall be permitted to be installed in accordance with the applicable provisions of Article 426.*

(3) Kitchens

## 210.9 Circuits Derived from Autotransformers.

Branch circuits shall not be derived from autotransformers unless the circuit supplied has a grounded conductor that is electrically connected to a grounded conductor of the system supplying the autotransformer.

*Exception No. 1: An autotransformer shall be permitted without the connection to a grounded conductor where transforming from a nominal 208 volts to a nominal 240-volt supply or similarly from 240 volts to 208 volts.*

*Exception No. 2: In industrial occupancies, where conditions of maintenance and supervision ensure that only qualified persons service the installation, autotransformers shall be permitted to supply nominal 600-volt loads from nominal 480-volt systems, and 480-volt loads from nominal 600-volt systems, without the connection to a similar grounded conductor.*

## 210.10 Ungrounded Conductors Tapped from Grounded Systems.

Two-wire dc circuits and ac circuits of two or more ungrounded conductors shall be permitted to be tapped from the ungrounded conductors of circuits that have a grounded neutral conductor. Switching devices in each tapped circuit shall have a pole in each ungrounded conductor. All poles of multipole switching devices shall manually switch together where such switching devices also serve as a disconnecting means as required by the following:

(1) 410.48 for double-pole switched lampholders
(2) 410.54(B) for electric-discharge lamp auxiliary equipment switching devices
(3) 422.31(B) for an appliance
(4) 424.20 for a fixed electric space-heating unit
(5) 426.51 for electric deicing and snow-melting equipment
(6) 430.85 for a motor controller
(7) 430.103 for a motor

**210.11 Branch Circuits Required.** Branch circuits for lighting and for appliances, including motor-operated appliances, shall be provided to supply the loads computed in accordance with 220.3. In addition, branch circuits shall be provided for specific loads not covered by 220.3 where required elsewhere in this *Code* and for dwelling unit loads as specified in 210.11(C).

**(A) Number of Branch Circuits.** The minimum number of branch circuits shall be determined from the total computed load and the size or rating of the circuits used. In all installations, the number of circuits shall be sufficient to supply the load served. In no case shall the load on any circuit exceed the maximum specified by 220.4.

**(B) Load Evenly Proportioned Among Branch Circuits.** Where the load is computed on a volt-amperes/square meter or square foot basis, the wiring system up to and including the branch-circuit panelboard(s) shall be provided to serve not less than the calculated load. This load shall be evenly proportioned among multioutlet branch circuits within the panelboard(s). Branch-circuit overcurrent devices and circuits shall only be required to be installed to serve the connected load.

**(C) Dwelling Units.**

**(1) Small-Appliance Branch Circuits.** In addition to the number of branch circuits required by other parts of this section, two or more 20-ampere small-appliance branch circuits shall be provided for all receptacle outlets specified by 210.52(B).

**(2) Laundry Branch Circuits.** In addition to the number of branch circuits required by other parts of this section, at least one additional 20-ampere branch circuit shall be provided to supply the laundry receptacle outlet(s) required by 210.52(F). This circuit shall have no other outlets.

**(3) Bathroom Branch Circuits.** In addition to the number of branch circuits required by other parts of this section, at least one 20-ampere branch circuit shall be provided to supply the bathroom receptacle outlet(s). Such circuits shall have no other outlets.

*Exception: Where the 20-ampere circuit supplies a single bathroom, outlets for other equipment within the same bathroom shall be permitted to be supplied in accordance with 210.23(A).*

FPN: See Examples D1(A), D1(B), D2(B), and D4(A) in Annex D.

**210.12 Arc-Fault Circuit-Interrupter Protection.**

**(A) Definition.** An *arc-fault circuit interrupter* is a device intended to provide protection from the effects of arc faults by recognizing characteristics unique to arcing and by functioning to de-energize the circuit when an arc fault is detected.

**(B) Dwelling Unit Bedrooms.** All branch circuits that supply 125-volt, single-phase, 15- and 20-ampere outlets installed in dwelling unit bedrooms shall be protected by an arc-fault circuit interrupter listed to provide protection of the entire branch circuit.

**II. Branch-Circuit Ratings**

**210.19 Conductors — Minimum Ampacity and Size.**

**(A) Branch Circuits Not More Than 600 Volts.**

**(1) General.** Branch-circuit conductors shall have an ampacity not less than the maximum load to be served. Where a branch circuit supplies continuous loads or any combination of continuous and noncontinuous loads, the minimum branch-circuit conductor size, before the application of any adjustment or correction factors, shall have an allowable ampacity not less than the noncontinuous load plus 125 percent of the continuous load.

*Exception: Where the assembly, including the overcurrent devices protecting the branch circuit(s), is listed for operation at 100 percent of its rating, the allowable ampacity of the branch circuit conductors shall be permitted to be not less than the sum of the continuous load plus the noncontinuous load.*

FPN No. 1: See 310.15 for ampacity ratings of conductors.

FPN No. 2: See Part II of Article 430 for minimum rating of motor branch-circuit conductors.

FPN No. 3: See 310.10 for temperature limitation of conductors.

FPN No. 4: Conductors for branch circuits as defined in Article 100, sized to prevent a voltage drop exceeding 3 percent at the farthest outlet of power, heating, and lighting loads, or combinations of such loads, and where the maximum total voltage drop on both feeders and branch circuits to the farthest outlet does not exceed 5 percent, provide reasonable efficiency of operation. See 215.2 for voltage drop on feeder conductors.

**(2) Multioutlet Branch Circuits.** Conductors of branch circuits supplying more than one receptacle for cord-and-plug-connected portable loads shall have an ampacity of not less than the rating of the branch circuit.

**(3) Household Ranges and Cooking Appliances.** Branch-circuit conductors supplying household ranges, wall-mounted ovens, counter-mounted cooking units, and other household cooking appliances shall have an ampacity not less than the rating of the branch circuit and not less than the maximum load to be served. For ranges of 8¾ kW or more rating, the minimum branch-circuit rating shall be 40 amperes.

*Exception No. 1: Tap conductors supplying electric ranges, wall-mounted electric ovens, and counter-mounted electric cooking units from a 50-ampere branch circuit shall have an ampacity of not less than 20 and shall be sufficient for the load to be served. The taps shall not be longer than necessary for servicing the appliance.*

*Exception No. 2: The neutral conductor of a 3-wire branch circuit supplying a household electric range, a wall-mounted oven, or a counter-mounted cooking unit shall be permitted to be smaller than the ungrounded conductors where the maximum demand of a range of 8¾ kW or more rating has been computed according to Column C of Table 220.19, but shall have an ampacity of not less than 70 percent of the branch-circuit rating and shall not be smaller than 10 AWG.*

**(4) Other Loads.** Branch-circuit conductors that supply loads other than those specified in 210.2 and other than cooking appliances as covered in 210.19(A)(3) shall have an ampacity sufficient for the loads served and shall not be smaller than 14 AWG.

*Exception No. 1: Tap conductors shall have an ampacity sufficient for the load served. In addition, they shall have an ampacity of not less than 15 for circuits rated less than 40 amperes and not less than 20 for circuits rated at 40 or 50 amperes and only where these tap conductors supply any of the following loads:*

*(a) Individual lampholders or luminaires (fixtures) with taps extending not longer than 450 mm (18 in.) beyond any portion of the lampholder or luminaire (fixture).*
*(b) A fixture having tap conductors as provided in 410.67.*
*(c) Individual outlets, other than receptacle outlets, with taps not over 450 mm (18 in.) long.*
*(d) Infrared lamp industrial heating appliances.*
*(e) Nonheating leads of deicing and snow-melting cables and mats.*

*Exception No. 2: Fixture wires and flexible cords shall be permitted to be smaller than 14 AWG as permitted by 240.5.*

**(B) Branch Circuits Over 600 Volts.** The ampacity of conductors shall be in accordance with 310.15 and 310.60 as applicable. Branch-circuit conductors over 600 volts shall be sized in accordance with 210.19(B)(1) or (B)(2).

**(1) General.** The ampacity of branch-circuit conductors shall not be less than 125 percent of the designed potential load of utilization equipment that will be operated simultaneously.

**(2) Supervised Installations.** For supervised installations, branch-circuit conductor sizing shall be permitted to be determined by qualified persons under engineering supervision. Supervised installations are defined as those portions of a facility where all of the following conditions are met:

(1) Conditions of design and installation are provided under engineering supervision.
(2) Qualified persons with documented training and experience in over 600-volt systems provide maintenance, monitoring, and servicing of the system.

**210.20 Overcurrent Protection.** Branch-circuit conductors and equipment shall be protected by overcurrent protective devices that have a rating or setting that complies with 210.20(A) through (D).

**(A) Continuous and Noncontinuous Loads.** Where a branch circuit supplies continuous loads or any combination of continuous and noncontinuous loads, the rating of the overcurrent device shall not be less than the noncontinuous load plus 125 percent of the continuous load.

*Exception: Where the assembly, including the overcurrent devices protecting the branch circuit(s), is listed for operation at 100 percent of its rating, the ampere rating of the overcurrent device shall be permitted to be not less than the sum of the continuous load plus the noncontinuous load.*

**(B) Conductor Protection.** Conductors shall be protected in accordance with 240.4. Flexible cords and fixture wires shall be protected in accordance with 240.5.

**(C) Equipment.** The rating or setting of the overcurrent protective device shall not exceed that specified in the applicable articles referenced in 240.3 for equipment.

**(D) Outlet Devices.** The rating or setting shall not exceed that specified in 210.21 for outlet devices.

**210.21 Outlet Devices.** Outlet devices shall have an ampere rating that is not less than the load to be served and shall comply with 210.21(A) and (B).

**(A) Lampholders.** Where connected to a branch circuit having a rating in excess of 20 amperes, lampholders shall be of the heavy-duty type. A heavy-duty lampholder shall have a rating of not less than 660 watts if of the admedium type and not less than 750 watts if of any other type.

**(B) Receptacles.**

**(1) Single Receptacle on an Individual Branch Circuit.** A single receptacle installed on an individual branch circuit shall have an ampere rating not less than that of the branch circuit.

*Exception No. 1: A receptacle installed in accordance with 430.81(C).*

*Exception No. 2: A receptacle installed exclusively for the use of a cord-and-plug-connected arc welder shall be permitted to have an ampere rating not less than the minimum branch-circuit conductor ampacity determined by 630.11(A) for arc welders.*

FPN: See definition of *receptacle* in Article 100.

**(2) Total Cord-and-Plug-Connected Load.** Where connected to a branch circuit supplying two or more receptacles or outlets, a receptacle shall not supply a total cord-and-plug-connected load in excess of the maximum specified in Table 210.21(B)(2).

**Table 210.21(B)(2) Maximum Cord-and-Plug-Connected Load to Receptacle**

| Circuit Rating (Amperes) | Receptacle Rating (Amperes) | Maximum Load (Amperes) |
|---|---|---|
| 15 or 20 | 15 | 12 |
| 20 | 20 | 16 |
| 30 | 30 | 24 |

**(3) Receptacle Ratings.** Where connected to a branch circuit supplying two or more receptacles or outlets, receptacle ratings shall conform to the values listed in Table 210.21(B)(3), or where larger than 50 amperes, the receptacle rating shall not be less than the branch-circuit rating.

*Exception No. 1: Receptacles for one or more cord-and-plug-connected arc welders shall be permitted to have ampere ratings not less than the minimum branch-circuit conductor ampacity permitted by 630.11(A) or (B) as applicable for arc welders.*

*Exception No. 2: The ampere rating of a receptacle installed for electric discharge lighting shall be permitted to be based on 410.30(C).*

**(4) Range Receptacle Rating.** The ampere rating of a range receptacle shall be permitted to be based on a single range demand load as specified in Table 220.19.

**Table 210.21(B)(3) Receptacle Ratings for Various Size Circuits**

| Circuit Rating (Amperes) | Receptacle Rating (Amperes) |
|---|---|
| 15 | Not over 15 |
| 20 | 15 or 20 |
| 30 | 30 |
| 40 | 40 or 50 |
| 50 | 50 |

**210.23 Permissible Loads.** In no case shall the load exceed the branch-circuit ampere rating. An individual branch circuit shall be permitted to supply any load for which it is rated. A branch circuit supplying two or more outlets or receptacles shall supply only the loads specified according to its size as specified in 210.23(A) through (D) and as summarized in 210.24 and Table 210.24.

**(A) 15- and 20-Ampere Branch Circuits.** A 15- or 20-ampere branch circuit shall be permitted to supply lighting units or other utilization equipment, or a combination of both, and shall comply with 210.23(A)(1) and (A)(2).

*Exception: The small appliance branch circuits, laundry branch circuits, and bathroom branch circuits required in a dwelling unit(s) by 210.11(C)(1), (2), and (3) shall supply only the receptacle outlets specified in that section.*

**(1) Cord-and-Plug-Connected Equipment.** The rating of any one cord-and-plug-connected utilization equipment shall not exceed 80 percent of the branch-circuit ampere rating.

**(2) Utilization Equipment Fastened in Place.** The total rating of utilization equipment fastened in place, other than luminaires (lighting fixtures), shall not exceed 50 percent of the branch-circuit ampere rating where lighting units, cord-and-plug-connected utilization equipment not fastened in place, or both, are also supplied.

**(B) 30-Ampere Branch Circuits.** A 30-ampere branch circuit shall be permitted to supply fixed lighting units with heavy-duty lampholders in other than a dwelling unit(s) or utilization equipment in any occupancy. A rating of any one cord-and-plug-connected utilization equipment shall not exceed 80 percent of the branch-circuit ampere rating.

**(C) 40- and 50-Ampere Branch Circuits.** A 40- or 50-ampere branch circuit shall be permitted to supply cooking appliances that are fastened in place in any occupancy. In other than dwelling units, such circuits shall be permitted to supply fixed lighting units with heavy-duty lampholders, infrared heating units, or other utilization equipment.

**(D) Branch Circuits Larger Than 50 Amperes.** Branch circuits larger than 50 amperes shall supply only nonlighting outlet loads.

**210.24 Branch-Circuit Requirements — Summary.** The requirements for circuits that have two or more outlets or receptacles, other than the receptacle circuits of 210.11(C)(1) and (2), are summarized in Table 210.24. This table provides only a summary of minimum requirements. See 210.19, 210.20, and 210.21 for the specific requirements applying to branch circuits.

**210.25 Common Area Branch Circuits.** Branch circuits in dwelling units shall supply only loads within that dwelling unit or loads associated only with that dwelling unit. Branch circuits required for the purpose of lighting, central alarm, signal, communications, or other needs for public or common areas of a two-family or multifamily dwelling shall not be supplied from equipment that supplies an individual dwelling unit.

## III. Required Outlets

**210.50 General.** Receptacle outlets shall be installed as specified in 210.52 through 210.63.

**(A) Cord Pendants.** A cord connector that is supplied by a permanently connected cord pendant shall be considered a receptacle outlet.

**(B) Cord Connections.** A receptacle outlet shall be installed wherever flexible cords with attachment plugs are used. Where flexible cords are permitted to be permanently connected, receptacles shall be permitted to be omitted for such cords.

**(C) Appliance Outlets.** Appliance receptacle outlets installed in a dwelling unit for specific appliances, such as laundry equipment, shall be installed within 1.8 m (6 ft) of the intended location of the appliance.

**210.52 Dwelling Unit Receptacle Outlets.** This section provides requirements for 125-volt, 15- and 20-ampere receptacle outlets. Receptacle outlets required by this section shall be in addition to any receptacle that is part of a luminaire (lighting fixture) or appliance, located within cabinets or cupboards, or located more than 1.7 m (5½ ft) above the floor.

Permanently installed electric baseboard heaters equipped with factory-installed receptacle outlets or outlets provided as a separate assembly by the manufacturer shall be permitted as the required outlet or outlets for the wall space utilized by such permanently installed heaters. Such receptacle outlets shall not be connected to the heater circuits.

FPN: Listed baseboard heaters include instructions that may not permit their installation below receptacle outlets.

**(A) General Provisions.** In every kitchen, family room, dining room, living room, parlor, library, den, sunroom, bedroom, recreation room, or similar room or area of dwelling units, receptacle outlets shall be installed in accordance with the general provisions specified in 210.52(A)(1) through (A)(3).

**(1) Spacing.** Receptacles shall be installed so that no point measured horizontally along the floor line in any wall space is more than 1.8 m (6 ft) from a receptacle outlet.

**(2) Wall Space.** As used in this section, a wall space shall include the following:

(1) Any space 600 mm (2 ft) or more in width (including space measured around corners) and unbroken along the floor line by doorways, fireplaces, and similar openings
(2) The space occupied by fixed panels in exterior walls, excluding sliding panels

**Table 210.24 Summary of Branch-Circuit Requirements**

| Circuit Rating | 15 A | 20 A | 30 A | 40 A | 50 A |
|---|---|---|---|---|---|
| Conductors (min. size): | | | | | |
|   Circuit wires[1] | 14 | 12 | 10 | 8 | 6 |
|   Taps | 14 | 14 | 14 | 12 | 12 |
|   Fixture wires and cords — See 240.5 | | | | | |
| **Overcurrent Protection** | **15 A** | **20 A** | **30 A** | **40 A** | **50 A** |
| Outlet devices: | | | | | |
|   Lampholders permitted | Any type | Any type | Heavy duty | Heavy duty | Heavy duty |
|   Receptacle rating[2] | 15 max. A | 15 or 20 A | 30 A | 40 or 50 A | 50 A |
| **Maximum Load** | **15 A** | **20 A** | **30 A** | **40 A** | **50 A** |
| Permissible load | See 210.23(A) | See 210.23(A) | See 210.23(B) | See 210.23(C) | See 210.23(C) |

[1]These gauges are for copper conductors.
[2]For receptacle rating of cord-connected electric-discharge luminaires (lighting fixtures), see 410.30(C).

(3) The space afforded by fixed room dividers such as free-standing bar-type counters or railings

**(3) Floor Receptacles.** Receptacle outlets in floors shall not be counted as part of the required number of receptacle outlets unless located within 450 mm (18 in.) of the wall.

**(B) Small Appliances.**

**(1) Receptacle Outlets Served.** In the kitchen, pantry, breakfast room, dining room, or similar area of a dwelling unit, the two or more 20-ampere small-appliance branch circuits required by 210.11(C)(1) shall serve all receptacle outlets covered by 210.52(A) and (C) and receptacle outlets for refrigeration equipment.

*Exception No. 1: In addition to the required receptacles specified by 210.52, switched receptacles supplied from a general-purpose branch circuit as defined in 210.70(A)(1), Exception No. 1, shall be permitted.*

*Exception No. 2: The receptacle outlet for refrigeration equipment shall be permitted to be supplied from an individual branch circuit rated 15 amperes or greater.*

**(2) No Other Outlets.** The two or more small-appliance branch circuits specified in 210.52(B)(1) shall have no other outlets.

*Exception No. 1: A receptacle installed solely for the electrical supply to and support of an electric clock in any of the rooms specified in 210.52(B)(1).*

*Exception No. 2: Receptacles installed to provide power for supplemental equipment and lighting on gas-fired ranges, ovens, or counter-mounted cooking units.*

**(3) Kitchen Receptacle Requirements.** Receptacles installed in a kitchen to serve countertop surfaces shall be supplied by not fewer than two small-appliance branch circuits, either or both of which shall also be permitted to supply receptacle outlets in the same kitchen and in other rooms specified in 210.52(B)(1). Additional small-appliance branch circuits shall be permitted to supply receptacle outlets in the kitchen and other rooms specified in 210.52(B)(1). No small-appliance branch circuit shall serve more than one kitchen.

**(C) Countertops.** In kitchens and dining rooms of dwelling units, receptacle outlets for counter spaces shall be installed in accordance with 210.52(C)(1) through (5).

**(1) Wall Counter Spaces.** A receptacle outlet shall be installed at each wall counter space that is 300 mm (12 in.) or wider. Receptacle outlets shall be installed so that no point along the wall line is more than 600 mm (24 in.) measured horizontally from a receptacle outlet in that space.

**(2) Island Counter Spaces.** At least one receptacle outlet shall be installed at each island counter space with a long dimension of 600 mm (24 in.) or greater and a short dimension of 300 mm (12 in.) or greater.

**(3) Peninsular Counter Spaces.** At least one receptacle outlet shall be installed at each peninsular counter space with a long dimension of 600 mm (24 in.) or greater and a short dimension of 300 mm (12 in.) or greater. A peninsular countertop is measured from the connecting edge.

**(4) Separate Spaces.** Countertop spaces separated by range tops, refrigerators, or sinks shall be considered as separate countertop spaces in applying the requirements of 210.52(C)(1), (2), and (3).

**(5) Receptacle Outlet Location.** Receptacle outlets shall be located above, but not more than 500 mm (20 in.) above, the countertop. Receptacle outlets rendered not readily accessible by appliances fastened in place, appliance garages, or appliances occupying dedicated space shall not be considered as these required outlets.

*Exception: To comply with the conditions specified in (a) or (b), receptacle outlets shall be permitted to be mounted not more than 300 mm (12 in.) below the countertop. Receptacles mounted below a countertop in accordance with this exception shall not be located where the countertop extends more than 150 mm (6 in.) beyond its support base.*

*(a) Construction for the physically impaired.*
*(b) On island and peninsular countertops where the countertop is flat across its entire surface (no backsplashes, dividers, etc.) and there are no means to mount a receptacle within 500 mm (20 in.) above the countertop, such as an overhead cabinet.*

**(D) Bathrooms.** In dwelling units, at least one wall receptacle outlet shall be installed in bathrooms within 900 mm (3 ft) of the outside edge of each basin. The receptacle outlet shall be located on a wall or partition that is adjacent to the basin or basin countertop.

**(E) Outdoor Outlets.** For a one-family dwelling and each unit of a two-family dwelling that is at grade level, at least one receptacle outlet accessible at grade level and not more than 2.0 m (6½ ft) above grade shall be installed at the front and back of the dwelling. See 210.8(A)(3).

**(F) Laundry Areas.** In dwelling units, at least one receptacle outlet shall be installed for the laundry.

*Exception No. 1: In a dwelling unit that is an apartment or living area in a multifamily building where laundry facilities are provided on the premises and are available to all building occupants, a laundry receptacle shall not be required.*

*Exception No. 2: In other than one-family dwellings where laundry facilities are not to be installed or permitted, a laundry receptacle shall not be required.*

**(G) Basements and Garages.** For a one-family dwelling, at least one receptacle outlet, in addition to any provided for laundry equipment, shall be installed in each basement and in each attached garage, and in each detached garage with electric power. See 210.8(A)(2) and (A)(5). Where a portion of the basement is finished into one or more habitable rooms, each separate unfinished portion shall have a receptacle outlet installed in accordance with this section.

**(H) Hallways.** In dwelling units, hallways of 3.0 m (10 ft) or more in length shall have at least one receptacle outlet.

As used in this subsection, the hall length shall be considered the length along the centerline of the hall without passing through a doorway.

### 210.60 Guest Rooms.

**(A) General.** Guest rooms in hotels, motels, and similar occupancies shall have receptacle outlets installed in accordance with 210.52(A) and 210.52(D). Guest rooms meeting the definition of a dwelling unit shall have receptacle outlets installed in accordance with all of the applicable rules in 210.52.

**(B) Receptacle Placement.** In applying the provisions of 210.52(A), the total number of receptacle outlets shall not be less than the minimum number that would comply with the provisions of that section. These receptacle outlets shall be permitted to be located conveniently for permanent furniture layout. At least two receptacle outlets shall be readily accessible. Where receptacles are installed behind the bed, the receptacle shall be located to prevent the bed from contacting any attachment plug that may be installed, or the receptacle shall be provided with a suitable guard.

**210.62 Show Windows.** At least one receptacle outlet shall be installed directly above a show window for each 3.7 linear m (12 ft) or major fraction thereof of show window area measured horizontally at its maximum width.

**210.63 Heating, Air-Conditioning, and Refrigeration Equipment Outlet.** A 125-volt, single-phase, 15- or 20-ampere-rated receptacle outlet shall be installed at an accessible location for the servicing of heating, air-conditioning, and refrigeration equipment. The receptacle shall be located on the same level and within 7.5 m (25 ft) of the heating, air-conditioning, and refrigeration equipment. The receptacle outlet shall not be connected to the load side of the equipment disconnecting means.

> FPN: See 210.8 for ground-fault circuit-interrupter requirements.

**210.70 Lighting Outlets Required.** Lighting outlets shall be installed where specified in 210.70(A), (B), and (C).

**(A) Dwelling Units.** In dwelling units, lighting outlets shall be installed in accordance with 210.70(A)(1), (2), and (3).

**(1) Habitable Rooms.** At least one wall switch-controlled lighting outlet shall be installed in every habitable room and bathroom.

*Exception No. 1: In other than kitchens and bathrooms, one or more receptacles controlled by a wall switch shall be permitted in lieu of lighting outlets.*

*Exception No. 2: Lighting outlets shall be permitted to be controlled by occupancy sensors that are (1) in addition to wall switches or (2) located at a customary wall switch location and equipped with a manual override that will allow the sensor to function as a wall switch.*

**(2) Additional Locations.** Additional lighting outlets shall be installed in accordance with (a), (b), and (c).

(a) At least one wall switch-controlled lighting outlet shall be installed in hallways, stairways, attached garages, and detached garages with electric power.

(b) For dwelling units, attached garages, and detached garages with electric power, at least one wall switch-controlled lighting outlet shall be installed to provide illumination on the exterior side of outdoor entrances or exits with grade level access. A vehicle door in a garage shall not be considered as an outdoor entrance or exit.

(c) Where one or more lighting outlet(s) are installed for interior stairways, there shall be a wall switch at each floor level, and landing level that includes an entry way, to control the lighting outlet(s) where the stairway between floor levels has six risers or more.

*Exception to (a), (b), and (c): In hallways, stairways, and at outdoor entrances, remote, central, or automatic control of lighting shall be permitted.*

**(3) Storage or Equipment Spaces.** For attics, underfloor spaces, utility rooms, and basements, at least one lighting outlet containing a switch or controlled by a wall switch shall be installed where these spaces are used for storage or contain equipment requiring servicing. At least one point of control shall be at the usual point of entry to these spaces. The lighting outlet shall be provided at or near the equipment requiring servicing.

**(B) Guest Rooms.** At least one wall switch–controlled lighting outlet or wall switch–controlled receptacle shall be installed in guest rooms in hotels, motels, or similar occupancies.

**(C) Other Than Dwelling Units.** For attics and underfloor spaces containing equipment requiring servicing, such as heating, air-conditioning, and refrigeration equipment, at least one lighting outlet containing a switch or controlled by a wall switch shall be installed in such spaces. At least one point of control shall be at the usual point of entry to these spaces. The lighting outlet shall be provided at or near the equipment requiring servicing.

# ARTICLE 215
# Feeders

**215.1 Scope.** This article covers the installation requirements, overcurrent protection requirements, minimum size, and ampacity of conductors for feeders supplying branch-circuit loads as computed in accordance with Article 220.

*Exception: Feeders for electrolytic cells as covered in 668.3(C)(1) and (4).*

## 215.2 Minimum Rating and Size.

**(A) Feeders Not More Than 600 Volts.**

**(1) General.** Feeder conductors shall have an ampacity not less than required to supply the load as computed in Parts II, III, and IV of Article 220. The minimum feeder-circuit conductor size, before the application of any adjustment or correction factors, shall have an allowable ampacity not less than the noncontinuous load plus 125 percent of the continuous load.

*Exception: Where the assembly, including the overcurrent devices protecting the feeder(s), is listed for operation at 100 percent of its rating, the allowable ampacity of the feeder conductors shall be permitted to be not less than the sum of the continuous load plus the noncontinuous load.*

Additional minimum sizes shall be as specified in (2), (3), and (4) under the conditions stipulated.

**(2) For Specified Circuits.** The ampacity of feeder conductors shall not be less than 30 amperes where the load supplied consists of any of the following number and types of circuits:

(1) Two or more 2-wire branch circuits supplied by a 2-wire feeder
(2) More than two 2-wire branch circuits supplied by a 3-wire feeder
(3) Two or more 3-wire branch circuits supplied by a 3-wire feeder
(4) Two or more 4-wire branch circuits supplied by a 3-phase, 4-wire feeder

**(3) Ampacity Relative to Service-Entrance Conductors.** The feeder conductor ampacity shall not be less than that of the service-entrance conductors where the feeder conduc-

tors carry the total load supplied by service-entrance conductors with an ampacity of 55 amperes or less.

**(4) Individual Dwelling Unit or Mobile Home Conductors.** Feeder conductors for individual dwelling units or mobile homes need not be larger than service-entrance conductors. Paragraph 310.15(B)(6) shall be permitted to be used for conductor size.

FPN No. 1: See Examples D1 through D10 in Annex D.

FPN No. 2: Conductors for feeders as defined in Article 100, sized to prevent a voltage drop exceeding 3 percent at the farthest outlet of power, heating, and lighting loads, or combinations of such loads, and where the maximum total voltage drop on both feeders and branch circuits to the farthest outlet does not exceed 5 percent, will provide reasonable efficiency of operation.

FPN No. 3: See 210.19(A), FPN No. 4, for voltage drop for branch circuits.

**(B) Feeders Over 600 Volts.** The ampacity of conductors shall be in accordance with 310.15 and 310.60 as applicable. Feeder conductors over 600 volts shall be sized in accordance with 215.2(B)(1), (2), or (3).

**(1) Feeders Supplying Transformers.** The ampacity of feeder conductors shall not be less than the sum of the nameplate ratings of the transformers supplied when only transformers are supplied.

**(2) Feeders Supplying Transformers and Utilization Equipment.** The ampacity of feeders supplying a combination of transformers and utilization equipment shall not be less than the sum of the nameplate ratings of the transformers and 125 percent of the designed potential load of the utilization equipment that will be operated simultaneously.

**(3) Supervised Installations.** For supervised installations, feeder conductor sizing shall be permitted to be determined by qualified persons under engineering supervision. Supervised installations are defined as those portions of a facility where all of the following conditions are met:

(1) Conditions of design and installation are provided under engineering supervision.
(2) Qualified persons with documented training and experience in over 600-volt systems provide maintenance, monitoring, and servicing of the system.

**215.3 Overcurrent Protection.** Feeders shall be protected against overcurrent in accordance with the provisions of Part I of Article 240. Where a feeder supplies continuous loads or any combination of continuous and noncontinuous loads, the rating of the overcurrent device shall not be less than the noncontinuous load plus 125 percent of the continuous load.

*Exception No. 1: Where the assembly, including the overcurrent devices protecting the feeder(s), is listed for operation at 100 percent of its rating, the ampere rating of the overcurrent device shall be permitted to be not less than the sum of the continuous load plus the noncontinuous load.*

*Exception No. 2: Overcurrent protection for feeders over 600 volts, nominal, shall comply with Part IX of Article 240.*

### 215.4 Feeders with Common Neutral.

**(A) Feeders with Common Neutral.** Two or three sets of 3-wire feeders or two sets of 4-wire or 5-wire feeders shall be permitted to utilize a common neutral.

**(B) In Metal Raceway or Enclosure.** Where installed in a metal raceway or other metal enclosure, all conductors of all feeders using a common neutral shall be enclosed within the same raceway or other enclosure as required in 300.20.

**215.5 Diagrams of Feeders.** If required by the authority having jurisdiction, a diagram showing feeder details shall be provided prior to the installation of the feeders. Such a diagram shall show the area in square feet of the building or other structure supplied by each feeder, the total computed load before applying demand factors, the demand factors used, the computed load after applying demand factors, and the size and type of conductors to be used.

**215.6 Feeder Conductor Grounding Means.** Where a feeder supplies branch circuits in which equipment grounding conductors are required, the feeder shall include or provide a grounding means, in accordance with the provisions of 250.134, to which the equipment grounding conductors of the branch circuits shall be connected.

**215.7 Ungrounded Conductors Tapped from Grounded Systems.** Two-wire dc circuits and ac circuits of two or more ungrounded conductors shall be permitted to be tapped from the ungrounded conductors of circuits having a grounded neutral conductor. Switching devices in each tapped circuit shall have a pole in each ungrounded conductor.

**215.8 Means of Identifying Conductor with the Higher Voltage to Ground.** On a 4-wire, delta-connected secondary where the midpoint of one phase winding is grounded to supply lighting and similar loads, the phase conductor having the higher voltage to ground shall be identified by an outer finish that is orange in color or by tagging or other effective means. Such identification shall be placed at each point where a connection is made if the grounded conductor is also present.

**215.9 Ground-Fault Circuit-Interrupter Protection for Personnel.** Feeders supplying 15- and 20-ampere receptacle branch circuits shall be permitted to be protected by a ground-fault circuit interrupter in lieu of the provisions for such interrupters as specified in 210.8 and Article 527.

**215.10 Ground-Fault Protection of Equipment.** Each feeder disconnect rated 1000 amperes or more and installed on solidly grounded wye electrical systems of more than 150 volts to ground, but not exceeding 600 volts phase-to-phase, shall be provided with ground-fault protection of equipment in accordance with the provisions of 230.95.

*Exception No. 1: The provisions of this section shall not apply to a disconnecting means for a continuous industrial process where a nonorderly shutdown will introduce additional or increased hazards.*

*Exception No. 2: The provisions of this section shall not apply to fire pumps.*

*Exception No. 3: The provisions of this section shall not apply if ground-fault protection of equipment is provided on the supply side of the feeder.*

**215.11 Circuits Derived from Autotransformers.** Feeders shall not be derived from autotransformers unless the system supplied has a grounded conductor that is electrically connected to a grounded conductor of the system supplying the autotransformer.

*Exception No. 1: An autotransformer shall be permitted without the connection to a grounded conductor where transforming from a nominal 208 volts to a nominal 240-volt supply or similarly from 240 volts to 208 volts.*

*Exception No. 2: In industrial occupancies, where conditions of maintenance and supervision ensure that only qualified persons service the installation, autotransformers shall be permitted to supply nominal 600-volt loads from nominal 480-volt systems, and 480-volt loads from nominal 600-volt systems, without the connection to a similar grounded conductor.*

## ARTICLE 220
## Branch-Circuit, Feeder, and Service Calculations

### I. General

**220.1 Scope.** This article provides requirements for computing branch-circuit, feeder, and service loads.

*Exception: Branch-circuit and feeder calculations for electrolytic cells as covered in 668.3(C)(1) and (4).*

### 220.2 Computations.

**(A) Voltages.** Unless other voltages are specified, for purposes of computing branch-circuit and feeder loads, nominal system voltages of 120, 120/240, 208Y/120, 240, 347, 480Y/277, 480, 600Y/347, and 600 volts shall be used.

**(B) Fractions of an Ampere.** Where computations result in a fraction of an ampere that is less than 0.5, such fractions shall be permitted to be dropped.

**220.3 Computation of Branch Circuit Loads.** Branch-circuit loads shall be computed as shown in 220.3(A) through (C).

**(A) Lighting Load for Specified Occupancies.** A unit load of not less than that specified in Table 220.3(A) for occupancies specified therein shall constitute the minimum lighting load. The floor area for each floor shall be computed from the outside dimensions of the building, dwelling unit, or other area involved. For dwelling units, the computed floor area shall not include open porches, garages, or unused or unfinished spaces not adaptable for future use.

> FPN: The unit values herein are based on minimum load conditions and 100 percent power factor and may not provide sufficient capacity for the installation contemplated.

**(B) Other Loads — All Occupancies.** In all occupancies, the minimum load for each outlet for general-use receptacles and outlets not used for general illumination shall not be less than that computed in 220.3(B)(1) through (11), the loads shown being based on nominal branch-circuit voltages.

*Exception: The loads of outlets serving switchboards and switching frames in telephone exchanges shall be waived from the computations.*

**(1) Specific Appliances or Loads.** An outlet for a specific appliance or other load not covered in (2) through (11) shall be computed based on the ampere rating of the appliance or load served.

**(2) Electric Dryers and Household Electric Cooking Appliances.** Load computations shall be permitted as specified in 220.18 for electric dryers and in 220.19 for electric ranges and other cooking appliances.

**(3) Motor Loads.** Outlets for motor loads shall be computed in accordance with the requirements in 430.22, 430.24, and 440.6.

**(4) Recessed Luminaires (Lighting Fixtures).** An outlet supplying recessed luminaire(s) [lighting fixture(s)] shall be computed based on the maximum volt-ampere rating of the equipment and lamps for which the luminaire(s) [fixture(s)] is rated.

**(5) Heavy-Duty Lampholders.** Outlets for heavy-duty lampholders shall be computed at a minimum of 600 volt-amperes.

**Table 220.3(A)  General Lighting Loads by Occupancy**

| Type of Occupancy | Unit Load | |
|---|---|---|
| | Volt-Amperes per Square Meter | Volt-Amperes per Square Foot |
| Armories and auditoriums | 11 | 1 |
| Banks | 39[b] | 3½[b] |
| Barber shops and beauty parlors | 33 | 3 |
| Churches | 11 | 1 |
| Clubs | 22 | 2 |
| Court rooms | 22 | 2 |
| Dwelling units[a] | 33 | 3 |
| Garages — commercial (storage) | 6 | ½ |
| Hospitals | 22 | 2 |
| Hotels and motels, including apartment houses without provision for cooking by tenants[a] | 22 | 2 |
| Industrial commercial (loft) buildings | 22 | 2 |
| Lodge rooms | 17 | 1½ |
| Office buildings | 39 | 3½[b] |
| Restaurants | 22 | 2 |
| Schools | 33 | 3 |
| Stores | 33 | 3 |
| Warehouses (storage) | 3 | ¼ |
| In any of the preceding occupancies except one-family dwellings and individual dwelling units of two-family and multifamily dwellings: | | |
| Assembly halls and auditoriums | 11 | 1 |
| Halls, corridors, closets, stairways | 6 | ½ |
| Storage spaces | 3 | ¼ |

[a]See 220.3(B)(10).
[b]In addition, a unit load of 11 volt-amperes/m² or 1 volt-ampere/ft² shall be included for general-purpose receptacle outlets where the actual number of general-purpose receptacle outlets is unknown.

**(6) Sign and Outline Lighting.** Sign and outline lighting outlets shall be computed at a minimum of 1200 volt-amperes for each required branch circuit specified in 600.5(A).

**(7) Show Windows.** Show windows shall be computed in accordance with either of the following:

(1) The unit load per outlet as required in other provisions of this section
(2) At 200 volt-amperes per 300 mm (1 ft) of show window

**(8) Fixed Multioutlet Assemblies.** Fixed multioutlet assemblies used in other than dwelling units or the guest rooms of hotels or motels shall be computed in accordance with (1) or (2). For the purposes of this section, the computation shall be permitted to be based on the portion that contains receptacle outlets.

(1) Where appliances are unlikely to be used simultaneously, each 1.5 m (5 ft) or fraction thereof of each separate and continuous length shall be considered as one outlet of not less than 180 volt-amperes.

(2) Where appliances are likely to be used simultaneously, each 300 mm (1 ft) or fraction thereof shall be considered as an outlet of not less than 180 volt-amperes.

**(9) Receptacle Outlets.** Except as covered in 220.3(B)(10), receptacle outlets shall be computed at not less than 180 volt-amperes for each single or for each multiple receptacle on one yoke. A single piece of equipment consisting of a multiple receptacle comprised of four or more receptacles shall be computed at not less than 90 volt-amperes per receptacle.

This provision shall not be applicable to the receptacle outlets specified in 210.11(C)(1) and (2).

**(10) Dwelling Occupancies.** In one-family, two-family, and multifamily dwellings and in guest rooms of hotels and motels, the outlets specified in (1), (2), and (3) are included in the general lighting load calculations of 220.3(A). No additional load calculations shall be required for such outlets.

(1) All general-use receptacle outlets of 20-ampere rating or less, including receptacles connected to the circuits in 210.11(C)(3)

(2) The receptacle outlets specified in 210.52(E) and (G)

(3) The lighting outlets specified in 210.70(A) and (B)

**(11) Other Outlets.** Other outlets not covered in 220.3(B)(1) through (10) shall be computed based on 180 volt-amperes per outlet.

**(C) Loads for Additions to Existing Installations.**

**(1) Dwelling Units.** Loads added to an existing dwelling unit(s) shall comply with the following as applicable:

(1) Loads for structural additions to an existing dwelling unit or for a previously unwired portion of an existing dwelling unit, either of which exceeds 46.5 m² (500 ft²), shall be computed in accordance with 220.3(A) and (B).

(2) Loads for new circuits or extended circuits in previously wired dwelling units shall be computed in accordance with either 220.3(A) or (B), as applicable.

**(2) Other Than Dwelling Units.** Loads for new circuits or extended circuits in other than dwelling units shall be computed in accordance with either 220.3(A) or (B), as applicable.

**220.4 Maximum Loads.** The total load shall not exceed the rating of the branch circuit, and it shall not exceed the maximum loads specified in 220.4(A) through (C) under the conditions specified therein.

**(A) Motor-Operated and Combination Loads.** Where a circuit supplies only motor-operated loads, Article 430 shall apply. Where a circuit supplies only air-conditioning equipment, refrigerating equipment, or both, Article 440 shall apply. For circuits supplying loads consisting of motor-operated utilization equipment that is fastened in place and has a motor larger than ⅛ hp in combination with other loads, the total computed load shall be based on 125 percent of the largest motor load plus the sum of the other loads.

**(B) Inductive Lighting Loads.** For circuits supplying lighting units that have ballasts, transformers, or autotransformers, the computed load shall be based on the total ampere ratings of such units and not on the total watts of the lamps.

**(C) Range Loads.** It shall be permissible to apply demand factors for range loads in accordance with Table 220.19, including Note 4.

## II. Feeders and Services

**220.10 General.** The computed load of a feeder or service shall not be less than the sum of the loads on the branch circuits supplied, as determined by Part I of this article, after any applicable demand factors permitted by Parts II, III, or IV have been applied.

> FPN: See Examples D1(A) through D10 in Annex D. See 220.4(B) for the maximum load in amperes permitted for lighting units operating at less than 100 percent power factor.

**220.11 General Lighting.** The demand factors specified in Table 220.11 shall apply to that portion of the total branch-circuit load computed for general illumination. They shall not be applied in determining the number of branch circuits for general illumination.

**220.12 Show-Window and Track Lighting.**

**(A) Show Windows.** For show-window lighting, a load of not less than 660 volt-amperes/linear meter or 200 volt-amperes/linear foot shall be included for a show window, measured horizontally along its base.

> FPN: See 220.3(B)(7) for branch circuits supplying show windows.

**Table 220.11 Lighting Load Demand Factors**

| Type of Occupancy | Portion of Lighting Load to Which Demand Factor Applies (Volt-Amperes) | Demand Factor (Percent) |
|---|---|---|
| Dwelling units | First 3000 or less at<br>From 3001 to 120,000 at<br>Remainder over 120,000 at | 100<br>35<br>25 |
| Hospitals* | First 50,000 or less at<br>Remainder over 50,000 at | 40<br>20 |
| Hotels and motels, including apartment houses without provision for cooking by tenants* | First 20,000 or less at<br>From 20,001 to 100,000 at<br>Remainder over 100,000 at | 50<br>40<br>30 |
| Warehouses (storage) | First 12,500 or less at<br>Remainder over 12,500 at | 100<br>50 |
| All others | Total volt-amperes | 100 |

*The demand factors of this table shall not apply to the computed load of feeders or services supplying areas in hospitals, hotels, and motels where the entire lighting is likely to be used at one time, as in operating rooms, ballrooms, or dining rooms.

**(B) Track Lighting.** For track lighting in other than dwelling units or guest rooms of hotels or motels, an additional load of 150 volt-amperes shall be included for every 600 mm (2 ft) of lighting track or fraction thereof. Where multicircuit track is installed, the load shall be considered to be divided equally between the track circuits.

**220.13 Receptacle Loads — Nondwelling Units.** In other than dwelling units, receptacle loads computed at not more than 180 volt-amperes per outlet in accordance with 220.3(B)(9) and fixed multioutlet assemblies computed in accordance with 220.3(B)(8) shall be permitted to be added to the lighting loads and made subject to the demand factors given in Table 220.11, or they shall be permitted to be made subject to the demand factors given in Table 220.13.

**Table 220.13 Demand Factors for Nondwelling Receptacle Loads**

| Portion of Receptacle Load to Which Demand Factor Applies (Volt-Amperes) | Demand Factor (Percent) |
|---|---|
| First 10 kVA or less at | 100 |
| Remainder over 10 kVA at | 50 |

**220.14 Motors.** Motor loads shall be computed in accordance with 430.24, 430.25, and 430.26 and with 440.6 for hermetic refrigerant motor compressors.

**220.15 Fixed Electric Space Heating.** Fixed electric space heating loads shall be computed at 100 percent of the total connected load; however, in no case shall a feeder or service load current rating be less than the rating of the largest branch circuit supplied.

*Exception: Where reduced loading of the conductors results from units operating on duty-cycle, intermittently, or from all units not operating at the same time, the authority having jurisdiction may grant permission for feeder and service conductors to have an ampacity less than 100 percent, provided the conductors have an ampacity for the load so determined.*

**220.16 Small Appliance and Laundry Loads — Dwelling Unit.**

**(A) Small Appliance Circuit Load.** In each dwelling unit, the load shall be computed at 1500 volt-amperes for each 2-wire small-appliance branch circuit required by 210.11(C)(1). Where the load is subdivided through two or more feeders, the computed load for each shall include not less than 1500 volt-amperes for each 2-wire small-appliance branch circuit. These loads shall be permitted to be included with the general lighting load and subjected to the demand factors provided in Table 220.11.

*Exception: The individual branch circuit permitted by 210.52(B)(1), Exception No. 2, shall be permitted to be excluded from the calculation required by 220.16.*

**(B) Laundry Circuit Load.** A load of not less than 1500 volt-amperes shall be included for each 2-wire laundry branch circuit installed as required by 210.11(C)(2). This load shall be permitted to be included with the general lighting load and subjected to the demand factors provided in Table 220.11.

**220.17 Appliance Load — Dwelling Unit(s).** It shall be permissible to apply a demand factor of 75 percent to the nameplate rating load of four or more appliances fastened in place, other than electric ranges, clothes dryers, space-heating equipment, or air-conditioning equipment, that are served by the same feeder or service in a one-family, two-family, or multifamily dwelling.

**220.18 Electric Clothes Dryers — Dwelling Unit(s).** The load for household electric clothes dryers in a dwelling unit(s) shall be 5000 watts (volt-amperes) or the nameplate rating, whichever is larger, for each dryer served. The use of the demand factors in Table 220.18 shall be permitted. Where two or more single-phase dryers are supplied by a 3-phase, 4-wire feeder or service, the total load shall be computed on the basis of twice the maximum number connected between any two phases.

**Table 220.18 Demand Factors for Household Electric Clothes Dryers**

| Number of Dryers | Demand Factor (Percent) |
|---|---|
| 1–4 | 100% |
| 5 | 85% |
| 6 | 75% |
| 7 | 65% |
| 8 | 60% |
| 9 | 55% |
| 10 | 50% |
| 11 | 47% |
| 12–22 | % = 47 – (number of dryers – 11) |
| 23 | 35% |
| 24–42 | % = 35 – [0.5 × (number of dryers – 23)] |
| 43 and over | 25% |

**220.19 Electric Ranges and Other Cooking Appliances — Dwelling Unit(s).** The demand load for household electric ranges, wall-mounted ovens, counter-mounted cooking units, and other household cooking appliances individually rated in excess of 1¾ kW shall be permitted to be computed in accordance with Table 220.19. Kilovolt-amperes (kVA) shall be considered equivalent to

**Table 220.19 Demand Loads for Household Electric Ranges, Wall-Mounted Ovens, Counter-Mounted Cooking Units, and Other Household Cooking Appliances over 1¾ kW Rating (Column C to be used in all cases except as otherwise permitted in Note 3.)**

| Number of Appliances | Demand Factor (Percent) (See Notes) | | Column C Maximum Demand (kW) (See Notes) (Not over 12 kW Rating) |
|---|---|---|---|
| | Column A (Less than 3½ kW Rating) | Column B (3½ kW to 8¾ kW Rating) | |
| 1 | 80 | 80 | 8 |
| 2 | 75 | 65 | 11 |
| 3 | 70 | 55 | 14 |
| 4 | 66 | 50 | 17 |
| 5 | 62 | 45 | 20 |
| 6 | 59 | 43 | 21 |
| 7 | 56 | 40 | 23 |
| 8 | 53 | 36 | 23 |
| 9 | 51 | 35 | 24 |
| 10 | 49 | 34 | 25 |
| 11 | 47 | 32 | 26 |
| 12 | 45 | 32 | 27 |
| 13 | 43 | 32 | 28 |

**Table 220.19**  *Continued*

| Number of Appliances | Demand Factor (Percent) (See Notes) | | Column C Maximum Demand (kW) (See Notes) (Not over 12 kW Rating) |
|---|---|---|---|
| | Column A (Less than 3½ kW Rating) | Column B (3½ kW to 8¾ kW Rating) | |
| 14 | 41 | 32 | 29 |
| 15 | 40 | 32 | 30 |
| 16 | 39 | 28 | 31 |
| 17 | 38 | 28 | 32 |
| 18 | 37 | 28 | 33 |
| 19 | 36 | 28 | 34 |
| 20 | 35 | 28 | 35 |
| 21 | 34 | 26 | 36 |
| 22 | 33 | 26 | 37 |
| 23 | 32 | 26 | 38 |
| 24 | 31 | 26 | 39 |
| 25 | 30 | 26 | 40 |
| 26–30 | 30 | 24 | 15 kW + 1 kW for each range |
| 31–40 | 30 | 22 | |
| 41–50 | 30 | 20 | 25 kW + ¾ kW for each range |
| 51–60 | 30 | 18 | |
| 61 and over | 30 | 16 | |

1. Over 12 kW through 27 kW ranges all of same rating. For ranges individually rated more than 12 kW but not more than 27 kW, the maximum demand in Column C shall be increased 5 percent for each additional kilowatt of rating or major fraction thereof by which the rating of individual ranges exceeds 12 kW.

2. Over 8¾ kW through 27 kW ranges of unequal ratings. For ranges individually rated more than 8¾ kW and of different ratings, but none exceeding 27 kW, an average value of rating shall be computed by adding together the ratings of all ranges to obtain the total connected load (using 12 kW for any range rated less than 12 kW) and dividing by the total number of ranges. Then the maximum demand in Column C shall be increased 5 percent for each kilowatt or major fraction thereof by which this average value exceeds 12 kW.

3. Over 1¾ kW through 8¾ kW. In lieu of the method provided in Column C, it shall be permissible to add the nameplate ratings of all household cooking appliances rated more than 1¾ kW but not more than 8¾ kW and multiply the sum by the demand factors specified in Column A or B for the given number of appliances. Where the rating of cooking appliances falls under both Column A and Column B, the demand factors for each column shall be applied to the appliances for that column, and the results added together.

4. Branch-Circuit Load. It shall be permissible to compute the branch-circuit load for one range in accordance with Table 220.19. The branch-circuit load for one wall-mounted oven or one counter-mounted cooking unit shall be the nameplate rating of the appliance. The branch-circuit load for a counter-mounted cooking unit and not more than two wall-mounted ovens, all supplied from a single branch circuit and located in the same room, shall be computed by adding the nameplate rating of the individual appliances and treating this total as equivalent to one range.

5. This table also applies to household cooking appliances rated over 1¾ kW and used in instructional programs.

kilowatts (kW) for loads computed under this section.

Where two or more single-phase ranges are supplied by a 3-phase, 4-wire feeder or service, the total load shall be computed on the basis of twice the maximum number connected between any two phases.

FPN No. 1: See Example D5(A) in Annex D.

FPN No. 2: See Table 220.20 for commercial cooking equipment.

FPN No. 3: See the examples in Annex D.

## 220.20 Kitchen Equipment — Other Than Dwelling Unit(s).
It shall be permissible to compute the load for commercial electric cooking equipment, dishwasher booster heaters, water heaters, and other kitchen equipment in accordance with Table 220.20. These demand factors shall be applied to all equipment that has either thermostatic control or intermittent use as kitchen equipment. They shall not apply to space-heating, ventilating, or air-conditioning equipment.

However, in no case shall the feeder or service demand be less than the sum of the largest two kitchen equipment loads.

**Table 220.20 Demand Factors for Kitchen Equipment — Other Than Dwelling Unit(s)**

| Number of Units of Equipment | Demand Factor (Percent) |
|:---:|:---:|
| 1 | 100 |
| 2 | 100 |
| 3 | 90 |
| 4 | 80 |
| 5 | 70 |
| 6 and over | 65 |

## 220.21 Noncoincident Loads.
Where it is unlikely that two or more noncoincident loads will be in use simultaneously, it shall be permissible to use only the largest load(s) that will be used at one time, in computing the total load of a feeder or service.

## 220.22 Feeder or Service Neutral Load.
The feeder or service neutral load shall be the maximum unbalance of the load determined by this article. The maximum unbalanced load shall be the maximum net computed load between the neutral and any one ungrounded conductor, except that the load thus obtained shall be multiplied by 140 percent for 3-wire, 2-phase or 5-wire, 2-phase systems. For a feeder or service supplying household electric ranges, wall-mounted ovens, counter-mounted cooking units, and electric dryers, the maximum unbalanced load shall be considered as 70 percent of the load on the ungrounded conductors, as determined in accordance with Table 220.19 for ranges and

Table 220.18 for dryers. For 3-wire dc or single-phase ac; 4-wire, 3-phase; 3-wire, 2-phase; or 5-wire, 2-phase systems, a further demand factor of 70 percent shall be permitted for that portion of the unbalanced load in excess of 200 amperes. There shall be no reduction of the neutral capacity for that portion of the load that consists of nonlinear loads supplied from a 4-wire, wye-connected, 3-phase system. There shall be no reduction in the capacity of the grounded conductor of a 3-wire circuit consisting of two phase wires and the neutral of a 4-wire, 3-phase, wye-connected system.

FPN No. 1: See Examples D1(A), D1(B), D2(B), D4(A), and D5(A) in Annex D.

FPN No. 2: A 3-phase, 4-wire, wye-connected power system used to supply power to nonlinear loads may necessitate that the power system design allow for the possibility of high harmonic neutral currents.

## III. Optional Calculations for Computing Feeder and Service Loads

## 220.30 Optional Calculation — Dwelling Unit.

**(A) Feeder and Service Load.** For a dwelling unit having the total connected load served by a single 3-wire, 120/240-volt or 208Y/120-volt set of service or feeder conductors with an ampacity of 100 or greater, it shall be permissible to compute the feeder and service loads in accordance with this section instead of the method specified in Part II of this article. The calculated load shall be the result of adding the loads from 220.30(B) and (C). Feeder and service-entrance conductors whose demand load is determined by this optional calculation shall be permitted to have the neutral load determined by 220.22.

**(B) General Loads.** The general calculated load shall be not less than 100 percent of the first 10 kVA plus 40 percent of the remainder of the following loads:

(1) 1500 volt-amperes for each 2-wire, 20-ampere small-appliance branch circuit and each laundry branch circuit specified in 220.16.

(2) 33 volt-amperes/m$^2$ or 3 volt-amperes/ft$^2$ for general lighting and general-use receptacles. The floor area for each floor shall be computed from the outside dimensions of the dwelling unit. The computed floor area shall not include open porches, garages, or unused or unfinished spaces not adaptable for future use.

(3) The nameplate rating of all appliances that are fastened in place, permanently connected, or located to be on a specific circuit, ranges, wall-mounted ovens, counter-mounted cooking units, clothes dryers, and water heaters.

(4) The nameplate ampere or kVA rating of all motors and of all low-power-factor loads.

**(C) Heating and Air-Conditioning Load.** The largest of the following six selections (load in kVA) shall be included:

(1) 100 percent of the nameplate rating(s) of the air conditioning and cooling.
(2) 100 percent of the nameplate ratings of the heat pump compressors and supplemental heating unless the controller prevents the compressor and supplemental heating from operating at the same time.
(3) 100 percent of the nameplate ratings of electric thermal storage and other heating systems where the usual load is expected to be continuous at the full nameplate value. Systems qualifying under this selection shall not be calculated under any other selection in 220.30(C).
(4) 65 percent of the nameplate rating(s) of the central electric space heating, including integral supplemental heating in heat pumps where the controller prevents the compressor and supplemental heating from operating at the same time.
(5) 65 percent of the nameplate rating(s) of electric space heating if less than four separately controlled units.
(6) 40 percent of the nameplate rating(s) of electric space heating if four or more separately controlled units.

**220.31 Optional Calculations for Additional Loads in an Existing Dwelling Unit.** This section shall be permitted to be used to determine if the existing service or feeder is of sufficient capacity to serve additional loads. Where the dwelling unit is served by a 120/240-volt or 208Y/120-volt, 3-wire service, it shall be permissible to compute the total load in accordance with 220.31(A) or (B).

**(A) Where Additional Air-Conditioning Equipment or Electric Space-Heating Equipment Is Not to Be Installed.** The following formula shall be used for existing and additional new loads.

| Load (kVa) | Percent of Load |
|---|---|
| First 8 kVA of load at | 100 |
| Remainder of load at | 40 |

Load calculations shall include the following:

(1) General lighting and general-use receptacles at 33 volt-amperes/m$^2$ or 3 volt-amperes/ft$^2$ as determined by 220.3(A)
(2) 1500 volt-amperes for each 2-wire, 20-ampere small-appliance branch circuit and each laundry branch circuit specified in 220.16
(3) Household range(s), wall-mounted oven(s), and counter-mounted cooking unit(s)
(4) All other appliances that are permanently connected, fastened in place, or connected to a dedicated circuit, at nameplate rating

**(B) Where Additional Air-Conditioning Equipment or Electric Space-Heating Equipment Is to Be Installed.** The following formula shall be used for existing and additional new loads. The larger connected load of air-conditioning or space-heating, but not both, shall be used.

| | |
|---|---|
| Air-conditioning equipment | 100 |
| Central electric space heating | 100 |
| Less than four separately controlled space-heating units | 100 |
| First 8 kVA of all other loads | 100 |
| Remainder of all other loads | 40 |

Other loads shall include the following:

(1) General lighting and general-use receptacles at 33 volt-amperes/m$^2$ or 3 volt-amperes/ft$^2$ as determined by 220.3(A)
(2) 1500 volt-amperes for each 2-wire, 20-ampere small-appliance branch circuit and each laundry branch circuit specified in 220.16
(3) Household range(s), wall-mounted oven(s), and counter-mounted cooking unit(s)
(4) All other appliances that are permanently connected, fastened in place, or connected to a dedicated circuit, including four or more separately controlled space-heating units, at nameplate rating

**220.32 Optional Calculation — Multifamily Dwelling.**

**(A) Feeder or Service Load.** It shall be permissible to compute the load of a feeder or service that supplies more than two dwelling units of a multifamily dwelling in accordance with Table 220.32 instead of Part II of this article where all the following conditions are met:

(1) No dwelling unit is supplied by more than one feeder.
(2) Each dwelling unit is equipped with electric cooking equipment.

*Exception: When the computed load for multifamily dwellings without electric cooking in Part II of this article exceeds that computed under Part III for the identical load plus electric cooking (based on 8 kW per unit), the lesser of the two loads shall be permitted to be used.*

(3) Each dwelling unit is equipped with either electric space heating, air conditioning, or both. Feeders and service conductors whose demand load is determined by this optional calculation shall be permitted to have the neutral load determined by 220.22.

**(B) House Loads.** House loads shall be computed in accordance with Part II of this article and shall be in addition to the dwelling unit loads computed in accordance with Table 220.32.

**(C) Connected Loads.** The computed load to which the demand factors of Table 220.32 apply shall include the following:

**Table 220.32 Optional Calculations — Demand Factors for Three or More Multifamily Dwelling Units**

| Number of Dwelling Units | Demand Factor (Percent) |
|---|---|
| 3–5 | 45 |
| 6–7 | 44 |
| 8–10 | 43 |
| 11 | 42 |
| 12–13 | 41 |
| 14–15 | 40 |
| 16–17 | 39 |
| 18–20 | 38 |
| 21 | 37 |
| 22–23 | 36 |
| 24–25 | 35 |
| 26–27 | 34 |
| 28–30 | 33 |
| 31 | 32 |
| 32–33 | 31 |
| 34–36 | 30 |
| 37–38 | 29 |
| 39–42 | 28 |
| 43–45 | 27 |
| 46–50 | 26 |
| 51–55 | 25 |
| 56–61 | 24 |
| 62 and over | 23 |

(1) 1500 volt-amperes for each 2-wire, 20-ampere small-appliance branch circuit and each laundry branch circuit specified in 220.16.

(2) 33 volt-amperes/m$^2$ or 3 volt-amperes/ft$^2$ for general lighting and general-use receptacles.

(3) The nameplate rating of all appliances that are fastened in place, permanently connected or located to be on a specific circuit, ranges, wall-mounted ovens, counter-mounted cooking units, clothes dryers, water heaters, and space heaters. If water heater elements are interlocked so that all elements cannot be used at the same time, the maximum possible load shall be considered the nameplate load.

(4) The nameplate ampere or kilovolt-ampere rating of all motors and of all low-power-factor loads.

(5) The larger of the air-conditioning load or the space-heating load.

**220.33 Optional Calculation — Two Dwelling Units.** Where two dwelling units are supplied by a single feeder and the computed load under Part II of this article exceeds that for three identical units computed under 220.32, the lesser of the two loads shall be permitted to be used.

**220.34 Optional Method — Schools.** The calculation of a feeder or service load for schools shall be permitted in accordance with Table 220.34 in lieu of Part II of this article where equipped with electric space heating, air con-

ditioning, or both. The connected load to which the demand factors of Table 220.34 apply shall include all of the interior and exterior lighting, power, water heating, cooking, other loads, and the larger of the air-conditioning load or space-heating load within the building or structure.

Feeders and service-entrance conductors whose demand load is determined by this optional calculation shall be permitted to have the neutral load determined by 220.22. Where the building or structure load is calculated by this optional method, feeders within the building or structure shall have ampacity as permitted in Part II of this article; however, the ampacity of an individual feeder shall not be required to be larger than the ampacity for the entire building.

This section shall not apply to portable classroom buildings.

**Table 220.34 Optional Method — Demand Factors for Feeders and Service-Entrance Conductors for Schools**

| Connected Load | Demand Factor (Percent) |
|---|---|
| First 33 VA/m$^2$ (3 VA/ft$^2$) at | 100 |
| Plus | |
| Over 33 to 220 VA/m$^2$ (3 to 20 VA/ft$^2$) at | 75 |
| Plus | |
| Remainder over 220 VA/m$^2$ (20 VA/ft$^2$) at | 25 |

**220.35 Optional Calculations for Determining Existing Loads.** The calculation of a feeder or service load for existing installations shall be permitted to use actual maximum demand to determine the existing load under the following conditions:

(1) The maximum demand data is available for a 1-year period.

*Exception: If the maximum demand data for a 1-year period is not available, the calculated load shall be permitted to be based on the maximum demand (measure of average power demand over a 15-minute period) continuously recorded over a minimum 30-day period using a recording ammeter or power meter connected to the highest loaded phase of the feeder or service, based on the initial loading at the start of the recording. The recording shall reflect the maximum demand of the feeder or service by being taken when the building or space is occupied and shall include by measurement or calculation the larger of the heating or cooling equipment load, and other loads that may be periodic in nature due to seasonal or similar conditions.*

(2) The maximum demand at 125 percent plus the new load does not exceed the ampacity of the feeder or rating of the service.

**Table 220.36 Optional Method — Permitted Load Calculations for Service and Feeder Conductors for New Restaurants**

| Total Connected Load (kVA) | All Electric Restaurant Calculated Loads (kVA) | Not All Electric Restaurant Calculated Loads (kVA) |
|---|---|---|
| 0–200 | 80% | 100% |
| 201–325 | 10% (amount over 200) + 160.0 | 50% (amount over 200) + 200.0 |
| 326–800 | 50% (amount over 325) + 172.5 | 45% (amount over 325) + 262.5 |
| Over 800 | 50% (amount over 800) + 410.0 | 20% (amount over 800) + 476.3 |

Note: Add all electrical loads, including both heating and cooling loads, to compute the total connected load. Select the one demand factor that applies from the table, and multiply the total connected load by this single demand factor.

(3) The feeder has overcurrent protection in accordance with 240.4, and the service has overload protection in accordance with 230.90.

**220.36 Optional Calculation — New Restaurants.** Calculation of a service or feeder load, where the feeder serves the total load, for a new restaurant shall be permitted in accordance with Table 220.36 in lieu of Part II of this article.

The overload protection of the service conductors shall be in accordance with 230.90 and 240.4.

Feeder conductors shall not be required to be of greater ampacity than the service conductors.

Service or feeder conductors whose demand load is determined by this optional calculation shall be permitted to have the neutral load determined by 220.22.

## IV. Method for Computing Farm Loads

**220.40 Farm Loads — Buildings and Other Loads.**

**(A) Dwelling Unit.** The feeder or service load of a farm dwelling unit shall be computed in accordance with the provisions for dwellings in Part II or III of this article. Where the dwelling has electric heat and the farm has electric grain-drying systems, Part III of this article shall not be used to compute the dwelling load where the dwelling and farm load are supplied by a common service.

**(B) Other Than Dwelling Unit.** Where a feeder or service supplies a farm building or other load having two or more separate branch circuits, the load for feeders, service conductors, and service equipment shall be computed in accordance with demand factors not less than indicated in Table 220.40.

**220.41 Farm Loads — Total.** Where supplied by a common service, the total load of the farm for service conductors and service equipment shall be computed in accordance with the farm dwelling unit load and demand factors specified in Table 220.41. Where there is equipment in two or more farm equipment buildings or for loads having the same function, such loads shall be computed in accordance

with Table 220.40 and shall be permitted to be combined as a single load in Table 220.41 for computing the total load.

**Table 220.40 Method for Computing Farm Loads for Other Than Dwelling Unit**

| Ampere Load at 240 Volts Maximum | Demand Factor (Percent) |
|---|---|
| Loads expected to operate without diversity, but not less than 125 percent full-load current of the largest motor and not less than the first 60 amperes of load | 100 |
| Next 60 amperes of all other loads | 50 |
| Remainder of other load | 25 |

**Table 220.41 Method for Computing Total Farm Load**

| Individual Loads Computed in Accordance with Table 220.40 | Demand Factor (Percent) |
|---|---|
| Largest load | 100 |
| Second largest load | 75 |
| Third largest load | 65 |
| Remaining loads | 50 |

Note: To this total load, add the load of the farm dwelling unit computed in accordance with Part II or III of this article. Where the dwelling has electric heat and the farm has electric grain-drying systems, Part III of this article shall not be used to compute the dwelling load.

## ARTICLE 225
## Outside Branch Circuits and Feeders

**225.1 Scope.** This article covers requirements for outside branch circuits and feeders run on or between buildings, structures, or poles on the premises; and electric equipment and wiring for the supply of utilization equipment that is located on or attached to the outside of buildings, structures, or poles.

> FPN: For additional information on wiring over 600 volts, see ANSI C2-1997, *National Electrical Safety Code.*

**225.2 Other Articles.** Application of other articles, including additional requirements to specific cases of equipment and conductors, is shown in Table 225.2.

**Table 225.2 Other Articles**

| Equipment/Conductors | Article |
| --- | --- |
| Branch circuits | 210 |
| Class 1, Class 2, and Class 3 remote-control, signaling, and power-limited circuits | 725 |
| Communications circuits | 800 |
| Community antenna television and radio distribution systems | 820 |
| Conductors for general wiring | 310 |
| Electrically driven or controlled irrigation machines | 675 |
| Electric signs and outline lighting | 600 |
| Feeders | 215 |
| Fire alarm systems | 760 |
| Fixed outdoor electric deicing and snow-melting equipment | 426 |
| Floating buildings | 553 |
| Grounding | 250 |
| Hazardous (classified) locations | 500 |
| Hazardous (classified) locations — specific | 510 |
| Marinas and boatyards | 555 |
| Messenger supported wiring | 396 |
| Open wiring on insulators | 398 |
| Over 600 volts, general | 490 |
| Overcurrent protection | 240 |
| Radio and television equipment | 810 |
| Services | 230 |
| Solar photovoltaic systems | 690 |
| Swimming pools, fountains, and similar installations | 680 |
| Use and identification of grounded conductors | 200 |

## I. General

**225.3 Calculation of Loads 600 Volts, Nominal, or Less.**

**(A) Branch Circuits.** The load on outdoor branch circuits shall be as determined by 220.3.

**(B) Feeders.** The load on outdoor feeders shall be as determined by Part II of Article 220.

**225.4 Conductor Covering.** Where within 3.0 m (10 ft) of any building or structure other than supporting poles or towers, open individual (aerial) overhead conductors shall be insulated or covered. Conductors in cables or raceways,

except Type MI cable, shall be of the rubber-covered type or thermoplastic type and, in wet locations, shall comply with 310.8. Conductors for festoon lighting shall be of the rubber-covered or thermoplastic type.

*Exception: Equipment grounding conductors and grounded circuit conductors shall be permitted to be bare or covered as specifically permitted elsewhere in this Code.*

**225.5 Size of Conductors 600 Volts, Nominal, or Less.** The ampacity of outdoor branch-circuit and feeder conductors shall be in accordance with 310.15 based on loads as determined under 220.3 and Part II of Article 220.

**225.6 Conductor Size and Support.**

**(A) Overhead Spans.** Open individual conductors shall not be smaller than the following:

(1) For 600 volts, nominal, or less, 10 AWG copper or 8 AWG aluminum for spans up to 15 m (50 ft) in length and 8 AWG copper or 6 AWG aluminum for a longer span, unless supported by a messenger wire
(2) For over 600 volts, nominal, 6 AWG copper or 4 AWG aluminum where open individual conductors and 8 AWG copper or 6 AWG aluminum where in cable

**(B) Festoon Lighting.** Overhead conductors for festoon lighting shall not be smaller than 12 AWG unless the conductors are supported by messenger wires. In all spans exceeding 12 m (40 ft), the conductors shall be supported by messenger wire. The messenger wire shall be supported by strain insulators. Conductors or messenger wires shall not be attached to any fire escape, downspout, or plumbing equipment.

**225.7 Lighting Equipment Installed Outdoors.**

**(A) General.** For the supply of lighting equipment installed outdoors, the branch circuits shall comply with Article 210 and 225.7(B) through (D).

**(B) Common Neutral.** The ampacity of the neutral conductor shall not be less than the maximum net computed load current between the neutral and all ungrounded conductors connected to any one phase of the circuit.

**(C) 277 Volts to Ground.** Circuits exceeding 120 volts, nominal, between conductors and not exceeding 277 volts, nominal, to ground shall be permitted to supply luminaires (lighting fixtures) for illumination of outdoor areas of industrial establishments, office buildings, schools, stores, and other commercial or public buildings where the luminaires (fixtures) are not less than 900 mm (3 ft) from windows, platforms, fire escapes, and the like.

**(D) 600 Volts Between Conductors.** Circuits exceeding 277 volts, nominal, to ground and not exceeding 600 volts, nominal, between conductors shall be permitted to supply the auxiliary equipment of electric-discharge lamps in accordance with 210.6(D)(1).

**225.9 Overcurrent Protection.** Overcurrent protection shall be in accordance with 210.20 for branch circuits and Article 240 for feeders.

**225.10 Wiring on Buildings.** The installation of outside wiring on surfaces of buildings shall be permitted for circuits of not over 600 volts, nominal, as open wiring on insulators, as multiconductor cable, as Type MC cable, as Type MI cable, as messenger supported wiring, in rigid metal conduit, in intermediate metal conduit, in rigid non-metallic conduit, in cable trays, as cablebus, in wireways, in auxiliary gutters, in electrical metallic tubing, in flexible metal conduit, in liquidtight flexible metal conduit, in liquidtight flexible nonmetallic conduit, and in busways. Circuits of over 600 volts, nominal, shall be installed as provided in 300.37. Circuits for signs and outline lighting shall be installed in accordance with Article 600.

**225.11 Circuit Exits and Entrances.** Where outside branch and feeder circuits leave or enter a building, the requirements of 230.52 and 230.54 shall apply.

**225.12 Open-Conductor Supports.** Open conductors shall be supported on glass or porcelain knobs, racks, brackets, or strain insulators.

**225.14 Open-Conductor Spacings.**

**(A) 600 Volts, Nominal, or Less.** Conductors of 600 volts, nominal, or less, shall comply with the spacings provided in Table 230.51(C).

**(B) Over 600 Volts, Nominal.** Conductors of over 600 volts, nominal, shall comply with the spacings provided in 110.36 and 490.24.

**(C) Separation from Other Circuits.** Open conductors shall be separated from open conductors of other circuits or systems by not less than 100 mm (4 in.).

**(D) Conductors on Poles.** Conductors on poles shall have a separation of not less than 300 mm (1 ft) where not placed on racks or brackets. Conductors supported on poles shall provide a horizontal climbing space not less than the following:

(1) Power conductors below communications conductors — 750 mm (30 in.)
(2) Power conductors alone or above communications conductors:

a. 300 volts or less — 600 mm (24 in.)
b. Over 300 volts — 750 mm (30 in.)

(3) Communications conductors below power conductors — same as power conductors
(4) Communications conductors alone — no requirement

**225.15 Supports over Buildings.** Supports over a building shall be in accordance with 230.29.

**225.16 Point of Attachment to Buildings.** The point of attachment to a building shall be in accordance with 230.26.

**225.17 Means of Attachment to Buildings.** The means of attachment to a building shall be in accordance with 230.27.

**225.18 Clearance from Ground.** Overhead spans of open conductors and open multiconductor cables of not over 600 volts, nominal, shall conform to the following:

(1) 3.0 m (10 ft) — above finished grade, sidewalks, or from any platform or projection from which they might be reached where the voltage does not exceed 150 volts to ground and accessible to pedestrians only
(2) 3.7 m (12 ft) — over residential property and driveways, and those commercial areas not subject to truck traffic where the voltage does not exceed 300 volts to ground
(3) 4.5 m (15 ft) — for those areas listed in the 3.7-m (12-ft) classification where the voltage exceeds 300 volts to ground
(4) 5.5 m (18 ft) — over public streets, alleys, roads, parking areas subject to truck traffic, driveways on other than residential property, and other land traversed by vehicles, such as cultivated, grazing, forest, and orchard.

**225.19 Clearances from Buildings for Conductors of Not Over 600 Volts, Nominal.**

**(A) Above Roofs.** Overhead spans of open conductors and open multiconductor cables shall have a vertical clearance of not less than 2.5 m (8 ft) above the roof surface. The vertical clearance above the roof level shall be maintained for a distance not less than 900 mm (3 ft) in all directions from the edge of the roof.

*Exception No. 1: The area above a roof surface subject to pedestrian or vehicular traffic shall have a vertical clearance from the roof surface in accordance with the clearance requirements of 225.18.*

*Exception No. 2: Where the voltage between conductors does not exceed 300, and the roof has a slope of 100 mm*

*(4 in.) in 300 mm (12 in.) or greater, a reduction in clearance to 900 mm (3 ft) shall be permitted.*

*Exception No. 3: Where the voltage between conductors does not exceed 300, a reduction in clearance above only the overhanging portion of the roof to not less than 450 mm (18 in.) shall be permitted if (1) not more than 1.8 m (6 ft) of the conductors, 1.2 m (4 ft) horizontally, pass above the roof overhang and (2) they are terminated at a through-the-roof raceway or approved support.*

*Exception No. 4: The requirement for maintaining the vertical clearance 900 mm (3 ft) from the edge of the roof shall not apply to the final conductor span where the conductors are attached to the side of a building.*

**(B) From Nonbuilding or Nonbridge Structures.** From signs, chimneys, radio and television antennas, tanks, and other nonbuilding or nonbridge structures, clearances — vertical, diagonal, and horizontal — shall not be less than 900 mm (3 ft).

**(C) Horizontal Clearances.** Clearances shall not be less than 900 mm (3 ft).

**(D) Final Spans.** Final spans of feeders or branch circuits shall comply with 225.19(D)(1), (2), and (3).

**(1) Clearance from Windows.** Final spans to the building they supply, or from which they are fed, shall be permitted to be attached to the building, but they shall be kept not less than 900 mm (3 ft) from windows that are designed to be opened, and from doors, porches, balconies, ladders, stairs, fire escapes, or similar locations.

*Exception: Conductors run above the top level of a window shall be permitted to be less than the 900-mm (3-ft) requirement.*

**(2) Vertical Clearance.** The vertical clearance of final spans above, or within 900 mm (3 ft) measured horizontally of, platforms, projections, or surfaces from which they might be reached shall be maintained in accordance with 225.18.

**(3) Building Openings.** The overhead branch-circuit and feeder conductors shall not be installed beneath openings through which materials may be moved, such as openings in farm and commercial buildings, and shall not be installed where they obstruct entrance to these buildings' openings.

**(E) Zone for Fire Ladders.** Where buildings exceed three stories or 15 m (50 ft) in height, overhead lines shall be arranged, where practicable, so that a clear space (or zone) at least 1.8 m (6 ft) wide will be left either adjacent to the buildings or beginning not over 2.5 m (8 ft) from them to facilitate the raising of ladders when necessary for fire fighting.

**225.20 Mechanical Protection of Conductors.** Mechanical protection of conductors on buildings, structures, or poles shall be as provided for services in 230.50.

**225.21 Multiconductor Cables on Exterior Surfaces of Buildings.** Supports for multiconductor cables on exterior surfaces of buildings shall be as provided in 230.51.

**225.22 Raceways on Exterior Surfaces of Buildings or Other Structures.** Raceways on exterior surfaces of buildings or other structures shall be raintight and arranged to drain.

*Exception: Flexible metal conduit, where permitted in 348.12(1), shall not be required to be raintight.*

**225.24 Outdoor Lampholders.** Where outdoor lampholders are attached as pendants, the connections to the circuit wires shall be staggered. Where such lampholders have terminals of a type that puncture the insulation and make contact with the conductors, they shall be attached only to conductors of the stranded type.

**225.25 Location of Outdoor Lamps.** Locations of lamps for outdoor lighting shall be below all energized conductors, transformers, or other electric utilization equipment, unless

(1) Clearances or other safeguards are provided for relamping operations, or
(2) Equipment is controlled by a disconnecting means that can be locked in the open position.

**225.26 Vegetation as Support.** Vegetation such as trees shall not be used for support of overhead conductor spans.

## II. More Than One Building or Other Structure

**225.30 Number of Supplies.** Where more than one building or other structure is on the same property and under single management, each additional building or other structure served that is on the load side of the service disconnecting means shall be supplied by one feeder or branch circuit unless permitted in 225.30(A) through (E). For the purpose of this section, a multiwire branch circuit shall be considered a single circuit.

**(A) Special Conditions.** Additional feeders or branch circuits shall be permitted to supply the following:

(1) Fire pumps
(2) Emergency systems
(3) Legally required standby systems
(4) Optional standby systems
(5) Parallel power production systems

**(B) Special Occupancies.** By special permission, additional feeders or branch circuits shall be permitted for the following:

(1) Multiple-occupancy buildings where there is no space available for supply equipment accessible to all occupants, or
(2) A single building or other structure sufficiently large to make two or more supplies necessary.

**(C) Capacity Requirements.** Additional feeders or branch circuits shall be permitted where the capacity requirements are in excess of 2000 amperes at a supply voltage of 600 volts or less.

**(D) Different Characteristics.** Additional feeders or branch circuits shall be permitted for different voltages, frequencies, or phases or for different uses, such as control of outside lighting from multiple locations.

**(E) Documented Switching Procedures.** Additional feeders or branch circuits shall be permitted to supply installations under single management where documented safe switching procedures are established and maintained for disconnection.

**225.31 Disconnecting Means.** Means shall be provided for disconnecting all ungrounded conductors that supply or pass through the building or structure.

**225.32 Location.** The disconnecting means shall be installed either inside or outside of the building or structure served or where the conductors pass through the building or structure. The disconnecting means shall be at a readily accessible location nearest the point of entrance of the conductors. For the purposes of this section, the requirements in 230.6 shall be permitted to be utilized.

*Exception No. 1: For installations under single management, where documented safe switching procedures are established and maintained for disconnection, and where the installation is monitored by qualified individuals, the disconnecting means shall be permitted to be located elsewhere on the premises.*

*Exception No. 2: For buildings or other structures qualifying under the provisions of Article 685, the disconnecting means shall be permitted to be located elsewhere on the premises.*

*Exception No. 3: For towers or poles used as lighting standards, the disconnecting means shall be permitted to be located elsewhere on the premises.*

*Exception No. 4: For poles or similar structures used only for support of signs installed in accordance with Article 600, the disconnecting means shall be permitted to be located elsewhere on the premises.*

**225.33 Maximum Number of Disconnects.**

**(A) General.** The disconnecting means for each supply permitted by 225.30 shall consist of not more than six switches or six circuit breakers mounted in a single enclosure, in a group of separate enclosures, or in or on a switchboard. There shall be no more than six disconnects per supply grouped in any one location.

*Exception: For the purposes of this section, disconnecting means used solely for the control circuit of the ground-fault protection system, or the control circuit of the power-operated supply disconnecting means, installed as part of the listed equipment, shall not be considered a supply disconnecting means.*

**(B) Single-Pole Units.** Two or three single-pole switches or breakers capable of individual operation shall be permitted on multiwire circuits, one pole for each ungrounded conductor, as one multipole disconnect, provided they are equipped with handle ties or a master handle to disconnect all ungrounded conductors with no more than six operations of the hand.

**225.34 Grouping of Disconnects.**

**(A) General.** The two to six disconnects as permitted in 225.33 shall be grouped. Each disconnect shall be marked to indicate the load served.

*Exception: One of the two to six disconnecting means permitted in 225.33, where used only for a water pump also intended to provide fire protection, shall be permitted to be located remote from the other disconnecting means.*

**(B) Additional Disconnecting Means.** The one or more additional disconnecting means for fire pumps or for emergency, legally required standby or optional standby system permitted by 225.30 shall be installed sufficiently remote from the one to six disconnecting means for normal supply to minimize the possibility of simultaneous interruption of supply.

**225.35 Access to Occupants.** In a multiple-occupancy building, each occupant shall have access to the occupant's supply disconnecting means.

*Exception: In a multiple-occupancy building where electric supply and electrical maintenance are provided by the building management and where these are under continuous building management supervision, the supply disconnecting means supplying more than one occupancy shall be permitted to be accessible to authorized management personnel only.*

**225.36 Suitable for Service Equipment.** The disconnecting means specified in 225.31 shall be suitable for use as service equipment.

*Exception: For garages and outbuildings on residential property, a snap switch or a set of 3-way or 4-way snap switches shall be permitted as the disconnecting means.*

**225.37 Identification.** Where a building or structure has any combination of feeders, branch circuits, or services passing through it or supplying it, a permanent plaque or directory shall be installed at each feeder and branch-circuit disconnect location denoting all other services, feeders, or branch circuits supplying that building or structure or passing through that building or structure and the area served by each.

*Exception No. 1: A plaque or directory shall not be required for large-capacity multibuilding industrial installations under single management, where it is ensured that disconnection can be accomplished by establishing and maintaining safe switching procedures.*

*Exception No. 2: This identification shall not be required for branch circuits installed from a dwelling unit to a second building or structure.*

**225.38 Disconnect Construction.** Disconnecting means shall meet the requirements of 225.38(A) through (D).

*Exception: For garages and outbuildings on residential property, snap switches or sets of 3-way or 4-way snap switches shall be permitted as the disconnecting means.*

**(A) Manually or Power Operable.** The disconnecting means shall consist of either (1) a manually operable switch or a circuit breaker equipped with a handle or other suitable operating means or (2) a power-operable switch or circuit breaker, provided the switch or circuit breaker can be opened by hand in the event of a power failure.

**(B) Simultaneous Opening of Poles.** Each building or structure disconnecting means shall simultaneously disconnect all ungrounded supply conductors that it controls from the building or structure wiring system.

**(C) Disconnection of Grounded Conductor.** Where the building or structure disconnecting means does not disconnect the grounded conductor from the grounded conductors in the building or structure wiring, other means shall be provided for this purpose at the location of disconnecting means. A terminal or bus to which all grounded conductors can be attached by means of pressure connectors shall be permitted for this purpose.

In a multisection switchboard, disconnects for the grounded conductor shall be permitted to be in any of the switchboard, provided any such switchboard is marked.

**(D) Indicating.** The building or structure disconnecting means shall plainly indicate whether it is in the open or closed position.

**225.39 Rating of Disconnect.** The feeder or branch-circuit disconnecting means shall have a rating of not less than the load to be carried, determined in accordance with Article 220. In no case shall the rating be lower than specified in 225.39(A), (B), (C), or (D).

**(A) One-Circuit Installation.** For installations to supply only limited loads of a single branch circuit, the branch circuit disconnecting means shall have a rating of not less than 15 amperes.

**(B) Two-Circuit Installations.** For installations consisting of not more than two 2-wire branch circuits, the feeder or branch-circuit disconnecting means shall have a rating of not less than 30 amperes.

**(C) One-Family Dwelling.** For a one-family dwelling, the feeder disconnecting means shall have a rating of not less than 100 amperes, 3-wire.

**(D) All Others.** For all other installations, the feeder or branch-circuit disconnecting means shall have a rating of not less than 60 amperes.

**225.40 Access to Overcurrent Protective Devices.** Where a feeder overcurrent device is not readily accessible, branch-circuit overcurrent devices shall be installed on the load side, shall be mounted in a readily accessible location, and shall be of a lower ampere rating than the feeder overcurrent device.

### III. Over 600 Volts

**225.50 Sizing of Conductors.** The sizing of conductors over 600 volts shall be in accordance with 210.19(B) for branch circuits and 215.2(B) for feeders.

**225.51 Isolating Switches.** Where oil switches or air, oil, vacuum, or sulfur hexafluoride circuit breakers constitute a building disconnecting means, an isolating switch with visible break contacts and meeting the requirements of 230.204(B), (C), and (D) shall be installed on the supply side of the disconnecting means and all associated equipment.

*Exception: The isolating switch shall not be required where the disconnecting means is mounted on removable truck panels or metal-enclosed switchgear units that cannot be opened unless the circuit is disconnected and that, when removed from the normal operating position, automatically disconnect the circuit breaker or switch from all energized parts.*

**225.52 Location.** A building or structure disconnecting means shall be located in accordance with 225.31, or it shall be electrically operated by a similarly located remote-control device.

**225.53 Type.** Each building or structure disconnect shall simultaneously disconnect all ungrounded supply conductors it controls and shall have a fault-closing rating not less than the maximum available short-circuit current available at its supply terminals.

Where fused switches or separately mounted fuses are installed, the fuse characteristics shall be permitted to contribute to the fault closing rating of the disconnecting means.

**225.60 Clearances over Roadways, Walkways, Rail, Water, and Open Land.**

**(A) 22 kV Nominal to Ground or Less.** The clearances over roadways, walkways, rail, water, and open land for conductors and live parts up to 22 kV nominal to ground or less shall be not less than the values shown in Table 225.60.

**(B) Over 22 kV Nominal to Ground.** Clearances for the categories shown in Table 225.60 shall be increased by 10 mm (0.4 in.) per kV above 22,000 volts.

**(C) Special Cases.** For special cases, such as where crossings will be made over lakes, rivers, or areas using large vehicles such as mining operations, specific designs shall be engineered considering the special circumstances and shall be approved by the authority having jurisdiction.

FPN: For additional information, see ANSI C2-1997, *National Electrical Safety Code.*

**Table 225.60 Clearances over Roadways, Walkways, Rail, Water, and Open Land**

| Location | Clearance | |
|---|---|---|
| | m | ft |
| Open land subject to vehicles, cultivation, or grazing | 5.6 | 18.5 |
| Roadways, driveways, parking lots, and alleys | 5.6 | 18.5 |
| Walkways | 4.1 | 13.5 |
| Rails | 8.1 | 26.5 |
| Spaces and ways for pedestrians and restricted traffic | 4.4 | 14.5 |
| Water areas not suitable for boating | 5.2 | 17 |

**225.61 Clearances over Buildings and Other Structures.**

**(A) 22 kV Nominal to Ground or Less.** The clearances over buildings and other structures for conductors and live parts up to 22 kV, nominal, to ground or less shall be not less than the values shown in Table 225.61.

**(B) Over 22 kV Nominal to Ground.** Clearances for the categories shown in Table 225.61 shall be increased by 10 mm (0.4 in.) per kV above 22,000 volts.

FPN: For additional information see ANSI C2-1997, *National Electrical Safety Code.*

**Table 225.61 Clearances over Buildings and Other Structures**

| Clearance from Conductors or Live Parts from: | Horizontal | | Vertical | |
|---|---|---|---|---|
| | m | ft | m | ft |
| Building walls, projections, and windows | 2.3 | 7.5 | — | — |
| Balconies, catwalks, and similar areas accessible to people | 2.3 | 7.5 | 4.1 | 13.5 |
| Over or under roofs or projections not readily accessible to people | — | — | 3.8 | 12.5 |
| Over roofs accessible to vehicles but not trucks | — | — | 4.1 | 13.5 |
| Over roofs accessible to trucks | — | — | 5.6 | 18.5 |
| Other structures | 2.3 | 7.5 | — | — |

## ARTICLE 230
## Services

**230.1 Scope.** This article covers service conductors and equipment for control and protection of services and their installation requirements.

FPN: See Figure 230.1.

### I. General

**230.2 Number of Services.** A building or other structure served shall be supplied by only one service unless permitted in 230.2(A) through (D). For the purpose of 230.40, Exception No. 2 only, underground sets of conductors,

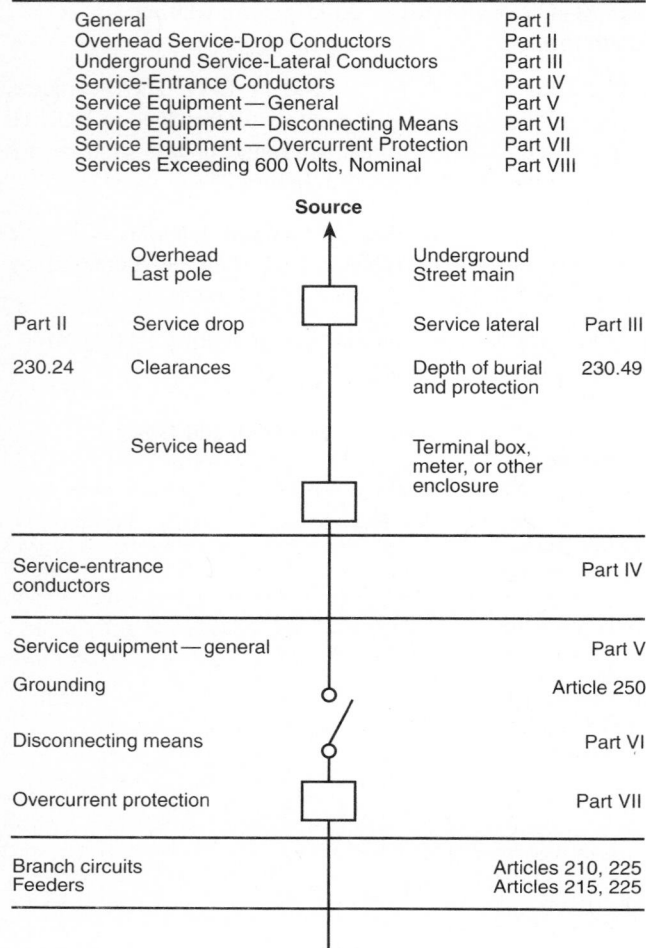

| General | Part I |
| Overhead Service-Drop Conductors | Part II |
| Underground Service-Lateral Conductors | Part III |
| Service-Entrance Conductors | Part IV |
| Service Equipment — General | Part V |
| Service Equipment — Disconnecting Means | Part VI |
| Service Equipment — Overcurrent Protection | Part VII |
| Services Exceeding 600 Volts, Nominal | Part VIII |

**Source**

| | Overhead Last pole | | Underground Street main | |
| Part II | Service drop | | Service lateral | Part III |
| 230.24 | Clearances | | Depth of burial and protection | 230.49 |
| | Service head | | Terminal box, meter, or other enclosure | |

| Service-entrance conductors | Part IV |
| Service equipment — general | Part V |
| Grounding | Article 250 |
| Disconnecting means | Part VI |
| Overcurrent protection | Part VII |
| Branch circuits | Articles 210, 225 |
| Feeders | Articles 215, 225 |

**Figure 230.1 Services.**

1/0 AWG and larger, running to the same location and connected together at their supply end but not connected together at their load end shall be considered to be supplying one service.

**(A) Special Conditions.** Additional services shall be permitted to supply the following:

(1) Fire pumps
(2) Emergency systems
(3) Legally required standby systems
(4) Optional standby systems
(5) Parallel power production systems

**(B) Special Occupancies.** By special permission, additional services shall be permitted for the following:

(1) Multiple-occupancy buildings where there is no available space for service equipment accessible to all occupants, or
(2) A single building or other structure sufficiently large to make two or more services necessary

**(C) Capacity Requirements.** Additional services shall be permitted under any of the following:

(1) Where the capacity requirements are in excess of 2000 amperes at a supply voltage of 600 volts or less
(2) Where the load requirements of a single-phase installation are greater than the serving agency normally supplies through one service
(3) By special permission

**(D) Different Characteristics.** Additional services shall be permitted for different voltages, frequencies, or phases, or for different uses, such as for different rate schedules.

**(E) Identification.** Where a building or structure is supplied by more than one service, or any combination of branch circuits, feeders, and services, a permanent plaque or directory shall be installed at each service disconnect location denoting all other services, feeders, and branch circuits supplying that building or structure and the area served by each. See 225.37.

**230.3 One Building or Other Structure Not to Be Supplied Through Another.** Service conductors supplying a building or other structure shall not pass through the interior of another building or other structure.

**230.6 Conductors Considered Outside the Building.** Conductors shall be considered outside of a building or other structure under any of the following conditions:

(1) Where installed under not less than 50 mm (2 in.) of concrete beneath a building or other structure
(2) Where installed within a building or other structure in a raceway that is encased in concrete or brick not less than 50 mm (2 in.) thick
(3) Where installed in any vault that meets the construction requirements of Article 450, Part III
(4) Where installed in conduit and under not less than 450 mm (18 in.) of earth beneath a building or other structure

**230.7 Other Conductors in Raceway or Cable.** Conductors other than service conductors shall not be installed in the same service raceway or service cable.

*Exception No. 1: Grounding conductors and bonding jumpers.*

*Exception No. 2: Load management control conductors having overcurrent protection.*

**230.8 Raceway Seal.** Where a service raceway enters a building or structure from an underground distribution system, it shall be sealed in accordance with 300.5(G). Spare or unused raceways shall also be sealed. Sealants shall be identified for use with the cable insulation, shield, or other components.

**230.9 Clearance from Building Openings.** Service conductors and final spans shall comply with 230.9(A), (B), and (C).

**(A) Clearance from Windows.** Service conductors installed as open conductors or multiconductor cable without an overall outer jacket shall have a clearance of not less than 900 mm (3 ft) from windows that are designed to be opened, doors, porches, balconies, ladders, stairs, fire escapes, or similar locations.

*Exception: Conductors run above the top level of a window shall be permitted to be less than the 900-mm (3-ft) requirement.*

**(B) Vertical Clearance.** The vertical clearance of final spans above, or within 900 mm (3 ft) measured horizontally of, platforms, projections, or surfaces from which they might be reached shall be maintained in accordance with 230.24(B).

**(C) Building Openings.** Overhead service conductors shall not be installed beneath openings through which materials may be moved, such as openings in farm and commercial buildings, and shall not be installed where they obstruct entrance to these building openings.

**230.10 Vegetation as Support.** Vegetation such as trees shall not be used for support of overhead service conductors.

## II. Overhead Service-Drop Conductors

**230.22 Insulation or Covering.** Individual conductors shall be insulated or covered.

*Exception: The grounded conductor of a multiconductor cable shall be permitted to be bare.*

### 230.23 Size and Rating.

**(A) General.** Conductors shall have sufficient ampacity to carry the current for the load as computed in accordance with Article 220 and shall have adequate mechanical strength.

**(B) Minimum Size.** The conductors shall not be smaller than 8 AWG copper or 6 AWG aluminum or copper-clad aluminum.

*Exception: Conductors supplying only limited loads of a single branch circuit — such as small polyphase power, controlled water heaters, and similar loads — shall not be smaller than 12 AWG hard-drawn copper or equivalent.*

**(C) Grounded Conductors.** The grounded conductor shall not be less than the minimum size as required by 250.24(B).

**230.24 Clearances.** Service-drop conductors shall not be readily accessible and shall comply with 230.24(A) through (D) for services not over 600 volts, nominal.

**(A) Above Roofs.** Conductors shall have a vertical clearance of not less than 2.5 m (8 ft) above the roof surface. The vertical clearance above the roof level shall be maintained for a distance of not less than 900 mm (3 ft) in all directions from the edge of the roof.

*Exception No. 1: The area above a roof surface subject to pedestrian or vehicular traffic shall have a vertical clearance from the roof surface in accordance with the clearance requirements of 230.24(B).*

*Exception No. 2: Where the voltage between conductors does not exceed 300 and the roof has a slope of 100 mm (4 in.) in 300 mm (12 in.), or greater, a reduction in clearance to 900 mm (3 ft) shall be permitted.*

*Exception No. 3: Where the voltage between conductors does not exceed 300, a reduction in clearance above only the overhanging portion of the roof to not less than 450 mm (18 in.) shall be permitted if (1) not more than 1.8 m (6 ft) of service-drop conductors, 1.2 m (4 ft) horizontally, pass above the roof overhang, and (2) they are terminated at a through-the-roof raceway or approved support.*

FPN: See 230.28 for mast supports.

*Exception No. 4: The requirement for maintaining the vertical clearance 900 mm (3 ft) from the edge of the roof shall not apply to the final conductor span where the service drop is attached to the side of a building.*

**(B) Vertical Clearance from Ground.** Service-drop conductors, where not in excess of 600 volts, nominal, shall have the following minimum clearance from final grade:

(1) 3.0 m (10 ft) — at the electric service entrance to buildings, also at the lowest point of the drip loop of the building electric entrance, and above areas or sidewalks accessible only to pedestrians, measured from final grade or other accessible surface only for service-drop cables supported on and cabled together with a grounded bare messenger where the voltage does not exceed 150 volts to ground

(2) 3.7 m (12 ft) — over residential property and driveways, and those commercial areas not subject to truck traffic where the voltage does not exceed 300 volts to ground

(3) 4.5 m (15 ft) — for those areas listed in the 3.7 m (12 ft) classification where the voltage exceeds 300 volts to ground

(4) 5.5 m (18 ft) — over public streets, alleys, roads, parking areas subject to truck traffic, driveways on other than residential property, and other land such as cultivated, grazing, forest, and orchard

**(C) Clearance from Building Openings.** See 230.9.

**(D) Clearance from Swimming Pools.** See 680.8.

**230.26 Point of Attachment.** The point of attachment of the service-drop conductors to a building or other structure shall provide the minimum clearances as specified in 230.24. In no case shall this point of attachment be less than 3.0 m (10 ft) above finished grade.

**230.27 Means of Attachment.** Multiconductor cables used for service drops shall be attached to buildings or other structures by fittings identified for use with service conductors. Open conductors shall be attached to fittings identified for use with service conductors or to noncombustible, nonabsorbent insulators securely attached to the building or other structure.

**230.28 Service Masts as Supports.** Where a service mast is used for the support of service-drop conductors, it shall be of adequate strength or be supported by braces or guys to withstand safely the strain imposed by the service drop. Where raceway-type service masts are used, all raceway fittings shall be identified for use with service masts. Only power service-drop conductors shall be permitted to be attached to a service mast.

**230.29 Supports over Buildings.** Service-drop conductors passing over a roof shall be securely supported by substantial structures. Where practicable, such supports shall be independent of the building.

### III. Underground Service-Lateral Conductors

**230.30 Insulation.** Service-lateral conductors shall be insulated for the applied voltage.

*Exception: A grounded conductor shall be permitted to be uninsulated as follows:*

*(a) Bare copper used in a raceway.*
*(b) Bare copper for direct burial where bare copper is judged to be suitable for the soil conditions.*
*(c) Bare copper for direct burial without regard to soil conditions where part of a cable assembly identified for underground use.*
*(d) Aluminum or copper-clad aluminum without individual insulation or covering where part of a cable assembly identified for underground use in a raceway or for direct burial.*

**230.31 Size and Rating.**

**(A) General.** Service-lateral conductors shall have sufficient ampacity to carry the current for the load as computed in accordance with Article 220 and shall have adequate mechanical strength.

**(B) Minimum Size.** The conductors shall not be smaller than 8 AWG copper or 6 AWG aluminum or copper-clad aluminum.

*Exception: Conductors supplying only limited loads of a single branch circuit — such as small polyphase power, controlled water heaters, and similar loads — shall not be smaller than 12 AWG copper or 10 AWG aluminum or copper-clad aluminum.*

**(C) Grounded Conductors.** The grounded conductor shall not be less than the minimum size required by 250.24(B).

**230.32 Protection Against Damage.** Underground service-lateral conductors shall be protected against damage in accordance with 300.5. Service-lateral conductors entering a building shall be installed in accordance with 230.6 or protected by a raceway wiring method identified in 230.43.

**230.33 Spliced Conductors.** Service-lateral conductors shall be permitted to be spliced or tapped in accordance with 110.14, 300.5(E), 300.13, and 300.15.

### IV. Service-Entrance Conductors

**230.40 Number of Service-Entrance Conductor Sets.** Each service drop or lateral shall supply only one set of service-entrance conductors.

*Exception No. 1: A building with one or more than one occupancy shall be permitted to have one set of service-entrance conductors for each service of different characteristics, as defined in 230.2(D), run to each occupancy or group of occupancies.*

*Exception No. 2: Where two to six service disconnecting means in separate enclosures are grouped at one location and supply separate loads from one service drop or lateral, one set of service-entrance conductors shall be permitted to supply each or several such service equipment enclosures.*

*Exception No. 3: A single-family dwelling unit and a separate structure shall be permitted to have one set of service-entrance conductors run to each from a single service drop or lateral.*

*Exception No. 4: A two-family dwelling or a multifamily dwelling shall be permitted to have one set of service-entrance conductors installed to supply the circuits covered in 210.25.*

*Exception No. 5: One set of service-entrance conductors connected to the supply side of the normal service disconnecting means shall be permitted to supply each or several systems covered by 230.82(4) or (5).*

**230.41 Insulation of Service-Entrance Conductors.** Service-entrance conductors entering or on the exterior of buildings or other structures shall be insulated.

*Exception: A grounded conductor shall be permitted to be uninsulated as follows:*

(a) *Bare copper used in a raceway or part of a service cable assembly.*

(b) *Bare copper for direct burial where bare copper is judged to be suitable for the soil conditions.*

(c) *Bare copper for direct burial without regard to soil conditions where part of a cable assembly identified for underground use.*

(d) *Aluminum or copper-clad aluminum without individual insulation or covering where part of a cable assembly or identified for underground use in a raceway, or for direct burial.*

(e) *Bare conductors used in an auxiliary gutter.*

**230.42 Minimum Size and Rating.**

**(A) General.** The ampacity of the service-entrance conductors before the application of any adjustment or correction factors shall not be less than either (1) or (2). Loads shall be determined in accordance with Article 220. Ampacity shall be determined from 310.15. The maximum allowable current of busways shall be that value for which the busway has been listed or labeled.

(1) The sum of the noncontinuous loads plus 125 percent of continuous loads

(2) The sum of the noncontinuous load plus the continuous load if the service-entrance conductors terminate in an overcurrent device where both the overcurrent device and its assembly are listed for operation at 100 percent of their rating

**(B) Specific Installations.** In addition to the requirements of 230.42(A), the minimum ampacity for ungrounded conductors for specific installations shall not be less than the rating of the service disconnecting means specified in 230.79(A) through (D).

**(C) Grounded Conductors.** The grounded conductor shall not be less than the minimum size as required by 250.24(B).

**230.43 Wiring Methods for 600 Volts, Nominal, or Less.** Service-entrance conductors shall be installed in accordance with the applicable requirements of this *Code* covering the type of wiring method used and shall be limited to the following methods:

(1) Open wiring on insulators
(2) Type IGS cable
(3) Rigid metal conduit
(4) Intermediate metal conduit
(5) Electrical metallic tubing
(6) Electrical nonmetallic tubing (ENT)
(7) Service-entrance cables
(8) Wireways
(9) Busways
(10) Auxiliary gutters
(11) Rigid nonmetallic conduit
(12) Cablebus
(13) Type MC cable
(14) Mineral-insulated, metal-sheathed cable
(15) Flexible metal conduit not over 1.8 m (6 ft) long or liquidtight flexible metal conduit not over 1.8 m (6 ft) long between raceways, or between raceway and service equipment, with equipment bonding jumper routed with the flexible metal conduit or the liquidtight flexible metal conduit according to the provisions of 250.102(A), (B), (C), and (E)
(16) Liquidtight flexible nonmetallic conduit

**230.44 Cable Trays.** Cable tray systems shall be permitted to support cable used as service-entrance conductors.

**230.46 Spliced Conductors.** Service-entrance conductors shall be permitted to be spliced or tapped in accordance with 110.14, 300.5(E), 300.13, and 300.15.

**230.49 Protection Against Physical Damage — Underground.** Underground service-entrance conductors shall be protected against physical damage in accordance with 300.5.

**230.50 Protection of Open Conductors and Cables Against Damage — Above Ground.** Service-entrance conductors installed above ground shall be protected against physical damage as specified in 230.50(A) or (B).

**(A) Service Cables.** Service cables, where subject to physical damage, shall be protected by any of the following:

(1) Rigid metal conduit
(2) Intermediate metal conduit
(3) Schedule 80 rigid nonmetallic conduit
(4) Electrical metallic tubing
(5) Other approved means

**(B) Other Than Service Cable.** Individual open conductors and cables other than service cables shall not be installed within 3.0 m (10 ft) of grade level or where exposed to physical damage.

*Exception: Type MI and Type MC cable shall be permitted within 3.0 m (10 ft) of grade level where not exposed to physical damage or where protected in accordance with 300.5(D).*

**230.51 Mounting Supports.** Cables or individual open service conductors shall be supported as specified in 230.51(A), (B), or (C).

**(A) Service Cables.** Service cables shall be supported by straps or other approved means within 300 mm (12 in.) of every service head, gooseneck, or connection to a raceway or enclosure and at intervals not exceeding 750 mm (30 in.).

**(B) Other Cables.** Cables that are not approved for mounting in contact with a building or other structure shall be mounted on insulating supports installed at intervals not exceeding 4.5 m (15 ft) and in a manner that will maintain a clearance of not less than 50 mm (2 in.) from the surface over which they pass.

**(C) Individual Open Conductors.** Individual open conductors shall be installed in accordance with Table 230.51(C). Where exposed to the weather, the conductors shall be mounted on insulators or on insulating supports attached to racks, brackets, or other approved means. Where not exposed to the weather, the conductors shall be mounted on glass or porcelain knobs.

**230.52 Individual Conductors Entering Buildings or Other Structures.** Where individual open conductors enter a building or other structure, they shall enter through roof bushings or through the wall in an upward slant through individual, noncombustible, nonabsorbent insulating tubes. Drip loops shall be formed on the conductors before they enter the tubes.

**230.53 Raceways to Drain.** Where exposed to the weather, raceways enclosing service-entrance conductors shall be raintight and arranged to drain. Where embedded in masonry, raceways shall be arranged to drain.

*Exception: As permitted in 348.12(1).*

**230.54 Overhead Service Locations.**

**(A) Raintight Service Head.** Service raceways shall be equipped with a raintight service head at the point of connection to service-drop conductors.

**(B) Service Cable Equipped with Raintight Service Head or Gooseneck.** Service cables shall be equipped with a raintight service head.

*Exception: Type SE cable shall be permitted to be formed in a gooseneck and taped with a self-sealing weather-resistant thermoplastic.*

**(C) Service Heads Above Service-Drop Attachment.** Service heads and goosenecks in service-entrance cables shall be located above the point of attachment of the service-drop conductors to the building or other structure.

*Exception: Where it is impracticable to locate the service head above the point of attachment, the service head location shall be permitted not farther than 600 mm (24 in.) from the point of attachment.*

**(D) Secured.** Service cables shall be held securely in place.

**(E) Separately Bushed Openings.** Service heads shall have conductors of different potential brought out through separately bushed openings.

*Exception: For jacketed multiconductor service cable without splice.*

**(F) Drip Loops.** Drip loops shall be formed on individual conductors. To prevent the entrance of moisture, service-entrance conductors shall be connected to the service-drop conductors either (1) below the level of the service head or (2) below the level of the termination of the service-entrance cable sheath.

**Table 230.51(C) Supports**

| | Maximum Distance Between Supports | | Minimum Clearance | | | |
| | | | Between Conductors | | From Surface | |
| Maximum Volts | m | ft | mm | in. | mm | in. |
|---|---|---|---|---|---|---|
| 600 | 2.7 | 9 | 150 | 6 | 50 | 2 |
| 600 | 4.5 | 15 | 300 | 12 | 50 | 2 |
| 300 | 1.4 | 4½ | 75 | 3 | 50 | 2 |
| 600* | 1.4* | 4½* | 65* | 2½* | 25* | 1* |

*Where not exposed to weather.

**(G) Arranged That Water Will Not Enter Service Raceway or Equipment.** Service-drop conductors and service-entrance conductors shall be arranged so that water will not enter service raceway or equipment.

**230.56 Service Conductor with the Higher Voltage to Ground.** On a 4-wire, delta-connected service where the midpoint of one phase winding is grounded, the service conductor having the higher phase voltage to ground shall be durably and permanently marked by an outer finish that is orange in color, or by other effective means, at each termination or junction point.

## V. Service Equipment — General

**230.62 Service Equipment — Enclosed or Guarded.** Energized parts of service equipment shall be enclosed as specified in 230.62(A) or guarded as specified in 230.62(B).

**(A) Enclosed.** Energized parts shall be enclosed so that they will not be exposed to accidental contact or shall be guarded as in 230.62(B).

**(B) Guarded.** Energized parts that are not enclosed shall be installed on a switchboard, panelboard, or control board and guarded in accordance with 110.18 and 110.27. Where energized parts are guarded as provided in 110.27(A)(1) and (2), a means for locking or sealing doors providing access to energized parts shall be provided.

**230.66 Marking.** Service equipment rated at 600 volts or less shall be marked to identify it as being suitable for use as service equipment. Individual meter socket enclosures shall not be considered service equipment.

## VI. Service Equipment — Disconnecting Means

**230.70 General.** Means shall be provided to disconnect all conductors in a building or other structure from the service-entrance conductors.

**(A) Location.** The service disconnecting means shall be installed in accordance with 230.70(A)(1), (2), and (3).

**(1) Readily Accessible Location.** The service disconnecting means shall be installed at a readily accessible location either outside of a building or structure or inside nearest the point of entrance of the service conductors.

**(2) Bathrooms.** Service disconnecting means shall not be installed in bathrooms.

**(3) Remote Control.** Where a remote control device(s) is used to actuate the service disconnecting means, the service disconnecting means shall be located in accordance with 230.70(A)(1).

**(B) Marking.** Each service disconnect shall be permanently marked to identify it as a service disconnect.

**(C) Suitable for Use.** Each service disconnecting means shall be suitable for the prevailing conditions. Service equipment installed in hazardous (classified) locations shall comply with the requirements of Articles 500 through 517.

**230.71 Maximum Number of Disconnects.**

**(A) General.** The service disconnecting means for each service permitted by 230.2, or for each set of service-entrance conductors permitted by 230.40, Exception Nos. 1, 3, 4, or 5, shall consist of not more than six switches or sets of circuit breakers, or a combination of not more than six switches and sets of circuit breakers, mounted in a single enclosure, in a group of separate enclosures, or in or on a switchboard. There shall be no more than six sets of disconnects per service grouped in any one location. For the purpose of this section, disconnecting means used solely for power monitoring equipment, or the control circuit of the ground-fault protection system or power-operable service disconnecting means, installed as part of the listed equipment, shall not be considered a service disconnecting means.

**(B) Single-Pole Units.** Two or three single-pole switches or breakers, capable of individual operation, shall be permitted on multiwire circuits, one pole for each ungrounded conductor, as one multipole disconnect, provided they are equipped with handle ties or a master handle to disconnect all conductors of the service with no more than six operations of the hand.

> FPN: See 408.16(A) for service equipment in panelboards, and see 430.95 for service equipment in motor control centers.

**230.72 Grouping of Disconnects.**

**(A) General.** The two to six disconnects as permitted in 230.71 shall be grouped. Each disconnect shall be marked to indicate the load served.

*Exception: One of the two to six service disconnecting means permitted in 230.71, where used only for a water pump also intended to provide fire protection, shall be permitted to be located remote from the other disconnecting means.*

**(B) Additional Service Disconnecting Means.** The one or more additional service disconnecting means for fire pumps, for legally required standby, or for optional standby services permitted by 230.2 shall be installed remote from the one to six service disconnecting means for normal service to minimize the possibility of simultaneous interruption of supply.

**(C) Access to Occupants.** In a multiple-occupancy building, each occupant shall have access to the occupant's service disconnecting means.

*Exception: In a multiple-occupancy building where electric service and electrical maintenance are provided by the building management and where these are under continuous building management supervision, the service disconnecting means supplying more than one occupancy shall be permitted to be accessible to authorized management personnel only.*

**230.74 Simultaneous Opening of Poles.** Each service disconnect shall simultaneously disconnect all ungrounded service conductors that it controls from the premises wiring system.

**230.75 Disconnection of Grounded Conductor.** Where the service disconnecting means does not disconnect the grounded conductor from the premises wiring, other means shall be provided for this purpose in the service equipment. A terminal or bus to which all grounded conductors can be attached by means of pressure connectors shall be permitted for this purpose. In a multisection switchboard, disconnects for the grounded conductor shall be permitted to be in any section of the switchboard, provided any such switchboard section is marked.

**230.76 Manually or Power Operable.** The service disconnecting means for ungrounded service conductors shall consist of either (1) a manually operable switch or circuit breaker equipped with a handle or other suitable operating means or (2) a power-operated switch or circuit breaker, provided the switch or circuit breaker can be opened by hand in the event of a power supply failure.

**230.77 Indicating.** The service disconnecting means shall plainly indicate whether it is in the open or closed position.

**230.79 Rating of Service Disconnecting Means.** The service disconnecting means shall have a rating not less than the load to be carried, determined in accordance with Article 220. In no case shall the rating be lower than specified in 230.79(A), (B), (C), or (D).

**(A) One-Circuit Installation.** For installations to supply only limited loads of a single branch circuit, the service disconnecting means shall have a rating of not less than 15 amperes.

**(B) Two-Circuit Installations.** For installations consisting of not more than two 2-wire branch circuits, the service

disconnecting means shall have a rating of not less than 30 amperes.

**(C) One-Family Dwelling.** For a one-family dwelling, the service disconnecting means shall have a rating of not less than 100 amperes, 3-wire.

**(D) All Others.** For all other installations, the service disconnecting means shall have a rating of not less than 60 amperes.

**230.80 Combined Rating of Disconnects.** Where the service disconnecting means consists of more than one switch or circuit breaker, as permitted by 230.71, the combined ratings of all the switches or circuit breakers used shall not be less than the rating required by 230.79.

**230.81 Connection to Terminals.** The service conductors shall be connected to the service disconnecting means by pressure connectors, clamps, or other approved means. Connections that depend on solder shall not be used.

**230.82 Equipment Connected to the Supply Side of Service Disconnect.** Only the following equipment shall be permitted to be connected to the supply side of the service disconnecting means:

(1) Cable limiters or other current-limiting devices.
(2) Meters, meter sockets, or meter disconnect switches nominally rated not in excess of 600 volts, provided all metal housings and service enclosures are grounded.
(3) Instrument transformers (current and voltage), high-impedance shunts, load management devices, and surge arresters.
(4) Taps used only to supply load management devices, circuits for standby power systems, fire pump equipment, and fire and sprinkler alarms, if provided with service equipment and installed in accordance with requirements for service-entrance conductors.
(5) Solar photovoltaic systems, fuel cell systems, or interconnected electric power production sources.
(6) Control circuits for power-operable service disconnecting means, if suitable overcurrent protection and disconnecting means are provided.
(7) Ground-fault protection systems where installed as part of listed equipment, if suitable overcurrent protection and disconnecting means are provided.

## VII. Service Equipment — Overcurrent Protection

**230.90 Where Required.** Each ungrounded service conductor shall have overload protection.

**(A) Ungrounded Conductor.** Such protection shall be provided by an overcurrent device in series with each ungrounded service conductor that has a rating or setting not higher than the allowable ampacity of the conductor. A set of fuses shall be considered all the fuses required to protect all the ungrounded conductors of a circuit. Single-pole circuit breakers, grouped in accordance with 230.71(B), shall be considered as one protective device.

*Exception No. 1: For motor-starting currents, ratings that conform with 430.52, 430.62, and 430.63 shall be permitted.*

*Exception No. 2: Fuses and circuit breakers with a rating or setting that conform with 240.4(B) or (C) and 240.6 shall be permitted.*

*Exception No. 3: Two to six circuit breakers or sets of fuses shall be permitted as the overcurrent device to provide the overload protection. The sum of the ratings of the circuit breakers or fuses shall be permitted to exceed the ampacity of the service conductors, provided the calculated load does not exceed the ampacity of the service conductors.*

*Exception No. 4: Overload protection for fire pump supply conductors shall conform with 695.4(B)(1).*

*Exception No. 5: Overload protection for 120/240-volt, 3-wire, single-phase dwelling services shall be permitted in accordance with the requirements of 310.15(B)(6).*

**(B) Not in Grounded Conductor.** No overcurrent device shall be inserted in a grounded service conductor except a circuit breaker that simultaneously opens all conductors of the circuit.

**230.91 Location.** The service overcurrent device shall be an integral part of the service disconnecting means or shall be located immediately adjacent thereto.

**230.92 Locked Service Overcurrent Devices.** Where the service overcurrent devices are locked or sealed or are not readily accessible to the occupant, branch-circuit overcurrent devices shall be installed on the load side, shall be mounted in a readily accessible location, and shall be of lower ampere rating than the service overcurrent device.

**230.93 Protection of Specific Circuits.** Where necessary to prevent tampering, an automatic overcurrent device that protects service conductors supplying only a specific load, such as a water heater, shall be permitted to be locked or sealed where located so as to be accessible.

**230.94 Relative Location of Overcurrent Device and Other Service Equipment.** The overcurrent device shall protect all circuits and devices.

*Exception No. 1: The service switch shall be permitted on the supply side.*

*Exception No. 2: High-impedance shunt circuits, surge arresters, surge-protective capacitors, and instrument transformers (current and voltage) shall be permitted to be connected and installed on the supply side of the service disconnecting means as permitted in 230.82.*

*Exception No. 3: Circuits for load management devices shall be permitted to be connected on the supply side of the service overcurrent device where separately provided with overcurrent protection.*

*Exception No. 4: Circuits used only for the operation of fire alarm, other protective signaling systems, or the supply to fire pump equipment shall be permitted to be connected on the supply side of the service overcurrent device where separately provided with overcurrent protection.*

*Exception No. 5: Meters nominally rated not in excess of 600 volts, provided all metal housings and service enclosures are grounded in accordance with Article 250.*

*Exception No. 6: Where service equipment is power operable, the control circuit shall be permitted to be connected ahead of the service equipment if suitable overcurrent protection and disconnecting means are provided.*

**230.95 Ground-Fault Protection of Equipment.** Ground-fault protection of equipment shall be provided for solidly grounded wye electrical services of more than 150 volts to ground but not exceeding 600 volts phase-to-phase for each service disconnect rated 1000 amperes or more.

The rating of the service disconnect shall be considered to be the rating of the largest fuse that can be installed or the highest continuous current trip setting for which the actual overcurrent device installed in a circuit breaker is rated or can be adjusted.

**Solidly Grounded — Definition.** Connection of the grounded conductor to ground without inserting any resistor or impedance device.

*Exception No. 1: The ground-fault protection provisions of this section shall not apply to a service disconnect for a continuous industrial process where a nonorderly shutdown will introduce additional or increased hazards.*

*Exception No. 2: The ground-fault protection provisions of this section shall not apply to fire pumps.*

**(A) Setting.** The ground-fault protection system shall operate to cause the service disconnect to open all ungrounded conductors of the faulted circuit. The maximum setting of the ground-fault protection shall be 1200 amperes, and the maximum time delay shall be one second for ground-fault currents equal to or greater than 3000 amperes.

**(B) Fuses.** If a switch and fuse combination is used, the fuses employed shall be capable of interrupting any current higher than the interrupting capacity of the switch during a time that the ground-fault protective system will not cause the switch to open.

**(C) Performance Testing.** The ground-fault protection system shall be performance tested when first installed on site. The test shall be conducted in accordance with instructions that shall be provided with the equipment. A written record of this test shall be made and shall be available to the authority having jurisdiction.

> FPN No. 1: Ground-fault protection that functions to open the service disconnect affords no protection from faults on the line side of the protective element. It serves only to limit damage to conductors and equipment on the load side in the event of an arcing ground fault on the load side of the protective element.

> FPN No. 2: This added protective equipment at the service equipment may make it necessary to review the overall wiring system for proper selective overcurrent protection coordination. Additional installations of ground-fault protective equipment may be needed on feeders and branch circuits where maximum continuity of electrical service is necessary.

> FPN No. 3: Where ground-fault protection is provided for the service disconnect and interconnection is made with another supply system by a transfer device, means or devices may be needed to ensure proper ground-fault sensing by the ground-fault protection equipment.

## VIII. Services Exceeding 600 Volts, Nominal

**230.200 General.** Service conductors and equipment used on circuits exceeding 600 volts, nominal, shall comply with all the applicable provisions of the preceding sections of this article and with the following sections, which supplement or modify the preceding sections. In no case shall the provisions of Part VIII apply to equipment on the supply side of the service point.

> FPN: For clearances of conductors of over 600 volts, nominal, see ANSI C2-1997, *National Electrical Safety Code.*

**230.202 Service-Entrance Conductors.** Service-entrance conductors to buildings or enclosures shall be installed to conform to 230.202(A) and (B).

**(A) Conductor Size.** Service-entrance conductors shall not be smaller than 6 AWG unless in multiconductor cable. Multiconductor cable shall not be smaller than 8 AWG.

**(B) Wiring Methods.** Service-entrance conductors shall be installed by one of the wiring methods covered in 300.37 and 300.50.

**230.204 Isolating Switches.**

**(A) Where Required.** Where oil switches or air, oil, vacuum, or sulfur hexafluoride circuit breakers constitute the service disconnecting means, an isolating switch with visible break contacts shall be installed on the supply side of the disconnecting means and all associated service equipment.

*Exception: An isolating switch shall not be required where the circuit breaker or switch is mounted on removable truck panels or metal-enclosed switchgear units, that*

*(a) Cannot be opened unless the circuit is disconnected, and*

*(b) Where all energized parts are automatically disconnected when the circuit breaker or switch is removed from the normal operating position*

**(B) Fuses as Isolating Switch.** Where fuses are of the type that can be operated as a disconnecting switch, a set of such fuses shall be permitted as the isolating switch.

**(C) Accessible to Qualified Persons Only.** The isolating switch shall be accessible to qualified persons only.

**(D) Grounding Connection.** Isolating switches shall be provided with a means for readily connecting the load side conductors to ground when disconnected from the source of supply.

A means for grounding the load side conductors shall not be required for any duplicate isolating switch installed and maintained by the electric supply company.

**230.205 Disconnecting Means.**

**(A) Location.** The service disconnecting means shall be located in accordance with 230.70.

**(B) Type.** Each service disconnect shall simultaneously disconnect all ungrounded service conductors that it controls and shall have a fault-closing rating that is not less than the maximum short-circuit current available at its supply terminals.

Where fused switches or separately mounted fuses are installed, the fuse characteristics shall be permitted to contribute to the fault-closing rating of the disconnecting means.

**(C) Remote Control.** For multibuilding, industrial installations under single management, the service disconnecting means shall be permitted to be located at a separate building or structure. In such cases, the service disconnecting means shall be permitted to be electrically operated by a readily accessible, remote-control device.

**230.206 Overcurrent Devices as Disconnecting Means.** Where the circuit breaker or alternative for it, as specified in 230.208 for service overcurrent devices, meets the requirements specified in 230.205, they shall constitute the service disconnecting means.

**230.208 Protection Requirements.** A short-circuit protective device shall be provided on the load side of, or as an integral part of, the service disconnect, and shall protect all ungrounded conductors that it supplies. The protective device shall be capable of detecting and interrupting all values of current, in excess of its trip setting or melting point, that can occur at its location. A fuse rated in continuous amperes not to exceed three times the ampacity of the conductor, or a circuit breaker with a trip setting of not more than six times the ampacity of the conductors, shall be considered as providing the required short-circuit protection.

> FPN: See Tables 310.67 through 310.86 for ampacities of conductors rated 2001 volts and above.

Overcurrent devices shall conform to 230.208(A) and (B).

**(A) Equipment Type.** Equipment used to protect service-entrance conductors shall meet the requirements of Article 490, Part II.

**(B) Enclosed Overcurrent Devices.** The restriction to 80 percent of the rating for an enclosed overcurrent device for continuous loads shall not apply to overcurrent devices installed in systems operating at over 600 volts.

**230.209 Surge Arresters (Lightning Arresters).** Surge arresters installed in accordance with the requirements of Article 280 shall be permitted on each ungrounded overhead service conductor.

**230.210 Service Equipment — General Provisions.** Service equipment, including instrument transformers, shall conform to Article 490, Part I.

**230.211 Metal-Enclosed Switchgear.** Metal-enclosed switchgear shall consist of a substantial metal structure and a sheet metal enclosure. Where installed over a combustible floor, suitable protection thereto shall be provided.

**230.212 Over 35,000 Volts.** Where the voltage exceeds 35,000 volts between conductors that enter a building, they shall terminate in a metal-enclosed switchgear compartment or a vault conforming to the requirements of 450.41 through 450.48.

## ARTICLE 240
## Overcurrent Protection

### I. General

**240.1 Scope.** Parts I through VII of this article provide the general requirements for overcurrent protection and overcurrent protective devices not more than 600 volts, nominal. Part VIII covers overcurrent protection for those portions of supervised industrial installations operating at voltages of not more than 600 volts, nominal. Part IX covers overcurrent protection over 600 volts, nominal.

> FPN: Overcurrent protection for conductors and equipment is provided to open the circuit if the current reaches a value that will cause an excessive or dangerous temperature in conductors or conductor insulation. See also 110.9 for requirements for interrupting ratings and 110.10 for requirements for protection against fault currents.

**240.2 Definitions.**

**Coordination.** The proper localization of a fault condition to restrict outages to the equipment affected, accomplished by the choice of selective fault-protective devices.

**Current-Limiting Overcurrent Protective Device.** A device that, when interrupting currents in its current-limiting range, reduces the current flowing in the faulted circuit to a magnitude substantially less than that obtainable in the same circuit if the device were replaced with a solid conductor having comparable impedance.

**Supervised Industrial Installation.** For the purposes of Part VIII, the industrial portions of a facility where all of the following conditions are met:

(1) Conditions of maintenance and engineering supervision ensure that only qualified persons monitor and service the system.
(2) The premises wiring system has 2500 kVA or greater of load used in industrial process(es), manufacturing activities, or both, as calculated in accordance with Article 220.
(3) The premises has at least one service that is more than 150 volts to ground and more than 300 volts phase-to-phase.

This definition excludes installations in buildings used by the industrial facility for offices, warehouses, garages, machine shops, and recreational facilities that are not an integral part of the industrial plant, substation, or control center.

**Tap Conductors.** As used in this article, a tap conductor is defined as a conductor, other than a service conductor, that has overcurrent protection ahead of its point of supply that exceeds the value permitted for similar conductors that are protected as described elsewhere in 240.4.

**240.3 Other Articles.** Equipment shall be protected against overcurrent in accordance with the article in this *Code* that covers the type of equipment specified in Table 240.3.

**Table 240.3 Other Articles**

| Equipment | Article |
|---|---|
| Air-conditioning and refrigerating equipment | 440 |
| Appliances | 422 |
| Audio signal processing, amplification, and reproduction equipment | 640 |
| Branch circuits | 210 |
| Busways | 368 |
| Capacitors | 460 |
| Class 1, Class 2, and Class 3 remote-control, signaling, and power-limited circuits | 725 |
| Closed-loop and programmed power distribution | 780 |
| Cranes and hoists | 610 |
| Electric signs and outline lighting | 600 |
| Electric welders | 630 |
| Electrolytic cells | 668 |
| Elevators, dumbwaiters, escalators, moving walks, wheelchair lifts, and stairway chair lifts | 620 |
| Emergency systems | 700 |
| Fire alarm systems | 760 |
| Fire pumps | 695 |
| Fixed electric heating equipment for pipelines and vessels | 427 |
| Fixed electric space-heating equipment | 424 |
| Fixed outdoor electric deicing and snow-melting equipment | 426 |
| Generators | 445 |
| Health care facilities | 517 |
| Induction and dielectric heating equipment | 665 |
| Industrial machinery | 670 |
| Luminaires (lighting fixtures), lampholders, and lamps | 410 |
| Motion picture and television studios and similar locations | 530 |
| Motors, motor circuits, and controllers | 430 |
| Phase converters | 455 |
| Pipe organs | 650 |
| Places of assembly | 518 |
| Receptacles | 406 |
| Services | 230 |
| Solar photovoltaic systems | 690 |
| Switchboards and panelboards | 408 |
| Theaters, audience areas of motion picture and television studios, and similar locations | 520 |
| Transformers and transformer vaults | 450 |
| X-ray equipment | 660 |

**240.4 Protection of Conductors.** Conductors, other than flexible cords, flexible cables, and fixture wires, shall be protected against overcurrent in accordance with their am-

pacities specified in 310.15, unless otherwise permitted or required in 240.4(A) through (G).

**(A) Power Loss Hazard.** Conductor overload protection shall not be required where the interruption of the circuit would create a hazard, such as in a material-handling magnet circuit or fire pump circuit. Short-circuit protection shall be provided.

FPN: See NFPA 20-1999, *Standard for the Installation of Stationary Pumps for Fire Protection.*

**(B) Devices Rated 800 Amperes or Less.** The next higher standard overcurrent device rating (above the ampacity of the conductors being protected) shall be permitted to be used, provided all of the following conditions are met:

(1) The conductors being protected are not part of a multioutlet branch circuit supplying receptacles for cord-and-plug-connected portable loads.
(2) The ampacity of the conductors does not correspond with the standard ampere rating of a fuse or a circuit breaker without overload trip adjustments above its rating (but that shall be permitted to have other trip or rating adjustments).
(3) The next higher standard rating selected does not exceed 800 amperes.

**(C) Devices Rated Over 800 Amperes.** Where the overcurrent device is rated over 800 amperes, the ampacity of the conductors it protects shall be equal to or greater than the rating of the overcurrent device defined in 240.6.

**(D) Small Conductors.** Unless specifically permitted in 240.4(E) through (G), the overcurrent protection shall not exceed 15 amperes for 14 AWG, 20 amperes for 12 AWG, and 30 amperes for 10 AWG copper; or 15 amperes for 12 AWG and 25 amperes for 10 AWG aluminum and copper-clad aluminum after any correction factors for ambient temperature and number of conductors have been applied.

**(E) Tap Conductors.** Tap conductors shall be permitted to be protected against overcurrent in accordance with 210.19(A)(3) and (4), 240.5(B)(2), 240.21, 368.11, 368.12, and 430.53(D).

**(F) Transformer Secondary Conductors.** Single-phase (other than 2-wire) and multiphase (other than delta-delta, 3-wire) transformer secondary conductors shall not be considered to be protected by the primary overcurrent protective device. Conductors supplied by the secondary side of a single-phase transformer having a 2-wire (single-voltage) secondary, or a three-phase, delta-delta connected transformer having a 3-wire (single-voltage) secondary, shall be permitted to be protected by overcurrent protection provided on the primary (supply) side of the transformer, provided this protection is in accordance with 450.3 and does not exceed the value determined by multiplying the secondary conductor ampacity by the secondary to primary transformer voltage ratio.

**(G) Overcurrent Protection for Specific Conductor Applications.** Overcurrent protection for the specific conductors shall be permitted to be provided as referenced in Table 240.4(G).

**Table 240.4(G) Specific Conductor Applications**

| Conductor | Article | Section |
|---|---|---|
| Air-conditioning and refrigeration equipment circuit conductors | 440, Parts III, VI | |
| Capacitor circuit conductors | 460 | 460.8(B) and 460.25(A)–(D) |
| Control and instrumentation circuit conductors (Type ITC) | 727 | 727.9 |
| Electric welder circuit conductors | 630 | 630.12 and 630.32 |
| Fire alarm system circuit conductors | 760 | 760.23, 760.24, 760.41, and Chapter 9, Tables 12(A) and 12(B) |
| Motor-operated appliance circuit conductors | 422, Part II | |
| Motor and motor-control circuit conductors | 430, Parts III, IV, V, VI, VII | |
| Phase converter supply conductors | 455 | 455.7 |
| Remote-control, signaling, and power- limited circuit conductors | 725 | 725.23, 725.24, 725.41, and Chapter 9, Tables 11(A) and 11(B) |
| Secondary tie conductors | 450 | 450.6 |

**240.5 Protection of Flexible Cords, Flexible Cables, and Fixture Wires.** Flexible cord and flexible cable, including tinsel cord and extension cords, and fixture wires shall be protected against overcurrent by either 240.5(A) or (B).

**(A) Ampacities.** Flexible cord and flexible cable shall be protected by an overcurrent device in accordance with their ampacity as specified in Tables 400.5(A) and 400.5(B). Fixture wire shall be protected against overcurrent in accordance with its ampacity as specified in Table 402.5. Supplementary overcurrent protection, as in 240.10, shall be permitted to be an acceptable means for providing this protection.

**(B) Branch Circuit Overcurrent Device.** Flexible cord shall be protected where supplied by a branch circuit in accordance with one of the methods described in 240.5(B)(1), (2), or (3).

**(1) Supply Cord of Listed Appliance or Portable Lamps.** Where flexible cord or tinsel cord is approved for and used with a specific listed appliance or portable lamp, it shall be permitted to be supplied by a branch circuit of Article 210 in accordance with the following:

(1) 20-ampere circuits — tinsel cord or 18 AWG cord and larger
(2) 30-ampere circuits — 16 AWG cord and larger
(3) 40-ampere circuits — cord of 20-ampere capacity and over
(4) 50-ampere circuits — cord of 20-ampere capacity and over

**(2) Fixture Wire.** Fixture wire shall be permitted to be tapped to the branch circuit conductor of a branch circuit of Article 210 in accordance with the following:

(1) 20-ampere circuits — 18 AWG, up to 15 m (50 ft) of run length
(2) 20-ampere circuits — 16 AWG, up to 30 m (100 ft) of run length
(3) 20-ampere circuits — 14 AWG and larger
(4) 30-ampere circuits — 14 AWG and larger
(5) 40-ampere circuits — 12 AWG and larger
(6) 50-ampere circuits — 12 AWG and larger

**(3) Extension Cord Sets.** Flexible cord used in listed extension cord sets, or in extension cords made with separately listed and installed components, shall be permitted to be supplied by a branch circuit of Article 210 in accordance with the following:

20-ampere circuits — 16 AWG and larger

**240.6 Standard Ampere Ratings.**

**(A) Fuses and Fixed-Trip Circuit Breakers.** The standard ampere ratings for fuses and inverse time circuit breakers shall be considered 15, 20, 25, 30, 35, 40, 45, 50, 60, 70, 80, 90, 100, 110, 125, 150, 175, 200, 225, 250, 300, 350, 400, 450, 500, 600, 700, 800, 1000, 1200, 1600, 2000, 2500, 3000, 4000, 5000, and 6000 amperes. Additional standard ampere ratings for fuses shall be 1, 3, 6, 10, and 601. The use of fuses and inverse time circuit breakers with nonstandard ampere ratings shall be permitted.

**(B) Adjustable-Trip Circuit Breakers.** The rating of adjustable-trip circuit breakers having external means for adjusting the current setting (long-time pickup setting), not meeting the requirements of 240.6(C), shall be the maximum setting possible.

**(C) Restricted Access Adjustable-Trip Circuit Breakers.** A circuit breaker(s) that has restricted access to the adjusting means shall be permitted to have an ampere rating(s) that is equal to the adjusted current setting (long-time pickup setting). Restricted access shall be defined as located behind one of the following:

(1) Removable and sealable covers over the adjusting means
(2) Bolted equipment enclosure doors
(3) Locked doors accessible only to qualified personnel

**240.8 Fuses or Circuit Breakers in Parallel.** Fuses and circuit breakers shall be permitted to be connected in parallel where they are factory assembled in parallel and listed as a unit. Individual fuses, circuit breakers, or combinations thereof shall not otherwise be connected in parallel.

**240.9 Thermal Devices.** Thermal relays and other devices not designed to open short circuits or ground faults shall not be used for the protection of conductors against overcurrent due to short circuits or ground faults, but the use of such devices shall be permitted to protect motor branch-circuit conductors from overload if protected in accordance with 430.40.

**240.10 Supplementary Overcurrent Protection.** Where supplementary overcurrent protection is used for luminaires (lighting fixtures), appliances, and other equipment or for internal circuits and components of equipment, it shall not be used as a substitute for branch-circuit overcurrent devices or in place of the branch-circuit protection specified in Article 210. Supplementary overcurrent devices shall not be required to be readily accessible.

**240.12 Electrical System Coordination.** Where an orderly shutdown is required to minimize the hazard(s) to personnel and equipment, a system of coordination based on the following two conditions shall be permitted:

(1) Coordinated short-circuit protection
(2) Overload indication based on monitoring systems or devices

> FPN: The monitoring system may cause the condition to go to alarm, allowing corrective action or an orderly shutdown, thereby minimizing personnel hazard and equipment damage.

**240.13 Ground-Fault Protection of Equipment.** Ground-fault protection of equipment shall be provided in accordance with the provisions of 230.95 for solidly grounded wye electrical systems of more than 150 volts to ground but not exceeding 600 volts phase-to-phase for each individual device used as a building or structure main disconnecting means rated 1000 amperes or more.

The provisions of this section shall not apply to the disconnecting means for the following:

(1) Continuous industrial processes where a nonorderly shutdown will introduce additional or increased hazards
(2) Installations where ground-fault protection is provided by other requirements for services or feeders
(3) Fire pumps installed in accordance with Article 695

## II. Location

**240.20 Ungrounded Conductors.**

**(A) Overcurrent Device Required.** A fuse or an overcurrent trip unit of a circuit breaker shall be connected in series with each ungrounded conductor. A combination of a current transformer and overcurrent relay shall be considered equivalent to an overcurrent trip unit.

> FPN: For motor circuits, see Parts III, IV, V, and X of Article 430.

**(B) Circuit Breaker as Overcurrent Device.** Circuit breakers shall open all ungrounded conductors of the circuit unless otherwise permitted in 240.20(B)(1), (B)(2), and (B)(3).

**(1) Multiwire Branch Circuit.** Except where limited by 210.4(B), individual single-pole circuit breakers, with or without approved handle ties, shall be permitted as the protection for each ungrounded conductor of multiwire branch circuits that serve only single-phase line-to-neutral loads.

**(2) Grounded Single-Phase and 3-wire dc Circuits.** In grounded systems, individual single-pole circuit breakers with approved handle ties shall be permitted as the protection for each ungrounded conductor for line-to-line connected loads for single-phase circuits or 3-wire, direct-current circuits.

**(3) 3-Phase and 2-Phase Systems.** For line-to-line loads in 4-wire, 3-phase systems or 5-wire, 2-phase systems having a grounded neutral and no conductor operating at a voltage greater than permitted in 210.6, individual single-pole circuit breakers with approved handle ties shall be permitted as the protection for each ungrounded conductor.

**(C) Closed-Loop Power Distribution Systems.** Listed devices that provide equivalent overcurrent protection in closed-loop power distribution systems shall be permitted as a substitute for fuses or circuit breakers.

**240.21 Location in Circuit.** Overcurrent protection shall be provided in each ungrounded circuit conductor and shall be located at the point where the conductors receive their supply except as specified in 240.21(A) through (G). No conductor supplied under the provisions of 240.21(A) through (G) shall supply another conductor under those provisions, except through an overcurrent protective device meeting the requirements of 240.4.

**(A) Branch-Circuit Conductors.** Branch-circuit tap conductors meeting the requirements specified in 210.19 shall be permitted to have overcurrent protection located as specified in that section.

**(B) Feeder Taps.** Conductors shall be permitted to be tapped, without overcurrent protection at the tap, to a feeder as specified in 240.21(B)(1) through (5).

**(1) Taps Not Over 3 m (10 ft) Long.** Where the length of the tap conductors does not exceed 3 m (10 ft) and the tap conductors comply with all of the following:

(1) The ampacity of the tap conductors is

   a. Not less than the combined computed loads on the circuits supplied by the tap conductors, and

   b. Not less than the rating of the device supplied by the tap conductors or not less than the rating of the overcurrent-protective device at the termination of the tap conductors.

(2) The tap conductors do not extend beyond the switchboard, panelboard, disconnecting means, or control devices they supply.

(3) Except at the point of connection to the feeder, the tap conductors are enclosed in a raceway, which shall extend from the tap to the enclosure of an enclosed switchboard, panelboard, or control devices, or to the back of an open switchboard.

(4) For field installations where the tap conductors leave the enclosure or vault in which the tap is made, the rating of the overcurrent device on the line side of the tap conductors shall not exceed 10 times the ampacity of the tap conductor.

> FPN: For overcurrent protection requirements for lighting and appliance branch-circuit panelboards and certain power panelboards, see 408.16(A), (B), and (E).

**(2) Taps Not Over 7.5 m (25 ft) Long.** Where the length of the tap conductors does not exceed 7.5 m (25 ft) and the tap conductors comply with all the following:

(1) The ampacity of the tap conductors is not less than one-third of the rating of the overcurrent device protecting the feeder conductors.

(2) The tap conductors terminate in a single circuit breaker or a single set of fuses that will limit the load to the ampacity of the tap conductors. This device shall be permitted to supply any number of additional overcurrent devices on its load side.

(3) The tap conductors are suitably protected from physical damage or are enclosed in a raceway.

**(3) Taps Supplying a Transformer [Primary Plus Secondary Not Over 7.5 m (25 ft) Long].** Where the tap conductors supply a transformer and comply with all the following conditions:

(1) The conductors supplying the primary of a transformer have an ampacity at least one-third the rating of the overcurrent device protecting the feeder conductors.

(2) The conductors supplied by the secondary of the transformer shall have an ampacity that, when multiplied by the ratio of the secondary-to-primary voltage, is at least one-third of the rating of the overcurrent device protecting the feeder conductors.

(3) The total length of one primary plus one secondary conductor, excluding any portion of the primary conductor that is protected at its ampacity, is not over 7.5 m (25 ft).

(4) The primary and secondary conductors are suitably protected from physical damage.

(5) The secondary conductors terminate in a single circuit breaker or set of fuses that limit the load current to not more than the conductor ampacity that is permitted by 310.15.

**(4) Taps Over 7.5 m (25 ft) Long.** Where the feeder is in a high bay manufacturing building over 11 m (35 ft) high at walls and the installation complies with all the following conditions:

(1) Conditions of maintenance and supervision ensure that only qualified persons service the systems.

(2) The tap conductors are not over 7.5 m (25 ft) long horizontally and not over 30 m (100 ft) total length.

(3) The ampacity of the tap conductors is not less than one-third the rating of the overcurrent device protecting the feeder conductors.

(4) The tap conductors terminate at a single circuit breaker or a single set of fuses that limit the load to the ampacity of the tap conductors. This single overcurrent device shall be permitted to supply any number of additional overcurrent devices on its load side.

(5) The tap conductors are suitably protected from physical damage or are enclosed in a raceway.

(6) The tap conductors are continuous from end-to-end and contain no splices.

(7) The tap conductors are sized 6 AWG copper or 4 AWG aluminum or larger.

(8) The tap conductors do not penetrate walls, floors, or ceilings.

(9) The tap is made no less than 9 m (30 ft) from the floor.

**(5) Outside Taps of Unlimited Length.** Where the conductors are located outdoors of a building or structure, except at the point of load termination, and comply with all of the following conditions:

(1) The conductors are suitably protected from physical damage.

(2) The conductors terminate at a single circuit breaker or a single set of fuses that limit the load to the ampacity of the conductors. This single overcurrent device shall be permitted to supply any number of additional overcurrent devices on its load side.

(3) The overcurrent device for the conductors is an integral part of a disconnecting means or shall be located immediately adjacent thereto.

·(4) The disconnecting means for the conductors is installed at a readily accessible location complying with one of the following:

   a. Outside of a building or structure

   b. Inside, nearest the point of entrance of the conductors

   c. Where installed in accordance with 230.6, nearest the point of entrance of the conductors

**(C) Transformer Secondary Conductors.** Conductors shall be permitted to be connected to a transformer secondary, without overcurrent protection at the secondary, as specified in 240.21(C)(1) through (6).

FPN: For overcurrent protection requirements for transformers, see 450.3.

**(1) Protection by Primary Overcurrent Device.** Conductors supplied by the secondary side of a single-phase transformer having a 2-wire (single-voltage) secondary, or a three-phase, delta-delta connected transformer having a 3-wire (single-voltage) secondary, shall be permitted to be protected by overcurrent protection provided on the primary (supply) side of the transformer, provided this protection is in accordance with 450.3 and does not exceed the value determined by multiplying the secondary conductor ampacity by the secondary to primary transformer voltage ratio.

Single-phase (other than 2-wire) and multiphase (other than delta-delta, 3-wire) transformer secondary conductors are not considered to be protected by the primary overcurrent protective device.

**(2) Transformer Secondary Conductors Not Over 3 m (10 ft) Long.** Where the length of secondary conductor does not exceed 3 m (10 ft) and complies with all of the following:

(1) The ampacity of the secondary conductors is

   a. Not less than the combined computed loads on the circuits supplied by the secondary conductors, and

   b. Not less than the rating of the device supplied by the secondary conductors or not less than the rating of the overcurrent-protective device at the termination of the secondary conductors

(2) The secondary conductors do not extend beyond the switchboard, panelboard, disconnecting means, or control devices they supply.

(3) The secondary conductors are enclosed in a raceway, which shall extend from the transformer to the enclosure of an enclosed switchboard, panelboard, or control devices or to the back of an open switchboard.

FPN: For overcurrent protection requirements for lighting and appliance branch-circuit panelboards and certain power panelboards, see 408.16(A), (B), and (E).

**(3) Industrial Installation Secondary Conductors Not Over 7.5 m (25 ft) Long.** For industrial installations only, where the length of the secondary conductors does not exceed 7.5 m (25 ft) and complies with all of the following:

(1) The ampacity of the secondary conductors is not less than the secondary current rating of the transformer, and the sum of the ratings of the overcurrent devices does not exceed the ampacity of the secondary conductors.

(2) All overcurrent devices are grouped.

(3) The secondary conductors are suitably protected from physical damage.

**(4) Outside Secondary of Building or Structure Conductors.** Where the conductors are located outdoors of a building or structure, except at the point of load termination, and comply with all of the following conditions:

(1) The conductors are suitably protected from physical damage.

(2) The conductors terminate at a single circuit breaker or a single set of fuses that limit the load to the ampacity of the conductors. This single overcurrent device shall be permitted to supply any number of additional overcurrent devices on its load side.

(3) The overcurrent device for the conductors is an integral part of a disconnecting means or shall be located immediately adjacent thereto.

(4) The disconnecting means for the conductors is installed at a readily accessible location complying with one of the following:

   a. Outside of a building or structure

   b. Inside, nearest the point of entrance of the conductors

   c. Where installed in accordance with 230.6, nearest the point of entrance of the conductors

**(5) Secondary Conductors from a Feeder Tapped Transformer.** Transformer secondary conductors installed in accordance with 240.21(B)(3) shall be permitted to have overcurrent protection as specified in that section.

**(6) Secondary Conductors Not Over 7.5 m (25 ft) Long.** Where the length of secondary conductor does not exceed 7.5 m (25 ft) and complies with all of the following:

(1) The secondary conductors shall have an ampacity that, when multiplied by the ratio of the secondary-to-primary voltage, is at least one-third of the rating of the overcurrent device protecting the primary of the transformer.

(2) The secondary conductors terminate in a single circuit breaker or set of fuses that limit the load current to not more than the conductor ampacity that is permitted by 310.15.

(3) The secondary conductors are suitably protected from physical damage.

**(D) Service Conductors.** Service-entrance conductors shall be permitted to be protected by overcurrent devices in accordance with 230.91.

**(E) Busway Taps.** Busways and busway taps shall be permitted to be protected against overcurrent in accordance with 368.10 through 368.13.

**(F) Motor Circuit Taps.** Motor-feeder and branch-circuit conductors shall be permitted to be protected against overcurrent in accordance with 430.28 and 430.53, respectively.

**(G) Conductors from Generator Terminals.** Conductors from generator terminals that meet the size requirement in 445.13 shall be permitted to be protected against overload by the generator overload protective device(s) required by 445.12.

**240.22 Grounded Conductor.** No overcurrent device shall be connected in series with any conductor that is intentionally grounded, unless one of the following two conditions is met:

(1) The overcurrent device opens all conductors of the circuit, including the grounded conductor, and is designed so that no pole can operate independently.

(2) Where required by 430.36 or 430.37 for motor overload protection.

**240.23 Change in Size of Grounded Conductor.** Where a change occurs in the size of the ungrounded conductor, a similar change shall be permitted to be made in the size of the grounded conductor.

**240.24 Location in or on Premises.**

**(A) Accessibility.** Overcurrent devices shall be readily accessible unless one of the following applies:

(1) For busways, as provided in 368.12.

(2) For supplementary overcurrent protection, as described in 240.10.

(3) For overcurrent devices, as described in 225.40 and 230.92.

(4) For overcurrent devices adjacent to utilization equipment that they supply, access shall be permitted to be by portable means.

**(B) Occupancy.** Each occupant shall have ready access to all overcurrent devices protecting the conductors supplying that occupancy.

*Exception No. 1: Where electric service and electrical maintenance are provided by the building management and where these are under continuous building management supervision, the service overcurrent devices and feeder overcurrent devices supplying more than one occupancy shall be permitted to be accessible to only authorized management personnel in the following:*

*(a) Multiple-occupancy buildings*

*(b) Guest rooms of hotels and motels that are intended for transient occupancy*

*Exception No. 2: Where electric service and electrical maintenance are provided by the building management and where these are under continuous building management supervision, the branch circuit overcurrent devices supplying any guest rooms shall be permitted to be accessible to only authorized management personnel for guest rooms of hotels and motels that are intended for transient occupancy.*

**(C) Not Exposed to Physical Damage.** Overcurrent devices shall be located where they will not be exposed to physical damage.

FPN: See 110.11, Deteriorating Agents.

**(D) Not in Vicinity of Easily Ignitible Material.** Overcurrent devices shall not be located in the vicinity of easily ignitible material, such as in clothes closets.

**(E) Not Located in Bathrooms.** In dwelling units and guest rooms of hotels and motels, overcurrent devices, other than supplementary overcurrent protection, shall not be located in bathrooms as defined in Article 100.

**III. Enclosures**

**240.30 General.**

**(A) Protection from Physical Damage.** Overcurrent devices shall be protected from physical damage by one of the following:

(1) Installation in enclosures, cabinets, cutout boxes, or equipment assemblies

(2) Mounting on open-type switchboards, panelboards, or control boards that are in rooms or enclosures free from dampness and easily ignitible material and are accessible only to qualified personnel

**(B) Operating Handle.** The operating handle of a circuit breaker shall be permitted to be accessible without opening a door or cover.

**240.32 Damp or Wet Locations.** Enclosures for overcurrent devices in damp or wet locations shall comply with 312.2(A).

**240.33 Vertical Position.** Enclosures for overcurrent devices shall be mounted in a vertical position unless that is shown to be impracticable. Circuit breaker enclosures shall be permitted to be installed horizontally where the circuit breaker is installed in accordance with 240.81. Listed busway plug-in units shall be permitted to be mounted in orientations corresponding to the busway mounting position.

## IV. Disconnecting and Guarding

**240.40 Disconnecting Means for Fuses.** A disconnecting means shall be provided on the supply side of all fuses in circuits over 150 volts to ground and cartridge fuses in circuits of any voltage where accessible to other than qualified persons so that each individual circuit containing fuses can be independently disconnected from the source of power. A current-limiting device without a disconnecting means shall be permitted on the supply side of the service disconnecting means as permitted by 230.82. A single disconnecting means shall be permitted on the supply side of more than one set of fuses as permitted by 430.112, Exception, for group operation of motors and 424.22(C) for fixed electric space-heating equipment.

**240.41 Arcing or Suddenly Moving Parts.** Arcing or suddenly moving parts shall comply with 240.41(A) and (B).

**(A) Location.** Fuses and circuit breakers shall be located or shielded so that persons will not be burned or otherwise injured by their operation.

**(B) Suddenly Moving Parts.** Handles or levers of circuit breakers, and similar parts that may move suddenly in such a way that persons in the vicinity are likely to be injured by being struck by them, shall be guarded or isolated.

## V. Plug Fuses, Fuseholders, and Adapters

### 240.50 General.

**(A) Maximum Voltage.** Plug fuses shall be permitted to be used in the following circuits:

(1) Circuits not exceeding 125 volts between conductors
(2) Circuits supplied by a system having a grounded neutral where the line-to-neutral voltage does not exceed 150 volts

**(B) Marking.** Each fuse, fuseholder, and adapter shall be marked with its ampere rating.

**(C) Hexagonal Configuration.** Plug fuses of 15-ampere and lower rating shall be identified by a hexagonal configuration of the window, cap, or other prominent part to distinguish them from fuses of higher ampere ratings.

**(D) No Energized Parts.** Plug fuses, fuseholders, and adapters shall have no exposed energized parts after fuses or fuses and adapters have been installed.

**(E) Screw Shell.** The screw shell of a plug-type fuseholder shall be connected to the load side of the circuit.

### 240.51 Edison-Base Fuses.

**(A) Classification.** Plug fuses of the Edison-base type shall be classified at not over 125 volts and 30 amperes and below.

**(B) Replacement Only.** Plug fuses of the Edison-base type shall be used only for replacements in existing installations where there is no evidence of overfusing or tampering.

**240.52 Edison-Base Fuseholders.** Fuseholders of the Edison-base type shall be installed only where they are made to accept Type S fuses by the use of adapters.

**240.53 Type S Fuses.** Type S fuses shall be of the plug type and shall comply with 240.53(A) and (B).

**(A) Classification.** Type S fuses shall be classified at not over 125 volts and 0 to 15 amperes, 16 to 20 amperes, and 21 to 30 amperes.

**(B) Noninterchangeable.** Type S fuses of an ampere classification as specified in 240.53(A) shall not be interchangeable with a lower ampere classification. They shall be designed so that they cannot be used in any fuseholder other than a Type S fuseholder or a fuseholder with a Type S adapter inserted.

### 240.54 Type S Fuses, Adapters, and Fuseholders.

**(A) To Fit Edison-Base Fuseholders.** Type S adapters shall fit Edison-base fuseholders.

**(B) To Fit Type S Fuses Only.** Type S fuseholders and adapters shall be designed so that either the fuseholder itself or the fuseholder with a Type S adapter inserted cannot be used for any fuse other than a Type S fuse.

**(C) Nonremovable.** Type S adapters shall be designed so that once inserted in a fuseholder, they cannot be removed.

**(D) Nontamperable.** Type S fuses, fuseholders, and adapters shall be designed so that tampering or shunting (bridging) would be difficult.

**(E) Interchangeability.** Dimensions of Type S fuses, fuseholders, and adapters shall be standardized to permit interchangeability regardless of the manufacturer.

## VI. Cartridge Fuses and Fuseholders

### 240.60 General.

**(A) Maximum Voltage — 300-Volt Type.** Cartridge fuses and fuseholders of the 300-volt type shall be permitted to be used in the following circuits:

(1) Circuits not exceeding 300 volts between conductors
(2) Single-phase line-to-neutral circuits supplied from a 3-phase, 4-wire, solidly grounded neutral source where the line-to-neutral voltage does not exceed 300 volts

**(B) Noninterchangeable — 0–6000-Ampere Cartridge Fuseholders.** Fuseholders shall be designed so that it will be difficult to put a fuse of any given class into a fuseholder that is designed for a current lower, or voltage higher, than that of the class to which the fuse belongs. Fuseholders for current-limiting fuses shall not permit insertion of fuses that are not current-limiting.

**(C) Marking.** Fuses shall be plainly marked, either by printing on the fuse barrel or by a label attached to the barrel showing the following:

(1) Ampere rating
(2) Voltage rating
(3) Interrupting rating where other than 10,000 amperes
(4) Current limiting where applicable
(5) The name or trademark of the manufacturer

The interrupting rating shall not be required to be marked on fuses used for supplementary protection.

**240.61 Classification.** Cartridge fuses and fuseholders shall be classified according to voltage and amperage ranges. Fuses rated 600 volts, nominal, or less shall be permitted to be used for voltages at or below their ratings.

## VII. Circuit Breakers

**240.80 Method of Operation.** Circuit breakers shall be trip free and capable of being closed and opened by manual operation. Their normal method of operation by other than manual means, such as electrical or pneumatic, shall be permitted if means for manual operation are also provided.

**240.81 Indicating.** Circuit breakers shall clearly indicate whether they are in the open "off" or closed "on" position.

Where circuit breaker handles are operated vertically rather than rotationally or horizontally, the "up" position of the handle shall be the "on" position.

**240.82 Nontamperable.** A circuit breaker shall be of such design that any alteration of its trip point (calibration) or the time required for its operation requires dismantling of the device or breaking of a seal for other than intended adjustments.

### 240.83 Marking.

**(A) Durable and Visible.** Circuit breakers shall be marked with their ampere rating in a manner that will be durable and visible after installation. Such marking shall be permitted to be made visible by removal of a trim or cover.

**(B) Location.** Circuit breakers rated at 100 amperes or less and 600 volts or less shall have the ampere rating molded, stamped, etched, or similarly marked into their handles or escutcheon areas.

**(C) Interrupting Rating.** Every circuit breaker having an interrupting rating other than 5000 amperes shall have its interrupting rating shown on the circuit breaker. The interrupting rating shall not be required to be marked on circuit breakers used for supplementary protection.

**(D) Used as Switches.** Circuit breakers used as switches in 120-volt and 277-volt fluorescent lighting circuits shall be listed and shall be marked SWD or HID. Circuit breakers used as switches in high-intensity discharge lighting circuits shall be listed and shall be marked as HID.

**(E) Voltage Marking.** Circuit breakers shall be marked with a voltage rating not less than the nominal system voltage that is indicative of their capability to interrupt fault currents between phases or phase to ground.

**240.85 Applications.** A circuit breaker with a straight voltage rating, such as 240V or 480V, shall be permitted to be applied in a circuit in which the nominal voltage between any two conductors does not exceed the circuit breaker's voltage rating. A two-pole circuit breaker shall not be used for protecting a 3-phase, corner-grounded delta circuit unless the circuit breaker is marked 1φ–3φ to indicate such suitability.

A circuit breaker with a slash rating, such as 120/240V or 480Y/277V, shall be permitted to be applied in a solidly grounded circuit where the nominal voltage of any conductor to ground does not exceed the lower of the two values of the circuit breaker's voltage rating and the nominal voltage between any two conductors does not exceed the higher value of the circuit breaker's voltage rating.

FPN: Proper application of molded case circuit breakers on 3-phase systems, other than solidly grounded wye, particularly on corner grounded delta systems, considers the circuit breakers' individual pole-interrupting capability.

**240.86 Series Ratings.** Where a circuit breaker is used on a circuit having an available fault current higher than its marked interrupting rating by being connected on the load side of an acceptable overcurrent protective device having the higher rating, 240.86(A) and (B) shall apply.

**(A) Marking.** The additional series combination interrupting rating shall be marked on the end use equipment, such as switchboards and panelboards.

**(B) Motor Contribution.** Series ratings shall not be used where

(1) Motors are connected on the load side of the higher-rated overcurrent device and on the line side of the lower-rated overcurrent device, and

(2) The sum of the motor full-load currents exceeds 1 percent of the interrupting rating of the lower-rated circuit breaker.

## VIII. Supervised Industrial Installations

**240.90 General.** Overcurrent protection in areas of supervised industrial installations shall comply with all of the other applicable provisions of this article, except as provided in Part VIII. The provisions of Part VIII shall only be permitted to apply to those portions of the electrical system in the supervised industrial installation used exclusively for manufacturing or process control activities.

**240.92 Location in Circuit.** An overcurrent device shall be connected in each ungrounded circuit conductor as required in 240.92(A) through (D).

**(A) Feeder and Branch-Circuit Conductors.** Feeder and branch-circuit conductors shall be protected at the point the conductors receive their supply as permitted in 240.21 or as otherwise permitted in 240.92(B), (C), or (D).

**(B) Transformer Secondary Conductors of Separately Derived Systems.** Conductors shall be permitted to be connected to a transformer secondary of a separately derived system, without overcurrent protection at the connection, where the conditions of 240.92(B)(1), (2), and (3) are met.

**(1) Short-Circuit and Ground-Fault Protection.** The conductors shall be protected from short-circuit and ground-fault conditions by complying with one of the following conditions:

(1) The length of the secondary conductors does not exceed 30 m (100 ft) and the transformer primary overcurrent device has a rating or setting that does not exceed 150 percent of the value determined by multiplying the secondary conductor ampacity by the secondary-to-primary transformer voltage ratio.

(2) The conductors are protected by a differential relay with a trip setting equal to or less than the conductor ampacity.

(3) The conductors shall be considered to be protected if calculations, made under engineering supervision, determine that the system overcurrent devices will protect the conductors within recognized time vs. current limits for all short-circuit and ground-fault conditions.

**(2) Overload Protection.** The conductors shall be protected against overload conditions by complying with one of the following:

(1) The conductors terminate in a single overcurrent device that will limit the load to the conductor ampacity.

(2) The sum of the overcurrent devices at the conductor termination limits the load to the conductor ampacity. The overcurrent devices shall consist of not more than six circuit breakers or sets of fuses, mounted in a single enclosure, in a group of separate enclosures, or in or on a switchboard. There shall be no more than six overcurrent devices grouped in any one location.

(3) Overcurrent relaying is connected [with a current transformer(s), if needed] to sense all of the secondary conductor current and limit the load to the conductor ampacity by opening upstream or downstream devices.

(4) Conductors shall be considered to be protected if calculations, made under engineering supervision, determine that the system overcurrent devices will protect the conductors from overload conditions.

**(3) Physical Protection.** The secondary conductors shall be suitably protected from physical damage.

**(C) Outside Feeder Taps.** Outside conductors shall be permitted to be tapped to a feeder or to be connected at a transformer secondary, without overcurrent protection at the tap or connection, where all the following conditions are met:

(1) The conductors are suitably protected from physical damage.

(2) The sum of the overcurrent devices at the conductor termination limits the load to the conductor ampacity. The overcurrent devices shall consist of not more than six circuit breakers or sets of fuses mounted in a single enclosure, in a group of separate enclosures, or in or on a switchboard. There shall be no more than six overcurrent devices grouped in any one location.

(3) The tap conductors are installed outdoors of a building or structure except at the point of load termination.

(4) The overcurrent device for the conductors is an integral part of a disconnecting means or shall be located immediately adjacent thereto.

(5) The disconnecting means for the conductors is installed at a readily accessible location complying with one of the following:

    a. Outside of a building or structure

    b. Inside, nearest the point of entrance of the conductors

    c. Where installed in accordance with 230.6, nearest the point of entrance of the conductors

**(D) Protection by Primary Overcurrent Device.** Conductors supplied by the secondary side of a transformer shall be permitted to be protected by overcurrent protection provided on the primary (supply) side of the transformer, provided the primary device time–current protection characteristic, multiplied by the maximum effective primary-to-secondary transformer voltage ratio, effectively protects the secondary conductors.

## IX. Overcurrent Protection Over 600 Volts, Nominal

### 240.100 Feeders and Branch Circuits.

**(A) Location and Type of Protection.** Feeder and branch-circuit conductors shall have overcurrent protection in each ungrounded conductor located at the point where the conductor receives its supply or at an alternative location in the circuit when designed under engineering supervision that includes but is not limited to considering the appropriate fault studies and time–current coordination analysis of the protective devices and the conductor damage curves. The overcurrent protection shall be permitted to be provided by either 240.100(A)(1) or (A)(2).

**(1) Overcurrent Relays and Current Transformers.** Circuit breakers used for overcurrent protection of 3-phase circuits shall have a minimum of three overcurrent relay elements operated from three current transformers. The separate overcurrent relay elements (or protective functions) shall be permitted to be part of a single electronic protective relay unit.

On 3-phase, 3-wire circuits, an overcurrent relay element in the residual circuit of the current transformers shall be permitted to replace one of the phase relay elements.

An overcurrent relay element, operated from a current transformer that links all phases of a 3-phase, 3-wire circuit, shall be permitted to replace the residual relay element and one of the phase-conductor current transformers. Where the neutral is not regrounded on the load side of the circuit as permitted in 250.184(B), the current transformer shall be permitted to link all 3-phase conductors and the grounded circuit conductor (neutral).

**(2) Fuses.** A fuse shall be connected in series with each ungrounded conductor.

**(B) Protective Devices.** The protective device(s) shall be capable of detecting and interrupting all values of current that can occur at their location in excess of their trip-setting or melting point.

**(C) Conductor Protection.** The operating time of the protective device, the available short-circuit current, and the conductor used shall be coordinated to prevent damaging or dangerous temperatures in conductors or conductor insulation under short-circuit conditions.

### 240.101 Additional Requirements for Feeders.

**(A) Rating or Setting of Overcurrent Protective Devices.** The continuous ampere rating of a fuse shall not exceed three times the ampacity of the conductors. The long-time trip element setting of a breaker or the minimum trip setting of an electronically actuated fuse shall not exceed six times the ampacity of the conductor. For fire pumps, conductors shall be permitted to be protected for overcurrent in accordance with 695.4(B).

**(B) Feeder Taps.** Conductors tapped to a feeder shall be permitted to be protected by the feeder overcurrent device where that overcurrent device also protects the tap conductor.

# ARTICLE 250
# Grounding

## I. General

**250.1 Scope.** This article covers general requirements for grounding and bonding of electrical installations, and specific requirements in (1) through (6).

(1) Systems, circuits, and equipment required, permitted, or not permitted to be grounded

(2) Circuit conductor to be grounded on grounded systems

(3) Location of grounding connections

(4) Types and sizes of grounding and bonding conductors and electrodes

(5) Methods of grounding and bonding

(6) Conditions under which guards, isolation, or insulation may be substituted for grounding

## 250.2 Definitions.

**Effective Ground-Fault Current Path.** An intentionally constructed, permanent, low-impedance electrically conductive path designed and intended to carry current under ground-fault conditions from the point of a ground fault on a wiring system to the electrical supply source.

**Ground Fault.** An unintentional, electrically conducting connection between an ungrounded conductor of an electrical circuit and the normally non–current-carrying conductors, metallic enclosures, metallic raceways, metallic equipment, or earth.

**Ground-Fault Current Path.** An electrically conductive path from the point of a ground fault on a wiring system through normally non–current-carrying conductors, equipment, or the earth to the electrical supply source.

> FPN: Examples of ground-fault current paths could consist of any combination of equipment grounding conductors, metallic raceways, metallic cable sheaths, electrical equipment, and any other electrically conductive material such as metal water and gas piping, steel framing members, stucco mesh, metal ducting, reinforcing steel, shields of communications cables, and the earth itself.

**250.3 Application of Other Articles.** In other articles applying to particular cases of installation of conductors and equipment, there are requirements identified in Table 250.3 that are in addition to, or modifications of, those of this article.

**Table 250.3 Additional Grounding Requirements**

| Conductor/Equipment | Article | Section |
|---|---|---|
| Agricultural buildings | | 547.9 and 547.10 |
| Audio signal processing, amplification, and reproduction equipment | | 640.7 |
| Branch circuits | | 210.5, 210.6, 406.3 |
| Cablebus | | 370.9 |
| Capacitors | | 460.10, 460.27 |
| Circuits and equipment operating at less than 50 volts | 720 | |
| Class 1, Class 2, and Class 3 remote-control, signaling, and power-limited circuits | | 725.9 |
| Closed-loop and programmed power distribution | | 780.3 |
| Communications circuits | 800 | |
| Community antenna television and radio distribution systems | | 820.33, 820.40, 820.41 |
| Conductors for general wiring | 310 | |
| Cranes and hoists | 610 | |
| Electrically driven or controlled irrigation machines | | 675.11(C), 675.12, 675.13, 675.14, 675.15 |
| Electric signs and outline lighting | 600 | |
| Electrolytic cells | 668 | |
| Elevators, dumbwaiters, escalators, moving walks, wheelchair lifts, and stairway chair lifts | 620 | |

**Table 250.3** *Continued*

| Conductor/Equipment | Article | Section |
|---|---|---|
| Fire alarm systems | | 760.9 |
| Fixed electric heating equipment for pipelines and vessels | | 427.29, 427.48 |
| Fixed outdoor electric deicing and snow-melting equipment | | 426.27 |
| Flexible cords and cables | | 400.22, 400.23 |
| Floating buildings | | 553.8, 553.10, 553.11 |
| Grounding-type receptacles, adapters, cord connectors, and attachment plugs | | 406.9 |
| Hazardous (classified) locations | 500–517 | |
| Health care facilities | 517 | |
| Induction and dielectric heating equipment | 665 | |
| Industrial machinery | 670 | |
| Information technology equipment | | 645.15 |
| Intrinsically safe systems | | 504.50 |
| Luminaires (lighting fixtures) and lighting equipment | | 410.17, 410.18, 410.20, 410.21, 410.105(B) |
| Luminaires (fixtures), lampholders, lamps, and receptacles | 410 | |
| Marinas and boatyards | | 555.15 |
| Mobile homes and mobile home park | 550 | |
| Motion picture and television studios and similar locations | | 530.20, 530.66 |
| Motors, motor circuits, and controllers | 430 | |
| Outlet, device, pull and junction boxes, conduit bodies and fittings | | 314.4, 314.25 |
| Over 600 volts, nominal, underground wiring methods | | 300.50(B) |
| Panelboards | | 408.20 |
| Pipe organs | 650 | |
| Radio and television equipment | 810 | |
| Receptacles and cord connectors | | 406.3 |
| Recreational vehicles and recreational vehicle parks | 551 | |
| Services | 230 | |
| Solar photovoltaic systems | | 690.41, 690.42, 690.43, 690.45, 690.47 |
| Swimming pools, fountains, and similar installations | 680 | |

**Table 250.3** *Continued*

| Conductor/Equipment | Article | Section |
|---|---|---|
| Switchboards and panelboards | | 408.3(D) |
| Switches | | 404.12 |
| Theaters, audience areas of motion picture and television studios, and similar locations | | 520.81 |
| Transformers and transformer vaults | | 450.10 |
| Use and identification of grounded conductors | 200 | |
| X-ray equipment | 660 | 517.78 |

**250.4 General Requirements for Grounding and Bonding.** The following general requirements identify what grounding and bonding of electrical systems are required to accomplish. The prescriptive methods contained in Article 250 shall be followed to comply with the performance requirements of this section.

**(A) Grounded Systems.**

**(1) Electrical System Grounding.** Electrical systems that are grounded shall be connected to earth in a manner that will limit the voltage imposed by lightning, line surges, or unintentional contact with higher-voltage lines and that will stabilize the voltage to earth during normal operation.

**(2) Grounding of Electrical Equipment.** Non–current-carrying conductive materials enclosing electrical conductors or equipment, or forming part of such equipment, shall be connected to earth so as to limit the voltage to ground on these materials.

**(3) Bonding of Electrical Equipment.** Non–current-carrying conductive materials enclosing electrical conductors or equipment, or forming part of such equipment, shall be connected together and to the electrical supply source in a manner that establishes an effective ground-fault current path.

**(4) Bonding of Electrically Conductive Materials and Other Equipment.** Electrically conductive materials that are likely to become energized shall be connected together and to the electrical supply source in a manner that establishes an effective ground-fault current path.

**(5) Effective Ground-Fault Current Path.** Electrical equipment and wiring and other electrically conductive material likely to become energized shall be installed in a manner that creates a permanent, low-impedance circuit capable of safely carrying the maximum ground-fault current likely to be imposed on it from any point on the wiring system where a ground fault may occur to the electrical supply source. The earth shall not be used as the sole equipment grounding conductor or effective ground-fault current path.

**(B) Ungrounded Systems.**

**(1) Grounding Electrical Equipment.** Non–current-carrying conductive materials enclosing electrical conductors or equipment, or forming part of such equipment, shall be connected to earth in a manner that will limit the voltage imposed by lightning or unintentional contact with higher-voltage lines and limit the voltage to ground on these materials.

**(2) Bonding of Electrical Equipment.** Non–current-carrying conductive materials enclosing electrical conductors or equipment, or forming part of such equipment, shall be connected together and to the supply system grounded equipment in a manner that creates a permanent, low-impedance path for ground-fault current that is capable of carrying the maximum fault current likely to be imposed on it.

**(3) Bonding of Electrically Conductive Materials and Other Equipment.** Electrically conductive materials that are likely to become energized shall be connected together and to the supply system grounded equipment in a manner that creates a permanent, low-impedance path for ground-fault current that is capable of carrying the maximum fault current likely to be imposed on it.

**(4) Path for Fault Current.** Electrical equipment, wiring, and other electrically conductive material likely to become energized shall be installed in a manner that creates a permanent, low-impedance circuit from any point on the wiring system to the electrical supply source to facilitate the operation of overcurrent devices should a second fault occur on the wiring system. The earth shall not be used as the sole equipment grounding conductor or effective fault-current path.

> FPN No. 1: A second fault that occurs through the equipment enclosures and bonding is considered a ground fault.

> FPN No. 2: See Figure 250.4 for information on the organization of Article 250.

**250.6 Objectionable Current over Grounding Conductors.**

**(A) Arrangement to Prevent Objectionable Current.** The grounding of electrical systems, circuit conductors, surge arresters, and conductive non–current-carrying materials and equipment shall be installed and arranged in a manner that will prevent objectionable current over the grounding conductors or grounding paths.

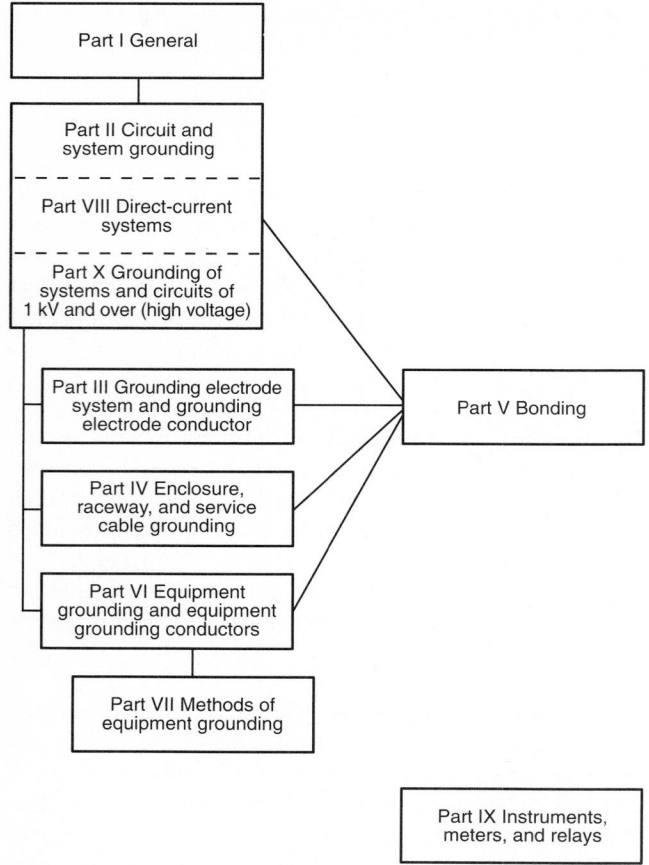

**Figure 250.4 Grounding.**

**(B) Alterations to Stop Objectionable Current.** If the use of multiple grounding connections results in objectionable current, one or more of the following alterations shall be permitted to be made, provided that the requirements of 250.4(A)(5) or 250.4(B)(4) are met:

(1) Discontinue one or more but not all of such grounding connections.
(2) Change the locations of the grounding connections.
(3) Interrupt the continuity of the conductor or conductive path interconnecting the grounding connections.
(4) Take other suitable remedial and approved action.

**(C) Temporary Currents Not Classified as Objectionable Currents.** Temporary currents resulting from accidental conditions, such as ground-fault currents, that occur only while the grounding conductors are performing their intended protective functions shall not be classified as objectionable current for the purposes specified in 250.6(A) and (B).

**(D) Limitations to Permissible Alterations.** The provisions of this section shall not be considered as permitting electronic equipment from being operated on ac systems or branch circuits that are not grounded as required by this article. Currents that introduce noise or data errors in electronic equipment shall not be considered the objectionable currents addressed in this section.

**(E) Isolation of Objectionable Direct-Current Ground Currents.** Where isolation of objectionable dc ground currents from cathodic protection systems is required, a listed ac coupling/dc isolating device shall be permitted in the equipment grounding path to provide an effective return path for ac ground-fault current while blocking dc current.

**250.8 Connection of Grounding and Bonding Equipment.** Grounding conductors and bonding jumpers shall be connected by exothermic welding, listed pressure connectors, listed clamps, or other listed means. Connection devices or fittings that depend solely on solder shall not be used. Sheet metal screws shall not be used to connect grounding conductors to enclosures.

**250.10 Protection of Ground Clamps and Fittings.** Ground clamps or other fittings shall be approved for general use without protection or shall be protected from physical damage as indicated in (1) or (2).

(1) In installations where they are not likely to be damaged
(2) Where enclosed in metal, wood, or equivalent protective covering

**250.12 Clean Surfaces.** Nonconductive coatings (such as paint, lacquer, and enamel) on equipment to be grounded shall be removed from threads and other contact surfaces to ensure good electrical continuity or be connected by means of fittings designed so as to make such removal unnecessary.

**II. Circuit and System Grounding**

**250.20 Alternating-Current Circuits and Systems to Be Grounded.** Alternating-current circuits and systems shall be grounded as provided for in 250.20(A), (B), (C), or (D). Other circuits and systems shall be permitted to be grounded. If such systems are grounded, they shall comply with the applicable provisions of this article.

> FPN: An example of a system permitted to be grounded is a corner-grounded delta transformer connection. See 250.26(4) for conductor to be grounded.

**(A) Alternating-Current Circuits of Less Than 50 Volts.** Alternating-current circuits of less than 50 volts shall be grounded under any of the following conditions:

(1) Where supplied by transformers, if the transformer supply system exceeds 150 volts to ground

(2) Where supplied by transformers, if the transformer supply system is ungrounded

(3) Where installed as overhead conductors outside of buildings

**(B) Alternating-Current Systems of 50 Volts to 1000 Volts.** Alternating-current systems of 50 volts to 1000 volts that supply premises wiring and premises wiring systems shall be grounded under any of the following conditions:

(1) Where the system can be grounded so that the maximum voltage to ground on the ungrounded conductors does not exceed 150 volts

(2) Where the system is 3-phase, 4-wire, wye connected in which the neutral is used as a circuit conductor

(3) Where the system is 3-phase, 4-wire, delta connected in which the midpoint of one phase winding is used as a circuit conductor

**(C) Alternating-Current Systems of 1 kV and Over.** Alternating-current systems supplying mobile or portable equipment shall be grounded as specified in 250.188. Where supplying other than mobile or portable equipment, such systems shall be permitted to be grounded.

**(D) Separately Derived Systems.** Separately derived systems, as covered in 250.20(A) or (B), shall be grounded as specified in 250.30.

> FPN No. 1: An alternate ac power source such as an on-site generator is not a separately derived system if the neutral is solidly interconnected to a service-supplied system neutral.

> FPN No. 2: For systems that are not separately derived and are not required to be grounded as specified in 250.30, see 445.13 for minimum size of conductors that must carry fault current.

**250.21 Alternating-Current Systems of 50 Volts to 1000 Volts Not Required to Be Grounded.** The following ac systems of 50 volts to 1000 volts shall be permitted to be grounded but shall not be required to be grounded:

(1) Electric systems used exclusively to supply industrial electric furnaces for melting, refining, tempering, and the like

(2) Separately derived systems used exclusively for rectifiers that supply only adjustable-speed industrial drives

(3) Separately derived systems supplied by transformers that have a primary voltage rating less than 1000 volts, provided that all the following conditions are met:

    a. The system is used exclusively for control circuits.

    b. The conditions of maintenance and supervision ensure that only qualified persons service the installation.

    c. Continuity of control power is required.

    d. Ground detectors are installed on the control system.

(4) High-impedance grounded neutral systems as specified in 250.36

(5) Other systems that are not required to be grounded in accordance with the requirements of 250.20(B)

**250.22 Circuits Not to Be Grounded.** The following circuits shall not be grounded:

(1) Cranes (circuits for electric cranes operating over combustible fibers in Class III locations, as provided in 503.13)

(2) Health care facilities (circuits as provided in Article 517)

(3) Electrolytic cells (circuits as provided in Article 668)

(4) Lighting systems [secondary circuits as provided in 411.5(A)]

**250.24 Grounding Service-Supplied Alternating-Current Systems.**

**(A) System Grounding Connections.** A premises wiring system supplied by a grounded ac service shall have a grounding electrode conductor connected to the grounded service conductor, at each service, in accordance with 250.24(A)(1) through (A)(5).

**(1) General.** The connection shall be made at any accessible point from the load end of the service drop or service lateral to and including the terminal or bus to which the grounded service conductor is connected at the service disconnecting means.

> FPN: See definitions of *Service Drop* and *Service Lateral* in Article 100.

**(2) Outdoor Transformer.** Where the transformer supplying the service is located outside the building, at least one additional grounding connection shall be made from the grounded service conductor to a grounding electrode, either at the transformer or elsewhere outside the building.

*Exception: The additional grounding connection shall not be made on high-impedance grounded neutral systems. The system shall meet the requirements of 250.36.*

**(3) Dual Fed Services.** For services that are dual fed (double ended) in a common enclosure or grouped together in separate enclosures and employing a secondary tie, a single grounding electrode connection to the tie point of the grounded circuit conductors from each power source shall be permitted.

**(4) Main Bonding Jumper as Wire or Busbar.** Where the main bonding jumper specified in 250.28 is a wire or busbar and is installed from the neutral bar or bus to the equipment grounding terminal bar or bus in the service equipment, the grounding electrode conductor shall be permitted to be connected to the equipment grounding terminal bar or bus to which the main bonding jumper is connected.

**(5) Load-Side Grounding Connections.** A grounding connection shall not be made to any grounded circuit conductor on the load side of the service disconnecting means except as otherwise permitted in this article.

> FPN: See 250.30(A) for separately derived systems, 250.32 for connections at separate buildings or structures, and 250.142 for use of the grounded circuit conductor for grounding equipment.

**(B) Grounded Conductor Brought to Service Equipment.** Where an ac system operating at less than 1000 volts is grounded at any point, the grounded conductor(s) shall be run to each service disconnecting means and shall be bonded to each disconnecting means enclosure. The grounded conductor(s) shall be installed in accordance with 250.24(B)(1) through (B)(3).

*Exception: Where more than one service disconnecting means are located in an assembly listed for use as service equipment, it shall be permitted to run the grounded conductor(s) to the assembly, and the conductor(s) shall be bonded to the assembly enclosure.*

**(1) Routing and Sizing.** This conductor shall be routed with the phase conductors and shall not be smaller than the required grounding electrode conductor specified in Table 250.66 but shall not be required to be larger than the largest ungrounded service-entrance phase conductor. In addition, for service-entrance phase conductors larger than 1100 kcmil copper or 1750 kcmil aluminum, the grounded conductor shall not be smaller than 12½ percent of the area of the largest service-entrance phase conductor. The grounded service entrance conductor of a 3-phase, 3-wire delta service shall have an ampacity not less than the ungrounded conductors.

**(2) Parallel Conductors.** Where the service-entrance phase conductors are installed in parallel, the size of the grounded conductor shall be based on the total circular mil area of the parallel conductors as indicated in this section. Where installed in two or more raceways, the size of the grounded conductor in each raceway shall be based on the size of the ungrounded service-entrance conductor in the raceway but not smaller than 1/0 AWG.

> FPN: See 310.4 for grounded conductors connected in parallel.

**(3) High Impedance.** The grounded conductor on a high-impedance grounded neutral system shall be grounded in accordance with 250.36.

**(C) Grounding Electrode Conductor.** A grounding electrode conductor shall be used to connect the equipment grounding conductors, the service-equipment enclosures, and, where the system is grounded, the grounded service conductor to the grounding electrode(s) required by Part III of this article.

High-impedance grounded neutral system connections shall be made as covered in 250.36.

> FPN: See 250.24(A) for ac system grounding connections.

**(D) Ungrounded System Grounding Connections.** A premises wiring system that is supplied by an ac service that is ungrounded shall have, at each service, a grounding electrode conductor connected to the grounding electrode(s) required by Part III of this article. The grounding electrode conductor shall be connected to a metal enclosure of the service conductors at any accessible point from the load end of the service drop or service lateral to the service disconnecting means.

**250.26 Conductor to Be Grounded — Alternating-Current Systems.** For ac premises wiring systems, the conductor to be grounded shall be as specified in the following:

(1) Single-phase, 2-wire — one conductor
(2) Single-phase, 3-wire — the neutral conductor
(3) Multiphase systems having one wire common to all phases — the common conductor
(4) Multiphase systems where one phase is grounded — one phase conductor
(5) Multiphase systems in which one phase is used as in (2) — the neutral conductor

**250.28 Main Bonding Jumper.** For a grounded system, an unspliced main bonding jumper shall be used to connect the equipment grounding conductor(s) and the service-disconnect enclosure to the grounded conductor of the system within the enclosure for each service disconnect.

*Exception No. 1: Where more than one service disconnecting means is located in an assembly listed for use as service equipment, an unspliced main bonding jumper shall bond the grounded conductor(s) to the assembly enclosure.*

*Exception No. 2: Impedance grounded neutral systems shall be permitted to be connected as provided in 250.36 and 250.186.*

**(A) Material.** Main bonding jumpers shall be of copper or other corrosion-resistant material. A main bonding jumper shall be a wire, bus, screw, or similar suitable conductor.

**(B) Construction.** Where a main bonding jumper is a screw only, the screw shall be identified with a green finish that shall be visible with the screw installed.

**(C) Attachment.** Main bonding jumpers shall be attached in the manner specified by the applicable provisions of 250.8.

**(D) Size.** The main bonding jumper shall not be smaller than the sizes shown in Table 250.66 for grounding electrode conductors. Where the service-entrance phase conductors are larger than 1100 kcmil copper or 1750 kcmil aluminum, the bonding jumper shall have an area that is not less than 12½ percent of the area of the largest phase conductor except that, where the phase conductors and the bonding jumper are of different materials (copper or aluminum), the minimum size of the bonding jumper shall be based on the assumed use of phase conductors of the same material as the bonding jumper and with an ampacity equivalent to that of the installed phase conductors.

### 250.30 Grounding Separately Derived Alternating-Current Systems.

**(A) Grounded Systems.** A separately derived ac system that is grounded shall comply with 250.30(A)(1) through (6).

*Exception: High-impedance grounded neutral system grounding connection requirements shall not be required to comply with 250.30(A)(1) and (2) and shall be made as specified in 250.36 and 250.186.*

**(1) Bonding Jumper.** A bonding jumper in compliance with 250.28(A) through (D) that is sized for the derived phase conductors shall be used to connect the equipment grounding conductors of the separately derived system to the grounded conductor. Except as permitted by 250.24(A)(3), this connection shall be made at any point on the separately derived system from the source to the first system disconnecting means or overcurrent device, or it shall be made at the source of a separately derived system that has no disconnecting means or overcurrent devices. The point of connection shall be the same as the grounding electrode conductor as required in 250.30(A)(2).

*Exception No. 1: A bonding jumper at both the source and the first disconnecting means shall be permitted where doing so does not establish a parallel path for the grounded circuit conductor. Where a grounded conductor is used in this manner, it shall not be smaller than the size specified for the bonding jumper but shall not be required to be larger than the ungrounded conductor(s). For the purposes of this exception, connection through the earth shall not be considered as providing a parallel path.*

*Exception No. 2: The size of the bonding jumper for a system that supplies a Class 1, Class 2, or Class 3 circuit, and is derived from a transformer rated not more than 1000 volt-amperes, shall not be smaller than the derived phase conductors and shall not be smaller than 14 AWG copper or 12 AWG aluminum.*

**(2) Grounding Electrode Conductor.** The grounding electrode conductor shall be installed in accordance with

(a) or (b). Where taps are connected to a common grounding electrode conductor, the installation shall comply with 250.30(A)(3).

(a) Single Separately Derived System. A grounding electrode conductor for a single separately derived system shall be sized in accordance with 250.66 for the derived phase conductors and shall be used to connect the grounded conductor of the derived system to the grounding electrode as specified in 250.30(A)(4). Except as permitted by 250.24(A)(3) or (A)(4), this connection shall be made at the same point on the separately derived system where the bonding jumper is installed.

*Exception: A grounding electrode conductor shall not be required for a system that supplies a Class 1, Class 2, or Class 3 circuit and is derived from a transformer rated not more than 1000 volt-amperes, provided the system grounded conductor is bonded to the transformer frame or enclosure by a jumper sized in accordance with 250.30(A)(1), Exception No. 2, and the transformer frame or enclosure is grounded by one of the means specified in 250.134.*

(b) Multiple Separately Derived Systems. Where more than one separately derived system is connected to a common grounding electrode conductor as provided in 250.30(A)(3), the common grounding electrode conductor shall be sized in accordance with 250.66, based on the total area of the largest derived phase conductor from each separately derived system.

**(3) Grounding Electrode Conductor Taps.** It shall be permissible to connect taps from a separately derived system to a common grounding electrode conductor. Each tap conductor shall connect the grounded conductor of the separately derived system to the common grounding electrode conductor.

(a) Tap Conductor Size. Each tap conductor shall be sized in accordance with 250.66 for the derived phase conductors of the separately derived system it serves.

(b) Connections. All connections shall be made at an accessible location by an irreversible compression connector listed for the purpose, listed connections to copper busbars not less than 6 mm × 50 mm (¼ in. × 2 in.), or by the exothermic welding process. The tap conductors shall be connected to the common grounding electrode conductor as specified in 250.30(A)(2)(b) in such a manner that the common grounding electrode conductor remains without a splice or joint.

(c) Installation. The common grounding electrode conductor and the taps to each separately derived system shall comply with 250.64(A), (B), (C), and (E).

(d) Bonding. Where exposed structural steel that is interconnected to form the building frame or interior metal piping exists in the area served by the separately derived system, it shall be bonded to the grounding electrode conductor in accordance with 250.104.

**(4) Grounding Electrode.** The grounding electrode shall be as near as practicable to and preferably in the same area as the grounding electrode conductor connection to the system. The grounding electrode shall be the nearest one of the following:

(1) An effectively grounded structural metal member of the structure
(2) An effectively grounded metal water pipe within 1.5 m (5 ft) from the point of entrance into the building

*Exception: In industrial and commercial buildings where conditions of maintenance and supervision ensure that only qualified persons service the installation and the entire length of the interior metal water pipe that is being used for the grounding electrode is exposed, the connection shall be permitted at any point on the water pipe system.*

(3) Other electrodes as specified by 250.52 where the electrodes specified by 250.30(A)(4)(1) or (A)(4)(2) are not available

*Exception to (1), (2), and (3):Where a separately derived system originates in listed equipment suitable for use as service equipment, the grounding electrode used for the service or feeder shall be permitted as the grounding electrode for the separately derived system, provided the grounding electrode conductor from the service or feeder to the grounding electrode is of sufficient size for the separately derived system. Where the equipment ground bus internal to the service equipment is not smaller than the required grounding electrode conductor, the grounding electrode connection for the separately derived system shall be permitted to be made to the bus.*

> FPN: See 250.104(A)(4) for bonding requirements of interior metal water piping in the area served by separately derived systems.

**(5) Equipment Bonding Jumper Size.** Where a bonding jumper is run with the derived phase conductors from the source of a separately derived system to the first disconnecting means, it shall be sized in accordance with 250.28(A) through (D), based on the size of the derived phase conductors.

**(6) Grounded Conductor.** Where a grounded conductor is installed and the bonding jumper is not located at the source of the separately derived system, the following shall apply:

(a) Routing and Sizing. This conductor shall be routed with the derived phase conductors and shall not be smaller than the required grounding electrode conductor specified in Table 250.66, but shall not be required to be larger than the largest ungrounded derived phase conductor. In addi-

tion, for phase conductors larger than 1100 kcmil copper or 1750 kcmil aluminum, the grounded conductor shall not be smaller than 12½ percent of the area of the largest derived phase conductor. The grounded conductor of a 3-phase, 3-wire delta system shall have an ampacity not less than the ungrounded conductors.

(b) Parallel Conductors. Where the derived phase conductors are installed in parallel, the size of the grounded conductor shall be based on the total circular mil area of the parallel conductors as indicated in this section. Where installed in two or more raceways, the size of the grounded conductor in each raceway shall be based on the size of the ungrounded conductors in the raceway but not smaller than 1/0 AWG.

> FPN: See 310.4 for grounded conductors connected in parallel.

(c) High Impedance. The grounded conductor on a high-impedance grounded neutral system shall be grounded in accordance with 250.36.

**(B) Ungrounded Systems.** The equipment of an ungrounded separately derived system shall be grounded as specified in 250.30(B)(1) and (2).

**(1) Grounding Electrode Conductor.** A grounding electrode conductor, sized in accordance with 250.66 for the derived phase conductors, shall be used to connect the metal enclosures of the derived system to the grounding electrode as specified in 250.30(B)(2). This connection shall be made at any point on the separately derived system from the source to the first system disconnecting means.

**(2) Grounding Electrode.** Except as permitted by 250.34 for portable and vehicle-mounted generators, the grounding electrode shall comply with 250.30(A)(4).

### 250.32 Two or More Buildings or Structures Supplied from a Common Service.

**(A) Grounding Electrode.** Where two or more buildings or structures are supplied from a common ac service by a feeder(s) or branch circuit(s), the grounding electrode(s) required in Part III of this article at each building or structure shall be connected in the manner specified in 250.32(B) or (C). Where there are no existing grounding electrodes, the grounding electrode(s) required in Part III of this article shall be installed.

*Exception: A grounding electrode at separate buildings or structures shall not be required where only one branch circuit supplies the building or structure and the branch circuit includes an equipment grounding conductor for grounding the conductive non–current-carrying parts of all equipment.*

**(B) Grounded Systems.** For a grounded system at the separate building or structure, the connection to the grounding electrode and grounding or bonding of equipment, structures, or frames required to be grounded or bonded shall comply with either 250.32(B)(1) or (2).

**(1) Equipment Grounding Conductor.** An equipment grounding conductor as described in 250.118 shall be run with the supply conductors and connected to the building or structure disconnecting means and to the grounding electrode(s). The equipment grounding conductor shall be used for grounding or bonding of equipment, structures, or frames required to be grounded or bonded. The equipment grounding conductor shall be sized in accordance with 250.122. Any installed grounded conductor shall not be connected to the equipment grounding conductor or to the grounding electrode(s).

**(2) Grounded Conductor.** Where (1) an equipment grounding conductor is not run with the supply to the building or structure, (2) there are no continuous metallic paths bonded to the grounding system in both buildings or structures involved, and (3) ground-fault protection of equipment has not been installed on the common ac service, the grounded circuit conductor run with the supply to the building or structure shall be connected to the building or structure disconnecting means and to the grounding electrode(s) and shall be used for grounding or bonding of equipment, structures, or frames required to be grounded or bonded. The size of the grounded conductor shall not be smaller than the larger of

(1) That required by 220.22
(2) That required by 250.122

**(C) Ungrounded Systems.** The grounding electrode(s) shall be connected to the building or structure disconnecting means.

**(D) Disconnecting Means Located in Separate Building or Structure on the Same Premises.** Where one or more disconnecting means supply one or more additional buildings or structures under single management, and where these disconnecting means are located remote from those buildings or structures in accordance with the provisions of 225.32, Exception Nos. 1 and 2, all of the following conditions shall be met:

(1) The connection of the grounded circuit conductor to the grounding electrode at a separate building or structure shall not be made.
(2) An equipment grounding conductor for grounding any non–current-carrying equipment, interior metal piping systems, and building or structural metal frames is run with the circuit conductors to a separate building or structure and bonded to existing grounding electrode(s)

required in Part III of this article, or, where there are no existing electrodes, the grounding electrode(s) required in Part III of this article shall be installed where a separate building or structure is supplied by more than one branch circuit.
(3) Bonding the equipment grounding conductor to the grounding electrode at a separate building or structure shall be made in a junction box, panelboard, or similar enclosure located immediately inside or outside the separate building or structure.

**(E) Grounding Electrode Conductor.** The size of the grounding electrode conductor to the grounding electrode(s) shall not be smaller than given in 250.66, based on the largest ungrounded supply conductor. The installation shall comply with Part III of this article.

**250.34 Portable and Vehicle-Mounted Generators.**

**(A) Portable Generators.** The frame of a portable generator shall not be required to be grounded and shall be permitted to serve as the grounding electrode for a system supplied by the generator under the following conditions:

(1) The generator supplies only equipment mounted on the generator, cord-and-plug-connected equipment through receptacles mounted on the generator, or both, and
(2) The non–current-carrying metal parts of equipment and the equipment grounding conductor terminals of the receptacles are bonded to the generator frame.

**(B) Vehicle-Mounted Generators.** The frame of a vehicle shall be permitted to serve as the grounding electrode for a system supplied by a generator located on the vehicle under the following conditions:

(1) The frame of the generator is bonded to the vehicle frame, and
(2) The generator supplies only equipment located on the vehicle or cord-and-plug-connected equipment through receptacles mounted on the vehicle, or both equipment located on the vehicle and cord-and-plug-connected equipment through receptacles mounted on the vehicle or on the generator, and
(3) The non–current-carrying metal parts of equipment and the equipment grounding conductor terminals of the receptacles are bonded to the generator frame, and
(4) The system complies with all other provisions of this article.

**(C) Grounded Conductor Bonding.** A system conductor that is required to be grounded by 250.26 shall be bonded to the generator frame where the generator is a component of a separately derived system.

FPN: For grounding portable generators supplying fixed wiring systems, see 250.20(D).

**250.36 High-Impedance Grounded Neutral Systems.** High-impedance grounded neutral systems in which a grounding impedance, usually a resistor, limits the ground-fault current to a low value shall be permitted for 3-phase ac systems of 480 volts to 1000 volts where all the following conditions are met:

(1) The conditions of maintenance and supervision ensure that only qualified persons service the installation.
(2) Continuity of power is required.
(3) Ground detectors are installed on the system.
(4) Line-to-neutral loads are not served.

High-impedance grounded neutral systems shall comply with the provisions of 250.36(A) through (G).

**(A) Grounding Impedance Location.** The grounding impedance shall be installed between the grounding electrode conductor and the system neutral. Where a neutral is not available, the grounding impedance shall be installed between the grounding electrode conductor and the neutral derived from a grounding transformer.

**(B) Neutral Conductor.** The neutral conductor from the neutral point of the transformer or generator to its connection point to the grounding impedance shall be fully insulated.

The neutral conductor shall have an ampacity of not less than the maximum current rating of the grounding impedance. In no case shall the neutral conductor be smaller than 8 AWG copper or 6 AWG aluminum or copper-clad aluminum.

**(C) System Neutral Connection.** The system neutral conductor shall not be connected to ground except through the grounding impedance.

> FPN: The impedance is normally selected to limit the ground-fault current to a value slightly greater than or equal to the capacitive charging current of the system. This value of impedance will also limit transient overvoltages to safe values. For guidance, refer to criteria for limiting transient overvoltages in ANSI/IEEE 142-1991, *Recommended Practice for Grounding of Industrial and Commercial Power Systems.*

**(D) Neutral Conductor Routing.** The conductor connecting the neutral point of the transformer or generator to the grounding impedance shall be permitted to be installed in a separate raceway. It shall not be required to run this conductor with the phase conductors to the first system disconnecting means or overcurrent device.

**(E) Equipment Bonding Jumper.** The equipment bonding jumper (the connection between the equipment grounding conductors and the grounding impedance) shall be an unspliced conductor run from the first system disconnecting means or overcurrent device to the grounded side of the grounding impedance.

**(F) Grounding Electrode Conductor Location.** The grounding electrode conductor shall be attached at any point from the grounded side of the grounding impedance to the equipment grounding connection at the service equipment or first system disconnecting means.

**(G) Equipment Bonding Jumper Size.** The equipment bonding jumper shall be sized in accordance with (1) or (2).

(1) Where the grounding electrode conductor connection is made at the grounding impedance, the equipment bonding jumper shall be sized in accordance with 250.66, based on the size of the service entrance conductors for a service or the derived phase conductors for a separately derived system.
(2) Where the grounding electrode conductor is connected at the first system disconnecting means or overcurrent device, the equipment bonding jumper shall be sized the same as the neutral conductor in 250.36(B).

## III. Grounding Electrode System and Grounding Electrode Conductor

**250.50 Grounding Electrode System.** If available on the premises at each building or structure served, each item in 250.52(A)(1) through (A)(6) shall be bonded together to form the grounding electrode system. Where none of these electrodes are available, one or more of the electrodes specified in 250.52(A)(4) through (A)(7) shall be installed and used.

**250.52 Grounding Electrodes.**

**(A) Electrodes Permitted for Grounding.**

**(1) Metal Underground Water Pipe.** A metal underground water pipe in direct contact with the earth for 3.0 m (10 ft) or more (including any metal well casing effectively bonded to the pipe) and electrically continuous (or made electrically continuous by bonding around insulating joints or insulating pipe) to the points of connection of the grounding electrode conductor and the bonding conductors. Interior metal water piping located more than 1.52 m (5 ft) from the point of entrance to the building shall not be used as a part of the grounding electrode system or as a conductor to interconnect electrodes that are part of the grounding electrode system.

*Exception: In industrial and commercial buildings or structures where conditions of maintenance and supervision ensure that only qualified persons service the installation, interior metal water piping located more than 1.52 m (5 ft) from the point of entrance to the building shall be permitted as a part of the grounding electrode system or as a conductor to interconnect electrodes that are part of the grounding electrode system, provided that the entire length, other than short sections passing perpendicular through walls, floors, or ceilings, of the interior metal water pipe that is being used for the conductor is exposed.*

**(2) Metal Frame of the Building or Structure.** The metal frame of the building or structure, where effectively grounded.

**(3) Concrete-Encased Electrode.** An electrode encased by at least 50 mm (2 in.) of concrete, located within and near the bottom of a concrete foundation or footing that is in direct contact with the earth, consisting of at least 6.0 m (20 ft) of one or more bare or zinc galvanized or other electrically conductive coated steel reinforcing bars or rods of not less than 13 mm (½ in.) in diameter, or consisting of at least 6.0 m (20 ft) of bare copper conductor not smaller than 4 AWG. Reinforcing bars shall be permitted to be bonded together by the usual steel tie wires or other effective means.

**(4) Ground Ring.** A ground ring encircling the building or structure, in direct contact with the earth, consisting of at least 6.0 m (20 ft) of bare copper conductor not smaller than 2 AWG.

**(5) Rod and Pipe Electrodes.** Rod and pipe electrodes shall not be less than 2.5 m (8 ft) in length and shall consist of the following materials.

(a) Electrodes of pipe or conduit shall not be smaller than metric designator 21 (trade size ¾) and, where of iron or steel, shall have the outer surface galvanized or otherwise metal-coated for corrosion protection.

(b) Electrodes of rods of iron or steel shall be at least 15.87 mm (⅝ in.) in diameter. Stainless steel rods less than 16 mm (⅝ in.) in diameter, nonferrous rods, or their equivalent shall be listed and shall not be less than 13 mm (½ in.) in diameter.

**(6) Plate Electrodes.** Each plate electrode shall expose not less than 0.186 m² (2 ft²) of surface to exterior soil. Electrodes of iron or steel plates shall be at least 6.4 mm (¼ in.) in thickness. Electrodes of nonferrous metal shall be at least 1.5 mm (0.06 in.) in thickness.

**(7) Other Local Metal Underground Systems or Structures.** Other local metal underground systems or structures such as piping systems and underground tanks.

**(B) Electrodes Not Permitted for Grounding.** The following shall not be used as grounding electrodes:

(1) Metal underground gas piping system
(2) Aluminum electrodes

## 250.53 Grounding Electrode System Installation.

FPN: See 547.9 and 547.10 for special grounding and bonding requirements for agricultural buildings.

**(A) Rod, Pipe, and Plate Electrodes.** Where practicable, rod, pipe, and plate electrodes shall be embedded below permanent moisture level. Rod, pipe, and plate electrodes shall be free from nonconductive coatings such as paint or enamel.

**(B) Electrode Spacing.** Where more than one of the electrodes of the type specified in 250.52(A)(5) or (A)(6) are used, each electrode of one grounding system (including that used for air terminals) shall not be less than 1.83 m (6 ft) from any other electrode of another grounding system. Two or more grounding electrodes that are effectively bonded together shall be considered a single grounding electrode system.

**(C) Bonding Jumper.** The bonding jumper(s) used to connect the grounding electrodes together to form the grounding electrode system shall be installed in accordance with 250.64(A), (B), and (E), shall be sized in accordance with 250.66, and shall be connected in the manner specified in 250.70.

**(D) Metal Underground Water Pipe.** Where used as a grounding electrode, metal underground water pipe shall meet the requirements of 250.53(D)(1) and (D)(2).

**(1) Continuity.** Continuity of the grounding path or the bonding connection to interior piping shall not rely on water meters or filtering devices and similar equipment.

**(2) Supplemental Electrode Required.** A metal underground water pipe shall be supplemented by an additional electrode of a type specified in 250.52(A)(2) through (A)(7). Where the supplemental electrode is a rod, pipe, or plate type, it shall comply with 250.56. The supplemental electrode shall be permitted to be bonded to the grounding electrode conductor, the grounded service-entrance conductor, the nonflexible grounded service raceway, or any grounded service enclosure.

*Exception: The supplemental electrode shall be permitted to be bonded to the interior metal water piping at any convenient point as covered in 250.52(A)(1), Exception.*

**(E) Supplemental Electrode Bonding Connection Size.** Where the supplemental electrode is a rod, pipe, or plate electrode, that portion of the bonding jumper that is the sole connection to the supplemental grounding electrode shall not be required to be larger than 6 AWG copper wire or 4 AWG aluminum wire.

**(F) Ground Ring.** The ground ring shall be buried at a depth below the earth's surface of not less than 750 mm (30 in.).

**(G) Rod and Pipe Electrodes.** The electrode shall be installed such that at least 2.44 m (8 ft) of length is in contact with the soil. It shall be driven to a depth of not less than 2.44 m (8 ft) except that, where rock bottom is encountered, the electrode shall be driven at an oblique angle not to exceed 45 degrees from the vertical or, where rock bottom is encountered at an angle up to 45 degrees, the electrode shall be permitted to be buried in a trench that is at least 750 mm (30 in.) deep. The upper end of the electrode shall be flush with or below ground level unless the aboveground end and the grounding electrode conductor attachment are protected against physical damage as specified in 250.10.

**(H) Plate Electrode.** Plate electrodes shall be installed not less than 750 mm (30 in.) below the surface of the earth.

**250.54 Supplementary Grounding Electrodes.**
Supplementary grounding electrodes shall be permitted to be connected to the equipment grounding conductors specified in 250.118 and shall not be required to comply with the electrode bonding requirements of 250.50 or 250.53(C) or the resistance requirements of 250.56, but the earth shall not be used as the sole equipment grounding conductor.

**250.56 Resistance of Rod, Pipe, and Plate Electrodes.** A single electrode consisting of a rod, pipe, or plate that does not have a resistance to ground of 25 ohms or less shall be augmented by one additional electrode of any of the types specified by 250.52(A)(2) through (A)(7). Where multiple rod, pipe, or plate electrodes are installed to meet the requirements of this section, they shall not be less than 1.8 m (6 ft) apart.

> FPN: The paralleling efficiency of rods longer than 2.5 m (8 ft) is improved by spacing greater than 1.8 m (6 ft).

**250.58 Common Grounding Electrode.** Where an ac system is connected to a grounding electrode in or at a building as specified in 250.24 and 250.32, the same electrode shall be used to ground conductor enclosures and equipment in or on that building. Where separate services supply a building and are required to be connected to a grounding electrode, the same grounding electrode shall be used.

Two or more grounding electrodes that are effectively bonded together shall be considered as a single grounding electrode system in this sense.

**250.60 Use of Air Terminals.** Air terminal conductors and driven pipes, rods, or plate electrodes used for grounding air terminals shall not be used in lieu of the grounding electrodes required by 250.50 for grounding wiring systems and equipment. This provision shall not prohibit the required bonding together of grounding electrodes of different systems.

> FPN No. 1: See 250.106 for spacing from air terminals. See 800.40(D), 810.21(J), and 820.40(D) for bonding of electrodes.

> FPN No. 2: Bonding together of all separate grounding electrodes will limit potential differences between them and between their associated wiring systems.

**250.62 Grounding Electrode Conductor Material.** The grounding electrode conductor shall be of copper, aluminum, or copper-clad aluminum. The material selected shall be resistant to any corrosive condition existing at the installation or shall be suitably protected against corrosion. The conductor shall be solid or stranded, insulated, covered, or bare.

**250.64 Grounding Electrode Conductor Installation.** Grounding electrode conductors shall be installed as specified in 250.64(A) through (F).

**(A) Aluminum or Copper-Clad Aluminum Conductors.** Bare aluminum or copper-clad aluminum grounding conductors shall not be used where in direct contact with masonry or the earth or where subject to corrosive conditions. Where used outside, aluminum or copper-clad aluminum grounding conductors shall not be terminated within 450 mm (18 in.) of the earth.

**(B) Securing and Protection from Physical Damage.** A grounding electrode conductor or its enclosure shall be securely fastened to the surface on which it is carried. A 4 AWG copper or aluminum or larger conductor shall be protected if exposed to severe physical damage. A 6 AWG grounding conductor that is free from exposure to physical damage shall be permitted to be run along the surface of the building construction without metal covering or protection where it is securely fastened to the construction; otherwise, it shall be in rigid metal conduit, intermediate metal conduit, rigid nonmetallic conduit, electrical metallic tubing, or cable armor. Grounding conductors smaller than 6 AWG shall be in rigid metal conduit, intermediate metal conduit, rigid nonmetallic conduit, electrical metallic tubing, or cable armor.

**(C) Continuous.** The grounding electrode conductor shall be installed in one continuous length without a splice or joint, unless spliced only by irreversible compression-type connectors listed for the purpose or by the exothermic welding process.

*Exception: Sections of busbars shall be permitted to be connected together to form a grounding electrode conductor.*

**(D) Grounding Electrode Conductor Taps.** Where a service consists of more than a single enclosure as permitted in 230.40, Exception No. 2, it shall be permitted to connect taps to the grounding electrode conductor. Each such tap conductor shall extend to the inside of each such enclosure. The grounding electrode conductor shall be sized in accordance with 250.66, but the tap conductors shall be permitted to be sized in accordance with the grounding electrode conductors specified in 250.66 for the largest conductor serving the respective enclosures. The tap conductors shall be connected to the grounding electrode conductor in such a manner that the grounding electrode conductor remains without a splice.

**(E) Enclosures for Grounding Electrode Conductors.** Metal enclosures for grounding electrode conductors shall be electrically continuous from the point of attachment to cabinets or equipment to the grounding electrode and shall be securely fastened to the ground clamp or fitting. Metal enclosures that are not physically continuous from cabinet or equipment to the grounding electrode shall be made electrically continuous by bonding each end to the grounding electrode conductor. Where a raceway is used as protection for a grounding electrode conductor, the installation shall comply with the requirements of the appropriate raceway article.

**(F) To Electrode(s).** A grounding electrode conductor shall be permitted to be run to any convenient grounding electrode available in the grounding electrode system or to one or more grounding electrode(s) individually. The grounding electrode conductor shall be sized for the largest grounding electrode conductor required among all the electrodes connected to it.

**250.66 Size of Alternating-Current Grounding Electrode Conductor.** The size of the grounding electrode conductor of a grounded or ungrounded ac system shall not be less than given in Table 250.66, except as permitted in 250.66(A) through (C).

> FPN: See 250.24(B) for size of ac system conductor brought to service equipment.

**(A) Connections to Rod, Pipe, or Plate Electrodes.** Where the grounding electrode conductor is connected to rod, pipe, or plate electrodes as permitted in 250.52(A)(5) or 250.52(A)(6), that portion of the conductor that is the sole connection to the grounding electrode shall not be required to be larger than 6 AWG copper wire or 4 AWG aluminum wire.

**(B) Connections to Concrete-Encased Electrodes.** Where the grounding electrode conductor is connected to a concrete-encased electrode as permitted in 250.52(A)(3), that portion of the conductor that is the sole connection to the grounding electrode shall not be required to be larger than 4 AWG copper wire.

**Table 250.66 Grounding Electrode Conductor for Alternating-Current Systems**

| Size of Largest Ungrounded Service-Entrance Conductor or Equivalent Area for Parallel Conductors[a] (AWG/kcmil) | | Size of Grounding Electrode Conductor (AWG/kcmil) | |
|---|---|---|---|
| Copper | Aluminum or Copper-Clad Aluminum | Copper | Aluminum or Copper-Clad Aluminum[b] |
| 2 or smaller | 1/0 or smaller | 8 | 6 |
| 1 or 1/0 | 2/0 or 3/0 | 6 | 4 |
| 2/0 or 3/0 | 4/0 or 250 | 4 | 2 |
| Over 3/0 through 350 | Over 250 through 500 | 2 | 1/0 |
| Over 350 through 600 | Over 500 through 900 | 1/0 | 3/0 |
| Over 600 through 1100 | Over 900 through 1750 | 2/0 | 4/0 |
| Over 1100 | Over 1750 | 3/0 | 250 |

Notes:
1. Where multiple sets of service-entrance conductors are used as permitted in 230.40, Exception No. 2, the equivalent size of the largest service-entrance conductor shall be determined by the largest sum of the areas of the corresponding conductors of each set.
2. Where there are no service-entrance conductors, the grounding electrode conductor size shall be determined by the equivalent size of the largest service-entrance conductor required for the load to be served.
[a]This table also applies to the derived conductors of separately derived ac systems.
[b]See installation restrictions in 250.64(A).

**(C) Connections to Ground Rings.** Where the grounding electrode conductor is connected to a ground ring as permitted in 250.52(A)(4), that portion of the conductor that is the sole connection to the grounding electrode shall not be required to be larger than the conductor used for the ground ring.

**250.68 Grounding Electrode Conductor and Bonding Jumper Connection to Grounding Electrodes.**

**(A) Accessibility.** The connection of a grounding electrode conductor or bonding jumper to a grounding electrode shall be accessible.

*Exception: An encased or buried connection to a concrete-encased, driven, or buried grounding electrode shall not be required to be accessible.*

**(B) Effective Grounding Path.** The connection of a grounding electrode conductor or bonding jumper to a grounding electrode shall be made in a manner that will ensure a permanent and effective grounding path. Where necessary to ensure the grounding path for a metal piping system used as a grounding electrode, effective bonding shall be provided around insulated joints and around any equipment likely to be disconnected for repairs or replacement. Bonding conductors shall be of sufficient length to permit removal of such equipment while retaining the integrity of the bond.

**250.70 Methods of Grounding and Bonding Conductor Connection to Electrodes.** The grounding or bonding conductor shall be connected to the grounding electrode by exothermic welding, listed lugs, listed pressure connectors, listed clamps, or other listed means. Connections depending on solder shall not be used. Ground clamps shall be listed for the materials of the grounding electrode and the grounding electrode conductor and, where used on pipe, rod, or other buried electrodes, shall also be listed for direct soil burial or concrete encasement. Not more than one conductor shall be connected to the grounding electrode by a single clamp or fitting unless the clamp or fitting is listed for multiple conductors. One of the following methods shall be used:

(1) A pipe fitting, pipe plug, or other approved device screwed into a pipe or pipe fitting
(2) A listed bolted clamp of cast bronze or brass, or plain or malleable iron
(3) For indoor telecommunications purposes only, a listed sheet metal strap-type ground clamp having a rigid metal base that seats on the electrode and having a strap of such material and dimensions that it is not likely to stretch during or after installation
(4) An equally substantial approved means

## IV. Enclosure, Raceway, and Service Cable Grounding

**250.80 Service Raceways and Enclosures.** Metal enclosures and raceways for service conductors and equipment shall be grounded.

*Exception: A metal elbow that is installed in an underground installation of rigid nonmetallic conduit and is isolated from possible contact by a minimum cover of 450 mm (18 in.) to any part of the elbow shall not be required to be grounded.*

**250.84 Underground Service Cable or Conduit.**

**(A) Underground Service Cable.** The sheath or armor of a continuous underground metal-sheathed service cable system that is metallically connected to the underground system shall not be required to be grounded at the building. The sheath or armor shall be permitted to be insulated from the interior conduit or piping.

**(B) Underground Service Conduit Containing Cable.** An underground service conduit that contains a metal-sheathed cable bonded to the underground system shall not be required to be grounded at the building. The sheath or armor shall be permitted to be insulated from the interior conduit or piping.

**250.86 Other Conductor Enclosures and Raceways.** Except as permitted by 250.112(I), metal enclosures and raceways for other than service conductors shall be grounded.

*Exception No. 1: Metal enclosures and raceways for conductors added to existing installations of open wire, knob and tube wiring, and nonmetallic-sheathed cable shall not be required to be grounded where these enclosures or wiring methods*

*(a) Do not provide an equipment ground;*
*(b) Are in runs of less than 7.5 m (25 ft);*
*(c) Are free from probable contact with ground, grounded metal, metal lath, or other conductive material; and*
*(d) Are guarded against contact by persons.*

*Exception No. 2: Short sections of metal enclosures or raceways used to provide support or protection of cable assemblies from physical damage shall not be required to be grounded.*

*Exception No. 3: A metal elbow shall not be required to be grounded where it is installed in a nonmetallic raceway and is isolated from possible contact by a minimum cover of 450 mm (18 in.) to any part of the elbow or is encased in not less than 50 mm (2 in.) of concrete.*

## V. Bonding

**250.90 General.** Bonding shall be provided where necessary to ensure electrical continuity and the capacity to conduct safely any fault current likely to be imposed.

**250.92 Services.**

**(A) Bonding of Services.** The non–current-carrying metal parts of equipment indicated in 250.92(A)(1), (2), and (3) shall be effectively bonded together.

(1) The service raceways, cable trays, cablebus framework, auxiliary gutters, or service cable armor or sheath except as permitted in 250.84.
(2) All service enclosures containing service conductors, including meter fittings, boxes, or the like, interposed in the service raceway or armor.
(3) Any metallic raceway or armor enclosing a grounding electrode conductor as specified in 250.64(B). Bonding shall apply at each end and to all intervening raceways, boxes, and enclosures between the service equipment and the grounding electrode.

**(B) Method of Bonding at the Service.** Electrical continuity at service equipment, service raceways, and service conductor enclosures shall be ensured by one of the following methods:

(1) Bonding equipment to the grounded service conductor in a manner provided in 250.8
(2) Connections utilizing threaded couplings or threaded bosses on enclosures where made up wrenchtight
(3) Threadless couplings and connectors where made up tight for metal raceways and metal-clad cables
(4) Other approved devices, such as bonding-type locknuts and bushings

Bonding jumpers meeting the other requirements of this article shall be used around concentric or eccentric knockouts that are punched or otherwise formed so as to impair the electrical connection to ground. Standard locknuts or bushings shall not be the sole means for the bonding required by this section.

**250.94 Bonding for Other Systems.** An accessible means external to enclosures for connecting intersystem bonding and grounding conductors shall be provided at the service equipment and at the disconnecting means for any additional buildings or structures by at least one of the following means:

(1) Exposed nonflexible metallic raceways
(2) Exposed grounding electrode conductor
(3) Approved means for the external connection of a copper or other corrosion-resistant bonding or grounding conductor to the grounded raceway or equipment

FPN No. 1: A 6 AWG copper conductor with one end bonded to the grounded nonflexible metallic raceway or equipment and with 150 mm (6 in.) or more of the other end made accessible on the outside wall is an example of the approved means covered in 250.94(3).

FPN No. 2: See 800.40, 810.21, and 820.40 for bonding and grounding requirements for communications circuits, radio and television equipment, and CATV circuits.

**250.96 Bonding Other Enclosures.**

**(A) General.** Metal raceways, cable trays, cable armor, cable sheath, enclosures, frames, fittings, and other metal non–current-carrying parts that are to serve as grounding conductors, with or without the use of supplementary equipment grounding conductors, shall be effectively bonded where necessary to ensure electrical continuity and the capacity to conduct safely any fault current likely to be imposed on them. Any nonconductive paint, enamel, or similar coating shall be removed at threads, contact points, and contact surfaces or be connected by means of fittings designed so as to make such removal unnecessary.

**(B) Isolated Grounding Circuits.** Where required for the reduction of electrical noise (electromagnetic interference) on the grounding circuit, an equipment enclosure supplied by a branch circuit shall be permitted to be isolated from a raceway containing circuits supplying only that equipment by one or more listed nonmetallic raceway fittings located at the point of attachment of the raceway to the equipment enclosure. The metal raceway shall comply with provisions of this article and shall be supplemented by an internal insulated equipment grounding conductor installed in accordance with 250.146(D) to ground the equipment enclosure.

FPN: Use of an isolated equipment grounding conductor does not relieve the requirement for grounding the raceway system.

**250.97 Bonding for Over 250 Volts.** For circuits of over 250 volts to ground, the electrical continuity of metal raceways and cables with metal sheaths that contain any conductor other than service conductors shall be ensured by one or more of the methods specified for services in 250.92(B), except for (1).

*Exception: Where oversized, concentric, or eccentric knockouts are not encountered, or where a box or enclosure with concentric or eccentric knockouts is listed for the purpose, the following methods shall be permitted:*

(a) *Threadless couplings and connectors for cables with metal sheaths*
(b) *Two locknuts, on rigid metal conduit or intermediate metal conduit, one inside and one outside of boxes and cabinets*
(c) *Fittings with shoulders that seat firmly against the box or cabinet, such as electrical metallic tubing connectors, flexible metal conduit connectors, and cable connectors, with one locknut on the inside of boxes and cabinets*
(d) *Listed fittings that are identified for the purpose*

**250.98 Bonding Loosely Jointed Metal Raceways.** Expansion fittings and telescoping sections of metal raceways shall be made electrically continuous by equipment bonding jumpers or other means.

**250.100 Bonding in Hazardous (Classified) Locations.** Regardless of the voltage of the electrical system, the electrical continuity of non–current-carrying metal parts of equipment, raceways, and other enclosures in any hazardous (classified) location as defined in Article 500 shall be ensured by any of the methods specified for services in 250.92(B) that are approved for the wiring method used.

**250.102 Equipment Bonding Jumpers.**

**(A) Material.** Equipment bonding jumpers shall be of copper or other corrosion-resistant material. A bonding jumper shall be a wire, bus, screw, or similar suitable conductor.

**(B) Attachment.** Equipment bonding jumpers shall be attached in the manner specified by the applicable provisions of 250.8 for circuits and equipment and by 250.70 for grounding electrodes.

**(C) Size — Equipment Bonding Jumper on Supply Side of Service.** The bonding jumper shall not be smaller than the sizes shown in Table 250.66 for grounding electrode conductors. Where the service-entrance phase conductors are larger than 1100 kcmil copper or 1750 kcmil aluminum, the bonding jumper shall have an area not less than 12½ percent of the area of the largest phase conductor except that, where the phase conductors and the bonding jumper are of different materials (copper or aluminum), the minimum size of the bonding jumper shall be based on the assumed use of phase conductors of the same material as the bonding jumper and with an ampacity equivalent to that of the installed phase conductors. Where the service-entrance conductors are paralleled in two or more raceways or cables, the equipment bonding jumper, where routed with the raceways or cables, shall be run in parallel. The size of the bonding jumper for each raceway or cable shall be based on the size of the service-entrance conductors in each raceway or cable.

The bonding jumper for a grounding electrode conductor raceway or cable armor as covered in 250.64(E) shall be the same size or larger than the required enclosed grounding electrode conductor.

**(D) Size — Equipment Bonding Jumper on Load Side of Service.** The equipment bonding jumper on the load side of the service overcurrent devices shall be sized, as a minimum, in accordance with the sizes listed in Table 250.122, but shall not be required to be larger than the largest ungrounded circuit conductors supplying the equipment and shall not be smaller than 14 AWG.

A single common continuous equipment bonding jumper shall be permitted to bond two or more raceways or cables where the bonding jumper is sized in accordance with Table 250.122 for the largest overcurrent device supplying circuits therein.

**(E) Installation.** The equipment bonding jumper shall be permitted to be installed inside or outside of a raceway or enclosure. Where installed on the outside, the length of the equipment bonding jumper shall not exceed 1.8 m (6 ft) and shall be routed with the raceway or enclosure. Where installed inside of a raceway, the equipment bonding jumper shall comply with the requirements of 250.119 and 250.148.

*Exception: An equipment bonding jumper longer than 1.8 m (6 ft) shall be permitted at outside pole locations for the purpose of bonding or grounding isolated sections of metal raceways or elbows installed in exposed risers of metal conduit or other metal raceway.*

**250.104 Bonding of Piping Systems and Exposed Structural Steel.**

**(A) Metal Water Piping.** The metal water piping system shall be bonded as required in (1), (2), (3), or (4) of this section. The bonding jumper(s) shall be installed in accordance with 250.64(A), (B), and (E). The points of attachment of the bonding jumper(s) shall be accessible.

**(1) General.** Metal water piping system(s) installed in or attached to a building or structure shall be bonded to the service equipment enclosure, the grounded conductor at the service, the grounding electrode conductor where of sufficient size, or to the one or more grounding electrodes used. The bonding jumper(s) shall be sized in accordance with Table 250.66 except as permitted in 250.104(A)(2) and (A)(3).

**(2) Buildings of Multiple Occupancy.** In buildings of multiple occupancy where the metal water piping system(s) installed in or attached to a building or structure for the individual occupancies is metallically isolated from all other occupancies by use of nonmetallic water piping, the metal water piping system(s) for each occupancy shall be permitted to be bonded to the equipment grounding terminal of the panelboard or switchboard enclosure (other than service equipment) supplying that occupancy. The bonding jumper shall be sized in accordance with Table 250.122.

**(3) Multiple Buildings or Structures Supplied from a Common Service.** The metal water piping system(s) installed in or attached to a building or structure shall be bonded to the building or structure disconnecting means enclosure where located at the building or structure, to the equipment grounding conductor run with the supply conductors, or to the one or more grounding electrodes used. The bonding jumper(s) shall be sized in accordance with 250.66, based on the size of the feeder or branch circuit conductors that supply the building. The bonding jumper shall not be required to be larger than the largest ungrounded feeder or branch circuit conductor supplying the building.

**(4) Separately Derived Systems.** The grounded conductor of each separately derived system shall be bonded to the nearest available point of the interior metal water piping system(s) in the area served by each separately derived system. This connection shall be made at the same point on the separately derived system where the grounding electrode conductor is connected. Each bonding jumper shall be sized in accordance with Table 250.66.

*Exception: A separate water piping bonding jumper shall not be required where the effectively grounded metal frame of a building or structure is used as the grounding electrode for a separately derived system and is bonded to the metal-*

*lic water piping in the area served by the separately derived system.*

**(B) Other Metal Piping.** Where installed in or attached to a building or structure, metal piping system(s), including gas piping, that may become energized shall be bonded to the service equipment enclosure, the grounded conductor at the service, the grounding electrode conductor where of sufficient size, or to the one or more grounding electrodes used. The bonding jumper(s) shall be sized in accordance with 250.122 using the rating of the circuit that may energize the piping system(s). The equipment grounding conductor for the circuit that may energize the piping shall be permitted to serve as the bonding means. The points of attachment of the bonding jumper(s) shall be accessible.

> FPN: Bonding all piping and metal air ducts within the premises will provide additional safety.

**(C) Structural Steel.** Exposed structural steel that is interconnected to form a steel building frame and is not intentionally grounded and may become energized shall be bonded to the service equipment enclosure, the grounded conductor at the service, the grounding electrode conductor where of sufficient size, or the one or more grounding electrodes used. The bonding jumper(s) shall be sized in accordance with Table 250.66 and installed in accordance with 250.64(A), (B), and (E). The points of attachment of the bonding jumper(s) shall be accessible.

**250.106 Lightning Protection Systems.** The lightning protection system ground terminals shall be bonded to the building or structure grounding electrode system.

> FPN No. 1: See 250.60 for use of air terminals. For further information, see NFPA 780-1997, *Standard for the Installation of Lightning Protection Systems*, which contains detailed information on grounding, bonding, and spacing from lightning protection systems.

> FPN No. 2: Metal raceways, enclosures, frames, and other non–current-carrying metal parts of electric equipment installed on a building equipped with a lightning protection system may require bonding or spacing from the lightning protection conductors in accordance with NFPA 780-1997, *Standard for the Installation of Lightning Protection Systems*. Separation from lightning protection conductors is typically 1.8 m (6 ft) through air or 900 mm (3 ft) through dense materials such as concrete, brick, or wood.

## VI. Equipment Grounding and Equipment Grounding Conductors

**250.110 Equipment Fastened in Place or Connected by Permanent Wiring Methods (Fixed).** Exposed non–current-carrying metal parts of fixed equipment likely to become energized shall be grounded under any of the following conditions:

(1) Where within 2.5 m (8 ft) vertically or 1.5 m (5 ft) horizontally of ground or grounded metal objects and subject to contact by persons
(2) Where located in a wet or damp location and not isolated
(3) Where in electrical contact with metal
(4) Where in a hazardous (classified) location as covered by Articles 500 through 517
(5) Where supplied by a metal-clad, metal-sheathed, metal-raceway, or other wiring method that provides an equipment ground, except as permitted by 250.86, Exception No. 2, for short sections of metal enclosures
(6) Where equipment operates with any terminal at over 150 volts to ground

*Exception No. 1: Metal frames of electrically heated appliances, exempted by special permission, in which case the frames shall be permanently and effectively insulated from ground.*

*Exception No. 2: Distribution apparatus, such as transformer and capacitor cases, mounted on wooden poles, at a height exceeding 2.5 m (8 ft) above ground or grade level.*

*Exception No. 3: Listed equipment protected by a system of double insulation, or its equivalent, shall not be required to be grounded. Where such a system is employed, the equipment shall be distinctively marked.*

**250.112 Fastened in Place or Connected by Permanent Wiring Methods (Fixed) — Specific.** Exposed, non–current-carrying metal parts of the kinds of equipment described in 250.112(A) through (K), and non–current-carrying metal parts of equipment and enclosures described in 250.112(L) and (M), shall be grounded regardless of voltage.

**(A) Switchboard Frames and Structures.** Switchboard frames and structures supporting switching equipment, except frames of 2-wire dc switchboards where effectively insulated from ground.

**(B) Pipe Organs.** Generator and motor frames in an electrically operated pipe organ, unless effectively insulated from ground and the motor driving it.

**(C) Motor Frames.** Motor frames, as provided by 430.142.

**(D) Enclosures for Motor Controllers.** Enclosures for motor controllers unless attached to ungrounded portable equipment.

**(E) Elevators and Cranes.** Electric equipment for elevators and cranes.

**(F) Garages, Theaters, and Motion Picture Studios.** Electric equipment in commercial garages, theaters, and motion picture studios, except pendant lampholders supplied by circuits not over 150 volts to ground.

**(G) Electric Signs.** Electric signs, outline lighting, and associated equipment as provided in Article 600.

**(H) Motion Picture Projection Equipment.** Motion picture projection equipment.

**(I) Power-Limited Remote-Control, Signaling, and Fire Alarm Circuits.** Equipment supplied by Class 1 power-limited circuits and Class 1, Class 2, and Class 3 remote-control and signaling circuits, and by fire alarm circuits, shall be grounded where system grounding is required by Part II or Part VIII of this article.

**(J) Luminaires (Lighting Fixtures).** Luminaires (lighting fixtures) as provided in Part V of Article 410.

**(K) Skid Mounted Equipment.** Permanently mounted electrical equipment and skids shall be grounded with an equipment bonding jumper sized as required by 250.122.

**(L) Motor-Operated Water Pumps.** Motor-operated water pumps, including the submersible type.

**(M) Metal Well Casings.** Where a submersible pump is used in a metal well casing, the well casing shall be bonded to the pump circuit equipment grounding conductor.

**250.114 Equipment Connected by Cord and Plug.** Under any of the conditions described in (1) through (4), exposed non–current-carrying metal parts of cord-and-plug-connected equipment likely to become energized shall be grounded.

*Exception: Listed tools, listed appliances, and listed equipment covered in (2) through (4) shall not be required to be grounded where protected by a system of double insulation or its equivalent. Double insulated equipment shall be distinctively marked.*

(1) In hazardous (classified) locations (see Articles 500 through 517)

(2) Where operated at over 150 volts to ground

*Exception No. 1: Motors, where guarded, shall not be required to be grounded.*

*Exception No. 2: Metal frames of electrically heated appliances, exempted by special permission, shall not be required to be grounded, in which case the frames shall be permanently and effectively insulated from ground.*

(3) In residential occupancies:
   a. Refrigerators, freezers, and air conditioners
   b. Clothes-washing, clothes-drying, dish-washing machines; kitchen waste disposers; information tech-

nology equipment; sump pumps and electrical aquarium equipment
   c. Hand-held motor-operated tools, stationary and fixed motor-operated tools, light industrial motor-operated tools
   d. Motor-operated appliances of the following types: hedge clippers, lawn mowers, snow blowers, and wet scrubbers
   e. Portable handlamps

(4) In other than residential occupancies:
   a. Refrigerators, freezers, and air conditioners
   b. Clothes-washing, clothes-drying, dish-washing machines; information technology equipment; sump pumps and electrical aquarium equipment
   c. Hand-held motor-operated tools, stationary and fixed motor-operated tools, light industrial motor-operated tools
   d. Motor-operated appliances of the following types: hedge clippers, lawn mowers, snow blowers, and wet scrubbers
   e. Portable handlamps
   f. Cord-and-plug-connected appliances used in damp or wet locations or by persons standing on the ground or on metal floors or working inside of metal tanks or boilers
   g. Tools likely to be used in wet or conductive locations

*Exception: Tools and portable handlamps likely to be used in wet or conductive locations shall not be required to be grounded where supplied through an isolating transformer with an ungrounded secondary of not over 50 volts.*

**250.116 Nonelectric Equipment.** The metal parts of non-electric equipment described in this section shall be grounded.

(1) Frames and tracks of electrically operated cranes and hoists
(2) Frames of nonelectrically driven elevator cars to which electric conductors are attached
(3) Hand-operated metal shifting ropes or cables of electric elevators

   FPN: Where extensive metal in or on buildings may become energized and is subject to personal contact, adequate bonding and grounding will provide additional safety.

**250.118 Types of Equipment Grounding Conductors.** The equipment grounding conductor run with or enclosing the circuit conductors shall be one or more or a combination of the following:

(1) A copper, aluminum, or copper-clad aluminum conductor. This conductor shall be solid or stranded; insulated, covered, or bare; and in the form of a wire or a busbar of any shape.

(2) Rigid metal conduit.

(3) Intermediate metal conduit.

(4) Electrical metallic tubing.

(5) Flexible metal conduit where both the conduit and fittings are listed for grounding.

(6) Listed flexible metal conduit that is not listed for grounding, meeting all the following conditions:

   a. The conduit is terminated in fittings listed for grounding.

   b. The circuit conductors contained in the conduit are protected by overcurrent devices rated at 20 amperes or less.

   c. The combined length of flexible metal conduit and flexible metallic tubing and liquidtight flexible metal conduit in the same ground return path does not exceed 1.8 m (6 ft).

   d. The conduit is not installed for flexibility.

(7) Listed liquidtight flexible metal conduit meeting all the following conditions:

   a. The conduit is terminated in fittings listed for grounding.

   b. For metric designators 12 through 16 (trade sizes ⅜ through ½), the circuit conductors contained in the conduit are protected by overcurrent devices rated at 20 amperes or less.

   c. For metric designators 21 through 35 (trade sizes ¾ through 1¼), the circuit conductors contained in the conduit are protected by overcurrent devices rated not more than 60 amperes and there is no flexible metal conduit, flexible metallic tubing, or liquidtight flexible metal conduit in trade sizes metric designators 12 through 16 (trade sizes ⅜ through ½) in the grounding path.

   d. The combined length of flexible metal conduit and flexible metallic tubing and liquidtight flexible metal conduit in the same ground return path does not exceed 1.8 m (6 ft).

   e. The conduit is not installed for flexibility.

(8) Flexible metallic tubing where the tubing is terminated in fittings listed for grounding and meeting the following conditions:

   a. The circuit conductors contained in the tubing are protected by overcurrent devices rated at 20 amperes or less.

   b. The combined length of flexible metal conduit and flexible metallic tubing and liquidtight flexible metal conduit in the same ground return path does not exceed 1.8 m (6 ft).

(9) Armor of Type AC cable as provided in 320.108.

(10) The copper sheath of mineral-insulated, metal-sheathed cable.

(11) Type MC cable where listed and identified for grounding in accordance with the following:

   a. The combined metallic sheath and grounding conductor of interlocked metal tape–type MC cable

   b. The metallic sheath or the combined metallic sheath and grounding conductors of the smooth or corrugated tube type MC cable

(12) Cable trays as permitted in 392.3(C) and 392.7.

(13) Cablebus framework as permitted in 370.3.

(14) Other electrically continuous metal raceways and auxiliary gutters listed for grounding.

**250.119 Identification of Equipment Grounding Conductors.** Unless required elsewhere in this *Code,* equipment grounding conductors shall be permitted to be bare, covered, or insulated. Individually covered or insulated equipment grounding conductors shall have a continuous outer finish that is either green or green with one or more yellow stripes except as permitted in this section.

**(A) Conductors Larger Than 6 AWG.** An insulated or covered conductor larger than 6 AWG copper or aluminum shall be permitted, at the time of installation, to be permanently identified as an equipment grounding conductor at each end and at every point where the conductor is accessible. Identification shall encircle the conductor and shall be accomplished by one of the following:

(1) Stripping the insulation or covering from the entire exposed length

(2) Coloring the exposed insulation or covering green

(3) Marking the exposed insulation or covering with green tape or green adhesive labels

**(B) Multiconductor Cable.** Where the conditions of maintenance and supervision ensure that only qualified persons service the installation, one or more insulated conductors in a multiconductor cable, at the time of installation, shall be permitted to be permanently identified as equipment grounding conductors at each end and at every point where the conductors are accessible by one of the following means:

(1) Stripping the insulation from the entire exposed length

(2) Coloring the exposed insulation green

(3) Marking the exposed insulation with green tape or green adhesive labels

**(C) Flexible Cord.** An uninsulated equipment grounding conductor shall be permitted, but, if individually covered, the covering shall have a continuous outer finish that is either green or green with one or more yellow stripes.

**250.120 Equipment Grounding Conductor Installation.** An equipment grounding conductor shall be installed in accordance with 250.120(A), (B), and (C).

**(A) Raceway, Cable Trays, Cable Armor, Cablebus, or Cable Sheaths.** Where it consists of a raceway, cable tray, cable armor, cablebus framework, or cable sheath or where it is a wire within a raceway or cable, it shall be installed in accordance with the applicable provisions in this *Code* using fittings for joints and terminations approved for use with the type raceway or cable used. All connections, joints, and fittings shall be made tight using suitable tools.

**(B) Aluminum and Copper-Clad Aluminum Conductors.** Equipment grounding conductors of bare or insulated aluminum or copper-clad aluminum shall be permitted. Bare conductors shall not come in direct contact with masonry or the earth or where subject to corrosive conditions. Aluminum or copper-clad aluminum conductors shall not be terminated within 450 mm (18 in.) of the earth.

**(C) Equipment Grounding Conductors Smaller Than 6 AWG.** Equipment grounding conductors smaller than 6 AWG shall be protected from physical damage by a raceway or cable armor except where run in hollow spaces of walls or partitions, where not subject to physical damage, or where protected from physical damage.

## 250.122 Size of Equipment Grounding Conductors.

**(A) General.** Copper, aluminum, or copper-clad aluminum equipment grounding conductors of the wire type shall not be smaller than shown in Table 250.122 but shall not be required to be larger than the circuit conductors supplying the equipment. Where a raceway or a cable armor or sheath is used as the equipment grounding conductor, as provided in 250.118 and 250.134(A), it shall comply with 250.4(A)(5) or 250.4(B)(4).

**(B) Increased in Size.** Where ungrounded conductors are increased in size, equipment grounding conductors, where installed, shall be increased in size proportionately according to circular mil area of the ungrounded conductors.

**(C) Multiple Circuits.** Where a single equipment grounding conductor is run with multiple circuits in the same raceway or cable, it shall be sized for the largest overcurrent device protecting conductors in the raceway or cable.

**(D) Motor Circuits.** Where the overcurrent device consists of an instantaneous trip circuit breaker or a motor short-circuit protector, as allowed in 430.52, the equipment grounding conductor size shall be permitted to be based on the rating of the motor overload protective device but not less than the size shown in Table 250.122.

**(E) Flexible Cord and Fixture Wire.** Equipment grounding conductors that are part of flexible cords or used with fixture wires in accordance with 240.5 shall be not smaller than 18 AWG copper and not smaller than the circuit conductors.

**(F) Conductors in Parallel.** Where conductors are run in parallel in multiple raceways or cables as permitted in 310.4, the equipment grounding conductors, where used, shall be run in parallel in each raceway or cable. One of the methods in 250.122(F)(1) or (2) shall be used to ensure the equipment grounding conductors are protected.

**(1)** Each parallel equipment grounding conductor shall be sized on the basis of the ampere rating of the overcurrent device protecting the circuit conductors in the raceway or cable in accordance with Table 250.122.

**(2)** Where ground-fault protection of equipment is installed, each parallel equipment grounding conductor in a multiconductor cable shall be permitted to be sized in accordance with Table 250.122 on the basis of the trip rating of the ground-fault protection where the following conditions are met:

(1) Conditions of maintenance and supervision ensure that only qualified persons will service the installation.

**Table 250.122 Minimum Size Equipment Grounding Conductors for Grounding Raceway and Equipment**

| Rating or Setting of Automatic Overcurrent Device in Circuit Ahead of Equipment, Conduit, etc., Not Exceeding (Amperes) | Size (AWG or kcmil) | |
|---|---|---|
| | Copper | Aluminum or Copper-Clad Aluminum* |
| 15 | 14 | 12 |
| 20 | 12 | 10 |
| 30 | 10 | 8 |
| 40 | 10 | 8 |
| 60 | 10 | 8 |
| 100 | 8 | 6 |
| 200 | 6 | 4 |
| 300 | 4 | 2 |
| 400 | 3 | 1 |
| 500 | 2 | 1/0 |
| 600 | 1 | 2/0 |
| 800 | 1/0 | 3/0 |
| 1000 | 2/0 | 4/0 |
| 1200 | 3/0 | 250 |
| 1600 | 4/0 | 350 |
| 2000 | 250 | 400 |
| 2500 | 350 | 600 |
| 3000 | 400 | 600 |
| 4000 | 500 | 800 |
| 5000 | 700 | 1200 |
| 6000 | 800 | 1200 |

Note: Where necessary to comply with 250.4(A)(5) or 250.4(B)(4), the equipment grounding conductor shall be sized larger than given in this table.

*See installation restrictions in 250.120.

(2) The ground-fault protection equipment is set to trip at not more than the ampacity of a single ungrounded conductor of one of the cables in parallel.

(3) The ground-fault protection is listed for the purpose.

**250.124 Equipment Grounding Conductor Continuity.**

**(A) Separable Connections.** Separable connections such as those provided in drawout equipment or attachment plugs and mating connectors and receptacles shall provide for first-make, last-break of the equipment grounding conductor. First-make, last-break shall not be required where interlocked equipment, plugs, receptacles, and connectors preclude energization without grounding continuity.

**(B) Switches.** No automatic cutout or switch shall be placed in the equipment grounding conductor of a premises wiring system unless the opening of the cutout or switch disconnects all sources of energy.

**250.126 Identification of Wiring Device Terminals.** The terminal for the connection of the equipment grounding conductor shall be identified by one of the following:

(1) A green, not readily removable terminal screw with a hexagonal head.

(2) A green, hexagonal, not readily removable terminal nut.

(3) A green pressure wire connector. If the terminal for the grounding conductor is not visible, the conductor entrance hole shall be marked with the word *green* or *ground*, the letters *G* or *GR* or the grounding symbol shown in Figure 250.126, or otherwise identified by a distinctive green color. If the terminal for the equipment grounding conductor is readily removable, the area adjacent to the terminal shall be similarly marked.

**Figure 250.126 Grounding symbol.**

## VII. Methods of Equipment Grounding

**250.130 Equipment Grounding Conductor Connections.** Equipment grounding conductor connections at the source of separately derived systems shall be made in accordance with 250.30(A)(1). Equipment grounding conductor connections at service equipment shall be made as indicated in 250.130(A) or (B). For replacement of non–grounding-type receptacles with grounding-type receptacles and for branch-circuit extensions only in existing installations that do not have an equipment grounding conductor in the branch circuit, connections shall be permitted as indicated in 250.130(C).

**(A) For Grounded Systems.** The connection shall be made by bonding the equipment grounding conductor to the grounded service conductor and the grounding electrode conductor.

**(B) For Ungrounded Systems.** The connection shall be made by bonding the equipment grounding conductor to the grounding electrode conductor.

**(C) Nongrounding Receptacle Replacement or Branch Circuit Extensions.** The equipment grounding conductor of a grounding-type receptacle or a branch-circuit extension shall be permitted to be connected to any of the following:

(1) Any accessible point on the grounding electrode system as described in 250.50

(2) Any accessible point on the grounding electrode conductor

(3) The equipment grounding terminal bar within the enclosure where the branch circuit for the receptacle or branch circuit originates

(4) For grounded systems, the grounded service conductor within the service equipment enclosure

(5) For ungrounded systems, the grounding terminal bar within the service equipment enclosure

FPN: See 406.3(D) for the use of a ground-fault circuit-interrupting type of receptacle.

**250.132 Short Sections of Raceway.** Isolated sections of metal raceway or cable armor, where required to be grounded, shall be grounded in accordance with 250.134.

**250.134 Equipment Fastened in Place or Connected by Permanent Wiring Methods (Fixed) — Grounding.** Unless grounded by connection to the grounded circuit conductor as permitted by 250.32, 250.140, and 250.142, non–current-carrying metal parts of equipment, raceways, and other enclosures, if grounded, shall be grounded by one of the following methods.

**(A) Equipment Grounding Conductor Types.** By any of the equipment grounding conductors permitted by 250.118.

**(B) With Circuit Conductors.** By an equipment grounding conductor contained within the same raceway, cable, or otherwise run with the circuit conductors.

*Exception No. 1: As provided in 250.130(C), the equipment grounding conductor shall be permitted to be run separately from the circuit conductors.*

*Exception No. 2: For dc circuits, the equipment grounding conductor shall be permitted to be run separately from the circuit conductors.*

FPN No. 1: See 250.102 and 250.168 for equipment bonding jumper requirements.

FPN No. 2: See 400.7 for use of cords for fixed equipment.

## 250.136 Equipment Considered Effectively Grounded.
Under the conditions specified in 250.136(A) and (B), the non–current-carrying metal parts of the equipment shall be considered effectively grounded.

**(A) Equipment Secured to Grounded Metal Supports.** Electrical equipment secured to and in electrical contact with a metal rack or structure provided for its support and grounded by one of the means indicated in 250.134. The structural metal frame of a building shall not be used as the required equipment grounding conductor for ac equipment.

**(B) Metal Car Frames.** Metal car frames supported by metal hoisting cables attached to or running over metal sheaves or drums of elevator machines that are grounded by one of the methods indicated in 250.134.

## 250.138 Cord-and-Plug-Connected Equipment.
Non–current-carrying metal parts of cord-and-plug-connected equipment, if grounded, shall be grounded by one of the methods in 250.138(A) or (B).

**(A) By Means of an Equipment Grounding Conductor.** By means of an equipment grounding conductor run with the power supply conductors in a cable assembly or flexible cord properly terminated in a grounding-type attachment plug with one fixed grounding contact.

*Exception: The grounding contacting pole of grounding-type plug-in ground-fault circuit interrupters shall be permitted to be of the movable, self-restoring type on circuits operating at not over 150 volts between any two conductors or over 150 volts between any conductor and ground.*

**(B) By Means of a Separate Flexible Wire or Strap.** By means of a separate flexible wire or strap, insulated or bare, protected as well as practicable against physical damage, where part of equipment.

## 250.140 Frames of Ranges and Clothes Dryers.
This section shall apply to existing branch-circuit installations only. New branch-circuit installations shall comply with 250.134 and 250.138. Frames of electric ranges, wall-mounted ovens, counter-mounted cooking units, clothes dryers, and outlet or junction boxes that are part of the circuit for these appliances shall be grounded in the manner specified by 250.134 or 250.138; or, except for mobile homes and recreational vehicles, shall be permitted to be grounded to the grounded circuit conductor if all the following conditions are met.

(1) The supply circuit is 120/240-volt, single-phase, 3-wire; or 208Y/120-volt derived from a 3-phase, 4-wire, wye-connected system.

(2) The grounded conductor is not smaller than 10 AWG copper or 8 AWG aluminum.

(3) The grounded conductor is insulated, or the grounded conductor is uninsulated and part of a Type SE service-entrance cable and the branch circuit originates at the service equipment.

(4) Grounding contacts of receptacles furnished as part of the equipment are bonded to the equipment.

## 250.142 Use of Grounded Circuit Conductor for Grounding Equipment.

**(A) Supply-Side Equipment.** A grounded circuit conductor shall be permitted to ground non–current-carrying metal parts of equipment, raceways, and other enclosures at any of the following locations:

(1) On the supply side or within the enclosure of the ac service-disconnecting means

(2) On the supply side or within the enclosure of the main disconnecting means for separate buildings as provided in 250.32(B)

(3) On the supply side or within the enclosure of the main disconnecting means or overcurrent devices of a separately derived system where permitted by 250.30(A)(1)

**(B) Load-Side Equipment.** Except as permitted in 250.30(A)(1) and 250.32(B), a grounded circuit conductor shall not be used for grounding non–current-carrying metal parts of equipment on the load side of the service disconnecting means or on the load side of a separately derived system disconnecting means or the overcurrent devices for a separately derived system not having a main disconnecting means.

*Exception No. 1: The frames of ranges, wall-mounted ovens, counter-mounted cooking units, and clothes dryers under the conditions permitted for existing installations by 250.140 shall be permitted to be grounded by a grounded circuit conductor.*

*Exception No. 2: It shall be permissible to ground meter enclosures by connection to the grounded circuit conductor on the load side of the service disconnect if*

*(a) No service ground-fault protection is installed, and*

*(b) All meter enclosures are located near the service disconnecting means, and*

*(c) The size of the grounded circuit conductor is not smaller than the size specified in Table 250.122 for equipment grounding conductors.*

*Exception No. 3: Direct-current systems shall be permitted to be grounded on the load side of the disconnecting means or overcurrent device in accordance with 250.164.*

*Exception No. 4: Electrode-type boilers operating at over 600 volts shall be grounded as required in 490.72(E)(1) and 490.74.*

**250.144 Multiple Circuit Connections.** Where equipment is required to be grounded and is supplied by separate connection to more than one circuit or grounded premises wiring system, a means for grounding shall be provided for each such connection as specified in 250.134 and 250.138.

**250.146 Connecting Receptacle Grounding Terminal to Box.** An equipment bonding jumper shall be used to connect the grounding terminal of a grounding-type receptacle to a grounded box unless grounded as in 250.146(A) through (D).

**(A) Surface Mounted Box.** Where the box is mounted on the surface, direct metal-to-metal contact between the device yoke and the box shall be permitted to ground the receptacle to the box. This provision shall not apply to cover-mounted receptacles unless the box and cover combination are listed as providing satisfactory ground continuity between the box and the receptacle.

**(B) Contact Devices or Yokes.** Contact devices or yokes designed and listed for the purpose shall be permitted in conjunction with the supporting screws to establish the grounding circuit between the device yoke and flush-type boxes.

**(C) Floor Boxes.** Floor boxes designed for and listed as providing satisfactory ground continuity between the box and the device shall be permitted.

**(D) Isolated Receptacles.** Where required for the reduction of electrical noise (electromagnetic interference) on the grounding circuit, a receptacle in which the grounding terminal is purposely insulated from the receptacle mounting means shall be permitted. The receptacle grounding terminal shall be grounded by an insulated equipment grounding conductor run with the circuit conductors. This grounding conductor shall be permitted to pass through one or more panelboards without connection to the panelboard grounding terminal as permitted in 408.20, Exception, so as to terminate within the same building or structure directly at an equipment grounding conductor terminal of the applicable derived system or service.

> FPN: Use of an isolated equipment grounding conductor does not relieve the requirement for grounding the raceway system and outlet box.

**250.148 Continuity and Attachment of Equipment Grounding Conductors to Boxes.** Where circuit conductors are spliced within a box, or terminated on equipment within or supported by a box, any separate equipment grounding conductors associated with those circuit conductors shall be spliced or joined within the box or to the box with devices suitable for the use. Connections depending solely on solder shall not be used. Splices shall be made in accordance with 110.14(B) except that insulation shall not be required. The arrangement of grounding connections shall be such that the disconnection or the removal of a receptacle, luminaire (fixture), or other device fed from the box will not interfere with or interrupt the grounding continuity.

*Exception: The equipment grounding conductor permitted in 250.146(D) shall not be required to be connected to the other equipment grounding conductors or to the box.*

**(A) Metal Boxes.** A connection shall be made between the one or more equipment grounding conductors and a metal box by means of a grounding screw that shall be used for no other purpose or a listed grounding device.

**(B) Nonmetallic Boxes.** One or more equipment grounding conductors brought into a nonmetallic outlet box shall be arranged so that a connection can be made to any fitting or device in that box requiring grounding.

## VIII. Direct-Current Systems

**250.160 General.** Direct-current systems shall comply with Part VIII and other sections of Article 250 not specifically intended for ac systems.

**250.162 Direct-Current Circuits and Systems to Be Grounded.** Direct-current circuits and systems shall be grounded as provided for in 250.162(A) and (B).

**(A) Two-Wire, Direct-Current Systems.** A 2-wire, dc system supplying premises wiring and operating at greater than 50 volts but not greater than 300 volts shall be grounded.

*Exception No. 1: A system equipped with a ground detector and supplying only industrial equipment in limited areas shall not be required to be grounded.*

*Exception No. 2: A rectifier-derived dc system supplied from an ac system complying with 250.20 shall not be required to be grounded.*

*Exception No. 3: Direct-current fire alarm circuits having a maximum current of 0.030 amperes as specified in Article 760, Part III, shall not be required to be grounded.*

**(B) Three-Wire, Direct-Current Systems.** The neutral conductor of all 3-wire, dc systems supplying premises wiring shall be grounded.

**250.164 Point of Connection for Direct-Current Systems.**

**(A) Off-Premises Source.** Direct-current systems to be grounded and supplied from an off-premises source shall have the grounding connection made at one or more supply stations. A grounding connection shall not be made at individual services or at any point on the premises wiring.

**(B) On-Premises Source.** Where the dc system source is located on the premises, a grounding connection shall be made at one of the following:

(1) The source
(2) The first system disconnection means or overcurrent device
(3) By other means that accomplish equivalent system protection and that utilize equipment listed and identified for the use

**250.166 Size of Direct-Current Grounding Electrode Conductor.** The size of the grounding electrode conductor for a dc system shall be as specified in 250.166(A) through (E).

**(A) Not Smaller Than the Neutral Conductor.** Where the dc system consists of a 3-wire balancer set or a balancer winding with overcurrent protection as provided in 445.12(D), the grounding electrode conductor shall not be smaller than the neutral conductor and not smaller than 8 AWG copper or 6 AWG aluminum.

**(B) Not Smaller Than the Largest Conductor.** Where the dc system is other than as in 250.166(A), the grounding electrode conductor shall not be smaller than the largest conductor supplied by the system, and not smaller than 8 AWG copper or 6 AWG aluminum.

**(C) Connected to Rod, Pipe, or Plate Electrodes.** Where connected to rod, pipe, or plate electrodes as in 250.52(A)(5) or 250.52(A)(6), that portion of the grounding electrode conductor that is the sole connection to the grounding electrode shall not be required to be larger than 6 AWG copper wire or 4 AWG aluminum wire.

**(D) Connected to a Concrete-Encased Electrode.** Where connected to a concrete-encased electrode as in 250.52(A)(3), that portion of the grounding electrode conductor that is the sole connection to the grounding electrode shall not be required to be larger than 4 AWG copper wire.

**(E) Connected to a Ground Ring.** Where connected to a ground ring as in 250.52(A)(4), that portion of the grounding electrode conductor that is the sole connection to the grounding electrode shall not be required to be larger than the conductor used for the ground ring.

**250.168 Direct-Current Bonding Jumper.** For dc systems, the size of the bonding jumper shall not be smaller than the system grounding conductor specified in 250.166.

**250.169 Ungrounded Direct-Current Separately Derived Systems.** Except as otherwise permitted in 250.34 for portable and vehicle-mounted generators, an ungrounded dc separately derived system supplied from a stand-alone power source (such as an engine–generator set) shall have a grounding electrode conductor connected to an electrode that complies with Part III to provide for grounding of metal enclosures, raceways, cables, and exposed non–current-carrying metal parts of equipment. The grounding electrode conductor connection shall be to the metal enclosure at any point on the separately derived system from the source to the first system disconnecting means or overcurrent device, or it shall be made at the source of a separately derived system that has no disconnecting means or overcurrent devices.

The size of the grounding electrode conductor shall be in accordance with 250.166.

## IX. Instruments, Meters, and Relays

**250.170 Instrument Transformer Circuits.** Secondary circuits of current and potential instrument transformers shall be grounded where the primary windings are connected to circuits of 300 volts or more to ground and, where on switchboards, shall be grounded irrespective of voltage.

*Exception: Circuits where the primary windings are connected to circuits of less than 1000 volts with no live parts or wiring exposed or accessible to other than qualified persons.*

**250.172 Instrument Transformer Cases.** Cases or frames of instrument transformers shall be grounded where accessible to other than qualified persons.

*Exception: Cases or frames of current transformers, the primaries of which are not over 150 volts to ground and that are used exclusively to supply current to meters.*

**250.174 Cases of Instruments, Meters, and Relays Operating at Less Than 1000 Volts.** Instruments, meters, and relays operating with windings or working parts at less than 1000 volts shall be grounded as specified in 250.174(A), (B), or (C).

**(A) Not on Switchboards.** Instruments, meters, and relays not located on switchboards, operating with windings or working parts at 300 volts or more to ground, and accessible to other than qualified persons, shall have the cases and other exposed metal parts grounded.

**(B) On Dead-Front Switchboards.** Instruments, meters, and relays (whether operated from current and potential transformers or connected directly in the circuit) on switchboards having no live parts on the front of the panels shall have the cases grounded.

**(C) On Live-Front Switchboards.** Instruments, meters, and relays (whether operated from current and potential transformers or connected directly in the circuit) on switch-boards having exposed live parts on the front of panels shall not have their cases grounded. Mats of insulating rubber or other suitable floor insulation shall be provided for the operator where the voltage to ground exceeds 150.

**250.176 Cases of Instruments, Meters, and Relays — Operating Voltage 1 kV and Over.** Where instruments, meters, and relays have current-carrying parts of 1 kV and over to ground, they shall be isolated by elevation or protected by suitable barriers, grounded metal, or insulating covers or guards. Their cases shall not be grounded.

*Exception: Cases of electrostatic ground detectors where the internal ground segments of the instrument are connected to the instrument case and grounded and the ground detector is isolated by elevation.*

**250.178 Instrument Grounding Conductor.** The grounding conductor for secondary circuits of instrument transformers and for instrument cases shall not be smaller than 12 AWG copper or 10 AWG aluminum. Cases of instrument transformers, instruments, meters, and relays that are mounted directly on grounded metal surfaces of enclosures or grounded metal switchboard panels shall be considered to be grounded, and no additional grounding conductor shall be required.

## X. Grounding of Systems and Circuits of 1 kV and Over (High Voltage)

**250.180 General.** Where high-voltage systems are grounded, they shall comply with all applicable provisions of the preceding sections of this article and with 250.182 through 250.190, which supplement and modify the preceding sections.

**250.182 Derived Neutral Systems.** A system neutral derived from a grounding transformer shall be permitted to be used for grounding high-voltage systems.

**250.184 Solidly Grounded Neutral Systems.**

**(A) Neutral Conductor.** The minimum insulation level for neutral conductors of solidly grounded systems shall be 600 volts.

*Exception No. 1: Bare copper conductors shall be permitted to be used for the neutral of service entrances and the neutral of direct-buried portions of feeders.*

*Exception No. 2: Bare conductors shall be permitted for the neutral of overhead portions installed outdoors.*

FPN: See 225.4 for conductor covering where within 3.0 m (10 ft) of any building or other structure.

**(B) Multiple Grounding.** The neutral of a solidly grounded neutral system shall be permitted to be grounded at more than one point. Grounding shall be permitted at one or more of the following locations:

(1) Transformers supplying conductors to a building or other structure
(2) Underground circuits where the neutral is exposed
(3) Overhead circuits installed outdoors

**(C) Neutral Grounding Conductor.** The neutral grounding conductor shall be permitted to be a bare conductor if isolated from phase conductors and protected from physical damage.

**(D) Multigrounded Neutral Conductor.** Where a multigrounded neutral system is used, the following shall apply:

(1) The multigrounded neutral conductor shall be of sufficient ampacity for the load imposed on the conductor but not less than 33⅓ percent of the ampacity of the phase conductors.

*Exception: In industrial and commercial premises under engineering supervision, it shall be permissible to size the ampacity of the neutral conductor to not less than 20 percent of the ampacity of the phase conductor.*

(2) The multigrounded neutral conductor shall be grounded at each transformer and at other additional locations by connection to a made or existing electrode.

(3) At least one grounding electrode shall be installed and connected to the multigrounded neutral circuit conductor every 400 m (1300 ft).

(4) The maximum distance between any two adjacent electrodes shall not be more than 400 m (1300 ft).

(5) In a multigrounded shielded cable system, the shielding shall be grounded at each cable joint that is exposed to personnel contact.

**250.186 Impedance Grounded Neutral Systems.** Impedance grounded neutral systems in which a grounding impedance, usually a resistor, limits the ground-fault current, shall be permitted where all of the following conditions are met.

(1) The conditions of maintenance and supervision ensure that only qualified persons will service the installation.
(2) Ground detectors are installed on the system.
(3) Line-to-neutral loads are not served.

Impedance grounded neutral systems shall comply with the provisions of 250.186(A) through (D).

**(A) Location.** The grounding impedance shall be inserted in the grounding conductor between the grounding electrode of the supply system and the neutral point of the supply transformer or generator.

**(B) Identified and Insulated.** The neutral conductor of an impedance grounded neutral system shall be identified, as well as fully insulated with the same insulation as the phase conductors.

**(C) System Neutral Connection.** The system neutral shall not be connected to ground, except through the neutral grounding impedance.

**(D) Equipment Grounding Conductors.** Equipment grounding conductors shall be permitted to be bare and shall be electrically connected to the ground bus and grounding electrode conductor.

**250.188 Grounding of Systems Supplying Portable or Mobile Equipment.** Systems supplying portable or mobile high-voltage equipment, other than substations installed on a temporary basis, shall comply with 250.188(A) through (F).

**(A) Portable or Mobile Equipment.** Portable or mobile high-voltage equipment shall be supplied from a system having its neutral grounded through an impedance. Where a delta-connected high-voltage system is used to supply portable or mobile equipment, a system neutral shall be derived.

**(B) Exposed Non–Current-Carrying Metal Parts.** Exposed non–current-carrying metal parts of portable or mobile equipment shall be connected by an equipment grounding conductor to the point at which the system neutral impedance is grounded.

**(C) Ground-Fault Current.** The voltage developed between the portable or mobile equipment frame and ground by the flow of maximum ground-fault current shall not exceed 100 volts.

**(D) Ground-Fault Detection and Relaying.** Ground-fault detection and relaying shall be provided to automatically de-energize any high-voltage system component that has developed a ground fault. The continuity of the equipment grounding conductor shall be continuously monitored so as to de-energize automatically the high-voltage circuit to the portable or mobile equipment upon loss of continuity of the equipment grounding conductor.

**(E) Isolation.** The grounding electrode to which the portable or mobile equipment system neutral impedance is connected shall be isolated from and separated in the ground by at least 6.0 m (20 ft) from any other system or equipment grounding electrode, and there shall be no direct connection between the grounding electrodes, such as buried pipe and fence, and so forth.

**(F) Trailing Cable and Couplers.** High-voltage trailing cable and couplers for interconnection of portable or mobile equipment shall meet the requirements of Part III of Article 400 for cables and 490.55 for couplers.

**250.190 Grounding of Equipment.** All non–current-carrying metal parts of fixed, portable, and mobile equipment and associated fences, housings, enclosures, and supporting structures shall be grounded.

*Exception: Where isolated from ground and located so as to prevent any person who can make contact with ground from contacting such metal parts when the equipment is energized.*

Grounding conductors not an integral part of a cable assembly shall not be smaller than 6 AWG copper or 4 AWG aluminum.

FPN: See 250.110, Exception No. 2, for pole-mounted distribution apparatus.

# ARTICLE 280
# Surge Arresters

## I. General

**280.1 Scope.** This article covers general requirements, installation requirements, and connection requirements for surge arresters installed on premises wiring systems.

**280.2 Definition.**

**Surge Arrester.** A protective device for limiting surge voltages by discharging or bypassing surge current, and it also prevents continued flow of follow current while remaining capable of repeating these functions.

**280.3 Number Required.** Where used at a point on a circuit, a surge arrester shall be connected to each ungrounded conductor. A single installation of such surge arresters shall be permitted to protect a number of interconnected circuits, provided that no circuit is exposed to surges while disconnected from the surge arresters.

**280.4 Surge Arrester Selection.**

**(A) Circuits of Less Than 1000 Volts.** The rating of the surge arrester shall be equal to or greater than the maximum continuous phase-to-ground power frequency voltage available at the point of application.

Surge arresters installed on circuits of less than 1000 volts shall be listed for the purpose.

**(B) Circuits of 1 kV and Over — Silicon Carbide Types.** The rating of a silicon carbide-type surge arrester shall be not less than 125 percent of the maximum continuous phase-to-ground voltage available at the point of application.

FPN No. 1: For further information on surge arresters, see ANSI/IEEE C62.1-1989, *Standard for Gapped Silicon-Carbide Surge Arresters for AC Power Circuits*; ANSI/IEEE C62.2-1987, *Guide for the Application of Gapped Silicon-Carbide Surge Arresters for Alternating-Current Systems*; ANSI/IEEE C62.11-1993, *Standard for Metal-Oxide Surge Arresters for Alternating-Current Power Circuits*; and ANSI/IEEE C62.22-1991, *Guide for the Application of Metal-Oxide Surge Arresters for Alternating-Current Systems*.

FPN No. 2: The selection of a properly rated metal oxide arrester is based on considerations of maximum continuous operating voltage and the magnitude and duration of over-voltages at the arrester location as affected by phase-to-ground faults, system grounding techniques, switching surges, and other causes. See the manufacturer's application rules for selection of the specific arrester to be used at a particular location.

## II. Installation

**280.11 Location.** Surge arresters shall be permitted to be located indoors or outdoors. Surge arresters shall be made inaccessible to unqualified persons, unless listed for installation in accessible locations.

**280.12 Routing of Surge Arrester Connections.** The conductor used to connect the surge arrester to line or bus and to ground shall not be any longer than necessary and shall avoid unnecessary bends.

## III. Connecting Surge Arresters

**280.21 Installed at Services of Less Than 1000 Volts.** Line and ground connecting conductors shall not be smaller than 14 AWG copper or 12 AWG aluminum. The arrester grounding conductor shall be connected to one of the following:

(1) Grounded service conductor
(2) Grounding electrode conductor
(3) Grounding electrode for the service
(4) Equipment grounding terminal in the service equipment

**280.22 Installed on the Load Side Services of Less Than 1000 Volts.** Line and ground connecting conductors shall not be smaller than 14 AWG copper or 12 AWG aluminum. A surge arrester shall be permitted to be connected between any two conductors — ungrounded conductor(s), grounded conductor, grounding conductor. The grounded conductor and the grounding conductor shall be interconnected only by the normal operation of the surge arrester during a surge.

**280.23 Circuits of 1 kV and Over — Surge-Arrester Conductors.** The conductor between the surge arrester and the line and the surge arrester and the grounding connection shall not be smaller than 6 AWG copper or aluminum.

**280.24 Circuits of 1 kV and Over — Interconnections.** The grounding conductor of a surge arrester protecting a transformer that supplies a secondary distribution system shall be interconnected as specified in 280.24(A), (B), or (C).

**(A) Metallic Interconnections.** A metallic interconnection shall be made to the secondary grounded circuit conductor or the secondary circuit grounding conductor provided that, in addition to the direct grounding connection at the surge arrester, the following occurs:

(1) The grounded conductor of the secondary has elsewhere a grounding connection to a continuous metal underground water piping system. However, in urban water-pipe areas where there are at least four water-pipe connections on the neutral and not fewer than four such connections in each mile of neutral, the metallic interconnection shall be permitted to be made to the secondary neutral with omission of the direct grounding connection at the surge arrester.
(2) The grounded conductor of the secondary system is a part of a multiground neutral system of which the primary neutral has at least four ground connections in each mile of line in addition to a ground at each service.

**(B) Through Spark Gap or Device.** Where the surge arrester grounding conductor is not connected as in 280.24(A) or where the secondary is not grounded as in 280.24(A) but is otherwise grounded as in 250.52, an interconnection shall be made through a spark gap or listed device as follows:

(1) For ungrounded or unigrounded primary systems, the spark gap or listed device shall have a 60-Hz breakdown voltage of at least twice the primary circuit voltage but not necessarily more than 10 kV, and there shall be at least one other ground on the grounded conductor of the secondary that is not less than 6.0 m (20 ft) distant from the surge arrester grounding electrode.
(2) For multigrounded neutral primary systems, the spark gap or listed device shall have a 60-Hz breakdown of not more than 3 kV, and there shall be at least one other ground on the grounded conductor of the secondary that is not less than 6.0 m (20 ft) distant from the surge arrester grounding electrode.

**(C) By Special Permission.** An interconnection of the surge arrester ground and the secondary neutral, other than as provided in 280.24(A) or (B), shall be permitted to be made only by special permission.

**280.25 Grounding.** Except as indicated in this article, surge arrester grounding connections shall be made as specified in Article 250. Grounding conductors shall not be run in metal enclosures unless bonded to both ends of such enclosure.

## ARTICLE 285
## Transient Voltage Surge Suppressors: TVSSs

### I. General

**285.1 Scope.** This article covers general requirements, installation requirements, and connection requirements for transient voltage surge suppressors (TVSS) permanently installed on premises wiring systems.

**285.2 Definition.**

**Transient Voltage Surge Suppressor (TVSS).** A protective device for limiting transient voltages by diverting or limiting surge current; it also prevents continued flow of follow current while remaining capable of repeating these functions.

**285.3 Uses Not Permitted.** A TVSS shall not be used in the following:

(1) Circuits exceeding 600 volts
(2) Ungrounded electrical systems as permitted in 250.21
(3) Where the rating of the TVSS is less than the maximum continuous phase-to-ground power frequency voltage available at the point of application

> FPN: For further information on TVSSs, see NEMA LS 1-1992, *Standard for Low Voltage Surge Suppression Devices*. The selection of a properly rated TVSS is based on criteria such as maximum continuous operating voltage, the magnitude and duration of overvoltages at the suppressor location as affected by phase-to-ground faults, system grounding techniques, and switching surges.

**285.4 Number Required.** Where used at a point on a circuit, the TVSS shall be connected to each ungrounded conductor.

**285.5 Listing.** A TVSS shall be a listed device.

**285.6 Short Circuit Current Rating.** The TVSS shall be marked with a short circuit current rating and shall not be installed at a point on the system where the available fault current is in excess of that rating. This marking requirement shall not apply to receptacles.

### II. Installation

**285.11 Location.** TVSSs shall be permitted to be located indoors or outdoors and shall be made inaccessible to unqualified persons, unless listed for installation in accessible locations.

**285.12 Routing of Connections.** The conductors used to connect the the TVSS to the line or bus and to ground shall not be any longer than necessary and shall avoid unnecessary bends.

### III. Connecting Transient Voltage Surge Suppressors

**285.21 Connection.** Where a TVSS is installed, it shall be connected as follows.

**(A)) Location.**

**(1)) Service Supplied Building or Structure.** The transient voltage surge suppressor shall be connected on the load side of a service disconnect overcurrent device required in 230.91.

**(2)) Feeder Supplied Building or Structure.** The transient voltage surge suppressor shall be connected on the load side of the first overcurrent device at the building or structure.

*Exception to (1) and (2): Where the TVSS is also listed as a surge arrester, the connection shall be as permitted by Article 280.*

**(3)) Separately Derived System.** The TVSS shall be connected on the load side of the first overcurrent device in a separately derived system.

**(B)) Conductor Size.** Line and ground connecting conductors shall not be smaller than 14 AWG copper or 12 AWG aluminum.

**(C)) Connection Between Conductors.** A TVSS shall be permitted to be connected between any two conductors — ungrounded conductor(s), grounded conductor, grounding conductor. The grounded conductor and the grounding conductor shall be interconnected only by the normal operation of the TVSS during a surge.

**285.25 Grounding.** Grounding conductors shall not be run in metal enclosures unless bonded to both ends of such enclosure.

## Chapter 3   Wiring Methods and Materials

### ARTICLE 300
### Wiring Methods

## I. General Requirements

### 300.1 Scope.

**(A) All Wiring Installations.** This article covers wiring methods for all wiring installations unless modified by other articles.

**(B) Integral Parts of Equipment.** The provisions of this article are not intended to apply to the conductors that form an integral part of equipment, such as motors, controllers, motor control centers, or factory assembled control equipment or listed utilization equipment.

**(C) Metric Designators and Trade Sizes.** Metric designators and trade sizes for conduit, tubing, and associated fittings and accessories shall be as designated in Table 300.1(C).

**Table 300.1(C)  Metric Designator and Trade Sizes**

| Metric Designator | Trade Size |
|---|---|
| 12 | ⅜ |
| 16 | ½ |
| 21 | ¾ |
| 27 | 1 |
| 35 | 1¼ |
| 41 | 1½ |
| 53 | 2 |
| 63 | 2½ |
| 78 | 3 |
| 91 | 3½ |
| 103 | 4 |
| 129 | 5 |
| 155 | 6 |

Note: The metric designators and trade sizes are for identification purposes only and are not actual dimensions.

### 300.2 Limitations.

**(A) Voltage.** Wiring methods specified in Chapter 3 shall be used for 600 volts, nominal, or less where not specifically limited in some section of Chapter 3. They shall be permitted for over 600 volts, nominal, where specifically permitted elsewhere in this *Code*.

**(B) Temperature.** Temperature limitation of conductors shall be in accordance with 310.10.

### 300.3 Conductors.

**(A) Single Conductors.** Single conductors specified in Table 310.13 shall only be installed where part of a recognized wiring method of Chapter 3.

**(B) Conductors of the Same Circuit.** All conductors of the same circuit and, where used, the grounded conductor and all equipment grounding conductors and bonding conductors shall be contained within the same raceway, auxiliary gutter, cable tray, cablebus assembly, trench, cable, or cord, unless otherwise permitted in accordance with 300.3(B)(1) through (4).

**(1) Paralleled Installations.** Conductors shall be permitted to be run in parallel in accordance with the provisions of 310.4. The requirement to run all circuit conductors within the same raceway, auxiliary gutter, cable tray, trench, cable, or cord shall apply separately to each portion of the paralleled installation, and the equipment grounding conductors shall comply with the provisions of 250.122. Parallel runs in cable tray shall comply with the provisions of 392.8(D).

*Exception: Conductors installed in nonmetallic raceways run underground shall be permitted to be arranged as isolated phase installations. The raceways shall be installed in close proximity, and the conductors shall comply with the provisions of 300.20(B).*

**(2) Grounding and Bonding Conductors.** Equipment grounding conductors shall be permitted to be installed outside a raceway or cable assembly where in accordance with the provisions of 250.130(C) for certain existing installations or in accordance with 250.134(B), Exception No. 2, for dc circuits. Equipment bonding conductors shall be permitted to be installed on the outside of raceways in accordance with 250.102(E).

**(3) Nonferrous Wiring Methods.** Conductors in wiring methods with a nonmetallic or other nonmagnetic sheath, where run in different raceways, auxiliary gutters, cable trays, trenches, cables, or cords, shall comply with the provisions of 300.20(B). Conductors in single-conductor Type MI cable with a nonmagnetic sheath shall comply with the provisions of 332.31. Conductors of single-conductor–type MC cable with a nonmagnetic sheath shall comply with the provisions of 330.31, 330.116, and 300.20(B).

**(4) Enclosures.** Where an auxiliary gutter runs between a column-width panelboard and a pull box, and the pull box includes neutral terminations, the neutral conductors of circuits supplied from the panelboard shall be permitted to originate in the pull box.

**(C) Conductors of Different Systems.**

**(1) 600 Volts, Nominal, or Less.** Conductors of circuits rated 600 volts, nominal, or less, ac circuits, and dc circuits shall be permitted to occupy the same equipment wiring enclosure, cable, or raceway. All conductors shall have an insulation rating equal to at least the maximum circuit voltage applied to any conductor within the enclosure, cable, or raceway.

*Exception: For solar photovoltaic systems in accordance with 690.4(B).*

FPN: See 725.55(A) for Class 2 and Class 3 circuit conductors.

**(2) Over 600 Volts, Nominal.** Conductors of circuits rated over 600 volts, nominal, shall not occupy the same equipment wiring enclosure, cable, or raceway with conductors of circuits rated 600 volts, nominal, or less unless otherwise permitted in (a) through (e).

(a) Secondary wiring to electric-discharge lamps of 1000 volts or less, if insulated for the secondary voltage involved, shall be permitted to occupy the same luminaire (fixture), sign, or outline lighting enclosure as the branch-circuit conductors.

(b) Primary leads of electric-discharge lamp ballasts, insulated for the primary voltage of the ballast, where contained within the individual wiring enclosure, shall be permitted to occupy the same luminaire (fixture), sign, or outline lighting enclosure as the branch-circuit conductors.

(c) Excitation, control, relay, and ammeter conductors used in connection with any individual motor or starter shall be permitted to occupy the same enclosure as the motor-circuit conductors.

(d) In motors, switchgear and control assemblies, and similar equipment, conductors of different voltage ratings shall be permitted.

(e) In manholes, if the conductors of each system are permanently and effectively separated from the conductors of the other systems and securely fastened to racks, insulators, or other approved supports, conductors of different voltage ratings shall be permitted.

Conductors having nonshielded insulation and operating at different voltage levels shall not occupy the same enclosure, cable, or raceway.

**300.4 Protection Against Physical Damage.** Where subject to physical damage, conductors shall be adequately protected.

**(A) Cables and Raceways Through Wood Members.**

**(1) Bored Holes.** In both exposed and concealed locations, where a cable- or raceway-type wiring method is installed through bored holes in joists, rafters, or wood members, holes shall be bored so that the edge of the hole is not less than 32 mm (1¼ in.) from the nearest edge of the wood member. Where this distance cannot be maintained, the cable or raceway shall be protected from penetration by screws or nails by a steel plate or bushing, at least 1.6 mm (¹⁄₁₆ in.) thick, and of appropriate length and width installed to cover the area of the wiring.

*Exception: Steel plates shall not be required to protect rigid metal conduit, intermediate metal conduit, rigid nonmetallic conduit, or electrical metallic tubing.*

**(2) Notches in Wood.** Where there is no objection because of weakening the building structure, in both exposed and concealed locations, cables or raceways shall be permitted to be laid in notches in wood studs, joists, rafters, or other wood members where the cable or raceway at those points is protected against nails or screws by a steel plate at least 1.6 mm (¹⁄₁₆ in.) thick installed before the building finish is applied.

*Exception: Steel plates shall not be required to protect rigid metal conduit, intermediate metal conduit, rigid nonmetallic conduit, or electrical metallic tubing.*

**(B) Nonmetallic-Sheathed Cables and Electrical Nonmetallic Tubing Through Metal Framing Members.**

**(1) Nonmetallic-Sheathed Cable.** In both exposed and concealed locations where nonmetallic-sheathed cables pass through either factory or field punched, cut, or drilled slots or holes in metal members, the cable shall be protected by listed bushings or listed grommets covering all metal edges that are securely fastened in the opening prior to installation of the cable.

**(2) Nonmetallic-Sheathed Cable and Electrical Nonmetallic Tubing.** Where nails or screws are likely to penetrate nonmetallic-sheathed cable or electrical nonmetallic tubing, a steel sleeve, steel plate, or steel clip not less than 1.6 mm (¹⁄₁₆ in.) in thickness shall be used to protect the cable or tubing.

**(C) Cables Through Spaces Behind Panels Designed to Allow Access.** Cables or raceway-type wiring methods, installed behind panels designed to allow access, shall be supported according to their applicable articles.

**(D) Cables and Raceways Parallel to Framing Members.** In both exposed and concealed locations, where a cable- or raceway-type wiring method is installed parallel to framing members, such as joists, rafters, or studs, the cable or raceway shall be installed and supported so that the nearest outside surface of the cable or raceway is not less than 32 mm (1¼ in.) from the nearest edge of the framing member where nails or screws are likely to penetrate.

Where this distance cannot be maintained, the cable or raceway shall be protected from penetration by nails or screws by a steel plate, sleeve, or equivalent at least 1.6 mm ($\frac{1}{16}$ in.) thick.

*Exception No. 1: Steel plates, sleeves, or the equivalent shall not be required to protect rigid metal conduit, intermediate metal conduit, rigid nonmetallic conduit, or electrical metallic tubing.*

*Exception No. 2: For concealed work in finished buildings, or finished panels for prefabricated buildings where such supporting is impracticable, it shall be permissible to fish the cables between access points.*

**(E) Cables and Raceways Installed in Shallow Grooves.** Cable- or raceway-type wiring methods installed in a groove, to be covered by wallboard, siding, paneling, carpeting, or similar finish, shall be protected by 1.6 mm ($\frac{1}{16}$ in.) thick steel plate, sleeve, or equivalent or by not less than 32 mm ($1\frac{1}{4}$ in.) free space for the full length of the groove in which the cable or raceway is installed.

*Exception: Steel plates, sleeves, or the equivalent shall not be required to protect rigid metal conduit, intermediate metal conduit, rigid nonmetallic conduit, or electrical metallic tubing.*

**(F) Insulated Fittings.** Where raceways containing ungrounded conductors 4 AWG or larger enter a cabinet, box enclosure, or raceway, the conductors shall be protected by a substantial fitting providing a smoothly rounded insulating surface, unless the conductors are separated from the fitting or raceway by substantial insulating material that is securely fastened in place.

*Exception: Where threaded hubs or bosses that are an integral part of a cabinet, box enclosure, or raceway provide a smoothly rounded or flared entry for conductors.*

Conduit bushings constructed wholly of insulating material shall not be used to secure a fitting or raceway. The insulating fitting or insulating material shall have a temperature rating not less than the insulation temperature rating of the installed conductors.

**300.5 Underground Installations.**

**(A) Minimum Cover Requirements.** Direct-buried cable or conduit or other raceways shall be installed to meet the minimum cover requirements of Table 300.5.

**(B) Grounding.** All underground installations shall be grounded and bonded in accordance with Article 250.

**(C) Underground Cables Under Buildings.** Underground cable installed under a building shall be in a raceway that is extended beyond the outside walls of the building.

**(D) Protection from Damage.** Direct-buried conductors and cables shall be protected from damage in accordance with (1) through (5).

**(1) Emerging from Grade.** Direct-buried conductors and enclosures emerging from grade shall be protected by enclosures or raceways extending from the minimum cover distance required by 300.5(A) below grade to a point at least 2.5 m (8 ft) above finished grade. In no case shall the protection be required to exceed 450 mm (18 in.) below finished grade.

**(2) Conductors Entering Buildings.** Conductors entering a building shall be protected to the point of entrance.

**(3) Service Conductors.** Underground service conductors that are not encased in concrete and that are buried 450 mm (18 in.) or more below grade shall have their location identified by a warning ribbon that is placed in the trench at least 300 mm (12 in.) above the underground installation.

**(4) Enclosure or Raceway Damage.** Where the enclosure or raceway is subject to physical damage, the conductors shall be installed in rigid metal conduit, intermediate metal conduit, Schedule 80 rigid nonmetallic conduit, or equivalent.

**(5) Listing.** Cables and insulated conductors installed in enclosures or raceways in underground installations shall be listed for use in wet locations.

**(E) Splices and Taps.** Direct-buried conductors or cables shall be permitted to be spliced or tapped without the use of splice boxes. The splices or taps shall be made in accordance with 110.14(B).

**(F) Backfill.** Backfill that contains large rocks, paving materials, cinders, large or sharply angular substances, or corrosive material shall not be placed in an excavation where materials may damage raceways, cables, or other substructures or prevent adequate compaction of fill or contribute to corrosion of raceways, cables, or other substructures.

Where necessary to prevent physical damage to the raceway or cable, protection shall be provided in the form of granular or selected material, suitable running boards, suitable sleeves, or other approved means.

**(G) Raceway Seals.** Conduits or raceways through which moisture may contact energized live parts shall be sealed or plugged at either or both ends.

FPN: Presence of hazardous gases or vapors may also necessitate sealing of underground conduits or raceways entering buildings.

**(H) Bushing.** A bushing, or terminal fitting, with an integral bushed opening shall be used at the end of a conduit or other raceway that terminates underground where the con-

| Table 300.5 Minimum Cover Requirements, 0 to 600 Volts, Nominal, Burial in Millimeters (Inches) |

| Location of Wiring Method or Circuit | Column 1 Direct Burial Cables or Conductors | | Column 2 Rigid Metal Conduit or Intermediate Metal Conduit | | Column 3 Nonmetallic Raceways Listed for Direct Burial Without Concrete Encasement or Other Approved Raceways | | Column 4 Residential Branch Circuits Rated 120 Volts or Less with GFCI Protection and Maximum Overcurrent Protection of 20 Amperes | | Column 5 Circuits for Control of Irrigation and Landscape Lighting Limited to Not More Than 30 Volts and Installed with Type UF or in Other Identified Cable or Raceway | |
|---|---|---|---|---|---|---|---|---|---|---|
| | mm | in. | mm | in. | mm | in. | mm | in. | mm | in. |
| All locations not specified below | 600 | 24 | 150 | 6 | 450 | 18 | 300 | 12 | 150 | 6 |
| In trench below 50-mm (2-in.) thick concrete or equivalent | 450 | 18 | 150 | 6 | 300 | 12 | 150 | 6 | 150 | 6 |
| Under a building | 0 (in raceway only) | 0 | 0 | 0 | 0 | 0 | 0 (in raceway only) | 0 | 0 (in raceway only) | 0 |
| Under minimum of 102-mm (4-in.) thick concrete exterior slab with no vehicular traffic and the slab extending not less than 152 mm (6 in.) beyond the underground installation | 450 | 18 | 100 | 4 | 100 | 4 | 150 (direct burial) 100 (in raceway) | 6 4 | 150 | 6 |
| Under streets, highways, roads, alleys, driveways, and parking lots | 600 | 24 | 600 | 24 | 600 | 24 | 600 | 24 | 600 | 24 |
| One- and two-family dwelling driveways and outdoor parking areas, and used only for dwelling-related purposes | 450 | 18 | 450 | 18 | 450 | 18 | 300 | 12 | 450 | 18 |
| In or under airport runways, including adjacent areas where trespassing prohibited | 450 | 18 | 450 | 18 | 450 | 18 | 450 | 18 | 450 | 18 |

Notes:
1. Cover is defined as the shortest distance in millimeters (inches) measured between a point on the top surface of any direct-buried conductor, cable, conduit, or other raceway and the top surface of finished grade, concrete, or similar cover.
2. Raceways approved for burial only where concrete encased shall require concrete envelope not less than 50 mm (2 in.) thick.
3. Lesser depths shall be permitted where cables and conductors rise for terminations or splices or where access is otherwise required.

4. Where one of the wiring method types listed in Columns 1–3 is used for one of the circuit types in Columns 4 and 5, the shallower depth of burial shall be permitted.
5. Where solid rock prevents compliance with the cover depths specified in this table, the wiring shall be installed in metal or nonmetallic raceway permitted for direct burial. The raceways shall be covered by a minimum of 50 mm (2 in.) of concrete extending down to rock.

ductors or cables emerge as a direct burial wiring method. A seal incorporating the physical protection characteristics of a bushing shall be permitted to be used in lieu of a bushing.

**(I) Conductors of the Same Circuit.** All conductors of the same circuit and, where used, the grounded conductor and all equipment grounding conductors shall be installed in the same raceway or cable or shall be installed in close proximity in the same trench.

*Exception No. 1: Conductors in parallel in raceways or cables shall be permitted, but each raceway or cable shall contain all conductors of the same circuit including grounding conductors.*

*Exception No. 2: Isolated phase, polarity, grounded conductor, and equipment grounding and bonding conductor installations shall be permitted in nonmetallic raceways or cables with a nonmetallic covering or nonmagnetic sheath in close proximity where conductors are paralleled as permitted in 310.4, and where the conditions of 300.20(B) are met.*

**(J) Ground Movement.** Where direct-buried conductors, raceways, or cables are subject to movement by settlement or frost, direct-buried conductors, raceways, or cables shall be arranged to prevent damage to the enclosed conductors or to equipment connected to the raceways.

> FPN: This section recognizes "S" loops in underground direct burial to raceway transitions, expansion fittings in raceway risers to fixed equipment, and, generally, the provision of flexible connections to equipment subject to settlement or frost heaves.

**(K) Directional Boring.** Cables or raceways installed using directional boring equipment shall be approved for the purpose.

**300.6 Protection Against Corrosion.** Metal raceways, cable trays, cablebus, auxiliary gutters, cable armor, boxes, cable sheathing, cabinets, elbows, couplings, fittings, supports, and support hardware shall be of materials suitable for the environment in which they are to be installed.

**(A) General.** Ferrous raceways, cable trays, cablebus, auxiliary gutters, cable armor, boxes, cable sheathing, cabinets, metal elbows, couplings, fittings, supports, and support hardware shall be suitably protected against corrosion inside and outside (except threads at joints) by a coating of approved corrosion-resistant material such as zinc, cadmium, or enamel. Where protected from corrosion solely by enamel, they shall not be used outdoors or in wet locations as described in 300.6(C). Where boxes or cabinets have an approved system of organic coatings and are marked "Raintight," "Rainproof," or "Outdoor Type," they shall be permitted outdoors. Where corrosion protection is

necessary and the conduit is threaded in the field, the threads shall be coated with an approved electrically conductive, corrosion-resistant compound.

**(B) In Concrete or in Direct Contact with the Earth.** Ferrous or nonferrous metal raceways, cable armor, boxes, cable sheathing, cabinets, elbows, couplings, fittings, supports, and support hardware shall be permitted to be installed in concrete or in direct contact with the earth, or in areas subject to severe corrosive influences where made of material judged suitable for the condition, or where provided with corrosion protection approved for the condition.

**(C) Indoor Wet Locations.** In portions of dairy processing facilities, laundries, canneries, and other indoor wet locations, and in locations where walls are frequently washed or where there are surfaces of absorbent materials, such as damp paper or wood, the entire wiring system, where installed exposed, including all boxes, fittings, conduits, and cable used therewith, shall be mounted so that there is at least a 6-mm (¼-in.) airspace between it and the wall or supporting surface.

*Exception: Nonmetallic raceways, boxes, and fittings shall be permitted to be installed without the airspace on a concrete, masonry, tile, or similar surface.*

> FPN: In general, areas where acids and alkali chemicals are handled and stored may present such corrosive conditions, particularly when wet or damp. Severe corrosive conditions may also be present in portions of meatpacking plants, tanneries, glue houses, and some stables; in installations immediately adjacent to a seashore and swimming pool areas; in areas where chemical deicers are used; and in storage cellars or rooms for hides, casings, fertilizer, salt, and bulk chemicals.

**300.7 Raceways Exposed to Different Temperatures.**

**(A) Sealing.** Where portions of a cable, raceway, or sleeve are known to be subjected to different temperatures and where condensation is known to be a problem, as in cold storage areas of buildings or where passing from the interior to the exterior of a building, the raceway or sleeve shall be filled with an approved material to prevent the circulation of warm air to a colder section of the raceway or sleeve. An explosionproof seal shall not be required for this purpose.

**(B) Expansion Fittings.** Raceways shall be provided with expansion fittings where necessary to compensate for thermal expansion and contraction.

> FPN: Table 352.44(A) provides the expansion information for polyvinyl chloride (PVC). A nominal number for steel conduit can be determined by multiplying the expansion length in this table by 0.20. The coefficient of expansion for steel electrical metallic tubing, intermediate metal conduit, and rigid conduit is $11.70 \times 10^{-6}$(0.0000117 mm per mm of

conduit for each °C in temperature change) [6.50 × 10⁻⁶ (0.0000065 in. per inch of conduit for each °F in temperature change)].

**300.8 Installation of Conductors with Other Systems.** Raceways or cable trays containing electric conductors shall not contain any pipe, tube, or equal for steam, water, air, gas, drainage, or any service other than electrical.

**300.10 Electrical Continuity of Metal Raceways and Enclosures.** Metal raceways, cable armor, and other metal enclosures for conductors shall be metallically joined together into a continuous electric conductor and shall be connected to all boxes, fittings, and cabinets so as to provide effective electrical continuity. Unless specifically permitted elsewhere in this *Code*, raceways and cable assemblies shall be mechanically secured to boxes, fittings, cabinets, and other enclosures.

*Exception No. 1: Short sections of raceways used to provide support or protection of cable assemblies from physical damage shall not be required to be made electrically continuous.*

*Exception No. 2: Equipment enclosures to be isolated, as permitted by 250.96(B), shall not be required to be metallically joined to the metal raceway.*

**300.11 Securing and Supporting.**

**(A) Secured in Place.** Raceways, cable assemblies, boxes, cabinets, and fittings shall be securely fastened in place. Support wires that do not provide secure support shall not be permitted as the sole support. Support wires and associated fittings that provide secure support and that are installed in addition to the ceiling grid support wires shall be permitted as the sole support. Where independent support wires are used, they shall be secured at both ends. Cables and raceways shall not be supported by ceiling grids.

**(1) Fire-Rated Assemblies.** Wiring located within the cavity of a fire-rated floor–ceiling or roof–ceiling assembly shall not be secured to, or supported by, the ceiling assembly, including the ceiling support wires. An independent means of secure support shall be provided. Where independent support wires are used, they shall be distinguishable by color, tagging, or other effective means from those that are part of the fire-rated design.

*Exception: The ceiling support system shall be permitted to support wiring and equipment that have been tested as part of the fire-rated assembly.*

FPN: One method of determining fire rating is testing in accordance with NFPA 251-1999, *Standard Methods of Tests of Fire Endurance of Building Construction and Materials.*

**(2) Non–Fire-Rated Assemblies.** Wiring located within the cavity of a non–fire-rated floor–ceiling or roof–ceiling assembly shall not be secured to, or supported by, the ceiling assembly, including the ceiling support wires. An independent means of secure support shall be provided.

*Exception: The ceiling support system shall be permitted to support branch-circuit wiring and associated equipment where installed in accordance with the ceiling system manufacturer's instructions.*

**(B) Raceways Used as Means of Support.** Raceways shall only be used as a means of support for other raceways, cables, or nonelectric equipment under the following conditions:

(1) Where the raceway or means of support is identified for the purpose; or
(2) Where the raceway contains power supply conductors for electrically controlled equipment and is used to support Class 2 circuit conductors or cables that are solely for the purpose of connection to the equipment control circuits; or
(3) Where the raceway is used to support boxes or conduit bodies in accordance with 314.23 or to support luminaires (fixtures) in accordance with 410.16(F)

**(C) Cables Not Used as Means of Support.** Cable wiring methods shall not be used as a means of support for other cables, raceways, or nonelectrical equipment.

**300.12 Mechanical Continuity — Raceways and Cables.** Metal or nonmetallic raceways, cable armors, and cable sheaths shall be continuous between cabinets, boxes, fittings, or other enclosures or outlets.

*Exception: Short sections of raceways used to provide support or protection of cable assemblies from physical damage shall not be required to be mechanically continuous.*

**300.13 Mechanical and Electrical Continuity — Conductors.**

**(A) General.** Conductors in raceways shall be continuous between outlets, boxes, devices, and so forth. There shall be no splice or tap within a raceway unless permitted by 300.15; 368.8(A); 376.56; 378.56; 384.56; 386.56; 388.56; or 390.6.

**(B) Device Removal.** In multiwire branch circuits, the continuity of a grounded conductor shall not depend on device connections such as lampholders, receptacles, and so forth, where the removal of such devices would interrupt the continuity.

**300.14 Length of Free Conductors at Outlets, Junctions, and Switch Points.** At least 150 mm (6 in.) of free conductor, measured from the point in the box where it emerges from its raceway or cable sheath, shall be left at each outlet, junction, and switch point for splices or the connection of luminaires (fixtures) or devices. Where the opening to an outlet, junction, or switch point is less than 200 mm (8 in.) in any dimension, each conductor shall be long enough to extend at least 75 mm (3 in.) outside the opening.

*Exception: Conductors that are not spliced or terminated at the outlet, junction, or switch point shall not be required to comply with 300.14.*

**300.15 Boxes, Conduit Bodies, or Fittings — Where Required.** A box shall be installed at each outlet and switch point for concealed knob-and-tube wiring.

Fittings and connectors shall be used only with the specific wiring methods for which they are designed and listed.

Where the wiring method is conduit, tubing, Type AC cable, Type MC cable, Type MI cable, nonmetallic-sheathed cable, or other cables, a box or conduit body complying with Article 314 shall be installed at each conductor splice point, outlet point, switch point, junction point, termination point, or pull point, unless otherwise permitted in 300.15(A) through (M).

**(A) Wiring Methods with Interior Access.** A box or conduit body shall not be required for each splice, junction, switch, pull, termination, or outlet points in wiring methods with removable covers, such as wireways, multioutlet assemblies, auxiliary gutters, and surface raceways. The covers shall be accessible after installation.

**(B) Equipment.** An integral junction box or wiring compartment as part of approved equipment shall be permitted in lieu of a box.

**(C) Protection.** A box or conduit body shall not be required where cables enter or exit from conduit or tubing that is used to provide cable support or protection against physical damage. A fitting shall be provided on the end(s) of the conduit or tubing to protect the cable from abrasion.

**(D) Type MI Cable.** A box or conduit body shall not be required where accessible fittings are used for straight-through splices in mineral-insulated metal-sheathed cable.

**(E) Integral Enclosure.** A wiring device with integral enclosure identified for the use, having brackets that securely fasten the device to walls or ceilings of conventional on-site frame construction, for use with nonmetallic-sheathed cable, shall be permitted in lieu of a box or conduit body.

FPN: See 334.30(C); 545.10; 550.15(I); 551.47(E), Exception No. 1; and 552.48(E), Exception No. 1.

**(F) Fitting.** A fitting identified for the use shall be permitted in lieu of a box or conduit body where conductors are not spliced or terminated within the fitting. The fitting shall be accessible after installation.

**(G) Direct-Buried Conductors.** As permitted in 300.5(E), a box or conduit body shall not be required for splices and taps in direct-buried conductors and cables.

**(H) Insulated Devices.** As permitted in 334.40(B), a box or conduit body shall not be required for insulated devices supplied by nonmetallic-sheathed cable.

**(I) Enclosures.** A box or conduit body shall not be required where a splice, switch, terminal, or pull point is in a cabinet or cutout box, in an enclosure for a switch or overcurrent device as permitted in 312.8, in a motor controller as permitted in 430.10(A), or in a motor control center.

**(J) Luminaires (Fixtures).** A box or conduit body shall not be required where a luminaire (fixture) is used as a raceway as permitted in 410.31 and 410.32.

**(K) Embedded.** A box or conduit body shall not be required for splices where conductors are embedded as permitted in 424.40, 424.41(D), 426.22(B), 426.24(A), and 427.19(A).

**(L) Manholes.** Where accessible only to qualified persons, a box or conduit body shall not be required for conductors in manholes, except where connecting to electrical equipment. The installation shall comply with the provisions of Part IV of Article 314.

**(M) Closed Loop.** A box shall not be required with a closed-loop power distribution system where a device identified and listed as suitable for installation without a box is used.

**300.16 Raceway or Cable to Open or Concealed Wiring.**

**(A) Box or Fitting.** A box or terminal fitting having a separately bushed hole for each conductor shall be used wherever a change is made from conduit, electrical metallic tubing, electrical nonmetallic tubing, nonmetallic-sheathed cable, Type AC cable, Type MC cable, or mineral-insulated, metal-sheathed cable and surface raceway wiring to open wiring or to concealed knob-and-tube wiring. A fitting used for this purpose shall contain no taps or splices and shall not be used at luminaire (fixture) outlets.

**(B) Bushing.** A bushing shall be permitted in lieu of a box or terminal where the conductors emerge from a raceway and enter or terminate at equipment, such as open switchboards, unenclosed control equipment, or similar equipment. The bushing shall be of the insulating type for other than lead-sheathed conductors.

**300.17 Number and Size of Conductors in Raceway.** The number and size of conductors in any raceway shall not be more than will permit dissipation of the heat and ready installation or withdrawal of the conductors without damage to the conductors or to their insulation.

> FPN: See the following sections of this *Code*: intermediate metal conduit, 342.22; rigid metal conduit, 344.22; flexible metal conduit, 348.22; liquidtight flexible metal conduit, 350.22; rigid nonmetallic conduit, 352.22; liquidtight nonmetallic flexible conduit, 356.22; electrical metallic tubing, 358.22; flexible metallic tubing, 360.22; electrical nonmetallic tubing, 362.22; cellular concrete floor raceways, 372.11; cellular metal floor raceways, 374.5; metal wireways, 376.22; nonmetallic wireways, 378.22; surface metal raceways, 386.22; surface nonmetallic raceways 388.22; underfloor raceways, 390.5; fixture wire, 402.7; theaters, 520.6; signs, 600.31(C); elevators, 620.33; audio signal processing, amplification, and reproduction equipment, 640.23(A) and 640.24; Class 1, Class 2, and Class 3 circuits, Article 725; fire alarm circuits, Article 760; and optical fiber cables and raceways, Article 770.

**300.18 Raceway Installations.**

**(A) Complete Runs.** Raceways, other than busways or exposed raceways having hinged or removable covers, shall be installed complete between outlet, junction, or splicing points prior to the installation of conductors. Where required to facilitate the installation of utilization equipment, the raceway shall be permitted to be initially installed without a terminating connection at the equipment. Prewired raceway assemblies shall be permitted only where specifically permitted in this *Code* for the applicable wiring method.

**(B) Welding.** Metal raceways shall not be supported, terminated, or connected by welding to the raceway unless specifically designed to be or otherwise specifically permitted to be in this *Code*.

**300.19 Supporting Conductors in Vertical Raceways.**

**(A) Spacing Intervals — Maximum.** Conductors in vertical raceways shall be supported if the vertical rise exceeds the values in Table 300.19(A). One cable support shall be provided at the top of the vertical raceway or as close to the top as practical. Intermediate supports shall be provided as necessary to limit supported conductor lengths to not greater than those values specified in Table 300.19(A).

*Exception: Steel wire armor cable shall be supported at the top of the riser with a cable support that clamps the steel wire armor. A safety device shall be permitted at the lower end of the riser to hold the cable in the event there is slippage of the cable in the wire-armored cable support. Additional wedge-type supports shall be permitted to relieve the strain on the equipment terminals caused by expansion of the cable under load.*

**(B) Support Methods.** One of the following methods of support shall be used.

(1) By clamping devices constructed of or employing insulating wedges inserted in the ends of the raceways. Where clamping of insulation does not adequately support the cable, the conductor also shall be clamped.

(2) By inserting boxes at the required intervals in which insulating supports are installed and secured in a satisfactory manner to withstand the weight of the conductors attached thereto, the boxes being provided with covers.

(3) In junction boxes, by deflecting the cables not less than 90 degrees and carrying them horizontally to a distance not less than twice the diameter of the cable, the cables being carried on two or more insulating supports and additionally secured thereto by tie wires if desired. Where this method is used, cables shall be supported at intervals not greater than 20 percent of those mentioned in the preceding tabulation.

(4) By a method of equal effectiveness.

**300.20 Induced Currents in Metal Enclosures or Metal Raceways.**

**(A) Conductors Grouped Together.** Where conductors carrying alternating current are installed in metal enclo-

**Table 300.19(A) Spacings for Conductor Supports**

| Size of Wire | Support of Conductors in Vertical Raceways | Aluminum or Copper-Clad Aluminum | | Copper | |
|---|---|---|---|---|---|
| | | m | ft | m | ft |
| 18 AWG through 8 AWG | Not greater than | 30 | 100 | 30 | 100 |
| 6 AWG through 1/0 AWG | Not greater than | 60 | 200 | 30 | 100 |
| 2/0 AWG through 4/0 AWG | Not greater than | 55 | 180 | 25 | 80 |
| Over 4/0 AWG through 350 kcmil | Not greater than | 41 | 135 | 18 | 60 |
| Over 350 kcmil through 500 kcmil | Not greater than | 36 | 120 | 15 | 50 |
| Over 500 kcmil through 750 kcmil | Not greater than | 28 | 95 | 12 | 40 |
| Over 750 kcmil | Not greater than | 26 | 85 | 11 | 35 |

sures or metal raceways, they shall be arranged so as to avoid heating the surrounding metal by induction. To accomplish this, all phase conductors and, where used, the grounded conductor and all equipment grounding conductors shall be grouped together.

*Exception No. 1: Equipment grounding conductors for certain existing installations shall be permitted to be installed separate from their associated circuit conductors where run in accordance with the provisions of 250.130(C).*

*Exception No. 2: A single conductor shall be permitted to be installed in a ferromagnetic enclosure and used for skin-effect heating in accordance with the provisions of 426.42 and 427.47.*

**(B) Individual Conductors.** Where a single conductor carrying alternating current passes through metal with magnetic properties, the inductive effect shall be minimized by (1) cutting slots in the metal between the individual holes through which the individual conductors pass or (2) passing all the conductors in the circuit through an insulating wall sufficiently large for all of the conductors of the circuit.

*Exception: In the case of circuits supplying vacuum or electric-discharge lighting systems or signs or X-ray apparatus, the currents carried by the conductors are so small that the inductive heating effect can be ignored where these conductors are placed in metal enclosures or pass through metal.*

> FPN: Because aluminum is not a magnetic metal, there will be no heating due to hysteresis; however, induced currents will be present. They will not be of sufficient magnitude to require grouping of conductors or special treatment in passing conductors through aluminum wall sections.

**300.21 Spread of Fire or Products of Combustion.** Electrical installations in hollow spaces, vertical shafts, and ventilation or air-handling ducts shall be made so that the possible spread of fire or products of combustion will not be substantially increased. Openings around electrical penetrations through fire-resistant-rated walls, partitions, floors, or ceilings shall be firestopped using approved methods to maintain the fire resistance rating.

> FPN: Directories of electrical construction materials published by qualified testing laboratories contain many listing installation restrictions necessary to maintain the fire-resistive rating of assemblies where penetrations or openings are made. Building codes also contain restrictions on membrane penetrations on opposite sides of a fire-resistance–rated wall assembly. An example is the 600-mm (24-in.) minimum horizontal separation that usually applies between boxes installed on opposite sides of the wall. Assistance in complying with 300.21 can be found in building codes, fire resistance directories, and product listings.

**300.22 Wiring in Ducts, Plenums, and Other Air-Handling Spaces.** The provisions of this section apply to the installation and uses of electric wiring and equipment in ducts, plenums, and other air-handling spaces.

> FPN: See Article 424, Part VI, for duct heaters.

**(A) Ducts for Dust, Loose Stock, or Vapor Removal.** No wiring systems of any type shall be installed in ducts used to transport dust, loose stock, or flammable vapors. No wiring system of any type shall be installed in any duct, or shaft containing only such ducts, used for vapor removal or for ventilation of commercial-type cooking equipment.

**(B) Ducts or Plenums Used for Environmental Air.** Only wiring methods consisting of Type MI cable, Type MC cable employing a smooth or corrugated impervious metal sheath without an overall nonmetallic covering, electrical metallic tubing, flexible metallic tubing, intermediate metal conduit, or rigid metal conduit without an overall nonmetallic covering shall be installed in ducts or plenums specifically fabricated to transport environmental air. Flexible metal conduit and liquidtight flexible metal conduit shall be permitted, in lengths not to exceed 1.2 m (4 ft), to connect physically adjustable equipment and devices permitted to be in these ducts and plenum chambers. The connectors used with flexible metal conduit shall effectively close any openings in the connection. Equipment and devices shall be permitted within such ducts or plenum chambers only if necessary for their direct action upon, or sensing of, the contained air. Where equipment or devices are installed and illumination is necessary to facilitate maintenance and repair, enclosed gasketed-type luminaires (fixtures) shall be permitted.

**(C) Other Space Used for Environmental Air.** This section applies to space used for environmental air-handling purposes other than ducts and plenums as specified in 300.22(A) and (B). It does not include habitable rooms or areas of buildings, the prime purpose of which is not air handling.

> FPN: The space over a hung ceiling used for environmental air-handling purposes is an example of the type of other space to which this section applies.

*Exception: This section shall not apply to the joist or stud spaces of dwelling units where the wiring passes through such spaces perpendicular to the long dimension of such spaces.*

**(1) Wiring Methods.** The wiring methods for such other space shall be limited to totally enclosed, nonventilated, insulated busway having no provisions for plug-in connections, Type MI cable, Type MC cable without an overall nonmetallic covering, Type AC cable, or other factory-assembled multiconductor control or power cable that is specifically listed for the use, or listed prefabricated cable assemblies of metallic manufactured wiring systems without nonmetallic sheath. Other types of cables and conductors shall be installed in electrical metallic tubing, flexible metallic tubing, intermediate metal conduit, rigid metal conduit without an overall nonmetallic covering, flexible metal conduit, or, where accessible, surface metal raceway or metal wireway with metal covers or solid bottom metal cable tray with solid metal covers.

**(2) Equipment.** Electrical equipment with a metal enclosure, or with a nonmetallic enclosure listed for the use and having adequate fire-resistant and low-smoke-producing characteristics, and associated wiring material suitable for the ambient temperature shall be permitted to be installed in such other space unless prohibited elsewhere in this *Code*.

*Exception: Integral fan systems shall be permitted where specifically identified for such use.*

**(D) Information Technology Equipment.** Electric wiring in air-handling areas beneath raised floors for information technology equipment shall be permitted in accordance with Article 645.

**300.23 Panels Designed to Allow Access.** Cables, raceways, and equipment installed behind panels designed to allow access, including suspended ceiling panels, shall be arranged and secured so as to allow the removal of panels and access to the equipment.

## II. Requirements for Over 600 Volts, Nominal

**300.31 Covers Required.** Suitable covers shall be installed on all boxes, fittings, and similar enclosures to prevent accidental contact with energized parts or physical damage to parts or insulation.

**300.32 Conductors of Different Systems.** See 300.3(C)(2).

**300.34 Conductor Bending Radius.** The conductor shall not be bent to a radius less than 8 times the overall diameter for nonshielded conductors or 12 times the diameter for shielded or lead-covered conductors during or after installation. For multiconductor or multiplexed single conductor cables having individually shielded conductors, the minimum bending radius is 12 times the diameter of the individually shielded conductors or 7 times the overall diameter, whichever is greater.

**300.35 Protection Against Induction Heating.** Metallic raceways and associated conductors shall be arranged so as to avoid heating of the raceway in accordance with the applicable provisions of 300.20.

**300.37 Aboveground Wiring Methods.** Aboveground conductors shall be installed in rigid metal conduit, in in-termediate metal conduit, in electrical metallic tubing, in rigid nonmetallic conduit, in cable trays, as busways, as cablebus, in other identified raceways, or as open runs of metal-clad cable suitable for the use and purpose. In locations accessible to qualified persons only, open runs of Type MV cables, bare conductors, and bare busbars shall also be permitted. Busbars shall be permitted to be either copper or aluminum.

**300.39 Braid-Covered Insulated Conductors — Open Installation.** Open runs of braid-covered insulated conductors shall have a flame-retardant braid. If the conductors used do not have this protection, a flame-retardant saturant shall be applied to the braid covering after installation. This treated braid covering shall be stripped back a safe distance at conductor terminals, according to the operating voltage. This distance shall not be less than 25 mm (1 in.) for each kilovolt of the conductor-to-ground voltage of the circuit, where practicable.

**300.40 Insulation Shielding.** Metallic and semiconducting insulation shielding components of shielded cables shall be removed for a distance dependent on the circuit voltage and insulation. Stress reduction means shall be provided at all terminations of factory-applied shielding.

Metallic shielding components such as tapes, wires, or braids, or combinations thereof, and their associated conducting or semiconducting components shall be grounded.

**300.42 Moisture or Mechanical Protection for Metal-Sheathed Cables.** Where cable conductors emerge from a metal sheath and where protection against moisture or physical damage is necessary, the insulation of the conductors shall be protected by a cable sheath terminating device.

**300.50 Underground Installations.**

**(A) General.** Underground conductors shall be identified for the voltage and conditions under which they are installed. Direct burial cables shall comply with the provisions of 310.7. Underground cables shall be installed in accordance with 300.50(A)(1) or (2), and the installation shall meet the depth requirements of Table 300.50.

*Exception No. 1: Areas subject to vehicular traffic, such as thoroughfares or commercial parking areas, shall have a minimum cover of 600 mm (24 in.).*

*Exception No. 2: The minimum cover requirements for other than rigid metal conduit and intermediate metal conduit shall be permitted to be reduced 150 mm (6 in.) for each 50 mm (2 in.) of concrete or equivalent protection placed in the trench over the underground installation.*

*Exception No. 3: The minimum cover requirements shall not apply to conduits or other raceways that are located*

**Table 300.50 Minimum Cover Requirements**

| Circuit Voltage | Direct-Buried Cables | | Rigid Nonmetallic Conduit Approved for Direct Burial* | | Rigid Metal Conduit and Intermediate Metal Conduit | |
|---|---|---|---|---|---|---|
| | mm | in. | mm | in. | mm | in. |
| Over 600 V through 22 kV | 750 | 30 | 450 | 18 | 150 | 6 |
| Over 22 kV through 40 kV | 900 | 36 | 600 | 24 | 150 | 6 |
| Over 40 kV | 1000 | 42 | 750 | 30 | 150 | 6 |

Note: *Cover* is defined as the shortest distance in millimeters measured between a point on the top surface of any direct-buried conductor, cable, conduit, or other raceway and the top surface of finished grade, concrete, or similar cover.

*Listed by a qualified testing agency as suitable for direct burial without encasement. All other nonmetallic systems shall require 50 mm (2 in.) of concrete or equivalent above conduit in addition to above depth.

*under a building or exterior concrete slab not less than 100 mm (4 in.) in thickness and extending not less than 150 mm (6 in.) beyond the underground installation. A warning ribbon or other effective means suitable for the conditions shall be placed above the underground installation.*

*Exception No. 4: Lesser depths shall be permitted where cables and conductors rise for terminations or splices or where access is otherwise required.*

*Exception No. 5: In airport runways, including adjacent defined areas where trespass is prohibited, cable shall be permitted to be buried not less than 450 mm (18 in.) deep and without raceways, concrete enclosement, or equivalent.*

*Exception No. 6: Raceways installed in solid rock shall be permitted to be buried at lesser depth where covered by 50 mm (2 in.) of concrete, which shall be permitted to extend to the rock surface.*

**(1) Shielded Cables and Nonshielded Cables in Metal-Sheathed Cable Assemblies.** Underground cables, including nonshielded, Type MC and moisture-impervious metal sheath cables, shall have those sheaths grounded through an effective grounding path meeting the requirements of 250.4(A)(5) or 250.4(B)(4). They shall be direct buried or installed in raceways identified for the use.

**(2) Other Nonshielded Cables.** Other nonshielded cables not covered in 300.50(A)(1) shall be installed in rigid metal conduit, intermediate metal conduit, or rigid nonmetallic conduit encased in not less than 75 mm (3 in.) of concrete.

**(B) Protection from Damage.** Conductors emerging from the ground shall be enclosed in listed raceways. Raceways installed on poles shall be of rigid metal conduit, intermediate metal conduit, PVC Schedule 80, or equivalent, extending from the minimum cover depth specified in Table 300.50 to a point 2.5 m (8 ft) above finished grade. Conductors entering a building shall be protected by an approved enclosure or raceway from the minimum cover depth to the point of entrance. Where direct-buried conductors, raceways, or cables are subject to movement by settlement or frost, they shall be installed to prevent damage to the enclosed conductors or to the equipment connected to the raceways. Metallic enclosures shall be grounded.

**(C) Splices.** Direct burial cables shall be permitted to be spliced or tapped without the use of splice boxes, provided they are installed using materials suitable for the application. The taps and splices shall be watertight and protected from mechanical damage. Where cables are shielded, the shielding shall be continuous across the splice or tap.

*Exception: At splices of an engineered cabling system, metallic shields of direct-buried single-conductor cables with maintained spacing between phases shall be permitted to be interrupted and overlapped. Where shields are interrupted and overlapped, each shield section shall be grounded at one point.*

**(D) Backfill.** Backfill containing large rocks, paving materials, cinders, large or sharply angular substances, or corrosive materials shall not be placed in an excavation where materials can damage raceways, cables, or other substructures, or prevent adequate compaction of fill, or contribute to corrosion of raceways, cables, or other substructures.

Protection in the form of granular or selected material or suitable sleeves shall be provided to prevent physical damage to the raceway or cable.

**(E) Raceway Seal.** Where a raceway enters from an underground system, the end within the building shall be sealed with an identified compound so as to prevent the entrance of moisture or gases, or it shall be so arranged to prevent moisture from contacting live parts.

## ARTICLE 310
## Conductors for General Wiring

**310.1 Scope.** This article covers general requirements for conductors and their type designations, insulations, markings, mechanical strengths, ampacity ratings, and uses. These requirements do not apply to conductors that form an integral part of equipment, such as motors, motor controllers, and similar equipment, or to conductors specifically provided for elsewhere in this *Code.*

> FPN: For flexible cords and cables, see Article 400. For fixture wires, see Article 402.

### 310.2 Conductors.

**(A) Insulated.** Conductors shall be insulated.

*Exception: Where covered or bare conductors are specifically permitted elsewhere in this Code.*

> FPN: See 250.184 for insulation of neutral conductors of a solidly grounded high-voltage system.

**(B) Conductor Material.** Conductors in this article shall be of aluminum, copper-clad aluminum, or copper unless otherwise specified.

### 310.3 Stranded Conductors. Where installed in raceways, conductors of size 8 AWG and larger shall be stranded.

*Exception: As permitted or required elsewhere in this Code.*

### 310.4 Conductors in Parallel. Aluminum, copper-clad aluminum, or copper conductors of size 1/0 AWG and larger, comprising each phase, neutral, or grounded circuit conductor, shall be permitted to be connected in parallel (electrically joined at both ends to form a single conductor).

*Exception No. 1: As permitted in 620.12(A)(1) .*

*Exception No. 2: Conductors in sizes smaller than 1/0 AWG shall be permitted to be run in parallel to supply control power to indicating instruments, contactors, relays, solenoids, and similar control devices provided*

*(a) They are contained within the same raceway or cable,*
*(b) The ampacity of each individual conductor is sufficient to carry the entire load current shared by the parallel conductors, and*
*(c) The overcurrent protection is such that the ampacity of each individual conductor will not be exceeded if one or more of the parallel conductors become inadvertently disconnected.*

*Exception No. 3: Conductors in sizes smaller than 1/0 AWG shall be permitted to be run in parallel for frequencies of 360 Hz and higher where conditions (a), (b), and (c) of Exception No. 2 are met.*

*Exception No. 4: Under engineering supervision, grounded neutral conductors in sizes 2 AWG and larger shall be permitted to be run in parallel for existing installations.*

> FPN: Exception No. 4 can be used to alleviate overheating of neutral conductors in existing installations due to high content of triplen harmonic currents.

The paralleled conductors in each phase, neutral, or grounded circuit conductor shall

(1) Be the same length
(2) Have the same conductor material
(3) Be the same size in circular mil area
(4) Have the same insulation type
(5) Be terminated in the same manner

Where run in separate raceways or cables, the raceways or cables shall have the same physical characteristics. Conductors of one phase, neutral, or grounded circuit conductor shall not be required to have the same physical characteristics as those of another phase, neutral, or grounded circuit conductor to achieve balance.

> FPN: Differences in inductive reactance and unequal division of current can be minimized by choice of materials, methods of construction, and orientation of conductors.

Where equipment grounding conductors are used with conductors in parallel, they shall comply with the requirements of this section except that they shall be sized in accordance with 250.122.

Conductors installed in parallel shall comply with the provisions of 310.15(B)(2)(a).

### 310.5 Minimum Size of Conductors. The minimum size of conductors shall be as shown in Table 310.5.

**Table 310.5 Minimum Size of Conductors**

| Conductor Voltage Rating (Volts) | Minimum Conductor Size (AWG) | |
| --- | --- | --- |
| | Copper | Aluminum or Copper-Clad Aluminum |
| 0–2000 | 14 | 12 |
| 2001–8000 | 8 | 8 |
| 8001–15,000 | 2 | 2 |
| 15,001–28,000 | 1 | 1 |
| 28,001–35,000 | 1/0 | 1/0 |

*Exception No. 1:  For flexible cords as permitted by 400.12.*

*Exception No. 2:  For fixture wire as permitted by 402.6.*

*Exception No. 3:  For motors rated 1 hp or less as permitted by 430.22(F).*

*Exception No. 4:  For cranes and hoists as permitted by 610.14.*

*Exception No. 5:  For elevator control and signaling circuits as permitted by 620.12.*

*Exception No. 6:  For Class 1, Class 2, and Class 3 circuits as permitted by 725.27(A) and 725.51, Exception.*

*Exception No. 7: For fire alarm circuits as permitted by 760.27(A), 760.51, Exception, and 760.71(B).*

*Exception No. 8: For motor-control circuits as permitted by 430.72.*

*Exception No. 9: For control and instrumentation circuits as permitted by 727.6.*

*Exception No. 10: For electric signs and outline lighting as permitted in 600.31(B) and 600.32(B).*

**310.6 Shielding.** Solid dielectric insulated conductors operated above 2000 volts in permanent installations shall have ozone-resistant insulation and shall be shielded. All metallic insulation shields shall be grounded through an effective grounding path meeting the requirements of 250.4(A)(5) or 250.4(B)(4). Shielding shall be for the purpose of confining the voltage stresses to the insulation.

*Exception: Nonshielded insulated conductors listed by a qualified testing laboratory shall be permitted for use up to 8000 volts under the following conditions:*

*(a) Conductors shall have insulation resistant to electric discharge and surface tracking, or the insulated conductor(s) shall be covered with a material resistant to ozone, electric discharge, and surface tracking.*

*(b) Where used in wet locations, the insulated conductor(s) shall have an overall nonmetallic jacket or a continuous metallic sheath.*

*(c) Where operated at 5001 to 8000 volts, the insulated conductor(s) shall have a nonmetallic jacket over the insulation. The insulation shall have a specific inductive capacity not greater than 3.6, and the jacket shall have a specific inductive capacity not greater than 10 and not less than 6.*

*(d) Insulation and jacket thicknesses shall be in accordance with Table 310.63.*

**310.7 Direct Burial Conductors.** Conductors used for direct burial applications shall be of a type identified for such use.

Cables rated above 2000 volts shall be shielded.

*Exception: Nonshielded multiconductor cables rated 2001–5000 volts shall be permitted if the cable has an overall metallic sheath or armor.*

The metallic shield, sheath, or armor shall be grounded through an effective grounding path meeting the requirements of 250.4(A)(5) or 250.4(B)(4).

FPN No. 1: See 300.5 for installation requirements for conductors rated 600 volts or less.

FPN No. 2: See 300.50 for installation requirements for conductors rated over 600 volts.

**310.8 Locations.**

**(A) Dry Locations.** Insulated conductors and cables used in dry locations shall be any of the types identified in this *Code.*

**(B) Dry and Damp Locations.** Insulated conductors and cables used in dry and damp locations shall be Types FEP, FEPB, MTW, PFA, RHH, RHW, RHW-2, SA, THHN, THW, THW-2, THHW, THHW-2, THWN, THWN-2, TW, XHH, XHHW, XHHW-2, Z, or ZW.

**(C) Wet Locations.** Insulated conductors and cables used in wet locations shall be

(1) Moisture-impervious metal-sheathed;
(2) Types MTW, RHW, RHW-2, TW, THW, THW-2, THHW, THHW-2, THWN, THWN-2, XHHW, XHHW-2, ZW; or
(3) Of a type listed for use in wet locations.

**(D) Locations Exposed to Direct Sunlight.** Insulated conductors and cables used where exposed to direct rays of the sun shall be of a type listed for sunlight resistance or listed and marked "sunlight resistant."

**310.9 Corrosive Conditions.** Conductors exposed to oils, greases, vapors, gases, fumes, liquids, or other substances having a deleterious effect on the conductor or insulation shall be of a type suitable for the application.

**310.10 Temperature Limitation of Conductors.** No conductor shall be used in such a manner that its operating temperature exceeds that designated for the type of insulated conductor involved. In no case shall conductors be associated together in such a way with respect to type of circuit, the wiring method employed, or the number of conductors that the limiting temperature of any conductor is exceeded.

FPN: The temperature rating of a conductor (see Table 310.13 and Table 310.61) is the maximum temperature, at any location along its length, that the conductor can withstand over a prolonged time period without serious degradation. The allowable ampacity tables, the ampacity tables of Article 310 and the ampacity tables of Annex B, the correction factors at the bottom of these tables, and the notes to the tables provide guidance for coordinating conductor sizes, types, allowable ampacities, ampacities, ambient temperatures, and number of associated conductors.

The principal determinants of operating temperature are as follows:

(1) Ambient temperature — ambient temperature may vary along the conductor length as well as from time to time.
(2) Heat generated internally in the conductor as the result of load current flow, including fundamental and harmonic currents.

(3) The rate at which generated heat dissipates into the ambient medium. Thermal insulation that covers or surrounds conductors affects the rate of heat dissipation.

(4) Adjacent load-carrying conductors — adjacent conductors have the dual effect of raising the ambient temperature and impeding heat dissipation.

## 310.11 Marking.

**(A) Required Information.** All conductors and cables shall be marked to indicate the following information, using the applicable method described in 310.11(B):

(1) The maximum rated voltage.

(2) The proper type letter or letters for the type of wire or cable as specified elsewhere in this *Code*.

(3) The manufacturer's name, trademark, or other distinctive marking by which the organization responsible for the product can be readily identified.

(4) The AWG size or circular mil area.

> FPN: See Conductor Properties, Table 8 of Chapter 9 for conductor area expressed in SI units for conductor sizes specified in AWG or circular mil area.

(5) Cable assemblies where the neutral conductor is smaller than the ungrounded conductors shall be so marked.

**(B) Method of Marking.**

**(1) Surface Marking.** The following conductors and cables shall be durably marked on the surface. The AWG size or circular mil area shall be repeated at intervals not exceeding 610 mm (24 in.). All other markings shall be repeated at intervals not exceeding 1.0 m (40 in.).

(1) Single- and multiconductor rubber- and thermoplastic-insulated wire and cable

(2) Nonmetallic-sheathed cable

(3) Service-entrance cable

(4) Underground feeder and branch-circuit cable

(5) Tray cable

(6) Irrigation cable

(7) Power-limited tray cable

(8) Instrumentation tray cable

**(2) Marker Tape.** Metal-covered multiconductor cables shall employ a marker tape located within the cable and running for its complete length.

*Exception No. 1: Mineral-insulated, metal-sheathed cable.*

*Exception No. 2: Type AC cable.*

*Exception No. 3: The information required in 310.11(A) shall be permitted to be durably marked on the outer nonmetallic covering of Type MC, Type ITC, or Type PLTC cables at intervals not exceeding 1.0 m (40 in.).*

*Exception No. 4: The information required in 310.11(A) shall be permitted to be durably marked on a nonmetallic*

*covering under the metallic sheath of Type ITC or Type PLTC cable at intervals not exceeding 1.0 m (40 in.).*

> FPN: Included in the group of metal-covered cables are Type AC cable (Article 320), Type MC cable (Article 330), and lead-sheathed cable.

**(3) Tag Marking.** The following conductors and cables shall be marked by means of a printed tag attached to the coil, reel, or carton:

(1) Mineral-insulated, metal-sheathed cable

(2) Switchboard wires

(3) Metal-covered, single-conductor cables

(4) Type AC cable

**(4) Optional Marking of Wire Size.** The information required in 310.11(A)(4) shall be permitted to be marked on the surface of the individual insulated conductors for the following multiconductor cables:

(1) Type MC cable

(2) Tray cable

(3) Irrigation cable

(4) Power-limited tray cable

(5) Power-limited fire alarm cable

(6) Instrumentation tray cable

**(C) Suffixes to Designate Number of Conductors.** A type letter or letters used alone shall indicate a single insulated conductor. The letter suffixes shall be indicated as follows:

(1) D — For two insulated conductors laid parallel within an outer nonmetallic covering

(2) M — For an assembly of two or more insulated conductors twisted spirally within an outer nonmetallic covering

**(D) Optional Markings.** All conductors and cables contained in Chapter 3 shall be permitted to be surface marked to indicate special characteristics of the cable materials. These markings include, but are not limited to, markings for limited smoke, sunlight resistant, and so forth.

## 310.12 Conductor Identification.

**(A) Grounded Conductors.** Insulated or covered grounded conductors shall be identified in accordance with 200.6.

**(B) Equipment Grounding Conductors.** Equipment grounding conductors shall be in accordance with 250.119.

**(C) Ungrounded Conductors.** Conductors that are intended for use as ungrounded conductors, whether used as a single conductor or in multiconductor cables, shall be finished to be clearly distinguishable from grounded and grounding conductors. Distinguishing markings shall not conflict in any manner with the surface markings required by 310.11(B)(1).

*Exception: Conductor identification shall be permitted in accordance with 200.7.*

### 310.13 Conductor Constructions and Applications.

Insulated conductors shall comply with the applicable provisions of one or more of the following: Tables 310.13, 310.61, 310.62, 310.63, and 310.64.

These conductors shall be permitted for use in any of the wiring methods recognized in Chapter 3 and as specified in their respective tables.

FPN: Thermoplastic insulation may stiffen at temperatures lower than minus 10°C (plus 14°F). Thermoplastic insulation may also be deformed at normal temperatures where subjected to pressure, such as at points of support. Thermoplastic insulation, where used on dc circuits in wet locations, may result in electroendosmosis between conductor and insulation.

**Table 310.13  Conductor Application and Insulations**

| Trade Name | Type Letter | Maximum Operating Temperature | Application Provisions | Insulation | Thickness of Insulation | | | | Outer Covering[1] |
|---|---|---|---|---|---|---|---|---|---|
| | | | | | AWG or kcmil | mm | | Mils | |
| Fluorinated ethylene propylene | FEP or FEPB | 90°C 194°F | Dry and damp locations | Fluorinated ethylene propylene | 14–10 8–2 | 0.51 0.76 | | 20 30 | None |
| | | | | Fluorinated ethylene propylene | 14–8 | 0.36 | | 14 | Glass braid |
| | | 200°C 392°F | Dry locations — special applications[2] | | 6–2 | 0.36 | | 14 | Glass or other suitable braid material |
| Mineral insulation (metal sheathed) | MI | 90°C 194°F | Dry and wet locations | Magnesium oxide | 18–16[3] 16–10 9–4 3–500 | 0.58 0.91 1.27 1.40 | | 23 36 50 55 | Copper or alloy steel |
| | | 250°C 482°F | For special applications[2] | | | | | | |
| Moisture-, heat-, and oil-resistant thermoplastic | MTW | 60°C 140°F | Machine tool wiring in wet locations as permitted in NFPA 79 (See Article 670.) Machine tool wiring in dry locations as permitted in NFPA 79 (See Article 670.) | Flame-retardant moisture-, heat-, and oil-resistant thermoplastic | 22–12 10 8 6 4–2 1–4/0 213–500 501–1000 | (A) 0.76 0.76 1.14 1.52 1.52 2.03 2.41 2.79 | (B) 0.38 0.51 0.76 0.76 1.02 1.27 1.52 1.78 | (A) 30 30 45 60 60 80 95 110 | (B) 15 20 30 30 40 50 60 70 | (A) None (B) Nylon jacket or equivalent |
| | | 90°C 194°F | | | | | | | |
| Paper | | 85°C 185°F | For underground service conductors, or by special permission | Paper | | | | | Lead sheath |
| Perfluoro-alkoxy | PFA | 90°C 194°F | Dry and damp locations | Perfluoro-alkoxy | 14–10 8–2 1–4/0 | 0.51 0.76 1.14 | | 20 30 45 | None |
| | | 200°C 392°F | Dry locations — special applications[2] | | | | | | |

**Table 310.13**  *Continued*

| Trade Name | Type Letter | Maximum Operating Temperature | Application Provisions | Insulation | Thickness of Insulation | | | Outer Covering[1] |
|---|---|---|---|---|---|---|---|---|
| | | | | | AWG or kcmil | mm | Mils | |
| Perfluoro-alkoxy | PFAH | 250°C 482°F | Dry locations only. Only for leads within apparatus or within raceways connected to apparatus (nickel or nickel-coated copper only) | Perfluoro-alkoxy | 14–10 8–2 1–4/0 | 0.51 0.76 1.14 | 20 30 45 | None |
| Thermoset | RHH | 90°C 194°F | Dry and damp locations | | 14-10 8–2 1–4/0 213–500 501–1000 1001–2000 For 601–2000, *see* Table 310.62. | 1.14 1.52 2.03 2.41 2.79 3.18 | 45 60 80 95 110 125 | Moisture-resistant, flame-retardant, nonmetallic covering[1] |
| Moisture-resistant thermoset | RHW [4] | 75°C 167°F | Dry and wet locations | Flame-retardant, moisture-resistant thermo-set | 14–10 8–2 1–4/0 213–500 501–1000 1001–2000 For 601–2000, *see* Table 310.62. | 1.14 1.52 2.03 2.41 2.79 3.18 | 45 60 80 95 110 125 | Moisture-resistant, flame-retardant, nonmetallic covering[5] |
| Moisture-resistant thermoset | RHW-2 | 90°C 194°F | Dry and wet locations | Flame-retardant moisture-resistant thermo-set | 14–10 8–2 1–4/0 213–500 501–1000 1001–2000 For 601–2000, see Table 310.62. | 1.14 1.52 2.03 2.41 2.79 3.18 | 45 60 80 95 110 125 | Moisture-resistant, flame-retardant, nonmetallic covering[5] |
| Silicone | SA | 90°C 194°F / 200°C 392°F | Dry and damp locations / For special application [2] | Silicone rubber | 14–10 8–2 1–4/0 213–500 501–1000 1001–2000 | 1.14 1.52 2.03 2.41 2.79 3.18 | 45 60 80 95 110 125 | Glass or other suitable braid material |
| Thermoset | SIS | 90°C 194°F | Switchboard wiring only | Flame-retardant thermoset | 14–10 8–2 1–4/0 | 0.76 1.14 2.41 | 30 45 95 | None |

**Table 310.13** *Continued*

| Trade Name | Type Letter | Maximum Operating Temperature | Application Provisions | Insulation | Thickness of Insulation | | | Outer Covering[1] |
|---|---|---|---|---|---|---|---|---|
| | | | | | AWG or kcmil | mm | Mils | |
| Thermoplastic and fibrous outer braid | TBS | 90°C 194°F | Switchboard wiring only | Thermoplastic | 14–10<br>8<br>6–2<br>1–4/0 | 0.76<br>1.14<br>1.52<br>2.03 | 30<br>45<br>60<br>80 | Flame-retardant, nonmetallic covering |
| Extended polytetra-fluoro-ethylene | TFE | 250°C 482°F | Dry locations only. Only for leads within apparatus or within raceways connected to apparatus, or as open wiring (nickel or nickel-coated copper only) | Extruded polytetra-fluoro-ethylene | 14–10<br>8–2<br>1–4/0 | 0.51<br>0.76<br>1.14 | 20<br>30<br>45 | None |
| Heatresistant thermoplastic | THHN | 90°C 194°F | Dry and damp locations | Flame-retardant, heat-resistant thermoplastic | 14–12<br>10<br>8–6<br>4–2<br>1–4/0<br>250–500<br>501–1000 | 0.38<br>0.51<br>0.76<br>1.02<br>1.27<br>1.52<br>1.78 | 15<br>20<br>30<br>40<br>50<br>60<br>70 | Nylon jacket or equivalent |
| Moisture- and heat-resistant thermoplastic | THHW | 75°C 167°F 90°C 194°F | Wet location<br><br>Dry location | Flame-retardant, moisture- and heat-resistant thermoplastic | 14–10<br>8<br>6–2<br>1–4/0<br>213–500<br>501–1000 | 0.76<br>1.14<br>1.52<br>2.03<br>2.41<br>2.79 | 30<br>45<br>60<br>80<br>95<br>110 | None |
| Moisture- and heat-resistant thermoplastic | THW [4] | 75°C 167°F<br><br>90°C 194°F | Dry and wet locations<br><br>Special applications within electric discharge lighting equipment. Limited to 1000 open-circuit volts or less. (size 14-8 only as permitted in 410.33) | Flame-retardant, moisture- and heat-resistant thermoplastic | 14–10<br>8<br>6–2<br>1–4/0<br>213–500<br>501–1000<br>1001–2000 | 0.76<br>1.14<br>1.52<br>2.03<br>2.41<br>2.79<br>3.18 | 30<br>45<br>60<br>80<br>95<br>110<br>125 | None |
| Moisture- and heat-resistant thermoplastic | THWN[4] | 75°C 167°F | Dry and wet locations | Flame-retardant, moisture- and heat-resistant thermoplastic | 14–12<br>10<br>8–6<br>4–2<br>1–4/0<br>250–500<br>501–1000 | 0.38<br>0.51<br>0.76<br>1.02<br>1.27<br>1.52<br>1.78 | 15<br>20<br>30<br>40<br>50<br>60<br>70 | Nylon jacket or equivalent |

**Table 310.13** *Continued*

| Trade Name | Type Letter | Maximum Operating Temperature | Application Provisions | Insulation | Thickness of Insulation | | | Outer Covering[1] |
|---|---|---|---|---|---|---|---|---|
| | | | | | AWG or kcmil | mm | Mils | |
| Moisture-resistant thermo-plastic | TW | 60°C 140°F | Dry and wet locations | Flame-retardant, moisture-resistant thermo-plastic | 14–10<br>8<br>6–2<br>1–4/0<br>213–500<br>501–1000<br>1001–2000 | 0.76<br>1.14<br>1.52<br>2.03<br>2.41<br>2.79<br>3.18 | 30<br>45<br>60<br>80<br>95<br>110<br>125 | None |
| Underground feeder and branch-circuit cable — single conductor (For Type UF cable employing more than one conductor, *see* Articles 339, 340.) | UF | 60°C 140°F 75°C 167°F [7] | See Article 340. | Moisture-resistant<br><br>Moisture-and heat-resistant | 14–10<br>8–2<br>1–4/0 | 1.52<br>2.03<br>2.41 | 60[6]<br>80[6]<br>95[6] | Integral with insulation |
| Underground service-entrance cable — single conductor (For Type USE cable employing more than one conductor, *see* Article 338.) | USE[4] | 75°C 167°F | See Article 338. | Heat- and moisture-resistant | 14–10<br>8–2<br>1–4/0<br>213–500<br>501–1000<br>1001–2000 | 1.14<br>1.52<br>2.03<br>2.41<br>2.79<br>3.18 | 45<br>60<br>80<br>95[8]<br>110<br>125 | Moisture-resistant nonmetallic covering (See 338.2.) |
| Thermoset | XHH | 90°C 194°F | Dry and damp locations | Flame-retardant thermoset | 14–10<br>8–2<br>1–4/0<br>213–500<br>501–1000<br>1001–2000 | 0.76<br>1.14<br>1.40<br>1.65<br>2.03<br>2.41 | 30<br>45<br>55<br>65<br>80<br>95 | None |
| Moisture-resistant thermoset | XHHW [4] | 90°C 194°F<br><br>75°C 167°F | Dry and damp locations<br><br>Wet locations | Flame-retardant, moisture-resistant thermoset | 14–10<br>8–2<br>1–4/0<br>213–500<br>501–1000<br>1001–2000 | 0.76<br>1.14<br>1.40<br>1.65<br>2.03<br>2.41 | 30<br>45<br>55<br>65<br>80<br>95 | None |

**Table 310.13**  *Continued*

| Trade Name | Type Letter | Maximum Operating Temperature | Application Provisions | Insulation | Thickness of Insulation | | | Outer Covering[1] |
|---|---|---|---|---|---|---|---|---|
| | | | | | AWG or kcmil | mm | Mils | |
| Moisture-resistant thermoset | XHHW-2 | 90°C 194°F | Dry and wet locations | Flame-retardant, moisture-resistant thermoset | 14–10 8–2 1–4/0 213–500 501–1000 1001–2000 | 0.76 1.14 1.40 1.65 2.03 2.41 | 30 45 55 65 80 95 | None |
| Modified ethylene tetra-fluoro-ethylene | Z | 90°C 194°F  150°C 302°F | Dry and damp locations  Dry locations — special applications[2] | Modified ethylene tetra-fluoro-ethylene | 14–12 10 8–4 3–1 1/0–4/0 | 0.38 0.51 0.64 0.89 1.14 | 15 20 25 35 45 | None |
| Modified ethylene tetra-fluoro-ethylene | ZW[4] | 75°C 167°F  90°C 194°F  150°C 302°F | Wet locations  Dry and damp locations  Dry locations — special applications[2] | Modified ethylene tetra-fluoro-ethylene | 14–10 8–2 | 0.76 1.14 | 30 45 | None |

[1] Some insulations do not require an outer covering.

[2] Where design conditions require maximum conductor operating temperatures above 90°C (194°F).

[3] For signaling circuits permitting 300-volt insulation.

[4] Listed wire types designated with the suffix "2," such as RHW-2, shall be permitted to be used at a continuous 90°C (194°F) operating temperature, wet or dry.

[5] Some rubber insulations do not require an outer covering.

[6] Includes integral jacket.

[7] For ampacity limitation, see 340.80.

[8] Insulation thickness shall be permitted to be 2.03 mm (80 mils) for listed Type USE conductors that have been subjected to special investigations. The nonmetallic covering over individual rubber-covered conductors of aluminum-sheathed cable and of lead-sheathed or multiconductor cable shall not be required to be flame retardant.
For Type MC cable, see 330.104. For nonmetallic-sheathed cable, *see* Article 334, Part III. For Type UF cable, *see* Article 340, Part III.

**310.14 Aluminum Conductor Material.** Solid aluminum conductors 8, 10, and 12 AWG shall be made of an AA-8000 series electrical grade aluminum alloy conductor material. Stranded aluminum conductors 8 AWG through 1000 kcmil marked as Type RHH, RHW, XHHW, THW, THHW, THWN, THHN, service-entrance Type SE Style U and SE Style R shall be made of an AA-8000 series electrical grade aluminum alloy conductor material.

**310.15 Ampacities for Conductors Rated 0–2000 Volts.**

**(A) General.**

**(1) Tables or Engineering Supervision.** Ampacities for conductors shall be permitted to be determined by tables or under engineering supervision, as provided in 310.15(B) and (C).

> FPN No. 1: Ampacities provided by this section do not take voltage drop into consideration. See 210.19(A), FPN No. 4, for branch circuits and 215.2(D), FPN No. 2, for feeders.

> FPN No. 2: For the allowable ampacities of Type MTW wire, see Table 11 in NFPA 79-1997, *Electrical Standard for Industrial Machinery*.

**(2) Selection of Ampacity.** Where more than one calculated or tabulated ampacity could apply for a given circuit length, the lowest value shall be used.

*Exception: Where two different ampacities apply to adjacent portions of a circuit, the higher ampacity shall be permitted to be used beyond the point of transition, a distance equal to 3.0 m (10 ft) or 10 percent of the circuit length figured at the higher ampacity, whichever is less.*

> FPN: See 110.14(C) for conductor temperature limitations due to termination provisions.

**(B) Tables.** Ampacities for conductors rated 0 to 2000 volts shall be as specified in the Allowable Ampacity Tables 310.16 through 310.19 and Ampacity Tables 310.20 through 310.23 as modified by (1) through (6).

> FPN: Tables 310.16 through 310.19 are application tables for use in determining conductor sizes on loads calculated in accordance with Article 220. Allowable ampacities result from consideration of one or more of the following:
>
> (1) Temperature compatibility with connected equipment, especially the connection points.
> (2) Coordination with circuit and system overcurrent protection.
> (3) Compliance with the requirements of product listings or certifications. See 110.3(B).
> (4) Preservation of the safety benefits of established industry practices and standardized procedures.

**(1) General.** For explanation of type letters used in tables and for recognized sizes of conductors for the various conductor insulations, see 310.13. For installation requirements, see 310.1 through 310.10 and the various articles of this *Code*. For flexible cords, see Tables, 400.4, 400.5(A), and 400.5(B).

**(2) Adjustment Factors.**

(a) More Than Three Current-Carrying Conductors in a Raceway or Cable. Where the number of current-carrying conductors in a raceway or cable exceeds three, or where single conductors or multiconductor cables are stacked or bundled longer than 600 mm (24 in.) without maintaining spacing and are not installed in raceways, the allowable ampacity of each conductor shall be reduced as shown in Table 310.15(B)(2)(a).

> FPN: See Annex B, Table B.310.11, for adjustment factors for more than three current-carrying conductors in a raceway or cable with load diversity.

*Exception No. 1: Where conductors of different systems, as provided in 300.3, are installed in a common raceway or cable, the derating factors shown in Table 310.15(B)(2)(a) shall apply to the number of power and lighting conductors only (Articles 210, 215, 220, and 230).*

*Exception No. 2: For conductors installed in cable trays, the provisions of 392.11 shall apply.*

*Exception No. 3: Derating factors shall not apply to conductors in nipples having a length not exceeding 600 mm (24 in.).*

*Exception No. 4: Derating factors shall not apply to underground conductors entering or leaving an outdoor trench if those conductors have physical protection in the form of rigid metal conduit, intermediate metal conduit, or rigid nonmetallic conduit having a length not exceeding 3.05 m (10 ft) and if the number of conductors does not exceed four.*

*Exception No. 5: Adjustment factors shall not apply to Type AC cable or to Type MC cable without an overall outer jacket under the following conditions:*

*(a) Each cable has not more than three current-carrying conductors.*
*(b) The conductors are 12 AWG copper.*
*(c) Not more than 20 current-carrying conductors are bundled, stacked, or supported on "bridle rings."*

*A 60 percent adjustment factor shall be applied where the current-carrying conductors in these cables that are stacked or bundled longer than 600 mm (24 in.) without maintaining spacing exceeds 20.*

(b) More Than One Conduit, Tube, or Raceway. Spacing between conduits, tubing, or raceways shall be maintained.

**Table 310.15(B)(2)(a) Adjustment Factors for More Than Three Current-Carrying Conductors in a Raceway or Cable**

| Number of Current-Carrying Conductors | Percent of Values in Tables 310.16 through 310.19 as Adjusted for Ambient Temperature if Necessary |
|---|---|
| 4–6 | 80 |
| 7–9 | 70 |
| 10–20 | 50 |
| 21–30 | 45 |
| 31–40 | 40 |
| 41 and above | 35 |

**(3) Bare or Covered Conductors.** Where bare or covered conductors are used with insulated conductors, their allowable ampacities shall be limited to those permitted for the adjacent insulated conductors.

**(4) Neutral Conductor.**

(a) A neutral conductor that carries only the unbalanced current from other conductors of the same circuit shall not be required to be counted when applying the provisions of 310.15(B)(2)(a).

(b)   In a 3-wire circuit consisting of two phase wires and the neutral of a 4-wire, 3-phase, wye-connected system, a common conductor carries approximately the same current as the line-to-neutral load currents of the other conductors and shall be counted when applying the provisions of 310.15(B)(2)(a).

(c)   On a 4-wire, 3-phase wye circuit where the major portion of the load consists of nonlinear loads, harmonic currents are present in the neutral conductor; the neutral shall therefore be considered a current-carrying conductor.

**(5) Grounding or Bonding Conductor.** A grounding or bonding conductor shall not be counted when applying the provisions of 310.15(B)(2)(a).

**(6) 120/240-Volt, 3-Wire, Single-Phase Dwelling Services and Feeders.** For dwelling units, conductors, as listed in Table 310.15(B)(6), shall be permitted as 120/240-volt, 3-wire, single-phase service-entrance conductors, service lateral conductors, and feeder conductors that serve as the main power feeder to a dwelling unit and are installed in raceway or cable with or without an equipment grounding conductor. For application of this section, the main power feeder shall be the feeder(s) between the main disconnect and the lighting and appliance branch-circuit panelboard(s). The feeder conductors to a dwelling unit shall not be required to be larger than their service-entrance conductors. The grounded conductor shall be permitted to be smaller than the ungrounded conductors, provided the requirements of 215.2, 220.22, and 230.42 are met.

**(C) Engineering Supervision.** Under engineering supervision, conductor ampacities shall be permitted to be calculated by means of the following general formula:

**Table 310.15(B)(6) Conductor Types and Sizes for 120/240-Volt, 3-Wire, Single-Phase Dwelling Services and Feeders. Conductor Types RHH, RHW, RHW-2, THHN, THHW, THW, THW-2, THWN, THWN-2, XHHW, XHHW-2, SE, USE, USE-2**

| Conductor (AWG or kcmil) | | |
|---|---|---|
| Copper | Aluminum or Copper-Clad Aluminum | Service or Feeder Rating (Amperes) |
| 4 | 2 | 100 |
| 3 | 1 | 110 |
| 2 | 1/0 | 125 |
| 1 | 2/0 | 150 |
| 1/0 | 3/0 | 175 |
| 2/0 | 4/0 | 200 |
| 3/0 | 250 | 225 |
| 4/0 | 300 | 250 |
| 250 | 350 | 300 |
| 350 | 500 | 350 |
| 400 | 600 | 400 |

$$I = \sqrt{\frac{TC - (TA + \Delta TD)}{RDC(1 + YC)RCA}}$$

where:

$TC$ = conductor temperature in degrees Celsius (°C)

$TA$ = ambient temperature in degrees Celsius (°C)

$\Delta TD$ = dielectric loss temperature rise

$RDC$ = dc resistance of conductor at temperature $TC$

$YC$ = component ac resistance resulting from skin effect and proximity effect

$RCA$ = effective thermal resistance between conductor and surrounding ambient

FPN:  See Annex B for examples of formula applications.

**Table 310.16 Allowable Ampacities of Insulated Conductors Rated 0 Through 2000 Volts, 60°C Through 90°C (140°F Through 194°F), Not More Than Three Current-Carrying Conductors in Raceway, Cable, or Earth (Directly Buried), Based on Ambient Temperature of 30°C (86°F)**

| Size AWG or kcmil | Temperature Rating of Conductor (See Table 310.13.) | | | | | | Size AWG or kcmil |
| --- | --- | --- | --- | --- | --- | --- | --- |
| | 60°C (140°F) | 75°C (167°F) | 90°C (194°F) | 60°C (140°F) | 75°C (167°F) | 90°C (194°F) | |
| | Types TW, UF | Types RHW, THHW, THW, THWN, XHHW, USE, ZW | Types TBS, SA, SIS, FEP, FEPB, MI, RHH, RHW-2, THHN, THHW, THW-2, THWN-2, USE-2, XHH, XHHW, XHHW-2, ZW-2 | Types TW, UF | Types RHW, THHW, THW, THWN, XHHW, USE | Types TBS, SA, SIS, THHN, THHW, THW-2, THWN-2, RHH, RHW-2, USE-2, XHH, XHHW, XHHW-2, ZW-2 | |
| | COPPER | | | ALUMINUM OR COPPER-CLAD ALUMINUM | | | |
| 18 | — | — | 14 | — | — | — | — |
| 16 | — | — | 18 | — | — | — | — |
| 14* | 20 | 20 | 25 | — | — | — | — |
| 12* | 25 | 25 | 30 | 20 | 20 | 25 | 12* |
| 10* | 30 | 35 | 40 | 25 | 30 | 35 | 10* |
| 8 | 40 | 50 | 55 | 30 | 40 | 45 | 8 |
| 6 | 55 | 65 | 75 | 40 | 50 | 60 | 6 |
| 4 | 70 | 85 | 95 | 55 | 65 | 75 | 4 |
| 3 | 85 | 100 | 110 | 65 | 75 | 85 | 3 |
| 2 | 95 | 115 | 130 | 75 | 90 | 100 | 2 |
| 1 | 110 | 130 | 150 | 85 | 100 | 115 | 1 |
| 1/0 | 125 | 150 | 170 | 100 | 120 | 135 | 1/0 |
| 2/0 | 145 | 175 | 195 | 115 | 135 | 150 | 2/0 |
| 3/0 | 165 | 200 | 225 | 130 | 155 | 175 | 3/0 |
| 4/0 | 195 | 230 | 260 | 150 | 180 | 205 | 4/0 |
| 250 | 215 | 255 | 290 | 170 | 205 | 230 | 250 |
| 300 | 240 | 285 | 320 | 190 | 230 | 255 | 300 |
| 350 | 260 | 310 | 350 | 210 | 250 | 280 | 350 |
| 400 | 280 | 335 | 380 | 225 | 270 | 305 | 400 |
| 500 | 320 | 380 | 430 | 260 | 310 | 350 | 500 |
| 600 | 355 | 420 | 475 | 285 | 340 | 385 | 600 |
| 700 | 385 | 460 | 520 | 310 | 375 | 420 | 700 |
| 750 | 400 | 475 | 535 | 320 | 385 | 435 | 750 |
| 800 | 410 | 490 | 555 | 330 | 395 | 450 | 800 |
| 900 | 435 | 520 | 585 | 355 | 425 | 480 | 900 |
| 1000 | 455 | 545 | 615 | 375 | 445 | 500 | 1000 |
| 1250 | 495 | 590 | 665 | 405 | 485 | 545 | 1250 |
| 1500 | 520 | 625 | 705 | 435 | 520 | 585 | 1500 |
| 1750 | 545 | 650 | 735 | 455 | 545 | 615 | 1750 |
| 2000 | 560 | 665 | 750 | 470 | 560 | 630 | 2000 |

**CORRECTION FACTORS**

| Ambient Temp. (°C) | For ambient temperatures other than 30°C (86°F), multiply the allowable ampacities shown above by the appropriate factor shown below. | | | | | | Ambient Temp. (°F) |
| --- | --- | --- | --- | --- | --- | --- | --- |
| 21–25 | 1.08 | 1.05 | 1.04 | 1.08 | 1.05 | 1.04 | 70–77 |
| 26–30 | 1.00 | 1.00 | 1.00 | 1.00 | 1.00 | 1.00 | 78–86 |
| 31–35 | 0.91 | 0.94 | 0.96 | 0.91 | 0.94 | 0.96 | 87–95 |
| 36–40 | 0.82 | 0.88 | 0.91 | 0.82 | 0.88 | 0.91 | 96–104 |
| 41–45 | 0.71 | 0.82 | 0.87 | 0.71 | 0.82 | 0.87 | 105–113 |
| 46–50 | 0.58 | 0.75 | 0.82 | 0.58 | 0.75 | 0.82 | 114–122 |
| 51–55 | 0.41 | 0.67 | 0.76 | 0.41 | 0.67 | 0.76 | 123–131 |
| 56–60 | — | 0.58 | 0.71 | — | 0.58 | 0.71 | 132–140 |
| 61–70 | — | 0.33 | 0.58 | — | 0.33 | 0.58 | 141–158 |
| 71–80 | — | — | 0.41 | — | — | 0.41 | 159–176 |

* See 240.4(D).

**Table 310.17 Allowable Ampacities of Single-Insulated Conductors Rated 0 Through 2000 Volts in Free Air, Based on Ambient Air Temperature of 30°C (86°F)**

| Size AWG or kcmil | Temperature Rating of Conductor (See Table 310.13.) | | | | | | Size AWG or kcmil |
|---|---|---|---|---|---|---|---|
| | 60°C (140°F) | 75°C (167°F) | 90°C (194°F) | 60°C (140°F) | 75°C (167°F) | 90°C (194°F) | |
| | Types TW, UF | Types RHW, THHW, THW, THWN, XHHW, ZW | Types TBS, SA, SIS, FEP, FEPB, MI, RHH, RHW-2, THHN, THHW, THW-2, THWN-2, USE-2, XHH, XHHW, XHHW-2, ZW-2 | Types TW, UF | Types RHW, THHW, THW, THWN, XHHW | Types TBS, SA, SIS, THHN, THHW, THW-2, THWN-2, RHH, RHW-2, USE-2, XHH, XHHW, XHHW-2, ZW-2 | |
| | COPPER | | | ALUMINUM OR COPPER-CLAD ALUMINUM | | | |
| 18 | — | — | 18 | — | — | — | — |
| 16 | — | — | 24 | — | — | — | — |
| 14* | 25 | 30 | 35 | — | — | — | — |
| 12* | 30 | 35 | 40 | 25 | 30 | 35 | 12* |
| 10* | 40 | 50 | 55 | 35 | 40 | 40 | 10* |
| 8 | 60 | 70 | 80 | 45 | 55 | 60 | 8 |
| 6 | 80 | 95 | 105 | 60 | 75 | 80 | 6 |
| 4 | 105 | 125 | 140 | 80 | 100 | 110 | 4 |
| 3 | 120 | 145 | 165 | 95 | 115 | 130 | 3 |
| 2 | 140 | 170 | 190 | 110 | 135 | 150 | 2 |
| 1 | 165 | 195 | 220 | 130 | 155 | 175 | 1 |
| 1/0 | 195 | 230 | 260 | 150 | 180 | 205 | 1/0 |
| 2/0 | 225 | 265 | 300 | 175 | 210 | 235 | 2/0 |
| 3/0 | 260 | 310 | 350 | 200 | 240 | 275 | 3/0 |
| 4/0 | 300 | 360 | 405 | 235 | 280 | 315 | 4/0 |
| 250 | 340 | 405 | 455 | 265 | 315 | 355 | 250 |
| 300 | 375 | 445 | 505 | 290 | 350 | 395 | 300 |
| 350 | 420 | 505 | 570 | 330 | 395 | 445 | 350 |
| 400 | 455 | 545 | 615 | 355 | 425 | 480 | 400 |
| 500 | 515 | 620 | 700 | 405 | 485 | 545 | 500 |
| 600 | 575 | 690 | 780 | 455 | 540 | 615 | 600 |
| 700 | 630 | 755 | 855 | 500 | 595 | 675 | 700 |
| 750 | 655 | 785 | 885 | 515 | 620 | 700 | 750 |
| 800 | 680 | 815 | 920 | 535 | 645 | 725 | 800 |
| 900 | 730 | 870 | 985 | 580 | 700 | 785 | 900 |
| 1000 | 780 | 935 | 1055 | 625 | 750 | 845 | 1000 |
| 1250 | 890 | 1065 | 1200 | 710 | 855 | 960 | 1250 |
| 1500 | 980 | 1175 | 1325 | 795 | 950 | 1075 | 1500 |
| 1750 | 1070 | 1280 | 1445 | 875 | 1050 | 1185 | 1750 |
| 2000 | 1155 | 1385 | 1560 | 960 | 1150 | 1335 | 2000 |

**Table 310.17**   *Continued*

| | Temperature Rating of Conductor (See Table 310.13.) | | | | | | |
|---|---|---|---|---|---|---|---|
| | 60°C (140°F) | 75°C (167°F) | 90°C (194°F) | 60°C (140°F) | 75°C (167°F) | 90°C (194°F) | |
| | Types TW, UF | Types RHW, THHW, THW, THWN, XHHW, ZW | Types TBS, SA, SIS, FEP, FEPB, MI, RHH, RHW-2, THHN, THHW, THW-2, THWN-2, USE-2, XHH, XHHW, XHHW-2, ZW-2 | Types TW, UF | Types RHW, THHW, THW, THWN, XHHW | Types TBS, SA, SIS, THHN, THHW, THW-2, THWN-2, RHH, RHW-2, USE-2, XHH, XHHW, XHHW-2, ZW-2 | |
| Size AWG or kcmil | COPPER | | | ALUMINUM OR COPPER-CLAD ALUMINUM | | | Size AWG or kcmil |
| | CORRECTION FACTORS | | | | | | |
| Ambient Temp. (°C) | For ambient temperatures other than 30°C (86°F), multiply the allowable ampacities shown above by the appropriate factor shown below. | | | | | | Ambient Temp. (°F) |
| 21–25 | 1.08 | 1.05 | 1.04 | 1.08 | 1.05 | 1.04 | 70–77 |
| 26–30 | 1.00 | 1.00 | 1.00 | 1.00 | 1.00 | 1.00 | 78–86 |
| 31–35 | 0.91 | 0.94 | 0.96 | 0.91 | 0.94 | 0.96 | 87–95 |
| 36–40 | 0.82 | 0.88 | 0.91 | 0.82 | 0.88 | 0.91 | 96–104 |
| 41–45 | 0.71 | 0.82 | 0.87 | 0.71 | 0.82 | 0.87 | 105–113 |
| 46–50 | 0.58 | 0.75 | 0.82 | 0.58 | 0.75 | 0.82 | 114–122 |
| 51–55 | 0.41 | 0.67 | 0.76 | 0.41 | 0.67 | 0.76 | 123–131 |
| 56–60 | — | 0.58 | 0.71 | — | 0.58 | 0.71 | 132–140 |
| 61–70 | — | 0.33 | 0.58 | — | 0.33 | 0.58 | 141–158 |
| 71–80 | — | — | 0.41 | — | — | 0.41 | 159–176 |

* See 240.4(D).

**Table 310.18 Allowable Ampacities of Insulated Conductors Rated 0 Through 2000 Volts, 150°C Through 250°C (302°F Through 482°F). Not More Than Three Current-Carrying Conductors in Raceway or Cable, Based on Ambient Air Temperature of 40°C (104°F)**

| Size AWG or kcmil | Temperature Rating of Conductor (See Table 310.13.) | | | | Size AWG or kcmil |
|---|---|---|---|---|---|
| | 150°C (302°F) | 200°C (392°F) | 250°C (482°F) | 150°C (302°F) | |
| | Type Z | Types FEP, FEPB, PFA | Types PFAH, TFE | Type Z | |
| | COPPER | | NICKEL OR NICKEL-COATED COPPER | ALUMINUM OR COPPER-CLAD ALUMINUM | |
| 14 | 34 | 36 | 39 | — | 14 |
| 12 | 43 | 45 | 54 | 30 | 12 |
| 10 | 55 | 60 | 73 | 44 | 10 |
| 8 | 76 | 83 | 93 | 57 | 8 |
| 6 | 96 | 110 | 117 | 75 | 6 |
| 4 | 120 | 125 | 148 | 94 | 4 |
| 3 | 143 | 152 | 166 | 109 | 3 |
| 2 | 160 | 171 | 191 | 124 | 2 |
| 1 | 186 | 197 | 215 | 145 | 1 |
| 1/0 | 215 | 229 | 244 | 169 | 1/0 |
| 2/0 | 251 | 260 | 273 | 198 | 2/0 |
| 3/0 | 288 | 297 | 308 | 227 | 3/0 |
| 4/0 | 332 | 346 | 361 | 260 | 4/0 |

**CORRECTION FACTORS**

| Ambient Temp. (°C) | For ambient temperatures other than 40°C (104°F), multiply the allowable ampacities shown above by the appropriate factor shown below. | | | | Ambient Temp. (°F) |
|---|---|---|---|---|---|
| 41–50 | 0.95 | 0.97 | 0.98 | 0.95 | 105–122 |
| 51–60 | 0.90 | 0.94 | 0.95 | 0.90 | 123–140 |
| 61–70 | 0.85 | 0.90 | 0.93 | 0.85 | 141–158 |
| 71–80 | 0.80 | 0.87 | 0.90 | 0.80 | 159–176 |
| 81–90 | 0.74 | 0.83 | 0.87 | 0.74 | 177–194 |
| 91–100 | 0.67 | 0.79 | 0.85 | 0.67 | 195–212 |
| 101–120 | 0.52 | 0.71 | 0.79 | 0.52 | 213–248 |
| 121–140 | 0.30 | 0.61 | 0.72 | 0.30 | 249–284 |
| 141–160 | — | 0.50 | 0.65 | — | 285–320 |
| 161–180 | — | 0.35 | 0.58 | — | 321–356 |
| 181–200 | — | — | 0.49 | — | 357–392 |
| 201–225 | — | — | 0.35 | — | 393–437 |

**Table 310.19 Allowable Ampacities of Single-Insulated Conductors, Rated 0 Through 2000 Volts, 150°C Through 250°C (302°F Through 482°F), in Free Air, Based on Ambient Air Temperature of 40°C (104°F)**

| Size AWG or kcmil | Temperature Rating of Conductor (See Table 310.13.) | | | | Size AWG or kcmil |
| | 150°C (302°F) | 200°C (392°F) | 250°C (482°F) | 150°C (302°F) | |
| | Type Z | Types FEP, FEPB, PFA | Types PFAH, TFE | Type Z | |
| | COPPER | | NICKEL, OR NICKEL-COATED COPPER | ALUMINUM OR COPPER-CLAD ALUMINUM | |
| 14 | 46 | 54 | 59 | — | 14 |
| 12 | 60 | 68 | 78 | 47 | 12 |
| 10 | 80 | 90 | 107 | 63 | 10 |
| 8 | 106 | 124 | 142 | 83 | 8 |
| 6 | 155 | 165 | 205 | 112 | 6 |
| 4 | 190 | 220 | 278 | 148 | 4 |
| 3 | 214 | 252 | 327 | 170 | 3 |
| 2 | 255 | 293 | 381 | 198 | 2 |
| 1 | 293 | 344 | 440 | 228 | 1 |
| 1/0 | 339 | 399 | 532 | 263 | 1/0 |
| 2/0 | 390 | 467 | 591 | 305 | 2/0 |
| 3/0 | 451 | 546 | 708 | 351 | 3/0 |
| 4/0 | 529 | 629 | 830 | 411 | 4/0 |

**CORRECTION FACTORS**

| Ambient Temp. (°C) | For ambient temperatures other than 40°C (104°F), multiply the allowable ampacities shown above by the appropriate factor shown below. | | | | Ambient Temp. (°F) |
| 41–50 | 0.95 | 0.97 | 0.98 | 0.95 | 105–122 |
| 51–60 | 0.90 | 0.94 | 0.95 | 0.90 | 123–140 |
| 61–70 | 0.85 | 0.90 | 0.93 | 0.85 | 141–158 |
| 71–80 | 0.80 | 0.87 | 0.90 | 0.80 | 159–176 |
| 81–90 | 0.74 | 0.83 | 0.87 | 0.74 | 177–194 |
| 91–100 | 0.67 | 0.79 | 0.85 | 0.67 | 195–212 |
| 101–120 | 0.52 | 0.71 | 0.79 | 0.52 | 213–248 |
| 121–140 | 0.30 | 0.61 | 0.72 | 0.30 | 249–284 |
| 141–160 | — | 0.50 | 0.65 | — | 285–320 |
| 161–180 | — | 0.35 | 0.58 | — | 321–356 |
| 181–200 | — | — | 0.49 | — | 357–392 |
| 201–225 | — | — | 0.35 | — | 393–437 |

**Table 310.20 Ampacities of Not More Than Three Single Insulated Conductors, Rated 0 Through 2000 Volts, Supported on a Messenger, Based on Ambient Air Temperature of 40°C (104°F)**

| Size AWG or kcmil | Temperature Rating of Conductor (See Table 310.13.) | | | | Size AWG or kcmil |
| --- | --- | --- | --- | --- | --- |
| | 75°C (167°F) | 90°C (194°F) | 75°C (167°F) | 90°C (194°F) | |
| | Types RHW, THHW, THW, THWN, XHHW, ZW | Types MI, THHN, THHW, THW-2, THWN-2, RHH, RHW-2, USE-2, XHHW, XHHW-2, ZW-2 | Types RHW, THW, THWN, THHW, XHHW | Types THHN, THHW, RHH, XHHW, RHW-2, XHHW-2, THW-2, THWN-2, USE-2, ZW-2 | |
| | COPPER | | ALUMINUM OR COPPER-CLAD ALUMINUM | | |
| 8 | 57 | 66 | 44 | 51 | 8 |
| 6 | 76 | 89 | 59 | 69 | 6 |
| 4 | 101 | 117 | 78 | 91 | 4 |
| 3 | 118 | 138 | 92 | 107 | 3 |
| 2 | 135 | 158 | 106 | 123 | 2 |
| 1 | 158 | 185 | 123 | 144 | 1 |
| 1/0 | 183 | 214 | 143 | 167 | 1/0 |
| 2/0 | 212 | 247 | 165 | 193 | 2/0 |
| 3/0 | 245 | 287 | 192 | 224 | 3/0 |
| 4/0 | 287 | 335 | 224 | 262 | 4/0 |
| 250 | 320 | 374 | 251 | 292 | 250 |
| 300 | 359 | 419 | 282 | 328 | 300 |
| 350 | 397 | 464 | 312 | 364 | 350 |
| 400 | 430 | 503 | 339 | 395 | 400 |
| 500 | 496 | 580 | 392 | 458 | 500 |
| 600 | 553 | 647 | 440 | 514 | 600 |
| 700 | 610 | 714 | 488 | 570 | 700 |
| 750 | 638 | 747 | 512 | 598 | 750 |
| 800 | 660 | 773 | 532 | 622 | 800 |
| 900 | 704 | 826 | 572 | 669 | 900 |
| 1000 | 748 | 879 | 612 | 716 | 1000 |

**CORRECTION FACTORS**

| Ambient Temp. (°C) | For ambient temperatures other than 40°C (104°F), multiply the allowable ampacities shown above by the appropriate factor shown below. | | | | Ambient Temp. (°F) |
| --- | --- | --- | --- | --- | --- |
| 21–25 | 1.20 | 1.14 | 1.20 | 1.14 | 70–77 |
| 26–30 | 1.13 | 1.10 | 1.13 | 1.10 | 79–86 |
| 31–35 | 1.07 | 1.05 | 1.07 | 1.05 | 88–95 |
| 36–40 | 1.00 | 1.00 | 1.00 | 1.00 | 97–104 |
| 41–45 | 0.93 | 0.95 | 0.93 | 0.95 | 106–113 |
| 46–50 | 0.85 | 0.89 | 0.85 | 0.89 | 115–122 |
| 51–55 | 0.76 | 0.84 | 0.76 | 0.84 | 124–131 |
| 56–60 | 0.65 | 0.77 | 0.65 | 0.77 | 133–140 |
| 61–70 | 0.38 | 0.63 | 0.38 | 0.63 | 142–158 |
| 71–80 | — | 0.45 | — | 0.45 | 160–176 |

**Table 310.21 Ampacities of Bare or Covered Conductors in Free Air, Based on 40°C (104°F) Ambient, 80°C (176°F) Total Conductor Temperature, 610 mm/sec (2 ft/sec) Wind Velocity**

| Copper Conductors | | | | AAC Aluminum Conductors | | | |
| Bare | | Covered | | Bare | | Covered | |
| AWG or kcmil | Amperes | AWG or kcmil | Amperes | AWG or kcmil | Amperes | AWG or kcmil | Amperes |
|---|---|---|---|---|---|---|---|
| 8 | 98 | 8 | 103 | 8 | 76 | 8 | 80 |
| 6 | 124 | 6 | 130 | 6 | 96 | 6 | 101 |
| 4 | 155 | 4 | 163 | 4 | 121 | 4 | 127 |
| 2 | 209 | 2 | 219 | 2 | 163 | 2 | 171 |
| 1/0 | 282 | 1/0 | 297 | 1/0 | 220 | 1/0 | 231 |
| 2/0 | 329 | 2/0 | 344 | 2/0 | 255 | 2/0 | 268 |
| 3/0 | 382 | 3/0 | 401 | 3/0 | 297 | 3/0 | 312 |
| 4/0 | 444 | 4/0 | 466 | 4/0 | 346 | 4/0 | 364 |
| 250 | 494 | 250 | 519 | 266.8 | 403 | 266.8 | 423 |
| 300 | 556 | 300 | 584 | 336.4 | 468 | 336.4 | 492 |
| 500 | 773 | 500 | 812 | 397.5 | 522 | 397.5 | 548 |
| 750 | 1000 | 750 | 1050 | 477.0 | 588 | 477.0 | 617 |
| 1000 | 1193 | 1000 | 1253 | 556.5 | 650 | 556.5 | 682 |
| — | — | — | — | 636.0 | 709 | 636.0 | 744 |
| — | — | — | — | 795.0 | 819 | 795.0 | 860 |
| — | — | — | — | 954.0 | 920 | — | — |
| — | — | — | — | 1033.5 | 968 | 1033.5 | 1017 |
| — | — | — | — | 1272 | 1103 | 1272 | 1201 |
| — | — | — | — | 1590 | 1267 | 1590 | 1381 |
| — | — | — | — | 2000 | 1454 | 2000 | 1527 |

## 310.60 Conductors Rated 2001 to 35,000 Volts.

### (A) Definitions.

**Electrical Ducts.** As used in Article 310, electrical ducts shall include any of the electrical conduits recognized in Chapter 3 as suitable for use underground; other raceways round in cross section, listed for underground use, and embedded in earth or concrete.

**Thermal Resistivity.** As used in this *Code*, the heat transfer capability through a substance by conduction. It is the reciprocal of thermal conductivity and is designated Rho and expressed in the units °C-cm/watt.

### (B) Ampacities of Conductors Rated 2001 to 35,000 Volts. Ampacities for solid dielectric-insulated conductors shall be permitted to be determined by tables or under engineering supervision, as provided in 310.60(C) and (D).

**(1) Selection of Ampacity.** Where more than one calculated or tabulated ampacity could apply for a given circuit length, the lowest value shall be used.

*Exception: Where two different ampacities apply to adjacent portions of a circuit, the higher ampacity shall be permitted to be used beyond the point of transition, a distance equal to 3.0 m (10 ft) or 10 percent of the circuit length figured at the higher ampacity, whichever is less.*

FPN: See 110.40 for conductor temperature limitations due to termination provisions.

**(C) Tables.** Ampacities for conductors rated 2001 to 35,000 volts shall be as specified in the Ampacity Tables 310.67 through 310.86. Ampacities at ambient temperatures other than those shown in the tables shall be determined by the formula in 310.60(C)(4).

FPN No. 1: For ampacities calculated in accordance with 310.60(B), reference IEEE 835-1994 (IPCEA Pub. No. P-46-426), *Standard Power Cable Ampacity Tables,* and the references therein for availability of all factors and constants.

FPN No. 2: Ampacities provided by this section do not take voltage drop into consideration. See 210.19(A), FPN No. 4, for branch circuits and 215.2(D), FPN No. 2, for feeders.

**(1) Grounded Shields.** Ampacities shown in Tables 310.69, 310.70, 310.81, and 310.82 are for cable with shields grounded at one point only. Where shields are grounded at more than one point, ampacities shall be adjusted to take into consideration the heating due to shield currents.

**(2) Burial Depth of Underground Circuits.** Where the burial depth of direct burial or electrical duct bank circuits is modified from the values shown in a figure or table, ampacities shall be permitted to be modified as indicated in (a) and (b).

(a) Where burial depths are increased in part(s) of an electrical duct run, no decrease in ampacity of the conductors is needed, provided the total length of parts of the duct run increased in depth is less than 25 percent of the total run length.

(b) Where burial depths are deeper than shown in a specific underground ampacity table or figure, an ampacity derating factor of 6 percent per 300-mm (1-ft) increase in depth for all values of rho shall be permitted.

No rating change is needed where the burial depth is decreased.

**(3) Electrical Ducts in Figure 310.60.** At locations where electrical ducts enter equipment enclosures from underground, spacing between such ducts, as shown in Figure 310.60, shall be permitted to be reduced without requiring the ampacity of conductors therein to be reduced.

**(4) Ambients Not in Tables.** Ampacities at ambient temperatures other than those shown in the tables shall be determined by means of the following formula:

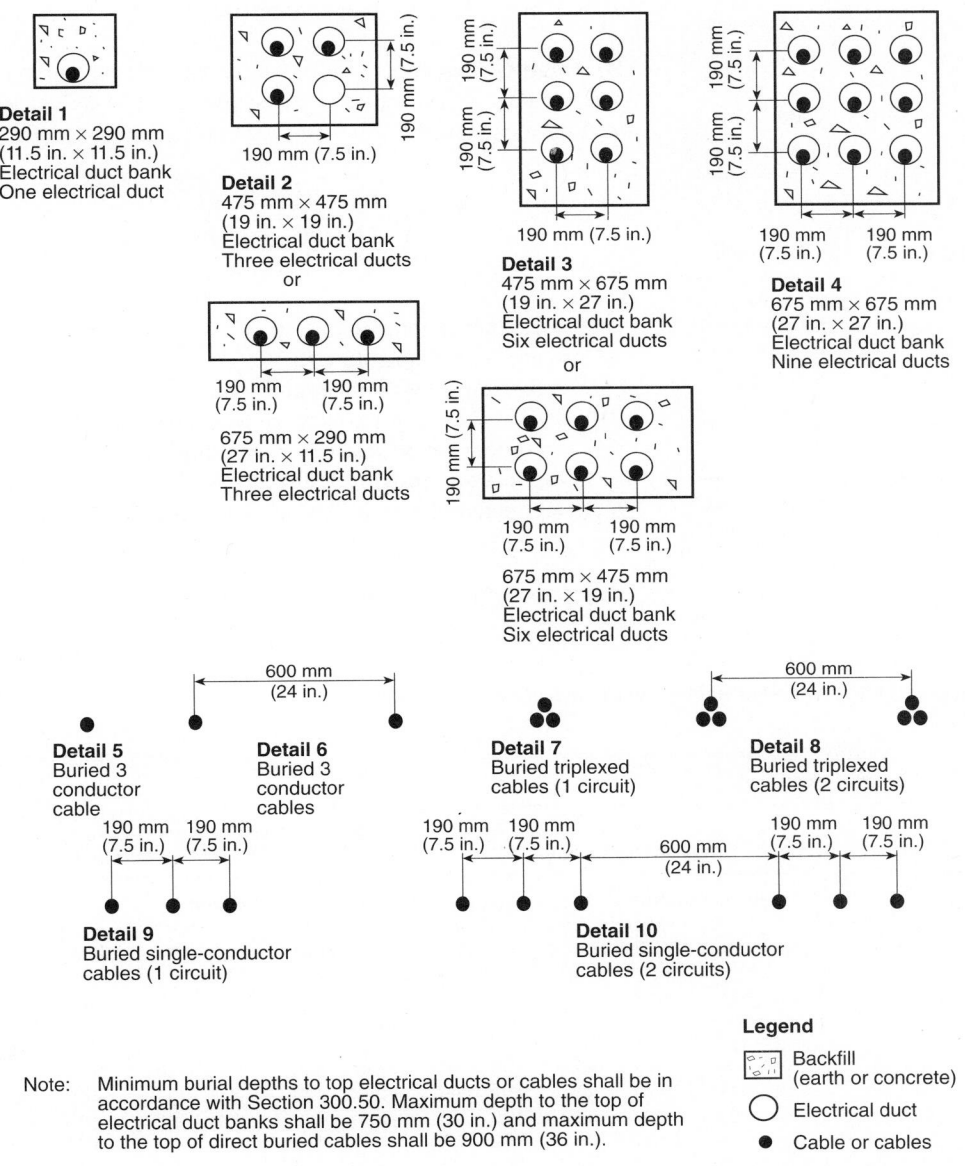

Note: Minimum burial depths to top electrical ducts or cables shall be in accordance with Section 300.50. Maximum depth to the top of electrical duct banks shall be 750 mm (30 in.) and maximum depth to the top of direct buried cables shall be 900 mm (36 in.).

**Figure 310.60 Cable installation dimensions for use with Tables 310.77 through 310.86.**

**Table 310.61 Conductor Application and Insulation**

| Trade Name | Type Letter | Maximum Operating Temperature | Application Provision | Insulation | Outer Covering |
|---|---|---|---|---|---|
| Medium voltage solid dielectric | MV-90 MV–105* | 90°C 105°C | Dry or wet locations rated 2001 volts and higher | Thermoplastic or thermosetting | Jacket, sheath, or armor |

*Where design conditions require maximum conductor temperatures above 90°C.

**Table 310.62 Thickness of Insulation for 601- to 2000-Volt Nonshielded Types RHH and RHW**

| Conductor Size (AWG or kcmil) | Column A[1] mm | Column A[1] mils | Column B[2] mm | Column B[2] mils |
|---|---|---|---|---|
| 14–10 | 2.03 | 80 | 1.52 | 60 |
| 8 | 2.03 | 80 | 1.78 | 70 |
| 6–2 | 2.41 | 95 | 1.78 | 70 |
| 1–2/0 | 2.79 | 110 | 2.29 | 90 |
| 3/0–4/0 | 2.79 | 110 | 2.29 | 90 |
| 213–500 | 3.18 | 125 | 2.67 | 105 |
| 501–1000 | 3.56 | 140 | 3.05 | 120 |

[1]Column A insulations are limited to natural, SBR, and butyl rubbers.
[2]Column B insulations are materials such as cross-linked polyethylene, ethylene propylene rubber, and composites thereof.

$$I_2 = I_1 \sqrt{\frac{TC - TA_2 - \Delta TD}{TC - TA_1 - \Delta TD}}$$

where:

$I_1$ = ampacity from tables at ambient $TA_1$

$I_2$ = ampacity at desired ambient $TA_2$

$TC$ = conductor temperature in degrees Celsius (°C)

$TA_1$ = surrounding ambient from tables in degrees Celsius (°C)

$TA_2$ = desired ambient in degrees Celsius (°C)

$\Delta TD$ = dielectric loss temperature rise

**(D) Engineering Supervision.** Under engineering supervision, conductor ampacities shall be permitted to be calculated by means of the following general formula:

$$I = \sqrt{\frac{TC - (TA + \Delta TD)}{RDC(1 + YC)RCA}}$$

where:

$TC$ = conductor temperature in °C

$TA$ = ambient temperature in °C

$\Delta TD$ = dielectric loss temperature rise

$RDC$ = dc resistance of conductor at temperature $TC$

$YC$ = component ac resistance resulting from skin effect and proximity effect

$RCA$ = effective thermal resistance between conductor and surrounding ambient

FPN: See Annex B for examples of formula applications.

**Table 310.63 Thickness of Insulation and Jacket for Nonshielded Solid Dielectric Insulated Conductors Rated 2001 to 8000 Volts**

| Conductor Size (AWG or kcmil) | 2001–5000 Volts Dry Locations, Single Conductor Without Jacket Insulation mm | mils | With Jacket Insulation mm | mils | With Jacket Jacket mm | mils | Wet or Dry Locations Single Conductor Insulation mm | mils | Single Conductor Jacket mm | mils | Multi-conductor Insulation* mm | mils | 5001–8000 Volts 100 Percent Insulation Level Wet or Dry Locations Single Conductor Insulation mm | mils | Single Conductor Jacket mm | mils | Multi-conductor Insulation* mm | mils |
|---|---|---|---|---|---|---|---|---|---|---|---|---|---|---|---|---|---|---|
| 8 | 2.79 | 110 | 2.29 | 90 | 0.76 | 30 | 3.18 | 125 | 2.03 | 80 | 2.29 | 90 | 4.57 | 180 | 2.03 | 80 | 4.57 | 180 |
| 6 | 2.79 | 110 | 2.29 | 90 | 0.76 | 30 | 3.18 | 125 | 2.03 | 80 | 2.29 | 90 | 4.57 | 180 | 2.03 | 80 | 4.57 | 180 |
| 4–2 | 2.79 | 110 | 2.29 | 90 | 1.14 | 45 | 3.18 | 125 | 2.03 | 80 | 2.29 | 90 | 4.57 | 180 | 2.41 | 95 | 4.57 | 180 |
| 1–2/0 | 2.79 | 110 | 2.29 | 90 | 1.14 | 45 | 3.18 | 125 | 2.03 | 80 | 2.29 | 90 | 4.57 | 180 | 2.41 | 95 | 4.57 | 180 |
| 3/0–4/0 | 2.79 | 110 | 2.29 | 90 | 1.65 | 65 | 3.18 | 125 | 2.41 | 95 | 2.29 | 90 | 4.57 | 180 | 2.79 | 110 | 4.57 | 180 |
| 213–500 | 3.05 | 120 | 2.29 | 90 | 1.65 | 65 | 3.56 | 140 | 2.79 | 110 | 2.29 | 90 | 5.33 | 210 | 2.79 | 110 | 5.33 | 210 |
| 501–750 | 3.30 | 130 | 2.29 | 90 | 1.65 | 65 | 3.94 | 155 | 3.18 | 125 | 2.29 | 90 | 5.97 | 235 | 3.18 | 125 | 5.97 | 235 |
| 751–1000 | 3.30 | 130 | 2.29 | 90 | 1.65 | 65 | 3.94 | 155 | 3.18 | 125 | 2.29 | 90 | 6.35 | 250 | 3.56 | 140 | 6.35 | 250 |

*Under a common overall covering such as a jacket, sheath, or armor.

**Table 310.64 Thickness of Insulation for Shielded Solid Dielectric Insulated Conductors Rated 2001 to 35,000 Volts**

| Conductor Size (AWG or kcmil) | 2001–5000 Volts | | 5001–8000 Volts | | | | 8001–15,000 Volts | | | | 15,001–25,000 Volts | | | | 25,001–28,000 Volts | | | | 28,001–35,000 Volts | | | |
|---|---|---|---|---|---|---|---|---|---|---|---|---|---|---|---|---|---|---|---|---|---|---|
| | | | 100 Percent Insulation Level 1 | | 133 Percent Insulation Level 2 | | 100 Percent Insulation Level 1 | | 133 Percent Insulation Level 2 | | 100 Percent Insulation Level 1 | | 133 Percent Insulation Level 2 | | 100 Percent Insulation Level 1 | | 133 Percent Insulation Level 2 | | 100 Percent Insulation Level 1 | | 133 Percent Insulation Level 2 | |
| | mm | mils | mm | mils | mm | mils | mm | mils | mm | mils | mm | mils | mm | mils | mm | mils | mm | mils | mm | mils | mm | mils |
| 8 | 2.29 | 90 | — | — | — | — | — | — | — | — | — | — | — | — | — | — | — | — | — | — | — | — |
| 6–4 | 2.29 | 90 | 2.92 | 115 | 3.56 | 140 | — | — | — | — | — | — | — | — | — | — | — | — | — | — | — | — |
| 2 | 2.29 | 90 | 2.92 | 115 | 3.56 | 140 | 4.45 | 175 | 5.46 | 215 | — | — | — | — | — | — | — | — | — | — | — | — |
| 1 | 2.29 | 90 | 2.92 | 115 | 3.56 | 140 | 4.45 | 175 | 5.46 | 215 | 6.60 | 260 | 8.76 | 345 | 7.11 | 280 | 8.76 | 345 | — | — | — | — |
| 1/0–2000 | 2.29 | 90 | 2.92 | 115 | 3.56 | 140 | 4.45 | 175 | 5.46 | 215 | 6.60 | 260 | 8.76 | 345 | 7.11 | 280 | 8.76 | 345 | 8.76 | 345 | 10.67 | 420 |

[1]**100 Percent Insulation Level.** Cables in this category shall be permitted to be applied where the system is provided with relay protection such that ground faults will be cleared as rapidly as possible but, in any case, within 1 minute. While these cables are applicable to the great majority of cable installations that are on grounded systems, they shall be permitted to be used also on other systems for which the application of cables is acceptable, provided the above clearing requirements are met in completely de-energizing the faulted section.

[2]**133 Percent Insulation Level.** This insulation level corresponds to that formerly designated for ungrounded systems. Cables in this category shall be permitted to be applied in situations where the clearing time requirements of the 100 percent level category cannot be met and yet there is adequate assurance that the faulted section will be de-energized in a time not exceeding 1 hour. Also, they shall be permitted to be used where additional insulation strength over the 100 percent level category is desirable.

**Table 310.67 Ampacities of Insulated Single Copper Conductor Cables Triplexed in Air Based on Conductor Temperatures of 90°C (194°F) and 105°C (221°F) and Ambient Air Temperature of 40°C (104°F)**

| Conductor Size (AWG or kcmil) | 2001–5000 Volts Ampacity | | 5001–35,000 Volts Ampacity | |
|---|---|---|---|---|
| | 90°C (194°F) Type MV-90 | 105°C (221°F) Type MV-105 | 90°C (194°F) Type MV-90 | 105°C (221°F) Type MV-105 |
| 8 | 65 | 74 | — | — |
| 6 | 90 | 99 | 100 | 110 |
| 4 | 120 | 130 | 130 | 140 |
| 2 | 160 | 175 | 170 | 195 |
| 1 | 185 | 205 | 195 | 225 |
| 1/0 | 215 | 240 | 225 | 255 |
| 2/0 | 250 | 275 | 260 | 295 |
| 3/0 | 290 | 320 | 300 | 340 |
| 4/0 | 335 | 375 | 345 | 390 |
| 250 | 375 | 415 | 380 | 430 |
| 350 | 465 | 515 | 470 | 525 |
| 500 | 580 | 645 | 580 | 650 |
| 750 | 750 | 835 | 730 | 820 |
| 1000 | 880 | 980 | 850 | 950 |

**Table 310.68 Ampacities of Insulated Single Aluminum Conductor Cables Triplexed in Air Based on Conductor Temperatures of 90°C (194°F) and 105°C (221°F) and Ambient Air Temperature of 40°C (104°F)**

| Conductor Size (AWG or kcmil) | 2001–5000 Volts Ampacity | | 5001–35,000 Volts Ampacity | |
|---|---|---|---|---|
| | 90°C (194°F) Type MV-90 | 105°C (221°F) Type MV-105 | 90°C (194°F) Type MV-90 | 105°C (221°F) Type MV-105 |
| 8 | 50 | 57 | — | — |
| 6 | 70 | 77 | 75 | 84 |
| 4 | 90 | 100 | 100 | 110 |
| 2 | 125 | 135 | 130 | 150 |
| 1 | 145 | 160 | 150 | 175 |
| 1/0 | 170 | 185 | 175 | 200 |
| 2/0 | 195 | 215 | 200 | 230 |
| 3/0 | 225 | 250 | 230 | 265 |
| 4/0 | 265 | 290 | 270 | 305 |
| 250 | 295 | 325 | 300 | 335 |
| 350 | 365 | 405 | 370 | 415 |
| 500 | 460 | 510 | 460 | 515 |
| 750 | 600 | 665 | 590 | 660 |
| 1000 | 715 | 800 | 700 | 780 |

**Table 310.69 Ampacities of Insulated Single Copper Conductor Isolated in Air Based on Conductor Temperatures of 90°C (194°F) and 105°C (221°F) and Ambient Air Temperature of 40°C (104°F)**

| Conductor Size (AWG or kcmil) | Temperature Rating of Conductor (See Table 310.61.) | | | | | |
| | 2001–5000 Volts Ampacity | | 5001–15,000 Volts Ampacity | | 15,001–35,000 Volts Ampacity | |
| | 90°C (194°F) Type MV-90 | 105°C (221°F) Type MV-105 | 90°C (194°F) Type MV-90 | 105°C (221°F) Type MV-105 | 90°C (194°F) Type MV-90 | 105°C (221°F) Type MV-105 |
|---|---|---|---|---|---|---|
| 8 | 83 | 93 | — | — | — | — |
| 6 | 110 | 120 | 110 | 125 | — | — |
| 4 | 145 | 160 | 150 | 165 | — | — |
| 2 | 190 | 215 | 195 | 215 | — | — |
| 1 | 225 | 250 | 225 | 250 | 225 | 250 |
| 1/0 | 260 | 290 | 260 | 290 | 260 | 290 |
| 2/0 | 300 | 330 | 300 | 335 | 300 | 330 |
| 3/0 | 345 | 385 | 345 | 385 | 345 | 380 |
| 4/0 | 400 | 445 | 400 | 445 | 395 | 445 |
| 250 | 445 | 495 | 445 | 495 | 440 | 490 |
| 350 | 550 | 615 | 550 | 610 | 545 | 605 |
| 500 | 695 | 775 | 685 | 765 | 680 | 755 |
| 750 | 900 | 1000 | 885 | 990 | 870 | 970 |
| 1000 | 1075 | 1200 | 1060 | 1185 | 1040 | 1160 |
| 1250 | 1230 | 1370 | 1210 | 1350 | 1185 | 1320 |
| 1500 | 1365 | 1525 | 1345 | 1500 | 1315 | 1465 |
| 1750 | 1495 | 1665 | 1470 | 1640 | 1430 | 1595 |
| 2000 | 1605 | 1790 | 1575 | 1755 | 1535 | 1710 |

**Table 310.70 Ampacities of Insulated Single Aluminum Conductor Isolated in Air Based on Conductor Temperatures of 90°C (194°F) and 105°C (221°F) and Ambient Air Temperature of 40°C (104°F)**

| Conductor Size (AWG or kcmil) | Temperature Rating of Conductor (See Table 310.61.) | | | | | |
| | 2001–5000 Volts Ampacity | | 5001–15,000 Volts Ampacity | | 15,001–35,000 Volts Ampacity | |
| | 90°C (194°F) Type MV-90 | 105°C (221°F) Type MV-105 | 90°C (194°F) Type MV-90 | 105°C (221°F) Type MV-105 | 90°C (194°F) Type MV-90 | 105°C (221°F) Type MV-105 |
|---|---|---|---|---|---|---|
| 8 | 64 | 71 | — | — | — | — |
| 6 | 85 | 95 | 87 | 97 | — | — |
| 4 | 115 | 125 | 115 | 130 | — | — |
| 2 | 150 | 165 | 150 | 170 | — | — |
| 1 | 175 | 195 | 175 | 195 | 175 | 195 |
| 1/0 | 200 | 225 | 200 | 225 | 200 | 225 |
| 2/0 | 230 | 260 | 235 | 260 | 230 | 260 |
| 3/0 | 270 | 300 | 270 | 300 | 270 | 300 |
| 4/0 | 310 | 350 | 310 | 350 | 310 | 345 |
| 250 | 345 | 385 | 345 | 385 | 345 | 380 |
| 350 | 430 | 480 | 430 | 480 | 430 | 475 |
| 500 | 545 | 605 | 535 | 600 | 530 | 590 |
| 750 | 710 | 790 | 700 | 780 | 685 | 765 |
| 1000 | 855 | 950 | 840 | 940 | 825 | 920 |
| 1250 | 980 | 1095 | 970 | 1080 | 950 | 1055 |
| 1500 | 1105 | 1230 | 1085 | 1215 | 1060 | 1180 |
| 1750 | 1215 | 1355 | 1195 | 1335 | 1165 | 1300 |
| 2000 | 1320 | 1475 | 1295 | 1445 | 1265 | 1410 |

**Table 310.71 Ampacities of an Insulated Three-Conductor Copper Cable Isolated in Air Based on Conductor Temperatures of 90°C (194°F) and 105°C (221°F) and Ambient Air Temperature of 40°C (104°F)**

| Conductor Size (AWG or kcmil) | Temperature Rating of Conductor (See Table 310.61.) | | | |
| | 2001–5000 Volts Ampacity | | 5001–35,000 Volts Ampacity | |
| | 90°C (194°F) Type MV-90 | 105°C (221°F) Type MV-105 | 90°C (194°F) Type MV-90 | 105°C (221°F) Type MV-105 |
|---|---|---|---|---|
| 8 | 59 | 66 | — | — |
| 6 | 79 | 88 | 93 | 105 |
| 4 | 105 | 115 | 120 | 135 |
| 2 | 140 | 154 | 165 | 185 |
| 1 | 160 | 180 | 185 | 210 |
| 1/0 | 185 | 205 | 215 | 240 |
| 2/0 | 215 | 240 | 245 | 275 |
| 3/0 | 250 | 280 | 285 | 315 |
| 4/0 | 285 | 320 | 325 | 360 |
| 250 | 320 | 355 | 360 | 400 |
| 350 | 395 | 440 | 435 | 490 |
| 500 | 485 | 545 | 535 | 600 |
| 750 | 615 | 685 | 670 | 745 |
| 1000 | 705 | 790 | 770 | 860 |

**Table 310.72 Ampacities of an Insulated Three-Conductor Aluminum Cable Isolated in Air Based on Conductor Temperatures of 90°C (194°F) and 105°C (221°F) and Ambient Air Temperature of 40°C (104°F)**

| Conductor Size (AWG or kcmil) | Temperature Rating of Conductor (See Table 310.61.) | | | |
| | 2001–5000 Volts Ampacity | | 5001–35,000 Volts Ampacity | |
| | 90°C (194°F) Type MV-90 | 105°C (221°F) Type MV-105 | 90°C (194°F) Type MV-90 | 105°C (221°F) Type MV-105 |
|---|---|---|---|---|
| 8 | 46 | 51 | — | — |
| 6 | 61 | 68 | 72 | 80 |
| 4 | 81 | 90 | 95 | 105 |
| 2 | 110 | 120 | 125 | 145 |
| 1 | 125 | 140 | 145 | 165 |
| 1/0 | 145 | 160 | 170 | 185 |
| 2/0 | 170 | 185 | 190 | 215 |
| 3/0 | 195 | 215 | 220 | 245 |
| 4/0 | 225 | 250 | 255 | 285 |
| 250 | 250 | 280 | 280 | 315 |
| 350 | 310 | 345 | 345 | 385 |
| 500 | 385 | 430 | 425 | 475 |
| 750 | 495 | 550 | 540 | 600 |
| 1000 | 585 | 650 | 635 | 705 |

**Table 310.73 Ampacities of an Insulated Triplexed or Three Single-Conductor Copper Cables in Isolated Conduit in Air Based on Conductor Temperatures of 90°C (194°F) and 105°C (221°F) and Ambient Air Temperature of 40°C (104°F)**

| | Temperature Rating of Conductor (See Table 310.61.) | | | |
|---|---|---|---|---|
| | 2001–5000 Volts Ampacity | | 5001–35,000 Volts Ampacity | |
| Conductor Size (AWG or kcmil) | 90°C (194°F) Type MV-90 | 105°C (221°F) Type MV-105 | 90°C (194°F) Type MV-90 | 105°C (221°F) Type MV-105 |
| 8 | 55 | 61 | — | — |
| 6 | 75 | 84 | 83 | 93 |
| 4 | 97 | 110 | 110 | 120 |
| 2 | 130 | 145 | 150 | 165 |
| 1 | 155 | 175 | 170 | 190 |
| 1/0 | 180 | 200 | 195 | 215 |
| 2/0 | 205 | 225 | 225 | 255 |
| 3/0 | 240 | 270 | 260 | 290 |
| 4/0 | 280 | 305 | 295 | 330 |
| 250 | 315 | 355 | 330 | 365 |
| 350 | 385 | 430 | 395 | 440 |
| 500 | 475 | 530 | 480 | 535 |
| 750 | 600 | 665 | 585 | 655 |
| 1000 | 690 | 770 | 675 | 755 |

**Table 310.74 Ampacities of an Insulated Triplexed or Three Single-Conductor Aluminum Cables in Isolated Conduit in Air Based on Conductor Temperatures of 90°C (194°F) and 105°C (221°F) and Ambient Air Temperature of 40°C (104°F)**

| | Temperature Rating of Conductor (See Table 310.61.) | | | |
|---|---|---|---|---|
| | 2001–5000 Volts Ampacity | | 5001–35,000 Volts Ampacity | |
| Conductor Size (AWG or kcmil) | 90°C (194°F) Type MV-90 | 105°C (221°F) Type MV-105 | 90°C (194°F) Type MV-90 | 105°C (221°F) Type MV-105 |
| 8 | 43 | 48 | — | — |
| 6 | 58 | 65 | 65 | 72 |
| 4 | 76 | 85 | 84 | 94 |
| 2 | 100 | 115 | 115 | 130 |
| 1 | 120 | 135 | 130 | 150 |
| 1/0 | 140 | 155 | 150 | 170 |
| 2/0 | 160 | 175 | 175 | 200 |
| 3/0 | 190 | 210 | 200 | 225 |
| 4/0 | 215 | 240 | 230 | 260 |
| 250 | 250 | 280 | 255 | 290 |
| 350 | 305 | 340 | 310 | 350 |
| 500 | 380 | 425 | 385 | 430 |
| 750 | 490 | 545 | 485 | 540 |
| 1000 | 580 | 645 | 565 | 640 |

**Table 310.75 Ampacities of an Insulated Three-Conductor Copper Cable in Isolated Conduit in Air Based on Conductor Temperatures of 90°C (194°F) and 105°C (221°F) and Ambient Air Temperature of 40°C (104°F)**

| | Temperature Rating of Conductor (See Table 310.61.) | | | |
|---|---|---|---|---|
| | 2001–5000 Volts Ampacity | | 5001–35,000 Volts Ampacity | |
| Conductor Size (AWG or kcmil) | 90°C (194°F) Type MV-90 | 105°C (221°F) Type MV-105 | 90°C (194°F) Type MV-90 | 105°C (221°F) Type MV-105 |
| 8 | 52 | 58 | — | — |
| 6 | 69 | 77 | 83 | 92 |
| 4 | 91 | 100 | 105 | 120 |
| 2 | 125 | 135 | 145 | 165 |
| 1 | 140 | 155 | 165 | 185 |
| 1/0 | 165 | 185 | 195 | 215 |
| 2/0 | 190 | 210 | 220 | 245 |
| 3/0 | 220 | 245 | 250 | 280 |
| 4/0 | 255 | 285 | 290 | 320 |
| 250 | 280 | 315 | 315 | 350 |
| 350 | 350 | 390 | 385 | 430 |
| 500 | 425 | 475 | 470 | 525 |
| 750 | 525 | 585 | 570 | 635 |
| 1000 | 590 | 660 | 650 | 725 |

**Table 310.76 Ampacities of an Insulated Three-Conductor Aluminum Cable in Isolated Conduit in Air Based on Conductor Temperatures of 90°C (194°F) and 105°C (221°F) and Ambient Air Temperature of 40°C (104°F)**

| | Temperature Rating of Conductor (See Table 310.61.) | | | |
|---|---|---|---|---|
| | 2001–5000 Volts Ampacity | | 5001–35,000 Volts Ampacity | |
| Conductor Size (AWG or kcmil) | 90°C (194°F) Type MV-90 | 105°C (221°F) Type MV-105 | 90°C (194°F) Type MV-90 | 105°C (221°F) Type MV-105 |
| 8 | 41 | 46 | — | — |
| 6 | 53 | 59 | 64 | 71 |
| 4 | 71 | 79 | 84 | 94 |
| 2 | 96 | 105 | 115 | 125 |
| 1 | 110 | 125 | 130 | 145 |
| 1/0 | 130 | 145 | 150 | 170 |
| 2/0 | 150 | 165 | 170 | 190 |
| 3/0 | 170 | 190 | 195 | 220 |
| 4/0 | 200 | 225 | 225 | 255 |
| 250 | 220 | 245 | 250 | 280 |
| 350 | 275 | 305 | 305 | 340 |
| 500 | 340 | 380 | 380 | 425 |
| 750 | 430 | 480 | 470 | 520 |
| 1000 | 505 | 560 | 550 | 615 |

**Table 310.77 Ampacities of Three Single-Insulated Copper Conductors in Underground Electrical Ducts (Three Conductors per Electrical Duct) Based on Ambient Earth Temperature of 20°C (68°F), Electrical Duct Arrangement per Figure 310.60, 100 Percent Load Factor, Thermal Resistance (RHO) of 90, Conductor Temperatures of 90°C (194°F) and 105°C (221°F)**

| | Temperature Rating of Conductor (See Table 310.61.) | | | |
| | 2001–5000 Volts Ampacity | | 5001–35,000 Volts Ampacity | |
| Conductor Size (AWG or kcmil) | 90°C (194°F) Type MV-90 | 105°C (221°F) Type MV-105 | 90°C (194°F) Type MV-90 | 105°C (221°F) Type MV-105 |
|---|---|---|---|---|
| **One Circuit (See Figure 310.60, Detail 1.)** | | | | |
| 8 | 64 | 69 | — | — |
| 6 | 85 | 92 | 90 | 97 |
| 4 | 110 | 120 | 115 | 125 |
| 2 | 145 | 155 | 155 | 165 |
| 1 | 170 | 180 | 175 | 185 |
| 1/0 | 195 | 210 | 200 | 215 |
| 2/0 | 220 | 235 | 230 | 245 |
| 3/0 | 250 | 270 | 260 | 275 |
| 4/0 | 290 | 310 | 295 | 315 |
| 250 | 320 | 345 | 325 | 345 |
| 350 | 385 | 415 | 390 | 415 |
| 500 | 470 | 505 | 465 | 500 |
| 750 | 585 | 630 | 565 | 610 |
| 1000 | 670 | 720 | 640 | 690 |
| **Three Circuits (See Figure 310.60, Detail 2.)** | | | | |
| 8 | 56 | 60 | — | — |
| 6 | 73 | 79 | 77 | 83 |
| 4 | 95 | 100 | 99 | 105 |
| 2 | 125 | 130 | 130 | 135 |
| 1 | 140 | 150 | 145 | 155 |
| 1/0 | 160 | 175 | 165 | 175 |
| 2/0 | 185 | 195 | 185 | 200 |
| 3/0 | 210 | 225 | 210 | 225 |
| 4/0 | 235 | 255 | 240 | 255 |
| 250 | 260 | 280 | 260 | 280 |
| 350 | 315 | 335 | 310 | 330 |
| 500 | 375 | 405 | 370 | 395 |
| 750 | 460 | 495 | 440 | 475 |
| 1000 | 525 | 565 | 495 | 535 |
| **Six Circuits (See Figure 310.60, Detail 3.)** | | | | |
| 8 | 48 | 52 | — | — |
| 6 | 62 | 67 | 64 | 68 |
| 4 | 80 | 86 | 82 | 88 |
| 2 | 105 | 110 | 105 | 115 |
| 1 | 115 | 125 | 120 | 125 |
| 1/0 | 135 | 145 | 135 | 145 |
| 2/0 | 150 | 160 | 150 | 165 |
| 3/0 | 170 | 185 | 170 | 185 |
| 4/0 | 195 | 210 | 190 | 205 |
| 250 | 210 | 225 | 210 | 225 |
| 350 | 250 | 270 | 245 | 265 |
| 500 | 300 | 325 | 290 | 310 |
| 750 | 365 | 395 | 350 | 375 |
| 1000 | 410 | 445 | 390 | 415 |

**Table 310.78 Ampacities of Three Single-Insulated Aluminum Conductors in Underground Electrical Ducts (Three Conductors per Electrical Duct) Based on Ambient Earth Temperature of 20°C (68°F), Electrical Duct Arrangement per Figure 310.60, 100 Percent Load Factor, Thermal Resistance (RHO) of 90, Conductor Temperatures of 90°C (194°F) and 105°C (221°F)**

| | Temperature Rating of Conductor (See Table 310.61.) | | | |
| | 2001–5000 Volts Ampacity | | 5001–35,000 Volts Ampacity | |
| Conductor Size (AWG or kcmil) | 90°C (194°F) Type MV-90 | 105°C (221°F) Type MV-105 | 90°C (194°F) Type MV-90 | 105°C (221°F) Type MV-105 |
|---|---|---|---|---|
| **One Circuit (See Figure 310.60, Detail 1.)** | | | | |
| 8 | 50 | 54 | — | — |
| 6 | 66 | 71 | 70 | 75 |
| 4 | 86 | 93 | 91 | 98 |
| 2 | 115 | 125 | 120 | 130 |
| 1 | 130 | 140 | 135 | 145 |
| 1/0 | 150 | 160 | 155 | 165 |
| 2/0 | 170 | 185 | 175 | 190 |
| 3/0 | 195 | 210 | 200 | 215 |
| 4/0 | 225 | 245 | 230 | 245 |
| 250 | 250 | 270 | 250 | 270 |
| 350 | 305 | 325 | 305 | 330 |
| 500 | 370 | 400 | 370 | 400 |
| 750 | 470 | 505 | 455 | 490 |
| 1000 | 545 | 590 | 525 | 565 |
| **Three Circuits (See Figure 310.60, Detail 2.)** | | | | |
| 8 | 44 | 47 | — | — |
| 6 | 57 | 61 | 60 | 65 |
| 4 | 74 | 80 | 77 | 83 |
| 2 | 96 | 105 | 100 | 105 |
| 1 | 110 | 120 | 110 | 120 |
| 1/0 | 125 | 135 | 125 | 140 |
| 2/0 | 145 | 155 | 145 | 155 |
| 3/0 | 160 | 175 | 165 | 175 |
| 4/0 | 185 | 200 | 185 | 200 |
| 250 | 205 | 220 | 200 | 220 |
| 350 | 245 | 265 | 245 | 260 |
| 500 | 295 | 320 | 290 | 315 |
| 750 | 370 | 395 | 355 | 385 |
| 1000 | 425 | 460 | 405 | 440 |
| **Six Circuits (See Figure 310.60, Detail 3.)** | | | | |
| 8 | 38 | 41 | — | — |
| 6 | 48 | 52 | 50 | 54 |
| 4 | 62 | 67 | 64 | 69 |
| 2 | 80 | 86 | 80 | 88 |
| 1 | 91 | 98 | 90 | 99 |
| 1/0 | 105 | 110 | 105 | 110 |
| 2/0 | 115 | 125 | 115 | 125 |
| 3/0 | 135 | 145 | 130 | 145 |
| 4/0 | 150 | 165 | 150 | 160 |
| 250 | 165 | 180 | 165 | 175 |
| 350 | 195 | 210 | 195 | 210 |
| 500 | 240 | 255 | 230 | 250 |
| 750 | 290 | 315 | 280 | 305 |
| 1000 | 335 | 360 | 320 | 345 |

**Table 310.79 Ampacities of Three Insulated Copper Conductors Cabled Within an Overall Covering (Three-Conductor Cable) in Underground Electrical Ducts (One Cable per Electrical Duct) Based on Ambient Earth Temperature of 20°C (68°F), Electrical Duct Arrangement per Figure 310.60, 100 Percent Load Factor, Thermal Resistance (RHO) of 90, Conductor Temperatures of 90°C (194°F) and 105°C (221°C)**

| | Temperature Rating of Conductor (See Table 310.61.) | | | |
| | 2001–5000 Volts Ampacity | | 5001–35,000 Volts Ampacity | |
| Conductor Size (AWG or kcmil) | 90°C (194°F) Type MV-90 | 105°C (221°F) Type MV-105 | 90°C (194°F) Type MV-90 | 105°C (221°F) Type MV-105 |
|---|---|---|---|---|
| **One Circuit (See Figure 310.60, Detail 1.)** | | | | |
| 8 | 59 | 64 | — | — |
| 6 | 78 | 84 | 88 | 95 |
| 4 | 100 | 110 | 115 | 125 |
| 2 | 135 | 145 | 150 | 160 |
| 1 | 155 | 165 | 170 | 185 |
| 1/0 | 175 | 190 | 195 | 210 |
| 2/0 | 200 | 220 | 220 | 235 |
| 3/0 | 230 | 250 | 250 | 270 |
| 4/0 | 265 | 285 | 285 | 305 |
| 250 | 290 | 315 | 310 | 335 |
| 350 | 355 | 380 | 375 | 400 |
| 500 | 430 | 460 | 450 | 485 |
| 750 | 530 | 570 | 545 | 585 |
| 1000 | 600 | 645 | 615 | 660 |
| **Three Circuits (See Figure 310.60, Detail 2.)** | | | | |
| 8 | 53 | 57 | — | — |
| 6 | 69 | 74 | 75 | 81 |
| 4 | 89 | 96 | 97 | 105 |
| 2 | 115 | 125 | 125 | 135 |
| 1 | 135 | 145 | 140 | 155 |
| 1/0 | 150 | 165 | 160 | 175 |
| 2/0 | 170 | 185 | 185 | 195 |
| 3/0 | 195 | 210 | 205 | 220 |
| 4/0 | 225 | 240 | 230 | 250 |
| 250 | 245 | 265 | 255 | 270 |
| 350 | 295 | 315 | 305 | 325 |
| 500 | 355 | 380 | 360 | 385 |
| 750 | 430 | 465 | 430 | 465 |
| 1000 | 485 | 520 | 485 | 515 |
| **Six Circuits (See Figure 310.60, Detail 3.)** | | | | |
| 8 | 46 | 50 | — | — |
| 6 | 60 | 65 | 63 | 68 |
| 4 | 77 | 83 | 81 | 87 |
| 2 | 98 | 105 | 105 | 110 |
| 1 | 110 | 120 | 115 | 125 |
| 1/0 | 125 | 135 | 130 | 145 |
| 2/0 | 145 | 155 | 150 | 160 |
| 3/0 | 165 | 175 | 170 | 180 |
| 4/0 | 185 | 200 | 190 | 200 |
| 250 | 200 | 220 | 205 | 220 |
| 350 | 240 | 270 | 245 | 275 |
| 500 | 290 | 310 | 290 | 305 |
| 750 | 350 | 375 | 340 | 365 |
| 1000 | 390 | 420 | 380 | 405 |

**Table 310.80 Ampacities of Three Insulated Aluminum Conductors Cabled Within an Overall Covering (Three-Conductor Cable) in Underground Electrical Ducts (One Cable per Electrical Duct) Based on Ambient Earth Temperature of 20°C (68°F), Electrical Duct Arrangement per Figure 310.60, 100 Percent Load Factor, Thermal Resistance (RHO) of 90, Conductor Temperatures of 90°C (194°F) and 105°C (221°C)**

| | Temperature Rating of Conductor (See Table 310.61.) | | | |
| | 2001–5000 Volts Ampacity | | 5001–35,000 Volts Ampacity | |
| Conductor Size (AWG or kcmil) | 90°C (194°F) Type MV-90 | 105°C (221°F) Type MV-105 | 90°C (194°F) Type MV-90 | 105°C (221°F) Type MV-105 |
|---|---|---|---|---|
| **One Circuit (See Figure 310.60, Detail 1.)** | | | | |
| 8 | 46 | 50 | — | — |
| 6 | 61 | 66 | 69 | 74 |
| 4 | 80 | 86 | 89 | 96 |
| 2 | 105 | 110 | 115 | 125 |
| 1 | 120 | 130 | 135 | 145 |
| 1/0 | 140 | 150 | 150 | 165 |
| 2/0 | 160 | 170 | 170 | 185 |
| 3/0 | 180 | 195 | 195 | 210 |
| 4/0 | 205 | 220 | 220 | 240 |
| 250 | 230 | 245 | 245 | 265 |
| 350 | 280 | 310 | 295 | 315 |
| 500 | 340 | 365 | 355 | 385 |
| 750 | 425 | 460 | 440 | 475 |
| 1000 | 495 | 535 | 510 | 545 |
| **Three Circuits (See Figure 310.60, Detail 2.)** | | | | |
| 8 | 41 | 44 | — | — |
| 6 | 54 | 58 | 59 | 64 |
| 4 | 70 | 75 | 75 | 81 |
| 2 | 90 | 97 | 100 | 105 |
| 1 | 105 | 110 | 110 | 120 |
| 1/0 | 120 | 125 | 125 | 135 |
| 2/0 | 135 | 145 | 140 | 155 |
| 3/0 | 155 | 165 | 160 | 175 |
| 4/0 | 175 | 185 | 180 | 195 |
| 250 | 190 | 205 | 200 | 215 |
| 350 | 230 | 250 | 240 | 255 |
| 500 | 280 | 300 | 285 | 305 |
| 750 | 345 | 375 | 350 | 375 |
| 1000 | 400 | 430 | 400 | 430 |
| **Six Circuits (See Figure 310.60, Detail 3.)** | | | | |
| 8 | 36 | 39 | — | — |
| 6 | 46 | 50 | 49 | 53 |
| 4 | 60 | 65 | 63 | 68 |
| 2 | 77 | 83 | 80 | 86 |
| 1 | 87 | 94 | 90 | 98 |
| 1/0 | 99 | 105 | 105 | 110 |
| 2/0 | 110 | 120 | 115 | 125 |
| 3/0 | 130 | 140 | 130 | 140 |
| 4/0 | 145 | 155 | 150 | 160 |
| 250 | 160 | 170 | 160 | 170 |
| 350 | 190 | 205 | 190 | 205 |
| 500 | 230 | 245 | 230 | 245 |
| 750 | 280 | 305 | 275 | 295 |
| 1000 | 320 | 345 | 315 | 335 |

**Table 310.81 Ampacities of Single Insulated Copper Conductors Directly Buried in Earth Based on Ambient Earth Temperature of 20°C (68°F), Arrangement per Figure 310.60, 100 Percent Load Factor, Thermal Resistance (RHO) of 90, Conductor Temperatures of 90°C (194°F) and 105°C (221°C)**

| Conductor Size (AWG or kcmil) | Temperature Rating of Conductor (See Table 310.61.) | | | |
| | 2001–5000 Volts Ampacity | | 5001–35,000 Volts Ampacity | |
| | 90°C (194°F) Type MV-90 | 105°C (221°F) Type MV-105 | 90°C (194°F) Type MV-90 | 105°C (221°F) Type MV-105 |
|---|---|---|---|---|
| **One Circuit, Three Conductors (See Figure 310.60, Detail 9.)** | | | | |
| 8 | 110 | 115 | — | — |
| 6 | 140 | 150 | 130 | 140 |
| 4 | 180 | 195 | 170 | 180 |
| 2 | 230 | 250 | 210 | 225 |
| 1 | 260 | 280 | 240 | 260 |
| 1/0 | 295 | 320 | 275 | 295 |
| 2/0 | 335 | 365 | 310 | 335 |
| 3/0 | 385 | 415 | 355 | 380 |
| 4/0 | 435 | 465 | 405 | 435 |
| 250 | 470 | 510 | 440 | 475 |
| 350 | 570 | 615 | 535 | 575 |
| 500 | 690 | 745 | 650 | 700 |
| 750 | 845 | 910 | 805 | 865 |
| 1000 | 980 | 1055 | 930 | 1005 |
| **Two Circuits, Six Conductors (See Figure 310.60, Detail 10.)** | | | | |
| 8 | 100 | 110 | — | — |
| 6 | 130 | 140 | 120 | 130 |
| 4 | 165 | 180 | 160 | 170 |
| 2 | 215 | 230 | 195 | 210 |
| 1 | 240 | 260 | 225 | 240 |
| 1/0 | 275 | 295 | 255 | 275 |
| 2/0 | 310 | 335 | 290 | 315 |
| 3/0 | 355 | 380 | 330 | 355 |
| 4/0 | 400 | 430 | 375 | 405 |
| 250 | 435 | 470 | 410 | 440 |
| 350 | 520 | 560 | 495 | 530 |
| 500 | 630 | 680 | 600 | 645 |
| 750 | 775 | 835 | 740 | 795 |
| 1000 | 890 | 960 | 855 | 920 |

**Table 310.82 Ampacities of Single Insulated Aluminum Conductors Directly Buried in Earth Based on Ambient Earth Temperature of 20°C (68°F), Arrangement per Figure 310.60, 100 Percent Load Factor, Thermal Resistance (RHO) of 90, Conductor Temperatures of 90°C (194°F) and 105°C (221°F)**

| Conductor Size (AWG or kcmil) | Temperature Rating of Conductor (See Table 310.61) | | | |
| | 2001–5000 Volts Ampacity | | 5001–35,000 Volts Ampacity | |
| | 90°C (194°F) Type MV-90 | 105°C (221°F) Type MV-105 | 90°C (194°F) Type MV-90 | 105°C (221°F) Type MV-105 |
|---|---|---|---|---|
| **One Circuit, Three Conductors (See Figure 310.60, Detail 9.)** | | | | |
| 8 | 85 | 90 | — | — |
| 6 | 110 | 115 | 100 | 110 |
| 4 | 140 | 150 | 130 | 140 |
| 2 | 180 | 195 | 165 | 175 |
| 1 | 205 | 220 | 185 | 200 |
| 1/0 | 230 | 250 | 215 | 230 |
| 2/0 | 265 | 285 | 245 | 260 |
| 3/0 | 300 | 320 | 275 | 295 |
| 4/0 | 340 | 365 | 315 | 340 |
| 250 | 370 | 395 | 345 | 370 |
| 350 | 445 | 480 | 415 | 450 |
| 500 | 540 | 580 | 510 | 545 |
| 750 | 665 | 720 | 635 | 680 |
| 1000 | 780 | 840 | 740 | 795 |
| **Two Circuits, Six Conductors (See Figure 310.60, Detail 10.)** | | | | |
| 8 | 80 | 85 | — | — |
| 6 | 100 | 110 | 95 | 100 |
| 4 | 130 | 140 | 125 | 130 |
| 2 | 165 | 180 | 155 | 165 |
| 1 | 190 | 200 | 175 | 190 |
| 1/0 | 215 | 230 | 200 | 215 |
| 2/0 | 245 | 260 | 225 | 245 |
| 3/0 | 275 | 295 | 255 | 275 |
| 4/0 | 310 | 335 | 290 | 315 |
| 250 | 340 | 365 | 320 | 345 |
| 350 | 410 | 440 | 385 | 415 |
| 500 | 495 | 530 | 470 | 505 |
| 750 | 610 | 655 | 580 | 625 |
| 1000 | 710 | 765 | 680 | 730 |

**Table 310.83 Ampacities of Three Insulated Copper Conductors Cabled Within an Overall Covering (Three-Conductor Cable), Directly Buried in Earth Based on Ambient Earth Temperature of 20°C (68°F), Arrangement per Figure 310.60, 100 Percent Load Factor, Thermal Resistance (RHO) of 90, Conductor Temperatures of 90°C (194°F) and 105°C (221°F)**

| Conductor Size (AWG or kcmil) | Temperature Rating of Conductor (See Table 310.61.) | | | |
| --- | --- | --- | --- | --- |
| | 2001–5000 Volts Ampacity | | 5001–35,000 Volts Ampacity | |
| | 90°C (194°F) Type MV-90 | 105°C (221°F) Type MV-105 | 90°C (194°F) Type MV-90 | 105°C (221°F) Type MV-105 |
| **One Circuit (See Figure 310.60, Detail 5.)** | | | | |
| 8 | 85 | 89 | — | — |
| 6 | 105 | 115 | 115 | 120 |
| 4 | 135 | 150 | 145 | 155 |
| 2 | 180 | 190 | 185 | 200 |
| 1 | 200 | 215 | 210 | 225 |
| 1/0 | 230 | 245 | 240 | 255 |
| 2/0 | 260 | 280 | 270 | 290 |
| 3/0 | 295 | 320 | 305 | 330 |
| 4/0 | 335 | 360 | 350 | 375 |
| 250 | 365 | 395 | 380 | 410 |
| 350 | 440 | 475 | 460 | 495 |
| 500 | 530 | 570 | 550 | 590 |
| 750 | 650 | 700 | 665 | 720 |
| 1000 | 730 | 785 | 750 | 810 |
| **Two Circuits (See Figure 310.60, Detail 6.)** | | | | |
| 8 | 80 | 84 | — | — |
| 6 | 100 | 105 | 105 | 115 |
| 4 | 130 | 140 | 135 | 145 |
| 2 | 165 | 180 | 170 | 185 |
| 1 | 185 | 200 | 195 | 210 |
| 1/0 | 215 | 230 | 220 | 235 |
| 2/0 | 240 | 260 | 250 | 270 |
| 3/0 | 275 | 295 | 280 | 305 |
| 4/0 | 310 | 335 | 320 | 345 |
| 250 | 340 | 365 | 350 | 375 |
| 350 | 410 | 440 | 420 | 450 |
| 500 | 490 | 525 | 500 | 535 |
| 750 | 595 | 640 | 605 | 650 |
| 1000 | 665 | 715 | 675 | 730 |

**Table 310.84 Ampacities of Three Insulated Aluminum Conductors Cabled Within an Overall Covering (Three-Conductor Cable), Directly Buried in Earth Based on Ambient Earth Temperature of 20°C (68°F), Arrangement per Figure 310.60, 100 Percent Load Factor, Thermal Resistance (RHO) of 90, Conductor Temperatures of 90°C (194°F) and 105°C (221°F)**

| Conductor Size (AWG or kcmil) | Temperature Rating of Conductor (See Table 310.61.) | | | |
| --- | --- | --- | --- | --- |
| | 2001–5000 Volts Ampacity | | 5001–35,000 Volts Ampacity | |
| | 90°C (194°F) Type MV-90 | 105°C (221°F) Type MV-105 | 90°C (194°F) Type MV-90 | 105°C (221°F) Type MV-105 |
| **One Circuit (See Figure 310.60, Detail 5.)** | | | | |
| 8 | 65 | 70 | — | — |
| 6 | 80 | 88 | 90 | 95 |
| 4 | 105 | 115 | 115 | 125 |
| 2 | 140 | 150 | 145 | 155 |
| 1 | 155 | 170 | 165 | 175 |
| 1/0 | 180 | 190 | 185 | 200 |
| 2/0 | 205 | 220 | 210 | 225 |
| 3/0 | 230 | 250 | 240 | 260 |
| 4/0 | 260 | 280 | 270 | 295 |
| 250 | 285 | 310 | 300 | 320 |
| 350 | 345 | 375 | 360 | 390 |
| 500 | 420 | 450 | 435 | 470 |
| 750 | 520 | 560 | 540 | 580 |
| 1000 | 600 | 650 | 620 | 665 |
| **Two Circuits (See Figure 310.60, Detail 6.)** | | | | |
| 8 | 60 | 66 | — | — |
| 6 | 75 | 83 | 80 | 95 |
| 4 | 100 | 110 | 105 | 115 |
| 2 | 130 | 140 | 135 | 145 |
| 1 | 145 | 155 | 150 | 165 |
| 1/0 | 165 | 180 | 170 | 185 |
| 2/0 | 190 | 205 | 195 | 210 |
| 3/0 | 215 | 230 | 220 | 240 |
| 4/0 | 245 | 260 | 250 | 270 |
| 250 | 265 | 285 | 275 | 295 |
| 350 | 320 | 345 | 330 | 355 |
| 500 | 385 | 415 | 395 | 425 |
| 750 | 480 | 515 | 485 | 525 |
| 1000 | 550 | 590 | 560 | 600 |

**Table 310.85 Ampacities of Three Triplexed Single Insulated Copper Conductors Directly Buried in Earth Based on Ambient Earth Temperature of 20°C (68°F), Arrangement per Figure 310.60, 100 Percent Load Factor, Thermal Resistance (RHO) of 90, Conductor Temperatures 90°C (194°F) and 105°C (221°F)**

| | Temperature Rating of Conductor (See Table 310.61.) | | | |
| | 2001–5000 Volts Ampacity | | 5001–35,000 Volts Ampacity | |
| Conductor Size (AWG or kcmil) | 90°C (194°F) Type MV-90 | 105°C (221°F) Type MV-105 | 90°C (194°F) Type MV-90 | 105°C (221°F) Type MV-105 |
|---|---|---|---|---|
| **One Circuit, Three Conductors (See Figure 310.60, Detail 7.)** | | | | |
| 8 | 90 | 95 | — | — |
| 6 | 120 | 130 | 115 | 120 |
| 4 | 150 | 165 | 150 | 160 |
| 2 | 195 | 205 | 190 | 205 |
| 1 | 225 | 240 | 215 | 230 |
| 1/0 | 255 | 270 | 245 | 260 |
| 2/0 | 290 | 310 | 275 | 295 |
| 3/0 | 330 | 360 | 315 | 340 |
| 4/0 | 375 | 405 | 360 | 385 |
| 250 | 410 | 445 | 390 | 410 |
| 350 | 490 | 580 | 470 | 505 |
| 500 | 590 | 635 | 565 | 605 |
| 750 | 725 | 780 | 685 | 740 |
| 1000 | 825 | 885 | 770 | 830 |
| **Two Circuits, Six Conductors (See Figure 310.60, Detail 8.)** | | | | |
| 8 | 85 | 90 | — | — |
| 6 | 110 | 115 | 105 | 115 |
| 4 | 140 | 150 | 140 | 150 |
| 2 | 180 | 195 | 175 | 190 |
| 1 | 205 | 220 | 200 | 215 |
| 1/0 | 235 | 250 | 225 | 240 |
| 2/0 | 265 | 285 | 255 | 275 |
| 3/0 | 300 | 320 | 290 | 315 |
| 4/0 | 340 | 365 | 325 | 350 |
| 250 | 370 | 395 | 355 | 380 |
| 350 | 445 | 480 | 425 | 455 |
| 500 | 535 | 575 | 510 | 545 |
| 750 | 650 | 700 | 615 | 660 |
| 1000 | 740 | 795 | 690 | 745 |

**Table 310.86 Ampacities of Three Triplexed Single Insulated Aluminum Conductors Directly Buried in Earth Based on Ambient Earth Temperature of 20°C (68°F), Arrangement per Figure 310.60, 100 Percent Load Factor, Thermal Resistance (RHO) of 90, Conductor Temperatures 90°C (194°F) and 105°C (221°F)**

| | Temperature Rating of Conductor (See Table 310.61.) | | | |
| | 2001–5000 Volts Ampacity | | 5001–35,000 Volts Ampacity | |
| Conductor Size (AWG or kcmil) | 90°C (194°F) Type MV-90 | 105°C (221°F) Type MV-105 | 90°C (194°F) Type MV-90 | 105°C (221°F) Type MV-105 |
|---|---|---|---|---|
| **One Circuit, Three Conductors (See Figure 310.60, Detail 7.)** | | | | |
| 8 | 70 | 75 | — | — |
| 6 | 90 | 100 | 90 | 95 |
| 4 | 120 | 130 | 115 | 125 |
| 2 | 155 | 165 | 145 | 155 |
| 1 | 175 | 190 | 165 | 175 |
| 1/0 | 200 | 210 | 190 | 205 |
| 2/0 | 225 | 240 | 215 | 230 |
| 3/0 | 255 | 275 | 245 | 265 |
| 4/0 | 290 | 310 | 280 | 305 |
| 250 | 320 | 350 | 305 | 325 |
| 350 | 385 | 420 | 370 | 400 |
| 500 | 465 | 500 | 445 | 480 |
| 750 | 580 | 625 | 550 | 590 |
| 1000 | 670 | 725 | 635 | 680 |
| **Two Circuits, Six Conductors (See Figure 310.60, Detail 8.)** | | | | |
| 8 | 65 | 70 | — | — |
| 6 | 85 | 95 | 85 | 90 |
| 4 | 110 | 120 | 105 | 115 |
| 2 | 140 | 150 | 135 | 145 |
| 1 | 160 | 170 | 155 | 170 |
| 1/0 | 180 | 195 | 175 | 190 |
| 2/0 | 205 | 220 | 200 | 215 |
| 3/0 | 235 | 250 | 225 | 245 |
| 4/0 | 265 | 285 | 255 | 275 |
| 250 | 290 | 310 | 280 | 300 |
| 350 | 350 | 375 | 335 | 360 |
| 500 | 420 | 455 | 405 | 435 |
| 750 | 520 | 560 | 485 | 525 |
| 1000 | 600 | 645 | 565 | 605 |

# ARTICLE 312
## Cabinets, Cutout Boxes, and Meter Socket Enclosures

**312.1 Scope.** This article covers the installation and construction specifications of cabinets, cutout boxes, and meter socket enclosures.

## I. Installation

### 312.2 Damp, Wet, or Hazardous (Classified) Locations.

**(A) Damp and Wet Locations.** In damp or wet locations, surface-type enclosures within the scope of this article shall be placed or equipped so as to prevent moisture or water from entering and accumulating within the cabinet or cutout box, and shall be mounted so there is at least 6 mm (¼ in.) airspace between the enclosure and the wall or other supporting surface. Enclosures installed in wet locations shall be weatherproof.

*Exception: Nonmetallic enclosures shall be permitted to be installed without the airspace on a concrete, masonry, tile, or similar surface.*

FPN: For protection against corrosion, see 300.6.

**(B) Hazardous (Classified) Locations.** Installations in hazardous (classified) locations shall conform to Articles 500 through 517.

**312.3 Position in Wall.** In walls of concrete, tile, or other noncombustible material, cabinets shall be installed so that the front edge of the cabinet is not set back of the finished surface more than 6 mm (¼ in.). In walls constructed of wood or other combustible material, cabinets shall be flush with the finished surface or project therefrom.

**312.5 Cabinets, Cutout Boxes, and Meter Socket Enclosures.** Conductors entering enclosures within the scope of this article shall be protected from abrasion and shall comply with 312.5(A) through (C).

**(A) Openings to Be Closed.** Openings through which conductors enter shall be adequately closed.

**(B) Metal Cabinets, Cutout Boxes, and Meter Socket Enclosures.** Where metal enclosures within the scope of this article are installed with open wiring or concealed knob-and-tube wiring, conductors shall enter through insulating bushings or, in dry locations, through flexible tubing extending from the last insulating support and firmly secured to the enclosure.

**(C) Cables.** Where cable is used, each cable shall be secured to the cabinet, cutout box, or meter socket enclosure.

*Exception: Cables with entirely nonmetallic sheaths shall be permitted to enter the top of a surface-mounted enclosure through one or more nonflexible raceways not less than 450 mm (18 in.) or more than 3.0 m (10 ft) in length, provided all the following conditions are met:*

*(a) Each cable is fastened within 300 mm (12 in.), measured along the sheath, of the outer end of the raceway.*
*(b) The raceway extends directly above the enclosure and does not penetrate a structural ceiling.*
*(c) A fitting is provided on each end of the raceway to protect the cable(s) from abrasion and the fittings remain accessible after installation.*
*(d) The raceway is sealed or plugged at the outer end using approved means so as to prevent access to the enclosure through the raceway.*
*(e) The cable sheath is continuous through the raceway and extends into the enclosure beyond the fitting not less than 6 mm (¼ in.).*
*(f) The raceway is fastened at its outer end and at other points in accordance with the applicable article.*
*(g) Where installed as conduit or tubing, the allowable cable fill does not exceed that permitted for complete conduit or tubing systems by Table 1 of Chapter 9 of this Code and all applicable notes thereto.*

FPN: See Table 1 in Chapter 9, including Note 9, for allowable cable fill in circular raceways. See 310.15(B)(2)(a) for required ampacity reductions for multiple cables installed in a common raceway.

**312.6 Deflection of Conductors.** Conductors at terminals or conductors entering or leaving cabinets or cutout boxes and the like shall comply with 312.6(A) through (C).

*Exception: Wire-bending space in enclosures for motor controllers with provisions for one or two wires per terminal shall comply with 430.10(B).*

**(A) Width of Wiring Gutters.** Conductors shall not be deflected within a cabinet or cutout box unless a gutter having a width in accordance with Table 312.6(A) is provided. Conductors in parallel in accordance with 310.4 shall be judged on the basis of the number of conductors in parallel.

**(B) Wire-Bending Space at Terminals.** Wire-bending space at each terminal shall be provided in accordance with 312.6(B)(1) or (2).

**(1) Conductors Not Entering or Leaving Opposite Wall.** Table 312.6(A) shall apply where the conductor does not enter or leave the enclosure through the wall opposite its terminal.

**Table 312.6(A) Minimum Wire-Bending Space at Terminals and Minimum Width of Wiring Gutters**

| Wire Size (AWG or kcmil) | Wires per Terminal | | | | | | | | | |
|---|---|---|---|---|---|---|---|---|---|---|
| | 1 | | 2 | | 3 | | 4 | | 5 | |
| | mm | in. | mm | in. | mm | in. | mm | in. | mm | in. |
| 14–10 | Not specified | | — | — | — | — | — | — | — | — |
| 8–6 | 38.1 | 1½ | — | — | — | — | — | — | — | — |
| 4–3 | 50.8 | 2 | — | — | — | — | — | — | — | — |
| 2 | 63.5 | 2½ | — | — | — | — | — | — | — | — |
| 1 | 76.2 | 3 | — | — | — | — | — | — | — | — |
| 1/0–2/0 | 88.9 | 3½ | — | — | — | — | — | — | — | — |
| 3/0–4/0 | 102 | 4 | 127 | 5 | — | — | — | — | — | — |
| 250 | 114 | 4½ | 152 | 6 | 203 | 8 | — | — | — | — |
| 300–350 | 127 | 5 | 152 | 6 | 203 | 8 | 254 | 10 | — | — |
| 400–500 | 152 | 6 | 203 | 8 | 254 | 10 | 305 | 12 | — | — |
| 600–700 | 203 | 8 | 203 | 8 | 254 | 10 | 305 | 12 | 356 | 14 |
| 750–900 | 203 | 8 | 254 | 10 | 305 | 12 | 356 | 14 | 406 | 16 |
| 1000–1250 | 254 | 10 | 305 | 12 | 356 | 14 | 406 | 16 | 457 | 18 |
| 1500–2000 | 305 | 12 | — | — | — | — | — | — | — | — |

Note: Bending space at terminals shall be measured in a straight line from the end of the lug or wire connector (in the direction that the wire leaves the terminal) to the wall, barrier, or obstruction.

**(2) Conductors Entering or Leaving Opposite Wall.** Table 312.6(B) shall apply where the conductor does enter or leave the enclosure through the wall opposite its terminal.

*Exception No. 1: Where the distance between the wall and its terminal is in accordance with Table 312.6(A), a conductor shall be permitted to enter or leave an enclosure through the wall opposite its terminal, provided the conductor enters or leaves the enclosure where the gutter joins an adjacent gutter that has a width that conforms to Table 312.6(B) for the conductor.*

*Exception No. 2: A conductor not larger than 350 kcmil shall be permitted to enter or leave an enclosure containing only a meter socket(s) through the wall opposite its terminal, provided the distance between the terminal and the opposite wall is not less than that specified in Table 312.6(A) and the terminal is a lay-in type where the terminal is either of the following:*

(a) *Directed toward the opening in the enclosure and within a 45 degree angle of directly facing the enclosure wall*

(b) *Directly facing the enclosure wall and offset not greater than 50 percent of the bending space specified in Table 312.6(A)*

FPN: *Offset* is the distance measured along the enclosure wall from the axis of the centerline of the terminal to a line passing through the center of the opening in the enclosure.

**(C) Conductors 4 AWG or Larger.** Installation shall comply with 300.4(F).

**312.7 Space in Enclosures.** Cabinets and cutout boxes shall have sufficient space to accommodate all conductors installed in them without crowding.

**312.8 Enclosures for Switches or Overcurrent Devices.** Enclosures for switches or overcurrent devices shall not be used as junction boxes, auxiliary gutters, or raceways for conductors feeding through or tapping off to other switches or overcurrent devices, unless adequate space for this purpose is provided. The conductors shall not fill the wiring space at any cross section to more than 40 percent of the cross-sectional area of the space, and the conductors, splices, and taps shall not fill the wiring space at any cross section to more than 75 percent of the cross-sectional area of that space.

**312.9 Side or Back Wiring Spaces or Gutters.** Cabinets and cutout boxes shall be provided with back-wiring spaces, gutters, or wiring compartments as required by 312.11(C) and (D).

## II. Construction Specifications

**312.10 Material.** Cabinets, cutout boxes, and meter socket enclosures shall comply with 312.10(A) through (C).

**(A) Metal Cabinets and Cutout Boxes.** Metal enclosures within the scope of this article shall be protected both inside and outside against corrosion.

FPN: For information on protection against corrosion, see 300.6.

**Table 312.6(B)  Minimum Wire-Bending Space at Terminals**

| Wire Size (AWG or kcmil) | | Wires per Terminal | | | | | | | |
| --- | --- | --- | --- | --- | --- | --- | --- | --- | --- |
| | | 1 | | 2 | | 3 | | 4 or More | |
| All Other Conductors | Compact Stranded AA-8000 Aluminum Alloy Conductors (See Note 3.) | mm | in. | mm | in. | mm | in. | mm | in. |
| 14–10 | 12–8 | Not specified | | — | — | — | — | — | — |
| 8 | 6 | 38.1 | 1½ | — | — | — | — | — | — |
| 6 | 4 | 50.8 | 2 | — | — | — | — | — | — |
| 4 | 2 | 76.2 | 3 | — | — | — | — | — | — |
| 3 | 1 | 76.2 | 3 | — | — | — | — | — | — |
| 2 | 1/0 | 88.9 | 3½ | — | — | — | — | — | — |
| 1 | 2/0 | 114 | 4½ | — | — | — | — | — | — |
| 1/0 | 3/0 | 140 | 5½ | 140 | 5½ | 178 | 7 | — | — |
| 2/0 | 4/0 | 152 | 6 | 152 | 6 | 190 | 7½ | — | — |
| 3/0 | 250 | 165[a] | 6½[a] | 165[a] | 6½[a] | 203 | 8 | — | — |
| 4/0 | 300 | 178[b] | 7[b] | 190[c] | 7½[c] | 216[a] | 8½[a] | — | — |
| 250 | 350 | 216[d] | 8½[d] | 229[d] | 8½[d] | 254 | 9[b] | 254 | 10 |
| 300 | 400 | 254[e] | 10[e] | 254[b] | 10[d] | 279[b] | 11[b] | 305 | 12 |
| 350 | 500 | 305[e] | 12[e] | 305[e] | 12[e] | 330[e] | 13[e] | 356[d] | 14[d] |
| 400 | 600 | 330[e] | 13[e] | 330[e] | 13[e] | 356[e] | 14[e] | 381[e] | 15[e] |
| 500 | 700–750 | 356[e] | 14[e] | 356[e] | 14[e] | 381[e] | 15[e] | 406[e] | 16[e] |
| 600 | 800–900 | 381[e] | 15[e] | 406[e] | 16[e] | 457[e] | 18[e] | 483[e] | 19[e] |
| 700 | 1000 | 406[e] | 16[e] | 457[e] | 18[e] | 508[e] | 20[e] | 559[e] | 22[e] |
| 750 | — | 432[e] | 17[e] | 483[e] | 19[e] | 559[e] | 22[e] | 610[e] | 24[e] |
| 800 | — | 457 | 18 | 508 | 20 | 559 | 22 | 610 | 24 |
| 900 | — | 483 | 19 | 559 | 22 | 610 | 24 | 610 | 24 |
| 1000 | — | 508 | 20 | — | — | — | — | — | — |
| 1250 | — | 559 | 22 | — | — | — | — | — | — |
| 1500 | — | 610 | 24 | — | — | — | — | — | — |
| 1750 | — | 610 | 24 | — | — | — | — | — | — |
| 2000 | — | 610 | 24 | — | — | — | — | — | — |

1. Bending space at terminals shall be measured in a straight line from the end of the lug or wire connector in a direction perpendicular to the enclosure wall.

2. For removable and lay-in wire terminals intended for only one wire, bending space shall be permitted to be reduced by the following number of millimeters (inches):

[a] 12.7 mm (½ in.)     [d] 50.8 mm (2 in.)
[b] 25.4 mm (1 in.)     [e] 76.2 mm (3 in.)
[c] 38.1 mm (1½ in.)

3. This column shall be permitted to determine the required wire-bending space for compact stranded aluminum conductors in sizes up to 1000 kcmil and manufactured using AA-8000 series electrical grade aluminum alloy conductor material in accordance with 310.14.

**(B) Strength.** The design and construction of enclosures within the scope of this article shall be such as to secure ample strength and rigidity. If constructed of sheet steel, the metal thickness shall not be less than 1.35 mm (0.053 in.) uncoated.

**(C) Nonmetallic Cabinets.** Nonmetallic cabinets shall be listed or they shall be submitted for approval prior to installation.

**312.11 Spacing.** The spacing within cabinets and cutout boxes shall comply with 312.11(A) through (D).

**(A) General.** Spacing within cabinets and cutout boxes shall be sufficient to provide ample room for the distribution of wires and cables placed in them and for a separation between metal parts of devices and apparatus mounted within them as follows.

**(1) Base.** Other than at points of support, there shall be an airspace of at least 1.59 mm (0.0625 in.) between the base of the device and the wall of any metal cabinet or cutout box in which the device is mounted.

**(2) Doors.** There shall be an airspace of at least 25.4 mm (1.00 in.) between any live metal part, including live metal parts of enclosed fuses, and the door.

*Exception: Where the door is lined with an approved insulating material or is of a thickness of metal not less than 2.36 mm (0.093 in.) uncoated, the airspace shall not be less than 12.7 mm (0.500 in.).*

**(3) Live Parts.** There shall be an airspace of at least 12.7 mm (0.500 in.) between the walls, back, gutter partition, if of metal, or door of any cabinet or cutout box and the nearest exposed current-carrying part of devices mounted within the cabinet where the voltage does not exceed 250. This spacing shall be increased to at least 25.4 mm (1.00 in.) for voltages of 251 to 600, nominal.

*Exception: Where the conditions in 312.11(A)(2), Exception, are met, the airspace for nominal voltages from 251 to 600 shall be permitted to be not less than 12.7 mm (0.500 in.).*

**(B) Switch Clearance.** Cabinets and cutout boxes shall be deep enough to allow the closing of the doors when 30-ampere branch-circuit panelboard switches are in any position, when combination cutout switches are in any position, or when other single-throw switches are opened as far as their construction permits.

**(C) Wiring Space.** Cabinets and cutout boxes that contain devices or apparatus connected within the cabinet or box to more than eight conductors, including those of branch circuits, meter loops, feeder circuits, power circuits, and similar circuits, but not including the supply circuit or a continuation thereof, shall have back-wiring spaces or one or more side-wiring spaces, side gutters, or wiring compartments.

**(D) Wiring Space — Enclosure.** Side-wiring spaces, side gutters, or side-wiring compartments of cabinets and cutout boxes shall be made tight enclosures by means of covers, barriers, or partitions extending from the bases of the devices contained in the cabinet, to the door, frame, or sides of the cabinet.

*Exception: Side-wiring spaces, side gutters, and side-wiring compartments of cabinets shall not be required to be made tight enclosures where those side spaces contain only conductors that enter the cabinet directly opposite to the devices where they terminate.*

Partially enclosed back-wiring spaces shall be provided with covers to complete the enclosure. Wiring spaces that are required by 312.11(C) and are exposed when doors are open shall be provided with covers to complete the enclosure. Where adequate space is provided for feed-through conductors and for splices as required in 312.8, additional barriers shall not be required.

## ARTICLE 314
## Outlet, Device, Pull, and Junction Boxes; Conduit Bodies; Fittings; and Manholes

### I. Scope and General

**314.1 Scope.** This article covers the installation and use of all boxes and conduit bodies used as outlet, device, junction, or pull boxes, depending on their use, and manholes and other electric enclosures intended for personnel entry. Cast, sheet metal, nonmetallic, and other boxes such as FS, FD, and larger boxes are not classified as conduit bodies. This article also includes installation requirements for fittings used to join raceways and to connect raceways and cables to boxes and conduit bodies.

**314.2 Round Boxes.** Round boxes shall not be used where conduits or connectors requiring the use of locknuts or bushings are to be connected to the side of the box.

**314.3 Nonmetallic Boxes.** Nonmetallic boxes shall be permitted only with open wiring on insulators, concealed knob-and-tube wiring, cabled wiring methods with entirely nonmetallic sheaths, flexible cords, and nonmetallic raceways.

*Exception No. 1: Where internal bonding means are provided between all entries, nonmetallic boxes shall be permitted to be used with metal raceways or metal-armored cables.*

*Exception No. 2: Where integral bonding means with a provision for attaching an equipment bonding jumper inside the box are provided between all threaded entries in nonmetallic boxes listed for the purpose, nonmetallic boxes shall be permitted to be used with metal raceways or metal-armored cables.*

**314.4 Metal Boxes.** All metal boxes shall be grounded in accordance with the provisions of Article 250.

**314.5 Short-Radius Conduit Bodies.** Conduit bodies such as capped elbows and service-entrance elbows that enclose conductors 6 AWG or smaller, and are only intended to enable the installation of the raceway and the contained conductors, shall not contain splices, taps, or devices and shall be of sufficient size to provide free space for all conductors enclosed in the conduit body.

## II. Installation

### 314.15 Damp, Wet, or Hazardous (Classified) Locations.

**(A) Damp or Wet Locations.** In damp or wet locations, boxes, conduit bodies, and fittings shall be placed or equipped so as to prevent moisture from entering or accumulating within the box, conduit body, or fitting. Boxes, conduit bodies, and fittings installed in wet locations shall be listed for use in wet locations.

> FPN No. 1: For boxes in floors, see 314.27(C).
>
> FPN No. 2: For protection against corrosion, see 300.6.

**(B) Hazardous (Classified) Locations.** Installations in hazardous (classified) locations shall conform to Articles 500 through 517.

### 314.16 Number of Conductors in Outlet, Device, and Junction Boxes, and Conduit Bodies. Boxes and conduit bodies shall be of sufficient size to provide free space for all enclosed conductors. In no case shall the volume of the box, as calculated in 314.16(A), be less than the fill calculation as calculated in 314.16(B). The minimum volume for conduit bodies shall be as calculated in 314.16(C).

The provisions of this section shall not apply to terminal housings supplied with motors.

> FPN: For volume requirements of motor terminal housings, see 430.12.

Boxes and conduit bodies enclosing conductors 4 AWG or larger shall also comply with the provisions of 314.28.

**(A) Box Volume Calculations.** The volume of a wiring enclosure (box) shall be the total volume of the assembled sections, and, where used, the space provided by plaster rings, domed covers, extension rings, and so forth, that are marked with their volume or are made from boxes the dimensions of which are listed in Table 314.16(A).

**(1) Standard Boxes.** The volumes of standard boxes that are not marked with their volume shall be as given in Table 314.16(A).

**(2) Other Boxes.** Boxes 1650 cm$^3$ (100 in.$^3$) or less, other than those described in Table 314.16(A), and nonmetallic boxes shall be durably and legibly marked by the manufacturer with their volume. Boxes described in Table 314.16(A) that have a volume larger than is designated in the table shall be permitted to have their volume marked as required by this section.

**(B) Box Fill Calculations.** The volumes in paragraphs 314.16(B)(1) through (5), as applicable, shall be added together. No allowance shall be required for small fittings such as locknuts and bushings.

**(1) Conductor Fill.** Each conductor that originates outside the box and terminates or is spliced within the box shall be counted once, and each conductor that passes through the box without splice or termination shall be counted once. The conductor fill shall be computed using Table 314.16(B). A conductor, no part of which leaves the box, shall not be counted.

*Exception: An equipment grounding conductor or conductors or not over four fixture wires smaller than 14 AWG, or both, shall be permitted to be omitted from the calculations where they enter a box from a domed luminaire (fixture) or similar canopy and terminate within that box.*

**(2) Clamp Fill.** Where one or more internal cable clamps, whether factory or field supplied, are present in the box, a single volume allowance in accordance with Table 314.16(B) shall be made based on the largest conductor present in the box. No allowance shall be required for a cable connector with its clamping mechanism outside the box.

**(3) Support Fittings Fill.** Where one or more luminaire (fixture) studs or hickeys are present in the box, a single volume allowance in accordance with Table 314.16(B) shall be made for each type of fitting based on the largest conductor present in the box.

**(4) Device or Equipment Fill.** For each yoke or strap containing one or more devices or equipment, a double volume allowance in accordance with Table 314.16(B) shall be made for each yoke or strap based on the largest conductor connected to a device(s) or equipment supported by that yoke or strap.

**(5) Equipment Grounding Conductor Fill.** Where one or more equipment grounding conductors or equipment bonding jumpers enter a box, a single volume allowance in accordance with Table 314.16(B) shall be made based on the largest equipment grounding conductor or equipment bonding jumper present in the box. Where an additional set of equipment grounding conductors, as permitted by 250.146(D), is present in the box, an additional volume allowance shall be made based on the largest equipment grounding conductor in the additional set.

**(C) Conduit Bodies.**

**(1) General.** Conduit bodies enclosing 6 AWG conductors or smaller, other than short-radius conduit bodies as described in 314.5, shall have a cross-sectional area not less than twice the cross-sectional area of the largest conduit or

**Table 314.16(A)  Metal Boxes**

| Box Trade Size | | | Minimum Volume | | Maximum Number of Conductors* | | | | | | |
|---|---|---|---|---|---|---|---|---|---|---|---|
| mm | in. | | cm³ | in.³ | 18 | 16 | 14 | 12 | 10 | 8 | 6 |
| 100 × 32 | (4 × 1¼) | round/octagonal | 205 | 12.5 | 8 | 7 | 6 | 5 | 5 | 5 | 2 |
| 100 × 38 | (4 × 1½) | round/octagonal | 254 | 15.5 | 10 | 8 | 7 | 6 | 6 | 5 | 3 |
| 100 × 54 | (4 × 2⅛) | round/octagonal | 353 | 21.5 | 14 | 12 | 10 | 9 | 8 | 7 | 4 |
| 100 × 32 | (4 × 1¼) | square | 295 | 18.0 | 12 | 10 | 9 | 8 | 7 | 6 | 3 |
| 100 × 38 | (4 × 1½) | square | 344 | 21.0 | 14 | 12 | 10 | 9 | 8 | 7 | 4 |
| 100 × 54 | (4 × 2⅛) | square | 497 | 30.3 | 20 | 17 | 15 | 13 | 12 | 10 | 6 |
| 120 × 32 | (4¹¹⁄₁₆ × 1¼) | square | 418 | 25.5 | 17 | 14 | 12 | 11 | 10 | 8 | 5 |
| 120 × 38 | (4¹¹⁄₁₆ × 1½) | square | 484 | 29.5 | 19 | 16 | 14 | 13 | 11 | 9 | 5 |
| 120 × 54 | (4¹¹⁄₁₆ × 2⅛) | square | 689 | 42.0 | 28 | 24 | 21 | 18 | 16 | 14 | 8 |
| 75 × 50 × 38 | (3 × 2 × 1½) | device | 123 | 7.5 | 5 | 4 | 3 | 3 | 3 | 2 | 1 |
| 75 × 50 × 50 | (3 × 2 × 2) | device | 164 | 10.0 | 6 | 5 | 5 | 4 | 4 | 3 | 2 |
| 75 × 50 × 57 | (3 × 2 × 2¼) | device | 172 | 10.5 | 7 | 6 | 5 | 4 | 4 | 3 | 2 |
| 75 × 50 × 65 | (3 × 2 × 2½) | device | 205 | 12.5 | 8 | 7 | 6 | 5 | 5 | 4 | 2 |
| 75 × 50 × 70 | (3 × 2 × 2¾) | device | 230 | 14.0 | 9 | 8 | 7 | 6 | 5 | 4 | 2 |
| 75 × 50 × 90 | (3 × 2 × 3½) | device | 295 | 18.0 | 12 | 10 | 9 | 8 | 7 | 6 | 3 |
| 100 × 54 × 38 | (4 × 2⅛ × 1½) | device | 169 | 10.3 | 6 | 5 | 5 | 4 | 4 | 3 | 2 |
| 100 × 54 × 48 | (4 × 2⅛ × 1⅞) | device | 213 | 13.0 | 8 | 7 | 6 | 5 | 5 | 4 | 2 |
| 100 × 54 × 54 | (4 × 2⅛ × 2⅛) | device | 238 | 14.5 | 9 | 8 | 7 | 6 | 5 | 4 | 2 |
| 95 × 50 × 65 | (3¾ × 2 × 2½) | masonry box/gang | 230 | 14.0 | 9 | 8 | 7 | 6 | 5 | 4 | 2 |
| 95 × 50 × 90 | (3¾ × 2 × 3½) | masonry box/gang | 344 | 21.0 | 14 | 12 | 10 | 9 | 8 | 7 | 2 |
| min. 44.5 depth | FS — single cover/gang (1¾) | | 221 | 13.5 | 9 | 7 | 6 | 6 | 5 | 4 | 2 |
| min. 60.3 depth | FD — single cover/gang (2⅜) | | 295 | 18.0 | 12 | 10 | 9 | 8 | 7 | 6 | 3 |
| min. 44.5 depth | FS — multiple cover/gang (1¾) | | 295 | 18.0 | 12 | 10 | 9 | 8 | 7 | 6 | 3 |
| min. 60.3 depth | FD — multiple cover/gang (2⅜) | | 395 | 24.0 | 16 | 13 | 12 | 10 | 9 | 8 | 4 |

*Where no volume allowances are required by 314.16(B)(2) through 314.16(B)(5).

**Table 314.16(B)  Volume Allowance Required per Conductor**

| Size of Conductor (AWG) | Free Space Within Box for Each Conductor | |
|---|---|---|
| | cm³ | in.³ |
| 18 | 24.6 | 1.50 |
| 16 | 28.7 | 1.75 |
| 14 | 32.8 | 2.00 |
| 12 | 36.9 | 2.25 |
| 10 | 41.0 | 2.50 |
| 8 | 49.2 | 3.00 |
| 6 | 81.9 | 5.00 |

tubing to which it is attached. The maximum number of conductors permitted shall be the maximum number permitted by Table 1 of Chapter 9 for the conduit or tubing to which it is attached.

**(2) With Splices, Taps, or Devices.** Only those conduit bodies that are durably and legibly marked by the manufacturer with their volume shall be permitted to contain splices, taps, or devices. The maximum number of conductors shall be computed in accordance with 314.16(B). Conduit bodies shall be supported in a rigid and secure manner.

**314.17 Conductors Entering Boxes, Conduit Bodies, or Fittings.** Conductors entering boxes, conduit bodies, or fittings shall be protected from abrasion and shall comply with 314.17(A) through (D).

**(A) Openings to Be Closed.** Openings through which conductors enter shall be adequately closed.

**(B) Metal Boxes and Conduit Bodies.** Where metal boxes or conduit bodies are installed with open wiring or concealed knob-and-tube wiring, conductors shall enter through insulating bushings or, in dry locations, through flexible tubing extending from the last insulating support to not less than 6 mm (¼ in.) inside the box and beyond any cable clamps. Except as provided in 300.15(C), the wiring shall be firmly secured to the box or conduit body. Where raceway or cable is installed with metal boxes or conduit bodies, the raceway or cable shall be secured to such boxes and conduit bodies.

**(C) Nonmetallic Boxes and Conduit Bodies.** Nonmetallic boxes and conduit bodies shall be suitable for the lowest temperature-rated conductor entering the box. Where nonmetallic boxes and conduit bodies are used with open wiring or concealed knob-and-tube wiring, the conductors shall enter the box through individual holes. Where flexible tubing is used to enclose the conductors, the tubing shall extend from the last insulating support to not less than 6 mm (¼ in.) inside the box and beyond any cable clamp. Where nonmetallic-sheathed cable or multiconductor Type UF cable is used, the sheath shall extend not less than 6 mm (¼ in.) inside the box and beyond any cable clamp. In all instances, all permitted wiring methods shall be secured to the boxes.

*Exception: Where nonmetallic-sheathed cable or multiconductor Type UF cable is used with single gang boxes not larger than a nominal size 57 mm × 100 mm (2¼ in. × 4 in.) mounted in walls or ceilings, and where the cable is fastened within 200 mm (8 in.) of the box measured along the sheath and where the sheath extends through a cable knockout not less than 6 mm (¼ in.), securing the cable to the box shall not be required. Multiple cable entries shall be permitted in a single cable knockout opening.*

**(D) Conductors 4 AWG or Larger.** Installation shall comply with 300.4(F).

> FPN: See 110.12(A) for requirements on closing unused cable and raceway knockout openings.

**314.19 Boxes Enclosing Flush Devices.** Boxes used to enclose flush devices shall be of such design that the devices will be completely enclosed on back and sides and substantial support for the devices will be provided. Screws for supporting the box shall not be used in attachment of the device contained therein.

**314.20 In Wall or Ceiling.** In walls or ceilings with a surface of concrete, tile, gypsum, plaster, or other noncombustible material, boxes shall be installed so that the front edge of the box will not be set back of the finished surface more than 6 mm (¼ in.).

In walls and ceilings constructed of wood or other combustible surface material, boxes shall be flush with the finished surface or project therefrom.

**314.21 Repairing Plaster and Drywall or Plasterboard.** Plaster, drywall, or plasterboard surfaces that are broken or incomplete shall be repaired so there will be no gaps or open spaces greater than 3 mm (⅛ in.) at the edge of the box or fitting.

**314.22 Exposed Surface Extensions.** Surface extensions from a flush-mounted box shall be made by mounting and mechanically securing an extension ring over the flush box. Equipment grounding and bonding shall be in accordance with Article 250.

*Exception: A surface extension shall be permitted to be made from the cover of a flush-mounted box where the cover is designed so it is unlikely to fall off or be removed if its securing means becomes loose. The wiring method shall be flexible for a length sufficient to permit removal of the cover and provide access to the box interior, and arranged so that any bonding or grounding continuity is independent of the connection between the box and cover.*

**314.23 Supports.** Enclosures within the scope of this article shall be supported in accordance with one or more of the provisions in 314.23(A) through (H).

**(A) Surface Mounting.** An enclosure mounted on a building or other surface shall be rigidly and securely fastened in place. If the surface does not provide rigid and secure support, additional support in accordance with other provisions of this section shall be provided.

**(B) Structural Mounting.** An enclosure supported from a structural member of a building or from grade shall be rigidly supported either directly or by using a metal, polymeric, or wood brace.

**(1) Nails and Screws.** Nails and screws, where used as a fastening means, shall be attached by using brackets on the outside of the enclosure, or they shall pass through the interior within 6 mm (¼ in.) of the back or ends of the enclosure.

**(2) Braces.** Metal braces shall be protected against corrosion and formed from metal that is not less than 0.51 mm (0.020 in.) thick uncoated. Wood braces shall have a cross section not less than nominal 25 mm × 50 mm (1 in. × 2 in.). Wood braces in wet locations shall be treated for the conditions. Polymeric braces shall be identified as being suitable for the use.

**(C) Mounting in Finished Surfaces.** An enclosure mounted in a finished surface shall be rigidly secured thereto by clamps, anchors, or fittings identified for the application.

**(D) Suspended Ceilings.** An enclosure mounted to structural or supporting elements of a suspended ceiling shall be not more than 1650 cm³ (100 in.³) in size and shall be securely fastened in place in accordance with either (D)(1) or (D)(2).

**(1) Framing Members.** An enclosure shall be fastened to the framing members by mechanical means such as bolts, screws, or rivets, or by the use of clips or other securing means identified for use with the type of ceiling framing member(s) and enclosure(s) employed. The framing members shall be adequately supported and securely fastened to each other and to the building structure.

**(2) Support Wires.** The installation shall comply with the provisions of 300.11(A). The enclosure shall be secured, using methods identified for the purpose, to ceiling support wire(s), including any additional support wire(s) installed for that purpose. Support wire(s) used for enclosure support shall be fastened at each end so as to be taut within the ceiling cavity.

**(E) Raceway Supported Enclosure, Without Devices, Luminaires (Fixtures), or Lampholders.** An enclosure that does not contain a device(s) other than splicing devices or support a luminaire(s) [fixture(s)], lampholder, or other equipment and is supported by entering raceways shall not exceed 1650 cm³ (100 in.³) in size. It shall have threaded entries or have hubs identified for the purpose. It shall be supported by two or more conduits threaded wrenchtight into the enclosure or hubs. Each conduit shall be secured within 900 mm (3 ft) of the enclosure, or within 450 mm (18 in.) of the enclosure if all conduit entries are on the same side.

*Exception: Rigid metal, intermediate metal, or rigid non-metallic conduit or electrical metallic tubing shall be permitted to support a conduit body of any size, including a conduit body constructed with only one conduit entry, provided the trade size of the conduit body is not larger than the largest trade size of the conduit or electrical metallic tubing.*

**(F) Raceway Supported Enclosures, with Devices, Luminaires (Fixtures), or Lampholders.** An enclosure that contains a device(s) or supports a luminaire(s) [fixture(s)], lampholder, or other equipment and is supported by entering raceways shall not exceed 1650 cm³ (100 in.³) in size. It shall have threaded entries or have hubs identified for the purpose. It shall be supported by two or more conduits threaded wrenchtight into the enclosure or hubs. Each conduit shall be secured within 450 mm (18 in.) of the enclosure.

*Exception No. 1: Rigid metal or intermediate metal conduit shall be permitted to support a conduit body of any size, including a conduit body constructed with only one conduit entry, provided the trade size of the conduit body is not larger than the largest trade size of the conduit.*

*Exception No. 2: An unbroken length(s) of rigid or intermediate metal conduit shall be permitted to support a box used for luminaire (fixture) or lampholder support, or to support a wiring enclosure that is an integral part of a luminaire (fixture) and used in lieu of a box in accordance with 300.15(B), where all of the following conditions are met.*

(a) *The conduit is securely fastened at a point so that the length of conduit beyond the last point of conduit support does not exceed 900 mm (3 ft).*

(b) *The unbroken conduit length before the last point of conduit support is 300 mm (12 in.) or greater, and that portion of the conduit is securely fastened at some point not less than 300 mm (12 in.) from its last point of support.*

(c) *Where accessible to unqualified persons, the luminaire (fixture) or lampholder, measured to its lowest point, is at least 2.5 m (8 ft) above grade or standing area and at least 900 mm (3 ft) measured horizontally to the 2.5 m (8 ft) elevation from windows, doors, porches, fire escapes, or similar locations.*

(d) *A luminaire (fixture) supported by a single conduit does not exceed 300 mm (12 in.) in any direction from the point of conduit entry.*

(e) *The weight supported by any single conduit does not exceed 9 kg (20 lb).*

(f) *At the luminaire (fixture) or lampholder end, the conduit(s) is threaded wrenchtight into the box, conduit body, or integral wiring enclosure, or into hubs identified for the purpose. Where a box or conduit body is used for support, the luminaire (fixture) shall be secured directly to the box or conduit body, or through a threaded conduit nipple not over 75 mm (3 in.) long.*

**(G) Enclosures in Concrete or Masonry.** An enclosure supported by embedment shall be identified as suitably protected from corrosion and securely embedded in concrete or masonry.

**(H) Pendant Boxes.** An enclosure supported by a pendant shall comply with 314.23(H)(1) or (2).

**(1) Flexible Cord.** A box shall be supported from a multiconductor cord or cable in an approved manner that protects the conductors against strain, such as a strain-relief connector threaded into a box with a hub.

**(2) Conduit.** A box supporting lampholders or luminaires (lighting fixtures), or wiring enclosures within luminaires (fixtures) used in lieu of boxes in accordance with 300.15(B), shall be supported by rigid or intermediate metal conduit stems. For stems longer than 450 mm (18 in.), the stems shall be connected to the wiring system with flexible fittings suitable for the location. At the luminaire (fixture) end, the conduit(s) shall be threaded wrenchtight into the box or wiring enclosure, or into hubs identified for the purpose.

Where supported by only a single conduit, the threaded joints shall be prevented from loosening by the use of set-screws or other effective means, or the luminaire (fixture), at any point, shall be at least 2.5 m (8 ft) above grade or standing area and at least 900 mm (3 ft) measured horizontally to the 2.5 m (8 ft) elevation from windows, doors, porches, fire escapes, or similar locations. A luminaire (fixture) supported by a single conduit shall not exceed 300 mm (12 in.) in any horizontal direction from the point of conduit entry.

**314.24 Depth of Outlet Boxes.** No box shall have an internal depth of less than 12.7 mm (½ in.). Boxes intended to enclose flush devices shall have an internal depth of not less than 23.8 mm ($^{15}$⁄$_{16}$ in.).

**314.25 Covers and Canopies.** In completed installations, each box shall have a cover, faceplate, lampholder, or luminaire (fixture) canopy, except where the installation complies with 410.14(B).

**(A) Nonmetallic or Metal Covers and Plates.** Nonmetallic or metal covers and plates shall be permitted. Where metal covers or plates are used, they shall comply with the grounding requirements of 250.110.

> FPN: For additional grounding requirements, see 410.18(A) for metal luminaire (fixture) canopies, and 404.12 and 406.5(B) for metal faceplates.

**(B) Exposed Combustible Wall or Ceiling Finish.** Where a luminaire (fixture) canopy or pan is used, any combustible wall or ceiling finish exposed between the edge of the canopy or pan and the outlet box shall be covered with noncombustible material.

**(C) Flexible Cord Pendants.** Covers of outlet boxes and conduit bodies having holes through which flexible cord pendants pass shall be provided with bushings designed for the purpose or shall have smooth, well-rounded surfaces on which the cords may bear. So-called hard rubber or composition bushings shall not be used.

**314.27 Outlet Boxes.**

**(A) Boxes at Luminaire (Lighting Fixture) Outlets.** Boxes used at luminaire (lighting fixture) or lampholder outlets shall be designed for the purpose. At every outlet used exclusively for lighting, the box shall be designed or installed so that a luminaire (lighting fixture) may be attached.

*Exception: A wall-mounted luminaire (fixture) weighing not more than 3 kg (6 lb) shall be permitted to be supported on other boxes or plaster rings that are secured to other boxes, provided the luminaire (fixture) or its supporting yoke is secured to the box with no fewer than two No. 6 or larger screws.*

**(B) Maximum Luminaire (Fixture) Weight.** Outlet boxes or fittings installed as required by 314.23 shall be permitted to support luminaires (lighting fixtures) weighing 23 kg (50 lb) or less. A luminaire (lighting fixture) that weighs more than 23 kg (50 lb) shall be supported independently of the outlet box unless the outlet box is listed for the weight to be supported.

**(C) Floor Boxes.** Boxes listed specifically for this application shall be used for receptacles located in the floor.

*Exception: Where the authority having jurisdiction judges them free from likely exposure to physical damage, moisture, and dirt, boxes located in elevated floors of show windows and similar locations shall be permitted to be other than those listed for floor applications. Receptacles and covers shall be listed as an assembly for this type of location.*

**(D) Boxes at Ceiling-Suspended (Paddle) Fan Outlets.** Where a box is used as the sole support of a ceiling-suspended (paddle) fan, the box shall be listed for the application and for the weight of the fan to be supported. The installation shall comply with 422.18.

**314.28 Pull and Junction Boxes and Conduit Bodies.** Boxes and conduit bodies used as pull or junction boxes shall comply with 314.28(A) through (D).

*Exception: Terminal housings supplied with motors shall comply with the provisions of 430.12.*

**(A) Minimum Size.** For raceways containing conductors of 4 AWG or larger, and for cables containing conductors of 4 AWG or larger, the minimum dimensions of pull or junction boxes installed in a raceway or cable run shall comply with the following. Where an enclosure dimension is to be calculated based on the diameter of entering raceways, the diameter shall be the metric designator (trade size) expressed in the units of measurement employed.

**(1) Straight Pulls.** In straight pulls, the length of the box shall not be less than eight times the metric designator (trade size) of the largest raceway.

**(2) Angle or U Pulls.** Where splices or where angle or U pulls are made, the distance between each raceway entry inside the box and the opposite wall of the box shall not be less than six times the metric designator (trade size) of the largest raceway in a row. This distance shall be increased for additional entries by the amount of the sum of the diameters of all other raceway entries in the same row on the same wall of the box. Each row shall be calculated individually, and the single row that provides the maximum distance shall be used.

*Exception: Where a raceway or cable entry is in the wall of a box or conduit body opposite a removable cover, the distance from that wall to the cover shall be permitted to comply with the distance required for one wire per terminal in Table 312.6(A).*

The distance between raceway entries enclosing the same conductor shall not be less than six times the metric designator (trade size) of the larger raceway.

When transposing cable size into raceway size in 314.28(A)(1) and (A)(2), the minimum metric designator (trade size) raceway required for the number and size of conductors in the cable shall be used.

**(3) Smaller Dimensions.** Boxes or conduit bodies of dimensions less than those required in 314.28(A)(1) and (A)(2) shall be permitted for installations of combinations of conductors that are less than the maximum conduit or tubing fill (of conduits or tubing being used) permitted by Table 1 of Chapter 9, provided the box or conduit body has been listed for and is permanently marked with the maximum number and maximum size of conductors permitted.

**(B) Conductors in Pull or Junction Boxes.** In pull boxes or junction boxes having any dimension over 1.8 m (6 ft), all conductors shall be cabled or racked up in an approved manner.

**(C) Covers.** All pull boxes, junction boxes, and conduit bodies shall be provided with covers compatible with the box or conduit body construction and suitable for the conditions of use. Where metal covers are used, they shall comply with the grounding requirements of 250.110. An extension from the cover of an exposed box shall comply with 314.22, Exception.

**(D) Permanent Barriers.** Where permanent barriers are installed in a box, each section shall be considered as a separate box.

**314.29 Boxes and Conduit Bodies to Be Accessible.** Boxes and conduit bodies shall be installed so that the wiring contained in them can be rendered accessible without removing any part of the building or, in underground circuits, without excavating sidewalks, paving, earth, or other substance that is to be used to establish the finished grade.

*Exception: Listed boxes shall be permitted where covered by gravel, light aggregate, or noncohesive granulated soil if their location is effectively identified and accessible for excavation.*

## III. Construction Specifications

### 314.40 Metal Boxes, Conduit Bodies, and Fittings.

**(A) Corrosion Resistant.** Metal boxes, conduit bodies, and fittings shall be corrosion resistant or shall be well-galvanized, enameled, or otherwise properly coated inside and out to prevent corrosion.

> FPN: See 300.6 for limitation in the use of boxes and fittings protected from corrosion solely by enamel.

**(B) Thickness of Metal.** Sheet steel boxes not over 1650 cm$^3$ (100 in.$^3$) in size shall be made from steel not less than 1.59 mm (0.0625 in.) thick. The wall of a malleable iron box or conduit body and a die-cast or permanent-mold cast aluminum, brass, bronze, or zinc box or conduit body shall not be less than 2.38 mm ($\frac{3}{32}$ in.) thick. Other cast metal boxes or conduit bodies shall have a wall thickness not less than 3.17 mm ($\frac{1}{8}$ in.).

*Exception No. 1: Listed boxes and conduit bodies shown to have equivalent strength and characteristics shall be permitted to be made of thinner or other metals.*

*Exception No. 2: The walls of listed short radius conduit bodies, as covered in 314.5, shall be permitted to be made of thinner metal.*

**(C) Metal Boxes Over 1650 cm$^3$ (100 in.$^3$).** Metal boxes over 1650 cm$^3$ (100 in.$^3$) in size shall be constructed so as to be of ample strength and rigidity. If of sheet steel, the metal thickness shall not be less than 1.35 mm (0.053 in.) uncoated.

**(D) Grounding Provisions.** A means shall be provided in each metal box for the connection of an equipment grounding conductor. The means shall be permitted to be a tapped hole or equivalent.

**314.41 Covers.** Metal covers shall be of the same material as the box or conduit body with which they are used, or they shall be lined with firmly attached insulating material that is not less than 0.79 mm ($\frac{1}{32}$ in.) thick, or they shall be listed for the purpose. Metal covers shall be the same thickness as the boxes or conduit body for which they are used, or they shall be listed for the purpose. Covers of porcelain or other approved insulating materials shall be permitted if of such form and thickness as to afford the required protection and strength.

**314.42 Bushings.** Covers of outlet boxes and conduit bodies having holes through which flexible cord pendants may pass shall be provided with approved bushings or shall have smooth, well-rounded surfaces on which the cord may bear. Where individual conductors pass through a metal cover, a separate hole equipped with a bushing of suitable insulating material shall be provided for each conductor.

Such separate holes shall be connected by a slot as required by 300.20.

**314.43 Nonmetallic Boxes.** Provisions for supports or other mounting means for nonmetallic boxes shall be outside of the box, or the box shall be constructed so as to prevent contact between the conductors in the box and the supporting screws.

**314.44 Marking.** All boxes and conduit bodies, covers, extension rings, plaster rings, and the like shall be durably and legibly marked with the manufacturer's name or trademark.

## IV. Manholes and Other Electric Enclosures Intended for Personnel Entry

**314.50 General.** Electric enclosures intended for personnel entry and specifically fabricated for this purpose shall be of sufficient size to provide safe work space about electric equipment with live parts that is likely to require examination, adjustment, servicing, or maintenance while energized. They shall have sufficient size to permit ready installation or withdrawal of the conductors employed without damage to the conductors or to their insulation. They shall comply with the provisions of this part.

*Exception: Where electric enclosures covered by Part IV of this article are part of an industrial wiring system operating under conditions of maintenance and supervision that ensure only qualified persons monitor and supervise the system, they shall be permitted to be designed and installed in accordance with appropriate engineering practice. If required by the authority having jurisdiction, design documentation shall be provided.*

**314.51 Strength.** Manholes, vaults, and their means of access shall be designed under qualified engineering supervision and shall withstand all loads likely to be imposed on the structures.

> FPN: See ANSI C2-1997, *National Electrical Safety Code*, for additional information on the loading that can be expected to bear on underground enclosures.

**314.52 Cabling Work Space.** A clear work space not less than 900 mm (3 ft) wide shall be provided where cables are located on both sides, and not less than 750 mm (2½ ft) where cables are only on one side. The vertical headroom shall not be less than 1.8 m (6 ft) unless the opening is within 300 mm (1 ft), measured horizontally, of the adjacent interior side wall of the enclosure.

*Exception: A manhole containing only one or more of the following shall be permitted to have one of the horizontal work space dimensions reduced to 600 mm (2 ft) where the other horizontal clear work space is increased so the sum of the two dimensions is not less than 1.8 m (6 ft):*

*(a) Optical fiber cables as covered in Article 770.*
*(b) Power-limited fire alarm circuits supplied in accordance with 760.41.*
*(c) Class 2 or Class 3 remote-control and signaling circuits, or both, supplied in accordance with 725.41.*

**314.53 Equipment Work Space.** Where electric equipment with live parts that is likely to require examination, adjustment, servicing, or maintenance while energized is installed in a manhole, vault, or other enclosure designed for personnel access, the work space and associated requirements in 110.26 shall be met for installations operating at 600 volts or less. Where the installation is over 600 volts, the work space and associated requirements in 110.34 shall be met. A manhole access cover that weighs over 45 kg (100 lb) shall be considered as meeting the requirements of 110.34(C).

**314.54 Bending Space for Conductors.** Bending space for conductors operating at 600 volts or below shall be provided in accordance with the requirements of 314.28(A). Conductors operating over 600 volts shall be provided with bending space in accordance with 314.71(A) and 314.71(B), as applicable. All conductors shall be cabled, racked up, or arranged in an approved manner that provides ready and safe access for persons to enter for installation and maintenance.

*Exception: Where 314.71(B) applies, each row or column of ducts on one wall of the enclosure shall be calculated individually, and the single row or column that provides the maximum distance shall be used.*

**314.55 Access to Manholes.**

**(A) Dimensions.** Rectangular access openings shall not be less than 650 mm × 550 mm (26 in. × 22 in.). Round access openings in a manhole shall not be less than 650 mm (26 in.) in diameter.

*Exception: A manhole that has a fixed ladder that does not obstruct the opening or that contains only one or more of the following shall be permitted to reduce the minimum cover diameter to 600 mm (2 ft):*

*(a) Optical fiber cables as covered in Article 770.*
*(b) Power-limited fire alarm circuits supplied in accordance with 760.41.*
*(c) Class 2 or Class 3 remote-control and signaling circuits, or both, supplied in accordance with 725.41.*

**(B) Obstructions.** Manhole openings shall be free of protrusions that could injure personnel or prevent ready egress.

**(C) Location.** Manhole openings for personnel shall be located where they are not directly above electric equipment or conductors in the enclosure. Where this is not practicable, either a protective barrier or a fixed ladder shall be provided.

**(D) Covers.** Covers shall be over 45 kg (100 lb) or otherwise designed to require the use of tools to open. They shall be designed or restrained so they cannot fall into the manhole or protrude sufficiently to contact electrical conductors or equipment within the manhole.

**(E) Marking.** Manhole covers shall have an identifying mark or logo that prominently indicates their function, such as "electric."

**314.56 Access to Vaults and Tunnels.**

**(A) Location.** Access openings for personnel shall be located where they are not directly above electric equipment or conductors in the enclosure. Other openings shall be permitted over equipment to facilitate installation, maintenance, or replacement of equipment.

**(B) Locks.** In addition to compliance with the requirements of 110.34(C), if applicable, access openings for personnel shall be arranged so that a person on the inside can exit when the access door is locked from the outside, or in the case of normally locking by padlock, the locking arrangement shall be such that the padlock can be closed on the locking system to prevent locking from the outside.

**314.57 Ventilation.** Where manholes, tunnels, and vaults have communicating openings into enclosed areas used by the public, ventilation to open air shall be provided wherever practicable.

**314.58 Guarding.** Where conductors or equipment, or both, could be contacted by objects falling or being pushed through a ventilating grating, both conductors and live parts shall be protected in accordance with the requirements of 110.27(A)(2) or 110.31(B)(1), depending on the voltage.

**314.59 Fixed Ladders.** Fixed ladders shall be corrosion resistant.

## V. Pull and Junction Boxes for Use on Systems Over 600 Volts, Nominal

**314.70 General.** Where pull and junction boxes are used on systems over 600 volts, the installation shall comply with the provisions of Part V and also with the following general provisions of this article:

(1) In Part I, 314.2, 314.3, and 314.4

(2) In Part II, 314.15; 314.17; 314.20; 314.23(A), (B), or (G); 314.28(B); and 314.29

(3) In Part III, 314.40(A) and (C) and 314.41

**314.71 Size of Pull and Junction Boxes.** Pull and junction boxes shall provide adequate space and dimensions for the installation of conductors, and they shall comply with the specific requirements of this section.

*Exception: Terminal housings supplied with motors shall comply with the provisions of 430.12.*

**(A) For Straight Pulls.** The length of the box shall not be less than 48 times the outside diameter, over sheath, of the largest shielded or lead-covered conductor or cable entering the box. The length shall not be less than 32 times the outside diameter of the largest nonshielded conductor or cable.

**(B) For Angle or U Pulls.**

**(1) Distance to Opposite Wall.** The distance between each cable or conductor entry inside the box and the opposite wall of the box shall not be less than 36 times the outside diameter, over sheath, of the largest cable or conductor. This distance shall be increased for additional entries by the amount of the sum of the outside diameters, over sheath, of all other cables or conductor entries through the same wall of the box.

*Exception No. 1: Where a conductor or cable entry is in the wall of a box opposite a removable cover, the distance from that wall to the cover shall be permitted to be not less than the bending radius for the conductors as provided in 300.34.*

*Exception No. 2: Where cables are nonshielded and not lead covered, the distance of 36 times the outside diameter shall be permitted to be reduced to 24 times the outside diameter.*

**(2) Distance Between Entry and Exit.** The distance between a cable or conductor entry and its exit from the box shall not be less than 36 times the outside diameter, over sheath, of that cable or conductor.

*Exception: Where cables are nonshielded and not lead covered, the distance of 36 times the outside diameter shall be permitted to be reduced to 24 times the outside diameter.*

**(C) Removable Sides.** One or more sides of any pull box shall be removable.

**314.72 Construction and Installation Requirements.**

**(A) Corrosion Protection.** Boxes shall be made of material inherently resistant to corrosion or shall be suitably protected, both internally and externally, by enameling, galvanizing, plating, or other means.

**(B) Passing Through Partitions.** Suitable bushings, shields, or fittings having smooth, rounded edges shall be provided where conductors or cables pass through partitions and at other locations where necessary.

**(C) Complete Enclosure.** Boxes shall provide a complete enclosure for the contained conductors or cables.

**(D) Wiring Is Accessible.** Boxes shall be installed so that the wiring is accessible without removing any part of the building. Working space shall be provided in accordance with 110.34.

**(E) Suitable Covers.** Boxes shall be closed by suitable covers securely fastened in place. Underground box covers that weigh over 45 kg (100 lb) shall be considered meeting this requirement. Covers for boxes shall be permanently marked "DANGER — HIGH VOLTAGE — KEEP OUT." The marking shall be on the outside of the box cover and shall be readily visible. Letters shall be block type and at least 13 mm (½ in.) in height.

**(F) Suitable for Expected Handling.** Boxes and their covers shall be capable of withstanding the handling to which they may likely be subjected.

## ARTICLE 320
## Armored Cable: Type AC

### I. General

**320.1 Scope.** This article covers the use, installation, and construction specifications for armored cable, Type AC.

**320.2 Definition.**

**Armored Cable, Type AC.** A fabricated assembly of insulated conductors in a flexible metallic enclosure. See 320.100.

### II. Installation

**320.10 Uses Permitted.** Where not subject to physical damage, Type AC cable shall be permitted as follows:

(1) In both exposed and concealed work
(2) In cable trays where identified for such use
(3) In dry locations
(4) Embedded in plaster finish on brick or other masonry, except in damp or wet locations

(5) To be run or fished in the air voids of masonry block or tile walls where such walls are not exposed or subject to excessive moisture or dampness

**320.12 Uses Not Permitted.** Type AC cable shall not be used as follows:

(1) In theaters and similar locations, except where permitted in 518.4
(2) In motion picture studios
(3) In hazardous (classified) locations except where permitted in

    a. 501.4(B), Exception
    b. 502.4(B), Exception No. 1
    c. 504.20

(4) Where exposed to corrosive fumes or vapors
(5) In storage battery rooms
(6) In hoistways, or on elevators or escalators, except where permitted in 620.21
(7) In commercial garages where prohibited in 511.4 and 511.7

**320.15 Exposed Work.** Exposed runs of cable, except as provided in 300.11(A), shall closely follow the surface of the building finish or of running boards. Exposed runs shall also be permitted to be installed on the underside of joists where supported at each joist and located so as not to be subject to physical damage.

**320.17 Through or Parallel to Framing Members.** Type AC cable shall be protected in accordance with 300.4 where installed through or parallel to framing members.

**320.23 In Accessible Attics.** Type AC cables in accessible attics or roof spaces shall be installed as specified in 320.23(A) and (B).

**(A) Where Run Across the Top of Floor Joists.** Where run across the top of floor joists, or within 2.1 m (7 ft) of floor or floor joists across the face of rafters or studding, in attics and roof spaces that are accessible, the cable shall be protected by substantial guard strips that are at least as high as the cable. Where this space is not accessible by permanent stairs or ladders, protection shall only be required within 1.8 m (6 ft) of the nearest edge of the scuttle hole or attic entrance.

**(B) Cable Installed Parallel to Framing Members.** Where the cable is installed parallel to the sides of rafters, studs, or floor joists, neither guard strips nor running boards shall be required, and the installation shall also comply with 300.4(D).

**320.24 Bending Radius.** Bends in Type AC cable shall be made so that the cable will not be damaged. The radius of the curve of the inner edge of any bend shall not be less than five times the diameter of the Type AC cable.

**320.30 Securing and Supporting.** Type AC cable shall be secured by staples, cable ties, straps, hangers, or similar fittings designed and installed so as not to damage the cable at intervals not exceeding 1.4 m (4½ ft) and within 300 mm (12 in.) of every outlet box, junction box, cabinet, or fitting.

**(A) Horizontal Runs Through Holes and Notches.** In other than vertical runs, cables installed in accordance with 300.4 shall be considered supported and secured where such support does not exceed 1.4-m (4½-ft) intervals and the armored cable is securely fastened in place by an approved means within 300 mm (12 in.) of each box, cabinet, conduit body, or other armored cable termination.

**(B) Unsupported Cables.** Type AC cable shall be permitted to be unsupported where the cable:

(1) Is fished between access points, where concealed in finished buildings or structures and supporting is impracticable; or

(2) Is not more than 600 mm (2 ft) in length at terminals where flexibility is necessary; or

(3) Is not more than 1.8 m (6 ft) from the last point of support for connections within an accessible ceiling to luminaire(s) [(lighting fixture(s)] or equipment.

**(C) Cable Trays.** Type AC cable installed in cable trays shall comply with 392.8(B).

**320.40 Boxes and Fittings.** At all points where the armor of AC cable terminates, a fitting shall be provided to protect wires from abrasion, unless the design of the outlet boxes or fittings is such as to afford equivalent protection, and, in addition, an insulating bushing or its equivalent protection shall be provided between the conductors and the armor. The connector or clamp by which the Type AC cable is fastened to boxes or cabinets shall be of such design that the insulating bushing or its equivalent will be visible for inspection. Where change is made from Type AC cable to other cable or raceway wiring methods, a box, fitting, or conduit body shall be installed at junction points as required in 300.15.

**320.80 Ampacity.** The ampacity shall be determined by 310.15.

**(A) Thermal Insulation.** Armored cable installed in thermal insulation shall have conductors rated at 90°C (194°F). The ampacity of cable installed in these applications shall be that of 60°C (140°F) conductors.

**(B) Cable Tray.** The ampacity of Type AC cable installed in cable tray shall be determined in accordance with 392.11.

## III. Construction Specifications

**320.100 Construction.** Type AC cable shall have an armor of flexible metal tape and shall have an internal bonding strip of copper or aluminum in intimate contact with the armor for its entire length.

**320.104 Conductors.** Insulated conductors shall be of a type listed in Table 310.13 or those identified for use in this cable. In addition, the conductors shall have an overall moisture-resistant and fire-retardant fibrous covering. For Type ACT, a moisture-resistant fibrous covering shall be required only on the individual conductors.

**320.108 Equipment Grounding.** Type AC cable shall provide an adequate path for equipment grounding as required by 250.4(A)(5) or 250.4(B)(4).

**320.120 Marking.** The cable shall be marked in accordance with 310.11, except that Type AC shall have ready identification of the manufacturer by distinctive external markings on the cable sheath throughout its entire length.

# ARTICLE 322
# Flat Cable Assemblies: Type FC

## I. General

**322.1 Scope.** This article covers the use, installation, and construction specifications for flat cable assemblies, Type FC.

**322.2 Definition.**

**Flat Cable Assembly, Type FC.** An assembly of parallel conductors formed integrally with an insulating material web specifically designed for field installation in surface metal raceway.

## II. Installation

**322.10 Uses Permitted.** Flat cable assemblies shall be permitted only as follows:

(1) As branch circuits to supply suitable tap devices for lighting, small appliances, or small power loads. The rating of the branch circuit shall not exceed 30 amperes.

(2) Where installed for exposed work.

(3) In locations where they will not be subjected to physical damage. Where a flat cable assembly is installed less than 2.5 m (8 ft) above the floor or fixed working

platform, it shall be protected by a cover identified for the use.

(4) In surface metal raceways identified for the use. The channel portion of the surface metal raceway systems shall be installed as complete systems before the flat cable assemblies are pulled into the raceways.

**322.12 Uses Not Permitted.** Flat cable assemblies shall not be used as follows:

(1) Where subject to corrosive vapors unless suitable for the application
(2) In hoistways or on elevators or escalators
(3) In any hazardous (classified) location
(4) Outdoors or in wet or damp locations unless identified for the use

**322.30 Securing and Supporting.** The flat cable assemblies shall be supported by means of their special design features, within the surface metal raceways.

The surface metal raceways shall be supported as required for the specific raceway to be installed.

**322.40 Boxes and Fittings.**

**(A) Dead Ends.** Each flat cable assembly dead end shall be terminated in an end-cap device identified for the use.

The dead-end fitting for the enclosing surface metal raceway shall be identified for the use.

**(B) Luminaire (Fixture) Hangers.** Luminaire (fixture) hangers installed with the flat cable assemblies shall be identified for the use.

**(C) Fittings.** Fittings to be installed with flat cable assemblies shall be designed and installed to prevent physical damage to the cable assemblies.

**(D) Extensions.** All extensions from flat cable assemblies shall be made by approved wiring methods, within the junction boxes, installed at either end of the flat cable assembly runs.

**322.56 Splices and Taps.**

**(A) Splices.** Splices shall be made in listed junction boxes.

**(B) Taps.** Taps shall be made between any phase conductor and the grounded conductor or any other phase conductor by means of devices and fittings identified for the use. Tap devices shall be rated at not less than 15 amperes, or more than 300 volts to ground, and they shall be color-coded in accordance with the requirements of 322.120(C).

**III. Construction**

**322.100 Construction.** Flat cable assemblies shall consist of two, three, four, or five conductors.

**322.104 Conductors.** Flat cable assemblies shall have conductors of 10 AWG special stranded copper wires.

**322.112 Insulation.** The entire flat cable assembly shall be formed to provide a suitable insulation covering all the conductors and using one of the materials recognized in Table 310.13 for general branch-circuit wiring.

**322.120 Marking.**

**(A) Temperature Rating.** In addition to the provisions of 310.11, Type FC cable shall have the temperature rating durably marked on the surface at intervals not exceeding 600 mm (24 in.).

**(B) Identification of Grounded Conductor.** The grounded conductor shall be identified throughout its length by means of a distinctive and durable white or gray marking.

> FPN: The color gray may have been used in the past as an ungrounded conductor. Care should be taken when working on existing systems.

**(C) Terminal Block Identification.** Terminal blocks identified for the use shall have distinctive and durable markings for color or word coding. The grounded conductor section shall have a white marking or other suitable designation. The next adjacent section of the terminal block shall have a black marking or other suitable designation. The next section shall have a red marking or other suitable designation. The final or outer section, opposite the grounded conductor section of the terminal block, shall have a blue marking or other suitable designation.

# ARTICLE 324
# Flat Conductor Cable: Type FCC

**I. General**

**324.1 Scope.** This article covers a field-installed wiring system for branch circuits incorporating Type FCC cable and associated accessories as defined by the article. The wiring system is designed for installation under carpet squares.

**324.2 Definitions.**

**Bottom Shield.** A protective layer that is installed between the floor and Type FCC flat conductor cable to protect the cable from physical damage and may or may not be incorporated as an integral part of the cable.

**Cable Connector.** A connector designed to join Type FCC cables without using a junction box.

**FCC System.** A complete wiring system for branch circuits that is designed for installation under carpet squares. The FCC system includes Type FCC cable and associated shielding, connectors, terminators, adapters, boxes, and receptacles.

**Insulating End.** An insulator designed to electrically insulate the end of a Type FCC cable.

**Metal Shield Connections.** Means of connection designed to electrically and mechanically connect a metal shield to another metal shield, to a receptacle housing or self-contained device, or to a transition assembly.

**Top Shield.** A grounded metal shield covering under-carpet components of the FCC system for the purposes of providing protection against physical damage.

**Transition Assembly.** An assembly to facilitate connection of the FCC system to other wiring systems, incorporating (1) a means of electrical interconnection and (2) a suitable box or covering for providing electrical safety and protection against physical damage.

**Type FCC Cable.** Three or more flat copper conductors placed edge-to-edge and separated and enclosed within an insulating assembly.

## II. Installation

### 324.10 Uses Permitted.

**(A) Branch Circuits.** Use of FCC systems shall be permitted both for general-purpose and appliance branch circuits and for individual branch circuits.

**(B) Branch-Circuit Ratings.**

**(1) Voltage.** Voltage between ungrounded conductors shall not exceed 300 volts. Voltage between ungrounded conductors and the grounded conductor shall not exceed 150 volts.

**(2) Current.** General-purpose and appliance branch circuits shall have ratings not exceeding 20 amperes. Individual branch circuits shall have ratings not exceeding 30 amperes.

**(C) Floors.** Use of FCC systems shall be permitted on hard, sound, smooth, continuous floor surfaces made of concrete, ceramic, or composition flooring, wood, and similar materials.

**(D) Walls.** Use of FCC systems shall be permitted on wall surfaces in surface metal raceways.

**(E) Damp Locations.** Use of FCC systems in damp locations shall be permitted.

**(F) Heated Floors.** Materials used for floors heated in excess of 30°C (86°F) shall be identified as suitable for use at these temperatures.

**(G) System Height.** Any portion of an FCC system with a height above floor level exceeding 2.3 mm (0.090 in.) shall be tapered or feathered at the edges to floor level.

**(H) Coverings.** Floor-mounted Type FCC cable, cable connectors, and insulating ends shall be covered with carpet squares not larger than 914 mm (36 in.) square. Those carpet squares that are adhered to the floor shall be attached with release-type adhesives.

**(I) Corrosion Resistance.** Metal components of the system shall be either corrosion resistant, coated with corrosion-resistant materials, or insulated from contact with corrosive substances.

**(J) Metal-Shield Connectors.** Metal shields shall be connected to each other and to boxes, receptacle housings, self-contained devices, and transition assemblies using metal-shield connectors.

### 324.12 Uses Not Permitted. FCC systems shall not be used:

(1) Outdoors or in wet locations
(2) Where subject to corrosive vapors
(3) In any hazardous (classified) location
(4) In residential, school, and hospital buildings

### 324.18 Crossings. 
Crossings of more than two Type FCC cable runs shall not be permitted at any one point. Crossings of a Type FCC cable over or under a flat communications or signal cable shall be permitted. In each case, a grounded layer of metal shielding shall separate the two cables, and crossings of more than two flat cables shall not be permitted at any one point.

### 324.30 Securing and Supporting. 
All FCC system components shall be firmly anchored to the floor or wall using an adhesive or mechanical anchoring system identified for this use. Floors shall be prepared to ensure adherence of the FCC system to the floor until the carpet squares are placed.

### 324.40 Boxes and Fittings.

**(A) Cable Connections and Insulating Ends.** All Type FCC cable connections shall use connectors identified for their use, installed such that electrical continuity, insulation, and sealing against dampness and liquid spillage are provided. All bare cable ends shall be insulated and sealed against dampness and liquid spillage using listed insulating ends.

**(B) Polarization of Connections.** All receptacles and connections shall be constructed and installed so as to maintain proper polarization of the system.

**(C) Shields.**

**(1) Top Shield.** A metal top shield shall be installed over all floor-mounted Type FCC cable, connectors, and insulating ends. The top shield shall completely cover all cable runs, corners, connectors, and ends.

**(2) Bottom Shield.** A bottom shield shall be installed beneath all Type FCC cable, connectors, and insulating ends.

**(D) Connection to Other Systems.** Power feed, grounding connection, and shield system connection between the FCC system and other wiring systems shall be accomplished in a transition assembly identified for this use.

**324.42 Devices.**

**(A) Receptacles.** All receptacles, receptacle housings, and self-contained devices used with the FCC system shall be identified for this use and shall be connected to the Type FCC cable and metal shields. Connection from any grounding conductor of the Type FCC cable shall be made to the shield system at each receptacle.

**(B) Receptacles and Housings.** Receptacle housings and self-contained devices designed either for floor mounting or for in-wall or on-wall mounting shall be permitted for use with the FCC system. Receptacle housings and self-contained devices shall incorporate means for facilitating entry and termination of Type FCC cable and for electrically connecting the housing or device with the metal shield. Receptacles and self-contained devices shall comply with 406.3. Power and communications outlets installed together in common housing shall be permitted in accordance with 800.52(A)(1)(c), Exception No. 2.

**324.56 Splices and Taps.**

**(A) FCC Systems Alterations.** Alterations to FCC systems shall be permitted. New cable connectors shall be used at new connection points to make alterations. It shall be permitted to leave unused cable runs and associated cable connectors in place and energized. All cable ends shall be covered with insulating ends.

**(B) Transition Assemblies.** All transition assemblies shall be identified for their use. Each assembly shall incorporate means for facilitating entry of the Type FCC cable into the assembly, for connecting the Type FCC cable to grounded conductors, and for electrically connecting the assembly to the metal cable shields and to equipment grounding conductors.

**324.60 Grounding.** All metal shields, boxes, receptacle housings, and self-contained devices shall be electrically continuous to the equipment grounding conductor of the supplying branch circuit. All such electrical connections shall be made with connectors identified for this use. The electrical resistivity of such shield system shall not be more than that of one conductor of the Type FCC cable used in the installation.

**III. Construction**

**324.100 Construction.**

**(A) Type FCC Cable.** Type FCC cable shall be listed for use with the FCC system and shall consist of three, four, or five flat copper conductors, one of which shall be an equipment grounding conductor.

**(B) Shields.**

**(1) Materials and Dimensions.** All top and bottom shields shall be of designs and materials identified for their use. Top shields shall be metal. Both metallic and nonmetallic materials shall be permitted for bottom shields.

**(2) Resistivity.** Metal shields shall have cross-sectional areas that provide for electrical resistivity of not more than that of one conductor of the Type FCC cable used in the installation.

**324.112 Insulation.** The insulating material of the cable shall be moisture resistant and flame retardant. All insulating materials in the FCC systems shall be identified for their use.

**324.120 Markings.**

**(A) Cable Marking.** Type FCC cable shall be clearly and durably marked on both sides at intervals of not more than 610 mm (24 in.) with the information required by 310.11(A) and with the following additional information:

(1) Material of conductors
(2) Maximum temperature rating
(3) Ampacity

**(B) Conductor Identification.** Conductors shall be clearly and durably identified on both sides throughout their length as specified in 310.12.

# ARTICLE 326
## Integrated Gas Spacer Cable: Type IGS

## I. General

**326.1 Scope.** This article covers the use, installation, and construction specifications for integrated gas spacer cable, Type IGS.

**326.2 Definition.**

**Integrated Gas Spacer Cable, Type IGS.** A factory assembly of one or more conductors, each individually insulated and enclosed in a loose fit, nonmetallic flexible conduit as an integrated gas spacer cable rated 0 through 600 volts.

## II. Installation

**326.10 Uses Permitted.** Type IGS cable shall be permitted for use under ground, including direct burial in the earth, as the following:

(1) Service-entrance conductors
(2) Feeder or branch-circuit conductors

**326.12 Uses Not Permitted.** Type IGS cable shall not be used as interior wiring or be exposed in contact with buildings.

**326.24 Bending Radius.** Where the coilable nonmetallic conduit and cable is bent for installation purposes or is flexed or bent during shipment or installation, the radii of bends measured to the inside of the bend shall not be less than specified in Table 326.24.

**Table 326.24 Minimum Radii of Bends**

| Conduit Size | | Minimum Radii | |
|---|---|---|---|
| Metric Designator | Trade Size | mm | in. |
| 53 | 2 | 600 | 24 |
| 78 | 3 | 900 | 35 |
| 103 | 4 | 1150 | 45 |

**326.26 Bends.** A run of Type IGS cable between pull boxes or terminations shall not contain more than the equivalent of four quarter bends (360 degrees total), including those bends located immediately at the pull box or terminations.

**326.40 Fittings.** Terminations and splices for Type IGS cable shall be identified as a type that is suitable for maintaining the gas pressure within the conduit. A valve and cap shall be provided for each length of the cable and conduit to check the gas pressure or to inject gas into the conduit.

**326.80 Ampacity.** The ampacity of Type IGS cable shall not exceed the values shown in Table 326.80.

**Table 326.80 Ampacity of Type IGS Cable**

| Size (kcmil) | Amperes | Size (kcmil) | Amperes |
|---|---|---|---|
| 250 | 119 | 2500 | 376 |
| 500 | 168 | 3000 | 412 |
| 750 | 206 | 3250 | 429 |
| 1000 | 238 | 3500 | 445 |
| 1250 | 266 | 3750 | 461 |
| 1500 | 292 | 4000 | 476 |
| 1750 | 344 | 4250 | 491 |
| 2000 | 336 | 4500 | 505 |
| 2250 | 357 | 4750 | 519 |

## III. Construction Specifications

**326.104 Conductors.** The conductors shall be solid aluminum rods, laid parallel, consisting of one to nineteen 12.7 mm ($\frac{1}{2}$ in.) diameter rods. The minimum conductor size shall be 250 kcmil, and the maximum size shall be 4750 kcmil.

**326.112 Insulation.** The insulation shall be dry kraft paper tapes and a pressurized sulfur hexafluoride gas (SF6), both approved for electrical use. The nominal gas pressure shall be 138 kPa gauge (20 pounds per square inch gauge). The thickness of the paper spacer shall be as specified in Table 326.112.

**Table 326.112 Paper Spacer Thickness**

| Size (kcmil) | Thickness | |
|---|---|---|
| | mm | in. |
| 250–1000 | 1.02 | 0.040 |
| 1250–4750 | 1.52 | 0.060 |

**326.116 Conduit.** The conduit shall be a medium density polyethylene identified as suitable for use with natural gas rated pipe in metric designator 53, 78, or 103 (trade size 2, 3, or 4). The percent fill dimensions for the conduit are shown in Table 326.116.

The size of the conduit permitted for each conductor size shall be calculated for a percent fill not to exceed those found in Table 1, Chapter 9.

**326.120 Marking.** The cable shall be marked in accordance with 310.11(A), 310.11(B)(1), and 310.11(D).

**Table 326.116 Conduit Dimensions**

| Conduit Size | | Actual Outside Diameter | | Actual Inside Diameter | |
|---|---|---|---|---|---|
| Metric Designator | Trade Size | mm | in. | mm | in. |
| 53 | 2 | 60 | 2.375 | 49.46 | 1.947 |
| 78 | 3 | 89 | 3.500 | 73.30 | 2.886 |
| 103 | 4 | 114 | 4.500 | 94.23 | 3.710 |

# ARTICLE 328
# Medium Voltage Cable: Type MV

## I. General

**328.1 Scope.** This article covers the use, installation, and construction specifications for medium voltage cable, Type MV.

## 328.2 Definition.

**Medium Voltage Cable, Type MV.** A single or multiconductor solid dielectric insulated cable rated 2001 volts or higher.

## II. Installation

**328.10 Uses Permitted.** Type MV cables shall be permitted for use on power systems rated up to 35,000 volts, nominal, as follows:

(1) In wet or dry locations
(2) In raceways
(3) In cable trays as specified in 392.3(B)(1)
(4) Direct buried in accordance with 300.50
(5) In messenger-supported wiring

**328.12 Uses Not Permitted.** Type MV cable shall not be used unless identified for the use as follows:

(1) Where exposed to direct sunlight
(2) In cable trays

**328.80 Ampacity.** The ampacity of Type MV cable shall be determined in accordance with 310.60. The ampacity of Type MV cable installed in cable tray shall be determined in accordance with 392.13.

## III. Construction Specifications

**328.100 Construction.** Type MV cables shall have copper, aluminum, or copper-clad aluminum conductors and shall be constructed in accordance with Article 310.

**328.120 Marking.** Medium voltage cable shall be marked as required in 310.11.

# ARTICLE 330
# Metal-Clad Cable: Type MC

## I. General

**330.1 Scope.** This article covers the use, installation, and construction specifications of metal-clad cable, Type MC.

## 330.2 Definition.

**Metal Clad Cable, Type MC.** A factory assembly of one or more insulated circuit conductors with or without optical fiber members enclosed in an armor of interlocking metal tape, or a smooth or corrugated metallic sheath.

## II. Installation

### 330.10 Uses Permitted.

**(A) General Uses.** Where not subject to physical damage, Type MC cables shall be permitted as follows:

(1) For services, feeders, and branch circuits
(2) For power, lighting, control, and signal circuits
(3) Indoors or outdoors
(4) Where exposed or concealed
(5) Direct buried where identified for such use
(6) In cable tray
(7) In any raceway
(8) As open runs of cable
(9) As aerial cable on a messenger
(10) In hazardous (classified) locations as permitted in Articles 501, 502, 503, 504, and 505
(11) In dry locations and embedded in plaster finish on brick or other masonry except in damp or wet locations
(12) In wet locations where any of the following conditions are met:
   a. The metallic covering is impervious to moisture.
   b. A lead sheath or moisture-impervious jacket is provided under the metal covering.
   c. The insulated conductors under the metallic covering are listed for use in wet locations.
(13) Where single-conductor cables are used, all phase conductors and, where used, the neutral conductor shall be grouped together to minimize induced voltage on the sheath.

**(B) Specific Uses.** Type MC cable shall be installed in compliance with Articles 300, 490, 725, and 770.52 as applicable and in accordance with 330.10(B)(1) through (B)(4).

**(1) Cable Tray.** Type MC cable installed in cable tray shall comply with Article 392.

**(2) Direct Buried.** Direct-buried cable shall comply with 300.5 or 300.50, as appropriate.

**(3) Installed as Service-Entrance Cable.** Type MC cable installed as service-entrance cable shall comply with Article 230.

**(4) Installed Outside of Buildings or as Aerial Cable.** Type MC cable installed outside of buildings or as aerial cable shall comply with Article 225 and Article 396.

**330.12 Uses Not Permitted.** Type MC cable shall not be used where exposed to the following destructive corrosive conditions, unless the metallic sheath is suitable for the conditions or is protected by material suitable for the conditions:

(1) Direct burial in the earth
(2) In concrete
(3) Where exposed to cinder fills, strong chlorides, caustic alkalis, or vapors of chlorine or of hydrochloric acids

**330.17 Through or Parallel to Framing Members.** Type MC cable shall be protected in accordance with 300.4 where installed through or parallel to framing members.

**330.23 In Accessible Attics.** The installation of Type MC cable in accessible attics or roof spaces shall also comply with 320.23.

**330.24 Bending Radius.** Bends in Type MC cable shall be made so that the cable will not be damaged. The radius of the curve of the inner edge of any bend shall not be less than shown in 330.24(A) through (C).

**(A) Smooth Sheath.**

(1) Ten times the external diameter of the metallic sheath for cable not more than 19 mm (¾ in.) in external diameter
(2) Twelve times the external diameter of the metallic sheath for cable more than 19 mm (¾ in.) but not more than 38 mm (1½ in.) in external diameter
(3) Fifteen times the external diameter of the metallic sheath for cable more than 38 mm (1½ in.) in external diameter

**(B) Interlocked-Type Armor or Corrugated Sheath.** Seven times the external diameter of the metallic sheath.

**(C) Shielded Conductors.** Twelve times the overall diameter of one of the individual conductors or seven times the overall diameter of the multiconductor cable, whichever is greater.

**330.30 Securing and Supporting.** Type MC cable shall be supported and secured at intervals not exceeding 1.8 m (6 ft).

**(A) Horizontal Runs Through Holes and Notches.** In other than vertical runs, cables installed in accordance with 300.4 shall be considered supported and secured where such support does not exceed 1.8-m (6-ft) intervals.

**(B) Unsupported Cables.** Type MC cable shall be permitted to be unsupported where the cable:

(1) Is fished between access points, where concealed in finished buildings or structures and supporting is impracticable
(2) Is not more than 1.8 m (6 ft) from the last point of support for connections within an accessible ceiling to luminaire(s) [lighting fixture(s)] or equipment

**(C) At Terminations.** Cables containing four or fewer conductors, sized no larger than 10 AWG, shall be secured within 300 mm (12 in.) of every box, cabinet, fitting, or other cable termination.

**330.31 Single Conductors.** Where single-conductor cables with a nonferrous armor or sheath are used, the installation shall comply with 300.20.

**330.40 Boxes and Fitting.** Fittings used for connecting Type MC cable to boxes, cabinets, or other equipment shall be listed and identified for such use.

**330.80 Ampacity.** The ampacity of Type MC cable shall be determined in accordance with 310.15 or 310.60 for 14 AWG and larger conductors and in accordance with Table 402.5 for 18 AWG and 16 AWG conductors. The installation shall not exceed the temperature ratings of terminations and equipment.

**(A) Type MC Cable Installed in Cable Tray.** The ampacities for Type MC cable installed in cable tray shall be determined in accordance with 392.11 and 392.13.

**(B) Single Type MC Conductors Grouped Together.** Where single Type MC conductors are grouped together in a triangular or square configuration and installed on a messenger or as open runs with a maintained free airspace of not less than 2.15 times one conductor diameter (2.15 × O.D.) of the largest conductor contained within the configuration and adjacent conductor configurations or cables, the ampacity of the conductors shall not exceed the allowable ampacities of:

(1) Table 310.20 for conductors rated 0 through 2000 volts
(2) Tables 310.67 and 310.68 for conductors rated over 2000 volts

## III. Construction Specifications

**330.104 Conductors.** The conductors shall be of copper, aluminum, or copper-clad aluminum, solid or stranded. The minimum conductor size shall be 18 AWG copper and 12 AWG aluminum or copper-clad aluminum.

**330.108 Equipment Grounding.** Type MC cable shall provide an adequate path for equipment grounding as required by Article 250.

**330.112 Insulation.** The insulated conductors shall comply with 330.112(A) or (B).

**(A) 600 Volts.** Insulated conductors in sizes 18 AWG and 16 AWG shall be of a type listed in Table 402.3, with a maximum operating temperature not less than 90°C (194°F) and as permitted by 725.27. Conductors larger than 16 AWG shall be of a type listed in Table 310.13 or of a type identified for use in Type MC cable.

**(B) Over 600 Volts.** Insulated conductors shall be of a type listed in Tables 310.61 through 310.64.

**330.116 Sheath.** The metallic covering shall be one of the following types: smooth metallic sheath, corrugated metallic sheath, interlocking metal tape armor. The metallic sheath shall be continuous and close fitting. A nonmagnetic sheath or armor shall be used on single conductor Type MC. Supplemental protection of an outer covering of corrosion-resistant material shall be permitted and shall be required where such protection is needed. The sheath shall not be used as a current-carrying conductor.

FPN: See 300.6 for protection against corrosion.

# ARTICLE 332
## Mineral-Insulated, Metal-Sheathed Cable: Type MI

## I. General

**332.1 Scope.** This article covers the use, installation, and construction specifications for mineral-insulated, metal-sheathed cable, Type MI.

## 332.2 Definition.

**Mineral-Insulated, Metal-Sheathed Cable, Type MI.** A factory assembly of one or more conductors insulated with a highly compressed refractory mineral insulation and enclosed in a liquidtight and gastight continuous copper or alloy steel sheath.

## II. Installation

**332.10 Uses Permitted.** Type MI cable shall be permitted as follows:

(1) For services, feeders, and branch circuits
(2) For power, lighting, control, and signal circuits
(3) In dry, wet, or continuously moist locations
(4) Indoors or outdoors
(5) Where exposed or concealed
(6) Embedded in plaster, concrete, fill, or other masonry, whether above or below grade
(7) In any hazardous (classified) location
(8) Where exposed to oil and gasoline
(9) Where exposed to corrosive conditions not deteriorating to its sheath
(10) In underground runs where suitably protected against physical damage and corrosive conditions

**332.12 Uses Not Permitted.** Type MI cable shall not be used where exposed to conditions that are destructive and corrosive to the metallic sheath unless additionally protected by materials suitable for the conditions.

**332.17 Through or Parallel to Framing Members.** Type MI cable shall be protected in accordance with 300.4 where installed through or parallel to framing members.

**332.24 Bending Radius.** Bends in Type MI cable shall be made so that the cable will not be damaged. The radius of the inner edge of any bend shall not be less than shown as follows:

(1) Five times the external diameter of the metallic sheath for cable not more than 19 mm (¾ in.) in external diameter
(2) Ten times the external diameter of the metallic sheath for cable greater than 19 mm (¾ in.) but not more than 25 mm (1 in.) in external diameter

**332.30 Securing and Supporting.** Type MI cable shall be supported securely at intervals not exceeding 1.8 m (6 ft) by straps, staples, hangers, or similar fittings designed and installed so as not to damage the cable.

**(A) Horizontal Runs Through Holes and Notches.** In other than vertical runs, cables installed in accordance with

300.4 shall be considered supported and secured where such support does not exceed 1.8-m (6-ft) intervals.

**(B) Unsupported Cable.** Type MI cable shall be permitted to be unsupported where the cable is fished.

**(C) Cable Trays.** Type MI cable installed in cable trays shall comply with 392.8(B).

**332.31 Single Conductors.** Where single-conductor cables are used, all phase conductors and, where used, the neutral conductor shall be grouped together to minimize induced voltage on the sheath.

**332.40 Boxes and Fittings.**

**(A) Fittings.** Fittings used for connecting Type MI cable to boxes, cabinets, or other equipment shall be identified for such use.

**(B) Terminal Seals.** Where Type MI cable terminates, an end seal fitting shall be installed immediately after stripping to prevent the entrance of moisture into the insulation. The conductors extending beyond the sheath shall be individually provided with an insulating material.

**332.80 Ampacity.** The ampacity of Type MI cable shall be determined in accordance with 310.15. The conductor temperature at the end seal fitting shall not exceed the temperature rating of the listed end seal fitting, and the installation shall not exceed the temperature ratings of terminations or equipment.

**(A) Type MI Cable Installed in Cable Tray.** The ampacities for Type MI cable installed in cable tray shall be determined in accordance with 392.11.

**(B) Single Type MI Conductors Grouped Together.** Where single Type MI conductors are grouped together in a triangular or square configuration, as required by 332.31, and installed on a messenger or as open runs with a maintained free air space of not less than 2.15 times one conductor diameter (2.15 × O.D.) of the largest conductor contained within the configuration and adjacent conductor configurations or cables, the ampacity of the conductors shall not exceed the allowable ampacities of Table 310.17.

### III. Construction Specifications

**332.104 Conductors.** Type MI cable conductors shall be of solid copper, nickel, or nickel-coated copper with a resistance corresponding to standard AWG and kcmil sizes.

**332.108 Equipment Grounding.** Where the outer sheath is made of copper, it shall provide an adequate path for

equipment grounding purposes. Where made of steel, an equipment grounding conductor shall be provided.

**332.112 Insulation.** The conductor insulation in Type MI cable shall be a highly compressed refractory mineral that provides proper spacing for all conductors.

**332.116 Sheath.** The outer sheath shall be of a continuous construction to provide mechanical protection and moisture seal.

## ARTICLE 334
## Nonmetallic-Sheathed Cable:
## Types NM, NMC, and NMS

### I. General

**334.1 Scope.** This article covers the use, installation, and construction specifications of nonmetallic-sheathed cable.

**334.2 Definition.**

**Nonmetallic-Sheathed Cable.** A factory assembly of two or more insulated conductors having an outer sheath of nonmetallic material.

**334.6 Listed.** Type NM, Type NMC, and Type NMS cables shall be listed.

### II. Installation

**334.10 Uses Permitted.** Type NM, Type NMC, and Type NMS cables shall be permitted to be used in the following:

(1) One- and two-family dwellings.
(2) Multifamily dwellings permitted to be of Types III, IV, and V construction except as prohibited in 334.12.
(3) Other structures permitted to be of Types III, IV, and V construction except as prohibited in 334.12. Cables shall be concealed within walls, floors, or ceilings that provide a thermal barrier of material that has at least a 15-minute finish rating as identified in listings of fire-rated assemblies.

> FPN No. 1: Building constructions are defined in NFPA 220-1999, *Standard on Types of Building Construction*, or the applicable building code, or both.

> FPN No. 2: See Annex E for determination of building types [NFPA 220, Table 3-1].

(4) Cable trays, where the cables are identified for the use.

> FPN: See 310.10 for temperature limitation of conductors.

**(A) Type NM.** Type NM cable shall be permitted as follows:

(1) For both exposed and concealed work in normally dry locations except as prohibited in 334.10(3).

(2) To be installed or fished in air voids in masonry block or tile walls

**(B) Type NMC.** Type NMC cable shall be permitted as follows:

(1) For both exposed and concealed work in dry, moist, damp, or corrosive locations, except as prohibited in 334.10(3)

(2) In outside and inside walls of masonry block or tile

(3) In a shallow chase in masonry, concrete, or adobe protected against nails or screws by a steel plate at least 1.59 mm (1/16 in.) thick and covered with plaster, adobe, or similar finish

**(C) Type NMS.** Type NMS cable shall be permitted as follows:

(1) For both exposed and concealed work in normally dry locations except as prohibited in 334.10(3)

(2) To be installed or fished in air voids in masonry block or tile walls

(3) To be used as permitted in Article 780

**334.12 Uses Not Permitted.**

**(A) Types NM, NMC, and NMS.** Types NM, NMC, and NMS cables shall not be used as follows:

(1) As open runs in dropped or suspended ceilings in other than one- and two-family and multifamily dwellings.

(2) As service-entrance cable.

(3) In commercial garages having hazardous (classified) locations as defined in 511.3.

(4) In theaters and similar locations, except where permitted in 518.4.

(5) In motion picture studios.

(6) In storage battery rooms.

(7) In hoistways or on elevators or escalators.

(8) Embedded in poured cement, concrete, or aggregate.

(9) In hazardous (classified) locations, except where permitted in the following:

   a. 501.4(B), Exception

   b. 502.4(B), Exception No. 1

   c. 504.20

(10) Types NM and NMS. Types NM and NMS cable shall not be used as follows:

   a. Where exposed to corrosive fumes or vapors

   b. Where embedded in masonry, concrete, adobe, fill, or plaster

   c. In a shallow chase in masonry, concrete, or adobe and covered with plaster, adobe, or similar finish

   d. Where exposed or subject to excessive moisture or dampness

**334.15 Exposed Work.** In exposed work, except as provided in 300.11(A), the cable shall be installed as specified in 334.15(A) through (C).

**(A) To Follow Surface.** The cable shall closely follow the surface of the building finish or of running boards.

**(B) Protection from Physical Damage.** The cable shall be protected from physical damage where necessary by conduit, electrical metallic tubing, Schedule 80 PVC rigid nonmetallic conduit, pipe, guard strips, listed surface metal or nonmetallic raceway, or other means. Where passing through a floor, the cable shall be enclosed in rigid metal conduit, intermediate metal conduit, electrical metallic tubing, Schedule 80 PVC rigid nonmetallic conduit, listed surface metal or nonmetallic raceway, or other metal pipe extending at least 150 mm (6 in.) above the floor.

**(C) In Unfinished Basements.** Where the cable is run at angles with joists in unfinished basements, it shall be permissible to secure cables not smaller than two 6 AWG or three 8 AWG conductors directly to the lower edges of the joists. Smaller cables shall be run either through bored holes in joists or on running boards.

**334.17 Through or Parallel to Framing Members.** Types NM, NMC, or NMS cable shall be protected in accordance with 300.4 where installed through or parallel to framing members. Grommets used as required in 300.4(B)(1) shall remain in place and be listed for the purpose of cable protection.

**334.23 In Accessible Attics.** The installation of cable in accessible attics or roof spaces shall also comply with 320.23.

**334.24 Bending Radius.** Bends in Types NM, NMC, and NMS cable shall be made so that the cable will not be damaged. The radius of the curve of the inner edge of any bend during or after installation shall not be less than five times the diameter of the cable.

**334.30 Securing and Supporting.** Nonmetallic-sheathed cable shall be secured by staples, cable ties, straps, hangers, or similar fittings designed and installed so as not to damage the cable at intervals not exceeding 1.4 m (4½ ft) and within 300 mm (12 in.) of every cabinet, box, or fitting. Flat cables shall not be stapled on edge.

**(A) Horizontal Runs through Holes and Notches.** In other than vertical runs, cables installed in accordance with 300.4 shall be considered supported and secured where such support does not exceed 1.4-m (4½-ft) intervals and the nonmetallic-sheathed cable is securely fastened in place by an approved means within 300 mm (12 in.) of each box, cabinet, conduit body, or other nonmetallic-sheathed cable termination.

> FPN: See 314.17(C) for support where nonmetallic boxes are used.

**(B) Unsupported Cables.** Nonmetallic-sheathed cable shall be permitted to be unsupported where the cable:

(1) Is fished between access points, where concealed in finished buildings or finished panels for prefabricated buildings and supporting is impracticable

(2) Is not more than 1.4 m (4½ ft) from the last point of support for connections within an accessible ceiling to luminaire(s) [lighting fixture(s)] or equipment

**(C) Wiring Device Without a Separate Outlet Box.** A wiring device identified for the use, without a separate outlet box, incorporating an integral cable clamp shall be permitted where the cable is secured in place at intervals not exceeding 1.4 m (4½ ft) and within 300 mm (12 in.) from the wiring device wall opening, and there shall be at least a 300 mm (12 in.) loop of unbroken cable or 150 mm (6 in.) of a cable end available on the interior side of the finished wall to permit replacement.

### 334.40 Boxes and Fittings.

**(A) Boxes of Insulating Material.** Nonmetallic outlet boxes shall be permitted as provided in 314.3.

**(B) Devices of Insulating Material.** Switch, outlet, and tap devices of insulating material shall be permitted to be used without boxes in exposed cable wiring and for rewiring in existing buildings where the cable is concealed and fished. Openings in such devices shall form a close fit around the outer covering of the cable, and the device shall fully enclose the part of the cable from which any part of the covering has been removed. Where connections to conductors are by binding-screw terminals, there shall be available as many terminals as conductors.

**(C) Devices with Integral Enclosures.** Wiring devices with integral enclosures identified for such use shall be permitted as provided in 300.15(E).

### 334.80 Ampacity.
The ampacity of Types NM, NMC, and NMS cable shall be determined in accordance with 310.15. The ampacity shall be in accordance with the 60°C (140°F) conductor temperature rating. The 90°C (194°F) rating shall be permitted to be used for ampacity derating purposes, provided the final derated ampacity does not exceed that for a 60°C (140°F) rated conductor. The ampacity of

Types NM, NMC, and NMS cable installed in cable tray shall be determined in accordance with 392.11.

### III. Construction Specifications

**334.100 Construction.** The outer cable sheath of nonmetallic-sheathed cable shall be a nonmetallic material.

**334.104 Conductors.** The insulated power conductors shall be sizes 14 AWG through 2 AWG with copper conductors or sizes 12 AWG through 2 AWG with aluminum or copper-clad aluminum conductors. The signaling conductors shall comply with 780.5.

**334.108 Equipment Grounding.** In addition to the insulated conductors, the cable shall be permitted to have an insulated or bare conductor for equipment grounding purposes only. Where provided, the grounding conductor shall be sized in accordance with Article 250.

**334.112 Insulation.** The insulated power conductors shall be one of the types listed in Table 310.13 that is suitable for branch circuit wiring or one that is identified for use in these cables. Conductor insulation shall be rated at 90°C (194°F).

> FPN: Types NM, NMC, and NMS cable identified by the markings NM-B, NMC-B, and NMS-B meet this requirement.

**334.116 Sheath.** The outer sheath of nonmetallic-sheathed cable shall comply with 334.116(A), (B), and (C).

**(A) Type NM.** The overall covering shall be flame retardant and moisture resistant.

**(B) Type NMC.** The overall covering shall be flame retardant, moisture resistant, fungus resistant, and corrosion resistant.

**(C) Type NMS.** The overall covering shall be flame retardant and moisture resistant. The sheath shall be applied so as to separate the power conductors from the communications and signaling conductors. The signaling conductors shall be permitted to be shielded. An optional outer jacket shall be permitted.

> FPN: For composite optical cable, see 770.5 and 770.52.

## ARTICLE 336
## Power and Control Tray Cable: Type TC

### I. General

**336.1 Scope.** This article covers the use, installation, and construction specifications for power and control tray cable, Type TC.

## 336.2 Definition.

**Power and Control Tray Cable, Type TC.** A factory assembly of two or more insulated conductors, with or without associated bare or covered grounding conductors, under a nonmetallic jacket, for installation in cable trays, in raceways, or where supported by a messenger wire.

## II. Installation

**336.10 Uses Permitted.** Type TC tray cable shall be permitted to be used in the following:

(1) For power, lighting, control, and signal circuits.
(2) In cable trays, or in raceways, or where supported in outdoor locations by a messenger wire.
(3) In cable trays in hazardous (classified) locations as permitted in Articles 392, 501, 502, 504, and 505 in industrial establishments where the conditions of maintenance and supervision ensure that only qualified persons service the installation.
(4) For Class I circuits as permitted in Article 725.
(5) For non–power-limited fire alarm circuits if conductors comply with the requirements of 760.27.
(6) In industrial establishments where the conditions of maintenance and supervision ensure that only qualified persons service the installation, and where the cable is continuously supported and protected against physical damage using mechanical protection, such as struts, angles, or channel, Type TC tray cable that complies with the crush and impact requirements of Type MC cable and is identified for such use shall be permitted between a cable tray and the utilization equipment or device. The cable shall be secured at intervals not exceeding 1.8 m (6 ft). Equipment grounding for the utilization equipment shall be provided by an equipment grounding conductor within the cable.
(7) Where installed in wet locations, Type TC cable shall also be resistant to moisture and corrosive agents.

FPN: See 310.10 for temperature limitation of conductors.

**336.12 Uses Not Permitted.** Type TC tray cable shall not be used in the following:

(1) Installed where it will be exposed to physical damage
(2) Installed as open cable on brackets or cleats, except as permitted in 340.10(6)
(3) Used where exposed to direct rays of the sun, unless identified as sunlight resistant
(4) Direct buried, unless identified for such use

**336.24 Bending Radius.** Bends in Type TC cable shall be made so as not to damage the cable. For Type TC cable without metal shielding, the minimum bending radius shall be as follows:

(1) Four times the overall diameter for cables 25 mm (1 in.) or less in diameter
(2) Five times the overall diameter for cables larger than 25 mm (1 in.) but not more than 50 mm (2 in.) in diameter
(3) Six times the overall diameter for cables larger than 50 mm (2 in.) in diameter

Type TC cables with metallic shielding shall have a minimum bending radius of not less than 12 times the cable overall diameter.

**336.80 Ampacity.** The ampacity of Type TC tray cable shall be determined in accordance with 392.11 for 14 AWG and larger conductors, in accordance with 402.5 for 18 AWG through 16 AWG conductors where installed in cable tray, and in accordance with 310.15 where installed in a raceway or as messenger supported wiring.

## III. Construction Specifications

**336.100 Construction.** A metallic sheath or armor as defined in 330.116 shall not be permitted either under or over the nonmetallic jacket. Metallic shield(s) shall be permitted over groups of conductors, under the outer jacket, or both.

**336.104 Conductors.** The insulated conductors of Type TC tray cable shall be in sizes 18 AWG through 1000 kcmil copper and sizes 12 AWG through 1000 kcmil aluminum or copper-clad aluminum. Insulated conductors of sizes 14 AWG and larger copper and sizes 12 AWG and larger aluminum or copper-clad aluminum shall be one of the types listed in Table 310.13 or Table 310.62 that is suitable for branch circuit and feeder circuits or one that is identified for such use.

**(A) Fire Alarm Systems.** Where used for fire alarm systems, conductors shall also be in accordance with 760.27.

**(B) Thermocouple Circuits.** Conductors in Type TC cables used for thermocouple circuits in accordance with Article 725 shall also be permitted to be any of the materials used for thermocouple extension wire.

**(C) Class I Circuit Conductors.** Insulated conductors of 18 AWG and 16 AWG copper shall also be in accordance with 725.27.

**336.116 Jacket.** The outer jacket shall be a flame-retardant, nonmetallic material.

**336.120 Marking.** There shall be no voltage marking on a Type TC cable employing thermocouple extension wire.

# ARTICLE 338
## Service-Entrance Cable:
## Types SE and USE

## I. General

**338.1 Scope.** This article covers the use, installation, and construction specifications of service-entrance cable.

## 338.2 Definitions.

**Service-Entrance Cable.** A single conductor or multiconductor assembly provided with or without an overall covering, primarily used for services, and of the following types:

*Type SE.* Service-entrance cable having a flame-retardant, moisture-resistant covering.

*Type USE.* Service-entrance cable, identified for underground use, having a moisture-resistant covering, but not required to have a flame-retardant covering.

## II. Installation

## 338.10 Uses Permitted.

**(A) Service-Entrance Conductors.** Service-entrance cable used as service-entrance conductors shall be installed as required by Article 230.

Type USE used for service laterals shall be permitted to emerge from the ground outside at terminations in meter bases or other enclosures where protected in accordance with 300.5(D).

**(B) Branch Circuits or Feeders.**

**(1) Grounded Conductor Insulated.** Type SE service-entrance cables shall be permitted in wiring systems where all of the circuit conductors of the cable are of the rubber-covered or thermoplastic type.

**(2) Grounded Conductor Not Insulated.** Type SE service-entrance cable shall be permitted for use where the insulated conductors are used for circuit wiring and the uninsulated conductor is used only for equipment grounding purposes.

*Exception: Uninsulated conductors shall be permitted as a grounded conductor in accordance with 250.140.*

**(3) Temperature Limitations.** Type SE service-entrance cable used to supply appliances shall not be subject to conductor temperatures in excess of the temperature specified for the type of insulation involved.

**(4) Installation Methods for Branch Circuits and Feeders.**

(a) Interior Installations. In addition to the provisions of this article, Type SE service-entrance cable used for interior wiring shall comply with the installation requirements of Parts I and II of Article 334, excluding 334.80.

FPN: See 310.10 for temperature limitation of conductors.

(b) Exterior Installations. In addition to the provisions of this article, service-entrance cable used for feeders or branch circuits, where installed as exterior wiring, shall be installed as required by Article 225. The cable shall be supported in accordance with 334.30, unless used as messenger-supported wiring as allowed by Article 396.

Type USE cable shall be installed outside in accordance with the provisions of Article 340. Where Type USE cable emerges from the ground at terminations, it shall be protected in accordance with 300.5(D).

Multiconductor service-entrance cable shall be permitted to be installed as messenger-supported wiring in accordance with Articles 225 and 396.

**338.24 Bending Radius.** Bends in Types USE and SE cable shall be made so that the cable will not be damaged. The radius of the curve of the inner edge of any bend, during or after installation, shall not be less than five times the diameter of the cable.

## III. Construction

**338.100 Construction.** Cabled, single-conductor, Type USE constructions recognized for underground use shall be permitted to have a bare copper conductor cabled with the assembly. Type USE single, parallel, or cabled conductor assemblies recognized for underground use shall be permitted to have a bare copper concentric conductor applied. These constructions shall not require an outer overall covering.

FPN: See 230.41, Exception, item (b), for directly buried, uninsulated service-entrance conductors.

Type SE or USE cable containing two or more conductors shall be permitted to have one conductor uninsulated.

**338.120 Marking.** Service-entrance cable shall be marked as required in 310.11. Cable with the neutral conductor smaller than the ungrounded conductors shall be so marked.

# ARTICLE 340
# Underground Feeder and Branch-Circuit Cable: Type UF

## I. General

**340.1 Scope.** This article covers the use, installation, and construction specifications for underground feeder and branch-circuit cable, Type UF.

## 340.2 Definition.

**Underground Feeder and Branch-Circuit Cable, Type UF.** A listed factory assembly of one or more insulated conductors with an integral or an overall covering of non-metallic material suitable for direct burial in the earth.

## II. Installation

**340.10 Uses Permitted.** Type UF cable shall be permitted as follows:

(1) For use underground, including direct burial in the earth. For underground requirements, see 300.5
(2) As single-conductor cables. Where installed as single-conductor cables, all conductors of the feeder grounded conductor or branch circuit, including the grounded conductor and equipment grounding conductor, if any, shall be installed in accordance with 300.3.
(3) For wiring in wet, dry, or corrosive locations under the recognized wiring methods of this *Code*.
(4) Installed as nonmetallic-sheathed cable. Where so installed, the installation and conductor requirements shall comply with the provisions of Article 334 and shall be of the multiconductor type.
(5) For solar photovoltaic systems in accordance with 690.31.
(6) As single-conductor cables as the nonheating leads for heating cables as provided in 424.43.
(7) Supported by cable trays. Type UF cable supported by cable trays shall be of the multiconductor type.

FPN: See 310.10 for temperature limitation of conductors.

**340.12 Uses Not Permitted.** Type UF cable shall not be used as follows:

(1) As service-entrance cable
(2) In commercial garages
(3) In theaters and similar locations
(4) In motion picture studios
(5) In storage battery rooms
(6) In hoistways, or on elevators or escalators
(7) In hazardous (classified) locations

(8) Embedded in poured cement, concrete, or aggregate, except where embedded in plaster as nonheating leads where permitted in 424.43
(9) Where exposed to direct rays of the sun, unless identified as sunlight resistant
(10) Where subject to physical damage
(11) As overhead cable, except where installed as messenger-supported wiring in accordance with Article 396

**340.24 Bending Radius.** Bends in Type UF cable shall be made so that the cable shall not be damaged. The radius of the curve of the inner edge of any bend shall not be less than five times the diameter of the cable.

**340.80 Ampacity.** The ampacity of Type UF cable shall be that of 60°C (140°F) conductors in accordance with 310.15.

## III. Construction Specifications

**340.104 Conductors.** The conductors shall be sizes 14 AWG copper or 12 AWG aluminum or copper-clad aluminum through 4/0 AWG.

**340.108 Equipment Grounding.** In addition to the insulated conductors, the cable shall be permitted to have an insulated or bare conductor for equipment grounding purposes only.

**340.112 Insulation.** The conductors of Type UF shall be one of the moisture-resistant types listed in Table 310.13 that is suitable for branch-circuit wiring or one that is identified for such use.

**340.116 Sheath.** The overall covering shall be flame retardant; moisture, fungus, and corrosion resistant; and suitable for direct burial in the earth.

# ARTICLE 342
# Intermediate Metal Conduit: Type IMC

## I. General

**342.1 Scope.** This article covers the use, installation, and construction specifications for intermediate metal conduit (IMC) and associated fittings.

## 342.2 Definition.

**Intermediate Metal Conduit (IMC).** A steel threadable raceway of circular cross section designed for the physical protection and routing of conductors and cables and for use as an equipment grounding conductor when installed with its integral or associated coupling and appropriate fittings.

**342.6 Listing Requirements.** IMC, factory elbows and couplings, and associated fittings shall be listed.

## II. Installation

### 342.10 Uses Permitted.

**(A) All Atmospheric Conditions and Occupancies.** Use of IMC shall be permitted under all atmospheric conditions and occupancies.

**(B) Corrosion Environments.** IMC, elbows, couplings, and fittings shall be permitted to be installed in concrete, in direct contact with the earth, or in areas subject to severe corrosive influences where protected by corrosion protection and judged suitable for the condition.

**(C) Cinder Fill.** IMC shall be permitted to be installed in or under cinder fill where subject to permanent moisture where protected on all sides by a layer of noncinder concrete not less than 50 mm (2 in.) thick; where the conduit is not less than 450 mm (18 in.) under the fill; or where protected by corrosion protection and judged suitable for the condition.

**(D) Wet Locations.** All supports, bolts, straps, screws, and so forth, shall be of corrosion-resistant materials or protected against corrosion by corrosion-resistant materials.

> FPN: See 300.6 for protection against corrosion.

**342.14 Dissimilar Metals.** Where practicable, dissimilar metals in contact anywhere in the system shall be avoided to eliminate the possibility of galvanic action.

Aluminum fittings and enclosures shall be permitted to be used with IMC.

### 342.20 Size.

**(A) Minimum.** IMC smaller than metric designator 16 (trade size ½) shall not be used.

**(B) Maximum.** IMC larger than metric designator 103 (trade size 4) shall not be used.

> FPN: See 300.1(C) for the metric designators and trade sizes. These are for identification purposes only and do not relate to actual dimensions.

**342.22 Number of Conductors.** The number of conductors shall not exceed that permitted by the percentage fill specified in Table 1, Chapter 9.

Cables shall be permitted to be installed where such use is permitted by the respective cable articles. The number of cables shall not exceed the allowable percentage fill specified in Table 1, Chapter 9.

**342.24 Bends — How Made.** Bends of IMC shall be made so that the conduit will not be damaged and so that the internal diameter of the conduit will not be effectively reduced. The radius of the curve of any field bend to the centerline of the conduit shall not be less than indicated in Table 344.24.

**342.26 Bends — Number in One Run.** There shall not be more than the equivalent of four quarter bends (360 degrees total) between pull points, for example, conduit bodies and boxes.

**342.28 Reaming and Threading.** All cut ends shall be reamed or otherwise finished to remove rough edges. Where conduit is threaded in the field, a standard cutting die with a taper of 1 in 16 (¾ in. taper per foot) shall be used.

> FPN: See ANSI/ASME B.1.20.1-1983, *Standard for Pipe Threads, General Purpose (Inch).*

**342.30 Securing and Supporting.** IMC shall be installed as a complete system as provided in Article 300 and shall be securely fastened in place and supported in accordance with 342.30(A) and (B).

**(A) Securely Fastened.** Each IMC shall be securely fastened within 900 mm (3 ft) of each outlet box, junction box, device box, cabinet, conduit body, or other conduit termination. Fastening shall be permitted to be increased to a distance of 1.5 m (5 ft) where structural members do not readily permit fastening within 900 mm (3 ft). Where approved, conduit shall not be required to be securely fastened within 900 mm (3 ft) of the service head for above-the-roof termination of a mast.

**(B) Supports.** IMC shall be supported in accordance with one of the following:

(1) Conduit shall be supported at intervals not exceeding 3 m (10 ft).
(2) The distance between supports for straight runs of conduit shall be permitted in accordance with Table 344.30(B)(2), provided the conduit is made up with threaded couplings and such supports prevent transmission of stresses to termination where conduit is deflected between supports.
(3) Exposed vertical risers from industrial machinery or fixed equipment shall be permitted to be supported at intervals not exceeding 6 m (20 ft), if the conduit is made up with threaded couplings, the conduit is firmly supported at the top and bottom of the riser, and no other means of intermediate support is readily available.

(4) Horizontal runs of IMC supported by openings through framing members at intervals not exceeding 3 m (10 ft) and securely fastened within 900 mm (3 ft) of termination points shall be permitted.

**342.42 Couplings and Connectors.**

**(A) Threadless.** Threadless couplings and connectors used with conduit shall be made tight. Where buried in masonry or concrete, they shall be the concretetight type. Where installed in wet locations, they shall be the raintight type. Threadless couplings and connectors shall not be used on threaded conduit ends unless listed for the purpose.

**(B) Running Threads.** Running threads shall not be used on conduit for connection at couplings.

**342.46 Bushings.** Where a conduit enters a box, fitting, or other enclosure, a bushing shall be provided to protect the wire from abrasion unless the design of the box, fitting, or enclosure is such as to afford equivalent protection.

FPN: See 300.4(F) for the protection of conductors 4 AWG and larger at bushings.

**342.56 Splices and Taps.** Splices and taps shall be made in accordance with 300.15.

**342.60 Grounding.** IMC shall be permitted as an equipment grounding conductor.

### III. Construction Specifications

**342.120 Marking.** Each length shall be clearly and durably marked at least every 1.5 m (5 ft) with the letters IMC. Each length shall be marked as required in 110.21.

**342.130 Standard Lengths.** The standard length of IMC shall be 3.05 m (10 ft), including an attached coupling, and each end shall be threaded. Longer or shorter lengths with or without coupling and threaded or unthreaded shall be permitted.

## ARTICLE 344
## Rigid Metal Conduit: Type RMC

### I. General

**344.1 Scope.** This article covers the use, installation, and construction specifications for rigid metal conduit (RMC) and associated fittings.

**344.2 Definition.**

**Rigid Metal Conduit (RMC).** A threadable raceway of circular cross section designed for the physical protection and routing of conductors and cables and for use as an equipment grounding conductor when installed with its integral or associated coupling and appropriate fittings. RMC is generally made of steel (ferrous) with protective coatings or aluminum (nonferrous). Special use types are silicon bronze and stainless steel.

**344.6 Listing Requirements.** RMC, factory elbows and couplings, and associated fittings shall be listed.

### II. Installation

**344.10 Uses Permitted.**

**(A) All Atmospheric Conditions and Occupancies.** Use of RMC shall be permitted under all atmospheric conditions and occupancies. Ferrous raceways and fittings protected from corrosion solely by enamel shall be permitted only indoors and in occupancies not subject to severe corrosive influences.

**(B) Corrosion Environments.** RMC, elbows, couplings, and fittings shall be permitted to be installed in concrete, in direct contact with the earth, or in areas subject to severe corrosive influences where protected by corrosion protection and judged suitable for the condition.

**(C) Cinder Fill.** RMC shall be permitted to be installed in or under cinder fill where subject to permanent moisture where protected on all sides by a layer of noncinder concrete not less than 50 mm (2 in.) thick; where the conduit is not less than 450 mm (18 in.) under the fill; or where protected by corrosion protection and judged suitable for the condition.

**(D) Wet Locations.** All supports, bolts, straps, screws, and so forth, shall be of corrosion-resistant materials or protected against corrosion by corrosion-resistant materials.

FPN: See 300.6 for protection against corrosion.

**344.14 Dissimilar Metals.** Where practicable, dissimilar metals in contact anywhere in the system shall be avoided to eliminate the possibility of galvanic action. Aluminum fittings and enclosures shall be permitted to be used with steel RMC, and steel fittings and enclosures shall be permitted to be used with aluminum RMC where not subject to severe corrosive influences.

**344.20 Size.**

**(A) Minimum.** RMC smaller than metric designator 16 (trade size ½) shall not be used.

*Exception: For enclosing the leads of motors as permitted in 430.145(B).*

**(B) Maximum.** RMC larger than metric designator 155 (trade size 6) shall not be used.

> FPN: See 300.1(C) for the metric designators and trade sizes. These are for identification purposes only and do not relate to actual dimensions.

**344.22 Number of Conductors.** The number of conductors or cables shall not exceed that permitted by the percentage fill specified in Table 1, Chapter 9.

Cables shall be permitted to be installed where such use is permitted by the respective cable articles. The number of cables shall not exceed the allowable percentage fill specified in Table 1, Chapter 9.

**344.24 Bends — How Made.** Bends of RMC shall be made so that the conduit is not damaged and the internal diameter of the conduit is not effectively reduced. The radius of the curve of any field bend to the centerline of the conduit shall not be less than indicated in Table 344.24.

**Table 344.24 Radius of Conduit Bends**

| Conduit Size | | One Shot and Full Shoe Benders | | Other Bends | |
|---|---|---|---|---|---|
| Metric Designator | Trade Size | mm | in. | mm | in. |
| 16 | ½ | 101.6 | 4 | 101.6 | 4 |
| 21 | ¾ | 114.3 | 4½ | 127 | 5 |
| 27 | 1 | 146.05 | 5¾ | 152.4 | 6 |
| 35 | 1¼ | 184.15 | 7¼ | 203.2 | 8 |
| 41 | 1½ | 209.55 | 8¼ | 254 | 10 |
| 53 | 2 | 241.3 | 9½ | 304.8 | 12 |
| 63 | 2½ | 266.7 | 10½ | 381 | 15 |
| 78 | 3 | 330.2 | 13 | 457.2 | 18 |
| 91 | 3½ | 381 | 15 | 533.4 | 21 |
| 103 | 4 | 406.4 | 16 | 609.6 | 24 |
| 129 | 5 | 609.6 | 24 | 762 | 30 |
| 155 | 6 | 762 | 30 | 914.4 | 36 |

**344.26 Bends — Number in One Run.** There shall not be more than the equivalent of four quarter bends (360 degrees total) between pull points, for example, conduit bodies and boxes.

**344.28 Reaming and Threading.** All cut ends shall be reamed or otherwise finished to remove rough edges. Where conduit is threaded in the field, a standard cutting die with a 1 in 16 taper (¾-in. taper per foot) shall be used.

> FPN: See ANSI/ASME B.1.20.1-1983, *Standard for Pipe Threads, General Purpose (Inch)*.

**344.30 Securing and Supporting.** RMC shall be installed as a complete system as provided in Article 300 and shall be securely fastened in place and supported in accordance with 344.30(A) and (B).

**(A) Securely Fastened.** RMC shall be securely fastened within 900 mm (3 ft) of each outlet box, junction box, device box, cabinet, conduit body, or other conduit termination. Fastening shall be permitted to be increased to a distance of 1.5 m (5 ft) where structural members do not readily permit fastening within 900 mm (3 ft). Where approved, conduit shall not be required to be securely fastened within 900 mm (3 ft) of the service head for above-the-roof termination of a mast.

**(B) Supports.** RMC shall be supported in accordance with one of the following.

(1) Conduit shall be supported at intervals not exceeding 3 m (10 ft)

(2) The distance between supports for straight runs of conduit shall be permitted in accordance with Table 346.30(B)(2), provided the conduit is made up with threaded couplings, and such supports prevent transmission of stresses to termination where conduit is deflected between supports.

(3) Exposed vertical risers from industrial machinery or fixed equipment shall be permitted to be supported at intervals not exceeding 6 m (20 ft), if the conduit is made up with threaded couplings, the conduit is firmly supported at the top and bottom of the riser, and no other means of intermediate support is readily available.

(4) Horizontal runs of RMC supported by openings through framing members at intervals not exceeding 3 m (10 ft) and securely fastened within 900 mm (3 ft) of termination points shall be permitted.

**Table 344.30(B)(2) Supports for Rigid Metal Conduit**

| Conduit Size | | Maximum Distance Between Rigid Metal Conduit Supports | |
|---|---|---|---|
| Metric Designator | Trade Size | m | ft |
| 16 – 21 | ½ – ¾ | 3.0 | 10 |
| 27 | 1 | 3.7 | 12 |
| 35 – 41 | 1¼ – 1½ | 4.3 | 14 |
| 53 – 63 | 2 – 2½ | 4.9 | 16 |
| 78 and larger | 3 and larger | 6.1 | 20 |

**344.42 Couplings and Connectors.**

**(A) Threadless.** Threadless couplings and connectors used with conduit shall be made tight. Where buried in masonry or concrete, they shall be the concretetight type. Where installed in wet locations, they shall be the raintight type. Threadless couplings and connectors shall not be used on threaded conduit ends unless listed for the purpose.

**(B) Running Threads.** Running threads shall not be used on conduit for connection at couplings.

**344.46 Bushings.** Where a conduit enters a box, fitting, or other enclosure, a bushing shall be provided to protect the wire from abrasion unless the design of the box, fitting, or enclosure is such as to afford equivalent protection.

> FPN: See 300.4(F) for the protection of conductors sizes 4 AWG and larger at bushings.

**344.56 Splices and Taps.** Splices and taps shall be made in accordance with 300.15.

**344.60 Grounding.** RMC shall be permitted as an equipment grounding conductor.

### III. Construction Specifications

**344.120 Marking.** Each length shall be clearly and durably identified in every 3 m (10 ft) as required in the first sentence of 110.21. Nonferrous conduit of corrosion-resistant material shall have suitable markings.

**344.130 Standard Lengths.** The standard length of RMC shall be 3.05 m (10 ft), including an attached coupling, and each end shall be threaded. Longer or shorter lengths with or without coupling and threaded or unthreaded shall be permitted.

## ARTICLE 348
## Flexible Metal Conduit: Type FMC

### I. General

**348.1 Scope.** This article covers the use, installation, and construction specifications for flexible metal conduit (FMC) and associated fittings.

**348.2 Definition.**

**Flexible Metal Conduit (FMC).** A raceway of circular cross section made of helically wound, formed, interlocked metal strip.

**348.6 Listing Requirements.** FMC and associated fittings shall be listed.

### II. Installation

**348.10 Uses Permitted.** FMC shall be permitted to be used in exposed and concealed locations.

**348.12 Uses Not Permitted.** FMC shall not be used in the following:

(1) In wet locations unless the conductors are approved for the specific conditions and the installation is such that liquid is not likely to enter raceways or enclosures to which the conduit is connected
(2) In hoistways, other than as permitted in 620.21(A)(1)
(3) In storage battery rooms
(4) In any hazardous (classified) location other than as permitted in 501.4(B) and 504.20
(5) Where exposed to materials having a deteriorating effect on the installed conductors, such as oil or gasoline
(6) Underground or embedded in poured concrete or aggregate
(7) Where subject to physical damage

**348.20 Size.**

**(A) Minimum.** FMC less than metric designator 16 (trade size ½) shall not be used unless permitted in 348.20(A)(1) through (5) for metric designator 12 (trade size ⅜).

(1) For enclosing the leads of motors as permitted in 430.145(B)
(2) In lengths not in excess of 1.8 m (6 ft) for any of the following uses:
  a. For utilization equipment
  b. As part of a listed assembly
  c. For tap connections to luminaires (lighting fixtures) as permitted in 410.67(C)
(3) For manufactured wiring systems as permitted in 604.6(A)
(4) In hoistways as permitted in 620.21(A)(1)
(5) As part of a listed assembly to connect wired luminaire (fixture) sections as permitted in 410.77(C)

**(B) Maximum.** FMC larger than metric designator 103 (trade size 4) shall not be used.

> FPN: See 300.1(C) for the metric designators and trade sizes. These are for identification purposes only and do not relate to actual dimensions.

**348.22 Number of Conductors.** The number of conductors shall not exceed that permitted by the percentage fill specified in Table 1, Chapter 9, or as permitted in Table 348.22 for metric designator 12 (trade size ⅜).

Cables shall be permitted to be installed where such use is permitted by the respective cable articles. The number of cables shall not exceed the allowable percentage fill specified in Table 1, Chapter 9.

**Table 348.22 Maximum Number of Insulated Conductors in Metric Designator 12 (Trade Size ⅜) Flexible Metal Conduit***

| Size (AWG) | Types RFH-2, SF-2 | | Types TF, XHHW, TW | | Types TFN, THHN, THWN | | Types FEP, FEBP, PF, PGF | |
| --- | --- | --- | --- | --- | --- | --- | --- | --- |
| | Fittings Inside Conduit | Fittings Outside Conduit | Fittings Inside Conduit | Fittings Outside Conduit | Fittings Inside Conduit | Fittings Outside Conduit | Fittings Inside Conduit | Fittings Outside Conduit |
| 18 | 2 | 3 | 3 | 5 | 5 | 8 | 5 | 8 |
| 16 | 1 | 2 | 3 | 4 | 4 | 6 | 4 | 6 |
| 14 | 1 | 2 | 2 | 3 | 3 | 4 | 3 | 4 |
| 12 | — | — | 1 | 2 | 2 | 3 | 2 | 3 |
| 10 | — | — | 1 | 1 | 1 | 1 | 1 | 2 |

*In addition, one covered or bare equipment grounding conductor of the same size shall be permitted.

**348.24 Bends —How Made.** Bends in conduit shall be made so that the conduit is not damaged and the internal diameter of the conduit is not effectively reduced. Bends shall be permitted to be made manually without auxiliary equipment. The radius of the curve to the centerline of any bend shall not be less than shown in Table 344.24 using the column "Other Bends."

**348.26 Bends — Number in One Run.** There shall not be more than the equivalent of four quarter bends (360 degrees total) between pull points, for example, conduit bodies and boxes.

**348.28 Trimming.** All cut ends shall be trimmed or otherwise finished to remove rough edges, except where fittings that thread into the convolutions are used.

**348.30 Securing and Supporting.** FMC shall be securely fastened in place and supported in accordance with 348.30(A) and (B).

**(A) Securely Fastened.** FMC shall be securely fastened in place by an approved means within 300 mm (12 in.) of each box, cabinet, conduit body, or other conduit termination and shall be supported and secured at intervals not to exceed 1.4 m (4½ ft).

*Exception No. 1: Where FMC is fished.*

*Exception No. 2: Lengths not exceeding 900 mm (3 ft) at terminals where flexibility is required.*

*Exception No. 3: Lengths not exceeding 1.8 m (6 ft) from a luminaire (fixture) terminal connection for tap connections to luminaires (light fixtures) as permitted in 410.67(C).*

**(B) Supports.** Horizontal runs of flexible metal conduit FMC supported by openings through framing members at intervals not greater than 1.4 m (4½ ft) and securely fastened within 300 mm (12 in.) of termination points shall be permitted.

**348.42 Couplings and Connectors.** Angle connectors shall not be used for concealed raceway installations.

**348.56 Splices and Taps.** Splices and taps shall be made in accordance with 300.15.

**348.60 Grounding and Bonding.** Where used to connect equipment where flexibility is required, an equipment grounding conductor shall be installed.

Where required or installed, equipment grounding conductors shall be installed in accordance with 250.134(B).

Where required or installed, equipment bonding jumpers shall be installed in accordance with 250.102.

## ARTICLE 350
## Liquidtight Flexible Metal Conduit: Type LFMC

### I. General

**350.1 Scope.** This article covers the use, installation, and construction specifications for liquidtight flexible metal conduit (LFMC) and associated fittings.

**350.2 Definition.**

**Liquidtight Flexible Metal Conduit (LFMC).** A raceway of circular cross section having an outer liquidtight, non-metallic, sunlight-resistant jacket over an inner flexible metal core with associated couplings, connectors, and fittings for the installation of electric conductors.

**350.6 Listing Requirements.** LFMC and associated fittings shall be listed.

## II. Installation

**350.10 Uses Permitted.** LFMC shall be permitted to be used in exposed or concealed locations as follows:

(1) Where conditions of installation, operation, or maintenance require flexibility or protection from liquids, vapors, or solids
(2) As permitted by 501.4(B), 502.4, 503.3, and 504.20 and in other hazardous (classified) locations where specifically approved, and by 553.7(B)
(3) For direct burial where listed and marked for the purpose

**350.12 Uses Not Permitted.** LFMC shall not be used as follows:

(1) Where subject to physical damage
(2) Where any combination of ambient and conductor temperature produces an operating temperature in excess of that for which the material is approved

**350.20 Size.**

**(A) Minimum.** LFMC smaller than metric designator 16 (trade size ½) shall not be used.

*Exception: LFMC of metric designator 12 (trade size ⅜) shall be permitted as covered in 348.20(A).*

**(B) Maximum.** The maximum size of LFMC shall be metric designator 103 (trade size 4).

> FPN: See 300.1(C) for the metric designators and trade sizes. These are for identification purposes only and do not relate to actual dimensions.

**350.22 Number of Conductors or Cables.**

**(A) Metric Designators 16 through 103 (Trade Sizes ½ through 4).** The number of conductors shall not exceed that permitted by the percentage fill specified in Table 1, Chapter 9.

Cables shall be permitted to be installed where such use is permitted by the respective cable articles. The number of cables shall not exceed the allowable percentage fill specified in Table 1, Chapter 9.

**(B) Metric Designator 12 (Trade Size ⅜).** The number of conductors shall not exceed that permitted in Table 348.22, "Fittings Outside Conduit" columns.

**350.24 Bends — How Made.** Bends in conduit shall be made so that the conduit will not be damaged and the internal diameter of the conduit will not be effectively reduced. Bends shall be permitted to be made manually without auxiliary equipment. The radius of the curve to the centerline of any bend shall not be less than shown in Table 344.24 using the column "Other Bends."

**350.26 Bends — Number in One Run.** There shall not be more than the equivalent of four quarter bends (360 degrees total) between pull points, for example, conduit bodies and boxes.

**350.30 Securing and Supporting.** LFMC shall be securely fastened in place and supported in accordance with 350.30(A) and (B).

**(A) Securely Fastened.** LFMC shall be securely fastened in place by an approved means within 300 mm (12 in.) of each box, cabinet, conduit body, or other conduit termination and shall be supported and secured at intervals not to exceed 1.4 m (4½ ft).

*Exception No. 1: Where LFMC is fished.*

*Exception No. 2: Lengths not exceeding 900 mm (3 ft) at terminals where flexibility is necessary.*

*Exception No. 3: Lengths not exceeding 1.8 m (6 ft) from a luminaire (fixture) terminal connection for tap conductors to luminaires (lighting fixtures), as permitted in 410.67(C).*

**(B) Supports.** Horizontal runs of LFMC supported by openings through framing members at intervals not greater than 1.4 m (4½ ft) and securely fastened within 300 mm (12 in.) of termination points shall be permitted.

**350.42 Couplings and Connectors.** Angle connectors shall not be used for concealed raceway installations.

**350.56 Splices and Taps.** Splices and taps shall be made in accordance with 300.15.

**350.60 Grounding and Bonding.** Where used to connect equipment where flexibility is required, an equipment grounding conductor shall be installed.

Where required or installed, equipment grounding conductors shall be installed in accordance with 250.134(B).

Where required or installed, equipment bonding jumpers shall be installed in accordance with 250.102.

> FPN: See 501.16(B), 502.16(B), and 503.16(B) for types of equipment grounding conductors.

## III. Construction Specifications

**350.120 Marking.** LFMC shall be marked according to 110.21. The trade size and other information required by the listing shall also be marked on the conduit. Conduit suitable for direct burial shall be so marked.

# ARTICLE 352
## Rigid Nonmetallic Conduit: Type RNC

## I. General

**352.1 Scope.** This article covers the use, installation, and construction specifications for rigid nonmetallic conduit (RNC) and associated fittings.

**352.2 Definition.**

**Rigid Nonmetallic Conduit (RNC).** A nonmetallic raceway of circular cross section, with integral or associated couplings, connectors, and fittings for the installation of electrical conductors.

**352.6 Listing Requirements.** RNC, factory elbows, and associated fittings shall be listed.

## II. Installation

**352.10 Uses Permitted.** The use of RNC shall be permitted under the following conditions.

> FPN: Extreme cold may cause some nonmetallic conduits to become brittle and therefore more susceptible to damage from physical contact.

**(A) Concealed.** In walls, floors, and ceilings.

**(B) Corrosive Influences.** In locations subject to severe corrosive influences as covered in 300.6 and where subject to chemicals for which the materials are specifically approved.

**(C) Cinders.** In cinder fill.

**(D) Wet Locations.** In portions of dairies, laundries, canneries, or other wet locations and in locations where walls are frequently washed, the entire conduit system including boxes and fittings used therewith shall be installed and equipped so as to prevent water from entering the conduit. All supports, bolts, straps, screws, and so forth, shall be of corrosion-resistant materials or be protected against corrosion by approved corrosion-resistant materials.

**(E) Dry and Damp Locations.** In dry and damp locations not prohibited by 352.12.

**(F) Exposed.** For exposed work where not subject to physical damage if identified for such use.

**(G) Underground Installations.** For underground installations, see 300.5 and 300.50. Conduits listed for the purpose shall be permitted to be installed underground in continuous lengths from a reel.

**(H) Support of Conduit Bodies.** Rigid nonmetallic conduit shall be permitted to support nonmetallic conduit bodies not larger than the largest trade size of an entering raceway. The conduit bodies shall not contain devices or support luminaires (fixtures) or other equipment.

**352.12 Uses Not Permitted.** RNC shall not be used in the following locations.

**(A) Hazardous (Classified) Locations.**

(1) In hazardous (classified) locations, except as permitted in 503.3(A), 504.20, 514.8, and 515.8
(2) In Class I, Division 2 locations, except as permitted in 501.4(B), Exception

**(B) Support of Luminaires (Fixtures).** For the support of luminaires (fixtures) or other equipment not described in 352.10(H).

**(C) Physical Damage.** Where subject to physical damage unless identified for such use.

**(D) Ambient Temperatures.** Where subject to ambient temperatures in excess of 50°C (122°F) unless listed otherwise.

**(E) Insulation Temperature Limitations.** For conductors whose insulation temperature limitations would exceed those for which the conduit is listed.

**(F) Theaters and Similar Locations.** In theaters and similar locations, except as provided in Articles 518 and 520.

**352.20 Size.**

**(A) Minimum.** RNC smaller than metric designator 16 (trade size ½) shall not be used.

**(B) Maximum.** RNC larger than metric designator 155 (trade size 6) shall not be used.

> FPN: The trade sizes and metric designators are for identification purposes only and do not relate to actual dimensions. See 300.1(C).

**352.22 Number of Conductors.** The number of conductors shall not exceed that permitted by the percentage fill specified in Table 1, Chapter 9.

Cables shall be permitted to be installed where such use is permitted by the respective cable articles. The number of cables shall not exceed the allowable percentage fill specified in Table 1, Chapter 9.

**352.24 Bends — How Made.** Bends shall be made so that the conduit will not be damaged and the internal diameter of the conduit will not be effectively reduced. Field bends shall be made only with bending equipment identified for the purpose. The radius of the curve to the centerline of such bends shall not be less than shown in Table 344.24, column "Other Bends."

**352.26 Bends — Number in One Run.** There shall not be more than the equivalent of four quarter bends (360 degrees total) between pull points, for example, conduit bodies and boxes.

**352.28 Trimming.** All cut ends shall be trimmed inside and outside to remove rough edges.

**352.30 Securing and Supporting.** RNC shall be installed as a complete system as provided in 300.18 and shall be fastened so that movement from thermal expansion or contraction is permitted. RNC shall be securely fastened and supported in accordance with 352.30(A) and (B).

**(A) Securely Fastened.** RNC shall be securely fastened within 900 mm (3 ft) of each outlet box, junction box, device box, conduit body, or other conduit termination. Conduit listed for securing at other than 900 mm (3 ft) shall be permitted to be installed in accordance with the listing.

**(B) Supports.** RNC shall be supported as required in Table 352.30(B). Conduit listed for support at spacings other than as shown in Table 352.30(B) shall be permitted to be installed in accordance with the listing. Horizontal runs of RNC supported by openings through framing members at intervals not exceeding those in Table 352.30(B) and securely fastened within 900 mm (3 ft) of termination points shall be permitted.

**Table 352.30(B) Support of Rigid Nonmetallic Conduit (RNC)**

| Conduit Size | | Maximum Spacing Between Supports | |
|---|---|---|---|
| Metric Designator | Trade Size | mm or m | ft |
| 16–27 | ½–1 | 900 mm | 3 |
| 35–53 | 1¼–2 | 1.5 m | 5 |
| 63–78 | 2½–3 | 1.8 m | 6 |
| 91–129 | 3½–5 | 2.1 m | 7 |
| 155 | 6 | 2.5 m | 8 |

**352.44 Expansion Fittings.** Expansion fittings for RNC shall be provided to compensate for thermal expansion and contraction where the length change, in accordance with Table 352.44(A) or (B), is expected to be 6 mm (¼ in.)or greater in a straight run between securely mounted items such as boxes, cabinets, elbows, or other conduit terminations.

**352.46 Bushings.** Where a conduit enters a box, fitting, or other enclosure, a bushing or adapter shall be provided to protect the wire from abrasion unless the box, fitting, or enclosure design provides equivalent protection.

FPN: See 300.4(F) for the protection of conductors 4 AWG and larger at bushings.

**352.48 Joints.** All joints between lengths of conduit, and between conduit and couplings, fittings, and boxes, shall be made by an approved method.

**352.56 Splices and Taps.** Splices and taps shall be made in accordance with 300.15.

**352.60 Grounding.** Where equipment grounding is required by Article 250, a separate equipment grounding conductor shall be installed in the conduit.

*Exception No. 1: As permitted in 250.134(B), Exception No. 2, for dc circuits and 250.134(B), Exception No. 1, for separately run equipment grounding conductors.*

*Exception No. 2: Where the grounded conductor is used to ground equipment as permitted in 250.142.*

### III. Construction Specifications

**352.100 Construction.** RNC and fittings shall be composed of suitable nonmetallic material that is resistant to moisture and chemical atmospheres. For use above ground, it shall also be flame retardant, resistant to impact and crushing, resistant to distortion from heat under conditions likely to be encountered in service, and resistant to low temperature and sunlight effects. For use underground, the material shall be acceptably resistant to moisture and corrosive agents and shall be of sufficient strength to withstand abuse, such as by impact and crushing, in handling and during installation. Where intended for direct burial, without encasement in concrete, the material shall also be capable of withstanding continued loading that is likely to be encountered after installation.

**352.120 Marking.** Each length of RNC shall be clearly and durably marked at least every 3 m (10 ft) as required in the first sentence of 110.21. The type of material shall also be included in the marking unless it is visually identifiable. For conduit recognized for use above ground, these markings shall be permanent. For conduit limited to underground use only, these markings shall be sufficiently durable to remain legible until the material is installed. Conduit shall be permitted to be surface marked to indicate special characteristics of the material.

FPN: Examples of these markings include but are not limited to "limited smoke" and "sunlight resistant."

**Table 352.44(A) Expansion Characteristics of PVC Rigid Nonmetallic Conduit Coefficient of Thermal Expansion = 6.084 × 10$^{-5}$ mm/mm/°C (3.38 × 10$^{-5}$ in./in./°F)**

| Temperature Change (°C) | Length Change of PVC Conduit (mm/m) | Temperature Change (°F) | Length Change of PVC Conduit (in./100 ft) | Temperature Change (°F) | Length Change of PVC Conduit (in./100 ft) |
|---|---|---|---|---|---|
| 5 | 0.30 | 5 | 0.20 | 105 | 4.26 |
| 10 | 0.61 | 10 | 0.41 | 110 | 4.46 |
| 15 | 0.91 | 15 | 0.61 | 115 | 4.66 |
| 20 | 1.22 | 20 | 0.81 | 120 | 4.87 |
| 25 | 1.52 | 25 | 1.01 | 125 | 5.07 |
| 30 | 1.83 | 30 | 1.22 | 130 | 5.27 |
| 35 | 2.13 | 35 | 1.42 | 135 | 5.48 |
| 40 | 2.43 | 40 | 1.62 | 140 | 5.68 |
| 45 | 2.74 | 45 | 1.83 | 145 | 5.88 |
| 50 | 3.04 | 50 | 2.03 | 150 | 6.08 |
| 55 | 3.35 | 55 | 2.23 | 155 | 6.29 |
| 60 | 3.65 | 60 | 2.43 | 160 | 6.49 |
| 65 | 3.95 | 65 | 2.64 | 165 | 6.69 |
| 70 | 4.26 | 70 | 2.84 | 170 | 6.90 |
| 75 | 4.56 | 75 | 3.04 | 175 | 7.10 |
| 80 | 4.87 | 80 | 3.24 | 180 | 7.30 |
| 85 | 5.17 | 85 | 3.45 | 185 | 7.50 |
| 90 | 5.48 | 90 | 3.65 | 190 | 7.71 |
| 95 | 5.78 | 95 | 3.85 | 195 | 7.91 |
| 100 | 6.08 | 100 | 4.06 | 200 | 8.11 |

**Table 352.44(B) Expansion Characteristics of Reinforced Thermosetting Resin Conduit (RTRC) Coefficient of Thermal Expansion = 2.7 × 10$^{-5}$ mm/mm/°C (1.5 × 10$^{-5}$ in./in./°F)**

| Temperature Change (°C) | Length Change of RTRC Conduit (mm/m) | Temperature Change (°F) | Length Change of RTRC Conduit (in./100 ft) | Temperature Change (°F) | Length Change of RTRC Conduit (in./100 ft) |
|---|---|---|---|---|---|
| 5 | 0.14 | 5 | 0.09 | 105 | 1.89 |
| 10 | 0.27 | 10 | 0.18 | 110 | 1.98 |
| 15 | 0.41 | 15 | 0.27 | 115 | 2.07 |
| 20 | 0.54 | 20 | 0.36 | 120 | 2.16 |
| 25 | 0.68 | 25 | 0.45 | 125 | 2.25 |
| 30 | 0.81 | 30 | 0.54 | 130 | 2.34 |
| 35 | 0.95 | 35 | 0.63 | 135 | 2.43 |
| 40 | 1.08 | 40 | 0.72 | 140 | 2.52 |
| 45 | 1.22 | 45 | 0.81 | 145 | 2.61 |
| 50 | 1.35 | 50 | 0.90 | 150 | 2.70 |
| 55 | 1.49 | 55 | 0.99 | 155 | 2.79 |
| 60 | 1.62 | 60 | 1.08 | 160 | 2.88 |
| 65 | 1.76 | 65 | 1.17 | 165 | 2.97 |
| 70 | 1.89 | 70 | 1.26 | 170 | 3.06 |
| 75 | 2.03 | 75 | 1.35 | 175 | 3.15 |
| 80 | 2.16 | 80 | 1.44 | 180 | 3.24 |
| 85 | 2.30 | 85 | 1.53 | 185 | 3.33 |
| 90 | 2.43 | 90 | 1.62 | 190 | 3.42 |
| 95 | 2.57 | 95 | 1.71 | 195 | 3.51 |
| 100 | 2.70 | 100 | 1.80 | 200 | 3.60 |

# ARTICLE 354
# Nonmetallic Underground Conduit with Conductors: Type NUCC

## I. General

**354.1 Scope.** This article covers the use, installation, and construction specifications for nonmetallic underground conduit with conductors (NUCC).

**354.2 Definition.**

**Nonmetallic Underground Conduit with Conductors (NUCC).** A factory assembly of conductors or cables inside a nonmetallic, smooth wall conduit with a circular cross section.

**354.6 Listing Requirements.** NUCC and associated fittings shall be listed.

## II. Installation

**354.10 Uses Permitted.** The use of NUCC and fittings shall be permitted in the following:

(1) For direct burial underground installation (For minimum cover requirements, see Table 300.5 and Table 300.50 under Rigid Nonmetallic Conduit.)
(2) Encased or embedded in concrete
(3) In cinder fill
(4) In underground locations subject to severe corrosive influences as covered in 300.6 and where subject to chemicals for which the assembly is specifically approved

**354.12 Uses Not Permitted.** NUCC shall not be used in the following:

(1) In exposed locations
(2) Inside buildings

*Exception: The conductor or the cable portion of the assembly, where suitable, shall be permitted to extend within the building for termination purposes in accordance with 300.3.*

(3) In hazardous (classified) locations except as permitted by 503.3(A), 504.20, 514.8, and 515.8, and in Class I, Division 2 locations as permitted in 501.4(B)(3)

**354.20 Size.**

**(A) Minimum.** NUCC smaller than metric designator 16 (trade size ½) shall not be used.

**(B) Maximum.** NUCC larger than metric designator 103 (trade size 4) shall not be used.

FPN: See 300.1(C) for the metric designators and trade sizes. These are for identification purposes only and do not relate to actual dimensions.

**354.22 Number of Conductors.** The number of conductors or cables shall not exceed that permitted by the percentage fill in Table 1, Chapter 9.

**354.24 Bends — How Made.** Bends shall be manually made so that the conduit will not be damaged and the internal diameter of the conduit will not be effectively reduced. The radius of the curve of the centerline of such bends shall not be less than shown in Table 354.24.

**Table 354.24 Minimum Bending Radius for Nonmetallic Underground Conduit with Conductors (NUCC)**

| Conduit Size | | Minimum Bending Radius | |
|---|---|---|---|
| Metric Designator | Trade Size | mm | in. |
| 16 | ½ | 250 | 10 |
| 21 | ¾ | 300 | 12 |
| 27 | 1 | 350 | 14 |
| 35 | 1¼ | 450 | 18 |
| 41 | 1½ | 500 | 20 |
| 53 | 2 | 650 | 26 |
| 63 | 2½ | 900 | 36 |
| 78 | 3 | 1200 | 48 |
| 103 | 4 | 1500 | 60 |

**354.26 Bends — Number in One Run.** There shall not be more than the equivalent of four quarter bends (360 degrees total) between termination points.

**354.28 Trimming.** For termination, the conduit shall be trimmed away from the conductors or cables using an approved method that will not damage the conductor or cable insulation or jacket. All conduit ends shall be trimmed inside and out to remove rough edges.

**354.46 Bushings.** Where the NUCC enters a box, fitting, or other enclosure, a bushing or adapter shall be provided to protect the conductor or cable from abrasion unless the design of the box, fitting, or enclosure provides equivalent protection.

FPN: See 300.4(F) for the protection of conductors size 4 AWG or larger.

**354.48 Joints.** All joints between conduit, fittings, and boxes shall be made by an approved method.

**354.50 Conductor Terminations.** All terminations between the conductors or cables and equipment shall be made by an approved method for that type of conductor or cable.

**354.56 Splices and Taps.** Splices and taps shall be made in junction boxes or other enclosures.

**354.60 Grounding.** Where equipment grounding is required by Article 250, an assembly containing a separate equipment grounding conductor shall be used.

## III. Construction Specifications

**354.100 Construction.**

**(A) General.** NUCC is an assembly that is provided in continuous lengths shipped in a coil, reel, or carton.

**(B) Nonmetallic Underground Conduit.** The nonmetallic underground conduit shall be listed and composed of a material that is resistant to moisture and corrosive agents. It shall also be capable of being supplied on reels without damage or distortion and shall be of sufficient strength to withstand abuse, such as impact or crushing, in handling and during installation without damage to conduit or conductors.

**(C) Conductors and Cables.** Conductors and cables used in NUCC shall be listed and shall comply with 310.8(C). Conductors of different systems shall be installed in accordance with 300.3(C).

**(D) Conductor Fill.** The maximum number of conductors or cables in NUCC shall not exceed that permitted by the percentage fill in Table 1, Chapter 9.

**354.120 Marking.** NUCC shall be clearly and durably marked at least every 3.05 m (10 ft) as required by 110.21. The type of conduit material shall also be included in the marking.

Identification of conductors or cables used in the assembly shall be provided on a tag attached to each end of the assembly or to the side of a reel. Enclosed conductors or cables shall be marked in accordance with 310.11.

## ARTICLE 356
### Liquidtight Flexible Nonmetallic Conduit: Type LFNC

## I. General

**356.1 Scope.** This article covers the use, installation, and construction specifications for liquidtight flexible nonmetallic conduit (LFNC) and associated fittings.

**356.2 Definition.**

**Liquidtight Flexible Nonmetallic Conduit (LFNC).** A raceway of circular cross section of various types as follows:

(1) A smooth seamless inner core and cover bonded together and having one or more reinforcement layers between the core and covers, designated as Type LFNC-A

(2) A smooth inner surface with integral reinforcement within the conduit wall, designated as Type LFNC-B

(3) A corrugated internal and external surface without integral reinforcement within the conduit wall, designated as LFNC-C.

LFNC is flame resistant and with fittings and is approved for the installation of electrical conductors.

FPN: FNMC is an alternative designation for LFNC.

**356.6 Listing Requirements.** LFNC and associated fittings shall be listed.

## II. Installation

**356.10 Uses Permitted.** LFNC shall be permitted to be used in exposed or concealed locations for the following purposes:

FPN: Extreme cold may cause some types of nonmetallic conduits to become brittle and therefore more susceptible to damage from physical contact.

(1) Where flexibility is required for installation, operation, or maintenance

(2) Where protection of the contained conductors is required from vapors, liquids, or solids

(3) For outdoor locations where listed and marked as suitable for the purpose

(4) For direct burial where listed and marked for the purpose

(5) Type LFNC-B shall be permitted to be installed in lengths longer than 1.8 m (6 ft) where secured in accordance with 356.30

(6) Type LFNC-B as a listed manufactured prewired assembly, metric designator 16 through 27 (trade size ½ through 1) conduit

**356.12 Uses Not Permitted.** LFNC shall not be used as follows:

(1) Where subject to physical damage

(2) Where any combination of ambient and conductor temperatures is in excess of that for which the LFNC is approved

(3) In lengths longer than 1.8 m (6 ft), except as permitted by 356.10(5) or where a longer length is approved as essential for a required degree of flexibility

(4) Where voltage of the contained conductors is in excess of 600 volts, nominal

## 356.20 Size.

**(A) Minimum.** LFNC smaller than metric designator 16 (trade size ½) shall not be used unless permitted in 356.20(A)(1) through (3) for metric designator 12 (trade size ⅜).

(1) For enclosing the leads of motors as permitted in 430.145(B)
(2) In lengths not exceeding 1.8 m (6 ft ) as part of a listed assembly for tap connections to luminaires (lighting fixtures) as required in 410.67(C), or for utilization equipment
(3) For electric sign conductors in accordance with 600.32(A)

**(B) Maximum.** LFNC larger than metric designator 103 (trade size 4) shall not be used.

> FPN: See 300.1(C) for the metric designators and trade sizes. These are for identification purposes only and do not relate to actual dimensions.

**356.22 Number of Conductors.** The number of conductors shall not exceed that permitted by the percentage fill specified in Table 1, Chapter 9.

Cables shall be permitted to be installed where such use is permitted by the respective cable articles. The number of cables shall not exceed the allowable percentage fill specified in Table 1, Chapter 9.

**356.24 Bends — How Made.** Bends in conduit shall be made so that the conduit is not damaged and the internal diameter of the conduit is not effectively reduced. Bends shall be permitted to be made manually without auxiliary equipment. The radius of the curve to the centerline of any bend shall not be less than shown in Table 344.24 using the column "Other Bends."

**356.26 Bends — Number in One Run.** There shall not be more than the equivalent of four quarter bends (360 degrees total) between pull points, for example, conduit bodies and boxes.

**356.28 Trimming.** All cut ends of conduit shall be trimmed inside and outside to remove rough edges.

**356.30 Securing and Supporting.** Type LFNC-B shall be securely fastened and supported in accordance with one of the following:

(1) The conduit shall be securely fastened at intervals not exceeding 900 mm (3 ft) and within 300 mm (12 in.) on each side of every outlet box, junction box, cabinet, or fitting.
(2) Securing or supporting of the conduit shall not be required where it is fished, installed in lengths not exceeding 900 mm (3 ft) at terminals where flexibility is required, or installed in lengths not exceeding 1.8 m (6 ft) from a luminaire (fixture) terminal connection for tap conductors to luminaires (lighting fixtures) permitted in 410.67(C).
(3) Horizontal runs of LFNC supported by openings through framing members at intervals not exceeding 900 mm (3 ft) and securely fastened within 300 mm (12 in.) of termination points shall be permitted.

**356.42 Couplings and Connectors.** Angle connectors shall not be used for concealed raceway installations.

**356.56 Splices and Taps.** Splices and taps shall be made in accordance with 300.15.

**356.60 Grounding and Bonding.** Where used to connect equipment where flexibility is required, an equipment grounding conductor shall be installed.

Where required or installed, equipment grounding conductors shall be installed in accordance with 250.134(B).

Where required or installed, equipment bonding jumpers shall be installed in accordance with 250.102.

## III. Construction Specifications

**356.100 Construction.** LFNC-B as a prewired manufactured assembly shall be provided in continuous lengths capable of being shipped in a coil, reel, or carton without damage.

**356.120 Marking.** LFNC shall be marked at least every 600 mm (2 ft) in accordance with 110.21. The marking shall include a type designation in accordance with 356.2 and the trade size. Conduit that is intended for outdoor use or direct burial shall be marked.

The type, size, and quantity of conductors used in prewired manufactured assemblies shall be identified by means of a printed tag or label attached to each end of the manufactured assembly and either the carton, coil, or reel. The enclosed conductors shall be marked in accordance with 310.11.

## ARTICLE 358
## Electrical Metallic Tubing: Type EMT

### I. General

**358.1 Scope.** This article covers the use, installation, and construction specifications for electrical metallic tubing (EMT) and associated fittings.

**358.2 Definition.**

**Electrical Metallic Tubing (EMT).** An unthreaded thin-wall raceway of circular cross section designed for the physical protection and routing of conductors and cables and for use as an equipment grounding conductor when installed utilizing appropriate fittings. EMT is generally made of steel (ferrous) with protective coatings or aluminum (nonferrous).

**358.6 Listing Requirements.** EMT, factory elbows, and associated fittings shall be listed.

### II. Installation

**358.10 Uses Permitted.**

**(A) Exposed and Concealed.** The use of EMT shall be permitted for both exposed and concealed work.

**(B) Corrosion Protection.** Ferrous or nonferrous EMT, elbows, couplings, and fittings shall be permitted to be installed in concrete, in direct contact with the earth, or in areas subject to severe corrosive influences where protected by corrosion protection and judged suitable for the condition.

**(C) Wet Locations.** All supports, bolts, straps, screws, and so forth shall be of corrosion-resistant materials or protected against corrosion by corrosion-resistant materials.

FPN: See 300.6 for protection against corrosion.

**358.12 Uses Not Permitted.** EMT shall not be used under the following conditions:

(1) Where, during installation or afterward, it will be subject to severe physical damage
(2) Where protected from corrosion solely by enamel
(3) In cinder concrete or cinder fill where subject to permanent moisture unless protected on all sides by a layer of noncinder concrete at least 50 mm (2 in.) thick or unless the tubing is at least 450 mm (18 in.) under the fill
(4) In any hazardous (classified) location except as permitted by 502.4, 503.3, and 504.20

(5) For the support of luminaires (fixtures) or other equipment except conduit bodies no larger than the largest trade size of the tubing
(6) Where practicable, dissimilar metals in contact anywhere in the system shall be avoided to eliminate the possibility of galvanic action

*Exception: Aluminum fittings and enclosures shall be permitted to be used with steel EMT where not subject to severe corrosive influences.*

**358.20 Size.**

**(A) Minimum.** EMT smaller than metric designator 16 (trade size ½) shall not be used.

*Exception: For enclosing the leads of motors as permitted in 430.145(B).*

**(B) Maximum.** The maximum size of EMT shall be metric designator 103 (trade size 4).

FPN: See 300.1(C) for the metric designators and trade sizes. These are for identification purposes only and do not relate to actual dimensions.

**358.22 Number of Conductors.** The number of conductors shall not exceed that permitted by the percentage fill specified in Table 1, Chapter 9.

Cables shall be permitted to be installed where such use is permitted by the respective cable articles. The number of cables shall not exceed the allowable percentage fill specified in Table 1, Chapter 9.

**358.24 Bends — How Made.** Bends shall be made so that the tubing is not damaged and the internal diameter of the tubing is not effectively reduced. The radius of the curve of any field bend to the centerline of the conduit shall not be less than shown in Table 344.24 for one-shot and full shoe benders.

**358.26 Bends — Number in One Run.** There shall not be more than the equivalent of four quarter bends (360 degrees total) between pull points, for example, conduit bodies and boxes.

**358.28 Reaming and Threading.**

**(A) Reaming.** All cut ends of EMT shall be reamed or otherwise finished to remove rough edges.

**(B) Threading.** EMT shall not be threaded.

*Exception: EMT with factory threaded integral couplings complying with 358.100.*

**358.30 Securing and Supporting.** EMT shall be installed as a complete system as provided in Article 300 and shall be securely fastened in place and supported in accordance with 358.30(A) and (B).

**(A) Securely Fastened.** EMT shall be securely fastened in place at least every 3 m (10 ft). In addition, each EMT run between termination points shall be securely fastened within 900 mm (3 ft) of each outlet box, junction box, device box, cabinet, conduit body, or other tubing termination.

*Exception No. 1: Fastening of unbroken lengths shall be permitted to be increased to a distance of 1.5 m (5 ft) where structural members do not readily permit fastening within 900 mm (3 ft).*

*Exception No. 2: For concealed work in finished buildings or prefinished wall panels where such securing is impracticable, unbroken lengths (without coupling) of EMT shall be permitted to be fished.*

**(B) Supports.** Horizontal runs of EMT supported by openings through framing members at intervals not greater than 3 m (10 ft) and securely fastened within 900 mm (3 ft) of termination points shall be permitted.

**358.42 Couplings and Connectors.** Couplings and connectors used with EMT shall be made up tight. Where buried in masonry or concrete, they shall be concretetight type. Where installed in wet locations, they shall be of the raintight type.

**358.56 Splices and Taps.** Splices and taps shall be made in accordance with 300.15.

**358.60 Grounding.** EMT shall be permitted as an equipment grounding conductor.

### III. Construction Specifications

**358.100 Construction.** Factory-threaded integral couplings shall be permitted. Where EMT with a threaded integral coupling is used, threads for both the tubing and coupling shall be factory-made. The coupling and EMT threads shall be designed so as to prevent bending of the tubing at any part of the thread.

**358.120 Marking.** EMT shall be clearly and durably marked at least every 3 m (10 ft) as required in the first sentence of 110.21.

---

## ARTICLE 360
## Flexible Metallic Tubing: Type FMT

### I. General

**360.1 Scope.** This article covers the use, installation, and construction specifications for flexible metallic tubing (FMT) and associated fittings.

**360.2 Definition.**

**Flexible Metallic Tubing (FMT).** A raceway that is circular in cross section, flexible, metallic, and liquidtight without a nonmetallic jacket.

**360.6 Listing Requirements.** FMT and associated fittings shall be listed.

### II. Installation

**360.10 Uses Permitted.** FMT shall be permitted to be used for branch circuits as follows:

(1) In dry locations
(2) Where concealed
(3) In accessible locations
(4) For system voltages of 1000 volts maximum

**360.12 Uses Not Permitted.** FMT shall not be used as follows:

(1) In hoistways
(2) In storage battery rooms
(3) In hazardous (classified) locations unless otherwise permitted under other articles in this *Code*
(4) Under ground for direct earth burial, or embedded in poured concrete or aggregate
(5) Where subject to physical damage
(6) In lengths over 1.8 m (6 ft)

**360.20 Size.**

**(A) Minimum.** FMT smaller than metric designator 16 (trade size ½) shall not be used.

*Exception No. 1: FMT of metric designator 12 (trade size ⅜) shall be permitted to be installed in accordance with 300.22(B) and (C).*

*Exception No. 2: FMT of metric designator 12 (trade size ⅜) shall be permitted in lengths not in excess of 1.8 m (6 ft) as part of an approved assembly or for luminaires (lighting fixtures). See 410.67(C).*

**(B) Maximum.** The maximum size of FMT shall be metric designator 21 (trade size ¾).

FPN: See 300.1(C) for the metric designators and trade sizes. These are for identification purposes only and do not relate to actual dimensions.

## 360.22 Number of Conductors.

**(A) FMT — Metric Designators 16 and 21 (Trade Sizes ½ and ¾).** The number of conductors in metric designators 16 (trade size ½) and 21 (trade size ¾) shall not exceed that permitted by the percentage fill specified in Table 1, Chapter 9.

Cables shall be permitted to be installed where such use is permitted by the respective cable articles. The number of cables shall not exceed the allowable percentage fill specified in Table 1, Chapter 9.

**(B) FMT — Metric Designator 12 (Trade Size ⅜).** The number of conductors in metric designator 12 (trade size ⅜) shall not exceed that permitted in Table 348.22.

## 360.24 Bends.

**(A) Infrequent Flexing Use.** Where FMT may be infrequently flexed in service after installation, the radii of bends measured to the inside of the bend shall not be less than specified in Table 360.24(A).

**Table 360.24(A) Minimum Radii for Flexing Use**

| Metric Designator | Trade Size | Minimum Radii for Flexing Use | |
|---|---|---|---|
| | | mm | in. |
| 12 | ⅜ | 25.4 | 10 |
| 16 | ½ | 317.5 | 12½ |
| 21 | ¾ | 444.5 | 17½ |

**(B) Fixed Bends.** Where FMT is bent for installation purposes and is not flexed or bent as required by use after installation, the radii of bends measured to the inside of the bend shall not be less than specified in Table 360.24(B).

**Table 360.24(B) Minimum Radii for Fixed Bends**

| Metric Designator | Trade Size | Minimum Radii for Fixed Bends | |
|---|---|---|---|
| | | mm | in. |
| 12 | ⅜ | 88.9 | 3½ |
| 16 | ½ | 101.6 | 4 |
| 21 | ¾ | 127.0 | 5 |

**360.40 Boxes and Fittings.** Fittings shall effectively close any openings in the connection.

**360.56 Splices and Taps.** Splices and taps shall be made in accordance with 300.15.

**360.60 Grounding.** FMT shall be permitted as an equipment grounding conductor where installed in accordance with 250.118(8).

### III. Construction Specifications

**360.120 Marking.** FMT shall be marked according to 110.21.

## ARTICLE 362
## Electrical Nonmetallic Tubing: Type ENT

### I. General

**362.1 Scope.** This article covers the use, installation, and construction specifications for electrical nonmetallic tubing (ENT) and associated fittings.

### 362.2 Definition.

**Electrical Nonmetallic Tubing (ENT).** A nonmetallic pliable corrugated raceway of circular cross section with integral or associated couplings, connectors, and fittings for the installation of electric conductors. ENT is composed of a material that is resistant to moisture and chemical atmospheres and is flame retardant.

A pliable raceway is a raceway that can be bent by hand with a reasonable force, but without other assistance.

**362.6 Listing Requirements.** ENT and associated fittings shall be listed.

### II. Installation

**362.10 Uses Permitted.** For the purpose of this article, the first floor of a building shall be that floor that has 50 percent or more of the exterior wall surface area level with or above finished grade. One additional level that is the first level and not designed for human habitation and used only for vehicle parking, storage, or similar use shall be permitted. The use of ENT and fittings shall be permitted in the following:

(1) In any building not exceeding three floors above grade

     a. For exposed work, where not prohibited by 362.12
     b. Concealed within walls, floors, and ceilings

(2) In any building exceeding three floors above grade, ENT shall be concealed within walls, floors, and ceil-

ings where the walls, floors, and ceilings provide a thermal barrier of material that has at least a 15-minute finish rating as identified in listings of fire-rated assemblies. The 15-minute-finish-rated thermal barrier shall be permitted to be used for combustible or noncombustible walls, floors, and ceilings.

*Exception: Where a fire sprinkler system(s) is installed in accordance with NFPA 13-1999, Standard for the Installation of Sprinkler Systems, on all floors, ENT is permitted to be used within walls, floors, and ceilings, exposed or concealed, in buildings exceeding three floors above grade.*

FPN: A finish rating is established for assemblies containing combustible (wood) supports. The finish rating is defined as the time at which the wood stud or wood joist reaches an average temperature rise of 121°C (250°F) or an individual temperature of 163°C (325°F) as measured on the plane of the wood nearest the fire. A finish rating is not intended to represent a rating for a membrane ceiling.

(3) In locations subject to severe corrosive influences as covered in 300.6 and where subject to chemicals for which the materials are specifically approved.

(4) In concealed, dry, and damp locations not prohibited by 362.12.

(5) Above suspended ceilings where the suspended ceilings provide a thermal barrier of material that has at least a 15-minute finish rating as identified in listings of fire-rated assemblies, except as permitted in 362.10(1)(a).

*Exception: Where a fire sprinkler system(s) is installed in accordance with NFPA 13-1999, Standard for the Installation of Sprinkler Systems, on all floors, ENT is permitted to be used within walls, floors, and ceilings, exposed or concealed, in buildings exceeding three floors above grade.*

(6) Encased in poured concrete, or embedded in a concrete slab on grade where ENT is placed on sand or approved screenings, provided fittings identified for this purpose are used for connections.

(7) For wet locations indoors as permitted in this section or in a concrete slab on or below grade, with fittings listed for the purpose.

(8) Metric designator 16 through 27 (trade size ½ through 1) as listed manufactured prewired assembly.

FPN: Extreme cold may cause some types of nonmetallic conduits to become brittle and therefore more susceptible to damage from physical contact.

**362.12 Uses Not Permitted.** ENT shall not be used in the following:

(1) In hazardous (classified) locations, except as permitted by 504.20 and 505.15(A)(1)

(2) For the support of luminaires (fixtures) and other equipment

(3) Where subject to ambient temperatures in excess of 50°C (122°F) unless listed otherwise

(4) For conductors whose insulation temperature limitations would exceed those for which the tubing is listed

(5) For direct earth burial

(6) Where the voltage is over 600 volts

(7) In exposed locations, except as permitted by 362.10(1), 362.10(5), and 362.10(7)

(8) In theaters and similar locations, except as provided in Articles 518 and 520

(9) Where exposed to the direct rays of the sun, unless identified as sunlight resistant

(10) Where subject to physical damage

**362.20 Size.**

**(A) Minimum.** ENT smaller than metric designator 16 (trade size ½) shall not be used.

**(B) Maximum.** ENT larger than metric designator 53 (trade size 2) shall not be used.

FPN: See 300.1(C) for the metric designators and trade sizes. These are for identification purposes only and do not relate to actual dimensions.

**362.22 Number of Conductors.** The number of conductors shall not exceed that permitted by the percentage fill in Table 1, Chapter 9.

Cables shall be permitted to be installed where such use is permitted by the respective cable articles. The number of cables shall not exceed the allowable percentage fill specified in Table 1, Chapter 9.

**362.24 Bends — How Made.** Bends shall be made so that the tubing will not be damaged and that the internal diameter of the tubing will not be effectively reduced. Bends shall be permitted to be made manually without auxiliary equipment, and the radius of the curve to the centerline of such bends shall not be less than shown in Table 344.24 using the column "Other Bends."

**362.26 Bends — Number in One Run.** There shall not be more than the equivalent of four quarter bends (360 degrees total) between pull points, for example, conduit bodies and boxes.

**362.28 Trimming.** All cut ends shall be trimmed inside and outside to remove rough edges.

**362.30 Securing and Supporting.** ENT shall be installed as a complete system as provided in Article 300 and shall be securely fastened in place and supported in accordance with 362.30(A) and (B).

**(A) Securely Fastened.** ENT shall be securely fastened at intervals not exceeding 900 mm (3 ft). In addition, ENT shall be securely fastened in place within 900 mm (3 ft) of each outlet box, device box, junction box, cabinet, or fitting where it terminates.

*Exception: Lengths not exceeding a distance of 1.8 m (6 ft) from a luminaire (fixture) terminal connection for tap connections to lighting luminaires (fixtures) shall be permitted without being secured.*

**(B) Supports.** Horizontal runs of ENT supported by openings in framing members at intervals not exceeding 900 mm (3 ft) and securely fastened within 900 mm (3 ft) of termination points shall be permitted.

**362.46 Bushings.** Where a tubing enters a box, fitting, or other enclosure, a bushing or adapter shall be provided to protect the wire from abrasion unless the box, fitting, or enclosure design provides equivalent protection.

> FPN: See 300.4(F) for the protection of conductors size 4 AWG or larger.

**362.48 Joints.** All joints between lengths of tubing and between tubing and couplings, fittings, and boxes shall be by an approved method.

**362.56 Splices and Taps.** Splices and taps shall be made only in accordance with 300.15.

> FPN: See Article 314 for rules on the installation and use of boxes and conduit bodies.

**362.60 Grounding.** Where equipment grounding is required by Article 250, a separate equipment grounding conductor shall be installed in the raceway.

### III. Construction Specifications

**362.100 Construction.** ENT shall be made of material that does not exceed the ignitibility, flammability, smoke generation, and toxicity characteristics of rigid (nonplasticized) polyvinyl chloride.

ENT, as a prewired manufactured assembly, shall be provided in continuous lengths capable of being shipped in a coil, reel, or carton without damage.

**362.120 Marking.** ENT shall be clearly and durably marked at least every 3 m (10 ft) as required in the first sentence of 110.21. The type of material shall also be included in the marking. Marking for limited smoke shall be permitted on the tubing that has limited smoke-producing characteristics.

The type, size, and quantity of conductors used in prewired manufactured assemblies shall be identified by means of a printed tag or label attached to each end of the manufactured assembly and either the carton, coil, or reel. The enclosed conductors shall be marked in accordance with 310.11.

---

# ARTICLE 366
## Auxiliary Gutters

**366.1 Scope.** This article covers the use, installation and construction requirements of metal auxiliary gutters and nonmetallic auxiliary gutters and associated fittings.

**366.2 Use.** Auxiliary gutters shall be permitted to supplement wiring spaces at meter centers, distribution centers, switchboards, and similar points of wiring systems and may enclose conductors or busbars but shall not be used to enclose switches, overcurrent devices, appliances, or other similar equipment.

**366.3 Extension Beyond Equipment.** An auxiliary gutter shall not extend a greater distance than 9 m (30 ft) beyond the equipment that it supplements.

*Exception: As permitted in 620.35 for elevators, an auxiliary gutter shall be permitted to extend a distance greater than 9 m (30 ft) beyond the equipment that it supplements.*

> FPN: For wireways, see Articles 376 and 378. For busways, see Article 368.

**366.4 Supports.**

**(A) Sheet Metal Auxiliary Gutters.** Sheet metal auxiliary gutters shall be supported throughout their entire length at intervals not exceeding 1.5 m (5 ft).

**(B) Nonmetallic Auxiliary Gutters.** Nonmetallic auxiliary gutters shall be supported at intervals not to exceed 900 mm (3 ft) and at each end or joint, unless listed for other support intervals. In no case shall the distance between supports exceed 3 m (10 ft).

**366.5 Covers.** Covers shall be securely fastened to the gutter.

**366.6 Number of Conductors.**

**(A) Sheet Metal Auxiliary Gutters.** The sum of the cross-sectional areas of all contained conductors at any cross section of a sheet metal auxiliary gutter shall not exceed 20 percent of the interior cross-sectional area of the sheet metal auxiliary gutter. The derating factors in 310.15(B)(2)(a) shall be applied only where the number of current-carrying conductors, including neutral conductors classified as current-carrying under the provisions of 310.15(B)(4), exceeds 30. Conductors for signaling circuits or controller conductors between a motor and its starter and used only for starting duty shall not be considered as current-carrying conductors.

**(B) Nonmetallic Auxiliary Gutters.** The sum of cross-sectional areas of all contained conductors at any cross section of the nonmetallic auxiliary gutter shall not exceed 20 percent of the interior cross-sectional area of the non-metallic auxiliary gutter.

### 366.7 Ampacity of Conductors.

**(A) Sheet Metal Auxiliary Gutters.** Where the number of current-carrying conductors contained in the sheet metal auxiliary gutter is 30 or less, the correction factors specified in 310.15(B)(2)(a) shall not apply. The current carried continuously in bare copper bars in sheet metal auxiliary gutters shall not exceed 1.55 amperes/mm² (1000 amperes/in.²) of cross section of the conductor. For aluminum bars, the current carried continuously shall not exceed 1.09 amperes/mm² (700 amperes/in.²) of cross section of the conductor.

**(B) Nonmetallic Auxiliary Gutters.** The derating factors specified in 310.15(B)(2)(a) shall be applicable to the current-carrying conductors in the nonmetallic auxiliary gutter.

### 366.8 Clearance of Bare Live Parts.
Bare conductors shall be securely and rigidly supported so that the minimum clearance between bare current-carrying metal parts of different potential mounted on the same surface will not be less than 50 mm (2 in.), nor less than 25 mm (1 in.) for parts that are held free in the air. A clearance not less than 25 mm (1 in.) shall be secured between bare current-carrying metal parts and any metal surface. Adequate provisions shall be made for the expansion and contraction of busbars.

### 366.9 Splices and Taps.
Splices and taps shall comply with 366.9(A) through (D).

**(A) Within Gutters.** Splices or taps shall be permitted within gutters where they are accessible by means of removable covers or doors. The conductors, including splices and taps, shall not fill the gutter to more than 75 percent of its area.

**(B) Bare Conductors.** Taps from bare conductors shall leave the gutter opposite their terminal connections, and conductors shall not be brought in contact with uninsulated current-carrying parts of different potential.

**(C) Suitably Identified.** All taps shall be suitably identified at the gutter as to the circuit or equipment that they supply.

**(D) Overcurrent Protection.** Tap connections from conductors in auxiliary gutters shall be provided with overcurrent protection as required in 240.21.

### 366.10 Construction and Installation.
Auxiliary gutters shall comply with 366.10(A) through (F).

**(A) Electrical and Mechanical Continuity.** Gutters shall be constructed and installed so that adequate electrical and mechanical continuity of the complete system is secured.

**(B) Substantial Construction.** Gutters shall be of substantial construction and shall provide a complete enclosure for the contained conductors. All surfaces, both interior and exterior, shall be suitably protected from corrosion. Corner joints shall be made tight, and where the assembly is held together by rivets, bolts, or screws, such fasteners shall be spaced not more than 300 mm (12 in.) apart.

**(C) Smooth Rounded Edges.** Suitable bushings, shields, or fittings having smooth, rounded edges shall be provided where conductors pass between gutters, through partitions, around bends, between gutters and cabinets or junction boxes, and at other locations where necessary to prevent abrasion of the insulation of the conductors.

**(D) Deflected Insulated Conductors.** Where insulated conductors are deflected within an auxiliary gutter, either at the ends or where conduits, fittings, or other raceways or cables enter or leave the gutter, or where the direction of the gutter is deflected greater than 30 degrees, dimensions corresponding to 312.6 shall apply.

**(E) Indoor and Outdoor Use.**

**(1) Sheet Metal Auxiliary Gutters.** Sheet metal auxiliary gutters installed in wet locations shall be suitable for such locations.

**(2) Nonmetallic Auxiliary Gutters.**

(a) Nonmetallic auxiliary gutters installed outdoors shall comply with the following:

(1) Be listed and marked as suitable for exposure to sunlight
(2) Be listed and marked as suitable for use in wet locations
(3) Be listed for the maximum ambient temperature of the installation, and marked for the installed conductor insulation temperature rating
(4) Have expansion fittings installed where the expected length change due to expansion and contraction due to temperature change is more than 6 mm (0.25 in.)

(b) Nonmetallic auxiliary gutters installed indoors shall comply with the following:

(1) Be listed for the maximum ambient temperature of the installation and marked for the installed conductor insulation temperature rating

(2) Have expansion fittings installed where expected length change, due to expansion and contraction due to temperature change, is more than 6 mm (0.25 in.)

FPN: Extreme cold may cause nonmetallic auxiliary gutter to become brittle and therefore more susceptible to damage from physical contact.

**(F) Grounding.** Grounding shall be in accordance with the provisions of Article 250.

# ARTICLE 368
# Busways

## I. General Requirements

**368.1 Scope.** This article covers service-entrance, feeder, and branch-circuit busways and associated fittings.

**368.2 Definition.**

**Busway.** A grounded metal enclosure containing factory-mounted, bare or insulated conductors, which are usually copper or aluminum bars, rods, or tubes.

FPN: For cablebus, refer to Article 370.

**368.4 Use.**

**(A) Uses Permitted.** Busways shall be permitted to be installed where they are located as follows:

(1) Located in the open and are visible, except as permitted in 368.6, or

(2) Installed behind access panels, provided the busways are totally enclosed, of nonventilating-type construction, and installed so that the joints between sections and at fittings are accessible for maintenance purposes. Where installed behind access panels, means of access shall be provided, and the following conditions shall be met:

    a. The space behind the access panels shall not be used for air-handling purposes, or

    b. Where the space behind the access panels is used for environmental air, other than ducts and plenums, there shall be no provisions for plug-in connections, and the conductors shall be insulated.

**(B) Uses Not Permitted.** Busways shall not be installed as follows:

(1) Where subject to severe physical damage or corrosive vapors

(2) In hoistways

(3) In any hazardous (classified) location, unless specifically approved for such use

FPN: See 501.4(B).

(4) Outdoors or in wet or damp locations unless identified for such use

Lighting busway and trolley busway shall not be installed less than 2.5 m (8 ft) above the floor or working platform unless provided with a cover identified for the purpose.

**368.5 Support.** Busways shall be securely supported at intervals not exceeding 1.5 m (5 ft) unless otherwise designed and marked.

**368.6 Through Walls and Floors.**

**(A) Walls.** Unbroken lengths of busway shall be permitted to be extended through dry walls.

**(B) Floors.** Floor penetrations shall comply with (1) and (2):

(1) Busways shall be permitted to be extended vertically through dry floors if totally enclosed (unventilated) where passing through and for a minimum distance of 1.8 m (6 ft) above the floor to provide adequate protection from physical damage.

(2) In other than industrial establishments, where a vertical riser penetrates two or more dry floors, a minimum 100 mm (4 in.) high curb shall be installed around all floor openings for riser busways to prevent liquids from entering the opening. The curb shall be installed within 300 mm (12 in.) of the floor opening. Electrical equipment shall be located so that it will not be damaged by liquids that are retained by the curb.

FPN: See 300.21 for information concerning the spread of fire or products of combustion.

**368.7 Dead Ends.** A dead end of a busway shall be closed.

**368.8 Branches from Busways.** Branches from busways shall be permitted to be made in accordance with 368.8(A), (B), and (C).

**(A) General.** Branches from busways shall be made in accordance with Articles 320, 330, 332, 342, 344, 348, 350, 352, 356, 358, 362, 368, 384, 386, and 388. Where a separate equipment grounding conductor is used, connection of the equipment grounding conductor to the busway shall comply with 250.8 and 250.12.

**(B) Cord and Cable Assemblies.** Suitable cord and cable assemblies approved for extra-hard usage or hard usage and listed bus drop cable shall be permitted as branches from busways for the connection of portable equipment or the connection of stationary equipment to facilitate their interchange in accordance with 400.7 and 400.8 and the following conditions:

(1) The cord or cable shall be attached to the building by an approved means.
(2) The length of the cord or cable from a busway plug-in device to a suitable tension take-up support device shall not exceed 1.8 m (6 ft).

*Exception: In industrial establishments only, where the conditions of maintenance and supervision ensure that only qualified persons service the installation, lengths exceeding 1.8 m (6 ft) shall be permitted between the busway plug-in device and the tension take-up support device where the cord or cable is supported at intervals not exceeding 2.5 m (8 ft).*

(3) The cord or cable shall be installed as a vertical riser from the tension take-up support device to the equipment served.
(4) Strain relief cable grips shall be provided for the cord or cable at the busway plug-in device and equipment terminations.

**(C) Branches from Trolley-Type Busways.** Suitable cord and cable assemblies approved for extra-hard usage or hard usage and listed bus drop cable shall be permitted as branches from trolley-type busways for the connection of movable equipment in accordance with 400.7 and 400.8.

**368.9 Overcurrent Protection.** Overcurrent protection shall be provided in accordance with 368.10 through 368.13.

**368.10 Rating of Overcurrent Protection — Feeders.** A busway shall be protected against overcurrent in accordance with the allowable current rating of the busway.

*Exception No. 1: The applicable provisions of 240.4 shall be permitted.*

*Exception No. 2: Where used as transformer secondary ties, the provisions of 450.6(A)(3) shall be permitted.*

**368.11 Reduction in Ampacity Size of Busway.** Overcurrent protection shall be required where busways are reduced in ampacity.

*Exception: For industrial establishments only, omission of overcurrent protection shall be permitted at points where busways are reduced in ampacity, provided that the length of the busway having the smaller ampacity does not exceed 15 m (50 ft) and has an ampacity at least equal to one-third*

*the rating or setting of the overcurrent device next back on the line, and provided that such busway is free from contact with combustible material.*

**368.12 Feeder or Branch Circuits.** Where a busway is used as a feeder, devices or plug-in connections for tapping off feeder or branch circuits from the busway shall contain the overcurrent devices required for the protection of the feeder or branch circuits. The plug-in device shall consist of an externally operable circuit breaker or an externally operable fusible switch. Where such devices are mounted out of reach and contain disconnecting means, suitable means such as ropes, chains, or sticks shall be provided for operating the disconnecting means from the floor.

*Exception No. 1: As permitted in 240.21.*

*Exception No. 2: For fixed or semifixed luminaires (lighting fixtures), where the branch-circuit overcurrent device is part of the luminaire (fixture) cord plug on cord-connected luminaires (fixtures).*

*Exception No. 3: Where luminaires (fixtures) without cords are plugged directly into the busway and the overcurrent device is mounted on the luminaire (fixture).*

**368.13 Rating of Overcurrent Protection — Branch Circuits.** A busway used as a branch circuit shall be protected against overcurrent in accordance with 210.20. Where so used, the circuit shall comply with the applicable requirements of Articles 210, 430, and 440.

**368.15 Marking.** Busways shall be marked with the voltage and current rating for which they are designed, and with the manufacturer's name or trademark in such manner as to be visible after installation.

**II. Requirements for Over 600 Volts, Nominal**

**368.21 Identification.** Each bus run shall be provided with a permanent nameplate on which the following information shall be provided:

(1) Rated voltage
(2) Rated continuous current; if bus is forced-cooled, both the normal forced-cooled rating and the self-cooled (not forced-cooled) rating for the same temperature rise shall be given
(3) Rated frequency
(4) Rated impulse withstand voltage
(5) Rated 60-Hz withstand voltage (dry)
(6) Rated momentary current
(7) Manufacturer's name or trademark

FPN: See ANSI C37.23-1987 (R1991), *Guide for Metal-Enclosed Bus and Calculating Losses in Isolated-Phase*

*Bus,* for construction and testing requirements for metal-enclosed buses.

**368.22 Grounding.** Metal-enclosed bus shall be grounded in accordance with Article 250.

**368.23 Adjacent and Supporting Structures.** Metal-enclosed busways shall be installed so that temperature rise from induced circulating currents in any adjacent metallic parts will not be hazardous to personnel or constitute a fire hazard.

**368.24 Neutral.** Neutral bus, where required, shall be sized to carry all neutral load current, including harmonic currents, and shall have adequate momentary and short-circuit rating consistent with system requirements.

**368.25 Barriers and Seals.** Bus runs that have sections located both inside and outside of buildings shall have a vapor seal at the building wall to prevent interchange of air between indoor and outdoor sections.

*Exception: Vapor seals shall not be required in forced-cooled bus.*

Fire barriers shall be provided where fire walls, floors, or ceilings are penetrated.

FPN: See 300.21 for information concerning the spread of fire or products of combustion.

**368.26 Drain Facilities.** Drain plugs, filter drains, or similar methods shall be provided to remove condensed moisture from low points in bus run.

**368.27 Ventilated Bus Enclosures.** Ventilated bus enclosures shall be installed in accordance with Article 110, Part III, and 490.24.

**368.28 Terminations and Connections.** Where bus enclosures terminate at machines cooled by flammable gas, seal-off bushings, baffles, or other means shall be provided to prevent accumulation of flammable gas in the bus enclosures.

Flexible or expansion connections shall be provided in long, straight runs of bus to allow for temperature expansion or contraction, or where the bus run crosses building vibration insulation joints.

All conductor termination and connection hardware shall be accessible for installation, connection, and maintenance.

**368.29 Switches.** Switching devices or disconnecting links provided in the bus run shall have the same momentary rating as the bus. Disconnecting links shall be plainly marked to be removable only when bus is de-energized. Switching devices that are not load-break shall be interlocked to prevent operation under load, and disconnecting link enclosures shall be interlocked to prevent access to energized parts.

**368.30 Wiring 600 Volts or Less, Nominal.** Secondary control devices and wiring that are provided as part of the metal-enclosed bus run shall be insulated by fire-retardant barriers from all primary circuit elements with the exception of short lengths of wire, such as at instrument transformer terminals.

## ARTICLE 370
## Cablebus

**370.1 Scope.** This article covers the use and installation requirements of cablebus and associated fittings.

**370.2 Definition.**

**Cablebus.** An assembly of insulated conductors with fittings and conductor terminations in a completely enclosed, ventilated protective metal housing. Cablebus is ordinarily assembled at the point of installation from the components furnished or specified by the manufacturer in accordance with instructions for the specific job. This assembly is designed to carry fault current and to withstand the magnetic forces of such current.

**370.3 Use.** Approved cablebus shall be permitted at any voltage or current for which spaced conductors are rated and shall be installed for exposed work only, except as permitted in 370.6. Cablebus installed outdoors or in corrosive, wet, or damp locations shall be identified for such use. Cablebus shall be not be installed in hoistways or hazardous (classified) locations unless specifically approved for such use. Cablebus shall be permitted to be used for branch circuits, feeders, and services.

Cablebus framework, where bonded as required by Article 250, shall be permitted as the equipment grounding conductor for branch circuits and feeders.

**370.4 Conductors.**

**(A) Types of Conductors.** The current-carrying conductors in cablebus shall have an insulation rating of 75°C (167°F) or higher of an approved type and suitable for the application in accordance with Articles 310 and 490.

**(B) Ampacity of Conductors.** The ampacity of conductors in cablebus shall be in accordance with Tables 310.17 and 310.19, or with Tables 310.69 and 310.70 for installations over 600 volts.

**(C) Size and Number of Conductors.** The size and number of conductors shall be that for which the cablebus is designed, and in no case smaller than 1/0 AWG.

**(D) Conductor Supports.** The insulated conductors shall be supported on blocks or other mounting means designed for the purpose.

The individual conductors in a cablebus shall be supported at intervals not greater than 900 mm (3 ft) for horizontal runs and 450 mm (1½ ft) for vertical runs. Vertical and horizontal spacing between supported conductors shall not be less than one conductor diameter at the points of support.

**370.5 Overcurrent Protection.** Cablebus shall be protected against overcurrent in accordance with the allowable ampacity of the cablebus conductors in accordance with 240.4.

*Exception: Overcurrent protection shall be permitted in accordance with 240.100 and 240.101 for over 600 volts, nominal.*

**370.6 Support and Extension Through Walls and Floors.**

**(A) Support.** Cablebus shall be securely supported at intervals not exceeding 3.7 m (12 ft).

*Exception: Where spans longer than 3.7 m (12 ft) are required, the structure shall be specifically designed for the required span length.*

**(B) Transversely Routed.** Cablebus shall be permitted to extend transversely through partitions or walls, other than fire walls, provided the section within the wall is continuous, protected against physical damage, and unventilated.

**(C) Through Dry Floors and Platforms.** Except where firestops are required, cablebus shall be permitted to extend vertically through dry floors and platforms, provided the cablebus is totally enclosed at the point where it passes through the floor or platform and for a distance of 1.8 m (6 ft) above the floor or platform.

**(D) Through Floors and Platforms in Wet Locations.** Except where firestops are required, cablebus shall be permitted to extend vertically through floors and platforms in wet locations where (1) there are curbs or other suitable means to prevent waterflow through the floor or platform opening, and (2) where the cablebus is totally enclosed at the point where it passes through the floor or platform and for a distance of 1.8 m (6 ft) above the floor or platform.

**370.7 Fittings.** A cablebus system shall include approved fittings for the following:

(1) Changes in horizontal or vertical direction of the run
(2) Dead ends
(3) Terminations in or on connected apparatus or equipment or the enclosures for such equipment
(4) Additional physical protection where required, such as guards where subject to severe physical damage

**370.8 Conductor Terminations.** Approved terminating means shall be used for connections to cablebus conductors.

**370.9 Grounding.** A cablebus installation shall be grounded and bonded in accordance with Article 250, excluding 250.86, Exception No. 2.

**370.10 Marking.** Each section of cablebus shall be marked with the manufacturer's name or trade designation and the maximum diameter, number, voltage rating, and ampacity of the conductors to be installed. Markings shall be located so as to be visible after installation.

## ARTICLE 372
## Cellular Concrete Floor Raceways

**372.1 Scope.** This article covers cellular concrete floor raceways, the hollow spaces in floors constructed of precast cellular concrete slabs, together with suitable metal fittings designed to provide access to the floor cells.

**372.2 Definitions.**

**Cell.** A single, enclosed tubular space in a floor made of precast cellular concrete slabs, the direction of the cell being parallel to the direction of the floor member.

**Header.** Transverse metal raceways for electric conductors, providing access to predetermined cells of a precast cellular concrete floor, thereby permitting the installation of electric conductors from a distribution center to the floor cells.

**372.3 Other Articles.** Cellular concrete floor raceways shall comply with the applicable provisions of Article 300.

**372.4 Uses Not Permitted.** Conductors shall not be installed in precast cellular concrete floor raceways as follows:

(1) Where subject to corrosive vapor
(2) In any hazardous (classified) locations except as permitted by 504.20, and in Class I, Division 2 locations as permitted in 501.4(B)(3)

(3) In commercial garages, other than for supplying ceiling outlets or extensions to the area below the floor but not above

FPN: See 300.8 for installation of conductors with other systems.

**372.5 Header.** The header shall be installed in a straight line at right angles to the cells. The header shall be mechanically secured to the top of the precast cellular concrete floor. The end joints shall be closed by a metal closure fitting and sealed against the entrance of concrete. The header shall be electrically continuous throughout its entire length and shall be electrically bonded to the enclosure of the distribution center.

**372.6 Connection to Cabinets and Other Enclosures.** Connections from headers to cabinets and other enclosures shall be made by means of listed metal raceways and listed fittings.

**372.7 Junction Boxes.** Junction boxes shall be leveled to the floor grade and sealed against the free entrance of water or concrete. Junction boxes shall be of metal and shall be mechanically and electrically continuous with the header.

**372.8 Markers.** A suitable number of markers shall be installed for the future location of cells.

**372.9 Inserts.** Inserts shall be leveled and sealed against the entrance of concrete. Inserts shall be of metal and shall be fitted with grounded-type receptacles. A grounding conductor shall connect the insert receptacles to a positive ground connection provided on the header. Where cutting through the cell wall for setting inserts or other purposes (such as providing access openings between header and cells), chips and other dirt shall not be allowed to remain in the raceway, and the tool used shall be designed so as to prevent the tool from entering the cell and damaging the conductors.

**372.10 Size of Conductors.** No conductor larger than 1/0 AWG shall be installed, except by special permission.

**372.11 Maximum Number of Conductors.** The combined cross-sectional area of all conductors or cables shall not exceed 40 percent of the cross-sectional area of the cell or header.

**372.12 Splices and Taps.** Splices and taps shall be made only in header access units or junction boxes.

For the purposes of this section, so-called loop wiring (continuous unbroken conductor connecting the individual outlets) shall not be considered to be a splice or tap.

**372.13 Discontinued Outlets.** When an outlet is abandoned, discontinued, or removed, the sections of circuit conductors supplying the outlet shall be removed from the raceway. No splices or reinsulated conductors, such as would be the case of abandoned outlets on loop wiring, shall be allowed in raceways.

## ARTICLE 374
## Cellular Metal Floor Raceways

**374.1 Scope.** This article covers the use and installation requirements for cellular metal floor raceways.

**374.2 Definitions.**

**Cellular Metal Floor Raceway.** The hollow spaces of cellular metal floors, together with suitable fittings, that may be approved as enclosures for electric conductors.

*Cell.* A single, enclosed tubular space in a cellular metal floor member, the axis of the cell being parallel to the axis of the metal floor member.

*Header.* A transverse raceway for electric conductors, providing access to predetermined cells of a cellular metal floor, thereby permitting the installation of electric conductors from a distribution center to the cells.

**374.3 Uses Not Permitted.** Conductors shall not be installed in cellular metal floor raceways as follows:

(1) Where subject to corrosive vapor
(2) In any hazardous (classified) location except as permitted by 504.20, and in Class I, Division 2 locations as permitted in 501.4(B)(3)
(3) In commercial garages, other than for supplying ceiling outlets or extensions to the area below the floor but not above

FPN: See 300.8 for installation of conductors with other systems.

### I. Installation

**374.4 Size of Conductors.** No conductor larger than 1/0 AWG shall be installed, except by special permission.

**374.5 Maximum Number of Conductors in Raceway.** The combined cross-sectional area of all conductors or cables shall not exceed 40 percent of the interior cross-sectional area of the cell or header.

**374.6 Splices and Taps.** Splices and taps shall be made only in header access units or junction boxes.

For the purposes of this section, so-called loop wiring (continuous unbroken conductor connecting the individual outlets) shall not be considered to be a splice or tap.

**374.7 Discontinued Outlets.** When an outlet is abandoned, discontinued, or removed, the sections of circuit conductors supplying the outlet shall be removed from the raceway. No splices or reinsulated conductors, such as would be the case with abandoned outlets on loop wiring, shall be allowed in raceways.

**374.8 Markers.** A suitable number of markers shall be installed for locating cells in the future.

**374.9 Junction Boxes.** Junction boxes shall be leveled to the floor grade and sealed against the free entrance of water or concrete. Junction boxes used with these raceways shall be of metal and shall be electrically continuous with the raceway.

**374.10 Inserts.** Inserts shall be leveled to the floor grade and sealed against the entrance of concrete. Inserts shall be of metal and shall be electrically continuous with the raceway. In cutting through the cell wall and setting inserts, chips and other dirt shall not be allowed to remain in the raceway, and tools shall be used that are designed to prevent the tool from entering the cell and damaging the conductors.

**374.11 Connection to Cabinets and Extensions from Cells.** Connections between raceways and distribution centers and wall outlets shall be made by means of flexible metal conduit where not installed in concrete, rigid metal conduit, intermediate metal conduit, electrical metallic tubing, or approved fittings. Where there are provisions for the termination of an equipment grounding conductor, nonmetallic conduit, electrical nonmetallic tubing, or liquidtight flexible nonmetallic conduit where not installed in concrete shall be permitted.

## II. Construction Specifications

**374.12 General.** Cellular metal floor raceways shall be constructed so that adequate electrical and mechanical continuity of the complete system will be secured. They shall provide a complete enclosure for the conductors. The interior surfaces shall be free from burrs and sharp edges, and surfaces over which conductors are drawn shall be smooth. Suitable bushings or fittings having smooth rounded edges shall be provided where conductors pass.

## ARTICLE 376
## Metal Wireways

### I. General

**376.1 Scope.** This article covers the use, installation, and construction specifications for metal wireways and associated fittings.

**376.2 Definition.**

**Metal Wireways.** Sheet metal troughs with hinged or removable covers for housing and protecting electric wires and cable and in which conductors are laid in place after the wireway has been installed as a complete system.

### II. Installation

**376.10 Uses Permitted.** The use of metal wireways shall be permitted in the following:

(1) For exposed work
(2) In concealed spaces as permitted in 376.10(4)
(3) In hazardous (classified) locations as permitted by 501.4(B) for Class I, Division 2 locations; 502.4(B) for Class II, Division 2 locations; and 504.20 for intrinsically safe wiring. Where installed in wet locations, wireways shall be listed for the purpose.
(4) As extensions to pass transversely through walls if the length passing through the wall is unbroken. Access to the conductors shall be maintained on both sides of the wall.

**376.12 Uses Not Permitted.** Metal wireways shall not be used in the following:

(1) Where subject to severe physical damage
(2) Where subject to severe corrosive environments

**376.21 Size of Conductors.** No conductor larger than that for which the wireway is designed shall be installed in any wireway.

**376.22 Number of Conductors.** The sum of the cross-sectional areas of all contained conductors at any cross section of a wireway shall not exceed 20 percent of the interior cross-sectional area of the wireway. The derating factors in 310.15(B)(2)(a) shall be applied only where the number of current-carrying conductors, including neutral conductors classified as current-carrying under the provisions of 310.15(B)(4), exceeds 30. Conductors for signaling circuits or controller conductors between a motor and its starter and used only for starting duty shall not be considered as current-carrying conductors.

**376.23 Insulated Conductors.** Insulated conductors installed in a metallic wireway shall comply with 376.23(A) and (B).

**(A) Deflected Insulated Conductors.** Where insulated conductors are deflected within a metallic wireway, either at the ends or where conduits, fittings, or other raceways or cables enter or leave the metallic wireway, or where the direction of the metallic wireway is deflected greater than 30 degrees, dimensions corresponding to 312.6(A) shall apply.

**(B) Metallic Wireways Used as Pullboxes.** Where insulated conductors 4 AWG or larger are pulled through a wireway, the distance between raceway and cable entries enclosing the same conductor shall not be less than that required in 314.28(A)(1) for straight pulls and 314.28(A)(2) for angle pulls.

**376.30 Securing and Supporting.** Metal wireways shall be supported in accordance with 376.30(A) and (B).

**(A) Horizontal Support.** Wireways shall be supported where run horizontally at each end and at intervals not to exceed 1.5 m (5 ft) or for individual lengths longer than 1.5 m (5 ft) at each end or joint, unless listed for other support intervals. The distance between supports shall not exceed 3 m (10 ft).

**(B) Vertical Support.** Vertical runs of wireways shall be securely supported at intervals not exceeding 4.5 m (15 ft) and shall not have more than one joint between supports. Adjoining wireway sections shall be securely fastened together to provide a rigid joint.

**376.56 Splices and Taps.** Splices and taps shall be permitted within a wireway provided they are accessible. The conductors, including splices and taps, shall not fill the wireway to more than 75 percent of its area at that point.

**376.58 Dead Ends.** Dead ends of metal wireways shall be closed.

**376.70 Extensions from Metal Wireways.** Extensions from wireways shall be made with cord pendants installed in accordance with 400.10 or any wiring method in Chapter 3 that includes a means for equipment grounding. Where a separate equipment grounding conductor is employed, connection of the equipment grounding conductors in the wiring method to the wireway shall comply with 250.8 and 250.12.

## III. Construction Specifications

**376.120 Marking.** Metal wireways shall be marked so that their manufacturer's name or trademark will be visible after installation.

## ARTICLE 378
## Nonmetallic Wireways

### I. General

**378.1 Scope.** This article covers the use, installation, and construction specifications for nonmetallic wireways and associated fittings.

**378.2 Definition.**

**Nonmetallic Wireways.** Flame retardant, nonmetallic troughs with removable covers for housing and protecting electric wires and cables in which conductors are laid in place after the wireway has been installed as a complete system.

**378.3 Other Articles.** Installations of nonmetallic wireways shall comply with the applicable provisions of Article 300.

**378.6 Listing Requirements.** Nonmetallic wireways and associated fittings shall be listed.

### II. Installation

**378.10 Uses Permitted.** The use of nonmetallic wireways shall be permitted in the following:

(1) Only for exposed work, except as permitted in 378.10(4).
(2) Where subject to corrosive environments where identified for the use.
(3) In wet locations where listed for the purpose.

> FPN: Extreme cold may cause nonmetallic wireways to become brittle and therefore more susceptible to damage from physical contact.

(4) As extensions to pass transversely through walls if the length passing through the wall is unbroken. Access to the conductors shall be maintained on both sides of the wall.

**378.12 Uses Not Permitted.** Nonmetallic wireways shall not be used in the following:

(1) Where subject to physical damage
(2) In any hazardous (classified) location, except as permitted in 504.20
(3) Where exposed to sunlight unless listed and marked as suitable for the purpose
(4) Where subject to ambient temperatures other than those for which nonmetallic wireway is listed

(5) For conductors whose insulation temperature limitations would exceed those for which the nonmetallic wireway is listed

**378.21 Size of Conductors.** No conductor larger than that for which the nonmetallic wireway is designed shall be installed in any nonmetallic wireway.

**378.22 Number of Conductors.** The sum of cross-sectional areas of all contained conductors at any cross section of the nonmetallic wireway shall not exceed 20 percent of the interior cross-sectional area of the nonmetallic wireway. Conductors for signaling circuits or controller conductors between a motor and its starter and used only for starting duty shall not be considered as current-carrying conductors.

The derating factors specified in 310.15(B)(2)(a) shall be applicable to the current-carrying conductors up to and including the 20 percent fill specified above.

**378.23 Insulated Conductors.** Insulated conductors installed in a nonmetallic wireway shall comply with 378.23(A) and (B).

**(A) Deflected Insulated Conductors.** Where insulated conductors are deflected within a nonmetallic wireway, either at the ends or where conduits, fittings, or other raceways or cables enter or leave the nonmetallic wireway, or where the direction of the nonmetallic wireway is deflected greater than 30 degrees, dimensions corresponding to 312.6(A) shall apply.

**(B) Nonmetallic Wireways Used as Pull Boxes.** Where insulated conductors 4 AWG or larger are pulled through a wireway, the distance between raceway and cable entries enclosing the same conductor shall not be less than that required in 314.28(A)(1) for straight pulls and in 314.28(A)(2) for angle pulls.

**378.30 Securing and Supporting.** Nonmetallic wireway shall be supported in accordance with 378.30(A) and (B).

**(A) Horizontal Support.** Nonmetallic wireways shall be supported where run horizontally at intervals not to exceed 900 mm (3 ft), and at each end or joint, unless listed for other support intervals. In no case shall the distance between supports exceed 3 m (10 ft).

**(B) Vertical Support.** Vertical runs of nonmetallic wireway shall be securely supported at intervals not exceeding 1.2 m (4 ft), unless listed for other support intervals, and shall not have more than one joint between supports. Adjoining nonmetallic wireway sections shall be securely fastened together to provide a rigid joint.

**378.44 Expansion Fittings.** Expansion fittings for nonmetallic wireway shall be provided to compensate for thermal expansion and contraction where the length change is expected to be 6 mm (0.25 in.) or greater in a straight run.

FPN: See Table 352.44(A) for expansion characteristics of PVC rigid nonmetallic conduit. The expansion characteristics of PVC nonmetallic wireway are identical.

**378.56 Splices and Taps.** Splices and taps shall be permitted within a nonmetallic wireway, provided they are accessible. The conductors, including splices and taps, shall not fill the nonmetallic wireway to more than 75 percent of its area at that point.

**378.58 Dead Ends.** Dead ends of nonmetallic wireway shall be closed using listed fittings.

**378.60 Grounding.** Where equipment grounding is required by Article 250, a separate equipment grounding conductor shall be installed in the nonmetallic wireway. A separate equipment grounding conductor shall not be required where the grounded conductor is used to ground equipment as permitted in 250.142.

**378.70 Extensions from Nonmetallic Wireways.** Extensions from nonmetallic wireway shall be made with cord pendants or any wiring method of Chapter 3. A separate equipment grounding conductor shall be installed in, or an equipment grounding connection shall be made to, any of the wiring methods used for the extension.

### III. Construction Specifications

**378.120 Marking.** Nonmetallic wireways shall be marked so that the manufacturer's name or trademark and interior cross-sectional area in square inches shall be visible after installation. Marking for limited smoke shall be permitted on the nonmetallic wireways that have limited smoke-producing characteristics.

## ARTICLE 380
## Multioutlet Assembly

**380.1 Scope.** This article covers the use and installation requirements for multioutlet assemblies.

**380.2 Use.**

**(A) Permitted.** The use of a multioutlet assembly shall be permitted in dry locations.

**(B) Not Permitted.** A multioutlet assembly shall not be installed as follows:

(1) Where concealed, except that it shall be permissible to surround the back and sides of a metal multioutlet assembly by the building finish or recess a nonmetallic multioutlet assembly in a baseboard
(2) Where subject to severe physical damage
(3) Where the voltage is 300 volts or more between conductors unless the assembly is of metal having a thickness of not less than 1.02 mm (0.040 in.)
(4) Where subject to corrosive vapors
(5) In hoistways
(6) In any hazardous (classified) locations except Class I, Division 2 locations as permitted in 501.4(B)(3)

**380.3 Metal Multioutlet Assembly Through Dry Partitions.** It shall be permissible to extend a metal multioutlet assembly through (not run within) dry partitions if arrangements are made for removing the cap or cover on all exposed portions and no outlet is located within the partitions.

# ARTICLE 382
# Nonmetallic Extensions

## I. General

**382.1 Scope.** This article covers the use, installation, and construction specifications for nonmetallic extensions.

**382.2 Definition.**

**Nonmetallic Extension.** An assembly of two insulated conductors within a nonmetallic jacket or an extruded thermoplastic covering. The classification includes surface extensions intended for mounting directly on the surface of walls or ceilings.

## II. Installation

**382.10 Uses Permitted.** Nonmetallic extensions shall be permitted only where all the conditions in 382.10(A), (B), and (C) are met.

**(A) From an Existing Outlet.** The extension is from an existing outlet on a 15- or 20-ampere branch circuit.

**(B) Exposed and in a Dry Location.** The extension is run exposed and in a dry location.

**(C) Residential or Offices.** For nonmetallic surface extensions mounted directly on the surface of walls or ceilings,

the building is occupied for residential or office purposes and does not exceed three floors above grade.

FPN No. 1: See 310.10 for temperature limitation of conductors.

FPN No. 2: See 362.10 for definition of *first floor.*

**382.12 Uses Not Permitted.** Nonmetallic extensions shall not be used as follows:

(1) In unfinished basements, attics, or roof spaces
(2) Where the voltage between conductors exceeds 150 volts for nonmetallic surface extension and 300 volts for aerial cable
(3) Where subject to corrosive vapors
(4) Where run through a floor or partition, or outside the room in which it originates

**382.15 Exposed.** One or more extensions shall be permitted to be run in any direction from an existing outlet, but not on the floor or within 50 mm (2 in.) from the floor.

**382.26 Bends.** A bend that reduces the normal spacing between the conductors shall be covered with a cap to protect the assembly from physical damage.

**382.30 Securing and Supporting.** Nonmetallic surface extensions shall be secured in place by approved means at intervals not exceeding 200 mm (8 in.), with an allowance for 300 mm (12 in.) to the first fastening where the connection to the supplying outlet is by means of an attachment plug. There shall be at least one fastening between each two adjacent outlets supplied. An extension shall be attached to only woodwork or plaster finish and shall not be in contact with any metal work or other conductive material other than with metal plates on receptacles.

**382.40 Boxes and Fittings.** Each run shall terminate in a fitting that covers the end of the assembly. All fittings and devices shall be of a type identified for the use.

**382.56 Splices and Taps.** Extensions shall consist of a continuous unbroken length of the assembly, without splices, and without exposed conductors between fittings. Taps shall be permitted where approved fittings completely covering the tap connections are used. Aerial cable and its tap connectors shall be provided with an approved means for polarization. Receptacle-type tap connectors shall be of the locking type.

# ARTICLE 384
# Strut-Type Channel Raceway

## I. General

**384.1 Scope.** This article covers the use, installation, and construction specifications of strut-type channel raceway.

**384.2 Definition.**

**Strut-Type Channel Raceway.** A metallic raceway that is intended to be mounted to the surface of or suspended from a structure, with associated accessories for the installation of electrical conductors.

**384.6 Listing Requirements.** Strut-type channel raceways, closure strips, and accessories shall be listed and identified for such use.

## II. Installation

**384.10 Uses Permitted.** The use of strut-type channel raceways shall be permitted in the following:

(1) Where exposed.
(2) In dry locations.
(3) In locations subject to corrosive vapors where protected by finishes judged suitable for the condition.
(4) Where the voltage is 600 volts or less.
(5) As power poles.
(6) In Class I, Division 2 hazardous (classified) locations as permitted in 501.4(B)(3).
(7) As extensions of unbroken lengths through walls, partitions, and floors where closure strips are removable from either side and the portion within the wall, partition, or floor remains covered.
(8) Ferrous channel raceways and fittings protected from corrosion solely by enamel shall be permitted only indoors.

**384.12 Uses Not Permitted.** Strut type channel raceways shall not be used as follows:

(1) Where concealed.
(2) Ferrous channel raceways and fittings protected from corrosion solely by enamel shall not be permitted where subject to severe corrosive influences.

**384.21 Size of Conductors.** No conductor larger than that for which the raceway is listed shall be installed in strut-type channel raceways.

**384.22 Number of Conductors.** The number of conductors permitted in strut-type channel raceways shall not ex-

**Table 384.22 Channel Size and Inside Diameter Area**

| Size Channel | Area | | 40% Area[*] | | 25% Area[**] | |
|---|---|---|---|---|---|---|
| | in.² | mm² | in.² | mm² | in.² | mm² |
| 1⅝ × ¹³⁄₁₆ | 0.887 | 572 | 0.355 | 229 | 0.222 | 143 |
| 1⅝ × 1 | 1.151 | 743 | 0.460 | 297 | 0.288 | 186 |
| 1⅝ × 1⅜ | 1.677 | 1076 | 0.671 | 433 | 0.419 | 270 |
| 1⅝ × 1⅝ | 2.028 | 1308 | 0.811 | 523 | 0.507 | 327 |
| 1⅝ × 2⁷⁄₁₆ | 3.169 | 2045 | 1.267 | 817 | 0.792 | 511 |
| 1⅝ × 3¼ | 4.308 | 2780 | 1.723 | 1112 | 1.077 | 695 |
| 1½ × ¾ | 0.849 | 548 | 0.340 | 219 | 0.212 | 137 |
| 1½ × 1½ | 1.828 | 1179 | 0.731 | 472 | 0.457 | 295 |
| 1½ × 1⅞ | 2.301 | 1485 | 0.920 | 594 | 0.575 | 371 |
| 1½ × 3 | 3.854 | 2487 | 1.542 | 995 | 0.964 | 622 |

[*]Raceways with external joiners shall use a 40 percent wire fill calculation to determine the number of conductors permitted.
[**]Raceways with internal joiners shall use a 25 percent wire fill calculation to determine the number of conductors permitted.

ceed the percentage fill using Table 384.22 and applicable outside diameter (O.D.) dimensions of specific types and sizes of wire given in the tables in Chapter 9.

The derating factors of 310.15(B)(2)(a) shall not apply to conductors installed in strut-type channel raceways where all of the following conditions are met:

(1) The cross-sectional area of the raceway exceeds 2500 mm² (4 in.²).
(2) The current-carrying conductors do not exceed 30 in number.
(3) The sum of the cross-sectional areas of all contained conductors does not exceed 20 percent of the interior cross-sectional area of the strut-type channel raceways. Formula for wire fill:

$$n = \frac{ca}{wa}$$

where:
$n$ = number of wires
$ca$ = channel area in square inches
$wa$ = wire area

**384.30 Securing and Supporting.**

**(A) Surface Mount.** A surface mount strut-type channel raceway shall be secured to the mounting surface with retention straps external to the channel at intervals not exceeding 3 m (10 ft) and within 900 mm (3 ft) of each outlet box, cabinet, junction box, or other channel raceway termination.

**(B) Suspension Mount.** Strut-type channel raceways shall be permitted to be suspension mounted in air with approved appropriate methods designed for the purpose at intervals not to exceed 3 m (10 ft) and within 900 mm (3 ft) of channel raceway terminations and ends.

**384.56 Splices and Taps.** Splices and taps shall be permitted in raceways that are accessible after installation by having a removable cover. The conductors, including splices and taps, shall not fill the raceway to more than 75 percent of its area at that point. All splices and taps shall be made by approved methods.

**384.60 Grounding.** Strut-type channel raceway enclosures providing a transition to or from other wiring methods shall have a means for connecting an equipment grounding conductor. Strut-type channel raceways shall be permitted as an equipment grounding conductor in accordance with 250.118(14). Where a snap-fit metal cover for strut-type channel raceways is used to achieve electrical continuity in accordance with the listing, this cover shall not be permitted as the means for providing electrical continuity for a receptacle mounted in the cover.

### III. Construction Specifications

**384.100 Construction.** Strut-type channel raceways and their accessories shall be of a construction that distinguishes them from other raceways. Raceways and their elbows, couplings, and other fittings shall be designed so that the sections can be electrically and mechanically coupled together and installed without subjecting the wires to abrasion. They shall comply with 384.100(A), (B), and (C).

**(A) Material.** Raceways and accessories shall be formed of steel, stainless steel, or aluminum.

**(B) Corrosion Protection.** Steel raceways and accessories shall be protected against corrosion by galvanizing or an organic coating.

> FPN: Enamel and PVC coatings are examples of organic coatings that provide corrosion protection.

**(C) Cover.** Covers of strut-type channel raceway shall be either metallic or nonmetallic.

**384.120 Marking.** Each length of strut-type channel raceways shall be clearly and durably identified as required in the first sentence of 110.21.

## ARTICLE 386
## Surface Metal Raceways

### I. General

**386.1 Scope.** This article covers the use, installation, and construction specifications for surface metal raceways and associated fittings.

**386.2 Definition.**

**Surface Metal Raceway.** A metallic raceway that is intended to be mounted to the surface of a structure, with associated couplings, connectors, boxes, and fittings for the installation of electrical conductors.

**386.6 Listing Requirements.** Surface metal raceway and associated fittings shall be listed.

### II. Installation

**386.10 Uses Permitted.** The use of surface metal raceways shall be permitted in the following:

(1) In dry locations.
(2) In Class I, Division 2 hazardous (classified) locations as permitted in 501.4(B)(3).
(3) Under raised floors, as permitted in 645.5(D)(2).
(4) Extension through walls and floors. Surface metal raceway shall be permitted to pass transversely through dry walls, dry partitions, and dry floors if the length passing through is unbroken. Access to the conductors shall be maintained on both sides of the wall, partition, or floor.

**386.12 Uses Not Permitted.** Surface metal raceways shall not be used in the following:

(1) Where subject to severe physical damage, unless otherwise approved
(2) Where the voltage is 300 volts or more between conductors, unless the metal has a thickness of not less than 1.02 mm (0.040 in.) nominal
(3) Where subject to corrosive vapors
(4) In hoistways
(5) Where concealed, except as permitted in 386.10(4)

**386.21 Size of Conductors.** No conductor larger than that for which the raceway is designed shall be installed in surface metal raceway.

**386.22 Number of Conductors or Cables.** The number of conductors or cables installed in surface metal raceway shall not be greater than the number for which the raceway is designed. Cables shall be permitted to be installed where such use is permitted by the respective cable articles.

The derating factors of 310.15(B)(2)(a) shall not apply to conductors installed in surface metal raceways where all of the following conditions are met:

(1) The cross-sectional area of the raceway exceeds 2500 mm$^2$ (4 in.$^2$)
(2) The current-carrying conductors do not exceed 30 in number

(3) The sum of the cross-sectional areas of all contained conductors does not exceed 20 percent of the interior cross-sectional area of the surface metal raceway

**386.56 Splices and Taps.** Splices and taps shall be permitted in surface metal raceways having a removable cover that is accessible after installation. The conductors, including splices and taps, shall not fill the raceway to more than 75 percent of its area at that point. Splices and taps in surface metal raceways without removable covers shall be made only in junction boxes. All splices and taps shall be made by approved methods.

Taps of Type FC cable installed in surface metal raceway shall be made in accordance with 322.56(B).

**386.60 Grounding.** Surface metal raceway enclosures providing a transition from other wiring methods shall have a means for connecting an equipment grounding conductor.

**386.70 Combination Raceways.** When combination surface metal raceways are used both for signaling and for lighting and power circuits, the different systems shall be run in separate compartments identified by sharply contrasting colors of the interior finish, and the same relative position of compartments shall be maintained throughout the premises.

### III. Construction Specifications

**386.100 Construction.** Surface metal raceways shall be of such construction as will distinguish them from other raceways. Surface metal raceways and their elbows, couplings, and similar fittings shall be designed so that the sections can be electrically and mechanically coupled together and installed without subjecting the wires to abrasion.

Where covers and accessories of nonmetallic materials are used on surface metal raceways, they shall be identified for such use.

## ARTICLE 388
## Surface Nonmetallic Raceways

### I. General

**388.1 Scope.** This article covers the use, installation, and construction specifications for surface nonmetallic raceways and associated fittings.

**388.2 Definition.**

**Surface Nonmetallic Raceway.** A nonmetallic raceway that is intended to be mounted to the surface of a structure, with associated couplings, connectors, boxes, and fittings for the installation of electrical conductors.

**388.6 Listing Requirements.** Surface nonmetallic raceway and associated fittings shall be listed.

### II. Installation

**388.10 Uses Permitted.** Surface nonmetallic raceway shall be permitted as follows:

(1) The use of surface nonmetallic raceways shall be permitted in dry locations.
(2) Extension through walls and floors shall be permitted. Surface nonmetallic raceway shall be permitted to pass transversely through dry walls, dry partitions, and dry floors if the length passing through is unbroken. Access to the conductors shall be maintained on both sides of the wall, partition, or floor.

**388.12 Uses Not Permitted.** Surface nonmetallic raceways shall not be used in the following:

(1) Where concealed, except as permitted in 388.10(2)
(2) Where subject to severe physical damage
(3) Where the voltage is 300 volts or more between conductors, unless listed for higher voltage
(4) In hoistways
(5) In any hazardous (classified) location except Class I, Division 2 locations as permitted in 501.4(B)(3)
(6) Where subject to ambient temperatures exceeding those for which the nonmetallic raceway is listed
(7) For conductors whose insulation temperature limitations would exceed those for which the nonmetallic raceway is listed

**388.21 Size of Conductors.** No conductor larger than that for which the raceway is designed shall be installed in surface nonmetallic raceway.

**388.22 Number of Conductors or Cables.** The number of conductors or cables installed in surface nonmetallic raceway shall not be greater than the number for which the raceway is designed. Cables shall be permitted to be installed where such use is permitted by the respective cable articles.

**388.56 Splices and Taps.** Splices and taps shall be permitted in surface nonmetallic raceways having a removable cover that is accessible after installation. The conductors, including splices and taps, shall not fill the raceway to more than 75 percent of its area at that point. Splices and taps in surface nonmetallic raceways without removable covers shall be made only in junction boxes. All splices and taps shall be made by approved methods.

**388.60 Grounding.** Where equipment grounding is required by Article 250, a separate equipment grounding conductor shall be installed in the raceway.

**388.70 Combination Raceways.** When combination surface nonmetallic raceways are used both for signaling and for lighting and power circuits, the different systems shall be run in separate compartments identified by sharply contrasting colors of the interior finish.

### III. Construction Specifications

**388.100 Construction.** Surface nonmetallic raceways shall be of such construction as will distinguish them from other raceways. Surface nonmetallic raceways and their elbows, couplings, and similar fittings shall be designed so that the sections can be mechanically coupled together and installed without subjecting the wires to abrasion.

Surface nonmetallic raceways and fittings are made of suitable nonmetallic material that is resistant to moisture and chemical atmospheres. It shall also be flame retardant, resistant to impact and crushing, resistant to distortion from heat under conditions likely to be encountered in service, and resistant to low-temperature effects.

**388.120 Marking.** Surface nonmetallic raceways that have limited smoke-producing characteristics shall be permitted to be so identified.

## ARTICLE 390
## Underfloor Raceways

**390.1 Scope.** This article covers the use and installation requirements for underfloor raceways.

**390.2 Use.**

**(A) Permitted.** The installation of underfloor raceways shall be permitted beneath the surface of concrete or other flooring material or in office occupancies where laid flush with the concrete floor and covered with linoleum or equivalent floor covering.

**(B) Not Permitted.** Underfloor raceways shall not be installed (1) where subject to corrosive vapors or (2) in any hazardous (classified) locations, except as permitted by 504.20 and in Class I, Division 2 locations as permitted in 501.4(B)(3). Unless made of a material judged suitable for the condition or unless corrosion protection approved for the condition is provided, ferrous or nonferrous metal underfloor raceways, junction boxes, and fittings shall not be installed in concrete or in areas subject to severe corrosive influences.

**390.3 Covering.** Raceway coverings shall comply with 390.3(A) through (D).

**(A) Raceways Not Over 100 mm (4 in.) Wide.** Half-round and flat-top raceways not over 100 mm (4 in.) in width shall have not less than 20 mm (¾ in.) of concrete or wood above the raceway.

*Exception: As permitted in 390.3(C) and (D) for flat-top raceways.*

**(B) Raceways Over 100 mm (4 in.) Wide But Not Over 200 mm (8 in.) Wide.** Flat-top raceways over 100 mm (4 in.) but not over 200 mm (8 in.) wide with a minimum of 25 mm (1 in.) spacing between raceways shall be covered with concrete to a depth of not less than 25 mm (1 in.). Raceways spaced less than 25 mm (1 in.) apart shall be covered with concrete to a depth of 38 mm (1½ in.).

**(C) Trench-Type Raceways Flush with Concrete.** Trench-type flush raceways with removable covers shall be permitted to be laid flush with the floor surface. Such approved raceways shall be designed so that the cover plates provide adequate mechanical protection and rigidity equivalent to junction box covers.

**(D) Other Raceways Flush with Concrete.** In office occupancies, approved metal flat-top raceways, if not over 100 mm (4 in.) in width, shall be permitted to be laid flush with the concrete floor surface, provided they are covered with substantial linoleum that is not less than 1.6 mm (¹⁄₁₆ in.) thick or with equivalent floor covering. Where more than one and not more than three single raceways are each installed flush with the concrete, they shall be contiguous with each other and joined to form a rigid assembly.

**390.4 Size of Conductors.** No conductor larger than that for which the raceway is designed shall be installed in underfloor raceways.

**390.5 Maximum Number of Conductors in Raceway.** The combined cross-sectional area of all conductors or cables shall not exceed 40 percent of the interior cross-sectional area of the raceway.

**390.6 Splices and Taps.** Splices and taps shall be made only in junction boxes.

For the purposes of this section, so-called loop wiring (continuous, unbroken conductor connecting the individual outlets) shall not be considered to be a splice or tap.

*Exception: Splices and taps shall be permitted in trench-type flush raceway having a removable cover that is acces-*

*sible after installation. The conductors, including splices and taps, shall not fill more than 75 percent of the raceway area at that point.*

**390.7 Discontinued Outlets.** When an outlet is abandoned, discontinued, or removed, the sections of circuit conductors supplying the outlet shall be removed from the raceway. No splices or reinsulated conductors, such as would be the case with abandoned outlets on loop wiring, shall be allowed in raceways.

**390.8 Laid in Straight Lines.** Underfloor raceways shall be laid so that a straight line from the center of one junction box to the center of the next junction box coincides with the centerline of the raceway system. Raceways shall be firmly held in place to prevent disturbing this alignment during construction.

**390.9 Markers at Ends.** A suitable marker shall be installed at or near each end of each straight run of raceways to locate the last insert.

**390.10 Dead Ends.** Dead ends of raceways shall be closed.

**390.13 Junction Boxes.** Junction boxes shall be leveled to the floor grade and sealed to prevent the free entrance of water or concrete. Junction boxes used with metal raceways shall be metal and shall be electrically continuous with the raceways.

**390.14 Inserts.** Inserts shall be leveled and sealed to prevent the entrance of concrete. Inserts used with metal raceways shall be metal and shall be electrically continuous with the raceway. Inserts set in or on fiber raceways before the floor is laid shall be mechanically secured to the raceway. Inserts set in fiber raceways after the floor is laid shall be screwed into the raceway. When cutting through the raceway wall and setting inserts, chips and other dirt shall not be allowed to remain in the raceway, and tools shall be used that are designed so as to prevent the tool from entering the raceway and damaging conductors that may be in place.

**390.15 Connections to Cabinets and Wall Outlets.** Connections from underfloor raceways to distribution centers and wall outlets shall be made by approved fittings or by any of the wiring methods in Chapter 3, where installed in accordance with the provisions of the respective articles.

## ARTICLE 392
## Cable Trays

**392.1 Scope.** This article covers cable tray systems, including ladder, ventilated trough, ventilated channel, solid bottom, and other similar structures.

> FPN: For further information on cable trays, see NEMA–VE 1, 1998-*Metal Cable Tray Systems*; NEMA–VE 2-1996, *Metal Cable Tray Installation Guidelines*; and NEMA–FG-, 1998, *Nonmetallic Cable Tray Systems.*

**392.2 Definition.**

**Cable Tray System.** A unit or assembly of units or sections and associated fittings forming a structural system used to securely fasten or support cables and raceways.

**392.3 Uses Permitted.** Cable tray shall be permitted to be used as a support system for services, feeders, branch circuits, communications circuits, control circuits, and signaling circuits. Cable tray installations shall not be limited to industrial establishments. Where exposed to direct rays of the sun, insulated conductors and jacketed cables shall be identified as being sunlight resistant. Cable trays and their associated fittings shall be identified for the intended use.

**(A) Wiring Methods.** The wiring methods in Table 392.3(A) shall be permitted to be installed in cable tray systems under the conditions described in their respective articles and sections.

**(B) In Industrial Establishments.** The wiring methods in Table 392.3(A) shall be permitted to be used in any industrial establishment under the conditions described in their respective articles. In industrial establishments only, where conditions of maintenance and supervision ensure that only qualified persons service the installed cable tray system, any of the cables in 392.3(B)(1) and (2) shall be permitted to be installed in ladder, ventilated trough, solid bottom, or ventilated channel cable trays.

**(1) Single Conductors.** Single-conductor cables shall be permitted to be installed in accordance with the following:

(a) Single-conductor cable shall be 1/0 AWG or larger and shall be of a type listed and marked on the surface for use in cable trays. Where 1/0 AWG through 4/0 AWG single-conductor cables are installed in ladder cable tray, the maximum allowable rung spacing for the ladder cable tray shall be 230 mm (9 in.).

(b) Welding cables shall comply with the provisions of Article 630, Part IV.

**Table 392.3(A) Wiring Methods**

| Wiring Method | Article | Section |
|---|---|---|
| Armored cable | 320 | |
| Communication raceways | 800 | |
| Electrical metallic tubing | 358 | |
| Electrical nonmetallic tubing | 362 | |
| Fire alarm cables | 760 | |
| Flexible metal conduit | 348 | |
| Flexible metallic tubing | 360 | |
| Instrumentation tray cable | 727 | |
| Intermediate metal conduit | 342 | |
| Liquidtight flexible metal conduit | 350 | |
| Liquidtight flexible nonmetallic conduit | 356 | |
| Metal-clad cable | 330 | |
| Mineral-insulated, metal-sheathed cable | 332 | |
| Multiconductor service-entrance cable | 338 | |
| Multiconductor underground feeder and branch-circuit cable | 340 | |
| Multipurpose and communications cables | 800 | |
| Nonmetallic-sheathed cable | 334 | |
| Power and control tray cable | 336 | |
| Power-limited tray cable | | 725.61(C) and 725.71(F) |
| Optical fiber cables | 770 | |
| Optical fiber raceways | 770 | |
| Other factory-assembled, multiconductor control, signal, or power cables that are specifically approved for installation in cable trays | | |
| Rigid metal conduit | 344 | |
| Rigid nonmetallic conduit | 352 | |

(c) Single conductors used as equipment grounding conductors shall be insulated, covered, or bare, and they shall be 4 AWG or larger.

**(2) Medium Voltage.** Single- and multiconductor medium voltage cables shall be Type MV cable (Article 328). Single conductors shall be installed in accordance with 392.3(B)(1).

**(C) Equipment Grounding Conductors.** Metallic cable trays shall be permitted to be used as equipment grounding conductors where continuous maintenance and supervision ensure that qualified persons service the installed cable tray system and the cable tray complies with provisions of 392.7.

**(D) Hazardous (Classified) Locations.** Cable trays in hazardous (classified) locations shall contain only the cable types permitted in 501.4, 502.4, 503.3, 504.20, and 505.15.

**(E) Nonmetallic Cable Tray.** In addition to the uses permitted elsewhere in Article 392, nonmetallic cable tray

shall be permitted in corrosive areas and in areas requiring voltage isolation.

**392.4 Uses Not Permitted.** Cable tray systems shall not be used in hoistways or where subject to severe physical damage. Cable tray systems shall not be used in environmental airspaces, except as permitted in 300.22, to support wiring methods recognized for use in such spaces.

**392.5 Construction Specifications.**

**(A) Strength and Rigidity.** Cable trays shall have suitable strength and rigidity to provide adequate support for all contained wiring.

**(B) Smooth Edges.** Cable trays shall not have sharp edges, burrs, or projections that could damage the insulation or jackets of the wiring.

**(C) Corrosion Protection.** Cable tray systems shall be corrosion resistant. If made of ferrous material, the system shall be protected from corrosion as required by 300.6.

**(D) Side Rails.** Cable trays shall have side rails or equivalent structural members.

**(E) Fittings.** Cable trays shall include fittings or other suitable means for changes in direction and elevation of runs.

**(F) Nonmetallic Cable Tray.** Nonmetallic cable trays shall be made of flame-retardant material.

**392.6 Installation.**

**(A) Complete System.** Cable trays shall be installed as a complete system. Field bends or modifications shall be made so that the electrical continuity of the cable tray system and support for the cables is maintained. Cable tray systems shall be permitted to have mechanically discontinuous segments between cable tray runs or between cable tray runs and equipment. The system shall provide for the support of the cables in accordance with their corresponding articles.

Where cable trays support individual conductors and where the conductors pass from one cable tray to another, or from a cable tray to raceway(s) or from a cable tray to equipment where the conductors are terminated, the distance between cable trays or between the cable tray and the raceway(s) or the equipment shall not exceed 1.8 m (6 ft). The conductors shall be secured to the cable tray(s) at the transition, and they shall be protected, by guarding or by location, from physical damage.

A bonding jumper sized in accordance with 250.102 shall connect the two sections of cable tray, or the cable tray and the raceway or equipment. Bonding shall be in accordance with 250.96.

**(B) Completed Before Installation.** Each run of cable tray shall be completed before the installation of cables.

**(C) Supports.** Supports shall be provided to prevent stress on cables where they enter raceways or other enclosures from cable tray systems.

Cable trays shall be supported at intervals in accordance with the installation instructions.

**(D) Covers.** In portions of runs where additional protection is required, covers or enclosures providing the required protection shall be of a material that is compatible with the cable tray.

**(E) Multiconductor Cables Rated 600 Volts or Less.** Multiconductor cables rated 600 volts or less shall be permitted to be installed in the same cable tray.

**(F) Cables Rated Over 600 Volts.** Cables rated over 600 volts and those rated 600 volts or less installed in the same cable tray shall comply with either of the following:

(1) The cables rated over 600 volts are Type MC.
(2) The cables rated over 600 volts are separated from the cables rated 600 volts or less by a solid fixed barrier of a material compatible with the cable tray.

**(G) Through Partitions and Walls.** Cable trays shall be permitted to extend transversely through partitions and walls or vertically through platforms and floors in wet or dry locations where the installations, complete with installed cables, are made in accordance with the requirements of 300.21.

**(H) Exposed and Accessible.** Cable trays shall be exposed and accessible except as permitted by 392.6(G).

**(I) Adequate Access.** Sufficient space shall be provided and maintained about cable trays to permit adequate access for installing and maintaining the cables.

**(J) Raceways, Cables, Boxes, and Conduit Bodies Supported from Cable Tray Systems.** In industrial facilities where conditions of maintenance and supervision ensure that only qualified persons service the installation and where the cable tray systems are designed and installed to support the load, such systems shall be permitted to support raceways and cables, and boxes and conduit bodies covered in 314.1. For raceways terminating at the tray, a listed cable tray clamp or adapter shall be used to securely fasten the raceway to the cable tray system. Additional supporting and securing of the raceway shall be in accordance with the requirements of the appropriate raceway article.

For raceways or cables running parallel to and attached to the bottom or side of a cable tray system, fastening and supporting shall be in accordance with the requirements of the appropriate raceway or cable article.

For boxes and conduit bodies attached to the bottom or side of a cable tray system, fastening and supporting shall be in accordance with the requirements of 314.23.

## 392.7 Grounding.

**(A) Metallic Cable Trays.** Metallic cable trays that support electrical conductors shall be grounded as required for conductor enclosures in Article 250.

**(B) Steel or Aluminum Cable Tray Systems.** Steel or aluminum cable tray systems shall be permitted to be used as equipment grounding conductors, provided that all the following requirements are met:

(1) The cable tray sections and fittings shall be identified for grounding purposes.
(2) The minimum cross-sectional area of cable trays shall conform to the requirements in Table 392.7(B).
(3) All cable tray sections and fittings shall be legibly and durably marked to show the cross-sectional area of metal in channel cable trays, or cable trays of one-piece construction, and the total cross-sectional area of both side rails for ladder or trough cable trays.
(4) Cable tray sections, fittings, and connected raceways shall be bonded in accordance with 250.96 using bolted mechanical connectors or bonding jumpers sized and installed in accordance with 250.102.

## 392.8 Cable Installation.

**(A) Cable Splices.** Cable splices made and insulated by approved methods shall be permitted to be located within a cable tray, provided they are accessible and do not project above the side rails.

**(B) Fastened Securely.** In other than horizontal runs, the cables shall be fastened securely to transverse members of the cable trays.

**(C) Bushed Conduit and Tubing.** A box shall not be required where cables or conductors are installed in bushed conduit and tubing used for support or for protection against physical damage.

**(D) Connected in Parallel.** Where single conductor cables comprising each phase or neutral of a circuit are connected in parallel as permitted in 310.4, the conductors shall be installed in groups consisting of not more than one conductor per phase or neutral to prevent current unbalance in the paralleled conductors due to inductive reactance.

Single conductors shall be securely bound in circuit groups to prevent excessive movement due to fault-current magnetic forces unless single conductors are cabled together, such as triplexed assemblies.

**(E) Single Conductors.** Where any of the single conductors installed in ladder or ventilated trough cable trays are

**Table 392.7(B) Metal Area Requirements for Cable Trays Used as Equipment Grounding Conductor**

| Maximum Fuse Ampere Rating, Circuit Breaker Ampere Trip Setting, or Circuit Breaker Protective Relay Ampere Trip Setting for Ground-Fault Protection of Any Cable Circuit in the Cable Tray System | Minimum Cross-Sectional Area of Metal[a] | | | |
|---|---|---|---|---|
| | Steel Cable Trays | | Aluminum Cable Trays | |
| | mm² | in.² | mm² | in.² |
| 60 | 129 | 0.20 | 129 | 0.20 |
| 100 | 258 | 0.40 | 129 | 0.20 |
| 200 | 451.5 | 0.70 | 129 | 0.20 |
| 400 | 645 | 1.00 | 258 | 0.40 |
| 600 | 967.5 | 1.50[b] | 258 | 0.40 |
| 1000 | — | — | 387 | 0.60 |
| 1200 | — | — | 645 | 1.00 |
| 1600 | — | — | 967.5 | 1.50 |
| 2000 | — | — | 1290 | 2.00[b] |

[a]Total cross-sectional area of both side rails for ladder or trough cable trays; or the minimum cross-sectional area of metal in channel cable trays or cable trays of one-piece construction.

[b]Steel cable trays shall not be used as equipment grounding conductors for circuits with ground-fault protection above 600 amperes. Aluminum cable trays shall not be used as equipment grounding conductors for circuits with ground-fault protection above 2000 amperes.

1/0 through 4/0 AWG, all single conductors shall be installed in a single layer. Conductors that are bound together to comprise each circuit group shall be permitted to be installed in other than a single layer.

**392.9 Number of Multiconductor Cables, Rated 2000 Volts or Less, in Cable Trays.** The number of multiconductor cables, rated 2000 volts or less, permitted in a single cable tray shall not exceed the requirements of this section. The conductor sizes herein apply to both aluminum and copper conductors.

**(A) Any Mixture of Cables.** Where ladder or ventilated trough cable trays contain multiconductor power or lighting cables, or any mixture of multiconductor power, lighting, control, and signal cables, the maximum number of cables shall conform to the following:

(1) Where all of the cables are 4/0 AWG or larger, the sum of the diameters of all cables shall not exceed the cable tray width, and the cables shall be installed in a single layer.

(2) Where all of the cables are smaller than 4/0 AWG, the sum of the cross-sectional areas of all cables shall not exceed the maximum allowable cable fill area in Column 1 of Table 392.9 for the appropriate cable tray width.

(3) Where 4/0 AWG or larger cables are installed in the same cable tray with cables smaller than 4/0 AWG, the sum of the cross-sectional areas of all cables smaller than 4/0 AWG shall not exceed the maximum allowable fill area resulting from the computation in Column 2 of Table 392.9 for the appropriate cable tray width. The 4/0 AWG and larger cables shall be installed in a single layer, and no other cables shall be placed on them.

**(B) Multiconductor Control and/or Signal Cables Only.** Where a ladder or ventilated trough cable tray having a usable inside depth of 150 mm (6 in.) or less contains multiconductor control and/or signal cables only, the sum of the cross-sectional areas of all cables at any cross section shall not exceed 50 percent of the interior cross-sectional area of the cable tray. A depth of 150 mm (6 in.) shall be used to compute the allowable interior cross-sectional area of any cable tray that has a usable inside depth of more than 150 mm (6 in.).

**(C) Solid Bottom Cable Trays Containing Any Mixture.** Where solid bottom cable trays contain multiconductor power or lighting cables, or any mixture of multiconductor power, lighting, control, and signal cables, the maximum number of cables shall conform to the following:

(1) Where all of the cables are 4/0 AWG or larger, the sum of the diameters of all cables shall not exceed 90 percent of the cable tray width, and the cables shall be installed in a single layer.

(2) Where all of the cables are smaller than 4/0 AWG, the sum of the cross-sectional areas of all cables shall not exceed the maximum allowable cable fill area in Column 3 of Table 392.9 for the appropriate cable tray width.

(3) Where 4/0 AWG or larger cables are installed in the same cable tray with cables smaller than 4/0 AWG, the sum of the cross-sectional areas of all cables smaller than 4/0 AWG shall not exceed the maximum allowable fill area resulting from the computation in Column 4 of Table 392.9 for the appropriate cable tray width. The 4/0 AWG and larger cables shall be installed in a single layer, and no other cables shall be placed on them.

**(D) Solid Bottom Cable Tray — Multiconductor Control and/or Signal Cables Only.** Where a solid bottom cable tray having a usable inside depth of 150 mm (6 in.) or less contains multiconductor control and/or signal cables only, the sum of the cross-sectional areas of all cables at any cross section shall not exceed 40 percent of the interior cross-sectional area of the cable tray. A depth of 150 mm (6 in.) shall be used to compute the allowable interior cross-sectional area of any cable tray that has a usable inside depth of more than 150 mm (6 in.).

**Table 392.9 Allowable Cable Fill Area for Multiconductor Cables in Ladder, Ventilated Trough, or Solid Bottom Cable Trays for Cables Rated 2000 Volts or Less**

| Inside Width of Cable Tray | | Maximum Allowable Fill Area for Multiconductor Cables | | | | | | | |
| --- | --- | --- | --- | --- | --- | --- | --- | --- | --- |
| | | Ladder or Ventilated Trough Cable Trays, 392.9(A) | | | | Solid Bottom Cable Trays, 392.9(C) | | | |
| | | Column 1 Applicable for 392.9(A)(2) Only | | Column 2[a] Applicable for 392.9(A)(3) Only | | Column 3 Applicable for 392.9(C)(2) Only | | Column 4[a] Applicable for 392.9(C)(3) Only | |
| mm | in. | mm² | in.² | mm² | in.² | mm² | in.² | mm² | in.² |
| 150 | 6.0 | 4,500 | 7.0 | 4,500 – (1.2 Sd)[b] | 7 – (1.2 Sd)[b] | 3,500 | 5.5 | 3,500 – Sd[b] | 5.5 – Sdb |
| 225 | 9.0 | 6,800 | 10.5 | 6,800 – (1.2 Sd) | 10.5 – (1.2 Sd) | 5,100 | 8.0 | 5,100 – Sd | 8.0 – Sd |
| 300 | 12.0 | 9,000 | 14.0 | 9,000 – (1.2 Sd) | 14 – (1.2 Sd) | 7,100 | 11.0 | 7,100 – Sd | 11.0 – Sd |
| 450 | 18.0 | 13,500 | 21.0 | 13,500 – (1.2 Sd) | 21 – (1.2 Sd) | 10,600 | 16.5 | 10,600 – Sd | 16.5 – Sd |
| 600 | 24.0 | 18,000 | 28.0 | 18,000 – (1.2 Sd) | 28 – (1.2 Sd) | 14,200 | 22.0 | 14,200 – Sd | 22.0 – Sd |
| 750 | 30.0 | 22,500 | 35.0 | 22,500 – (1.2 Sd) | 35 – (1.2 Sd) | 17,700 | 27.5 | 17,700 – Sd | 27.5 – Sd |
| 900 | 36.0 | 27,000 | 42.0 | 27,000 – (1.2 Sd) | 42 – (1.2 Sd) | 21,300 | 33.0 | 21,300 – Sd | 33.0 – Sd |

[a]The maximum allowable fill areas in Columns 2 and 4 shall be computed. For example, the maximum allowable fill in mm² for a 150-mm wide cable tray in Column 2 shall be 4500 minus (1.2 multiplied by Sd) [the maximum allowable fill, in square inches, for a 6-in. wide cable tray in Column 2 shall be 7 minus (1.2 multiplied by Sd)].
[b]The term Sd in Columns 2 and 4 is equal to the sum of the diameters, in mm, of all cables 107.2 mm (in inches, of all 4/0 AWG) and larger multiconductor cables in the same cable tray with smaller cables.

**(E) Ventilated Channel Cable Trays.** Where ventilated channel cable trays contain multiconductor cables of any type, the following shall apply:

(1) Where only one multiconductor cable is installed, the cross-sectional area shall not exceed the value specified in Column 1 of Table 392.9(E).

(2) Where more than one multiconductor cable is installed, the sum of the cross-sectional area of all cables shall not exceed the value specified in Column 2 of Table 392.9(E).

**(F) Solid Channel Cable Trays.** Where solid channel cable trays contain multiconductor cables of any type, the following shall apply:

(1) Where only one multiconductor cable is installed, the cross-sectional area of the cable shall not exceed the value specified in Column 1 of Table 392.9(F).

(2) Where more than one multiconductor cable is installed, the sum of the cross-sectional area of all cable shall not exceed the value specified in Column 2 of Table 392.9(F).

**392.10 Number of Single-Conductor Cables, Rated 2000 Volts or Less, in Cable Trays.** The number of single-conductor cables, rated 2000 volts or less, permitted in a single cable tray section shall not exceed the requirements of this section. The single conductors, or conductor assemblies, shall be evenly distributed across the cable tray. The conductor sizes herein apply to both aluminum and copper conductors.

**Table 392.9(E) Allowable Cable Fill Area for Multiconductor Cables in Ventilated Channel Cable Trays for Cables Rated 2000 Volts or Less**

| Inside Width of Cable Tray | | Maximum Allowable Fill Area for Multiconductor Cables | | | |
| --- | --- | --- | --- | --- | --- |
| | | Column 1 One Cable | | Column 2 More Than One Cable | |
| mm | in. | mm² | in.² | mm² | in.² |
| 75 | 3 | 1500 | 2.3 | 850 | 1.3 |
| 100 | 4 | 2900 | 4.5 | 1600 | 2.5 |
| 150 | 6 | 4500 | 7.0 | 2450 | 3.8 |

**Table 392.9(F) Allowable Cable Fill Area for Multiconductor Cables in Solid Channel Cable Trays for Cables Rated 2000 Volts or Less**

| Inside Width of Cable Tray | | Column 1 One Cable | | Column 2 More Than One Cable | |
| --- | --- | --- | --- | --- | --- |
| mm | in. | mm² | in.² | mm² | in.² |
| 50 | 2 | 850 | 1.3 | 500 | 0.8 |
| 75 | 3 | 1300 | 2.0 | 700 | 1.1 |
| 100 | 4 | 2400 | 3.7 | 1400 | 2.1 |
| 150 | 6 | 3600 | 5.5 | 2100 | 3.2 |

**(A) Ladder or Ventilated Trough Cable Trays.** Where ladder or ventilated trough cable trays contain single-conductor cables, the maximum number of single conductors shall conform to the following:

(1) Where all of the cables are 1000 kcmil or larger, the sum of the diameters of all single-conductor cables shall not exceed the cable tray width.

(2) Where all of the cables are from 250 kcmil up to 1000 kcmil, the sum of the cross-sectional areas of all single-conductor cables shall not exceed the maximum allowable cable fill area in Column 1 of Table 392.10(A) for the appropriate cable tray width.

(3) Where 1000 kcmil or larger single-conductor cables are installed in the same cable tray with single-conductor cables smaller than 1000 kcmil, the sum of the cross-sectional areas of all cables smaller than 1000 kcmil shall not exceed the maximum allowable fill area resulting from the computation in Column 2 of Table 392.10(A) for the appropriate cable tray width.

(4) Where any of the single conductor cables are 1/0 through 4/0 AWG, the sum of the diameters of all single conductor cables shall not exceed the cable tray width.

**(B) Ventilated Channel Cable Trays.** Where 50 mm (2 in.), 75 mm (3 in.), 100 mm (4 in.), or 150 mm (6 in.) wide ventilated channel cable trays contain single-conductor cables, the sum of the diameters of all single conductors shall not exceed the inside width of the channel.

**392.11 Ampacity of Cables, Rated 2000 Volts or Less, in Cable Trays.**

**(A) Multiconductor Cables.** The allowable ampacity of multiconductor cables, nominally rated 2000 volts or less, installed according to the requirements of 392.9 shall be as given in Tables 310.16 and 310.18, subject to the provisions of (1), (2), (3), and 310.15(A)(2).

(1) The derating factors of 310.15(B)(2)(a) shall apply only to multiconductor cables with more than three current-carrying conductors. Derating shall be limited to the number of current-carrying conductors in the cable and not to the number of conductors in the cable tray.

(2) Where cable trays are continuously covered for more than 1.8 m (6 ft) with solid unventilated covers, not over 95 percent of the allowable ampacities of Tables 310.16 and 310.18 shall be permitted for multiconductor cables.

(3) Where multiconductor cables are installed in a single layer in uncovered trays, with a maintained spacing of not less than one cable diameter between cables, the ampacity shall not exceed the allowable ambient temperature-corrected ampacities of multiconductor cables, with not more than three insulated conductors rated 0 through 2000 volts in free air, in accordance with 310.15(C).

FPN: See Table B.310.3.

**(B) Single-Conductor Cables.** The allowable ampacity of single-conductor cables shall be as permitted by 310.15(A)(2). The derating factors of 310.15(B)(2)(a) shall not apply to the ampacity of cables in cable trays. The ampacity of single-conductor cables, or single conductors cabled together (triplexed, quadruplexed, etc.), nominally rated 2000 volts or less shall comply with the following:

(1) Where installed according to the requirements of 392.10, the ampacities for 600 kcmil and larger single-

**Table 392.10(A) Allowable Cable Fill Area for Single-Conductor Cables in Ladder or Ventilated Trough Cable Trays for Cables Rated 2000 Volts or Less**

| Inside Width of Cable Tray | | Maximum Allowable Fill Area for Single-Conductor Cables in Ladder or Ventilated Trough Cable Trays | | | |
|---|---|---|---|---|---|
| | | Column 1 Applicable for 392.10(A)(2) Only | | Column 2[a] Applicable for 392.10(A)(3) Only | |
| mm | in. | mm² | in.² | mm² | in.² |
| 150 | 6 | 4,200 | 6.5 | 4,200–(1.1Sd)[b] | 6.5–(1.1Sd)[b] |
| 225 | 9 | 6,100 | 9.5 | 6,100–(1.1 Sd) | 9.5–(1.1 Sd) |
| 300 | 12 | 8,400 | 13.0 | 8,400–(1.1 Sd) | 13.0–(1.1 Sd) |
| 450 | 18 | 12,600 | 19.5 | 12,600–(1.1 Sd) | 19.5–(1.1 Sd) |
| 600 | 24 | 16,800 | 26.0 | 16,800–(1.1 Sd) | 26.0–(1.1 Sd) |
| 750 | 30 | 21,000 | 32.5 | 21,000–(1.1 Sd) | 32.5–(1.1 Sd) |
| 900 | 36 | 25,200 | 39.0 | 25,200–(1.1 Sd) | 39.0–(1.1 Sd) |

[a]The maximum allowable fill areas in Column 2 shall be computed. For example, the maximum allowable fill, in mm² for a 150 mm wide cable tray in Column 2 shall be 4192.5 minus (1.1 multiplied by Sd) [the maximum allowable fill, in square inches, for a 6-in. wide cable tray in Column 2 shall be 6.5 minus (1.1 multiplied by Sd)].

[b]The term Sd in Column 2 is equal to the sum of the diameters, in mm, of all cables 507 mm² (in inches, of all 1000 kcmil) and larger single-conductor cables in the same ladder or ventilated trough cable tray with small cables.

conductor cables in uncovered cable trays shall not exceed 75 percent of the allowable ampacities in Tables 310.17 and 310.19. Where cable trays are continuously covered for more than 1.8 m (6 ft) with solid unventilated covers, the ampacities for 600 kcmil and larger cables shall not exceed 70 percent of the allowable ampacities in Tables 310.17 and 310.19.

(2) Where installed according to the requirements of 392.10, the ampacities for 1/0 AWG through 500 kcmil single-conductor cables in uncovered cable trays shall not exceed 65 percent of the allowable ampacities in Tables 310.17 and 310.19. Where cable trays are continuously covered for more than 1.8 m (6 ft) with solid unventilated covers, the ampacities for 1/0 AWG through 500 kcmil cables shall not exceed 60 percent of the allowable ampacities in Tables 310.17 and 310.19.

(3) Where single conductors are installed in a single layer in uncovered cable trays, with a maintained space of not less than one cable diameter between individual conductors, the ampacity of 1/0 AWG and larger cables shall not exceed the allowable ampacities in Tables 310.17 and 310.19.

(4) Where single conductors are installed in a triangular or square configuration in uncovered cable trays, with a maintained free airspace of not less than 2.15 times one conductor diameter (2.15 × O.D.) of the largest conductor contained within the configuration and adjacent conductor configurations or cables, the ampacity of 1/0 AWG and larger cables shall not exceed the allowable ampacities of two or three single insulated conductors rated 0 through 2000 volts supported on a messenger in accordance with 310.15(B).

FPN: See Table 310.20.

**392.12 Number of Type MV and Type MC Cables (2001 Volts or Over) in Cable Trays.** The number of cables rated 2001 volts or over permitted in a single cable tray shall not exceed the requirements of this section.

The sum of the diameters of single-conductor and multiconductor cables shall not exceed the cable tray width, and the cables shall be installed in a single layer. Where single conductor cables are triplexed, quadruplexed, or bound together in circuit groups, the sum of the diameters of the single conductors shall not exceed the cable tray width, and these groups shall be installed in single layer arrangement.

**392.13 Ampacity of Type MV and Type MC Cables (2001 Volts or Over) in Cable Trays.** The ampacity of cables, rated 2001 volts, nominal, or over, installed according to 392.12 shall not exceed the requirements of this section.

**(A) Multiconductor Cables (2001 Volts or Over).** The allowable ampacity of multiconductor cables shall be as given in Tables 310.75 and 310.76, subject to the following provisions:

(1) Where cable trays are continuously covered for more than 1.8 m (6 ft) with solid unventilated covers, not more than 95 percent of the allowable ampacities of Tables 310.75 and 310.76 shall be permitted for multiconductor cables.

(2) Where multiconductor cables are installed in a single layer in uncovered cable trays, with maintained spacing of not less than one cable diameter between cables, the ampacity shall not exceed the allowable ampacities of Tables 310.71 and 310.72.

**(B) Single-Conductor Cables (2001 Volts or Over).** The ampacity of single-conductor cables, or single conductors cabled together (triplexed, quadruplexed, etc.), shall comply with the following:

(1) The ampacities for 1/0 AWG and larger single-conductor cables in uncovered cable trays shall not exceed 75 percent of the allowable ampacities in Tables 310.69 and 310.70. Where the cable trays are covered for more than 1.8 m (6 ft) with solid unventilated covers, the ampacities for 1/0 AWG and larger single-conductor cables shall not exceed 70 percent of the allowable ampacities in Tables 310.69 and 310.70.

(2) Where single-conductor cables are installed in a single layer in uncovered cable trays, with a maintained space of not less than one cable diameter between individual conductors, the ampacity of 1/0 AWG and larger cables shall not exceed the allowable ampacities in Tables 310.69 and 310.70.

(3) Where single conductors are installed in a triangular or square configuration in uncovered cable trays, with a maintained free air space of not less than 2.15 times the diameter (2.15 × O.D.) of the largest conductor contained within the configuration and adjacent conductor configurations or cables, the ampacity of 1/0 AWG and larger cables shall not exceed the allowable ampacities in Tables 310.67 and 310.68.

# ARTICLE 394
# Concealed Knob-and-Tube Wiring

## I. General

**394.1 Scope.** This article covers the use, installation, and construction specifications of concealed knob-and-tube wiring.

## 394.2. Definition.

**Concealed Knob-and-Tube Wiring.** A wiring method using knobs, tubes, and flexible nonmetallic tubing for the protection and support of single insulated conductors.

## II. Installation

**394.10 Uses Permitted.** Concealed knob-and-tube wiring shall be permitted to be installed in the hollow spaces of walls and ceilings or in unfinished attics and roof spaces as provided in 394.23 only as follows:

(1) For extensions of existing installations
(2) Elsewhere by special permission

**394.12 Uses Not Permitted.** Concealed knob-and-tube wiring shall not be used in the following:

(1) Commercial garages
(2) Theaters and similar locations
(3) Motion picture studios
(4) Hazardous (classified) locations
(5) Hollow spaces of walls, ceilings, and attics where such spaces are insulated by loose, rolled, or foamed-in-place insulating material that envelops the conductors

**394.17 Through or Parallel to Framing Members.** Conductors shall comply with 398.17 where passing through holes in structural members. Where passing through wood cross members in plastered partitions, conductors shall be protected by noncombustible, nonabsorbent, insulating tubes extending not less than 75 mm (3 in.) beyond the wood member.

## 394.19 Clearances.

**(A) General.** A clearance of not less than 75 mm (3 in.) shall be maintained between conductors and a clearance of not less than 25 mm (1 in.) between the conductor and the surface over which it passes.

**(B) Limited Conductor Space.** Where space is too limited to provide these minimum clearances, such as at meters, panelboards, outlets, and switch points, the individual conductors shall be enclosed in flexible nonmetallic tubing, which shall be continuous in length between the last support and the enclosure or terminal point.

**(C) Clearance from Piping, Exposed Conductors, and So Forth.** Conductors shall comply with 398.19 for clearances from other exposed conductors, piping, and so forth.

**394.23 In Accessible Attics.** Conductors in unfinished attics and roof spaces shall comply with 394.23(A) or (B).

FPN: See 310.10 for temperature limitation of conductors.

**(A) Accessible by Stairway or Permanent Ladder.** Conductors shall be installed along the side of or through bored holes in floor joists, studs, or rafters. Where run through bored holes, conductors in the joists and in studs or rafters to a height of not less than 2.1 m (7 ft) above the floor or floor joists shall be protected by substantial running boards extending not less than 25 mm (1 in.) on each side of the conductors. Running boards shall be securely fastened in place. Running boards and guard strips shall not be required where conductors are installed along the sides of joists, studs, or rafters.

**(B) Not Accessible by Stairway or Permanent Ladder.** Conductors shall be installed along the sides of or through bored holes in floor joists, studs, or rafters.

*Exception: In buildings completed before the wiring is installed, attic and roof spaces that are not accessible by stairway or permanent ladder and have headroom at all points less than 900 mm (3 ft), the wiring shall be permitted to be installed on the edges of rafters or joists facing the attic or roof space.*

## 394.30 Securing and Supporting.

**(A) Supporting.** Conductors shall be rigidly supported on noncombustible, nonabsorbent insulating materials and shall not contact any other objects. Supports shall be installed as follows:

(1) Within 150 mm (6 in.) of each side of each tap or splice, and
(2) At intervals not exceeding 1.4 m (4½ ft).

Where it is impracticable to provide supports, conductors shall be permitted to be fished through hollow spaces in dry locations, provided each conductor is individually enclosed in flexible nonmetallic tubing that is in continuous lengths between supports, between boxes, or between a support and a box.

**(B) Securing.** Where solid knobs are used, conductors shall be securely tied thereto by tie wires having insulation equivalent to that of the conductor.

**394.42 Devices.** Switches shall comply with 404.4 and 404.10(B).

**394.56 Splices and Taps.** Splices shall be soldered unless approved splicing devices are used. In-line or strain splices shall not be used.

## III. Construction Specifications

**394.104 Conductors.** Conductors shall be of a type specified by Article 310.

# ARTICLE 396
## Messenger Supported Wiring

### I. General

**396.1 Scope.** This article covers the use, installation, and construction specifications for messenger supported wiring.

**396.2 Definition.**

**Messenger Supported Wiring.** An exposed wiring support system using a messenger wire to support insulated conductors by any one of the following:

(1) A messenger with rings and saddles for conductor support
(2) A messenger with a field-installed lashing material for conductor support
(3) Factory-assembled aerial cable
(4) Multiplex cables utilizing a bare conductor, factory assembled and twisted with one or more insulated conductors, such as duplex, triplex, or quadruplex type of construction

### II. Installation

**396.10 Uses Permitted.**

**(A) Cable Types.** The cable types in Table 396.10(A) shall be permitted to be installed in messenger supported wiring under the conditions described in the article or section referenced for each.

**Table 396.10(A) Cable Types**

| Cable Type | Section | Article |
|---|---|---|
| Metal-clad cable | | 330 |
| Mineral-insulated, metal-sheathed cable | | 332 |
| Multiconductor service-entrance cable | | 338 |
| Multiconductor underground feeder and branch-circuit cable | | 340 |
| Other factory-assembled, multiconductor control, signal, or power cables that are identified for the use | | |
| Power and control tray cable | | 336 |
| Power-limited tray cable | 725.61(C) and 725.71(E) | |

**(B) In Industrial Establishments.** In industrial establishments only, where conditions of maintenance and supervision ensure that only qualified persons service the installed messenger supported wiring, the following shall be permitted:

(1) Any of the conductor types shown in Table 310.13 or Table 310.62
(2) MV cable

Where exposed to weather, conductors shall be listed for use in wet locations. Where exposed to direct rays of the sun, conductors or cables shall be sunlight resistant.

**(C) Hazardous (Classified) Locations.** Messenger supported wiring shall be permitted to be used in hazardous (classified) locations where the contained cables are permitted for such use in 501.4, 502.4, 503.3, and 504.20.

**396.12 Uses Not Permitted.** Messenger supported wiring shall not be used in hoistways or where subject to physical damage.

**396.30 Messenger Support.** The messenger shall be supported at dead ends and at intermediate locations so as to eliminate tension on the conductors. The conductors shall not be permitted to come into contact with the messenger supports or any structural members, walls, or pipes.

**396.56 Conductor Splices and Taps.** Conductor splices and taps made and insulated by approved methods shall be permitted in messenger supported wiring.

**396.60 Grounding.** The messenger shall be grounded as required by 250.80 and 250.86 for enclosure grounding.

# ARTICLE 398
## Open Wiring on Insulators

### I. General

**398.1 Scope.** This article covers the use, installation, and construction specifications of open wiring on insulators.

**398.2 Definition.**

**Open Wiring on Insulators.** An exposed wiring method using cleats, knobs, tubes, and flexible tubing for the protection and support of single insulated conductors run in or on buildings.

### II. Installation

**398.10 Uses Permitted.** Open wiring on insulators shall be permitted only for industrial or agricultural establishments on systems of 600 volts, nominal, or less, as follows:

(1) Indoors or outdoors
(2) In wet or dry locations
(3) Where subject to corrosive vapors
(4) For services

**398.12 Uses Not Permitted.** Open wiring on insulators shall not be installed where concealed by the building structure.

**398.15 Exposed Work.**

**(A) Dry Locations.** In dry locations, where not exposed to severe physical damage, conductors shall be permitted to be separately enclosed in flexible nonmetallic tubing. The tubing shall be in continuous lengths not exceeding 4.5 m (15 ft) and secured to the surface by straps at intervals not exceeding 1.4 m (4½ ft).

**(B) Entering Spaces Subject to Dampness, Wetness, or Corrosive Vapors.** Conductors entering or leaving locations subject to dampness, wetness, or corrosive vapors shall have drip loops formed on them and shall then pass upward and inward from the outside of the buildings, or from the damp, wet, or corrosive location, through noncombustible, nonabsorbent insulating tubes.

> FPN: See 230.52 for individual conductors entering buildings or other structures.

**(C) Exposed to Physical Damage.** Conductors within 2.1 m (7 ft) from the floor shall be considered exposed to physical damage. Where open conductors cross ceiling joists and wall studs and are exposed to physical damage, they shall be protected by one of the following methods:

(1) Guard strips not less than 25 mm (1 in.) nominal in thickness and at least as high as the insulating supports, placed on each side of and close to the wiring.
(2) A substantial running board at least 13 mm (½ in.) thick in back of the conductors with side protections. Running boards shall extend at least 25 mm (1 in.) outside the conductors, but not more than 50 mm (2 in.), and the protecting sides shall be at least 50 mm (2 in.) high and at least 25 mm (1 in.) nominal in thickness.
(3) Boxing made in accordance with (C)(1) or (C)(2) and furnished with a cover kept at least 25 mm (1 in.) away from the conductors within. Where protecting vertical conductors on side walls, the boxing shall be closed at the top and the holes through which the conductors pass shall be bushed.
(4) Rigid metal conduit, intermediate metal conduit, rigid nonmetallic conduit, or electrical metallic tubing, in which case the rules of Articles 342, 344, 352, or 358 shall apply; or by metal piping, in which case the conductors shall be encased in continuous lengths of approved flexible tubing.

**398.17 Through or Parallel to Framing Members.** Open conductors shall be separated from contact with walls, floors, wood cross members, or partitions through which they pass by tubes or bushings of noncombustible, nonabsorbent insulating material. Where the bushing is shorter than the hole, a waterproof sleeve of noninductive material shall be inserted in the hole and an insulating bushing slipped into the sleeve at each end in such a manner as to keep the conductors absolutely out of contact with the sleeve. Each conductor shall be carried through a separate tube or sleeve.

> FPN: See 310.10 for temperature limitation of conductors.

**398.19 Clearances.** Open conductors shall be separated at least 50 mm (2 in.) from metal raceways, piping, or other conducting material, and from any exposed lighting, power, or signaling conductor, or shall be separated therefrom by a continuous and firmly fixed nonconductor in addition to the insulation of the conductor. Where any insulating tube is used, it shall be secured at the ends. Where practicable, conductors shall pass over rather than under any piping subject to leakage or accumulations of moisture.

**398.23 In Accessible Attics.** Conductors in unfinished attics and roof spaces shall comply with 398.23(A) or (B).

**(A) Accessible by Stairway or Permanent Ladder.** Conductors shall be installed along the side of or through bored holes in floor joists, studs, or rafters. Where run through bored holes, conductors in the joists and in studs or rafters to a height of not less than 2.1 m (7 ft) above the floor or floor joists shall be protected by substantial running boards extending not less than 25 mm (1 in.) on each side of the conductors. Running boards shall be securely fastened in place. Running boards and guard strips shall not be required for conductors installed along the sides of joists, studs, or rafters.

**(B) Not Accessible by Stairway or Permanent Ladder.** Conductors shall be installed along the sides of or through bored holes in floor joists, studs, or rafters.

*Exception: In buildings completed before the wiring is installed, attic and roof spaces that are not accessible by stairway or permanent ladder and have headroom at all points less than 900 mm (3 ft), the wiring shall be permitted to be installed on the edges of rafters or joists facing the attic or roof space.*

**398.30 Securing and Supporting.**

**(A) Conductor Sizes Smaller Than 8 AWG.** Conductors smaller than 8 AWG shall be rigidly supported on noncombustible, nonabsorbent insulating materials and shall not contact any other objects. Supports shall be installed as follows:

(1) Within 150 mm (6 in.) from a tap or splice
(2) Within 300 mm (12 in.) of a dead-end connection to a lampholder or receptacle
(3) At intervals not exceeding 1.4 m (4½ ft) and at closer intervals sufficient to provide adequate support where likely to be disturbed

**(B) Conductor Sizes 8 AWG and Larger.** Supports for conductors 8 AWG or larger installed across open spaces shall be permitted up to 4.5 m (15 ft) apart if noncombustible, nonabsorbent insulating spacers are used at least every 1.4 m (4½ ft) to maintain at least 65 mm (2½ in.) between conductors.

Where not likely to be disturbed in buildings of mill construction, 8 AWG and larger conductors shall be permitted to be run across open spaces if supported from each wood cross member on approved insulators maintaining 150 mm (6 in.) between conductors.

**(C) Industrial Establishments.** In industrial establishments only, where conditions of maintenance and supervision ensure that only qualified persons service the system,

conductors of sizes 250 kcmil and larger shall be permitted to be run across open spaces where supported at intervals up to 9.0 m (30 ft) apart.

**(D) Mounting of Conductor Supports.** Where nails are used to mount knobs, they shall not be smaller than ten-penny. Where screws are used to mount knobs, or where nails or screws are used to mount cleats, they shall be of a length sufficient to penetrate the wood to a depth equal to at least one-half the height of the knob and the full thickness of the cleat. Cushion washers shall be used with nails.

**(E) Tie Wires.** 8 AWG or larger conductors supported on solid knobs shall be securely tied thereto by tie wires having an insulation equivalent to that of the conductor.

**398.42 Devices.** Surface-type snap switches shall be mounted in accordance with 404.10(A), and boxes shall not be required. Other type switches shall be installed in accordance with 404.4.

### III. Construction Specifications

**398.104 Conductors.** Conductors shall be of a type specified by Article 310.

# Chapter 4 Equipment for General Use

## ARTICLE 400
## Flexible Cords and Cables

### I. General

**400.1 Scope.** This article covers general requirements, applications, and construction specifications for flexible cords and flexible cables.

**400.2 Other Articles.** Flexible cords and flexible cables shall comply with this article and with the applicable provisions of other articles of this *Code*.

**400.3 Suitability.** Flexible cords and cables and their associated fittings shall be suitable for the conditions of use and location.

**400.4 Types.** Flexible cords and flexible cables shall conform to the description in Table 400.4. Types of flexible cords and flexible cables other than those listed in the table shall be the subject of special investigation.

**Table 400.4 Flexible Cords and Cables (See 400.4.)**

| Trade Name | Type Letter | Voltage | AWG or kcmil | Number of Conductors | Insulation | Nominal Insulation Thickness[1] | | | Braid on Each Conductor | Outer Covering | Use | | |
|---|---|---|---|---|---|---|---|---|---|---|---|---|---|
| | | | | | | AWG or kcmil | mm | mils | | | | | |
| Lamp cord | C | 300 600 | 18–16 14–10 | 2 or more | Thermoset or thermoplastic | 18–16 14–10 | 0.76 1.14 | 30 45 | Cotton | None | Pendant or portable | Dry locations | Not hard usage |
| Elevator cable | E See Note 5. See Note 9. See Note 10. | 300 or 600 | 20–2 | 2 or more | Thermoset | 20–16 14–12 12–10 8–2 | 0.51 0.76 1.14 1.52 | 20 30 45 60 | Cotton | Three cotton, Outer one flame-retardant & moisture-resistant. See Note 3. | Elevator lighting and control | Unclassified locations | |
| | | | | | | 20–16 14–12 12–10 8–2 | 0.51 0.76 1.14 1.52 | 20 30 45 60 | Flexible nylon jacket | | | | |
| Elevator cable | EO See Note 5. See Note 10. | 300 or 600 | 20–2 | 2 or more | Thermoset | 20–16 14–12 12–10 8–2 | 0.51 0.76 1.14 1.52 | 20 30 45 60 | Cotton | Outer one Three cotton, flame-retardant & moisture-resistant. See Note 3. | Elevator lighting and control | Unclassified locations | |
| | | | | | | | | | | One cotton and a neoprene jacket. See Note 3. | | Hazardous (classified) locations | |
| Elevator cable | ET See Note 5. See Note 10. | 300 or 600 | 20–2 | 2 or more | Thermoplastic | 20–16 14–12 12–10 8–2 | 0.51 0.76 1.14 1.52 | 20 30 45 60 | Rayon | Three cotton or equivalent. Outer one flame-retardant & moisture-resistant. See Note 3. | Unclassified locations | | |
| | ETLB See Note 5. See Note 10. | 300 or 600 | | | | | | | None | | | | |
| | ETP See Note 5. See Note 10. | 300 or 600 | | | | | | | Rayon | Thermoplastic | Hazardous (classified) locations | | |
| | ETT See Note 5. See Note 10. | 300 or 600 | | | | | | | None | One cotton or equivalent and a thermoplastic jacket | | | |
| Portable power cable | G | 2000 | 12–500 | 2–6 plus grounding conductor(s) | Thermoset | 12–2 1–4/0 250–500 | 1.52 2.03 2.41 | 60 80 95 | | Oil-resistant thermoset | Portable and extra hard usage | | |

**Table 400.4** *Continued*

| Trade Name | Type Letter | Voltage | AWG or kcmil | Number of Conductors | Insulation | Nominal Insulation Thickness[1] AWG or kcmil | mm | mils | Braid on Each Conductor | Outer Covering | Use | | |
|---|---|---|---|---|---|---|---|---|---|---|---|---|---|
| | G-GC | 2000 | 12–500 | 3–6 plus grounding conductors and 1 ground check conductor | Thermoset | 12–2 1–4/0 250–500 | 1.52 2.03 2.41 | 60 80 95 | | Oil-resistant thermoset | | | |
| Heater cord | HPD | 300 | 18–12 | 2, 3, or 4 | Thermoset | 18–16 14–12 | 0.38 0.76 | 15 30 | None | Cotton or rayon | Portable heaters | Dry locations | Not hard usage |
| Parallel heater cord | HPN See Note 6. | 300 | 18–12 | 2 or 3 | Oil-resistant thermoset | 18–16 14–12 | 1.14 1.52 2.41 | 45 60 95 | None | Oil-resistant thermoset | Portable | Damp locations | Not hard usage |
| Thermoset jacketed heater cords | HSJ | 300 | 18–12 | 2, 3, or 4 | Thermoset | 18–16 | 0.76 | 30 | None | Cotton and Thermoset | Portable or portable heater | Damp locations | Hard usage |
| | HSJO | 300 | 18–12 | | Oil-resistant thermoset | 14–12 | 1.14 | 45 | | Cotton and oil-resistant thermoset | | | |
| | HSJOO | 300 | 18–12 | | | | | | | | | | |
| Non-integral parallel cords | NISP-1 See Note 6. | 300 | 20–18 | 2 or 3 | Thermoset | 20–18 | 0.38 | 15 | None | Thermoset | Pendant or portable | Damp locations | Not hard usage |
| | NISP-2 See Note 6. | 300 | 18–16 | | | 18–16 | 0.76 | 30 | | | | | |
| | NISPE-1 See Note 6. | 300 | 20–18 | | Thermoplastic elastomer | 20–18 | 0.38 | 15 | | Thermoplastic elastomer | | | |
| | NISPE-2 See Note 6. | 300 | 18–16 | | | 18–16 | 0.76 | 30 | | | | | |
| | NISPT-1 See Note 6. | 300 | 20–18 | | Thermoplastic | 20–18 | 0.38 | 15 | | Thermoplastic | | | |
| | NISPT-2 See Note 6. | 300 | 18–16 | | | 18–16 | 0.76 | 30 | | | | | |
| Twisted portable cord | PD | 300 600 | 18–16 14–10 | 2 or more | Thermoset or thermoplastic | 18–16 14–10 | 0.76 1.14 | 30 45 | Cotton | Cotton or rayon | Pendant or portable | Dry locations | Not hard usage |
| Portable power cable | PPE | 2000 | 12–500 | 1–6 plus optional grounding conductor(s) | Thermoplastic elastomer | 12–2 1–4/0 250–500 | 1.52 2.03 2.41 | 60 80 95 | | Oil-resistant thermoplastic elastomer | Portable, extra hard usage | | |
| Hard service cord | S See Note 4. | 600 | 18–12 | 2 or more | Thermoset | 18–16 14–10 8–2 | 0.76 1.14 1.52 | 30 45 60 | None | Thermoset | Pendant or portable | Damp locations | Extra hard usage |
| Flexible stage and lighting power cable | SC | 600 | 8–250 | 1 or more | | 8–2 1–4/0 250 | 1.52 2.03 2.41 | 60 80 95 | | Thermoset[2] | Portable, extra hard usage | | |
| | SCE | 600 | | | Thermoplastic elastomer | | | | | Thermoplastic elastomer[2] | | | |
| | SCT | 600 | | | Thermoplastic | | | | | Thermoplastic[2] | | | |
| Hard service cord | SE See Note 4. | 600 | 18–2 | 2 or more | Thermoplastic elastomer | 18–16 14–10 8–2 | 0.76 1.14 1.52 | 30 45 60 | None | Thermoplastic elastomer | Pendant or portable | Damp locations | Extra hard usage |
| | SEW See Note 4. See Note 13. | 600 | | | | | | | | | | Damp and wet locations | |

**Table 400.4** *Continued*

| Trade Name | Type Letter | Voltage | AWG or kcmil | Number of Conductors | Insulation | Nominal Insulation Thickness[1] | | | Braid on Each Conductor | Outer Covering | | Use | |
|---|---|---|---|---|---|---|---|---|---|---|---|---|---|
| | | | | | | AWG or kcmil | mm | mils | | | | | |
| | SEO See Note 4. | 600 | | | | | | | | Oil-resistant thermoplastic elastomer | | Damp locations | |
| | SEOW See Note 4. See Note 13. | 600 | | | | | | | | | | Damp and wet locations | |
| | SEOO See Note 4. | 600 | | | Oil-resistant thermoplastic elastomer | | | | | | | Damp locations | |
| | SEOOW See Note 4. See Note 13. | 600 | | | | | | | | | | Damp and wet locations | |
| Junior hard service cord | SJ | 300 | 18–10 | 2–6 | Thermoset | 18–12 | 0.76 | 30 | None | Thermoset | Pendant or portable | Damp locations | Hard usage |
| | SJE | 300 | | | Thermoplastic elastomer | | | | | Thermoplastic elastomer | | | |
| | SJEW See Note 13. | 300 | | | | | | | | | | Damp and wet locations | |
| | SJEO | 300 | | | | | | | | Oil-resistant thermoplastic elastomer | | Damp locations | |
| | SJEOW See Note 13. | 300 | | | | | | | | | | Damp and wet locations | |
| | SJEOO | 300 | | | Oil-resistant thermoplastic elastomer | | | | | | | Damp locations | |
| | SJEOOW See Note 13. | 300 | | | | | | | | | | Damp and wet locations | |
| | SJO | 300 | | | Thermoset | | | | | Oil-resistant thermoset | | Damp locations | |
| | SJOW See Note 13. | 300 | | | | | | | | | | Damp and wet locations | |
| | SJOO | 300 | | | Oil-resistant thermoset | | | | | | | Damp locations | |
| | SJOOW See Note 13. | 300 | | | | | | | | | | Damp and wet locations | |
| | SJT | 300 | | | Thermoplastic | 10 | 1.14 | 45 | | Thermoplastic | | Damp locations | |
| | SJTW See Note 13. | 300 | | | | | | | | | | Damp and wet locations | |
| | SJTO | 300 | | | Thermoplastic | 18–12 | 0.76 | 30 | | Oil-resistant thermoplastic | | Damp locations | |
| | SJTOW See Note 13. | 300 | | | | | | | | | | Damp and wet locations | |
| | SJTOO | 300 | | | Oil-resistant thermoplastic | | | | | | | Damp locations | |
| | SJTOOW See Note 13. | 300 | | | | | | | | | | Damp and wet locations | |

Table 400.4  *Continued*

| Trade Name | Type Letter | Voltage | AWG or kcmil | Number of Conductors | Insulation | Nominal Insulation Thickness[1] | | | Braid on Each Conductor | Outer Covering | Use | | |
|---|---|---|---|---|---|---|---|---|---|---|---|---|---|
| | | | | | | AWG or kcmil | mm | mils | | | | | |
| Hard service cord | SO See Note 4. | 600 | 18–2 | 2 or more | Thermoset | 18–16 | 0.76 | 30 | | Oil-resistant thermoset | Pendant or portable | Damp locations | Extra hard usage |
| | SOW See Note 4. See Note 13. | 600 | | | | | | | | | | Damp and wet locations | |
| | SOO See Note 4. | 600 | | | Oil-resistant thermoset | 14–10 8–2 | 1.14 1.52 | 45 60 | | | | Damp locations | |
| | SOOW See Note 4. See Note 13. | 600 | | | | | | | | | | Damp and wet locations | |
| All thermoset parallel cord | SP-1 See Note 6. | 300 | 20–18 | 2 or 3 | Thermoset | 20–18 | 0.76 | 30 | None | None | Pendant or portable | Damp locations | Not hard usage |
| | SP-2 See Note 6. | 300 | 18–16 | | | 18-16 | 1.14 | 45 | | | | | |
| | SP-3 See Note 6. | 300 | 18–10 | | | 18–16 14 12 10 | 1.52 2.03 2.41 2.80 | 60 80 95 110 | | | Refrigerators, room air conditioners, and as permitted in 422.16(B) | | |
| All elastomer (thermoplastic) parallel cord | SPE-1 See Note 6. | 300 | 20-18 | 2 or 3 | Thermoplastic elastomer | 20–18 | 0.76 | 30 | None | None | Pendant or portable | Damp locations | Not Hard usage |
| | SPE-2 See Note 6. | 300 | 18–16 | | | 18–16 | 1.14 | 45 | | | | | |
| | SPE-3 See Note 6. | 300 | 18–10 | | | 18–16 14 12 10 | 1.52 2.03 2.41 2.80 | 60 80 95 110 | | | Refrigerators, room air conditioners, and as permitted in 422.16(B) | | |
| All plastic parallel cord | SPT-1 See Note 6. | 300 | 20–18 | 2 or 3 | Thermoplastic | 20–18 | 0.76 | 30 | None | None | Pendant or portable | Damp locations | Not hard usage |
| | SPT-1W See Note 6. See Note 13. | 300 | | | | | | | | | | Damp and wet locations | |
| | SPT-2 See Note 6. | 300 | 18–16 | | | 18–16 | 1.14 | 45 | | | | Damp locations | |
| | SPT-2W See Note 6. See Note 13. | 300 | | | | | | | | | | Damp and wet locations | |
| | SPT-3 See Note 6. | 300 | 18–10 | | | 18–16 14 12 10 | 1.52 2.03 2.41 2.80 | 60 80 95 110 | | | Refrigerators, room air conditioners, and as permitted in 422.16(B) | Damp locations | Not hard usage |
| Range, dryer cable | SRD | 300 | 10–4 | 3 or 4 | Thermoset | 10–4 | 1.14 | 45 | None | Thermoset | Portable | Damp locations | Ranges, dryers |
| | SRDE | 300 | 10–4 | 3 or 4 | Thermoplastic elastomer | | | | None | Thermoplastic elastomer | | | |
| | SRDT | 300 | 10–4 | 3 or 4 | Thermoplastic | | | | None | Thermoplastic | | | |

Table 400.4 *Continued*

| Trade Name | Type Letter | Voltage | AWG or kcmil | Number of Conductors | Insulation | Nominal Insulation Thickness[1] | | | Braid on Each Conductor | Outer Covering | Use | | |
|---|---|---|---|---|---|---|---|---|---|---|---|---|---|
| | | | | | | AWG or kcmil | mm | mils | | | | | |
| Hard service cord | ST See Note 4. | 600 | 18–2 | 2 or more | Thermoplastic | 18–16 14–10 8–2 | 0.76 1.14 1.52 | 30 45 60 | None | Thermoplastic | Pendant or portable | Damp locations | Extra hard usage |
| | STW See Note 4. See Note 13. | 600 | | | | | | | | | | Damp and wet locations | |
| | STO See Note 4. | 600 | | | | | | | | Oil-resistant thermoplastic | | Damp locations | |
| | STOW See Note 4. See Note 13. | 600 | | | | | | | | | | Damp and wet locations | |
| | STOO See Note 4. | 600 | | | | Oil-resistant thermoplastic | | | | | | Damp locations | |
| | STOOW See Note 4. See Note 13. | 600 | | | | | | | | | | Damp and wet locations | |
| Vacuum cleaner cord | SV See Note 6. | 300 | 18–16 | 2 or 3 | Thermoset | 18–16 | 0.38 | 15 | None | Thermoset | Pendant or portable | Damp locations | Not hard usage |
| | SVE See Note 6. | 300 | | | Thermoplastic elastomer | | | | | Thermoplastic elastomer | | | |
| | SVEO See Note 6. | 300 | | | | | | | | Oil-resistant thermoplastic elastomer | | | |
| | SVEOO See Note 6. | 300 | | | Oil-resistant thermoplas tic elastomer | | | | | | | | |
| | SVO | 300 | | | Thermoset | | | | | Oil-resistant thermoset | | | |
| | SVOO | 300 | | | Oil-resistant thermoset | | | | | Oil-resistant thermoset | | | |
| | SVT See Note 6. | 300 | | | Thermoplastic | | | | | Thermoplastic | | | |
| | SVTO See Note 6. | 300 | | | Thermoplastic | | | | | Oil-resistant thermoplastic | | | |
| | SVTOO | 300 | | | Oil-resistant thermoplastic | | | | | | | | |
| Parallel tinsel cord | TPT See Note 2. | 300 | 27 | 2 | Thermoplastic | 27 | 0.76 | 30 | None | Thermoplastic | Attached to an appliance | Damp locations | Not hard usage |
| Jacketed tinsel cord | TST See Note 2. | 300 | 27 | 2 | Thermoplastic | 27 | 0.38 | 15 | None | Thermoplastic | Attached to an appliance | Damp locations | Not Hard Usage |
| Portable power-cable | W | 2000 | 12–500 501–1000 | 1–6 1 | Thermoset | 12–2 1–4/0 250–500 501–1000 | 1.52 2.03 2.41 2.80 | 60 80 95 110 | | Oil-resistant thermoset | Portable, extra hard usage | | |
| Electric vehicle cable | EV | 600 | 18–500 See Note 11. | 2 or more plus grounding conductor(s), plus optional hybrid data, signal communications, and optical fiber cables | Thermoset with optional nylon See Note 12. | 18–16 14–10 8–2 1–4/0 250–500 | 0.76 (0.51) 1.14 (0.76) 1.52 (1.14) 2.03 (1.52) 2.41 (1.90) | 30 (20) 45 (30) 60 (45) 80 (60) 95 (75) See Note 12. | Optional | Thermoset | Electric vehicle charging | Wet locations | Extra hard usage |
| | EVJ | 300 | 18-12 See Note 11. | | | 18–12 | 0.76 (0.51) | 30 (20) See Note 12. | | | | | Hard usage |

**Table 400.4**  *Continued*

| Trade Name | Type Letter | Voltage | AWG or kcmil | Number of Conductors | Insulation | Nominal Insulation Thickness[1] | | | Braid on Each Conductor | Outer Covering | Use | | |
|---|---|---|---|---|---|---|---|---|---|---|---|---|---|
| | | | | | | AWG or kcmil | mm | mils | | | | | |
| | EVE | 600 | 18–500 See Note 11. | 2 or more plus grounding conductor(s), plus optional hybrid data, signal communications, and optical fiber cables | Thermoplastic elastomer with optional nylon See Note 12. | 18–16 14–10 8–2 1–4/0 250–500 | 0.76 (0.51) 1.14 (0.76) 1.52 (1.14) 2.03 (1.52) 2.41 (1.90) | 30 (20) 45 (30) 60 (45) 80 (60) 95 (75) See Note 12. | | Thermoplastic elastomer | | | Extra hard usage |
| | EVJE | 300 | 18–12 See Note 11. | | | 18–12 | 0.76 (0.51) | 30 (20) See Note 12. | | | | | Hard usage |
| | EVT | 600 | 18–500 See Note 11. | 2 or more plus grounding conductor(s), plus optional hybrid data, signal communications, and optical fiber cables | Thermoplastic with optional nylon See Note 12. | 18–16 14–10 8–2 1–4/0 250–500 | 0.76 (0.51) 1.14 (0.76) 1.52 (1.14) 2.03 (1.52) 2.41 (1.90) | 30 (20) 45 (30) 60 (45) 80 (60) 95 (75) See Note 12. | Optional | Thermoplastic | Electric vehicle charging | Wet Locations | Extra hard usage |
| | EVJT | 300 | 18–12 See Note 11. | | | 18–12 | 0.76 (0.51) | 30 (20) See Note 12. | | | | | Hard usage |

*See Note 8.

**The required outer covering on some single conductor cables may be integral with the insulation.

Notes:

1. All types listed in Table 400.4 shall have individual conductors twisted together except for Types HPN, SP-1, SP-2, SP-3, SPE-1, SPE-2, SPE-3, SPT-1, SPT-2, SPT-3, TPT, NISP-1, NISP-2, NISPT-1, NISPT-2, NISPE-1, NISPE-2, and three-conductor parallel versions of SRD, SRDE, and SRDT.

2. Types TPT and TST shall be permitted in lengths not exceeding 2.5 m (8 ft) where attached directly, or by means of a special type of plug, to a portable appliance rated at 50 watts or less and of such nature that extreme flexibility of the cord is essential.

3. Rubber-filled or varnished cambric tapes shall be permitted as a substitute for the inner braids.

4. Types G, G-GC, S, SC, SCE, SCT, SE, SEO, SEOO, SO, SOO, ST, STO, STOO, PPE, and W shall be permitted for use on theater stages, in garages, and elsewhere where flexible cords are permitted by this *Code*.

5. Elevator traveling cables for operating control and signal circuits shall contain nonmetallic fillers as necessary to maintain concentricity. Cables shall have steel supporting members as required for suspension by 620.41. In locations subject to excessive moisture or corrosive vapors or gases, supporting members of other materials shall be permitted. Where steel supporting members are used, they shall run straight through the center of the cable assembly and shall not be cabled with the copper strands of any conductor.

In addition to conductors used for control and signaling circuits, Types E, EO, ET, ETLB, ETP, and ETT elevator cables shall be permitted to incorporate in the construction, one or more 20 AWG telephone conductor pairs, one or more coaxial cables, or one or more optical fibers. The 20 AWG conductor pairs shall be permitted to be covered with suitable shielding for telephone, audio, or higher frequency communications circuits; the coaxial cables consist of a center conductor, insulation, and shield for use in video or other radio frequency communications circuits. The optical fiber shall be suitably covered with flame-retardant thermoplastic of thickness not less than specified for the other conductors of the particular type of cable. Metallic shields shall have their own protective covering. Where used, these components shall be permitted to be incorporated in any layer of the cable assembly but shall not run straight through the center.

6. The third conductor in these cables shall be used for equipment grounding purpose only. The insulation of the grounding conductor for Types SPE-1, SPE-2, SPE-3, SPT-1, SPT-2, SPT-3, NISPT-1, NISPT-2, NISPE-1, and NISPE-2 shall be permitted to be thermoset polymer.

7. The individual conductors of all cords, except those of heat-resistant cords, shall have a thermoset or thermoplastic insulation, except that the equipment grounding conductor where used shall be in accordance with 400.23(B).

8. Where the voltage between any two conductors exceeds 300, but does not exceed 600, flexible cord of 10 AWG and smaller shall have thermoset or thermoplastic insulation on the individual conductors at least 1.14 mm (45 mils) in thickness, unless Type S, SE, SEO, SEOO, SO, SOO, ST, STO, or STOO cord is used.

9. Insulations and outer coverings that meet the requirements as flame retardant, limited smoke, and are so listed, shall be permitted to be marked for limited smoke after the code type designation.

10. Elevator cables in sizes 20 AWG through 14 AWG are rated 300 volts, and sizes 10 through 2 are rated 600 volts. 12 AWG is rated 300 volts with a 0.76-mm (30-mil) insulation thickness and 600 volts with a 1.14-mm (45-mil) insulation thickness.

11. Conductor size for Types EV, EVJ, EVE, EVJE, EVT, and EVJT cables apply to nonpower-limited circuits only. Conductors of power-limited (data, signal, or communications) circuits may extend beyond the stated AWG size range. All conductors shall be insulated for the same cable voltage rating.

12. Insulation thickness for Types EV, EVJ, EVEJE, EVT, and EVJT cables of nylon construction is indicated in parentheses.

13. Cords that comply with the requirements for outdoor cords and are so listed shall be permitted to be designated as weather and water resistant with the suffix "W" after the code type designation. Cords with the "W" suffix are suitable for use in wet locations.

**400.5 Ampacities for Flexible Cords and Cables.** Table 400.5(A) provides the allowable ampacities, and Table 400.5(B) provides the ampacities for flexible cords and cables with not more than three current-carrying conductors. These tables shall be used in conjunction with applicable end-use product standards to ensure selection of the proper size and type. If the number of current-carrying conductors exceeds three, the allowable ampacity or the ampacity of each conductor shall be reduced from the 3-conductor rating as shown in Table 400.5.

**Table 400.5 Adjustment Factors for More Than Three Current-Carrying Conductors in a Flexible Cord or Cable**

| Number of Conductors | Percent of Value in Tables 400.5(A) and 400.5(B) |
|---|---|
| 4 – 6 | 80 |
| 7 – 9 | 70 |
| 10 – 20 | 50 |
| 21 – 30 | 45 |
| 31 – 40 | 40 |
| 41 and above | 35 |

**Ultimate Insulation Temperature.** In no case shall conductors be associated together in such a way with respect to the kind of circuit, the wiring method used, or the number of conductors such that the limiting temperature of the conductors is exceeded.

A neutral conductor that carries only the unbalanced current from other conductors of the same circuit shall not be required to meet the requirements of a current-carrying conductor.

In a 3-wire circuit consisting of two phase wires and the neutral of a 4-wire, 3-phase, wye-connected system, a common conductor carries approximately the same current as the line-to-neutral currents of the other conductors and shall be considered to be a current-carrying conductor.

On a 4-wire, 3-phase, wye circuit where the major portion of the load consists of nonlinear loads, there are harmonic currents present in the neutral conductor and the neutral shall be considered to be a current-carrying conductor.

An equipment grounding conductor shall not be considered a current-carrying conductor.

Where a single conductor is used for both equipment grounding and to carry unbalanced current from other conductors, as provided for in 250.140 for electric ranges and electric clothes dryers, it shall not be considered as a current-carrying conductor.

*Exception: For other loading conditions, adjustment factors shall be permitted to be calculated under 310.15(C).*

FPN: See Annex B, Table B.310.11, for adjustment factors for more than three current-carrying conductors in a raceway or cable with load diversity.

**400.6 Markings.**

**(A) Standard Markings.** Flexible cords and cables shall be marked by means of a printed tag attached to the coil reel or carton. The tag shall contain the information required in 310.11(A). Types S, SC, SCE, SCT, SE, SEO,

**Table 400.5(A) Allowable Ampacity for Flexible Cords and Cables [Based on Ambient Temperature of 30°C (86°F). See 400.13 and Table 400.4.]**

| Size (AWG) | Thermoplastic Types TPT, TST | Thermoset Types C, E, EO, PD, S, SJ, SJO, SJOW, SJOO, SJOOW, SO, SOW, SOO, SOOW, SP-1, SP-2, SP-3, SRD, SV, SVO, SVOO / Thermoplastic Types ET, ETLB, ETP, ETT, SE, SEW, SEO, SEOW, SEOOW, SJE, SJEW, SJEO, SJEOW, SJEOOW, SJT, SJTW, SJTO, SJTOW, SJTOO, SJTOOW, SPE-1, SPE-2, SPE-3, SPT-1, SPT-1W, SPT-2, SPT-2W, SPT-3, ST, SRDE, SRDT, STO, STOW, STOO, STOOW, SVE, SVEO, SVT, SVTO, SVTOO | | Types HPD, HPN, HSJ, HSJO, HSJOO |
|---|---|---|---|---|
| | | A+ | B+ | |
| 27* | 0.5 | — | — | — |
| 20 | — | 5** | *** | — |
| 18 | — | 7 | 10 | 10 |
| 17 | — | — | 12 | 13 |
| 16 | — | 10 | 13 | 15 |
| 15 | — | — | — | 17 |
| 14 | — | 15 | 18 | 20 |
| 12 | — | 20 | 25 | 30 |
| 10 | — | 25 | 30 | 35 |
| 8 | — | 35 | 40 | — |
| 6 | — | 45 | 55 | — |
| 4 | — | 60 | 70 | — |
| 2 | — | 80 | 95 | — |

*Tinsel cord.
**Elevator cables only.
***7 amperes for elevator cables only; 2 amperes for other types.
+The allowable currents under subheading A apply to 3-conductor cords and other multiconductor cords connected to utilization equipment so that only 3 conductors are current-carrying. The allowable currents under subheading B apply to 2-conductor cords and other multiconductor cords connected to utilization equipment so that only 2 conductors are current carrying.

**Table 400.5(B) Ampacity of Cable Types SC, SCE, SCT, PPE, G, G-GC, and W. [Based on Ambient Temperature of 30°C (86°F). See Table 400.4.] Temperature Rating of Cable.**

| Size (AWG or kcmil) | 60°C (140°F) | | | 75°C (167°F) | | | 90°C (194°F) | | |
|---|---|---|---|---|---|---|---|---|---|
| | D[1] | E[2] | F[3] | D[1] | E[2] | F[3] | D[1] | E[2] | F[3] |
| 12 | — | 31 | 26 | — | 37 | 31 | — | 42 | 35 |
| 10 | — | 44 | 37 | — | 52 | 43 | — | 59 | 49 |
| 8 | 60 | 55 | 48 | 70 | 65 | 57 | 80 | 74 | 65 |
| 6 | 80 | 72 | 63 | 95 | 88 | 77 | 105 | 99 | 87 |
| 4 | 105 | 96 | 84 | 125 | 115 | 101 | 140 | 130 | 114 |
| 3 | 120 | 113 | 99 | 145 | 135 | 118 | 165 | 152 | 133 |
| 2 | 140 | 128 | 112 | 170 | 152 | 133 | 190 | 174 | 152 |
| 1 | 165 | 150 | 131 | 195 | 178 | 156 | 220 | 202 | 177 |
| 1/0 | 195 | 173 | 151 | 230 | 207 | 181 | 260 | 234 | 205 |
| 2/0 | 225 | 199 | 174 | 265 | 238 | 208 | 300 | 271 | 237 |
| 3/0 | 260 | 230 | 201 | 310 | 275 | 241 | 350 | 313 | 274 |
| 4/0 | 300 | 265 | 232 | 360 | 317 | 277 | 405 | 361 | 316 |
| 250 | 340 | 296 | 259 | 405 | 354 | 310 | 455 | 402 | 352 |
| 300 | 375 | 330 | 289 | 445 | 395 | 346 | 505 | 449 | 393 |
| 350 | 420 | 363 | 318 | 505 | 435 | 381 | 570 | 495 | 433 |
| 400 | 455 | 392 | 343 | 545 | 469 | 410 | 615 | 535 | 468 |
| 500 | 515 | 448 | 392 | 620 | 537 | 470 | 700 | 613 | 536 |
| 600 | 575 | — | — | 690 | — | — | 780 | — | — |
| 700 | 630 | — | — | 755 | — | — | 855 | — | — |
| 750 | 655 | — | — | 785 | — | — | 885 | — | — |
| 800 | 680 | — | — | 815 | — | — | 920 | — | — |
| 900 | 730 | — | — | 870 | — | — | 985 | — | — |
| 1000 | 780 | — | — | 935 | — | — | 1055 | — | — |

[1]The ampacities under subheading D shall be permitted for single-conductor Types SC, SCE, SCT, PPE, and W cable only where the individual conductors are not installed in raceways and are not in physical contact with each other except in lengths not to exceed 600 mm (24 in.) where passing through the wall of an enclosure.

[2]The ampacities under subheading E apply to two-conductor cables and other multiconductor cables connected to utilization equipment so that only two conductors are current carrying.

[3]The ampacities under subheading F apply to three-conductor cables and other multiconductor cables connected to utilization equipment so that only three conductors are current carrying.

SEOO, SJ, SJE, SJEO, SJEOO, SJO, SJT, SJTO, SJTOO, SO, SOO, ST, STO, STOO, SEW, SEOW, SEOOW, SJEW, SJEOW, SJEOOW, SJOW, SJTW, SJTOW, SJTOOW, SOW, SOOW, STW, STOW, and STOOW flexible cords and G, G-GC, PPE, and W flexible cables shall be durably marked on the surface at intervals not exceeding 610 mm (24 in.) with the type designation, size, and number of conductors.

**(B) Optional Markings.** Flexible cords and cable types listed in Table 400.4 shall be permitted to be surface marked to indicate special characteristics of the cable materials. These markings include, but are not limited to, markings for limited smoke, sunlight resistance, and so forth.

### 400.7 Uses Permitted.

**(A) Uses.** Flexible cords and cables shall be used only for the following:

(1) Pendants

(2) Wiring of luminaires (fixtures)

(3) Connection of portable lamps, portable and mobile signs, or appliances

(4) Elevator cables

(5) Wiring of cranes and hoists

(6) Connection of utilization equipment to facilitate frequent interchange

(7) Prevention of the transmission of noise or vibration

(8) Appliances where the fastening means and mechanical connections are specifically designed to permit ready removal for maintenance and repair, and the appliance is intended or identified for flexible cord connection

(9) Data processing cables as permitted by 645.5

(10) Connection of moving parts

(11) Temporary wiring as permitted in 527.4(B) and 527.4(C)

**(B) Attachment Plugs.** Where used as permitted in 400.7(A)(3), (A)(6), and (A)(8), each flexible cord shall be equipped with an attachment plug and shall be energized from a receptacle outlet.

*Exception: As permitted in 368.8.*

**400.8 Uses Not Permitted.** Unless specifically permitted in 400.7, flexible cords and cables shall not be used for the following:

(1) As a substitute for the fixed wiring of a structure
(2) Where run through holes in walls, structural ceilings, suspended ceilings, dropped ceilings, or floors
(3) Where run through doorways, windows, or similar openings
(4) Where attached to building surfaces

*Exception: Flexible cord and cable shall be permitted to be attached to building surfaces in accordance with the provisions of 368.8.*

(5) Where concealed by walls, floors, or ceilings or located above suspended or dropped ceilings
(6) Where installed in raceways, except as otherwise permitted in this *Code*

**400.9 Splices.** Flexible cord shall be used only in continuous lengths without splice or tap where initially installed in applications permitted by 400.7(A). The repair of hard-service cord and junior hard-service cord (see Trade Name column in Table 400.4) 14 AWG and larger shall be permitted if conductors are spliced in accordance with 110.14(B) and the completed splice retains the insulation, outer sheath properties, and usage characteristics of the cord being spliced.

**400.10 Pull at Joints and Terminals.** Flexible cords and cables shall be connected to devices and to fittings so that tension is not transmitted to joints or terminals.

*Exception: Listed portable single pole devices that are intended to accommodate such tension at their terminals shall be permitted to be used with single-conductor flexible cable.*

> FPN: Some methods of preventing pull on a cord from being transmitted to joints or terminals are knotting the cord, winding with tape, and fittings designed for the purpose.

**400.11 In Show Windows and Show Cases.** Flexible cords used in show windows and show cases shall be Type S, SE, SEO, SEOO, SJ, SJE, SJEO, SJEOO, SJO, SJOO, SJT, SJTO, SJTOO, SO, SOO, ST, STO, STOO, SEW, SEOW, SEOOW, SJEW, SJEOW, SJEOOW, SJOW, SJOOW, SJTW, SJTOW, SJTOOW, SOW, SOOW, STW, STOW, or STOOW.

*Exception No. 1: For the wiring of chain-supported luminaires (lighting fixtures).*

*Exception No. 2: As supply cords for portable lamps and other merchandise being displayed or exhibited.*

**400.12 Minimum Size.** The individual conductors of a flexible cord or cable shall not be smaller than the sizes in Table 400.4.

*Exception: The size of the insulated ground-check conductor of Type G-GC cables shall be not smaller than 10 AWG.*

**400.13 Overcurrent Protection.** Flexible cords not smaller than 18 AWG, and tinsel cords or cords having equivalent characteristics of smaller size approved for use with specific appliances, shall be considered as protected against overcurrent by the overcurrent devices described in 240.5.

**400.14 Protection from Damage.** Flexible cords and cables shall be protected by bushings or fittings where passing through holes in covers, outlet boxes, or similar enclosures.

## II. Construction Specifications

**400.20 Labels.** Flexible cords shall be examined and tested at the factory and labeled before shipment.

**400.21 Nominal Insulation Thickness.** The nominal thickness of insulation for conductors of flexible cords and cables shall not be less than specified in Table 400.4.

*Exception: The nominal insulation thickness for the ground-check conductors of Type G-GC cables shall not be less than 1.14 mm (45 mils) for 8 AWG and not less than 0.76 mm (30 mils) for 10 AWG.*

**400.22 Grounded-Conductor Identification.** One conductor of flexible cords that is intended to be used as a grounded circuit conductor shall have a continuous marker that readily distinguishes it from the other conductor or conductors. The identification shall consist of one of the methods indicated in 400.22(A) through (F).

**(A) Colored Braid.** A braid finished to show a white or gray color and the braid on the other conductor or conductors finished to show a readily distinguishable solid color or colors.

**(B) Tracer in Braid.** A tracer in a braid of any color contrasting with that of the braid and no tracer in the braid of the other conductor or conductors. No tracer shall be used in the braid of any conductor of a flexible cord that contains a conductor having a braid finished to show white or gray.

*Exception: In the case of Types C and PD and cords having the braids on the individual conductors finished to show white or gray. In such cords, the identifying marker shall be permitted to consist of the solid white or gray finish on one conductor, provided there is a colored tracer in the braid of each other conductor.*

**(C) Colored Insulation.** A white or gray insulation on one conductor and insulation of a readily distinguishable color or colors on the other conductor or conductors for cords having no braids on the individual conductors.

For jacketed cords furnished with appliances, one conductor having its insulation colored light blue, with the other conductors having their insulation of a readily distinguishable color other than white or gray.

*Exception: Cords that have insulation on the individual conductors integral with the jacket.*

The insulation shall be permitted to be covered with an outer finish to provide the desired color.

**(D) Colored Separator.** A white or gray separator on one conductor and a separator of a readily distinguishable solid color on the other conductor or conductors of cords having insulation on the individual conductors integral with the jacket.

**(E) Tinned Conductors.** One conductor having the individual strands tinned and the other conductor or conductors having the individual strands untinned for cords having insulation on the individual conductors integral with the jacket.

**(F) Surface Marking.** One or more stripes, ridges, or grooves located on the exterior of the cord so as to identify one conductor for cords having insulation on the individual conductors integral with the jacket.

**400.23 Equipment Grounding Conductor Identification.** A conductor intended to be used as an equipment grounding conductor shall have a continuous identifying marker readily distinguishing it from the other conductor or conductors. Conductors having a continuous green color or a continuous green color with one or more yellow stripes shall not be used for other than equipment grounding purposes. The identifying marker shall consist of one of the methods in 400.23(A) or (B).

**(A) Colored Braid.** A braid finished to show a continuous green color or a continuous green color with one or more yellow stripes.

**(B) Colored Insulation or Covering.** For cords having no braids on the individual conductors, an insulation of a con-

tinuous green color or a continuous green color with one or more yellow stripes.

**400.24 Attachment Plugs.** Where a flexible cord is provided with an equipment grounding conductor and equipped with an attachment plug, the attachment plug shall comply with 250.138(A) and (B).

### III. Portable Cables Over 600 Volts, Nominal

**400.30 Scope.** This part applies to multiconductor portable cables used to connect mobile equipment and machinery.

**400.31 Construction.**

**(A) Conductors.** The conductors shall be 8 AWG copper or larger and shall employ flexible stranding.

*Exception: The size of the insulated ground-check conductor of Type G-GC cables shall be not smaller than 10 AWG.*

**(B) Shields.** Cables operated at over 2000 volts shall be shielded. Shielding shall be for the purpose of confining the voltage stresses to the insulation.

**(C) Equipment Grounding Conductor(s).** An equipment grounding conductor(s) shall be provided. The total area shall not be less than that of the size of the equipment grounding conductor required in 250.122.

**400.32 Shielding.** All shields shall be grounded.

**400.33 Grounding.** Grounding conductors shall be connected in accordance with Part V of Article 250.

**400.34 Minimum Bending Radii.** The minimum bending radii for portable cables during installation and handling in service shall be adequate to prevent damage to the cable.

**400.35 Fittings.** Connectors used to connect lengths of cable in a run shall be of a type that lock firmly together. Provisions shall be made to prevent opening or closing these connectors while energized. Suitable means shall be used to eliminate tension at connectors and terminations.

**400.36 Splices and Terminations.** Portable cables shall not contain splices unless the splices are of the permanent molded, vulcanized types in accordance with 110.14(B). Terminations on portable cables rated over 600 volts, nominal, shall be accessible only to authorized and qualified personnel.

# ARTICLE 402
# Fixture Wires

**402.1 Scope.** This article covers general requirements and construction specifications for fixture wires.

**402.2 Other Articles.** Fixture wires shall comply with this article and also with the applicable provisions of other articles of this *Code*.

FPN: For application in luminaires (lighting fixtures), see Article 410.

**402.3 Types.** Fixture wires shall be of a type listed in Table 402.3, and they shall comply with all requirements of that table. The fixture wires listed in Table 402.3 are all suitable for service at 600 volts, nominal, unless otherwise specified.

FPN: Thermoplastic insulation may stiffen at temperatures colder than −10°C (+14°F), requiring that care be exercised during installation at such temperatures. Thermoplastic insulation may also be deformed at normal temperatures where subjected to pressure, requiring that care be exercised during installation and at points of support.

**Table 402.3 Fixture Wires**

| Name | Type Letter | Insulation | Thickness of Insulation | | | Outer Covering | Maximum Operating Temperature | Application Provisions |
| --- | --- | --- | --- | --- | --- | --- | --- | --- |
| | | | AWG | mm | mils | | | |
| Heat-resistant rubber-covered fixture wire — flexible stranding | FFH-2 | Heat-resistant rubber Cross-linked synthetic polymer | 18–16 18–16 | 0.76 0.76 | 30 30 | Nonmetallic covering | 75°C 167°F | Fixture wiring |
| ECTFE — solid or 7-strand | HF | Ethylene chlorotrifluoroethylene | 18–14 | 0.38 | 15 | None | 150°C 302°F | Fixture wiring |
| ECTFE — flexible stranding | HFF | Ethylene chlorotrifluoroethylene | 18–14 | 0.38 | 15 | None | 150°C 302°F | Fixture wiring |
| Tape insulated fixture wire — solid or 7-strand | KF-1 | Aromatic polyimide tape | 18–10 | 0.14 | 5.5 | None | 200°C 392°F | Fixture wiring — limited to 300 volts |
| | KF-2 | Aromatic polyimide tape | 18–10 | 0.21 | 8.4 | None | 200°C 392°F | Fixture wiring |
| Tape insulated fixture wire — flexible stranding | KFF-1 | Aromatic polyimide tape | 18–10 | 0.14 | 5.5 | None | 200°C 392°F | Fixture wiring — limited to 300 volts |
| | KFF-2 | Aromatic polyimide tape | 18–10 | 0.21 | 8.4 | None | 200°C 392°F | Fixture wiring |
| Perfluoro-alkoxy — solid or 7-strand (nickel or nickel-coated copper) | PAF | Perfluoro-alkoxy | 18–14 | 0.51 | 20 | None | 250°C 482°F | Fixture wiring (nickel or nickel-coated copper) |
| Perfluoro-alkoxy — flexible stranding | PAFF | Perfluoro-alkoxy | 18–14 | 0.51 | 20 | None | 150°C 302°F | Fixture wiring |

**Table 402.3** *Continued*

| Name | Type Letter | Insulation | Thickness of Insulation | | | Outer Covering | Maximum Operating Temperature | Application Provisions |
|------|-------------|------------|------|-----|------|----------------|-------------------------------|------------------------|
| | | | AWG | mm | mils | | | |
| Fluorinated ethylene propylene fixture wire — solid or 7-strand | PF | Fluorinated ethylene propylene | 18–14 | 0.51 | 20 | None | 200°C 392°F | Fixture wiring |
| Fluorinated ethylene propylene fixture wire — flexible stranding | PFF | Fluorinated ethylene propylene | 18–14 | 0.51 | 20 | None | 150°C 302°F | Fixture wiring |
| Fluorinated ethylene propylene fixture wire — solid or 7-strand | PGF | Fluorinated ethylene propylene | 18–14 | 0.36 | 14 | Glass braid | 200°C 392°F | Fixture wiring |
| Fluorinated ethylene propylene fixture wire — flexible stranding | PGFF | Fluorinated ethylene propylene | 18–14 | 0.36 | 14 | Glass braid | 150°C 302°F | Fixture wiring |
| Extruded polytetra-fluoroethylene — solid or 7-strand (nickel or nickel-coated copper) | PTF | Extruded polytetra-fluoroethyl-ene | 18–14 | 0.51 | 20 | None | 250°C 482°F | Fixture wiring (nickel or nickel-coated copper) |
| Extruded polytetra-fluoroethylene — flexible stranding 26-36 (AWG silver or nickel-coated copper) | PTFF | Extruded polytetra-fluoroethyl-ene | 18–14 | 0.51 | 20 | None | 150°C 302°F | Fixture wiring (silver or nickel-coated copper) |
| Heat-resistant rubber-covered fixture wire — solid or 7-strand | RFH-1 | Heat-resis-tant rubber | 18 | 0.38 | 15 | Nonmetallic covering | 75°C 167°F | Fixture wiring — limited to 300 volts |
| | RFH-2 | Heat-resis-tant rubber Cross-linked synthetic polymer | 18–16 | 0.76 | 30 | None or nonmetallic covering | 75°C 167°F | Fixture wiring |
| Heat-resistant cross-linked synthetic polymer-insulated fixture wire — solid or stranded | RFHH-2* | Cross-linked synthetic polymer | 18–16 | 0.76 | 30 | None or nonmetallic covering | 90°C 194°F | Fixture wiring — multi-conductor cable |
| | RFHH-3* | | 18–16 | 1.14 | 45 | | | |
| Silicone insulated fixture wire — solid or 7-strand | SF-1 | Silicone rubber | 18 | 0.38 | 15 | Nonmetallic covering | 200°C 392°F | Fixture wiring — limited to 300 volts |
| | SF-2 | Silicone rubber | 18–12 10 | 0.76 1.14 | 30 45 | Nonmetallic covering | 200°C 392°F | Fixture wiring |

**Table 402.3**  *Continued*

| Name | Type Letter | Insulation | Thickness of Insulation | | | Outer Covering | Maximum Operating Temperature | Application Provisions |
|---|---|---|---|---|---|---|---|---|
| | | | AWG | mm | mils | | | |
| Silicone insulated fixture wire — flexible stranding | SFF-1 | Silicone rubber | 18 | 0.38 | 15 | Nonmetallic covering | 150°C 302°F | Fixture wiring — limited to 300 volts |
| | SFF-2 | Silicone rubber | 18–12 10 | 0.76 1.14 | 30 45 | Nonmetallic covering | 150°C 302°F | Fixture wiring |
| Thermoplastic covered fixture wire — solid or 7-strand | TF* | Thermoplastic | 18–16 | 0.76 | 30 | None | 60°C 140°F | Fixture wiring |
| Thermoplastic covered fixture wire — flexible stranding | TFF* | Thermoplastic | 18–16 | 0.76 | 30 | None | 60°C 140°F | Fixture wiring |
| Heat-resistant thermoplastic covered fixture wire — solid or 7-strand | TFN* | Thermoplastic | 18–16 | 0.38 | 15 | Nylon-jacketed or equivalent | 90°C 194°F | Fixture wiring |
| Heat-resistant thermoplastic covered fixture wire — flexible stranded | TFFN* | Thermoplastic | 18–16 | 0.38 | 15 | Nylon-jacketed or equivalent | 90°C 194°F | Fixture wiring |
| Cross-linked polyolefin insulated fixture wire — solid or 7-strand | XF* | Cross-linked polyolefin | 18–14 12-10 | 0.76 1.14 | 30 45 | None | 150°C 302°F | Fixture wiring — limited to 300 volts |
| Cross-linked polyolefin insulated fixture wire — flexible stranded | XFF* | Cross-linked polyolefin | 18–14 12–10 | 0.76 1.14 | 30 45 | None | 150°C 302°F | Fixture wiring — limited to 300 volts |
| Modified ETFE — solid or 7-strand | ZF | Modified ethylene tetrafluoroethylene | 18–14 | 0.38 | 15 | None | 150°C 302°F | Fixture wiring |
| Flexible stranding | ZFF | Modified ethylene tetrafluoroethylene | 18–14 | 0.38 | 15 | None | 150°C 302°F | Fixture wiring |
| High temp. modified ETFE — solid or 7-strand | ZHF | Modified ethylene tetrafluoroethylene | 18–14 | 0.38 | 15 | None | 200°C 392°F | Fixture wiring |

*Insulations and outer coverings that meet the requirements of flame retardant, limited smoke, and are so listed shall be permitted to be marked for limited smoke after the *Code* type designation.

**402.5 Allowable Ampacities for Fixture Wires.** The allowable ampacity of fixture wire shall be as specified in Table 402.5.

**Table 402.5 Allowable Ampacity for Fixture Wires**

| Size (AWG) | Allowable Ampacity |
|:---:|:---:|
| 18 | 6 |
| 16 | 8 |
| 14 | 17 |
| 12 | 23 |
| 10 | 28 |

No conductor shall be used under such conditions that its operating temperature exceeds the temperature specified in Table 402.3 for the type of insulation involved.

FPN: See 310.10 for temperature limitation of conductors.

**402.6 Minimum Size.** Fixture wires shall not be smaller than 18 AWG.

**402.7 Number of Conductors in Conduit or Tubing.** The number of fixture wires permitted in a single conduit or tubing shall not exceed the percentage fill specified in Table 1, Chapter 9.

**402.8 Grounded Conductor Identification.** One conductor of fixture wires that is intended to be used as a grounded conductor shall be identified by means of stripes or by the means described in 400.22(A) through (E).

**402.9 Marking.**

**(A) Method of Marking.** Thermoplastic insulated fixture wire shall be durably marked on the surface at intervals not exceeding 610 mm (24 in.). All other fixture wire shall be marked by means of a printed tag attached to the coil, reel, or carton.

**(B) Optional Marking.** Fixture wire types listed in Table 402.3 shall be permitted to be surface marked to indicate special characteristics of the cable materials. These markings include, but are not limited to, markings for limited smoke, sunlight resistance, and so forth.

**402.10 Uses Permitted.** Fixture wires shall be permitted (1) for installation in luminaires (lighting fixtures) and in similar equipment where enclosed or protected and not subject to bending or twisting in use, or (2) for connecting luminaires (lighting fixtures) to the branch-circuit conductors supplying the luminaires (fixtures).

**402.11 Uses Not Permitted.** Fixture wires shall not be used as branch-circuit conductors.

**402.12 Overcurrent Protection.** Overcurrent protection for fixture wires shall be as specified in 240.5.

# ARTICLE 404
## Switches

### I. Installation

**404.1 Scope.** The provisions of this article shall apply to all switches, switching devices, and circuit breakers where used as switches.

**404.2 Switch Connections.**

**(A) Three-Way and Four-Way Switches.** Three-way and four-way switches shall be wired so that all switching is done only in the ungrounded circuit conductor. Where in metal raceways or metal-armored cables, wiring between switches and outlets shall be in accordance with 300.20(A).

*Exception: Switch loops shall not require a grounded conductor.*

**(B) Grounded Conductors.** Switches or circuit breakers shall not disconnect the grounded conductor of a circuit.

*Exception: A switch or circuit breaker shall be permitted to disconnect a grounded circuit conductor where all circuit conductors are disconnected simultaneously, or where the device is arranged so that the grounded conductor cannot be disconnected until all the ungrounded conductors of the circuit have been disconnected.*

**404.3 Enclosure.**

**(A) General.** Switches and circuit breakers shall be of the externally operable type mounted in an enclosure listed for the intended use. The minimum wire-bending space at terminals and minimum gutter space provided in switch enclosures shall be as required in 312.6.

*Exception No. 1: Pendant- and surface-type snap switches and knife switches mounted on an open-face switchboard or panelboard shall be permitted without enclosures.*

*Exception No. 2: Switches and circuit breakers installed in accordance with 110.27(A)(1), (2), (3), or (4) shall be permitted without enclosures.*

**(B) Used as a Raceway.** Enclosures shall not be used as junction boxes, auxiliary gutters, or raceways for conductors feeding through or tapping off to other switches or overcurrent devices, unless the enclosure complies with 312.8.

**404.4 Wet Locations.** A switch or circuit breaker in a wet location or outside of a building shall be enclosed in a weatherproof enclosure or cabinet that shall comply with 312.2(A). Switches shall not be installed within wet locations in tub or shower spaces unless installed as part of a listed tub or shower assembly.

**404.5 Time Switches, Flashers, and Similar Devices.** Time switches, flashers, and similar devices shall be of the enclosed type or shall be mounted in cabinets or boxes or equipment enclosures. Energized parts shall be barriered to prevent operator exposure when making manual adjustments or switching.

*Exception: Devices mounted so they are accessible only to qualified persons shall be permitted without barriers, provided they are located within an enclosure such that any energized parts within 152 mm (6.0 in.) of the manual adjustment or switch are covered by suitable barriers.*

**404.6 Position and Connection of Switches.**

**(A) Single-Throw Knife Switches.** Single-throw knife switches shall be placed so that gravity will not tend to close them. Single-throw knife switches, approved for use in the inverted position, shall be provided with a locking device that ensures that the blades remain in the open position when so set.

**(B) Double-Throw Knife Switches.** Double-throw knife switches shall be permitted to be mounted so that the throw is either vertical or horizontal. Where the throw is vertical, a locking device shall be provided to hold the blades in the open position when so set.

**(C) Connection of Switches.** Single-throw knife switches and switches with butt contacts shall be connected so that their blades are de-energized when the switch is in the open position. Bolted pressure contact switches shall have barriers that prevent inadvertent contact with energized blades. Single-throw knife switches, bolted pressure contact switches, molded case switches, switches with butt contacts, and circuit breakers used as switches shall be connected so that the terminals supplying the load are de-energized when the switch is in the open position.

*Exception: The blades and terminals supplying the load of a switch shall be permitted to be energized when the switch is in the open position where the switch is connected to circuits or equipment inherently capable of providing a backfeed source of power. For such installations, a permanent sign shall be installed on the switch enclosure or immediately adjacent to open switches with the following words or equivalent: WARNING — LOAD SIDE TERMINALS MAY BE ENERGIZED BY BACKFEED.*

**404.7 Indicating.** General-use and motor-circuit switches, circuit breakers, and molded case switches, where mounted in an enclosure as described in 404.3, shall clearly indicate whether they are in the open (off) or closed (on) position.

Where these switch or circuit breaker handles are operated vertically rather than rotationally or horizontally, the up position of the handle shall be the (on) position.

*Exception: Vertically operated double-throw switches shall be permitted to be in the closed (on) position with the handle in either the up or down position.*

**404.8 Accessibility and Grouping.**

**(A) Location.** All switches and circuit breakers used as switches shall be located so that they may be operated from a readily accessible place. They shall be installed so that the center of the grip of the operating handle of the switch or circuit breaker, when in its highest position, is not more than 2.0 m (6 ft 7 in.) above the floor or working platform.

*Exception No. 1: On busway installations, fused switches and circuit breakers shall be permitted to be located at the same level as the busway. Suitable means shall be provided to operate the handle of the device from the floor.*

*Exception No. 2: Switches and circuit breakers installed adjacent to motors, appliances, or other equipment that they supply shall be permitted to be located higher than specified in the foregoing and to be accessible by portable means.*

*Exception No. 3: Hookstick operable isolating switches shall be permitted at greater heights.*

**(B) Voltage Between Adjacent Devices.** A snap switch shall not be grouped or ganged in enclosures with other snap switches, receptacles, or similar devices, unless they are arranged so that the voltage between adjacent devices does not exceed 300 volts, or unless they are installed in enclosures equipped with permanently installed barriers between adjacent devices.

**404.9 Provisions for General-Use Snap Switches.**

**(A) Faceplates.** Faceplates provided for snap switches mounted in boxes and other enclosures shall be installed so as to completely cover the opening and, where the switch is flush mounted, seat against the finished surface.

**(B) Grounding.** Snap switches, including dimmer and similar control switches, shall be effectively grounded and shall provide a means to ground metal faceplates, whether or not a metal faceplate is installed. Snap switches shall be considered effectively grounded if either of the following conditions is met.

(1) The switch is mounted with metal screws to a metal box or to a nonmetallic box with integral means for grounding devices.
(2) An equipment grounding conductor or equipment bonding jumper is connected to an equipment grounding termination of the snap switch.

*Exception to (B): Where no grounding means exists within the snap-switch enclosure or where the wiring method does not include or provide an equipment ground, a snap switch without a grounding connection shall be permitted for replacement purposes only. A snap switch wired under the provisions of this exception and located within reach of earth, grade conducting floors, or other conducting surfaces shall be provided with a faceplate of nonconducting, noncombustible material.*

**(C) Construction.** Metal faceplates shall be of ferrous metal not less than 0.76 mm (0.030 in.) in thickness or of nonferrous metal not less than 1.02 mm (0.040 in.) in thickness. Faceplates of insulating material shall be noncombustible and not less than 2.54 mm (0.010 in.) in thickness, but they shall be permitted to be less than 2.54 mm (0.010 in.) in thickness if formed or reinforced to provide adequate mechanical strength.

**404.10 Mounting of Snap Switches.**

**(A) Surface-Type.** Snap switches used with open wiring on insulators shall be mounted on insulating material that separates the conductors at least 13 mm (½ in.) from the surface wired over.

**(B) Box Mounted.** Flush-type snap switches mounted in boxes that are set back of the wall surface as permitted in 314.20 shall be installed so that the extension plaster ears are seated against the surface of the wall. Flush-type snap switches mounted in boxes that are flush with the wall surface or project from it shall be installed so that the mounting yoke or strap of the switch is seated against the box.

**404.11 Circuit Breakers as Switches.** A hand-operable circuit breaker equipped with a lever or handle, or a power-operated circuit breaker capable of being opened by hand in the event of a power failure, shall be permitted to serve as a switch if it has the required number of poles.

FPN: See the provisions contained in 240.81 and 240.83.

**404.12 Grounding of Enclosures.** Metal enclosures for switches or circuit breakers shall be grounded as specified in Article 250. Where nonmetallic enclosures are used with metal raceways or metal-armored cables, provision shall be made for grounding continuity.

Except as covered in 404.9(B), Exception, nonmetallic boxes for switches shall be installed with a wiring method that provides or includes an equipment ground.

**404.13 Knife Switches.**

**(A) Isolating Switches.** Knife switches rated at over 1200 amperes at 250 volts or less, and at over 600 amperes at 251 to 600 volts, shall be used only as isolating switches and shall not be opened under load.

**(B) To Interrupt Currents.** To interrupt currents over 1200 amperes at 250 volts, nominal, or less, or over 600 amperes at 251 to 600 volts, nominal, a circuit breaker or a switch of special design listed for such purpose shall be used.

**(C) General-Use Switches.** Knife switches of ratings less than specified in 404.13(A) and (B) shall be considered general-use switches.

FPN: See definition of *General-Use Switch* in Article 100.

**(D) Motor-Circuit Switches.** Motor-circuit switches shall be permitted to be of the knife-switch type.

FPN: See definition of a *Motor-Circuit Switch* in Article 100.

**404.14 Rating and Use of Snap Switches.** Snap switches shall be used within their ratings and as indicated in 404.14(A) through (D).

FPN No. 1: For switches on signs and outline lighting, see 600.6.

FPN No. 2: For switches controlling motors, see 430.83, 430.109, and 430.110.

**(A) Alternating Current General-Use Snap Switch.** A form of general-use snap switch suitable only for use on ac circuits for controlling the following:

(1) Resistive and inductive loads, including electric-discharge lamps, not exceeding the ampere rating of the switch at the voltage involved
(2) Tungsten-filament lamp loads not exceeding the ampere rating of the switch at 120 volts
(3) Motor loads not exceeding 80 percent of the ampere rating of the switch at its rated voltage

**(B) Alternating-Current or Direct-Current General-Use Snap Switch.** A form of general-use snap switch suitable for use on either ac or dc circuits for controlling the following:

(1) Resistive loads not exceeding the ampere rating of the switch at the voltage applied.

(2) Inductive loads not exceeding 50 percent of the ampere rating of the switch at the applied voltage. Switches rated in horsepower are suitable for controlling motor loads within their rating at the voltage applied.

(3) Tungsten-filament lamp loads not exceeding the ampere rating of the switch at the applied voltage if T-rated.

**(C) CO/ALR Snap Switches.** Snap switches rated 20 amperes or less directly connected to aluminum conductors shall be listed and marked CO/ALR.

**(D) Alternating-Current Specific-Use Snap Switches Rated for 347 Volts.** Snap switches rated 347 volts ac shall be listed and shall be used only for controlling the following.

**(1) Noninductive Loads.** Noninductive loads other than tungsten-filament lamps not exceeding the ampere and voltage ratings of the switch.

**(2) Inductive Loads.** Inductive loads not exceeding the ampere and voltage ratings of the switch. Where particular load characteristics or limitations are specified as a condition of the listing, those restrictions shall be observed regardless of the ampere rating of the load.

The ampere rating of the switch shall not be less than 15 amperes at a voltage rating of 347 volts ac. Flush-type snap switches rated 347 volts ac shall not be readily interchangeable in box mounting with switches identified in 404.14(A) and (B).

**(E) Dimmer Switches.** General-use dimmer switches shall be used only to control permanently installed incandescent luminaires (lighting fixtures) unless listed for the control of other loads and installed accordingly.

## II. Construction Specifications

### 404.15 Marking.

**(A) Ratings.** Switches shall be marked with the current, voltage, and, if horsepower rated, the maximum rating for which they are designed.

**(B) Off Indication.** Where in the off position, a switching device with a marked OFF position shall completely disconnect all ungrounded conductors to the load it controls.

**404.16 600-Volt Knife Switches.** Auxiliary contacts of a renewable or quick-break type or the equivalent shall be provided on all knife switches rated 600 volts and designed for use in breaking current over 200 amperes.

**404.17 Fused Switches.** A fused switch shall not have fuses in parallel except as permitted in 240.8.

**404.18 Wire-Bending Space.** The wire-bending space required by 404.3 shall meet Table 312.6(B) spacings to the enclosure wall opposite the line and load terminals.

## ARTICLE 406
## Receptacles, Cord Connectors, and Attachment Plugs (Caps)

**406.1 Scope.** This article covers the rating, type, and installation of receptacles, cord connectors, and attachment plugs (cord caps).

### 406.2 Receptacle Rating and Type.

**(A) Receptacles.** Receptacles shall be listed for the purpose and marked with the manufacturer's name or identification and voltage and ampere ratings.

**(B) Rating.** Receptacles and cord connectors shall be rated not less than 15 amperes, 125 volts, or 15 amperes, 250 volts, and shall be of a type not suitable for use as lampholders.

> FPN: See 210.21(B) for receptacle ratings where installed on branch circuits.

**(C) Receptacles for Aluminum Conductors.** Receptacles rated 20 amperes or less and designed for the direct connection of aluminum conductors shall be marked CO/ALR.

**(D) Isolated Ground Receptacles.** Receptacles incorporating an isolated grounding connection intended for the reduction of electrical noise (electromagnetic interference) as permitted in 250.146(D) shall be identified by an orange triangle located on the face of the receptacle.

**(1)** Receptacles so identified shall be used only with grounding conductors that are isolated in accordance with 250.146(D).

**(2)** Isolated ground receptacles installed in nonmetallic boxes shall be covered with a nonmetallic faceplate.

*Exception: Where an isolated ground receptacle is installed in a nonmetallic box, a metal faceplate shall be permitted if the box contains a feature or accessory that permits the effective grounding of the faceplate.*

**406.3 General Installation Requirements.** Receptacle outlets shall be located in branch circuits in accordance with Part III of Article 210. General installation requirements shall be in accordance with 406.3(A) through (F).

**(A) Grounding Type.** Receptacles installed on 15- and 20-ampere branch circuits shall be of the grounding type. Grounding-type receptacles shall be installed only on circuits of the voltage class and current for which they are rated, except as provided in Tables 210.21(B)(2) and (B)(3).

*Exception: Nongrounding-type receptacles installed in accordance with 406.3(D).*

**(B) To Be Grounded.** Receptacles and cord connectors that have grounding contacts shall have those contacts effectively grounded.

*Exception No. 1: Receptacles mounted on portable and vehicle-mounted generators in accordance with 250.34.*

*Exception No. 2: Replacement receptacles as permitted by 406.3(D).*

**(C) Methods of Grounding.** The grounding contacts of receptacles and cord connectors shall be grounded by connection to the equipment grounding conductor of the circuit supplying the receptacle or cord connector.

> FPN: For installation requirements for the reduction of electrical noise, see 250.146(D).

The branch-circuit wiring method shall include or provide an equipment-grounding conductor to which the grounding contacts of the receptacle or cord connector shall be connected.

> FPN No. 1: 250.118 describes acceptable grounding means.

> FPN No. 2: For extensions of existing branch circuits, see 250.130.

**(D) Replacements.** Replacement of receptacles shall comply with 406.3(D)(1), (2), and (3) as applicable.

**(1) Grounding-Type Receptacles.** Where a grounding means exists in the receptacle enclosure or a grounding conductor is installed in accordance with 250.130(C), grounding-type receptacles shall be used and shall be connected to the grounding conductor in accordance with 406.3(C) or 250.130(C).

**(2) Ground-Fault Circuit Interrupters.** Ground-fault circuit-interrupter protected receptacles shall be provided where replacements are made at receptacle outlets that are required to be so protected elsewhere in this *Code*.

**(3) Nongrounding-Type Receptacles.** Where grounding means does not exist in the receptacle enclosure, the installation shall comply with (a), (b), or (c).

(a) A nongrounding-type receptacle(s) shall be permitted to be replaced with another nongrounding-type receptacle(s).

(b) A nongrounding-type receptacle(s) shall be permitted to be replaced with a ground-fault circuit interrupter-type of receptacle(s). These receptacles shall be marked "No Equipment Ground." An equipment grounding conductor shall not be connected from the ground-fault circuit-interrupter-type receptacle to any outlet supplied from the ground-fault circuit-interrupter receptacle.

(c) A nongrounding-type receptacle(s) shall be permitted to be replaced with a grounding-type receptacle(s) where supplied through a ground-fault circuit interrupter. Grounding-type receptacles supplied through the ground-fault circuit interrupter shall be marked "GFCI Protected" and "No Equipment Ground." An equipment grounding conductor shall not be connected between the grounding-type receptacles.

**(E) Cord-and-Plug-Connected Equipment.** The installation of grounding-type receptacles shall not be used as a requirement that all cord-and-plug-connected equipment be of the grounded type.

> FPN: See 250.114 for types of cord-and-plug-connected equipment to be grounded.

**(F) Noninterchangeable Types.** Receptacles connected to circuits that have different voltages, frequencies, or types of current (ac or dc) on the same premises shall be of such design that the attachment plugs used on these circuits are not interchangeable.

**406.4 Receptacle Mounting.** Receptacles shall be mounted in boxes or assemblies designed for the purpose, and such boxes or assemblies shall be securely fastened in place.

**(A) Boxes That Are Set Back.** Receptacles mounted in boxes that are set back of the wall surface, as permitted in 314.20, shall be installed so that the mounting yoke or strap of the receptacle is held rigidly at the surface of the wall.

**(B) Boxes That Are Flush.** Receptacles mounted in boxes that are flush with the wall surface or project therefrom shall be installed so that the mounting yoke or strap of the receptacle is held rigidly against the box or raised box cover.

**(C) Receptacles Mounted on Covers.** Receptacles mounted to and supported by a cover shall be held rigidly against the cover by more than one screw or shall be a device assembly or box cover listed and identified for securing by a single screw.

**(D) Position of Receptacle Faces.** After installation, receptacle faces shall be flush with or project from faceplates of insulating material and shall project a minimum of 0.4 mm (0.015 in.) from metal faceplates.

**(E) Receptacles in Countertops and Similar Work Surfaces in Dwelling Units.** Receptacles shall not be installed in a face-up position in countertops or similar work surfaces.

**(F) Exposed Terminals.** Receptacles shall be enclosed so that live wiring terminals are not exposed to contact.

**406.5 Receptacle Faceplates (Cover Plates).** Receptacle faceplates shall be installed so as to completely cover the opening and seat against the mounting surface.

**(A)** Metal faceplates shall be of ferrous metal not less than 0.76 mm (0.030 in.) in thickness or of nonferrous metal not less than 1.02 mm (0.040 in.) in thickness.

**(B)** Metal faceplates shall be grounded.

**(C)** Faceplates of insulating material shall be noncombustible and not less than 2.54 mm (0.10 in.) in thickness but shall be permitted to be less than 2.54 mm (0.10 in.) in thickness if formed or reinforced to provide adequate mechanical strength.

**406.6 Attachment Plugs.** All attachment plugs and cord connectors shall be listed for the purpose and marked with the manufacturer's name or identification and voltage and ampere ratings.

**(A)** Attachment plugs and cord connectors shall be constructed so that there are no exposed current-carrying parts except the prongs, blades, or pins. The cover for wire terminations shall be a part that is essential for the operation of an attachment plug or connector (dead-front construction).

**(B)** Attachment plugs shall be installed so that their prongs, blades, or pins are not energized unless inserted into an energized receptacle. No receptacle shall be installed so as to require an energized attachment plug as its source of supply.

**(C) Attachment Plug Ejector Mechanisms.** Attachment plug ejector mechanisms shall not adversely affect engagement of the blades of the attachment plug with the contacts of the receptacle.

**406.7 Noninterchangeability.** Receptacles, cord connectors, and attachment plugs shall be constructed so that receptacle or cord connectors do not accept an attachment plug with a different voltage or current rating from that for which the device is intended. However, a 20-ampere T-slot receptacle or cord connector shall be permitted to accept a 15-ampere attachment plug of the same voltage rating. Non–grounding-type receptacles and connectors shall not accept grounding-type attachment plugs.

**406.8 Receptacles in Damp or Wet Locations.**

**(A) Damp Locations.** A receptacle installed outdoors in a location protected from the weather or in other damp locations shall have an enclosure for the receptacle that is weatherproof when the receptacle is covered (attachment plug cap not inserted and receptacle covers closed).

An installation suitable for wet locations shall also be considered suitable for damp locations.

A receptacle shall be considered to be in a location protected from the weather where located under roofed open porches, canopies, marquees, and the like, and will not be subjected to a beating rain or water runoff.

**(B) Wet Locations.**

**(1) 15- and 20-Ampere Outdoor Receptacles.** 15- and 20-ampere, 125- and 250-volt receptacles installed outdoors in a wet location shall have an enclosure that is weatherproof whether or not the attachment plug cap is inserted.

**(2) Other Receptacles.** All other receptacles installed in a wet location shall comply with (a) or (b):

(a)   A receptacle installed in a wet location where the product intended to be plugged into it is not attended while in use (e.g., sprinkler system controller, landscape lighting, holiday lights, and so forth) shall have an enclosure that is weatherproof with the attachment plug cap inserted or removed.

(b)   A receptacle installed in a wet location where the product intended to be plugged into it will be attended while in use (e.g., portable tools, and so forth) shall have an enclosure that is weatherproof when the attachment plug is removed.

**(C) Bathtub and Shower Space.** A receptacle shall not be installed within a bathtub or shower space.

**(D) Protection for Floor Receptacles.** Standpipes of floor receptacles shall allow floor-cleaning equipment to be operated without damage to receptacles.

**(E) Flush Mounting with Faceplate.** The enclosure for a receptacle installed in an outlet box flush-mounted on a wall surface shall be made weatherproof by means of a weatherproof faceplate assembly that provides a watertight connection between the plate and the wall surface.

**406.9 Grounding-Type Receptacles, Adapters, Cord Connectors, and Attachment Plugs.**

**(A) Grounding Poles.** Grounding-type receptacles, cord connectors, and attachment plugs shall be provided with one fixed grounding pole in addition to the circuit poles. The grounding contacting pole of grounding-type plug-in ground-fault circuit interrupters shall be permitted to be of the movable, self-restoring type on circuits operating at not over 150 volts between any two conductors or any conductor and ground.

**(B) Grounding-Pole Identification.** Grounding-type receptacles, adapters, cord connections, and attachment plugs shall have a means for connection of a grounding conductor to the grounding pole.

A terminal for connection to the grounding pole shall be designated by one of the following:

(1) A green-colored hexagonal-headed or -shaped terminal screw or nut, not readily removable.

(2) A green-colored pressure wire connector body (a wire barrel).

(3) A similar green-colored connection device, in the case of adapters. The grounding terminal of a grounding adapter shall be a green-colored rigid ear, lug, or similar device. The grounding connection shall be designed so that it cannot make contact with current-carrying parts of the receptacle, adapter, or attachment plug. The adapter shall be polarized.

(4) If the terminal for the equipment grounding conductor is not visible, the conductor entrance hole shall be marked with the word *green* or *ground*, the letters G or GR, or the grounding symbol, as shown in Figure 406.9(B)(4), or otherwise identified by a distinctive green color. If the terminal for the equipment grounding conductor is readily removable, the area adjacent to the terminal shall be similarly marked.

**Figure 406.9(B)(4)   Grounding symbol.**

**(C) Grounding Terminal Use.** A grounding terminal or grounding-type device shall not be used for purposes other than grounding.

**(D) Grounding-Pole Requirements.** Grounding-type attachment plugs and mating cord connectors and receptacles shall be designed so that the grounding connection is made before the current-carrying connections. Grounding-type devices shall be designed so grounding poles of attachment plugs cannot be brought into contact with current-carrying parts of receptacles or cord connectors.

**(E) Use.** Grounding-type attachment plugs shall be used only with a cord having an equipment grounding conductor.

FPN: See 200.10(B) for identification of grounded conductor terminals.

**406.10 Connecting Receptacle Grounding Terminal to Box.** The connection of the receptacle grounding terminal shall comply with 250.146.

## ARTICLE 408
## Switchboards and Panelboards

### I. General

**408.1 Scope.** This article covers the following:

(1) All switchboards, panelboards, and distribution boards installed for the control of light and power circuits

(2) Battery-charging panels supplied from light or power circuits

**408.2 Other Articles.** Switches, circuit breakers, and overcurrent devices used on switchboards, panelboards, and distribution boards, and their enclosures, shall comply with this article and also with the requirements of Articles 240, 250, 312, 314, 404, and other articles that apply. Switchboards and panelboards in hazardous (classified) locations shall comply with the requirements of Articles 500 through 517.

**408.3 Support and Arrangement of Busbars and Conductors.**

**(A) Conductors and Busbars on a Switchboard or Panelboard.** Conductors and busbars on a switchboard or panelboard shall comply with 408.3(A)(1), (2), and (3) as applicable.

**(1) Location.** Conductors and busbars shall be located so as to be free from physical damage and shall be held firmly in place.

**(2) Service Switchboards.** Barriers shall be placed in all service switchboards such that no uninsulated, ungrounded service busbar or service terminal is exposed to inadvertent contact by persons or maintenance equipment while servicing load terminations.

**(3) Same Vertical Section.** Other than the required interconnections and control wiring, only those conductors that are intended for termination in a vertical section of a switchboard shall be located in that section.

*Exception: Conductors shall be permitted to travel horizontally through vertical sections of switchboards where such conductors are isolated from busbars by a barrier.*

**(B) Overheating and Inductive Effects.** The arrangement of busbars and conductors shall be such as to avoid overheating due to inductive effects.

**(C) Used as Service Equipment.** Each switchboard or panelboard, if used as service equipment, shall be provided with a main bonding jumper sized in accordance with 250.28(D) or the equivalent placed within the panelboard or one of the sections of the switchboard for connecting the grounded service conductor on its supply side to the switchboard or panelboard frame. All sections of a switchboard shall be bonded together using an equipment grounding conductor sized in accordance with Table 250.122.

*Exception: Switchboards and panelboards used as service equipment on high-impedance grounded-neutral systems in accordance with 250.36 shall not be required to be provided with a main bonding jumper.*

**(D) Terminals.** In switchboards and panelboards, load terminals for field wiring, including grounded circuit conductor load terminals and connections to the ground bus for load equipment grounding conductors, shall be located so that it is not necessary to reach across or beyond an uninsulated ungrounded line bus in order to make connections.

**(E) Phase Arrangement.** The phase arrangement on 3-phase buses shall be A, B, C from front to back, top to bottom, or left to right, as viewed from the front of the switchboard or panelboard. The B phase shall be that phase having the higher voltage to ground on 3-phase, 4-wire, delta-connected systems. Other busbar arrangements shall be permitted for additions to existing installations and shall be marked.

*Exception: Equipment within the same single section or multisection switchboard or panelboard as the meter on 3-phase, 4-wire, delta-connected systems shall be permitted to have the same phase configuration as the metering equipment.*

> FPN: See 110.15 for requirements on marking the busbar or phase conductor having the higher voltage to ground where supplied from a 4-wire, delta-connected system.

**(F) Minimum Wire-Bending Space.** The minimum wire-bending space at terminals and minimum gutter space provided in panelboards and switchboards shall be as required in 312.6.

**408.4 Circuit Directory.** All circuits and circuit modifications shall be legibly identified as to purpose or use on a circuit directory located on the face or inside of the panel door in the case of a panelboard, and at each switch on a switchboard.

## II. Switchboards

**408.5 Location of Switchboards.** Switchboards that have any exposed live parts shall be located in permanently dry locations and then only where under competent supervision and accessible only to qualified persons. Switchboards shall be located so that the probability of damage from equipment or processes is reduced to a minimum.

**408.6 Switchboards in Damp or Wet Locations.** Switchboards in damp or wet locations shall be installed to comply with 312.2(A).

**408.7 Location Relative to Easily Ignitible Material.** Switchboards shall be placed so as to reduce to a minimum the probability of communicating fire to adjacent combustible materials. Where installed over a combustible floor, suitable protection thereto shall be provided.

**408.8 Clearances.**

**(A) From Ceiling.** For other than a totally enclosed switchboard, a space not less than 900 mm (3 ft) shall be provided between the top of the switchboard and any combustible ceiling, unless a noncombustible shield is provided between the switchboard and the ceiling.

**(B) Around Switchboards.** Clearances around switchboards shall comply with the provisions of 110.26.

**408.9 Conductor Insulation.** An insulated conductor used within a switchboard shall be listed, shall be flame retardant, and shall be rated not less than the voltage applied to it and not less than the voltage applied to other conductors or busbars with which it may come in contact.

**408.10 Clearance for Conductors Entering Bus Enclosures.** Where conduits or other raceways enter a switchboard, floor-standing panelboard, or similar enclosure at the bottom, sufficient space shall be provided to permit installation of conductors in the enclosure. The wiring space shall not be less than shown in Table 408.10 where the conduit or raceways enter or leave the enclosure below the busbars, their supports, or other obstructions. The conduit or raceways, including their end fittings, shall not rise more than 75 mm (3 in.) above the bottom of the enclosure.

**Table 408.10 Clearance for Conductors Entering Bus Enclosures**

| | Minimum Spacing Between Bottom of Enclosure and Busbars, Their Supports, or Other Obstructions | |
|---|---|---|
| Conductor | mm | in. |
| Insulated busbars, their supports, or other obstructions | 200 | 8 |
| Noninsulated busbars | 250 | 10 |

**408.12 Grounding of Instruments, Relays, Meters, and Instrument Transformers on Switchboards.** Instruments, relays, meters, and instrument transformers located on switchboards shall be grounded as specified in 250.170 through 250.178.

## III. Panelboards

**408.13 General.** All panelboards shall have a rating not less than the minimum feeder capacity required for the load computed in accordance with Article 220. Panelboards shall be durably marked by the manufacturer with the voltage and the current rating and the number of phases for which they are designed and with the manufacturer's name or trademark in such a manner so as to be visible after installation, without disturbing the interior parts or wiring.

FPN: See 110.22 for additional requirements.

**408.14 Classification of Panelboards.** Panelboards shall be classified for the purposes of this article as either lighting and appliance branch-circuit panelboards or power panelboards, based on their content. A lighting and appliance branch circuit is a branch circuit that has a connection to the neutral of the panelboard and that has overcurrent protection of 30 amperes or less in one or more conductors.

**(A) Lighting and Appliance Branch-Circuit Panelboard.** A lighting and appliance branch-circuit panelboard is one having more than 10 percent of its overcurrent devices protecting lighting and appliance branch circuits.

**(B) Power Panelboard.** A power panelboard is one having 10 percent or fewer of its overcurrent devices protecting lighting and appliance branch circuits.

**408.15 Number of Overcurrent Devices on One Panelboard.** Not more than 42 overcurrent devices (other than those provided for in the mains) of a lighting and appliance branch-circuit panelboard shall be installed in any one cabinet or cutout box.

A lighting and appliance branch-circuit panelboard shall be provided with physical means to prevent the installation of more overcurrent devices than that number for which the panelboard was designed, rated, and approved.

For the purposes of this article, a 2-pole circuit breaker shall be considered two overcurrent devices; a 3-pole circuit breaker shall be considered three overcurrent devices.

**408.16 Overcurrent Protection.**

**(A) Lighting and Appliance Branch-Circuit Panelboard Individually Protected.** Each lighting and appliance branch-circuit panelboard shall be individually protected on the supply side by not more than two main circuit breakers or two sets of fuses having a combined rating not greater than that of the panelboard.

*Exception No. 1: Individual protection for a lighting and appliance panelboard shall not be required if the panelboard feeder has overcurrent protection not greater than the rating of the panelboard.*

*Exception No. 2: For existing installations, individual protection for lighting and appliance branch-circuit panelboards shall not be required where such panelboards are used as service equipment in supplying an individual residential occupancy.*

**(B) Power Panelboard Protection.** In addition to the requirements of 408.13, a power panelboard with supply conductors that include a neutral and having more than 10 percent of its overcurrent devices protecting branch circuits rated 30 amperes or less shall be protected by an overcurrent protective device having a rating not greater than that of the panelboard. The overcurrent protective device shall be located within or at any point on the supply side of the panelboard.

*Exception: This individual protection shall not be required for a power panelboard used as service equipment with multiple disconnecting means in accordance with 230.71.*

**(C) Snap Switches Rated at 30 Amperes or Less.** Panelboards equipped with snap switches rated at 30 amperes or less shall have overcurrent protection not in excess of 200 amperes.

**(D) Supplied Through a Transformer.** Where a panelboard is supplied through a transformer, the overcurrent protection in 408.16(A), (B), and (C) shall be located on the secondary side of the transformer.

*Exception: A panelboard supplied by the secondary side of a transformer shall be considered as protected by the overcurrent protection provided on the primary side of the transformer where that protection is in accordance with 240.21(C)(1).*

**(E) Delta Breakers.** A 3-phase disconnect or overcurrent device shall not be connected to the bus of any panelboard that has less than 3-phase buses. Delta breakers shall not be installed in panelboards.

**(F) Back-Fed Devices.** Plug-in-type overcurrent protection devices or plug-in type-main lug assemblies that are backfed and used to terminate field-installed ungrounded supply conductors shall be secured in place by an additional fastener that requires other than a pull to release the device from the mounting means on the panel.

**408.17 Panelboards in Damp or Wet Locations.** Panelboards in damp or wet locations shall be installed to comply with 312.2(A).

**408.18 Enclosure.** Panelboards shall be mounted in cabinets, cutout boxes, or enclosures designed for the purpose and shall be dead-front.

*Exception: Panelboards other than of the dead-front, externally operable type shall be permitted where accessible only to qualified persons.*

**408.19 Relative Arrangement of Switches and Fuses.** In panelboards, fuses of any type shall be installed on the load side of any switches.

*Exception: Fuses installed as part of service equipment in accordance with the provisions of 230.94 shall be permitted on the line side of the service switch.*

**408.20 Grounding of Panelboards.** Panelboard cabinets and panelboard frames, if of metal, shall be in physical contact with each other and shall be grounded. Where the panelboard is used with nonmetallic raceway or cable or where separate grounding conductors are provided, a terminal bar for the grounding conductors shall be secured inside the cabinet. The terminal bar shall be bonded to the cabinet and panelboard frame, if of metal; otherwise it shall be connected to the grounding conductor that is run with the conductors feeding the panelboard.

*Exception: Where an isolated equipment grounding conductor is provided as permitted by 250.146(D), the insulated equipment grounding conductor that is run with the circuit conductors shall be permitted to pass through the panelboard without being connected to the panelboard's equipment grounding terminal bar.*

Grounding conductors shall not be connected to a terminal bar provided for grounded conductors (may be a neutral) unless the bar is identified for the purpose and is located where interconnection between equipment grounding conductors and grounded circuit conductors is permitted or required by Article 250.

**408.21 Grounded Conductor Terminations.** Each grounded conductor shall terminate within the panelboard in an individual terminal that is not also used for another conductor.

*Exception: Grounded conductors of circuits with parallel conductors shall be permitted to terminate in a single terminal if the terminal is identified for connection of more than one conductor.*

## IV. Construction Specifications

**408.30 Panels.** The panels of switchboards shall be made of moisture-resistant, noncombustible material.

**408.31 Busbars.** Insulated or bare busbars shall be rigidly mounted.

**408.32 Protection of Instrument Circuits.** Instruments, pilot lights, potential transformers, and other switchboard

devices with potential coils shall be supplied by a circuit that is protected by standard overcurrent devices rated 15 amperes or less.

*Exception No. 1: Overcurrent devices rated more than 15 amperes shall be permitted where the interruption of the circuit could create a hazard. Short-circuit protection shall be provided.*

*Exception No. 2: For ratings of 2 amperes or less, special types of enclosed fuses shall be permitted.*

**408.33 Component Parts.** Switches, fuses, and fuseholders used on panelboards shall comply with the applicable requirements of Articles 240 and 404.

**408.35 Wire-Bending Space in Panelboards.** The enclosure for a panelboard shall have the top and bottom wire-bending space sized in accordance with Table 312.6(B) for the largest conductor entering or leaving the enclosure. Side wire-bending space shall be in accordance with Table 312.6(A) for the largest conductor to be terminated in that space.

*Exception No. 1: Either the top or bottom wire-bending space shall be permitted to be sized in accordance with Table 312.6(A) for a lighting and appliance branch-circuit panelboard rated 225 amperes or less.*

*Exception No. 2: Either the top or bottom wire-bending space for any panelboard shall be permitted to be sized in accordance with Table 312.6(A) where at least one side wire-bending space is sized in accordance with Table 312.6(B) for the largest conductor to be terminated in any side wire-bending space.*

*Exception No. 3: The top and bottom wire-bending space shall be permitted to be sized in accordance with Table 312.6(A) spacings if the panelboard is designed and constructed for wiring using only one single 90 degree bend for each conductor, including the grounded circuit conductor, and the wiring diagram shows and specifies the method of wiring that shall be used.*

*Exception No. 4: Either the top or the bottom wire-bending space, but not both, shall be permitted to be sized in accordance with Table 312.6(A) where there are no conductors terminated in that space.*

**408.36 Minimum Spacings.** The distance between bare metal parts, busbars, and so forth shall not be less than specified in Table 408.36.

Where close proximity does not cause excessive heating, parts of the same polarity at switches, enclosed fuses, and so forth shall be permitted to be placed as close together as convenience in handling will allow.

*Exception: The distance shall be permitted to be less than that specified in Table 408.36 at circuit breakers and*

**Table 408.36 Minimum Spacings Between Bare Metal Parts**

| Voltage | Opposite Polarity Where Mounted on the Same Surface | | Opposite Polarity Where Held Free in Air | | Live Parts to Ground* | |
|---|---|---|---|---|---|---|
| | mm | in. | mm | in. | mm | in. |
| Not over 125 volts, nominal | 19.1 | ¾ | 12.7 | ½ | 12.7 | ½ |
| Not over 250 volts, nominal | 31.8 | 1¼ | 19.1 | ¾ | 12.7 | ½ |
| Not over 600 volts, nominal | 50.8 | 2 | 25.4 | 1 | 25.4 | 1 |

*For spacing between live parts and doors of cabinets, see 312.11(A)(1), (2), and (3).

*switches and in listed components installed in switchboards and panelboards.*

# ARTICLE 410
## Luminaires (Lighting Fixtures), Lampholders, and Lamps

## I. General

**410.1 Scope.** This article covers luminaires (lighting fixtures), lampholders, pendants, incandescent filament lamps, arc lamps, electric-discharge lamps, the wiring and equipment forming part of such lamps, luminaires (fixtures), and lighting installations.

**410.2 Application of Other Articles.** Equipment for use in hazardous (classified) locations shall conform to Articles 500 through 517. Lighting systems operating at 30 volts or less shall conform to Article 411. Arc lamps used in theaters shall comply with 520.61, and arc lamps used in projection machines shall comply with 540.20. Arc lamps used on constant-current systems shall comply with the general requirements of Article 490.

**410.3 Live Parts.** Luminaires (fixtures), lampholders, and lamps shall have no live parts normally exposed to contact. Exposed accessible terminals in lampholders and switches shall not be installed in metal luminaire (fixture) canopies or in open bases of portable table or floor lamps.

*Exception: Cleat-type lampholders located at least 2.5 m (8 ft) above the floor shall be permitted to have exposed terminals.*

## II. Luminaire (Fixture) Locations

**410.4 Luminaires (Fixtures) in Specific Locations.**

**(A) Wet and Damp Locations.** Luminaires (fixtures) installed in wet or damp locations shall be installed so that water cannot enter or accumulate in wiring compartments, lampholders, or other electrical parts. All luminaires (fixtures) installed in wet locations shall be marked, "Suitable for Wet Locations." All luminaires (fixtures) installed in damp locations shall be marked, "Suitable for Wet Locations" or "Suitable for Damp Locations."

**(B) Corrosive Locations.** Luminaires (fixtures) installed in corrosive locations shall be of a type suitable for such locations.

**(C) In Ducts or Hoods.** Luminaires (fixtures) shall be permitted to be installed in commercial cooking hoods where all of the following conditions are met:

(1) The luminaire (fixture) shall be identified for use within commercial cooking hoods and installed so that the temperature limits of the materials used are not exceeded.
(2) The luminaire (fixture) shall be constructed so that all exhaust vapors, grease, oil, or cooking vapors are excluded from the lamp and wiring compartment. Diffusers shall be resistant to thermal shock.
(3) Parts of the luminaire (fixture) exposed within the hood shall be corrosion resistant or protected against corrosion, and the surface shall be smooth so as not to collect deposits and to facilitate cleaning.
(4) Wiring methods and materials supplying the luminaire(s) [fixture(s)] shall not be exposed within the cooking hood.

FPN: See 110.11 for conductors and equipment exposed to deteriorating agents.

**(D) Bathtub and Shower Areas.** No parts of cord-connected luminaires (fixtures), hanging luminaires (fixtures), lighting track, pendants, or ceiling-suspended (paddle) fans shall be located within a zone measured 900 mm (3 ft) horizontally and 2.5 m (8 ft) vertically from the top of the bathtub rim or shower stall threshold. This zone is all encompassing and includes the zone directly over the tub or shower stall.

**410.5 Luminaires (Fixtures) Near Combustible Material.** Luminaires (fixtures) shall be constructed, installed, or equipped with shades or guards so that combustible material is not subjected to temperatures in excess of 90°C (194°F).

**410.6 Luminaires (Fixtures) Over Combustible Material.** Lampholders installed over highly combustible material shall be of the unswitched type. Unless an individual switch is provided for each luminaire (fixture), lampholders shall be located at least 2.5 m (8 ft) above the floor or shall be located or guarded so that the lamps cannot be readily removed or damaged.

**410.7 Luminaires (Fixtures) in Show Windows.** Chain-supported luminaires (fixtures) used in a show window shall be permitted to be externally wired. No other externally wired luminaires (fixtures) shall be used.

**410.8 Luminaires (Fixtures) in Clothes Closets.**

**(A) Definition.**

**Storage Space.** The volume bounded by the sides and back closet walls and planes extending from the closet floor vertically to a height of 1.8 m (6 ft) or the highest clothes-hanging rod and parallel to the walls at a horizontal distance of 600 mm (24 in.) from the sides and back of the closet walls, respectively, and continuing vertically to the closet ceiling parallel to the walls at a horizontal distance of 300 mm (12 in.) or the width of the shelf, whichever is greater; for a closet that permits access to both sides of a hanging rod, this space includes the volume below the highest rod extending 300 mm (12 in.) on either side of the rod on a plane horizontal to the floor extending the entire length of the rod.

FPN: See Figure 410.8.

**(B) Luminaire (Fixture) Types Permitted.** Listed luminaires (fixtures) of the following types shall be permitted to be installed in a closet:

(1) A surface-mounted or recessed incandescent luminaire (fixture) with a completely enclosed lamp

(2) A surface-mounted or recessed fluorescent luminaire (fixture)

**(C) Luminaire (Fixture) Types Not Permitted.** Incandescent luminaires (fixtures) with open or partially enclosed lamps and pendant luminaires (fixtures) or lampholders shall not be permitted.

**(D) Location.** Luminaires (fixtures) in clothes closets shall be permitted to be installed as follows:

(1) Surface-mounted incandescent luminaires (fixtures) installed on the wall above the door or on the ceiling, provided there is a minimum clearance of 300 mm

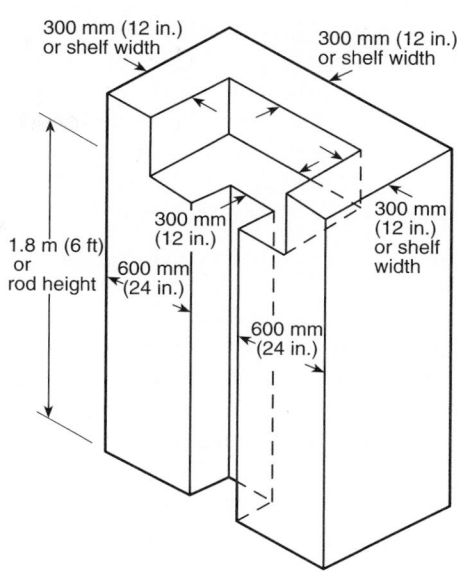

**Figure 410.8 Closet storage space.**

(12 in.) between the luminaire (fixture) and the nearest point of a storage space

(2) Surface-mounted fluorescent luminaires (fixtures) installed on the wall above the door or on the ceiling, provided there is a minimum clearance of 150 mm (6 in.) between the luminaire (fixture) and the nearest point of a storage space

(3) Recessed incandescent luminaires (fixtures) with a completely enclosed lamp installed in the wall or the ceiling, provided there is a minimum clearance of 150 mm (6 in.) between the luminaire (fixture) and the nearest point of a storage space

(4) Recessed fluorescent luminaires (fixtures) installed in the wall or the ceiling, provided there is a minimum clearance of 150 mm (6 in.) between the luminaire (fixture) and the nearest point of a storage space

**410.9 Space for Cove Lighting.** Coves shall have adequate space and shall be located so that lamps and equipment can be properly installed and maintained.

**III. Provisions at Luminaire (Fixture) Outlet Boxes, Canopies, and Pans**

**410.10 Space for Conductors.** Canopies and outlet boxes taken together shall provide adequate space so that luminaire (fixture) conductors and their connecting devices can be properly installed.

**410.11 Temperature Limit of Conductors in Outlet Boxes.** Luminaires (fixtures) shall be of such construction or installed so that the conductors in outlet boxes shall not be subjected to temperatures greater than that for which the conductors are rated.

Branch-circuit wiring, other than 2-wire or multiwire branch circuits supplying power to luminaires (fixtures) connected together, shall not be passed through an outlet box that is an integral part of a luminaire (fixture) unless the luminaire (fixture) is identified for through-wiring.

FPN: See 410.32 for wiring supplying power to fixtures connected together.

**410.12 Outlet Boxes to Be Covered.** In a completed installation, each outlet box shall be provided with a cover unless covered by means of a luminaire (fixture) canopy, lampholder, receptacle, or similar device.

**410.13 Covering of Combustible Material at Outlet Boxes.** Any combustible wall or ceiling finish exposed between the edge of a luminaire (fixture) canopy or pan and an outlet box shall be covered with noncombustible material.

**410.14 Connection of Electric-Discharge Luminaires (Lighting Fixtures).**

**(A) Independent of the Outlet Box.** Electric-discharge luminaires (lighting fixtures) supported independently of the outlet box shall be connected to the branch circuit through metal raceway, nonmetallic raceway, Type MC cable, Type AC cable, Type MI cable, nonmetallic sheathed cable, or by flexible cord as permitted in 410.30(B) or (C).

**(B) Access to Boxes.** Electric-discharge luminaires (lighting fixtures) surface mounted over concealed outlet, pull, or junction boxes shall be installed with suitable openings in back of the fixture to provide access to the boxes.

## IV. Luminaire (Fixture) Supports

**410.15 Supports.**

**(A) General.** Luminaires (fixtures) and lampholders shall be securely supported. A luminaire (fixture) that weighs more than 3 kg (6 lb) or exceeds 400 mm (16 in.) in any dimension shall not be supported by the screw shell of a lampholder.

**(B) Metal Poles Supporting Luminaires (Lighting Fixtures).** Metal poles shall be permitted to be used to support luminaires (lighting fixtures) and as a raceway to enclose supply conductors, provided the following conditions are met:

(1) A metal pole shall have a handhole not less than 50 mm × 100 mm (2 in. × 4 in.) with a raintight cover to provide access to the supply terminations within the pole or pole base.

*Exception No. 1: No handhole shall be required in a pole 2.5 m (8 ft) or less in height above grade where the supply wiring method continues without splice or pull point, and where the interior of the pole and any splices are accessible by removing the luminaire (fixture).*

*Exception No. 2: No handhole shall be required in a metal pole 6.0 m (20 ft) or less in height above grade that is provided with a hinged base.*

(2) Where raceway risers or cable is not installed within the pole, a threaded fitting or nipple shall be brazed or welded to the pole opposite the handhole for the supply connection.

(3) A metal pole shall be provided with a grounding terminal.

　　a. A pole with a handhole shall have the grounding terminal accessible from the handhole.

　　b. A pole with a hinged base shall have the grounding terminal accessible within the base.

*Exception: No grounding terminal shall be required in a pole 2.5 m (8 ft) or less in height above grade where the supply wiring method continues without splice or pull, and where the interior of the pole and any splices are accessible by removing the luminaire (fixture).*

(4) A pole with a hinged base shall have the hinged base and pole bonded together.

(5) Metal raceways or other equipment grounding conductors shall be bonded to the pole with an equipment grounding conductor recognized by 250.118 and sized in accordance with 250.122.

(6) Conductors in vertical metal poles used as raceway shall be supported as provided in 300.19.

**410.16 Means of Support.**

**(A) Outlet Boxes.** Outlet boxes or fittings installed as required by 314.23 shall be permitted to support luminaires (fixtures).

**(B) Inspection.** Luminaires (fixtures) shall be installed so that the connections between the luminaire (fixture) conductors and the circuit conductors can be inspected without requiring the disconnection of any part of the wiring unless the luminaires (fixtures) are connected by attachment plugs and receptacles.

**(C) Suspended Ceilings.** Framing members of suspended ceiling systems used to support luminaires (fixtures) shall be securely fastened to each other and shall be securely attached to the building structure at appropriate intervals. Luminaires (fixtures) shall be securely fastened to the ceiling framing member by mechanical means such as bolts, screws, or rivets. Listed clips identified for use with the type of ceiling framing member(s) and luminaire(s) [fixture(s)] shall also be permitted.

**(D) Luminaire (Fixture) Studs.** Luminaire (fixture) studs that are not a part of outlet boxes, hickeys, tripods, and crowfeet shall be made of steel, malleable iron, or other material suitable for the application.

**(E) Insulating Joints.** Insulating joints that are not designed to be mounted with screws or bolts shall have an exterior metal casing, insulated from both screw connections.

**(F) Raceway Fittings.** Raceway fittings used to support a luminaire(s) [lighting fixture(s)] shall be capable of supporting the weight of the complete fixture assembly and lamp(s).

**(G) Busways.** Luminaires (fixtures) shall be permitted to be connected to busways in accordance with 368.12.

**(H) Trees.** Outdoor luminaires (lighting fixtures) and associated equipment shall be permitted to be supported by trees.

> FPN No. 1: See 225.26 for restrictions for support of overhead conductors.

> FPN No. 2: See 300.5(D) for protection of conductors.

## V. Grounding

**410.17 General.** Luminaires (fixtures) and lighting equipment shall be grounded as required in Article 250 and Part V of this article.

**410.18 Exposed Luminaire (Fixture) Parts.**

**(A) Exposed Conductive Parts.** Exposed metal parts shall be grounded or insulated from ground and other conducting surfaces or be inaccessible to unqualified personnel. Lamp tie wires, mounting screws, clips, and decorative bands on glass spaced at least 38 mm (1½ in.) from lamp terminals shall not be required to be grounded.

**(B) Made of Insulating Material.** Luminaires (fixtures) directly wired or attached to outlets supplied by a wiring method that does not provide a ready means for grounding shall be made of insulating material and shall have no exposed conductive parts.

*Exception: Replacement luminaires (fixtures) shall be permitted to connect an equipment grounding conductor from the outlet in compliance with 250.130(C). The luminaire (fixture) shall then be grounded in accordance wtih 410.18(A).*

**410.20 Equipment Grounding Conductor Attachment.** Luminaires (fixtures) with exposed metal parts shall be provided with a means for connecting an equipment grounding conductor for such luminaires (fixtures).

**410.21 Methods of Grounding.** Luminaires (fixtures) and equipment shall be considered grounded where mechanically connected to an equipment grounding conductor as specified in 250.118 and sized in accordance with 250.122.

## VI. Wiring of Luminaires (Fixtures)

**410.22 Luminaire (Fixture) Wiring — General.** Wiring on or within fixtures shall be neatly arranged and shall not be exposed to physical damage. Excess wiring shall be avoided. Conductors shall be arranged so that they are not subjected to temperatures above those for which they are rated.

**410.23 Polarization of Luminaires (Fixtures).** Luminaires (fixtures) shall be wired so that the screw shells of lampholders are connected to the same luminaire (fixture) or circuit conductor or terminal. The grounded conductor, where connected to a screw-shell lampholder, shall be connected to the screw shell.

**410.24 Conductor Insulation.** Luminaires (fixtures) shall be wired with conductors having insulation suitable for the environmental conditions, current, voltage, and temperature to which the conductors will be subjected.

> FPN: For ampacity of luminaire (fixture) wire, maximum operating temperature, voltage limitations, minimum wire size, and so forth, see Article 402.

**410.27 Pendant Conductors for Incandescent Filament Lamps.**

**(A) Support.** Pendant lampholders with permanently attached leads, where used for other than festoon wiring, shall be hung from separate stranded rubber-covered conductors that are soldered directly to the circuit conductors but supported independently thereof.

**(B) Size.** Unless part of listed decorative lighting assemblies, pendant conductors shall not be smaller than 14 AWG for mogul-base or medium-base screw-shell lampholders or smaller than 18 AWG for intermediate or candelabra-base lampholders.

**(C) Twisted or Cabled.** Pendant conductors longer than 900 mm (3 ft) shall be twisted together where not cabled in a listed assembly.

**410.28 Protection of Conductors and Insulation.**

**(A) Properly Secured.** Conductors shall be secured in a manner that does not tend to cut or abrade the insulation.

**(B) Protection Through Metal.** Conductor insulation shall be protected from abrasion where it passes through metal.

**(C) Luminaire (Fixture) Stems.** Splices and taps shall not be located within luminaire (fixture) arms or stems.

**(D) Splices and Taps.** No unnecessary splices or taps shall be made within or on a luminaire (fixture).

FPN: For approved means of making connections, see 110.14.

**(E) Stranding.** Stranded conductors shall be used for wiring on luminaire (fixture) chains and on other movable or flexible parts.

**(F) Tension.** Conductors shall be arranged so that the weight of the luminaire (fixture) or movable parts does not put tension on the conductors.

**410.29 Cord-Connected Showcases.** Individual showcases, other than fixed, shall be permitted to be connected by flexible cord to permanently installed receptacles, and groups of not more than six such showcases shall be permitted to be coupled together by flexible cord and separable locking-type connectors with one of the group connected by flexible cord to a permanently installed receptacle.

The installation shall comply with 410.29(A) through (E).

**(A) Cord Requirements.** Flexible cord shall be of the hard-service type, having conductors not smaller than the branch-circuit conductors, having ampacity at least equal to the branch-circuit overcurrent device, and having an equipment grounding conductor.

FPN: See Table 250.122 for size of equipment grounding conductor.

**(B) Receptacles, Connectors, and Attachment Plugs.** Receptacles, connectors, and attachment plugs shall be of a listed grounding type rated 15 or 20 amperes.

**(C) Support.** Flexible cords shall be secured to the undersides of showcases so that

(1) Wiring is not exposed to mechanical damage.
(2) A separation between cases not in excess of 50 mm (2 in.), or more than 300 mm (12 in.) between the first case and the supply receptacle, is ensured.
(3) The free lead at the end of a group of showcases has a female fitting not extending beyond the case.

**(D) No Other Equipment.** Equipment other than showcases shall not be electrically connected to showcases.

**(E) Secondary Circuit(s).** Where showcases are cord-connected, the secondary circuit(s) of each electric-discharge lighting ballast shall be limited to one showcase.

**410.30 Cord-Connected Lampholders and Luminaires (Fixtures).**

**(A) Lampholders.** Where a metal lampholder is attached to a flexible cord, the inlet shall be equipped with an insulating bushing that, if threaded, is not smaller than metric designator 12 (trade size ⅜) pipe size. The cord hole shall be of a size appropriate for the cord, and all burrs and fins shall be removed in order to provide a smooth bearing surface for the cord.

Bushing having holes 7 mm (⁹⁄₃₂ in.) in diameter shall be permitted for use with plain pendant cord and holes 11 mm (¹³⁄₃₂ in.) in diameter with reinforced cord.

**(B) Adjustable Luminaires (Fixtures).** Luminaires (fixtures) that require adjusting or aiming after installation shall not be required to be equipped with an attachment plug or cord connector, provided the exposed cord is of the hard-usage or extra-hard-usage type and is not longer than that required for maximum adjustment. The cord shall not be subject to strain or physical damage.

**(C) Electric-Discharge Luminaires (Fixtures).**

**(1)** A listed luminaire (fixture) or a listed assembly shall be permitted to be cord connected if the following conditions apply:

(1) The luminaire (fixture) is located directly below the outlet box or busway.
(2) The flexible cord meets all the following:
     a. Is visible for its entire length outside the luminaire (fixture)
     b. Is not subject to strain or physical damage
     c. Is terminated in a grounding-type attachment plug cap or busway plug or has a luminaire (fixture) assembly with a strain relief and canopy

**(2)** Electric-discharge luminaires (lighting fixtures) provided with mogul-base, screw-shell lampholders shall be permitted to be connected to branch circuits of 50 amperes or less by cords complying with 240.5. Receptacles and attachment plugs shall be permitted to be of lower ampere rating than the branch circuit but not less than 125 percent of the luminaire (fixture) full-load current.

**(3)** Electric-discharge luminaires (lighting fixtures) equipped with a flanged surface inlet shall be permitted to be supplied by cord pendants equipped with cord connectors. Inlets and connectors shall be permitted to be of lower ampere rating than the branch circuit but not less than 125 percent of the luminaire (fixture) load current.

**410.31 Luminaires (Fixtures) as Raceways.** Luminaires (fixtures) shall not be used as a raceway for circuit conductors unless listed and marked for use as a raceway.

**410.32 Wiring Supplying Luminaires (Fixtures) Connected Together.** Luminaires (fixtures) designed for end-to-end connection to form a continuous assembly, or luminaires (fixtures) connected together by recognized wiring methods, shall be permitted to contain the conductors of a 2-wire branch circuit, or one multiwire branch circuit, supplying the connected luminaires (fixtures) and need not be listed as a raceway. One additional 2-wire branch circuit separately supplying one or more of the connected luminaires (fixtures) shall also be permitted.

FPN: See Article 100 for the definition of *Multiwire Branch Circuit.*

### 410.33 Branch Circuit Conductors and Ballasts.

Branch-circuit conductors within 75 mm (3 in.) of a ballast shall have an insulation temperature rating not lower than 90°C (194°F) unless supplying a luminaire (fixture) listed and marked as suitable for a different insulation temperature.

## VII. Construction of Luminaires (Fixtures)

### 410.34 Combustible Shades and Enclosures. Adequate airspace shall be provided between lamps and shades or other enclosures of combustible material.

### 410.35 Luminaire (Fixture) Rating.

**(A) Marking.** All luminaires (fixtures) shall be marked with the maximum lamp wattage or electrical rating, manufacturer's name, trademark, or other suitable means of identification. A luminaire (fixture) requiring supply wire rated higher than 60°C (140°F) shall be marked in letters not smaller than 6 mm (¼ in.) high, prominently displayed on the luminaire (fixture) and shipping carton or equivalent.

**(B) Electrical Rating.** The electrical rating shall include the voltage and frequency and shall indicate the current rating of the unit, including the ballast, transformer, or autotransformer.

### 410.36 Design and Material. Luminaires (fixtures) shall be constructed of metal, wood, or other material suitable for the application and shall be designed and assembled so as to secure requisite mechanical strength and rigidity. Wiring compartments, including their entrances, shall be such that conductors may be drawn in and withdrawn without physical damage.

### 410.37 Nonmetallic Luminaires (Fixtures). When luminaire (fixture) wiring compartments are constructed from combustible material, armored or lead-covered conductors with suitable fittings shall be used or the wiring compartment shall be lined with metal.

### 410.38 Mechanical Strength.

**(A) Tubing for Arms.** Tubing used for arms and stems where provided with cut threads shall not be less than 1.02 mm (0.040 in.) in thickness and where provided with

rolled (pressed) threads shall not be less than 0.64 mm (0.025 in.) in thickness. Arms and other parts shall be fastened to prevent turning.

**(B) Metal Canopies.** Metal canopies supporting lampholders, shades, and so forth exceeding 4 kg (8 lb), or incorporating attachment-plug receptacles, shall not be less than 0.51 mm (0.020 in.) in thickness. Other canopies shall not be less than 0.41 mm (0.016 in.) if made of steel and not less than 0.51 mm (0.020 in.) if of other metals.

**(C) Canopy Switches.** Pull-type canopy switches shall not be inserted in the rims of metal canopies that are less than 0.64 mm (0.025 in.) in thickness unless the rims are reinforced by the turning of a bead or the equivalent. Pull-type canopy switches, whether mounted in the rims or elsewhere in sheet metal canopies, shall not be located more than 90 mm (3½ in.) from the center of the canopy. Double set-screws, double canopy rings, a screw ring, or equal method shall be used where the canopy supports a pull-type switch or pendant receptacle.

The thickness requirements in the preceding paragraph shall apply to measurements made on finished (formed) canopies.

### 410.39 Wiring Space. Bodies of luminaires (fixtures), including portable lamps, shall provide ample space for splices and taps and for the installation of devices, if any. Splice compartments shall be of nonabsorbent, noncombustible material.

### 410.42 Portable Lamps.

**(A) General.** Portable lamps shall be wired with flexible cord recognized by 400.4 and an attachment plug of the polarized or grounding type. Where used with Edison-base lampholders, the grounded conductor shall be identified and attached to the screw shell and the identified blade of the attachment plug.

**(B) Portable Handlamps.** In addition to the provisions of 410.42(A), portable handlamps shall comply with the following.

(1) Metal shell, paper-lined lampholders shall not be used.
(2) Handlamps shall be equipped with a handle of molded composition or other insulating material.
(3) Handlamps shall be equipped with a substantial guard attached to the lampholder or handle.
(4) Metallic guards shall be grounded by means of an equipment grounding conductor run with circuit conductors within the power-supply cord.
(5) Portable handlamps shall not be required to be grounded where supplied through an isolating transformer with an ungrounded secondary of not over 50 volts.

**410.44 Cord Bushings.** A bushing or the equivalent shall be provided where flexible cord enters the base or stem of a portable lamp. The bushing shall be of insulating material unless a jacketed type of cord is used.

**410.45 Tests.** All wiring shall be free from short circuits and grounds and shall be tested for these defects prior to being connected to the circuit.

**410.46 Live Parts.** Exposed live parts within porcelain luminaires (fixtures) shall be suitably recessed and located so as to make it improbable that wires come in contact with them. There shall be a spacing of at least 13 mm (½ in.) between live parts and the mounting plane of the luminaire (fixture).

## VIII. Installation of Lampholders

**410.47 Screw-Shell Type.** Lampholders of the screw-shell type shall be installed for use as lampholders only. Where supplied by a circuit having a grounded conductor, the grounded conductor shall be connected to the screw shell.

**410.48 Double-Pole Switched Lampholders.** Where supplied by the ungrounded conductors of a circuit, the switching device of lampholders of the switched type shall simultaneously disconnect both conductors of the circuit.

**410.49 Lampholders in Wet or Damp Locations.** Lampholders installed in wet or damp locations shall be of the weatherproof type.

## IX. Construction of Lampholders

**410.50 Insulation.** The outer metal shell and the cap shall be lined with insulating material that prevents the shell and cap from becoming a part of the circuit. The lining shall not extend beyond the metal shell more than 3 mm (⅛ in.) but shall prevent any current-carrying part of the lamp base from being exposed when a lamp is in the lampholding device.

**410.52 Switched Lampholders.** Switched lampholders shall be of such construction that the switching mechanism interrupts the electrical connection to the center contact. The switching mechanism shall also be permitted to interrupt the electrical connection to the screw shell if the connection to the center contact is simultaneously interrupted.

## X. Lamps and Auxiliary Equipment

**410.53 Bases, Incandescent Lamps.** An incandescent lamp for general use on lighting branch circuits shall not be equipped with a medium base if rated over 300 watts, or

with a mogul base if rated over 1500 watts. Special bases or other devices shall be used for over 1500 watts.

**410.54 Electric-Discharge Lamp Auxiliary Equipment.**

**(A) Enclosures.** Auxiliary equipment for electric-discharge lamps shall be enclosed in noncombustible cases and treated as sources of heat.

**(B) Switching.** Where supplied by the ungrounded conductors of a circuit, the switching device of auxiliary equipment shall simultaneously disconnect all conductors.

## XI. Special Provisions for Flush and Recessed Luminaires (Fixtures)

**410.64 General.** Luminaires (fixtures) installed in recessed cavities in walls or ceilings shall comply with 410.65 through 410.72.

**410.65 Temperature.**

**(A) Combustible Material.** Luminaires (fixtures) shall be installed so that adjacent combustible material will not be subjected to temperatures in excess of 90°C (194°F).

**(B) Fire-Resistant Construction.** Where a luminaire (fixture) is recessed in fire-resistant material in a building of fire-resistant construction, a temperature higher than 90°C (194°F) but not higher than 150°C (302°F) shall be considered acceptable if the luminaire (fixture) is plainly marked that it is listed for that service.

**(C) Recessed Incandescent Luminaires (Fixtures).** Incandescent luminaires (fixtures) shall have thermal protection and shall be identified as thermally protected.

*Exception No. 1: Thermal protection shall not be required in a recessed luminaire (fixture) identified for use and installed in poured concrete.*

*Exception No. 2: Thermal protection shall not be required in a recessed luminaire (fixture) whose design, construction, and thermal performance characteristics are equivalent to a thermally protected luminaire (fixture) and are identified as inherently protected.*

**410.66 Clearance and Installation.**

**(A) Clearance.**

**(1) Non-Type IC.** A recessed luminaire (fixture) that is not identified for contact with insulation shall have all recessed parts spaced not less than 13 mm (½ in.) from combustible materials. The points of support and the trim finishing off the opening in the ceiling or wall surface shall be permitted to be in contact with combustible materials.

**(2) Type IC.** A recessed luminaire (fixture) that is identified for contact with insulation, Type IC, shall be permitted to be in contact with combustible materials at recessed parts, points of support, and portions passing through or finishing off the opening in the building structure.

**(B) Installation.** Thermal insulation shall not be installed above a recessed luminaire (fixture) or within 75 mm (3 in.) of the recessed luminaire's (fixture's) enclosure, wiring compartment, or ballast unless it is identified for contact with insulation, Type IC.

### 410.67 Wiring.

**(A) General.** Conductors that have insulation suitable for the temperature encountered shall be used.

**(B) Circuit Conductors.** Branch-circuit conductors that have an insulation suitable for the temperature encountered shall be permitted to terminate in the luminaire (fixture).

**(C) Tap Conductors.** Tap conductors of a type suitable for the temperature encountered shall be permitted to run from the luminaire (fixture) terminal connection to an outlet box placed at least 300 mm (1 ft) from the luminaire (fixture). Such tap conductors shall be in suitable raceway or Type AC or MC cable of at least 450 mm (18 in.) but not more than 1.8 m (6 ft) in length.

## XII. Construction of Flush and Recessed Luminaires (Fixtures)

**410.68 Temperature.** Luminaires (fixtures) shall be constructed so that adjacent combustible material is not subject to temperatures in excess of 90°C (194°F).

**410.70 Lamp Wattage Marking.** Incandescent lamp luminaires (fixtures) shall be marked to indicate the maximum allowable wattage of lamps. The markings shall be permanently installed, in letters at least 6 mm (¼ in.) high, and shall be located where visible during relamping.

**410.71 Solder Prohibited.** No solder shall be used in the construction of a luminaire (fixture) box.

**410.72 Lampholders.** Lampholders of the screw-shell type shall be of porcelain or other suitable insulating materials. Where used, cements shall be of the high-heat type.

## XIII. Special Provisions for Electric-Discharge Lighting Systems of 1000 Volts or Less

### 410.73 General.

**(A) Open-Circuit Voltage of 1000 Volts or Less.** Equipment for use with electric-discharge lighting systems and designed for an open-circuit voltage of 1000 volts or less shall be of a type intended for such service.

**(B) Considered as Energized.** The terminals of an electric-discharge lamp shall be considered as energized where any lamp terminal is connected to a circuit of over 300 volts.

**(C) Transformers of the Oil-Filled Type.** Transformers of the oil-filled type shall not be used.

**(D) Additional Requirements.** In addition to complying with the general requirements for luminaires (lighting fixtures), such equipment shall comply with Part XIII of this article.

**(E) Thermal Protection — Fluorescent Luminaires (Fixtures).**

**(1) Integral Thermal Protection.** The ballast of a fluorescent luminaire (fixture) installed indoors shall have integral thermal protection. Replacement ballasts shall also have thermal protection integral with the ballast.

**(2) Simple Reactance Ballasts.** A simple reactance ballast in a fluorescent luminaire (fixture) with straight tubular lamps shall not be required to be thermally protected.

**(3) Exit Fixtures.** A ballast in a fluorescent exit luminaire (fixture) shall not have thermal protection.

**(4) Emergency Egress Luminaires (Fixtures).** A ballast in a fluorescent luminaire (fixture) that is used for egress lighting and energized only during an emergency shall not have thermal protection.

**(F) High-Intensity Discharge Luminaires (Fixtures).**

**(1) Recessed.** Recessed high-intensity luminaires (fixtures) designed to be installed in wall or ceiling cavities shall have thermal protection and be identified as thermally protected.

**(2) Inherently Protected.** Thermal protection shall not be required in a recessed high-intensity luminaire (fixture) whose design, construction, and thermal performance characteristics are equivalent to a thermally protected luminaire (fixture) and are identified as inherently protected.

**(3) Installed in Poured Concrete.** Thermal protection shall not be required in a recessed high-intensity discharge luminaire (fixture) identified for use and installed in poured concrete.

**(4) Recessed Remote Ballasts.** A recessed remote ballast for a high-intensity discharge luminaire (fixture) shall have thermal protection that is integral with the ballast and be identified as thermally protected.

**410.74 Direct-Current Equipment.** Luminaires (fixtures) installed on dc circuits shall be equipped with auxiliary equipment and resistors designed for dc operation. The luminaires (fixtures) shall be marked for dc operation.

**410.75 Open-Circuit Voltage Exceeding 300 Volts.** Equipment having an open-circuit voltage exceeding 300 volts shall not be installed in dwelling occupancies unless such equipment is designed so that there will be no exposed live parts when lamps are being inserted, are in place, or are being removed.

**410.76 Luminaire (Fixture) Mounting.**

**(A) Exposed Ballasts.** Luminaires (fixtures) that have exposed ballasts or transformers shall be installed so that such ballasts or transformers will not be in contact with combustible material.

**(B) Combustible Low-Density Cellulose Fiberboard.** Where a surface-mounted luminaire (fixture) containing a ballast is to be installed on combustible low-density cellulose fiberboard, it shall be listed for this condition or shall be spaced not less than 38 mm (1½ in.) from the surface of the fiberboard. Where such luminaires (fixtures) are partially or wholly recessed, the provisions of 410.64 through 410.72 shall apply.

> FPN: Combustible low-density cellulose fiberboard includes sheets, panels, and tiles that have a density of 320 kg/m³ (20 lb/ft³) or less and that are formed of bonded plant fiber material but does not include solid or laminated wood or fiberboard that has a density in excess of 320 kg/m³ (20 lb/ft³) or is a material that has been integrally treated with fire-retarding chemicals to the degree that the flame spread in any plane of the material will not exceed 25, determined in accordance with tests for surface burning characteristics of building materials. See ANSI/ASTM E84-1997, *Test Method for Surface Burning Characteristics of Building Materials.*

**410.77 Equipment Not Integral with Luminaire (Fixture).**

**(A) Metal Cabinets.** Auxiliary equipment, including reactors, capacitors, resistors, and similar equipment, where not installed as part of a luminaire (lighting fixture) assembly, shall be enclosed in accessible, permanently installed metal cabinets.

**(B) Separate Mounting.** Separately mounted ballasts that are intended for direct connection to a wiring system shall not be required to be separately enclosed.

**(C) Wired Luminaire (Fixture) Sections.** Wired luminaire (fixture) sections are paired, with a ballast(s) supplying a lamp or lamps in both. For interconnection between paired units, it shall be permissible to use metric designator 12 (trade size ⅜) flexible metal conduit in lengths not exceeding 7.5 m (25 ft), in conformance with Article 348. Luminaire (fixture) wire operating at line voltage, supplying only the ballast(s) of one of the paired luminaires (fixtures), shall be permitted in the same raceway as the lamp supply wires of the paired luminaires (fixtures).

**410.78 Autotransformers.** An autotransformer that is used to raise the voltage to more than 300 volts, as part of a ballast for supplying lighting units, shall be supplied only by a grounded system.

**410.79 Switches.** Snap switches shall comply with 404.14.

**XIV. Special Provisions for Electric-Discharge Lighting Systems of More Than 1000 Volts**

**410.80 General.**

**(A) Listing.** Electric-discharge lighting systems with an open-circuit voltage exceeding 1000 volts shall be listed and installed in conformance with that listing.

**(B) Dwelling Occupancies.** Equipment that has an open-circuit voltage exceeding 1000 volts shall not be installed in or on dwelling occupancies.

**(C) Live Parts.** The terminal of an electric-discharge lamp shall be considered as a live part.

**(D) Additional Requirements.** In addition to complying with the general requirements for luminaires (lighting fixtures), such equipment shall comply with Part XIV of this article.

> FPN: For signs and outline lighting, see Article 600.

**410.81 Control.**

**(A) Disconnection.** Luminaires (fixtures) or lamp installation shall be controlled either singly or in groups by an externally operable switch or circuit breaker that opens all ungrounded primary conductors.

**(B) Within Sight or Locked Type.** The switch or circuit breaker shall be located within sight from the luminaires (fixtures) or lamps, or it shall be permitted elsewhere if it is provided with a means for locking in the open position.

**410.82 Lamp Terminals and Lampholders.** Parts that must be removed for lamp replacement shall be hinged or held captive. Lamps or lampholders shall be designed so that there are no exposed live parts when lamps are being inserted or removed.

## 410.83 Transformers.

**(A) Type.** Transformers shall be enclosed, identified for the use, and listed.

**(B) Voltage.** The secondary-circuit voltage shall not exceed 15,000 volts, nominal, under any load condition. The voltage to ground of any ouput terminals of the secondary circuit shall not exceed 7500 volts, under any load conditions.

**(C) Rating.** Transformers shall have a secondary short-circuit current rating of not more than 150 mA if the open-circuit voltage is over 7500 volts, and not more than 300 mA if the open-circuit voltage rating is 7500 volts or less.

**(D) Secondary Connections.** Secondary circuit outputs shall not be connected in parallel or in series.

## 410.84 Transformer Locations.

**(A) Accessible.** Transformers shall be accessible after installation.

**(B) Secondary Conductors.** Transformers shall be installed as near to the lamps as practicable to keep the secondary conductors as short as possible.

**(C) Adjacent to Combustible Materials.** Transformers shall be located so that adjacent combustible materials are not subjected to temperatures in excess of 90°C (194°F).

## 410.85 Exposure to Damage. Lamps shall not be located where normally exposed to physical damage.

**410.86 Marking.** Each luminaire (fixture) or each secondary circuit of tubing having an open-circuit voltage of over 1000 volts shall have a clearly legible marking in letters not less than 6 mm (¼ in.) high reading "Caution . . . volts." The voltage indicated shall be the rated open-circuit voltage.

**410.87 Switches.** Snap switches shall comply with 404.4.

## XV. Lighting Track

### 410.100 Definition.

**Lighting Track.** A manufactured assembly designed to support and energize luminaires (lighting fixtures) that are capable of being readily repositioned on the track. Its length may be altered by the addition or subtraction of sections of track.

### 410.101 Installation.

**(A) Lighting Track.** Lighting track shall be permanently installed and permanently connected to a branch circuit. Only lighting track fittings shall be installed on lighting track. Lighting track fittings shall not be equipped with general-purpose receptacles.

**(B) Connected Load.** The connected load on lighting track shall not exceed the rating of the track. Lighting track shall be supplied by a branch circuit having a rating not more than that of the track.

**(C) Locations Not Permitted.** Lighting track shall not be installed in the following locations:

(1) Where likely to be subjected to physical damage
(2) In wet or damp locations
(3) Where subject to corrosive vapors
(4) In storage battery rooms
(5) In hazardous (classified) locations
(6) Where concealed
(7) Where extended through walls or partitions
(8) Less than 1.5 m (5 ft) above the finished floor except where protected from physical damage or track operating at less than 30 volts rms open-circuit voltage
(9) Within the zone measured 900 mm (3 ft) horizontally and 2.5 m (8 ft) vertically from the top of the bathtub rim

**(D) Support.** Fittings identified for use on lighting track shall be designed specifically for the track on which they are to be installed. They shall be securely fastened to the track, shall maintain polarization and grounding, and shall be designed to be suspended directly from the track.

**410.103 Heavy-Duty Lighting Track.** Heavy-duty lighting track is lighting track identified for use exceeding 20 amperes. Each fitting attached to a heavy-duty lighting track shall have individual overcurrent protection.

**410.104 Fastening.** Lighting track shall be securely mounted so that each fastening will be suitable for supporting the maximum weight of luminaires (fixtures) that can be installed. Unless identified for supports at greater intervals, a single section 1.2 m (4 ft) or shorter in length shall have two supports, and, where installed in a continuous row, each individual section of not more than 1.2 m (4 ft) in length shall have one additional support.

### 410.105 Construction Requirements.

**(A) Construction.** The housing for the lighting track system shall be of substantial construction to maintain rigidity. The conductors shall be installed within the track housing, permitting insertion of a luminaire (fixture), and designed to prevent tampering and accidental contact with live parts. Components of lighting track systems of different voltages shall not be interchangeable. The track conductors shall be a minimum 12 AWG or equal and shall be copper. The track system ends shall be insulated and capped.

**(B) Grounding.** Lighting track shall be grounded in accordance with Article 250, and the track sections shall be securely coupled to maintain continuity of the circuitry, polarization, and grounding throughout.

# ARTICLE 411
# Lighting Systems Operating at 30 Volts or Less

**411.1 Scope.** This article covers lighting systems operating at 30 volts or less and their associated components.

**411.2 Definition.**

**Lighting Systems Operating at 30 Volts or Less.** A lighting system consisting of an isolating power supply operating at 30 volts (42.4 volts peak) or less, under any load condition, with one or more secondary circuits, each limited to 25 amperes maximum, supplying luminaires (lighting fixtures) and associated equipment identified for the use.

**411.3 Listing Required.** Lighting systems operating at 30 volts or less shall be listed for the purpose.

**411.4 Locations Not Permitted.** Lighting systems operating at 30 volts or less shall not be installed (1) where concealed or extended through a building wall, unless using a wiring method specified in Chapter 3, or (2) within 3.0 m (10 ft) of pools, spas, fountains, or similar locations, except as permitted by Article 680.

**411.5 Secondary Circuits.**

**(A) Grounding.** Secondary circuits shall not be grounded.

**(B) Isolation.** The secondary circuit shall be insulated from the branch circuit by an isolating transformer.

**(C) Bare Conductors.** Exposed bare conductors and current-carrying parts shall be permitted. Bare conductors shall not be installed less than 2.1 m (7 ft) above the finished floor, unless specifically listed for a lower installation height.

**411.6 Branch Circuit.** Lighting systems operating at 30 volts or less shall be supplied from a maximum 20-ampere branch circuit.

**411.7 Hazardous (Classified) Locations.** Where installed in hazardous (classified) locations, these systems shall conform with Articles 500 through 517 in addition to this article.

# ARTICLE 422
# Appliances

## I. General

**422.1 Scope.** This article covers electric appliances used in any occupancy.

**422.3 Other Articles.** Appliances for use in hazardous (classified) locations shall comply with Articles 500 through 517.

The requirements of Article 430 shall apply to the installation of motor-operated appliances, and the requirements of Article 440 shall apply to the installation of appliances containing a hermetic refrigerant motor-compressor(s), except as specifically amended in this article.

**422.4 Live Parts.** Appliances shall have no live parts normally exposed to contact other than those parts functioning as open-resistance heating elements, such as the heating element of a toaster, which are necessarily exposed.

## II. Installation

**422.10 Branch-Circuit Rating.** This section specifies the ratings of branch circuits capable of carrying appliance current without overheating under the conditions specified.

**(A) Individual Circuits.** The rating of an individual branch circuit shall not be less than the marked rating of the appliance or the marked rating of an appliance having combined loads as provided in 422.62.

The rating of an individual branch circuit for motor-operated appliances not having a marked rating shall be in accordance with Part II of Article 430.

The branch-circuit rating for an appliance that is continuously loaded, other than a motor-operated appliance, shall not be less than 125 percent of the marked rating, or not less than 100 percent of the marked rating if the branch-circuit device and its assembly are listed for continuous loading at 100 percent of its rating.

Branch circuits for household cooking appliances shall be permitted to be in accordance with Table 220.19.

**(B) Circuits Supplying Two or More Loads.** For branch circuits supplying appliance and other loads, the rating shall be determined in accordance with 210.23.

**422.11 Overcurrent Protection.** Appliances shall be protected against overcurrent in accordance with 422.11(A) through (G) and 422.10.

**(A) Branch-Circuit Overcurrent Protection.** Branch circuits shall be protected in accordance with 240.4.

If a protective device rating is marked on an appliance, the branch-circuit overcurrent device rating shall not exceed the protective device rating marked on the appliance.

**(B) Household-Type Appliance with Surface Heating Elements.** A household-type appliance with surface heating elements having a maximum demand of more than 60 amperes computed in accordance with Table 220.19 shall have its power supply subdivided into two or more circuits, each of which shall be provided with overcurrent protection rated at not over 50 amperes.

**(C) Infrared Lamp Commercial and Industrial Heating Appliances.** Infrared lamp commercial and industrial heating appliances shall have overcurrent protection not exceeding 50 amperes.

**(D) Open-Coil or Exposed Sheathed-Coil Types of Surface Heating Elements in Commercial-Type Heating Appliances.** Open-coil or exposed sheathed-coil types of surface heating elements in commercial-type heating appliances shall be protected by overcurrent protective devices rated at not over 50 amperes.

**(E) Single Nonmotor-Operated Appliance.** If the branch circuit supplies a single non–motor-operated appliance, the rating of overcurrent protection shall

(1) Not exceed that marked on the appliance;
(2) If the overcurrent protection rating is not marked and the appliance is rated 13.3 amperes or less, not exceed 20 amperes; or
(3) If the overcurrent protection rating is not marked and the appliance is rated over 13.3 amperes, not exceed 150 percent of the appliance rated current. Where 150 percent of the appliance rating does not correspond to a standard overcurrent device ampere rating, the next higher standard rating shall be permitted.

**(F) Electric Heating Appliances Employing Resistance-Type Heating Elements Rated More Than 48 Amperes.**

**(1) Electric Heating Appliances.** Electric heating appliances employing resistance-type heating elements rated more than 48 amperes, other than household appliances with surface heating elements covered by 422.11(B), and commercial-type heating appliances covered by 422.11(D), shall have the heating elements subdivided. Each subdivided load shall not exceed 48 amperes and shall be protected at not more than 60 amperes.

These supplementary overcurrent protective devices shall be (1) factory-installed within or on the heater enclosure or provided as a separate assembly by the heater manufacturer; (2) accessible; and (3) suitable for branch-circuit protection.

The main conductors supplying these overcurrent protective devices shall be considered branch-circuit conductors.

**(2) Commercial Kitchen and Cooking Appliances.** Commercial kitchen and cooking appliances using sheathed-type heating elements not covered in 422.11(D) shall be permitted to be subdivided into circuits not exceeding 120 amperes and protected at not more than 150 amperes where one of the following is met:

(1) Elements are integral with and enclosed within a cooking surface.
(2) Elements are completely contained within an enclosure identified as suitable for this use.
(3) Elements are contained within an ASME-rated and stamped vessel.

**(3) Water Heaters and Steam Boilers.** Water heaters and steam boilers employing resistance-type immersion electric heating elements contained in an ASME-rated and stamped vessel or listed instantaneous water heaters shall be permitted to be subdivided into circuits not exceeding 120 amperes and protected at not more than 150 amperes.

**(G) Motor-Operated Appliances.** Motors of motor-operated appliances shall be provided with overload protection in accordance with Part III of Article 430. Hermetic refrigerant motor-compressors in air-conditioning or refrigerating equipment shall be provided with overload protection in accordance with Part VI of Article 440. Where appliance overcurrent protective devices that are separate from the appliance are required, data for selection of these devices shall be marked on the appliance. The minimum marking shall be that specified in 430.7 and 440.4.

**422.12 Central Heating Equipment.** Central heating equipment other than fixed electric space-heating equipment shall be supplied by an individual branch circuit.

*Exception: Auxiliary equipment, such as a pump, valve, humidifier, or electrostatic air cleaner directly associated with the heating equipment, shall be permitted to be connected to the same branch circuit.*

**422.13 Storage-Type Water Heaters.** A branch circuit supplying a fixed storage-type water heater that has a capacity of 450 L (120 gal) or less shall have a rating not less than 125 percent of the nameplate rating of the water heater.

FPN: For branch-circuit rating, see 422.10.

**422.14 Infrared Lamp Industrial Heating Appliances.** Infrared industrial heating appliance lampholders shall be permitted to be connected to any of the branch circuits in

Article 210 and, in industrial occupancies, shall be permitted to be operated in series on circuits of over 150 volts to ground, provided the voltage rating of the lampholders is not less than the circuit voltage.

Each section, panel, or strip carrying a number of infrared lampholders (including the internal wiring of such section, panel, or strip) shall be considered an appliance. The terminal connection block of each such assembly shall be considered an individual outlet.

## 422.15 Central Vacuum Outlet Assemblies.

**(A)** Listed central vacuum outlet assemblies shall be permitted to be connected to a branch circuit in accordance with 210.23(A).

**(B)** The ampacity of the connecting conductors shall not be less than the ampacity of the branch circuit conductors to which they are connected.

**(C)** An equipment grounding conductor shall be used where the central vacuum outlet assembly has accessible non–current-carrying metal parts.

## 422.16 Flexible Cords.

**(A) General.** Flexible cord shall be permitted (1) for the connection of appliances to facilitate their frequent interchange or to prevent the transmission of noise or vibration or (2) to facilitate the removal or disconnection of appliances that are fastened in place, where the fastening means and mechanical connections are specifically designed to permit ready removal for maintenance or repair and the appliance is intended or identified for flexible cord connection.

**(B) Specific Appliances.**

**(1) Electrically Operated Kitchen Waste Disposers.** Electrically operated kitchen waste disposers shall be permitted to be cord-and-plug connected with a flexible cord identified as suitable for the purpose in the installation instructions of the appliance manufacturer, where all of the following conditions are met.

(1) The flexible cord shall be terminated with a grounding type attachment plug.

*Exception: A listed kitchen waste disposer distinctly marked to identify it as protected by a system of double insulation, or its equivalent, shall not be required to be terminated with a grounding-type attachment plug.*

(2) The length of the cord shall not be less than 450 mm (18 in.) and not over 900 mm (36 in.).

(3) Receptacles shall be located to avoid physical damage to the flexible cord.

(4) The receptacle shall be accessible.

**(2) Built-in Dishwashers and Trash Compactors.** Built-in dishwashers and trash compactors shall be permitted to be cord-and-plug connected with a flexible cord identified as suitable for the purpose in the installation instructions of the appliance manufacturer where all of the following conditions are met.

(1) The flexible cord shall be terminated with a grounding-type attachment plug.

*Exception: A listed dishwasher or trash compactor distinctly marked to identify it as protected by a system of double insulation, or its equivalent, shall not be required to be terminated with a grounding-type attachment plug.*

(2) The length of the cord shall be 0.9 m to 1.2 m (3 ft to 4 ft) measured from the face of the attachment plug to the plane of the rear of the appliance.

(3) Receptacles shall be located to avoid physical damage to the flexible cord.

(4) The receptacle shall be located in the space occupied by the appliance or adjacent thereto.

(5) The receptacle shall be accessible.

**(3) Wall-Mounted Ovens and Counter-Mounted Cooking Units.** Wall-mounted ovens and counter-mounted cooking units complete with provisions for mounting and for making electrical connections shall be permitted to be permanently connected or, only for ease in servicing or for installation, cord-and-plug connected.

A separable connector or a plug and receptacle combination in the supply line to an oven or cooking unit shall be approved for the temperature of the space in which it is located.

## 422.17 Protection of Combustible Material.
Each electrically heated appliance that is intended by size, weight, and service to be located in a fixed position shall be placed so as to provide ample protection between the appliance and adjacent combustible material.

## 422.18 Support of Ceiling-Suspended (Paddle) Fans.

**(A) Ceiling-Suspended (Paddle) Fans 16 kg (35 lb) or Less.** Ceiling-suspended (paddle) fans that do not exceed 16 kg (35 lb) in weight, with or without accessories, shall be permitted to be supported by outlet boxes identified for such use and supported in accordance with 314.23 and 314.27.

**(B) Ceiling-Suspended (Paddle) Fans Exceeding 16 kg (35 lb).** Ceiling-suspended (paddle) fans exceeding 16 kg (35 lb) in weight, with or without accessories, shall be supported independently of the outlet box. See 314.23.

*Exception: Listed outlet boxes or outlet box systems that are identified for the purpose shall be permitted to support ceiling-suspended fans, with or without accessories, that weigh 32 kg (70 lb) or less.*

**422.20 Other Installation Methods.** Appliances employing methods of installation other than covered by this article shall be permitted to be used only by special permission.

### III. Disconnecting Means

**422.30 General.** A means shall be provided to disconnect each appliance from all ungrounded conductors in accordance with the following sections of Part III. If an appliance is supplied by more than one source, the disconnecting means shall be grouped and identified.

**422.31 Disconnection of Permanently Connected Appliances.**

**(A) Rated at Not Over 300 Volt-Amperes or ⅛ Horsepower.** For permanently connected appliances rated at not over 300 volt-amperes or ⅛ hp, the branch-circuit overcurrent device shall be permitted to serve as the disconnecting means.

**(B) Appliances Rated Over 300 Volt-Amperes or ⅛ Horsepower.** For permanently connected appliances rated over 300 volt-amperes or ⅛ hp, the branch-circuit switch or circuit breaker shall be permitted to serve as the disconnecting means where the switch or circuit breaker is within sight from the appliance or is capable of being locked in the open position.

> FPN: For appliances employing unit switches, see 422.34.

**422.32 Disconnecting Means for Motor-Driven Appliance.** If a switch or circuit breaker serves as the disconnecting means for a permanently connected motor-driven appliance of more than ⅛ hp, it shall be located within sight from the motor controller and shall comply with Part IX of Article 430.

*Exception: If a motor-driven appliance of more than ⅛ hp is provided with a unit switch that complies with 422.34(A), (B), (C), or (D), the switch or circuit breaker serving as the other disconnecting means shall be permitted to be out of sight from the motor controller.*

**422.33 Disconnection of Cord-and-Plug-Connected Appliances.**

**(A) Separable Connector or an Attachment Plug and Receptacle.** For cord-and-plug-connected appliances, an accessible separable connector or an accessible plug and receptacle shall be permitted to serve as the disconnecting means. Where the separable connector or plug and recep-

tacle are not accessible, cord-and-plug-connected appliances shall be provided with disconnecting means in accordance with 422.31.

**(B) Connection at the Rear Base of a Range.** For cord-and-plug-connected household electric ranges, an attachment plug and receptacle connection at the rear base of a range, if it is accessible from the front by removal of a drawer, shall be considered as meeting the intent of 422.33(A).

**(C) Rating.** The rating of a receptacle or of a separable connector shall not be less than the rating of any appliance connected thereto.

*Exception: Demand factors authorized elsewhere in this Code shall be permitted to be applied to the rating of a receptacle or of a separable connector.*

**422.34 Unit Switch(es) as Disconnecting Means.** A unit switch(es) with a marked-off position that is a part of an appliance and disconnects all ungrounded conductors shall be permitted as the disconnecting means required by this article where other means for disconnection are provided in the following types of occupancies.

**(A) Multifamily Dwellings.** In multifamily dwellings, the other disconnecting means shall be within the dwelling unit, or on the same floor as the dwelling unit in which the appliance is installed, and shall be permitted to control lamps and other appliances.

**(B) Two-Family Dwellings.** In two-family dwellings, the other disconnecting means shall be permitted either inside or outside of the dwelling unit in which the appliance is installed. In this case, an individual switch or circuit breaker for the dwelling unit shall be permitted and shall also be permitted to control lamps and other appliances.

**(C) One-Family Dwellings.** In one-family dwellings, the service disconnecting means shall be permitted to be the other disconnecting means.

**(D) Other Occupancies.** In other occupancies, the branch-circuit switch or circuit breaker, where readily accessible for servicing of the appliance, shall be permitted as the other disconnecting means.

**422.35 Switch and Circuit Breaker to Be Indicating.** Switches and circuit breakers used as disconnecting means shall be of the indicating type.

### IV. Construction

**422.40 Polarity in Cord-and-Plug-Connected Appliances.** If the appliance is provided with a manually operated, line-connected, single-pole switch for appliance on–off operation, an Edison-base lampholder, or a 15- or 20-ampere receptacle, the attachment plug shall be of the polarized or grounding type.

A 2-wire, nonpolarized attachment plug shall be permitted to be used on a listed double-insulated shaver.

FPN: For polarity of Edison-base lampholders, see 410.42(A).

**422.41 Cord-and-Plug-Connected Appliances Subject to Immersion.** Cord-and-plug-connected portable, free-standing hydromassage units and hand-held hair dryers shall be constructed to provide protection for personnel against electrocution when immersed while in the "on" or "off" position.

**422.42 Signals for Heated Appliances.** In other than dwelling-type occupancies, each electrically heated appliance or group of appliances intended to be applied to combustible material shall be provided with a signal or an integral temperature-limiting device.

**422.43 Flexible Cords.**

**(A) Heater Cords.** All cord-and-plug-connected smoothing irons and electrically heated appliances that are rated at more than 50 watts and produce temperatures in excess of 121°C (250°F) on surfaces with which the cord is likely to be in contact shall be provided with one of the types of approved heater cords listed in Table 400.4.

**(B) Other Heating Appliances.** All other cord-and-plug-connected electrically heated appliances shall be connected with one of the approved types of cord listed in Table 400.4, selected in accordance with the usage specified in that table.

**422.44 Cord-and-Plug-Connected Immersion Heaters.** Electric heaters of the cord-and-plug-connected immersion type shall be constructed and installed so that current-carrying parts are effectively insulated from electrical contact with the substance in which they are immersed.

**422.45 Stands for Cord-and-Plug-Connected Appliances.** Each smoothing iron and other cord-and-plug-connected electrically heated appliance intended to be applied to combustible material shall be equipped with an approved stand, which shall be permitted to be a separate piece of equipment or a part of the appliance.

**422.46 Flatirons.** Electrically heated smoothing irons shall be equipped with an identified temperature-limiting means.

**422.47 Water Heater Controls.** All storage or instantaneous-type water heaters shall be equipped with a temperature-limiting means in addition to its control thermostat to disconnect all ungrounded conductors. Such means shall be as follows:

(1) Installed to sense maximum water temperature; and
(2) Either a trip-free, manually reset type or a type having a replacement element. Such water heaters shall be marked to require the installation of a temperature and pressure relief valve.

*Exception No. 1: Storage water heaters that are identified as being suitable for use with supply water temperature of 82°C (180°F) or above and a capacity of 60 kW or above.*

*Exception No. 2: Instantaneous-type water heaters that are identified as being suitable for such use, with a capacity of 4 L (1 gal) or less.*

FPN: See ANSI Z21.22-1999/CSA 4.4-M99, *Relief Valves for Hot Water Supply Systems.*

**422.48 Infrared Lamp Industrial Heating Appliances.**

**(A) 300 Watts or Less.** Infrared heating lamps rated at 300 watts or less shall be permitted with lampholders of the medium-base, unswitched porcelain type or other types identified as suitable for use with infrared heating lamps rated 300 watts or less.

**(B) Over 300 Watts.** Screw-shell lampholders shall not be used with infrared lamps rated over 300 watts, unless the lampholders are identified as being suitable for use with infrared heating lamps rated over 300 watts.

**422.49 High-Pressure Spray Washers.** All single-phase cord-and-plug-connected high-pressure spray washing machines rated at 250 volts or less shall be provided with factory-installed ground-fault circuit-interrupter protection for personnel. The ground-fault circuit interrupter shall be an integral part of the attachment plug or shall be located in the supply cord within 300 mm (12 in.) of the attachment plug.

**422.50 Cord-and-Plug-Connected Pipe Heating Assemblies.** Cord-and-plug-connected pipe heating assemblies intended to prevent freezing of piping shall be listed.

**V. Marking**

**422.60 Nameplate.**

**(A) Nameplate Marking.** Each electric appliance shall be provided with a nameplate giving the identifying name and the rating in volts and amperes, or in volts and watts. If the appliance is to be used on a specific frequency or frequencies, it shall be so marked.

Where motor overload protection external to the appliance is required, the appliance shall be so marked.

FPN: See 422.11 for overcurrent protection requirements.

**(B) To Be Visible.** Marking shall be located so as to be visible or easily accessible after installation.

**422.61 Marking of Heating Elements.** All heating elements that are rated over one ampere, replaceable in the field, and a part of an appliance shall be legibly marked with the ratings in volts and amperes, or in volts and watts, or with the manufacturer's part number.

**422.62 Appliances Consisting of Motors and Other Loads.**

**(A) Nameplate Horsepower Markings.** Where a motor-operated appliance nameplate includes a horsepower rating, that rating shall not be less than the horsepower rating on the motor nameplate. Where an appliance consists of multiple motors, or one or more motors and other loads, the nameplate value shall not be less than the equivalent horsepower of the combined loads, calculated in accordance with 430.110(C)(1).

**(B) Additional Nameplate Markings.** Appliances, other than those factory-equipped with cords and attachment plugs and with nameplates in compliance with 422.60, shall be marked in accordance with 422.62(B)(1) or (2).

**(1) Marking.** In addition to the marking required in 422.60, the marking on an appliance consisting of a motor with other load(s) or motors with or without other load(s) shall specify the minimum supply circuit conductor ampacity and the maximum rating of the circuit overcurrent protective device. This requirement shall not apply to an appliance with a nameplate in compliance with 422.60 where both the minimum supply circuit conductor ampacity and maximum rating of the circuit overcurrent protective device are not more than 15 amperes.

**(2) Alternate Marking Method.** An alternative marking method shall be permitted to specify the rating of the largest motor in volts and amperes, and the additional load(s) in volts and amperes, or volts and watts in addition to the marking required in 422.60. The ampere rating of a motor ⅛ horsepower or less or a nonmotor load 1 ampere or less shall be permitted to be omitted unless such loads constitute the principal load.

## ARTICLE 424
## Fixed Electric Space-Heating Equipment

### I. General

**424.1 Scope.** This article covers fixed electric equipment used for space heating. For the purpose of this article, heating equipment shall include heating cable, unit heaters, boilers, central systems, or other approved fixed electric space-heating equipment. This article shall not apply to process heating and room air conditioning.

**424.2 Other Articles.** All requirements of this *Code* shall apply where applicable. Fixed electric space-heating equipment for use in hazardous (classified) locations shall comply with Articles 500 through 517. Fixed electric space-heating equipment incorporating a hermetic refrigerant motor-compressor shall also comply with Article 440.

**424.3 Branch Circuits.**

**(A) Branch-Circuit Requirements.** Individual branch circuits shall be permitted to supply any size fixed electric space-heating equipment.

Branch circuits supplying two or more outlets for fixed electric space-heating equipment shall be rated 15, 20, 25, or 30 amperes. In other than residential occupancies, fixed infrared heating equipment shall be permitted to be supplied from branch circuits rated not over 50 amperes.

**(B) Branch-Circuit Sizing.** The ampacity of the branch-circuit conductors and the rating or setting of overcurrent protective devices supplying fixed electric space-heating equipment consisting of resistance elements with or without a motor shall not be less than 125 percent of the total load of the motors and the heaters. The rating or setting of overcurrent protective devices shall be permitted in accordance with 240.4(B). A contactor, thermostat, relay, or similar device, listed for continuous operation at 100 percent of its rating, shall be permitted to supply its full-rated load as provided in 210.19(A), Exception.

The size of the branch-circuit conductors and overcurrent protective devices supplying fixed electric space-heating equipment, including a hermetic refrigerant motor-compressor with or without resistance units, shall be computed in accordance with 440.34 and 440.35. The provisions of this section shall not apply to conductors that form an integral part of approved fixed electric space-heating equipment.

### II. Installation

**424.9 General.** All fixed electric space-heating equipment shall be installed in an approved manner.

Permanently installed electric baseboard heaters equipped with factory-installed receptacle outlets, or outlets provided as a separate listed assembly, shall be permitted in lieu of a receptacle outlet(s) that is required by 210.50(B). Such receptacle outlets shall not be connected to the heater circuits.

> FPN: Listed baseboard heaters include instructions that may not permit their installation below receptacle outlets.

**424.10 Special Permission.** Fixed electric space-heating equipment and systems installed by methods other than covered by this article shall be permitted only by special permission.

**424.11 Supply Conductors.** Fixed electric space-heating equipment requiring supply conductors with over 60°C insulation shall be clearly and permanently marked. This marking shall be plainly visible after installation and shall be permitted to be adjacent to the field connection box.

**424.12 Locations.**

**(A) Exposed to Physical Damage.** Where subject to physical damage, fixed electric space-heating equipment shall be protected in an approved manner.

**(B) Damp or Wet Locations.** Heaters and related equipment installed in damp or wet locations shall be approved for such locations and shall be constructed and installed so that water or other liquids cannot enter or accumulate in or on wired sections, electrical components, or ductwork.

> FPN No. 1: See 110.11 for equipment exposed to deteriorating agents.

> FPN No. 2: See 680.27(C) for pool deck areas.

**424.13 Spacing from Combustible Materials.** Fixed electric space-heating equipment shall be installed to provide the required spacing between the equipment and adjacent combustible material, unless it has been found to be acceptable where installed in direct contact with combustible material.

## III. Control and Protection of Fixed Electric Space-Heating Equipment

**424.19 Disconnecting Means.** Means shall be provided to disconnect the heater, motor controller(s), and supplementary overcurrent protective device(s) of all fixed electric space-heating equipment from all ungrounded conductors. Where heating equipment is supplied by more than one source, the disconnecting means shall be grouped and marked.

**(A) Heating Equipment with Supplementary Overcurrent Protection.** The disconnecting means for fixed electric space-heating equipment with supplementary overcurrent protection shall be within sight from the supplementary overcurrent protective device(s), on the supply side of these devices, if fuses, and, in addition, shall comply with either 424.19(A)(1) or (2).

**(1) Heater Containing No Motor Rated Over ⅛ Horsepower.** The above disconnecting means or unit switches complying with 424.19(C) shall be permitted to serve as the required disconnecting means for both the motor controller(s) and heater under either item (1) or (2):

(1) The disconnecting means provided is also within sight from the motor controller(s) and the heater.
(2) The disconnecting means provided shall be capable of being locked in the open position.

**(2) Heater Containing a Motor(s) Rated Over ⅛ Horsepower.** The above disconnecting means shall be permitted to serve as the required disconnecting means for both the motor controller(s) and heater by one of the means specified in items (1) through (4):

(1) Where the disconnecting means is also in sight from the motor controller(s) and the heater.
(2) Where the disconnecting means is not within sight from the heater, a separate disconnecting means shall be installed, or the disconnecting means shall be capable of being locked in the open position, or unit switches complying with 424.19(C) shall be permitted.
(3) Where the disconnecting means is not within sight from the motor controller location, a disconnecting means complying with 430.102 shall be provided.
(4) Where the motor is not in sight from the motor controller location, 430.102(B) shall apply.

**(B) Heating Equipment Without Supplementary Overcurrent Protection.**

**(1) Without Motor or with Motor Not Over ⅛ Horsepower.** For fixed electric space-heating equipment without a motor rated over ⅛ hp, the branch-circuit switch or circuit breaker shall be permitted to serve as the disconnecting means where the switch or circuit breaker is within sight from the heater or is capable of being locked in the open position.

**(2) Over ⅛ Horsepower.** For motor-driven electric space-heating equipment with a motor rated over ⅛ hp, a disconnecting means shall be located within sight from the motor controller or shall be permitted to comply with the requirements in 424.19(A)(2).

**(C) Unit Switch(es) as Disconnecting Means.** A unit switch(es) with a marked "off" position that is part of a fixed heater and disconnects all ungrounded conductors shall be permitted as the disconnecting means required by this article where other means for disconnection are provided in the types of occupancies in 424.19(C)(1) through (C)(4).

**(1) Multifamily Dwellings.** In multifamily dwellings, the other disconnecting means shall be within the dwelling unit, or on the same floor as the dwelling unit in which the fixed heater is installed, and shall also be permitted to control lamps and appliances.

**(2) Two-Family Dwellings.** In two-family dwellings, the other disconnecting means shall be permitted either inside or outside of the dwelling unit in which the fixed heater is installed. In this case, an individual switch or circuit breaker for the dwelling unit shall be permitted and shall also be permitted to control lamps and appliances.

**(3) One-Family Dwellings.** In one-family dwellings, the service disconnecting means shall be permitted to be the other disconnecting means.

**(4) Other Occupancies.** In other occupancies, the branch-circuit switch or circuit breaker, where readily accessible for servicing of the fixed heater, shall be permitted as the other disconnecting means.

### 424.20 Thermostatically Controlled Switching Devices.

**(A) Serving as Both Controllers and Disconnecting Means.** Thermostatically controlled switching devices and combination thermostats and manually controlled switches shall be permitted to serve as both controllers and disconnecting means, provided all of the following conditions are met:

(1) Provided with a marked "off" position
(2) Directly open all ungrounded conductors when manually placed in the "off" position
(3) Designed so that the circuit cannot be energized automatically after the device has been manually placed in the "off" position
(4) Located as specified in 424.19

**(B) Thermostats That Do Not Directly Interrupt All Ungrounded Conductors.** Thermostats that do not directly interrupt all ungrounded conductors and thermostats that operate remote-control circuits shall not be required to meet the requirements of 424.20(A). These devices shall not be permitted as the disconnecting means.

### 424.21 Switch and Circuit Breaker to Be Indicating.
Switches and circuit breakers used as disconnecting means shall be of the indicating type.

### 424.22 Overcurrent Protection.

**(A) Branch-Circuit Devices.** Electric space-heating equipment, other than such motor-operated equipment as required by Articles 430 and 440 to have additional overcurrent protection, shall be permitted to be protected against overcurrent where supplied by one of the branch circuits in Article 210.

**(B) Resistance Elements.** Resistance-type heating elements in electric space-heating equipment shall be protected at not more than 60 amperes. Equipment rated more than 48 amperes and employing such elements shall have the heating elements subdivided, and each subdivided load shall not exceed 48 amperes. Where a subdivided load is less than 48 amperes, the rating of the supplementary overcurrent protective device shall comply with 424.3(B). A boiler employing resistance-type immersion heating elements contained in an ASME rated and stamped vessel shall be permitted to comply with 424.72(A).

**(C) Overcurrent Protective Devices.** The supplementary overcurrent protective devices for the subdivided loads specified in 424.22(B) shall be (1) factory-installed within or on the heater enclosure or supplied for use with the heater as a separate assembly by the heater manufacturer; (2) accessible, but shall not be required to be readily accessible; and (3) suitable for branch-circuit protection.

FPN: See 240.10.

Where cartridge fuses are used to provide this overcurrent protection, a single disconnecting means shall be permitted to be used for the several subdivided loads.

FPN No. 1: For supplementary overcurrent protection, see 240.10.

FPN No. 2: For disconnecting means for cartridge fuses in circuits of any voltage, see 240.40.

**(D) Branch-Circuit Conductors.** The conductors supplying the supplementary overcurrent protective devices shall be considered branch-circuit conductors.

Where the heaters are rated 50 kW or more, the conductors supplying the supplementary overcurrent protective devices specified in 424.22(C) shall be permitted to be sized at not less than 100 percent of the nameplate rating of the heater, provided all of the following conditions are met:

(1) The heater is marked with a minimum conductor size.
(2) The conductors are not smaller than the marked minimum size.
(3) A temperature-actuated device controls the cyclic operation of the equipment.

**(E) Conductors for Subdivided Loads.** Field-wired conductors between the heater and the supplementary overcurrent protective devices shall be sized at not less than

125 percent of the load served. The supplementary overcurrent protective devices specified in 424.22(C) shall protect these conductors in accordance with 240.4.

Where the heaters are rated 50 kW or more, the ampacity of field-wired conductors between the heater and the supplementary overcurrent protective devices shall be permitted to be not less than 100 percent of the load of their respective subdivided circuits, provided all of the following conditions are met:

(1) The heater is marked with a minimum conductor size.
(2) The conductors are not smaller than the marked minimum size.
(3) A temperature-activated device controls the cyclic operation of the equipment.

## IV. Marking of Heating Equipment

### 424.28 Nameplate.

**(A) Marking Required.** Each unit of fixed electric space-heating equipment shall be provided with a nameplate giving the identifying name and the normal rating in volts and watts or in volts and amperes.

Electric space-heating equipment intended for use on alternating current only or direct current only shall be marked to so indicate. The marking of equipment consisting of motors over $\frac{1}{8}$ hp and other loads shall specify the rating of the motor in volts, amperes, and frequency, and the heating load in volts and watts or in volts and amperes.

**(B) Location.** This nameplate shall be located so as to be visible or easily accessible after installation.

### 424.29 Marking of Heating Elements.
All heating elements that are replaceable in the field and are a part of an electric heater shall be legibly marked with the ratings in volts and watts or in volts and amperes.

## V. Electric Space-Heating Cables

### 424.34 Heating Cable Construction.
Heating cables shall be furnished complete with factory-assembled nonheating leads at least 2.1 m (7 ft) in length.

### 424.35 Marking of Heating Cables.
Each unit shall be marked with the identifying name or identification symbol, catalog number, and ratings in volts and watts or in volts and amperes.

Each unit length of heating cable shall have a permanent legible marking on each nonheating lead located within 75 mm (3 in.) of the terminal end. The lead wire shall have the following color identification to indicate the circuit voltage on which it is to be used:

(1) 120 volt, nominal — yellow
(2) 208 volt, nominal — blue
(3) 240 volt, nominal — red
(4) 277 volt, nominal — brown
(5) 480 volt, nominal — orange

### 424.36 Clearances of Wiring in Ceilings.
Wiring located above heated ceilings shall be spaced not less than 50 mm (2 in.) above the heated ceiling and shall be considered as operating at an ambient temperature of 50°C (122°F). The ampacity of conductors shall be computed on the basis of the correction factors shown in the 0–2000 volt ampacity tables of Article 310. If this wiring is located above thermal insulation having a minimum thickness of 50 mm (2 in.), the wiring shall not require correction for temperature.

### 424.37 Location of Branch-Circuit and Feeder Wiring in Exterior Walls.
Wiring methods shall comply with Article 300 and 310.10.

### 424.38 Area Restrictions.

**(A) Shall Not Extend Beyond the Room or Area.** Heating cables shall not extend beyond the room or area in which they originate.

**(B) Uses Prohibited.** Heating cables shall not be installed in the following:

(1) In closets
(2) Over walls
(3) Over partitions that extend to the ceiling, unless they are isolated single runs of embedded cable
(4) Over cabinets whose clearance from the ceiling is less than the minimum horizontal dimension of the cabinet to the nearest cabinet edge that is open to the room or area

**(C) In Closet Ceilings as Low-Temperature Heat Sources to Control Relative Humidity.** The provisions of 424.38(B) shall not prevent the use of cable in closet ceilings as low-temperature heat sources to control relative humidity, provided they are used only in those portions of the ceiling that are unobstructed to the floor by shelves or other permanent luminaires (fixtures).

### 424.39 Clearance from Other Objects and Openings.
Heating elements of cables shall be separated at least 200 mm (8 in.) from the edge of outlet boxes and junction boxes that are to be used for mounting surface luminaires (lighting fixtures). A clearance of not less than 50 mm (2 in.) shall be provided from recessed luminaires (fixtures) and their trims, ventilating openings, and other such openings in room surfaces. Sufficient area shall be provided to ensure that no heating cable will be covered by any surface-mounted units.

**424.40 Splices.** Embedded cables shall be spliced only where necessary and only by approved means, and in no case shall the length of the heating cable be altered.

**424.41 Installation of Heating Cables on Dry Board, in Plaster, and on Concrete Ceilings.**

**(A) In Walls.** Cables shall not be installed in walls unless it is necessary for an isolated single run of cable to be installed down a vertical surface to reach a dropped ceiling.

**(B) Adjacent Runs.** Adjacent runs of cable not exceeding 9 watts/m (2¾ watts/ft) shall not be installed less than 38 mm (1½ in.) on centers.

**(C) Surfaces to Be Applied.** Heating cables shall be applied only to gypsum board, plaster lath, or other fire-resistant material. With metal lath or other electrically conductive surfaces, a coat of plaster shall be applied to completely separate the metal lath or conductive surface from the cable.

FPN: See also 424.41(F).

**(D) Splices.** All heating cables, the splice between the heating cable and nonheating leads, and 75-mm (3-in.) minimum of the nonheating lead at the splice shall be embedded in plaster or dry board in the same manner as the heating cable.

**(E) Ceiling Surface.** The entire ceiling surface shall have a finish of thermally noninsulating sand plaster that has a nominal thickness of 13 mm (½ in.), or other noninsulating material identified as suitable for this use and applied according to specified thickness and directions.

**(F) Secured.** Cables shall be secured by means of approved stapling, tape, plaster, nonmetallic spreaders, or other approved means either at intervals not exceeding 400 mm (16 in.) or at intervals not exceeding 1.8 m (6 ft) for cables identified for such use. Staples or metal fasteners that straddle the cable shall not be used with metal lath or other electrically conductive surfaces.

**(G) Dry Board Installations.** In dry board installations, the entire ceiling below the heating cable shall be covered with gypsum board not exceeding 13 mm (½ in.) thickness. The void between the upper layer of gypsum board, plaster lath, or other fire-resistant material and the surface layer of gypsum board shall be completely filled with thermally conductive, nonshrinking plaster or other approved material or equivalent thermal conductivity.

**(H) Free from Contact with Conductive Surfaces.** Cables shall be kept free from contact with metal or other electrically conductive surfaces.

**(I) Joists.** In dry board applications, cable shall be installed parallel to the joist, leaving a clear space centered under the joist of 65 mm (2½ in.) (width) between centers of adjacent runs of cable. A surface layer of gypsum board shall be mounted so that the nails or other fasteners do not pierce the heating cable.

**(J) Crossing Joists.** Cables shall cross joists only at the ends of the room unless the cable is required to cross joists elsewhere in order to satisfy the manufacturer's instructions that the installer avoid placing the cable too close to ceiling penetrations and luminaires (lighting fixtures).

**424.42 Finished Ceilings.** Finished ceilings shall not be covered with decorative panels or beams constructed of materials that have thermal insulating properties, such as wood, fiber, or plastic. Finished ceilings shall be permitted to be covered with paint, wallpaper, or other approved surface finishes.

**424.43 Installation of Nonheating Leads of Cables.**

**(A) Free Nonheating Leads.** Free nonheating leads of cables shall be installed in accordance with approved wiring methods from the junction box to a location within the ceiling. Such installations shall be permitted to be single conductors in approved raceways, single or multiconductor Type UF, Type NMC, Type MI, or other approved conductors.

**(B) Leads in Junction Box.** Not less than 150 mm (6 in.) of free nonheating lead shall be within the junction box. The marking of the leads shall be visible in the junction box.

**(C) Excess Leads.** Excess leads of heating cables shall not be cut but shall be secured to the underside of the ceiling and embedded in plaster or other approved material, leaving only a length sufficient to reach the junction box with not less than 150 mm (6 in.) of free lead within the box.

**424.44 Installation of Cables in Concrete or Poured Masonry Floors.**

**(A) Watts per Linear Foot.** Constant wattage heating cables shall not exceed 54 watts/linear meter (16½ watts/linear foot) of cable.

**(B) Spacing Between Adjacent Runs.** The spacing between adjacent runs of cable shall not be less than 25 mm (1 in.) on centers.

**(C) Secured in Place.** Cables shall be secured in place by nonmetallic frames or spreaders or other approved means while the concrete or other finish is applied.

Cables shall not be installed where they bridge expansion joints unless protected from expansion and contraction.

**(D) Spacings Between Heating Cable and Metal Embedded in the Floor.** Spacings shall be maintained between the heating cable and metal embedded in the floor, unless the cable is a grounded metal-clad cable.

**(E) Leads Protected.** Leads shall be protected where they leave the floor by rigid metal conduit, intermediate metal conduit, rigid nonmetallic conduit, electrical metallic tubing, or by other approved means.

**(F) Bushings or Approved Fittings.** Bushings or approved fittings shall be used where the leads emerge within the floor slab.

**(G) Ground-Fault Circuit-Interrupter Protection for Heated Floors of Bathrooms, and in Hydromassage Bathtub, Spa, and Hot Tub Locations.** Ground-fault circuit-interrupter protection for personnel shall be provided for electrically heated floors in bathrooms, and in hydromassage bathtub, spa, and hot tub locations.

**424.45 Inspection and Tests.** Cable installations shall be made with due care to prevent damage to the cable assembly and shall be inspected and approved before cables are covered or concealed.

## VI. Duct Heaters

**424.57 General.** Part VI shall apply to any heater mounted in the airstream of a forced-air system where the air-moving unit is not provided as an integral part of the equipment.

**424.58 Identification.** Heaters installed in an air duct shall be identified as suitable for the installation.

**424.59 Airflow.** Means shall be provided to ensure uniform and adequate airflow over the face of the heater in accordance with the manufacturer's instructions.

FPN: Heaters installed within 1.2 m (4 ft) of the outlet of an air-moving device, heat pump, air conditioner, elbows, baffle plates, or other obstructions in ductwork may require turning vanes, pressure plates, or other devices on the inlet side of the duct heater to ensure an even distribution of air over the face of the heater.

**424.60 Elevated Inlet Temperature.** Duct heaters intended for use with elevated inlet air temperature shall be identified as suitable for use at the elevated temperatures.

**424.61 Installation of Duct Heaters with Heat Pumps and Air Conditioners.** Heat pumps and air conditioners having duct heaters closer than 1.2 m (4 ft) to the heat pump or air conditioner shall have both the duct heater and heat pump or air conditioner identified as suitable for such installation and so marked.

**424.62 Condensation.** Duct heaters used with air conditioners or other air-cooling equipment that could result in condensation of moisture shall be identified as suitable for use with air conditioners.

**424.63 Fan Circuit Interlock.** Means shall be provided to ensure that the fan circuit is energized when any heater circuit is energized. However, time- or temperature-controlled delay in energizing the fan motor shall be permitted.

**424.64 Limit Controls.** Each duct heater shall be provided with an approved, integral, automatic-reset temperature-limiting control or controllers to de-energize the circuit or circuits.

In addition, an integral independent supplementary control or controllers shall be provided in each duct heater that disconnects a sufficient number of conductors to interrupt current flow. This device shall be manually resettable or replaceable.

**424.65 Location of Disconnecting Means.** Duct heater controller equipment shall be either accessible with the disconnecting means installed at or within sight from the controller or as permitted by 424.19(A).

**424.66 Installation.** Duct heaters shall be installed in accordance with the manufacturer's instructions in such a manner that operation will not create a hazard to persons or property. Furthermore, duct heaters shall be located with respect to building construction and other equipment so as to permit access to the heater. Sufficient clearance shall be maintained to permit replacement of controls and heating elements and for adjusting and cleaning of controls and other parts requiring such attention. See 110.26.

FPN: For additional installation information, see NFPA 90A-1999, *Standard for the Installation of Air Conditioning and Ventilating Systems*, and NFPA 90B-1999, *Standard for the Installation of Warm Air Heating and Air Conditioning Systems*.

## VII. Resistance-Type Boilers

**424.70 Scope.** The provisions in Part VII of this article shall apply to boilers employing resistance-type heating elements. Electrode-type boilers shall not be considered as employing resistance-type heating elements. See Part VIII of this article.

**424.71 Identification.** Resistance-type boilers shall be identified as suitable for the installation.

## 424.72 Overcurrent Protection.

**(A) Boiler Employing Resistance-Type Immersion Heating Elements in an ASME Rated and Stamped Vessel.** A boiler employing resistance-type immersion heating elements contained in an ASME rated and stamped vessel shall have the heating elements protected at not more than 150 amperes. Such a boiler rated more than 120 amperes shall have the heating elements subdivided into loads not exceeding 120 amperes.

Where a subdivided load is less than 120 amperes, the rating of the overcurrent protective device shall comply with 424.3(B).

**(B) Boiler Employing Resistance-Type Heating Elements Rated More Than 48 Amperes and Not Contained in an ASME Rated and Stamped Vessel.** A boiler employing resistance-type heating elements not contained in an ASME rated and stamped vessel shall have the heating elements protected at not more than 60 amperes. Such a boiler rated more than 48 amperes shall have the heating elements subdivided into loads not exceeding 48 amperes.

Where a subdivided load is less than 48 amperes, the rating of the overcurrent protective device shall comply with 424.3(B).

**(C) Supplementary Overcurrent Protective Devices.** The supplementary overcurrent protective devices for the subdivided loads as required by 424.72(A) and (B) shall be as follows:

(1) Factory-installed within or on the boiler enclosure or provided as a separate assembly by the boiler manufacturer
(2) Accessible, but need not be readily accessible
(3) Suitable for branch-circuit protection

Where cartridge fuses are used to provide this overcurrent protection, a single disconnecting means shall be permitted for the several subdivided circuits. See 240.40.

**(D) Conductors Supplying Supplementary Overcurrent Protective Devices.** The conductors supplying these supplementary overcurrent protective devices shall be considered branch-circuit conductors.

Where the heaters are rated 50 kW or more, the conductors supplying the overcurrent protective device specified in 424.72(C) shall be permitted to be sized at not less than 100 percent of the nameplate rating of the heater, provided all of the following conditions are met:

(1) The heater is marked with a minimum conductor size.
(2) The conductors are not smaller than the marked minimum size.
(3) A temperature- or pressure-actuated device controls the cyclic operation of the equipment.

**(E) Conductors for Subdivided Loads.** Field-wired conductors between the heater and the supplementary overcurrent protective devices shall be sized at not less than 125 percent of the load served. The supplementary overcurrent protective devices specified in 424.72(C) shall protect these conductors in accordance with 240.4.

Where the heaters are rated 50 kW or more, the ampacity of field-wired conductors between the heater and the supplementary overcurrent protective devices shall be permitted to be not less than 100 percent of the load of their respective subdivided circuits, provided all of the following conditions are met:

(1) The heater is marked with a minimum conductor size.
(2) The conductors are not smaller than the marked minimum size.
(3) A temperature-activated device controls the cyclic operation of the equipment.

**424.73 Overtemperature Limit Control.** Each boiler designed so that in normal operation there is no change in state of the heat transfer medium shall be equipped with a temperature-sensitive limiting means. It shall be installed to limit maximum liquid temperature and shall directly or indirectly disconnect all ungrounded conductors to the heating elements. Such means shall be in addition to a temperature regulating system and other devices protecting the tank against excessive pressure.

**424.74 Overpressure Limit Control.** Each boiler designed so that in normal operation there is a change in state of the heat transfer medium from liquid to vapor shall be equipped with a pressure-sensitive limiting means. It shall be installed to limit maximum pressure and shall directly or indirectly disconnect all ungrounded conductors to the heating elements. Such means shall be in addition to a pressure regulating system and other devices protecting the tank against excessive pressure.

## VIII. Electrode-Type Boilers

**424.80 Scope.** The provisions in Part VIII of this article shall apply to boilers for operation at 600 volts, nominal, or less, in which heat is generated by the passage of current between electrodes through the liquid being heated.

FPN: For over 600 volts, see Part V of Article 490.

**424.81 Identification.** Electrode-type boilers shall be identified as suitable for the installation.

**424.82 Branch-Circuit Requirements.** The size of branch-circuit conductors and overcurrent protective devices shall be calculated on the basis of 125 percent of the total load (motors not included). A contactor, relay, or other device, approved for continuous operation at 100 percent of its rating, shall be permitted to supply its full-rated load. See 210.19(A), Exception. The provisions of this section shall not apply to conductors that form an integral part of an approved boiler.

Where an electrode boiler is rated 50 kW or more, the conductors supplying the boiler electrode(s) shall be permitted to be sized at not less than 100 percent of the nameplate rating of the electrode boiler, provided all the following conditions are met:

(1) The electrode boiler is marked with a minimum conductor size.
(2) The conductors are not smaller than the marked minimum size.
(3) A temperature- or pressure-actuated device controls the cyclic operation of the equipment.

**424.83 Overtemperature Limit Control.** Each boiler designed so that in normal operation there is no change in state of the heat transfer medium shall be equipped with a temperature-sensitive limiting means. It shall be installed to limit maximum liquid temperature and shall directly or indirectly interrupt all current flow through the electrodes. Such means shall be in addition to the temperature regulating system and other devices protecting the tank against excessive pressure.

**424.84 Overpressure Limit Control.** Each boiler designed so that in normal operation there is a change in state of the heat transfer medium from liquid to vapor shall be equipped with a pressure-sensitive limiting means. It shall be installed to limit maximum pressure and shall directly or indirectly interrupt all current flow through the electrodes. Such means shall be in addition to a pressure regulating system and other devices protecting the tank against excessive pressure.

**424.85 Grounding.** For those boilers designed such that fault currents do not pass through the pressure vessel, and the pressure vessel is electrically isolated from the electrodes, all exposed non–current-carrying metal parts, including the pressure vessel, supply, and return connecting piping, shall be grounded in accordance with Article 250.

For all other designs, the pressure vessel containing the electrodes shall be isolated and electrically insulated from ground.

**424.86 Markings.** All electrode-type boilers shall be marked to show the following:

(1) The manufacturer's name
(2) The normal rating in volts, amperes, and kilowatts
(3) The electrical supply required specifying frequency, number of phases, and number of wires

(4) The marking "Electrode-Type Boiler"
(5) A warning marking, "All Power Supplies Shall Be Disconnected Before Servicing, Including Servicing the Pressure Vessel"

The nameplate shall be located so as to be visible after installation.

## IX. Electric Radiant Heating Panels and Heating Panel Sets

**424.90 Scope.** The provisions of Part IX of this article shall apply to radiant heating panels and heating panel sets.

**424.91 Definitions.**

**Heating Panel.** A complete assembly provided with a junction box or a length of flexible conduit for connection to a branch circuit.

**Heating Panel Set.** A rigid or nonrigid assembly provided with nonheating leads or a terminal junction assembly identified as being suitable for connection to a wiring system.

**424.92 Markings.**

**(A)** Markings shall be permanent and in a location that is visible prior to application of panel finish.

**(B)** Each unit shall be identified as suitable for the installation.

**(C)** Each unit shall be marked with the identifying name or identification symbol, catalog number, and rating in volts and watts or in volts and amperes.

**(D)** The manufacturers of heating panels or heating panel sets shall provide marking labels that indicate that the space-heating installation incorporates heating panels or heating panel sets and instructions that the labels shall be affixed to the panelboards to identify which branch circuits supply the circuits to those space-heating installations. If the heating panels and heating panel set installations are visible and distinguishable after installation, the labels shall not be required to be provided and affixed to the panelboards.

**424.93 Installation.**

**(A) General.**

**(1) Manufacturer's Instructions.** Heating panels and heating panel sets shall be installed in accordance with the manufacturer's instructions.

**(2) Locations Not Permitted.** The heating portion shall not

(1) Be installed in or behind surfaces where subject to physical damage.

(2) Be run through or above walls, partitions, cupboards, or similar portions of structures that extend to the ceiling.

(3) Be run in or through thermal insulation, but shall be permitted to be in contact with the surface of thermal insulation.

**(3) Separation From Outlets for Luminaires (Lighting Fixtures).** Edges of panels and panel sets shall be separated by not less than 200 mm (8 in.) from the edges of any outlet boxes and junction boxes that are to be used for mounting surface luminaires (lighting fixtures). A clearance of not less than 50 mm (2 in.) shall be provided from recessed luminaires (fixtures) and their trims, ventilating openings, and other such openings in room surfaces, unless the heating panels and panel sets are listed and marked for lesser clearances, in which case they shall be permitted to be installed at the marked clearances. Sufficient area shall be provided to ensure that no heating panel or heating panel set is to be covered by any surface-mounted units.

**(4) Surfaces Covering Heating Panels.** After the heating panels or heating panel sets are installed and inspected, it shall be permitted to install a surface that has been identified by the manufacturer's instructions as being suitable for the installation. The surface shall be secured so that the nails or other fastenings do not pierce the heating panels or heating panel sets.

**(5) Surface Coverings.** Surfaces permitted by 424.93(A)(4) shall be permitted to be covered with paint, wallpaper, or other approved surfaces identified in the manufacturer's instructions as being suitable.

**(B) Heating Panel Sets.**

**(1) Mounting Location.** Heating panel sets shall be permitted to be secured to the lower face of joists or mounted in between joists, headers, or nailing strips.

**(2) Parallel to Joists or Nailing Strips.** Heating panel sets shall be installed parallel to joists or nailing strips.

**(3) Installation of Nails, Staples, or Other Fasteners.** Nailing or stapling of heating panel sets shall be done only through the unheated portions provided for this purpose. Heating panel sets shall not be cut through or nailed through any point closer than 6 mm (¼ in.) to the element. Nails, staples, or other fasteners shall not be used where they penetrate current-carrying parts.

**(4) Installed as Complete Unit.** Heating panel sets shall be installed as complete units unless identified as suitable for field cutting in an approved manner.

**424.94 Clearances of Wiring in Ceilings.** Wiring located above heated ceilings shall be spaced not less than 50 mm (2 in.) above the heated ceiling and shall be considered as operating at an ambient of 50°C (122°F). The ampacity shall be computed on the basis of the correction factors given in the 0–2000 volt ampacity tables of Article 310. If this wiring is located above thermal insulations having a minimum thickness of 50 mm (2 in.), the wiring shall not require correction for temperature.

**424.95 Location of Branch-Circuit and Feeder Wiring in Walls.**

**(A) Exterior Walls.** Wiring methods shall comply with Article 300 and 310.10.

**(B) Interior Walls.** Any wiring behind heating panels or heating panel sets located in interior walls or partitions shall be considered as operating at an ambient temperature of 40°C (104°F), and the ampacity shall be computed on the basis of the correction factors given in the 0–2000 volt ampacity tables of Article 310.

**424.96 Connection to Branch-Circuit Conductors.**

**(A) General.** Heating panels or heating panel sets assembled together in the field to form a heating installation in one room or area shall be connected in accordance with the manufacturer's instructions.

**(B) Heating Panels.** Heating panels shall be connected to branch-circuit wiring by an approved wiring method.

**(C) Heating Panel Sets.**

**(1) Connection to Branch Circuit Wiring.** Heating panel sets shall be connected to branch-circuit wiring by a method identified as being suitable for the purpose.

**(2) Panel Sets with Terminal Junction Assembly.** A heating panel set provided with terminal junction assembly shall be permitted to have the nonheating leads attached at the time of installation in accordance with the manufacturer's instructions.

**424.97 Nonheating Leads.** Excess nonheating leads of heating panels or heating panel sets shall be permitted to be cut to the required length. They shall meet the installation requirements of the wiring method employed in accordance with 424.96. Nonheating leads shall be an integral part of a heating panel and a heating panel set and shall not be subjected to the ampacity requirements of 424.3(B) for branch circuits.

**424.98 Installation in Concrete or Poured Masonry.**

**(A) Maximum Heated Area.** Heating panels or heating panel sets shall not exceed 355 watts/m$^2$ (33 watts/ft$^2$) of heated area.

**(B) Secured in Place and Identified as Suitable.** Heating panels or heating panel sets shall be secured in place by means specified in the manufacturer's instructions and identified as suitable for the installation.

**(C) Expansion Joints.** Heating panels or heating panel sets shall not be installed where they bridge expansion joints unless provision is made for expansion and contraction.

**(D) Spacings.** Spacings shall be maintained between heating panels or heating panel sets and metal embedded in the floor. Grounded metal-clad heating panels shall be permitted to be in contact with metal embedded in the floor.

**(E) Protection of Leads.** Leads shall be protected where they leave the floor by rigid metal conduit, intermediate metal conduit, rigid nonmetallic conduit, or electrical metallic tubing, or by other approved means.

**(F) Bushings or Fittings Required.** Bushings or approved fittings shall be used where the leads emerge within the floor slabs.

### 424.99 Installation Under Floor Covering.

**(A) Identification.** Heating panels or heating panel sets for installation under floor covering shall be identified as suitable for installation under floor covering.

**(B) Maximum Heated Area.** Heating panels or panel sets, installed under floor covering, shall not exceed 160 watts/m$^2$ (15 watts/ft$^2$) of heated area.

**(C) Installation.** Listed heating panels or panel sets, if installed under floor covering, shall be installed on floor surfaces that are smooth and flat in accordance with the manufacturer's instructions and shall also comply with 424.99(C)(1) through (C) (5).

**(1) Expansion Joints.** Heating panels or heating panel sets shall not be installed where they bridge expansion joints unless protected from expansion and contraction.

**(2) Connection to Conductors.** Heating panels and heating panel sets shall be connected to branch-circuit and supply wiring by wiring methods recognized in Chapter 3.

**(3) Anchoring.** Heating panels and heating panel sets shall be firmly anchored to the floor using an adhesive or anchoring system identified for this use.

**(4) Coverings.** After heating panels or heating panel sets are installed and inspected, they shall be permitted to be covered by a floor covering that has been identified by the manufacturer as being suitable for the installation. The covering shall be secured to the heating panel or heating panel

sets with release-type adhesives or by means identified for this use.

**(5) Fault Protection.** A device to open all ungrounded conductors supplying the heating panels or heating panel sets, provided by the manufacturer, shall function when a low- or high-resistance line-to-line, line-to-grounded conductor, or line-to-ground fault occurs, such as the result of a penetration of the element or element assembly.

> FPN: An integral grounding shield may be required to provide this protection.

## ARTICLE 426
## Fixed Outdoor Electric Deicing and Snow-Melting Equipment

### I. General

**426.1 Scope.** The requirements of this article shall apply to electrically energized heating systems and the installation of these systems.

**(A) Embedded.** Embedded in driveways, walks, steps, and other areas.

**(B) Exposed.** Exposed on drainage systems, bridge structures, roofs, and other structures.

**426.2 Definitions.** For the purpose of this article:

**Heating System.** A complete system consisting of components such as heating elements, fastening devices, nonheating circuit wiring, leads, temperature controllers, safety signs, junction boxes, raceways, and fittings.

**Impedance Heating System.** A system in which heat is generated in a pipe or rod, or combination of pipes and rods, by causing current to flow through the pipe or rod by direct connection to an ac voltage source from a dual-winding transformer. The pipe or rod shall be permitted to be embedded in the surface to be heated, or constitute the exposed components to be heated.

**Resistance Heating Element.** A specific separate element to generate heat that is embedded in or fastened to the surface to be heated.

> FPN: Tubular heaters, strip heaters, heating cable, heating tape, and heating panels are examples of resistance heaters.

**Skin-Effect Heating System.** A system in which heat is generated on the inner surface of a ferromagnetic envelope embedded in or fastened to the surface to be heated.

FPN: Typically, an electrically insulated conductor is routed through and connected to the envelope at the other end. The envelope and the electrically insulated conductor are connected to an ac voltage source from a dual-winding transformer.

**426.3 Application of Other Articles.** All requirements of this *Code* shall apply except as specifically amended in this article. Cord-and-plug-connected fixed outdoor electric deicing and snow-melting equipment intended for specific use and identified as suitable for this use shall be installed according to Article 422. Fixed outdoor electric deicing and snow-melting equipment for use in hazardous (classified) locations shall comply with Articles 500 through 516.

**426.4 Branch-Circuit Sizing.** The ampacity of branch-circuit conductors and the rating or setting of overcurrent protective devices supplying fixed outdoor electric deicing and snow-melting equipment shall not be less than 125 percent of the total load of the heaters. The rating or setting of overcurrent protective devices shall be permitted in accordance with 240.4(B).

## II. Installation

**426.10 General.** Equipment for outdoor electric deicing and snow melting shall be identified as being suitable for the following:

(1) The chemical, thermal, and physical environment
(2) Installation in accordance with the manufacturer's drawings and instructions

**426.11 Use.** Electrical heating equipment shall be installed in such a manner as to be afforded protection from physical damage.

**426.12 Thermal Protection.** External surfaces of outdoor electric deicing and snow-melting equipment that operate at temperatures exceeding 60°C (140°F) shall be physically guarded, isolated, or thermally insulated to protect against contact by personnel in the area.

**426.13 Identification.** The presence of outdoor electric deicing and snow-melting equipment shall be evident by the posting of appropriate caution signs or markings where clearly visible.

**426.14 Special Permission.** Fixed outdoor deicing and snow-melting equipment employing methods of construction or installation other than covered by this article shall be permitted only by special permission.

## III. Resistance Heating Elements

**426.20 Embedded Deicing and Snow-Melting Equipment.**

**(A) Watt Density.** Panels or units shall not exceed 1300 watts/m$^2$ (120 watts/ft$^2$) of heated area.

**(B) Spacing.** The spacing between adjacent cable runs is dependent upon the rating of the cable and shall be not less than 25 mm (1 in.) on centers.

**(C) Cover.** Units, panels, or cables shall be installed as follows:

(1) On a substantial asphalt or masonry base at least 50 mm (2 in.) thick and have at least 38 mm (1½ in.) of asphalt or masonry applied over the units, panels, or cables; or
(2) They shall be permitted to be installed over other approved bases and embedded within 90 mm (3½ in.) of masonry or asphalt but not less than 38 mm (1½ in.) from the top surface; or
(3) Equipment that has been specially investigated for other forms of installation shall be installed only in the manner for which it has been investigated.

**(D) Secured.** Cables, units, and panels shall be secured in place by frames or spreaders or other approved means while the masonry or asphalt finish is applied.

**(E) Expansion and Contraction.** Cables, units, and panels shall not be installed where they bridge expansion joints unless provision is made for expansion and contraction.

**426.21 Exposed Deicing and Snow-Melting Equipment.**

**(A) Secured.** Heating element assemblies shall be secured to the surface being heated by approved means.

**(B) Overtemperature.** Where the heating element is not in direct contact with the surface being heated, the design of the heater assembly shall be such that its temperature limitations shall not be exceeded.

**(C) Expansion and Contraction.** Heating elements and assemblies shall not be installed where they bridge expansion joints unless provision is made for expansion and contraction.

**(D) Flexural Capability.** Where installed on flexible structures, the heating elements and assemblies shall have a flexural capability that is compatible with the structure.

**426.22 Installation of Nonheating Leads for Embedded Equipment.**

**(A) Grounding Sheath or Braid.** Nonheating leads having a grounding sheath or braid shall be permitted to be embedded in the masonry or asphalt in the same manner as the heating cable without additional physical protection.

**(B) Raceways.** All but 25 mm to 150 mm (1 in. to 6 in.) of nonheating leads of Type TW and other approved types not having a grounding sheath shall be enclosed in a rigid conduit, electrical metallic tubing, intermediate metal conduit, or other raceways within asphalt or masonry; and the distance from the factory splice to raceway shall not be less than 25 mm (1 in.) or more than 150 mm (6 in.).

**(C) Bushings.** Insulating bushings shall be used in the asphalt or masonry where leads enter conduit or tubing.

**(D) Expansion and Contraction.** Leads shall be protected in expansion joints and where they emerge from masonry or asphalt by rigid conduit, electrical metallic tubing, intermediate metal conduit, other raceways, or other approved means.

**(E) Leads in Junction Boxes.** Not less than 150 mm (6 in.) of free nonheating lead shall be within the junction box.

### 426.23 Installation of Nonheating Leads for Exposed Equipment.

**(A) Nonheating Leads.** Power supply nonheating leads (cold leads) for resistance elements shall be suitable for the temperature encountered. Preassembled nonheating leads on approved heaters shall be permitted to be shortened if the markings specified in 426.25 are retained. Not less than 150 mm (6 in.) of nonheating leads shall be provided within the junction box.

**(B) Protection.** Nonheating power supply leads shall be enclosed in a rigid conduit, intermediate metal conduit, electrical metallic tubing, or other approved means.

### 426.24 Electrical Connection.

**(A) Heating Element Connections.** Electrical connections, other than factory connections of heating elements to nonheating elements embedded in masonry or asphalt or on exposed surfaces, shall be made with insulated connectors identified for the use.

**(B) Circuit Connections.** Splices and terminations at the end of the nonheating leads, other than the heating element end, shall be installed in a box or fitting in accordance with 110.14 and 300.15.

### 426.25 Marking.
Each factory-assembled heating unit shall be legibly marked within 75 mm (3 in.) of each end of the nonheating leads with the permanent identification symbol, catalog number, and ratings in volts and watts or in volts and amperes.

### 426.26 Corrosion Protection.
Ferrous and nonferrous metal raceways, cable armor, cable sheaths, boxes, fittings, supports, and support hardware shall be permitted to be installed in concrete or in direct contact with the earth, or in areas subject to severe corrosive influences, where made of material suitable for the condition, or where provided with corrosion protection identified as suitable for the condition.

### 426.27 Grounding Braid or Sheath.
Grounding means, such as copper braid, metal sheath, or other approved means, shall be provided as part of the heated section of the cable, panel, or unit.

### 426.28 Equipment Protection.
Ground-fault protection of equipment shall be provided for fixed outdoor electric deicing and snow-melting equipment, except for equipment that employs mineral-insulated, metal-sheathed cable embedded in a noncombustible medium.

## IV. Impedance Heating

### 426.30 Personnel Protection.
Exposed elements of impedance heating systems shall be physically guarded, isolated, or thermally insulated with a weatherproof jacket to protect against contact by personnel in the area.

### 426.31 Isolation Transformer.
A dual-winding transformer with a grounded shield between the primary and secondary windings shall be used to isolate the distribution system from the heating system.

### 426.32 Voltage Limitations.
Unless protected by ground-fault circuit-interrupter protection for personnel, the secondary winding of the isolation transformer connected to the impedance heating elements shall not have an output voltage greater than 30 volts ac.

Where ground-fault circuit-interrupter protection for personnel is provided, the voltage shall be permitted to be greater than 30 but not more than 80 volts.

### 426.33 Induced Currents.
All current-carrying components shall be installed in accordance with 300.20.

### 426.34 Grounding.
An impedance heating system that is operating at a voltage greater than 30 but not more than 80 shall be grounded at a designated point(s).

## V. Skin-Effect Heating

### 426.40 Conductor Ampacity.
The current through the electrically insulated conductor inside the ferromagnetic envelope shall be permitted to exceed the ampacity values shown in Article 310, provided it is identified as suitable for this use.

**426.41 Pull Boxes.** Where pull boxes are used, they shall be accessible without excavation by location in suitable vaults or above grade. Outdoor pull boxes shall be of watertight construction.

**426.42 Single Conductor in Enclosure.** The provisions of 300.20 shall not apply to the installation of a single conductor in a ferromagnetic envelope (metal enclosure).

**426.43 Corrosion Protection.** Ferromagnetic envelopes, ferrous or nonferrous metal raceways, boxes, fittings, supports, and support hardware shall be permitted to be installed in concrete or in direct contact with the earth, or in areas subjected to severe corrosive influences, where made of material suitable for the condition, or where provided with corrosion protection identified as suitable for the condition. Corrosion protection shall maintain the original wall thickness of the ferromagnetic envelope.

**426.44 Grounding.** The ferromagnetic envelope shall be grounded at both ends; and, in addition, it shall be permitted to be grounded at intermediate points as required by its design.

The provisions of 250.30 shall not apply to the installation of skin-effect heating systems.

FPN: For grounding methods, see Article 250.

## VI. Control and Protection

**426.50 Disconnecting Means.**

**(A) Disconnection.** All fixed outdoor deicing and snow-melting equipment shall be provided with a means for disconnection from all ungrounded conductors. Where readily accessible to the user of the equipment, the branch-circuit switch or circuit breaker shall be permitted to serve as the disconnecting means. Switches used as the disconnecting means shall be of the indicating type.

**(B) Cord-and-Plug-Connected Equipment.** The factory-installed attachment plug of cord-and-plug-connected equipment rated 20 amperes or less and 150 volts or less to ground shall be permitted to be the disconnecting means.

**426.51 Controllers.**

**(A) Temperature Controller with "Off" Position.** Temperature controlled switching devices that indicate an "off" position and that interrupt line current shall open all ungrounded conductors when the control device is in the "off" position. These devices shall not be permitted to serve as the disconnecting means unless provided with a positive lockout in the "off" position.

**(B) Temperature Controller Without "Off" Position.** Temperature controlled switching devices that do not have an "off" position shall not be required to open all ungrounded conductors and shall not be permitted to serve as the disconnecting means.

**(C) Remote Temperature Controller.** Remote controlled temperature-actuated devices shall not be required to meet the requirements of 426.51(A). These devices shall not be permitted to serve as the disconnecting means.

**(D) Combined Switching Devices.** Switching devices consisting of combined temperature-actuated devices and manually controlled switches that serve both as the controller and the disconnecting means shall comply with all of the following conditions:

(1) Open all ungrounded conductors when manually placed in the "off" position
(2) Be so designed that the circuit cannot be energized automatically if the device has been manually placed in the "off" position
(3) Be provided with a positive lockout in the "off" position

**426.52 Overcurrent Protection.** Fixed outdoor electric deicing and snow-melting equipment shall be permitted to be protected against overcurrent where supplied by a branch circuit as specified in 426.4.

**426.54 Cord-and-Plug-Connected Deicing and Snow-Melting Equipment.** Cord-and-plug-connected deicing and snow-melting equipment shall be listed.

## ARTICLE 427
## Fixed Electric Heating Equipment for Pipelines and Vessels

## I. General

**427.1 Scope.** The requirements of this article shall apply to electrically energized heating systems and the installation of these systems used with pipelines or vessels or both.

**427.2 Definitions.**

**Impedance Heating System.** A system in which heat is generated in a pipeline or vessel wall by causing current to flow through the pipeline or vessel wall by direct connection to an ac voltage source from a dual-winding transformer.

**Induction Heating System.** A system in which heat is generated in a pipeline or vessel wall by inducing current and hysteresis effect in the pipeline or vessel wall from an external isolated ac field source.

**Integrated Heating System.** A complete system consisting of components such as pipelines, vessels, heating elements, heat transfer medium, thermal insulation, moisture barrier, nonheating leads, temperature controllers, safety signs, junction boxes, raceways, and fittings.

**Pipeline.** A length of pipe including pumps, valves, flanges, control devices, strainers, and/or similar equipment for conveying fluids.

**Resistance Heating Element.** A specific separate element to generate heat that is applied to the pipeline or vessel externally or internally.

> FPN: Tubular heaters, strip heaters, heating cable, heating tape, heating blankets, and immersion heaters are examples of resistance heaters.

**Skin-Effect Heating System.** A system in which heat is generated on the inner surface of a ferromagnetic envelope attached to a pipeline or vessel, or both.

> FPN: Typically, an electrically insulated conductor is routed through and connected to the envelope at the other end. The envelope and the electrically insulated conductor are connected to an ac voltage source from a dual-winding transformer.

**Vessel.** A container such as a barrel, drum, or tank for holding fluids or other material.

**427.3 Application of Other Articles.** All requirements of this *Code* shall apply except as specifically amended in this article. Cord-connected pipe heating assemblies intended for specific use and identified as suitable for this use shall be installed according to Article 422. Fixed electric pipeline and vessel heating equipment for use in hazardous (classified) locations shall comply with Articles 500 through 516.

**427.4 Branch-Circuit Sizing.** The ampacity of branch-circuit conductors and the rating or setting of overcurrent protective devices that supply fixed electric heating equipment for pipelines and vessels shall be not less than 125 percent of the total load of the heaters. The rating or setting of overcurrent protective devices shall be permitted in accordance with 240.4(B).

## II. Installation

**427.10 General.** Equipment for pipeline and vessel electrical heating shall be identified as being suitable for (1) the chemical, thermal, and physical environment and (2) instal-

lation in accordance with the manufacturer's drawings and instructions.

**427.11 Use.** Electrical heating equipment shall be installed in such a manner as to be afforded protection from physical damage.

**427.12 Thermal Protection.** External surfaces of pipeline and vessel heating equipment that operate at temperatures exceeding 60°C (140°F) shall be physically guarded, isolated, or thermally insulated to protect against contact by personnel in the area.

**427.13 Identification.** The presence of electrically heated pipelines, vessels, or both, shall be evident by the posting of appropriate caution signs or markings at frequent intervals along the pipeline or vessel.

## III. Resistance Heating Elements

**427.14 Secured.** Heating element assemblies shall be secured to the surface being heated by means other than the thermal insulation.

**427.15 Not in Direct Contact.** Where the heating element is not in direct contact with the pipeline or vessel being heated, means shall be provided to prevent overtemperature of the heating element unless the design of the heater assembly is such that its temperature limitations will not be exceeded.

**427.16 Expansion and Contraction.** Heating elements and assemblies shall not be installed where they bridge expansion joints unless provisions are made for expansion and contraction.

**427.17 Flexural Capability.** Where installed on flexible pipelines, the heating elements and assemblies shall have a flexural capability that is compatible with the pipeline.

**427.18 Power Supply Leads.**

**(A) Nonheating Leads.** Power supply nonheating leads (cold leads) for resistance elements shall be suitable for the temperature encountered. Preassembled nonheating leads on approved heaters shall be permitted to be shortened if the markings specified in 427.20 are retained. Not less than 150 mm (6 in.) of nonheating leads shall be provided within the junction box.

**(B) Power Supply Leads Protection.** Nonheating power supply leads shall be protected where they emerge from electrically heated pipeline or vessel heating units by rigid metal conduit, intermediate metal conduit, electrical metallic tubing, or other raceways identified as suitable for the application.

**(C) Interconnecting Leads.** Interconnecting nonheating leads connecting portions of the heating system shall be permitted to be covered by thermal insulation in the same manner as the heaters.

**427.19 Electrical Connections.**

**(A) Nonheating Interconnections.** Nonheating interconnections, where required under thermal insulation, shall be made with insulated connectors identified as suitable for this use.

**(B) Circuit Connections.** Splices and terminations outside the thermal insulation shall be installed in a box or fitting in accordance with 110.14 and 300.15.

**427.20 Marking.** Each factory-assembled heating unit shall be legibly marked within 75 mm (3 in.) of each end of the nonheating leads with the permanent identification symbol, catalog number, and ratings in volts and watts or in volts and amperes.

**427.22 Equipment Protection.** Ground-fault protection of equipment shall be provided for electric heat tracing and heating panels. This requirement shall not apply in industrial establishments where there is alarm indication of ground faults and

(1) Conditions of maintenance and supervision ensure that only qualified persons service the installed systems.
(2) Continued circuit operation is necessary for safe operation of equipment or processes.

**427.23 Grounded Conductive Covering.** Electric heating equipment shall be listed and have a grounded conductive covering in accordance with 427.23(A) or (B). The conductive covering shall provide an effective ground path for equipment protection.

**(A) Heating Wires or Cables.** Heating wires or cables shall have a grounded conductive covering that surrounds the heating element and bus wires, if any, and their electrical insulation.

**(B) Heating Panels.** Heating panels shall have a grounded conductive covering over the heating element and its electrical insulation on the side opposite the side attached to the surface to be heated.

## IV. Impedance Heating

**427.25 Personnel Protection.** All accessible external surfaces of the pipeline, vessel, or both, being heated shall be

physically guarded, isolated, or thermally insulated (with a weatherproof jacket for outside installations) to protect against contact by personnel in the area.

**427.26 Isolation Transformer.** A dual-winding transformer with a grounded shield between the primary and secondary windings shall be used to isolate the distribution system from the heating system.

**427.27 Voltage Limitations.** Unless protected by ground-fault circuit-interrupter protection for personnel, the secondary winding of the isolation transformer connected to the pipeline or vessel being heated shall not have an output voltage greater than 30 volts ac.

Where ground-fault circuit-interrupter protection for personnel is provided, the voltage shall be permitted to be greater than 30 but not more than 80 volts.

**427.28 Induced Currents.** All current-carrying components shall be installed in accordance with 300.20.

**427.29 Grounding.** The pipeline, vessel, or both, being heated that is operating at a voltage greater than 30 but not more than 80 shall be grounded at designated points.

**427.30 Secondary Conductor Sizing.** The ampacity of the conductors connected to the secondary of the transformer shall be at least 100 percent of the total load of the heater.

## V. Induction Heating

**427.35 Scope.** This part covers the installation of line frequency induction heating equipment and accessories for pipelines and vessels.

FPN: See Article 665 for other applications.

**427.36 Personnel Protection.** Induction coils that operate or may operate at a voltage greater than 30 volts ac shall be enclosed in a nonmetallic or split metallic enclosure, isolated, or made inaccessible by location to protect personnel in the area.

**427.37 Induced Current.** Induction coils shall be prevented from inducing circulating currents in surrounding metallic equipment, supports, or structures by shielding, isolation, or insulation of the current paths. Stray current paths shall be bonded to prevent arcing.

## VI. Skin-Effect Heating

**427.45 Conductor Ampacity.** The ampacity of the electrically insulated conductor inside the ferromagnetic envelope shall be permitted to exceed the values given in Article 310, provided it is identified as suitable for this use.

**427.46 Pull Boxes.** Pull boxes for pulling the electrically insulated conductor in the ferromagnetic envelope shall be permitted to be buried under the thermal insulation, provided their locations are indicated by permanent markings on the insulation jacket surface and on drawings. For outdoor installations, pull boxes shall be of watertight construction.

**427.47 Single Conductor in Enclosure.** The provisions of 300.20 shall not apply to the installation of a single conductor in a ferromagnetic envelope (metal enclosure).

**427.48 Grounding.** The ferromagnetic envelope shall be grounded at both ends, and, in addition, it shall be permitted to be grounded at intermediate points as required by its design. The ferromagnetic envelope shall be bonded at all joints to ensure electrical continuity.

The provisions of 250.30 shall not apply to the installation of skin-effect heating systems.

FPN: See Article 250 for grounding methods.

## VII. Control and Protection

### 427.55 Disconnecting Means.

**(A) Switch or Circuit Breaker.** Means shall be provided to disconnect all fixed electric pipeline or vessel heating equipment from all ungrounded conductors. The branch-circuit switch or circuit breaker, where readily accessible to the user of the equipment, shall be permitted to serve as the disconnecting means. The disconnecting means shall be of the indicating type and shall be provided with a positive lockout in the "off" position.

**(B) Cord-and-Plug-Connected Equipment.** The factory-installed attachment plug of cord-and-plug-connected equipment rated 20 amperes or less and 150 volts or less to ground shall be permitted to be the disconnecting means.

### 427.56 Controls.

**(A) Temperature Control with "Off" Position.** Temperature controlled switching devices that indicate an "off" position and that interrupt line current shall open all ungrounded conductors when the control device is in this "off" position. These devices shall not be permitted to serve as the disconnecting means unless provided with a positive lockout in the "off" position.

**(B) Temperature Control Without "Off" Position.** Temperature controlled switching devices that do not have an "off" position shall not be required to open all ungrounded conductors and shall not be permitted to serve as the disconnecting means.

**(C) Remote Temperature Controller.** Remote controlled temperature-actuated devices shall not be required to meet the requirements of 427.56(A) and (B). These devices shall not be permitted to serve as the disconnecting means.

**(D) Combined Switching Devices.** Switching devices consisting of combined temperature-actuated devices and manually controlled switches that serve both as the controllers and the disconnecting means shall comply with all the following conditions:

(1) Open all ungrounded conductors when manually placed in the "off" position
(2) Be designed so that the circuit cannot be energized automatically if the device has been manually placed in the "off" position
(3) Be provided with a positive lockout in the "off" position

**427.57 Overcurrent Protection.** Heating equipment shall be considered as protected against overcurrent where supplied by a branch circuit as specified in 427.4.

## ARTICLE 430
## Motors, Motor Circuits, and Controllers

## I. General

**430.1 Scope.** This article covers motors, motor branch-circuit and feeder conductors and their protection, motor overload protection, motor control circuits, motor controllers, and motor control centers.

FPN No. 1: Installation requirements for motor control centers are covered in 110.26(F). Air conditioning and refrigerating equipment are covered in Article 440.

FPN No. 2: Figure 430.1 is for information only.

**430.2 Adjustable-Speed Drive Systems.** The incoming branch circuit or feeder to power conversion equipment included as a part of an adjustable-speed drive system shall be based on the rated input to the power conversion equipment. Where the power conversion equipment is marked to indicate that overload protection is included, additional overload protection shall not be required.

The disconnecting means shall be permitted to be in the incoming line to the conversion equipment and shall have a rating not less than 115 percent of the rated input current of the conversion unit.

| | |
|---|---|
| General, 430.1 through 430.18 | Part I |
| Motor Circuit Conductors, 430.21 through 430.29 | Part II |
| Motor and Branch-Circuit Overload Protection, 430.31 through 430.44 | Part III |
| Motor Branch-Circuit Short-Circuit and Ground-Fault Protection, 430.51 through 430.58 | Part IV |
| Motor Feeder Short-Circuit and Ground-Fault Protection, 430.61 through 430.63 | Part V |
| Motor Control Circuits, 430.71 through 430.74 | Part VI |
| Motor Controllers, 430.81 through 430.91 | Part VII |
| Motor Control Centers, 430.92 through 430.98 | Part VIII |
| Disconnecting Means, 430.101 through 430.113 | Part IX |
| Over 600 Volts, Nominal, 430.121 through 430.127 | Part X |
| Protection of Live Parts—All Voltages, 430.131 through 430.133 | Part XI |
| Grounding—All Voltages, 430.141 through 430.145 | Part XII |
| Tables, Tables 430.147 through 430.151(B) | Part XIII |

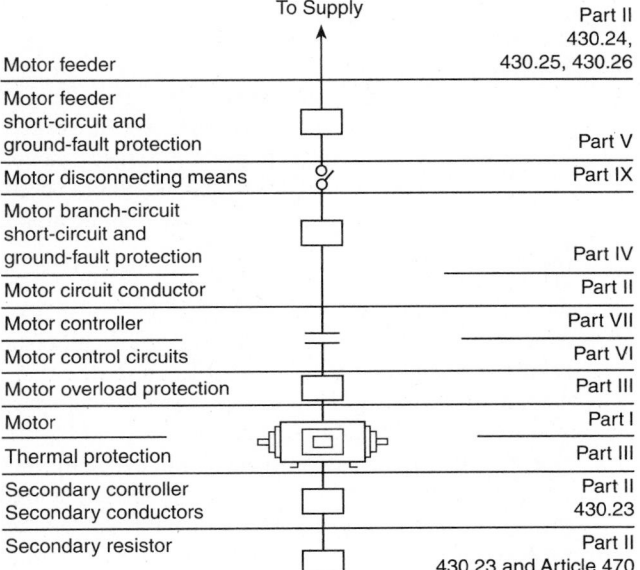

| | |
|---|---|
| To Supply | Part II |
| | 430.24, |
| Motor feeder | 430.25, 430.26 |
| Motor feeder short-circuit and ground-fault protection | Part V |
| Motor disconnecting means | Part IX |
| Motor branch-circuit short-circuit and ground-fault protection | Part IV |
| Motor circuit conductor | Part II |
| Motor controller | Part VII |
| Motor control circuits | Part VI |
| Motor overload protection | Part III |
| Motor | Part I |
| Thermal protection | Part III |
| Secondary controller | Part II |
| Secondary conductors | 430.23 |
| Secondary resistor | Part II |
| | 430.23 and Article 470 |

**Figure 430.1 Article 430 contents.**

FPN: Electrical resonance can result from the interaction of the nonsinusoidal currents from this type of load with power factor correction capacitors.

**430.3 Part-Winding Motors.** A part-winding start induction or synchronous motor is one that is arranged for starting by first energizing part of its primary (armature) winding and, subsequently, energizing the remainder of this winding in one or more steps. A standard part-winding start induction motor is arranged so that one-half of its primary winding can be energized initially, and, subsequently, the remaining half can be energized, both halves then carrying equal current. A hermetic refrigerant compressor motor shall not be considered a standard part-winding start induction motor.

Where separate overload devices are used with a standard part-winding start induction motor, each half of the motor winding shall be individually protected in accordance with 430.32 and 430.37 with a trip current one-half that specified.

Each motor-winding connection shall have branch-circuit short-circuit and ground-fault protection rated at not more than one-half that specified by 430.52.

*Exception: A short-circuit and ground-fault protective device shall be permitted for both windings if the device will allow the motor to start. Where time-delay (dual-element) fuses are used, they shall be permitted to have a rating not exceeding 150 percent of the motor full-load current.*

**430.5 Other Articles.** Motors and controllers shall also comply with the applicable provisions of Table 430.5.

**Table 430.5 Other Articles**

| Equipment/Occupancy | Article | Section |
|---|---|---|
| Air-conditioning and refrigerating equipment | 440 | |
| Capacitors | | 460.8, 460.9 |
| Commercial garages; aircraft hangars; motor fuel dispensing facilities; bulk storage plants; spray application, dipping, and coating processes; and inhalation anesthetizing locations | 511, 513, 514, 515, 516, and 517 Part IV | |
| Cranes and hoists | 610 | |
| Electrically driven or controlled irrigation machines | 675 | |
| Elevators, dumbwaiters, escalators, moving walks, wheelchair lifts, and stairway chair lifts | 620 | |
| Fire pumps | 695 | |
| Hazardous (classified) locations | 500–503 and 505 | |
| Industrial machinery | 670 | |
| Motion picture projectors | | 540.11 and 540.20 |
| Motion picture and television studios and similar locations | 530 | |
| Resistors and reactors | 470 | |
| Theaters, audience areas of motion picture and television studios, and similar locations | | 520.48 |
| Transformers and transformer vaults | 450 | |

**430.6 Ampacity and Motor Rating Determination.** The size of conductors supplying equipment covered by Article 430 shall be selected from the allowable ampacity tables in accordance with 310.15(B) or shall be calculated in accordance with 310.15(C). Where flexible cord is used, the size of the conductor shall be selected in accordance with 400.5.

The required ampacity and motor ratings shall be determined as specified in 430.6(A), (B), and (C).

**(A) General Motor Applications.** For general motor applications, current ratings shall be determined based on (1) and (2).

**(1) Table Values.** The values given in Tables 430.147, 430.148, 430.149, and 430.150, including notes, shall be used to determine the ampacity of conductors or ampere ratings of switches, branch-circuit short-circuit and ground-fault protection, instead of the actual current rating marked on the motor nameplate. Where a motor is marked in amperes, but not horsepower, the horsepower rating shall be assumed to be that corresponding to the value given in Tables 430.147, 430.148, 430.149, and 430.150, interpolated if necessary.

*Exception No. 1: Multispeed motors shall be in accordance with 430.22(A) and 430.52.*

*Exception No. 2: For equipment that employs a shaded-pole or permanent-split capacitor-type fan or blower motor that is marked with the motor type, the full load current for such motor marked on the nameplate of the equipment in which the fan or blower motor is employed shall be used instead of the horsepower rating to determine the ampacity or rating of the disconnecting means, the branch-circuit conductors, the controller, the branch-circuit short-circuit and ground-fault protection, and the separate overload protection. This marking on the equipment nameplate shall not be less than the current marked on the fan or blower motor nameplate.*

*Exception No. 3: For a listed motor-operated appliance that is marked with both motor horsepower and full-load current, the motor full-load current marked on the nameplate of the appliance shall be used instead of the horsepower rating on the appliance nameplate to determine the ampacity or rating of the disconnecting means, the branch-circuit conductors, the controller, the branch-circuit short-circuit and ground-fault protection, and any separate overload protection.*

**(2) Nameplate Values.** Separate motor overload protection shall be based on the motor nameplate current rating.

**(B) Torque Motors.** For torque motors, the rated current shall be locked-rotor current, and this nameplate current shall be used to determine the ampacity of the branch-circuit conductors covered in 430.22 and 430.24, the ampere rating of the motor overload protection, and the ampere rating of motor branch-circuit short-circuit and ground-fault protection in accordance with 430.52(B).

FPN: For motor controllers and disconnecting means, see 430.83(D) and 430.110.

**(C) Alternating-Current Adjustable Voltage Motors.** For motors used in alternating-current, adjustable voltage, variable torque drive systems, the ampacity of conductors, or ampere ratings of switches, branch-circuit short-circuit and ground-fault protection, and so forth, shall be based on the maximum operating current marked on the motor or control nameplate, or both. If the maximum operating current does not appear on the nameplate, the ampacity determination shall be based on 150 percent of the values given in Tables 430.149 and 430.150.

### 430.7 Marking on Motors and Multimotor Equipment.

**(A) Usual Motor Applications.** A motor shall be marked with the following information.

(1) Manufacturer's name.
(2) Rated volts and full-load amperes. For a multispeed motor, full-load amperes for each speed, except shaded-pole and permanent-split capacitor motors where amperes are required only for maximum speed.
(3) Rated frequency and number of phases if an ac motor.
(4) Rated full-load speed.
(5) Rated temperature rise or the insulation system class and rated ambient temperature.
(6) Time rating. The time rating shall be 5, 15, 30, or 60 minutes, or continuous.
(7) Rated horsepower if ⅛ hp or more. For a multispeed motor ⅛ hp or more, rated horsepower for each speed, except shaded-pole and permanent-split capacitor motors ⅛ hp or more where rated horsepower is required only for maximum speed. Motors of arc welders are not required to be marked with the horsepower rating.
(8) Code letter or locked-rotor amperes if an alternating-current motor rated ½ hp or more. On polyphase wound-rotor motors, the code letter shall be omitted.

FPN: See 430.7(B).

(9) Design letter for design B, C, D, or E motors.

FPN: Motor design letter definitions are found in ANSI/NEMA MG 1-1993, *Motors and Generators, Part 1, Definitions,* and in IEEE 100-1996, *Standard Dictionary of Electrical and Electronic Terms.*

(10) Secondary volts and full-load amperes if a wound-rotor induction motor.
(11) Field current and voltage for dc excited synchronous motors.
(12) Winding — straight shunt, stabilized shunt, compound, or series, if a dc motor. Fractional horsepower dc motors 175 mm (7 in.) or less in diameter shall not be required to be marked.
(13) A motor provided with a thermal protector complying with 430.32(A)(2) or (B)(2) shall be marked "Thermally Protected." Thermally protected motors rated 100 watts or less and complying with 430.32(B)(2)

shall be permitted to use the abbreviated marking "T.P."

(14) A motor complying with 430.32(B)(4) shall be marked "Impedance Protected." Impedance protected motors rated 100 watts or less and complying with 430.32(B)(4) shall be permitted to use the abbreviated marking "Z.P."

**(B) Locked-Rotor Indicating Code Letters.** Code letters marked on motor nameplates to show motor input with locked rotor shall be in accordance with Table 430.7(B).

The code letter indicating motor input with locked rotor shall be in an individual block on the nameplate, properly designated.

**Table 430.7(B) Locked-Rotor Indicating Code Letters**

| Code Letter | Kilovolt-Amperes per Horsepower with Locked Rotor |
|---|---|
| A | 0 – 3.14 |
| B | 3.15 – 3.54 |
| C | 3.55 – 3.99 |
| D | 4.0 – 4.49 |
| E | 4.5 – 4.99 |
| F | 5.0 – 5.59 |
| G | 5.6 – 6.29 |
| H | 6.3 – 7.09 |
| J | 7.1 – 7.99 |
| K | 8.0 – 8.99 |
| L | 9.0 – 9.99 |
| M | 10.0 – 11.19 |
| N | 11.2 – 12.49 |
| P | 12.5 – 13.99 |
| R | 14.0 – 15.99 |
| S | 16.0 – 17.99 |
| T | 18.0 – 19.99 |
| U | 20.0 – 22.39 |
| V | 22.4 and up |

**(1) Multispeed Motors.** Multispeed motors shall be marked with the code letter designating the locked-rotor kilovolt-ampere (kVA) per horsepower for the highest speed at which the motor can be started.

*Exception: Constant horsepower multispeed motors shall be marked with the code letter giving the highest locked-rotor kilovolt-ampere (kVA) per horsepower.*

**(2) Single-Speed Motors.** Single-speed motors starting on wye connection and running on delta connections shall be marked with a code letter corresponding to the locked-rotor kilovolt-ampere (kVA) per horsepower for the wye connection.

**(3) Dual-Voltage Motors.** Dual-voltage motors that have a different locked-rotor kilovolt-ampere (kVA) per horse-power on the two voltages shall be marked with the code letter for the voltage giving the highest locked-rotor kilovolt-ampere (kVA) per horsepower.

**(4) 50/60 Hz Motors.** Motors with 50- and 60-Hz ratings shall be marked with a code letter designating the locked-rotor kilovolt-ampere (kVA) per horsepower on 60 Hz.

**(5) Part Winding Motors.** Part-winding start motors shall be marked with a code letter designating the locked-rotor kilovolt-ampere (kVA) per horsepower that is based on the locked-rotor current for the full winding of the motor.

**(C) Torque Motors.** Torque motors are rated for operation at standstill and shall be marked in accordance with 430.7(A), except that locked-rotor torque shall replace horsepower.

**(D) Multimotor and Combination-Load Equipment.**

**(1) Factory-Wired.** Multimotor and combination-load equipment shall be provided with a visible nameplate marked with the manufacturer's name, the rating in volts, frequency, number of phases, minimum supply circuit conductor ampacity, and the maximum ampere rating of the circuit short-circuit and ground-fault protective device. The conductor ampacity shall be computed in accordance with 430.24 and counting all of the motors and other loads that will be operated at the same time. The short-circuit and ground-fault protective device rating shall not exceed the value computed in accordance with 430.53. Multimotor equipment for use on two or more circuits shall be marked with the preceding information for each circuit.

**(2) Not Factory-Wired.** Where the equipment is not factory-wired and the individual nameplates of motors and other loads are visible after assembly of the equipment, the individual nameplates shall be permitted to serve as the required marking.

**430.8 Marking on Controllers.** A controller shall be marked with the manufacturer's name or identification, the voltage, the current or horsepower rating, and such other necessary data to properly indicate the motors for which it is suitable. A controller that includes motor overload protection suitable for group motor application shall be marked with the motor overload protection and the maximum branch-circuit short-circuit and ground-fault protection for such applications.

Combination controllers that employ adjustable instantaneous trip circuit breakers shall be clearly marked to indicate the ampere settings of the adjustable trip element.

Where a controller is built-in as an integral part of a motor or of a motor-generator set, individual marking of the controller shall not be required if the necessary data are on the nameplate. For controllers that are an integral part of equipment approved as a unit, the above marking shall be permitted on the equipment nameplate.

> FPN: See 110.10 for information on circuit impedance and other characteristics.

### 430.9 Terminals.

**(A) Markings.** Terminals of motors and controllers shall be suitably marked or colored where necessary to indicate the proper connections.

**(B) Conductors.** Motor controllers and terminals of control circuit devices shall be connected with copper conductors unless identified for use with a different conductor.

**(C) Torque Requirements.** Control circuit devices with screw-type pressure terminals used with 14 AWG or smaller copper conductors shall be torqued to a minimum of 0.8 N•m (7 lb-in.) unless identified for a different torque value.

### 430.10 Wiring Space in Enclosures.

**(A) General.** Enclosures for motor controllers and disconnecting means shall not be used as junction boxes, auxiliary gutters, or raceways for conductors feeding through or tapping off to the other apparatus unless designs are employed that provide adequate space for this purpose.

> FPN: See 312.8 for switch and overcurrent-device enclosures.

**(B) Wire-Bending Space in Enclosures.** Minimum wire-bending space within the enclosures for motor controllers shall be in accordance with Table 430.10(B) where measured in a straight line from the end of the lug or wire connector (in the direction the wire leaves the terminal) to the wall or barrier. Where alternate wire termination means are substituted for that supplied by the manufacturer of the controller, they shall be of a type identified by the manufacturer for use with the controller and shall not reduce the minimum wire-bending space.

### 430.11 Protection Against Liquids.
Suitable guards or enclosures shall be provided to protect exposed current-carrying parts of motors and the insulation of motor leads where installed directly under equipment, or in other locations where dripping or spraying oil, water, or other injurious liquid may occur, unless the motor is designed for the existing conditions.

### 430.12 Motor Terminal Housings.

**(A) Material.** Where motors are provided with terminal housings, the housings shall be of metal and of substantial construction.

**Table 430.10(B) Minimum Wire-Bending Space at the Terminals of Enclosed Motor Controllers**

| Size of Wire (AWG or kcmil) | Wires per Terminal[*] | | | |
| | 1 | | 2 | |
| | mm | in. | mm | in. |
|---|---|---|---|---|
| 14–10 | Not specified | | — | — |
| 8–6 | 38 | 1½ | — | — |
| 4–3 | 50 | 2 | — | — |
| 2 | 65 | 2½ | — | — |
| 1 | 75 | 3 | — | — |
| 1/0 | 125 | 5 | 125 | 5 |
| 2/0 | 150 | 6 | 150 | 6 |
| 3/0–4/0 | 175 | 7 | 175 | 7 |
| 250 | 200 | 8 | 200 | 8 |
| 300 | 250 | 10 | 250 | 10 |
| 350–500 | 300 | 12 | 300 | 12 |
| 600–700 | 350 | 14 | 400 | 16 |
| 750–900 | 450 | 18 | 475 | 19 |

[*]Where provision for three or more wires per terminal exists, the minimum wire-bending space shall be in accordance with the requirements of Article 312.

*Exception: In other than hazardous (classified) locations, substantial, nonmetallic, nonburning housings shall be permitted, provided an internal grounding means between the motor frame and the equipment grounding connection is incorporated within the housing.*

**(B) Dimensions and Space — Wire-to-Wire Connections.** Where these terminal housings enclose wire-to-wire connections, they shall have minimum dimensions and usable volumes in accordance with Table 430.12(B).

**(C) Dimensions and Space — Fixed Terminal Connections.** Where these terminal housings enclose rigidly mounted motor terminals, the terminal housing shall be of sufficient size to provide minimum terminal spacings and usable volumes in accordance with Tables 430.12(C)(1) and 430.12(C)(2).

**Table 430.12(B) Terminal Housings — Wire-to-Wire Connections Motors 275 mm (11 in.) in Diameter or Less**

| Horsepower | Cover Opening Minimum Dimension | | Usable Volume Minimum | |
| | mm | in. | cm³ | in.³ |
|---|---|---|---|---|
| 1 and smaller[a] | 41 | 1⅝ | 170 | 10.5 |
| 1½, 2, and 3[b] | 45 | 1¾ | 275 | 16.8 |
| 5 and 7½ | 50 | 2 | 365 | 22.4 |
| 10 and 15 | 65 | 2½ | 595 | 36.4 |

### Motors Over 275 mm (11 in.) in Diameter — Alternating-Current Motors

| Maximum Full Load Current for 3-Phase Motors with Maximum of 12 Leads (Amperes) | Terminal Box Cover Opening Minimum Dimension | | Usable Volume Minimum | | Typical Maximum Horsepower 3-Phase | |
|---|---|---|---|---|---|---|
| | mm | in. | cm³ | in.³ | 230 Volt | 460 Volt |
| 45 | 65 | 2.5 | 595 | 36.4 | 15 | 30 |
| 70 | 84 | 3.3 | 1,265 | 77 | 25 | 50 |
| 110 | 100 | 4.0 | 2,295 | 140 | 40 | 75 |
| 160 | 125 | 5.0 | 4,135 | 252 | 60 | 125 |
| 250 | 150 | 6.0 | 7,380 | 450 | 100 | 200 |
| 400 | 175 | 7.0 | 13,775 | 840 | 150 | 300 |
| 600 | 200 | 8.0 | 25,255 | 1540 | 250 | 500 |

### Direct-Current Motors

| Maximum Full-Load Current for Motors with Maximum of 6 Leads (Amperes) | Terminal Box Minimum Dimensions | | Usable Volume Minimum | |
|---|---|---|---|---|
| | mm | in. | cm³ | in.³ |
| 68 | 65 | 2.5 | 425 | 26 |
| 105 | 84 | 3.3 | 900 | 55 |
| 165 | 100 | 4.0 | 1,640 | 100 |
| 240 | 125 | 5.0 | 2,950 | 180 |
| 375 | 150 | 6.0 | 5,410 | 330 |
| 600 | 175 | 7.0 | 9,840 | 600 |
| 900 | 200 | 8.0 | 18,040 | 1,100 |

Note: Auxiliary leads for such items as brakes, thermostats, space heaters, and exciting fields shall be permitted to be neglected if their current-carrying area does not exceed 25 percent of the current-carrying area of the machine power leads.

[a]For motors rated 1 hp and smaller and with the terminal housing partially or wholly integral with the frame or end shield, the volume of the terminal housing shall not be less than 18.0 cm³ (1.1 in.³) per wire-to-wire connection. The minimum cover opening dimension is not specified.

[b]For motors rated 1½, 2, and 3 hp and with the terminal housing partially or wholly integral with the frame or end shield, the volume of the terminal housing shall not be less than 23.0 cm³ (1.4 in.³) per wire-to-wire connection. The minimum cover opening dimension is not specified.

### Table 430.12(C)(1) Terminal Spacings — Fixed Terminals

| | Minimum Spacing | | | |
|---|---|---|---|---|
| | Between Line Terminals | | Between Line Terminals and Other Uninsulated Metal Parts | |
| Nominal Volts | mm | in. | mm | in. |
| 240 or less | 6 | ¼ | 6 | ¼ |
| Over 250 – 600 | 10 | ⅜ | 10 | ⅜ |

### Table 430.12(C)(2) Usable Volumes — Fixed Terminals

| Power-Supply Conductor Size (AWG) | Minimum Usable Volume per Power-Supply Conductor | |
|---|---|---|
| | cm³ | in.³ |
| 14 | 16 | 1 |
| 12 and 10 | 20 | 1¼ |
| 8 and 6 | 37 | 2¼ |

**(D) Large Wire or Factory Connections.** For motors with larger ratings, greater number of leads, or larger wire sizes, or where motors are installed as a part of factory-wired equipment, without additional connection being required at the motor terminal housing during equipment installation, the terminal housing shall be of ample size to make connections, but the foregoing provisions for the volumes of terminal housings shall not be considered applicable.

**(E) Equipment Grounding Connections.** A means for attachment of an equipment grounding conductor termination in accordance with 250.8 shall be provided at motor terminal housings for wire-to-wire connections or fixed terminal connections. The means for such connections shall be permitted to be located either inside or outside the motor terminal housing.

*Exception: Where a motor is installed as a part of factory-wired equipment that is required to be grounded and without additional connection being required at the motor terminal housing during equipment installation, a separate means for motor grounding at the motor terminal housing shall not be required.*

**430.13 Bushing.** Where wires pass through an opening in an enclosure, conduit box, or barrier, a bushing shall be used to protect the conductors from the edges of openings having sharp edges. The bushing shall have smooth, well-rounded surfaces where it may be in contact with the conductors. If used where oils, greases, or other contaminants may be present, the bushing shall be made of material not deleteriously affected.

FPN: For conductors exposed to deteriorating agents, see 310.9.

**430.14 Location of Motors.**

**(A) Ventilation and Maintenance.** Motors shall be located so that adequate ventilation is provided and so that maintenance, such as lubrication of bearings and replacing of brushes, can be readily accomplished.

*Exception: Ventilation shall not be required for submersible types of motors.*

**(B) Open Motors.** Open motors that have commutators or collector rings shall be located or protected so that sparks cannot reach adjacent combustible material.

*Exception: Installation of these motors on wooden floors or supports shall be permitted.*

**430.16 Exposure to Dust Accumulations.** In locations where dust or flying material collects on or in motors in such quantities as to seriously interfere with the ventilation or cooling of motors and thereby cause dangerous temperatures, suitable types of enclosed motors that do not overheat under the prevailing conditions shall be used.

> FPN: Especially severe conditions may require the use of enclosed pipe-ventilated motors, or enclosure in separate dusttight rooms, properly ventilated from a source of clean air.

**430.17 Highest Rated or Smallest Rated Motor.** In determining compliance with 430.24, 430.53(B), and 430.53(C), the highest rated or smallest rated motor shall be based on the rated full-load current as selected from Tables 430.147, 430.148, 430.149, and 430.150.

**430.18 Nominal Voltage of Rectifier Systems.** The nominal value of the ac voltage being rectified shall be used to determine the voltage of a rectifier derived system.

*Exception: The nominal dc voltage of the rectifier shall be used if it exceeds the peak value of the ac voltage being rectified.*

## II. Motor Circuit Conductors

**430.21 General.** Part II specifies ampacities of conductors that are capable of carrying the motor current without overheating under the conditions specified.

The provisions of Part II shall not apply to motor circuits rated over 600 volts, nominal.

The provisions of Articles 250, 300, and 310 shall not apply to conductors that form an integral part of equipment, such as motors, motor controllers, motor control centers, or other factory-assembled control equipment.

> FPN No. 1: See 300.1(B) and 310.1 for similar requirements.
>
> FPN No. 2: See 110.14(C) and 430.9(B) for equipment device terminal requirements.
>
> FPN No. 3: For over 600 volts, nominal, see Part XI.

**430.22 Single Motor.**

**(A) General.** Branch-circuit conductors that supply a single motor used in a continuous duty application shall have an ampacity of not less than 125 percent of the motor's full-load current rating as determined by 430.6(A)(1).

*Exception No. 1: For dc motors operating from a rectified single-phase power supply, the conductors between the field wiring terminals of the rectifier and the motor shall have an ampacity of not less than the following percent of the motor full-load current rating:*

(a) *Where a rectifier bridge of the single-phase half-wave type is used, 190 percent.*

(b) *Where a rectifier bridge of the single-phase full-wave type is used, 150 percent.*

*Exception No. 2: Circuit conductors supplying power conversion equipment included as part of an adjustable-speed drive system shall have an ampacity not less than 125 percent of the rated input to the power conversion equipment.*

**(B) Multispeed Motor.** For a multispeed motor, the selection of branch-circuit conductors on the line side of the controller shall be based on the highest of the full-load current ratings shown on the motor nameplate. The selection of branch-circuit conductors between the controller and the motor shall be based on the current rating of the winding(s) that the conductors energize.

**(C) Wye-Start, Delta-Run Motor.** For a wye-start, delta-run connected motor, the selection of branch-circuit conductors on the line side of the controller shall be based on the motor full-load current. The selection of conductors between the controller and the motor shall be based on 58 percent of the motor full-load current.

**(D) Part-Winding Motor.** For a part-winding connected motor, the selection of branch-circuit conductors on the line side of the controller shall be based on the motor full-load current. The selection of conductors between the controller and the motor shall be based on 50 percent of the motor full-load current.

**(E) Other Than Continuous Duty.** Conductors for a motor used in a short-time, intermittent, periodic, or varying duty application shall have an ampacity of not less than the percentage of the motor nameplate current rating shown in Table 430.22(E), unless the authority having jurisdiction grants special permission for conductors of lower ampacity.

**(F) Separate Terminal Enclosure.** The conductors between a stationary motor rated 1 hp or less and the separate terminal enclosure permitted in 430.145(B) shall be permitted to be smaller than 14 AWG but not smaller than 18 AWG, provided they have an ampacity as specified in 430.22(A).

**430.23 Wound-Rotor Secondary.**

**(A) Continuous Duty.** For continuous duty, the conductors connecting the secondary of a wound-rotor ac motor to its

**Table 430.22(E) Duty-Cycle Service**

| Classification of Service | Nameplate Current Rating Percentages | | | |
| --- | --- | --- | --- | --- |
| | 5-Minute Rated Motor | 15-Minute Rated Motor | 30- & 60-Minute Rated Motor | Continuous Rated Motor |
| Short-time duty operating valves, raising or lowering rolls, etc. | 110 | 120 | 150 | — |
| Intermittent duty freight and passenger elevators, tool heads, pumps, drawbridges, turntables, etc. (for arc welders, see 630.11) | 85 | 85 | 90 | 140 |
| Periodic duty rolls, ore- and coal-handling machines, etc. | 85 | 90 | 95 | 140 |
| Varying duty | 110 | 120 | 150 | 200 |

Note: Any motor application shall be considered as continuous duty unless the nature of the apparatus it drives is such that the motor will not operate continuously with load under any condition of use.

controller shall have an ampacity not less than 125 percent of the full-load secondary current of the motor.

**(B) Other Than Continuous Duty.** For other than continuous duty, these conductors shall have an ampacity, in percent of full-load secondary current, not less than that specified in Table 430.22(E).

**(C) Resistor Separate from Controller.** Where the secondary resistor is separate from the controller, the ampacity of the conductors between controller and resistor shall not be less than that shown in Table 430.23(C).

**Table 430.23(C) Secondary Conductor**

| Resistor Duty Classification | Ampacity of Conductor in Percent of Full-Load Secondary Current |
| --- | --- |
| Light starting duty | 35 |
| Heavy starting duty | 45 |
| Extra-heavy starting duty | 55 |
| Light intermittent duty | 65 |
| Medium intermittent duty | 75 |
| Heavy intermittent duty | 85 |
| Continuous duty | 110 |

**430.24 Several Motors or a Motor(s) and Other Load(s).** Conductors supplying several motors, or a mo-

tor(s) and other load(s), shall have an ampacity not less than 125 percent of the full-load current rating of the highest rated motor plus the sum of the full-load current ratings of all the other motors in the group, as determined by 430.6(A), plus the ampacity required for the other loads.

FPN: See Annex D, Example No. D8.

*Exception No. 1: Where one or more of the motors of the group are used for short-time, intermittent, periodic, or varying duty, the ampere rating of such motors to be used in the summation shall be determined in accordance with 430.22(E). For the highest rated motor, the greater of either the ampere rating from 430.22(E) or the largest continuous duty motor full-load current multiplied by 1.25 shall be used in the summation.*

*Exception No. 2: The ampacity of conductors supplying motor-operated fixed electric space-heating equipment shall conform with 424.3(B).*

*Exception No. 3: Where the circuitry is interlocked so as to prevent operation of selected motors or other loads at the same time, the conductor ampacity shall be permitted to be based on the summation of the currents of the motors and other loads to be operated at the same time that results in the highest total current.*

**430.25 Multimotor and Combination-Load Equipment.** The ampacity of the conductors supplying multimotor and combination-load equipment shall not be less than the minimum circuit ampacity marked on the equipment in accordance with 430.7(D). Where the equipment is not factory-wired and the individual nameplates are visible in accordance with 430.7(D)(2), the conductor ampacity shall be determined in accordance with 430.24.

**430.26 Feeder Demand Factor.** Where reduced heating of the conductors results from motors operating on duty-cycle, intermittently, or from all motors not operating at one time, the authority having jurisdiction may grant permission for feeder conductors to have an ampacity less than specified in 430.24, provided the conductors have sufficient ampacity for the maximum load determined in accordance with the sizes and number of motors supplied and the character of their loads and duties.

**430.27 Capacitors with Motors.** Where capacitors are installed in motor circuits, conductors shall comply with 460.8 and 460.9.

**430.28 Feeder Taps.** Feeder tap conductors shall have an ampacity not less than that required by Part II, shall terminate in a branch-circuit protective device, and, in addition, shall meet one of the following requirements:

(1) Be enclosed either by an enclosed controller or by a raceway, be not more than 3.0 m (10 ft) in length, and, for field installation, be protected by an overcurrent device on the line side of the tap conductor, the rating or setting of which shall not exceed 1000 percent of the tap conductor ampacity

(2) Have an ampacity of at least one-third that of the feeder conductors, be suitably protected from physical damage or enclosed in a raceway, and be not more than 7.5 m (25 ft) in length

(3) Have the same ampacity as the feeder conductors

*Exception: Feeder taps over 7.5 m (25 ft) long. In high-bay manufacturing buildings [over 11 m (35 ft) high at walls], where conditions of maintenance and supervision ensure that only qualified persons service the systems, conductors tapped to a feeder shall be permitted to be not over 7.5 m (25 ft) long horizontally and not over 30.0 m (100 ft) in total length where all of the following conditions are met:*

(a) *The ampacity of the tap conductors is not less than one-third that of the feeder conductors.*

(b) *The tap conductors terminate with a single circuit breaker or a single set of fuses conforming with (1) Part IV, where the load-side conductors are a branch circuit, or (2) Part V, where the load-side conductors are a feeder.*

(c) *The tap conductors are suitably protected from physical damage and are installed in raceways.*

(d) *The tap conductors are continuous from end-to-end and contain no splices.*

(e) *The tap conductors shall be 6 AWG copper or 4 AWG aluminum or larger.*

(f) *The tap conductors shall not penetrate walls, floors, or ceilings.*

(g) *The tap shall not be made less than 9.0 m (30 ft) from the floor.*

### 430.29 Constant Voltage Direct-Current Motors — Power Resistors.
Conductors connecting the motor controller to separately mounted power accelerating and dynamic braking resistors in the armature circuit shall have an ampacity not less than the value calculated from Table 430.29 using motor full-load current. If an armature shunt resistor is used, the power accelerating resistor conductor ampacity shall be calculated using the total of motor full-load current and armature shunt resistor current.

Armature shunt resistor conductors shall have an ampacity of not less than that calculated from Table 430.29 using rated shunt resistor current as full-load current.

### III. Motor and Branch-Circuit Overload Protection

### 430.31 General.
Part III specifies overload devices intended to protect motors, motor-control apparatus, and mo-

**Table 430.29 Conductor Rating Factors for Power Resistors**

| Time in Seconds | | Ampacity of Conductor in Percent of Full-Load Current |
|---|---|---|
| **On** | **Off** | |
| 5 | 75 | 35 |
| 10 | 70 | 45 |
| 15 | 75 | 55 |
| 15 | 45 | 65 |
| 15 | 30 | 75 |
| 15 | 15 | 85 |
| Continuous Duty | | 110 |

tor branch-circuit conductors against excessive heating due to motor overloads and failure to start.

Overload in electrical apparatus is an operating overcurrent that, when it persists for a sufficient length of time, would cause damage or dangerous overheating of the apparatus. It does not include short circuits or ground faults.

These provisions shall not be interpreted as requiring overload protection where it might introduce additional or increased hazards, as in the case of fire pumps.

FPN: For protection of fire pump supply conductors, see 695.6.

The provisions of Part III shall not apply to motor circuits rated over 600 volts, nominal.

FPN No. 1: For over 600 volts, nominal, see Part X.

FPN No. 2: See Annex D, Example No. D8.

### 430.32 Continuous-Duty Motors.

**(A) More Than 1 Horsepower.** Each continuous-duty motor rated more than 1 hp shall be protected against overload by one of the means in 430.32(A)(1) through (A)(4).

**(1) Separate Overload Device.** A separate overload device that is responsive to motor current. This device shall be selected to trip or shall be rated at no more than the following percent of the motor nameplate full-load current rating:

| | |
|---|---|
| Motors with a marked service factor 1.15 or greater | 125% |
| Motors with a marked temperature rise 40°C or less | 125% |
| All other motors | 115% |

Modification of this value shall be permitted as provided in 430.32(C). For a multispeed motor, each winding connection shall be considered separately.

Where a separate motor overload device is connected so that it does not carry the total current designated on the motor nameplate, such as for wye-delta starting, the proper percentage of nameplate current applying to the selection or setting of the overload device shall be clearly designated on the equipment, or the manufacturer's selection table shall take this into account.

FPN: Where power factor correction capacitors are installed on the load side of the motor overload device, see 460.9.

**(2) Thermal Protector.** A thermal protector integral with the motor, approved for use with the motor it protects on the basis that it will prevent dangerous overheating of the motor due to overload and failure to start. The ultimate trip current of a thermally protected motor shall not exceed the following percentage of motor full-load current given in Tables 430.148, 430.149, and 430.150:

| | |
|---|---|
| Motor full-load current 9 amperes or less | 170% |
| Motor full-load current from 9.1 to, and including, 20 amperes | 156% |
| Motor full-load current greater than 20 amperes | 140% |

If the motor current-interrupting device is separate from the motor and its control circuit is operated by a protective device integral with the motor, it shall be arranged so that the opening of the control circuit will result in interruption of current to the motor.

**(3) Integral with Motor.** A protective device integral with a motor that will protect the motor against damage due to failure to start shall be permitted if the motor is part of an approved assembly that does not normally subject the motor to overloads.

**(4) Larger Than 1500 Horsepower.** For motors larger than 1500 hp, a protective device having embedded temperature detectors that cause current to the motor to be interrupted when the motor attains a temperature rise greater than marked on the nameplate in an ambient temperature of 40°C.

**(B) One Horsepower or Less, Automatically Started.** Any motor of 1 hp or less that is started automatically shall be protected against overload by one of the following means.

**(1) Separate Overload Device.** A separate overload device that is responsive to motor current. This device shall be selected to trip or shall be rated at not more than the following percentage of the motor nameplate full-load current rating:

| | |
|---|---|
| Motors with a marked service factor 1.15 or greater | 125% |
| Motors with a marked temperature rise 40°C or less | 125% |
| All other motors | 115% |

For a multispeed motor, each winding connection shall be considered separately. Modification of this value shall be permitted as provided in 430.32(C).

**(2) Thermal Protector.** A thermal protector integral with the motor, approved for use with the motor that it protects on the basis that it will prevent dangerous overheating of the motor due to overload and failure to start. Where the motor current-interrupting device is separate from the motor and its control circuit is operated by a protective device integral with the motor, it shall be arranged so that the opening of the control circuit results in interruption of current to the motor.

**(3) Integral with Motor.** A protective device integral with a motor that protects the motor against damage due to failure to start shall be permitted (1) if the motor is part of an approved assembly that does not subject the motor to overloads, or (2) if the assembly is also equipped with other safety controls (such as the safety combustion controls on a domestic oil burner) that protect the motor against damage due to failure to start. Where the assembly has safety controls that protect the motor, it shall be so indicated on the nameplate of the assembly where it will be visible after installation.

**(4) Impedance-Protected.** In case the impedance of the motor windings is sufficient to prevent overheating due to failure to start, the motor shall be permitted to be protected as specified in 430.32(D)(1) for manually started motors if the motor is part of an approved assembly in which the motor will limit itself so that it will not be dangerously overheated.

FPN: Many ac motors of less than 1/20 hp, such as clock motors, series motors, and so forth, and also some larger motors such as torque motors, come within this classification. It does not include split-phase motors having automatic switches that disconnect the starting windings.

**(C) Selection of Overload Relay.** Where the sensing element or setting of the overload relay selected in accordance with 430.32(A)(1) and 430.32(B)(1) is not sufficient to start the motor or to carry the load, higher size sensing elements or incremental settings shall be permitted to be used, provided the trip current of the overload relay does not exceed the following percentage of motor nameplate full-load current rating:

| | |
|---|---|
| Motors with marked service factor 1.15 or greater | 140% |
| Motors with a marked temperature rise 40°C or less | 140% |
| All other motors | 130% |

If not shunted during the starting period of the motor as provided in 430.35, the overload device shall have sufficient time delay to permit the motor to start and accelerate its load.

FPN: A Class 20 or 30 overload relay will provide a longer motor acceleration time than a Class 10 or 20, respectively.

Use of a higher class overload relay may preclude the need for selection of a higher trip current.

**(D) One Horsepower or Less, Nonautomatically Started.**

**(1) Within Sight from Controller.** Each continuous-duty motor rated at 1 hp or less that is not permanently installed, is nonautomatically started, and is within sight from the controller location shall be permitted to be protected against overload by the branch-circuit short-circuit and ground-fault protective device. This branch-circuit protective device shall not be larger than that specified in Part IV of Article 430.

*Exception: Any such motor shall be permitted on a nominal 120-volt branch circuit protected at not over 20 amperes.*

**(2) Not Within Sight from Controller.** Any such motor that is not in sight from the controller location shall be protected as specified in 430.32(B). Any motor rated at 1 hp or less that is permanently installed shall be protected in accordance with 430.32(B).

**(E) Wound-Rotor Secondaries.** The secondary circuits of wound-rotor ac motors, including conductors, controllers, resistors, and so forth, shall be permitted to be protected against overload by the motor-overload device.

**430.33 Intermittent and Similar Duty.** A motor used for a condition of service that is inherently short-time, intermittent, periodic, or varying duty, as illustrated by Table 430.22(E), shall be permitted to be protected against overload by the branch-circuit short-circuit and ground-fault protective device, provided the protective device rating or setting does not exceed that specified in Table 430.52.

Any motor application shall be considered to be for continuous duty unless the nature of the apparatus it drives is such that the motor cannot operate continuously with load under any condition of use.

**430.35 Shunting During Starting Period.**

**(A) Nonautomatically Started.** For a nonautomatically started motor, the overload protection shall be permitted to be shunted or cut out of the circuit during the starting period of the motor if the device by which the overload protection is shunted or cut out cannot be left in the starting position and if fuses or inverse time circuit breakers rated or set at not over 400 percent of the full-load current of the motor are located in the circuit so as to be operative during the starting period of the motor.

**(B) Automatically Started.** The motor overload protection shall not be shunted or cut out during the starting period if the motor is automatically started.

*Exception: The motor overload protection shall be permitted to be shunted or cut out during the starting period on an automatically started motor where*

*(a) The motor starting period exceeds the time delay of available motor overload protective devices, and*
*(b) Listed means are provided to*
   *(1) Sense motor rotation and to automatically prevent the shunting or cutout in the event that the motor fails to start, and*
   *(2) Limit the time of overload protection shunting or cutout to less than the locked rotor time rating of the protected motor, and*
   *(3) Provide for shutdown and manual restart if motor running condition is not reached.*

**430.36 Fuses — In Which Conductor.** Where fuses are used for motor overload protection, a fuse shall be inserted in each ungrounded conductor and also in the grounded conductor if the supply system is 3-wire, 3-phase ac with one conductor grounded.

**430.37 Devices Other Than Fuses — In Which Conductor.** Where devices other than fuses are used for motor overload protection, Table 430.37 shall govern the minimum allowable number and location of overload units such as trip coils or relays.

**430.38 Number of Conductors Opened by Overload Device.** Motor overload devices, other than fuses or thermal protectors, shall simultaneously open a sufficient number of ungrounded conductors to interrupt current flow to the motor.

**430.39 Motor Controller as Overload Protection.** A motor controller shall also be permitted to serve as an overload device if the number of overload units complies with Table 430.37 and if these units are operative in both the starting and running position in the case of a dc motor, and in the running position in the case of an ac motor.

**430.40 Overload Relays.** Overload relays and other devices for motor overload protection that are not capable of opening short circuits or ground faults shall be protected by fuses or circuit breakers with ratings or settings in accordance with 430.52 or by a motor short-circuit protector in accordance with 430.52.

*Exception: Where approved for group installation and marked to indicate the maximum size of fuse or inverse time circuit breaker by which they must be protected, the overload devices shall be protected in accordance with this marking.*

FPN: For instantaneous trip circuit breakers or motor short-circuit protectors, see 430.52.

**Table 430.37 Overload Units**

| Kind of Motor | Supply System | Number and Location of Overload Units, Such as Trip Coils or Relays |
|---|---|---|
| 1-phase ac or dc | 2-wire, 1-phase ac or dc ungrounded | 1 in either conductor |
| 1-phase ac or dc | 2-wire, 1-phase ac or dc, one conductor grounded | 1 in ungrounded conductor |
| 1-phase ac or dc | 3-wire, 1-phase ac or dc, grounded neutral | 1 in either ungrounded conductor |
| 1-phase ac | Any 3-phase | 1 in ungrounded conductor |
| 2-phase ac | 3-wire, 2-phase ac, ungrounded | 2, one in each phase |
| 2-phase ac | 3-wire, 2-phase ac, one conductor grounded | 2 in ungrounded conductors |
| 2-phase ac | 4-wire, 2-phase ac, grounded or ungrounded | 2, one per phase in ungrounded conductors |
| 2-phase ac | Grounded neutral or 5-wire, 2-phase ac, ungrounded | 2, one per phase in any ungrounded phase wire |
| 3-phase ac | Any 3-phase | 3, one in each phase* |

*Exception: An overload unit in each phase shall not be required where overload protection is provided by other approved means.*

**430.42 Motors on General-Purpose Branch Circuits.** Overload protection for motors used on general-purpose branch circuits as permitted in Article 210 shall be provided as specified in 430.42(A), (B), (C), or (D).

**(A) Not Over 1 Horsepower.** One or more motors without individual overload protection shall be permitted to be connected to a general-purpose branch circuit only where the installation complies with the limiting conditions specified in 430.32(B) and (D) and 430.53(A)(1) and (A)(2).

**(B) Over 1 Horsepower.** Motors of ratings larger than specified in 430.53(A) shall be permitted to be connected to general-purpose branch circuits only where each motor is protected by overload protection selected to protect the motor as specified in 430.32. Both the controller and the motor overload device shall be approved for group installation

with the short-circuit and ground-fault protective device selected in accordance with 430.53.

**(C) Cord-and-Plug Connected.** Where a motor is connected to a branch circuit by means of an attachment plug and receptacle and individual overload protection is omitted as provided in 430.42(A), the rating of the attachment plug and receptacle shall not exceed 15 amperes at 125 volts or 250 volts. Where individual overload protection is required as provided in 430.42(B) for a motor or motor-operated appliance that is attached to the branch circuit through an attachment plug and receptacle, the overload device shall be an integral part of the motor or of the appliance. The rating of the attachment plug and receptacle shall determine the rating of the circuit to which the motor may be connected, as provided in Article 210.

**(D) Time Delay.** The branch-circuit short-circuit and ground-fault protective device protecting a circuit to which a motor or motor-operated appliance is connected shall have sufficient time delay to permit the motor to start and accelerate its load.

**430.43 Automatic Restarting.** A motor overload device that can restart a motor automatically after overload tripping shall not be installed unless approved for use with the motor it protects. A motor overload device that can restart a motor automatically after overload tripping shall not be installed if automatic restarting of the motor can result in injury to persons.

**430.44 Orderly Shutdown.** If immediate automatic shutdown of a motor by a motor overload protective device(s) would introduce additional or increased hazard(s) to a person(s) and continued motor operation is necessary for safe shutdown of equipment or process, a motor overload sensing device(s) conforming with the provisions of Part III of this article shall be permitted to be connected to a supervised alarm instead of causing immediate interruption of the motor circuit, so that corrective action or an orderly shutdown can be initiated.

## IV. Motor Branch-Circuit Short-Circuit and Ground-Fault Protection

**430.51 General.** Part IV specifies devices intended to protect the motor branch-circuit conductors, the motor control apparatus, and the motors against overcurrent due to short circuits or grounds. These rules add to or amend the provisions of Article 240. The devices specified in Part IV do not include the types of devices required by 210.8, 230.95, and 527.6.

The provisions of Part IV shall not apply to motor circuits rated over 600 volts, nominal.

FPN No. 1: For over 600 volts, nominal, see Part X.

FPN No. 2: See Annex D, Example D8.

### 430.52 Rating or Setting for Individual Motor Circuit.

**(A) General.** The motor branch-circuit short-circuit and ground-fault protective device shall comply with 430.52(B) and either 430.52(C) or (D), as applicable.

**(B) All Motors.** The motor branch-circuit short-circuit and ground-fault protective device shall be capable of carrying the starting current of the motor.

**(C) Rating or Setting.**

**(1) In Accordance with Table 430.52.** A protective device that has a rating or setting not exceeding the value calculated according to the values given in Table 430.52 shall be used.

*Exception No. 1: Where the values for branch-circuit short-circuit and ground-fault protective devices determined by Table 430.52 do not correspond to the standard sizes or ratings of fuses, nonadjustable circuit breakers, thermal protective devices, or possible settings of adjustable circuit breakers, a higher size, rating, or possible setting that does not exceed the next higher standard ampere rating shall be permitted.*

*Exception No. 2: Where the rating specified in Table 430.52, as modified by Exception No. 1, is not sufficient for the starting current of the motor:*

(a) *The rating of a nontime-delay fuse not exceeding 600 amperes or a time-delay Class CC fuse shall be permitted to be increased but shall in no case exceed 400 percent of the full-load current.*

(b) *The rating of a time-delay (dual-element) fuse shall be permitted to be increased but shall in no case exceed 225 percent of the full-load current.*

(c) *The rating of an inverse time circuit breaker shall be permitted to be increased but shall in no case exceed 400 percent for full-load currents of 100 amperes or less or 300 percent for full-load currents greater than 100 amperes.*

(d) *The rating of a fuse of 601–6000 ampere classification shall be permitted to be increased but shall in no case exceed 300 percent of the full-load current.*

FPN: See Annex D, Example D8, and Figure 430.1.

**(2) Overload Relay Table.** Where maximum branch-circuit short-circuit and ground-fault protective device ratings are shown in the manufacturer's overload relay table for use with a motor controller or are otherwise marked on

**Table 430.52 Maximum Rating or Setting of Motor Branch-Circuit Short-Circuit and Ground-Fault Protective Devices**

| | Percentage of Full-Load Current | | | |
| Type of Motor | Nontime Delay Fuse[1] | Dual Element (Time-Delay) Fuse[1] | Instantaneous Trip Breaker | Inverse Time Breaker[2] |
|---|---|---|---|---|
| Single-phase motors | 300 | 175 | 800 | 250 |
| AC polyphase motors other than wound-rotor | | | | |
| Squirrel cage — other than Design E or Design B energy efficient | 300 | 175 | 800 | 250 |
| Design E or Design B energy efficient | 300 | 175 | 1100 | 250 |
| Synchronous[3] | 300 | 175 | 800 | 250 |
| Wound rotor | 150 | 150 | 800 | 150 |
| Direct current (constant voltage) | 150 | 150 | 250 | 150 |

Note: For certain exceptions to the values specified, see 430.54.
[1]The values in the Nontime Delay Fuse column apply to Time-Delay Class CC fuses.
[2]The values given in the last column also cover the ratings of nonadjustable inverse time types of circuit breakers that may be modified as in 430.52(C), Exception No. 1 and No. 2.
[3]Synchronous motors of the low-torque, low-speed type (usually 450 rpm or lower), such as are used to drive reciprocating compressors, pumps, and so forth, that start unloaded, do not require a fuse rating or circuit-breaker setting in excess of 200 percent of full-load current.

the equipment, they shall not be exceeded even if higher values are allowed as shown above.

**(3) Instantaneous Trip Circuit Breaker.** An instantaneous trip circuit breaker shall be used only if adjustable and if part of a listed combination motor controller having coordinated motor overload and short-circuit and ground-fault protection in each conductor, and the setting is adjusted to no more than the value specified in Table 430.52.

FPN: For the purpose of this article, instantaneous-trip circuit breakers may include a damping means to accommodate a transient motor inrush current without nuisance tripping of the circuit breaker.

*Exception No. 1: Where the setting specified in Table 430.52 is not sufficient for the starting current of the motor, the setting of an instantaneous trip circuit breaker shall be permitted to be increased but shall in no case exceed*

*1300 percent of the motor full-load current for other than Design E motors or Design B energy efficient motors and no more than 1700 percent of full-load motor current for Design E motors or Design B energy efficient motors. Trip settings above 800 percent for other than Design E motors or Design B energy efficient motors and above 1100 percent for Design E motors or Design B energy efficient motors shall be permitted where the need has been demonstrated by engineering evaluation. In such cases, it shall not be necessary to first apply an instantaneous-trip circuit breaker at 800 percent or 1100 percent.*

> FPN: For additional information on the requirements for a motor to be classified "energy efficient," see NEMA Standards Publication No. MG1-1993, Revision, *Motors and Generators*, Part 12.59.

*Exception No. 2: Where the motor full-load current is 8 amperes or less, the setting of the instantaneous-trip circuit breaker with a continuous current rating of 15 amperes or less in a listed combination motor controller that provides coordinated motor branch-circuit overload and short-circuit and ground-fault protection shall be permitted to be increased to the value marked on the controller.*

**(4) Multispeed Motor.** For a multispeed motor, a single short-circuit and ground-fault protective device shall be permitted for two or more windings of the motor, provided the rating of the protective device does not exceed the above applicable percentage of the nameplate rating of the smallest winding protected.

*Exception: For a multispeed motor, a single short-circuit and ground-fault protective device shall be permitted to be used and sized according to the full-load current of the highest current winding, where all of the following conditions are met:*

*(a) Each winding is equipped with individual overload protection sized according to its full-load current.*

*(b) The branch-circuit conductors supplying each winding are sized according to the full-load current of the highest full-load current winding.*

*(c) The controller for each winding has a horsepower rating not less than that required for the winding having the highest horsepower rating.*

**(5) Power Electronic Devices.** Suitable fuses shall be permitted in lieu of devices listed in Table 430.52 for power electronic devices in a solid state motor controller system, provided that the marking for replacement fuses is provided adjacent to the fuses.

**(6) Self-Protected Combination Controller.** A listed self-protected combination controller shall be permitted in lieu of the devices specified in Table 430.52. Adjustable instantaneous-trip settings shall not exceed 1300 percent of full-load motor current for other than Design E motors or

Design B energy efficient motors and not more than 1700 percent of full-load motor current for Design E motors or Design B energy efficient motors.

**(7) Motor Short-Circuit Protector.** A motor short-circuit protector shall be permitted in lieu of devices listed in Table 430.52 if the motor short-circuit protector is part of a listed combination motor controller having coordinated motor overload protection and short-circuit and ground-fault protection in each conductor and it will open the circuit at currents exceeding 1300 percent of motor full-load current for other than Design E motors or Design B energy efficient motors and 1700 percent of motor full-load motor current for Design E motors or Design B energy efficient motors.

**(D) Torque Motors.** Torque motor branch circuits shall be protected at the motor nameplate current rating in accordance with 240.4(B).

**430.53 Several Motors or Loads on One Branch Circuit.** Two or more motors or one or more motors and other loads shall be permitted to be connected to the same branch circuit under conditions specified in 430.53(D) and in 430.53(A), (B), or (C).

**(A) Not Over 1 Horsepower.** Several motors, each not exceeding 1 hp in rating, shall be permitted on a nominal 120-volt branch circuit protected at not over 20 amperes or a branch circuit of 600 volts, nominal, or less, protected at not over 15 amperes, if all of the following conditions are met:

(1) The full-load rating of each motor does not exceed 6 amperes.

(2) The rating of the branch-circuit short-circuit and ground-fault protective device marked on any of the controllers is not exceeded.

(3) Individual overload protection conforms to 430.32.

**(B) If Smallest Rated Motor Protected.** If the branch-circuit short-circuit and ground-fault protective device is selected not to exceed that allowed by 430.52 for the smallest rated motor, two or more motors or one or more motors and other load(s), with each motor having individual overload protection, shall be permitted to be connected to a branch circuit where it can be determined that the branch-circuit short-circuit and ground-fault protective device will not open under the most severe normal conditions of service that might be encountered.

**(C) Other Group Installations.** Two or more motors of any rating or one or more motors and other load(s), with each motor having individual overload protection, shall be permitted to be connected to one branch circuit where the motor controller(s) and overload device(s) are (1) installed as a listed factory assembly and the motor branch-circuit short-circuit and ground-fault protective device either is provided as part of the assembly or is specified by a marking on the assembly, or (2) the motor branch-circuit short-circuit and ground-fault protective device, the motor controller(s), and overload device(s) are field-installed as separate assemblies listed for such use and provided with manufacturers' instructions for use with each other, and (3) all of the following conditions are complied with:

(1) Each motor overload device is listed for group installation with a specified maximum rating of fuse, inverse time circuit breaker, or both.

(2) Each motor controller is listed for group installation with a specified maximum rating of fuse, circuit breaker, or both.

(3) Each circuit breaker is one of the inverse time type and listed for group installation.

(4) The branch circuit shall be protected by fuses or inverse time circuit breakers having a rating not exceeding that specified in 430.52 for the highest rated motor connected to the branch circuit plus an amount equal to the sum of the full-load current ratings of all other motors and the ratings of other loads connected to the circuit. Where this calculation results in a rating less than the ampacity of the supply conductors, it shall be permitted to increase the maximum rating of the fuses or circuit breaker to a value not exceeding that permitted by 240.4(B).

(5) The branch-circuit fuses or inverse time circuit breakers are not larger than allowed by 430.40 for the overload relay protecting the smallest rated motor of the group.

> FPN: See 110.10 for circuit impedance and other characteristics.

**(D) Single Motor Taps.** For group installations described above, the conductors of any tap supplying a single motor shall not be required to have an individual branch-circuit short-circuit and ground-fault protective device, provided they comply with one of the following:

(1) No conductor to the motor shall have an ampacity less than that of the branch-circuit conductors.

(2) No conductor to the motor shall have an ampacity less than one-third that of the branch-circuit conductors, with a minimum in accordance with 430.22, the conductors to the motor overload device being not more than 7.5 m (25 ft) long and being protected from physical damage.

(3) Conductors from the branch-circuit short-circuit and ground-fault protective device to a listed manual motor controller additionally marked "Suitable for Tap Conductor Protection in Group Installations" shall be permitted to have an ampacity not less than 1/10 the rating or setting of the branch-circuit short-circuit and

ground-fault protective device. The conductors from the controller to the motor shall have an ampacity in accordance with 430.22. The conductors from the branch-circuit short-circuit and ground-fault protective device to the controller shall (1) be suitably protected from physical damage and enclosed either by an enclosed controller or by a raceway and shall be not more than 3 m (10 ft) long or (2) shall have an ampacity not less than that of the branch circuit conductors.

**430.54 Multimotor and Combination-Load Equipment.** The rating of the branch-circuit short-circuit and ground-fault protective device for multimotor and combination-load equipment shall not exceed the rating marked on the equipment in accordance with 430.7(D).

**430.55 Combined Overcurrent Protection.** Motor branch-circuit short-circuit and ground-fault protection and motor overload protection shall be permitted to be combined in a single protective device where the rating or setting of the device provides the overload protection specified in 430.32.

**430.56 Branch-Circuit Protective Devices — In Which Conductor.** Branch-circuit protective devices shall comply with the provisions of 240.20.

**430.57 Size of Fuseholder.** Where fuses are used for motor branch-circuit short-circuit and ground-fault protection, the fuseholders shall not be of a smaller size than required to accommodate the fuses specified by Table 430.52.

*Exception: Where fuses having time delay appropriate for the starting characteristics of the motor are used, it shall be permitted to use fuseholders sized to fit the fuses that are used.*

**430.58 Rating of Circuit Breaker.** A circuit breaker for motor branch-circuit short-circuit and ground-fault protection shall have a current rating in accordance with 430.52 and 430.110.

### V. Motor Feeder Short-Circuit and Ground-Fault Protection

**430.61 General.** Part V specifies protective devices intended to protect feeder conductors supplying motors against overcurrents due to short circuits or grounds.

> FPN: See Annex D, Example D8.

**430.62 Rating or Setting — Motor Load.**

**(A) Specific Load.** A feeder supplying a specific fixed motor load(s) and consisting of conductor sizes based on

430.24 shall be provided with a protective device having a rating or setting not greater than the largest rating or setting of the branch-circuit short-circuit and ground-fault protective device for any motor supplied by the feeder [based on the maximum permitted value for the specific type of a protective device in accordance with 430.52, or 440.22(A) for hermetic refrigerant motor-compressors], plus the sum of the full-load currents of the other motors of the group.

Where the same rating or setting of the branch-circuit short-circuit and ground-fault protective device is used on two or more of the branch circuits supplied by the feeder, one of the protective devices shall be considered the largest for the above calculations.

*Exception No. 1: Where one or more instantaneous trip circuit breakers or motor short-circuit protectors are used for motor branch-circuit short-circuit and ground-fault protection as permitted in 430.52(C), the procedure provided above for determining the maximum rating of the feeder protective device shall apply with the following provision: For the purpose of the calculation, each instantaneous trip circuit breaker or motor short-circuit protector shall be assumed to have a rating not exceeding the maximum percentage of motor full-load current permitted by Table 430.52 for the type of feeder protective device employed.*

*Exception No. 2: Where the feeder overcurrent protective device also provides overcurrent protection for a motor control center, the provisions of 430.94 shall apply.*

FPN: See Annex D, Example D8.

**(B) Other Installations.** Where feeder conductors have an ampacity greater than required by 430.24, the rating or setting of the feeder overcurrent protective device shall be permitted to be based on the ampacity of the feeder conductors.

**430.63 Rating or Setting — Power and Lighting Loads.** Where a feeder supplies a motor load and, in addition, a lighting or a lighting and appliance load, the feeder protective device shall have a rating sufficient to carry the lighting or lighting and appliance load, plus the following:

(1) For a single motor, the rating permitted by 430.52
(2) For a single hermetic refrigerant motor-compressor, the rating permitted by 440.22
(3) For two or more motors, the rating permitted by 430.62

*Exception: Where the feeder overcurrent device provides the overcurrent protection for a motor control center, the provisions of 430.94 shall apply.*

## VI. Motor Control Circuits

**430.71 General.** Part VI contains modifications of the general requirements and applies to the particular conditions of motor control circuits.

FPN: See 430.9(B) for equipment device terminal requirements.

**Definition: Motor Control Circuit.** The circuit of a control apparatus or system that carries the electric signals directing the performance of the controller but does not carry the main power current.

### 430.72 Overcurrent Protection.

**(A) General.** A motor control circuit tapped from the load side of a motor branch-circuit short-circuit and ground-fault protective device(s) and functioning to control the motor(s) connected to that branch circuit shall be protected against overcurrent in accordance with 430.72. Such a tapped control circuit shall not be considered to be a branch circuit and shall be permitted to be protected by either a supplementary or branch-circuit overcurrent protective device(s). A motor control circuit other than such a tapped control circuit shall be protected against overcurrent in accordance with 725.23 or the notes to Tables 11(A) and (B) in Chapter 9, as applicable.

**(B) Conductor Protection.** The overcurrent protection for conductors shall be provided as specified in 430.72(B)(1) or (B)(2).

*Exception No. 1: Where the opening of the control circuit would create a hazard as, for example, the control circuit of a fire pump motor, and the like, conductors of control circuits shall require only short-circuit and ground-fault protection and shall be permitted to be protected by the motor branch-circuit short-circuit and ground-fault protective device(s).*

*Exception No. 2: Conductors supplied by the secondary side of a single-phase transformer having only a two-wire (single-voltage) secondary shall be permitted to be protected by overcurrent protection provided on the primary (supply) side of the transformer, provided this protection does not exceed the value determined by multiplying the appropriate maximum rating of the overcurrent device for the secondary conductor from Table 430.72(B) by the secondary-to-primary voltage ratio. Transformer secondary conductors (other than two-wire) shall not be considered to be protected by the primary overcurrent protection.*

**(1) Separate Overcurrent Protection.** Where the motor branch-circuit short-circuit and ground-fault protective device does not provide protection in accordance with 430.72(B)(2), separate overcurrent protection shall be provided. The overcurrent protection shall not exceed the values specified in Column A of Table 430.72(B).

**(2) Branch-Circuit Overcurrent Protective Device.** Conductors shall be permitted to be protected by the motor branch-circuit short-circuit and ground-fault protective device and shall require only short-circuit and ground-fault protection. Where the conductors do not extend beyond the motor control equipment enclosure, the rating of the protective device(s) shall not exceed the value specified in Column B of Table 430.72(B). Where the conductors extend beyond the motor control equipment enclosure, the rating of the protective device(s) shall not exceed the value specified in Column C of Table 430.72(B).

**(C) Control Circuit Transformer.** Where a motor control circuit transformer is provided, the transformer shall be protected in accordance with 430.72(C)(1), (2), (3), (4), or (5).

*Exception: Overcurrent protection shall be omitted where the opening of the control circuit would create a hazard as, for example, the control circuit of a fire pump motor and the like.*

**(1) Compliance with Article 725.** Where the transformer supplies a Class 1 power-limited circuit, Class 2, or Class 3 remote-control circuit conforming with the requirements of Article 725, the protection shall comply with Article 725.

**(2) Compliance with Article 450.** Protection shall be permitted to be provided in accordance with 450.3.

**(3) Less Than 50 Volt-Amperes.** Control circuit transformers rated less than 50 volt-amperes (VA) and that are an integral part of the motor controller and located within the motor controller enclosure shall be permitted to be protected by primary overcurrent devices, impedance limiting means, or other inherent protective means.

**(4) Primary Less Than 2 Amperes.** Where the control circuit transformer rated primary current is less than 2 am-

peres, an overcurrent device rated or set at not more than 500 percent of the rated primary current shall be permitted in the primary circuit.

**(5) Other Means.** Protection shall be permitted to be provided by other approved means.

**430.73 Mechanical Protection of Conductor.** Where damage to a motor control circuit would constitute a hazard, all conductors of such a remote motor control circuit that are outside the control device itself shall be installed in a raceway or be otherwise suitably protected from physical damage.

Where one side of the motor control circuit is grounded, the motor control circuit shall be arranged so that an accidental ground in the control circuit remote from the motor controller will (1) not start the motor and (2) not bypass manually operated shutdown devices or automatic safety shutdown devices.

**430.74 Disconnection.**

**(A) General.** Motor control circuits shall be arranged so that they will be disconnected from all sources of supply when the disconnecting means is in the open position. The disconnecting means shall be permitted to consist of two or more separate devices, one of which disconnects the motor and the controller from the source(s) of power supply for the motor, and the other(s), the motor control circuit(s) from its power supply. Where separate devices are used, they shall be located immediately adjacent to each other.

*Exception No. 1: Where more than 12 motor control circuit conductors are required to be disconnected, the disconnecting means shall be permitted to be located other than im-*

**Table 430.72(B) Maximum Rating of Overcurrent Protective Device in Amperes**

| | | | Protection Provided by Motor Branch-Circuit Protective Device(s) | | | |
|---|---|---|---|---|---|---|
| | Column A<br>Separate Protection<br>Provided | | Column B<br>Conductors Within<br>Enclosure | | Column C<br>Conductors Extend Beyond<br>Enclosure | |
| Control Circuit<br>Conductor Size<br>(AWG) | Copper | Aluminum or<br>Copper-Clad<br>Aluminum | Copper | Aluminum or<br>Copper-Clad<br>Aluminum | Copper | Aluminum or<br>Copper-Clad<br>Aluminum |
| 18 | 7 | — | 25 | — | 7 | — |
| 16 | 10 | — | 40 | — | 10 | — |
| 14 | (Note 1) | — | 100 | — | 45 | — |
| 12 | (Note 1) | (Note 1) | 120 | 100 | 60 | 45 |
| 10 | (Note 1) | (Note 1) | 160 | 140 | 90 | 75 |
| Larger than 10 | (Note 1) | (Note 1) | (Note 2) | (Note 2) | (Note3) | (Note 3) |

Notes:
1. Value specified in 310.15 as applicable.

2. 400 percent of value specified in Table 310.17 for 60°C conductors.
3. 300 percent of value specified in Table 310.16 for 60°C conductors.

*mediately adjacent to each other where all of the following conditions are complied with.*

*(a) Access to energized parts is limited to qualified persons in accordance with Part XI of this article.*

*(b) A warning sign is permanently located on the outside of each equipment enclosure door or cover permitting access to the live parts in the motor control circuit(s), warning that motor control circuit disconnecting means are remotely located and specifying the location and identification of each disconnect. Where energized parts are not in an equipment enclosure as permitted by 430.132 and 430.133, an additional warning sign(s) shall be located where visible to persons who may be working in the area of the energized parts.*

*Exception No. 2: The motor control circuit disconnecting means shall be permitted to be remote from the motor controller power supply disconnecting means where the opening of one or more motor control circuit disconnect means may result in potentially unsafe conditions for personnel or property and the conditions of items (a) and (b) of Exception No. 1 are complied with.*

**(B) Control Transformer in Controller Enclosure.** Where a transformer or other device is used to obtain a reduced voltage for the motor control circuit and is located in the controller enclosure, such transformer or other device shall be connected to the load side of the disconnecting means for the motor control circuit.

## VII. Motor Controllers

**430.81 General.** Part VII is intended to require suitable controllers for all motors.

**(A) Definition.** For the definition of *Controller*, see Article 100. For the purpose of this article, a controller is any switch or device that is normally used to start and stop a motor by making and breaking the motor circuit current.

**(B) Stationary Motor of ⅛ Horsepower or Less.** For a stationary motor rated at ⅛ hp or less that is normally left running and is constructed so that it cannot be damaged by overload or failure to start, such as clock motors and the like, the branch-circuit protective device shall be permitted to serve as the controller.

**(C) Portable Motor of ⅓ Horsepower or Less.** For a portable motor rated at ⅓ hp or less, the controller shall be permitted to be an attachment plug and receptacle.

**430.82 Controller Design.**

**(A) Starting and Stopping.** Each controller shall be capable of starting and stopping the motor it controls and

shall be capable of interrupting the locked-rotor current of the motor.

**(B) Autotransformer.** An autotransformer starter shall provide an "off" position, a running position, and at least one starting position. It shall be designed so that it cannot rest in the starting position or in any position that will render the overload device in the circuit inoperative.

**(C) Rheostats.** Rheostats shall be in compliance with the following:

(1) Motor-starting rheostats shall be designed so that the contact arm cannot be left on intermediate segments. The point or plate on which the arm rests when in the starting position shall have no electrical connection with the resistor.

(2) Motor-starting rheostats for dc motors operated from a constant voltage supply shall be equipped with automatic devices that will interrupt the supply before the speed of the motor has fallen to less than one-third its normal rate.

**430.83 Ratings.** The controller shall have a rating as specified in 430.83(A), unless otherwise permitted in 430.83(B) or (C), or as specified in 430.83(D), under the conditions specified.

**(A) General.**

**(1) Horsepower Ratings.** Controllers, other than inverse time circuit breakers and molded case switches, shall have horsepower ratings at the application voltage not lower than the horsepower rating of the motor. A controller for a Design E motor rated more than 2 hp shall (1) be marked as rated for use with a Design E motor or (2) have a horsepower rating not less than 1.4 times the rating of a motor rated 3 through 100 hp or not less than 1.3 times the rating of a motor rated over 100 hp.

**(2) Circuit Breaker.** A branch-circuit inverse time circuit breaker rated in amperes shall be permitted as a controller for all motors, including Design E. Where this circuit breaker is also used for overload protection, it shall conform to the appropriate provisions of this article governing overload protection.

**(3) Molded Case Switch.** A molded case switch rated in amperes shall be permitted as a controller for all motors, including Design E.

**(B) Small Motors.** Devices as specified in 430.81(B) and (C) shall be permitted as a controller.

**(C) Stationary Motors of 2 Horsepower or Less.** For stationary motors rated at 2 hp or less and 300 volts or less, the controller shall be permitted to be either of the following:

(1) A general-use switch having an ampere rating not less than twice the full-load current rating of the motor

(2) On ac circuits, a general-use snap switch suitable only for use on ac (not general-use ac–dc snap switches) where the motor full-load current rating is not more than 80 percent of the ampere rating of the switch

**(D) Torque Motors.** For torque motors, the controller shall have a continuous-duty, full-load current rating not less than the nameplate current rating of the motor. For a motor controller rated in horsepower but not marked with the foregoing current rating, the equivalent current rating shall be determined from the horsepower rating by using Tables 430.147, 430.148, 430.149, or 430.150.

**(E) Voltage Rating.** A controller with a straight voltage rating, for example, 240 volts or 480 volts, shall be permitted to be applied in a circuit in which the nominal voltage between any two conductors does not exceed the controller's voltage rating. A controller with a slash rating, for example, 120/240 volts or 480/277 volts, shall only be applied in a circuit in which the nominal voltage to ground from any conductor does not exceed the lower of the two values of the controller's voltage rating and the nominal voltage between any two conductors does not exceed the higher value of the controller's voltage rating.

**430.84 Need Not Open All Conductors.** The controller shall not be required to open all conductors to the motor.

*Exception: Where the controller serves also as a disconnecting means, it shall open all ungrounded conductors to the motor as provided in 430.111.*

**430.85 In Grounded Conductors.** One pole of the controller shall be permitted to be placed in a permanently grounded conductor, provided the controller is designed so that the pole in the grounded conductor cannot be opened without simultaneously opening all conductors of the circuit.

**430.87 Number of Motors Served by Each Controller.** Each motor shall be provided with an individual controller.

*Exception: For motors rated 600 volts or less, a single controller rated at not less than the equivalent horsepower, as determined in accordance with 430.110(C)(1), of all the motors in the group shall be permitted to serve the group under any of the following conditions:*

*(a) Where a number of motors drive several parts of a single machine or piece of apparatus, such as metal and woodworking machines, cranes, hoists, and similar apparatus*

*(b) Where a group of motors is under the protection of one overcurrent device as permitted in 430.53(A)*

*(c) Where a group of motors is located in a single room within sight from the controller location*

**430.88 Adjustable-Speed Motors.** Adjustable-speed motors that are controlled by means of field regulation shall be equipped and connected so that they cannot be started under a weakened field.

*Exception: Starting under a weakened field shall be permitted where the motor is designed for such starting.*

**430.89 Speed Limitation.** Machines of the following types shall be provided with speed-limiting devices or other speed-limiting means:

(1) Separately excited dc motors
(2) Series motors
(3) Motor-generators and converters that can be driven at excessive speed from the dc end, as by a reversal of current or decrease in load

*Exception: Separate speed-limiting devices or means shall not be required under either of the following conditions:*

*(a) Where the inherent characteristics of the machines, the system, or the load and the mechanical connection thereto are such as to safely limit the speed*

*(b) Where the machine is always under the manual control of a qualified operator*

**430.90 Combination Fuseholder and Switch as Controller.** The rating of a combination fuseholder and switch used as a motor controller shall be such that the fuseholder will accommodate the size of the fuse specified in Part III of this article for motor overload protection.

*Exception: Where fuses having time delay appropriate for the starting characteristics of the motor are used, fuseholders of smaller size than specified in Part III of this article shall be permitted.*

**430.91 Motor Controller Enclosure Types.** Table 430.91 provides the basis for selecting enclosures for use in specific locations other than hazardous (classified) locations. The enclosures are not intended to protect against conditions such as condensation, icing, corrosion, or contamination that may occur within the enclosure or enter via the conduit or unsealed openings. These internal conditions shall require special consideration by the installer and user.

## VIII. Motor Control Centers

**430.92 General.** Part VIII covers motor control centers installed for the control of motors, lighting, and power circuits.

**Table 430.91 Motor Controller Enclosure Selection**

**For Outdoor Use**

| Provides a Degree of Protection Against the Following Environmental Conditions | Enclosure Type Number[1] | | | | | | | | | |
|---|---|---|---|---|---|---|---|---|---|---|
| | 3 | 3R | 3S | 3X | 3RX | 3SX | 4 | 4X | 6 | 6P |
| Incidental contact with the enclosed equipment | X | X | X | X | X | X | X | X | X | X |
| Rain, snow, and sleet | X | X | X | X | X | X | X | X | X | X |
| Sleet[2] | — | — | X | — | — | X | — | — | — | — |
| Windblown dust | X | — | X | X | — | X | X | X | X | X |
| Hosedown | — | — | — | — | — | — | X | X | X | X |
| Corrosive agents | — | — | — | X | X | X | — | X | — | X |
| Temporary submersion | — | — | — | — | — | — | — | — | X | X |
| Prolonged submersion | — | — | — | — | — | — | — | — | — | X |

**For Indoor Use**

| Provides a Degree of Protection Against the Following Environmental Conditions | Enclosure Type Number[1] | | | | | | | | | |
|---|---|---|---|---|---|---|---|---|---|---|
| | 1 | 2 | 4 | 4X | 5 | 6 | 6P | 12 | 12K | 13 |
| Incidental contact with the enclosed equipment | X | X | X | X | X | X | X | X | X | X |
| Falling dirt | X | X | X | X | X | X | X | X | X | X |
| Falling liquids and light splashing | — | X | X | X | X | X | X | X | X | X |
| Circulating dust, lint, fibers, and flyings | — | — | X | X | — | X | X | X | X | X |
| Settling airborne dust, lint, fibers, and flyings | — | — | X | X | X | X | X | X | X | X |
| Hosedown and splashing water | — | — | X | X | — | X | X | — | — | — |
| Oil and coolant seepage | — | — | — | — | — | — | — | X | X | X |
| Oil or coolant spraying and splashing | — | — | — | — | — | — | — | — | — | X |
| Corrosive agents | — | — | — | X | — | — | X | — | — | — |
| Temporary submersion | — | — | — | — | — | X | X | — | — | — |
| Prolonged submersion | — | — | — | — | — | — | X | — | — | — |

[1]Enclosure type number shall be marked on the motor controller enclosure.
[2]Mechanism shall be operable when ice covered.

**430.94 Overcurrent Protection.** Motor control centers shall be provided with overcurrent protection in accordance with Parts I, II, and IX of Article 240. The ampere rating or setting of the overcurrent protective device shall not exceed the rating of the common power bus. This protection shall be provided by (1) an overcurrent protective device located ahead of the motor control center or (2) a main overcurrent protective device located within the motor control center.

**430.95 Service-Entrance Equipment.** Where used as service equipment, each motor control center shall be provided with a single main disconnecting means to disconnect all ungrounded service conductors.

*Exception: A second service disconnect shall be permitted to supply additional equipment.*

Where a grounded conductor is provided, the motor control center shall be provided with a main bonding jumper, sized in accordance with 250.28(D), within one of the sections for connecting the grounded conductor, on its supply side, to the motor control center equipment ground bus.

*Exception: High-impedance grounded neutral systems shall be permitted to be connected as provided in 250.36.*

**430.96 Grounding.** Multisection motor control centers shall be bonded together with an equipment grounding conductor or an equivalent grounding bus sized in accordance with Table 250.122. Equipment grounding conductors shall terminate on this grounding bus or to a grounding termination point provided in a single-section motor control center.

### 430.97 Busbars and Conductors.

**(A) Support and Arrangement.** Busbars shall be protected from physical damage and be held firmly in place. Other than for required interconnections and control wiring, only those conductors that are intended for termination in a vertical section shall be located in that section.

*Exception: Conductors shall be permitted to travel horizontally through vertical sections where such conductors are isolated from the busbars by a barrier.*

**(B) Phase Arrangement.** The phase arrangement on 3-phase horizontal common power and vertical buses shall be A, B, C from front to back, top to bottom, or left to right, as viewed from the front of the motor control center. The B phase shall be that phase having the higher voltage to ground on 3-phase, 4-wire, delta-connected systems. Other busbar arrangements shall be permitted for additions to existing installations and shall be marked.

*Exception: Rear-mounted units connected to a vertical bus that is common to front-mounted units shall be permitted to have a C, B, A phase arrangement where properly identified.*

**(C) Minimum Wire-Bending Space.** The minimum wire-bending space at the motor control center terminals and minimum gutter space shall be as required in Article 312.

**(D) Spacings.** Spacings between motor control center bus terminals and other bare metal parts shall not be less than specified in Table 430.97.

**(E) Barriers.** Barriers shall be placed in all service-entrance motor control centers to isolate service busbars and terminals from the remainder of the motor control center.

### 430.98 Marking.

**(A) Motor Control Centers.** Motor control centers shall be marked according to 110.21, and such marking shall be plainly visible after installation. Marking shall also include common power bus current rating and motor control center short-circuit rating.

**(B) Motor Control Units.** Motor control units in a motor control center shall comply with 430.8.

## IX. Disconnecting Means

### 430.101 General. Part IX is intended to require disconnecting means capable of disconnecting motors and controllers from the circuit.

FPN No. 1: See Figure 430.1.

FPN No. 2: See 110.22 for identification of disconnecting means.

### 430.102 Location.

**(A) Controller.** An individual disconnecting means shall be provided for each controller and shall disconnect the controller. The disconnecting means shall be located in sight from the controller location.

*Exception No. 1: For motor circuits over 600 volts, nominal, a controller disconnecting means capable of being locked in the open position shall be permitted to be out of sight of the controller, provided the controller is marked with a warning label giving the location of the disconnecting means.*

*Exception No. 2: A single disconnecting means shall be permitted for a group of coordinated controllers that drive several parts of a single machine or piece of apparatus. The disconnecting means shall be located in sight from the controllers, and both the disconnecting means and the controllers shall be located in sight from the machine or apparatus.*

**(B) Motor.** A disconnecting means shall be located in sight from the motor location and the driven machinery location. The disconnecting means required in accordance with 430.102(A) shall be permitted to serve as the disconnecting

**Table 430.97 Minimum Spacing Between Bare Metal Parts**

| Nominal Voltage | Opposite Polarity Where Mounted on the Same Surface | | Opposite Polarity Where Held Free in Air | | Live Parts to Ground | |
|---|---|---|---|---|---|---|
| | mm | in. | mm | in. | mm | in. |
| Not over 125 volts, nominal | 19.1 | ¾ | 12.7 | ½ | 12.7 | ½ |
| Not over 250 volts, nominal | 31.8 | 1¼ | 19.1 | ¾ | 12.7 | ½ |
| Not over 600 volts, nominal | 50.8 | 2 | 25.4 | 1 | 25.4 | 1 |

means for the motor if it is located in sight from the motor location and the driven machinery location.

*Exception: The disconnecting means shall not be required to be in sight from the motor and the driven machinery location under either condition (a) or (b), provided the disconnecting means required in accordance with 430.102(A) is individually capable of being locked in the open position. The provision for locking or adding a lock to the disconnecting means shall be permanently installed on or at the switch or circuit breaker used as the disconnecting means.*

*(a) Where such a location of the disconnecting means is impracticable or introduces additional or increased hazards to persons or property*

*(b) In industrial installations, with written safety procedures, where conditions of maintenance and supervision ensure that only qualified persons service the equipment*

FPN No. 1: Some examples of increased or additional hazards include, but are not limited to, motors rated in excess of 100 hp, multimotor equipment, submersible motors, motors associated with variable frequency drives, and motors located in hazardous (classified) locations.

FPN No. 2: For information on lockout/tagout procedures, see NFPA 70E-2000, *Standard for Electrical Safety Requirements for Employee Workplaces.*

**430.103 Operation.** The disconnecting means shall open all ungrounded supply conductors and shall be designed so that no pole can be operated independently. The disconnecting means shall be permitted in the same enclosure with the controller.

FPN: See 430.113 for equipment receiving energy from more than one source.

**430.104 To Be Indicating.** The disconnecting means shall plainly indicate whether it is in the open (off) or closed (on) position.

**430.105 Grounded Conductors.** One pole of the disconnecting means shall be permitted to disconnect a permanently grounded conductor, provided the disconnecting means is designed so that the pole in the grounded conductor cannot be opened without simultaneously disconnecting all conductors of the circuit.

**430.107 Readily Accessible.** At least one of the disconnecting means shall be readily accessible.

**430.108 Every Switch.** Every disconnecting means in the motor circuit between the point of attachment to the feeder and the point of connection to the motor shall comply with the requirements of 430.109 and 430.110.

**430.109 Type.** The disconnecting means shall be a type specified in 430.109(A), unless otherwise permitted in 430.109(B) through (G), under the conditions specified.

**(A) General.**

**(1) Motor Circuit Switch.** A listed motor-circuit switch rated in horsepower. For Design E motors rated greater than 2 hp, the motor circuit switch shall be either (1) marked as rated for use with Design E motors or (2) have a horsepower rating not less than 1.4 times the rating of a motor rated 3–100 hp, or not less than 1.3 times the rating of a motor rated over 100 hp.

**(2) Molded Case Circuit Breaker.** A listed molded case circuit breaker.

**(3) Molded Case Switch.** A listed molded case switch.

**(4) Instantaneous Trip Circuit Breaker.** An instantaneous trip circuit breaker that is part of a listed combination motor controller.

**(5) Self-Protected Combination Controller.** Listed self-protected combination controller.

**(6) Manual Motor Controller.** Listed manual motor controllers additionally marked "Suitable as Motor Disconnect" shall be permitted as a disconnecting means where installed between the final motor branch-circuit short-circuit and ground-fault protective device and the motor.

**(B) Stationary Motors of ⅛ Horsepower or Less.** For stationary motors of ⅛ hp or less, the branch-circuit overcurrent device shall be permitted to serve as the disconnecting means.

**(C) Stationary Motors of 2 Horsepower or Less.** For stationary motors rated at 2 hp or less and 300 volts or less, the disconnecting means shall be permitted to be one of the devices specified in (1), (2), or (3):

(1) A general-use switch having an ampere rating not less than twice the full-load current rating of the motor

(2) On ac circuits, a general-use snap switch suitable only for use on ac (not general-use ac–dc snap switches) where the motor full-load current rating is not more than 80 percent of the ampere rating of the switch

(3) A listed manual motor controller having a horsepower rating not less than the rating of the motor and marked "Suitable as Motor Disconnect"

**(D) Autotransformer-Type Controlled Motors.** For motors of over 2 hp to and including 100 hp, the separate disconnecting means required for a motor with an autotransformer-type controller shall be permitted to be a general-use switch where all of the following provisions are met:

(1) The motor drives a generator that is provided with overload protection.

(2) The controller is capable of interrupting the locked-rotor current of the motors, is provided with a no voltage release, and is provided with running overload protection not exceeding 125 percent of the motor full-load current rating.

(3) Separate fuses or an inverse time circuit breaker rated or set at not more than 150 percent of the motor full-load current are provided in the motor branch circuit.

**(E) Isolating Switches.** For stationary motors rated at more than 40 hp dc or 100 hp ac, the disconnecting means shall be permitted to be a general-use or isolating switch where plainly marked "Do not operate under load."

**(F) Cord-and-Plug-Connected Motors.** For a cord-and-plug-connected motor, a horsepower-rated attachment plug and receptacle having ratings no less than the motor ratings shall be permitted to serve as the disconnecting means for other than a Design E motor and for a Design E motor rated 2 hp or less. For a Design E motor rated more than 2 hp, an attachment plug and receptacle used as the disconnecting means shall have a horsepower rating not less than 1.4 times the motor rating. A horsepower-rated attachment plug and receptacle shall not be required for a cord-and-plug-connected appliance in accordance with 422.33, a room air conditioner in accordance with 440.63, or a portable motor rated ⅓ hp or less.

**(G) Torque Motors.** For torque motors, the disconnecting means shall be permitted to be a general-use switch.

**430.110 Ampere Rating and Interrupting Capacity.**

**(A) General.** The disconnecting means for motor circuits rated 600 volts, nominal, or less shall have an ampere rating of at least 115 percent of the full-load current rating of the motor.

*Exception: A listed nonfused motor-circuit switch having a horsepower rating equal to or greater than the motor horsepower shall be permitted to have an ampere rating less than 115 percent of the full-load current rating of the motor.*

**(B) For Torque Motors.** Disconnecting means for a torque motor shall have an ampere rating of at least 115 percent of the motor nameplate current.

**(C) For Combination Loads.** Where two or more motors are used together or where one or more motors are used in combination with other loads, such as resistance heaters, and where the combined load may be simultaneous on a single disconnecting means, the ampere and horsepower ratings of the combined load shall be determined as follows.

**(1) Horsepower Rating.** The rating of the disconnecting means shall be determined from the sum of all currents, including resistance loads, at the full-load condition and also at the locked-rotor condition. The combined full-load current and the combined locked-rotor current so obtained shall be considered as a single motor for the purpose of this requirement as follows.

The full-load current equivalent to the horsepower rating of each motor shall be selected from Tables 430.147, 430.148, 430.149, or 430.150. These full-load currents shall be added to the rating in amperes of other loads to obtain an equivalent full-load current for the combined load.

The locked-rotor current equivalent to the horsepower rating of each motor shall be selected from Tables 430.151(A) or 430.151(B). The locked-rotor currents shall be added to the rating in amperes of other loads to obtain an equivalent locked-rotor current for the combined load. Where two or more motors or other loads cannot be started simultaneously, the largest sum of locked rotor currents of a motor or group of motors that can be started simultaneously and the full load currents of other concurrent loads shall be permitted to be used to determine the equivalent locked-rotor current for the simultaneous combined loads.

*Exception: Where part of the concurrent load is resistance load, and where the disconnecting means is a switch rated in horsepower and amperes, the switch used shall be permitted to have a horsepower rating that is not less than the combined load of the motor(s), if the ampere rating of the switch is not less than the locked-rotor current of the motor(s) plus the resistance load.*

**(2) Ampere Rating.** The ampere rating of the disconnecting means shall not be less than 115 percent of the sum of all currents at the full-load condition determined in accordance with 430.110(C)(1).

*Exception: A listed nonfused motor-circuit switch having a horsepower rating equal to or greater than the equivalent horsepower of the combined loads, determined in accordance with 430.110(C)(1), shall be permitted to have an ampere rating less than 115 percent of the sum of all currents at the full-load condition.*

**(3) Small Motors.** For small motors not covered by Tables 430.147, 430.148, 430.149, or 430.150, the locked-rotor current shall be assumed to be six times the full-load current.

**430.111 Switch or Circuit Breaker as Both Controller and Disconnecting Means.** A switch or circuit breaker shall be permitted to be used as both the controller and disconnecting means if it complies with 430.111(A) and is one of the types specified in 430.111(B).

**(A) General.** The switch or circuit breaker complies with the requirements for controllers specified in 430.83, opens all ungrounded conductors to the motor, and is protected by an overcurrent device in each ungrounded conductor (which shall be permitted to be the branch-circuit fuses). The overcurrent device protecting the controller shall be permitted to be part of the controller assembly or shall be permitted to be separate. An autotransformer-type controller shall be provided with a separate disconnecting means.

**(B) Type.** The device shall be one of the types specified in 430.111(B)(1), (2), or (3).

**(1) Air-Break Switch.** An air-break switch, operable directly by applying the hand to a lever or handle.

**(2) Inverse Time Circuit Breaker.** An inverse time circuit breaker operable directly by applying the hand to a lever or handle. The circuit breaker shall be permitted to be both power and manually operable.

**(3) Oil Switch.** An oil switch used on a circuit whose rating does not exceed 600 volts or 100 amperes, or by special permission on a circuit exceeding this capacity where under expert supervision. The oil switch shall be permitted to be both power and manually operable.

**430.112 Motors Served by Single Disconnecting Means.** Each motor shall be provided with an individual disconnecting means.

*Exception: A single disconnecting means shall be permitted to serve a group of motors under any one of the conditions of (a), (b), and (c). The single disconnecting means shall be rated in accordance with 430.110(C).*

*(a) Where a number of motors drive several parts of a single machine or piece of apparatus, such as metal and woodworking machines, cranes, and hoists.*
*(b) Where a group of motors is under the protection of one set of branch-circuit protective devices as permitted by 430.53(A).*
*(c) Where a group of motors is in a single room within sight from the location of the disconnecting means.*

**430.113 Energy from More Than One Source.** Motor and motor-operated equipment receiving electrical energy from more than one source shall be provided with disconnecting means from each source of electrical energy immediately adjacent to the equipment served. Each source shall be permitted to have a separate disconnecting means. Where multiple disconnecting means are provided, a permanent warning sign shall be provided on or adjacent to each disconnecting means.

*Exception No. 1: Where a motor receives electrical energy from more than one source, the disconnecting means for the main power supply to the motor shall not be required to be immediately adjacent to the motor, provided the controller disconnecting means is capable of being locked in the open position.*

*Exception No. 2: A separate disconnecting means shall not be required for a Class 2 remote-control circuit conforming with Article 725, rated not more than 30 volts, and that is isolated and ungrounded.*

## X. Over 600 Volts, Nominal

**430.121 General.** Part X recognizes the additional hazard due to the use of higher voltages. It adds to or amends the other provisions of this article.

**430.122 Marking on Controllers.** In addition to the marking required by 430.8, a controller shall be marked with the control voltage.

**430.123 Conductor Enclosures Adjacent to Motors.** Flexible metal conduit or liquidtight flexible metal conduit not exceeding 1.8 m (6 ft) in length shall be permitted to be employed for raceway connection to a motor terminal enclosure.

**430.124 Size of Conductors.** Conductors supplying motors shall have an ampacity not less than the current at which the motor overload protective device(s) is selected to trip.

**430.125 Motor-Circuit Overcurrent Protection.**

**(A) General.** Each motor circuit shall include coordinated protection to automatically interrupt overload and fault currents in the motor, the motor-circuit conductors, and the motor control apparatus.

*Exception: Where a motor is vital to operation of the plant and the motor should operate to failure if necessary to prevent a greater hazard to persons, the sensing device(s) shall be permitted to be connected to a supervised annunciator or alarm instead of interrupting the motor circuit.*

**(B) Overload Protection.**

**(1) Type of Overload Device.** Each motor shall be protected against dangerous heating due to motor overloads and failure to start by a thermal protector integral with the motor or external current-sensing devices, or both.

**(2) Wound-Rotor AC Motors.** The secondary circuits of wound-rotor ac motors including conductors, controllers, and resistors rated for the application shall be considered as protected against overcurrent by the motor overload protection means.

**(3) Operation.** Operation of the overload interrupting device shall simultaneously disconnect all ungrounded conductors.

**(4) Automatic Reset.** Overload sensing devices shall not automatically reset after trip unless resetting of the overload sensing device does not cause automatic restarting of the motor or there is no hazard to persons created by automatic restarting of the motor and its connected machinery.

**(C) Fault-Current Protection.**

**(1) Type of Protection.** Fault-current protection shall be provided in each motor circuit by one of the following means.

(a) A circuit breaker of suitable type and rating arranged so that it can be serviced without hazard. The circuit breaker shall simultaneously disconnect all ungrounded conductors. The circuit breaker shall be permitted to sense the fault current by means of integral or external sensing elements.

(b) Fuses of a suitable type and rating placed in each ungrounded conductor. Fuses shall be used with suitable disconnecting means, or they shall be of a type that can also serve as the disconnecting means. They shall be arranged so that they cannot be serviced while they are energized.

**(2) Reclosing.** Fault-current interrupting devices shall not automatically reclose the circuit.

*Exception: Automatic reclosing of a circuit shall be permitted where the circuit is exposed to transient faults and where such automatic reclosing does not create a hazard to persons.*

**(3) Combination Protection.** Overload protection and fault-current protection shall be permitted to be provided by the same device.

**430.126 Rating of Motor Control Apparatus.** The ultimate trip current of overcurrent (overload) relays or other motor-protective devices used shall not exceed 115 percent of the controller's continuous current rating. Where the motor branch-circuit disconnecting means is separate from the controller, the disconnecting means current rating shall not be less than the ultimate trip setting of the overcurrent relays in the circuit.

**430.127 Disconnecting Means.** The controller disconnecting means shall be capable of being locked in the open position.

**XI. Protection of Live Parts — All Voltages**

**430.131 General.** Part XI specifies that live parts shall be protected in a manner judged adequate for the hazard involved.

**430.132 Where Required.** Exposed live parts of motors and controllers operating at 50 volts or more between terminals shall be guarded against accidental contact by enclosure or by location as follows:

(1) By installation in a room or enclosure that is accessible only to qualified persons
(2) By installation on a suitable balcony, gallery, or platform, elevated and arranged so as to exclude unqualified persons
(3) By elevation 2.5 m (8 ft) or more above the floor

*Exception: Live parts of motors operating at more than 50 volts between terminals shall not require additional guarding for stationary motors that have commutators, collectors, and brush rigging located inside of motor-end brackets and not conductively connected to supply circuits operating at more than 150 volts to ground.*

**430.133 Guards for Attendants.** Where live parts of motors or controllers operating at over 150 volts to ground are guarded against accidental contact only by location as specified in 430.132, and where adjustment or other attendance may be necessary during the operation of the apparatus, suitable insulating mats or platforms shall be provided so that the attendant cannot readily touch live parts unless standing on the mats or platforms.

FPN: For working space, see 110.26 and 110.34.

**XII. Grounding — All Voltages**

**430.141 General.** Part XII specifies the grounding of exposed non–current-carrying metal parts, likely to become energized, of motor and controller frames to prevent a voltage above ground in the event of accidental contact between energized parts and frames. Insulation, isolation, or guarding are suitable alternatives to grounding of motors under certain conditions.

**430.142 Stationary Motors.** The frames of stationary motors shall be grounded under any of the following conditions:

(1) Where supplied by metal-enclosed wiring
(2) Where in a wet location and not isolated or guarded
(3) If in a hazardous (classified) location as covered in Articles 500 through 517
(4) If the motor operates with any terminal at over 150 volts to ground

Where the frame of the motor is not grounded, it shall be permanently and effectively insulated from the ground.

**430.143 Portable Motors.** The frames of portable motors that operate at over 150 volts to ground shall be guarded or grounded.

FPN No. 1: See 250.114(4) for grounding of portable appliances in other than residential occupancies.

FPN No. 2: See 250.119(B) for color of equipment grounding conductor.

**430.144 Controllers.** Controller enclosures shall be grounded regardless of voltage. Controller enclosures shall have means for attachment of an equipment grounding conductor termination in accordance with 250.8.

*Exception: Enclosures attached to ungrounded portable equipment shall not be required to be grounded.*

**430.145 Method of Grounding.** Where required, grounding shall be done in the manner specified in Part V of Article 250.

**(A) Grounding Through Terminal Housings.** Where the wiring to fixed motors is metal-enclosed cable or in metal raceways, junction boxes to house motor terminals shall be provided, and the armor of the cable or the metal raceways shall be connected to them in the manner specified in Article 250.

FPN: See 430.12(E) for equipment grounding connection means required at motor terminal housings.

**(B) Separation of Junction Box from Motor.** The junction box required by 430.145(A) shall be permitted to be separated from the motor by not more than 1.8 m (6 ft), provided the leads to the motor are Type AC cable or armored cord or are stranded leads enclosed in liquidtight flexible metal conduit, flexible metal conduit, intermediate metal conduit, rigid metal conduit, or electrical metallic tubing not smaller than metric designator 12 (trade size ⅜), the armor or raceway being connected both to the motor and to the box.

Liquidtight flexible nonmetallic conduit and rigid nonmetallic conduit shall be permitted to enclose the leads to the motor, provided the leads are stranded and the required equipment grounding conductor is connected to both the motor and to the box.

Where stranded leads are used, protected as specified above, each strand within the conductor shall be not larger than 10 AWG and shall comply with other requirements of this *Code* for conductors to be used in raceways.

**(C) Grounding of Controller-Mounted Devices.** Instrument transformer secondaries and exposed non–current-carrying metal or other conductive parts or cases of instrument transformers, meters, instruments, and relays shall be grounded as specified in 250.170 through 250.178.

## XIII. Tables

**Table 430.147 Full-Load Current in Amperes, Direct-Current Motors**
The following values of full-load currents[*] are for motors running at base speed.

| Horse-power | Armature Voltage Rating[*] | | | | | |
|---|---|---|---|---|---|---|
| | 90 Volts | 120 Volts | 180 Volts | 240 Volts | 500 Volts | 550 Volts |
| ¼ | 4.0 | 3.1 | 2.0 | 1.6 | — | — |
| ⅓ | 5.2 | 4.1 | 2.6 | 2.0 | — | — |
| ½ | 6.8 | 5.4 | 3.4 | 2.7 | — | — |
| ¾ | 9.6 | 7.6 | 4.8 | 3.8 | — | — |
| 1 | 12.2 | 9.5 | 6.1 | 4.7 | — | — |
| 1½ | — | 13.2 | 8.3 | 6.6 | — | — |
| 2 | — | 17 | 10.8 | 8.5 | — | — |
| 3 | — | 25 | 16 | 12.2 | — | — |
| 5 | — | 40 | 27 | 20 | — | — |
| 7½ | — | 58 | — | 29 | 13.6 | 12.2 |
| 10 | — | 76 | — | 38 | 18 | 16 |
| 15 | — | — | — | 55 | 27 | 24 |
| 20 | — | — | — | 72 | 34 | 31 |
| 25 | — | — | — | 89 | 43 | 38 |
| 30 | — | — | — | 106 | 51 | 46 |
| 40 | — | — | — | 140 | 67 | 61 |
| 50 | — | — | — | 173 | 83 | 75 |
| 60 | — | — | — | 206 | 99 | 90 |
| 75 | — | — | — | 255 | 123 | 111 |
| 100 | — | — | — | 341 | 164 | 148 |
| 125 | — | — | — | 425 | 205 | 185 |
| 150 | — | — | — | 506 | 246 | 222 |
| 200 | — | — | — | 675 | 330 | 294 |

[*]These are average dc quantities.

**Table 430.148 Full-Load Currents in Amperes, Single-Phase Alternating-Current Motors**
The following values of full-load currents are for motors running at usual speeds and motors with normal torque characteristics. Motors built for especially low speeds or high torques may have higher full-load currents, and multispeed motors will have full-load current varying with speed, in which case the nameplate current ratings shall be used.

The voltages listed are rated motor voltages. The currents listed shall be permitted for system voltage ranges of 110 to 120 and 220 to 240 volts.

| Horsepower | 115 Volts | 200 Volts | 208 Volts | 230 Volts |
|---|---|---|---|---|
| ⅙ | 4.4 | 2.5 | 2.4 | 2.2 |
| ¼ | 5.8 | 3.3 | 3.2 | 2.9 |
| ⅓ | 7.2 | 4.1 | 4.0 | 3.6 |
| ½ | 9.8 | 5.6 | 5.4 | 4.9 |
| ¾ | 13.8 | 7.9 | 7.6 | 6.9 |
| 1 | 16 | 9.2 | 8.8 | 8.0 |
| 1½ | 20 | 11.5 | 11.0 | 10 |
| 2 | 24 | 13.8 | 13.2 | 12 |
| 3 | 34 | 19.6 | 18.7 | 17 |
| 5 | 56 | 32.2 | 30.8 | 28 |
| 7½ | 80 | 46.0 | 44.0 | 40 |
| 10 | 100 | 57.5 | 55.0 | 50 |

**Table 430.149 Full-Load Current, Two-Phase Alternating-Current Motors (4-Wire)**

The following values of full-load current are for motors running at speeds usual for belted motors and motors with normal torque characteristics. Motors built for especially low speeds or high torques may require more running current, and multispeed motors will have full-load current varying with speed, in which case the nameplate current rating shall be used. Current in the common conductor of a 2-phase, 3-wire system will be 1.41 times the value given.

The voltages listed are rated motor voltages. The currents listed shall be permitted for system voltage ranges of 110 to 120, 220 to 240, 440 to 480, and 550 to 600 volts.

| | Induction-Type Squirrel Cage and Wound Rotor (Amperes) | | | | |
|---|---|---|---|---|---|
| Horsepower | 115 Volts | 230 Volts | 460 Volts | 575 Volts | 2300 Volts |
| ½ | 4.0 | 2.0 | 1.0 | 0.8 | — |
| ¾ | 4.8 | 2.4 | 1.2 | 1.0 | — |
| 1 | 6.4 | 3.2 | 1.6 | 1.3 | — |
| 1½ | 9.0 | 4.5 | 2.3 | 1.8 | — |
| 2 | 11.8 | 5.9 | 3.0 | 2.4 | — |
| 3 | — | 8.3 | 4.2 | 3.3 | — |
| 5 | — | 13.2 | 6.6 | 5.3 | — |
| 7½ | — | 19 | 9.0 | 8.0 | — |

**Table 430.149** *Continued*

| | Induction-Type Squirrel Cage and Wound Rotor (Amperes) | | | | |
|---|---|---|---|---|---|
| Horsepower | 115 Volts | 230 Volts | 460 Volts | 575 Volts | 2300 Volts |
| 10 | — | 24 | 12 | 10 | — |
| 15 | — | 36 | 18 | 14 | — |
| 20 | — | 47 | 23 | 19 | — |
| 25 | — | 59 | 29 | 24 | — |
| 30 | — | 69 | 35 | 28 | — |
| 40 | — | 90 | 45 | 36 | — |
| 50 | — | 113 | 56 | 45 | — |
| 60 | — | 133 | 67 | 53 | 14 |
| 75 | — | 166 | 83 | 66 | 18 |
| 100 | — | 218 | 109 | 87 | 23 |
| 125 | — | 270 | 135 | 108 | 28 |
| 150 | — | 312 | 156 | 125 | 32 |
| 200 | — | 416 | 208 | 167 | 43 |

**Table 430.150 Full-Load Current, Three-Phase Alternating-Current Motors**

The following values of full-load currents are typical for motors running at speeds usual for belted motors and motors with normal torque characteristics.

Motors built for low speeds (1200 rpm or less) or high torques may require more running current, and multispeed motors will have full-load current varying with speed. In these cases, the nameplate current rating shall be used.

The voltages listed are rated motor voltages. The currents listed shall be permitted for system voltage ranges of 110 to 120, 220 to 240, 440 to 480, and 550 to 600 volts.

| | Induction-Type Squirrel Cage and Wound Rotor (Amperes) | | | | | | | Synchronous-Type Unity Power Factor* (Amperes) | | | |
|---|---|---|---|---|---|---|---|---|---|---|---|
| Horsepower | 115 Volts | 200 Volts | 208 Volts | 230 Volts | 460 Volts | 575 Volts | 2300 Volts | 230 Volts | 460 Volts | 575 Volts | 2300 Volts |
| ½ | 4.4 | 2.5 | 2.4 | 2.2 | 1.1 | 0.9 | — | — | — | — | — |
| ¾ | 6.4 | 3.7 | 3.5 | 3.2 | 1.6 | 1.3 | — | — | — | — | — |
| 1 | 8.4 | 4.8 | 4.6 | 4.2 | 2.1 | 1.7 | — | — | — | — | — |
| 1½ | 12.0 | 6.9 | 6.6 | 6.0 | 3.0 | 2.4 | — | — | — | — | — |
| 2 | 13.6 | 7.8 | 7.5 | 6.8 | 3.4 | 2.7 | — | — | — | — | — |
| 3 | — | 11.0 | 10.6 | 9.6 | 4.8 | 3.9 | — | — | — | — | — |
| 5 | — | 17.5 | 16.7 | 15.2 | 7.6 | 6.1 | — | — | — | — | — |
| 7½ | — | 25.3 | 24.2 | 22 | 11 | 9 | — | — | — | — | — |
| 10 | — | 32.2 | 30.8 | 28 | 14 | 11 | — | — | — | — | — |
| 15 | — | 48.3 | 46.2 | 42 | 21 | 17 | — | — | — | — | — |
| 20 | — | 62.1 | 59.4 | 54 | 27 | 22 | — | — | — | — | — |
| 25 | — | 78.2 | 74.8 | 68 | 34 | 27 | — | 53 | 26 | 21 | — |
| 30 | — | 92 | 88 | 80 | 40 | 32 | — | 63 | 32 | 26 | — |
| 40 | — | 120 | 114 | 104 | 52 | 41 | — | 83 | 41 | 33 | — |
| 50 | — | 150 | 143 | 130 | 65 | 52 | — | 104 | 52 | 42 | — |
| 60 | — | 177 | 169 | 154 | 77 | 62 | 16 | 123 | 61 | 49 | 12 |
| 75 | — | 221 | 211 | 192 | 96 | 77 | 20 | 155 | 78 | 62 | 15 |
| 100 | — | 285 | 273 | 248 | 124 | 99 | 26 | 202 | 101 | 81 | 20 |
| 125 | — | 359 | 343 | 312 | 156 | 125 | 31 | 253 | 126 | 101 | 25 |
| 150 | — | 414 | 396 | 360 | 180 | 144 | 37 | 302 | 151 | 121 | 30 |
| 200 | — | 552 | 528 | 480 | 240 | 192 | 49 | 400 | 201 | 161 | 40 |
| 250 | — | — | — | — | 302 | 242 | 60 | — | — | — | — |
| 300 | — | — | — | — | 361 | 289 | 72 | — | — | — | — |
| 350 | — | — | — | — | 414 | 336 | 83 | — | — | — | — |
| 400 | — | — | — | — | 477 | 382 | 95 | — | — | — | — |
| 450 | — | — | — | — | 515 | 412 | 103 | — | — | — | — |
| 500 | — | — | — | — | 590 | 472 | 118 | — | — | — | — |

*For 90 and 80 percent power factor, the figures shall be multiplied by 1.1 and 1.25, respectively.

**Table 430.151(A) Conversion Table of Single-Phase Locked-Rotor Currents for Selection of Disconnecting Means and Controllers as Determined from Horsepower and Voltage Rating**

For use only with 430.110, 440.12, 440.41 and 455.8(C).

| Rated Horsepower | Maximum Locked-Rotor Current in Amperes, Single Phase | | |
|---|---|---|---|
| | 115 Volts | 208 Volts | 230 Volts |
| ½ | 58.8 | 32.5 | 29.4 |
| ¾ | 82.8 | 45.8 | 41.4 |
| 1 | 96 | 53 | 48 |
| 1½ | 120 | 66 | 60 |
| 2 | 144 | 80 | 72 |
| 3 | 204 | 113 | 102 |
| 5 | 336 | 186 | 168 |
| 7½ | 480 | 265 | 240 |
| 10 | 600 | 332 | 300 |

**Table 430.151(B) Conversion Table of Polyphase Design B, C, D, and E Maximum Locked-Rotor Currents for Selection of Disconnecting Means and Controllers as Determined from Horsepower and Voltage Rating and Design Letter**

For use only with 430.110, 440.12*, 440.41* and 455.8(C).

| Rated Horsepower | Maximum Motor Locked-Rotor Current in Amperes, Two- and Three-Phase, Design B, C, D, and E | | | | | | | | | | | |
|---|---|---|---|---|---|---|---|---|---|---|---|---|
| | 115 Volts | | 200 Volts | | 208 Volts | | 230 Volts | | 460 Volts | | 575 Volts | |
| | B, C, D | E | B, C, D | E | B, C, D | E | B, C, D | E | B, C, D | E | B, C, D | E |
| ½ | 40 | 40 | 23 | 23 | 22.1 | 22.1 | 20 | 20 | 10 | 10 | 8 | 8 |
| ¾ | 50 | 50 | 28.8 | 28.8 | 27.6 | 27.6 | 25 | 25 | 12.5 | 12.5 | 10 | 10 |
| 1 | 60 | 60 | 34.5 | 34.5 | 33 | 33 | 30 | 30 | 15 | 15 | 12 | 12 |
| 1½ | 80 | 80 | 46 | 46 | 44 | 44 | 40 | 40 | 20 | 20 | 16 | 16 |
| 2 | 100 | 100 | 57.5 | 57.5 | 55 | 55 | 50 | 50 | 25 | 25 | 20 | 20 |
| 3 | — | — | 73.6 | 84 | 71 | 81 | 64 | 73 | 32 | 36.5 | 25.6 | 29.2 |
| 5 | — | — | 105.8 | 140 | 102 | 135 | 92 | 122 | 46 | 61 | 36.8 | 48.8 |
| 7½ | — | — | 146 | 210 | 140 | 202 | 127 | 183 | 63.5 | 91.5 | 50.8 | 73.2 |
| 10 | — | — | 186.3 | 259 | 179 | 249 | 162 | 225 | 81 | 113 | 64.8 | 90 |
| 15 | — | — | 267 | 388 | 257 | 373 | 232 | 337 | 116 | 169 | 93 | 135 |
| 20 | — | — | 334 | 516 | 321 | 497 | 290 | 449 | 145 | 225 | 116 | 180 |
| 25 | — | — | 420 | 646 | 404 | 621 | 365 | 562 | 183 | 281 | 146 | 225 |
| 30 | — | — | 500 | 775 | 481 | 745 | 435 | 674 | 218 | 337 | 174 | 270 |
| 40 | — | — | 667 | 948 | 641 | 911 | 580 | 824 | 290 | 412 | 232 | 330 |
| 50 | — | — | 834 | 1185 | 802 | 1139 | 725 | 1030 | 363 | 515 | 290 | 412 |
| 60 | — | — | 1001 | 1421 | 962 | 1367 | 870 | 1236 | 435 | 618 | 348 | 494 |
| 75 | — | — | 1248 | 1777 | 1200 | 1708 | 1085 | 1545 | 543 | 773 | 434 | 618 |
| 100 | — | — | 1668 | 2154 | 1603 | 2071 | 1450 | 1873 | 725 | 937 | 580 | 749 |
| 125 | — | — | 2087 | 2692 | 2007 | 2589 | 1815 | 2341 | 908 | 1171 | 726 | 936 |
| 150 | — | — | 2496 | 3230 | 2400 | 3106 | 2170 | 2809 | 1085 | 1405 | 868 | 1124 |
| 200 | — | — | 3335 | 4307 | 3207 | 4141 | 2900 | 3745 | 1450 | 1873 | 1160 | 1498 |
| 250 | — | — | — | — | — | — | — | — | 1825 | 2344 | 1460 | 1875 |
| 300 | — | — | — | — | — | — | — | — | 2200 | 2809 | 1760 | 2247 |
| 350 | — | — | — | — | — | — | — | — | 2550 | 3277 | 2040 | 2622 |
| 400 | — | — | — | — | — | — | — | — | 2900 | 3745 | 2320 | 2996 |
| 450 | — | — | — | — | — | — | — | — | 3250 | 4214 | 2600 | 3371 |
| 500 | — | — | — | — | — | — | — | — | 3625 | 4682 | 2900 | 3746 |

*In determining compliance with 440.12 and 440.41, the values in the B, C, D columns shall be used.

# ARTICLE 440
## Air-Conditioning and Refrigerating Equipment

## I. General

**440.1 Scope.** The provisions of this article apply to electric motor-driven air-conditioning and refrigerating equipment and to the branch circuits and controllers for such equipment. It provides for the special considerations necessary for circuits supplying hermetic refrigerant motor-compressors and for any air-conditioning or refrigerating equipment that is supplied from a branch circuit that supplies a hermetic refrigerant motor-compressor.

## 440.2 Definitions.

**Branch-Circuit Selection Current.** The value in amperes to be used instead of the rated-load current in determining the ratings of motor branch-circuit conductors, disconnecting means, controllers, and branch-circuit short-circuit and ground-fault protective devices wherever the running overload protective device permits a sustained current greater than the specified percentage of the rated-load current. The value of branch-circuit selection current will always be equal to or greater than the marked rated-load current.

**Hermetic Refrigerant Motor-Compressor.** A combination consisting of a compressor and motor, both of which are enclosed in the same housing, with no external shaft or shaft seals, the motor operating in the refrigerant.

**Leakage Current Detection and Interruption (LCDI) Protection.** A device provided in a power supply cord or cord set that senses leakage current flowing between or from the cord conductors and interrupts the circuit at a predetermined level of leakage current.

**Rated-Load Current.** The rated-load current for a hermetic refrigerant motor-compressor is the current resulting when the motor-compressor is operated at the rated load, rated voltage, and rated frequency of the equipment it serves.

## 440.3 Other Articles.

**(A) Article 430.** These provisions are in addition to, or amendatory of, the provisions of Article 430 and other articles in this *Code*, which apply except as modified in this article.

**(B) Articles 422, 424, or 430.** The rules of Articles 422, 424, or 430, as applicable, shall apply to air-conditioning and refrigerating equipment that does not incorporate a hermetic refrigerant motor-compressor. This equipment includes devices that employ refrigeration compressors driven by conventional motors, furnaces with air-conditioning evaporator coils installed, fan-coil units, remote forced air-cooled condensers, remote commercial refrigerators, and so forth.

**(C) Article 422.** Equipment such as room air conditioners, household refrigerators and freezers, drinking water coolers, and beverage dispensers shall be considered appliances, and the provisions of Article 422 shall also apply.

**(D) Other Applicable Articles.** Hermetic refrigerant motor-compressors, circuits, controllers, and equipment shall also comply with the applicable provisions of Table 440.3(D).

**Table 440.3(D) Other Articles**

| Equipment/Occupancy | Article | Section |
|---|---|---|
| Capacitors | | 460.9 |
| Commercial garages, aircraft hangars, motor fuel dispensing facilities, bulk storage plants, spray application, dipping, and coating processes, and inhalation anesthetizing locations | 511, 513, 514, 515, 516, and 517 Part IV | |
| Hazardous (classified) locations | 500–503 and 505 | |
| Motion picture and television studios and similar locations | 530 | |
| Resistors and reactors | 470 | |

## 440.4 Marking on Hermetic Refrigerant Motor-Compressors and Equipment.

**(A) Hermetic Refrigerant Motor-Compressor Name plate.** A hermetic refrigerant motor-compressor shall be provided with a nameplate that shall indicate the manufacturer's name, trademark, or symbol; identifying designation; phase; voltage; and frequency. The rated-load current in amperes of the motor-compressor shall be marked by the equipment manufacturer on either or both the motor-compressor nameplate and the nameplate of the equipment in which the motor-compressor is used. The locked-rotor current of each single-phase motor-compressor having a rated-load current of more than 9 amperes at 115 volts, or more than 4.5 amperes at 230 volts, and each polyphase motor-compressor shall be marked on the motor-compressor nameplate. Where a thermal protector complying with 440.52(A)(2) and (B)(2) is used, the motor-compressor nameplate or the equipment nameplate shall be marked with the words "thermally protected." Where a protective system complying with 440.52(A)(4) and (B)(4) is used and is furnished with the equipment, the equipment nameplate shall be marked with the words, "thermally protected system." Where a protective system complying with 440.52(A)(4) and (B)(4) is specified, the equipment nameplate shall be appropriately marked.

**(B) Multimotor and Combination-Load Equipment.** Multimotor and combination-load equipment shall be provided with a visible nameplate marked with the maker's name, the rating in volts, frequency and number of phases, minimum supply circuit conductor ampacity, and the maximum rating of the branch-circuit short-circuit and ground-fault protective device. The ampacity shall be calculated by using Part IV and counting all the motors and other loads that will be operated at the same time. The branch-circuit short-circuit and ground-fault protective device rating shall not exceed the value calculated by using Part III. Multimotor or combination-load equipment for use on two or more circuits shall be marked with the above information for each circuit.

*Exception No. 1: Multimotor and combination-load equipment that is suitable under the provisions of this article for connection to a single 15- or 20-ampere, 120-volt, or a 15-ampere, 208- or 240-volt, single-phase branch circuit shall be permitted to be marked as a single load.*

*Exception No. 2: The minimum supply circuit conductor ampacity and the maximum rating of the branch-circuit short-circuit and ground-fault protective device shall not be required to be marked on a room air conditioner conforming with 440.62(A).*

**(C) Branch-Circuit Selection Current.** A hermetic refrigerant motor-compressor, or equipment containing such a compressor, having a protection system that is approved for use with the motor-compressor that it protects and that permits continuous current in excess of the specified percentage of nameplate rated-load current given in 440.52(B)(2) or (B)(4) shall also be marked with a branch-circuit selection current that complies with 440.52(B)(2) or (B)(4). This marking shall be provided by the equipment manufacturer and shall be on the nameplate(s) where the rated-load current(s) appears.

**440.5 Marking on Controllers.** A controller shall be marked with the manufacturer's name, trademark, or symbol; identifying designation; voltage; phase; full-load and locked-rotor current (or horsepower) rating; and such other data as may be needed to properly indicate the motor-compressor for which it is suitable.

**440.6 Ampacity and Rating.** The size of conductors for equipment covered by this article shall be selected from Tables 310.16 through 310.19 or calculated in accordance with 310.15 as applicable. The required ampacity of conductors and rating of equipment shall be determined according to 440.6(A) and (B).

**(A) Hermetic Refrigerant Motor-Compressor.** For a hermetic refrigerant motor-compressor, the rated-load current marked on the nameplate of the equipment in which the motor-compressor is employed shall be used in determining the rating or ampacity of the disconnecting means, the branch-circuit conductors, the controller, the branch-circuit short-circuit and ground-fault protection, and the separate motor overload protection. Where no rated-load current is shown on the equipment nameplate, the rated-load current shown on the compressor nameplate shall be used.

*Exception No. 1: Where so marked, the branch-circuit selection current shall be used instead of the rated-load current to determine the rating or ampacity of the disconnecting means, the branch-circuit conductors, the controller, and the branch-circuit short-circuit and ground-fault protection.*

*Exception No. 2: For cord-and-plug-connected equipment, the nameplate marking shall be used in accordance with 440.22(B), Exception No. 2.*

> FPN: For disconnecting means and controllers, see 440.12 and 440.41.

**(B) Multimotor Equipment.** For multimotor equipment employing a shaded-pole or permanent split-capacitor-type fan or blower motor, the full-load current for such motor marked on the nameplate of the equipment in which the fan or blower motor is employed shall be used instead of the horsepower rating to determine the ampacity or rating of the disconnecting means, the branch-circuit conductors, the controller, the branch-circuit short-circuit and ground-fault protection, and the separate overload protection. This marking on the equipment nameplate shall not be less than the current marked on the fan or blower motor nameplate.

**440.7 Highest Rated (Largest) Motor.** In determining compliance with this article and with 430.24, 430.53(B) and (C), and 430.62(A), the highest rated (largest) motor shall be considered to be the motor that has the highest rated-load current. Where two or more motors have the same highest rated-load current, only one of them shall be considered as the highest rated (largest) motor. For other than hermetic refrigerant motor-compressors, and fan or blower motors as covered in 440.6(B), the full-load current used to determine the highest rated motor shall be the equivalent value corresponding to the motor horsepower rating selected from Tables 430.148, 430.149, or 430.150.

*Exception: Where so marked, the branch-circuit selection current shall be used instead of the rated-load current in determining the highest rated (largest) motor-compressor.*

**440.8 Single Machine.** An air-conditioning or refrigerating system shall be considered to be a single machine under the provisions of 430.87, Exception, and 430.112, Exception. The motors shall be permitted to be located remotely from each other.

## II. Disconnecting Means

**440.11 General.** The provisions of Part II are intended to require disconnecting means capable of disconnecting air-conditioning and refrigerating equipment, including motor-compressors and controllers from the circuit conductors.

**440.12 Rating and Interrupting Capacity.**

**(A) Hermetic Refrigerant Motor-Compressor.** A disconnecting means serving a hermetic refrigerant motor-compressor shall be selected on the basis of the nameplate rated-load current or branch-circuit selection current, whichever is greater, and locked-rotor current, respectively, of the motor-compressor as follows.

**(1) Ampere Rating.** The ampere rating shall be at least 115 percent of the nameplate rated-load current or branch-circuit selection current, whichever is greater.

*Exception: A listed nonfused motor circuit switch having a horsepower rating not less than the equivalent horsepower determined in accordance with 440.12(A)(2) shall be permitted to have an ampere rating less than 115 percent of the specified current.*

**(2) Equivalent Horsepower.** To determine the equivalent horsepower in complying with the requirements of 430.109, the horsepower rating shall be selected from Tables 430.148, 430.149, or 430.150 corresponding to the rated-load current or branch-circuit selection current, whichever is greater, and also the horsepower rating from Tables 430.151(A) or 430.151(B) corresponding to the locked-rotor current. In case the nameplate rated-load current or branch-circuit selection current and locked-rotor current do not correspond to the currents shown in Tables 430.148, 430.149, 430.150, 430.151(A), or 430.151(B), the horsepower rating corresponding to the next higher value shall be selected. In case different horsepower ratings are obtained when applying these tables, a horsepower rating at least equal to the larger of the values obtained shall be selected.

**(B) Combination Loads.** Where the combined load of two or more hermetic refrigerant motor-compressors or one or more hermetic refrigerant motor-compressor with other motors or loads may be simultaneous on a single disconnecting means, the rating for the disconnecting means shall be determined in accordance with 440.12(B)(1) and (B)(2).

**(1) Horsepower Rating.** The horsepower rating of the disconnecting means shall be determined from the sum of all currents, including resistance loads, at the rated-load condition and also at the locked-rotor condition. The combined rated-load current and the combined locked-rotor current so obtained shall be considered as a single motor for the purpose of this requirement as follows:

(a) The full-load current equivalent to the horsepower rating of each motor, other than a hermetic refrigerant motor-compressor, and fan or blower motors as covered in 440.6(B) shall be selected from Tables 430.148, 430.149, or 430.150. These full-load currents shall be added to the motor-compressor rated-load current(s) or branch-circuit selection current(s), whichever is greater, and to the rating in amperes of other loads to obtain an equivalent full-load current for the combined load.

(b) The locked-rotor current equivalent to the horsepower rating of each motor, other than a hermetic refrigerant motor-compressor, shall be selected from Tables 430.151(A) or 430.151(B), and, for fan and blower motors of the shaded-pole or permanent split-capacitor type marked with the locked-rotor current, the marked value shall be used. The locked-rotor currents shall be added to the motor-compressor locked-rotor current(s) and to the rating in amperes of other loads to obtain an equivalent locked-rotor current for the combined load. Where two or more motors or other loads such as resistance heaters, or both, cannot be started simultaneously, appropriate combinations of locked-rotor and rated-load current or branch-circuit selection current, whichever is greater, shall be an acceptable means of determining the equivalent locked-rotor current for the simultaneous combined load.

*Exception: Where part of the concurrent load is a resistance load and the disconnecting means is a switch rated in horsepower and amperes, the switch used shall be permitted to have a horsepower rating not less than the combined load to the motor-compressor(s) and other motor(s) at the locked-rotor condition, if the ampere rating of the switch is not less than this locked-rotor load plus the resistance load.*

**(2) Full-Load Current Equivalent.** The ampere rating of the disconnecting means shall be at least 115 percent of the sum of all currents at the rated-load condition determined in accordance with 440.12(B)(1).

*Exception: A listed nonfused motor circuit switch having a horsepower rating not less than the equivalent horsepower determined by 440.12(B)(1) shall be permitted to have an ampere rating less than 115 percent of the sum of all currents.*

**(C) Small Motor-Compressors.** For small motor-compressors not having the locked-rotor current marked on the nameplate, or for small motors not covered by Tables

430.147, 430.148, 430.149, or 430.150, the locked-rotor current shall be assumed to be six times the rated-load current.

**(D) Every Switch.** Every disconnecting means in the refrigerant motor-compressor circuit between the point of attachment to the feeder and the point of connection to the refrigerant motor-compressor shall comply with the requirements of 440.12.

**(E) Disconnecting Means Rated in Excess of 100 Horsepower.** Where the rated-load or locked-rotor current as determined above would indicate a disconnecting means rated in excess of 100 hp, the provisions of 430.109(E) shall apply.

**440.13 Cord-Connected Equipment.** For cord-connected equipment such as room air conditioners, household refrigerators and freezers, drinking water coolers, and beverage dispensers, a separable connector or an attachment plug and receptacle shall be permitted to serve as the disconnecting means.

FPN: For room air conditioners, see 440.63.

**440.14 Location.** Disconnecting means shall be located within sight from and readily accessible from the air-conditioning or refrigerating equipment. The disconnecting means shall be permitted to be installed on or within the air-conditioning or refrigerating equipment.

The disconnecting means shall not be located on panels that are designed to allow access to the air-conditioning or refrigeration equipment.

*Exception No. 1: Where the disconnecting means provided in accordance with 430.102(A) is capable of being locked in the open position, and the refrigerating or air-conditioning equipment is essential to an industrial process in a facility where the conditions of maintenance and the supervision ensure that only qualified persons service the equipment, a disconnecting means within sight from the equipment shall not be required.*

*Exception No. 2: Where an attachment plug and receptacle serve as the disconnecting means in accordance with 440.13, their location shall be accessible but shall not be required to be readily accessible.*

FPN: See Parts VII and IX of Article 430 for additional requirements.

## III. Branch-Circuit Short-Circuit and Ground-Fault Protection

**440.21 General.** The provisions of Part III specify devices intended to protect the branch-circuit conductors, control apparatus, and motors in circuits supplying hermetic refrig-

erant motor-compressors against overcurrent due to short circuits and grounds. They are in addition to or amendatory of the provisions of Article 240.

**440.22 Application and Selection.**

**(A) Rating or Setting for Individual Motor Compressor.** The motor-compressor branch-circuit short-circuit and ground-fault protective device shall be capable of carrying the starting current of the motor. A protective device having a rating or setting not exceeding 175 percent of the motor-compressor rated-load current or branch-circuit selection current, whichever is greater, shall be permitted, provided that, where the protection specified is not sufficient for the starting current of the motor, the rating or setting shall be permitted to be increased but shall not exceed 225 percent of the motor rated-load current or branch-circuit selection current, whichever is greater.

*Exception: The rating of the branch-circuit short-circuit and ground-fault protective device shall not be required to be less than 15 amperes.*

**(B) Rating or Setting for Equipment.** The equipment branch-circuit short-circuit and ground-fault protective device shall be capable of carrying the starting current of the equipment. Where the hermetic refrigerant motor-compressor is the only load on the circuit, the protection shall conform with 440.22(A). Where the equipment incorporates more than one hermetic refrigerant motor-compressor or a hermetic refrigerant motor-compressor and other motors or other loads, the equipment short-circuit and ground-fault protection shall conform with 430.53 and the following.

**(1) Motor-Compressor Largest Load.** Where a hermetic refrigerant motor-compressor is the largest load connected to the circuit, the rating or setting of the branch-circuit short-circuit and ground-fault protective device shall not exceed the value specified in 440.22(A) for the largest motor-compressor plus the sum of the rated-load current or branch-circuit selection current, whichever is greater, of the other motor-compressor(s) and the ratings of the other loads supplied.

**(2) Motor-Compressor Not Largest Load.** Where a hermetic refrigerant motor-compressor is not the largest load connected to the circuit, the rating or setting of the branch-circuit short-circuit and ground-fault protective device shall not exceed a value equal to the sum of the rated-load current or branch-circuit selection current, whichever is greater, rating(s) for the motor-compressor(s) plus the value specified in 430.53(C)(4) where other motor loads are supplied, or the value specified in 240.4 where only nonmotor loads are supplied in addition to the motor-compressor(s).

*Exception No. 1: Equipment that starts and operates on a 15- or 20-ampere 120-volt, or 15-ampere 208- or 240-volt single-phase branch circuit, shall be permitted to be protected by the 15- or 20-ampere overcurrent device protecting the branch circuit, but if the maximum branch-circuit short-circuit and ground-fault protective device rating marked on the equipment is less than these values, the circuit protective device shall not exceed the value marked on the equipment nameplate.*

*Exception No. 2: The nameplate marking of cord-and-plug-connected equipment rated not greater than 250 volts, single-phase, such as household refrigerators and freezers, drinking water coolers, and beverage dispensers, shall be used in determining the branch-circuit requirements, and each unit shall be considered as a single motor unless the nameplate is marked otherwise.*

**(C) Protective Device Rating Not to Exceed the Manufacturer's Values.** Where maximum protective device ratings shown on a manufacturer's overload relay table for use with a motor controller are less than the rating or setting selected in accordance with 440.22(A) and (B), the protective device rating shall not exceed the manufacturer's values marked on the equipment.

## IV. Branch-Circuit Conductors

**440.31 General.** The provisions of Part IV and Article 310 specify ampacities of conductors required to carry the motor current without overheating under the conditions specified, except as modified in 440.6(A), Exception No. 1.

The provisions of these articles shall not apply to integral conductors of motors, motor controllers and the like, or to conductors that form an integral part of approved equipment.

> FPN: See 300.1(B) and 310.1 for similar requirements.

**440.32 Single Motor-Compressor.** Branch-circuit conductors supplying a single motor-compressor shall have an ampacity not less than 125 percent of either the motor-compressor rated-load current or the branch-circuit selection current, whichever is greater.

For a wye-start, delta-run connected motor-compressor, the selection of branch-circuit conductors between the controller and the motor-compressor shall be permitted to be based on 58 percent of either the motor-compressor rated-load current or the branch-circuit selection current, whichever is greater.

**440.33 Motor-Compressor(s) With or Without Additional Motor Loads.** Conductors supplying one or more motor-compressor(s) with or without an additional load(s) shall have an ampacity not less than the sum of the rated-load or branch-circuit selection current ratings, whichever

is larger, of all the motor-compressors plus the full-load currents of the other motors, plus 25 percent of the highest motor or motor-compressor rating in the group.

*Exception No. 1: Where the circuitry is interlocked so as to prevent the starting and running of a second motor-compressor or group of motor-compressors, the conductor size shall be determined from the largest motor-compressor or group of motor-compressors that is to be operated at a given time.*

*Exception No. 2: The branch circuit conductors for room air conditioners shall be in accordance with Part VII of Article 440.*

**440.34 Combination Load.** Conductors supplying a motor-compressor load in addition to a lighting or appliance load as computed from Article 220 and other applicable articles shall have an ampacity sufficient for the lighting or appliance load plus the required ampacity for the motor-compressor load determined in accordance with 440.33, or, for a single motor-compressor, in accordance with 440.32.

*Exception: Where the circuitry is interlocked so as to prevent simultaneous operation of the motor-compressor(s) and all other loads connected, the conductor size shall be determined from the largest size required for the motor-compressor(s) and other loads to be operated at a given time.*

**440.35 Multimotor and Combination-Load Equipment.** The ampacity of the conductors supplying multimotor and combination-load equipment shall not be less than the minimum circuit ampacity marked on the equipment in accordance with 440.4(B).

## V. Controllers for Motor-Compressors

### 440.41 Rating.

**(A) Motor-Compressor Controller.** A motor-compressor controller shall have both a continuous-duty full-load current rating and a locked-rotor current rating not less than the nameplate rated-load current or branch-circuit selection current, whichever is greater, and locked-rotor current, respectively, of the compressor. In case the motor controller is rated in horsepower but is without one or both of the foregoing current ratings, equivalent currents shall be determined from the ratings as follows. Tables 430.148, 430.149, and 430.150 shall be used to determine the equivalent full-load current rating. Tables 430.151(A) and 430.151(B) shall be used to determine the equivalent locked-rotor current ratings.

**(B) Controller Serving More Than One Load.** A controller serving more than one motor-compressor or a motor-compressor and other loads shall have a continuous-duty full-load current rating and a locked-rotor current rating not less than the combined load as determined in accordance with 440.12(B).

## VI. Motor-Compressor and Branch-Circuit Overload Protection

**440.51 General.** The provisions of Part VI specify devices intended to protect the motor-compressor, the motor-control apparatus, and the branch-circuit conductors against excessive heating due to motor overload and failure to start.

> FPN: See 240.4(G) for application of Parts III and VI of Article 440.

**440.52 Application and Selection.**

**(A) Protection of Motor-Compressor.** Each motor-compressor shall be protected against overload and failure to start by one of the following means:

(1) A separate overload relay that is responsive to motor-compressor current. This device shall be selected to trip at not more than 140 percent of the motor-compressor rated-load current.

(2) A thermal protector integral with the motor-compressor, approved for use with the motor-compressor that it protects on the basis that it will prevent dangerous overheating of the motor-compressor due to overload and failure to start. If the current-interrupting device is separate from the motor-compressor and its control circuit is operated by a protective device integral with the motor-compressor, it shall be arranged so that the opening of the control circuit will result in interruption of current to the motor-compressor.

(3) A fuse or inverse time circuit breaker responsive to motor current, which shall also be permitted to serve as the branch-circuit short-circuit and ground-fault protective device. This device shall be rated at not more than 125 percent of the motor-compressor rated-load current. It shall have sufficient time delay to permit the motor-compressor to start and accelerate its load. The equipment or the motor-compressor shall be marked with this maximum branch-circuit fuse or inverse time circuit breaker rating.

(4) A protective system, furnished or specified and approved for use with the motor-compressor that it protects on the basis that it will prevent dangerous overheating of the motor-compressor due to overload and failure to start. If the current-interrupting device is separate from the motor-compressor and its control circuit is operated by a protective device that is not integral with the current-interrupting device, it shall be arranged so that the opening of the control circuit will

result in interruption of current to the motor-compressor.

**(B) Protection of Motor-Compressor Control Apparatus and Branch-Circuit Conductors.** The motor-compressor controller(s), the disconnecting means, and the branch-circuit conductors shall be protected against overcurrent due to motor overload and failure to start by one of the following means, which shall be permitted to be the same device or system protecting the motor-compressor in accordance with 440.52(A):

*Exception: Overload protection of motor-compressors and equipment on 15- and 20-ampere, single-phase, branch circuits shall be permitted to be in accordance with 440.54 and 440.55.*

(1) An overload relay selected in accordance with 440.52(A)(1)

(2) A thermal protector applied in accordance with 440.52(A)(2), that will not permit a continuous current in excess of 156 percent of the marked rated-load current or branch-circuit selection current

(3) A fuse or inverse time circuit breaker selected in accordance with 440.52(A)(3)

(4) A protective system, in accordance with 440.52(A)(4), that will not permit a continuous current in excess of 156 percent of the marked rated-load current or branch-circuit selection current

**440.53 Overload Relays.** Overload relays and other devices for motor overload protection that are not capable of opening short circuits shall be protected by fuses or inverse time circuit breakers with ratings or settings in accordance with Part III unless approved for group installation or for part-winding motors and marked to indicate the maximum size of fuse or inverse time circuit breaker by which they shall be protected.

*Exception: The fuse or inverse time circuit breaker size marking shall be permitted on the nameplate of approved equipment in which the overload relay or other overload device is used.*

**440.54 Motor-Compressors and Equipment on 15- or 20-Ampere Branch Circuits — Not Cord-and-Attachment-Plug-Connected.** Overload protection for motor-compressors and equipment used on 15- or 20-ampere 120-volt, or 15-ampere 208- or 240-volt single-phase branch circuits as permitted in Article 210 shall be permitted as indicated in 440.54(A) and (B).

**(A) Overload Protection.** The motor-compressor shall be provided with overload protection selected as specified in 440.52(A). Both the controller and motor overload protective device shall be approved for installation with the short-circuit and ground-fault protective device for the branch circuit to which the equipment is connected.

**(B) Time Delay.** The short-circuit and ground-fault protective device protecting the branch circuit shall have sufficient time delay to permit the motor-compressor and other motors to start and accelerate their loads.

**440.55 Cord-and-Attachment-Plug-Connected Motor-Compressors and Equipment on 15- or 20-Ampere Branch Circuits.** Overload protection for motor-compressors and equipment that are cord-and-attachment-plug-connected and used on 15- or 20-ampere 120-volt, or 15-ampere 208- or 240-volt single-phase branch circuits as permitted in Article 210 shall be permitted as indicated in 440.55(A), (B), and (C).

**(A) Overload Protection.** The motor-compressor shall be provided with overload protection as specified in 440.52(A). Both the controller and the motor overload protective device shall be approved for installation with the short-circuit and ground-fault protective device for the branch circuit to which the equipment is connected.

**(B) Attachment Plug and Receptacle Rating.** The rating of the attachment plug and receptacle shall not exceed 20 amperes at 125 volts or 15 amperes at 250 volts.

**(C) Time Delay.** The short-circuit and ground-fault protective device protecting the branch circuit shall have sufficient time delay to permit the motor-compressor and other motors to start and accelerate their loads.

## VII. Provisions for Room Air Conditioners

**440.60 General.** The provisions of Part VII shall apply to electrically energized room air conditioners that control temperature and humidity. For the purpose of Part VII, a room air conditioner (with or without provisions for heating) shall be considered as an ac appliance of the air-cooled window, console, or in-wall type that is installed in the conditioned room and that incorporates a hermetic refrigerant motor-compressor(s). The provisions of Part VII cover equipment rated not over 250 volts, single phase, and such equipment shall be permitted to be cord-and-attachment-plug-connected.

A room air conditioner that is rated three phase or rated over 250 volts shall be directly connected to a wiring method recognized in Chapter 3, and provisions of Part VII shall not apply.

**440.61 Grounding.** Room air conditioners shall be grounded in accordance with 250.110, 250.112, and 250.114.

**440.62 Branch-Circuit Requirements.**

**(A) Room Air Conditioner as a Single Motor Unit.** A room air conditioner shall be considered as a single motor unit in determining its branch-circuit requirements where all the following conditions are met:

(1) It is cord-and-attachment-plug-connected.
(2) Its rating is not more than 40 amperes and 250 volts, single phase.
(3) Total rated-load current is shown on the room air-conditioner nameplate rather than individual motor currents.
(4) The rating of the branch-circuit short-circuit and ground-fault protective device does not exceed the ampacity of the branch-circuit conductors or the rating of the receptacle, whichever is less.

**(B) Where No Other Loads Are Supplied.** The total marked rating of a cord-and-attachment-plug-connected room air conditioner shall not exceed 80 percent of the rating of a branch circuit where no other loads are supplied.

**(C) Where Lighting Units or Other Appliances Are Also Supplied.** The total marked rating of a cord-and-attachment-plug-connected room air conditioner shall not exceed 50 percent of the rating of a branch circuit where lighting outlets, other appliances, or general-use receptacles are also supplied. Where the circuitry is interlocked to prevent simultaneous operation of the room air conditioner and energization of other outlets on the same branch circuit, a cord-and-attachment-plug-connected room air conditioner shall not exceed 80 percent of the branch-circuit rating.

**440.63 Disconnecting Means.** An attachment plug and receptacle shall be permitted to serve as the disconnecting means for a single-phase room air conditioner rated 250 volts or less if (1) the manual controls on the room air conditioner are readily accessible and located within 1.8 m (6 ft) of the floor or (2) an approved manually operable switch is installed in a readily accessible location within sight from the room air conditioner.

**440.64 Supply Cords.** Where a flexible cord is used to supply a room air conditioner, the length of such cord shall not exceed 3.0 m (10 ft) for a nominal, 120-volt rating or 1.8 m (6 ft) for a nominal, 208- or 240-volt rating.

**440.65 Leakage Current Detection and Interruption (LCDI) and Arc Fault Circuit Interrupter (AFCI).** Single-phase cord-and-plug-connected room air conditioners shall be provided with factory-installed LCDI or AFCI protection. The LCDI or AFCI protection shall be an integral part of the attachment plug or be located in the power supply cord within 300 mm (12 in.) of the attachment plug.

# ARTICLE 445
## Generators

**445.1 Scope.** This article covers the installation of generators.

**445.3 Other Articles.** Generators and their associated wiring and equipment shall also comply with the applicable provisions of Articles 695, 700, 701, 702, and 705.

**445.10 Location.** Generators shall be of a type suitable for the locations in which they are installed. They shall also meet the requirements for motors in 430.14. Generators installed in hazardous (classified) locations as described in Articles 500 through 503 and Article 505, or in other locations as described in Articles 510 through 517, and in Articles 518, 520, 525, 530, 665, and 695 shall also comply with the applicable provisions of those articles.

**445.11 Marking.** Each generator shall be provided with a nameplate giving the manufacturer's name, the rated frequency, power factor, number of phases if of alternating current, the rating in kilowatts or kilovolt amperes, the normal volts and amperes corresponding to the rating, rated revolutions per minute, insulation system class and rated ambient temperature or rated temperature rise, and time rating.

**445.12 Overcurrent Protection.**

**(A) Constant-Voltage Generators.** Constant-voltage generators, except ac generator exciters, shall be protected from overloads by inherent design, circuit breakers, fuses, or other acceptable overcurrent protective means suitable for the conditions of use.

**(B) Two-Wire Generators.** Two-wire, dc generators shall be permitted to have overcurrent protection in one conductor only if the overcurrent device is actuated by the entire current generated other than the current in the shunt field. The overcurrent device shall not open the shunt field.

**(C) 65 Volts or Less.** Generators operating at 65 volts or less and driven by individual motors shall be considered as protected by the overcurrent device protecting the motor if these devices will operate when the generators are delivering not more than 150 percent of their full-load rated current.

**(D) Balancer Sets.** Two-wire, dc generators used in conjunction with balancer sets to obtain neutrals for 3-wire systems shall be equipped with overcurrent devices that will disconnect the 3-wire system in case of excessive unbalancing of voltages or currents.

**(E) Three-Wire, Direct-Current Generators.** Three-wire, dc generators, whether compound or shunt wound, shall be equipped with overcurrent devices, one in each armature lead, and connected so as to be actuated by the entire current from the armature. Such overcurrent devices shall consist either of a double-pole, double-coil circuit breaker or of a 4-pole circuit breaker connected in the main and equalizer leads and tripped by two overcurrent devices, one in each armature lead. Such protective devices shall be interlocked so that no one pole can be opened without simultaneously disconnecting both leads of the armature from the system.

*Exception to (A) through (E): Where deemed by the authority having jurisdiction, a generator is vital to the operation of an electrical system and the generator should operate to failure to prevent a greater hazard to persons. The overload sensing device(s) shall be permitted to be connected to an annunciator or alarm supervised by authorized personnel instead of interrupting the generator circuit.*

**445.13 Ampacity of Conductors.** The ampacity of the conductors from the generator terminals to the first distribution device(s) containing overcurrent protection shall not be less than 115 percent of the nameplate current rating of the generator. It shall be permitted to size the neutral conductors in accordance with 220.22. Conductors that must carry ground-fault currents shall not be smaller than required by 250.24(B). Neutral conductors of dc generators that must carry ground-fault currents shall not be smaller than the minimum required size of the largest conductor.

*Exception: Where the design and operation of the generator prevent overloading, the ampacity of the conductors shall not be less than 100 percent of the nameplate current rating of the generator.*

**445.14 Protection of Live Parts.** Live parts of generators operated at more than 50 volts to ground shall not be exposed to accidental contact where accessible to unqualified persons.

**445.15 Guards for Attendants.** Where necessary for the safety of attendants, the requirements of 430.133 shall apply.

**445.16 Bushings.** Where wires pass through an opening in an enclosure, conduit box, or barrier, a bushing shall be used to protect the conductors from the edges of an opening having sharp edges. The bushing shall have smooth, well-rounded surfaces where it may be in contact with the conductors. If used where oils, grease, or other contaminants may be present, the bushing shall be made of a material not deleteriously affected.

**445.17 Generator Terminal Housings.** Generator terminal housings shall comply with 430.12. Where a horsepower rating is required to determine the required minimum size of the generator terminal housing, the full-load current of the generator shall be compared with comparable motors in Tables 430.147 through 430.150. The higher horsepower rating of Tables 430.147 and 430.150 shall be used whenever the generator selection is between two ratings.

**445.18 Disconnecting Means Required for Generators.** Generators shall be equipped with a disconnect by means of which the generator and all protective devices and control apparatus are able to be disconnected entirely from the circuits supplied by the generator except where:

(1) The driving means for the generator can be readily shut down; and

(2) The generator is not arranged to operate in parallel with another generator or other source of voltage.

# ARTICLE 450
# Transformers and Transformer Vaults (Including Secondary Ties)

**450.1 Scope.** This article covers the installation of all transformers.

*Exception No. 1: Current transformers.*

*Exception No. 2: Dry-type transformers that constitute a component part of other apparatus and comply with the requirements for such apparatus.*

*Exception No. 3: Transformers that are an integral part of an X-ray, high-frequency, or electrostatic-coating apparatus.*

*Exception No. 4: Transformers used with Class 2 and Class 3 circuits that comply with Article 725.*

*Exception No. 5: Transformers for sign and outline lighting that comply with Article 600.*

*Exception No. 6: Transformers for electric-discharge lighting that comply with Article 410.*

*Exception No. 7: Transformers used for power-limited fire alarm circuits that comply with Part III of Article 760.*

*Exception No. 8: Transformers used for research, development, or testing, where effective arrangements are provided to safeguard persons from contacting energized parts.*

This article covers the installation of transformers dedicated to supplying power to a fire pump installation as modified by Article 695.

This article also covers the installation of transformers in hazardous (classified) locations as modified by Articles 501 through 504.

## I. General Provisions

**450.2 Definition.** For the purpose of this article, the following definition shall apply.

**Transformer.** An individual transformer, single- or polyphase, identified by a single nameplate, unless otherwise indicated in this article.

**450.3 Overcurrent Protection.** Overcurrent protection of transformers shall comply with 450.3(A), (B), or (C). As used in this section, the word *transformer* shall mean a transformer or polyphase bank of two or more single-phase transformers operating as a unit.

> FPN No. 1: See 240.4, 240.21, 240.100, and 240.101 for overcurrent protection of conductors.

> FPN No. 2: Nonlinear loads can increase heat in a transformer without operating its overcurrent protective device.

**(A) Transformers Over 600 Volts, Nominal.** Overcurrent protection shall be provided in accordance with Table 450.3(A).

**(B) Transformers 600 Volts, Nominal, or Less.** Overcurrent protection shall be provided in accordance with Table 450.3(B).

*Exception: Where the transformer is installed as a motor-control circuit transformer in accordance with 430.72(C)(1) through (5).*

**(C) Voltage Transformers.** Voltage transformers installed indoors or enclosed shall be protected with primary fuses.

> FPN: For protection of instrument circuits including voltage transformers, see 408.32.

**450.4 Autotransformers 600 Volts, Nominal, or Less.**

**(A) Overcurrent Protection.** Each autotransformer 600 volts, nominal, or less shall be protected by an individual overcurrent device installed in series with each ungrounded input conductor. Such overcurrent device shall be rated or set at not more than 125 percent of the rated full-load input current of the autotransformer. Where this calculation does not correspond to a standard rating of a fuse or nonadjustable circuit breaker and the rated input current is 9 amperes or more, the next higher standard rating described in 240.6 shall be permitted. An overcurrent device shall not be installed in series with the shunt winding (the winding common to both the input and the output circuits) of the autotransformer between Points A and B as shown in Figure 450.4.

**Table 450.3(A) Maximum Rating or Setting of Overcurrent Protection for Transformers Over 600 Volts (as a Percentage of Transformer-Rated Current)**

| Location Limitations | Transformer Rated Impedance | Primary Protection Over 600 Volts | | Secondary Protection (See Note 2.) | | |
| | | | | Over 600 Volts | | 600 Volts or Below |
| | | Circuit Breaker (See Note 4.) | Fuse Rating | Circuit Breaker (See Note 4.) | Fuse Rating | Circuit Breaker or Fuse Rating |
|---|---|---|---|---|---|---|
| Any location | Not more than 6% | 600% (See Note 1.) | 300% (See Note 1.) | 300% (See Note 1.) | 250% (See Note 1.) | 125% (See Note 1.) |
| | More than 6% and not more than 10% | 400% (See Note 1.) | 300% (See Note 1.) | 250% (See Note 1.) | 225% (See Note 1.) | 125% (See Note 1.) |
| Supervised locations only (See Note 3.) | Any | 300% (See Note 1.) | 250% (See Note 1.) | Not required | Not required | Not required |
| | Not more than 6% | 600% | 300% | 300% (See Note 5.) | 250% (See Note 5.) | 250% (See Note 5.) |
| | More than 6% and not more than 10% | 400% | 300% | 250% (See Note 5.) | 225% (See Note 5.) | 250% (See Note 5.) |

Notes:

1. Where the required fuse rating or circuit breaker setting does not correspond to a standard rating or setting, a higher rating or setting that does not exceed the next higher standard rating or setting shall be permitted.

2. Where secondary overcurrent protection is required, the secondary overcurrent device shall be permitted to consist of not more than six circuit breakers or six sets of fuses grouped in one location. Where multiple overcurrent devices are utilized, the total of all the device ratings shall not exceed the allowed value of a single overcurrent device. If both circuit breakers and fuses are used as the overcurrent device, the total of the device ratings shall not exceed that allowed for fuses.

3. A supervised location is a location where conditions of maintenance and supervision ensure that only qualified persons monitor and service the transformer installation.

4. Electronically actuated fuses that may be set to open at a specific current shall be set in accordance with settings for circuit breakers.

5. A transformer equipped with a coordinated thermal overload protection by the manufacturer shall be permitted to have separate secondary protection omitted.

*Exception: Where the rated input current of the autotransformer is less than 9 amperes, an overcurrent device rated or set at not more than 167 percent of the input current shall be permitted.*

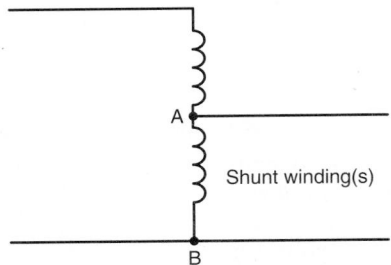

**Figure 450.4 Autotransformer.**

**(B) Transformer Field-Connected as an Auto-transformer.** A transformer field-connected as an autotransformer shall be identified for use at elevated voltage.

FPN: For information on permitted uses of autotransformers, see 210.9 and 215.11.

**450.5 Grounding Autotransformers.** Grounding autotransformers covered in this section are zigzag or T-connected transformers connected to 3-phase, 3-wire ungrounded systems for the purpose of creating a 3-phase, 4-wire distribution system or providing a neutral reference for grounding purposes. Such transformers shall have a continuous per-phase current rating and a continuous neutral current rating.

FPN: The phase current in a grounding autotransformer is one-third the neutral current.

**Table 450.3(B)  Maximum Rating or Setting of Overcurrent Protection for Transformers 600 Volts and Less (as a Percentage of Transformer-Rated Current)**

| Protection Method | Primary Protection | | | Secondary Protection (See Note 2.) | |
|---|---|---|---|---|---|
| | Currents of 9 Amperes or More | Currents Less Than 9 Amperes | Currents Less Than 2 Amperes | Currents of 9 Amperes or More | Currents Less Than 9 Amperes |
| Primary only protection | 125% (See Note 1.) | 167% | 300% | Not required | Not required |
| Primary and secondary protection | 250% (See Note 3.) | 250% (See Note 3.) | 250% (See Note 3.) | 125% (See Note 1.) | 167% |

Notes:

1. Where 125 percent of this current does not correspond to a standard rating of a fuse or nonadjustable circuit breaker, a higher rating that does not exceed the next higher standard rating shall be permitted.

2. Where secondary overcurrent protection is required, the secondary overcurrent device shall be permitted to consist of not more than six circuit breakers or six sets of fuses grouped in one location. Where multiple overcurrent devices are utilized, the total of all the device ratings shall not exceed the allowed value of a single overcurrent device. If both breakers and fuses are utilized as the overcurrent device, the total of the device ratings shall not exceed that allowed for fuses.

3. A transformer equipped with coordinated thermal overload protection by the manufacturer and arranged to interrupt the primary current shall be permitted to have primary overcurrent protection rated or set at a current value that is not more than six times the rated current of the transformer for transformers having not more than 6 percent impedance and not more than four times the rated current of the transformer for transformers having more than 6 percent but not more than 10 percent impedance.

**(A)  Three-Phase, 4-Wire System.** A grounding autotransformer used to create a 3-phase, 4-wire distribution system from a 3-phase, 3-wire ungrounded system shall conform to 450.5(A)(1) through (A)(4).

**(1)  Connections.** The transformer shall be directly connected to the ungrounded phase conductors and shall not be switched or provided with overcurrent protection that is independent of the main switch and common-trip overcurrent protection for the 3-phase, 4-wire system.

**(2)  Overcurrent Protection.** An overcurrent sensing device shall be provided that will cause the main switch or common-trip overcurrent protection referred to in 450.5 (A)(1) to open if the load on the autotransformer reaches or exceeds 125 percent of its continuous current per-phase or neutral rating. Delayed tripping for temporary overcurrents sensed at the autotransformer overcurrent device shall be permitted for the purpose of allowing proper operation of branch or feeder protective devices on the 4-wire system.

**(3)  Transformer Fault Sensing.** A fault-sensing system that causes the opening of a main switch or common-trip overcurrent device for the 3-phase, 4-wire system shall be provided to guard against single-phasing or internal faults.

> FPN: This can be accomplished by the use of two subtractive-connected donut-type current transformers installed to sense and signal when an unbalance occurs in the line current to the autotransformer of 50 percent or more of rated current.

**(4)  Rating.** The autotransformer shall have a continuous neutral-current rating that is sufficient to handle the maximum possible neutral unbalanced load current of the 4-wire system.

**(B)  Ground Reference for Fault Protection Devices.** A grounding autotransformer used to make available a specified magnitude of ground-fault current for operation of a ground-responsive protective device on a 3-phase, 3-wire ungrounded system shall conform to 450.5(B)(1) and (2).

**(1)  Rating.** The autotransformer shall have a continuous neutral-current rating sufficient for the specified ground-fault current.

**(2)  Overcurrent Protection.** An overcurrent protective device of adequate short-circuit rating that will open simultaneously all ungrounded conductors when it operates shall be applied in the grounding autotransformer branch circuit and shall be rated or set at a current not exceeding 125 percent of the autotransformer continuous per-phase current rating or 42 percent of the continuous-current rating of any series connected devices in the autotransformer neutral connection. Delayed tripping for temporary overcurrents to permit the proper operation of ground-responsive tripping devices on the main system shall be permitted but shall not exceed values that would be more than the short-time current rating of the grounding autotransformer or any series connected devices in the neutral connection thereto.

**(C)  Ground Reference for Damping Transitory Overvoltages.** A grounding autotransformer used to limit transitory overvoltages shall be of suitable rating and connected in accordance with 450.5(A)(1).

**450.6 Secondary Ties.** A secondary tie is a circuit operating at 600 volts, nominal, or less between phases that connects two power sources or power supply points, such as the secondaries of two transformers. The tie shall be permitted to consist of one or more conductors per phase.

As used in this section, the word *transformer* means a transformer or a bank of transformers operating as a unit.

**(A) Tie Circuits.** Tie circuits shall be provided with overcurrent protection at each end as required in Article 240.

Under the conditions described in 450.6(A)(1) and 450.6(A)(2), the overcurrent protection shall be permitted to be in accordance with 450.6(A)(3).

**(1) Loads at Transformer Supply Points Only.** Where all loads are connected at the transformer supply points at each end of the tie and overcurrent protection is not provided in accordance with Article 240, the rated ampacity of the tie shall not be less than 67 percent of the rated secondary current of the largest transformer connected to the secondary tie system.

**(2) Loads Connected Between Transformer Supply Points.** Where load is connected to the tie at any point between transformer supply points and overcurrent protection is not provided in accordance with Article 240, the rated ampacity of the tie shall not be less than 100 percent of the rated secondary current of the largest transformer connected to the secondary tie system.

*Exception: As otherwise provided in 450.6(A)(4).*

**(3) Tie Circuit Protection.** Under the conditions described in 450.6(A)(1) and (A)(2), both ends of each tie conductor shall be equipped with a protective device that opens at a predetermined temperature of the tie conductor under short-circuit conditions. This protection shall consist of one of the following: (1) a fusible link cable connector, terminal, or lug, commonly known as a limiter, each being of a size corresponding with that of the conductor and of construction and characteristics according to the operating voltage and the type of insulation on the tie conductors or (2) automatic circuit breakers actuated by devices having comparable current-time characteristics.

**(4) Interconnection of Phase Conductors Between Transformer Supply Points.** Where the tie consists of more than one conductor per phase, the conductors of each phase shall comply with one of the following provisions.

(a) *Interconnected.* The conductors shall be interconnected in order to establish a load supply point, and the protection specified in 450.6(A)(3) shall be provided in each tie conductor at this point.

(b) *Not Interconnected.* The loads shall be connected to one or more individual conductors of a paralleled conductor tie without interconnecting the conductors of each

phase and without the protection specified in 450.6(A)(3) at load connection points. Where this is done, the tie conductors of each phase shall have a combined capacity of not less than 133 percent of the rated secondary current of the largest transformer connected to the secondary tie system, the total load of such taps shall not exceed the rated secondary current of the largest transformer, and the loads shall be equally divided on each phase and on the individual conductors of each phase as far as practicable.

**(5) Tie Circuit Control.** Where the operating voltage exceeds 150 volts to ground, secondary ties provided with limiters shall have a switch at each end that, when open, de-energizes the associated tie conductors and limiters. The current rating of the switch shall not be less than the rated current of the conductors connected to the switch. It shall be capable of opening its rated current, and it shall be constructed so that it will not open under the magnetic forces resulting from short-circuit current.

**(B) Overcurrent Protection for Secondary Connections.** Where secondary ties are used, an overcurrent device rated or set at not more than 250 percent of the rated secondary current of the transformers shall be provided in the secondary connections of each transformer. In addition, an automatic circuit breaker actuated by a reverse-current relay set to open the circuit at not more than the rated secondary current of the transformer shall be provided in the secondary connection of each transformer.

**450.7 Parallel Operation.** Transformers shall be permitted to be operated in parallel and switched as a unit provided the overcurrent protection for each transformer meets the requirements of 450.3(A) for primary and secondary protective devices over 600 volts or 450.3(B) for primary and secondary protective devices 600 volts or less.

**450.8 Guarding.** Transformers shall be guarded as specified in 450.8(A) through (D).

**(A) Mechanical Protection.** Appropriate provisions shall be made to minimize the possibility of damage to transformers from external causes where the transformers are exposed to physical damage.

**(B) Case or Enclosure.** Dry-type transformers shall be provided with a noncombustible moisture-resistant case or enclosure that provides protection against the accidental insertion of foreign objects.

**(C) Exposed Energized Parts.** Switches or other equipment operating at 600 volts, nominal, or less and serving only equipment within a transformer enclosure shall be permitted to be installed in the transformer enclosure if accessible to qualified persons only. All energized parts shall be guarded in accordance with 110.27 and 110.34.

**(D) Voltage Warning.** The operating voltage of exposed live parts of transformer installations shall be indicated by signs or visible markings on the equipment or structures.

**450.9 Ventilation.** The ventilation shall be adequate to dispose of the transformer full-load losses without creating a temperature rise that is in excess of the transformer rating.

> FPN No. 1: See ANSI/IEEE C57.12.00-1993, *General Requirements for Liquid-Immersed Distribution, Power, and Regulating Transformers*, and ANSI/IEEE C57.12.01-1989, *General Requirements for Dry-Type Distribution and Power Transformers*.

> FPN No. 2: Additional losses may occur in some transformers where nonsinusoidal currents are present, resulting in increased heat in the transformer above its rating. See ANSI/IEEE C57.110-1993, *Recommended Practice for Establishing Transformer Capability When Supplying Nonsinusoidal Load Currents*, where transformers are utilized with nonlinear loads.

Transformers with ventilating openings shall be installed so that the ventilating openings are not blocked by walls or other obstructions. The required clearances shall be clearly marked on the transformer.

**450.10 Grounding.** Exposed non–current-carrying metal parts of transformer installations, including fences, guards, and so forth, shall be grounded where required under the conditions and in the manner specified for electric equipment and other exposed metal parts in Article 250.

**450.11 Marking.** Each transformer shall be provided with a nameplate giving the name of the manufacturer, rated kilovolt-amperes, frequency, primary and secondary voltage, impedance of transformers 25 kVA and larger, required clearances for transformers with ventilating openings, and the amount and kind of insulating liquid where used. In addition, the nameplate of each dry-type transformer shall include the temperature class for the insulation system.

**450.12 Terminal Wiring Space.** The minimum wire-bending space at fixed, 600-volt and below terminals of transformer line and load connections shall be as required in 312.6. Wiring space for pigtail connections shall conform to Table 314.16(B).

**450.13 Accessibility.** All transformers and transformer vaults shall be readily accessible to qualified personnel for inspection and maintenance or shall meet the requirements of 450.13(A) or (B).

**(A) Open Installations.** Dry-type transformers 600 volts, nominal, or less, located in the open on walls, columns, or structures, shall not be required to be readily accessible.

**(B) Hollow Space Installations.** Dry-type transformers 600 volts, nominal, or less and not exceeding 50 kVA shall be permitted in hollow spaces of buildings not permanently closed in by structure, provided they meet the ventilation requirements of 450.9 and separation from combustible materials requirements of 450.21(A). Transformers so installed shall not be required to be readily accessible.

## II. Specific Provisions Applicable to Different Types of Transformers

**450.21 Dry-Type Transformers Installed Indoors.**

**(A) Not Over 112½ kVA.** Dry-type transformers installed indoors and rated 112½ kVA or less shall have a separation of at least 305 mm (12 in.) from combustible material unless separated from the combustible material by a fire-resistant, heat-insulated barrier.

*Exception: This rule shall not apply to transformers rated for 600 volts, nominal, or less that are completely enclosed, with or without ventilating openings.*

**(B) Over 112½ kVA.** Individual dry-type transformers of more than 112½ kVA rating shall be installed in a transformer room of fire-resistant construction. Unless specified otherwise in this article, the term *fire resistant* means a construction having a minimum fire rating of 1 hour.

*Exception No. 1: Transformers with Class 155 or higher insulation systems and separated from combustible material by a fire-resistant, heat-insulating barrier or by not less than 1.83 m (6 ft) horizontally and 3.7 m (12 ft) vertically.*

*Exception No. 2: Transformers with Class 155 or higher insulation systems and completely enclosed except for ventilating openings.*

> FPN: See ANSI/ASTM E119-1995, *Method for Fire Tests of Building Construction and Materials*, and NFPA 251-1999, *Standard Methods of Tests of Fire Endurance of Building Construction and Materials*.

**(C) Over 35,000 Volts.** Dry-type transformers rated over 35,000 volts shall be installed in a vault complying with Part III of this article.

**450.22 Dry-Type Transformers Installed Outdoors.** Dry-type transformers installed outdoors shall have a weatherproof enclosure.

Transformers exceeding 112½ kVA shall not be located within 305 mm (12 in.) of combustible materials of buildings unless the transformer has Class 155 insulation systems or higher and is completely enclosed except for ventilating openings.

**450.23 Less-Flammable Liquid-Insulated Transformers.** Transformers insulated with listed less-flammable liquids that have a fire point of not less than 300°C shall be permitted to be installed in accordance with 450.23(A) or (B).

**(A) Indoor Installations.** Indoor installations shall be permitted in accordance with one of the following:

(1) In Type I or Type II buildings, in areas where all of the following requirements are met:
   a. The transformer is rated 35,000 volts or less.
   b. No combustible materials are stored.
   c. A liquid confinement area is provided.
   d. The installation complies with all restrictions provided for in the listing of the liquid.

(2) With an automatic fire extinguishing system and a liquid confinement area, provided the transformer is rated 35,000 volts or less

(3) In accordance with 450.26

**(B) Outdoor Installations.** Less-flammable liquid-filled transformers shall be permitted to be installed outdoors, attached to, adjacent to, or on the roof of buildings, where installed in accordance with (1) or (2):

(1) For Type I and Type II buildings, the installation shall comply with all restrictions provided for in the listing of the liquid.

   FPN: Installations adjacent to combustible material, fire escapes, or door and window openings may require additional safeguards such as those listed in 450.27.

(2) In accordance with 450.27.

   FPN No. 1: As used in this section, *Type I and Type II buildings* refers to Type I and Type II building construction as defined in NFPA 220-1999, *Standard on Types of Building Construction. Combustible materials* refers to those materials not classified as noncombustible or limited-combustible as defined in NFPA 220-1999, *Standard on Types of Building Construction.*

   FPN No. 2: See definition of *Listed* in Article 100.

**450.24 Nonflammable Fluid-Insulated Transformers.** Transformers insulated with a dielectric fluid identified as nonflammable shall be permitted to be installed indoors or outdoors. Such transformers installed indoors and rated over 35,000 volts shall be installed in a vault. Such transformers installed indoors shall be furnished with a liquid confinement area and a pressure-relief vent. The transformers shall be furnished with a means for absorbing any gases generated by arcing inside the tank, or the pressure-relief vent shall be connected to a chimney or flue that will carry such gases to an environmentally safe area.

   FPN: Safety may be increased if fire hazard analyses are performed for such transformer installations.

For the purposes of this section, a nonflammable dielectric fluid is one that does not have a flash point or fire point and is not flammable in air.

**450.25 Askarel-Insulated Transformers Installed Indoors.** Askarel-insulated transformers installed indoors and rated over 25 kVA shall be furnished with a pressure-relief vent. Where installed in a poorly ventilated place, they shall be furnished with a means for absorbing any gases generated by arcing inside the case, or the pressure-relief vent shall be connected to a chimney or flue that carries such gases outside the building. Askarel-insulated transformers rated over 35,000 volts shall be installed in a vault.

**450.26 Oil-Insulated Transformers Installed Indoors.** Oil-insulated transformers installed indoors shall be installed in a vault constructed as specified in Part III of this article.

*Exception No. 1: Where the total capacity does not exceed 112½ kVA, the vault specified in Part III of this article shall be permitted to be constructed of reinforced concrete that is not less than 100 mm (4 in.) thick.*

*Exception No. 2: Where the nominal voltage does not exceed 600, a vault shall not be required if suitable arrangements are made to prevent a transformer oil fire from igniting other materials and the total capacity in one location does not exceed 10 kVA in a section of the building classified as combustible or 75 kVA where the surrounding structure is classified as fire-resistant construction.*

*Exception No. 3: Electric furnace transformers that have a total rating not exceeding 75 kVA shall be permitted to be installed without a vault in a building or room of fire-resistant construction, provided suitable arrangements are made to prevent a transformer oil fire from spreading to other combustible material.*

*Exception No. 4: A transformer that has a total rating not exceeding 75 kVA and a supply voltage of 600 volts or less that is an integral part of charged-particle-accelerating equipment shall be permitted to be installed without a vault in a building or room of noncombustible or fire-resistant construction, provided suitable arrangements are made to prevent a transformer oil fire from spreading to other combustible material.*

*Exception No. 5: Transformers shall be permitted to be installed in a detached building that does not comply with Part III of this article if neither the building nor its contents present a fire hazard to any other building or property, and if the building is used only in supplying electric service and the interior is accessible only to qualified persons.*

*Exception No. 6: Oil-insulated transformers shall be permitted to be used without a vault in portable and mobile surface mining equipment (such as electric excavators) if each of the following conditions is met:*

(a) *Provision is made for draining leaking fluid to the ground.*

(b) *Safe egress is provided for personnel.*

(c) *A minimum 6-mm (¼-in.) steel barrier is provided for personnel protection.*

**450.27 Oil-Insulated Transformers Installed Outdoors.** Combustible material, combustible buildings, and parts of buildings, fire escapes, and door and window openings shall be safeguarded from fires originating in oil-insulated transformers installed on roofs, attached to or adjacent to a building or combustible material.

In cases where the transformer installation presents a fire hazard, one or more of the following safeguards shall be applied according to the degree of hazard involved:

(1) Space separations
(2) Fire-resistant barriers
(3) Automatic fire suppression systems
(4) Enclosures that confine the oil of a ruptured transformer tank

Oil enclosures shall be permitted to consist of fire-resistant dikes, curbed areas or basins, or trenches filled with coarse, crushed stone. Oil enclosures shall be provided with trapped drains where the exposure and the quantity of oil involved are such that removal of oil is important.

FPN: For additional information on transformers installed on poles or structures or under ground, see ANSI C2-1997, *National Electrical Safety Code.*

**450.28 Modification of Transformers.** When modifications are made to a transformer in an existing installation that change the type of the transformer with respect to Part II of this article, such transformer shall be marked to show the type of insulating liquid installed, and the modified transformer installation shall comply with the applicable requirements for that type of transformer.

## III. Transformer Vaults

**450.41 Location.** Vaults shall be located where they can be ventilated to the outside air without using flues or ducts wherever such an arrangement is practicable.

**450.42 Walls, Roofs, and Floors.** The walls and roofs of vaults shall be constructed of materials that have adequate structural strength for the conditions with a minimum fire resistance of 3 hours. The floors of vaults in contact with the earth shall be of concrete that is not less than 100 mm (4 in.) thick, but where the vault is constructed with a vacant space or other stories below it, the floor shall have

adequate structural strength for the load imposed thereon and a minimum fire resistance of 3 hours. For the purposes of this section, studs and wallboard construction shall not be acceptable.

*Exception: Where transformers are protected with automatic sprinkler, water spray, carbon dioxide, or halon, construction of 1-hour rating shall be permitted.*

FPN No. 1: For additional information, see ANSI/ASTM E119-1995, *Method for Fire Tests of Building Construction and Materials,* and NFPA 251-1999, *Standard Methods of Tests of Fire Endurance of Building Construction and Materials.*

FPN No. 2: A typical 3-hour construction is 150 mm (6 in.) thick reinforced concrete.

**450.43 Doorways.** Vault doorways shall be protected in accordance with 450.43(A), (B), and (C).

**(A) Type of Door.** Each doorway leading into a vault from the building interior shall be provided with a tight-fitting door that has a minimum fire rating of 3 hours. The authority having jurisdiction shall be permitted to require such a door for an exterior wall opening where conditions warrant.

*Exception: Where transformers are protected with automatic sprinkler, water spray, carbon dioxide, or halon, construction of 1-hour rating shall be permitted.*

FPN: For additional information, see NFPA 80-1999, *Standard for Fire Doors and Fire Windows.*

**(B) Sills.** A door sill or curb that is of sufficient height to confine the oil from the largest transformer within the vault shall be provided, and in no case shall the height be less than 100 mm (4 in.).

**(C) Locks.** Doors shall be equipped with locks, and doors shall be kept locked, access being allowed only to qualified persons. Personnel doors shall swing out and be equipped with panic bars, pressure plates, or other devices that are normally latched but open under simple pressure.

**450.45 Ventilation Openings.** Where required by 450.9, openings for ventilation shall be provided in accordance with 450.45(A) through (F).

**(A) Location.** Ventilation openings shall be located as far as possible from doors, windows, fire escapes, and combustible material.

**(B) Arrangement.** A vault ventilated by natural circulation of air shall be permitted to have roughly half of the total area of openings required for ventilation in one or more openings near the floor and the remainder in one or more openings in the roof or in the sidewalls near the roof, or all of the area required for ventilation shall be permitted in one or more openings in or near the roof.

**(C) Size.** For a vault ventilated by natural circulation of air to an outdoor area, the combined net area of all ventilating openings, after deducting the area occupied by screens, gratings, or louvers, shall not be less than 1900 mm$^2$ (3 in.$^2$) per kVA of transformer capacity in service, and in no case shall the net area be less than 0.1 m$^2$ (1 ft$^2$) for any capacity under 50 kVA.

**(D) Covering.** Ventilation openings shall be covered with durable gratings, screens, or louvers, according to the treatment required in order to avoid unsafe conditions.

**(E) Dampers.** All ventilation openings to the indoors shall be provided with automatic closing fire dampers that operate in response to a vault fire. Such dampers shall possess a standard fire rating of not less than 1½ hours.

FPN: See ANSI/UL 555-1995, *Standard for Fire Dampers*.

**(F) Ducts.** Ventilating ducts shall be constructed of fire-resistant material.

**450.46 Drainage.** Where practicable, vaults containing more than 100 kVA transformer capacity shall be provided with a drain or other means that will carry off any accumulation of oil or water in the vault unless local conditions make this impracticable. The floor shall be pitched to the drain where provided.

**450.47 Water Pipes and Accessories.** Any pipe or duct system foreign to the electrical installation shall not enter or pass through a transformer vault. Piping or other facilities provided for vault fire protection, or for transformer cooling, shall not be considered foreign to the electrical installation.

**450.48 Storage in Vaults.** Materials shall not be stored in transformer vaults.

# ARTICLE 455
# Phase Converters

## I. General

**455.1 Scope.** This article covers the installation and use of phase converters.

**455.2 Definitions.**

**Manufactured Phase.** The manufactured or derived phase originates at the phase converter and is not solidly connected to either of the single-phase input conductors.

**Phase Converter.** An electrical device that converts single-phase power to 3-phase electrical power.

FPN: Phase converters have characteristics that modify the starting torque and locked-rotor current of motors served, and consideration is required in selecting a phase converter for a specific load.

**Rotary-Phase Converter.** A device that consists of a rotary transformer and capacitor panel(s) that permits the operation of 3-phase loads from a single-phase supply.

**Static-Phase Converter.** A device without rotating parts, sized for a given 3-phase load to permit operation from a single-phase supply.

**455.3 Other Articles.** All applicable requirements of this *Code* shall apply to phase converters except as amended by this article.

**455.4 Marking.** Each phase converter shall be provided with a permanent nameplate indicating the following:

(1) Manufacturer's name
(2) Rated input and output voltages
(3) Frequency
(4) Rated single-phase input full-load amperes
(5) Rated minimum and maximum single load in kilovolt-amperes (kVA) or horsepower
(6) Maximum total load in kilovolt-amperes (kVA) or horsepower
(7) For a rotary-phase converter, 3-phase amperes at full load

**455.5 Equipment Grounding Connection.** A means for attachment of an equipment grounding conductor termination in accordance with 250.8 shall be provided.

**455.6 Conductors.**

**(A) Ampacity.** The ampacity of the single-phase supply conductors shall be determined by 455.6(A)(1) or (A)(2).

FPN: Single-phase conductors sized to prevent a voltage drop not exceeding 3 percent from the source of supply to the phase converter may help ensure proper starting and operation of motor loads.

**(1) Variable Loads.** Where the loads to be supplied are variable, the conductor ampacity shall not be less than 125 percent of the phase converter nameplate single-phase input full-load amperes.

**(2) Fixed Loads.** Where the phase converter supplies specific fixed loads, and the conductor ampacity is less than 125 percent of the phase converter nameplate single-phase input full-load amperes, the conductors shall have an ampacity not less than 250 percent of the sum of the full-load, 3-phase current rating of the motors and other loads served where the input and output voltages of the phase converter are identical. Where the input and output voltages of the phase converter are different, the current as determined by this section shall be multiplied by the ratio of output to input voltage.

**(B) Manufactured Phase Marking.** The manufactured phase conductors shall be identified in all accessible locations with a distinctive marking. The marking shall be consistent throughout the system and premises.

**455.7 Overcurrent Protection.** The single-phase supply conductors and phase converter shall be protected from overcurrent by 455.7(A) or (B). Where the required fuse rating or circuit breaker setting does not correspond to a standard rating or setting, the next higher standard rating or setting shall be permitted.

**(A) Variable Loads.** Where the loads to be supplied are variable, overcurrent protection shall be set at not more than 125 percent of the phase converter nameplate single-phase input full-load amperes.

**(B) Fixed Loads.** Where the phase converter supplies specific fixed loads and the conductors are sized in accordance with 455.6(A)(2), the conductors shall be protected in accordance with their ampacity. The overcurrent protection determined from this section shall not exceed 125 percent of the phase converter nameplate single-phase input amperes.

**455.8 Disconnecting Means.** Means shall be provided to disconnect simultaneously all ungrounded single-phase supply conductors to the phase converter.

**(A) Location.** The disconnecting means shall be readily accessible and located in sight from the phase converter.

**(B) Type.** The disconnecting means shall be a switch rated in horsepower, a circuit breaker, or a molded-case switch. Where only nonmotor loads are served, an ampere-rated switch shall be permitted.

**(C) Rating.** The ampere rating of the disconnecting means shall not be less than 115 percent of the rated maximum single-phase input full-load amperes or, for specific fixed loads, shall be permitted to be selected from 455.7(C)(1) or (C)(2).

**(1) Current Rated Disconnect.** The disconnecting means shall be a circuit breaker or molded-case switch with an ampere rating not less than 250 percent of the sum of the following:

(1) Full-load, 3-phase current ratings of the motors
(2) Other loads served

**(2) Horsepower Rated Disconnect.** The disconnecting means shall be a switch with a horsepower rating. The equivalent locked rotor current of the horsepower rating of the switch shall not be less than 200 percent of the sum of the following:

(1) Nonmotor loads
(2) The 3-phase, locked-rotor current of the largest motor as determined from Table 430.151(B)
(3) The full-load current of all other 3-phase motors operating at the same time

**(D) Voltage Ratios.** The calculations in 455.8(C) shall apply directly where the input and output voltages of the phase converter are identical. Where the input and output voltages of the phase converter are different, the current shall be multiplied by the ratio of the output to input voltage.

**455.9 Connection of Single-Phase Loads.** Where single-phase loads are connected on the load side of a phase converter, they shall not be connected to the manufactured phase.

**455.10 Terminal Housings.** A terminal housing shall be provided on a phase converter, and the terminal housing shall be in accordance with the provisions of 430.12.

## II. Specific Provisions Applicable to Different Types of Phase Converters

**455.20 Disconnecting Means.** The single-phase disconnecting means for the input of a static phase converter shall be permitted to serve as the disconnecting means for the phase converter and a single load if the load is within sight of the disconnecting means.

**455.21 Start-Up.** Power to the utilization equipment shall not be supplied until the rotary-phase converter has been started.

**455.22 Power Interruption.** Utilization equipment supplied by a rotary-phase converter shall be controlled in such a manner that power to the equipment will be disconnected in the event of a power interruption.

> FPN: Magnetic motor starters, magnetic contactors, and similar devices, with manual or time delay restarting for the load, provide restarting after power interruption.

**455.23 Capacitors.** Capacitors that are not an integral part of the rotary-phase conversion system but are installed for a motor load shall be connected to the line side of that motor overload protective device.

# ARTICLE 460
# Capacitors

**460.1 Scope.** This article covers the installation of capacitors on electric circuits.

Surge capacitors or capacitors included as a component part of other apparatus and conforming with the requirements of such apparatus are excluded from these requirements.

This article also covers the installation of capacitors in hazardous (classified) locations as modified by Articles 501 through 503.

**460.2 Enclosing and Guarding.**

**(A) Containing More Than 11 L (3 gal) of Flammable Liquid.** Capacitors containing more than 11 L (3 gal) of flammable liquid shall be enclosed in vaults or outdoor fenced enclosures complying with Article 110, Part III. This limit shall apply to any single unit in an installation of capacitors.

**(B) Accidental Contact.** Where capacitors are accessible to unauthorized and unqualified persons, they shall be enclosed, located, or guarded so that persons cannot come into accidental contact or bring conducting materials into accidental contact with exposed energized parts, terminals, or buses associated with them. However, no additional guarding is required for enclosures accessible only to authorized and qualified persons.

## I. 600 Volts, Nominal, and Under

**460.6 Discharge of Stored Energy.** Capacitors shall be provided with a means of discharging stored energy.

**(A) Time of Discharge.** The residual voltage of a capacitor shall be reduced to 50 volts, nominal, or less within 1 minute after the capacitor is disconnected from the source of supply.

**(B) Means of Discharge.** The discharge circuit shall be either permanently connected to the terminals of the capacitor or capacitor bank or provided with automatic means of connecting it to the terminals of the capacitor bank on removal of voltage from the line. Manual means of switching or connecting the discharge circuit shall not be used.

**460.8 Conductors.**

**(A) Ampacity.** The ampacity of capacitor circuit conductors shall not be less than 135 percent of the rated current of the capacitor. The ampacity of conductors that connect a capacitor to the terminals of a motor or to motor circuit conductors shall not be less than one-third the ampacity of the motor circuit conductors and in no case less than 135 percent of the rated current of the capacitor.

**(B) Overcurrent Protection.** An overcurrent device shall be provided in each ungrounded conductor for each capacitor bank. The rating or setting of the overcurrent device shall be as low as practicable.

*Exception: A separate overcurrent device shall not be required for a capacitor connected on the load side of a motor overload protective device.*

**(C) Disconnecting Means.** A disconnecting means shall be provided in each ungrounded conductor for each capacitor bank and shall meet the following requirements:

(1) The disconnecting means shall open all ungrounded conductors simultaneously.
(2) The disconnecting means shall be permitted to disconnect the capacitor from the line as a regular operating procedure.
(3) The rating of the disconnecting means shall not be less than 135 percent of the rated current of the capacitor.

*Exception: A separate disconnecting means shall not be required where a capacitor is connected on the load side of a motor controller.*

**460.9 Rating or Setting of Motor Overload Device.** Where a motor installation includes a capacitor connected on the load side of the motor overload device, the rating or setting of the motor overload device shall be based on the improved power factor of the motor circuit.

The effect of the capacitor shall be disregarded in determining the motor circuit conductor rating in accordance with 430.22.

**460.10 Grounding.** Capacitor cases shall be grounded in accordance with Article 250.

*Exception: Capacitor cases shall not be grounded where the capacitor units are supported on a structure designed to operate at other than ground potential.*

**460.12 Marking.** Each capacitor shall be provided with a nameplate giving the name of the manufacturer, rated voltage, frequency, kilovar or amperes, number of phases, and, if filled with a combustible liquid, the volume of liquid. Where filled with a nonflammable liquid, the nameplate shall so state. The nameplate shall also indicate whether a capacitor has a discharge device inside the case.

## II. Over 600 Volts, Nominal

**460.24 Switching.**

**(A) Load Current.** Group-operated switches shall be used for capacitor switching and shall be capable of the following:

(1) Carrying continuously not less than 135 percent of the rated current of the capacitor installation

(2) Interrupting the maximum continuous load current of each capacitor, capacitor bank, or capacitor installation that will be switched as a unit

(3) Withstanding the maximum inrush current, including contributions from adjacent capacitor installations

(4) Carrying currents due to faults on capacitor side of switch

**(B) Isolation.**

**(1) General.** A means shall be installed to isolate from all sources of voltage each capacitor, capacitor bank, or capacitor installation that will be removed from service as a unit. The isolating means shall provide a visible gap in the electrical circuit adequate for the operating voltage.

**(2) Isolating or Disconnecting Switches with No Interrupting Rating.** Isolating or disconnecting switches (with no interrupting rating) shall be interlocked with the load-interrupting device or shall be provided with prominently displayed caution signs in accordance with 490.22 to prevent switching load current.

**(C) Additional Requirements for Series Capacitors.** The proper switching sequence shall be ensured by use of one of the following:

(1) Mechanically sequenced isolating and bypass switches
(2) Interlocks
(3) Switching procedure prominently displayed at the switching location

**460.25 Overcurrent Protection.**

**(A) Provided to Detect and Interrupt Fault Current.** A means shall be provided to detect and interrupt fault current likely to cause dangerous pressure within an individual capacitor.

**(B) Single Pole or Multipole Devices.** Single-pole or multipole devices shall be permitted for this purpose.

**(C) Protected Individually or in Groups.** Capacitors shall be permitted to be protected individually or in groups.

**(D) Protective Devices Rated or Adjusted.** Protective devices for capacitors or capacitor equipment shall be rated or adjusted to operate within the limits of the safe zone for individual capacitors. If the protective devices are rated or adjusted to operate within the limits for Zone 1 or Zone 2, the capacitors shall be enclosed or isolated.

In no event shall the rating or adjustment of the protective devices exceed the maximum limit of Zone 2.

FPN: For definitions of *Safe Zone, Zone 1,* and *Zone 2,* see ANSI/IEEE 18-1992, *Shunt Power Capacitors.*

**460.26 Identification.** Each capacitor shall be provided with a permanent nameplate giving the manufacturer's name, rated voltage, frequency, kilovar or amperes, number of phases, and the volume of liquid identified as flammable, if such is the case.

**460.27 Grounding.** Capacitor neutrals and cases, if grounded, shall be grounded in accordance with Article 250.

*Exception: Where the capacitor units are supported on a structure that is designed to operate at other than ground potential.*

**460.28 Means for Discharge.**

**(A) Means to Reduce the Residual Voltage.** A means shall be provided to reduce the residual voltage of a capacitor to 50 volts or less within 5 minutes after the capacitor is disconnected from the source of supply.

**(B) Connection to Terminals.** A discharge circuit shall be either permanently connected to the terminals of the capacitor or provided with automatic means of connecting it to the terminals of the capacitor bank after disconnection of the capacitor from the source of supply. The windings of motors, transformers, or other equipment directly connected to capacitors without a switch or overcurrent device interposed shall meet the requirements of 460.28(A).

# ARTICLE 470
## Resistors and Reactors

**I. 600 Volts, Nominal, and Under**

**470.1 Scope.** This article covers the installation of separate resistors and reactors on electric circuits.

*Exception: Resistors and reactors that are component parts of other apparatus.*

This article also covers the installation of resistors and reactors in hazardous (classified) locations as modified by Articles 501 through 504.

**470.2 Location.** Resistors and reactors shall not be placed where exposed to physical damage.

**470.3 Space Separation.** A thermal barrier shall be required if the space between the resistors and reactors and any combustible material is less than 305 mm (12 in.).

**470.4 Conductor Insulation.** Insulated conductors used for connections between resistance elements and controllers shall be suitable for an operating temperature of not less than 90°C (194°F).

*Exception: Other conductor insulations shall be permitted for motor starting service.*

## II. Over 600 Volts, Nominal

**470.18 General.**

**(A) Protected Against Physical Damage.** Resistors and reactors shall be protected against physical damage.

**(B) Isolated by Enclosure or Elevation.** Resistors and reactors shall be isolated by enclosure or elevation to protect personnel from accidental contact with energized parts.

**(C) Combustible Materials.** Resistors and reactors shall not be installed in close enough proximity to combustible materials to constitute a fire hazard and shall have a clearance of not less than 305 mm (12 in.) from combustible materials.

**(D) Clearances.** Clearances from resistors and reactors to grounded surfaces shall be adequate for the voltage involved.

FPN: See Article 490.

**(E) Temperature Rise from Induced Circulating Currents.** Metallic enclosures of reactors and adjacent metal parts shall be installed so that the temperature rise from induced circulating currents is not hazardous to personnel or does not constitute a fire hazard.

**470.19 Grounding.** Resistor and reactor cases or enclosures shall be grounded in accordance with Article 250.

*Exception: Resistor or reactor cases or enclosures supported on a structure designed to operate at other than ground potential shall not be grounded.*

**470.20 Oil-Filled Reactors.** Installation of oil-filled reactors, in addition to the above requirements, shall comply with applicable requirements of Article 450.

## ARTICLE 480
## Storage Batteries

**480.1 Scope.** The provisions of this article shall apply to all stationary installations of storage batteries.

**480.2 Definitions.**

**Nominal Battery Voltage.** The voltage computed on the basis of 2 volts per cell for the lead-acid type and 1.2 volts per cell for the alkali type.

**Sealed Cell or Battery.** A sealed cell or battery is one that has no provision for the addition of water or electrolyte or for external measurement of electrolyte specific gravity. The individual cells shall be permitted to contain a venting arrangement as described in 480.10(B).

**Storage Battery.** A battery comprised of one or more rechargeable cells of the lead-acid, nickel-cadmium, or other rechargeable electrochemical types.

**480.3 Wiring and Equipment Supplied from Batteries.** Wiring and equipment supplied from storage batteries shall be subject to the requirements of this *Code* applying to wiring and equipment operating at the same voltage, unless otherwise permitted by 480.4.

**480.4 Overcurrent Protection for Prime Movers.** Overcurrent protection shall not be required for conductors from a battery rated less than 50 volts if the battery provides power for starting, ignition, or control of prime movers. Section 300.3 shall not apply to these conductors.

**480.5 Grounding.** The requirements of Article 250 shall apply.

**480.6 Insulation of Batteries Not Over 250 Volts.** This section shall apply to storage batteries having cells connected so as to operate at a nominal battery voltage of not over 250 volts.

**(A) Vented Lead-Acid Batteries.** Cells and multicompartment batteries with covers sealed to containers of nonconductive, heat-resistant material shall not require additional insulating support.

**(B) Vented Alkaline-Type Batteries.** Cells with covers sealed to jars of nonconductive, heat-resistant material shall require no additional insulation support. Cells in jars of conductive material shall be installed in trays of nonconductive material with not more than 20 cells (24 volts, nominal) in the series circuit in any one tray.

**(C) Rubber Jars.** Cells in rubber or composition containers shall require no additional insulating support where the total nominal voltage of all cells in series does not exceed 150 volts. Where the total voltage exceeds 150 volts, batteries shall be sectionalized into groups of 150 volts or less, and each group shall have the individual cells installed in trays or on racks.

**(D) Sealed Cells or Batteries.** Sealed cells and multicompartment sealed batteries constructed of nonconductive, heat-resistant material shall not require additional insulating support. Batteries constructed of a conducting container shall have insulating support if a voltage is present between the container and ground.

**480.7 Insulation of Batteries of Over 250 Volts.** The provisions of 480.6 shall apply to storage batteries having the cells connected so as to operate at a nominal voltage exceeding 250 volts, and, in addition, the provisions of this section shall also apply to such batteries. Cells shall be installed in groups having a total nominal voltage of not over 250 volts. Insulation, which can be air, shall be provided between groups and shall have a minimum separation between live battery parts of opposite polarity of 50 mm (2 in.) for battery voltages not exceeding 600 volts.

**480.8 Racks and Trays.** Racks and trays shall comply with 480.8(A) and (B).

**(A) Racks.** Racks, as required in this article, are rigid frames designed to support cells or trays. They shall be substantial and be made of one of the following:

(1) Metal, treated so as to be resistant to deteriorating action by the electrolyte and provided with nonconducting members directly supporting the cells or with continuous insulating material other than paint on conducting members
(2) Other construction such as fiberglass or other suitable nonconductive materials

**(B) Trays.** Trays are frames, such as crates or shallow boxes usually of wood or other nonconductive material, constructed or treated so as to be resistant to deteriorating action by the electrolyte.

**480.9 Battery Locations.** Battery locations shall conform to 480.9(A), (B) and (C).

**(A) Ventilation.** Provisions shall be made for sufficient diffusion and ventilation of the gases from the battery to prevent the accumulation of an explosive mixture.

**(B) Live Parts.** Guarding of live parts shall comply with 110.27.

**(C) Working Space.** Working space about the battery systems shall comply with 110.26. Working clearance shall be measured from the edge of the battery rack.

**480.10 Vents.**

**(A) Vented Cells.** Each vented cell shall be equipped with a flame arrester that is designed to prevent destruction of the cell due to ignition of gases within the cell by an external spark or flame under normal operating conditions.

**(B) Sealed Cells.** Sealed battery or cells shall be equipped with a pressure-release vent to prevent excessive accumulation of gas pressure, or the battery or cell shall be designed to prevent scatter of cell parts in event of a cell explosion.

## ARTICLE 490
## Equipment, Over 600 Volts, Nominal

### I. General

**490.1 Scope.** This article covers the general requirements for equipment operating at more than 600 volts, nominal.

> FPN No. 1: See NFPA 70E-2000, *Standard for Electrical Safety Requirements for Employee Workplaces*, for electrical safety requirements for employee workplaces.

> FPN No. 2: For further information on hazard signs and labels, see ANSI Z535-4, *Product Signs and Safety Labels*.

**490.2 Definition.**

**High Voltage.** For the purposes of this article, more than 600 volts, nominal.

**490.3 Oil-Filled Equipment.** Installation of electrical equipment, other than transformers covered in Article 450, containing more than 38 L (10 gal) of flammable oil per unit shall meet the requirements of Parts II and III of Article 450.

### II. Equipment — Specific Provisions

**490.21 Circuit-Interrupting Devices.**

**(A) Circuit Breakers.**

**(1) Location.**

(a) Circuit breakers installed indoors shall be mounted either in metal-enclosed units or fire-resistant cell-mounted units, or they shall be permitted to be open-mounted in locations accessible to qualified persons only.

(b) Circuit breakers used to control oil-filled transformers shall either be located outside the transformer vault or be capable of operation from outside the vault.

(c) Oil circuit breakers shall be arranged or located so that adjacent readily combustible structures or materials are safeguarded in an approved manner.

**(2) Operating Characteristics.** Circuit breakers shall have the following equipment or operating characteristics:

(1) An accessible mechanical or other approved means for manual tripping, independent of control power.
(2) Be release free (trip free).
(3) If capable of being opened or closed manually while energized, the main contacts shall operate independently of the speed of the manual operation.
(4) A mechanical position indicator at the circuit breaker to show the open or closed position of the main contacts.
(5) A means of indicating the open and closed position of the breaker at the point(s) from which they may be operated.

**(3) Nameplate.** A circuit breaker shall have a permanent and legible nameplate showing manufacturer's name or trademark, manufacturer's type or identification number, continuous current rating, interrupting rating in megavolt-amperes (MVA) or amperes, and maximum voltage rating. Modification of a circuit breaker affecting its rating(s) shall be accompanied by an appropriate change of nameplate information.

**(4) Rating.** Circuit breakers shall have the following ratings:

(1) The continuous current rating of a circuit breaker shall not be less than the maximum continuous current through the circuit breaker.
(2) The interrupting rating of a circuit breaker shall not be less than the maximum fault current the circuit breaker will be required to interrupt, including contributions from all connected sources of energy.
(3) The closing rating of a circuit breaker shall not be less than the maximum asymmetrical fault current into which the circuit breaker can be closed.
(4) The momentary rating of a circuit breaker shall not be less than the maximum asymmetrical fault current at the point of installation.
(5) The rated maximum voltage of a circuit breaker shall not be less than the maximum circuit voltage.

**(B) Power Fuses and Fuseholders.**

**(1) Use.** Where fuses are used to protect conductors and equipment, a fuse shall be placed in each ungrounded conductor. Two power fuses shall be permitted to be used in parallel to protect the same load if both fuses have identical ratings and both fuses are installed in an identified common mounting with electrical connections that will divide the current equally. Power fuses of the vented type shall not be used indoors, under ground, or in metal enclosures unless identified for the use.

**(2) Interrupting Rating.** The interrupting rating of power fuses shall not be less than the maximum fault current the fuse will be required to interrupt, including contributions from all connected sources of energy.

**(3) Voltage Rating.** The maximum voltage rating of power fuses shall not be less than the maximum circuit voltage. Fuses having a minimum recommended operating voltage shall not be applied below this voltage.

**(4) Identification of Fuse Mountings and Fuse Units.** Fuse mountings and fuse units shall have permanent and legible nameplates showing the manufacturer's type or designation, continuous current rating, interrupting current rating, and maximum voltage rating.

**(5) Fuses.** Fuses that expel flame in opening the circuit shall be designed or arranged so that they function properly without hazard to persons or property.

**(6) Fuseholders.** Fuseholders shall be designed or installed so that they are de-energized while a fuse is being replaced.

*Exception: Fuses and fuseholders designed to permit fuse replacement by qualified persons using equipment designed for the purpose without de-energizing the fuseholder shall be permitted.*

**(7) High-Voltage Fuses.** Metal-enclosed switchgear and substations that utilize high-voltage fuses shall be provided with a gang-operated disconnecting switch. Isolation of the fuses from the circuit shall be provided by either connecting a switch between the source and the fuses or providing roll-out switch and fuse-type construction. The switch shall be of the load-interrupter type, unless mechanically or electrically interlocked with a load-interrupting device arranged to reduce the load to the interrupting capability of the switch.

*Exception: More than one switch shall be permitted as the disconnecting means for one set of fuses where the switches are installed to provide connection to more than one set of supply conductors. The switches shall be mechanically or electrically interlocked to permit access to the fuses only when all switches are open. A conspicuous sign shall be placed at the fuses identifying the presence of more than one source.*

**(C) Distribution Cutouts and Fuse Links — Expulsion Type.**

**(1) Installation.** Cutouts shall be located so that they may be readily and safely operated and re-fused, and so that the exhaust of the fuses does not endanger persons. Distribution cutouts shall not be used indoors, underground, or in metal enclosures.

**(2) Operation.** Where fused cutouts are not suitable to interrupt the circuit manually while carrying full load, an approved means shall be installed to interrupt the entire load. Unless the fused cutouts are interlocked with the switch to prevent opening of the cutouts under load, a conspicuous sign shall be placed at such cutouts identifying that they shall not be operated under load.

**(3) Interrupting Rating.** The interrupting rating of distribution cutouts shall not be less than the maximum fault current the cutout is required to interrupt, including contributions from all connected sources of energy.

**(4) Voltage Rating.** The maximum voltage rating of cutouts shall not be less than the maximum circuit voltage.

**(5) Identification.** Distribution cutouts shall have on their body, door, or fuse tube a permanent and legible nameplate or identification showing the manufacturer's type or designation, continuous current rating, maximum voltage rating, and interrupting rating.

**(6) Fuse Links.** Fuse links shall have a permanent and legible identification showing continuous current rating and type.

**(7) Structure Mounted Outdoors.** The height of cutouts mounted outdoors on structures shall provide safe clearance between lowest energized parts (open or closed position) and standing surfaces, in accordance with 110.34(E).

**(D) Oil-Filled Cutouts.**

**(1) Continuous Current Rating.** The continuous current rating of oil-filled cutouts shall not be less than the maximum continuous current through the cutout.

**(2) Interrupting Rating.** The interrupting rating of oil-filled cutouts shall not be less than the maximum fault current the oil-filled cutout is required to interrupt, including contributions from all connected sources of energy.

**(3) Voltage Rating.** The maximum voltage rating of oil-filled cutouts shall not be less than the maximum circuit voltage.

**(4) Fault Closing Rating.** Oil-filled cutouts shall have a fault closing rating not less than the maximum asymmetrical fault current that can occur at the cutout location, unless suitable interlocks or operating procedures preclude the possibility of closing into a fault.

**(5) Identification.** Oil-filled cutouts shall have a permanent and legible nameplate showing the rated continuous current, rated maximum voltage, and rated interrupting current.

**(6) Fuse Links.** Fuse links shall have a permanent and legible identification showing the rated continuous current.

**(7) Location.** Cutouts shall be located so that they are readily and safely accessible for re-fusing, with the top of the cutout not over 1.5 m (5 ft) above the floor or platform.

**(8) Enclosure.** Suitable barriers or enclosures shall be provided to prevent contact with nonshielded cables or energized parts of oil-filled cutouts.

**(E) Load Interrupters.** Load-interrupter switches shall be permitted if suitable fuses or circuit breakers are used in conjunction with these devices to interrupt fault currents. Where these devices are used in combination, they shall be coordinated electrically so that they will safely withstand the effects of closing, carrying, or interrupting all possible currents up to the assigned maximum short-circuit rating.

Where more than one switch is installed with interconnected load terminals to provide for alternate connection to different supply conductors, each switch shall be provided with a conspicuous sign identifying this hazard.

**(1) Continuous Current Rating.** The continuous current rating of interrupter switches shall equal or exceed the maximum continuous current at the point of installation.

**(2) Voltage Rating.** The maximum voltage rating of interrupter switches shall equal or exceed the maximum circuit voltage.

**(3) Identification.** Interrupter switches shall have a permanent and legible nameplate including the following information: manufacturer's type or designation, continuous current rating, interrupting current rating, fault closing rating, maximum voltage rating.

**(4) Switching of Conductors.** The switching mechanism shall be arranged to be operated from a location where the operator is not exposed to energized parts and shall be arranged to open all ungrounded conductors of the circuit simultaneously with one operation. Switches shall be arranged to be locked in the open position. Metal-enclosed switches shall be operable from outside the enclosure.

**(5) Stored Energy for Opening.** The stored-energy operator shall be permitted to be left in the uncharged position after the switch has been closed if a single movement of the operating handle charges the operator and opens the switch.

**(6) Supply Terminals.** The supply terminals of fused interrupter switches shall be installed at the top of the switch enclosure, or, if the terminals are located elsewhere, the equipment shall have barriers installed so as to prevent persons from accidentally contacting energized parts or dropping tools or fuses into energized parts.

**490.22 Isolating Means.** Means shall be provided to completely isolate an item of equipment. The use of isolating switches shall not be required where there are other ways of de-energizing the equipment for inspection and repairs, such as draw-out-type metal-enclosed switchgear units and removable truck panels.

Isolating switches not interlocked with an approved circuit-interrupting device shall be provided with a sign warning against opening them under load.

A fuseholder and fuse, designed for the purpose, shall be permitted as an isolating switch.

**490.23 Voltage Regulators.** Proper switching sequence for regulators shall be ensured by use of one of the following:

(1) Mechanically sequenced regulator bypass switch(es)
(2) Mechanical interlocks
(3) Switching procedure prominently displayed at the switching location

**490.24 Minimum Space Separation.** In field-fabricated installations, the minimum air separation between bare live conductors and between such conductors and adjacent grounded surfaces shall not be less than the values given in Table 490.24. These values shall not apply to interior portions or exterior terminals of equipment designed, manufactured, and tested in accordance with accepted national standards.

## III. Equipment — Metal-Enclosed Power Switchgear and Industrial Control Assemblies

**490.30 General.** This part covers assemblies of metal-enclosed power switchgear and industrial control, including but not limited to switches, interrupting devices and their control, metering, protection and regulating equipment, where an integral part of the assembly, with associated interconnections and supporting structures. This part also includes metal-enclosed power switchgear assemblies that form a part of unit substations, power centers, or similar equipment.

**490.31 Arrangement of Devices in Assemblies.** Arrangement of devices in assemblies shall be such that individual components can safely perform their intended function without adversely affecting the safe operation of other components in the assembly.

**490.32 Guarding of High-Voltage Energized Parts Within a Compartment.** Where access for other than visual inspection is required to a compartment that contains energized high-voltage parts, barriers shall be provided to prevent accidental contact by persons, tools, or other equipment with energized parts. Exposed live parts shall only be permitted in compartments accessible to qualified persons.

**Table 490.24 Minimum Clearance of Live Parts***

| Nominal Voltage Rating (kV) | Impulse Withstand, B.I.L (kV) | | Minimum Clearance of Live Parts | | | | | | | |
| | | | Phase-to-Phase | | | | Phase-to-Ground | | | |
| | | | Indoors | | Outdoors | | Indoors | | Outdoors | |
| | Indoors | Outdoors | mm | in. | mm | in. | mm | in. | mm | in. |
|---|---|---|---|---|---|---|---|---|---|---|
| 2.4–4.16 | 60 | 95 | 115 | 4.5 | 180 | 7 | 80 | 3.0 | 155 | 6 |
| 7.2 | 75 | 95 | 140 | 5.5 | 180 | 7 | 105 | 4.0 | 155 | 6 |
| 13.8 | 95 | 110 | 195 | 7.5 | 305 | 12 | 130 | 5.0 | 180 | 7 |
| 14.4 | 110 | 110 | 230 | 9.0 | 305 | 12 | 170 | 6.5 | 180 | 7 |
| 23 | 125 | 150 | 270 | 10.5 | 385 | 15 | 190 | 7.5 | 255 | 10 |
| 34.5 | 150 | 150 | 320 | 12.5 | 385 | 15 | 245 | 9.5 | 255 | 10 |
| | 200 | 200 | 460 | 18.0 | 460 | 18 | 335 | 13.0 | 335 | 13 |
| 46 | — | 200 | — | — | 460 | 18 | — | — | 335 | 13 |
| | — | 250 | — | — | 535 | 21 | — | — | 435 | 17 |
| 69 | — | 250 | — | — | 535 | 21 | — | — | 435 | 17 |
| | — | 350 | — | — | 790 | 31 | — | — | 635 | 25 |
| 115 | — | 550 | — | — | 1350 | 53 | — | — | 1070 | 42 |
| 138 | — | 550 | — | — | 1350 | 53 | — | — | 1070 | 42 |
| | — | 650 | — | — | 1605 | 63 | — | — | 1270 | 50 |
| 161 | — | 650 | — | — | 1605 | 63 | — | — | 1270 | 50 |
| | — | 750 | — | — | 1830 | 72 | — | — | 1475 | 58 |
| 230 | — | 750 | — | — | 1830 | 72 | — | — | 1475 | 58 |
| | — | 900 | — | — | 2265 | 89 | — | — | 1805 | 71 |
| | — | 1050 | — | — | 2670 | 105 | — | — | 2110 | 83 |

*The values given are the minimum clearance for rigid parts and bare conductors under favorable service conditions. They shall be increased for conductor movement or under unfavorable service conditions or wherever space limitations permit. The selection of the associated impulse withstand voltage for a particular system voltage is determined by the characteristics of the surge protective equipment.

Fuses and fuseholders designed to enable future replacement without de-energizing the fuse holder shall only be permitted for use by qualified persons.

**490.33 Guarding of Low-Voltage Energized Parts Within a Compartment.** Energized bare parts mounted on doors shall be guarded where the door must be opened for maintenance of equipment or removal of draw-out equipment.

**490.34 Clearance for Cable Conductors Entering Enclosure.** The unobstructed space opposite terminals or opposite raceways or cables entering a switchgear or control assembly shall be adequate for the type of conductor and method of termination.

**490.35 Accessibility of Energized Parts.**

**(A) High-Voltage Equipment.** Doors that would provide unqualified persons access to high-voltage energized parts shall be locked.

**(B) Low-Voltage Control Equipment.** Low-voltage control equipment, relays, motors, and the like shall not be installed in compartments with exposed high-voltage energized parts or high-voltage wiring unless either of the following conditions is met:

(1) The access means is interlocked with the high-voltage switch or disconnecting means to prevent the access means from being opened or removed.
(2) The high-voltage switch or disconnecting means is in the isolating position.

**(C) High-Voltage Instruments or Control Transformers and Space Heaters.** High-voltage instrument or control transformers and space heaters shall be permitted to be installed in the high-voltage compartment without access restrictions beyond those that apply to the high-voltage compartment generally.

**490.36 Grounding.** Frames of switchgear and control assemblies shall be grounded.

**490.37 Grounding of Devices.** Devices with metal cases or frames, or both, such as instruments, relays, meters, and instrument and control transformers, located in or on switchgear or control, shall have the frame or case grounded.

**490.38 Door Stops and Cover Plates.** External hinged doors or covers shall be provided with stops to hold them in the open position. Cover plates intended to be removed for inspection of energized parts or wiring shall be equipped with lifting handles and shall not exceed 1.1 m² (12 ft²) in

area or 27 kg (60 lb) in weight, unless they are hinged and bolted or locked.

**490.39 Gas Discharge from Interrupting Devices.** Gas discharged during operating of interrupting devices shall be directed so as not to endanger personnel.

**490.40 Inspection Windows.** Windows intended for inspection of disconnecting switches or other devices shall be of suitable transparent material.

**490.41 Location of Devices.**

**(A) Control and Instrument Transfer Switch Handles or Push Buttons.** Control and instrument transfer switch handles or push buttons other than those covered in 490.41(B) shall be in a readily accessible location at an elevation of not over 2.0 m (78 in.).

*Exception: Operating handles requiring more than 23 kg (50 lb) of force shall be located no higher than 1.7 m (66 in.) in either the open or closed position.*

**(B) Infrequently Operated Devices.** Operating handles for infrequently operated devices, such as drawout fuses, fused potential or control transformers and their primary disconnects, and bus transfer switches, shall be permitted to be located where they are safely operable and serviceable from a portable platform.

**490.42 Interlocks — Interrupter Switches.** Interrupter switches equipped with stored energy mechanisms shall have mechanical interlocks to prevent access to the switch compartment unless the stored energy mechanism is in the discharged or blocked position.

**490.43 Stored Energy for Opening.** The stored energy operator shall be permitted to be left in the uncharged position after the switch has been closed if a single movement of the operating handle charges the operator and opens the switch.

**490.44 Fused Interrupter Switches.**

**(A) Supply Terminals.** The supply terminals of fused interrupter switches shall be installed at the top of the switch enclosure or, if the terminals are located elsewhere, the equipment shall have barriers installed so as to prevent persons from accidentally contacting energized parts or dropping tools or fuses into energized parts.

**(B) Backfeed.** Where fuses can be energized by backfeed, a sign shall be placed on the enclosure door identifying this hazard.

**(C) Switching Mechanism.** The switching mechanism shall be arranged to be operated from a location outside the enclosure where the operator is not exposed to energized parts and shall be arranged to open all ungrounded conductors of the circuit simultaneously with one operation. Switches shall be capable of being locked in the open position.

## 490.45 Circuit Breakers — Interlocks.

**(A) Circuit Breakers.** Circuit breakers equipped with stored energy mechanisms shall be designed to prevent the release of the stored energy unless the mechanism has been fully charged.

**(B) Mechanical Interlocks.** Mechanical interlocks shall be provided in the housing to prevent the complete withdrawal of the circuit breaker from the housing when the stored energy mechanism is in the fully charged position, unless a suitable device is provided to block the closing function of the circuit breaker before complete withdrawal.

## IV. Mobile and Portable Equipment

## 490.51 General.

**(A) Covered.** The provisions of this part shall apply to installations and use of high-voltage power distribution and utilization equipment that is portable, mobile, or both, such as substations and switch houses mounted on skids, trailers, or cars; mobile shovels; draglines; cranes; hoists; drills; dredges; compressors; pumps; conveyors; underground excavators; and the like.

**(B) Other Requirements.** The requirements of this part shall be additional to, or amendatory of, those prescribed in Articles 100 through 725 of this *Code*. Special attention shall be paid to Article 250.

**(C) Protection.** Adequate enclosures, guarding, or both, shall be provided to protect portable and mobile equipment from physical damage.

**(D) Disconnecting Means.** Disconnecting means shall be installed for mobile and portable high-voltage equipment according to the requirements of Part VIII of Article 230 and shall disconnect all ungrounded conductors.

**490.52 Overcurrent Protection.** Motors driving single or multiple dc generators supplying a system operating on a cyclic load basis do not require overload protection, provided that the thermal rating of the ac drive motor cannot be exceeded under any operating condition. The branch-circuit protective device(s) shall provide short-circuit and locked-rotor protection and shall be permitted to be external to the equipment.

**490.53 Enclosures.** All energized switching and control parts shall be enclosed in effectively grounded metal cabinets or enclosures. These cabinets or enclosures shall be marked "DANGER — HIGH VOLTAGE — KEEP OUT" and shall be locked so that only authorized and qualified persons can enter. Circuit breakers and protective equipment shall have the operating means projecting through the metal cabinet or enclosure so these units can be reset without opening locked doors. With doors closed, reasonable safe access for normal operation of these units shall be provided.

**490.54 Collector Rings.** The collector ring assemblies on revolving-type machines (shovels, draglines, etc.) shall be guarded to prevent accidental contact with energized parts by personnel on or off the machine.

**490.55 Power Cable Connections to Mobile Machines.** A metallic enclosure shall be provided on the mobile machine for enclosing the terminals of the power cable. The enclosure shall include provisions for a solid connection for the ground wire(s) terminal to effectively ground the machine frame. Ungrounded conductors shall be attached to insulators or be terminated in approved high-voltage cable couplers (which include ground wire connectors) of proper voltage and ampere rating. The method of cable termination used shall prevent any strain or pull on the cable from stressing the electrical connections. The enclosure shall have provision for locking so only authorized and qualified persons may open it and shall be marked

DANGER — HIGH VOLTAGE — KEEP OUT.

**490.56 High-Voltage Portable Cable for Main Power Supply.** Flexible high-voltage cable supplying power to portable or mobile equipment shall comply with Article 250 and Article 400, Part III.

## V. Electrode-Type Boilers

**490.70 General.** The provisions of this part shall apply to boilers operating over 600 volts, nominal, in which heat is generated by the passage of current between electrodes through the liquid being heated.

**490.71 Electric Supply System.** Electrode-type boilers shall be supplied only from a 3-phase, 4-wire solidly grounded wye system, or from isolating transformers arranged to provide such a system. Control circuit voltages shall not exceed 150 volts, shall be supplied from a grounded system, and shall have the controls in the ungrounded conductor.

## 490.72 Branch-Circuit Requirements.

**(A) Rating.** Each boiler shall be supplied from an individual branch circuit rated not less than 100 percent of the total load.

**(B) Common-Trip Fault-Interrupting Device.** The circuit shall be protected by a 3-phase, common-trip fault-interrupting device, which shall be permitted to automatically reclose the circuit upon removal of an overload condition but shall not reclose after a fault condition.

**(C) Phase-Fault Protection.** Phase-fault protection shall be provided in each phase, consisting of a separate phase-overcurrent relay connected to a separate current transformer in the phase.

**(D) Ground Current Detection.** Means shall be provided for detection of the sum of the neutral and ground currents and shall trip the circuit-interrupting device if the sum of those currents exceeds the greater of 5 amperes or 7½ percent of the boiler full-load current for 10 seconds or exceeds an instantaneous value of 25 percent of the boiler full-load current.

**(E) Grounded Neutral Conductor.** The grounded neutral conductor shall be as follows:

(1) Connected to the pressure vessel containing the electrodes

(2) Insulated for not less than 600 volts

(3) Have not less than the ampacity of the largest ungrounded branch-circuit conductor

(4) Installed with the ungrounded conductors in the same raceway, cable, or cable tray, or, where installed as open conductors, in close proximity to the ungrounded conductors

(5) Not used for any other circuit

**490.73 Pressure and Temperature Limit Control.** Each boiler shall be equipped with a means to limit the maximum temperature, pressure, or both, by directly or indirectly interrupting all current flow through the electrodes. Such means shall be in addition to the temperature, pressure, or both, regulating systems and pressure relief or safety valves.

**490.74 Grounding.** All exposed non–current-carrying metal parts of the boiler and associated exposed grounded structures or equipment shall be bonded to the pressure vessel or to the neutral conductor to which the vessel is connected in accordance with 250.102, except the ampacity of the bonding jumper shall not be less than the ampacity of the neutral conductor.

# Chapter 5  Special Occupancies

## ARTICLE 500
## Hazardous (Classified) Locations, Classes I, II, and III, Divisions 1 and 2

FPN:   Rules that are followed by a reference in brackets contain text that has been extracted from NFPA 497, *Recommended Practice for the Classification of Flammable Liquids, Gases, or Vapors and of Hazardous (Classified) Locations for Electrical Installations in Chemical Process Areas*, 1997 edition, and NFPA 499, *Recommended Practice for the Classification of Combustible Dusts and of Hazardous (Classified) Locations for Electrical Installation in Chemical Process Areas*, 1997 edition. Only editorial changes were made to the extracted text to make it consistent with this *Code*.

**500.1 Scope — Articles 500 Through 504.** Articles 500 through 504 cover the requirements for electrical and electronic equipment and wiring for all voltages in Class I, Divisions 1 and 2; Class II, Divisions 1 and 2; and Class III, Divisions 1 and 2 locations where fire or explosion hazards may exist due to flammable gases or vapors, flammable liquids, combustible dust, or ignitible fibers or flyings.

FPN:   For the requirements for electrical and electronic equipment and wiring for all voltages in Class I, Zone 0, Zone 1, and Zone 2 hazardous (classified) locations where fire or explosion hazards may exist due to flammable gases or vapors or flammable liquids, refer to Article 505.

**500.2 Definitions.** For purposes of Articles 500 through 504 and Articles 510 through 516, the following definitions apply.

**Associated Nonincendive Field Wiring Apparatus.** Apparatus in which the circuits are not necessarily nonincendive themselves but that affect the energy in nonincendive field wiring circuits and are relied upon to maintain nonincendive energy levels. Associated nonincendive field wiring apparatus may be either of the following:

(1) Electrical apparatus that has an alternative type of protection for use in the appropriate hazardous (classified) location
(2) Electrical apparatus not so protected that shall not be used in a hazardous (classified) location

FPN:   Associated nonincendive field wiring apparatus has designated associated nonincendive field wiring apparatus connections for nonincendive field wiring apparatus and may also have connections for other electrical apparatus.

**Combustible Gas Detection System.** A protection technique utilizing stationary gas detectors in industrial establishments.

**Control Drawing.** A drawing or other document provided by the manufacturer of the intrinsically safe or associated apparatus, or of the nonincendive field wiring apparatus or associated nonincendive field wiring apparatus, that details the allowed interconnections between the intrinsically safe and associated apparatus or between the nonincendive field wiring apparatus or associated nonincendive field wiring apparatus.

**Dust-Ignitionproof.** Equipment enclosed in a manner that excludes dusts and does not permit arcs, sparks, or heat otherwise generated or liberated inside of the enclosure to cause ignition of exterior accumulations or atmospheric suspensions of a specified dust on or in the vicinity of the enclosure.

FPN:   For further information on dust-ignitionproof enclosures, see Type 9 enclosure in ANSI/NEMA 250-1991, *Enclosures for Electrical Equipment*, and ANSI/UL 1203-1994, *Explosionproof and Dust-Ignitionproof Electrical Equipment for Hazardous (Classified) Locations*.

**Dusttight.** Enclosures constructed so that dust will not enter under specified test conditions.

FPN:   See ANSI/ISA 12.12.01-2000, *Nonincendive Electrical Equipment for Use in Class I and II, Division 2, and Class III, Divisions 1 and 2 Hazardous (Classified) Locations*, and UL 1604-1994, *Electrical Equipment for Use in Class I and II, Division 2 and Class III Hazardous (Classified) Locations*.

**Electrical and Electronic Equipment.** Materials, fittings, devices, appliances, and the like that are part of, or in connection with, an electrical installation.

FPN:   Portable or transportable equipment having self-contained power supplies, such as battery-operated equipment, could potentially become an ignition source in hazardous (classified) locations.

**Explosionproof Apparatus.** Apparatus enclosed in a case that is capable of withstanding an explosion of a specified gas or vapor that may occur within it and of preventing the ignition of a specified gas or vapor surrounding the enclosure by sparks, flashes, or explosion of the gas or vapor within, and that operates at such an external temperature that a surrounding flammable atmosphere will not be ignited thereby.

FPN:   For further information, see ANSI/UL 1203-1994, *Explosion-Proof and Dust-Ignition-Proof Electrical Equipment for Use in Hazardous (Classified) Locations*.

**Hermetically Sealed.** Equipment sealed against the entrance of an external atmosphere where the seal is made by fusion, for example, soldering, brazing, welding, or the fusion of glass to metal.

FPN: For further information, see ANSI/ISA 12.12.01-2000, *Nonincendive Electrical Equipment for Use in Class I and II, Division 2, and Class III, Division 1 and 2 Hazardous (Classified) Locations.*

**Nonincendive Circuit.** A circuit, other than field wiring, in which any arc or thermal effect produced under intended operating conditions of the equipment is not capable, under specified test conditions, of igniting the flammable gas–air, vapor–air, or dust–air mixture.

FPN: Conditions are described in ANSI/ISA 12.12.01-2000, *Nonincendive Electrical Equipment for Use in Class I and II, Division 2, and Class III, Divisions 1 and 2 Hazardous (Classified) Locations.*

**Nonincendive Component.** A component having contacts for making or breaking an incendive circuit and the contacting mechanism is constructed so that the component is incapable of igniting the specified flammable gas–air or vapor–air mixture. The housing of a nonincendive component is not intended to exclude the flammable atmosphere or contain an explosion.

FPN: For further information, see UL 1604-1994, *Electrical Equipment for Use in Class I and II, Division 2, and Class III Hazardous (Classified) Locations.*

**Nonincendive Equipment.** Equipment having electrical/electronic circuitry that is incapable, under normal operating conditions, of causing ignition of a specified flammable gas–air, vapor–air, or dust–air mixture due to arcing or thermal means.

FPN: For further information, see ANSI/ISA 12.12.01-2000, *Nonincendive Electrical Equipment for Use in Class I and II, Division 2, and Class III, Divisions 1 and 2 Hazardous (Classified) Locations.*

**Nonincendive Field Wiring.** Wiring that enters or leaves an equipment enclosure and, under normal operating conditions of the equipment, is not capable, due to arcing or thermal effects, of igniting the flammable gas–air, vapor–air, or dust–air mixture. Normal operation includes opening, shorting, or grounding the field wiring.

**Nonincendive Field Wiring Apparatus.** Apparatus intended to be connected to nonincendive field wiring.

FPN: For further information see ANSI/ISA 12.12.01-2000, *Nonincendive Electrical Equipment for Use in Class I and II, Division 2, and Class III, Divisions 1 and 2 Hazardous (Classified) Locations.*

**Oil Immersion.** Electrical equipment immersed in a protective liquid in such a way that an explosive atmosphere that may be above the liquid or outside the enclosure cannot be ignited.

FPN: For further information, see ANSI/UL 698-1995, *Industrial Control Equipment for Use in Hazardous (Classified) Locations.*

**Purged and Pressurized.** The process of supplying an enclosure with a protective gas at a sufficient flow and positive pressure to reduce the concentration of any flammable gas or vapor initially present to an acceptable level.

FPN: For further information, see ANSI/ NFPA 496-1998, *Purged and Pressurized Enclosures for Electrical Equipment.*

**Unclassified Locations.** Locations determined to be neither Class I, Division 1; Class I, Division 2; Class I, Zone 0; Class I, Zone 1; Class I, Zone 2; Class II, Division 1; Class II, Division 2; Class III, Division 1; Class III, Division 2; or any combination thereof.

**500.3 Other Articles.** Except as modified in Articles 500 through 504, all other applicable rules contained in this *Code* shall apply to electrical equipment and wiring installed in hazardous (classified) locations.

**500.4 General.**

**(A) Documentation.** All areas designated as hazardous (classified) locations shall be properly documented. This documentation shall be available to those authorized to design, install, inspect, maintain, or operate electrical equipment at the location.

**(B) Reference Standards.** Important information relating to topics covered in Chapter 5 may be found in other publications.

FPN No. 1: It is important that the authority having jurisdiction be familiar with recorded industrial experience as well as with the standards of the National Fire Protection Association (NFPA), the American Petroleum Institute (API), and the Instrumentation, Systems, and Automation Society (ISA) that may be of use in the classification of various locations, the determination of adequate ventilation, and the protection against static electricity and lightning hazards.

FPN No. 2: For further information on the classification of locations, see NFPA 30-2000, *Flammable and Combustible Liquids Code*; NFPA 32-2000, *Standard for Drycleaning Plants*; NFPA 33-2000, *Standard for Spray Application Using Flammable or Combustible Materials*; NFPA 34-2000, *Standard for Dipping and Coating Processes Using Flammable or Combustible Liquids*; NFPA 35-1999, *Standard for the Manufacture of Organic Coatings*; NFPA 36-2001, *Standard for Solvent Extraction Plants*; NFPA 45-2000, *Standard on Fire Protection for Laboratories Using Chemicals*; NFPA 50A-1999, *Standard for Gaseous Hydrogen Systems at Consumer Sites*; NFPA 50B-1999, *Standard for Liquefied Hydrogen Systems at Consumer Sites*; NFPA 58-2001, *Liquefied Petroleum Gas Code*; NFPA 59-2001, *Utility LP-Gas Plant Code*; NFPA 497-1997, *Recommended Practice for the Classification of Flammable Liquids, Gases, or Vapors and of Hazardous (Classified) Locations for Electrical Installations in Chemical Process Areas*; NFPA 499-1997, *Recommended Practice for the Classification of Combustible Dusts and of Hazardous (Classified)*

*Locations for Electrical Installations in Chemical Process Areas*; NFPA 820-1999, *Standard for Fire Protection in Wastewater Treatment and Collection Facilities*; ANSI/API RP500-1997, *Recommended Practice for Classification of Locations of Electrical Installations at Petroleum Facilities Classified as Class I, Division 1 and Division 2*; ISA 12.10-1988, *Area Classification In Hazardous (Classified) Dust Locations.*

FPN No. 3:  For further information on protection against static electricity and lightning hazards in hazardous (classified) locations, see NFPA 77-2000, *Recommended Practice on Static Electricity*; NFPA 780-1997, *Standard for the Installation of Lightning Protection Systems*; and API RP 2003-1998, *Protection Against Ignitions Arising Out of Static Lightning and Stray Currents.*

FPN No. 4:  For further information on ventilation, see NFPA 30-2000, *Flammable and Combustible Liquids Code*; and API RP 500-1997, *Recommended Practice for Classification of Locations for Electrical Installations at Petroleum Facilities Classified as Class I, Division 1 and Division 2.*

FPN No. 5:  For further information on electrical systems for hazardous (classified) locations on offshore oil- and gas-producing platforms, see ANSI/API RP 14F-1999, *Recommended Practice for Design and Installation of Electrical Systems for Fixed and Floating Offshore Petroleum Facilities for Unclassified and Class I, Division 1 and Division 2 Locations.*

## 500.5 Classifications of Locations.

**(A)  Classifications of Locations.** Locations shall be classified depending on the properties of the flammable vapors, liquids, or gases, or combustible dusts or fibers that may be present, and the likelihood that a flammable or combustible concentration or quantity is present. Where pyrophoric materials are the only materials used or handled, these locations shall not be classified. Each room, section, or area shall be considered individually in determining its classification.

FPN:  Through the exercise of ingenuity in the layout of electrical installations for hazardous (classified) locations, it is frequently possible to locate much of the equipment in a reduced level of classification or in an unclassified location and, thus, to reduce the amount of special equipment required.

Rooms and areas containing ammonia refrigeration systems that are equipped with adequate mechanical ventilation may be classified as "unclassified" locations.

FPN:  For further information regarding classification and ventilation of areas involving ammonia, see ANSI/ASHRAE 15-1994, *Safety Code for Mechanical Refrigeration*, and ANSI/CGA G2.1-1989, *Safety Requirements for the Storage and Handling of Anhydrous Ammonia.*

**(B)  Class I Locations.** Class I locations are those in which flammable gases or vapors are or may be present in the air in quantities sufficient to produce explosive or ignitible mixtures. Class I locations shall include those specified in 500.5(B)(1) and (B)(2).

**(1)  Class I, Division 1.** A Class I, Division 1 location is a location

(1) In which ignitible concentrations of flammable gases or vapors can exist under normal operating conditions, or

(2) In which ignitible concentrations of such gases or vapors may exist frequently because of repair or maintenance operations or because of leakage, or

(3) In which breakdown or faulty operation of equipment or processes might release ignitible concentrations of flammable gases or vapors and might also cause simultaneous failure of electrical equipment in such a way as to directly cause the electrical equipment to become a source of ignition.

FPN No. 1:  This classification usually includes the following locations:

(1) Where volatile flammable liquids or liquefied flammable gases are transferred from one container to another

(2) Interiors of spray booths and areas in the vicinity of spraying and painting operations where volatile flammable solvents are used

(3) Locations containing open tanks or vats of volatile flammable liquids

(4) Drying rooms or compartments for the evaporation of flammable solvents

(5) Locations containing fat- and oil-extraction equipment using volatile flammable solvents

(6) Portions of cleaning and dyeing plants where flammable liquids are used

(7) Gas generator rooms and other portions of gas manufacturing plants where flammable gas may escape

(8) Inadequately ventilated pump rooms for flammable gas or for volatile flammable liquids

(9) The interiors of refrigerators and freezers in which volatile flammable materials are stored in open, lightly stoppered, or easily ruptured containers

(10) All other locations where ignitible concentrations of flammable vapors or gases are likely to occur in the course of normal operations

FPN No. 2:  In some Division 1 locations, ignitible concentrations of flammable gases or vapors may be present continuously or for long periods of time. Examples include the following:

(1) The inside of inadequately vented enclosures containing instruments normally venting flammable gases or vapors to the interior of the enclosure

(2) The inside of vented tanks containing volatile flammable liquids

(3) The area between the inner and outer roof sections of a floating roof tank containing volatile flammable fluids

(4) Inadequately ventilated areas within spraying or coating operations using volatile flammable fluids

(5) The interior of an exhaust duct that is used to vent ignitible concentrations of gases or vapors

Experience has demonstrated the prudence of avoiding the installation of instrumentation or other electric equipment in these particular areas altogether or where it cannot be avoided because it is essential to the process and other locations are not feasible [see 500.5(A), FPN] using electric equipment or instrumentation approved for the specific application or consisting of intrinsically safe systems as described in Article 504.

**(2) Class I, Division 2.** A Class I, Division 2 location is a location

(1) In which volatile flammable liquids or flammable gases are handled, processed, or used, but in which the liquids, vapors, or gases will normally be confined within closed containers or closed systems from which they can escape only in case of accidental rupture or breakdown of such containers or systems or in case of abnormal operation of equipment, or

(2) In which ignitible concentrations of gases or vapors are normally prevented by positive mechanical ventilation, and which might become hazardous through failure or abnormal operation of the ventilating equipment, or

(3) That is adjacent to a Class I, Division 1 location, and to which ignitible concentrations of gases or vapors might occasionally be communicated unless such communication is prevented by adequate positive-pressure ventilation from a source of clean air and effective safeguards against ventilation failure are provided.

FPN No. 1:  This classification usually includes locations where volatile flammable liquids or flammable gases or vapors are used but that, in the judgment of the authority having jurisdiction, would become hazardous only in case of an accident or of some unusual operating condition. The quantity of flammable material that might escape in case of accident, the adequacy of ventilating equipment, the total area involved, and the record of the industry or business with respect to explosions or fires are all factors that merit consideration in determining the classification and extent of each location.

FPN No. 2:  Piping without valves, checks, meters, and similar devices would not ordinarily introduce a hazardous condition even though used for flammable liquids or gases. Depending on factors such as the quantity and size of the containers and ventilation, locations used for the storage of flammable liquids or liquefied or compressed gases in sealed containers may be considered either hazardous (classified) or unclassified locations. See NFPA 30-2000, *Flammable and Combustible Liquids Code*, and NFPA 58-2001, *Liquefied Petroleum Gas Code*.

**(C) Class II Locations.** Class II locations are those that are hazardous because of the presence of combustible dust. Class II locations shall include those specified in 500.5(C)(1) and (C)(2).

**(1) Class II, Division 1.** A Class II, Division 1 location is a location

(1) In which combustible dust is in the air under normal operating conditions in quantities sufficient to produce explosive or ignitible mixtures, or

(2) Where mechanical failure or abnormal operation of machinery or equipment might cause such explosive or ignitible mixtures to be produced, and might also provide a source of ignition through simultaneous failure of electric equipment, through operation of protection devices, or from other causes, or

(3) In which combustible dusts of an electrically conductive nature may be present in hazardous quantities.

FPN:  Combustible dusts that are electrically nonconductive include dusts produced in the handling and processing of grain and grain products, pulverized sugar and cocoa, dried egg and milk powders, pulverized spices, starch and pastes, potato and wood-flour, oil meal from beans and seed, dried hay, and other organic materials that may produce combustible dusts when processed or handled. Only Group E dusts are considered to be electrically conductive for classification purposes. Dusts containing magnesium or aluminum are particularly hazardous, and the use of extreme precaution is necessary to avoid ignition and explosion.

**(2) Class II, Division 2.** A Class II, Division 2 location is a location

(1) Where combustible dust is not normally in the air in quantities sufficient to produce explosive or ignitible mixtures, and dust accumulations are normally insufficient to interfere with the normal operation of electrical equipment or other apparatus, but combustible dust may be in suspension in the air as a result of infrequent malfunctioning of handling or processing equipment and

(2) Where combustible dust accumulations on, in, or in the vicinity of the electrical equipment may be sufficient to interfere with the safe dissipation of heat from electrical equipment or may be ignitible by abnormal operation or failure of electrical equipment.

FPN No. 1:  The quantity of combustible dust that may be present and the adequacy of dust removal systems are factors that merit consideration in determining the classification and may result in an unclassified area.

FPN No. 2:  Where products such as seed are handled in a manner that produces low quantities of dust, the amount of dust deposited may not warrant classification.

**(D) Class III Locations.** Class III locations are those that are hazardous because of the presence of easily ignitible fibers or flyings, but in which such fibers or flyings are not likely to be in suspension in the air in quantities sufficient to produce ignitible mixtures. Class III locations shall include those specified in 500.5(D)(1) and (D)(2).

**(1) Class III, Division 1.** A Class III, Division 1 location is a location in which easily ignitible fibers or materials producing combustible flyings are handled, manufactured, or used.

FPN No. 1: Such locations usually include some parts of rayon, cotton, and other textile mills; combustible fiber manufacturing and processing plants; cotton gins and cotton-seed mills; flax-processing plants; clothing manufacturing plants; woodworking plants; and establishments and industries involving similar hazardous processes or conditions.

FPN No. 2: Easily ignitible fibers and flyings include rayon, cotton (including cotton linters and cotton waste), sisal or henequen, istle, jute, hemp, tow, cocoa fiber, oakum, baled waste kapok, Spanish moss, excelsior, and other materials of similar nature.

**(2) Class III, Division 2.** A Class III, Division 2 location is a location in which easily ignitible fibers are stored or handled other than in the process of manufacture.

**500.6 Material Groups.** For purposes of testing, approval, and area classification, various air mixtures (not oxygen-enriched) shall be grouped in accordance with 500.6(A) and 500.6(B).

*Exception: Equipment identified for a specific gas, vapor, or dust.*

FPN: This grouping is based on the characteristics of the materials. Facilities are available for testing and identifying equipment for use in the various atmospheric groups.

**(A) Class I Group Classifications.** Class I groups shall be according to 500.6(A)(1) through (A)(4).

FPN No. 1: FPN Nos. 2 and 3 apply to 500.6(A).

FPN No. 2: The explosion characteristics of air mixtures of gases or vapors vary with the specific material involved. For Class I locations, Groups A, B, C, and D, the classification involves determinations of maximum explosion pressure and maximum safe clearance between parts of a clamped joint in an enclosure. It is necessary, therefore, that equipment be identified not only for class but also for the specific group of the gas or vapor that will be present.

FPN No. 3: Certain chemical atmospheres may have characteristics that require safeguards beyond those required for any of the Class I groups. Carbon disulfide is one of these chemicals because of its low ignition temperature [100°C (212°F)] and the small joint clearance permitted to arrest its flame.

**(1) Group A.** Acetylene. [NFPA 497, 1-3]

**(2) Group B.** Flammable gas, flammable liquid–produced vapor, or combustible liquid–produced vapor mixed with air that may burn or explode, having either a maximum experimental safe gap (MESG) value less than or equal to 0.45 mm or a minimum igniting current ratio (MIC ratio) less than or equal to 0.40. [NFPA 497, 1-3]

FPN: A typical Class I, Group B material is hydrogen.

*Exception No. 1: Group D equipment shall be permitted to be used for atmospheres containing butadiene, provided all conduit runs into explosionproof equipment are provided with explosionproof seals installed within 450 mm (18 in.) of the enclosure.*

*Exception No. 2: Group C equipment shall be permitted to be used for atmospheres containing allyl glycidyl ether, n-butyl glycidyl ether, ethylene oxide, propylene oxide, and acrolein, provided all conduit runs into explosionproof equipment are provided with explosionproof seals installed within 450 mm (18 in.) of the enclosure.*

**(3) Group C.** Flammable gas, flammable liquid–produced vapor, or combustible liquid–produced vapor mixed with air that may burn or explode, having either a maximum experimental safe gap (MESG) value greater than 0.45 mm and less than or equal to 0.75 mm, or a minimum igniting current ratio (MIC ratio) greater than 0.40 and less than or equal to 0.80. [NFPA 497, 1-3]

FPN: A typical Class I, Group C material is ethylene.

**(4) Group D.** Flammable gas, flammable liquid–produced vapor, or combustible liquid–produced vapor mixed with air that may burn or explode, having either a maximum experimental safe gap (MESG) value greater than 0.75 mm or a minimum igniting current ratio (MIC ratio) greater than 0.80. [NFPA 497, 1-3]

FPN No. 1: A typical Class I, Group D material is propane.

FPN No. 2: For classification of areas involving ammonia atmospheres, see ANSI/ASHRAE 15-1994, *Safety Code for Mechanical Refrigeration*, and ANSI/CGA G2.1-1989, *Safety Requirements for the Storage and Handling of Anhydrous Ammonia*.

**(B) Class II Group Classifications.** Class II groups shall be according to 500.6(B)(1) through (B)(3).

**(1) Group E.** Atmospheres containing combustible metal dusts, including aluminum, magnesium, and their commercial alloys, or other combustible dusts whose particle size, abrasiveness, and conductivity present similar hazards in the use of electrical equipment. [NFPA 499, 1-3]

FPN: Certain metal dusts may have characteristics that require safeguards beyond those required for atmospheres containing the dusts of aluminum, magnesium, and their commercial alloys. For example, zirconium, thorium, and uranium dusts have extremely low ignition temperatures [as low as 20°C (68°F)] and minimum ignition energies lower than any material classified in any of the Class I or Class II groups.

**(2) Group F.** Atmospheres containing combustible carbonaceous dusts that have more than 8 percent total entrapped volatiles (see ASTM D 3175-89, *Standard Test Method for Volatile Material in the Analysis Sample for Coal and Coke*, for coal and coke dusts) or that have been sensitized by other materials so that they present an explosion hazard. Coal, carbon black, charcoal, and coke dusts are examples of carbonaceous dusts. [NFPA 499, 1-3]

**(3) Group G.** Atmospheres containing combustible dusts not included in Group E or F, including flour, grain, wood, plastic, and chemicals.

FPN No. 1: For additional information on group classification of Class II materials, see NFPA 499-1997, *Recommended Practice for the Classification of Combustible Dusts and of Hazardous (Classified) Locations for Electrical Installations in Chemical Process Areas.*

FPN No. 2: The explosion characteristics of air mixtures of dust vary with the materials involved. For Class II locations, Groups E, F, and G, the classification involves the tightness of the joints of assembly and shaft openings to prevent the entrance of dust in the dust-ignitionproof enclosure, the blanketing effect of layers of dust on the equipment that may cause overheating, and the ignition temperature of the dust. It is necessary, therefore, that equipment be identified not only for the class, but also for the specific group of dust that will be present.

FPN No. 3: Certain dusts may require additional precautions due to chemical phenomena that can result in the generation of ignitible gases. See ANSI C2-1997, *National Electrical Safety Code*, Section 127A, Coal Handling Areas.

**500.7 Protection Techniques.** Section 500.7(A) through (L) shall be acceptable protection techniques for electrical and electronic equipment in hazardous (classified) locations.

**(A) Explosionproof Apparatus.** This protection technique shall be permitted for equipment in Class I, Division 1 or 2 locations.

**(B) Dust Ignitionproof.** This protection technique shall be permitted for equipment in Class II, Division 1 or 2 locations.

**(C) Dusttight.** This protection technique shall be permitted for equipment in Class II, Division 2 or Class III, Division 1 or 2 locations.

**(D) Purged and Pressurized.** This protection technique shall be permitted for equipment in any hazardous (classified) location for which it is identified.

**(E) Intrinsic Safety.** This protection technique shall be permitted for equipment in Class I, Division 1 or 2; or Class II, Division 1 or 2; or Class III, Division 1 or 2 locations. The provisions of Articles 501 through 503 and Articles 510 through 516 shall not be considered applicable to such installations, except as required by Article 504, and installation of intrinsically safe apparatus and wiring shall be in accordance with the requirements of Article 504.

**(F) Nonincendive Circuit.** This protection technique shall be permitted for equipment in Class I, Division 2; Class II, Division 2; or Class III, Division 1 or 2 locations.

**(G) Nonincendive Equipment.** This protection technique shall be permitted for equipment in Class I, Division 2; Class II, Division 2; or Class III, Division 1 or 2 locations.

**(H) Nonincendive Component.** This protection technique shall be permitted for equipment in Class I, Division 2; Class II, Division 2; or Class III, Division 1 or 2 locations.

**(I) Oil Immersion.** This protection technique shall be permitted for current-interrupting contacts in Class I, Division 2 locations as described in 501.6(B)(1)(2).

**(J) Hermetically Sealed.** This protection technique shall be permitted for equipment in Class I, Division 2; Class II, Division 2; or Class III, Division 1 or 2 locations.

**(K) Combustible Gas Detection System.** A combustible gas detection system shall be permitted as a means of protection in industrial establishments with restricted public access and where the conditions of maintenance and supervision ensure that only qualified persons service the installation. Gas detection equipment shall be listed for detection of the specific gas or vapor to be encountered. Where such a system is installed, equipment specified in 500.7(K)(1), (2), or (3) shall be permitted.

**(1) Inadequate Ventilation.** In a Class I, Division 1 location that is so classified due to inadequate ventilation, electrical equipment suitable for Class I, Division 2 locations shall be permitted.

**(2) Interior of a Building.** In a building located in, or with an opening into, a Class I, Division 2 location where the interior does not contain a source of flammable gas or vapor, electrical equipment for unclassified locations shall be permitted.

**(3) Interior of a Control Panel.** In the interior of a control panel containing instrumentation utilizing or measuring flammable liquids, gases, or vapors, electrical equipment suitable for Class I, Division 2 locations shall be permitted.

FPN No. 1: For further information, see ANSI/ISA-12.13.01, *Performance Requirements, Combustible Gas Detectors.*

FPN No. 2: For further information., see ANSI/API RP 500, *Recommended Practice for Classification of Locations for Electrical Installations at Petroleum Facilities Classified as Class I, Division I or Division 2.*

FPN No. 3: For further information, see ISA-RP12.13.02, *Installation, Operation, and Maintenance of Combustible Gas Detection Instruments.*

**(L) Other Protection Techniques.** Other protection techniques used in equipment identified for use in hazardous (classified) locations.

**500.8 Equipment.** Articles 500 through 504 require equipment construction and installation that ensure safe performance under conditions of proper use and maintenance.

FPN No. 1: It is important that inspection authorities and users exercise more than ordinary care with regard to installation and maintenance.

FPN No. 2: Low ambient conditions require special consideration. Explosionproof or dust-ignitionproof equipment may not be suitable for use at temperatures lower than −25°C (−13°F) unless they are identified for low-temperature service. However, at low ambient temperatures, flammable concentrations of vapors may not exist in a location classified Class I, Division 1 at normal ambient temperature.

**(A) Approval for Class and Properties.**

**(1)** Equipment shall be identified not only for the class of location but also for the explosive, combustible, or ignitible properties of the specific gas, vapor, dust, fiber, or flyings that will be present. In addition, Class I equipment shall not have any exposed surface that operates at a temperature in excess of the ignition temperature of the specific gas or vapor. Class II equipment shall not have an external temperature higher than that specified in 500.8(C)(2). Class III equipment shall not exceed the maximum surface temperatures specified in 503.1.

FPN: Luminaires (lighting fixtures) and other heat-producing apparatus, switches, circuit breakers, and plugs and receptacles are potential sources of ignition and are investigated for suitability in classified locations. Such types of equipment, as well as cable terminations for entry into explosionproof enclosures, are available as listed for Class I, Division 2 locations. Fixed wiring, however, may utilize wiring methods that are not evaluated with respect to classified locations. Wiring products such as cable, raceways, boxes, and fittings, therefore, are not marked as being suitable for Class I, Division 2 locations. Also see Exception No. 3 to 500.8(B).

Suitability of identified equipment shall be determined by any of the following:

(1) Equipment listing or labeling
(2) Evidence of equipment evaluation from a qualified testing laboratory or inspection agency concerned with product evaluation
(3) Evidence acceptable to the authority having jurisdiction such as a manufacturer's self-evaluation or an owner's engineering judgment.

**(2)** Equipment that has been identified for a Division 1 location shall be permitted in a Division 2 location of the same class and group.

**(3)** Where specifically permitted in Articles 501 through 503, general-purpose equipment or equipment in general-purpose enclosures shall be permitted to be installed in Division 2 locations if the equipment does not constitute a source of ignition under normal operating conditions.

**(4)** Equipment, regardless of the classification of the location in which it is installed, that depends on a single compression seal, diaphragm, or tube to prevent flammable or combustible fluids from entering the equipment shall be identified for a Class I, Division 2 location. Equipment installed in a Class I, Division 1 location shall be identified for the Class I, Division 1 location.

FPN: See 501.5(F)(3) for additional requirements.

**(5)** Unless otherwise specified, normal operating conditions for motors shall be assumed to be rated full-load steady conditions.

**(6)** Where flammable gases or combustible dusts are or may be present at the same time, the simultaneous presence of both shall be considered when determining the safe operating temperature of the electrical equipment.

FPN: The characteristics of various atmospheric mixtures of gases, vapors, and dusts depend on the specific material involved.

**(B) Marking.** Equipment shall be marked to show the class, group, and operating temperature or temperature class referenced to a 40°C ambient.

*Exception No. 1: Equipment of the non–heat-producing type, such as junction boxes, conduit, and fittings, and equipment of the heat-producing type having a maximum temperature not more than 100°C (212°F) shall not be required to have a marked operating temperature or temperature class.*

*Exception No. 2: Fixed luminaires (lighting fixtures) marked for use in Class I, Division 2 or Class II, Division 2 locations only shall not be required to be marked to indicate the group.*

*Exception No. 3: Fixed general-purpose equipment in Class I locations, other than fixed luminaires (lighting fixtures), that is acceptable for use in Class I, Division 2 locations shall not be required to be marked with the class, group, division, or operating temperature.*

*Exception No. 4: Fixed dusttight equipment other than fixed luminaires (lighting fixtures) that is acceptable for use in Class II, Division 2 and Class III locations shall not be required to be marked with the class, group, division, or operating temperature.*

*Exception No. 5:  Electric equipment suitable for ambient temperatures exceeding 40°C (104°F) shall be marked with both the maximum ambient temperature and the operating temperature or temperature class at that ambient temperature.*

FPN:  Equipment not marked to indicate a division, or marked "Division 1" or "Div. 1," is suitable for both Division 1 and 2 locations. Equipment marked "Division 2" or "Div. 2" is suitable for Division 2 locations only.

The temperature class, if provided, shall be indicated using the temperature class (T Codes) shown in Table 500.8(B). The temperature class (T Code) marked on equipment nameplates shall be in accordance with Table 500.8(B). Equipment for Class I and Class II shall be marked with the maximum safe operating temperature, as determined by simultaneous exposure to the combinations of Class I and Class II conditions.

FPN:  Since there is no consistent relationship between explosion properties and ignition temperature, the two are independent requirements.

**Table 500.8(B)  Classification of Maximum Surface Temperature**

| Maximum Temperature | | Temperature Class (T Code) |
|---|---|---|
| C° | F° | |
| 450 | 842 | T1 |
| 300 | 572 | T2 |
| 280 | 536 | T2A |
| 260 | 500 | T2B |
| 230 | 446 | T2C |
| 215 | 419 | T2D |
| 200 | 392 | T3 |
| 180 | 356 | T3A |
| 165 | 329 | T3B |
| 160 | 320 | T3C |
| 135 | 275 | T4 |
| 120 | 248 | T4A |
| 100 | 212 | T5 |
| 85 | 185 | T6 |

**(C)  Temperature.**

**(1)  Class I Temperature.** The temperature marking specified in 500.8(B) shall not exceed the ignition temperature of the specific gas or vapor to be encountered.

FPN:  For information regarding ignition temperatures of gases and vapors, see NFPA 497-1997, *Recommended Practice for the Classification of Flammable Liquids, Gases, or Vapors, and of Hazardous (Classified) Locations for Electrical Installations in Chemical Process Areas.*

**(2)  Class II Temperature.** The temperature marking specified in 500.8(B) shall be less than the ignition tem-perature of the specific dust to be encountered. For organic dusts that may dehydrate or carbonize, the temperature marking shall not exceed the lower of either the ignition temperature or 165°C (329°F).

FPN:  See NFPA 499-1997, *Recommended Practice for the Classification of Combustible Dusts and of Hazardous (Classified) Locations for Electrical Installations in Chemical Process Areas*, for minimum ignition temperatures of specific dusts.

The ignition temperature for which equipment was approved prior to this requirement shall be assumed to be as shown in Table 500.8(C)(2).

**Table 500.8(C)(2)  Class II Temperatures**

| Class II Group | Equipment Not Subject to Overloading | | Equipment (Such as Motors or Power Transformers) That May Be Overloaded | | | |
|---|---|---|---|---|---|---|
| | | | Normal Operation | | Abnormal Operation | |
| | °C | °F | °C | °F | °C | °F |
| E | 200 | 392 | 200 | 392 | 200 | 392 |
| F | 200 | 392 | 150 | 302 | 200 | 392 |
| G | 165 | 329 | 120 | 248 | 165 | 329 |

**(D)  Threading.** All threaded conduit or fittings referred to herein shall be threaded with a National (American) Standard Pipe Taper (NPT) standard conduit cutting die that provides a taper of 1 in 16 (¾-in. taper per foot). Such conduit shall be made wrenchtight to prevent sparking when fault current flows through the conduit system and to ensure the explosionproof or flameproof integrity of the conduit system where applicable. Equipment provided with threaded entries for field wiring connections shall be installed in accordance with 500.8(D)(1) or (D)(2).

**(1)  Equipment Provided with Threaded Entries for NPT Threaded Conduit or Fittings.** For equipment provided with threaded entries for NPT threaded conduit or fittings, listed conduit, conduit fittings, or cable fittings shall be used.

FPN:  Thread form specifications for NPT threads are located in ANSI/ASME B1.20.1-1983, *Pipe Threads, General Purpose (Inch).*

**(2)  Equipment Provided with Threaded Entries for Metric Threaded Conduit or Fittings.** For equipment with metric threaded entries, such entries shall be identified as being metric, or listed adapters to permit connection to conduit or NPT-threaded fittings shall be provided with the equipment. Adapters shall be used for connection to conduit or NPT-threaded fittings. Listed cable fittings that have metric threads shall be permitted to be used.

FPN: Threading specifications for metric threaded entries are located in ISO 965/1-1980, *Metric Screw Threads*, and ISO 965/3-1980, *Metric Screw Threads*.

**(E) Fiber Optic Cable Assembly.** Where a fiber optic cable assembly contains conductors that are capable of carrying current, the fiber optic cable assembly shall be installed in accordance with the requirements of Articles 500, 501, 502, or 503, as applicable.

**500.9 Specific Occupancies.** Articles 510 through 517 cover garages, aircraft hangars, motor fuel dispensing facilities, bulk storage plants, spray application, dipping and coating processes, and health care facilities.

# ARTICLE 501
# Class I Locations

**501.1 General.** The general rules of this *Code* shall apply to the electric wiring and equipment in locations classified as Class I in 500.5. Equipment listed and marked in accordance with 505.9(C)(2) for use in Class I, Zone 0, 1, or 2 locations shall be permitted in Class I, Division 2 locations for the same gas and with a suitable temperature class.

*Exception: As modified by this article.*

**501.2 Transformers and Capacitors.**

**(A) Class I, Division 1.** In Class I, Division 1 locations, transformers and capacitors shall comply with 501.2(A)(1) and (A)(2).

**(1) Containing Liquid That Will Burn.** Transformers and capacitors containing a liquid that will burn shall be installed only in vaults that comply with 450.41 through 450.48 and, in addition, with (a) through (d).

(a) There shall be no door or other communicating opening between the vault and the Division 1 location.

(b) Ample ventilation shall be provided for the continuous removal of flammable gases or vapors.

(c) Vent openings or ducts shall lead to a safe location outside of buildings.

(d) Vent ducts and openings shall be of sufficient area to relieve explosion pressures within the vault, and all portions of vent ducts within the buildings shall be of reinforced concrete construction.

**(2) Not Containing Liquid That Will Burn.** Transformers and capacitors that do not contain a liquid that will burn

shall be installed in vaults complying with 501.2(A)(1) or be approved for Class I locations.

**(B) Class I, Division 2.** In Class I, Division 2 locations, transformers and capacitors shall comply with 450.21 through 450.27.

**501.3 Meters, Instruments, and Relays.**

**(A) Class I, Division 1.** In Class I, Division 1 locations, meters, instruments, and relays, including kilowatt-hour meters, instrument transformers, resistors, rectifiers, and thermionic tubes, shall be provided with enclosures identified for Class I, Division 1 locations. Enclosures for Class I, Division 1 locations include explosionproof enclosures and purged and pressurized enclosures.

FPN: See NFPA 496-1998, *Standard for Purged and Pressurized Enclosures for Electrical Equipment.*

**(B) Class I, Division 2.** In Class I, Division 2 locations, meters, instruments, and relays shall comply with 501.3(B)(1) through (B)(6).

**(1) Contacts.** Switches, circuit breakers, and make-and-break contacts of pushbuttons, relays, alarm bells, and horns shall have enclosures identified for Class I, Division 1 locations in accordance with 501.3(A).

*Exception: General-purpose enclosures shall be permitted if current-interrupting contacts are*

(a) *Immersed in oil, or*
(b) *Enclosed within a chamber that is hermetically sealed against the entrance of gases or vapors, or*
(c) *In nonincendive circuits, or*
(d) *Part of a listed nonincendive component.*

**(2) Resistors and Similar Equipment.** Resistors, resistance devices, thermionic tubes, rectifiers, and similar equipment that are used in or in connection with meters, instruments, and relays shall comply with 501.3(A).

*Exception: General-purpose-type enclosures shall be permitted if such equipment is without make-and-break or sliding contacts [other than as provided in 501.3(B)(1)] and if the maximum operating temperature of any exposed surface will not exceed 80 percent of the ignition temperature in degrees Celsius of the gas or vapor involved or has been tested and found incapable of igniting the gas or vapor. This exception shall not apply to thermionic tubes.*

**(3) Without Make-or-Break Contacts.** Transformer windings, impedance coils, solenoids, and other windings that do not incorporate sliding or make-or-break contacts shall be provided with enclosures. General-purpose-type enclosures shall be permitted.

**(4) General-Purpose Assemblies.** Where an assembly is made up of components for which general-purpose enclosures are acceptable as provided in 501.3(B)(1), (B)(2), and (B)(3), a single general-purpose enclosure shall be acceptable for the assembly. Where such an assembly includes any of the equipment described in 501.3(B)(2), the maximum obtainable surface temperature of any component of the assembly shall be clearly and permanently indicated on the outside of the enclosure. Alternatively, equipment shall be permitted to be marked to indicate the temperature class for which it is suitable, using the temperature class (T Code) of Table 500.8(B).

**(5) Fuses.** Where general-purpose enclosures are permitted in 501.3(B)(1), (B)(2), (B)(3), and (B)(4), fuses for overcurrent protection of instrument circuits not subject to overloading in normal use shall be permitted to be mounted in general-purpose enclosures if each such fuse is preceded by a switch complying with 501.3(B)(1).

**(6) Connections.** To facilitate replacements, process control instruments shall be permitted to be connected through flexible cord, attachment plug, and receptacle, provided all of the following conditions apply:

(1) A switch complying with 501.3(B)(1) is provided so that the attachment plug is not depended on to interrupt current.
(2) The current does not exceed 3 amperes at 120 volts, nominal.
(3) The power-supply cord does not exceed 900 mm (3 ft), is of a type listed for extra-hard usage or for hard usage if protected by location, and is supplied through an attachment plug and receptacle of the locking and grounding type.
(4) Only necessary receptacles are provided.
(5) The receptacle carries a label warning against unplugging under load.

**501.4 Wiring Methods.** Wiring methods shall comply with 501.4(A) or (B).

**(A) Class I, Division 1.**

**(1) General.** In Class I, Division 1 locations, the wiring methods in (a) through (d) shall be permitted.

(a)   Threaded rigid metal conduit or threaded steel intermediate metal conduit. Threaded joints shall be made up with at least five threads fully engaged.

*Exception: Rigid nonmetallic conduit complying with Article 352 shall be permitted where encased in a concrete envelope a minimum of 50 mm (2 in.) thick and provided with not less than 600 mm (24 in.) of cover measured from the top of the conduit to grade. The concrete encasement shall be permitted to be omitted where subject to the pro-*

*visions of 511.4, Exception; 514.8, Exception No. 2; and 515.8(A). Threaded rigid metal conduit or threaded steel intermediate metal conduit shall be used for the last 600 mm (24 in.) of the underground run to emergence or to the point of connection to the aboveground raceway. An equipment grounding conductor shall be included to provide for electrical continuity of the raceway system and for grounding of non–current-carrying metal parts.*

(b)   Type MI cable with termination fittings listed for the location. Type MI cable shall be installed and supported in a manner to avoid tensile stress at the termination fittings.

(c)   In industrial establishments with restricted public access, where the conditions of maintenance and supervision ensure that only qualified persons service the installation, Type MC-HL cable, listed for use in Class I, Division 1 locations, with a gas/vaportight continuous corrugated metallic sheath, an overall jacket of suitable polymeric material, separate grounding conductors in accordance with 250.122, and provided with termination fittings listed for the application.

> FPN:   See 330.10 and 330.12 for restrictions on use of Type MC cable.

(d)   In industrial establishments with restricted public access, where the conditions of maintenance and supervision ensure that only qualified persons service the installation, Type ITC-HL cable, listed for use in Class I, Division 1 locations, with a gas/vaportight continuous corrugated metallic sheath, an overall jacket of suitable polymeric material and provided with termination fittings listed for the application.

**(2) Flexible Connections.** Where necessary to employ flexible connections, as at motor terminals, flexible fittings listed for Class I, Division 1 locations or flexible cord in accordance with the provisions of 501.11 shall be permitted.

**(3) Boxes, Fittings, and Joints.** All boxes, fittings, and joints shall be approved for Class I, Division 1.

**(B) Class I, Division 2.**

**(1) General.** In Class I, Division 2 locations, the following wiring methods shall be permitted:

(1) All wiring methods permitted in Article 501.4(A).
(2) Threaded rigid metal conduit, threaded steel intermediate metal conduit.
(3) Enclosed gasketed busways, enclosed gasketed wireways.
(4) Type PLTC cable in accordance with the provisions of Article 725, or in cable tray systems. PLTC shall be installed in a manner to avoid tensile stress at the termination fittings.

(5) Type ITC cable in cable trays, in raceways, supported by messenger wire, afforded mechanical protection and run as open wiring, or directly buried where the cable is listed for this use.

(6) Type MI, MC, MV, or TC cable with termination fittings, or in cable tray systems and installed in a manner to avoid tensile stress at the termination fittings.

**(2) Flexible Connections.** Where provision must be made for limited flexibility, flexible metal fittings, flexible metal conduit with listed fittings, liquidtight flexible metal conduit with listed fittings, liquidtight flexible nonmetallic conduit with listed fittings, or flexible cord listed for extra-hard usage and provided with listed bushed fittings shall be used. An additional conductor for grounding shall be included in the flexible cord.

FPN: See 501.16(B) for grounding requirements where flexible conduit is used.

**(3) Nonincendive Field Wiring.** Nonincendive field wiring shall be permitted using any of the wiring methods permitted for unclassified locations. Nonincendive field wiring systems shall be installed in accordance with the control drawing(s). Simple apparatus, not shown on the control drawing, shall be permitted in a nonincendive field wiring circuit, provided the simple apparatus does not interconnect the nonincendive field wiring circuit to any other circuit.

FPN: Simple apparatus is defined in 504.2.

Separate nonincendive field wiring circuits shall be installed in accordance with one of the following:

(1) In separate cables
(2) In multiconductor cables where the conductors of each circuit are within a grounded metal shield
(3) In multiconductor cables, where the conductors of each circuit have insulation with a minimum thickness of 0.25 mm (0.01 in.)

**(4) Boxes, Fittings, and Joints.** Boxes, fittings, and joints shall not be required to be explosionproof except as required by 501.3(B)(1), 501.6(B)(1), and 501.14(B)(1).

**501.5 Sealing and Drainage.** Seals in conduit and cable systems shall comply with 501.5(A) through (F). Sealing compound shall be used in Type MI cable termination fittings to exclude moisture and other fluids from the cable insulation.

FPN No. 1: Seals are provided in conduit and cable systems to minimize the passage of gases and vapors and prevent the passage of flames from one portion of the electrical installation to another through the conduit. Such communication through Type MI cable is inherently prevented by construction of the cable. Unless specifically designed and tested for the purpose, conduit and cable seals are not intended to prevent the passage of liquids, gases, or vapors at a continuous pressure differential across the seal. Even at differences in pressure across the seal equivalent to a few inches of water, there may be a slow passage of gas or vapor through a seal and through conductors passing through the seal. See 501.5(E)(2). Temperature extremes and highly corrosive liquids and vapors can affect the ability of seals to perform their intended function. See 501.5(C)(2).

FPN No. 2: Gas or vapor leakage and propagation of flames may occur through the interstices between the strands of standard stranded conductors larger than 2 AWG. Special conductor constructions, for example, compacted strands or sealing of the individual strands, are means of reducing leakage and preventing the propagation of flames.

**(A) Conduit Seals, Class I, Division 1.** In Class I, Division 1 locations, conduit seals shall be located in accordance with 501.5(A)(1) through (A)(4).

**(1) Entering Enclosures.** In each conduit entry into an explosionproof enclosure where either

(1) The enclosure contains apparatus, such as switches, circuit breakers, fuses, relays, or resistors, that may produce arcs, sparks, or high temperatures that are considered to be an ignition source in normal operation, or
(2) The entry is metric designator 53 (trade size 2) or larger and the enclosure contains terminals, splices, or taps.

For the purposes of this section, high temperatures shall be considered to be any temperatures exceeding 80 percent of the autoignition temperature in degrees Celsius of the gas or vapor involved.

*Exception to 501.5(A)(1)(1): Seals shall not be required for conduit entering an enclosure where such switches, circuit breakers, fuses, relays, or resistors are*

(a) *Enclosed within a chamber hermetically sealed against the entrance of gases or vapors, or*
(b) *Immersed in oil in accordance with 501.6(B)(1)(2), or*
(c) *Enclosed within a factory-sealed explosionproof chamber located within the enclosure, identified for the location, and marked "factory sealed" or equivalent, unless the enclosure entry is metric designator 53 (trade size 2) or larger.*
(d) *In nonincendive circuits.*

Factory-sealed enclosures shall not be considered to serve as a seal for another adjacent explosionproof enclosure that is required to have a conduit seal.

Conduit seals shall be installed within 450 mm (18 in.) from the enclosure. Only explosionproof unions, couplings, reducers, elbows, capped elbows, and conduit bodies similar to L, T, and Cross types that are not larger than the trade size of the conduit shall be permitted between the sealing fitting and the explosionproof enclosure.

**(2) Pressurized Enclosures.** In each conduit entry into a pressurized enclosure where the conduit is not pressurized as part of the protection system. Conduit seals shall be installed within 450 mm (18 in.) from the pressurized enclosure.

> FPN No. 1: Installing the seal as close as possible to the enclosure will reduce problems with purging the dead air-space in the pressurized conduit.

> FPN No. 2: For further information, see NFPA 496-1998, *Standard for Purged and Pressurized Enclosures for Electrical Equipment.*

**(3) Two or More Explosionproof Enclosures.** Where two or more explosionproof enclosures for which conduit seals are required under 501.5(A)(1) are connected by nipples or by runs of conduit not more than 900 mm (36 in.) long, a single conduit seal in each such nipple connection or run of conduit shall be considered sufficient if located not more than 450 mm (18 in.) from either enclosure.

**(4) Class I, Division 1 Boundary.** In each conduit run leaving a Class I, Division 1 location. The sealing fitting shall be permitted on either side of the boundary of such location within 3.05 m (10 ft) of the boundary and shall be designed and installed so as to minimize the amount of gas or vapor within the Division 1 portion of the conduit from being communicated to the conduit beyond the seal. Except for listed explosionproof reducers at the conduit seal, there shall be no union, coupling, box, or fitting between the conduit seal and the point at which the conduit leaves the Division 1 location.

*Exception No. 1: Metal conduit that contains no unions, couplings, boxes, or fittings and passes completely through a Class I, Division 1 location with no fittings less than 300 mm (12 in.) beyond each boundary shall not require a conduit seal if the termination points of the unbroken conduit are in unclassified locations.*

*Exception No. 2: For underground conduit installed in accordance with 300.5 where the boundary is beneath the ground, the sealing fitting shall be permitted to be installed after the conduit leaves the ground, but there shall be no union, coupling, box, or fitting, other than listed explosionproof reducers at the sealing fitting, in the conduit between the sealing fitting and the point at which the conduit leaves the ground.*

**(B) Conduit Seals, Class I, Division 2.** In Class I, Division 2 locations, conduit seals shall be located in accordance with 501.5(B)(1) and (B)(2).

**(1) Entering Enclosures.** For connections to enclosures that are required to be explosionproof, a conduit seal shall be provided in accordance with 501.5(A)(1)(1) and (A)(3). All portions of the conduit run or nipple between the seal and such enclosure shall comply with 501.4(A).

**(2) Class I, Division 2 Boundary.** In each conduit run passing from a Class I, Division 2 location into an unclassified location. The sealing fitting shall be permitted on either side of the boundary of such location within 3.05 m (10 ft) of the boundary and shall be designed and installed so as to minimize the amount of gas or vapor within the Division 2 portion of the conduit from being communicated to the conduit beyond the seal. Rigid metal conduit or threaded steel intermediate metal conduit shall be used between the sealing fitting and the point at which the conduit leaves the Division 2 location, and a threaded connection shall be used at the sealing fitting. Except for listed explosionproof reducers at the conduit seal, there shall be no union, coupling, box, or fitting between the conduit seal and the point at which the conduit leaves the Division 2 location.

*Exception No. 1: Metal conduit that contains no unions, couplings, boxes, or fittings and passes completely through a Class I, Division 2 location with no fittings less than 300 mm (12 in.) beyond each boundary shall not be required to be sealed if the termination points of the unbroken conduit are in unclassified locations.*

*Exception No. 2: Conduit systems terminating at an unclassified location where a wiring method transition is made to cable tray, cablebus, ventilated busway, Type MI cable, or open wiring shall not be required to be sealed where passing from the Class I, Division 2 location into the unclassified location. The unclassified location shall be outdoors or, if the conduit system is all in one room, it shall be permitted to be indoors. The conduits shall not terminate at an enclosure containing an ignition source in normal operation.*

*Exception No. 3: Conduit systems passing from an enclosure or room that is unclassified as a result of pressurization into a Class I, Division 2 location shall not require a seal at the boundary.*

> FPN: For further information, refer to NFPA 496-1998, *Standard for Purged and Pressurized Enclosures for Electrical Equipment.*

*Exception No. 4: Segments of aboveground conduit systems shall not be required to be sealed where passing from a Class I, Division 2 location into an unclassified location if the following conditions are met:*

*(a) No part of the conduit system segment passes through a Class I, Division 1 location where the conduit contains unions, couplings, boxes, or fittings within 300 mm (12 in.) of the Class I, Division 1 location; and*

*(b) The conduit system segment is located entirely in outdoor locations; and*

*(c) The conduit system segment is not directly connected to canned pumps, process or service connections for flow, pressure, or analysis measurement, and so forth, that*

*depend on a single compression seal, diaphragm, or tube to prevent flammable or combustible fluids from entering the conduit system; and*

*(d) The conduit system segment contains only threaded metal conduit, unions, couplings, conduit bodies, and fittings in the unclassified location; and*

*(e) The conduit system segment is sealed at its entry to each enclosure or fitting housing terminals, splices, or taps in Class I, Division 2 locations*

**(C) Class I, Divisions 1 and 2.** Where required, seals in Class I, Division 1 and 2 locations shall comply with 501.5(C)(1) through (C)(6).

**(1) Fittings.** Enclosures for connections or equipment shall be provided with an integral means for sealing, or sealing fittings listed for the location shall be used. Sealing fittings shall be listed for use with one or more specific compounds and shall be accessible.

**(2) Compound.** The compound shall provide a seal against passage of gas or vapors through the seal fitting, shall not be affected by the surrounding atmosphere or liquids, and shall not have a melting point of less than 93°C (200°F).

**(3) Thickness of Compounds.** In a completed seal, the minimum thickness of the sealing compound shall not be less than the trade size of the sealing fitting and, in no case, less than 16 mm (⅝ in.).

*Exception: Listed cable sealing fittings shall not be required to have a minimum thickness equal to the trade size of the fitting.*

**(4) Splices and Taps.** Splices and taps shall not be made in fittings intended only for sealing with compound, nor shall other fittings in which splices or taps are made be filled with compound.

**(5) Assemblies.** In an assembly where equipment that may produce arcs, sparks, or high temperatures is located in a compartment separate from the compartment containing splices or taps, and an integral seal is provided where conductors pass from one compartment to the other, the entire assembly shall be identified for the location. Seals in conduit connections to the compartment containing splices or taps shall be provided in Class I, Division 1 locations where required by 501.5(A)(1)(2).

**(6) Conductor Fill.** The cross-sectional area of the conductors permitted in a seal shall not exceed 25 percent of the cross-sectional area of a rigid metal conduit of the same trade size unless it is specifically identified for a higher percentage of fill.

**(D) Cable Seals, Class I, Division 1.** In Class I, Division 1 locations, cable seals shall be located according to 501.5(D)(1) through (D)(3).

**(1) At Terminations.** Cable shall be sealed at all terminations. The sealing fitting shall comply with 501.5(C). Multiconductor Type MC-HL cables with a gas/vaportight continuous corrugated metallic sheath and an overall jacket of suitable polymeric material shall be sealed with a listed fitting after removing the jacket and any other covering so that the sealing compound surrounds each individual insulated conductor in such a manner as to minimize the passage of gases and vapors.

*Exception: Shielded cables and twisted pair cables shall not require the removal of the shielding material or separation of the twisted pairs, provided the termination is by an approved means to minimize the entrance of gases or vapors and prevent propagation of flame into the cable core.*

**(2) Cables Capable of Transmitting Gases or Vapors.** Cables in conduit with a gas/vaportight continuous sheath capable of transmitting gases or vapors through the cable core shall be sealed in the Division 1 location after removing the jacket and any other coverings so that the sealing compound will surround each individual insulated conductor and the outer jacket.

*Exception: Multiconductor cables with a gas/vaportight continuous sheath capable of transmitting gases or vapors through the cable core shall be permitted to be considered as a single conductor by sealing the cable in the conduit within 450 mm (18 in.) of the enclosure and the cable end within the enclosure by an approved means to minimize the entrance of gases or vapors and prevent the propagation of flame into the cable core, or by other approved methods. For shielded cables and twisted pair cables, it shall not be required to remove the shielding material or separate the twisted pair.*

**(3) Cables Incapable of Transmitting Gases or Vapors.** Each multiconductor cable in conduit shall be considered as a single conductor if the cable is incapable of transmitting gases or vapors through the cable core. These cables shall be sealed in accordance with 501.5(A).

**(E) Cable Seals, Class I, Division 2.** In Class I, Division 2 locations, cable seals shall be located in accordance with 501.5(E)(1) through (E)(4).

**(1) Terminations.** Cables entering enclosures that are required to be explosionproof shall be sealed at the point of entrance. The sealing fitting shall comply with 501.5(B)(1). Multiconductor cables with a gas/vaportight continuous sheath capable of transmitting gases or vapors through the cable core shall be sealed in a listed fitting in the Division 2 location after removing the jacket and any other coverings so that the sealing compound surrounds each individual insulated conductor in such a manner as to minimize the passage of gases and vapors. Multiconductor cables in conduit shall be sealed as described in 501.5(D).

*Exception No. 1: Cables passing from an enclosure or room that is unclassified as a result of Type Z pressurization into a Class I, Division 2 location shall not require a seal at the boundary.*

*Exception No. 2: Shielded cables and twisted pair cables shall not require the removal of the shielding material or separation of the twisted pairs, provided the termination is by an approved means to minimize the entrance of gases or vapors and prevent propagation of flame into the cable core.*

**(2) Cables That Do Not Transmit Gases or Vapors.** Cables that have a gas/vaportight continuous sheath and do not transmit gases or vapors through the cable core in excess of the quantity permitted for seal fittings shall not be required to be sealed except as required in 501.5(E)(1). The minimum length of such cable run shall not be less than that length that limits gas or vapor flow through the cable core to the rate permitted for seal fittings [200 cm$^3$/hr (0.007 ft$^3$/hr) of air at a pressure of 1500 pascals (6 in. of water)].

    FPN No. 1: See ANSI/UL 886-1994, *Outlet Boxes and Fittings for Use in Hazardous (Classified) Locations.*

    FPN No. 2: The cable core does not include the interstices of the conductor strands.

**(3) Cables Capable of Transmitting Gases or Vapors.** Cables with a gas/vaportight continuous sheath capable of transmitting gases or vapors through the cable core shall not be required to be sealed except as required in 501.5(E)(1), unless the cable is attached to process equipment or devices that may cause a pressure in excess of 1500 pascals (6 in. of water) to be exerted at a cable end, in which case a seal, barrier, or other means shall be provided to prevent migration of flammables into an unclassified location.

*Exception: Cables with an unbroken gas/vaportight continuous sheath shall be permitted to pass through a Class I, Division 2 location without seals.*

**(4) Cables Without Gas/Vaportight Sheath.** Cables that do not have gas/vaportight continuous sheath shall be sealed at the boundary of the Division 2 and unclassified location in such a manner as to minimize the passage of gases or vapors into an unclassified location.

**(F) Drainage.**

**(1) Control Equipment.** Where there is a probability that liquid or other condensed vapor may be trapped within enclosures for control equipment or at any point in the raceway system, approved means shall be provided to pre-vent accumulation or to permit periodic draining of such liquid or condensed vapor.

**(2) Motors and Generators.** Where the authority having jurisdiction judges that there is a probability that liquid or condensed vapor may accumulate within motors or generators, joints and conduit systems shall be arranged to minimize the entrance of liquid. If means to prevent accumulation or to permit periodic draining are judged necessary, such means shall be provided at the time of manufacture and shall be considered an integral part of the machine.

**(3) Canned Pumps, Process, or Service Connections, etc.** For canned pumps, process, or service connections for flow, pressure, or analysis measurement, and so forth, that depend on a single compression seal, diaphragm, or tube to prevent flammable or combustible fluids from entering the electrical raceway or cable system capable of transmitting fluids, an additional approved seal, barrier, or other means shall be provided to prevent the flammable or combustible fluid from entering the raceway or cable system capable of transmitting fluids beyond the additional devices or means, if the primary seal fails. The additional approved seal or barrier and the interconnecting enclosure shall meet the temperature and pressure conditions to which they will be subjected upon failure of the primary seal, unless other approved means are provided to accomplish the purpose above. Drains, vents, or other devices shall be provided so that primary seal leakage will be obvious.

    FPN: See also the fine print notes to 501.5.

**501.6 Switches, Circuit Breakers, Motor Controllers, and Fuses.**

**(A) Class I, Division 1.** In Class I, Division 1 locations, switches, circuit breakers, motor controllers, and fuses, including pushbuttons, relays, and similar devices, shall be provided with enclosures, and the enclosure in each case, together with the enclosed apparatus, shall be identified as a complete assembly for use in Class I locations.

**(B) Class I, Division 2.** Switches, circuit breakers, motor controllers, and fuses in Class I, Division 2 locations shall comply with 501.6(B)(1) through (B)(4).

**(1) Type Required.** Circuit breakers, motor controllers, and switches intended to interrupt current in the normal performance of the function for which they are installed shall be provided with enclosures identified for Class I, Division 1 locations in accordance with 501.3(A), unless general-purpose enclosures are provided and any of the following apply:

(1) The interruption of current occurs within a chamber hermetically sealed against the entrance of gases and vapors.

(2) The current make-and-break contacts are oil-immersed and of the general-purpose type having a 50-mm (2-in.) minimum immersion for power contacts and a 25-mm (1-in.) minimum immersion for control contacts.

(3) The interruption of current occurs within a factory-sealed explosionproof chamber.

(4) The device is a solid state, switching control without contacts, where the surface temperature does not exceed 80 percent of the ignition temperature in degrees Celsius of the gas or vapor involved.

**(2) Isolating Switches.** Fused or unfused disconnect and isolating switches for transformers or capacitor banks that are not intended to interrupt current in the normal performance of the function for which they are installed shall be permitted to be installed in general-purpose enclosures.

**(3) Fuses.** For the protection of motors, appliances, and lamps, other than as provided in 501.6(B)(4), standard plug or cartridge fuses shall be permitted, provided they are placed within enclosures identified for the location; or fuses shall be permitted if they are within general-purpose enclosures, and if they are of a type in which the operating element is immersed in oil or other approved liquid, or the operating element is enclosed within a chamber hermetically sealed against the entrance of gases and vapors, or the fuse is a nonindicating, filled, current-limiting type.

**(4) Fuses Internal to Luminaires (Lighting Fixtures).** Listed cartridge fuses shall be permitted as supplementary protection within luminaires (lighting fixtures).

**501.7 Control Transformers and Resistors.** Transformers, impedance coils, and resistors used as, or in conjunction with, control equipment for motors, generators, and appliances shall comply with 501.7(A) and (B).

**(A) Class I, Division 1.** In Class I, Division 1 locations, transformers, impedance coils, and resistors, together with any switching mechanism associated with them, shall be provided with enclosures identified for Class I, Division 1 locations in accordance with 501.3(A).

**(B) Class I, Division 2.** In Class I, Division 2 locations, control transformers and resistors shall comply with 501.7(B)(1) through (B)(3).

**(1) Switching Mechanisms.** Switching mechanisms used in conjunction with transformers, impedance coils, and resistors shall comply with 501.6(B).

**(2) Coils and Windings.** Enclosures for windings of transformers, solenoids, or impedance coils shall be permitted to be of the general-purpose type.

**(3) Resistors.** Resistors shall be provided with enclosures; and the assembly shall be identified for Class I locations,

unless resistance is nonvariable and maximum operating temperature, in degrees Celsius, will not exceed 80 percent of the ignition temperature of the gas or vapor involved or has been tested and found incapable of igniting the gas or vapor.

**501.8 Motors and Generators.**

**(A) Class I, Division 1.** In Class I, Division 1 locations, motors, generators, and other rotating electric machinery shall be as follows:

(1) Identified for Class I, Division 1 locations; or

(2) Of the totally enclosed type supplied with positive-pressure ventilation from a source of clean air with discharge to a safe area, so arranged to prevent energizing of the machine until ventilation has been established and the enclosure has been purged with at least 10 volumes of air, and also arranged to automatically de-energize the equipment when the air supply fails; or

(3) Of the totally enclosed inert gas-filled type supplied with a suitable reliable source of inert gas for pressuring the enclosure, with devices provided to ensure a positive pressure in the enclosure and arranged to automatically de-energize the equipment when the gas supply fails; or

(4) Of a type designed to be submerged in a liquid that is flammable only when vaporized and mixed with air, or in a gas or vapor at a pressure greater than atmospheric and that is flammable only when mixed with air; and the machine is arranged so to prevent energizing it until it has been purged with the liquid or gas to exclude air, and also arranged to automatically de-energize the equipment when the supply of liquid or gas or vapor fails or the pressure is reduced to atmospheric.

Totally enclosed motors of the types specified in 501.8(A)(2) or (A)(3) shall have no external surface with an operating temperature in degrees Celsius in excess of 80 percent of the ignition temperature of the gas or vapor involved. Appropriate devices shall be provided to detect and automatically de-energize the motor or provide an adequate alarm if there is any increase in temperature of the motor beyond designed limits. Auxiliary equipment shall be of a type identified for the location in which it is installed.

FPN: See D 2155-69, ASTM Test Procedure.

**(B) Class I, Division 2.** In Class I, Division 2 locations, motors, generators, and other rotating electric machinery in which are employed sliding contacts, centrifugal or other types of switching mechanism (including motor overcurrent, overloading, and overtemperature devices), or integral resistance devices, either while starting or while running, shall be identified for Class I, Division 1 locations, unless such sliding contacts, switching mechanisms, and resistance devices are provided with enclosures identified for Class I, Division 2 locations in accordance with 501.3(B). The exposed surface of space heaters used to prevent condensation of moisture during shutdown periods shall not exceed 80 percent of the ignition temperature in degrees Celsius of the gas or vapor involved when operated at rated voltage, and the maximum surface temperature [based on a 40°C (104°F) ambient] shall be permanently marked on a visible nameplate mounted on the motor. Otherwise, space heaters shall be identified for Class I, Division 2 locations. In Class I, Division 2 locations, the installation of open or nonexplosionproof enclosed motors, such as squirrel-cage induction motors without brushes, switching mechanisms, or similar arc-producing devices that are not identified for use in a Class I, Division 2 location, shall be permitted.

FPN No. 1: It is important to consider the temperature of internal and external surfaces that may be exposed to the flammable atmosphere.

FPN No. 2: It is important to consider the risk of ignition due to currents arcing across discontinuities and overheating of parts in multisection enclosures of large motors and generators. Such motors and generators may need equipotential bonding jumpers across joints in the enclosure and from enclosure to ground. Where the presence of ignitible gases or vapors is suspected, clean-air purging may be needed immediately prior to and during start-up periods.

**501.9 Luminaires (Lighting Fixtures).** Luminaires (lighting fixtures) shall comply with 501.9(A) or (B).

**(A) Class I, Division 1.** In Class I, Division 1 locations, luminaires (lighting fixtures) shall comply with 501.9(A)(1) through (A)(4).

**(1) Luminaires (Lighting Fixtures).** Each luminaire (lighting fixture) shall be identified as a complete assembly for the Class I, Division 1 location and shall be clearly marked to indicate the maximum wattage of lamps for which it is identified. Luminaires (lighting fixtures) intended for portable use shall be specifically listed as a complete assembly for that use.

**(2) Physical Damage.** Each luminaire (lighting fixture) shall be protected against physical damage by a suitable guard or by location.

**(3) Pendant Luminaires (Lighting Fixtures).** Pendant luminaires (lighting fixtures) shall be suspended by and supplied through threaded rigid metal conduit stems or threaded steel intermediate conduit stems, and threaded joints shall be provided with set-screws or other effective means to prevent loosening. For stems longer than 300 mm (12 in.), permanent and effective bracing against lateral displacement shall be provided at a level not more than 300 mm (12 in.) above the lower end of the stem, or flexibility in the form of a fitting or flexible connector identi-

fied for the Class I, Division 1 location shall be provided not more than 300 mm (12 in.) from the point of attachment to the supporting box or fitting.

**(4) Supports.** Boxes, box assemblies, or fittings used for the support of luminaires (lighting fixtures) shall be identified for Class I locations.

**(B) Class I, Division 2.** In Class I, Division 2 locations, luminaires (lighting fixtures) shall comply with 501.9(B)(1) through (B)(5).

**(1) Portable Lighting Equipment.** Portable lighting equipment shall comply with 501.9(A)(1).

*Exception: Where portable lighting equipment is mounted on movable stands and is connected by flexible cords, as covered in 501.11, it shall be permitted, where mounted in any position, if it conforms to 501.9(B)(2).*

**(2) Fixed Luminaires (Lighting Fixtures).** Luminaires (lighting fixtures) for fixed lighting shall be protected from physical damage by suitable guards or by location. Where there is danger that falling sparks or hot metal from lamps or fixtures might ignite localized concentrations of flammable vapors or gases, suitable enclosures or other effective protective means shall be provided. Where lamps are of a size or type that may, under normal operating conditions, reach surface temperatures exceeding 80 percent of the ignition temperature in degrees Celsius of the gas or vapor involved, fixtures shall comply with 501.9(A)(1) or shall be of a type that has been tested in order to determine the marked operating temperature or temperature class (T Code).

**(3) Pendant Luminaires (Fixtures).** Pendant luminaires (lighting fixtures) shall be suspended by threaded rigid metal conduit stems, threaded steel intermediate metal conduit stems, or other approved means. For rigid stems longer than 300 mm (12 in.), permanent and effective bracing against lateral displacement shall be provided at a level not more than 300 mm (12 in.) above the lower end of the stem, or flexibility in the form of an identified fitting or flexible connector shall be provided not more than 300 mm (12 in.) from the point of attachment to the supporting box or fitting.

**(4) Switches.** Switches that are a part of an assembled fixture or of an individual lampholder shall comply with 501.6(B)(1).

**(5) Starting Equipment.** Starting and control equipment for electric-discharge lamps shall comply with 501.7(B).

*Exception: A thermal protector potted into a thermally protected fluorescent lamp ballast if the luminaire (lighting fixture) is identified for the location.*

**501.10 Utilization Equipment.**

**(A) Class I, Division 1.** In Class I, Division 1 locations, all utilization equipment shall be identified for Class I, Division 1 locations.

**(B) Class I, Division 2.** In Class I, Division 2 locations, all utilization equipment shall comply with 501.10(B)(1) through (B)(3).

**(1) Heaters.** Electrically heated utilization equipment shall conform with either item (1) or (2):

(1) The heater shall not exceed 80 percent of the ignition temperature in degrees Celsius of the gas or vapor involved on any surface that is exposed to the gas or vapor when continuously energized at the maximum rated ambient temperature. If a temperature controller is not provided, these conditions shall apply when the heater is operated at 120 percent of rated voltage.

*Exception No. 1: For motor-mounted anticondensation space heaters, see 501.8(B).*

*Exception No. 2: Where current-limiting device is applied to the circuit serving the heater to limit the current in the heater to a value less than that required to raise the heater surface temperature to 80 percent of the ignition temperature.*

(2) The heater shall be identified for Class I, Division 1 locations.

*Exception: Electrical resistance heat tracing identified for Class I, Division 2 locations.*

**(2) Motors.** Motors of motor-driven utilization equipment shall comply with 501.8(B).

**(3) Switches, Circuit Breakers, and Fuses.** Switches, circuit breakers, and fuses shall comply with 501.6(B).

**501.11 Flexible Cords, Class I, Divisions 1 and 2.** A flexible cord shall be permitted for connection between portable lighting equipment or other portable utilization equipment and the fixed portion of their supply circuit. Flexible cord shall also be permitted for that portion of the circuit where the fixed wiring methods of 501.4(A) cannot provide the necessary degree of movement for fixed and mobile electrical utilization equipment, in an industrial establishment where conditions of maintenance and engineering supervision ensure that only qualified persons install and service the installation, and the flexible cord is protected by location or by a suitable guard from damage. The length of the flexible cord shall be continuous. Where flexible cords are used, the cords shall be as follows:

(1) Of a type listed for extra-hard usage
(2) Contain, in addition to the conductors of the circuit, a grounding conductor complying with 400.23

(3) Connected to terminals or to supply conductors in an approved manner
(4) Supported by clamps or by other suitable means in such a manner that there is no tension on the terminal connections
(5) Provided with suitable seals where the flexible cord enters boxes, fittings, or enclosures of the explosion-proof type

*Exception: As provided in 501.3(B)(6) and 501.4(B).*

Electric submersible pumps with means for removal without entering the wet-pit shall be considered portable utilization equipment. The extension of the flexible cord within a suitable raceway between the wet-pit and the power source shall be permitted. Electric mixers intended for travel into and out of open-type mixing tanks or vats shall be considered portable utilization equipment.

FPN: See 501.13 for flexible cords exposed to liquids having a deleterious effect on the conductor insulation.

**501.12 Receptacles and Attachment Plugs, Class I, Divisions 1 and 2.** Receptacles and attachment plugs shall be of the type providing for connection to the grounding conductor of a flexible cord and shall be identified for the location.

*Exception: As provided in 501.3(B)(6).*

**501.13 Conductor Insulation, Class I, Divisions 1 and 2.** Where condensed vapors or liquids may collect on, or come in contact with, the insulation on conductors, such insulation shall be of a type identified for use under such conditions; or the insulation shall be protected by a sheath of lead or by other approved means.

**501.14 Signaling, Alarm, Remote-Control, and Communications Systems.**

**(A) Class I, Division 1.** In Class I, Division 1 locations, all apparatus and equipment of signaling, alarm, remote-control, and communications systems, regardless of voltage, shall be identified for Class I, Division 1 locations, and all wiring shall comply with 501.4(A) and 501.5(A) and (C).

**(B) Class I, Division 2.** In Class I, Division 2 locations, signaling, alarm, remote-control, and communications systems shall comply with 501.14(B)(1) through (B)(4).

**(1) Contacts.** Switches, circuit breakers, and make-and-break contacts of pushbuttons, relays, alarm bells, and horns shall have enclosures identified for Class I, Division 1 locations in accordance with 501.3(A).

*Exception: General-purpose enclosures shall be permitted if current-interrupting contacts are one of the following:*

(a) *Immersed in oil, or*

(b) *Enclosed within a chamber hermetically sealed against the entrance of gases or vapors, or*

(c) *In nonincendive circuits, or*

(d) *Part of a listed nonincendive component*

**(2) Resistors and Similar Equipment.** Resistors, resistance devices, thermionic tubes, rectifiers, and similar equipment shall comply with 501.3(B)(2).

**(3) Protectors.** Enclosures shall be provided for lightning protective devices and for fuses. Such enclosures shall be permitted to be of the general-purpose type.

**(4) Wiring and Sealing.** All wiring shall comply with 501.4(B) and 501.5(B) and (C).

**501.15 Live Parts, Class I, Divisions 1 and 2.** There shall be no exposed live parts.

**501.16 Grounding, Class I, Divisions 1 and 2.** Wiring and equipment in Class I, Division 1 and 2 locations shall be grounded as specified in Article 250 and with the requirements in 501.16(A) and (B).

**(A) Bonding.** The locknut-bushing and double-locknut types of contacts shall not be depended on for bonding purposes, but bonding jumpers with proper fittings or other approved means of bonding shall be used. Such means of bonding shall apply to all intervening raceways, fittings, boxes, enclosures, and so forth between Class I locations and the point of grounding for service equipment or point of grounding of a separately derived system.

*Exception: The specific bonding means shall only be required to the nearest point where the grounded circuit conductor and the grounding electrode are connected together on the line side of the building or structure disconnecting means as specified in 250.32(A), (B), and (C), provided the branch-circuit overcurrent protection is located on the load side of the disconnecting means.*

FPN: See 250.100 for additional bonding requirements in hazardous (classified) locations.

**(B) Types of Equipment Grounding Conductors.** Where flexible metal conduit or liquidtight flexible metal conduit is used as permitted in 501.4(B) and is to be relied on to complete a sole equipment grounding path, it shall be installed with internal or external bonding jumpers in parallel with each conduit and complying with 250.102.

*Exception: In Class I, Division 2 locations, the bonding jumper shall be permitted to be deleted where all the following conditions are met.*

(a) *Listed liquidtight flexible metal conduit 1.8 m (6 ft) or less in length, with fittings listed for grounding, is used.*

(b) *Overcurrent protection in the circuit is limited to 10 amperes or less.*

(c) *The load is not a power utilization load.*

**501.17 Surge Protection.**

**(A) Class I, Division 1.** Surge arresters, including their installation and connection, shall comply with Article 280. The surge arresters and capacitors shall be installed in enclosures identified for Class I, Division 1 locations. Surge-protective capacitors shall be of a type designed for specific duty.

**(B) Class I, Division 2.** Surge arresters shall be nonarcing, such as metal-oxide varistor (MOV), sealed type, and surge-protective capacitors shall be of a type designed for specific duty. Installation and connection shall comply with Article 280. Enclosures shall be permitted to be of the general-purpose type. Surge protection of types other than described above shall be installed in enclosures identified for Class I, Division 1 locations.

**501.18 Multiwire Branch Circuits.** In a Class I, Division 1 location, a multiwire branch circuit shall not be permitted.

*Exception: Where the disconnect device(s) for the circuit opens all ungrounded conductors of the multiwire circuit simultaneously.*

# ARTICLE 502
# Class II Locations

**502.1 General.** The general rules of this *Code* shall apply to the electric wiring and equipment in locations classified as Class II locations in 500.5(C).

*Exception: As modified by this article.*

Equipment installed in Class II locations shall be able to function at full rating without developing surface temperatures high enough to cause excessive dehydration or gradual carbonization of any organic dust deposits that may occur.

FPN: Dust that is carbonized or excessively dry is highly susceptible to spontaneous ignition.

Explosionproof equipment and wiring shall not be required and shall not be acceptable in Class II locations unless identified for such locations. Where Class II, Group E dusts are present in hazardous quantities, there are only Division 1 locations.

## 502.2 Transformers and Capacitors.

**(A) Class II, Division 1.** In Class II, Division 1 locations, transformers and capacitors shall comply with 502.2(A)(1) through (A)(3).

**(1) Containing Liquid That Will Burn.** Transformers and capacitors containing a liquid that will burn shall be installed only in vaults complying with 450.41 through 450.48, and, in addition, (a), (b), and (c) shall apply.

(a) Doors or other openings communicating with the Division 1 location shall have self-closing fire doors on both sides of the wall, and the doors shall be carefully fitted and provided with suitable seals (such as weather stripping) to minimize the entrance of dust into the vault.

(b) Vent openings and ducts shall communicate only with the outside air.

(c) Suitable pressure-relief openings communicating with the outside air shall be provided.

**(2) Not Containing Liquid That Will Burn.** Transformers and capacitors that do not contain a liquid that will burn shall be installed in vaults complying with 450.41 through 450.48 or be identified as a complete assembly, including terminal connections for Class II locations.

**(3) Metal Dusts.** No transformer or capacitor shall be installed in a location where dust from magnesium, aluminum, aluminum bronze powders, or other metals of similarly hazardous characteristics may be present.

**(B) Class II, Division 2.** In Class II, Division 2 locations, transformers and capacitors shall comply with 502.2(B)(1) through (B)(3).

**(1) Containing Liquid That Will Burn.** Transformers and capacitors containing a liquid that will burn shall be installed in vaults that comply with 450.41 through 450.48.

**(2) Containing Askarel.** Transformers containing askarel and rated in excess of 25 kVA shall be as follows:

(1) Provided with pressure-relief vents
(2) Provided with a means for absorbing any gases generated by arcing inside the case, or the pressure-relief vents shall be connected to a chimney or flue that will carry such gases outside the building
(3) Have an airspace of not less than 150 mm (6 in.) between the transformer cases and any adjacent combustible material

**(3) Dry-Type Transformers.** Dry-type transformers shall be installed in vaults or shall have their windings and terminal connections enclosed in tight metal housings without ventilating or other openings and shall operate at not over 600 volts, nominal.

## 502.4 Wiring Methods. Wiring methods shall comply with 502.4(A) or (B).

**(A) Class II, Division 1.**

**(1) General.** In Class II, Division 1 locations, the wiring methods in (a) through (e) shall be permitted.

(a) Threaded rigid metal conduit, or threaded steel intermediate metal conduit.

(b) Type MI cable with termination fittings listed for the location. Type MI cable shall be installed and supported in a manner to avoid tensile stress at the termination fittings.

(c) In industrial establishments with limited public access, where the conditions of maintenance and supervision ensure that only qualified persons service the installation, Type MC cable, listed for use in Class II, Division 1 locations, with a gas/vaportight continuous corrugated metallic sheath, an overall jacket of suitable polymeric material, separate grounding conductors in accordance with 250.122, and provided with termination fittings listed for the application, shall be permitted.

(d) Fittings and boxes shall be provided with threaded bosses for connection to conduit or cable terminations and shall be dusttight. Fittings and boxes in which taps, joints, or terminal connections are made, or that are used in Group E locations, shall be identified for Class II locations.

(e) Where necessary to employ flexible connections, dusttight flexible connectors, liquidtight flexible metal conduit with listed fittings, liquidtight flexible nonmetallic conduit with listed fittings, or flexible cord listed for extra-hard usage and provided with bushed fittings shall be used. Where flexible cords are used, they shall comply with 502.12. Where flexible connections are subject to oil or other corrosive conditions, the insulation of the conductors shall be of a type listed for the condition or shall be protected by means of a suitable sheath.

> FPN: See 502.16(B) for grounding requirements where flexible conduit is used.

**(B) Class II, Division 2.**

**(1) General.** In Class II, Division 2 locations, the following wiring methods shall be permitted:

(1) All wiring methods permitted in 502.4(A).
(2) Rigid metal conduit, intermediate metal conduit, electrical metallic tubing, dusttight wireways.
(3) Type MC or MI cable with listed termination fittings.
(4) Type PLTC in cable trays.
(5) Type ITC in cable trays.
(6) Type MC, MI, or TC cable installed in ladder, ventilated trough, or ventilated channel cable trays in a single layer, with a space not less than the larger cable diameter between the two adjacent cables, shall be the wiring method employed.

*Exception: Type MC cable listed for use in Class II, Division 1 locations shall be permitted to be installed without the above required spacings.*

**(2) Flexible Connections.** Where provision must be made for flexibility, 502.4(A)(1)(e) shall apply.

**(3) Nonincendive Field Wiring.** Nonincendive field wiring shall be permitted using any of the wiring methods permitted for unclassified locations. Nonincendive field wiring systems shall be installed in accordance with the control drawing(s). Simple apparatus, not shown on the control drawing, shall be permitted in a nonincendive field wiring circuit, provided the simple apparatus does not interconnect the nonincendive field wiring circuit to any other circuit.

FPN: Simple apparatus is defined in 504.2.

Separate nonincendive field wiring circuits shall be as follows:

(1) In separate cables, or
(2) In multiconductor cables where the conductors of each circuit are within a grounded metal shield, or
(3) In multiconductor cables where the conductors of each circuit have insulation with a minimum thickness of 0.25 mm (0.01 in.)

**(4) Boxes and Fittings.** All boxes and fittings shall be dusttight.

**502.5 Sealing, Class II, Divisions 1 and 2.** Where a raceway provides communication between an enclosure that is required to be dust-ignitionproof and one that is not, suitable means shall be provided to prevent the entrance of dust into the dust-ignitionproof enclosure through the raceway. One of the following means shall be permitted:

(1) A permanent and effective seal
(2) A horizontal raceway not less than 3.05 m (10 ft) long, or
(3) A vertical raceway not less than 1.5 m (5 ft) long and extending downward from the dust-ignitionproof enclosure

Where a raceway provides communication between an enclosure that is required to be dust-ignitionproof and an enclosure in an unclassified location, seals shall not be required.

Sealing fittings shall be accessible.

Seals shall not be required to be explosionproof.

FPN: Electrical sealing putty is a method of sealing.

**502.6 Switches, Circuit Breakers, Motor Controllers, and Fuses.**

**(A) Class II, Division 1.** In Class II, Division 1 locations, switches, circuit breakers, motor controllers, and fuses shall comply with 502.6(A)(1) through (A)(3).

**(1) Type Required.** Switches, circuit breakers, motor controllers, and fuses, including pushbuttons, relays, and similar devices that are intended to interrupt current during normal operation or that are installed where combustible dusts of an electrically conductive nature may be present, shall be provided with identified dust-ignitionproof enclosures.

**(2) Isolating Switches.** Disconnecting and isolating switches containing no fuses and not intended to interrupt current and not installed where dusts may be of an electrically conductive nature shall be provided with tight metal enclosures that shall be designed to minimize the entrance of dust and that shall (1) be equipped with telescoping or close-fitting covers or with other effective means to prevent the escape of sparks or burning material and (2) have no openings (such as holes for attachment screws) through which, after installation, sparks or burning material might escape or through which exterior accumulations of dust or adjacent combustible material might be ignited.

**(3) Metal Dusts.** In locations where dust from magnesium, aluminum, aluminum bronze powders, or other metals of similarly hazardous characteristics may be present, fuses, switches, motor controllers, and circuit breakers shall have enclosures identified for such locations.

**(B) Class II, Division 2.** In Class II, Division 2 locations, enclosures for fuses, switches, circuit breakers, and motor controllers, including pushbuttons, relays, and similar devices, shall be dusttight.

**502.7 Control Transformers and Resistors.**

**(A) Class II, Division 1.** In Class II, Division 1 locations, control transformers, solenoids, impedance coils, resistors, and any overcurrent devices or switching mechanisms associated with them shall have dust-ignitionproof enclosures identified for Class II locations. No control transformer, impedance coil, or resistor shall be installed in a location where dust from magnesium, aluminum, aluminum bronze powders, or other metals of similarly hazardous characteristics may be present unless provided with an enclosure identified for the specific location.

**(B) Class II, Division 2.** In Class II, Division 2 locations, transformers and resistors shall comply with 502.7(B)(1) through (B)(3).

**(1) Switching Mechanisms.** Switching mechanisms (including overcurrent devices) associated with control transformers, solenoids, impedance coils, and resistors shall be provided with dusttight enclosures.

**(2) Coils and Windings.** Where not located in the same enclosure with switching mechanisms, control transformers, solenoids, and impedance coils shall be provided with tight metal housings without ventilating openings.

**(3) Resistors.** Resistors and resistance devices shall have dust-ignitionproof enclosures identified for Class II locations.

*Exception: Where the maximum normal operating temperature of the resistor will not exceed 120°C (248°F), nonadjustable resistors or resistors that are part of an automatically timed starting sequence shall be permitted to have enclosures complying with 502.7(B)(2).*

### 502.8 Motors and Generators.

**(A) Class II, Division 1.** In Class II, Division 1 locations, motors, generators, and other rotating electrical machinery shall be

(1) Identified for Class II, Division 1 locations, or
(2) Totally enclosed pipe-ventilated, meeting temperature limitations in 502.1.

**(B) Class II, Division 2.** In Class II, Division 2 locations, motors, generators, and other rotating electrical equipment shall be totally enclosed nonventilated, totally enclosed pipe-ventilated, totally enclosed water-air-cooled, totally enclosed fan-cooled or dust-ignitionproof for which maximum full-load external temperature shall be in accordance with 500.8(C)(2) for normal operation when operating in free air (not dust blanketed) and shall have no external openings.

*Exception: If the authority having jurisdiction believes accumulations of nonconductive, nonabrasive dust will be moderate and if machines can be easily reached for routine cleaning and maintenance, the following shall be permitted to be installed:*

*(a) Standard open-type machines without sliding contacts, centrifugal or other types of switching mechanism (including motor overcurrent, overloading, and overtemperature devices), or integral resistance devices*
*(b) Standard open-type machines with such contacts, switching mechanisms, or resistance devices enclosed within dusttight housings without ventilating or other openings*
*(c) Self-cleaning textile motors of the squirrel-cage type*

### 502.9 Ventilating Piping.
Ventilating pipes for motors, generators, or other rotating electric machinery, or for enclosures for electric equipment, shall be of metal not less than 0.53 mm (0.021 in.) in thickness or of equally substantial noncombustible material and shall comply with all of the following:

(1) Lead directly to a source of clean air outside of buildings

(2) Be screened at the outer ends to prevent the entrance of small animals or birds
(3) Be protected against physical damage and against rusting or other corrosive influences

Ventilating pipes shall also comply with 502.9(A) and (B).

**(A) Class II, Division 1.** In Class II, Division 1 locations, ventilating pipes, including their connections to motors or to the dust-ignitionproof enclosures for other equipment, shall be dusttight throughout their length. For metal pipes, seams and joints shall comply with one of the following:

(1) Be riveted and soldered
(2) Be bolted and soldered
(3) Be welded
(4) Be rendered dusttight by some other equally effective means

**(B) Class II, Division 2.** In Class II, Division 2 locations, ventilating pipes and their connections shall be sufficiently tight to prevent the entrance of appreciable quantities of dust into the ventilated equipment or enclosure and to prevent the escape of sparks, flame, or burning material that might ignite dust accumulations or combustible material in the vicinity. For metal pipes, lock seams and riveted or welded joints shall be permitted; and tight-fitting slip joints shall be permitted where some flexibility is necessary, as at connections to motors.

### 502.10 Utilization Equipment.

**(A) Class II, Division 1.** In Class II, Division 1 locations, all utilization equipment shall be identified for Class II locations. Where dust from magnesium, aluminum, aluminum bronze powders, or other metals of similarly hazardous characteristics may be present, such equipment shall be identified for the specific location.

**(B) Class II, Division 2.** In Class II, Division 2 locations, all utilization equipment shall comply with 502.10(B)(1) through (B)(4).

**(1) Heaters.** Electrically heated utilization equipment shall be identified for Class II locations.

*Exception: Metal-enclosed radiant heating panel equipment shall be dusttight and marked in accordance with 500.8(B).*

**(2) Motors.** Motors of motor-driven utilization equipment shall comply with 502.8(B).

**(3) Switches, Circuit Breakers, and Fuses.** Enclosures for switches, circuit breakers, and fuses shall be dusttight.

**(4) Transformers, Solenoids, Impedance Coils, and Resistors.** Transformers, solenoids, impedance coils, and resistors shall comply with 502.7(B).

**502.11 Luminaires (Lighting Fixtures).** Luminaires (lighting fixtures) shall comply with 502.11(A) and (B).

**(A) Class II, Division 1.** In Class II, Division 1 locations, luminaires (lighting fixtures) for fixed and portable lighting shall comply with 502.11(A)(1) through (A)(4).

**(1) Fixtures.** Each luminaire (fixture) shall be identified for Class II locations and shall be clearly marked to indicate the maximum wattage of the lamp for which it is designed. In locations where dust from magnesium, aluminum, aluminum bronze powders, or other metals of similarly hazardous characteristics may be present, luminaires (fixtures) for fixed or portable lighting and all auxiliary equipment shall be identified for the specific location.

**(2) Physical Damage.** Each luminaire (fixture) shall be protected against physical damage by a suitable guard or by location.

**(3) Pendant Luminaires (Fixtures).** Pendant luminaires (fixtures) shall be suspended by threaded rigid metal conduit stems, threaded steel intermediate metal conduit stems, by chains with approved fittings, or by other approved means. For rigid stems longer than 300 mm (12 in.), permanent and effective bracing against lateral displacement shall be provided at a level not more than 300 mm (12 in.) above the lower end of the stem, or flexibility in the form of a fitting or a flexible connector listed for the location shall be provided not more than 300 mm (12 in.) from the point of attachment to the supporting box or fitting. Threaded joints shall be provided with set-screws or other effective means to prevent loosening. Where wiring between an outlet box or fitting and a pendant luminaire (fixture) is not enclosed in conduit, flexible cord listed for hard usage shall be used, and suitable seals shall be provided where the cord enters the luminaire (fixture) and the outlet box or fitting. Flexible cord shall not serve as the supporting means for a fixture.

**(4) Supports.** Boxes, box assemblies, or fittings used for the support of luminaires (lighting fixtures) shall be identified for Class II locations.

**(B) Class II, Division 2.** In Class II, Division 2 locations, luminaires (lighting fixtures) shall comply with 502.11(B)(1) through (B)(5).

**(1) Portable Lighting Equipment.** Portable lighting equipment shall be identified for Class II locations. They shall be clearly marked to indicate the maximum wattage of lamps for which they are designed.

**(2) Fixed Lighting.** Luminaires (lighting fixtures) for fixed lighting, where not of a type identified for Class II locations, shall provide enclosures for lamps and lampholders that shall be designed to minimize the deposit of dust on lamps and to prevent the escape of sparks, burning material, or hot metal. Each fixture shall be clearly marked to indicate the maximum wattage of the lamp that shall be permitted without exceeding an exposed surface temperature in accordance with 500.8(C)(2) under normal conditions of use.

**(3) Physical Damage.** Luminaires (lighting fixtures) for fixed lighting shall be protected from physical damage by suitable guards or by location.

**(4) Pendant Luminaires (Fixtures).** Pendant luminaires (fixtures) shall be suspended by threaded rigid metal conduit stems, threaded steel intermediate metal conduit stems, by chains with approved fittings, or by other approved means. For rigid stems longer than 300 mm (12 in.), permanent and effective bracing against lateral displacement shall be provided at a level not more than 300 mm (12 in.) above the lower end of the stem, or flexibility in the form of an identified fitting or a flexible connector shall be provided not more than 300 mm (12 in.) from the point of attachment to the supporting box or fitting. Where wiring between an outlet box or fitting and a pendant luminaire (fixture) is not enclosed in conduit, flexible cord listed for hard usage shall be used. Flexible cord shall not serve as the supporting means for a fixture.

**(5) Electric-Discharge Lamps.** Starting and control equipment for electric-discharge lamps shall comply with the requirements of 502.7(B).

**502.12 Flexible Cords — Class II, Divisions 1 and 2.** Flexible cords used in Class II locations shall comply with all of the following:

(1) Be of a type listed for extra-hard usage

*Exception: Flexible cord listed for hard usage as permitted by 502.11(A)(3) and (B)(4).*

(2) Contain, in addition to the conductors of the circuit, a grounding conductor complying with 400.23

(3) Be connected to terminals or to supply conductors in an approved manner

(4) Be supported by clamps or by other suitable means in such a manner that there will be no tension on the terminal connections

(5) Be provided with suitable seals to prevent the entrance of dust where the flexible cord enters boxes or fittings that are required to be dust-ignitionproof

## 502.13 Receptacles and Attachment Plugs.

**(A) Class II, Division 1.** In Class II, Division 1 locations, receptacles and attachment plugs shall be of the type providing for connection to the grounding conductor of the flexible cord and shall be identified for Class II locations.

**(B) Class II, Division 2.** In Class II, Division 2 locations, receptacles and attachment plugs shall be of the type that provides for connection to the grounding conductor of the flexible cord and shall be designed so that connection to the supply circuit cannot be made or broken while live parts are exposed.

## 502.14 Signaling, Alarm, Remote-Control, and Communications Systems; and Meters, Instruments, and Relays.

> FPN: See Article 800 for rules governing the installation of communications circuits.

**(A) Class II, Division 1.** In Class II, Division 1 locations, signaling, alarm, remote-control, and communications systems; and meters, instruments, and relays shall comply with 502.14(A)(1) through (A)(6).

**(1) Wiring Methods.** The wiring method shall comply with 502.4(A).

**(2) Contacts.** Switches, circuit breakers, relays, contactors, fuses and current-breaking contacts for bells, horns, howlers, sirens, and other devices in which sparks or arcs may be produced shall be provided with enclosures identified for a Class II location.

*Exception: Where current-breaking contacts are immersed in oil or where the interruption of current occurs within a chamber sealed against the entrance of dust, enclosures shall be permitted to be of the general-purpose type.*

**(3) Resistors and Similar Equipment.** Resistors, transformers, choke coils, rectifiers, thermionic tubes, and other heat-generating equipment shall be provided with enclosures identified for Class II locations.

*Exception: Where resistors or similar equipment are immersed in oil or enclosed in a chamber sealed against the entrance of dust, enclosures shall be permitted to be of the general-purpose type.*

**(4) Rotating Machinery.** Motors, generators, and other rotating electric machinery shall comply with 502.8(A).

**(5) Combustible, Electrically Conductive Dusts.** Where dusts are of a combustible, electrically conductive nature, all wiring and equipment shall be identified for Class II locations.

**(6) Metal Dusts.** Where dust from magnesium, aluminum, aluminum bronze powders, or other metals of similarly hazardous characteristics may be present, all apparatus and equipment shall be identified for the specific conditions.

**(B) Class II, Division 2.** In Class II, Division 2 locations, signaling, alarm, remote-control, and communications systems; and meters, instruments, and relays shall comply with the following.

**(1) Contacts.** Enclosures shall comply with 502.14(A), (2) or contacts shall have tight metal enclosures designed to minimize the entrance of dust and shall have telescoping or tight-fitting covers and no openings through which, after installation, sparks or burning material might escape.

*Exception: In nonincendive circuits, enclosures shall be permitted to be of the general-purpose type.*

**(2) Transformers and Similar Equipment.** The windings and terminal connections of transformers, choke coils, and similar equipment shall be provided with tight metal enclosures without ventilating openings.

**(3) Resistors and Similar Equipment.** Resistors, resistance devices, thermionic tubes, rectifiers, and similar equipment shall comply with 502.14(A)(3).

*Exception: Enclosures for thermionic tubes, nonadjustable resistors, or rectifiers for which maximum operating temperature will not exceed 120°C (248°F) shall be permitted to be of the general-purpose type.*

**(4) Rotating Machinery.** Motors, generators, and other rotating electric machinery shall comply with 502.8(B).

**(5) Wiring Methods.** The wiring method shall comply with 502.4(B).

## 502.15 Live Parts, Class II, Divisions 1 and 2. Live parts shall not be exposed.

## 502.16 Grounding, Class II, Divisions 1 and 2. Wiring and equipment in Class II, Divisions 1 and 2 locations shall be grounded as specified in Article 250 and with the requirements in 502.16(A) and (B).

**(A) Bonding.** The locknut-bushing and double-locknut types of contact shall not be depended on for bonding purposes, but bonding jumpers with proper fittings or other approved means of bonding shall be used. Such means of bonding shall apply to all intervening raceways, fittings, boxes, enclosures, and so forth, between Class II locations and the point of grounding for service equipment or point of grounding of a separately derived system.

*Exception: The specific bonding means shall only be required to the nearest point where the grounded circuit con-*

*ductor and the grounding electrode conductor are connected together on the line side of the building or structure disconnecting means as specified in 250.32(A), (B), and (C), if the branch-circuit overcurrent protection is located on the load side of the disconnecting means.*

FPN: See 250.100 for additional bonding requirements in hazardous (classified) locations.

**(B) Types of Equipment Grounding Conductors.** Where flexible conduit is used as permitted in 502.4, it shall be installed with internal or external bonding jumpers in parallel with each conduit and complying with 250.102.

*Exception: In Class II, Division 2 locations, the bonding jumper shall be permitted to be deleted where all the following conditions are met:*

(a) *Listed liquidtight flexible metal conduit 1.8 m (6 ft) or less in length, with fittings listed for grounding, is used.*
(b) *Overcurrent protection in the circuit is limited to 10 amperes or less.*
(c) *The load is not a power utilization load.*

**502.17 Surge Protection — Class II, Divisions 1 and 2.** Surge arresters, including their installation and connection, shall comply with Article 280. In addition, surge arresters, if installed in a Class II, Division 1 location, shall be in suitable enclosures. Surge-protective capacitors shall be of a type designed for specific duty.

**502.18 Multiwire Branch Circuits.** In a Class II, Division 1 location, a multiwire branch circuit shall not be permitted.

*Exception: Where the disconnect device(s) for the circuit opens all ungrounded conductors of the multiwire circuit simultaneously.*

## ARTICLE 503
## Class III Locations

**503.1 General.** The general rules of this *Code* shall apply to electric wiring and equipment in locations classified as Class III locations in 500.5(D).

*Exception: As modified by this article.*

Equipment installed in Class III locations shall be able to function at full rating without developing surface temperatures high enough to cause excessive dehydration or gradual carbonization of accumulated fibers or flyings. Organic material that is carbonized or excessively dry is highly susceptible to spontaneous ignition. The maximum

surface temperatures under operating conditions shall not exceed 165°C (329°F) for equipment that is not subject to overloading, and 120°C (248°F) for equipment (such as motors or power transformers) that may be overloaded.

FPN: For electric trucks, see NFPA 505-1999, *Fire Safety Standard for Powered Industrial Trucks Including Type Designations, Areas of Use, Conversions, Maintenance, and Operation.*

**503.2 Transformers and Capacitors — Class III, Divisions 1 and 2.** Transformers and capacitors shall comply with 502.2(B).

**503.3 Wiring Methods.** Wiring methods shall comply with 503.3(A) or (B).

**(A) Class III, Division 1.** In Class III, Division 1 locations, the wiring method shall be rigid metal conduit, rigid nonmetallic conduit, intermediate metal conduit, electrical metallic tubing, dusttight wireways, or Type MC or MI cable with listed termination fittings.

**(1) Boxes and Fittings.** All boxes and fittings shall be dusttight.

**(2) Flexible Connections.** Where necessary to employ flexible connections, dusttight flexible connectors, liquidtight flexible metal conduit with listed fittings, liquidtight flexible nonmetallic conduit with listed fittings, or flexible cord in conformance with 503.10 shall be used.

FPN: See 503.16(B) for grounding requirements where flexible conduit is used.

**(B) Class III, Division 2.** In Class III, Division 2 locations, the wiring method shall comply with 503.3(A).

*Exception: In sections, compartments, or areas used solely for storage and containing no machinery, open wiring on insulators shall be permitted where installed in accordance with Article 398, but only on condition that protection as required by 398.15(C) be provided where conductors are not run in roof spaces and are well out of reach of sources of physical damage.*

**503.4 Switches, Circuit Breakers, Motor Controllers, and Fuses — Class III, Divisions 1 and 2.** Switches, circuit breakers, motor controllers, and fuses, including pushbuttons, relays, and similar devices, shall be provided with dusttight enclosures.

**503.5 Control Transformers and Resistors — Class III, Divisions 1 and 2.** Transformers, impedance coils, and resistors used as or in conjunction with control equipment for motors, generators, and appliances shall be provided with dusttight enclosures complying with the temperature limitations in 503.1.

**503.6 Motors and Generators — Class III, Divisions 1 and 2.** In Class III, Divisions 1 and 2 locations, motors, generators, and other rotating machinery shall be totally enclosed nonventilated, totally enclosed pipe ventilated, or totally enclosed fan cooled.

*Exception: In locations where, in the judgment of the authority having jurisdiction, only moderate accumulations of lint or flyings are likely to collect on, in, or in the vicinity of a rotating electric machine and where such machine is readily accessible for routine cleaning and maintenance, one of the following shall be permitted:*

*(a) Self-cleaning textile motors of the squirrel-cage type*

*(b) Standard open-type machines without sliding contacts, centrifugal or other types of switching mechanisms, including motor overload devices*

*(c) Standard open-type machines having such contacts, switching mechanisms, or resistance devices enclosed within tight housings without ventilating or other openings*

**503.7 Ventilating Piping — Class III, Divisions 1 and 2.** Ventilating pipes for motors, generators, or other rotating electric machinery, or for enclosures for electric equipment, shall be of metal not less than 0.53 mm (0.021 in.) in thickness, or of equally substantial noncombustible material, and shall comply with the following:

(1) Lead directly to a source of clean air outside of buildings

(2) Be screened at the outer ends to prevent the entrance of small animals or birds

(3) Be protected against physical damage and against rusting or other corrosive influences

Ventilating pipes shall be sufficiently tight, including their connections, to prevent the entrance of appreciable quantities of fibers or flyings into the ventilated equipment or enclosure and to prevent the escape of sparks, flame, or burning material that might ignite accumulations of fibers or flyings or combustible material in the vicinity. For metal pipes, lock seams and riveted or welded joints shall be permitted; and tight-fitting slip joints shall be permitted where some flexibility is necessary, as at connections to motors.

**503.8 Utilization Equipment — Class III, Divisions 1 and 2.**

**(A) Heaters.** Electrically heated utilization equipment shall be identified for Class III locations.

**(B) Motors.** Motors of motor-driven utilization equipment shall comply with 503.6.

**(C) Switches, Circuit Breakers, Motor Controllers, and Fuses.** Switches, circuit breakers, motor controllers, and fuses shall comply with 503.4.

**503.9 Luminaires (Lighting Fixtures) — Class III, Divisions 1 and 2.**

**(A) Fixed Lighting.** Luminaires (lighting fixtures) for fixed lighting shall provide enclosures for lamps and lampholders that are designed to minimize entrance of fibers and flyings and to prevent the escape of sparks, burning material, or hot metal. Each luminaire (fixture) shall be clearly marked to show the maximum wattage of the lamps that shall be permitted without exceeding an exposed surface temperature of 165°C (329°F) under normal conditions of use.

**(B) Physical Damage.** A luminaire (fixture) that may be exposed to physical damage shall be protected by a suitable guard.

**(C) Pendant Luminaires (Fixtures).** Pendant luminaires (fixtures) shall be suspended by stems of threaded rigid metal conduit, threaded intermediate metal conduit, threaded metal tubing of equivalent thickness, or by chains with approved fittings. For stems longer than 300 mm (12 in.), permanent and effective bracing against lateral displacement shall be provided at a level not more than 300 mm (12 in.) above the lower end of the stem, or flexibility in the form of an identified fitting or a flexible connector shall be provided not more than 300 mm (12 in.) from the point of attachment to the supporting box or fitting.

**(D) Portable Lighting Equipment.** Portable lighting equipment shall be equipped with handles and protected with substantial guards. Lampholders shall be of the unswitched type with no provision for receiving attachment plugs. There shall be no exposed current-carrying metal parts, and all exposed non–current-carrying metal parts shall be grounded. In all other respects, portable lighting equipment shall comply with 503.9(A).

**503.10 Flexible Cords — Class III, Divisions 1 and 2.** Flexible cords shall comply with the following:

(1) Be of a type listed for extra-hard usage

(2) Contain, in addition to the conductors of the circuit, a grounding conductor complying with 400.23

(3) Be connected to terminals or to supply conductors in an approved manner

(4) Be supported by clamps or other suitable means in such a manner that there will be no tension on the terminal connections

(5) Be provided with suitable means to prevent the entrance of fibers or flyings where the cord enters boxes or fittings

**503.11 Receptacles and Attachment Plugs — Class III, Divisions 1 and 2.** Receptacles and attachment plugs shall be of the grounding type and shall be designed so as to minimize the accumulation or the entry of fibers or flyings, and shall prevent the escape of sparks or molten particles.

*Exception: In locations where, in the judgment of the authority having jurisdiction, only moderate accumulations of lint or flyings will be likely to collect in the vicinity of a receptacle, and where such receptacle is readily accessible for routine cleaning, general-purpose grounding-type receptacles mounted so as to minimize the entry of fibers or flyings shall be permitted.*

**503.12 Signaling, Alarm, Remote-Control, and Local Loudspeaker Intercommunications Systems — Class III, Divisions 1 and 2.** Signaling, alarm, remote-control, and local loudspeaker intercommunications systems shall comply with the requirements of Article 503 regarding wiring methods, switches, transformers, resistors, motors, luminaires (lighting fixtures), and related components.

**503.13 Electric Cranes, Hoists, and Similar Equipment — Class III, Divisions 1 and 2.** Where installed for operation over combustible fibers or accumulations of flyings, traveling cranes and hoists for material handling, traveling cleaners for textile machinery, and similar equipment shall comply with 503.13(A) through (D).

**(A) Power Supply.** Power supply to contact conductors shall be isolated from all other systems and shall be equipped with an acceptable ground detector that gives an alarm and automatically de-energizes the contact conductors in case of a fault to ground or gives a visual and audible alarm as long as power is supplied to the contact conductors and the ground fault remains.

**(B) Contact Conductors.** Contact conductors shall be located or guarded so as to be inaccessible to other than authorized persons and shall be protected against accidental contact with foreign objects.

**(C) Current Collectors.** Current collectors shall be arranged or guarded so as to confine normal sparking and prevent escape of sparks or hot particles. To reduce sparking, two or more separate surfaces of contact shall be provided for each contact conductor. Reliable means shall be provided to keep contact conductors and current collectors free of accumulations of lint or flyings.

**(D) Control Equipment.** Control equipment shall comply with 503.4 and 503.5.

**503.14 Storage Battery Charging Equipment — Class III, Divisions 1 and 2.** Storage battery charging equipment shall be located in separate rooms built or lined with substantial noncombustible materials. The rooms shall be constructed to prevent the entrance of ignitible amounts of flyings or lint and shall be well ventilated.

**503.15 Live Parts — Class III, Divisions 1 and 2.** Live parts shall not be exposed.

*Exception: As provided in 503.13.*

**503.16 Grounding — Class III, Divisions 1 and 2.** Wiring and equipment in Class III, Divisions 1 and 2 locations shall be grounded as specified in Article 250 and with the following additional requirements in 503.16(A) and (B).

**(A) Bonding.** The locknut-bushing and double-locknut types of contacts shall not be depended on for bonding purposes, but bonding jumpers with proper fittings or other approved means of bonding shall be used. Such means of bonding shall apply to all intervening raceways, fittings, boxes, enclosures, and so forth, between Class III locations and the point of grounding for service equipment or point of grounding of a separately derived system.

*Exception: The specific bonding means shall only be required to the nearest point where the grounded circuit conductor and the grounding electrode conductor are connected together on the line side of the building or structure disconnecting means as specified in 250.32(A), (B), and (C), if the branch-circuit overcurrent protection is located on the load side of the disconnecting means.*

> FPN: See 250.100 for additional bonding requirements in hazardous (classified) locations.

**(B) Types of Equipment Grounding Conductors.** Where flexible conduit is used as permitted in 503.3, it shall be installed with internal or external bonding jumpers in parallel with each conduit and complying with 250.102.

*Exception: In Class III, Division 1 and 2 locations, the bonding jumper shall be permitted to be deleted where all the following conditions are met:*

*(a) Listed liquidtight flexible metal 1.8 m (6 ft) or less in length, with fittings listed for grounding, is used.*
*(b) Overcurrent protection in the circuit is limited to 10 amperes or less.*
*(c) The load is not a power utilization load.*

## ARTICLE 504
## Intrinsically Safe Systems

**504.1 Scope.** This article covers the installation of intrinsically safe (I.S.) apparatus, wiring, and systems for Class I, II, and III locations.

FPN: For further information, see ANSI/ISA RP 12.6-1995, *Wiring Practices for Hazardous (Classified) Locations Instrumentation — Part 1: Intrinsic Safety.*

## 504.2 Definitions.

**Associated Apparatus.** Apparatus in which the circuits are not necessarily intrinsically safe themselves but that affect the energy in the intrinsically safe circuits and are relied on to maintain intrinsic safety. Associated apparatus may be either of the following:

(1) Electrical apparatus that has an alternative-type protection for use in the appropriate hazardous (classified) location, or

(2) Electrical apparatus not so protected that shall not be used within a hazardous (classified) location

FPN No. 1: Associated apparatus has identified intrinsically safe connections for intrinsically safe apparatus and also may have connections for nonintrinsically safe apparatus.

FPN No. 2: An example of associated apparatus is an intrinsic safety barrier, which is a network designed to limit the energy (voltage and current) available to the protected circuit in the hazardous (classified) location, under specified fault conditions.

**Control Drawing.** See definition in 500.2.

**Different Intrinsically Safe Circuits.** Intrinsically safe circuits in which the possible interconnections have not been evaluated and identified as intrinsically safe.

**Intrinsically Safe Apparatus.** Apparatus in which all the circuits are intrinsically safe.

**Intrinsically Safe Circuit.** A circuit in which any spark or thermal effect is incapable of causing ignition of a mixture of flammable or combustible material in air under prescribed test conditions.

FPN: Test conditions are described in ANSI/UL 913-1997, *Standard for Safety, Intrinsically Safe Apparatus and Associated Apparatus for Use in Class I, II, and III, Division 1, Hazardous (Classified) Locations.*

**Intrinsically Safe System.** An assembly of interconnected intrinsically safe apparatus, associated apparatus, and interconnecting cables in that those parts of the system that may be used in hazardous (classified) locations are intrinsically safe circuits.

FPN: An intrinsically safe system may include more than one intrinsically safe circuit.

**Simple Apparatus.** An electrical component or combination of components of simple construction with well-defined electrical parameters that does not generate more than 1.5 volts, 100 milliamps, and 25 milliwatts, or a passive component that does not dissipate more than 1.3 watts

and is compatible with the intrinsic safety of the circuit in which it is used.

FPN: The following apparatus are examples of simple apparatus:

(a) Passive components; for example, switches, junction boxes, resistance temperature devices, and simple semiconductor devices such as LEDs

(b) Sources of generated energy, for example, thermocouples and photocells, which do not generate more than 1.5 V, 100 mA, and 25 mW

## 504.3 Application of Other Articles.
Except as modified by this article, all applicable articles of this *Code* shall apply.

## 504.4 Equipment.
All intrinsically safe apparatus and associated apparatus shall be listed.

*Exception: Simple apparatus, as described on the control drawing, shall not be required to be listed.*

## 504.10 Equipment Installation.

**(A) Control Drawing.** Intrinsically safe apparatus, associated apparatus, and other equipment shall be installed in accordance with the control drawing(s).

*Exception: A simple apparatus that does not interconnect intrinsically safe circuits.*

FPN: The control drawing identification is marked on the apparatus.

**(B) Location.** Intrinsically safe apparatus shall be permitted to be installed in any hazardous (classified) locations for which it has been identified. General-purpose enclosures shall be permitted for intrinsically safe apparatus.

Associated apparatus shall be permitted to be installed in any hazardous (classified) location for which it has been identified or, if protected by other means, permitted by Articles 501 through 503 and Article 505.

## 504.20 Wiring Methods.
Intrinsically safe apparatus and wiring shall be permitted to be installed using any of the wiring methods suitable for unclassified locations, including Chapter 7 and Chapter 8. Sealing shall be as provided in 504.70, and separation shall be as provided in 504.30.

## 504.30 Separation of Intrinsically Safe Conductors.

**(A) From Nonintrinsically Safe Circuit Conductors.**

**(1) Open Wiring.** Conductors and cables of intrinsically safe circuits not in raceways or cable trays shall be separated at least 50 mm (2 in.) and secured from conductors and cables of any nonintrinsically safe circuits.

*Exception: Where either (1) all of the intrinsically safe circuit conductors are in Type MI or MC cables or (2) all of the nonintrinsically safe circuit conductors are in raceways or Type MI or MC cables where the sheathing or cladding is capable of carrying fault current to ground.*

**(2) In Raceways, Cable Trays, and Cables.** Conductors of intrinsically safe circuits shall not be placed in any raceway, cable tray, or cable with conductors of any nonintrinsically safe circuit.

*Exception No. 1: Where conductors of intrinsically safe circuits are separated from conductors of nonintrinsically safe circuits by a distance of at least 50 mm (2 in.) and secured, or by a grounded metal partition or an approved insulating partition.*

FPN: No. 20 gauge sheet metal partitions 0.91 mm (0.0359 in.) or thicker are generally considered acceptable.

*Exception No. 2: Where either (1) all of the intrinsically safe circuit conductors or (2) all of the nonintrinsically safe circuit conductors are in grounded metal-sheathed or metal-clad cables where the sheathing or cladding is capable of carrying fault current to ground.*

FPN: Cables meeting the requirements of Articles 330 and 332 are typical of those considered acceptable.

**(3) Within Enclosures.**

(a) Conductors of intrinsically safe circuits shall be separated at least 50 mm (2 in.) from conductors of any nonintrinsically safe circuits, or as specified in 504.30(A)(2).

(b) All conductors shall be secured so that any conductor that might come loose from a terminal cannot come in contact with another terminal.

FPN No. 1: The use of separate wiring compartments for the intrinsically safe and nonintrinsically safe terminals is the preferred method of complying with this requirement.

FPN No. 2: Physical barriers such as grounded metal partitions or approved insulating partitions or approved restricted access wiring ducts separated from other such ducts by at least 19 mm (¾ in.) can be used to help ensure the required separation of the wiring.

**(B) From Different Intrinsically Safe Circuit Conductors.** Different intrinsically safe circuits shall be in separate cables or shall be separated from each other by one of the following means:

(1) The conductors of each circuit are within a grounded metal shield.

(2) The conductors of each circuit have insulation with a minimum thickness of 0.25 mm (0.01 in.).

*Exception: Unless otherwise identified.*

**504.50 Grounding.**

**(A) Intrinsically Safe Apparatus, Associated Apparatus, and Raceways.** Intrinsically safe apparatus, associated apparatus, cable shields, enclosures, and raceways, if of metal, shall be grounded.

FPN: Supplementary bonding to the grounding electrode may be needed for some associated apparatus, for example, zener diode barriers, if specified in the control drawing. See ANSI/ISA RP 12.6-1995, *Wiring Practices for Hazardous (Classified) Locations Instrumentation Part 1: Intrinsic Safety.*

**(B) Connection to Grounding Electrodes.** Where connection to a grounding electrode is required, the grounding electrode shall be as specified in 250.52(A)(1), (2), (3), and (4) and shall comply with 250.30(A)(3). Section 250.52(A)(5), (6) and (7) shall not be used if electrodes specified in 250.52(A)(1), (2), (3) or (4) are available.

**(C) Shields.** Where shielded conductors or cables are used, shields shall be grounded.

*Exception: Where a shield is part of an intrinsically safe circuit.*

**504.60 Bonding.**

**(A) Hazardous Locations.** In hazardous (classified) locations, intrinsically safe apparatus shall be bonded in the hazardous (classified) location in accordance with 250.100.

**(B) Unclassified.** In unclassified or nonhazardous locations, where metal raceways are used for intrinsically safe system wiring in hazardous (classified) locations, associated apparatus shall be bonded in accordance with 501.16(A), 502.16(A), 503.16(A), or 505.25, as applicable.

**504.70 Sealing.** Conduits and cables that are required to be sealed by 501.5 and 502.5 shall be sealed to minimize the passage of gases, vapors, or dusts. Such seals shall not be required to be explosionproof or flameproof.

*Exception: Seals shall not be required for enclosures that contain only intrinsically safe apparatus, except as required by 501.5(F)(3).*

**504.80 Identification.** Labels required by this section shall be suitable for the environment where they are installed with consideration given to exposure to chemicals and sunlight.

**(A) Terminals.** Intrinsically safe circuits shall be identified at terminal and junction locations in a manner that will prevent unintentional interference with the circuits during testing and servicing.

**(B) Wiring.** Raceways, cable trays, and other wiring methods for intrinsically safe system wiring shall be identified with permanently affixed labels with the wording "Intrinsic Safety Wiring" or equivalent. The labels shall be located so as to be visible after installation and placed so that they may be readily traced through the entire length of the installation. Intrinsic safety circuit labels shall appear in every section of the wiring system that is separated by enclosures, walls, partitions, or floors. Spacing between labels shall not be more than 7.5 m (25 ft).

*Exception: Circuits run underground shall be permitted to be identified where they become accessible after emergence from the ground.*

FPN No. 1: Wiring methods permitted in unclassified locations may be used for intrinsically safe systems in hazardous (classified) locations. Without labels to identify the application of the wiring, enforcement authorities cannot determine that an installation is in compliance with the *Code*.

FPN No. 2: In unclassified locations, the identification is necessary to ensure that nonintrinsically safe wire will not be inadvertently added to existing raceways at a later date.

**(C) Color Coding.** Color coding shall be permitted to identify intrinsically safe conductors where they are colored light blue and where no other conductors colored light blue are used. Likewise, color coding shall be permitted to identify raceways, cable trays, and junction boxes where they are colored light blue and contain only intrinsically safe wiring.

# ARTICLE 505
# Class I, Zone 0, 1, and 2 Locations

FPN: Rules that are followed by a reference in brackets contain text that has been extracted from NFPA 497-1997, *Recommended Practice for the Classification of Flammable Liquids, Gases, or Vapors and of Hazardous (Classified) Locations for Electrical Installations in Chemical Process Areas.* Only editorial changes were made to the extracted text to make it consistent with this *Code*.

**505.1 Scope.** This article covers the requirements for the zone classification system as an alternative to the division classification system covered in Article 500 for electrical and electronic equipment and wiring for all voltages in Class I, Zone 0, Zone 1, and Zone 2 hazardous (classified) locations where fire or explosion hazards may exist due to flammable gases, vapors, or liquids.

FPN: For the requirements for electrical and electronic equipment and wiring for all voltages in Class I, Division 1 or Division 2; Class II, Division 1 or Division 2; and Class III, Division 1 or Division 2 hazardous (classified) locations where fire or explosion hazards may exist due to flammable gases or vapors, flammable liquids, or combustible dusts or fibers, refer to Articles 500 through 504.

**505.2 Definitions.** For purposes of this article, the following definitions apply.

**Combustible Gas Detection System.** A protection technique utilizing stationary gas detectors in industrial establishments.

**Electrical and Electronic Equipment.** Materials, fittings, devices, appliances, and the like that are part of, or in connection with, an electrical installation.

FPN: Portable or transportable equipment having self-contained power supplies, such as battery-operated equipment, could potentially become an ignition source in hazardous (classified) locations.

**Encapsulation "m."** Type of protection where electrical parts that could ignite an explosive atmosphere by either sparking or heating are enclosed in a compound in such a way that this explosive atmosphere cannot be ignited.

FPN: See ISA 12.23.01-1998, *Electrical Apparatus for Use in Class I, Zone 1 Hazardous (Classified) Locations, Type of Protection — Encapsulation "m"*; IEC 60079-18-1992, *Electrical Apparatus for Explosive Gas Atmospheres — Part 18: Encapsulation "m"* ; and ANSI/UL 2279-1997(Part 18), *Electrical Equipment for Use in Class I, Zone 0, 1, and 2 Hazardous (Classified) Locations.*

**Flameproof "d."** Type of protection where the enclosure will withstand an internal explosion of a flammable mixture that has penetrated into the interior, without suffering damage and without causing ignition, through any joints or structural openings in the enclosure, of an external explosive gas atmosphere consisting of one or more of the gases or vapors for which it is designed.

FPN: See ISA 12.22.01-1998, *Electrical Apparatus for Use in Class I, Zone 1 and 2 Hazardous (Classified) Locations, Type of Protection — Flameproof "d"*; IEC 60079-1-2000, *Electrical Apparatus for Explosive Gas Atmospheres, Part 1 — Construction and Verification Test of Flameproof Enclosures of Electrical Apparatus*; ANSI/UL 2279-1997 (Part 1), *Electrical Equipment for Use in Class I, Zone 0, 1, and 2 Hazardous (Classified) Locations.*

**Increased Safety "e."** Type of protection applied to electrical equipment that does not produce arcs or sparks in normal service and under specified abnormal conditions, in which additional measures are applied so as to give increased security against the possibility of excessive temperatures and of the occurrence of arcs and sparks.

FPN: See ISA — 12.16.01-1998, *Electrical Apparatus for Use in Class I, Zone 1 Hazardous (Classified) Locations, Type of Protection — Increased Safety "e"*; IEC 60079-7-1990, *Electrical Apparatus for Explosive Gas Atmospheres*

— *Part 7: Increased Safety "e"*, Amendment No. 1 (1991) and Amendment No. 2 (1993); and ANSI/UL 2279-1997(Part 7), *Electrical Equipment for Use in Class I, Zone 0, 1, and 2 Hazardous (Classified) Locations.*

**Intrinsic Safety "i."** Type of protection where any spark or thermal effect is incapable of causing ignition of a mixture of flammable or combustible material in air under prescribed test conditions.

> FPN No. 1: See ANSI/UL 913-1997, *Intrinsically Safe Apparatus and Associated Apparatus for Use in Class I, II, and III, Hazardous Locations*; ISA —12.02.01-1999, *Electrical Apparatus for Use in Class I, Zones 0, 1 and 2 Hazardous (Classified) Locations — Intrinsic Safety "i"*; IEC 60079-11-1999, *Electrical Apparatus for Explosive Gas Atmospheres — Part 11: Intrinsic Safety "i"*; and ANSI/UL 2279-1997 (Part 11), *Electrical Equipment for Use in Class I, Zone 0, 1, and 2 Hazardous (Classified) Locations.*

> FPN No. 2: Intrinsic safety is designated type of protection "ia" for use in Zone 0 locations. Intrinsic safety is designated type of protection "ib" for use in Zone 1 locations.

> FPN No. 3: Intrinsically safe associated apparatus, designated by [ia] or [ib], is connected to intrinsically safe apparatus ("ia" or "ib," respectively) but is located outside the hazardous (classified) location unless also protected by another type of protection (such as flameproof).

**Oil Immersion "o."** Type of protection where electrical equipment is immersed in a protective liquid in such a way that an explosive atmosphere that may be above the liquid or outside the enclosure cannot be ignited.

> FPN: See ISA 12.26.01-1998, *Electrical Apparatus for Use in Class I, Zone 1 Hazardous (Classified) Locations, Type of Protection — Oil-Immersion "o"*; IEC 60079-6-1995, *Electrical Apparatus for Explosive Gas Atmospheres, Part 6 — Oil-Immersion "o"* ; and ANSI/UL 2279-1997 (Part 6), *Electrical Equipment for Use in Class I, Zone 0, 1, and 2 Hazardous (Classified) Locations.*

**Powder Filling "q."** Type of protection where electrical parts capable of igniting an explosive atmosphere are fixed in position and completely surrounded by filling material (glass or quartz powder) to prevent the ignition of an external explosive atmosphere.

> FPN: See ISA-12.25.01-1996, *Electrical Apparatus for Use in Class I, Zone 1 Hazardous (Classified) Locations Type of Protection — Powder Filling "q"*; IEC 60079-5-1996, *Electrical Apparatus for Explosive Gas Atmospheres — Part 5: Powder Filling, Type of Protection "q"* ; and ANSI/UL 2279-1997 (Part 5), *Electrical Equipment for Use in Class I, Zone 0, 1, and 2 Hazardous (Classified) Locations.*

**Purged and Pressurized.** Type of protection for electrical equipment that uses the technique of guarding against the ingress of the external atmosphere, which may be explosive, into an enclosure by maintaining a protective gas therein at a pressure above that of the external atmosphere.

> FPN No. 1: See NFPA 496-1998, *Standard for Purged and Pressurized Enclosures for Electrical Equipment.*

> FPN No. 2: See IEC 60079-2-2000, *Electrical Apparatus for Explosive Gas Atmospheres — Part 2: Electrical Apparatus, Type of Protection "p"*; and IEC 60079-13-1982, *Electrical Apparatus for Explosive Gas Atmospheres — Part 13: Construction and Use of Rooms or Buildings Protected by Pressurization.*

**Type of Protection "n."** Type of protection where electrical equipment, in normal operation, is not capable of igniting a surrounding explosive gas atmosphere and a fault capable of causing ignition is not likely to occur.

> FPN: See IEC 60079-15-2000, *Electrical Apparatus for Explosive Gas Atmospheres, Part 15 — Electrical Apparatus with Type of Protection "n"*; and ANSI/UL 2279-1997 (Part 15), *Electrical Equipment for Use in Class I, Zone 0, 1, and 2 Hazardous (Classified) Locations.*

**Unclassified Locations.** Locations determined to be neither Class I, Division 1; Class I, Division 2; Class I, Zone 0; Class I, Zone 1; Class I, Zone 2; Class II, Division 1; Class II, Division 2; Class III, Division 1; Class III, Division 2; or any combination thereof.

**505.3 Other Articles.** All other applicable rules contained in this *Code* shall apply to electrical equipment and wiring installed in hazardous (classified) locations.

*Exception: As modified by Article 504 and this article.*

**505.4 General.**

**(A) Documentation for Industrial Occupancies.** All areas in industrial occupancies designated as hazardous (classified) locations shall be properly documented. This documentation shall be available to those authorized to design, install, inspect, maintain, or operate electrical equipment at the location.

> FPN: For examples of area classification drawings, see ANSI/API RP 505-1997, *Recommended Practice for Classification of Locations for Electrical Installations at Petroleum Facilities Classified as Class I, Zone 0, Zone 1, or Zone 2*; ISA RP12.24.01-1998, *Recommended Practice for Classification of Locations for Electrical Installations Classified as Class I, Zone 0, Zone 1, or Zone 2*; IEC 60079-10-1995, *Electrical Apparatus for Explosive Gas Atmospheres, Classification of Hazardous Areas*; and *Model Code of Safe Practice in the Petroleum Industry, Part 15: Area Classification Code for Petroleum Installations*, IP 15, The Institute of Petroleum, London.

**(B) Reference Standards.** Important information relating to topics covered in Chapter 5 may be found in other publications.

> FPN No. 1: It is important that the authority having jurisdiction be familiar with recorded industrial experience as well as with standards of the National Fire Protection Association (NFPA), the American Petroleum Institute (API),

Instrumentation, Systems, and Automation Society (ISA), and the International Electrotechnical Commission (IEC) that may be of use in the classification of various locations, the determination of adequate ventilation, and the protection against static electricity and lightning hazards.

FPN No. 2: For further information on the classification of locations, see ANSI/API RP 505-1997, *Recommended Practice for Classification of Locations for Electrical Installations at Petroleum Facilities Classified as Class I, Zone 0, Zone 1, or Zone 2*; ISA RP12.24.01-1998, *Recommended Practice for Classification of Locations for Electrical Installations Classified as Class I, Zone 0, Zone 1, or Zone 2*; IEC 60079-10-1995, *Electrical Apparatus for Explosive Gas Atmospheres, Classification of Hazardous Areas*; and *Model Code of Safe Practice in the Petroleum Industry, Part 15: Area Classification Code for Petroleum Installations*, IP 15, The Institute of Petroleum, London.

FPN No. 3: For further information on protection against static electricity and lightning hazards in hazardous (classified) locations, see NFPA 77-2000, *Recommended Practice on Static Electricity*; NFPA 780-1997, *Standard for the Installation of Lightning Protection Systems*; and API RP 2003-1998, *Protection Against Ignitions Arising Out of Static Lightning and Stray Currents*.

FPN No. 4: For further information on ventilation, see NFPA 30-2000, *Flammable and Combustible Liquids Code*, and ANSI/API RP 505-1997, *Recommended Practice for Classification of Locations for Electrical Installations at Petroleum Facilities Classified as Class I, Zone 0, Zone 1, or Zone 2*.

FPN No. 5: For further information on electrical systems for hazardous (classified) locations on offshore oil and gas producing platforms, see ANSI/API RP 14FZ-2000, *Recommended Practice for Design and Installation of Electrical Systems for Fixed and Floating Offshore Petroleum Facilities for Unclassified and Class I, Zone 0, Zone 1, and Zone 2 Locations*.

FPN No. 6: For further information on the installation of electrical equipment in hazardous (classified) locations in general, see IEC 60079-14-1996, *Electrical Apparatus for Explosive Gas Atmospheres — Part 14: Electrical Installations in Explosive Gas Atmospheres (Other Than Mines)*, and IEC 60079-16-1990, *Electrical Apparatus for Explosive Gas Atmospheres — Part 16: Artificial Ventilation for the Protection of Analyzer(s) Houses*.

FPN No. 7: For further information on application of electrical equipment in hazardous (classified) locations in general, see ISA 12.00.01-1999, *Electrical Apparatus for Use in Class I, Zones 0 and 1, Hazardous (Classified) Locations: General Requirements*; ISA 12.01.01-1999, *Definitions and Information Pertaining to Electrical Apparatus in Hazardous (Classified) Locations*; and ANSI/UL 2279-1997 (Part 0), *Electrical Equipment for Use in Class I, Zone 0, 1, and 2 Hazardous (Classified) Locations*.

## 505.5 Classifications of Locations.

**(A) Classification of Locations.** Locations shall be classified depending on the properties of the flammable vapors, liquids, or gases that may be present and the likelihood that a flammable or combustible concentration or quantity is present. Where pyrophoric materials are the only materials used or handled, these locations shall not be classified. Each room, section, or area shall be considered individually in determining its classification.

FPN No. 1: See 505.7 for restrictions on area classification.

FPN No. 2: Through the exercise of ingenuity in the layout of electrical installations for hazardous (classified) locations, it is frequently possible to locate much of the equipment in reduced level of classification or in an unclassified location and, thus, to reduce the amount of special equipment required.

Rooms and areas containing ammonia refrigeration systems that are equipped with adequate mechanical ventilation may be classified as "unclassified" locations.

FPN: For further information regarding classification and ventilation of areas involving ammonia, see ANSI/ASHRAE 15-1994, *Safety Code for Mechanical Refrigeration*; and ANSI/CGA G2.1-1989 (14-39), *Safety Requirements for the Storage and Handling of Anhydrous Ammonia*.

**(B) Class I, Zone 0, 1, and 2 Locations.** Class I, Zone 0, 1, and 2 locations are those in which flammable gases or vapors are or may be present in the air in quantities sufficient to produce explosive or ignitible mixtures. Class I, Zone 0, 1, and 2 locations shall include those specified in 505(B)(1), (B)(2), and (B)(3).

**(1) Class I, Zone 0.** A Class I, Zone 0 location is a location in which

(1) Ignitible concentrations of flammable gases or vapors are present continuously, or
(2) Ignitible concentrations of flammable gases or vapors are present for long periods of time.

FPN No. 1: As a guide in determining when flammable gases or vapors are present continuously or for long periods of time, refer to ANSI/API RP 505-1997, *Recommended Practice for Classification of Locations for Electrical Installations of Petroleum Facilities Classified as Class I, Zone 0, Zone 1 or Zone 2*; ISA 12.24.01-1998, *Recommended Practice for Classification of Locations for Electrical Installations Classified as Class I, Zone 0, Zone 1, or Zone 2*; IEC 60079-10-1995, *Electrical Apparatus for Explosive Gas Atmospheres, Classifications of Hazardous Areas*; and *Area Classification Code for Petroleum Installations, Model Code, Part 15*, Institute of Petroleum.

FPN No. 2: This classification includes locations inside vented tanks or vessels that contain volatile flammable liquids; inside inadequately vented spraying or coating enclosures, where volatile flammable solvents are used; between the inner and outer roof sections of a floating roof tank containing volatile flammable liquids; inside open vessels, tanks and pits containing volatile flammable liquids; the interior of an exhaust duct that is used to vent ignitible concentrations of gases or vapors; and inside inadequately ventilated enclosures that contain normally venting instruments utilizing or analyzing flammable fluids and venting to the inside of the enclosures.

FPN No. 3: It is not good practice to install electrical equipment in Zone 0 locations except when the equipment is essential to the process or when other locations are not feasible. [See 505.5(A) FPN No. 2.] If it is necessary to install electrical systems in a Zone 0 location, it is good practice to install intrinsically safe systems as described by Article 504.

**(2) Class I, Zone 1.** A Class I, Zone 1 location is a location

(1) In which ignitible concentrations of flammable gases or vapors are likely to exist under normal operating conditions; or

(2) In which ignitible concentrations of flammable gases or vapors may exist frequently because of repair or maintenance operations or because of leakage; or

(3) In which equipment is operated or processes are carried on, of such a nature that equipment breakdown or faulty operations could result in the release of ignitible concentrations of flammable gases or vapors and also cause simultaneous failure of electrical equipment in a mode to cause the electrical equipment to become a source of ignition; or

(4) That is adjacent to a Class I, Zone 0 location from which ignitible concentrations of vapors could be communicated, unless communication is prevented by adequate positive pressure ventilation from a source of clean air and effective safeguards against ventilation failure are provided.

FPN No. 1: Normal operation is considered the situation when plant equipment is operating within its design parameters. Minor releases of flammable material may be part of normal operations. Minor releases include the releases from mechanical packings on pumps. Failures that involve repair or shutdown (such as the breakdown of pump seals and flange gaskets, and spillage caused by accidents) are not considered normal operation.

FPN No. 2: This classification usually includes locations where volatile flammable liquids or liquefied flammable gases are transferred from one container to another. In areas in the vicinity of spraying and painting operations where flammable solvents are used; adequately ventilated drying rooms or compartments for evaporation of flammable solvents; adequately ventilated locations containing fat and oil extraction equipment using volatile flammable solvents; portions of cleaning and dyeing plants where volatile flammable liquids are used; adequately ventilated gas generator rooms and other portions of gas manufacturing plants where flammable gas may escape; inadequately ventilated pump rooms for flammable gas or for volatile flammable liquids; the interiors of refrigerators and freezers in which volatile flammable materials are stored in the open, lightly stoppered, or in easily ruptured containers; and other locations where ignitible concentrations of flammable vapors or gases are likely to occur in the course of normal operation but not classified Zone 0.

**(3) Class I, Zone 2.** A Class I, Zone 2 location is a location

(1) In which ignitible concentrations of flammable gases or vapors are not likely to occur in normal operation and, if they do occur, will exist only for a short period; or

(2) In which volatile flammable liquids, flammable gases, or flammable vapors are handled, processed, or used but in which the liquids, gases, or vapors normally are confined within closed containers of closed systems from which they can escape, only as a result of accidental rupture or breakdown of the containers or system, or as a result of the abnormal operation of the equipment with which the liquids or gases are handled, processed, or used; or

(3) In which ignitible concentrations of flammable gases or vapors normally are prevented by positive mechanical ventilation but which may become hazardous as a result of failure or abnormal operation of the ventilation equipment; or

(4) That is adjacent to a Class I, Zone 1 location, from which ignitible concentrations of flammable gases or vapors could be communicated, unless such communication is prevented by adequate positive-pressure ventilation from a source of clean air and effective safeguards against ventilation failure are provided.

FPN: The Zone 2 classification usually includes locations where volatile flammable liquids or flammable gases or vapors are used but which would become hazardous only in case of an accident or of some unusual operating condition.

**505.6 Material Groups.** For purposes of testing, approval, and area classification, various air mixtures (not oxygen enriched) shall be grouped as required in 505.6(A), (B), and (C).

FPN: Group I is intended for use in describing atmospheres that contain firedamp (a mixture of gases, composed mostly of methane, found underground, usually in mines). This *Code* does not apply to installations underground in mines. See 90.2(B).

Group II shall be subdivided into IIC, IIB, and IIA, as noted in 505.6(A), (B), and (C), according to the nature of the gas or vapor, for protection techniques "d," "ia," "ib," "[ia]," and "[ib]," and, where applicable, "n" and "o."

FPN No. 1: The gas and vapor subdivision as described above is based on the maximum experimental safe gap (MESG), minimum igniting current (MIC), or both. Test equipment for determining the MESG is described in IEC 60079-1A-1975, Amendment No. 1 (1993), *Construction and Verification Tests of Flameproof Enclosures of Electrical Apparatus*; and *UL Technical Report No. 58* (1993). The test equipment for determining MIC is described in IEC 60079-11-1999, *Electrical Apparatus for Explosive Gas Atmospheres — Part 11: Intrinsic Safety "i"*. The classification of gases or vapors according to their maximum experimental safe gaps and minimum igniting currents is described in IEC 60079-12-1978, *Classification of Mixtures of Gases or Vapours with Air According to Their Maximum Experimental Safe Gaps and Minimum Igniting Currents.*

FPN No. 2: Verification of electrical equipment utilizing protection techniques "e," "m," "p," and "q," due to design technique, does not require tests involving MESG or MIC. Therefore, Group II is not required to be subdivided for these protection techniques.

FPN No. 3: It is necessary that the meanings of the different equipment markings and Group II classifications be carefully observed to avoid confusion with Class I, Divisions 1 and 2, Groups A, B, C, and D.

Class I, Zone 0, 1, and 2, groups shall be as follows:

**(A) Group IIC.** Atmospheres containing acetylene, hydrogen, or flammable gas, flammable liquid-produced vapor, or combustible liquid-produced vapor mixed with air that may burn or explode, having either a maximum experimental safe gap (MESG) value less than or equal to 0.50 mm or minimum igniting current ratio (MIC ratio) less than or equal to 0.45. [NFPA 497, 1-3]

FPN: Group IIC is equivalent to a combination of Class I, Group A, and Class I, Group B, as described in 500.6(A)(1) and 500.6(A)(2).

**(B) Group IIB.** Atmospheres containing acetaldehyde ethylene, or flammable gas, flammable liquid-produced vapor, or combustible liquid-produced vapor mixed with air that may burn or explode, having either maximum experimental safe gap (MESG) values greater than 0.50 mm and less than or equal to 0.90 mm or minimum igniting current ratio (MIC ratio) greater than 0.45 and less than or equal to 0.80. [NFPA 497, 1-3]

FPN: Group IIB is equivalent to Class I, Group C as described in 500.6(A)(3).

**(C) Group IIA.** Atmospheres containing acetone, ammonia, ethyl alcohol, gasoline, methane, propane, or flammable gas, flammable liquid-produced vapor, or combustible liquid-produced vapor mixed with air that may burn or explode, having either a maximum experiment safe gap (MESG) value greater than 0.90 mm or minumum igniting current ratio (MIC ratio) greater than 0.80. [NFPA 497, 1-3]

FPN: Group IIA is equivalent to Class I, Group D as described in 500.6(A)(4).

**505.7 Special Precaution.** Article 505 requires equipment construction and installation that ensures safe performance under conditions of proper use and maintenance.

FPN No. 1: It is important that inspection authorities and users exercise more than ordinary care with regard to the installation and maintenance of electrical equipment in hazardous (classified) locations.

FPN No. 2: Low ambient conditions require special consideration. Electrical equipment depending on the protection techniques described by 505.8(A) may not be suitable for use at temperatures lower than −20°C (−4°F) unless they are identified for use at lower temperatures. However, at low ambient temperatures, flammable concentrations of vapors may not exist in a location classified Class I, Zones 0, 1, or 2 at normal ambient temperature.

**(A) Supervision of Work.** Classification of areas and selection of equipment and wiring methods shall be under the supervision of a qualified Registered Professional Engineer.

**(B) Dual Classification.** In instances of areas within the same facility classified separately, Class I, Zone 2 locations shall be permitted to abut, but not overlap, Class I, Division 2 locations. Class I, Zone 0 or Zone 1 locations shall not abut Class I, Division 1 or Division 2 locations.

**(C) Reclassification Permitted.** A Class I, Division 1 or Division 2 location shall be permitted to be reclassified as a Class I, Zone 0, Zone 1, or Zone 2 location, provided all of the space that is classified because of a single flammable gas or vapor source is reclassified under the requirements of this article.

**(D) Solid Obstacles.** Flameproof equipment with flanged joints shall not be installed such that the flange openings are closer than the distances shown in Table 505.7(D) to any solid obstacle that is not a part of the equipment (such as steelworks, walls, weather guards, mounting brackets, pipes, or other electrical equipment) unless the equipment is listed for a smaller distance of separation.

**Table 505.7(D) Minimum Distance of Obstructions from Flameproof "d" Flange Openings**

| | Minimum Distance | |
|---|---|---|
| **Gas Group** | **mm** | **in.** |
| IIC | 40 | 1³⁷/₆₄ |
| IIB | 30 | 1³/₁₆ |
| IIA | 10 | ²⁵/₆₄ |

**505.8 Protection Techniques.** Acceptable protection techniques for electrical and electronic equipment in hazardous (classified) locations shall be as described in 505.8(A) through (I).

FPN: For additional information, see ISA 12.00.01-1999, *Electrical Apparatus for Use in Class I, Zones 0 and 1 Hazardous (Classified) Locations, General Requirements;* ISA 12.01.01-1999, *Definitions and Information Pertaining to Electrical Apparatus in Hazardous (Classified) Locations;* ANSI/UL 2279, 1997, *Electrical Equipment for Use in Class I, Zone 0, 1, and 2 Hazardous (Classified) Loca-*

*tions*; and IEC 60079-0-1998, *Electrical Apparatus for Explosive Gas Atmospheres — Part 0: General Requirements.*

**(A) Flameproof "d".** This protection technique shall be permitted for equipment in Class I, Zone 1 or Zone 2 locations.

**(B) Purged and Pressurized.** This protection technique shall be permitted for equipment in those Class I, Zone 1 or Zone 2 locations for which it is identified.

**(C) Intrinsic Safety.** This protection technique shall be permitted for apparatus and associated apparatus in Class I, Zone 0, Zone 1, or Zone 2 locations for which it is listed.

**(D) Type of Protection "n".** This protection technique shall be permitted for equipment in Class I, Zone 2 locations. Type of protection "n" is further subdivided into nA, nC, and nR.

FPN: See Table 505.9(C)(2)(4) for the descriptions of subdivisions for type of protection "n."

**(E) Oil Immersion "o".** This protection technique shall be permitted for equipment in Class I, Zone 1 or Zone 2 locations.

**(F) Increased Safety "e".** This protection technique shall be permitted for equipment in Class I, Zone 1 or Zone 2 locations.

**(G) Encapsulation "m".** This protection technique shall be permitted for equipment in Class I, Zone 1 or Zone 2 locations.

**(H) Powder Filling "q".** This protection technique shall be permitted for equipment in Class I, Zone 1 or Zone 2 locations.

**(I) Combustible Gas Detection System.** A combustible gas detection system shall be permitted as a means of protection in industrial establishments with restricted public access and where the conditions of maintenance and supervision ensure that only qualified persons service the installation. Gas detection equipment shall be listed for detection of the specific gas or vapor to be encountered. Where such a system is installed, equipment specified in 505.8(I)(1), (2), or (3) shall be permitted.

**(1) Inadequate Ventilation.** In a Class I, Zone 1 location that is so classified due to inadequate ventilation, electrical equipment suitable for Class I, Zone 2 locations shall be permitted.

**(2) Interior of a Building.** In a building located in, or with an opening into, a Class I, Zone 2 location where the interior does not contain a source of flammable gas or vapor, electrical equipment for unclassified locations shall be permitted.

**(3) Interior of a Control Panel.** In the interior of a control panel containing instrumentation utilizing or measuring flammable liquids, gases, or vapors, electrical equipment suitable for Class I, Zone 2 locations shall be permitted.

FPN No. 1: For further information, see ANSI/ISA-12.13.01, *Performance Requirements, Combustible Gas Detectors.*

FPN No. 2: For further information, see ANSI/API RP 500, *Recommended Practice for Classification of Locations for Electrical Installations at Petroleum Facilities Classified as Class I, Division 1 or Division 2.*

FPN No. 3: For further information, see ISA-RP12.13.02, *Installation, Operation, and Maintenance of Combustible Gas Detection Instruments.*

**505.9 Equipment.**

**(A) Suitability.** Suitability of identified equipment shall be determined by one of the following:

(1) Equipment listing or labeling
(2) Evidence of equipment evaluation from a qualified testing laboratory or inspection agency concerned with product evaluation
(3) Evidence acceptable to the authority having jurisdiction such as a manufacturer's self-evaluation or an owner's engineering judgment

**(B) Listing.**

(1) Equipment that is listed for a Zone 0 location shall be permitted in a Zone 1 or Zone 2 location of the same gas or vapor. Equipment that is listed for a Zone 1 location shall be permitted in a Zone 2 location of the same gas or vapor.
(2) Equipment shall be permitted to be listed for a specific gas or vapor, specific mixtures of gases or vapors, or any specific combination of gases or vapors.

FPN: One common example is equipment marked for "IIB. + H2."

**(C) Marking.** Equipment shall be marked in accordance with 505.9(C)(1) or (2).

**(1) Division Equipment.** Equipment identified for Class I, Division 1 or Class I, Division 2 shall, in addition to being marked in accordance with 500.8(B), be permitted to be marked with the following:

(1) Class I, Zone 1 or Class I, Zone 2 (as applicable), and
(2) Applicable gas classification group(s) in accordance with Table 505.9(C), and
(3) Temperature classification in accordance with 505.9(D)(1).

**(2) Zone Equipment.** Equipment meeting one or more of the protection techniques described in 505.8 shall be marked with the following in the order shown:

**Table 505.9(C) Gas Classification Groups**

| Gas Group | Comment |
|-----------|---------|
| IIC | See 505.6(A)(1) |
| IIB | See 505.6(A)(2) |
| IIA | See 505.6(A)(3) |

(1) Class
(2) Zone
(3) Symbol "AEx"
(4) Protection technique(s) in accordance with Table 505.9(C)(2)(4)
(5) Applicable gas classification group(s) in accordance with Table 505.9(C)
(6) Temperature classification in accordance with 505.9(D)(1).

*Exception: Intrinsically safe associated apparatus shall be required to be marked only with (3), (4), and (5).*

**Table 505.9(C)(2)(4) Types of Protection Designation**

| Designation | Technique | Zone* |
|-------------|-----------|-------|
| d | Flameproof enclosure | 1 |
| e | Increased safety | 1 |
| ia | Intrinsic safety | 0 |
| ib | Intrinsic safety | 1 |
| [ia] | Intrinsically safe associated apparatus | Unclassified |
| [ib] | Intrinsically safe associated apparatus | Unclassified |
| m | Encapsulation | 1 |
| nA | Nonsparking equipment | 2 |
| nC | Sparking equipment in which the contacts are suitably protected other than by restricted breathing enclosure | 2 |
| nR | Restricted breathing enclosure | 2 |
| o | Oil immersion | 1 |
| p | Purged and pressurized | 1 or 2 |
| q | Powder filled | 1 |

*Does not address use where a combination of techniques is used.

Electrical equipment of types of protection "e," "m," "p," or "q," shall be marked Group II. Electrical equipment of types of protection "d," "ia," "ib," "[ia]," or "[ib]" shall be marked Group IIA, IIB, or IIC, or for a specific gas or vapor. Electrical equipment of types of protection "n" shall be marked Group II unless it contains enclosed-break devices, nonincendive components, or energy-limited equipment or circuits, in which case it shall be marked Group IIA, IIB, or IIC, or a specific gas or vapor. Electrical equip-

ment of other types of protection shall be marked Group II unless the type of protection utilized by the equipment requires that it shall be marked Group IIA, IIB, or IIC, or a specific gas or vapor.

FPN: An example of such a required marking is "Class I, Zone 0, AEx ia IIC T6." An explanation of the marking that is required is shown in FPN Figure 505.9(C)(2).

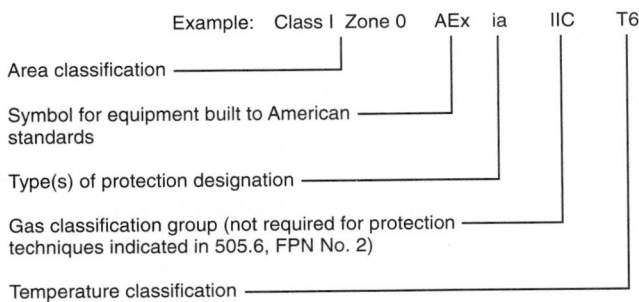

**FPN Figure 505.9(C)(2) Zone equipment marking.**

**(D) Class I Temperature.** The temperature marking specified below shall not exceed the ignition temperature of the specific gas or vapor to be encountered.

FPN: For information regarding ignition temperatures of gases and vapors, see NFPA 497-1997, *Recommended Practice for the Classification of Flammable Liquids, Gases, or Vapors and of Hazardous (Classified) Locations for Electrical Installations in Chemical Process Areas*; and IEC 60079-20-1996, *Electrical Apparatus for Explosive Gas Atmospheres, Data for Flammable Gases and Vapours, Relating to the Use of Electrical Apparatus.*

**(1) Temperature Classifications.** Equipment shall be marked to show the operating temperature or temperature class referenced to a 40°C (104°F) ambient. The temperature class, if provided, shall be indicated using the temperature class (T Code) shown in Table 505.9(D)(1).

**Table 505.9(D)(1) Classification of Maximum Surface Temperature for Group II Electrical Equipment**

| Temperature Class (T Code) | Maximum Surface Temperature (°C) |
|-----------------------------|----------------------------------|
| T1 | ≤450 |
| T2 | ≤300 |
| T3 | ≤200 |
| T4 | ≤135 |
| T5 | ≤100 |
| T6 | ≤85 |

Electrical equipment designed for use in the ambient temperature range between −20°C and +40°C shall require no additional ambient temperature marking.

Electrical equipment that is designed for use in a range of ambient temperatures other than −20°C and +40°C is considered to be special; and the ambient temperature range shall then be marked on the equipment, including either the symbol "Ta" or "Tamb" together with the special range of ambient temperatures. As an example, such a marking might be "−30°C ≤ Ta ≤ + 40°C."

Electrical equipment suitable for ambient temperatures exceeding 40°C (104°F) shall be marked with both the maximum ambient temperature and the operating temperature or temperature class at that ambient temperature.

*Exception No. 1: Equipment of the non–heat-producing type, such as conduit fittings, and equipment of the heat-producing type having a maximum temperature of not more than 100°C (212°F) shall not be required to have a marked operating temperature or temperature class.*

*Exception No. 2: Equipment identified for Class I, Division 1 or Division 2 locations as permitted by 505.20(B) and (C) shall be permitted to be marked in accordance with 500.8(B) and Table 500.8(B).*

**(E) Threading.** All threaded conduit referred to herein shall be threaded with a National (American) Standard Pipe Taper (NPT) standard conduit cutting die that provides a taper of 1 in 16 (¾-in. taper per foot). Such conduit shall be made wrenchtight to prevent sparking when fault current flows through the conduit system, and to ensure the explosionproof or flameproof integrity of the conduit system where applicable. Threaded joints shall be made up with at least five threads fully engaged for entries into flameproof or explosionproof equipment.

Equipment provided with threaded entries for field wiring connections shall be installed in accordance with 505.9(D)(1) or (2).

**(1) Equipment Provided with Threaded Entries for NPT Threaded Conduit or Fittings.** For equipment provided with threaded entries for NPT threaded conduit or fittings, listed conduit fittings, or cable fittings shall be used.

FPN: Thread form specifications for NPT threads are located in ANSI/ASME B1.20.1-1983, *Pipe Threads, General Purpose (Inch).*

**(2) Equipment Provided with Threaded Entries for Metric Threaded Conduit or Fittings.** For equipment with metric threaded entries, such entries shall be identified as being metric, or listed adapters to permit connection to conduit or NPT-threaded fittings shall be provided with the equipment. Adapters shall be used for connection to conduit or NPT-threaded fittings. Listed cable fittings that have metric threads shall be permitted to be used.

FPN: Threading specifications for metric threaded entries are located in ISO 965/1-1980, *Metric Screw Threads*; and ISO 965/3-1980, *Metric Screw Threads.*

**505.15 Wiring Methods.** Wiring methods shall maintain the integrity of protection techniques and shall comply with 505.15(A), (B), or (C).

**(A) Class I, Zone 0.** In Class I, Zone 0 locations, only intrinsically safe wiring methods in accordance with Article 504 shall be permitted.

FPN: Article 504 only includes protection technique "ia."

**(B) Class I, Zone 1.**

**(1) General.** In Class I, Zone 1 locations, the wiring methods in (a) through (e) shall be permitted.

(a) In industrial establishments with restricted public access, where the conditions of maintenance and supervision ensure that only qualified persons service the installation, and where the cable is not subject to physical damage, Type MC-HL cable listed for use in Class I, Zone 1 or Division 1 locations, with a gas/vaportight continuous corrugated metallic sheath, an overall jacket of suitable polymeric material, separate grounding conductors in accordance with 250.122, and provided with termination fittings listed for the application.

FPN: See 330.10 and 330.12 for restrictions on use of Type MC cable.

(b) In industrial establishments with restricted public access, where the conditions of maintenance and supervision ensure that only qualified persons service the installation, and where the cable is not subject to physical damage, Type ITC-HL cable, listed for use in Class I, Zone 1 or Division 1 locations, with a gas/vaportight continuous corrugated metallic sheath, an overall jacket of suitable polymeric material and provided with termination fittings listed for the application.

(c) Type MI cable with termination fittings listed for Class I, Zone 1 or Division 1 locations. Type MI cable shall be installed and supported in a manner to avoid tensile stress at the termination fittings.

(d) Threaded rigid metal conduit, or threaded steel intermediate metal conduit.

(e) Rigid nonmetallic conduit complying with Article 352 shall be permitted where encased in a concrete envelope a minimum of 50 mm (2 in.) thick and provided with not less than 600 mm (24 in.) of cover measured from the top of the conduit to grade. Threaded rigid metal conduit or threaded steel intermediate metal conduit shall be used for the last 600 mm (24 in.) of the underground run to emergence or to the point of connection to the aboveground raceway. An equipment grounding conductor shall be included to provide for electrical continuity of the raceway system and for grounding of non–current-carrying metal parts.

**(2) Flexible Connections.** Where necessary to employ flexible connections, flexible fittings listed for Class I, Zone 1 or Division 1 locations or flexible cord in accordance with the provisions of 505.17 shall be permitted.

**(C) Class I, Zone 2.**

**(1) General.** In Class I, Zone 2 locations, the wiring methods in (a) through (g) shall be permitted.

(a)  All wiring methods permitted by 505.15(B).

(b)  Types MI, MC, MV, or TC cable with termination fittings, or in cable tray systems and installed in a manner to avoid tensile stress at the termination fittings.

(c)  Type ITC cable in cable trays, in raceways, supported by messenger wire, where afforded mechanical protection and run as open wiring, or directly buried where the cable is listed for this use.

(d)  Type PLTC cable in accordance with the provisions of Article 725, or in cable tray systems. PLTC shall be installed in a manner to avoid tensile stress at the termination fittings.

(e)  Enclosed gasketed busways, enclosed gasketed wireways.

(f)  Threaded rigid metal conduit, threaded steel intermediate metal conduit.

(g)  Nonincendive field wiring shall be permitted using any of the wiring methods permitted for unclassified locations. Nonincendive field wiring systems shall be installed in accordance with the control drawing(s). Simple apparatus, not shown on the control drawing, shall be permitted in a nonincendive field wiring circuit, provided the simple apparatus does not interconnect the nonincendive field wiring circuit to any other circuit.

FPN:  Simple apparatus is defined in 504.2.

Separate nonincendive field wiring circuits shall be installed as follows:

(1) In separate cables, or
(2) In multiconductor cables where the conductors of each circuit are within a grounded metal shield, or
(3) In multiconductor cables where the conductors of each circuit have insulation with a minimum thickness of 0.25 mm (0.01 in.)

**(2) Flexible Connections.** Where provision must be made for limited flexibility, flexible metal fittings, flexible metal conduit with listed fittings, liquidtight flexible metal conduit with listed fittings, liquidtight flexible nonmetallic conduit with listed fittings, or flexible cord in accordance with the provisions of 505.17 shall be permitted.

FPN:  See 505.25(B) for grounding requirements where flexible conduit is used.

**505.16 Sealing and Drainage.** Seals in conduit and cable systems shall comply with 505.16(A) through (E). Sealing compound shall be used in Type MI cable termination fittings to exclude moisture and other fluids from the cable insulation.

FPN No. 1:  Seals are provided in conduit and cable systems to minimize the passage of gases and vapors and prevent the passage of flames from one portion of the electrical installation to another through the conduit. Such communication through Type MI cable is inherently prevented by construction of the cable. Unless specifically designed and tested for the purpose, conduit and cable seals are not intended to prevent the passage of liquids, gases, or vapors at a continuous pressure differential across the seal. Even at differences in pressure across the seal equivalent to a few inches of water, there may be a slow passage of gas or vapor through a seal, and through conductors passing through the seal. See 505.16(C)(2)(b). Temperature extremes and highly corrosive liquids and vapors can affect the ability of seals to perform their intended function. See 505.16(D)(2).

FPN No. 2:  Gas or vapor leakage and propagation of flames may occur through the interstices between the strands of standard stranded conductors larger than 2 AWG. Special conductor constructions, for example, compacted strands or sealing of the individual strands, are means of reducing leakage and preventing the propagation of flames.

**(A) Zone 0.** In Class I, Zone 0 locations, seals shall be located according to 505.16(A)(1), (A)(2), and (A)(3).

**(1) Conduit Seals.** Seals shall be provided within 3.05 m (10 ft) of where a conduit leaves a Zone 0 location. There shall be no unions, couplings, boxes, or fittings, except listed reducers at the seal, in the conduit run between the seal and the point at which the conduit leaves the location.

*Exception:  A rigid unbroken conduit that passes completely through the Zone 0 location with no fittings less than 300 mm (12 in.) beyond each boundary shall not be required to be sealed if the termination points of the unbroken conduit are in unclassified locations.*

**(2) Cable Seals.** Seals shall be provided on cables at the first point of termination after entry into the Zone 0 location.

**(3) Not Required to Be Explosionproof or Flameproof.** Seals shall not be required to be explosionproof or flameproof.

**(B) Zone 1.** In Class I, Zone 1 locations, seals shall be located in accordance with 505.16(B)(1) through (B)(8).

**(1) Type of Protection "d" or "e" Enclosures.** Conduit seals shall be provided for each conduit entering enclosures having type of protection "d" or "e."

*Exception:  Where the enclosure having type of protection "d" is marked to indicate that a seal is not required.*

**(2) Explosionproof Equipment.** Conduit seals shall be provided for each conduit entering explosionproof equipment according to (a), (b), and (c).

(a)   In each conduit entry into an explosionproof enclosure where either (1) the enclosure contains apparatus, such as switches, circuit breakers, fuses, relays, or resistors, that may produce arcs, sparks, or high temperatures that are considered to be an ignition source in normal operation, or (2) the entry is metric designator 53 (trade size 2) or larger and the enclosure contains terminals, splices, or taps. For the purposes of this section, high temperatures shall be considered to be any temperatures exceeding 80 percent of the autoignition temperature in degrees Celsius of the gas or vapor involved.

*Exception: Conduit entering an enclosure where such switches, circuit breakers, fuses, relays, or resistors are*

(a) *Enclosed within a chamber hermetically sealed against the entrance of gases or vapors, or*
(b) *Immersed in oil, or*
(c) *Enclosed within a factory-sealed explosionproof chamber located within the enclosure, identified for the location, and marked "factory sealed" or equivalent, unless the entry is metric designator 53 (trade size 2) or larger. Factory-sealed enclosures shall not be considered to serve as a seal for another adjacent explosionproof enclosure that is required to have a conduit seal.*

(b)   Conduit seals shall be installed within 450 mm (18 in.) from the enclosure. Only explosionproof unions, couplings, reducers, elbows, capped elbows, and conduit bodies similar to L, T, and Cross types that are not larger than the trade size of the conduit shall be permitted between the sealing fitting and the explosionproof enclosure.

(c)   Where two or more explosionproof enclosures for which conduit seals are required under 505.16(B)(2) are connected by nipples or by runs of conduit not more than 900 mm (36 in.) long, a single conduit seal in each such nipple connection or run of conduit shall be considered sufficient if located not more than 450 mm (18 in.) from either enclosure.

**(3) Pressurized Enclosures.** Conduit seals shall be provided in each conduit entry into a pressurized enclosure where the conduit is not pressurized as part of the protection system. Conduit seals shall be installed within 450 mm (18 in.) from the pressurized enclosure.

>   FPN No. 1:  Installing the seal as close as possible to the enclosure reduces problems with purging the dead airspace in the pressurized conduit.

>   FPN No. 2:  For further information, see NFPA 496-1998, *Standard for Purged and Pressurized Enclosures for Electrical Equipment.*

**(4) Class I, Zone 1 Boundary.** Conduit seals shall be provided in each conduit run leaving a Class I, Zone 1 loca-

tion. The sealing fitting shall be permitted on either side of the boundary of such location within 3.05 m (10 ft) of the boundary and shall be designed and installed so as to minimize the amount of gas or vapor within the Zone 1 portion of the conduit from being communicated to the conduit beyond the seal. Except for listed explosionproof reducers at the conduit seal, there shall be no union, coupling, box, or fitting between the conduit seal and the point at which the conduit leaves the Zone 1 location.

*Exception:   Metal conduit containing no unions, couplings, boxes, or fittings and passing completely through a Class I, Zone 1 location with no fittings less than 300 mm (12 in.) beyond each boundary shall not require a conduit seal if the termination points of the unbroken conduit are in unclassified locations.*

**(5) Cables Capable of Transmitting Gases or Vapors.** Conduits containing cables with a gas/vaportight continuous sheath capable of transmitting gases or vapors through the cable core shall be sealed in the Zone 1 location after removing the jacket and any other coverings so that the sealing compound surrounds each individual insulated conductor and the outer jacket.

*Exception:   Multiconductor cables with a gas/vaportight continuous sheath capable of transmitting gases or vapors through the cable core shall be permitted to be considered as a single conductor by sealing the cable in the conduit within 450 mm (18 in.) of the enclosure and the cable end within the enclosure by an approved means to minimize the entrance of gases or vapors and prevent the propagation of flame into the cable core, or by other approved methods. For shielded cables and twisted pair cables, it shall not be required to remove the shielding material or separate the twisted pair.*

**(6) Cables Incapable of Transmitting Gases or Vapors.** Each multiconductor cable in conduit shall be considered as a single conductor if the cable is incapable of transmitting gases or vapors through the cable core. These cables shall be sealed in accordance with 505.16(D).

**(7) Cables Entering Enclosures.** Cable seals shall be provided for each cable entering flameproof or explosionproof enclosures. The seal shall comply with 505.16(D).

**(8) Class I, Zone 1 Boundary.** Cables shall be sealed at the point at which they leave the Zone 1 location.

*Exception:   Where cable is sealed at the termination point.*

**(C) Zone 2.** In Class I, Zone 2 locations, seals shall be located in accordance with 505.16(C)(1) and (C)(2).

**(1)   Conduit Seals.**  Conduit seals shall be located in accordance with (a) and (b).

(a)   For connections to enclosures that are required to be flameproof or explosionproof, a conduit seal shall be provided in accordance with 505.16(B)(1) and 505.16(B)(2). All portions of the conduit run or nipple between the seal and such enclosure shall comply with 505.16(B).

(b)   In each conduit run passing from a Class I, Zone 2 location into an unclassified location. The sealing fitting shall be permitted on either side of the boundary of such location within 3.05 m (10 ft) of the boundary and shall be designed and installed so as to minimize the amount of gas or vapor within the Zone 2 portion of the conduit from being communicated to the conduit beyond the seal. Rigid metal conduit or threaded steel intermediate metal conduit shall be used between the sealing fitting and the point at which the conduit leaves the Zone 2 location, and a threaded connection shall be used at the sealing fitting. Except for listed explosionproof reducers at the conduit seal, there shall be no union, coupling, box, or fitting between the conduit seal and the point at which the conduit leaves the Zone 2 location.

*Exception No. 1: Metal conduit containing no unions, couplings, boxes, or fittings and passing completely through a Class I, Zone 2 location with no fittings less than 300 mm (12 in.) beyond each boundary shall not be required to be sealed if the termination points of the unbroken conduit are in unclassified locations.*

*Exception No. 2: Conduit systems terminating at an unclassified location where a wiring method transition is made to cable tray, cablebus, ventilated busway, Type MI cable, or open wiring shall not be required to be sealed where passing from the Class I, Zone 2 location into the unclassified location. The unclassified location shall be outdoors or, if the conduit system is all in one room, it shall be permitted to be indoors. The conduits shall not terminate at an enclosure containing an ignition source in normal operation.*

*Exception No. 3: Conduit systems passing from an enclosure or room that is unclassified as a result of pressurization into a Class I, Zone 2 location shall not require a seal at the boundary.*

FPN:  For further information, refer to NFPA 496-1998, *Standard for Purged and Pressurized Enclosures for Electrical Equipment.*

*Exception No. 4:  Segments of aboveground conduit systems shall not be required to be sealed where passing from a Class I, Zone 2 location into an unclassified location if all the following conditions are met:*

*(a) No part of the conduit system segment passes through a Class I, Zone 0 or Class I, Zone 1 location where the conduit contains unions, couplings, boxes, or fittings within 300 mm (12 in.) of the Class I, Zone 0 or Class I, Zone 1 location.*

*(b) The conduit system segment is located entirely in outdoor locations.*

*(c) The conduit system segment is not directly connected to canned pumps, process or service connections for flow, pressure, or analysis measurement, and so forth, that depend on a single compression seal, diaphragm, or tube to prevent flammable or combustible fluids from entering the conduit system.*

*(d) The conduit system segment contains only threaded metal conduit, unions, couplings, conduit bodies, and fittings in the unclassified location.*

*(e) The conduit system segment is sealed at its entry to each enclosure or fitting housing terminals, splices, or taps in Class I, Zone 2 locations.*

**(2)  Cable Seals.** Cable seals shall be located in accordance with (a), (b), and (c).

(a)   Explosionproof and Flameproof Enclosures. Cables entering enclosures required to be flameproof or explosionproof shall be sealed at the point of entrance. The seal shall comply with 505.16(D). Multiconductor cables with a gas/vaportight continuous sheath capable of transmitting gases or vapors through the cable core shall be sealed in the Zone 2 location after removing the jacket and any other coverings so that the sealing compound surrounds each individual insulated conductor in such a manner as to minimize the passage of gases and vapors. Multiconductor cables in conduit shall be sealed as described in 505.16(B)(4).

*Exception No. 1: Cables passing from an enclosure or room that is unclassified as a result of Type Z pressurization into a Class I, Zone 2 location shall not require a seal at the boundary.*

*Exception No. 2: Shielded cables and twisted pair cables shall not require the removal of the shielding material or separation of the twisted pairs, provided the termination is by an approved means to minimize the entrance of gases or vapors and prevent propagation of flame into the cable core.*

(b)   Cables That Will Not Transmit Gases or Vapors. Cables with a gas/vaportight continuous sheath and that will not transmit gases or vapors through the cable core in excess of the quantity permitted for seal fittings shall not be required to be sealed except as required in 505.16(C)(2)(a). The minimum length of such cable run shall not be less than the length that limits gas or vapor flow through the cable core to the rate permitted for seal fittings [200 cm$^3$/hr (0.007 ft$^3$/hr) of air at a pressure of 1500 pascals (6 in. of water)].

FPN No. 1:  See ANSI/UL 886-1994, *Outlet Boxes and Fittings for Use in Hazardous (Classified) Locations.*

FPN No. 2:  The cable core does not include the interstices of the conductor strands.

(c) Cables Capable of Transmitting Gases or Vapors. Cables with a gas/vaportight continuous sheath capable of transmitting gases or vapors through the cable core shall not be required to be sealed except as required in 505.16(C)(2)(a), unless the cable is attached to process equipment or devices that may cause a pressure in excess of 1500 pascals (6 in. of water) to be exerted at a cable end, in which case a seal, barrier, or other means shall be provided to prevent migration of flammables into an unclassified area.

*Exception: Cables with an unbroken gas/vaportight continuous sheath shall be permitted to pass through a Class I, Zone 2 location without seals.*

(d) Cables Without Gas/Vaportight Continuous Sheath. Cables that do not have gas/vaportight continuous sheath shall be sealed at the boundary of the Zone 2 and unclassified location in such a manner as to minimize the passage of gases or vapors into an unclassified location.

FPN: The cable sheath may be either metal or a nonmetallic material.

(D) Class I, Zones 0, 1, and 2. Where required, seals in Class I, Zones 0, 1, and 2 locations shall comply with 505.16(D)(1) through (D)(5).

(1) Fittings. Enclosures for connections or equipment shall be provided with an integral means for sealing, or sealing fittings listed for the location shall be used. Sealing fittings shall be listed for use with one or more specific compounds and shall be accessible.

(2) Compound. The compound shall provide a seal against passage of gas or vapors through the seal fitting, shall not be affected by the surrounding atmosphere or liquids, and shall not have a melting point less than 93°C (200°F).

(3) Thickness of Compounds. In a completed seal, the minimum thickness of the sealing compound shall not be less than the trade size of the sealing fitting and, in no case, less than 16 mm (⅝ in.).

*Exception: Listed cable sealing fittings shall not be required to have a minimum thickness equal to the trade size of the fitting.*

(4) Splices and Taps. Splices and taps shall not be made in fittings intended only for sealing with compound, nor shall other fittings in which splices or taps are made be filled with compound.

(5) Conductor Fill. The cross-sectional area of the conductors permitted in a seal shall not exceed 25 percent of the cross-sectional area of a rigid metal conduit of the same trade size unless it is specifically listed for a higher percentage of fill.

(E) Drainage.

(1) Control Equipment. Where there is a probability that liquid or other condensed vapor may be trapped within enclosures for control equipment or at any point in the raceway system, approved means shall be provided to prevent accumulation or to permit periodic draining of such liquid or condensed vapor.

(2) Motors and Generators. Where the authority having jurisdiction judges that there is a probability that liquid or condensed vapor may accumulate within motors or generators, joints and conduit systems shall be arranged to minimize entrance of liquid. If means to prevent accumulation or to permit periodic draining are judged necessary, such means shall be provided at the time of manufacture and shall be considered an integral part of the machine.

(3) Canned Pumps, Process or Service Connections, and So Forth. For canned pumps, process or service connections for flow, pressure, or analysis measurement, and so forth, that depend upon a single compression seal, diaphragm, or tube to prevent flammable or combustible fluids from entering the electrical conduit system, an additional approved seal, barrier, or other means shall be provided to prevent the flammable or combustible fluid from entering the conduit system beyond the additional devices or means if the primary seal fails.

The additional approved seal or barrier and the interconnecting enclosure shall meet the temperature and pressure conditions to which they will be subjected upon failure of the primary seal, unless other approved means are provided to accomplish the purpose in the preceding paragraph.

Drains, vents, or other devices shall be provided so that primary seal leakage is obvious.

505.17 Flexible Cords, Class I, Zones 1 and 2. A flexible cord shall be permitted for connection between portable lighting equipment or other portable utilization equipment and the fixed portion of their supply circuit. Flexible cord shall also be permitted for that portion of the circuit where the fixed wiring methods of 505.15(B) cannot provide the necessary degree of movement for fixed and mobile electrical utilization equipment, in an industrial establishment where conditions of maintenance and engineering supervision ensure that only qualified persons install and service the installation, and the flexible cord is protected by location or by a suitable guard from damage. The length of the flexible cord shall be continuous. Where flexible cords are used, the cords shall be as follows:

(1) Of a type listed for extra-hard usage;
(2) Contain, in addition to the conductors of the circuit, a grounding conductor complying with 400.23;

(3) Connected to terminals or to supply conductors in an approved manner;

(4) Be supported by clamps or by other suitable means in such a manner that there will be no tension on the terminal connections; and

(5) Be provided with listed seals where the flexible cord enters boxes, fittings, or enclosures that are required to be explosionproof or flameproof.

*Exception: As provided in 505.16.*

Electric submersible pumps with means for removal without entering the wet-pit shall be considered portable utilization equipment. The extension of the flexible cord within a suitable raceway between the wet-pit and the power source shall be permitted.

Electric mixers intended for travel into and out of open-type mixing tanks or vats shall be considered portable utilization equipment.

FPN: See 505.18 for flexible cords exposed to liquids having a deleterious effect on the conductor insulation.

**505.18 Conductors and Conductor Insulation.**

**(A) Conductors.** For type of protection "e," field wiring conductors shall be copper.

**(B) Conductor Insulation.** Where condensed vapors or liquids may collect on, or come in contact with, the insulation on conductors, such insulation shall be of a type identified for use under such conditions or the insulation shall be protected by a sheath of lead or by other approved means.

**505.19 Live Parts.** There shall be no exposed live parts.

**505.20 Equipment Requirements.**

**(A) Zone 0.** In Class I, Zone 0 locations, only equipment specifically listed and marked as suitable for the location shall be permitted.

*Exception: Intrinsically safe apparatus listed for use in Class I, Division 1 locations for the same gas, or as permitted by 505.9(B)(2), and with a suitable temperature class shall be permitted.*

**(B) Zone 1.** In Class I, Zone 1 locations, only equipment specifically listed and marked as suitable for the location shall be permitted.

*Exception No. 1: Equipment identified for use in Class I, Division 1 or listed for use in Class I, Zone 0 locations for the same gas, or as permitted by 505.9(B)(2), and with a suitable temperature class shall be permitted.*

*Exception No. 2: Equipment identified for Class I, Zone 1, or Zone 2 type of protection "p" shall be permitted.*

**(C) Zone 2.** In Class I, Zone 2 locations, only equipment specifically listed and marked as suitable for the location shall be permitted.

*Exception No. 1: Equipment listed for use in Class I, Zone 0 or Zone 1 locations for the same gas, or as permitted by 505.9(B)(2), and with a suitable temperature class, shall be permitted.*

*Exception No. 2: Equipment identified for Class I, Zone 1 or Zone 2 type of protection "p" shall be permitted.*

*Exception No. 3: Equipment identified for use in Class I, Division 1 or Division 2 locations for the same gas, or as permitted by 505.9(B)(2), and with a suitable temperature class shall be permitted.*

*Exception No. 4: In Class I, Zone 2 locations, the installation of open or nonexplosionproof or nonflameproof enclosed motors, such as squirrel-cage induction motors without brushes, switching mechanisms, or similar arc-producing devices that are not identified for use in a Class I, Zone 2 location shall be permitted.*

FPN No. 1: It is important to consider the temperature of internal and external surfaces that may be exposed to the flammable atmosphere.

FPN No. 2: It is important to consider the risk of ignition due to currents arcing across discontinuities and overheating of parts in multisection enclosures of large motors and generators. Such motors and generators may need equipotential bonding jumpers across joints in the enclosure and from enclosure to ground. Where the presence of ignitible gases or vapors is suspected, clean air purging may be needed immediately prior to and during start-up periods.

**(D) Manufacturer's Instructions.** Electrical equipment installed in hazardous (classified) locations shall be installed in accordance with the instructions (if any) provided by the manufacturer.

**505.21 Multiwire Branch Circuits.** In a Class I, Zone 1 location, a multiwire branch circuit shall not be permitted.

*Exception: Where the disconnect device(s) for the circuit opens all ungrounded conductors of the multiwire circuit simultaneously.*

**505.22 Increased Safety "e" Motors and Generators.** In Class I, Zone 1 locations, Increased Safety "e" motors and generators of all voltage ratings shall be listed for Class I, Zone 1 locations, and shall comply with the following:

(1) Motors shall be marked with the current ratio, $I_A/I_N$, and time, $t_E$.

(2) Motors shall have controllers marked with the model or identification number, output rating (horsepower or kilowatt), full-load amperes, starting current ratio ($I_A/I_N$), and time ($t_E$) of the motors that they are intended to protect; the controller marking shall also include the specific over-

load protection type (and setting, if applicable) that is listed with the motor or generator.

(3) Connections shall be made with the specific terminals listed with the motor or generator.

(4) Terminal housings shall be permitted to be of substantial, nonmetallic, nonburning material, provided an internal grounding means between the motor frame and the equipment grounding connection is incorporated within the housing.

(5) The provisions of Part III of Article 430 shall apply regardless of the voltage rating of the motor.

(6) The motors shall be protected against overload by a separate overload device that is responsive to motor current. This device shall be selected to trip or shall be rated in accordance with the listing of the motor and its overload protection.

(7) Sections 430.32(C) and 430.44 shall not apply to such motors.

(8) The motor overload protection shall not be shunted or cut out during the starting period.

**505.25 Grounding and Bonding.** Grounding and bonding shall comply with Article 250 and the requirements in 505.25(A) and (B).

**(A) Bonding.** The locknut-bushing and double-locknut types of contacts shall not be depended on for bonding purposes, but bonding jumpers with proper fittings or other approved means of bonding shall be used. Such means of bonding shall apply to all intervening raceways, fittings, boxes, enclosures, and so forth, between Class I locations and the point of grounding for service equipment or point of grounding of a separately derived system.

*Exception: The specific bonding means shall only be required to the nearest point where the grounded circuit conductor and the grounding electrode are connected together on the line side of the building or structure disconnecting means as specified in 250.32(A), (B), and (C), provided the branch-circuit overcurrent protection is located on the load side of the disconnecting means.*

FPN: See 250.100 for additional bonding requirements in hazardous (classified) locations.

**(B) Types of Equipment Grounding Conductors.** Where flexible metal conduit or liquidtight flexible metal conduit is used as permitted in 505.15(C) and is to be relied on to complete a sole equipment grounding path, it shall be installed with internal or external bonding jumpers in parallel with each conduit and complying with 250.102.

*Exception: In Class I, Zone 2 locations, the bonding jumper shall be permitted to be deleted where all the following conditions are met:*

*(a) Listed liquidtight flexible metal conduit 1.8 m (6 ft) or less in length, with fittings listed for grounding, is used.*

*(b) Overcurrent protection in the circuit is limited to 10 amperes or less.*

*(c) The load is not a power utilization load.*

# ARTICLE 510
# Hazardous (Classified) Locations — Specific

**510.1 Scope.** Articles 511 through 517 cover occupancies or parts of occupancies that are or may be hazardous because of atmospheric concentrations of flammable liquids, gases, or vapors, or because of deposits or accumulations of materials that may be readily ignitible.

**510.2 General.** The general rules of this *Code* and the provisions of Articles 500 through 504 shall apply to electric wiring and equipment in occupancies within the scope of Articles 511 through 517, except as such rules are modified in Articles 511 through 517. Where unusual conditions exist in a specific occupancy, the authority having jurisdiction shall judge with respect to the application of specific rules.

# ARTICLE 511
# Commercial Garages, Repair and Storage

FPN: Rules that are followed by a reference in brackets contain text that has been extracted from NFPA 88B-1997, *Standard for Repair Garages*. Only editorial changes were made to the extracted text to make it consistent with this *Code*.

**511.1 Scope.** These occupancies shall include locations used for service and repair operations in connection with self-propelled vehicles (including, but not limited to, passenger automobiles, buses, trucks, and tractors) in which volatile flammable liquids or flammable gases are used for fuel or power.

**511.3 Classifications of Locations.**

**(A) Unclassified Locations.** Parking garages used for parking or storage and where no repair work is done except exchange of parts and routine maintenance requiring no use of electrical equipment, open flame, welding, or the use of volatile flammable liquids are not classified.

The storage, handling, or dispensing into motor vehicles of alcohol-based windshield washer fluid in areas used for the service and repair operations of the vehicles shall not cause such areas to be classified as hazardous (classified) locations.

FPN No. 1: For further information, see NFPA 88A-1998, *Standard for Parking Structures* and NFPA 88B-1997, *Standard for Repair Garages.*

FPN No. 2: For further information, see 8.3.5 of NFPA 30A, *Code for Motor Fuel Dispensing Facilities and Repair Garages.*

**(B) Classified Locations.** Classification shall be in accordance with Article 500. Areas in which flammable fuel is transferred to vehicle fuel tanks shall also conform to Article 514.

**(1) Up to a Level of 450 mm (18 in.) Above the Floor.** For each floor, the entire area up to a level of 450 mm (18 in.) above the floor shall be considered to be a Class I, Division 2 location.

*Exception: Where the enforcing agency determines that there is mechanical ventilation providing a minimum of four air changes per hour or one cubic foot per minute of exchanged air for each square foot of floor area. Ventilation shall provide for air exchange across the entire floor area within 0.3 m (12 in.) of the floor.*

**(2) Within 450 mm (18 in.) of the Ceiling.** Where compressed natural gas (CNG) vehicles are repaired or stored, the area within 450 mm (18 in.) of the ceiling shall be classified as Class I, Division 2, except where ventilation of at least four air changes per hour is provided. [NFPA 88B, 3-1.1]

**(3) Any Pit or Depression Below Floor Level.** Any pit or depression below floor level shall be considered to be a Class I, Division 1 location and shall extend up to said floor level.

*Exception No. 1: Any pit or depression in which six air changes per hour are exhausted at the floor level of the pit shall be permitted to be judged by the enforcing agency to be a Class I, Division 2 location.*

*Exception No. 2: Lubrication and service rooms without dispensing shall be classified in accordance with Table 514.3(B)(1).*

**(4) Areas Adjacent to Defined Locations or with Positive-Pressure Ventilation.** Areas adjacent to defined locations in which flammable vapors are not likely to be released, such as stock rooms, switchboard rooms, and other similar locations, shall not be classified where mechanically ventilated at a rate of four or more air changes per hour or where effectively cut off by walls or partitions.

**(5) Adjacent Areas by Special Permission.** Adjacent areas that by reason of ventilation, air pressure differentials, or physical spacing are such that, in the opinion of the authority enforcing this *Code,* no ignition hazard exists, shall be unclassified.

**511.4 Wiring and Equipment in Class I Locations.**

**(A) Wiring Located in Class I Locations.** Within Class I locations as classified in 511.3, wiring shall conform to applicable provisions of Article 501.

**(1) Raceways.** Raceways embedded in a masonry wall or buried beneath a floor shall be considered to be within the Class I location above the floor if any connections or extensions lead into or through such areas.

*Exception: Rigid nonmetallic conduit that complies with Article 352 shall be permitted where buried under not less than 600 mm (24 in.) of cover. Where rigid nonmetallic conduit is used, threaded rigid metal conduit or threaded steel intermediate metal conduit shall be used for the last 600 mm (24 in.) of the underground run to emergence or to the point of connection to the aboveground raceway and an equipment grounding conductor shall be included to provide electrical continuity of the raceway system and for grounding of non–current-carrying metal parts.*

**(B) Equipment Located in Class I Locations.** Within Class I locations as defined in 511.3, equipment shall conform to applicable provisions of Article 501.

**(1) Fuel-Dispensing Units.** Where fuel-dispensing units (other than liquid petroleum gas, which is prohibited) are located within buildings, the requirements of Article 514 shall govern.

Where mechanical ventilation is provided in the dispensing area, the control shall be interlocked so that the dispenser cannot operate without ventilation as prescribed in 500.5(B)(2).

**(2) Portable Lighting Equipment.** Portable lighting equipment shall be equipped with handle, lampholder, hook, and substantial guard attached to the lampholder or handle. All exterior surfaces that might come in contact with battery terminals, wiring terminals, or other objects shall be of nonconducting material or shall be effectively protected with insulation. Lampholders shall be of an unswitched type and shall not provide means for plug-in of attachment plugs. The outer shell shall be of molded composition or other suitable material. Unless the lamp and its cord are supported or arranged in such a manner that they cannot be used in the locations classified in 511.3, they shall be of a type identified for Class I, Division 1 locations.

**511.7 Wiring and Equipment Installed Above Class I Locations.**

**(A) Wiring in Spaces Above Class I Locations.**

**(1) Fixed Wiring Above Class I Locations.** All fixed wiring above Class I locations shall be in metal raceways, rigid nonmetallic conduit, electrical nonmetallic tubing, flexible metal conduit, liquidtight flexible metal conduit, or liquidtight flexible nonmetallic conduit or shall be Type MC, MI, manufactured wiring systems, or PLTC cable in accordance with Article 725, or Type TC cable or Type ITC cable in accordance with Article 727. Cellular metal floor raceways or cellular concrete floor raceways shall be permitted to be used only for supplying ceiling outlets or extensions to the area below the floor, but such raceways shall have no connections leading into or through any Class I location above the floor.

**(2) Pendant.** For pendants, flexible cord suitable for the type of service and listed for hard usage shall be used.

**(B) Electrical Equipment Installed Above Class I Locations.**

**(1) Fixed Electrical Equipment.** Electrical equipment in a fixed position shall be located above the level of any defined Class I location or shall be identified for the location.

(a)  Arcing Equipment. Equipment that is less than 3.7 m (12 ft) above the floor level and that may produce arcs, sparks, or particles of hot metal, such as cutouts, switches, charging panels, generators, motors, or other equipment (excluding receptacles, lamps, and lampholders) having make-and-break or sliding contacts, shall be of the totally enclosed type or constructed so as to prevent the escape of sparks or hot metal particles.

(b)  Fixed Lighting. Lamps and lampholders for fixed lighting that is located over lanes through which vehicles are commonly driven or that may otherwise be exposed to physical damage shall be located not less than 3.7 m (12 ft) above floor level, unless of the totally enclosed type or constructed so as to prevent escape of sparks or hot metal particles.

**511.9 Sealing.** Seals conforming to the requirements of 501.5 and 501.5(B)(2) shall be provided and shall apply to horizontal as well as vertical boundaries of the defined Class I locations.

**511.10 Special Equipment.**

**(A) Battery Charging Equipment.** Battery chargers and their control equipment, and batteries being charged, shall not be located within locations classified in 511.3.

**(B) Electric Vehicle Charging Equipment.**

**(1) General.** All electrical equipment and wiring shall be installed in accordance with Article 625, except as noted in 511.10(B)(2) and (B)(3). Flexible cords shall be of a type identified for extra-hard usage.

**(2) Connector Location.** No connector shall be located within a Class I location as defined in 511.3.

**(3) Plug Connections to Vehicles.** Where the cord is suspended from overhead, it shall be arranged so that the lowest point of sag is at least 150 mm (6 in.) above the floor. Where an automatic arrangement is provided to pull both cord and plug beyond the range of physical damage, no additional connector shall be required in the cable or at the outlet.

**511.12 Ground-Fault Circuit-Interrupter Protection for Personnel.** All 125-volt, single-phase, 15- and 20-ampere receptacles installed in areas where electrical diagnostic equipment, electrical hand tools, or portable lighting equipment are to be used shall have ground-fault circuit-interrupter protection for personnel.

**511.16 Grounded and Grounding Requirements.**

**(A) General Grounding Requirements.** All metal raceways, the metal armor or metallic sheath on cables, and all non–current-carrying metal parts of fixed or portable electrical equipment, regardless of voltage, shall be grounded as provided in Article 250.

**(B) Supplying Circuits with Grounded and Grounding Conductors in Class I Locations.** Grounding in Class I locations shall comply with 501.16.

**(1) Circuits Supplying Portable Equipment or Pendants.** Where a circuit supplies portables or pendants and includes a grounded conductor as provided in Article 200, receptacles, attachment plugs, connectors, and similar devices shall be of the grounding type, and the grounded conductor of the flexible cord shall be connected to the screw shell of any lampholder or to the grounded terminal of any utilization equipment supplied.

**(2) Approved Means.** Approved means shall be provided for maintaining continuity of the grounding conductor between the fixed wiring system and the non–current-carrying metal portions of pendant luminaires (fixtures), portable lamps, and portable utilization equipment.

## ARTICLE 513
## Aircraft Hangars

**513.1 Scope.** This article shall apply to buildings or structures in any part of which aircraft containing Class I (flammable) liquids or Class II (combustible) liquids whose temperatures are above their flash points are housed or stored and in which aircraft might undergo service, repairs, or alterations. It shall not apply to locations used exclusively for aircraft that have never contained fuel or unfueled aircraft.

FPN No. 1: For definitions of aircraft hangar and unfueled aircraft, see NFPA 409-1995, *Standard on Aircraft Hangars.*

FPN No. 2: For further information on fuel classification see NFPA 30-2000, *Flammable and Combustible Liquids Code.*

**513.2 Definitions.** For the purpose of this article, the following definitions shall apply.

**Mobile Equipment.** Equipment with electric components suitable to be moved only with mechanical aids or is provided with wheels for movement by person(s) or powered devices.

**Portable Equipment.** Equipment with electric components suitable to be moved by a single person without mechanical aids.

**513.3 Classification of Locations.**

**(A) Below Floor Level.** Any pit or depression below the level of the hangar floor shall be classified as a Class I, Division 1 or Zone 1 location that shall extend up to said floor level.

**(B) Areas Not Cut Off or Ventilated.** The entire area of the hangar, including any adjacent and communicating areas not suitably cut off from the hangar, shall be classified as a Class I, Division 2 or Zone 2 location up to a level 450 mm (18 in.) above the floor.

**(C) Vicinity of Aircraft.** The area within 1.5 m (5 ft) horizontally from aircraft power plants or aircraft fuel tanks shall be classified as a Class I, Division 2 or Zone 2 location that shall extend upward from the floor to a level 1.5 m (5 ft) above the upper surface of wings and of engine enclosures.

**(D) Areas Suitably Cut Off and Ventilated.** Adjacent areas in which flammable liquids or vapors are not likely to be released, such as stock rooms, electrical control rooms, and other similar locations, shall not be classified where adequately ventilated and where effectively cut off from the hangar itself by walls or partitions.

**513.4 Wiring and Equipment in Class I Locations.**

**(A) General.** All wiring and equipment that is or may be installed or operated within any of the Class I locations defined in 513.3 shall comply with the applicable provi-

sions of Article 501 or Article 505 for the division or zone in which they are used.

Attachment plugs and receptacles in Class I locations shall be identified for Class I locations or shall be designed so that they cannot be energized while the connections are being made or broken.

**(B) Stanchions, Rostrums, and Docks.** Electric wiring, outlets, and equipment (including lamps) on or attached to stanchions, rostrums, or docks that are located or likely to be located in a Class I location, as defined in 513.3(C), shall comply with the applicable provisions of Article 501 or Article 505 for the division or zone in which they are used.

**513.7 Wiring and Equipment Not Installed in Class I Locations.**

**(A) Fixed Wiring.** All fixed wiring in a hangar but not installed in a Class I location as classified in 513.3 shall be installed in metal raceways or shall be Type MI, TC, or MC cable.

*Exception: Wiring in unclassified locations, as classified in 513.3(D), shall be of a type recognized in Chapter 3.*

**(B) Pendants.** For pendants, flexible cord suitable for the type of service and identified for hard usage or extra-hard usage shall be used. Each such cord shall include a separate equipment grounding conductor.

**(C) Arcing Equipment.** In locations above those described in 513.3, equipment that is less than 3.0 m (10 ft) above wings and engine enclosures of aircraft and that may produce arcs, sparks, or particles of hot metal, such as lamps and lampholders for fixed lighting, cutouts, switches, receptacles, charging panels, generators, motors, or other equipment having make-and-break or sliding contacts, shall be of the totally enclosed type or constructed so as to prevent the escape of sparks or hot metal particles.

*Exception: Equipment in areas described in 513.3(D) shall be permitted to be of the general-purpose type.*

**(D) Lampholders.** Lampholders of metal-shell, fiber-lined types shall not be used for fixed incandescent lighting.

**(E) Stanchions, Rostrums, or Docks.** Where stanchions, rostrums, or docks are not located or likely to be located in a Class I location, as defined in 513.3(C), wiring and equipment shall comply with 513.7, except that such wiring and equipment not more than 457 mm (18 in.) above the floor in any position shall comply with 513.4(B). Receptacles and attachment plugs shall be of a locking type that will not readily disconnect.

**(F) Mobile Stanchions.** Mobile stanchions with electric equipment complying with 513.7(E) shall carry at least one permanently affixed warning sign with the following words or equivalent:

WARNING
KEEP 5 FT CLEAR OF AIRCRAFT
ENGINES AND FUEL TANK AREAS

or

WARNING
KEEP 1.5 METERS CLEAR OF AIRCRAFT
ENGINES AND FUEL TANK AREAS

**513.8 Wiring and Equipment Embedded, Under Slab, or Under Ground.** All wiring installed in or under the hangar floor shall comply with the requirements for Class I, Division 1 locations. Where such wiring is located in vaults, pits, or ducts, adequate drainage shall be provided.

**513.9 Sealing.** Seals shall be provided in accordance with 501.5 and 505.16. Sealing requirements specified shall apply to horizontal as well as to vertical boundaries of the defined Class I locations. Raceways embedded in a concrete floor or buried beneath a floor shall be considered to be within the Class I location above the floor.

**513.10 Special Equipment.**

**(A) Aircraft Electrical Systems.**

**(1) De-energizing Aircraft Electrical Systems.** Aircraft electrical systems shall be de-energized when the aircraft is stored in a hangar and, whenever possible, while the aircraft is undergoing maintenance.

**(2) Aircraft Batteries.** Aircraft batteries shall not be charged where installed in an aircraft located inside or partially inside a hangar.

**(B) Aircraft Battery Charging and Equipment.** Battery chargers and their control equipment shall not be located or operated within any of the Class I locations defined in 513.3 and shall preferably be located in a separate building or in an area such as defined in 513.3(D). Mobile chargers shall carry at least one permanently affixed warning sign with the following words or equivalent:

WARNING
KEEP 5 FT CLEAR OF AIRCRAFT ENGINES
AND FUEL TANK AREAS

or

WARNING
KEEP 1.5 METERS CLEAR OF AIRCRAFT
ENGINES AND FUEL TANK AREAS

Tables, racks, trays, and wiring shall not be located within a Class I location and, in addition, shall comply with Article 480.

**(C) External Power Sources for Energizing Aircraft.**

**(1) Not Less Than 450 mm (18 in.) Above Floor.** Aircraft energizers shall be designed and mounted so that all electric equipment and fixed wiring will be at least 450 mm (18 in.) above floor level and shall not be operated in a Class I location as defined in 513.3(C).

**(2) Marking for Mobile Units.** Mobile energizers shall carry at least one permanently affixed warning sign with the following words or equivalent:

WARNING
KEEP 5 FT CLEAR OF AIRCRAFT ENGINES
AND FUEL TANK AREAS.

or

WARNING
KEEP 1.5 METERS CLEAR OF AIRCRAFT
ENGINES AND FUEL TANK AREAS

**(3) Cords.** Flexible cords for aircraft energizers and ground support equipment shall be identified for the type of service and extra-hard usage and shall include an equipment grounding conductor.

**(D) Mobile Servicing Equipment with Electric Components.**

**(1) General.** Mobile servicing equipment (such as vacuum cleaners, air compressors, air movers, etc.) having electric wiring and equipment not suitable for Class I, Division 2 or Zone 2 locations shall be designed and mounted so that all such fixed wiring and equipment will be at least 450 mm (18 in.) above the floor. Such mobile equipment shall not be operated within the Class I location defined in 513.3(C) and shall carry at least one permanently affixed warning sign with the following words or equivalent:

WARNING
KEEP 5 FT CLEAR OF AIRCRAFT ENGINES
AND FUEL TANK AREAS

or

WARNING
KEEP 1.5 METERS CLEAR OF AIRCRAFT
ENGINES AND FUEL TANK AREAS

**(2) Cords and Connectors.** Flexible cords for mobile equipment shall be suitable for the type of service and identified for extra-hard usage and shall include an equipment grounding conductor. Attachment plugs and receptacles shall be identified for the location in which they are installed and shall provide for connection of the equipment grounding conductor.

**(3) Restricted Use.** Equipment that is not identified as suitable for Class I, Division 2 locations shall not be operated in locations where maintenance operations likely to release flammable liquids or vapors are in progress.

**(E) Portable Equipment.**

**(1) Portable Lighting Equipment.** Portable lighting equipment that is used within a hangar shall be identified for the location in which they are used. For portable lamps, flexible cord suitable for the type of service and identified for extra-hard usage shall be used. Each such cord shall include a separate equipment grounding conductor.

**(2) Portable Utilization Equipment.** Portable utilization equipment that is or may be used within a hangar shall be of a type suitable for use in Class I, Division 2 or Zone 2 locations. For portable utilization equipment flexible cord suitable for the type of service and approved for extra-hard usage shall be used. Each such cord shall include a separate equipment grounding conductor.

## 513.16 Grounded and Grounding Requirements.

**(A) General Grounding Requirements.** All metal raceways, the metal armor or metallic sheath on cables, and all non–current-carrying metal parts of fixed or portable electrical equipment, regardless of voltage, shall be grounded as provided in Article 250. Grounding in Class I locations shall comply with 501.16 for Class I, Division 1 and 2 locations and 505.25 for Class I, Zone 0, 1, and 2 locations.

**(B) Supplying Circuits with Grounded and Grounding Conductors in Class I Locations.**

**(1) Circuits Supplying Portable Equipment or Pendants.** Where a circuit supplies portables or pendants and includes a grounded conductor as provided in Article 200, receptacles, attachment plugs, connectors, and similar devices shall be of the grounding type, and the grounded conductor of the flexible cord shall be connected to the screw shell of any lampholder or to the grounded terminal of any utilization equipment supplied.

**(2) Approved Means.** Approved means shall be provided for maintaining continuity of the grounding conductor between the fixed wiring system and the non–current-carrying metal portions of pendant luminaires (fixtures), portable lamps, and portable utilization equipment.

## ARTICLE 514
## Motor Fuel Dispensing Facilities

FPN: Rules that are followed by a reference in brackets contain text that has been extracted from NFPA 30A-2000, *Automotive and Marine Service Station Code*. Only editorial changes were made to the extracted text to make it consistent with this *Code*.

**514.1 Scope.** These occupancies shall include locations where gasoline or other volatile flammable liquids or liquefied flammable gases are transferred to the fuel tanks (including auxiliary fuel tanks) of self-propelled vehicles or approved containers.

FPN: For further information regarding safeguards for motor fuel dispensing facilities, see NFPA 30A-2000, *Code for Motor Fuel Dispensing Facilities and Repair Garages.*

**514.2 Definition.**

**Motor Fuel Dispensing Facility.** A location where gasoline or other volatile flammable liquids or liquefied flammable gases are transferred to the fuel tanks (including auxiliary fuel tanks) of self-propelled vehicles or approved containers.

FPN: Refer to Articles 510 and 511 with respect to electric wiring and equipment for other areas used as lubritoriums, service rooms, repair rooms, offices, salesrooms, compressor rooms, and similar locations.

**514.3 Classification of Locations.**

**(A) Unclassified Locations.** Where the authority having jurisdiction can satisfactorily determine that flammable liquids having a flash point below 38°C (100°F), such as gasoline, will not be handled, such location shall not be required to be classified.

**(B) Classified Locations.**

**(1) Class I Locations.** Table 514.3(B)(1) shall be applied where Class I liquids are stored, handled, or dispensed and shall be used to delineate and classify motor fuel dispensing facilities and commercial garages as defined in Article 511. Table 515.3 shall be used for the purpose of delineating and classifying aboveground tanks. A Class I location shall not extend beyond an unpierced wall, roof, or other solid partition. [NFPA 30A, 8.1, 8.3]

**(2) Compressed Natural Gas, Liquefied Natural Gas, and Liquefied Petroleum Gas Areas.** Table 514.3(B)(2) shall be used to delineate and classify areas where compressed natural gas (CNG), liquefied natural gas (LNG), or liquefied petroleum gas (LPG) are stored, handled, or dispensed. Where CNG or LNG dispensers are installed beneath a canopy or enclosure, either the canopy or enclosure shall be designed to prevent accumulation or entrapment of ignitible vapors, or all electrical equipment installed beneath the canopy or enclosure shall be suitable for Class I, Division 2 hazardous (classified) locations. Dispensing devices for liquefied petroleum gas shall be located not less than 1.5 m (5 ft) from any dispensing device for Class I liquids. [NFPA 30A, 12.1, 12.4, 12.5]

FPN No. 1: For information on area classification where liquefied petroleum gases are dispensed, see NFPA 58-2001, *Liquefied Petroleum Gas Code.*

FPN No. 2: For information on classified areas pertaining to LP-Gas systems other than residential or commercial, see NFPA 58-2001, *Liquefied Petroleum Gas Code* and NFPA 59-2001, *Utility LP-Gas Plant Code*.

FPN No. 3: See 555.21 for gasoline dispensing stations in marinas and boatyards.

**Table 514.3(B)(1) Class I Locations — Motor Fuel Dispensing Facilities and Commercial Garages**

| Location | Class I, Group D Division | Extent of Classified Location |
|---|---|---|
| **Underground Tank** | | |
| Fill opening | 1 | Any pit, box, or space below grade level, any part of which is within the Division 1 or Division 2, Zone 1 or Zone 2 classified location |
| | 2 | Up to 450 mm (18 in.) above grade level within a horizontal radius of 3.0 m (10 ft) from a loose fill connection and within a horizontal radius of 1.5 m (5 ft) from a tight fill connection |
| Vent — discharging upward | 1 | Within 900 mm (3 ft) of open end of vent, extending in all directions |
| | 2 | Space between 900 mm (3 ft) and 1.5 m (5 ft) of open end of vent, extending in all directions |
| **Dispensing Device**[1,4] (except overhead type)[2] | | |
| Pits | 1 | Any pit, box, or space below grade level, any part of which is within the Division 1 or Division 2, Zone 1 or Zone 2 classified location |
| Dispenser | | FPN: Space classification inside the dispenser enclosure is covered in ANSI/UL 87-1995, *Power Operated Dispensing Devices for Petroleum Products.* |
| | 2 | Within 450 mm (18 in.) horizontally in all directions extending to grade from the dispenser enclosure or that portion of the dispenser enclosure containing liquid-handling components FPN: Space classification inside the dispenser enclosure is covered in ANSI/UL 87-1995, *Power Operated Dispensing Devices for Petroleum Products.* |
| Outdoor | 2 | Up to 450 mm (18 in.) above grade level within 6.0 m (20 ft) horizontally of any edge of enclosure. |
| Indoor with mechanical ventilation | 2 | Up to 450 mm (18 in.) above grade or floor level within 6.0 m (20 ft) horizontally of any edge of enclosure |
| with gravity ventilation | 2 | Up to 450 mm (18 in.) above grade or floor level within 7.5 m (25 ft) horizontally of any edge of enclosure |
| **Dispensing Device**[4] | | |
| Overhead type [2] | 1 | The space within the dispenser enclosure, and all electrical equipment integral with the dispensing hose or nozzle |
| | 2 | A space extending 450 mm (18 in.) horizontally in all directions beyond the enclosure and extending to grade |
| | 2 | Up to 450 mm (18 in.) above grade level within 6.0 m (20 ft) horizontally measured from a point vertically below the edge of any dispenser enclosure |
| **Remote Pump — Outdoor** | 1 | Any pit, box, or space below grade level if any part is within a horizontal distance of 3.0 m (10 ft) from any edge of pump |
| | 2 | Within 900 mm (3 ft) of any edge of pump, extending in all directions. Also up to 450 mm (18 in.) above grade level within 3.0 m (10 ft) horizontally from any edge of pump |
| **Remote Pump — Indoor** | 1 | Entire space within any pit |
| | 2 | Within 1.5 m (5 ft) of any edge of pump, extending in all directions. Also up to 900 mm (3 ft) above grade level within 7.5 m (25 ft) horizontally from any edge of pump |
| **Lubrication or Service Room — with Dispensing** | 1 | Any pit within any unventilated space |
| | 2 | Any pit with ventilation |
| | 2 | Space up to 450 mm (18 in.) above floor or grade level and 900 mm (3 ft) horizontally from a lubrication pit |

**Table 514.3(B)(1)**  *Continued*

| Location | Class I, Group D Division | Extent of Classified Location |
|---|---|---|
| Dispenser for Class I liquids | 2 | Within 900 mm (3 ft) of any fill or dispensing point, extending in all directions |
| **Lubrication or Service Room — Without Dispensing** | 2 | Entire area within any pit used for lubrication or similar services where Class I liquids may be released |
| | 2 | Area up to 450 mm (18 in.) above any such pit and extending a distance of 900 mm (3 ft) horizontally from any edge of the pit |
| | 2 | Entire unventilated area within any pit, belowgrade area, or subfloor area |
| | 2 | Area up to 450 mm (18 in.) above any such unventilated pit, belowgrade work area, or subfloor work area and extending a distance of 900 mm (3 ft) horizontally from the edge of any such pit, belowgrade work area, or subfloor work area |
| | Unclassified | Any pit, belowgrade work area, or subfloor work area that is provided with exhaust ventilation at a rate of not less than 0.3 m$^3$/min/m$^2$ (1 cfm/ft$^2$) of floor area at all times that the building is occupied or when vehicles are parked in or over this area and where exhaust air is taken from a point within 300 mm (12 in.) of the floor of the pit, belowgrade work area, or subfloor work area |
| **Special Enclosure Inside Building[3]** | 1 | Entire enclosure |
| **Sales, Storage, and Rest Rooms** | Unclassified | If there is any opening to these rooms within the extent of a Division 1 location, the entire room shall be classified as Division 1 |
| **Vapor Processing Systems Pits** | 1 | Any pit, box, or space below grade level, any part of which is within a Division 1 or Division 2 classified location or that houses any equipment used to transport or process vapors |
| **Vapor Processing Equipment Located Within Protective Enclosures** FPN: See 10.1.7 of NFPA 30A-2000, *Code for Motor Fuel Dispensing Facilities and Repair Garages.* | 2 | Within any protective enclosure housing vapor processing equipment |
| **Vapor Processing Equipment Not Within Protective Enclosures** (excluding piping and combustion devices) | 2 | The space within 450 mm (18 in.) in all directions of equipment containing flammable vapor or liquid extending to grade level. Up to 450 mm (18 in.) above grade level within 3.0 m (10 ft) horizontally of the vapor processing equipment |
| **Equipment Enclosures** | 1 | Any space within the enclosure where vapor or liquid is present under normal operating conditions |
| **Vacuum-Assist Blowers** | 2 | The space within 450 mm (18 in.) in all directions extending to grade level. Up to 450 mm (18 in.) above grade level within 3.0 m (10 ft) horizontally |

[1]Refer to Figure 514.3 for an illustration of classified location around dispensing devices.
[2]Ceiling mounted hose reel.
[3]FPN: See 4.3.9 of NFPA 30A-2000, *Code for Motor Fuel Dispensing Facilities and Repair Garages.*
[4]FPN: Area classification inside the dispenser enclosure is covered in ANSI/UL 87-1995, *Power-Operated Dispensing Devices for Petroleum Products.*

[NFPA 30A, Table 8.3.1]

**Table 514.3(B)(2)  Electrical Equipment Classified Areas for Dispensing Devices**

| Dispensing Device | Extent of Classified Area | |
| --- | --- | --- |
| | Class I, Division 1 | Class I, Division 2 |
| Compressed Natural Gas | Entire space within the dispenser enclosure | 1.5 m (5 ft) in all directions from dispenser enclosure |
| Liquefied Natural Gas | Entire space within the dispenser enclosure and 1.5 m (5 ft) in all directions from the dispenser enclosure | From 1.5 m to 3.0 m (5 ft to 10 ft) in all directions from the dispenser enclosure |
| Liquefied Petroleum Gas | Entire space within the dispenser enclosure; 450 mm (18 in.) from the exterior surface of the dispenser enclosure to an elevation of 1.2 m (4 ft) above the base of the dispenser; the entire pit or open space beneath the dispenser and within 6.0 m (20 ft) horizontally from any edge of the dispenser when the pit or trench is not mechanically ventilated. | Up to 450 mm (18 in.) aboveground and within 6.0 m (20 ft) horizontally from any edge of the dispenser enclosure, including pits or trenches within this area when provided with adequate mechanical ventilation |

[NFPA 30A, Table 12.6.2]

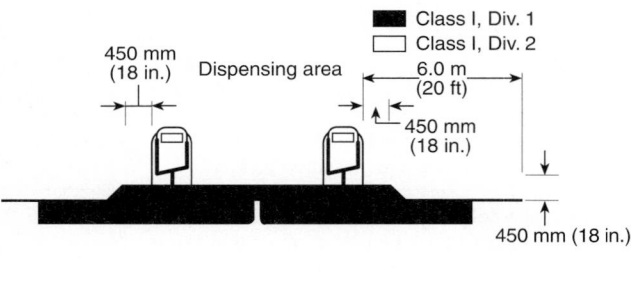

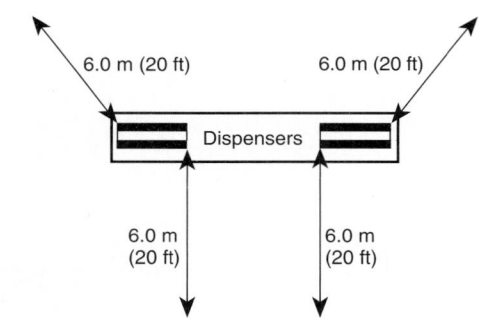

**Figure 514.3  Classified locations adjacent to dispensers as detailed in Table 514.3(B)(1). [NFPA 30A, Figure 8.3]**

**514.4  Wiring and Equipment Installed in Class I Locations.** All electrical equipment and wiring installed in Class I locations as classified in 514.3 shall comply with the applicable provisions of Article 501.

*Exception:  As permitted in 514.8.*

FPN:  For special requirements for conductor insulation, see 501.13.

**514.7  Wiring and Equipment Above Class I Locations.** Wiring and equipment above the Class I locations as classified in 514.3 shall comply with 511.7.

**514.8  Underground Wiring.** Underground wiring shall be installed in threaded rigid metal conduit or threaded steel intermediate metal conduit. Any portion of electrical wiring or equipment that is below the surface of a Class I, Division 1, or a Class I, Division 2 location [as classified in Table 514.3(B)(1) and Table 514.3(B)(2)] shall be considered to be in a Class I, Division 1, location that shall extend at least to the point of emergence above grade. Refer to Table 300.5.

*Exception No. 1:  Type MI cable shall be permitted where it is installed in accordance with Article 332.*

*Exception No. 2:  Rigid nonmetallic conduit complying with Article 352 shall be permitted where buried under not less than 600 mm (2 ft) of cover. Where rigid nonmetallic conduit is used, threaded rigid metal conduit or threaded steel intermediate metal conduit shall be used for the last 600 mm (2 ft) of the underground run to emergence or to the point of connection to the aboveground raceway, and an equipment grounding conductor shall be included to provide electrical continuity of the raceway system and for grounding of non–current-carrying metal parts.*

**514.9  Sealing.**

**(A)  At Dispenser.** A listed seal shall be provided in each conduit run entering or leaving a dispenser or any cavities or enclosures in direct communication therewith. The sealing fitting shall be the first fitting after the conduit emerges from the earth or concrete.

**(B) At Boundary.** Additional seals shall be provided in accordance with 501.5. Sections 501.5(A)(4) and (B)(2) shall apply to horizontal as well as to vertical boundaries of the defined Class I locations.

### 514.11 Circuit Disconnects.

**(A) General.** Each circuit leading to or through dispensing equipment, including equipment for remote pumping systems, shall be provided with a clearly identified and readily accessible switch or other acceptable means, located remote from the dispensing devices, to disconnect simultaneously from the source of supply, all conductors of the circuits, including the grounded conductor, if any.

Single-pole breakers utilizing handle ties shall not be permitted.

**(B) Attended Self-Service Motor Fuel Dispensing Facilities.** Emergency controls as specified in 514.11(A) shall be installed at a location acceptable to the authority having jurisdiction, but controls shall not be more than 30 m (100 ft) from dispensers. [NFPA 30A, 6.7.1]

**(C) Unattended Self-Service Motor Fuel Dispensing Facilities.** Emergency controls as specified in 514.11(A) shall be installed at a location acceptable to the authority having jurisdiction, but the control shall be more than 6 m (20 ft) but less than 30 m (100 ft) from the dispensers. Additional emergency controls shall be installed on each group of dispensers or the outdoor equipment used to control the dispensers. Emergency controls shall shut off all power to all dispensing equipment at the station. Controls shall be manually reset only in a manner approved by the authority having jurisdiction. [NFPA 30A, 6.7.2]

> FPN: For additional information, see 6.7.1 and 6.7.2 of NFPA 30A-2000, *Code For Motor Fuel Dispensing Facilities and Repair Garages.*

### 514.13 Provisions for Maintenance and Service of Dispensing Equipment. Each dispensing device shall be provided with a means to remove all external voltage sources, including feedback, during periods of maintenance and service of the dispensing equipment. The location of this means shall be permitted to be other than inside or adjacent to the dispensing device.

### 514.16 Grounding. All metal raceways, the metal armor or metallic sheath on cables, and all non–current-carrying metal parts of fixed portable electrical equipment, regardless of voltage, shall be grounded as provided in Article 250. Grounding in Class I locations shall comply with 501.16.

## ARTICLE 515
## Bulk Storage Plants

> FPN: Rules that are followed by a reference in brackets contain text that has been extracted from NFPA 30-2000, *Flammable and Combustible Liquids Code.* Only editorial changes were made to the extracted text to make it consistent with this *Code.*

**515.1 Scope.** This article covers a property or portion of a property where flammable liquids are received by tank vessel, pipelines, tank car, or tank vehicle and are stored or blended in bulk for the purpose of distributing such liquids by tank vessel, pipeline, tank car, tank vehicle, portable tank, or container.

**515.2 Definition.**

**Bulk Storage Plant.** That portion of a property where flammable liquids are received by tank vessel, pipelines, tank car, or tank vehicle, and are stored or blended in bulk for the purpose of distributing such liquids by tank vessel, pipelines, tank car, tank vehicle, portable tank, or container.

> FPN: For further information, see NFPA 30-2000, *Flammable and Combustible Liquids Code.*

**515.3 Class I Locations.** Table 515.3 shall be applied where Class I liquids are stored, handled, or dispensed and shall be used to delineate and classify bulk storage plants. The class location shall not extend beyond a floor, wall, roof, or other solid partition that has no communicating openings. [NFPA 30, 5.9.5.1, 5.9.5.3]

> FPN No. 1: The area classifications listed in Table 515.3 are based on the premise that the installation meets the applicable requirements of NFPA 30-2000, *Flammable and Combustible Liquids Code,* Chapter 5, in all respects. Should this not be the case, the authority having jurisdiction has the authority to classify the extent of the classified space.

> FPN No. 2: See 555.21 for gasoline dispensing stations in marinas and boatyards.

**515.4 Wiring and Equipment Located in Class I Locations.** All electrical wiring and equipment within the Class I locations defined in 515.2 shall comply with the applicable provisions of Article 501 or Article 505 for the division or zone in which they are used.

*Exception: As permitted in 515.8.*

## Table 515.3 Electrical Area Classifications

| Location | NEC Class I Division | Zone | Extent of Classified Area |
|---|---|---|---|
| Indoor equipment installed in accordance with 5.3 of NFPA 30 where flammable vapor–air mixtures can exist under normal operation | 1 | 0 | The entire area associated with such equipment where flammable gases or vapors are present continuously or for long periods of time |
| | 1 | 1 | Area within 1.5 m (5 ft) of any edge of such equipment, extending in all directions |
| | 2 | 2 | Area between 1.5 m and 2.5 m (5 ft and 8 ft) of any edge of such equipment, extending in all directions; also, space up to 900 mm (3 ft) above floor or grade level within 1.5 m to 7.5 m (5 ft to 25 ft) horizontally from any edge of such equipment[1] |
| Outdoor equipment of the type covered in 5.3 of NFPA 30 where flammable vapor–air mixtures may exist under normal operation | 1 | 0 | The entire area associated with such equipment where flammable gases or vapors are present continuously or for long periods of time |
| | 1 | 1 | Area within 900 mm (3 ft) of any edge of such equipment, extending in all directions |
| | 2 | 2 | Area between 900 mm (3 ft) and 2.5 m (8 ft) of any edge of such equipment, extending in all directions; also, space up to 900 mm (3 ft) above floor or grade level within 900 mm to 3.0 m (3 ft to 10 ft) horizontally from any edge of such equipment |
| Tank storage installations inside buildings | 1 | 1 | All equipment located below grade level |
| | 2 | 2 | Any equipment located at or above grade level |
| Tank – aboveground | 1 | 0 | Inside fixed roof tank |
| | 1 | 1 | Area inside dike where dike height is greater than the distance from the tank to the dike for more than 50 percent of the tank circumference |
| Shell, ends, or roof and dike area | 2 | 2 | Within 3.0 m (10 ft) from shell, ends, or roof of tank; also, area inside dike to level of top of tank |
| Vent | 1 | 0 | Area inside of vent piping or opening |
| | 1 | 1 | Within 1.5 m (5 ft) of open end of vent, extending in all directions |
| | 2 | 2 | Area between 1.5 m and 3.0 m (5 ft and 10 ft) from open end of vent, extending in all directions |
| Floating roof with fixed outer roof | 1 | 0 | Area between the floating and fixed roof sections and within the shell |
| Floating roof with no fixed outer roof | 1 | 1 | Area above the floating roof and within the shell |
| Underground tank fill opening | 1 | 1 | Any pit, or space below grade level, if any part is within a Division 1 or 2, or Zone 1 or 2, classified location |
| | 2 | 2 | Up to 450 mm (18 in.) above grade level within a horizontal radius of 3.0 m (10 ft) from a loose fill connection, and within a horizontal radius of 1.5 m (5 ft) from a tight fill connection |
| Vent – discharging upward | 1 | 0 | Area inside of vent piping or opening |
| | 1 | 1 | Within 900 mm (3 ft) of open end of vent, extending in all directions |
| | 2 | 2 | Area between 900 mm and 1.5 m (3 ft and 5 ft) of open end of vent, extending in all directions |

**Table 515.3** *Continued*

| Location | NEC Class I Division | Zone | Extent of Classified Area |
|---|---|---|---|
| Drum and container filling – outdoors or indoors | 1 | 0 | Area inside the drum or container |
| | 1 | 1 | Within 900 mm (3 ft) of vent and fill openings, extending in all directions |
| | 2 | 2 | Area between 900 mm and 1.5 m (3 ft and 5 ft) from vent or fill opening, extending in all directions; also, up to 450 mm (18 in.) above floor or grade level within a horizontal radius of 3.0 m (10 ft) from vent or fill opening |
| Pumps, bleeders, withdrawal fittings, Indoors | 2 | 2 | Within 1.5 m (5 ft) of any edge of such devices, extending in all directions; also, up to 900 mm (3 ft) above floor or grade level within 7.5 m (25 ft) horizontally from any edge of such devices |
| Outdoors | 2 | 2 | Within 900 mm (3 ft) of any edge of such devices, extending in all directions. Also, up to 450 mm (18 in.) above grade level within 3.0 m (10 ft) horizontally from any edge of such devices |
| Pits and sumps Without mechanical ventilation | 1 | 1 | Entire area within a pit or sump if any part is within a Division 1 or 2, or Zone 1 or 2, classified location |
| With adequate mechanical ventilation | 2 | 2 | Entire area within a pit or sump if any part is within a Division 1 or 2, or Zone 1 or 2, classified location |
| Containing valves, fittings, or piping, and not within a Division 1 or 2, or Zone 1 or 2, classified location | 2 | 2 | Entire pit or sump |
| Drainage ditches, separators, impounding basins Outdoors | 2 | 2 | Area up to 450 mm (18 in.) above ditch, separator, or basin; also, area up to 450 mm (18 in.) above grade within 4.5 m (15 ft) horizontally from any edge |
| Indoors | | | Same classified area as pits |
| Tank vehicle and tank car[2] loading through open dome | 1 | 0 | Area inside of the tank |
| | 1 | 1 | Within 900 mm (3 ft) of edge of dome, extending in all directions |
| | 2 | 2 | Area between 900 mm and 4.5 m (3 ft and 15 ft) from edge of dome, extending in all directions |
| Loading through bottom connections with atmospheric venting | 1 | 0 | Area inside of the tank |
| | 1 | 1 | Within 900 mm (3 ft) of point of venting to atmosphere, extending in all directions |
| | 2 | 2 | Area between 900 mm and 4.5 m (3 ft and 15 ft) from point of venting to atmosphere, extending in all directions; also, up to 450 mm (18 in.) above grade within a horizontal radius of 3.0 m (10 ft) from point of loading connection |
| Office and rest rooms | Ordinary | | If there is any opening to these rooms within the extent of an indoor classified location, the room shall be classified the same as if the wall, curb, or partition did not exist. |

**Table 515.3** *Continued*

| Location | NEC Class I Division | Zone | Extent of Classified Area |
|---|---|---|---|
| Loading through closed dome with atmospheric venting | 1 | 1 | Within 900 mm (3 ft) of open end of vent, extending in all directions |
|  | 2 | 2 | Area between 900 mm and 4.5 m (3 ft and 15 ft) from open end of vent, extending in all directions; also, within 900 mm (3 ft) of edge of dome, extending in all directions |
| Loading through closed dome with vapor control | 2 | 2 | Within 900 mm (3 ft) of point of connection of both fill and vapor lines extending in all directions |
| Bottom loading with vapor control or any bottom unloading | 2 | 2 | Within 900 mm (3 ft) of point of connections, extending in all directions; also up to 450 mm (18 in.) above grade within a horizontal radius of 3.0 m (10 ft) from point of connections |
| Storage and repair garage for tank vehicles | 1 | 1 | All pits or spaces below floor level |
|  | 2 | 2 | Area up to 450 mm (18 in.) above floor or grade level for entire storage or repair garage |
| Garages for other than tank vehicles | Ordinary |  | If there is any opening to these rooms within the extent of an outdoor classified location, the entire room shall be classified the same as the area classification at the point of the opening. |
| Outdoor drum storage | Ordinary |  |  |
| Inside rooms or storage lockers used for the storage of Class I liquids | 2 | 2 | Entire room |
| Indoor warehousing where there is no flammable liquid transfer | Ordinary |  | If there is any opening to these rooms within the extent of an indoor classified location, the room shall be classified the same as if the wall, curb, or partition did not exist. |
| Piers and wharves |  |  | See Figure 515.3. |

[1]The release of Class I liquids may generate vapors to the extent that the entire building, and possibly an area surrounding it, should be considered a Class I, Division 2 or Zone 2 location.

[2]When classifying extent of area, consideration shall be given to fact that tank cars or tank vehicles may be spotted at varying points. Therefore, the extremities of the loading or unloading positions shall be used. [NFPA 30: Table 5-9.5.3]

## 515.7 Wiring and Equipment Above Class I Locations.

**(A) Fixed Wiring.** All fixed wiring above Class I locations shall be in metal raceways or PVC Schedule 80 rigid nonmetallic conduit, or equivalent, or be Type MI, TC, or MC cable.

**(B) Fixed Equipment.** Fixed equipment that may produce arcs, sparks, or particles of hot metal, such as lamps and lampholders for fixed lighting, cutouts, switches, receptacles, motors, or other equipment having make-and-break or sliding contacts, shall be of the totally enclosed type or be constructed so as to prevent the escape of sparks or hot metal particles.

**(C) Portable Lamps or Other Utilization Equipment.** Portable lamps or other utilization equipment and their flexible cords shall comply with the provisions of Article 501 or Article 505 for the class of location above which they are connected or used.

## 515.8 Underground Wiring.

**(A) Wiring Method.** Underground wiring shall be installed in threaded rigid metal conduit or threaded steel intermediate metal conduit or, where buried under not less than 600 mm (2 ft) of cover, shall be permitted in rigid nonmetallic conduit or a listed cable. Where rigid nonmetallic conduit is used, threaded rigid metal conduit or threaded steel intermediate metal conduit shall be used for the last 600 mm (2 ft) of the conduit run to emergence or to the point of connection to the aboveground raceway. Where cable is used, it shall be enclosed in threaded rigid metal conduit or threaded steel intermediate metal conduit from the point of lowest buried cable level to the point of connection to the aboveground raceway.

**(B) Insulation.** Conductor insulation shall comply with 501.13.

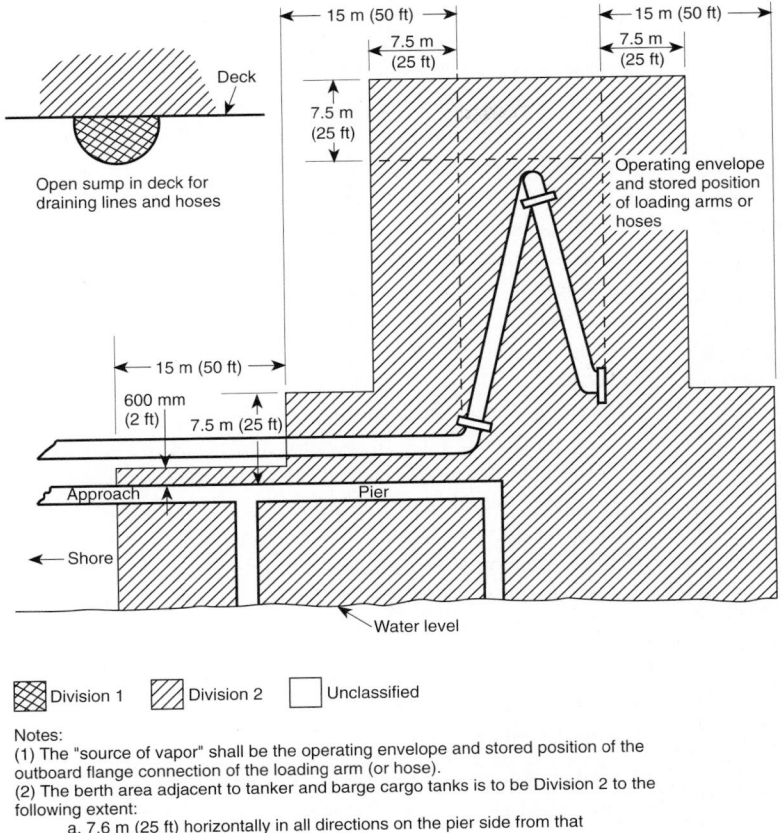

Division 1  Division 2  Unclassified

Notes:
(1) The "source of vapor" shall be the operating envelope and stored position of the outboard flange connection of the loading arm (or hose).
(2) The berth area adjacent to tanker and barge cargo tanks is to be Division 2 to the following extent:
    a. 7.6 m (25 ft) horizontally in all directions on the pier side from that portion of the hull containing cargo tanks
    b. From the water level to 7.6 m (25 ft) above the cargo tanks at their highest position
(3) Additional locations may have to be classified as required by the presence of other sources of flammable liquids on the berth, by Coast Guard, or other regulations.

**Figure 515.3 Marine terminal handling flammable liquids. [NFPA 30, Figure 5.7.16]**

**(C) Nonmetallic Wiring.** Where rigid nonmetallic conduit or cable with a nonmetallic sheath is used, an equipment grounding conductor shall be included to provide for electrical continuity of the raceway system and for grounding of non–current-carrying metal parts.

**515.9 Sealing.** Sealing requirements shall apply to horizontal as well as to vertical boundaries of the defined Class I locations. Buried raceways and cables under defined Class I locations shall be considered to be within a Class I, Division 1 or Zone 1 location.

**515.10 Special Equipment — Gasoline Dispensers.** Where gasoline or other volatile flammable liquids or liquefied flammable gases are dispensed at bulk stations, the applicable provisions of Article 514 shall apply.

**515.16 Grounding.** All metal raceways, the metal armor or metallic sheath on cables, and all non–current-carrying metal parts of fixed or portable electrical equipment, re-

gardless of voltage, shall be grounded as provided in Article 250. Grounding in Class I locations shall comply with 501.16 for Class I, Division 1 and 2 locations and Article 505.25 for Class I, Zone 0, 1, and 2 locations.

FPN: For information on grounding for static protection, see 5.6.3.4 and 5.6.3.5 of NFPA 30-2000, *Flammable and Combustible Liquids Code.*

## ARTICLE 516
## Spray Application, Dipping, and Coating Processes

FPN: Rules that are followed by a reference in brackets contain text that has been extracted from NFPA 33-2000, *Standard for Spray Application Using Flammable and Combustible Materials*, or NFPA 34-2000, *Standard for Dipping and Coating Processes Using Flammable or Com-*

*bustible Liquids.* Only editorial changes were made to the extracted text to make it consistent with this *Code.*

**516.1 Scope.** This article covers the regular or frequent application of flammable liquids, combustible liquids, and combustible powders by spray operations and the application of flammable liquids, or combustible liquids at temperatures above their flashpoint, by dipping, coating, or other means.

> FPN: For further information regarding safeguards for these processes, such as fire protection, posting of warning signs, and maintenance, see NFPA 33-2000, *Standard for Spray Application Using Flammable and Combustible Materials*, and NFPA 34-2000, *Standard for Dipping and Coating Processes Using Flammable or Combustible Liquids*. For additional information regarding ventilation, see NFPA 91-1999, *Standard for Exhaust Systems for Air Conveying of Vapors, Gases, Mists, and Noncombustible Particulate Solids.*

**516.2 Definitions.** For the purpose of this article, the following definitions shall apply.

**Spray Area.** Normally locations outside of buildings or localized operations within a larger room or space. Such are normally provided with some local vapor extraction/ventilation system. In automated operations, the area limits shall be the maximum area in the direct path of spray operations. In manual operations, the area limits shall be the maximum area of spray when aimed at 180 degrees to the application surface. [NFPA 33, 1.6, Definitions]

**Spray Booth.** An enclosure or insert within a larger room used for spray/coating/dipping applications. A spray booth may be fully enclosed or have open front or face and may include separate conveyor entrance and exit. The spray booth is provided with a dedicated ventilation exhaust but may draw supply air from the larger room or have a dedicated air supply. [NFPA 33, 1.6, Definitions]

**Spray Room.** A purposefully enclosed room built for spray/coating/dipping applications provided with dedicated ventilation supply and exhaust. Normally the room is configured to house the item to be painted, providing reasonable access around the item/process. Depending on the size of the item being painted, such rooms may actually be the entire building or the major portion thereof. [NFPA 33, 1.6, Definitions]

**516.3 Classification of Locations.** Classification is based on dangerous quantities of flammable vapors, combustible mists, residues, dusts, or deposits.

**(A) Class I or Class II, Division 1 Locations.** The following spaces shall be considered Class I or Class II, Division 1 locations, as applicable:

(1) The interior of spray booths and rooms except as specifically provided in 516.3(D).

(2) The interior of exhaust ducts.

(3) Any area in the direct path of spray operations.

(4) For dipping and coating operations, all space within a 1.5-m (5-ft) radial distance from the vapor sources extending from these surfaces to the floor. The vapor source shall be the liquid exposed in the process and the drainboard, and any dipped or coated object from which it is possible to measure vapor concentrations exceeding 25 percent of the lower flammable limit at a distance of 300 mm (1 ft), in any direction, from the object.

(5) Sumps, pits, or belowgrade channels within 7.5 m (25 ft) horizontally of a vapor source. If the sump, pit, or channel extends beyond 7.5 m (25 ft) from the vapor source, it shall be provided with a vapor stop or it shall be classified as Class I, Division 1 for its entire length.

(6) The interior of any enclosed dipping or coating process or apparatus. [NFPA 33, 1.6 Definitions; NFPA 34, 1.6, Definitions, 4.2.1, 4.2.2, 4.3.1]

**(B) Class I or Class II, Division 2 Locations.** The following spaces shall be considered Class I or Class II, Division 2 as applicable.

**(1) Open Spraying.** For open spraying, all space outside of but within 6 m (20 ft) horizontally and 3 m (10 ft) vertically of the Class I, Division 1 location as defined in 516.3(A), and not separated from it by partitions. See Figure 516.3(B)(1) .

**(2) Closed-Top, Open-Face, and Open-Front Spraying.** If spray application operations are conducted within a closed-top, open-face, or open-front booth or room, any electrical wiring or utilization equipment located outside of the booth or room but within the boundaries designated as Division 2 in Figure 516.3(B)(2) shall be suitable for Class I, Division 2 or Class II, Division 2 locations, whichever is applicable. The Class I, Division 2 or Class II, Division 2 locations shown in Figure 516.3(B)(2) shall extend from the edges of the open face or open front of the booth or room in accordance with the following:

(a)  If the exhaust ventilation system is interlocked with the spray application equipment, then the Division 2 location shall extend 1.5 m (5 ft) horizontally and 900 mm (3 ft) vertically from the open face or open front of the booth or room, as shown in Figure 516.3(B)(2) , top.

(b)  If the exhaust ventilation system is not interlocked with the spray application equipment, then the Division 2 location shall extend 3 m (10 ft) horizontally and 900 mm (3 ft) vertically from the open face or open front of the booth or room, as shown in Figure 516.3(B)(2), bottom.

For the purposes of this subsection, *interlocked* shall mean that the spray application equipment cannot be oper-

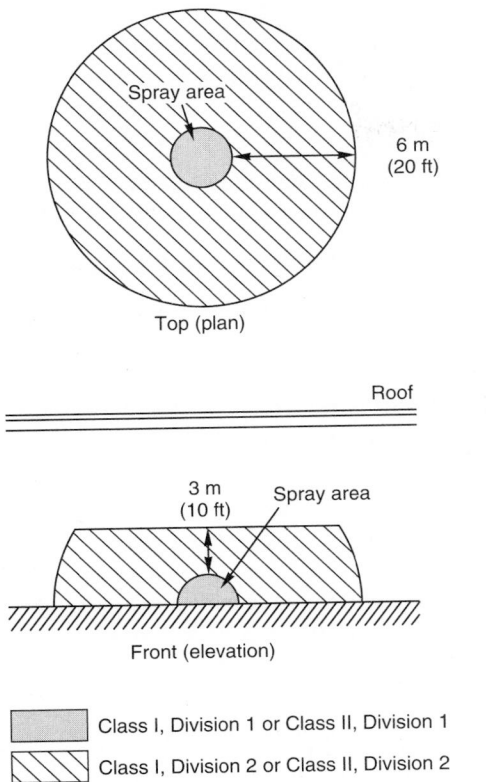

**Figure 516.3(B)(1)  Electrical area classification for open spray areas. [NFPA 33, Figure 4.3.1]**

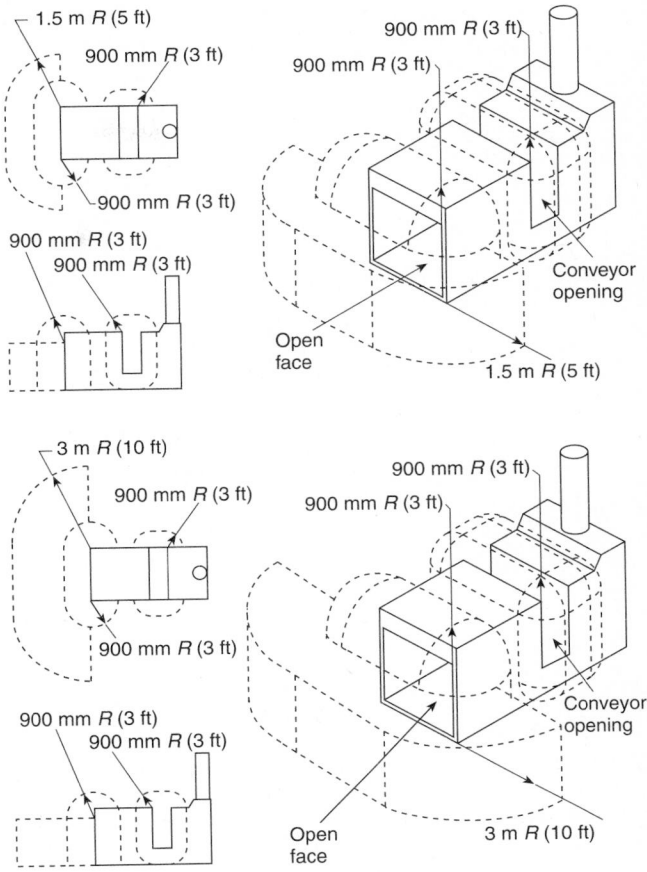

**Figure 516.3(B)(2)  Class I or Class II, Division 2 locations adjacent to a closed top, open face, or open front spray booth or room. [NFPA 33, Figures 4.3.2(a) and 4.3.2(b)]**

ated unless the exhaust ventilation system is operating and functioning properly and spray application is automatically stopped if the exhaust ventilation system fails.

**(3) Open-Top Spraying.** For spraying operations conducted within an open top spray booth, the space 900 mm (3 ft) vertically above the booth and within 900 mm (3 ft) of other booth openings shall be considered Class I or Class II, Division 2.

**(4) Enclosed Booths and Rooms.** For spraying operations confined to an enclosed spray booth or room, the space within 900 mm (3 ft) in all directions from any openings shall be considered Class I or Class II, Division 2 as shown in Figure 516.3B)(4) .

**(5) Dip Tanks and Drain Boards — Surrounding Space.** For dip tanks and drain boards, the 914-mm (3-ft) space surrounding the Class I, Division 1 location as defined in 516.3(A)(4) and as shown in Figure 516.3(B)(5).

**(6) Dip Tanks and Drain Boards — Space Above Floor.** For dip tanks and drain boards, the space 900 mm (3 ft) above the floor and extending 6 m (20 ft) horizontally in all directions from the Class I, Division 1 location.

*Exception:  This space shall not be required to be considered a hazardous (classified) location where the vapor source area is 0.46 m² (5 ft²) or less, and where the contents of the open tank trough or container do not exceed 19 L (5 gal). In addition, the vapor concentration during operation and shutdown periods shall not exceed 25 percent of the lower flammable limit outside the Class I location specified in 516.3(A)(4). [NFPA 33, 4.3.1, 4.3.2, 4.3.3, 4.3.4; NFPA 34, 4.2.3, 4.2.4]*

**(C) Enclosed Coating and Dipping Operations.** The space adjacent to an enclosed dipping or coating process or apparatus shall be considered unclassified.

*Exception:  The space within 900 mm (3 ft) in all directions from any opening in the enclosures shall be classified as Class I, Division 2. [NFPA 34, 4.3.2]*

**(D) Adjacent Locations.** Adjacent locations that are cut off from the defined Class I or Class II locations by tight partitions without communicating openings, and within which flammable vapors or combustible powders are not likely to be released, shall be unclassified.

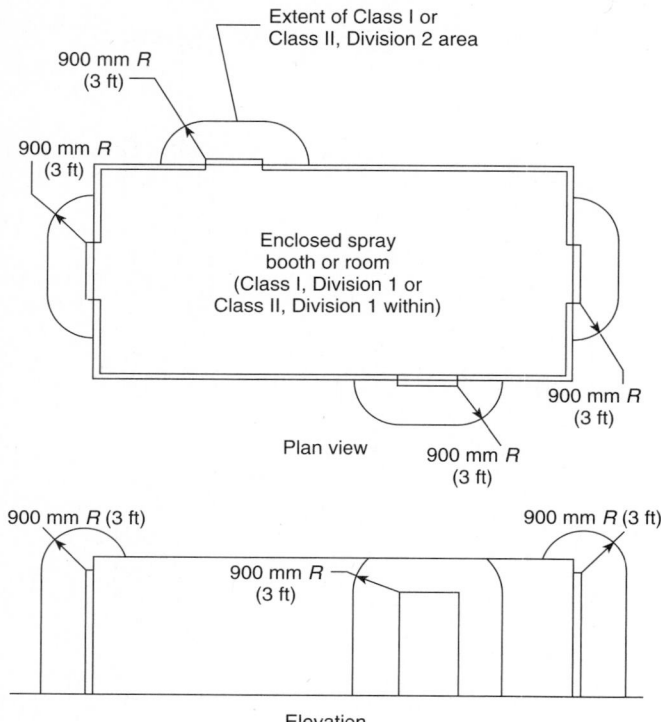

**Figure 516.3(B)(4)   Class I (or Class II), Division 2 locations adjacent to an enclosed spray booth or spray room. [NFPA 33, Figure 4.3.4]**

**(E) Unclassified Locations.** Locations using drying, curing, or fusion apparatus and provided with positive mechanical ventilation adequate to prevent accumulation of flammable concentrations of vapors, and provided with ef-

fective interlocks to de-energize all electrical equipment (other than equipment identified for Class I locations) in case the ventilating equipment is inoperative, shall be permitted to be unclassified where the authority having jurisdiction so judges.

> FPN:  For further information regarding safeguards, see NFPA 86-1999, *Standard for Ovens and Furnaces.*

### 516.4 Wiring and Equipment in Class I Locations.

**(A) Wiring and Equipment — Vapors.** All electric wiring and equipment within the Class I location (containing vapor only — not residues) defined in 516.3 shall comply with the applicable provisions of Article 501.

**(B) Wiring and Equipment — Vapors and Residues.** Unless specifically listed for locations containing deposits of dangerous quantities of flammable or combustible vapors, mists, residues, dusts, or deposits (as applicable), there shall be no electrical equipment in any spray area as herein defined whereon deposits of combustible residue may readily accumulate, except wiring in rigid metal conduit, intermediate metal conduit, Type MI cable, or in metal boxes or fittings containing no taps, splices, or terminal connections. [NFPA 33, 4.2]

**(C) Illumination.** Illumination of readily ignitible areas through panels of glass or other transparent or translucent material shall be permitted only if it complies with the following:

(1) Fixed lighting units are used as the source of illumination.
(2) The panel effectively isolates the Class I location from the area in which the lighting unit is located.

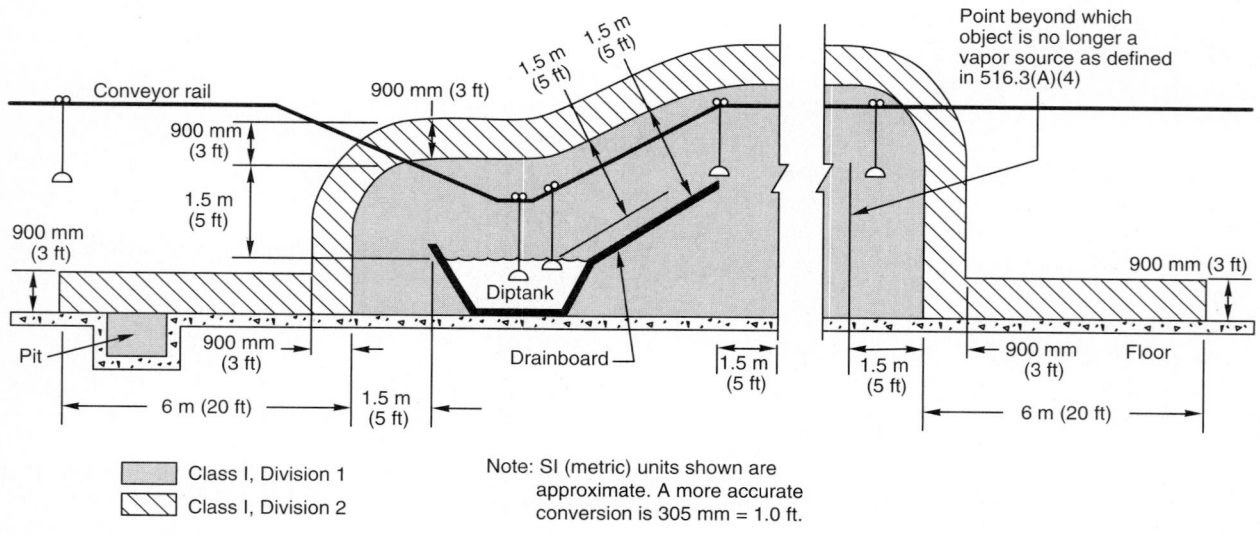

**Figure 516.3(B)(5)   Electrical area classification for open processes without vapor containment or ventilation. [NFPA 34, Figure 4.2(a)]**

(3) The lighting unit is identified for its specific location.

(4) The panel is of a material or is protected so that breakage is unlikely.

(5) The arrangement is such that normal accumulations of hazardous residue on the surface of the panel will not be raised to a dangerous temperature by radiation or conduction from the source of illumination.

**(D) Portable Equipment.** Portable electric lamps or other utilization equipment shall not be used in a spray area during spray operations.

*Exception No. 1: Where portable electric lamps are required for operations in spaces not readily illuminated by fixed lighting within the spraying area, they shall be of the type identified for Class I, Division 1 locations where readily ignitable residues may be present.*

*Exception No. 2: Where portable electric drying apparatus is used in automobile refinishing spray booths and the following requirements are met.*

*(a) The apparatus and its electrical connections are not located within the spray enclosure during spray operations.*

*(b) Electrical equipment within 450 mm (18 in.) of the floor is identified for Class I, Division 2 locations.*

*(c) All metallic parts of the drying apparatus are electrically bonded and grounded.*

*(d) Interlocks are provided to prevent the operation of spray equipment while drying apparatus is within the spray enclosure, to allow for a 3-minute purge of the enclosure before energizing the drying apparatus and to shut off drying apparatus on failure of ventilation system. [NFPA 33, 4.7]*

**(E) Electrostatic Equipment.** Electrostatic spraying or detearing equipment shall be installed and used only as provided in 516.10.

> FPN: For further information, see NFPA 33-2000, *Standard for Spray Application Using Flammable or Combustible Materials.*

## 516.7 Wiring and Equipment Not Within Class I and II Locations.

**(A) Wiring.** All fixed wiring above the Class I and II locations shall be in metal raceways, rigid nonmetallic conduit, or electrical nonmetallic tubing, or shall be Type MI, TC, or MC cable. Cellular metal floor raceways shall be permitted only for supplying ceiling outlets or extensions to the area below the floor of a Class I or II location, but such raceways shall have no connections leading into or through the Class I or II location above the floor unless suitable seals are provided.

**(B) Equipment.** Equipment that may produce arcs, sparks, or particles of hot metal, such as lamps and lampholders for fixed lighting, cutouts, switches, receptacles, motors, or other equipment having make-and-break or sliding contacts, where installed above a Class I or II location or above a location where freshly finished goods are handled, shall be of the totally enclosed type or be constructed so as to prevent the escape of sparks or hot metal particles.

## 516.10 Special Equipment.

**(A) Fixed Electrostatic Equipment.** This section shall apply to any equipment using electrostatically charged elements for the atomization, charging, and/or precipitation of hazardous materials for coatings on articles or for other similar purposes in which the charging or atomizing device is attached to a mechanical support or manipulator. This shall include robotic devices. This section shall not apply to devices that are held or manipulated by hand. Where robot or programming procedures involve manual manipulation of the robot arm while spraying with the high voltage on, the provisions of 516.10(B) shall apply. The installation of electrostatic spraying equipment shall comply with 516.10(A)(1) through (A)(10). Spray equipment shall be listed. All automatic electrostatic equipment systems shall comply with 516.4(A)(1) through (A)(9).

**(1) Power and Control Equipment.** Transformers, high-voltage supplies, control apparatus, and all other electric portions of the equipment shall be installed outside of the Class I location as defined in 516.3 or be of a type identified for the location.

*Exception: High-voltage grids, electrodes, electrostatic atomizing heads, and their connections shall be permitted within the Class I location.*

**(2) Electrostatic Equipment.** Electrodes and electrostatic atomizing heads shall be adequately supported in permanent locations and shall be effectively insulated from ground. Electrodes and electrostatic atomizing heads that are permanently attached to their bases, supports, reciprocators, or robots shall be deemed to comply with this section.

**(3) High-Voltage Leads.** High-voltage leads shall be properly insulated and protected from mechanical damage or exposure to destructive chemicals. Any exposed element at high voltage shall be effectively and permanently supported on suitable insulators and shall be effectively guarded against accidental contact or grounding.

**(4) Support of Goods.** Goods being coated using this process shall be supported on conveyors or hangers. The conveyors or hangers shall be arranged (1) to ensure that the parts being coated are electrically connected to ground with a resistance of 1 megohm or less and (2) to prevent parts from swinging.

**(5) Automatic Controls.** Electrostatic apparatus shall be equipped with automatic means that will rapidly de-energize the high-voltage elements under any of the following conditions:

(1) Stoppage of ventilating fans or failure of ventilating equipment from any cause

(2) Stoppage of the conveyor carrying goods through the high-voltage field unless stoppage is required by the spray process

(3) Occurrence of excessive current leakage at any point in the high-voltage system

(4) De-energizing the primary voltage input to the power supply

**(6) Grounding.** All electrically conductive objects in the spray area, except those objects required by the process to be at high voltage, shall be adequately grounded. This requirement shall apply to paint containers, wash cans, guards, hose connectors, brackets, and any other electrically conductive objects or devices in the area.

**(7) Isolation.** Safeguards such as adequate booths, fencing, railings, interlocks, or other means shall be placed about the equipment or incorporated therein so that they, either by their location, character, or both, ensure that a safe separation of the process is maintained.

**(8) Signs.** Signs shall be conspicuously posted to convey the following:

(1) Designate the process zone as dangerous with regard to fire and accident

(2) Identify the grounding requirements for all electrically conductive objects in the spray area

(3) Restrict access to qualified personnel only

**(9) Insulators.** All insulators shall be kept clean and dry.

**(10) Other Than Nonincendive Equipment.** Spray equipment that cannot be classified as nonincendive shall comply with (a) and (b).

(a)　Conveyors or hangers shall be arranged so as to maintain a safe distance of at least twice the sparking distance between goods being painted and electrodes, electrostatic atomizing heads, or charged conductors. Warnings defining this safe distance shall be posted.

(b)　The equipment shall provide an automatic means of rapidly de-energizing the high-voltage elements in the event the distance between the goods being painted and the electrodes or electrostatic atomizing heads falls below that specified in (a). [NFPA 33, Chapter 9]

**(B) Electrostatic Hand-Spraying Equipment.** This section shall apply to any equipment using electrostatically charged elements for the atomization, charging, and/or pre-cipitation of materials for coatings on articles, or for other similar purposes in which the atomizing device is hand-held or manipulated during the spraying operation. Electrostatic hand-spraying equipment and devices used in connection with paint-spraying operations shall be of listed types and shall comply with 516.10(B)(1) through (B)(5).

**(1) General.** The high-voltage circuits shall be designed so as not to produce a spark of sufficient intensity to ignite the most readily ignitible of those vapor–air mixtures likely to be encountered, or result in appreciable shock hazard upon coming in contact with a grounded object under all normal operating conditions. The electrostatically charged exposed elements of the handgun shall be capable of being energized only by an actuator that also controls the coating material supply.

**(2) Power Equipment.** Transformers, power packs, control apparatus, and all other electric portions of the equipment shall be located outside of the Class I location or be identified for the location.

*Exception: The handgun itself and its connections to the power supply shall be permitted within the Class I location.*

**(3) Handle.** The handle of the spraying gun shall be electrically connected to ground by a metallic connection and be constructed so that the operator in normal operating position is in intimate electrical contact with the grounded handle to prevent buildup of a static charge on the operator's body. Signs indicating the necessity for grounding other persons entering the spray area shall be conspicuously posted.

**(4) Electrostatic Equipment.** All electrically conductive objects in the spraying area shall be adequately grounded. This requirement shall apply to paint containers, wash cans, and any other electrical conductive objects or devices in the area. The equipment shall carry a prominent, permanently installed warning regarding the necessity for this grounding feature.

**(5) Support of Objects.** Objects being painted shall be maintained in metallic contact with the conveyor or other grounded support. Hooks shall be regularly cleaned to ensure adequate grounding of 1 megohm or less. Areas of contact shall be sharp points or knife edges where possible. Points of support of the object shall be concealed from random spray where feasible; and, where the objects being sprayed are supported from a conveyor, the point of attachment to the conveyor shall be located so as to not collect spray material during normal operation. [NFPA 33, Chapter 10]

**(C) Powder Coating.** This section shall apply to processes in which combustible dry powders are applied. The hazards associated with combustible dusts are present in such a process to a degree, depending on the chemical composition of the material, particle size, shape, and distribution.

**(1) Electric Equipment and Sources of Ignition.** Electric equipment and other sources of ignition shall comply with the requirements of Article 502. Portable electric lamps and other utilization equipment shall not be used within a Class II location during operation of the finishing processes. Where such lamps or utilization equipment are used during cleaning or repairing operations, they shall be of a type identified for Class II, Division 1 locations, and all exposed metal parts shall be effectively grounded.

*Exception: Where portable electric lamps are required for operations in spaces not readily illuminated by fixed lighting within the spraying area, they shall be of the type listed for Class II, Division 1 locations where readily ignitible residues may be present.*

**(2) Fixed Electrostatic Spraying Equipment.** The provisions of 516.10(A) and 516.10(C)(1) shall apply to fixed electrostatic spraying equipment.

**(3) Electrostatic Hand-Spraying Equipment.** The provisions of 516.10(B) and 516.10(C)(1) shall apply to electrostatic hand-spraying equipment.

**(4) Electrostatic Fluidized Beds.** Electrostatic fluidized beds and associated equipment shall be of identified types. The high-voltage circuits shall be designed so that any discharge produced when the charging electrodes of the bed are approached or contacted by a grounded object shall not be of sufficient intensity to ignite any powder–air mixture likely to be encountered or to result in an appreciable shock hazard.

(a) Transformers, power packs, control apparatus, and all other electric portions of the equipment shall be located outside the powder-coating area or shall otherwise comply with the requirements of 516.10(C)(1).

*Exception: The charging electrodes and their connections to the power supply shall be permitted within the powder-coating area.*

(b) All electrically conductive objects within the powder-coating area shall be adequately grounded. The powder-coating equipment shall carry a prominent, permanently installed warning regarding the necessity for grounding these objects.

(c) Objects being coated shall be maintained in electrical contact (less than 1 megohm) with the conveyor or other support in order to ensure proper grounding. Hangers shall be regularly cleaned to ensure effective electrical contact. Areas of electrical contact shall be sharp points or knife edges where possible.

(d) The electric equipment and compressed air supplies shall be interlocked with a ventilation system so that the equipment cannot be operated unless the ventilating fans are in operation. [NFPA 33, Chapter 13]

**516.16 Grounding.** All metal raceways, the metal armors or metallic sheath on cables, and all non–current-carrying metal parts of fixed or portable electrical equipment, regardless of voltage, shall be grounded as provided in Article 250. Grounding in Class I and Class II locations shall comply with 501.16 and 502.16, respectively.

# ARTICLE 517
# Health Care Facilities

FPN: Rules that are followed by a reference in brackets contain text that has been extracted from NFPA 99-1999, *Standard for Health Care Facilities*. Only editorial changes were made to the extracted text to make it consistent with this *Code*.

## I. General

**517.1 Scope.** The provisions of this article shall apply to electrical construction and installation criteria in health care facilities that provide services to human beings.

The requirements in Parts II and III not only apply to single-function buildings but are also intended to be individually applied to their respective forms of occupancy within a multifunction building (e.g., a doctor's examining room located within a limited care facility would be required to meet the provisions of 517.10).

FPN: For information concerning performance, maintenance, and testing criteria, refer to the appropriate health care facilities documents.

**517.2 Definitions.**

**Alternate Power Source.** One or more generator sets, or battery systems where permitted, intended to provide power during the interruption of the normal electrical services or the public utility electrical service intended to provide power during interruption of service normally provided by the generating facilities on the premises.

**Ambulatory Health Care Facility.** A building or part thereof used to provide services or treatment to four or more patients at the same time and meeting either (1) or (2).

(1) Those facilities that provide, on an outpatient basis, treatment for patients that would render them incapable of taking action for self-preservation under emergency

conditions without assistance from others, such as hemodialysis units or freestanding emergency medical units.

(2) Those facilities that provide, on an outpatient basis, surgical treatment requiring general anesthesia.

**Anesthetizing Location.** Any area of a facility that has been designated to be used for the administration of any flammable or nonflammable inhalation anesthetic agent in the course of examination or treatment, including the use of such agents for relative analgesia.

**Critical Branch.** A subsystem of the emergency system consisting of feeders and branch circuits supplying energy to task illumination, special power circuits, and selected receptacles serving areas and functions related to patient care, and which are connected to alternate power sources by one or more transfer switches during interruption of the normal power source.

**Electrical Life-Support Equipment.** Electrically powered equipment whose continuous operation is necessary to maintain a patient's life.

**Emergency System.** A system of circuits and equipment intended to supply alternate power to a limited number of prescribed functions vital to the protection of life and safety.

**Equipment System.** A system of circuits and equipment arranged for delayed, automatic, or manual connection to the alternate power source and that serves primarily 3-phase power equipment.

**Essential Electrical System.** A system comprised of alternate sources of power and all connected distribution systems and ancillary equipment, designed to ensure continuity of electrical power to designated areas and functions of a health care facility during disruption of normal power sources, and also designed to minimize disruption within the internal wiring system.

**Exposed Conductive Surfaces.** Those surfaces that are capable of carrying electric current and that are unprotected, unenclosed, or unguarded, permitting personal contact. Paint, anodizing, and similar coatings are not considered suitable insulation, unless they are listed for such use.

**Fault Hazard Current.** *See* Hazard Current.

**Flammable Anesthetics.** Gases or vapors, such as fluroxene, cyclopropane, divinyl ether, ethyl chloride, ethyl ether, and ethylene, which may form flammable or explosive mixtures with air, oxygen, or reducing gases such as nitrous oxide.

**Flammable Anesthetizing Location.** Any area of the facility that has been designated to be used for the adminis-

tration of any flammable inhalation anesthetic agents in the normal course of examination or treatment.

**Hazard Current.** For a given set of connections in an isolated power system, the total current that would flow through a low impedance if it were connected between either isolated conductor and ground.

*Fault Hazard Current.* The hazard current of a given isolated system with all devices connected except the line isolation monitor.

*Monitor Hazard Current.* The hazard current of the line isolation monitor alone.

*Total Hazard Current.* The hazard current of a given isolated system with all devices, including the line isolation monitor, connected.

**Health Care Facilities.** Buildings or portions of buildings in which medical, dental, psychiatric, nursing, obstetrical, or surgical care are provided. Health care facilities include, but are not limited to, hospitals, nursing homes, limited care facilities, clinics, medical and dental offices, and ambulatory care centers, whether permanent or movable.

**Hospital.** A building or part thereof used for the medical, psychiatric, obstetrical, or surgical care, on a 24-hour basis, of four or more inpatients. *Hospital*, wherever used in this *Code*, shall include general hospitals, mental hospitals, tuberculosis hospitals, children's hospitals, and any such facilities providing inpatient care.

**Isolated Power System.** A system comprising an isolating transformer or its equivalent, a line isolation monitor, and its ungrounded circuit conductors.

**Isolation Transformer.** A transformer of the multiple-winding type, with the primary and secondary windings physically separated, which inductively couples its secondary winding to the grounded feeder systems that energize its primary winding.

**Life Safety Branch.** A subsystem of the emergency system consisting of feeders and branch circuits, meeting the requirements of Article 700 and intended to provide adequate power needs to ensure safety to patients and personnel, and which are automatically connected to alternate power sources during interruption of the normal power source.

**Limited Care Facility.** A building or part thereof used on a 24-hour basis for the housing of four or more persons who are incapable of self-preservation because of age, physical limitation due to accident or illness, or mental limitations, such as mental retardation/developmental disability, mental illness, or chemical dependency.

**Line Isolation Monitor.** A test instrument designed to continually check the balanced and unbalanced impedance from each line of an isolated circuit to ground and equipped with a built-in test circuit to exercise the alarm without adding to the leakage current hazard.

**Monitor Hazard Current.** *See* Hazard Current.

**Nurses' Stations.** Areas intended to provide a center of nursing activity for a group of nurses serving bed patients, where the patient calls are received, nurses are dispatched, nurses' notes written, inpatient charts prepared, and medications prepared for distribution to patients. Where such activities are carried on in more than one location within a nursing unit, all such separate areas are considered a part of the nurses' station.

**Nursing Home.** A building or part thereof used for the lodging, boarding, and nursing care, on a 24-hour basis, of four or more persons who, because of mental or physical incapacity, may be unable to provide for their own needs and safety without the assistance of another person. *Nursing home*, wherever used in this *Code*, shall include nursing and convalescent homes, skilled nursing facilities, intermediate care facilities, and infirmaries of homes for the aged.

**Patient Bed Location.** The location of an inpatient sleeping bed; or the bed or procedure table used in a critical patient care area.

**Patient Care Area.** Any portion of a health care facility wherein patients are intended to be examined or treated. Areas of a health care facility in which patient care is administered are classified as general care areas or critical care areas, either of which may be classified as a wet location. The governing body of the facility designates these areas in accordance with the type of patient care anticipated and with the following definitions of the area classification.

> FPN: Business offices, corridors, lounges, day rooms, dining rooms, or similar areas typically are not classified as patient care areas.

*General Care Areas.* Patient bedrooms, examining rooms, treatment rooms, clinics, and similar areas in which it is intended that the patient will come in contact with ordinary appliances such as a nurse call system, electrical beds, examining lamps, telephone, and entertainment devices. In such areas, it may also be intended that patients be connected to electromedical devices (such as heating pads, electrocardiographs, drainage pumps, monitors, otoscopes, ophthalmoscopes, intravenous lines, etc.).

*Critical Care Areas.* Those special care units, intensive care units, coronary care units, angiography laboratories, cardiac catheterization laboratories, delivery rooms, operating rooms, and similar areas in which patients are intended to be subjected to invasive procedures and connected to line-operated, electromedical devices.

*Wet Locations.* Those patient care areas that are normally subject to wet conditions while patients are present. These include standing fluids on the floor or drenching of the work area, either of which condition is intimate to the patient or staff. Routine housekeeping procedures and incidental spillage of liquids do not define a wet location.

**Patient Equipment Grounding Point.** A jack or terminal bus that serves as the collection point for redundant grounding of electric appliances serving a patient vicinity or for grounding other items in order to eliminate electromagnetic interference problems.

**Patient Vicinity.** In an area in which patients are normally cared for, the *patient vicinity* is the space with surfaces likely to be contacted by the patient or an attendant who can touch the patient. Typically in a patient room, this encloses a space within the room not less than 1.8 m (6 ft) beyond the perimeter of the bed in its nominal location, and extending vertically not less than 2.3 m (7½ ft) above the floor.

**Psychiatric Hospital.** A building used exclusively for the psychiatric care, on a 24-hour basis, of four or more inpatients.

**Reference Grounding Point.** The ground bus of the panelboard or isolated power system panel supplying the patient care area.

**Relative Analgesia.** A state of sedation and partial block of pain perception produced in a patient by the inhalation of concentrations of nitrous oxide insufficient to produce loss of consciousness (conscious sedation).

**Selected Receptacles.** A minimum number of electric receptacles to accommodate appliances ordinarily required for local tasks or likely to be used in patient care emergencies.

**Task Illumination.** Provision for the minimum lighting required to carry out necessary tasks in the described areas, including safe access to supplies and equipment, and access to exits.

**Therapeutic High Frequency Diathermy Equipment.** Therapeutic high-frequency diathermy equipment is therapeutic induction and dielectric heating equipment.

**Total Hazard Current.** *See* Hazard Current.

**X-Ray Installations, Long-Time Rating.** A rating based on an operating interval of 5 minutes or longer.

**X-Ray Installations, Mobile.** X-ray equipment mounted on a permanent base with wheels, casters, or a combination of both to facilitate moving the equipment while completely assembled.

**X-Ray Installations, Momentary Rating.** A rating based on an operating interval that does not exceed 5 seconds.

**X-Ray Installations, Portable.** X-ray equipment designed to be hand carried.

**X-Ray Installations, Transportable.** X-ray equipment to be installed in a vehicle or that may be readily disassembled for transport in a vehicle.

## II. Wiring and Protection

### 517.10 Applicability.

**(A)** Part II shall apply to patient care areas of all health care facilities.

**(B)** Part II shall not apply to the following:

(1) Business offices, corridors, waiting rooms, and the like in clinics, medical and dental offices, and outpatient facilities

(2) Areas of nursing homes and limited care facilities wired in accordance with Chapters 1 through 4 of this *Code* where these areas are used exclusively as patient sleeping rooms

FPN: See NFPA *101®*-2000, *Life Safety Code®*.

### 517.11 General Installation — Construction Criteria. It is the purpose of this article to specify the installation criteria and wiring methods that minimize electrical hazards by the maintenance of adequately low potential differences only between exposed conductive surfaces that are likely to become energized and could be contacted by a patient.

FPN: In a health care facility, it is difficult to prevent the occurrence of a conductive or capacitive path from the patient's body to some grounded object, because that path may be established accidentally or through instrumentation directly connected to the patient. Other electrically conductive surfaces that may make an additional contact with the patient, or instruments that may be connected to the patient, then become possible sources of electric currents that can traverse the patient's body. The hazard is increased as more apparatus is associated with the patient, and, therefore, more intensive precautions are needed. Control of electric shock hazard requires the limitation of electric current that might flow in an electric circuit involving the patient's body by raising the resistance of the conductive circuit that includes the patient, or by insulating exposed surfaces that might become energized, in addition to reducing the potential difference that can appear between exposed conductive surfaces in the patient vicinity, or by combinations of these methods. A special problem is presented by the patient with an externalized direct conductive path to the heart muscle. The patient may be electrocuted at current levels so low that additional protection in the design of appliances, insulation of the catheter, and control of medical practice is required.

### 517.12 Wiring Methods. Except as modified in this article, wiring methods shall comply with the applicable requirements of Chapters 1 through 4 of this *Code*.

### 517.13 Grounding of Receptacles and Fixed Electric Equipment in Patient Care Areas. Wiring in patient care areas shall comply with 517.13(A) and (B).

**(A) Wiring Methods.** All branch circuits serving patient care areas shall be provided with a ground path for fault current by installation in a metal raceway system, or a cable armor or sheath assembly. The metal raceway system, or cable armor, or sheath assembly, shall itself qualify as an equipment grounding return path in accordance with 250.118. Type AC, Type MC, Type MI cables shall have an outer metal armor or sheath that is identified as an acceptable grounding return path.

**(B) Insulated Equipment Grounding Conductor.** In an area used for patient care, the grounding terminals of all receptacles and all non–current-carrying conductive surfaces of fixed electric equipment likely to become energized that are subject to personal contact, operating at over 100 volts, shall be grounded by an insulated copper conductor. The grounding conductor shall be sized in accordance with Table 250.122 and installed in metal raceways or metal-clad cables with the branch-circuit conductors supplying these receptacles or fixed equipment.

*Exception No. 1: Metal faceplates shall be permitted to be grounded by means of a metal mounting screw(s) securing the faceplate to a grounded outlet box or grounded wiring device.*

*Exception No. 2: Luminaires (light fixtures) more than 2.3 m (7½ ft) above the floor and switches located outside of the patient vicinity shall not be required to be grounded by an insulated equipment grounding conductor.*

### 517.14 Panelboard Bonding. The equipment grounding terminal buses of the normal and essential branch-circuit panelboards serving the same individual patient vicinity shall be bonded together with an insulated continuous copper conductor not smaller than 10 AWG. Where more than two panels serve the same location, this conductor shall be continuous from panel to panel but shall be permitted to be broken in order to terminate on the ground bus in each panel.

### 517.16 Receptacles with Insulated Grounding Terminals. Receptacles with insulated grounding terminals, as permitted in 250.146(D), shall be identified; such identification shall be visible after installation.

FPN: Caution is important in specifying such a system with receptacles having insulated grounding terminals, since the grounding impedance is controlled only by the grounding conductors and does not benefit functionally from any parallel grounding paths.

**517.17 Ground-Fault Protection.**

**(A) Feeders.** Where ground-fault protection is provided for operation of the service disconnecting means or feeder disconnecting means as specified by 230.95 or 215.10, an additional step of ground-fault protection shall be provided in the next level of feeder disconnecting means downstream toward the load. Such protection shall consist of overcurrent devices and current transformers or other equivalent protective equipment that shall cause the feeder disconnecting means to open.

The additional levels of ground-fault protection shall not be installed as follows:

(1) On the load side of an essential electrical system transfer switch

(2) Between the on-site generating unit(s) described in 517.35(B) and the essential electrical system transfer switch(es)

(3) On electrical systems that are not solidly grounded wye systems with greater than 150 volts to ground but not exceeding 600 volts phase-to-phase

**(B) Selectivity.** Ground-fault protection for operation of the service and feeder disconnecting means shall be fully selective such that the feeder device and not the service device shall open on ground faults on the load side of the feeder device. A six-cycle minimum separation between the service and feeder ground-fault tripping bands shall be provided. Operating time of the disconnecting devices shall be considered in selecting the time spread between these two bands to achieve 100 percent selectivity.

FPN: See 230.95, fine print note, for transfer of alternate source where ground-fault protection is applied.

**(C) Testing.** When equipment ground-fault protection is first installed, each level shall be performance tested to ensure compliance with 517.17(B).

**517.18 General Care Areas.**

**(A) Patient Bed Location.** Each patient bed location shall be supplied by at least two branch circuits, one from the emergency system and one from the normal system. All branch circuits from the normal system shall originate in the same panelboard.

*Exception No. 1: Branch circuits serving only special-purpose outlets or receptacles, such as portable X-ray outlets, shall not be required to be served from the same distribution panel or panels.*

*Exception No. 2: Requirements of 517.18(A) shall not apply to patient bed locations in clinics, medical and dental offices, and outpatient facilities; psychiatric, substance abuse, and rehabilitation hospitals; sleeping rooms of nurs-*

*ing homes and limited care facilities meeting the requirements of 517.10(B)(2).*

*Exception No. 3: A general care patient bed location served from two separate transfer switches on the emergency system shall not be required to have circuits from the normal system.*

**(B) Patient Bed Location Receptacles.** Each patient bed location shall be provided with a minimum of four receptacles. They shall be permitted to be of the single or duplex types or a combination of both. All receptacles, whether four or more, shall be listed "hospital grade" and so identified. Each receptacle shall be grounded by means of an insulated copper conductor sized in accordance with Table 250.122.

*Exception No. 1: Requirements of 517.18(B) shall not apply to psychiatric, substance abuse, and rehabilitation hospitals meeting the requirements of 517.10(B)(2).*

*Exception No. 2: Psychiatric security rooms shall not be required to have receptacle outlets installed in the room.*

FPN: It is not intended that there be a total, immediate replacement of existing non-hospital grade receptacles. It is intended, however, that non-hospital grade receptacles be replaced with hospital grade receptacles upon modification of use, renovation, or as existing receptacles need replacement.

**(C) Pediatric Locations.** Receptacles located within the patient care areas of pediatric wards, rooms, or areas shall be listed tamper resistant or shall employ a listed tamper resistant cover.

**517.19 Critical Care Areas.**

**(A) Patient Bed Location Branch Circuits.** Each patient bed location shall be supplied by at least two branch circuits, one or more from the emergency system and one or more circuits from the normal system. At least one branch circuit from the emergency system shall supply an outlet(s) only at that bed location. All branch circuits from the normal system shall be from a single panelboard. Emergency system receptacles shall be identified and shall also indicate the panelboard and circuit number supplying them.

*Exception No. 1: Branch circuits serving only special-purpose receptacles or equipment in critical care areas shall be permitted to be served by other panelboards.*

*Exception No. 2: Critical care locations served from two separate transfer switches on the emergency system shall not be required to have circuits from the normal system.*

**(B) Patient Bed Location Receptacles.**

**(1) Minimum Number and Supply.** Each patient bed location shall be provided with a minimum of six receptacles, at least one of which shall be connected to either of the following:

(1) The normal system branch circuit required in 517.19(A)
(2) An emergency system branch circuit supplied by a different transfer switch than the other receptacles at the same location

**(2) Receptacle Requirements.** The receptacles required in 517.19(B)(1) shall be permitted to be of the single or duplex types or a combination of both. All receptacles, whether six or more, shall be listed "hospital grade" and so identified. Each receptacle shall be grounded to the reference grounding point by means of an insulated copper equipment grounding conductor.

**(C) Patient Vicinity Grounding and Bonding (Optional).** A patient vicinity shall be permitted to have a patient equipment grounding point. The patient equipment grounding point, where supplied, shall be permitted to contain one or more jacks listed for the purpose. An equipment bonding jumper not smaller than 10 AWG shall be used to connect the grounding terminal of all grounding-type receptacles to the patient equipment grounding point. The bonding conductor shall be permitted to be arranged centrically or looped as convenient.

> FPN: Where there is no patient equipment grounding point, it is important that the distance between the reference grounding point and the patient vicinity be as short as possible to minimize any potential differences.

**(D) Panelboard Grounding.** Where a grounded electrical distribution system is used and metal feeder raceway or Type MC or MI cable is installed, grounding of a panelboard or switchboard shall be ensured by one of the following means at each termination or junction point of the raceway or Type MC or MI cable:

(1) A grounding bushing and a continuous copper bonding jumper, sized in accordance with 250.122, with the bonding jumper connected to the junction enclosure or the ground bus of the panel
(2) Connection of feeder raceways or Type MC or MI cable to threaded hubs or bosses on terminating enclosures
(3) Other approved devices such as bonding-type locknuts or bushings

**(E) Additional Protective Techniques in Critical Care Areas (Optional).** Isolated power systems shall be permitted to be used for critical care areas, and, if used, the isolated power system equipment shall be listed for the purpose and the system designed and installed so that it meets the provisions of and is in accordance with 517.160.

*Exception: The audible and visual indicators of the line isolation monitor shall be permitted to be located at the nursing station for the area being served.*

**(F) Isolated Power System Grounding.** Where an isolated ungrounded power source is used and limits the first-fault current to a low magnitude, the grounding conductor associated with the secondary circuit shall be permitted to be run outside of the enclosure of the power conductors in the same circuit.

> FPN: Although it is permitted to run the grounding conductor outside of the conduit, it is safer to run it with the power conductors to provide better protection in case of a second ground fault.

**(G) Special-Purpose Receptacle Grounding.** The equipment grounding conductor for special-purpose receptacles, such as the operation of mobile X-ray equipment, shall be extended to the reference grounding points of branch circuits for all locations likely to be served from such receptacles. Where such a circuit is served from an isolated ungrounded system, the grounding conductor shall not be required to be run with the power conductors; however, the equipment grounding terminal of the special-purpose receptacle shall be connected to the reference grounding point.

## 517.20 Wet Locations.

**(A) Receptacles and Fixed Equipment.** All receptacles and fixed equipment within the area of the wet location shall have ground-fault circuit-interrupter protection for personnel if interruption of power under fault conditions can be tolerated, or be served by an isolated power system if such interruption cannot be tolerated.

*Exception: Branch circuits supplying only listed, fixed, therapeutic and diagnostic equipment shall be permitted to be supplied from a normal grounded service, single- or 3-phase system, provided that*

*(a) Wiring for grounded and isolated circuits does not occupy the same raceway, and*
*(b) All conductive surfaces of the equipment are grounded.*

**(B) Isolated Power Systems.** Where an isolated power system is utilized, the equipment shall be listed for the purpose and installed so that it meets the provisions of and is in accordance with 517.160.

> FPN: For requirements for installation of therapeutic pools and tubs, see Part VI of Article 680.

## 517.21 Ground-Fault Circuit-Interrupter Protection for Personnel.
Ground-fault circuit-interrupter protection for personnel shall not be required for receptacles installed in those critical care areas where the toilet and basin are installed within the patient room.

## III. Essential Electrical System

**517.25 Scope.** The essential electrical system for these facilities shall comprise a system capable of supplying a limited amount of lighting and power service, which is considered essential for life safety and orderly cessation of procedures during the time normal electrical service is interrupted for any reason. This includes clinics, medical and dental offices, outpatient facilities, nursing homes, limited care facilities, hospitals, and other health care facilities serving patients.

FPN: For information as to the need for an essential electrical system, see NFPA 99-1999, *Standard for Health Care Facilities.*

### 517.30 Essential Electrical Systems for Hospitals.

**(A) Applicability.** The requirements of Part III, 517.30 through 517.35, shall apply to hospitals where an essential electrical system is required.

FPN No. 1: For performance, maintenance, and testing requirements of essential electrical systems in hospitals, see NFPA 99-1999, *Standard for Health Care Facilities.* For installation of centrifugal fire pumps, see NFPA 20-1999, *Standard for the Installation of Stationary Fire Pumps for Fire Protection.*

FPN No. 2: For additional information, see NFPA 99-1999, *Standard for Health Care Facilities.*

**(B) General.**

**(1) Separate Systems.** Essential electrical systems for hospitals shall be comprised of two separate systems capable of supplying a limited amount of lighting and power service, which is considered essential for life safety and effective hospital operation during the time the normal electrical service is interrupted for any reason. These two systems shall be the emergency system and the equipment system.

**(2) Emergency Systems.** The emergency system shall be limited to circuits essential to life safety and critical patient care. These are designated the life safety branch and the critical branch.

**(3) Equipment System.** The equipment system shall supply major electrical equipment necessary for patient care and basic hospital operation.

**(4) Transfer Switches.** The number of transfer switches to be used shall be based on reliability, design, and load considerations. Each branch of the emergency system and each equipment system shall have one or more transfer switches. One transfer switch shall be permitted to serve one or more branches or systems in a facility with a maximum demand on the essential electrical system of 150 kVA.

FPN No. 1: See NFPA 99-1999, *Standard for Health Care Facilities*: 3.4.3.2, Transfer Switch Operation Type I; 3.4.2.1.4, Automatic Transfer Switch Features; and 3.4.2.1.6, Nonautomatic Transfer Device Features.

FPN No. 2: See FPN Figure 517.30, No. 1.

FPN No. 3: See FPN Figure 517.30, No. 2.

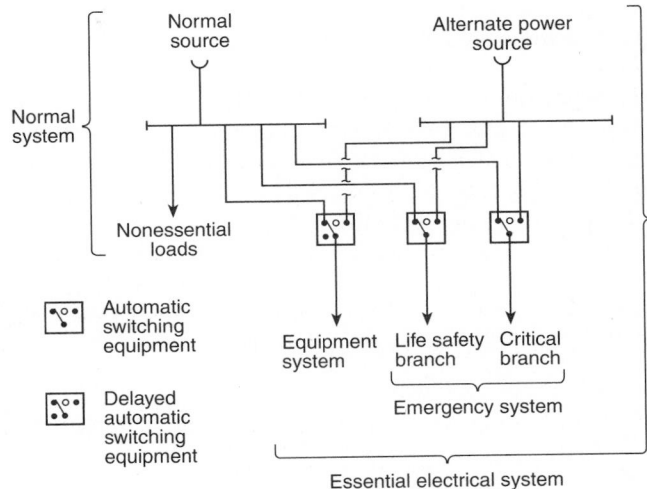

**FPN Figure 517.30, No. 1 Hospital — minimum requirement for transfer switch arrangement.**

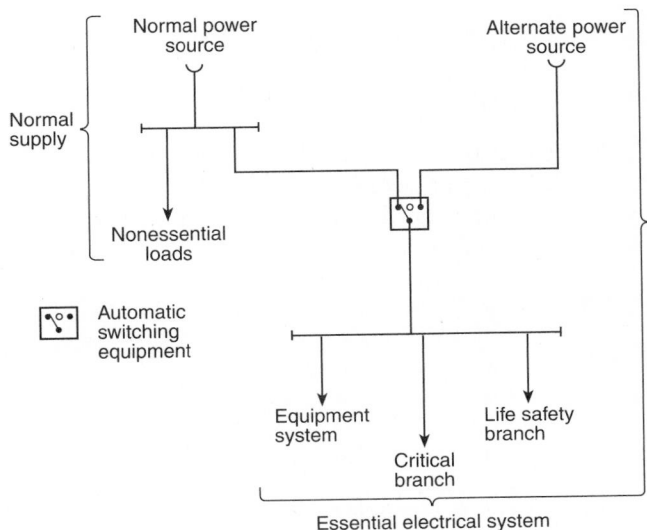

**FPN Figure 517.30, No. 2 Hospital — minimum requirement (150 kVA or less) for transfer switch arrangement.**

**(5) Other Loads.** Loads served by the generating equipment not specifically named in Article 517 shall be served by their own transfer switches such that these loads:

(1) Shall not be transferred if the transfer will overload the generating equipment.

(2) Shall be automatically shed upon generating equipment overloading.

**(6) Contiguous Facilities.** Hospital power sources and alternate power sources shall be permitted to serve the essential electrical systems of contiguous or same site facilities. [NFPA 99, 3.4.2.2.1, 12.3.3.2]

**(C) Wiring Requirements.**

**(1) Separation from Other Circuits.** The life safety branch and critical branch of the emergency system shall be kept entirely independent of all other wiring and equipment and shall not enter the same raceways, boxes, or cabinets with each other or other wiring.

Wiring of the life safety branch and the critical branch shall be permitted to occupy the same raceways, boxes, or cabinets of other circuits not part of the branch where such wiring is as follows:

(1) In transfer equipment enclosures, or
(2) In exit or emergency luminaires (lighting fixtures) supplied from two sources, or
(3) In a common junction box attached to exit or emergency luminaires (lighting fixtures) supplied from two sources, or
(4) For two or more emergency circuits supplied from the same branch

The wiring of the equipment system shall be permitted to occupy the same raceways, boxes, or cabinets of other circuits that are not part of the emergency system.

**(2) Isolated Power Systems.** Where isolated power systems are installed in any of the areas in 517.33(A)(1) and (A)(2), each system shall be supplied by an individual circuit serving no other load.

**(3) Mechanical Protection of the Emergency System.** The wiring of the emergency system of a hospital shall be mechanically protected by installation in nonflexible metal raceways, or shall be wired with Type MI cable.

*Exception No. 1: Flexible power cords of appliances, or other utilization equipment, connected to the emergency system shall not be required to be enclosed in raceways.*

*Exception No. 2: Secondary circuits of transformer-powered communications or signaling systems shall not be required to be enclosed in raceways unless otherwise specified by Chapters 7 or 8.*

*Exception No. 3: Schedule 80 rigid nonmetallic conduit shall be permitted if the branch circuits do not serve patient care areas and it is not prohibited elsewhere in this Code.*

*Exception No. 4: Where encased in not less than 50 mm (2 in.) of concrete, Schedule 40 rigid nonmetallic conduit or electrical nonmetallic tubing shall be permitted if the branch circuits do not serve patient care areas.*

*Exception No. 5: Flexible metal raceways and cable assemblies shall be permitted to be used in listed prefabricated medical headwalls, listed office furnishings, or where necessary for flexible connection to equipment.*

> FPN: See 517.13 for additional grounding requirements in patient care areas.

**(D) Capacity of Systems.** The essential electrical system shall have adequate capacity to meet the demand for the operation of all functions and equipment to be served by each system and branch.

Feeders shall be sized in accordance with Articles 215 and 220. The generator set(s) shall have sufficient capacity and proper rating to meet the demand produced by the load of the essential electrical system(s) at any given time.

Demand calculations for sizing of the generator set(s) shall be based on the following:

(1) Prudent demand factors and historical data, or
(2) Connected load, or
(3) Feeder calculation procedures described in Article 220, or
(4) Any combination of the above

The sizing requirements in 700.5 and 701.6 shall not apply to hospital generator set(s).

**(E) Receptacle Identification.** The cover plates for the electrical receptacles or the electrical receptacles themselves supplied from the emergency system shall have a distinctive color or marking so as to be readily identifiable. [NFPA 99, 3.4.2.2.4(b)2]

**517.31 Emergency System.** Those functions of patient care depending on lighting or appliances that are connected to the emergency system shall be divided into two mandatory branches: the life safety branch and the critical branch, described in 517.32 and 517.33.

The branches of the emergency system shall be installed and connected to the alternate power source so that all functions specified herein for the emergency system shall be automatically restored to operation within 10 seconds after interruption of the normal source. [NFPA 99, 3.4.2.2.2(a), 3.5.2.2.2]

**517.32 Life Safety Branch.** No function other than those listed in 517.32(A) through (G) shall be connected to the life safety branch. The life safety branch of the emergency system shall supply power for the following lighting, receptacles, and equipment.

**(A) Illumination of Means of Egress.** Illumination of means of egress, such as lighting required for corridors, passageways, stairways, and landings at exit doors, and all necessary ways of approach to exits. Switching arrangements to transfer patient corridor lighting in hospitals from general illumination circuits to night illumination circuits shall be permitted, provided only one of two circuits can be selected and both circuits cannot be extinguished at the same time.

FPN: See NFPA *101*-2000, *Life Safety Code*, 5.8 and 5.9.

**(B) Exit Signs.** Exit signs and exit directional signs.

FPN: See NFPA *101*-2000, *Life Safety Code*, 5.10.

**(C) Alarm and Alerting Systems.** Alarm and alerting systems including the following:

(1) Fire alarms

FPN: See NFPA *101*-2000, *Life Safety Code*, 7.6 and 12.3.4.

(2) Alarms required for systems used for the piping of nonflammable medical gases

FPN: See NFPA 99-1999, *Standard for Health Care Facilities*, 12.3.4.1.

**(D) Communications Systems.** Hospital communications systems, where used for issuing instructions during emergency conditions.

**(E) Generator Set Location.** Task illumination battery charger for emergency battery-powered lighting unit(s) and selected receptacles at the generator set location.

**(F) Elevators.** Elevator cab lighting, control, communications, and signal systems.

**(G) Automatic Doors.** Automatically operated doors used for building egress. [NFPA 99, 3.4.2.2.2(b)]

**517.33 Critical Branch.**

**(A) Task Illumination and Selected Receptacles.** The critical branch of the emergency system shall supply power for task illumination, fixed equipment, selected receptacles, and special power circuits serving the following areas and functions related to patient care:

(1) Critical care areas that utilize anesthetizing gases — task illumination, selected receptacles, and fixed equipment
(2) The isolated power systems in special environments
(3) Patient care areas — task illumination and selected receptacles in the following:

   a. Infant nurseries
   b. Medication preparation areas
   c. Pharmacy dispensing areas
   d. Selected acute nursing areas
   e. Psychiatric bed areas (omit receptacles)

   f. Ward treatment rooms
   g. Nurses' stations (unless adequately lighted by corridor luminaires)

(4) Additional specialized patient care task illumination and receptacles, where needed
(5) Nurse call systems
(6) Blood, bone, and tissue banks
(7) Telephone equipment rooms and closets
(8) Task illumination, selected receptacles, and selected power circuits for the following:

   a. General care beds (at least one duplex receptacle per patient bedroom)
   b. Angiographic labs
   c. Cardiac catheterization labs
   d. Coronary care units
   e. Hemodialysis rooms or areas
   f. Emergency room treatment areas (selected)
   g. Human physiology labs
   h. Intensive care units
   i. Postoperative recovery rooms (selected)

(9) Additional task illumination, receptacles, and selected power circuits needed for effective hospital operation. Single-phase fractional horsepower motors shall be permitted to be connected to the critical branch. [NFPA 99, 3.4.2.2.2(c)]

**(B) Subdivision of the Critical Branch.** It shall be permitted to subdivide the critical branch into two or more branches.

FPN: It is important to analyze the consequences of supplying an area with only critical care branch power when failure occurs between the area and the transfer switch. Some proportion of normal and critical power or critical power from separate transfer switches may be appropriate.

**517.34 Equipment System Connection to Alternate Power Source.** The equipment system shall be installed and connected to the alternate power source such that the equipment described in 517.34(A) is automatically restored to operation at appropriate time-lag intervals following the energizing of the emergency system. Its arrangement shall also provide for the subsequent connection of equipment described in 517.34(B). [NFPA 99, 3.4.2.2.3(b)]

*Exception: For essential electrical systems under 150 kVA, deletion of the time-lag intervals feature for delayed automatic connection to the equipment system shall be permitted.*

**(A) Equipment for Delayed Automatic Connection.** The following equipment shall be arranged for delayed automatic connection to the alternate power source.

(1) Central suction systems serving medical and surgical functions, including controls. Such suction systems shall be permitted on the critical branch.

(2) Sump pumps and other equipment required to operate for the safety of major apparatus, including associated control systems and alarms.

(3) Compressed air systems serving medical and surgical functions, including controls. Such air systems shall be permitted on the critical branch.

(4) Smoke control and stair pressurization systems, or both.

(5) Kitchen hood supply or exhaust systems, or both, if required to operate during a fire in or under the hood. [NFPA 99, 3.4.2.2.3(d)]

*Exception: Sequential delayed automatic connection to the alternate power source to prevent overloading the generator shall be permitted where engineering studies indicate it is necessary.*

**(B) Equipment for Delayed Automatic or Manual Connection.** The following equipment shall be arranged for either delayed automatic or manual connection to the alternate power source:

(1) Heating equipment to provide heating for operating, delivery, labor, recovery, intensive care, coronary care, nurseries, infection/isolation rooms, emergency treatment spaces, and general patient rooms and pressure maintenance (jockey or make-up) pump(s) for water-based fire protection systems.

*Exception: Heating of general patient rooms and infection/isolation rooms during disruption of the normal source shall not be required under any of the following conditions:*

*(a) The outside design temperature is higher than −6.7°C (20°F).*

*(b) The outside design temperature is lower than −6.7°C (20°F), and where a selected room(s) is provided for the needs of all confined patients, only such room(s) need be heated.*

*(c) The facility is served by a dual source of normal power.*

> FPN No. 1: The design temperature is based on the 97½ percent design value as shown in Chapter 24 of the ASHRAE *Handbook of Fundamentals* (1997).

> FPN No. 2: For a description of a dual source of normal power, see 517.35(C), FPN.

(2) An elevator(s) selected to provide service to patient, surgical, obstetrical, and ground floors during interruption of normal power. In instances where interruption of normal power would result in other elevators stopping between floors, throw-over facilities shall be provided to allow the temporary operation of any elevator for the release of patients or other persons who may be confined between floors.

(3) Supply, return, and exhaust ventilating systems for airborne infectious/isolation rooms, protective environ-ment rooms, exhaust fans for laboratory fume hoods, nuclear medicine areas where radioactive material is used, ethylene oxide evacuation and anesthesia evacuation. Where delayed automatic connection is not appropriate, such ventilation systems shall be permitted to be placed on the critical branch. [NFPA 99, 3.4.2.2.3(e)(4)]

(4) Hyperbaric facilities.

(5) Hypobaric facilities.

(6) Automatically operated doors.

(7) Minimal electrically heated autoclaving equipment shall be permitted to be arranged for either automatic or manual connection to the alternate source.

(8) Controls for equipment listed in 517.34.

(9) Other selected equipment shall be permitted to be served by the equipment system. [NFPA 99, 3.4.2.2.3(e)]

**517.35 Sources of Power.**

**(A) Two Independent Sources of Power.** Essential electrical systems shall have a minimum of two independent sources of power: a normal source generally supplying the entire electrical system and one or more alternate sources for use when the normal source is interrupted. [NFPA 99, 3.4.1.1.2]

**(B) Alternate Source of Power.** The alternate source of power shall be one of the following:

(1) Generator(s) driven by some form of prime mover(s) and located on the premises

(2) Another generating unit(s) where the normal source consists of a generating unit(s) located on the premises

(3) An external utility service when the normal source consists of a generating unit(s) located on the premises

**(C) Location of Essential Electrical System Components.** Careful consideration shall be given to the location of the spaces housing the components of the essential electrical system to minimize interruptions caused by natural forces common to the area (e.g., storms, floods, earthquakes, or hazards created by adjoining structures or activities). Consideration shall also be given to the possible interruption of normal electrical services resulting from similar causes as well as possible disruption of normal electrical service due to internal wiring and equipment failures.

> FPN: Facilities in which the normal source of power is supplied by two or more separate central station-fed services experience greater than normal electrical service reliability than those with only a single feed. Such a dual source of normal power consists of two or more electrical services fed from separate generator sets or a utility distribution network that has multiple power input sources and is arranged to provide mechanical and electrical separation so that a fault between the facility and the generating sources

is not likely to cause an interruption of more than one of the facility service feeders.

## 517.40 Essential Electrical Systems for Nursing Homes and Limited Care Facilities.

**(A) Applicability.** The requirements of Part III, 517.40(C) through 517.44, shall apply to nursing homes and limited care facilities.

*Exception: The requirements of Part III, 517.40(C) through 517.44, shall not apply to freestanding buildings used as nursing homes and limited care facilities, provided that the following apply:*

*(a) Admitting and discharge policies are maintained that preclude the provision of care for any patient or resident who may need to be sustained by electrical life-support equipment.*

*(b) No surgical treatment requiring general anesthesia is offered.*

*(c) An automatic battery-operated system(s) or equipment is provided that shall be effective for at least 1½ hours and is otherwise in accordance with 700.12 and that shall be capable of supplying lighting for exit lights, exit corridors, stairways, nursing stations, medical preparation areas, boiler rooms, and communications areas. This system shall also supply power to operate all alarm systems. [NFPA 99, 16.3.3.2 Exception, 17.3.3.2 Exception]*

FPN: See NFPA *101-2000, Life Safety Code.*

**(B) Inpatient Hospital Care Facilities.** Nursing homes and limited care facilities that provide inpatient hospital care shall comply with the requirements of Part III, 517.30 through 517.35.

**(C) Facilities Contiguous or Located on the Same Site with Hospitals.** Nursing homes and limited care facilities that are contiguous or located on the same site with a hospital shall be permitted to have their essential electrical systems supplied by that of the hospital.

FPN: For performance, maintenance, and testing requirements of essential electrical systems in nursing homes and limited care facilities, see NFPA 99-1999, *Standard for Health Care Facilities.*

## 517.41 Essential Electrical Systems.

**(A) General.** Essential electrical systems for nursing homes and limited care facilities shall be comprised of two separate branches capable of supplying a limited amount of lighting and power service, which is considered essential for the protection of life safety and effective operation of the institution during the time normal electrical service is interrupted for any reason. These two separate branches shall be the life safety branch and the critical branch. [NFPA 99, 3-5.2.2.1]

**(B) Transfer Switches.** The number of transfer switches to be used shall be based on reliability, design, and load considerations. Each branch of the essential electrical system shall be served by one or more transfer switches. One transfer switch shall be permitted to serve one or more branches or systems in a facility with a maximum demand on the essential electrical system of 150 kVa. [NFPA 99, 3.5.2.2.1]

FPN No. 1: See NFPA 99-1999, *Standard for Health Care Facilities,* 3.5.3.2, Transfer Switch Operation Type II; 3.4.2.1.4, Automatic Transfer Switch Features; and 3.4.2.1.6, Nonautomatic Transfer Device Features.

FPN No. 2: See FPN Figure 517.41, No. 1.

FPN No. 3: See FPN Figure 517.41, No. 2.

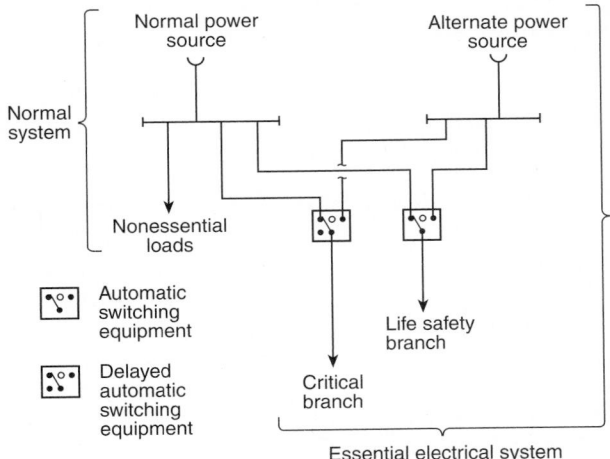

**FPN Figure 517.41, No. 1 Nursing home and limited health care facilities — minimum requirement for transfer switch arrangement.**

**(C) Capacity of System.** The essential electrical system shall have adequate capacity to meet the demand for the operation of all functions and equipment to be served by each branch at one time.

**(D) Separation from Other Circuits.** The life safety branch shall be kept entirely independent of all other wiring and equipment and shall not enter the same raceways, boxes, or cabinets with other wiring except as follows:

(1) In transfer switches
(2) In exit or emergency luminaires (lighting fixtures) supplied from two sources, or
(3) In a common junction box attached to exit or emergency luminaires (lighting fixtures) supplied from two sources

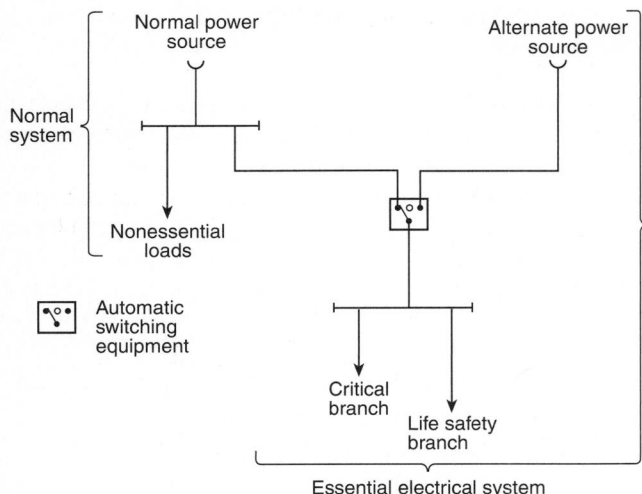

**FPN Figure 517.41, No. 2 Nursing home and limited health care facilities — minimum requirement (150 kVa or less) for transfer switch arrangement.**

The wiring of the critical branch shall be permitted to occupy the same raceways, boxes, or cabinets of other circuits that are not part of the life safety branch.

**(E) Receptacle Identification.** The cover plates for the electrical receptacles or the electrical receptacles themselves supplied from the emergency system shall have a distinctive color or marking so as to be readily identifiable. [NFPA 99, 3-5.2.2.4]

**517.42 Automatic Connection to Life Safety Branch.** The life safety branch shall be installed and connected to the alternate source of power so that all functions specified herein shall be automatically restored to operation within 10 seconds after the interruption of the normal source. No functions other than those listed in 517.42(A) through (G) shall be connected to the life safety branch. The life safety branch shall supply power for the following lighting, receptacles, and equipment.

> FPN: The life safety branch is called the emergency system in NFPA 99-1999, *Standard for Health Care Facilities.*

**(A) Illumination of Means of Egress.** Illumination of means of egress as is necessary for corridors, passageways, stairways, landings, and exit doors and all ways of approach to exits. Switching arrangement to transfer patient corridor lighting from general illumination circuits shall be permitted, providing only one of two circuits can be selected and both circuits cannot be extinguished at the same time.

> FPN: See NFPA *101*-2000, *Life Safety Code*, 5.8 and 5.9.

**(B) Exit Signs.** Exit signs and exit directional signs.

> FPN: See NFPA *101*-2000, *Life Safety Code*, 5.10.

**(C) Alarm and Alerting Systems.** Alarm and alerting systems, including the following:

(1) Fire alarms

> FPN: See NFPA *101*-2000, *Life Safety Code,*, 7.6 and 12.3.4.

(2) Alarms required for systems used for the piping of nonflammable medical gases

> FPN: See NFPA 99-1999, *Standard for Health Care Facilities,* 16.3.4.1.

**(D) Communications Systems.** Communications systems, where used for issuing instructions during emergency conditions.

**(E) Dining and Recreation Areas.** Sufficient lighting in dining and recreation areas to provide illumination to exit ways.

**(F) Generator Set Location.** Task illumination and selected receptacles in the generator set location.

**(G) Elevators.** Elevator cab lighting, control, communications, and signal systems. [NFPA 99, 3.5.2.2.2, 3.5.3.1]

**517.43 Connection to Critical Branch.** The critical branch shall be installed and connected to the alternate power source so that the equipment listed in 517.43(A) shall be automatically restored to operation at appropriate time-lag intervals following the restoration of the life safety branch to operation. Its arrangement shall also provide for the additional connection of equipment listed in 517.43(B) by either delayed automatic or manual operation.

*Exception:  For essential electrical systems under 150 kVA, deletion of the time-lag intervals feature for delayed automatic connection to the equipment system shall be permitted.*

**(A) Delayed Automatic Connection.** The following equipment shall be connected to the critical branch and shall be arranged for delayed automatic connection to the alternate power source:

(1) Patient care areas — task illumination and selected receptacles in the following:

    a. Medication preparation areas

    b. Pharmacy dispensing areas

    c. Nurses' stations (unless adequately lighted by corridor luminaires)

(2) Sump pumps and other equipment required to operate for the safety of major apparatus and associated control systems and alarms

(3) Smoke control and stair pressurization systems

(4) Kitchen hood supply and/or exhaust systems, if required to operate during a fire in or under the hood

(5) Supply, return and exhaust ventilating systems for airborne infectious isolation rooms

**(B) Delayed Automatic or Manual Connection.** The following equipment shall be connected to the critical branch and shall be arranged for either delayed automatic or manual connection to the alternate power source.

(1) Heating equipment to provide heating for patient rooms.

*Exception: Heating of general patient rooms during disruption of the normal source shall not be required under any of the following conditions:*

(a) *The outside design temperature is higher than −6.7°C (20°F), or*
(b) *The outside design temperature is lower than −6.7°C (20°F) and where a selected room(s) is provided for the needs of all confined patients, only such room(s) need be heated.*
(c) *The facility is served by a dual source of normal power as described in 517.44(C), FPN.*

FPN: The outside design temperature is based on the 97½ percent design values as shown in Chapter 24 of the ASHRAE *Handbook of Fundamentals* (1997).

(2) Elevator service — in instances where disruption of power would result in elevators stopping between floors, throw-over facilities shall be provided to allow the temporary operation of any elevator for the release of passengers. For elevator cab lighting, control, and signal system requirements, see 517.42(G).

(3) Additional illumination, receptacles, and equipment shall be permitted to be connected only to the critical branch. [NFPA 99, 3.5.2.2.3]

**517.44 Sources of Power.**

**(A) Two Independent Sources of Power.** Essential electrical systems shall have a minimum of two independent sources of power: a normal source generally supplying the entire electrical system and one or more alternate sources for use when the normal source is interrupted. [NFPA 99, 3.4.1.1.2, 3.5.1]

**(B) Alternate Source of Power.** The alternate source of power shall be a generator(s) driven by some form of prime mover(s) and located on the premises.

*Exception No. 1: Where the normal source consists of generating units on the premises, the alternate source shall be either another generator set or an external utility service.*

*Exception No. 2: Nursing homes or limited care facilities meeting the requirements of 517.40(A), Exception, shall be permitted to use a battery system or self-contained battery*

*integral with the equipment. [NFPA 99, 3.4.1.1.3, 3.5.1, 16.3.3.2.1, 17.3.3.2.1]*

**(C) Location of Essential Electrical System Components.** Careful consideration shall be given to the location of the spaces housing the components of the essential electrical system to minimize interruptions caused by natural forces common to the area (e.g., storms, floods, earthquakes, or hazards created by adjoining structures or activities). Consideration shall also be given to the possible interruption of normal electrical services resulting from similar causes as well as possible disruption of normal electrical service due to internal wiring and equipment failures.

FPN: Facilities in which the normal source of power is supplied by two or more separate central station-fed services experience greater than normal electrical service reliability than those with only a single feed. Such a dual source of normal power consists of two or more electrical services fed from separate generator sets or a utility distribution network that has multiple power input sources and is arranged to provide mechanical and electrical separation so that a fault between the facility and the generating sources will not likely cause an interruption of more than one of the facility service feeders.

**517.45 Essential Electrical Systems for Other Health Care Facilities.**

**(A) Essential Electrical Distribution.** The essential electrical distribution system shall be a battery or generator system. [NFPA 99, 13.3.3.2]

**(B) Electrical Life Support Equipment.** Where electrical life support equipment is required, the essential electrical distribution system shall be as described in 517.30 through 517.35. [NFPA 99, 13.3.3.2.1]

**(C) Critical Care Areas.** Where critical care areas are present, the essential electrical distribution system shall be as described in 517.30 through 517.35. [NFPA 99, 13.3.3.2.2]

**(D) Power Systems.** Battery systems shall be installed in accordance with the requirements of Article 700, and generator systems shall be as described in 517.30 through 517.35.

**IV. Inhalation Anesthetizing Locations**

FPN: For further information regarding safeguards for anesthetizing locations, see, NFPA 99-1999, *Standard for Health Care Facilities.*

**517.60 Anesthetizing Location Classification.**

FPN: If either of the anesthetizing locations in 517.60(A) or (B) is designated a wet location, refer to 517.20.

**(A) Hazardous (Classified) Location.**

**(1) Use Location.** In a location where flammable anesthetics are employed, the entire area shall be considered to be a Class I, Division 1 location that shall extend upward to a level 1.52 m (5 ft) above the floor. The remaining volume up to the structural ceiling is considered to be above a hazardous (classified) location. [NFPA 99, Annex 2, 2.1, 2.2]

**(2) Storage Location.** Any room or location in which flammable anesthetics or volatile flammable disinfecting agents are stored shall be considered to be a Class I, Division 1 location from floor to ceiling.

**(B) Other-Than-Hazardous (Classified) Location.** Any inhalation anesthetizing location designated for the exclusive use of nonflammable anesthetizing agents shall be considered to be an other-than-hazardous (classified) location.

**517.61 Wiring and Equipment.**

**(A) Within Hazardous (Classified) Anesthetizing Locations.**

**(1) Isolation.** Except as permitted in 517.160, each power circuit within, or partially within, a flammable anesthetizing location as referred to in 517.60 shall be isolated from any distribution system by the use of an isolated power system. [NFPA 99, Annex 2, 2.6.3.2]

**(2) Design and Installation.** Isolated power system equipment shall be listed for the purpose and the system designed and installed so that it meets the provisions and is in accordance with Part VII.

**(3) Equipment Operating at More Than 10 Volts.** In hazardous (classified) locations referred to in 517.60, all fixed wiring and equipment and all portable equipment, including lamps and other utilization equipment, operating at more than 10 volts between conductors shall comply with the requirements of 501.1 through 501.15 and 501.16(A) and (B) for Class I, Division 1 locations. All such equipment shall be specifically approved for the hazardous atmospheres involved. [NFPA 99, Annex 2, 2.2.1, 2.4.5, 2.4.6, 2.4.7]

**(4) Extent of Location.** Where a box, fitting, or enclosure is partially, but not entirely, within a hazardous (classified) location(s), the hazardous (classified) location(s) shall be considered to be extended to include the entire box, fitting, or enclosure.

**(5) Receptacles and Attachment Plugs.** Receptacles and attachment plugs in a hazardous (classified) location(s) shall be listed for use in Class I, Group C hazardous (classified) locations and shall have provision for the connection of a grounding conductor.

**(6) Flexible Cord Type.** Flexible cords used in hazardous (classified) locations for connection to portable utilization equipment, including lamps operating at more than 8 volts between conductors, shall be of a type approved for extra-hard usage in accordance with Table 400.4 and shall include an additional conductor for grounding.

**(7) Flexible Cord Storage.** A storage device for the flexible cord shall be provided and shall not subject the cord to bending at a radius of less than 75 mm (3 in.).

**(B) Above Hazardous (Classified) Anesthetizing Locations.**

**(1) Wiring Methods.** Wiring above a hazardous (classified) location referred to in 517.60 shall be installed in rigid metal conduit, electrical metallic tubing, intermediate metal conduit, Type MI cable, or Type MC cable that employs a continuous, gas/vaportight metal sheath.

**(2) Equipment Enclosure.** Installed equipment that may produce arcs, sparks, or particles of hot metal, such as lamps and lampholders for fixed lighting, cutouts, switches, generators, motors, or other equipment having make-and-break or sliding contacts, shall be of the totally enclosed type or be constructed so as to prevent escape of sparks or hot metal particles.

*Exception: Wall-mounted receptacles installed above the hazardous (classified) location in flammable anesthetizing locations shall not be required to be totally enclosed or have openings guarded or screened to prevent dispersion of particles.*

**(3) Luminaires (Lighting Fixtures).** Surgical and other luminaires (lighting fixtures) shall conform to 501.9(B).

*Exception No. 1: The surface temperature limitations set forth in 501.9(B)(2) shall not apply.*

*Exception No. 2: Integral or pendant switches that are located above and cannot be lowered into the hazardous (classified) location(s) shall not be required to be explosionproof.*

**(4) Seals.** Approved seals shall be provided in conformance with 501.5, and 501.5(A)(4) shall apply to horizontal as well as to vertical boundaries of the defined hazardous (classified) locations.

**(5) Receptacles and Attachment Plugs.** Receptacles and attachment plugs located above hazardous (classified) anesthetizing locations shall be listed for hospital use for services of prescribed voltage, frequency, rating, and number of conductors with provision for the connection of the grounding conductor. This requirement shall apply to attachment plugs and receptacles of the 2-pole, 3-wire grounding type for single-phase, 120-volt, nominal, ac service.

**(6) 250-Volt Receptacles and Attachment Plugs Rated 50 and 60 Amperes.** Receptacles and attachment plugs rated 250 volts, for connection of 50-ampere and 60-ampere ac medical equipment for use above hazardous (classified) locations shall be arranged so that the 60-ampere receptacle will accept either the 50-ampere or the 60-ampere plug. Fifty-ampere receptacles shall be designed so as not to accept the 60-ampere attachment plug. The attachment plugs shall be of the 2-pole, 3-wire design with a third contact connecting to the insulated (green or green with yellow stripe) equipment grounding conductor of the electrical system.

**(C) Other-Than-Hazardous (Classified) Anesthetizing Locations.**

**(1) Wiring Methods.** Wiring serving other-than-hazardous (classified) locations, as defined in 517.60, shall be installed in a metal raceway system or cable assembly. The metal raceway system or cable armor or sheath assembly shall qualify as an equipment grounding return path in accordance with 250.118. Type MC and Type MI cable shall have an outer metal armor or sheath that is identified as an acceptable grounding return path.

*Exception: Pendant receptacle constructions that employ at least Type SJO or equivalent flexible cords suspended not less than 1.8 m (6 ft) from the floor shall not be required to be installed in a metal raceway or cable assembly.*

**(2) Receptacles and Attachment Plugs.** Receptacles and attachment plugs installed and used in other-than-hazardous (classified) locations shall be listed for hospital use for services of prescribed voltage, frequency, rating, and number of conductors with provision for connection of the grounding conductor. This requirement shall apply to 2-pole, 3-wire grounding type for single-phase, 120-, 208-, or 240-volt, nominal, ac service.

**(3) 250-Volt Receptacles and Attachment Plugs Rated 50 and 60 Amperes.** Receptacles and attachment plugs rated 250 volts, for connection of 50-ampere and 60-ampere ac medical equipment for use in other-than-hazardous (classified) locations shall be arranged so that the 60-ampere receptacle will accept either the 50-ampere or the 60-ampere plug. Fifty-ampere receptacles shall be designed so as not to accept the 60-ampere attachment plug. The attachment plugs shall be of the 2-pole, 3-wire design with a third contact connecting to the insulated (green or green with yellow stripe) equipment grounding conductor of the electrical system.

**517.62 Grounding.** In any anesthetizing area, all metal raceways and metal-sheathed cables and all non–current-

carrying conductive portions of fixed electric equipment shall be grounded. Grounding in Class I locations shall comply with 501.16.

*Exception: Equipment operating at not more than 10 volts between conductors shall not be required to be grounded.*

**517.63 Grounded Power Systems in Anesthetizing Locations.**

**(A) Battery-Powered Emergency Lighting Units.** One or more battery-powered emergency lighting units shall be provided in accordance with 700.12(E).

**(B) Branch-Circuit Wiring.** Branch circuits supplying only listed, fixed, therapeutic and diagnostic equipment, permanently installed above the hazardous (classified) location and in other-than-hazardous (classified) locations, shall be permitted to be supplied from a normal grounded service, single- or three-phase system, provided the following apply:

(1) Wiring for grounded and isolated circuits does not occupy the same raceway or cable.
(2) All conductive surfaces of the equipment are grounded.
(3) Equipment (except enclosed X-ray tubes and the leads to the tubes) are located at least 2.5 m (8 ft) above the floor or outside the anesthetizing location.
(4) Switches for the grounded branch circuit are located outside the hazardous (classified) location.

*Exception: Sections 517.63(B)(3) and (B)(4) shall not apply in other-than-hazardous (classified) locations.*

**(C) Fixed Lighting Branch Circuits.** Branch circuits supplying only fixed lighting shall be permitted to be supplied by a normal grounded service, provided the following apply:

(1) Such luminaires (fixtures) are located at least 2.5 m (8 ft) above the floor.
(2) All conductive surfaces of luminaires (fixtures) are grounded.
(3) Wiring for circuits supplying power to luminaires (fixtures) does not occupy the same raceway or cable for circuits supplying isolated power.
(4) Switches are wall-mounted and located above hazardous (classified) locations.

*Exception: Sections 517.63(C)(1) and (C)(4) shall not apply in other-than-hazardous (classified) locations.*

**(D) Remote-Control Stations.** Wall-mounted remote-control stations for remote-control switches operating at 24 volts or less shall be permitted to be installed in any anesthetizing location.

**(E) Location of Isolated Power Systems.** An isolated power center listed for the purpose and its grounded primary feeder shall be permitted to be located in an anesthetizing location, provided it is installed above a hazardous (classified) location or in an other-than-hazardous (classified) location.

**(F) Circuits in Anesthetizing Locations.** Except as permitted above, each power circuit within, or partially within, a flammable anesthetizing location as referred to in 517.60 shall be isolated from any distribution system supplying other-than-anesthetizing locations.

### 517.64 Low-Voltage Equipment and Instruments.

**(A) Equipment Requirements.** Low-voltage equipment that is frequently in contact with the bodies of persons or has exposed current-carrying elements shall be as follows:

(1) Operate on an electrical potential of 10 volts or less, or
(2) Approved as intrinsically safe or double-insulated equipment, or
(3) Moisture resistant

**(B) Power Supplies.** Power shall be supplied to low-voltage equipment from the following:

(1) An individual portable isolating transformer (autotransformers shall not be used) connected to an isolated power circuit receptacle by means of an appropriate cord and attachment plug, or
(2) A common low-voltage isolating transformer installed in an other-than-hazardous (classified) location, or
(3) Individual dry-cell batteries, or
(4) Common batteries made up of storage cells located in an other-than-hazardous (classified) location

**(C) Isolated Circuits.** Isolating-type transformers for supplying low-voltage circuits shall have the following:

(1) Approved means for insulating the secondary circuit from the primary circuit, and
(2) The core and case grounded

**(D) Controls.** Resistance or impedance devices shall be permitted to control low-voltage equipment but shall not be used to limit the maximum available voltage to the equipment.

**(E) Battery-Powered Appliances.** Battery-powered appliances shall not be capable of being charged while in operation unless their charging circuitry incorporates an integral isolating-type transformer.

**(F) Receptacles or Attachment Plugs.** Any receptacle or attachment plug used on low-voltage circuits shall be of a type that does not permit interchangeable connection with circuits of higher voltage.

FPN: Any interruption of the circuit, even circuits as low as 10 volts, either by any switch or loose or defective con-

nections anywhere in the circuit, may produce a spark that is sufficient to ignite flammable anesthetic agents. See 7.5.1.2.3 of NFPA 99-1999, *Standard for Health Care Facilities.*

## V. X-Ray Installations

Nothing in this part shall be construed as specifying safeguards against the useful beam or stray X-ray radiation.

FPN No. 1: Radiation safety and performance requirements of several classes of X-ray equipment are regulated under Public Law 90-602 and are enforced by the Department of Health and Human Services.

FPN No. 2: In addition, information on radiation protection by the National Council on Radiation Protection and Measurements is published as *Reports of the National Council on Radiation Protection and Measurement.* These reports are obtainable from NCRP Publications, P.O. Box 30175, Washington, DC 20014.

### 517.71 Connection to Supply Circuit.

**(A) Fixed and Stationary Equipment.** Fixed and stationary X-ray equipment shall be connected to the power supply by means of a wiring method that meets the general requirements of this *Code.*

*Exception: Equipment properly supplied by a branch circuit rated at not over 30 amperes shall be permitted to be supplied through a suitable attachment plug and hard-service cable or cord.*

**(B) Portable, Mobile, and Transportable Equipment.** Individual branch circuits shall not be required for portable, mobile, and transportable medical X-ray equipment requiring a capacity of not over 60 amperes.

**(C) Over 600-Volt Supply.** Circuits and equipment operated on a supply circuit of over 600 volts shall comply with Article 490.

### 517.72 Disconnecting Means.

**(A) Capacity.** A disconnecting means of adequate capacity for at least 50 percent of the input required for the momentary rating or 100 percent of the input required for the long-time rating of the X-ray equipment, whichever is greater, shall be provided in the supply circuit.

**(B) Location.** The disconnecting means shall be operable from a location readily accessible from the X-ray control.

**(C) Portable Equipment.** For equipment connected to a 120-volt branch circuit of 30 amperes or less, a grounding-type attachment plug and receptacle of proper rating shall be permitted to serve as a disconnecting means.

**517.73 Rating of Supply Conductors and Overcurrent Protection.**

**(A) Diagnostic Equipment.**

**(1) Branch Circuits.** The ampacity of supply branch-circuit conductors and the current rating of overcurrent protective devices shall not be less than 50 percent of the momentary rating or 100 percent of the long-time rating, whichever is greater.

**(2) Feeders.** The ampacity of supply feeders and the current rating of overcurrent protective devices supplying two or more branch circuits supplying X-ray units shall not be less than 50 percent of the momentary demand rating of the largest unit plus 25 percent of the momentary demand rating of the next largest unit plus 10 percent of the momentary demand rating of each additional unit. Where simultaneous biplane examinations are undertaken with the X-ray units, the supply conductors and overcurrent protective devices shall be 100 percent of the momentary demand rating of each X-ray unit.

> FPN: The minimum conductor size for branch and feeder circuits is also governed by voltage regulation requirements. For a specific installation, the manufacturer usually specifies minimum distribution transformer and conductor sizes, rating of disconnecting means, and overcurrent protection.

**(B) Therapeutic Equipment.** The ampacity of conductors and rating of overcurrent protective devices shall not be less than 100 percent of the current rating of medical X-ray therapy equipment.

> FPN: The ampacity of the branch-circuit conductors and the ratings of disconnecting means and overcurrent protection for X-ray equipment are usually designated by the manufacturer for the specific installation.

**517.74 Control Circuit Conductors.**

**(A) Number of Conductors in Raceway.** The number of control circuit conductors installed in a raceway shall be determined in accordance with 300.17.

**(B) Minimum Size of Conductors.** Size 18 AWG or 16 AWG fixture wires as specified in 725.27 and flexible cords shall be permitted for the control and operating circuits of X-ray and auxiliary equipment where protected by not larger than 20-ampere overcurrent devices.

**517.75 Equipment Installations.** All equipment for new X-ray installations and all used or reconditioned X-ray equipment moved to and reinstalled at a new location shall be of an approved type.

**517.76 Transformers and Capacitors.** Transformers and capacitors that are part of X-ray equipment shall not be required to comply with Articles 450 and 460.

Capacitors shall be mounted within enclosures of insulating material or grounded metal.

**517.77 Installation of High-Tension X-Ray Cables.** Cables with grounded shields connecting X-ray tubes and image intensifiers shall be permitted to be installed in cable trays or cable troughs along with X-ray equipment control and power supply conductors without the need for barriers to separate the wiring.

**517.78 Guarding and Grounding.**

**(A) High-Voltage Parts.** All high-voltage parts, including X-ray tubes, shall be mounted within grounded enclosures. Air, oil, gas, or other suitable insulating media shall be used to insulate the high-voltage from the grounded enclosure. The connection from the high-voltage equipment to X-ray tubes and other high-voltage components shall be made with high-voltage shielded cables.

**(B) Low-Voltage Cables.** Low-voltage cables connecting to oil-filled units that are not completely sealed, such as transformers, condensers, oil coolers, and high-voltage switches, shall have insulation of the oil-resistant type.

**(C) Noncurrent–Carrying Metal Parts.** Noncurrent-carrying metal parts of X-ray and associated equipment (controls, tables, X-ray tube supports, transformer tanks, shielded cables, X-ray tube heads, etc.) shall be grounded in the manner specified in Article 250, as modified by 517.13(A) and (B).

**VI. Communications, Signaling Systems, Data Systems, Fire Alarm Systems, and Systems Less Than 120 Volts, Nominal**

**517.80 Patient Care Areas.** Equivalent insulation and isolation to that required for the electrical distribution systems in patient care areas shall be provided for communications, signaling systems, data system circuits, fire alarm systems, and systems less than 120 volts, nominal.

> FPN: An acceptable alternate means of providing isolation for patient/nurse call systems is by the use of nonelectrified signaling, communications, or control devices held by the patient or within reach of the patient.

**517.81 Other-Than-Patient-Care Areas.** In other-than-patient-care areas, installations shall be in accordance with the appropriate provisions of Articles 640, 725, 760, and 800.

**517.82 Signal Transmission Between Appliances.**

**(A) General.** Permanently installed signal cabling from an appliance in a patient location to remote appliances shall employ a signal transmission system that prevents hazardous grounding interconnection of the appliances.

> FPN: See 517.13(A) for additional grounding requirements in patient care areas.

**(B) Common Signal Grounding Wire.** Common signal grounding wires (i.e., the chassis ground for single-ended transmission) shall be permitted to be used between appliances all located within the patient vicinity, provided the appliances are served from the same reference grounding point.

## VII. Isolated Power Systems

### 517.160 Isolated Power Systems.

**(A) Installations.**

**(1) Isolated Power Circuits.** Each isolated power circuit shall be controlled by a switch that has a disconnecting pole in each isolated circuit conductor to simultaneously disconnect all power. Such isolation shall be accomplished by means of one or more transformers having no electrical connection between primary and secondary windings, by means of motor generator sets, or by means of suitably isolated batteries.

**(2) Circuit Characteristics.** Circuits supplying primaries of isolating transformers shall operate at not more than 600 volts between conductors and shall be provided with proper overcurrent protection. The secondary voltage of such transformers shall not exceed 600 volts between conductors of each circuit. All circuits supplied from such secondaries shall be ungrounded and shall have an approved overcurrent device of proper ratings in each conductor. Circuits supplied directly from batteries or from motor generator sets shall be ungrounded and shall be protected against overcurrent in the same manner as transformer-fed secondary circuits. If an electrostatic shield is present, it shall be connected to the reference grounding point. [NFPA 99, 3.3.2.2.1]

**(3) Equipment Location.** The isolating transformers, motor generator sets, batteries and battery chargers, and associated primary or secondary overcurrent devices shall not be installed in hazardous (classified) locations. The isolated secondary circuit wiring extending into a hazardous anesthetizing location shall be installed in accordance with 501.4.

**(4) Isolation Transformers.** An isolation transformer shall not serve more than one operating room except as covered in (a) and (b).

For purposes of this section, anesthetic induction rooms are considered part of the operating room or rooms served by the induction rooms.

(a) Induction Rooms. Where an induction room serves more than one operating room, the isolated circuits of the induction room shall be permitted to be supplied from the isolation transformer of any one of the operating rooms served by that induction room.

(b) Higher Voltages. Isolation transformers shall be permitted to serve single receptacles in several patient areas where the following apply:

(1) The receptacles are reserved for supplying power to equipment requiring 150 volts or higher, such as portable X-ray units.
(2) The receptacles and mating plugs are not interchangeable with the receptacles on the local isolated power system. [NFPA 99, 12.4.1.2.6(d), 12.4.1.2.6(e)]

**(5) Conductor Identification.** The isolated circuit conductors shall be identified as follows:

(1) Isolated Conductor No. 1 — Orange
(2) Isolated Conductor No. 2 — Brown

For 3-phase systems, the third conductor shall be identified as yellow. Where isolated circuit conductors supply 125-volt, single-phase, 15- and 20-ampere receptacles, the orange conductor(s) shall be connected to the terminal(s) on the receptacles that are identified in accordance with 200.10(B) for connection to the grounded circuit conductor.

**(6) Wire-Pulling Compounds.** Wire-pulling compounds that increase the dielectric constant shall not be used on the secondary conductors of the isolated power supply.

> FPN No. 1: It is desirable to limit the size of the isolation transformer to 10 kVA or less and to use conductor insulation with low leakage to meet impedance requirements.

> FPN No. 2: Minimizing the length of branch-circuit conductors and using conductor insulations with a dielectric constant less than 3.5 and insulation resistance constant greater than 6100 megohm-meters (20,000 megohm-ft) at 16°C (60°F) reduces leakage from line to ground, reducing the hazard current.

**(B) Line Isolation Monitor.**

**(1) Characteristics.** In addition to the usual control and overcurrent protective devices, each isolated power system shall be provided with a continually operating line isolation monitor that indicates total hazard current. The monitor shall be designed so that a green signal lamp, conspicuously visible to persons in each area served by the isolated power system, remains lighted when the system is adequately isolated from ground. An adjacent red signal lamp and an audible warning signal (remote if desired) shall be energized when the total hazard current (consisting of possible resistive and capacitive leakage currents) from either isolated conductor to ground reaches a threshold value of 5 mA under nominal line voltage conditions. The line monitor shall not alarm for a fault hazard of less than 3.7 mA or for a total hazard current of less than 5 mA.

*Exception: A system shall be permitted to be designed to operate at a lower threshold value of total hazard current. A line isolation monitor for such a system shall be permitted to be approved with the provision that the fault hazard current shall be permitted to be reduced but not to less than 35 percent of the corresponding threshold value of the total hazard current, and the monitor hazard current is to be correspondingly reduced to not more than 50 percent of the alarm threshold value of the total hazard current.*

**(2) Impedance.** The line isolation monitor shall be designed to have sufficient internal impedance such that, when properly connected to the isolated system, the maximum internal current that can flow through the line isolation monitor, when any point of the isolated system is grounded, shall be 1 mA.

*Exception: The line isolation monitor shall be permitted to be of the low-impedance type such that the current through the line isolation monitor, when any point of the isolated system is grounded, will not exceed twice the alarm threshold value for a period not exceeding 5 milliseconds.*

> FPN: Reduction of the monitor hazard current, provided this reduction results in an increased "not alarm" threshold value for the fault hazard current, will increase circuit capacity.

**(3) Ammeter.** An ammeter calibrated in the total hazard current of the system (contribution of the fault hazard current plus monitor hazard current) shall be mounted in a plainly visible place on the line isolation monitor with the "alarm on" zone at approximately the center of the scale.

*Exception: The line isolation monitor shall be permitted to be a composite unit, with a sensing section cabled to a separate display panel section on which the alarm or test functions are located.*

> FPN: It is desirable to locate the ammeter so that it is conspicuously visible to persons in the anesthetizing location.

# ARTICLE 518
# Places of Assembly

**518.1 Scope.** This article covers all buildings or portions of buildings or structures designed or intended for the assembly of 100 or more persons.

**518.2 General Classifications.**

**(A) Examples.** Places of assembly shall include but not be limited to the following:

| | |
|---|---|
| Armories | Courtrooms |
| Assembly halls | Dance halls |
| Auditoriums | Dining facilities |
| Auditoriums within | Exhibition halls |
|     Business establishments | Gymnasiums |
|     Mercantile establishments | Mortuary chapels |
|     Other occupancies | Multipurpose rooms |
|     Schools | Museums |
| Bowling lanes | Places of awaiting transportation |
| Church chapels | Pool rooms |
| Club rooms | Restaurants |
| Conference rooms | Skating rinks |

**(B) Multiple Occupancies.** Occupancy of any room or space for assembly purposes by less than 100 persons in a building of other occupancy, and incidental to such other occupancy, shall be classified as part of the other occupancy and subject to the provisions applicable thereto.

**(C) Theatrical Areas.** Where any such building structure, or portion thereof, contains a projection booth or stage platform or area for the presentation of theatrical or musical productions, either fixed or portable, the wiring for that area, including associated audience seating areas, and all equipment that is used in the referenced area, and portable equipment and wiring for use in the production that will not be connected to permanently installed wiring, shall comply with Article 520.

> FPN: For methods of determining population capacity, see local building code or, in its absence, NFPA *101*-2000, *Life Safety Code.*

**518.3 Other Articles.**

**(A) Hazardous (Classified) Areas.** Electrical installations in hazardous (classified) areas located in places of assembly shall comply with Article 500.

**(B) Temporary Wiring.** In exhibition halls used for display booths, as in trade shows, the temporary wiring shall be installed in accordance with Article 527. Flexible cables and cords approved for hard or extra-hard usage shall be permitted to be laid on floors where protected from contact by the general public. The ground-fault circuit-interrupter requirements of 527.6 shall not apply.

*Exception: Where conditions of supervision and maintenance ensure that only qualified persons will service the installation, flexible cords or cables identified in Table 400.4 for hard usage or extra-hard usage shall be permitted in cable trays used only for temporary wiring. All cords or cables shall be installed in a single layer. A permanent sign shall be attached to the cable tray at intervals not to exceed 7.5 m (25 ft). The sign shall read*

*CABLE TRAY FOR TEMPORARY WIRING ONLY*

**(C) Emergency Systems.** Control of emergency systems shall comply with Article 700.

## 518.4 Wiring Methods.

**(A) General.** The fixed wiring methods shall be metal raceways, flexible metal raceways, nonmetallic raceways encased in not less than 50 mm (2 in.) of concrete, Type MI, MC, or AC cable containing an insulated equipment grounding conductor sized in accordance with Table 250.122.

*Exception: Fixed wiring methods shall be as provided in*

*(a) Audio signal processing, amplification, and reproduction equipment — Article 640*
*(b) Communications circuits — Article 800*
*(c) Class 2 and Class 3 remote-control and signaling circuits — Article 725*
*(d) Fire alarm circuits — Article 760*

**(B) Nonrated Construction.** Nonmetallic-sheathed cable, Type AC cable, electrical nonmetallic tubing, and rigid nonmetallic conduit shall be permitted to be installed in those buildings or portions thereof that are not required to be of fire-rated construction by the applicable building code.

FPN:   Fire-rated construction is the fire-resistive classification used in building codes.

**(C) Spaces with Finish Rating.** Electrical nonmetallic tubing and rigid nonmetallic conduit shall be permitted to be installed in club rooms, college and university classrooms, conference and meeting rooms in hotels or motels, courtrooms, drinking establishments, dining facilities, restaurants, mortuary chapels, museums, passenger stations and terminals of air, surface, underground, and marine public transportation facilities, libraries, and places of religious worship where the following apply:

(1) The electrical nonmetallic tubing or rigid nonmetallic conduit is installed concealed within walls, floors, and ceilings where the walls, floors, and ceilings provide a thermal barrier of material that has at least a 15-minute finish rating as identified in listings of fire-rated assemblies.
(2) The electrical nonmetallic tubing or rigid nonmetallic conduit is installed above suspended ceilings where the suspended ceilings provide a thermal barrier of material that has at least a 15-minute finish rating as identified in listings of fire-rated assemblies.

Electrical nonmetallic tubing and rigid nonmetallic conduit are not recognized for use in other space used for environmental air in accordance with 300.22(C).

FPN:   A finish rating is established for assemblies containing combustible (wood) supports. The finish rating is defined as the time at which the wood stud or wood joist reaches an average temperature rise of 121°C (250°F) or an individual temperature rise of 163°C (325°F) as measured on the plane of the wood nearest the fire. A finish rating is not intended to represent a rating for a membrane ceiling.

**518.5 Supply.** Portable switchboards and portable power distribution equipment shall be supplied only from listed power outlets of sufficient voltage and ampere rating. Such power outlets shall be protected by overcurrent devices. Such overcurrent devices and power outlets shall not be accessible to the general public. Provisions for connection of an equipment grounding conductor shall be provided. The neutral of feeders supplying solid-state, 3-phase, 4-wire dimmer systems shall be considered a current-carrying conductor.

## ARTICLE 520
## Theaters, Audience Areas of Motion Picture and Television Studios, Performance Areas, and Similar Locations

### I. General

**520.1 Scope.** This article covers all buildings or that part of a building or structure, indoor or outdoor, designed or used for presentation, dramatic, musical, motion picture projection, or similar purposes and to specific audience seating areas within motion picture or television studios.

### 520.2 Definitions.

**Border Light.** A permanently installed overhead strip light.

**Breakout Assembly.** An adapter used to connect a multipole connector containing two or more branch circuits to multiple individual branch-circuit connectors.

**Bundled.** Cables or conductors that are physically tied, wrapped, taped, or otherwise periodically bound together.

**Connector Strip.** A metal wireway containing pendant or flush receptacles.

**Drop Box.** A box containing pendant- or flush-mounted receptacles attached to a multiconductor cable via strain relief or a multipole connector.

**Footlight.** A border light installed on or in the stage.

**Grouped.** Cables or conductors positioned adjacent to one another but not in continuous contact with each other.

**Performance Area.** The stage and audience seating area associated with a temporary stage structure, whether indoors or outdoors, constructed of scaffolding, truss, platforms, or similar devices, that is used for the presentation of theatrical or musical productions or for public presentations.

**Portable Equipment.** Equipment fed with portable cords or cables intended to be moved from one place to another.

**Portable Power Distribution Unit.** A power distribution box containing receptacles and overcurrent devices.

**Proscenium.** The wall and arch that separates the stage from the auditorium (house).

**Stand Lamp (Work Light).** A portable stand that contains a general-purpose luminaire (lighting fixture) or lampholder with guard for the purpose of providing general illumination on the stage or in the auditorium.

**Strip Light.** A luminaire (lighting fixture) with multiple lamps arranged in a row.

**Two-Fer.** An adapter cable containing one male plug and two female cord connectors used to connect two loads to one branch circuit.

**520.3 Motion Picture Projectors.** Motion picture equipment and its installation and use shall comply with Article 540.

**520.4 Audio Signal Processing, Amplification, and Reproduction Equipment.** Audio signal processing, amplification, and reproduction equipment and its installation shall comply with Article 640.

**520.5 Wiring Methods.**

**(A) General.** The fixed wiring method shall be metal raceways, nonmetallic raceways encased in at least 50 mm (2 in.) of concrete, Type MI cable, MC cable, or AC cable containing an insulated equipment grounding conductor sized in accordance with Table 250.122.

*Exception: Fixed wiring methods shall be as provided in Article 640 for audio signal processing, amplification, and reproduction equipment, in Article 800 for communication circuits, in Article 725 for Class 2 and Class 3 remote-control and signaling circuits, and in Article 760 for fire alarm circuits.*

**(B) Portable Equipment.** The wiring for portable switchboards, stage set lighting, stage effects, and other wiring not fixed as to location shall be permitted with approved flexible cords and cables as provided elsewhere in Article 520. Fastening such cables and cords by uninsulated staples or nailing shall not be permitted.

**(C) Nonrated Construction.** Nonmetallic-sheathed cable, Type AC cable, electrical nonmetallic tubing, and rigid nonmetallic conduit shall be permitted to be installed in those buildings or portions thereof that are not required to be of fire-rated construction by the applicable building code.

**520.6 Number of Conductors in Raceway.** The number of conductors permitted in any metal conduit, rigid nonmetallic conduit as permitted in this article, or electrical metallic tubing for border or stage pocket circuits or for remote-control conductors shall not exceed the percentage fill shown in Table 1 of Chapter 9. Where contained within an auxiliary gutter or a wireway, the sum of the cross-sectional areas of all contained conductors at any cross section shall not exceed 20 percent of the interior cross-sectional area of the auxiliary gutter or wireway. The 30-conductor limitation of 366.6 and 376.22 shall not apply.

**520.7 Enclosing and Guarding Live Parts.** Live parts shall be enclosed or guarded to prevent accidental contact by persons and objects. All switches shall be of the externally operable type. Dimmers, including rheostats, shall be placed in cases or cabinets that enclose all live parts.

**520.8 Emergency Systems.** Control of emergency systems shall comply with Article 700.

**520.9 Branch Circuits.** A branch circuit of any size supplying one or more receptacles shall be permitted to supply stage set lighting. The voltage rating of the receptacles shall not be less than the circuit voltage. Receptacle ampere ratings and branch-circuit conductor ampacity shall not be less than the branch-circuit overcurrent device ampere rating. Table 210.21(B)(2) shall not apply.

**520.10 Portable Equipment.** Portable stage and studio lighting equipment and portable power distribution equipment shall be permitted for temporary use outdoors, provided the equipment is supervised by qualified personnel while energized and barriered from the general public.

**II. Fixed Stage Switchboards**

**520.21 Dead Front.** Stage switchboards shall be of the dead-front type and shall comply with Part IV of Article 384 unless approved based on suitability as a stage switchboard as determined by a qualified testing laboratory and recognized test standards and principles.

**520.22 Guarding Back of Switchboard.** Stage switchboards having exposed live parts on the back of such boards shall be enclosed by the building walls, wire mesh grills, or by other approved methods. The entrance to this enclosure shall be by means of a self-closing door.

**520.23 Control and Overcurrent Protection of Receptacle Circuits.** Means shall be provided at a stage-lighting switchboard to which load circuits are connected for overcurrent protection of stage-lighting branch circuits, including branch circuits supplying stage and auditorium receptacles used for cord- and plug-connected stage equipment. Where the stage switchboard contains dimmers to control nonstage lighting, the locating of the overcurrent protective devices for these branch circuits at the stage switchboard shall be permitted.

**520.24 Metal Hood.** A stage switchboard that is not completely enclosed dead-front and dead-rear or recessed into a wall shall be provided with a metal hood extending the full length of the board to protect all equipment on the board from falling objects.

**520.25 Dimmers.** Dimmers shall comply with 520.25(A) through (D).

**(A) Disconnection and Overcurrent Protection.** Where dimmers are installed in ungrounded conductors, each dimmer shall have overcurrent protection not greater than 125 percent of the dimmer rating and shall be disconnected from all ungrounded conductors when the master or individual switch or circuit breaker supplying such dimmer is in the open position.

**(B) Resistance- or Reactor-Type Dimmers.** Resistance- or series reactor-type dimmers shall be permitted to be placed in either the grounded or the ungrounded conductor of the circuit. Where designed to open either the supply circuit to the dimmer or the circuit controlled by it, the dimmer shall then comply with 404.1. Resistance- or reactor-type dimmers placed in the grounded neutral conductor of the circuit shall not open the circuit.

**(C) Autotransformer-Type Dimmers.** The circuit supplying an autotransformer-type dimmer shall not exceed 150 volts between conductors. The grounded conductor shall be common to the input and output circuits.

FPN:  See 210.9 for circuits derived from autotransformers.

**(D) Solid-State-Type Dimmers.** The circuit supplying a solid-state dimmer shall not exceed 150 volts between conductors unless the dimmer is listed specifically for higher voltage operation. Where a grounded conductor supplies a dimmer, it shall be common to the input and output circuits.

Dimmer chassis shall be connected to the equipment grounding conductor.

**520.26 Type of Switchboard.** A stage switchboard shall be either one or a combination of the types specified in 520.26(A), (B), and (C).

**(A) Manual.** Dimmers and switches are operated by handles mechanically linked to the control devices.

**(B) Remotely Controlled.** Devices are operated electrically from a pilot-type control console or panel. Pilot control panels either shall be part of the switchboard or shall be permitted to be at another location.

**(C) Intermediate.** A stage switchboard with circuit interconnections is a secondary switchboard (patch panel) or panelboard remote to the primary stage switchboard. It shall contain overcurrent protection. Where the required branch-circuit overcurrent protection is provided in the dimmer panel, it shall be permitted to be omitted from the intermediate switchboard.

**520.27 Stage Switchboard Feeders.**

**(A) Type of Feeder.** Feeders supplying stage switchboards shall be one of the types in 520.27(A)(1) through (A)(3).

**(1) Single Feeder.** A single feeder disconnected by a single disconnect device.

**(2) Multiple Feeders to Intermediate Stage Switchboard (Patch Panel).** Multiple feeders of unlimited quantity shall be permitted, provided that all multiple feeders are part of a single system. Where combined, neutral conductors in a given raceway shall be of sufficient ampacity to carry the maximum unbalanced current supplied by multiple feeder conductors in the same raceway, but they need not be greater than the ampacity of the neutral supplying the primary stage switchboard. Parallel neutral conductors shall comply with 310.4.

**(3) Separate Feeders to Single Primary Stage Switchboard (Dimmer Bank).** Installations with separate feeders to a single primary stage switchboard shall have a disconnecting means for each feeder. The primary stage switchboard shall have a permanent and obvious label stating the number and location of disconnecting means. If the disconnecting means are located in more than one distribution switchboard, the primary stage switchboard shall be provided with barriers to correspond with these multiple locations.

**(B) Neutral.** The neutral of feeders supplying solid-state, 3-phase, 4-wire dimming systems shall be considered a current-carrying conductor.

**(C) Supply Capacity.** For the purposes of computing supply capacity to switchboards, it shall be permissible to consider the maximum load that the switchboard is intended to control in a given installation, provided that the following apply:

(1) All feeders supplying the switchboard shall be protected by an overcurrent device with a rating not greater than the ampacity of the feeder.

(2) The opening of the overcurrent device shall not affect the proper operation of the egress or emergency lighting systems.

FPN: For computation of stage switchboard feeder loads, see 220.10.

## III. Fixed Stage Equipment Other Than Switchboards

### 520.41 Circuit Loads.

**(A) Circuits Rated 20 Amperes or Less.** Footlights, border lights, and proscenium sidelights shall be arranged so that no branch circuit supplying such equipment carries a load exceeding 20 amperes.

**(B) Circuits Rated Greater Than 20 Amperes.** Where heavy-duty lampholders only are used, such circuits shall be permitted to comply with Article 210 for circuits supplying heavy-duty lampholders.

### 520.42 Conductor Insulation.

Foot, border, proscenium, or portable strip lights and connector strips shall be wired with conductors that have insulation suitable for the temperature at which the conductors are operated, but not less than 125°C (257°F). The ampacity of the 125°C (257°F) conductors shall be that of 60°C (140°F) conductors. All drops from connector strips shall be 90°C (194°F) wire sized to the ampacity of 60°C (140°F) cords and cables with no more than 150 mm (6 in.) of conductor extending into the connector strip. Section 310.15(B)(2)(a) shall not apply.

FPN: See Table 310.13 for conductor types.

### 520.43 Footlights.

**(A) Metal Trough Construction.** Where metal trough construction is employed for footlights, the trough containing the circuit conductors shall be made of sheet metal not lighter than 0.81 mm (0.032 in.) and treated to prevent oxidation. Lampholder terminals shall be kept at least 13 mm (½ in.) from the metal of the trough. The circuit conductors shall be soldered to the lampholder terminals.

**(B) Other-Than-Metal Trough Construction.** Where the metal trough construction specified in 520.43(A) is not used, footlights shall consist of individual outlets with lampholders wired with rigid metal conduit, intermediate metal conduit, or flexible metal conduit, Type MC cable, or mineral-insulated, metal-sheathed cable. The circuit conductors shall be soldered to the lampholder terminals.

**(C) Disappearing Footlights.** Disappearing footlights shall be arranged so that the current supply is automatically disconnected when the footlights are replaced in the storage recesses designed for them.

### 520.44 Borders and Proscenium Sidelights.

**(A) General.** Borders and proscenium sidelights shall be as follows:

(1) Constructed as specified in 520.43

(2) Suitably stayed and supported

(3) Designed so that the flanges of the reflectors or other adequate guards protect the lamps from mechanical damage and from accidental contact with scenery or other combustible material

**(B) Cords and Cables for Border Lights.**

**(1) General.** Cords and cables for supply to border lights shall be listed for extra-hard usage. The cords and cables shall be suitably supported. Such cords and cables shall be employed only where flexible conductors are necessary. Ampacity of the conductors shall be as provided in 400.5.

**(2) Cords and Cables Not in Contact with Heat-Producing Equipment.** Listed multiconductor extra-hard-usage-type cords and cables not in direct contact with equipment containing heat-producing elements shall be permitted to have their ampacity determined by Table 520.44. Maximum load current in any conductor with an ampacity determined by Table 520.44 shall not exceed the values in Table 520.44.

### 520.45 Receptacles.

Receptacles for electrical equipment on stages shall be rated in amperes. Conductors supplying receptacles shall be in accordance with Articles 310 and 400.

### 520.46 Connector Strips, Drop Boxes, Floor Pockets, and Other Outlet Enclosures.

Receptacles for the connection of portable stage-lighting equipment shall be pendant or mounted in suitable pockets or enclosures and shall comply with 520.45. Supply cables for connector strips and drop boxes shall be as specified in 520.44(B).

### 520.47 Backstage Lamps (Bare Bulbs).

Lamps (bare bulbs) installed in backstage and ancillary areas where they can come in contact with scenery shall be located and guarded so as to be free from physical damage and shall

**Table 520.44 Ampacity of Listed Extra-Hard-Usage Cords and Cables with Temperature Ratings of 75°C (167°F) and 90°C (194°F)\* [Based on Ambient Temperature of 30°C (86°F)]**

| Size (AWG) | Temperature Rating of Cords and Cables | | Maximum Rating of Overcurrent Device |
| | 75°C (167°F) | 90°C (194°F) | |
| --- | --- | --- | --- |
| 14 | 24 | 28 | 15 |
| 12 | 32 | 35 | 20 |
| 10 | 41 | 47 | 25 |
| 8 | 57 | 65 | 35 |
| 6 | 77 | 87 | 45 |
| 4 | 101 | 114 | 60 |
| 2 | 133 | 152 | 80 |

\*Ampacity shown is the ampacity for multiconductor cords and cables where only three copper conductors are current-carrying as described in 400.5. If the number of current-carrying conductors in a cord or cable exceeds three and the load diversity factor is a minimum of 50 percent, the ampacity of each conductor shall be reduced as shown in the following table.

| Number of Conductors | Percent of Ampacity |
| --- | --- |
| 4–6 | 80 |
| 7–24 | 70 |
| 25–42 | 60 |
| 43 and above | 50 |

Note: Ultimate insulation temperature. In no case shall conductors be associated together in such a way with respect to the kind of circuit, the wiring method used, or the number of conductors such that the temperature limit of the conductors is exceeded.

A neutral conductor that carries only the unbalanced current from other conductors of the same circuit need not be considered as a current-carrying conductor.

In a 3-wire circuit consisting of two phase wires and the neutral of a 4-wire, 3-phase, wye-connected system, a common conductor carries approximately the same current as the line-to-neutral currents of the other conductors and shall be considered to be a current-carrying conductor.

On a 4-wire, 3-phase, wye circuit where the major portion of the load consists of nonlinear loads such as electric-discharge lighting, electronic computer/data processing, or similar equipment, there are harmonic currents present in the neutral conductor, and the neutral shall be considered to be a current-carrying conductor.

provide an air space of not less than 50 mm (2 in.) between such lamps and any combustible material.

*Exception: Decorative lamps installed in scenery shall not be considered to be backstage lamps for the purpose of this section.*

**520.48 Curtain Machines.** Curtain machines shall be listed.

**520.49 Smoke Ventilator Control.** Where stage smoke ventilators are released by an electrical device, the circuit operating the device shall be normally closed and shall be controlled by at least two externally operable switches, one

switch being placed at a readily accessible location on stage and the other where designated by the authority having jurisdiction. The device shall be designed for the full voltage of the circuit to which it is connected, no resistance being inserted. The device shall be located in the loft above the scenery and shall be enclosed in a suitable metal box having a tight, self-closing door.

## IV. Portable Switchboards on Stage

**520.50 Road Show Connection Panel (A Type of Patch Panel).** A panel designed to allow for road show connection of portable stage switchboards to fixed lighting outlets by means of permanently installed supplementary circuits. The panel, supplementary circuits, and outlets shall comply with 520.50(A) through (D).

**(A) Load Circuits.** Circuits shall terminate in grounding-type polarized inlets of current and voltage rating that match the fixed-load receptacle.

**(B) Circuit Transfer.** Circuits that are transferred between fixed and portable switchboards shall have all circuit conductors transferred simultaneously.

**(C) Overcurrent Protection.** The supply devices of these supplementary circuits shall be protected by branch-circuit overcurrent protective devices. The individual supplementary circuit, within the road show connection panel and theater shall be protected by branch-circuit overcurrent protective devices of suitable ampacity installed within the road show connection panel.

**(D) Enclosure.** Panel construction shall be in accordance with Article 408.

**520.51 Supply.** Portable switchboards shall be supplied only from power outlets of sufficient voltage and ampere rating. Such power outlets shall include only externally operable, enclosed fused switches or circuit breakers mounted on stage or at the permanent switchboard in locations readily accessible from the stage floor. Provisions for connection of an equipment grounding conductor shall be provided. The neutral of feeders supplying solid-state, 3-phase, 4-wire dimmer systems shall be considered a current-carrying conductor.

**520.52 Overcurrent Protection.** Circuits from portable switchboards directly supplying equipment containing incandescent lamps of not over 300 watts shall be protected by overcurrent protective devices having a rating or setting of not over 20 amperes. Circuits for lampholders over 300 watts shall be permitted where overcurrent protection complies with Article 210.

**520.53 Construction and Feeders.** Portable switchboards and feeders for use on stages shall comply with 520.53(A) through (P).

**(A) Enclosure.** Portable switchboards shall be placed within an enclosure of substantial construction, which shall be permitted to be arranged so that the enclosure is open during operation. Enclosures of wood shall be completely lined with sheet metal of not less than 0.51 mm (0.020 in.) and shall be well galvanized, enameled, or otherwise properly coated to prevent corrosion or be of a corrosion-resistant material.

**(B) Energized Parts.** There shall not be exposed energized parts within the enclosure.

**(C) Switches and Circuit Breakers.** All switches and circuit breakers shall be of the externally operable, enclosed type.

**(D) Circuit Protection.** Overcurrent devices shall be provided in each ungrounded conductor of every circuit supplied through the switchboard. Enclosures shall be provided for all overcurrent devices in addition to the switchboard enclosure.

**(E) Dimmers.** The terminals of dimmers shall be provided with enclosures, and dimmer faceplates shall be arranged so that accidental contact cannot be readily made with the faceplate contacts.

**(F) Interior Conductors.**

**(1) Type.** All conductors other than busbars within the switchboard enclosure shall be stranded. Conductors shall be approved for an operating temperature at least equal to the approved operating temperature of the dimming devices used in the switchboard and in no case less than the following:

(1) Resistance-type dimmers — 200°C (392°F); or
(2) Reactor-type, autotransformer, and solid-state dimmers — 125°C (257°F).

All control wiring shall comply with Article 725.

**(2) Protection.** Each conductor shall have an ampacity not less than the rating of the circuit breaker, switch, or fuse that it supplies. Circuit interrupting and bus bracing shall be in accordance with 110.9 and 110.10. The short-circuit current rating shall be marked on the switchboard.

Conductors shall be enclosed in metal wireways or shall be securely fastened in position and shall be bushed where they pass through metal.

**(G) Pilot Light.** A pilot light shall be provided within the enclosure and shall be connected to the circuit supplying the board so that the opening of the master switch does not cut off the supply to the lamp. This lamp shall be on an individual branch circuit having overcurrent protection rated or set at not over 15 amperes.

**(H) Supply Conductors.**

**(1) General.** The supply to a portable switchboard shall be by means of listed extra-hard usage cords or cables. The supply cords or cable shall terminate within the switchboard enclosure, in an externally operable fused master switch or circuit breaker or in a connector assembly identified for the purpose. The supply cords or cable (and connector assembly) shall have sufficient ampacity to carry the total load connected to the switchboard and shall be protected by overcurrent devices.

**(2) Single-Conductor Cables.** Single-conductor portable supply cable sets shall not be smaller than 2 AWG conductors. The equipment grounding conductor shall not be smaller than 6 AWG conductor. Single-conductor grounded neutral cables for a supply shall be sized as per 520.53(O)(2). Where single conductors are paralleled for increased ampacity, the paralleled conductors shall be of the same length and size. Single-conductor supply cables shall be grouped together but not bundled. The equipment grounding conductor shall be permitted to be of a different type, provided it meets the other requirements of this section, and it shall be permitted to be reduced in size as permitted by 250.122. Grounded (neutral) and equipment grounding conductors shall be identified in accordance with 200.6, 250.119, and 310.12. Grounded conductors shall be permitted to be identified by marking at least the first 150 mm (6 in.) from both ends of each length of conductor with white or gray. Equipment grounding conductors shall be permitted to be identified by marking at least the first 150 mm (6 in.) from both ends of each length of conductor with green or green with yellow stripes. Where more than one nominal voltage exists within the same premises, each ungrounded conductor shall be identified by system.

**(3) Supply Conductors Not Over 3.0 m (10 ft) Long.** Where supply conductors do not exceed 3.0 m (10 ft) in length between supply and switchboard or supply and a subsequent overcurrent device, the supply conductors shall be permitted to be reduced in size where all of the following conditions are met:

(1) The ampacity of the supply conductors shall be at least one-quarter of the ampacity of the supply overcurrent protection device.
(2) The supply conductors shall terminate in a single overcurrent protection device that will limit the load to the ampacity of the supply conductors. This single overcurrent device shall be permitted to supply additional overcurrent devices on its load side.
(3) The supply conductors shall not penetrate walls, floors, or ceilings or be run through doors or traffic areas. The

supply conductors shall be adequately protected from physical damage.

(4) The supply conductors shall be suitably terminated in an approved manner.

(5) Conductors shall be continuous without splices or connectors.

(6) Conductors shall not be bundled.

(7) Conductors shall be supported above the floor in an approved manner.

**(4) Supply Conductors Not Over 6.0 m (20 ft) Long.** Where supply conductors do not exceed 6.0 m (20 ft) in length between supply and switchboard or supply and a subsequent overcurrent protection device, the supply conductors shall be permitted to be reduced in size where all of the following conditions are met:

(1) The ampacity of the supply conductors shall be at least one-half of the ampacity of the supply overcurrent protection device.

(2) The supply conductors shall terminate in a single overcurrent protection device that limits the load to the ampacity of the supply conductors. This single overcurrent device shall be permitted to supply additional overcurrent devices on its load side.

(3) The supply conductors shall not penetrate walls, floors, or ceilings or be run through doors or traffic areas. The supply conductors shall be adequately protected from physical damage.

(4) The supply conductors shall be suitably terminated in an approved manner.

(5) The supply conductors shall be supported in an approved manner at least 2.1 m (7 ft) above the floor except at terminations.

(6) The supply conductors shall not be bundled.

(7) Tap conductors shall be in unbroken lengths.

**(5) Supply Conductors Not Reduced in Size.** Supply conductors not reduced in size under provisions of 520.53(H)(3) or 520.53(H)(4) shall be permitted to pass through holes in walls specifically designed for the purpose. If penetration is through the fire-resistant–rated wall, it shall be in accordance with 300.21.

**(I) Cable Arrangement.** Cables shall be protected by bushings where they pass through enclosures and shall be arranged so that tension on the cable is not transmitted to the connections. Where power conductors pass through metal, the requirements of 300.20 shall apply.

**(J) Number of Supply Interconnections.** Where connectors are used in a supply conductor, there shall be a maximum number of three interconnections (mated connector pairs) where the total length from supply to switchboard does not exceed 30 m (100 ft). In cases where the total length from supply to switchboard exceeds 30 m (100 ft), one additional interconnection shall be permitted for each additional 30 m (100 ft) of supply conductor.

**(K) Single-Pole Separable Connectors.** Where single-pole portable cable connectors are used, they shall be listed and of the locking type. Sections 400.10, 406.6, and 406.7 shall not apply to listed single-pole separable connectors and single-conductor cable assemblies utilizing listed single-pole separable connectors. Where paralleled sets of current-carrying, single-pole separable connectors are provided as input devices, they shall be prominently labeled with a warning indicating the presence of internal parallel connections. The use of single-pole separable connectors shall comply with at least one of the following conditions:

(1) Connection and disconnection of connectors are only possible where the supply connectors are interlocked to the source and it is not possible to connect or disconnect connectors when the supply is energized.

(2) Line connectors are of the listed sequential-interlocking type so that load connectors shall be connected in the following sequence:

   a. Equipment grounding conductor connection
   b. Grounded circuit conductor connection, if provided
   c. Ungrounded conductor connection, and that disconnection shall be in the reverse order

(3) A caution notice shall be provided adjacent to the line connectors indicating that plug connection shall be in the following order:

   a. Equipment grounding conductor connectors
   b. Grounded circuit conductor connectors, if provided
   c. Ungrounded conductor connectors, and that disconnection shall be in the reverse order

**(L) Protection of Supply Conductors and Connectors.** All supply conductors and connectors shall be protected against physical damage by an approved means. This protection shall not be required to be raceways.

**(M) Flanged Surface Inlets.** Flanged surface inlets (recessed plugs) that are used to accept the power shall be rated in amperes.

**(N) Terminals.** Terminals to which stage cables are connected shall be located so as to permit convenient access to the terminals.

**(O) Neutral.**

**(1) Neutral Terminal.** In portable switchboard equipment designed for use with 3-phase, 4-wire with ground supply, the supply neutral terminal, its associated busbar, or equivalent wiring, or both, shall have an ampacity equal to at least twice the ampacity of the largest ungrounded supply terminal.

*Exception: Where portable switchboard equipment is specifically constructed and identified to be internally converted in the field, in an approved manner, from use with a balanced 3-phase, 4-wire with ground supply to a balanced single-phase, 3-wire with ground supply, the supply neutral terminal and its associated busbar, equivalent wiring, or both, shall have an ampacity equal to at least that of the largest ungrounded single-phase supply terminal.*

**(2) Supply Neutral.** The power supply conductors for portable switchboards shall be sized considering the neutral as a current-carrying conductor. Where single-conductor feeder cables, not installed in raceways, are used on multiphase circuits, the grounded neutral conductor shall have an ampacity of at least 130 percent of the ungrounded circuit conductors feeding the portable switchboard.

**(P) Qualified Personnel.** The routing of portable supply conductors, the making and breaking of supply connectors and other supply connections, and the energization and de-energization of supply services shall be performed by qualified personnel, and portable switchboards shall be so marked, indicating this requirement in a permanent and conspicuous manner.

*Exception: A portable switchboard shall be permitted to be connected to a permanently installed supply receptacle by other than qualified personnel, provided that the supply receptacle is protected for its rated ampacity by an overcurrent device of not greater than 150 amperes, and where the receptacle, interconnection, and switchboard further*

*(a) Employ listed multipole connectors suitable for the purpose for every supply interconnection, and*
*(b) Prevent access to all supply connections by the general public, and*
*(c) Employ listed extra-hard usage multiconductor cords or cables with an ampacity suitable for the type of load and not less than the ampere rating of the connectors.*

## V. Portable Stage Equipment Other Than Switchboards

**520.61 Arc Lamps.** Arc lamps, including enclosed arc lamps and associated ballasts, shall be listed. Interconnecting cord sets and interconnecting cords and cables shall be extra-hard usage type and listed.

**520.62 Portable Power Distribution Units.** Portable power distribution units shall comply with 520.62(A) through (E).

**(A) Enclosure.** The construction shall be such that no current-carrying part will be exposed.

**(B) Receptacles and Overcurrent Protection.** Receptacles shall comply with 520.45 and shall have branch-circuit overcurrent protection in the box. Fuses and circuit breakers shall be protected against physical damage. Cords or cables supplying pendant receptacles shall be listed for extra-hard usage.

**(C) Busbars and Terminals.** Busbars shall have an ampacity equal to the sum of the ampere ratings of all the circuits connected to the busbar. Lugs shall be provided for the connection of the master cable.

**(D) Flanged Surface Inlets.** Flanged surface inlets (recessed plugs) that are used to accept the power shall be rated in amperes.

**(E) Cable Arrangement.** Cables shall be adequately protected where they pass through enclosures and be arranged so that tension on the cable is not transmitted to the terminations.

**520.63 Bracket Fixture Wiring.**

**(A) Bracket Wiring.** Brackets for use on scenery shall be wired internally, and the fixture stem shall be carried through to the back of the scenery where a bushing shall be placed on the end of the stem. Externally wired brackets or other fixtures shall be permitted where wired with cords designed for hard usage that extend through scenery and without joint or splice in canopy of fixture back and terminate in an approved-type stage connector located, where practical, within 450 mm (18 in.) of the fixture.

**(B) Mounting.** Fixtures shall be securely fastened in place.

**520.64 Portable Strips.** Portable strips shall be constructed in accordance with the requirements for border lights and proscenium sidelights in 520.44(A). The supply cable shall be protected by bushings where it passes through metal and shall be arranged so that tension on the cable will not be transmitted to the connections.

> FPN No. 1: See 520.42 for wiring of portable strips.
>
> FPN No. 2: See 520.68(A)(3) for insulation types required on single conductors.

**520.65 Festoons.** Joints in festoon wiring shall be staggered. Lamps enclosed in lanterns or similar devices of combustible material shall be equipped with guards.

**520.66 Special Effects.** Electrical devices used for simulating lightning, waterfalls, and the like shall be constructed and located so that flames, sparks, or hot particles cannot come in contact with combustible material.

**520.67 Multipole Branch-Circuit Cable Connectors.** Multipole branch-circuit cable connectors, male and female, for flexible conductors shall be constructed so that tension on the cord or cable is not transmitted to the connections. The female half shall be attached to the load end of the power supply cord or cable. The connector shall be rated in amperes and designed so that differently rated devices cannot be connected together; however, a 20-ampere T-slot receptacle shall be permitted to accept a 15-ampere attachment plug of the same voltage rating. Alternating-current multipole connectors shall be polarized and comply with 406.6 and 406.9.

FPN: See 400.10 for pull at terminals.

### 520.68 Conductors for Portables.

**(A) Conductor Type.**

**(1) General.** Flexible conductors, including cable extensions, used to supply portable stage equipment shall be listed extra-hard usage cords or cables.

**(2) Stand Lamps.** Reinforced cord shall be permitted to supply stand lamps where the cord is not subject to severe physical damage and is protected by an overcurrent device rated at not over 20 amperes.

**(3) High-Temperature Applications.** A special assembly of conductors in sleeving not longer than 1.0 m (3.3 ft) shall be permitted to be employed in lieu of flexible cord if the individual wires are stranded and rated not less than 125°C (257°F) and the outer sleeve is glass fiber with a wall thickness of at least 0.635 mm (0.025 in.).

Portable stage equipment requiring flexible supply conductors with a higher temperature rating where one end is permanently attached to the equipment shall be permitted to employ alternate, suitable conductors as determined by a qualified testing laboratory and recognized test standards.

**(4) Breakouts.** Listed, hard usage (junior hard service) cords shall be permitted in breakout assemblies where all of the following conditions are met:

(1) The cords are utilized to connect between a single multipole connector containing two or more branch circuits and multiple 2-pole, 3-wire connectors.
(2) The longest cord in the breakout assembly does not exceed 6.0 m (20 ft).
(3) The breakout assembly is protected from physical damage by attachment over its entire length to a pipe, truss, tower, scaffold, or other substantial support structure.
(4) All branch circuits feeding the breakout assembly are protected by overcurrent devices rated at not over 20 amperes.

**(B) Conductor Ampacity.** The ampacity of conductors shall be as given in 400.5, except multiconductor, listed, extra-hard usage portable cords that are not in direct contact with equipment containing heat-producing elements shall be permitted to have their ampacity determined by Table 520.44. Maximum load current in any conductor with an ampacity determined by Table 520.44 shall not exceed the values in Table 520.44.

*Exception: Where alternate conductors are allowed in 520.68(A)(3), their ampacity shall be as given in the appropriate table in this Code for the types of conductors employed.*

**520.69 Adapters.** Adapters, two-fers, and other single- and multiple-circuit outlet devices shall comply with 520.69(A), (B), and (C).

**(A) No Reduction in Current Rating.** Each receptacle and its corresponding cable shall have the same current and voltage rating as the plug supplying it. It shall not be utilized in a stage circuit with a greater current rating.

**(B) Connectors.** All connectors shall be wired in accordance with 520.67.

**(C) Conductor Type.** Conductors for adapters and two-fers shall be listed, extra-hard usage or listed, hard usage (junior hard service) cord. Hard usage (junior hard service) cord shall be restricted in overall length to 1.0 m (3.3 ft).

## VI. Dressing Rooms

**520.71 Pendant Lampholders.** Pendant lampholders shall not be installed in dressing rooms.

**520.72 Lamp Guards.** All exposed incandescent lamps in dressing rooms, where less than 2.5 m (8 ft) from the floor, shall be equipped with open-end guards riveted to the outlet box cover or otherwise sealed or locked in place.

**520.73 Switches Required.** All lights and any receptacles adjacent to the mirror(s) and above the dressing table counter(s) installed in dressing rooms shall be controlled by wall switches installed in the dressing room(s). Each switch controlling receptacles adjacent to the mirror(s) and above the dressing table counter(s) shall be provided with a pilot light located outside the dressing room, adjacent to the door to indicate when the receptacles are energized. Other outlets installed in the dressing room shall not be required to be switched.

## VII. Grounding

**520.81 Grounding.** All metal raceways and metal-sheathed cables shall be grounded. The metal frames and enclosures of all equipment, including border lights and portable luminaires (lighting fixtures), shall be grounded. Grounding, where used, shall be in accordance with Article 250.

## ARTICLE 525
## Carnivals, Circuses, Fairs, and Similar Events

### I. General Requirements

**525.1 Scope.** This article covers the installation of portable wiring and equipment for carnivals, circuses, fairs, and similar functions, including wiring in or on all structures.

### 525.3 Other Articles.

**(A) Portable Wiring and Equipment.** Wherever the requirements of other articles of this *Code* and Article 525 differ, the requirements of Article 525 shall apply to the portable wiring and equipment.

**(B) Permanent Structures.** Articles 518 and 520 shall apply to wiring in permanent structures.

**(C) Audio Signal Processing, Amplification, and Reproduction Equipment.** Article 640 shall apply to the wiring and installation of audio signal processing, amplification, and reproduction equipment.

**(D) Attractions Utilizing Pools, Fountains, and Similar Installations with Contained Volumes of Water.** This equipment shall be installed to comply with the applicable requirements of Article 680.

### 525.5 Overhead Conductor Clearances.

**(A) Vertical Clearances.** Conductors shall have a vertical clearance to ground in accordance with 225.18. These clearances shall apply only to wiring installed outside of tents and concessions.

**(B) Clearance to Rides and Attractions.** Amusement rides and amusement attractions shall be maintained not less than 4.5 m (15 ft) in any direction from overhead conductors operating at 600 volts or less, except for the conductors supplying the amusement ride or attraction. Amusement rides or attractions shall not be located under or within 4.5 m (15 ft) horizontally of conductors operating in excess of 600 volts.

**525.6 Protection of Electrical Equipment.** Electrical equipment and wiring methods in or on rides, concessions, or other units shall be provided with mechanical protection where such equipment or wiring methods are subject to physical damage.

### II. Power Sources

### 525.10 Separately Derived Systems.

**(A) Generators.** Generators shall comply with the requirements of Article 445.

**(B) Transformers.** Transformers shall comply with the applicable requirements of 240.4(A), (B)(3), and (C); 250.30; and Article 450.

**525.11 Services.** Services shall be installed in accordance with the applicable requirements of Article 230 and, in addition, shall comply with 525.11(A) and (B).

**(A) Guarding.** Service equipment shall not be installed in a location that is accessible to unqualified persons, unless the equipment is lockable.

**(B) Mounting and Location.** Service equipment shall be mounted on a solid backing and be installed so as to be protected from the weather, unless of weatherproof construction.

### III. Wiring Methods

### 525.20 Wiring Methods.

**(A) Type.** Where flexible cords or cables are used, they shall be listed for extra hard usage. Where flexible cords or cables are used and are not subject to physical damage, they shall be permitted to be listed for hard usage. Where used outdoors, flexible cords and cables shall also be listed for wet locations and shall be sunlight resistant. Extra-hard usage flexible cords or cables shall be permitted for use as permanent wiring on portable amusement rides and attractions where not subject to physical damage.

**(B) Single-Conductor.** Single-conductor cable shall be permitted only in sizes 2 AWG or larger.

**(C) Open Conductors.** Open conductors are prohibited except as part of a listed assembly or festoon lighting installed in accordance with Article 225.

**(D) Splices.** Flexible cords or cables shall be continuous without splice or tap between boxes or fittings.

**(E) Cord Connectors.** Cord connectors shall not be laid on the ground unless listed for wet locations. Connectors and cable connections shall not be placed in audience traffic paths or within areas accessible to the public unless guarded.

**(F) Support.** Wiring for an amusement ride, attraction, tent, or similar structure shall not be supported by any other ride or structure unless specifically designed for the purpose.

**(G) Protection.** Flexible cords or cables accessible to the public shall be arranged to minimize the tripping hazard and shall be permitted to be covered with nonconductive matting, provided that the matting does not constitute a greater tripping hazard than the uncovered cables. It shall be permitted to bury cables. The requirements of 300.5 shall not apply.

**(H) Boxes and Fittings.** A box or fitting shall be installed at each connection point, outlet, switchpoint, or junction point.

### 525.21 Rides, Tents and Concessions.

**(A) Disconnecting Means.** Each ride and concession shall be provided with a fused disconnect switch or circuit breaker located within sight and within 1.8 m (6 ft) of the operator's station. The disconnecting means shall be readily accessible to the operator, including when the ride is in operation. Where accessible to unqualified persons, the enclosure for the switch or circuit breaker shall be of the lockable type. A shunt trip device that opens the fused disconnect or circuit breaker when a switch located in the ride operator's console is closed shall be a permissible method of opening the circuit.

**(B) Portable Wiring Inside Tents and Concessions.** Electrical wiring for lighting, where installed inside of tents and concessions, shall be securely installed and, where subject to physical damage, shall be provided with mechanical protection. All lamps for general illumination shall be protected from accidental breakage by a suitable fixture or lampholder with a guard.

### 525.22 Portable Distribution or Termination Boxes.

Portable distribution or termination boxes shall comply with 525.22(A) through (D).

**(A) Construction.** Boxes shall be designed so that no live parts are exposed to accidental contact. Where installed outdoors, the box shall be of weatherproof construction and mounted so that the bottom of the enclosure is not less than 150 mm (6 in.) above the ground.

**(B) Busbars and Terminals.** Busbars shall have an ampere rating not less than the overcurrent device supplying the feeder supplying the box. Where conductors terminate directly on busbars, busbar connectors shall be provided.

**(C) Receptacles and Overcurrent Protection.** Receptacles shall have overcurrent protection installed within the box. The overcurrent protection shall not exceed the ampere rating of the receptacle, except as permitted in Article 430 for motor loads.

**(D) Single-Pole Connectors.** Where single-pole connectors are used, they shall comply with 530.22.

### 525.23 Ground-Fault Circuit-Interrupter (GFCI) Protection for Personnel.

**(A) General-Use 15- and 20-Ampere, 125-Volt Receptacles.** All 125-volt, single-phase, 15- and 20-ampere receptacle outlets that are in use by personnel shall have listed ground-fault circuit-interrupter protection for personnel. The ground-fault circuit interrupter shall be permitted to be an integral part of the attachment plug or located in the power-supply cord, within 300 mm (12 in.) of the attachment plug. For the purposes of this section, listed cord sets incorporating ground-fault circuit-interrupter protection for personnel shall be permitted. Egress lighting shall not be connected to the load side terminals of a ground-fault circuit-interrupter receptacle.

**(B) Appliance Receptacles.** Receptacles supplying items, such as cooking and refrigeration equipment, that are incompatible with ground-fault circuit-interrupter devices shall not be required to have ground-fault circuit-interrupter protection.

**(C) Other Receptacles.** Other receptacle outlets not covered in 525.23(A) or (B) shall be permitted to have ground-fault circuit-interrupter protection for personnel, or a written procedure shall be continuously enforced at the site by one or more designated persons to ensure the safety of equipment grounding conductors for all cord sets and receptacles, as described in 527.6(B)(2).

### IV. Grounding and Bonding

**525.30 Equipment Bonding.** The following equipment connected to the same source shall be bonded:

(1) Metal raceways and metal-sheathed cable
(2) Metal enclosures of electric equipment
(3) Metal frames and metal parts of rides, concessions, tents, trailers, trucks, or other equipment that contain or support electrical equipment

**525.31 Equipment Grounding.** All equipment requiring grounding shall be grounded by an equipment grounding conductor of a type and size recognized by 250.118 and installed in accordance with Article 250. The equipment grounding conductor shall be bonded to the system grounded conductor at the service disconnecting means or, in the case of a separately derived system such as a generator, at the generator or first disconnecting means supplied by the generator. The grounded circuit conductor shall not be connected to the equipment grounding conductor on the load side of the service disconnecting means or on the load side of a separately derived system disconnecting means.

**525.32 Grounding Conductor Continuity Assurance.** The continuity of the grounding conductor system used to reduce electrical shock hazards as required by 250.114, 250.138, 406.3(C), and 527.4(D) shall be verified each time that portable electrical equipment is connected.

# ARTICLE 527
# Temporary Installations

**527.1 Scope.** The provisions of this article apply to temporary electrical power and lighting installations.

## 527.2 All Wiring Installations.

**(A) Other Articles.** Except as specifically modified in this article, all other requirements of this *Code* for permanent wiring shall apply to temporary wiring installations.

**(B) Approval.** Temporary wiring methods shall be acceptable only if approved based on the conditions of use and any special requirements of the temporary installation.

## 527.3 Time Constraints.

**(A) During the Period of Construction.** Temporary electrical power and lighting installations shall be permitted during the period of construction, remodeling, maintenance, repair, or demolition of buildings, structures, equipment, or similar activities.

**(B) 90 Days.** Temporary electrical power and lighting installations shall be permitted for a period not to exceed 90 days for holiday decorative lighting and similar purposes.

**(C) Emergencies and Tests.** Temporary electrical power and lighting installations shall be permitted during emergencies and for tests, experiments, and developmental work.

**(D) Removal.** Temporary wiring shall be removed immediately upon completion of construction or purpose for which the wiring was installed.

## 527.4 General.

**(A) Services.** Services shall be installed in conformance with Article 230.

**(B) Feeders.** Feeders shall be protected as provided in Article 240. They shall originate in an approved distribution center. Conductors shall be permitted within cable assemblies or within multiconductor cords or cables of a type identified in Table 400.4 for hard usage or extra-hard usage. For the purpose of this section, Type NM and Type NMC cables shall be permitted to be used in any dwelling, building, or structure without any height limitation.

*Exception: Single insulated conductors shall be permitted where installed for the purpose(s) specified in 527.3(C), where accessible only to qualified persons.*

**(C) Branch Circuits.** All branch circuits shall originate in an approved power outlet or panelboard. Conductors shall be permitted within cable assemblies or within multiconductor cord or cable of a type identified in Table 400.4 for hard usage or extra-hard usage. All conductors shall be protected as provided in Article 240. For the purposes of this section, Type NM and Type NMC cables shall be permitted to be used in any dwelling, building, or structure without any height limitation.

*Exception: Branch circuits installed for the purposes specified in 527.3(B) or (C) shall be permitted to be run as single insulated conductors. Where the wiring is installed in accordance with 527.3(B), the voltage to ground shall not exceed 150 volts, the wiring shall not be subject to physical damage, and the conductors shall be supported on insulators at intervals of not more than 3.0 m (10 ft); or, for festoon lighting, the conductors shall be arranged so that excessive strain is not transmitted to the lampholders.*

**(D) Receptacles.** All receptacles shall be of the grounding type. Unless installed in a continuous grounded metal raceway or metal-covered cable, all branch circuits shall contain a separate equipment grounding conductor, and all receptacles shall be electrically connected to the equipment grounding conductors. Receptacles on construction sites shall not be installed on branch circuits that supply temporary lighting. Receptacles shall not be connected to the same ungrounded conductor of multiwire circuits that supply temporary lighting.

**(E) Disconnecting Means.** Suitable disconnecting switches or plug connectors shall be installed to permit the disconnection of all ungrounded conductors of each temporary circuit. Multiwire branch circuits shall be provided with a means to disconnect simultaneously all ungrounded conductors at the power outlet or panelboard where the branch circuit originated. Approved handle ties shall be permitted.

**(F) Lamp Protection.** All lamps for general illumination shall be protected from accidental contact or breakage by a suitable fixture or lampholder with a guard.

Brass shell, paper-lined sockets, or other metal-cased sockets shall not be used unless the shell is grounded.

**(G) Splices.** On construction sites, a box shall not be required for splices or junction connections where the circuit conductors are multiconductor cord or cable assemblies, provided that the equipment grounding continuity is maintained with or without the box. See 110.14(B) and 400.9. A box, conduit body, or terminal fitting having a separately bushed hole for each conductor shall be used wherever a change is made to a conduit or tubing system or a metal-sheathed cable system.

**(H) Protection from Accidental Damage.** Flexible cords and cables shall be protected from accidental damage. Sharp corners and projections shall be avoided. Where passing through doorways or other pinch points, protection shall be provided to avoid damage.

**(I) Termination(s) at Devices.** Flexible cords and cables entering enclosures containing devices requiring termination shall be secured to the box with fittings designed for the purpose.

**(J) Support.** Cable assemblies and flexible cords and cables shall be supported in place at intervals that ensure that they will be protected from physical damage. Support shall be in the form of staples, cable ties, straps, or similar type fittings installed so as not to cause damage. Vegetation shall not be used for support of overhead spans of branch circuits or feeders.

**527.6 Ground-Fault Protection for Personnel.** Ground-fault protection for personnel for all temporary wiring installations shall be provided to comply with 527.6(A) and (B). This section shall apply only to temporary wiring installations used to supply temporary power to equipment used by personnel during construction, remodeling, maintenance, repair, or demolition of buildings, structures, equipment, or similar activities.

**(A) Receptacle Outlets.** All 125-volt, single-phase, 15-, 20-, and 30-ampere receptacle outlets that are not a part of the permanent wiring of the building or structure and that are in use by personnel shall have ground-fault circuit interrupter protection for personnel. If a receptacle(s) is installed or exists as part of the permanent wiring of the building or structure and is used for temporary electric power, ground-fault circuit-interrupter protection for personnel shall be provided. For the purposes of this section, cord sets or devices incorporating listed ground-fault circuit interrupter protection for personnel identified for portable use shall be permitted.

*Exception: In industrial establishments only, where conditions of maintenance and supervision ensure that only qualified personnel are involved, an assured equipment grounding conductor program as specified in 527.6(B)(2) shall be permitted for only those receptacle outlets used to supply equipment that would create a greater hazard if power was interrupted or having a design that is not compatible with GFCI protection.*

**(B) Use of Other Outlets.** Receptacles other than 125-volt, single-phase, 15-, 20-, and 30-ampere receptacles shall have protection in accordance with (1) or, the assured equipment grounding conductor program in accordance with (2).

**(1) GFCI Protection.** Ground-fault circuit interrupter protection for personnel.

**(2) Assured Equipment Grounding Conductor Program.** A written assured equipment grounding conductor program continuously enforced at the site by one or more designated persons to ensure that equipment grounding conductors for all cord sets, receptacles that are not a part of the permanent wiring of the building or structure, and equipment connected by cord and plug are installed and maintained in accordance with the applicable requirements of 250.114, 250.138, 406.3(C), and 527.4(D).

(a) The following tests shall be performed on all cord sets, receptacles that are not part of the permanent wiring of the building or structure, and cord- and plug-connected equipment required to be grounded:

(1) All equipment grounding conductors shall be tested for continuity and shall be electrically continuous.
(2) Each receptacle and attachment plug shall be tested for correct attachment of the equipment grounding conductor. The equipment grounding conductor shall be connected to its proper terminal.
(3) All required tests shall be performed as follows:
   a. Before first use on site
   b. When there is evidence of damage
   c. Before equipment is returned to service following any repairs
   d. At intervals not exceeding 3 months

(b) The tests required in item (2)(a) shall be recorded and made available to the authority having jurisdiction.

**527.7 Guarding.** For wiring over 600 volts, nominal, suitable fencing, barriers, or other effective means shall be provided to limit access only to authorized and qualified personnel.

# ARTICLE 530
# Motion Picture and Television Studios and Similar Locations

## I. General

**530.1 Scope.** The requirements of this article shall apply to television studios and motion picture studios using either film or electronic cameras, except as provided in 520.1, and exchanges, factories, laboratories, stages, or a portion of the building in which film or tape more than 22 mm (⅞ in.) in width is exposed, developed, printed, cut, edited, rewound, repaired, or stored.

FPN: For methods of protecting against cellulose nitrate film hazards, see NFPA 40-1997, *Standard for the Storage and Handling of Cellulose Nitrate Motion Picture Film*.

## 530.2 Definitions.

**Alternating-Current Power Distribution Box (Alternating-Current Plugging Box, Scatter Box).** An ac distribution center or box that contains one or more grounding-type polarized receptacles that may contain overcurrent protection devices.

**Bull Switch.** An externally operated wall-mounted safety switch that may or may not contain overcurrent protection and is designed for the connection of portable cables and cords.

**Location (Shooting Location).** A place outside a motion picture studio where a production or part of it is filmed or recorded.

**Location Board (Deuce Board).** Portable equipment containing a lighting contactor or contactors and overcurrent protection designed for remote control of stage lighting.

**Motion Picture Studio (Lot).** A building or group of buildings and other structures designed, constructed, or permanently altered for use by the entertainment industry for the purpose of motion picture or television production.

**Portable Equipment.** Equipment intended to be moved from one place to another.

**Plugging Box.** A dc device consisting of one or more 2-pole, 2-wire, nonpolarized, nongrounding-type receptacles intended to be used on dc circuits only.

**Single-Pole Separable Connector.** A device that is installed at the ends of portable, flexible, single-conductor cable that is used to establish connection or disconnection between two cables or one cable and a single-pole, panel-mounted separable connector.

**Spider (Cable Splicing Block).** A device that contains busbars that are insulated from each other for the purpose of splicing or distributing power to portable cables and cords that are terminated with single-pole busbar connectors.

**Stage Effect (Special Effect).** An electrical or electromechanical piece of equipment used to simulate a distinctive visual or audible effect such as wind machines, lightning simulators, sunset projectors, and the like.

**Stage Property.** An article or object used as a visual element in a motion picture or television production, except painted backgrounds (scenery) and costumes.

**Stage Set.** A specific area set up with temporary scenery and properties designed and arranged for a particular scene in a motion picture or television production.

**Stand Lamp (Work Light).** A portable stand that contains a general-purpose luminaire (lighting fixture) or lampholder with guard for the purpose of providing general illumination in the studio or stage.

**Television Studio or Motion Picture Stage (Sound Stage).** A building or portion of a building usually insulated from the outside noise and natural light for use by the entertainment industry for the purpose of motion picture, television, or commercial production.

**530.6 Portable Equipment.** Portable stage and studio lighting equipment and portable power distribution equipment shall be permitted for temporary use outdoors if the equipment is supervised by qualified personnel while energized and barriered from the general public.

## II. Stage or Set

**530.11 Permanent Wiring.** The permanent wiring shall be Type MC cable, Type AC cable containing an insulated equipment grounding conductor sized in accordance with Table 250.122, Type MI cable, or in approved raceways.

*Exception: Communications circuits; audio signal processing, amplification, and reproduction circuits; Class 1, Class 2, and Class 3 remote-control or signaling circuits and power-limited fire alarm circuits shall be permitted to be wired in accordance with Articles 640, 725, 760, and 800.*

### 530.12 Portable Wiring.

**(A) Stage Set Wiring.** The wiring for stage set lighting and other supply wiring not fixed as to location shall be done with listed hard usage flexible cords and cables. Where subject to physical damage, such wiring shall be listed extra-hard usage flexible cords and cables. Splices or taps in cables shall be permitted if the total connected load does not exceed the maximum ampacity of the cable.

**(B) Stage Effects and Electrical Equipment Used as Stage Properties.** The wiring for stage effects and electrical equipment used as stage properties shall be permitted to be wired with single- or multiconductor listed flexible cords or cables if the conductors are protected from physical damage and secured to the scenery by approved cable ties or by insulated staples. Splices or taps shall be permitted where such are made with listed devices and the circuit is protected at not more than 20 amperes.

**(C) Other Electrical Equipment.** Cords and cables other than extra-hard usage, where supplied as a part of a listed assembly, shall be permitted.

**530.13 Stage Lighting and Effects Control.** Switches used for studio stage set lighting and effects (on the stages and lots and on location) shall be of the externally operable type. Where contactors are used as the disconnecting means for fuses, an individual externally operable switch, such as a tumbler switch, for the control of each contactor shall be located at a distance of not more than 1.8 m (6 ft) from the contactor, in addition to remote-control switches. A single externally operable switch shall be permitted to simultaneously disconnect all the contactors on any one location board, where located at a distance of not more than 1.8 m (6 ft) from the location board.

**530.14 Plugging Boxes.** Each receptacle of dc plugging boxes shall be rated at not less than 30 amperes.

**530.15 Enclosing and Guarding Live Parts.**

**(A) Live Parts.** Live parts shall be enclosed or guarded to prevent accidental contact by persons and objects.

**(B) Switches.** All switches shall be of the externally operable type.

**(C) Rheostats.** Rheostats shall be placed in approved cases or cabinets that enclose all live parts, having only the operating handles exposed.

**(D) Current-Carrying Parts.** Current-carrying parts of bull switches, location boards, spiders, and plugging boxes shall be enclosed, guarded, or located so that persons cannot accidentally come into contact with them or bring conductive material into contact with them.

**530.16 Portable Lamps.** Portable lamps and work lights shall be equipped with flexible cords, composition or metal-sheathed porcelain sockets, and substantial guards.

*Exception: Portable lamps used as properties in a motion picture set or television stage set, on a studio stage or lot, or on location shall not be considered to be portable lamps for the purpose of this section.*

**530.17 Portable Arc Lamps.**

**(A) Portable Carbon Arc Lamps.** Portable carbon arc lamps shall be substantially constructed. The arc shall be provided with an enclosure designed to retain sparks and carbons and to prevent persons or materials from coming into contact with the arc or bare live parts. The enclosures shall be ventilated. All switches shall be of the externally operable type.

**(B) Portable Noncarbon Arc Electric-Discharge Lamps.** Portable noncarbon arc lamps, including enclosed arc lamps, and associated ballasts shall be listed. Interconnecting cord sets and interconnecting cords and cables shall be extra-hard usage type and listed.

**530.18 Overcurrent Protection — General.** Automatic overcurrent protective devices (circuit breakers or fuses) for motion picture studio stage set lighting and the stage cables for such stage set lighting shall be as given in 530.18(A) through (G). The maximum ampacity allowed on a given conductor, cable, or cord size shall be as given in the applicable tables of Articles 310 and 400.

**(A) Stage Cables.** Stage cables for stage set lighting shall be protected by means of overcurrent devices set at not more than 400 percent of the ampacity given in the applicable tables of Articles 310 and 400.

**(B) Feeders.** In buildings used primarily for motion picture production, the feeders from the substations to the stages shall be protected by means of overcurrent devices (generally located in the substation) having a suitable ampere rating. The overcurrent devices shall be permitted to be multipole or single-pole gang operated. No pole shall be required in the neutral conductor. The overcurrent device setting for each feeder shall not exceed 400 percent of the ampacity of the feeder, as given in the applicable tables of Article 310.

**(C) Cable Protection.** Cables shall be protected by bushings where they pass through enclosures and shall be arranged so that tension on the cable is not transmitted to the connections. Where power conductors pass through metal, the requirements of 300.20 shall apply.

Portable feeder cables shall be permitted to temporarily penetrate fire-rated walls, floors, or ceilings provided that all of the following apply:

(1) The opening is of noncombustible material.
(2) When in use, the penetration is sealed with a temporary seal of a listed firestop material.
(3) When not in use, the opening shall be capped with a material of equivalent fire rating.

**(D) Location Boards.** Overcurrent protection (fuses or circuit breakers) shall be provided at the location boards. Fuses in the location boards shall have an ampere rating of not over 400 percent of the ampacity of the cables between the location boards and the plugging boxes.

**(E) Plugging Boxes.** Cables and cords supplied through plugging boxes shall be of copper. Cables and cords smaller than 8 AWG shall be attached to the plugging box by means of a plug containing two cartridge fuses or a 2-pole circuit breaker. The rating of the fuses or the setting of the circuit breaker shall not be over 400 percent of the rated ampacity of the cables or cords as given in the applicable tables of Articles 310 and 400. Plugging boxes shall not be permitted on ac systems.

**(F) Alternating-Current Power Distribution Boxes.** Alternating-current power distribution boxes used on sound stages and shooting locations shall contain connection receptacles of a polarized, grounding type.

**(G) Lighting.** Work lights, stand lamps, and luminaires (fixtures) rated 1000 watts or less and connected to dc plugging boxes shall be by means of plugs containing two cartridge fuses not larger than 20 amperes, or they shall be permitted to be connected to special outlets on circuits protected by fuses or circuit breakers rated at not over 20 amperes. Plug fuses shall not be used unless they are on the load side of the fuse or circuit breakers on the location boards.

**530.19 Sizing of Feeder Conductors for Television Studio Sets.**

**(A) General.** It shall be permissible to apply the demand factors listed in Table 530.19(A) to that portion of the maximum possible connected load for studio or stage set lighting for all permanently installed feeders between substations and stages and to all permanently installed feeders between the main stage switchboard and stage distribution centers or location boards.

**(B) Portable Feeders.** A demand factor of 50 percent of maximum possible connected load shall be permitted for all portable feeders.

**Table 530.19(A) Demand Factors for Stage Set Lighting**

| Portion of Stage Set Lighting Load to Which Demand Factor Applied (volt-amperes) | Feeder Demand Factor |
|---|---|
| First 50,000 or less at | 100% |
| From 50,001 to 100,000 at | 75% |
| From 100,001 to 200,000 at | 60% |
| Remaining over 200,000 at | 50% |

**530.20 Grounding.** Type MC cable, Type MI cable, metal raceways, and all non–current-carrying metal parts of appliances, devices, and equipment shall be grounded as specified in Article 250. This shall not apply to pendant and portable lamps, to stage lighting and stage sound equipment, or to other portable and special stage equipment operating at not over 150 volts dc to ground.

**530.21 Plugs and Receptacles.**

**(A) Rating.** Plugs and receptacles shall be rated in amperes. The voltage rating of the plugs and receptacles shall not be less than the circuit voltage. Plug and receptacle ampere ratings for ac circuits shall not be less than the feeder or branch-circuit overcurrent device ampere rating. Table 210.21(B)(2) shall not apply.

**(B) Interchangeability.** Plugs and receptacles used in portable professional motion picture and television equipment shall be permitted to be interchangeable for ac or dc use on the same premises provided they are listed for ac/dc use and marked in a suitable manner to identify the system to which they are connected.

**530.22 Single-Pole Separable Connectors.**

**(A) General.** Where ac single-pole portable cable connectors are used, they shall be listed and of the locking type. Sections 400.10, 406.6, and 406.7 shall not apply to listed single-pole separable connections and single-conductor cable assemblies utilizing listed single-pole separable connectors. Where paralleled sets of current-carrying single-pole separable connectors are provided as input devices, they shall be prominently labeled with a warning indicating the presence of internal parallel connections. The use of single-pole separable connectors shall comply with at least one of the following conditions:

(1) Connection and disconnection of connectors are only possible where the supply connectors are interlocked to the source and it is not possible to connect or disconnect connectors when the supply is energized.
(2) Line connectors are of the listed sequential-interlocking type so that load connectors shall be connected in the following sequence:
  a. Equipment grounding conductor connection
  b. Grounded circuit conductor connection, if provided
  c. Ungrounded conductor connection, and that disconnection shall be in the reverse order
(3) A caution notice shall be provided adjacent to the line connectors, indicating that plug connection shall be in the following order:
  a. Equipment grounding conductor connectors
  b. Grounded circuit-conductor connectors, if provided
  c. Ungrounded conductor connectors, and that disconnection shall be in the reverse order

**(B) Interchangeability.** Single-pole separable connectors used in portable professional motion picture and television equipment shall be permitted to be interchangeable for ac or dc use or for different current ratings on the same premises, provided they are listed for ac/dc use and marked in a suitable manner to identify the system to which they are connected.

**530.23 Branch Circuits.** A branch circuit of any size supplying one or more receptacles shall be permitted to supply stage set lighting loads.

**III. Dressing Rooms**

**530.31 Dressing Rooms.** Fixed wiring in dressing rooms shall be installed in accordance with the wiring methods covered in Chapter 3. Wiring for portable dressing rooms shall be approved.

## IV. Viewing, Cutting, and Patching Tables

**530.41 Lamps at Tables.** Only composition or metal-sheathed, porcelain, keyless lampholders equipped with suitable means to guard lamps from physical damage and from film and film scrap shall be used at patching, viewing, and cutting tables.

## V. Cellulose Nitrate Film Storage Vaults

**530.51 Lamps in Cellulose Nitrate Film Storage Vaults.** Lamps in cellulose nitrate film storage vaults shall be installed in rigid fixtures of the glass-enclosed and gasketed type. Lamps shall be controlled by a switch having a pole in each ungrounded conductor. This switch shall be located outside of the vault and provided with a pilot light to indicate whether the switch is on or off. This switch shall disconnect from all sources of supply all ungrounded conductors terminating in any outlet in the vault.

**530.52 Electrical Equipment in Cellulose Nitrate Film Storage Vaults.** Except as permitted in 530.51, no receptacles, outlets, heaters, portable lights, or other portable electric equipment shall be located in cellulose nitrate film storage vaults. Electric motors shall be permitted, provided they are listed for the application and comply with Article 500, Class I, Division 2.

## VI. Substations

**530.61 Substations.** Wiring and equipment of over 600 volts, nominal, shall comply with Article 490.

**530.62 Portable Substations.** Wiring and equipment in portable substations shall conform to the sections applying to installations in permanently fixed substations, but, due to the limited space available, the working spaces shall be permitted to be reduced, provided that the equipment shall be arranged so that the operator can work safely and so that other persons in the vicinity cannot accidentally come into contact with current-carrying parts or bring conducting objects into contact with them while they are energized.

**530.63 Overcurrent Protection of Direct-Current Generators.** Three-wire generators shall have overcurrent protection in accordance with 445.12(E).

**530.64 Direct-Current Switchboards.**

**(A) General.** Switchboards of not over 250 volts dc between conductors, where located in substations or switch-board rooms accessible to qualified persons only, shall not be required to be dead-front.

**(B) Circuit Breaker Frames.** Frames of dc circuit breakers installed on switchboards shall not be required to be grounded.

## ARTICLE 540
## Motion Picture Projection Rooms

### I. General

**540.1 Scope.** The provisions of this article apply to motion picture projection rooms, motion picture projectors, and associated equipment of the professional and nonprofessional types using incandescent, carbon arc, xenon, or other light source equipment that develops hazardous gases, dust, or radiation.

> FPN: For further information, see NFPA 40-1997, *Standard for the Storage and Handling of Cellulose Nitrate Motion Picture Film.*

**540.2 Definitions.**

**Nonprofessional Projector.** Nonprofessional projectors are those types other than as described in 540.2.

**Professional Projector.** A type of projector using 35- or 70-mm film that has a minimum width of 35 mm (1⅜ in.) and has on each edge 212 perforations per meter (5.4 perforations per inch), or a type using carbon arc, xenon, or other light source equipment that develops hazardous gases, dust, or radiation.

### II. Equipment and Projectors of the Professional Type

**540.10 Motion Picture Projection Room Required.** Every professional-type projector shall be located within a projection room. Every projection room shall be of permanent construction, approved for the type of building in which the projection room is located. All projection ports, spotlight ports, viewing ports, and similar openings shall be provided with glass or other approved material so as to completely close the opening. Such rooms shall not be considered as hazardous (classified) locations as defined in Article 500.

> FPN: For further information on protecting openings in projection rooms handling cellulose nitrate motion picture film, see NFPA *101*-2000, *Life Safety Code.*

**540.11 Location of Associated Electrical Equipment.**

**(A) Motor Generator Sets, Transformers, Rectifiers, Rheostats, and Similar Equipment.** Motor generator sets, transformers, rectifiers, rheostats, and similar equipment for the supply or control of current to projection or spotlight equipment shall, where nitrate film is used, be located in a separate room. Where placed in the projection room, they shall be located or guarded so that arcs or sparks cannot come in contact with film, and the commutator end or ends of motor generator sets shall comply with one of the conditions in 540.11(A)(1) through (A)(6).

**(1) Types.** Be of the totally enclosed, enclosed fan-cooled, or enclosed pipe-ventilated type.

**(2) Separate Rooms or Housings.** Be enclosed in separate rooms or housings built of noncombustible material constructed so as to exclude flyings or lint, and properly ventilated from a source of clean air.

**(3) Solid Metal Covers.** Have the brush or sliding-contact end of motor-generator enclosed by solid metal covers.

**(4) Tight Metal Housings.** Have brushes or sliding contacts enclosed in substantial, tight metal housings.

**(5) Upper and Lower Half Enclosures.** Have the upper half of the brush or sliding-contact end of the motor-generator enclosed by a wire screen or perforated metal and the lower half enclosed by solid metal covers.

**(6) Wire Screens or Perforated Metal.** Have wire screens or perforated metal placed at the commutator of brush ends. No dimension of any opening in the wire screen or perforated metal shall exceed 1.27 mm (0.05 in.), regardless of the shape of the opening and of the material used.

**(B) Switches, Overcurrent Devices, or Other Equipment.** Switches, overcurrent devices, or other equipment not normally required or used for projectors, sound reproduction, flood or other special effect lamps, or other equipment shall not be installed in projection rooms.

*Exception No. 1: In projection rooms approved for use only with cellulose acetate (safety) film, the installation of appurtenant electrical equipment used in conjunction with the operation of the projection equipment and the control of lights, curtains, and audio equipment, and so forth, shall be permitted. In such projection rooms, a sign reading "Safety Film Only Permitted in This Room" shall be posted on the outside of each projection room door and within the projection room itself in a conspicuous location.*

*Exception No. 2: Remote-control switches for the control of auditorium lights or switches for the control of motors operating curtains and masking of the motion picture screen shall be permitted to be installed in projection rooms.*

**(C) Emergency Systems.** Control of emergency systems shall comply with Article 700.

**540.12 Work Space.** Each motion picture projector, floodlight, spotlight, or similar equipment shall have clear working space not less than 750 mm (30 in.) wide on each side and at the rear thereof.

*Exception: One such space shall be permitted between adjacent pieces of equipment.*

**540.13 Conductor Size.** Conductors supplying outlets for arc and xenon projectors of the professional type shall not be smaller than 8 AWG and shall be of sufficient size for the projector employed. Conductors for incandescent-type projectors shall conform to normal wiring standards as provided in 210.24.

**540.14 Conductors on Lamps and Hot Equipment.** Insulated conductors having a rated operating temperature of not less than 200°C (392°F) shall be used on all lamps or other equipment where the ambient temperature at the conductors as installed will exceed 50°C (122°F).

**540.15 Flexible Cords.** Cords approved for hard usage, as provided in Table 400.4, shall be used on portable equipment.

**540.20 Approval.** Projectors and enclosures for arc, xenon and incandescent lamps and rectifiers, transformers, rheostats, and similar equipment shall be listed.

**540.21 Marking.** Projectors and other equipment shall be marked with the manufacturer's name or trademark and with the voltage and current for which they are designed in accordance with 110.21.

### III. Nonprofessional Projectors

**540.31 Motion Picture Projection Room Not Required.** Projectors of the nonprofessional or miniature type, where employing cellulose acetate (safety) film, shall be permitted to be operated without a projection room.

**540.32 Approval.** Projection equipment shall be listed.

### IV. Audio Signal Processing, Amplification, and Reproduction Equipment

**540.50 Audio Signal Processing, Amplification, and Reproduction Equipment.** Audio signal processing, amplification, and reproduction equipment shall be installed as provided in Article 640.

# ARTICLE 545
## Manufactured Buildings

**545.1 Scope.** This article covers requirements for a manufactured building and building components as herein defined.

**545.2 Other Articles.** Wherever the requirements of other articles of this *Code* and Article 545 differ, the requirements of Article 545 shall apply.

**545.3 Definitions.**

**Building Component.** Any subsystem, subassembly, or other system designed for use in or integral with or as part of a structure, which can include structural, electrical, mechanical, plumbing, and fire protection systems, and other systems affecting health and safety.

**Building System.** Plans, specifications, and documentation for a system of manufactured building or for a type or a system of building components, which can include structural, electrical, mechanical, plumbing, and fire protection systems, and other systems affecting health and safety, and including such variations thereof as are specifically permitted by regulation, and which variations are submitted as part of the building system or amendment thereto.

**Closed Construction.** Any building, building component, assembly, or system manufactured in such a manner that all concealed parts of processes of manufacture cannot be inspected before installation at the building site without disassembly, damage, or destruction.

**Manufactured Building.** Any building that is of closed construction and is made or assembled in manufacturing facilities on or off the building site for installation, or for assembly and installation on the building site, other than manufactured homes, mobile homes, park trailers, or recreational vehicles.

**545.4 Wiring Methods.**

**(A) Methods Permitted.** All raceway and cable wiring methods included in this *Code* and such other wiring systems specifically intended and listed for use in manufactured buildings shall be permitted with listed fittings and with fittings listed and identified for manufactured buildings.

**(B) Securing Cables.** In closed construction, cables shall be permitted to be secured only at cabinets, boxes, or fittings where 10 AWG or smaller conductors are used and protection against physical damage is provided.

**545.5 Supply Conductors.** Provisions shall be made to route the service-entrance, service-lateral, feeder, or branch-circuit supply to the service or building disconnecting means conductors.

**545.6 Installation of Service-Entrance Conductors.** Service-entrance conductors shall be installed after erection at the building site.

*Exception: Where point of attachment is known prior to manufacture.*

**545.7 Service Equipment.** Service equipment shall be installed in accordance with 230.70.

**545.8 Protection of Conductors and Equipment.** Protection shall be provided for exposed conductors and equipment during processes of manufacturing, packaging, in transit, and erection at the building site.

**545.9 Boxes.**

**(A) Other Dimensions.** Boxes of dimensions other than those required in Table 314.16(A) shall be permitted to be installed where tested, identified, and listed to applicable standards.

**(B) Not Over 1650 cm³ (100 in.³).** Any box not over 1650 cm³ (100 in.³) in size, intended for mounting in closed construction, shall be affixed with anchors or clamps so as to provide a rigid and secure installation.

**545.10 Receptacle or Switch with Integral Enclosure.** A receptacle or switch with integral enclosure and mounting means, where tested, identified, and listed to applicable standards, shall be permitted to be installed.

**545.11 Bonding and Grounding.** Prewired panels and building components shall provide for the bonding, or bonding and grounding, of all exposed metals likely to become energized, in accordance with Article 250, Parts V, VI, and VII.

**545.12 Grounding Electrode Conductor.** Provisions shall be made to route a grounding electrode conductor from the service, feeder, or branch-circuit supply to the point of attachment to the grounding electrode.

**545.13 Component Interconnections.** Fittings and connectors that are intended to be concealed at the time of on-site assembly, where tested, identified, and listed to applicable standards, shall be permitted for on-site interconnection of modules or other building components. Such fittings and connectors shall be equal to the wiring method employed in insulation, temperature rise, and fault-current withstand and shall be capable of enduring the vibration and minor relative motions occurring in the components of manufactured building.

# ARTICLE 547
## Agricultural Buildings

**547.1 Scope.** The provisions of this article shall apply to the following agricultural buildings or that part of a building or adjacent areas of similar or like nature as specified in 547.1(A) and (B).

**(A) Excessive Dust and Dust with Water.** Agricultural buildings where excessive dust and dust with water may accumulate, including all areas of poultry, livestock, and fish confinement systems, where litter dust or feed dust, including mineral feed particles, may accumulate.

**(B) Corrosive Atmosphere.** Agricultural buildings where a corrosive atmosphere exists. Such buildings include areas where the following conditions exist:

(1) Poultry and animal excrement may cause corrosive vapors.
(2) Corrosive particles may combine with water.
(3) The area is damp and wet by reason of periodic washing for cleaning and sanitizing with water and cleansing agents.
(4) Similar conditions exist.

**547.2 Definitions.**

**Distribution Point.** An electrical supply point from which service drops, service laterals, feeders, or branch circuits to agricultural buildings, associated farm dwelling(s), and associated buildings under single management are supplied.

FPN No. 1: Distribution points are also known as the center yard pole, meterpole or the common distribution point.

FPN No. 2: The service point as defined in Article 100 is typically at the distribution point.

**Equipotential Plane.** An area where wire mesh or other conductive elements are embedded in or placed under concrete, bonded to all metal structures and fixed nonelectrical equipment that may become energized, and connected to the electrical grounding system to prevent a difference in voltage from developing within the plane.

**547.3 Other Articles.** For agricultural buildings not having conditions as specified in 547.1, the electrical installa-

tions shall be made in accordance with the applicable articles in this *Code*.

**547.4 Surface Temperatures.** Electrical equipment or devices installed in accordance with the provisions of this article shall be installed in a manner such that they will function at full rating without developing surface temperatures in excess of the specified normal safe operating range of the equipment or device.

**547.5 Wiring Methods.**

**(A) Wiring Systems.** Types UF, NMC, copper SE cables, jacketed Type MC cable, rigid nonmetallic conduit, liquidtight flexible nonmetallic conduit, or other cables or raceways suitable for the location, with approved termination fittings, shall be the wiring methods employed. Article 398 and Article 502 wiring methods shall be permitted for areas described in 547.1(A).

FPN: See 300.7 and 352.44 for installation of raceway systems exposed to widely different temperatures.

**(B) Mounting.** All cables shall be secured within 200 mm (8 in.) of each cabinet, box, or fitting. The 6-mm (¼-in.) airspace required for nonmetallic boxes, fittings, conduit, and cables in 300.6(C) shall not be required in buildings covered by this article.

**(C) Equipment Enclosures, Boxes, Conduit Bodies, and Fittings.**

**(1) Excessive Dust.** Equipment enclosures, boxes, conduit bodies, and fittings installed in areas of buildings where excessive dust may be present shall be designed to minimize the entrance of dust and shall have no openings (such as holes for attachment screws) through which dust could enter the enclosure.

**(2) Damp or Wet Locations.** In damp or wet locations, equipment enclosures, boxes, conduit bodies, and fittings shall be placed or equipped so as to prevent moisture from entering or accumulating within the enclosure, box, conduit body, or fitting. In wet locations, including normally dry or damp locations where surfaces are periodically washed or sprayed with water, boxes, conduit bodies, and fittings shall be listed for use in wet locations and equipment enclosures shall be weatherproof.

**(3) Corrosive Atmosphere.** Where wet dust, excessive moisture, corrosive gases or vapors, or other corrosive conditions may be present, equipment enclosures, boxes, conduit bodies, and fittings shall have corrosion resistance properties suitable for the conditions.

FPN No. 1: See Table 430.91 for appropriate enclosure type designations.

FPN No. 2:  Aluminum and magnetic ferrous materials may corrode in agricultural environments.

**(D) Flexible Connections.** Where necessary to employ flexible connections, dusttight flexible connectors, liquidtight flexible conduit, or flexible cord listed and identified for hard usage shall be used. All connectors and fittings used shall be listed and identified for the purpose.

**(E) Physical Protection.** All electrical wiring and equipment subject to physical damage shall be protected.

**(F) Separate Equipment Grounding Conductor.** Non–current-carrying metal parts of equipment, raceways, and other enclosures, where required to be grounded, shall be grounded by a copper equipment grounding conductor installed between the equipment and the building disconnecting means. If installed underground, the equipment grounding conductor shall be insulated or covered.

**(G) Receptacles.** All 125-volt, single-phase, 15- and 20-ampere general-purpose receptacles installed in the following locations shall have ground-fault circuit-interrupter protection for personnel:

(1)  In areas having an equipotential plane
(2)  Outdoors
(3)  Damp or wet locations

**547.6 Switches, Receptacles, Circuit Breakers, Controllers, and Fuses.** Switches, including pushbuttons, relays, and similar devices, receptacles, circuit breakers, controllers, and fuses, shall be provided with enclosures as specified in 547.5(C).

**547.7 Motors.** Motors and other rotating electrical machinery shall be totally enclosed or designed so as to minimize the entrance of dust, moisture, or corrosive particles.

**547.8 Luminaires (Lighting Fixtures).** Luminaires (lighting fixtures) shall comply with the following.

**(A) Minimize the Entrance of Dust.** Luminaires (lighting fixtures) shall be installed to minimize the entrance of dust, foreign matter, moisture, and corrosive material.

**(B) Exposed to Physical Damage.** Any luminaire (lighting fixture) that may be exposed to physical damage shall be protected by a suitable guard.

**(C) Exposed to Water.** A luminaire (fixture) that may be exposed to water from condensation, building cleansing water, or solution shall be watertight.

**547.9 Electrical Supply to Building or Structures from a Distribution Point.**

**(A) Site-Isolating Device.** A disconnecting means shall be installed at the distribution point where two or more agri-cultural buildings, structures, associated farm dwelling(s), or other buildings are supplied from the distribution point. For the purposes of applying the requirements of this section, this disconnecting means shall be classified as a site-isolating device and shall have provisions for bonding the grounding electrode conductor to the grounded conductor.

**(1) Purpose.** The disconnecting means shall simultaneously interrupt all ungrounded conductors for the purposes of isolation, system maintenance, emergency disconnection, or connection of optional standby systems.

**(2) Series Disconnects.** An additional disconnecting means shall not be required where the serving utility provides a disconnecting means as part of their service requirements and this disconnecting means is accessible to the user and meets the requirements of this section.

**(3) Rating.** The disconnecting means shall be rated for the calculated load as determined by Part IV of Article 220.

**(4) Overcurrent.** The disconnecting means shall not be required to contain overload protection.

**(5) Accessibility.** Where not readily accessible, the disconnecting means shall be capable of operation from a readily accessible point.

**(6) Grounding.** The grounded conductor of the system shall be connected to a grounding electrode through a grounding electrode conductor at the disconnecting means.

**(B) Electrical Supply.** The buildings or structures shall be permitted to be supplied by either 547.9(B)(1) or (B)(2).

**(1) Building(s) or Structure(s).** Where the disconnecting means and overcurrent protection are located at the buildings or structures, the supply conductors shall be sized in accordance with Part IV of Article 220 and installed in accordance with the requirements of Part II of Article 225.

For each building or structure, the conditions in either (a) or (b) shall be permitted.

(a)  The grounded circuit conductor shall be permitted to be connected to the building disconnecting means and to the grounding electrode system of that building or structure where all the requirements of 250.32(B)(2) are met.

(b)  A separate equipment grounding conductor shall be run with the supply conductors to the building(s) or structure(s) and the following conditions shall be met:

(1)  The equipment grounding conductor is the same size as the largest supply conductor, if of the same material, or is adjusted in size in accordance with the equivalent size columns of Table 250.122 if of different materials.
(2)  The equipment grounding conductor is bonded to the grounded circuit conductor at the disconnecting means

enclosure at the distribution point or at the source of a separately derived system.

(3) A grounding electrode system is provided in accordance with Part III of Article 250 and connected to the equipment grounding conductor at the building(s) or structure(s) disconnecting means.

(4) The grounded circuit conductor is not connected to a grounding electrode or to any equipment grounding conductor on the load side of the distribution point.

**(2) Disconnecting Means and Overcurrent Protection at the Distribution Point.** Where the disconnecting means and overcurrent protection for each set of feeder conductors are located at the distribution point, feeders to building(s) or structure(s) shall meet the requirements of 250.32 and Article 225, Parts I and II.

> FPN: Methods to reduce neutral-to-earth voltages in livestock facilities include supplying buildings or structures with 4-wire, single-phase services, sizing of 3-wire service conductors to limit voltage drop to 2 percent, and connecting loads line-to-line.

**(C) Underground Equipment Grounding Conductors.** Where livestock is housed, any portion of the equipment grounding conductor run underground to the building or structure shall be insulated or covered copper.

**547.10 Equipotential Planes and Bonding of Equipotential Planes.** For the purposes of this section, the term *livestock* shall not include poultry.

**(A) Areas Requiring Equipotential Planes.** Equipotential planes shall be installed in all concrete floor confinement areas of livestock buildings that contain metallic equipment that is accessible to animals and likely to become energized. Outdoor confinement areas, such as feedlots, shall have equipotential planes installed around metallic equipment that is accessible to animals and likely to become energized. The equipotential plane shall encompass the area around the equipment where the animal stands while accessing the equipment.

**(B) Areas Not Requiring Equipotential Planes.** Equipotential planes shall not be required in dirt confinement areas containing metallic equipment that is accessible to animals and likely to become energized. All circuits providing electric power to equipment that is accessible to animals in dirt confinement areas shall have GFCI protection.

**(C) Bonding.** Equipotential planes shall be bonded to the electrical grounding system. The bonding conductor shall be copper, insulated, covered or bare, and not smaller and 8 AWG. The means of bonding to wire mesh or conductive elements shall be by pressure connectors or clamps of brass, copper, copper alloy, or an equally substantial approved means. Slatted floors that are supported by structures that are a part of an equipotential plane shall not require bonding.

> FPN No. 1: Methods to establish equipotential planes are described in American Society of Agricultural Engineers (ASAE) EP473-2001, *Equipotential Planes in Animal Containment Areas.*
>
> FPN No. 2: Low grounding electrode system resistances may reduce potential differences in livestock facilities.

## ARTICLE 550
## Mobile Homes, Manufactured Homes, and Mobile Home Parks

### I. General

**550.1 Scope.** The provisions of this article cover the electrical conductors and equipment installed within or on mobile and manufactured homes, the conductors that connect mobile and manufactured homes to a supply of electricity, and the installation of electrical wiring, luminaires (fixtures), equipment, and appurtenances related to electrical installations within a mobile home park up to the mobile home service-entrance conductors or, if none, the mobile home service equipment.

> FPN: For additional information on manufactured housing see NFPA 501-1999, *Standard on Manufactured Housing,* and Part 3280, *Manufactured Home Construction and Safety Standards,* of the Federal Department of Housing and Urban Development.

**550.2 Definitions.**

**Appliance, Fixed.** An appliance that is fastened or otherwise secured at a specific location.

**Appliance, Portable.** An appliance that is actually moved or can easily be moved from one place to another in normal use.

> FPN: For the purpose of this article, the following major appliances, other than built-in, are considered portable if cord connected: refrigerators, range equipment, clothes washers, dishwashers without booster heaters, or other similar appliances.

**Appliance, Stationary.** An appliance that is not easily moved from one place to another in normal use.

**Distribution Panelboard.** See definition of panelboard in Article 100.

**Feeder Assembly.** The overhead or under-chassis feeder conductors, including the grounding conductor, together with the necessary fittings and equipment or a power-supply cord listed for mobile home use, designed for the purpose of delivering energy from the source of electrical supply to the distribution panelboard within the mobile home.

**Laundry Area.** An area containing or designed to contain a laundry tray, clothes washer, or a clothes dryer.

**Manufactured Home.** A structure, transportable in one or more sections, that is 25 m (8 body ft) or more in width or 12 m (40 body ft) or more in length in the traveling mode or, when erected on site, is 30 m$^2$ (320 ft$^2$) or more; which is built on a chassis and designed to be used as a dwelling, with or without a permanent foundation, when connected to the required utilities, including the plumbing, heating, air conditioning, and electrical systems contained therein. Calculations used to determine the number of square meters (square feet) in a structure will be based on the structure's exterior dimensions, measured at the largest horizontal projections when erected on site. These dimensions include all expandable rooms, cabinets, and other projections containing interior space, but do not include inside bay windows.

For the purpose of this *Code* and unless otherwise indicated, the term *mobile home* includes manufactured homes.

> FPN No. 1: See the applicable building code for definition of the term *permanent foundation*.
>
> FPN No. 2: See Part 3280, *Manufactured Home Construction and Safety Standards*, of the Federal Department of Housing and Urban Development, for additional information on the definition.

**Mobile Home.** A factory-assembled structure or structures transportable in one or more sections that is built on a permanent chassis and designed to be used as a dwelling without a permanent foundation where connected to the required utilities and that includes the plumbing, heating, air-conditioning, and electric systems contained therein.

For the purpose of this *Code* and unless otherwise indicated, the term *mobile home* includes manufactured homes.

**Mobile Home Accessory Building or Structure.** Any awning, cabana, ramada, storage cabinet, carport, fence, windbreak, or porch established for the use of the occupant of the mobile home on a mobile home lot.

**Mobile Home Lot.** A designated portion of a mobile home park designed for the accommodation of one mobile home and its accessory buildings or structures for the exclusive use of its occupants.

**Mobile Home Park.** A contiguous parcel of land that is used for the accommodation of occupied mobile homes.

**Mobile Home Service Equipment.** The equipment containing the disconnecting means, overcurrent protective devices, and receptacles or other means for connecting a mobile home feeder assembly.

**Park Electrical Wiring Systems.** All of the electrical wiring, luminaires (fixtures), equipment, and appurtenances related to electrical installations within a mobile home park, including the mobile home service equipment.

**550.3 Other Articles.** Wherever the requirements of other articles of this *Code* and Article 550 differ, the requirements of Article 550 shall apply.

**550.4 General Requirements.**

**(A) Mobile Home Not Intended as a Dwelling Unit.** A mobile home not intended as a dwelling unit — for example, those equipped for sleeping purposes only, contractor's on-site offices, construction job dormitories, mobile studio dressing rooms, banks, clinics, mobile stores, or intended for the display or demonstration of merchandise or machinery — shall not be required to meet the provisions of this article pertaining to the number or capacity of circuits required. It shall, however, meet all other applicable requirements of this article if provided with an electrical installation intended to be energized from a 120-volt or 120/240-volt ac power supply system. Where different voltage is required by either design or available power supply system, adjustment shall be made in accordance with other articles and sections for the voltage used.

**(B) In Other Than Mobile Home Parks.** Mobile homes installed in other than mobile home parks shall comply with the provisions of this article.

**(C) Connection to Wiring System.** The provisions of this article shall apply to mobile homes intended for connection to a wiring system rated 120/240 volts, nominal, 3-wire ac, with grounded neutral.

**(D) Listed or Labeled.** All electrical materials, devices, appliances, fittings, and other equipment shall be listed or labeled by a qualified testing agency and shall be connected in an approved manner when installed.

## II. Mobile and Manufactured Homes

**550.10 Power Supply.**

**(A) Feeder.** The power supply to the mobile home shall be a feeder assembly consisting of not more than one listed 50-ampere mobile home power-supply cord with an integrally molded or securely attached plug cap or a permanently installed feeder.

*Exception No. 1: A mobile home that is factory equipped with gas or oil-fired central heating equipment and cooking appliances shall be permitted to be provided with a listed mobile home power-supply cord rated 40 amperes.*

*Exception No. 2: Manufactured homes constructed in accordance with 550.32(B).*

**(B) Power-Supply Cord.** If the mobile home has a power-supply cord, it shall be permanently attached to the distribution panelboard or to a junction box permanently connected to the distribution panelboard, with the free end terminating in an attachment plug cap.

Cords with adapters and pigtail ends, extension cords, and similar items shall not be attached to, or shipped with, a mobile home.

A suitable clamp or the equivalent shall be provided at the distribution panelboard knockout to afford strain relief for the cord to prevent strain from being transmitted to the terminals when the power-supply cord is handled in its intended manner.

The cord shall be a listed type with four conductors, one of which shall be identified by a continuous green color or a continuous green color with one or more yellow stripes for use as the grounding conductor.

**(C) Attachment Plug Cap.** The attachment plug cap shall be a 3-pole, 4-wire, grounding type, rated 50 amperes, 125/250 volts with a configuration as shown in Figure 550.10(C) and intended for use with the 50-ampere, 125/250-volt receptacle configuration shown in Figure 550.10(C). It shall be listed, by itself or as part of a power-supply cord assembly, for the purpose and shall be molded to or installed on the flexible cord so that it is secured tightly to the cord at the point where the cord enters the attachment plug cap. If a right-angle cap is used, the configuration shall be oriented so that the grounding member is farthest from the cord.

> FPN: Complete details of the 50-ampere plug and receptacle configuration can be found in the National Electrical Manufacturers Association *Standard for Dimensions of Attachment Plugs and Receptacles*, ANSI/NEMA WD 6-1989, Figure 14-50.

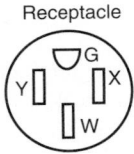  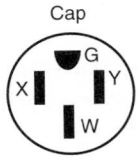

Receptacle          Cap

125/250-V, 50-A, 3-pole, 4-wire, grounding type

**Figure 550.10(C) 50-ampere, 125/250-volt receptacle and attachment plug cap configurations, 3-pole, 4-wire, grounding-types, used for mobile home supply cords and mobile home parks.**

**(D) Overall Length of a Power-Supply Cord.** The overall length of a power-supply cord, measured from the end of the cord, including bared leads, to the face of the attachment plug cap shall not be less than 6.4 m (21 ft) and shall not exceed 11 m (36½ ft). The length of the cord from the face of the attachment plug cap to the point where the cord enters the mobile home shall not be less than 6.0 m (20 ft).

**(E) Marking.** The power-supply cord shall bear the following marking:

FOR USE WITH MOBILE HOMES — 40 AMPERES.

or

FOR USE WITH MOBILE HOMES — 50 AMPERES.

**(F) Point of Entrance.** The point of entrance of the feeder assembly to the mobile home shall be in the exterior wall, floor, or roof.

**(G) Protected.** Where the cord passes through walls or floors, it shall be protected by means of conduits and bushings or equivalent. The cord shall be permitted to be installed within the mobile home walls, provided a continuous raceway having a maximum size of 32 mm (1¼ in.) is installed from the branch-circuit panelboard to the underside of the mobile home floor.

**(H) Protection Against Corrosion and Mechanical Damage.** Permanent provisions shall be made for the protection of the attachment plug cap of the power-supply cord and any connector cord assembly or receptacle against corrosion and mechanical damage if such devices are in an exterior location while the mobile home is in transit.

**(I) Mast Weatherhead or Raceway.** Where the calculated load exceeds 50 amperes or where a permanent feeder is used, the supply shall be by means of either of the following:

(1) One mast weatherhead installation, installed in accordance with Article 230, containing four continuous, insulated, color-coded feeder conductors, one of which shall be an equipment grounding conductor

(2) A metal raceway or rigid nonmetallic conduit from the disconnecting means in the mobile home to the underside of the mobile home, with provisions for the attachment to a suitable junction box or fitting to the raceway on the underside of the mobile home [with or without conductors as in 550.10(I)(1)]. The manufacturer shall provide written installation instructions stating the proper feeder conductor sizes for the raceway and the size of the junction box to be used.

**550.11 Disconnecting Means and Branch-Circuit Protective Equipment.** The branch-circuit equipment shall be permitted to be combined with the disconnecting means as a single assembly. Such a combination shall be permitted to be designated as a distribution panelboard. If a fused distribution panelboard is used, the maximum fuse size for the mains shall be plainly marked with lettering at least 6 mm (¼ in.) high and visible when fuses are changed.

Where plug fuses and fuseholders are used, they shall be tamper-resistant Type S, enclosed in dead-front fuse panelboards. Electrical distribution panelboards containing circuit breakers shall also be dead-front type.

FPN: See 110.22 concerning identification of each disconnecting means and each service, feeder, or branch circuit at the point where it originated and the type marking needed.

**(A) Disconnecting Means.** A single disconnecting means shall be provided in each mobile home consisting of a circuit breaker, or a switch and fuses and its accessories installed in a readily accessible location near the point of entrance of the supply cord or conductors into the mobile home. The main circuit breakers or fuses shall be plainly marked "Main." This equipment shall contain a solderless type of grounding connector or bar for the purposes of grounding, with sufficient terminals for all grounding conductors. The neutral bar termination of the grounded circuit conductors shall be insulated in accordance with 550.16(A). The disconnecting equipment shall have a rating suitable for the connected load. The distribution equipment, either circuit breaker or fused type, shall be located a minimum of 600 mm (24 in.) from the bottom of such equipment to the floor level of the mobile home.

FPN: See 550.20(B) for information on disconnecting means for branch circuits designed to energize heating or air-conditioning equipment, or both, located outside the mobile home, other than room air conditioners.

A distribution panelboard shall be rated not less than 50 amperes and employ a 2-pole circuit breaker rated 40 amperes for a 40-ampere supply cord, or 50 amperes for a 50-ampere supply cord. A distribution panelboard employing a disconnect switch and fuses shall be rated 60 amperes and shall employ a single 2-pole, 60-ampere fuseholder with 40- or 50-ampere main fuses for 40- or 50-ampere supply cords, respectively. The outside of the distribution panelboard shall be plainly marked with the fuse size.

The distribution panelboard shall be located in an accessible location but shall not be located in a bathroom or a clothes closet. A clear working space at least 750 mm (30 in.) wide and 750 mm (30 in.) in front of the distribution panelboard shall be provided. This space shall extend from the floor to the top of the distribution panelboard.

**(B) Branch-Circuit Protective Equipment.** Branch-circuit distribution equipment shall be installed in each mobile home and shall include overcurrent protection for each branch circuit consisting of either circuit breakers or fuses.

The branch-circuit overcurrent devices shall be rated as follows:

(1) Not more than the circuit conductors; and

(2) Not more than 150 percent of the rating of a single appliance rated 13.3 amperes or more that is supplied by an individual branch circuit; but

(3) Not more than the overcurrent protection size and of the type marked on the air conditioner or other motor-operated appliance.

**(C) Two-Pole Circuit Breakers.** Where circuit breakers are provided for branch-circuit protection, 240-volt circuits shall be protected by a 2-pole common or companion trip, or handle-tied paired circuit breakers.

**(D) Electrical Nameplates.** A metal nameplate on the outside adjacent to the feeder assembly entrance shall read:

THIS CONNECTION FOR 120/240-VOLT,
3-POLE, 4-WIRE, 60-HERTZ,
_____ AMPERE SUPPLY

The correct ampere rating shall be marked in the blank space.

*Exception: For manufactured homes, the manufacturer shall provide in its written installation instructions or in the data plate the minimum ampere rating of the feeder assembly or, where provided, the service entrance conductors intended for connection to the manufactured home. The rating provided shall not be less than the minimum load calculated in accordance with 550.18.*

**550.12 Branch Circuits.** The number of branch circuits required shall be determined in accordance with 550.12(A) through (E).

**(A) Lighting.** Based on 33 volt-amperes/m² (3 VA/ft²) times outside dimensions of the mobile home (coupler excluded) divided by 120 volts to determine the number of 15- or 20-ampere lighting area circuits, for example,

$$\frac{3 \times \text{length} \times \text{width}}{120 \times 15 \,(\text{or } 20)}$$

$$= \text{No. of 15 (or 20) ampere circuits}$$

**(B) Small Appliances.** For the small-appliance load in kitchens, pantries, dining rooms, and breakfast rooms, two or more 20-ampere appliance branch circuits, in addition to the number of branch circuits required by other parts of this section, shall be provided for all receptacle outlets required by 550.13(D) in these rooms. Such circuits shall have no other outlets.

*Exception No. 1: A receptacle installed solely for the electrical supply to and support of an electric clock in any of the rooms specified in (B).*

*Exception No. 2: Receptacles installed to provide power for supplemental equipment and lighting on gas-fired ranges, ovens, or counter-mounted cooking units.*

Countertop receptacle outlets installed in the kitchen shall be supplied by not less than two small-appliance branch circuits, either or both of which shall be permitted to supply receptacle outlets in the kitchen and other rooms specified above.

**(C) Laundry Area.** Where a laundry area is provided, a 20-ampere branch circuit shall be provided to supply the laundry receptacle outlet(s).

**(D) General Appliances.** (Including furnace, water heater, range, and central or room air conditioner, etc.). There shall be one or more circuits of adequate rating in accordance with the following:

FPN: For central air conditioning, see Article 440.

(1) The ampere rating of fixed appliances shall not be over 50 percent of the circuit rating if lighting outlets (receptacles, other than kitchen, dining area, and laundry, considered as lighting outlets) are on the same circuit.

(2) For fixed appliances on a circuit without lighting outlets, the sum of rated amperes shall not exceed the branch-circuit rating. Motor loads or other continuous duty loads shall not exceed 80 percent of the branch-circuit rating.

(3) The rating of a single cord-and-plug-connected appliance on a circuit having no other outlets shall not exceed 80 percent of the circuit rating.

(4) The rating of a range branch circuit shall be based on the range demand as specified for ranges in 550.18(B)(5).

**(E) Bathrooms.** Bathroom receptacle outlets shall be supplied by at least one 20-ampere branch circuit. Such circuits shall have no other outlets other than as provided for in 550.13(E)(2).

**550.13 Receptacle Outlets.**

**(A) Grounding-Type Receptacle Outlets.** All receptacle outlets shall comply with the following:

(1) Be of grounding type
(2) Be installed according to 406.3
(3) Except where supplying specific appliances, be 15- or 20-ampere, 125-volt, either single or duplex, and accept parallel-blade attachment plugs

**(B) Ground-Fault Circuit Interrupters (GFCI).** All 125-volt, single-phase, 15- and 20-ampere receptacle outlets installed outdoors, in compartments accessible from outside the unit, or in bathrooms, including receptacles in luminaires (light fixtures), shall have GFCI protection for personnel. GFCI protection for personnel shall be provided for receptacle outlets serving countertops in kitchens, and receptacle outlets located within 1.8 m (6 ft) of a wet bar sink.

*Exception: Receptacles installed for appliances in dedicated spaces, such as for dishwashers, disposals, refrigerators, freezers, and laundry equipment.*

Feeders supplying branch circuits shall be permitted to be protected by a ground-fault circuit-interrupter in lieu of the provision for such interrupters specified herein.

**(C) Cord-Connected Fixed Appliance.** A grounding-type receptacle outlet shall be provided for each cord-connected fixed appliance installed.

**(D) Receptacle Outlets Required.** Except in the bath, closet, and hall areas, receptacle outlets shall be installed at wall spaces 600 mm (2 ft) wide or more so that no point along the floor line is more than 1.8 m (6 ft) measured horizontally from an outlet in that space. In addition, a receptacle outlet shall be installed in the following locations:

(1) Over or adjacent to countertops in the kitchen [at least one on each side of the sink if countertops are on each side and are 300 mm (12 in.) or over in width].

(2) Adjacent to the refrigerator and freestanding gas-range space. A duplex receptacle shall be permitted to serve as the outlet for a countertop and a refrigerator.

(3) At countertop spaces for built-in vanities.

(4) At countertop spaces under wall-mounted cabinets.

(5) In the wall at the nearest point to where a bar-type counter attaches to the wall.

(6) In the wall at the nearest point to where a fixed room divider attaches to the wall.

(7) In laundry areas within 1.8 m (6 ft) of the intended location of the laundry appliance(s).

(8) At least one receptacle outlet located outdoors and accessible at grade level and not more than 2.0 m (6½ ft) above grade. A receptacle outlet located in a compartment accessible from the outside of the unit shall be considered an outdoor receptacle.

(9) At least one receptacle outlet shall be installed in bathrooms within 900 mm (36 in.) of the outside edge of each basin. The receptacle outlet shall be located above or adjacent to the basin location. This receptacle shall be in addition to any receptacle that is a part of a luminaire (fixture) or appliance. The receptacle shall not be enclosed within a bathroom cabinet or vanity.

**(E) Pipe Heating Cable(s) Outlet.** For the connection of pipe heating cable(s), a receptacle outlet shall be located on the underside of the unit as follows:

(1) Within 600 mm (2 ft) of the cold water inlet.

(2) Connected to an interior branch circuit, other than a small appliance branch circuit. It shall be permitted to use a bathroom receptacle circuit for this purpose.

(3) On a circuit where all of the outlets are on the load side of the ground-fault circuit-interrupter.

(4) This outlet shall not be considered as the receptacle required by 550.13(D)(8).

**(F) Receptacle Outlets Not Permitted.** Receptacle outlets shall not be permitted in the following locations:

(1) Receptacle outlets shall not be installed in or within reach [750 mm (30 in.)] of a shower or bathtub space.

(2) A receptacle shall not be installed in a face-up position in any countertop.

(3) Receptacle outlets shall not be installed above electric baseboard heaters, unless provided for in the listing or manufacturer's instructions.

**(G) Receptacle Outlets Not Required.** Receptacle outlets shall not be located in the following locations:

(1) In the wall space occupied by built-in kitchen or wardrobe cabinets

(2) In the wall space behind doors that can be opened fully against a wall surface

(3) In room dividers of the lattice type that are less than 2.5 m (8 ft) long, not solid, and within 150 mm (6 in.) of the floor

(4) In the wall space afforded by bar-type counters

**550.14 Luminaires (Fixtures) and Appliances.**

**(A) Fasten Appliances in Transit.** Means shall be provided to securely fasten appliances when the mobile home is in transit. (See 550.16 for provisions on grounding.)

**(B) Accessibility.** Every appliance shall be accessible for inspection, service, repair, or replacement without removal of permanent construction.

**(C) Pendants.** Listed pendant-type luminaires (fixtures) or pendant cords shall be permitted.

**(D) Bathtub and Shower Luminaires (Fixtures).** Where a luminaire (lighting fixture) is installed over a bathtub or in a shower stall, it shall be of the enclosed and gasketed type listed for wet locations.

**550.15 Wiring Methods and Materials.** Except as specifically limited in this section, the wiring methods and materials included in this *Code* shall be used in mobile homes. Aluminum conductors, aluminum alloy conductors, and aluminum core conductors such as copper-clad aluminum shall not be acceptable for use as branch-circuit wiring.

**(A) Nonmetallic Boxes.** Nonmetallic boxes shall be permitted only with nonmetallic cable or nonmetallic raceways.

**(B) Nonmetallic Cable Protection.** Nonmetallic cable located 380 mm (15 in.) or less above the floor, if exposed, shall be protected from physical damage by covering boards, guard strips, or raceways. Cable likely to be damaged by stowage shall be so protected in all cases.

**(C) Metal-Covered and Nonmetallic Cable Protection.** Metal-covered and nonmetallic cables shall be permitted to pass through the centers of the wide side of 2 by 4 studs. However, they shall be protected where they pass through 2 by 2 studs or at other studs or frames where the cable or armor would be less than 32 mm (1¼ in.) from the inside or outside surface of the studs where the wall covering materials are in contact with the studs. Steel plates on each side of the cable, or a tube, with not less than 1.35 mm (0.053 in.) wall thickness shall be required to protect the cable. These plates or tubes shall be securely held in place.

**(D) Metal Faceplates.** Where metal faceplates are used, they shall be effectively grounded.

**(E) Installation Requirements.** If a range, clothes dryer, or similar appliance is connected by metal-covered cable or flexible metal conduit, a length of not less than 900 mm (3 ft) of free cable or conduit shall be provided to permit moving the appliance. The cable or flexible metal conduit shall be secured to the wall. Type NM or Type SE cable shall not be used to connect a range or dryer. This shall not prohibit the use of Type NM or Type SE cable between the branch-circuit overcurrent-protective device and a junction box or range or dryer receptacle.

**(F) Raceways.** Where rigid metal conduit or intermediate metal conduit is terminated at an enclosure with a locknut and bushing connection, two locknuts shall be provided, one inside and one outside of the enclosure. Rigid nonmetallic conduit, electrical nonmetallic tubing, or surface raceway shall be permitted. All cut ends of conduit and tubing shall be reamed or otherwise finished to remove rough edges.

**(G) Switches.** Switches shall be rated as follows:

(1) For lighting circuits, switches shall be rated not less than 10 amperes, 120 to 125 volts, and in no case less than the connected load.

(2) For motors or other loads, switches shall have ampere or horsepower ratings, or both, adequate for loads controlled. (An ac general-use snap switch shall be permitted to control a motor 2 hp or less with full-load current not over 80 percent of the switch ampere rating.)

**(H) Under-Chassis Wiring (Exposed to Weather).** Where outdoor or under-chassis line-voltage (120 volts, nominal, or higher) wiring is exposed to moisture or physical damage, it shall be protected by rigid metal conduit or intermediate metal conduit. The conductors shall be suitable for wet locations.

*Exception: Electrical metallic tubing or rigid nonmetallic conduit shall be permitted where closely routed against frames and equipment enclosures.*

**(I) Boxes, Fittings, and Cabinets.** Boxes, fittings, and cabinets shall be securely fastened in place and shall be supported from a structural member of the home, either directly or by using a substantial brace.

*Exception: Snap-in-type boxes. Boxes provided with special wall or ceiling brackets and wiring devices with integral enclosures that securely fasten to walls or ceilings and are identified for the use shall be permitted without support from a structural member or brace. The testing and approval shall include the wall and ceiling construction systems for which the boxes and devices are intended to be used.*

**(J) Appliance Terminal Connections.** Appliances having branch-circuit terminal connections that operate at temperatures higher than 60°C (140°F) shall have circuit conductors as described in the following:

(1) Branch-circuit conductors having an insulation suitable for the temperature encountered shall be permitted to be run directly to the appliance.

(2) Conductors having an insulation suitable for the temperature encountered shall be run from the appliance terminal connection to a readily accessible outlet box placed at least 300 mm (1 ft) from the appliance. These conductors shall be in a suitable raceway or Type AC or MC cable of at least 450 mm (18 in.) but not more than 1.8 m (6 ft) in length.

**(K) Component Interconnections.** Fittings and connectors that are intended to be concealed at the time of assembly shall be listed and identified for the interconnection of building components. Such fittings and connectors shall be equal to the wiring method employed in insulation, temperature rise, and fault-current withstanding and shall be capable of enduring the vibration and shock occurring in mobile home transportation.

FPN: See 550.19 for interconnection of multiple section units.

**550.16 Grounding.** Grounding of both electrical and non-electrical metal parts in a mobile home shall be through connection to a grounding bus in the mobile home distribution panelboard. The grounding bus shall be grounded through the green-colored insulated conductor in the supply cord or the feeder wiring to the service ground in the service-entrance equipment located adjacent to the mobile home location. Neither the frame of the mobile home nor the frame of any appliance shall be connected to the grounded circuit conductor (neutral) in the mobile home. Where service equipment is installed in or on a manufac-

tured home as permitted in 550.32(B), the neutral conductors and the ground bus shall be permitted to be connected in the distribution panel.

**(A) Grounded (Neutral) Conductor.**

**(1) Insulated.** The grounded circuit conductor (neutral) shall be insulated from the grounding conductors and from equipment enclosures and other grounded parts. The grounded (neutral) circuit terminals in the distribution panelboard and in ranges, clothes dryers, counter-mounted cooking units, and wall-mounted ovens shall be insulated from the equipment enclosure. Bonding screws, straps, or buses in the distribution panelboard or in appliances shall be removed and discarded. Where service equipment is installed in or on a manufactured home as permitted in 550.32(B), the neutral conductors and the ground bus shall be permitted to be connected in the distribution panel.

**(2) Connections of Ranges and Clothes Dryers.** Connections of ranges and clothes dryers with 120/240-volt, 3-wire ratings shall be made with 4-conductor cord and 3-pole, 4-wire, grounding-type plugs or by Type AC cable, Type MC cable, or conductors enclosed in flexible metal conduit.

**(B) Equipment Grounding Means.**

**(1) Supply Cord or Permanent Feeder.** The green-colored insulated grounding wire in the supply cord or permanent feeder wiring shall be connected to the grounding bus in the distribution panelboard or disconnecting means.

**(2) Electrical System.** In the electrical system, all exposed metal parts, enclosures, frames, lamp fixture canopies, and so forth shall be effectively bonded to the grounding terminal or enclosure of the distribution panelboard.

**(3) Cord-Connected Appliances.** Cord-connected appliances, such as washing machines, clothes dryers, and refrigerators, and the electrical system of gas ranges and so forth, shall be grounded by means of a cord with grounding conductor and grounding-type attachment plug.

**(C) Bonding of Non–Current-Carrying Metal Parts.**

**(1) Exposed Non–Current-Carrying Metal Parts.** All exposed non–current-carrying metal parts that may become energized shall be effectively bonded to the grounding terminal or enclosure of the distribution panelboard. A bonding conductor shall be connected between the distribution panelboard and accessible terminal on the chassis.

**(2) Grounding Terminals.** Grounding terminals shall be of the solderless type and listed as pressure-terminal connectors recognized for the wire size used. The bonding conductor shall be solid or stranded, insulated or bare, and shall be 8 AWG copper minimum, or equivalent. The bonding conductor shall be routed so as not to be exposed to physical damage.

**(3) Metallic Piping and Ducts.** Metallic gas, water, and waste pipes and metallic air-circulating ducts shall be considered bonded if they are connected to the terminal on the chassis [see 550.16(C)(1)] by clamps, solderless connectors, or by suitable grounding-type straps.

**(4) Metallic Roof and Exterior Coverings.** Any metallic roof and exterior covering shall be considered bonded if the following conditions are met:

(1) The metal panels overlap one another and are securely attached to the wood or metal frame parts by metallic fasteners.
(2) The lower panel of the metallic exterior covering is secured by metallic fasteners at a cross member of the chassis by two metal straps per mobile home unit or section at opposite ends.

The bonding strap material shall be a minimum of 100 mm (4 in.) in width of material equivalent to the skin or a material of equal or better electrical conductivity. The straps shall be fastened with paint-penetrating fittings such as screws and starwashers or equivalent.

**550.17 Testing.**

**(A) Dielectric Strength Test.** The wiring of each mobile home shall be subjected to a 1-minute, 900-volt, dielectric strength test (with all switches closed) between live parts (including neutral) and the mobile home ground. Alternatively, the test shall be permitted to be performed at 1080 volts for 1 second. This test shall be performed after branch circuits are complete and after luminaires (fixtures) or appliances are installed.

*Exception: Listed luminaires (fixtures) or appliances shall not be required to withstand the dielectric strength test.*

**(B) Continuity and Operational Tests and Polarity Checks.** Each mobile home shall be subjected to all of the following:

(1) An electrical continuity test to ensure that all exposed electrically conductive parts are properly bonded
(2) An electrical operational test to demonstrate that all equipment, except water heaters and electric furnaces, is connected and in working order
(3) Electrical polarity checks of permanently wired equipment and receptacle outlets to determine that connections have been properly made

**550.18 Calculations.** The following method shall be employed in computing the supply-cord and distribution-panelboard load for each feeder assembly for each mobile home in lieu of the procedure shown in Article 220 and shall be based on a 3-wire, 120/240-volt supply with 120-volt loads balanced between the two legs of the 3-wire system.

**(A) Lighting, Small Appliance, and Laundry Load.**

**(1) Lighting Volt-Amperes.** Length times width of mobile home floor (outside dimensions) times 33 volt-amperes/$m^2$ (3 VA/$ft^2$), for example, length × width × 3 = lighting volt-amperes.

**(2) Small Appliance Volt-Amperes.** Number of circuits times 1500 volt-amperes for each 20-ampere appliance receptacle circuit (see definition of Appliance, Portable with note in 550.2), for example, number of circuits × 1500 = small appliance volt-amperes.

**(3) Laundry Area Circuit Volt-Amperes.** 1500 volt-amperes.

**(4) Total Volt-Amperes.** Lighting volt-amperes plus small appliance volt-amperes plus laundry area volt-amperes equals total volt-amperes.

**(5) Net Volt-Amperes.** First 3000 total volt-amperes at 100 percent plus remainder at 35 percent equals volt-amperes to be divided by 240 volts to obtain current (amperes) per leg.

**(B) Total Load for Determining Power Supply.** Total load for determining power supply is the sum of the following:

(1) Lighting and small appliance load as calculated in 550.18(A)(5).
(2) Nameplate amperes for motors and heater loads (exhaust fans, air conditioners, electric, gas, or oil heating). Omit smaller of the heating and cooling loads, except include blower motor if used as air-conditioner evaporator motor. Where an air conditioner is not installed and a 40-ampere power-supply cord is provided, allow 15 amperes per leg for air conditioning.
(3) Twenty-five percent of current of largest motor in (2).
(4) Total of nameplate amperes for waste disposer, dishwasher, water heater, clothes dryer, wall-mounted oven, cooking units. Where the number of these appliances exceeds three, use 75 percent of total.
(5) Derive amperes for freestanding range (as distinguished from separate ovens and cooking units) by dividing the following values by 240 volts.
(6) If outlets or circuits are provided for other than factory-installed appliances, include the anticipated load.

FPN: Refer to Annex D, Example D11, for an illustration of the application of this calculation.

| Nameplate Rating (watts) | Use (volt-amperes) |
|---|---|
| 0 – 10,000 | 80 percent of rating |
| Over 10,000 – 12,500 | 8,000 |
| Over 12,500 – 13,500 | 8,400 |
| Over 13,500 – 14,500 | 8,800 |
| Over 14,500 – 15,500 | 9,200 |
| Over 15,500 – 16,500 | 9,600 |
| Over 16,500 – 17,500 | 10,000 |

**(C) Optional Method of Calculation for Lighting and Appliance Load.** The optional method for calculating lighting and appliance load shown in 220.30 shall be permitted.

**550.19 Interconnection of Multiple-Section Mobile or Manufactured Home Units.**

**(A) Wiring Methods.** Approved and listed fixed-type wiring methods shall be used to join portions of a circuit that must be electrically joined and are located in adjacent sections after the home is installed on its support foundation. The circuit's junction shall be accessible for disassembly when the home is prepared for relocation.

FPN: See 550.15(K) for component interconnections.

**(B) Disconnecting Means.** Expandable or multi-unit manufactured homes not having permanently installed feeders that are to be moved from one location to another shall be permitted to have disconnecting means with branch-circuit protective equipment in each unit when so located that after assembly or joining together of units, the requirements of 550.10 will be met.

**550.20 Outdoor Outlets, Luminaires (Fixtures), Air-Cooling Equipment, and So Forth.**

**(A) Listed for Outdoor Use.** Outdoor luminaires (fixtures) and equipment shall be listed for outdoor use. Outdoor receptacle or convenience outlets shall be of a gasketed-cover type for use in wet locations. Where located on the underside of the home or located under roof extensions or similarly protected locations, outdoor luminaires (fixtures) and equipment shall be listed for use in damp locations.

**(B) Outside Heating Equipment, Air-Conditioning Equipment, or Both.** A mobile home provided with a branch circuit designed to energize outside heating equipment, air-conditioning equipment, or both, located outside the mobile home, other than room air conditioners, shall have such branch-circuit conductors terminate in a listed outlet box, or disconnecting means, located on the outside of the mobile home. A label shall be permanently affixed adjacent to the outlet box and shall contain the following information:

> THIS CONNECTION IS FOR HEATING AND/OR AIR-CONDITIONING EQUIPMENT. THE BRANCH CIRCUIT IS RATED AT NOT MORE THAN _____ AMPERES, AT _____ VOLTS, 60-HERTZ, _____ CONDUCTOR AMPACITY. A DISCONNECTING MEANS SHALL BE LOCATED WITHIN SIGHT OF THE EQUIPMENT.

The correct voltage and ampere rating shall be given. The tag shall be not less than 0.51 mm (0.020 in.) thick etched brass, stainless steel, anodized or alclad aluminum, or equivalent. The tag shall not be less than 75 mm by 45 mm (3 in. by 1¾ in.) minimum size.

**550.25 Arc-Fault Circuit-Interrupter Protection.**

**(A) Definition.** Arc-fault circuit interrupters are defined in 210.12(A).

**(B) Bedrooms of Mobile Homes and Manufactured Homes.** All branch circuits that supply 125-volt, single-phase, 15- and 20-ampere outlets installed in bedrooms of mobile homes and manufactured homes shall be protected by arc-fault circuit interrupter(s).

**III. Services and Feeders**

**550.30 Distribution System.** The mobile home park secondary electrical distribution system to mobile home lots shall be single-phase, 120/240 volts, nominal. For the purpose of Part III, where the park service exceeds 240 volts, nominal, transformers and secondary distribution panelboards shall be treated as services.

**550.31 Allowable Demand Factors.** Park electrical wiring systems shall be calculated (at 120/240 volts) on the larger of the following:

(1) 16,000 volt-amperes for each mobile home lot
(2) The load calculated in accordance with 550.18 for the largest typical mobile home that each lot will accept

It shall be permissible to compute the feeder or service load in accordance with Table 550.31. No demand factor shall be allowed for any other load, except as provided in this *Code*.

Service and feeder conductors to a mobile home in compliance with 310.15(B)(6) shall be permitted.

**550.32 Service Equipment.**

**(A) Mobile Home Service Equipment.** The mobile home service equipment shall be located adjacent to the mobile

**Table 550.31 Demand Factors for Services and Feeders**

| Number of Mobile Homes | Demand Factor (percent) |
|---|---|
| 1 | 100 |
| 2 | 55 |
| 3 | 44 |
| 4 | 39 |
| 5 | 33 |
| 6 | 29 |
| 7–9 | 28 |
| 10–12 | 27 |
| 13–15 | 26 |
| 16–21 | 25 |
| 22–40 | 24 |
| 41–60 | 23 |
| 61 and over | 22 |

home and not mounted in or on the mobile home. The service equipment shall be located in sight from and not more than 9.0 m (30 ft) from the exterior wall of the mobile home it serves. The service equipment shall be permitted to be located elsewhere on the premises, provided that a disconnecting means suitable for service equipment is located in sight from and not more than 9.0 m (30 ft) from the exterior wall of the mobile home it serves. Grounding at the disconnecting means shall be in accordance with 250.32.

**(B) Manufactured Home Service Equipment.** The manufactured home service equipment shall be permitted to be installed in or on a manufactured home, provided that all of the following conditions are met:

(1) The manufacturer shall include in its written installation instructions information indicating that the home shall be secured in place by an anchoring system or installed on and secured to a permanent foundation.

(2) The installation of the service equipment shall comply with Article 230.

(3) Means shall be provided for the connection of a grounding electrode conductor to the service equipment and routing it outside the structure.

(4) Bonding and grounding of the service shall be in accordance with Article 250.

(5) The manufacturer shall include in its written installation instructions one method of grounding the service equipment at the installation site. The instructions shall clearly state that other methods of grounding are found in Article 250.

(6) The minimum size grounding electrode conductor shall be specified in the instructions.

(7) A red warning label shall be mounted on or adjacent to the service equipment. The label shall state the following:

**WARNING
DO NOT PROVIDE ELECTRICAL POWER
UNTIL THE GROUNDING ELECTRODE(S)
IS INSTALLED AND CONNECTED
(SEE INSTALLATION INSTRUCTIONS).**

Where the service equipment is not installed in or on the unit, the installation shall comply with the other provisions of this section.

**(C) Rating.** Mobile home service equipment shall be rated at not less than 100 amperes at 120/240 volts, and provisions shall be made for connecting a mobile home feeder assembly by a permanent wiring method. Power outlets used as mobile home service equipment shall also be permitted to contain receptacles rated up to 50 amperes with appropriate overcurrent protection. Fifty-ampere receptacles shall conform to the configuration shown in Figure 550.10(C).

> FPN: Complete details of the 50-ampere plug and receptacle configuration can be found in ANSI/NEMA WD 6-1989, National Electrical Manufacturers Association *Standard for Wiring Devices — Dimensional Requirements*, Figure 14-50.

**(D) Additional Outside Electrical Equipment.** Means for connecting a mobile home accessory building or structure or additional electrical equipment located outside a mobile home by a fixed wiring method shall be provided in either the mobile home service equipment or the local external disconnecting means permitted in 550.32(A).

**(E) Additional Receptacles.** Additional receptacles shall be permitted for connection of electrical equipment located outside the mobile home, and all such 125-volt, single-phase, 15- and 20-ampere receptacles shall be protected by a listed ground-fault circuit interrupter.

**(F) Mounting Height.** Outdoor mobile home disconnecting means shall be installed so the bottom of the enclosure containing the disconnecting means is not less than 600 mm (2 ft) above finished grade or working platform. The disconnecting means shall be installed so that the center of the grip of the operating handle, when in the highest position, is not more than 2.0 m (6 ft 7 in.) above the finished grade or working platform.

**(G) Marking.** Where a 125/250-volt receptacle is used in mobile home service equipment, the service equipment shall be marked as follows:

**TURN DISCONNECTING SWITCH OR
CIRCUIT BREAKER OFF BEFORE INSERTING
OR REMOVING PLUG. PLUG MUST BE FULLY
INSERTED OR REMOVED.**

The marking shall be located on the service equipment adjacent to the receptacle outlet.

## 550.33 Feeder.

**(A) Feeder Conductors.** Feeder conductors shall consist of either a listed cord, factory installed in accordance with 550.10(B), or a permanently installed feeder consisting of four, insulated, color-coded conductors that shall be identified by the factory or field marking of the conductors in compliance with 310.12. Equipment grounding conductors shall not be identified by stripping the insulation.

*Exception: Where a feeder is installed between service equipment and a disconnecting means as covered in 550.32(A), it shall be permitted to omit the equipment grounding conductor where the grounded circuit conductor is grounded at the disconnecting means as required in 250.32(B).*

**(B) Adequate Feeder Capacity.** Mobile home and manufactured home lot feeder circuit conductors shall have adequate capacity for the loads supplied and shall be rated at not less than 100 amperes at 120/240 volts.

## ARTICLE 551
## Recreational Vehicles and Recreational Vehicle Parks

### I. General

**551.1 Scope.** The provisions of this article cover the electrical conductors and equipment installed within or on recreational vehicles, the conductors that connect recreational vehicles to a supply of electricity, and the installation of equipment and devices related to electrical installations within a recreational vehicle park.

**551.2 Definitions.** (See Article 100 for additional definitions.)

**Air-Conditioning or Comfort-Cooling Equipment.** All of that equipment intended or installed for the purpose of processing the treatment of air so as to control simultaneously its temperature, humidity, cleanliness, and distribution to meet the requirements of the conditioned space.

**Appliance, Fixed.** An appliance that is fastened or otherwise secured at a specific location.

**Appliance, Portable.** An appliance that is actually moved or can easily be moved from one place to another in normal use.

> FPN: For the purpose of this article, the following major appliances, other than built-in, are considered portable if cord connected: refrigerators, range equipment, clothes washers, dishwashers without booster heaters, or other similar appliances.

**Appliance, Stationary.** An appliance that is not easily moved from one place to another in normal use.

**Camping Trailer.** A vehicular portable unit mounted on wheels and constructed with collapsible partial side walls that fold for towing by another vehicle and unfold at the campsite to provide temporary living quarters for recreational, camping, or travel use. (*See* Recreational Vehicle.)

**Converter.** A device that changes electrical energy from one form to another, as from alternating current to direct current.

**Dead Front (as applied to switches, circuit breakers, switchboards, and distribution panelboards).** Designed, constructed, and installed so that no current-carrying parts are normally exposed on the front.

**Disconnecting Means.** The necessary equipment usually consisting of a circuit breaker or switch and fuses, and their accessories, located near the point of entrance of supply conductors in a recreational vehicle and intended to constitute the means of cutoff for the supply to that recreational vehicle.

**Distribution Panelboard.** A single panel or group of panel units designed for assembly in the form of a single panel, including buses, and with or without switches and/or automatic overcurrent-protective devices for the control of light, heat, or power circuits of small individual as well as aggregate capacity; designed to be placed in a cabinet or cutout box placed in or against a wall or partition and accessible only from the front.

**Frame.** Chassis rail and any welded addition thereto of metal thickness of 1.35 mm (0.053 in.) or greater.

**Low Voltage.** An electromotive force rated 24 volts, nominal, or less, supplied from a transformer, converter, or battery.

**Motor Home.** A vehicular unit designed to provide temporary living quarters for recreational, camping, or travel use built on or permanently attached to a self-propelled motor vehicle chassis or on a chassis cab or van that is an integral part of the completed vehicle. (*See* Recreational Vehicle.)

**Power-Supply Assembly.** The conductors, including ungrounded, grounded, and equipment grounding conductors, the connectors, attachment plug caps, and all other fittings, grommets, or devices installed for the purpose of delivering energy from the source of electrical supply to the distribution panel within the recreational vehicle.

**Recreational Vehicle.** A vehicular-type unit primarily designed as temporary living quarters for recreational, camping, or travel use, which either has its own motive power or is mounted on or drawn by another vehicle. The basic entities are travel trailer, camping trailer, truck camper, and motor home.

**Recreational Vehicle Park.** A plot of land upon which two or more recreational vehicle sites are located, established, or maintained for occupancy by recreational vehicles of the general public as temporary living quarters for recreation or vacation purposes.

**Recreational Vehicle Site.** A plot of ground within a recreational vehicle park set aside for the accommodation of a recreational vehicle on a temporary basis. It can be used as either a recreational vehicle site or as a camping unit site.

**Recreational Vehicle Site Feeder Circuit Conductors.** The conductors from the park service equipment to the recreational vehicle site supply equipment.

**Recreational Vehicle Site Supply Equipment.** The necessary equipment, usually a power outlet, consisting of a circuit breaker or switch and fuse and their accessories, located near the point of entrance of supply conductors to a recreational vehicle site and intended to constitute the disconnecting means for the supply to that site.

**Recreational Vehicle Stand.** That area of a recreational vehicle site intended for the placement of a recreational vehicle.

**Transformer.** A device that, when used, raises or lowers the voltage of alternating current of the original source.

**Travel Trailer.** A vehicular unit, mounted on wheels, designed to provide temporary living quarters for recreational, camping, or travel use, of such size or weight as not to require special highway movement permits when towed by a motorized vehicle, and of gross trailer area less than 30 m² (320 ft²). (*See* Recreational Vehicle.)

**Truck Camper.** A portable unit constructed to provide temporary living quarters for recreational, travel, or camping use, consisting of a roof, floor, and sides, designed to be loaded onto and unloaded from the bed of a pick-up truck. (*See* Recreational Vehicle.)

**551.3 Other Articles.** Wherever the requirements of other articles of this *Code* and Article 551 differ, the requirements of Article 551 shall apply.

**551.4 General Requirements.**

**(A) Not Covered.** A recreational vehicle not used for the purposes as defined in 551.2 shall not be required to meet the provisions of Part I pertaining to the number or capacity of circuits required. It shall, however, meet all other applicable requirements of this article if the recreational vehicle is provided with an electrical installation intended to be energized from a 120- or 120/240-volt, nominal, ac power-supply system.

**(B) Systems.** This article covers battery and other low-voltage power systems (24 volts or less), combination electrical systems, generator installations, and 120- or 120/240-volt, nominal, systems.

## II. Low-Voltage Systems

### 551.10 Low-Voltage Systems.

**(A) Low-Voltage Circuits.** Low-voltage circuits furnished and installed by the recreational vehicle manufacturer, other than automotive vehicle circuits or extensions thereof, are subject to this *Code*. Circuits supplying lights subject to federal or state regulations shall comply with applicable government regulations and this *Code*.

**(B) Low-Voltage Wiring.**

**(1) Material.** Copper conductors shall be used for low-voltage circuits.

*Exception: Metal chassis or frame shall be permitted as the return path to the source of supply.*

**(2) Conductor Types.** Conductors shall conform to the requirements for Type GXL, HDT, SGT, SGR, or Type SXL or shall have insulation in accordance with Table 310.13 or the equivalent. Conductor sizes 6 AWG through 18 AWG or SAE shall be listed. Single-wire, low-voltage conductors shall be of the stranded type.

> FPN: See SAE Standard J1128-1995 for Types GXL, HDT, and SXL, and SAE Standard J1127-1995 for Types SGT and SGR.

**(3) Marking.** All insulated low-voltage conductors shall be surface marked at intervals not greater than 1.2 m (4 ft) as follows:

(1) Listed conductors shall be marked as required by the listing agency.
(2) SAE conductors shall be marked with the name or logo of the manufacturer, specification designation, and wire gauge.
(3) Other conductors shall be marked with the name or logo of the manufacturer, temperature rating, wire gauge, conductor material, and insulation thickness.

**(4) Insulation Rating.** Conductors shall have a minimum insulation rating of 90°C (194°F) for interior installations and 125°C (257°F) for all engine compartment wiring or any under-chassis installations where conductors are located less than 450 mm (18 in.) from any component of an internal combustion engine exhaust system.

**(C) Low-Voltage Wiring Methods.**

**(1) Physical Protection.** Conductors shall be protected against physical damage and shall be secured. Where insulated conductors are clamped to the structure, the conductor insulation shall be supplemented by an additional wrap or layer of equivalent material, except that jacketed cables shall not be required to be so protected. Wiring shall be routed away from sharp edges, moving parts, or heat sources.

**(2) Splices.** Conductors shall be spliced or joined with splicing devices that provide a secure connection or by brazing, welding, or soldering with a fusible metal or alloy. Soldered splices shall first be spliced or joined so as to be mechanically and electrically secure without solder and then soldered. All splices, joints, and free ends of conductors shall be covered with an insulation equivalent to that on the conductors.

**(3) Separation.** Battery and dc circuits shall be physically separated by at least a 13-mm (½-in.) gap or other approved means from circuits of a different power source. Acceptable methods shall be by clamping, routing, or equivalent means that ensure permanent total separation. Where circuits of different power sources cross, the external jacket of the nonmetallic-sheathed cables shall be deemed adequate separation.

**(4) Ground Connections.** Ground connections to the chassis or frame shall be made in an accessible location and shall be mechanically secure. Ground connections shall be by means of copper conductors and copper or copper-alloy terminals of the solderless type identified for the size of wire used. The surface on which ground terminals make contact shall be cleaned and be free from oxide or paint or shall be electrically connected through the use of a cadmium, tin, or zinc-plated internal/external-toothed lockwasher or locking terminals. Ground terminal attaching screws, rivets or bolts, nuts, and lockwashers shall be cadmium, tin, or zinc-plated except rivets shall be permitted to be unanodized aluminum where attaching to aluminum structures.

The chassis-grounding terminal of the battery shall be bonded to the vehicle chassis with a minimum 8 AWG copper conductor. In the event the power lead from the battery exceeds 8 AWG, then the bonding conductor shall be of an equal size.

**(D) Battery Installations.** Storage batteries subject to the provisions of this *Code* shall be securely attached to the vehicle and installed in an area vaportight to the interior and ventilated directly to the exterior of the vehicle. Where batteries are installed in a compartment, the compartment shall be ventilated with openings having a minimum area of 1100 mm² (1.7 in.²) at both the top and at the bottom.

Where compartment doors are equipped for ventilation, the openings shall be within 50 mm (2 in.) of the top and bottom. Batteries shall not be installed in a compartment containing spark- or flame-producing equipment, except that they shall be permitted to be installed in the engine generator compartment if the only charging source is from the engine generator.

**(E) Overcurrent Protection.**

**(1) Rating.** Low-voltage circuit wiring shall be protected by overcurrent protective devices rated not in excess of the ampacity of copper conductors, in accordance with Table 551.10(E)(1).

**Table 551.10(E)(1) Low-Voltage Overcurrent Protection**

| Wire Size (AWG) | Ampacity | Wire Type |
|---|---|---|
| 18 | 6 | Stranded only |
| 16 | 8 | Stranded only |
| 14 | 15 | Stranded or solid |
| 12 | 20 | Stranded or solid |
| 10 | 30 | Stranded or solid |

**(2) Type.** Circuit breakers or fuses shall be of an approved type, including automotive types. Fuseholders shall be clearly marked with maximum fuse size, and both circuit breakers and fuses shall be protected against shorting and physical damage by a cover or equivalent means.

FPN: For further information, see ANSI/SAE J554-1987, *Standard for Electric Fuses (Cartridge Type)*; SAE J1284-1988, *Standard for Blade Type Electric Fuses,*; and UL 275-1993, *Standard for Automotive Glass Tube Fuses.*

**(3) Appliances.** DC appliances, such as pumps, compressors, heater blowers, and similar motor-driven appliances, shall be installed in accordance with the manufacturer's instructions.

Motors that are controlled by automatic switching or by latching-type manual switches shall be protected in accordance with 430.32(B).

**(4) Location.** The overcurrent protective device shall be installed in an accessible location on the vehicle within 450 mm (18 in.) of the point where the power supply connects to the vehicle circuits. If located outside the recreational vehicle, the device shall be protected against weather and physical damage.

*Exception: External low-voltage supply shall be permitted to have the overcurrent protective device within 450 mm (18 in.) after entering the vehicle or after leaving a metal raceway.*

**(F) Switches.** Switches shall have a dc rating not less than the connected load.

**(G) Luminaires (Lighting Fixtures).** All low-voltage interior luminaires (lighting fixtures) rated more than 4 watts, employing lamps rated more than 1.2 watts, shall be listed.

Ceiling luminaires (fixtures) in camping trailers shall be automatically de-energized by an interlock when folding down the trailer, or it shall be physically impossible to fold down the trailer unless the ceiling luminaire(s) (fixtures) are disconnected.

**(H) Cigarette Lighter Receptacles.** Twelve-volt receptacles that will accept and energize cigarette lighters shall be installed in a noncombustible outlet box, or the assembly shall be identified by the manufacturer of the product as thermally protected.

## III. Combination Electrical Systems

### 551.20 Combination Electrical Systems.

**(A) General.** Vehicle wiring suitable for connection to a battery or dc supply source shall be permitted to be connected to a 120-volt source, provided the entire wiring system and equipment are rated and installed in full conformity with Parts I, III, IV, V, and VI requirements of this article covering 120-volt electrical systems. Circuits fed from ac transformers shall not supply dc appliances.

**(B) Voltage Converters (120-Volt Alternating Current to Low-Voltage Direct Current).** The 120-volt ac side of the voltage converter shall be wired in full conformity with Parts I, III, IV, V, and VI requirements of this article for 120-volt electrical systems.

*Exception: Converters supplied as an integral part of a listed appliance shall not be subject to the above.*

All converters and transformers shall be listed for use in recreation vehicles and designed or equipped to provide over-temperature protection. To determine the converter rating, the following formula shall be applied to the total connected load, including average battery charging rate, of all 12-volt equipment:

> The first 20 amperes of load at 100 percent, plus
> The second 20 amperes of load at 50 percent, plus
> All load above 40 amperes at 25 percent

*Exception: A low-voltage appliance that is controlled by a momentary switch (normally open) that has no means for holding in the closed position or refrigerators with a 120-volt function shall not be considered as a connected load when determining the required converter rating. Momentarily energized appliances shall be limited to those used to prepare the vehicle for occupancy or travel.*

**(C) Bonding Voltage Converter Enclosures.** The non–current-carrying metal enclosure of the voltage converter shall be bonded to the frame of the vehicle with a minimum 8 AWG copper conductor. The voltage converter shall be provided with a separate chassis bonding conductor that shall not be used as a current-carrying conductor.

**(D) Dual-Voltage Fixtures, Including Luminaires or Appliances.** Fixtures, including luminaires, or appliances having both 120-volt and low-voltage connections shall be listed for dual voltage.

**(E) Autotransformers.** Autotransformers shall not be used.

**(F) Receptacles and Plug Caps.** Where a recreational vehicle is equipped with a 120-volt or 120/240-volt ac system, a low-voltage system, or both, receptacles and plug caps of the low-voltage system shall differ in configuration from those of the 120- or 120/240-volt system. Where a vehicle equipped with a battery or other low-voltage system has an external connection for low-voltage power, the connector shall have a configuration that will not accept 120-volt power.

## IV. Other Power Sources

### 551.30 Generator Installations.

**(A) Mounting.** Generators shall be mounted in such a manner as to be effectively bonded to the recreational vehicle chassis.

**(B) Generator Protection.** Equipment shall be installed to ensure that the current-carrying conductors from the engine generator and from an outside source are not connected to a vehicle circuit at the same time.

Receptacles used as disconnecting means shall be accessible (as applied to wiring methods) and capable of interrupting their rated current without hazard to the operator.

**(C) Installation of Storage Batteries and Generators.** Storage batteries and internal-combustion-driven generator units (subject to the provisions of this *Code*) shall be secured in place to avoid displacement from vibration and road shock.

**(D) Ventilation of Generator Compartments.** Compartments accommodating internal-combustion-driven generator units shall be provided with ventilation in accordance with instructions provided by the manufacturer of the generator unit.

FPN: For generator compartment construction requirements, see NFPA 1192-1999, *Standard on Recreational Vehicles.*

**(E) Supply Conductors.** The supply conductors from the engine generator to the first termination on the vehicle shall be of the stranded type and be installed in listed flexible conduit or listed liquidtight flexible conduit. The point of first termination shall be in one of the following:

(1) Panelboard
(2) Junction box with a blank cover
(3) Junction box with a receptacle
(4) Enclosed transfer switch
(5) Receptacle assembly listed in conjunction with the generator

The panelboard or junction box with a receptacle shall be installed within the vehicle's interior and within 450 mm (18 in.) of the compartment wall but not inside the compartment. If the generator is below the floor level and not in a compartment, the panelboard or junction box with receptacle shall be installed within the vehicle interior within 450 mm (18 in.) of the point of entry into the vehicle. A junction box with a blank cover shall be mounted on the compartment wall and shall be permitted inside or outside the compartment. A receptacle assembly listed in conjunction with the generator shall be mounted in accordance with its listing. If the generator is below floor level and not in a compartment, the junction box with blank cover shall be mounted either to any part of the generator supporting structure (but not to the generator) or to the vehicle floor within 450 mm (18 in.) of any point directly above the generator on either the inside or outside of the floor surface. Overcurrent protection in accordance with 240.4 shall be provided for supply conductors as an integral part of a listed generator or shall be located within 450 mm (18 in.) of their point of entry into the vehicle.

**551.31 Multiple Supply Source.**

**(A) Multiple Supply Sources.** Where a multiple supply system consisting of an alternate power source and a power-supply cord is installed, the feeder from the alternate power source shall be protected by an overcurrent-protective device. Installation shall be in accordance with 551.30(A) and (B) and 551.40.

**(B) Calculation of Loads.** Calculation of loads shall be in accordance with 551.42.

**(C) Multiple Supply Sources Capacity.** The multiple supply sources shall not be required to be of the same capacity.

**(D) Alternate Power Sources Exceeding 30 Amperes.** If an alternate power source exceeds 30 amperes, 120 volts, nominal, it shall be permissible to wire it as a 120-volt, nominal, system or a 120/240-volt, nominal, system, provided an overcurrent-protective device of the proper rating is installed in the feeder.

**(E) Power-Supply Assembly Not Less Than 30 Amperes.** The external power-supply assembly shall be permitted to be less than the calculated load but not less than 30 amperes and shall have overcurrent protection not greater than the capacity of the external power-supply assembly.

**551.32 Other Sources.** Other sources of ac power, such as inverters or motor generators, shall be listed for use in recreational vehicles and shall be installed in accordance with the terms of the listing. Other sources of ac power shall be wired in full conformity with the requirements in Parts I, III, IV, V, and VI of this article covering 120-volt electrical systems.

**551.33 Alternate Source Restriction.** Transfer equipment, if not integral with the listed power source, shall be installed to ensure that the current-carrying conductors from other sources of ac power and from an outside source are not connected to the vehicle circuit at the same time.

**V. Nominal 120-Volt or 120/240-Volt Systems**

**551.40 120-Volt or 120/240-Volt, Nominal, Systems.**

**(A) General Requirements.** The electrical equipment and material of recreational vehicles indicated for connection to a wiring system rated 120 volts, nominal, 2-wire with ground, or a wiring system rated 120/240 volts, nominal, 3-wire with ground, shall be listed and installed in accordance with the requirements of Parts I, III, IV, V, and VI of this article.

**(B) Materials and Equipment.** Electrical materials, devices, appliances, fittings, and other equipment installed in, intended for use in, or attached to the recreational vehicle shall be listed. All products shall be used only in the manner in which they have been tested and found suitable for the intended use.

**(C) Ground-Fault Circuit-Interrupter Protection.** The internal wiring of a recreational vehicle having only one 15- or 20-ampere branch circuit as permitted in 551.42(A) and (B) shall have ground-fault circuit-interrupter protection for personnel. The ground-fault circuit interrupter shall be installed at the point where the power supply assembly terminates within the recreational vehicle. Where a separable cord set is not employed, the ground-fault circuit interrupter shall be permitted to be an integral part of the attachment plug of the power supply assembly. The ground-fault circuit interrupter shall provide protection also under the conditions of an open grounded circuit conductor, interchanged circuit conductors, or both.

## 551.41 Receptacle Outlets Required.

**(A) Spacing.** Receptacle outlets shall be installed at wall spaces 600 mm (2 ft) wide or more so that no point along the floor line is more than 1.8 m (6 ft), measured horizontally, from an outlet in that space.

*Exception No. 1: Bath and hall areas.*

*Exception No. 2: Wall spaces occupied by kitchen cabinets, wardrobe cabinets, built-in furniture, behind doors that may open fully against a wall surface, or similar facilities.*

**(B) Location.** Receptacle outlets shall be installed as follows:

(1) Adjacent to countertops in the kitchen [at least one on each side of the sink if countertops are on each side and are 300 mm (12 in.) or over in width]
(2) Adjacent to the refrigerator and gas range space, except where a gas-fired refrigerator or cooking appliance, requiring no external electrical connection, is factory installed
(3) Adjacent to countertop spaces of 300 mm (12 in.) or more in width that cannot be reached from a receptacle required in 551.41(B)(1) by a cord of 1.8 m (6 ft) without crossing a traffic area, cooking appliance, or sink

**(C) Ground-Fault Circuit-Interrupter Protection.** Where provided, each 125-volt, single-phase, 15- or 20-ampere receptacle outlet shall have ground-fault circuit-interrupter protection for personnel in the following locations:

(1) Adjacent to a bathroom lavatory
(2) Where the receptacles are installed to serve the countertop surfaces and are within 1.8 m (6 ft) of any lavatory or sink

*Exception No. 1: Receptacles installed for appliances in dedicated spaces, such as for dishwashers, disposals, refrigerators, freezers, and laundry equipment.*

*Exception No. 2: Single receptacles for interior connections of expandable room sections.*

*Exception No. 3: De-energized receptacles that are within 1.8 m (6 ft) of any sink or lavatory due to the retraction of the expandable room section.*

(3) In the area occupied by a toilet, shower, tub, or any combination thereof
(4) On the exterior of the vehicle

*Exception: Receptacles that are located inside of an access panel that is installed on the exterior of the vehicle to supply power for an installed appliance shall not be required to have ground-fault circuit-interrupter protection.*

The receptacle outlet shall be permitted in a listed luminaire (lighting fixture). A receptacle outlet shall not be installed in a tub or combination tub–shower compartment.

**(D) Face-Up Position.** A receptacle shall not be installed in a face-up position in any countertop or similar horizontal surfaces within the living area.

**551.42 Branch Circuits Required.** Each recreational vehicle containing a 120-volt electrical system shall contain one of the following.

**(A) One 15-Ampere Circuit.** One 15-ampere circuit to supply lights, receptacle outlets, and fixed appliances. Such recreational vehicles shall be equipped with one 15-ampere switch and fuse or one 15-ampere circuit breaker.

**(B) One 20-Ampere Circuit.** One 20-ampere circuit to supply lights, receptacle outlets, and fixed appliances. Such recreational vehicles shall be equipped with one 20-ampere switch and fuse or one 20-ampere circuit breaker.

**(C) Two to Five 15- or 20-Ampere Circuits.** A maximum of five 15- or 20-ampere circuits to supply lights, receptacle outlets, and fixed appliances shall be permitted. Such recreational vehicles shall be equipped with a distribution panelboard rated at 120 volts maximum with a 30-ampere rated main power supply assembly. Not more than two 120-volt thermostatically controlled appliances (i.e., air conditioner and water heater) shall be installed in such systems unless appliance isolation switching, energy management systems, or similar methods are used.

*Exception: Additional 15- or 20-ampere circuits shall be permitted where a listed energy management system rated at 30-ampere maximum is employed within the system.*

> FPN: See 210.23(A) for permissible loads. See 551.45(C) for main disconnect and overcurrent protection requirements.

**(D) More Than Five Circuits Without a Listed Energy Management System.** A 50-ampere, 120/240-volt power-supply assembly shall be used where six or more circuits are employed. The load distribution shall ensure a reasonable current balance between phases.

## 551.43 Branch-Circuit Protection.

**(A) Rating.** The branch-circuit overcurrent devices shall be rated as follows:

(1) Not more than the circuit conductors, and
(2) Not more than 150 percent of the rating of a single appliance rated 13.3 amperes or more and supplied by an individual branch circuit, but

(3) Not more than the overcurrent protection size marked on an air conditioner or other motor-operated appliances

**(B) Protection for Smaller Conductors.** A 20-ampere fuse or circuit breaker shall be permitted for protection for fixtures, including a luminaire leads, cords, or small appliances, and 14 AWG tap conductors, not over 1.8 m (6 ft) long for recessed luminaires (lighting fixtures).

**(C) Fifteen-Ampere Receptacle Considered Protected by 20 Amperes.** If more than one receptacle or load is on a branch circuit, a 15-ampere receptacle shall be permitted to be protected by a 20-ampere fuse or circuit breaker.

**551.44 Power-Supply Assembly.** Each recreational vehicle shall have only one of the following main power-supply assemblies.

**(A) Fifteen-Ampere Main Power-Supply Assembly.** Recreational vehicles wired in accordance with 551.42(A) shall use a listed 15-ampere or larger main power-supply assembly.

**(B) Twenty-Ampere Main Power-Supply Assembly.** Recreational vehicles wired in accordance with 551.42(B) shall use a listed 20-ampere or larger main power-supply assembly.

**(C) Thirty-Ampere Main Power-Supply Assembly.** Recreational vehicles wired in accordance with 551.42(C) shall use a listed 30-ampere or larger main power-supply assembly.

**(D) Fifty-Ampere Power-Supply Assembly.** Recreational vehicles wired in accordance with 551.42(D) shall use a listed 50-ampere, 120/240-volt main power-supply assembly.

**551.45 Distribution Panelboard.**

**(A) Listed and Appropriately Rated.** A listed and appropriately rated distribution panelboard or other equipment specifically listed for the purpose shall be used. The grounded conductor termination bar shall be insulated from the enclosure as provided in 551.54(C). An equipment grounding terminal bar shall be attached inside the metal enclosure of the panelboard.

**(B) Location.** The distribution panelboard shall be installed in a readily accessible location. Working clearance for the panelboard shall be not less than 600 mm (24 in.) wide and 750 mm (30 in.) deep.

*Exception No. 1: Where the panelboard cover is exposed to the inside aisle space, one of the working clearance dimensions shall be permitted to be reduced to a minimum of 550 mm (22 in.). A panelboard is considered exposed where the panelboard cover is within 50 mm (2 in.) of the aisle's finished surface.*

*Exception No. 2: Compartment doors used for access to a generator shall be permitted to be equipped with a locking system.*

**(C) Dead-Front Type.** The distribution panelboard shall be of the dead-front type and shall consist of one or more circuit breakers or Type S fuseholders. A main disconnecting means shall be provided where fuses are used or where more than two circuit breakers are employed. A main overcurrent protective device not exceeding the power-supply assembly rating shall be provided where more than two branch circuits are employed.

**551.46 Means for Connecting to Power Supply.**

**(A) Assembly.** The power-supply assembly or assemblies shall be factory supplied or factory installed and be of one of the types specified herein.

**(1) Separable.** Where a separable power-supply assembly consisting of a cord with a female connector and molded attachment plug cap is provided, the vehicle shall be equipped with a permanently mounted, flanged surface inlet (male, recessed-type motor-base attachment plug) wired directly to the distribution panelboard by an approved wiring method. The attachment plug cap shall be of a listed type.

**(2) Permanently Connected.** Each power-supply assembly shall be connected directly to the terminals of the distribution panelboard or conductors within a junction box and provided with means to prevent strain from being transmitted to the terminals. The ampacity of the conductors between each junction box and the terminals of each distribution panelboard shall be at least equal to the ampacity of the power-supply cord. The supply end of the assembly shall be equipped with an attachment plug of the type described in 551.46(C). Where the cord passes through the walls or floors, it shall be protected by means of conduit and bushings or equivalent. The cord assembly shall have permanent provisions for protection against corrosion and mechanical damage while the vehicle is in transit.

**(B) Cord.** The cord exposed usable length shall be measured from the point of entrance to the recreational vehicle or the face of the flanged surface inlet (motor-base attachment plug) to the face of the attachment plug at the supply end.

The cord exposed usable length, measured to the point of entry on the vehicle exterior, shall be a minimum of 7.5 m (25 ft) where the point of entrance is at the side of the vehicle or shall be a minimum 9.0 m (30 ft) where the point of entrance is at the rear of the vehicle.

Where the cord entrance into the vehicle is more than 900 mm (3 ft) above the ground, the minimum cord lengths above shall be increased by the vertical distance of the cord entrance heights above 900 mm (3 ft).

FPN: See 551.46(E).

**(C) Attachment Plugs.**

**(1) Units with One 15-Ampere Branch Circuit.** Recreational vehicles having only one 15-ampere branch circuit as permitted by 551.42(A) shall have an attachment plug that shall be 2-pole, 3-wire grounding type, rated 15 amperes, 125 volts, conforming to the configuration shown in Figure 551.46(C).

FPN: Complete details of this configuration can be found in National Electrical Manufacturers Association's ANSI/NEMA WD 6-1989, *Standard for Dimensions of Attachment Plugs and Receptacle*, Figure 5.15.

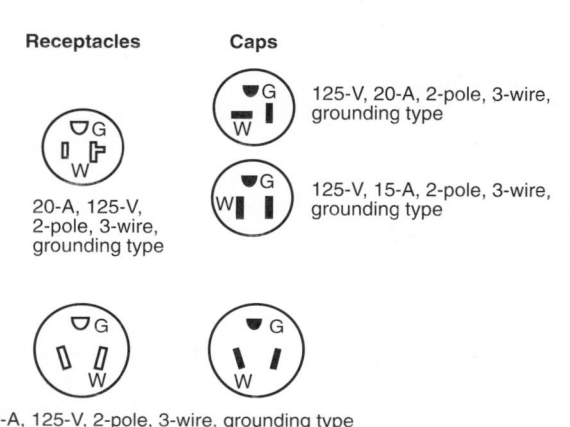

Figure 551.46(C)  Configurations for grounding-type receptacles and attachment plug caps used for recreational vehicle supply cords and recreational vehicle lots.

**(2) Units with One 20-Ampere Branch Circuit.** Recreational vehicles having only one 20-ampere branch circuit as permitted in 551.42(B) shall have an attachment plug that shall be 2-pole, 3-wire grounding type, rated 20 amperes, 125 volts, conforming to the configuration shown in Figure 551.46(C).

FPN: Complete details of this configuration can be found in ANSI/NEMA WD 6-1989, National Electrical Manufacturers Association's *Standard for Dimensions of Attachment Plugs and Receptacles*, Figure 5.20.

**(3) Units with Two to Five 15- or 20-Ampere Branch Circuits.** Recreational vehicles wired in accordance with 551.42(C) shall have an attachment plug that shall be 2-pole, 3-wire grounding type, rated 30 amperes, 125 volts, conforming to the configuration shown in Figure 551.46(C), intended for use with units rated at 30 amperes, 125 volts.

FPN: Complete details of this configuration can be found in ANSI/NEMA WD 6-1989, National Electrical Manufacturers Association's *Standard for Dimensions of Attachment Plugs and Receptacles*, Figure TT.

**(4) Units with 50-Ampere Power Supply Assembly.** Recreational vehicles having a power-supply assembly rated 50 amperes as permitted by 551.42(D) shall have a 3-pole, 4-wire grounding-type attachment plug rated 50 amperes, 125/250 volts, conforming to the configuration shown in Figure 551.46(C).

FPN: Complete details of this configuration can be found in ANSI/NEMA WD 6-1989, National Electrical Manufacturers Association's *Standard for Dimensions of Attachment Plugs and Receptacles*, Figure 14.50.

**(D) Labeling at Electrical Entrance.** Each recreational vehicle shall have permanently affixed to the exterior skin, at or near the point of entrance of the power-supply cord(s), a label 75 mm × 45 mm (3 in. × 1¾ in.) minimum size, made of etched, metal-stamped, or embossed brass, stainless steel, or anodized or alclad aluminum not less than 0.51 mm (0.020 in.) thick, or other suitable material [e.g., 0.13 mm (0.005 in.) thick plastic laminate] that reads, as appropriate, either

THIS CONNECTION IS FOR 110–125-VOLT AC,
60 HZ, _____ AMPERE SUPPLY.

or

THIS CONNECTION IS FOR 120/240-VOLT AC,
3-POLE, 4-WIRE,
60 HZ, _____ AMPERE SUPPLY.

The correct ampere rating shall be marked in the blank space.

**(E) Location.** The point of entrance of a power-supply assembly shall be located within 4.5 m (15 ft) of the rear, on the left (road) side or at the rear, left of the longitudinal center of the vehicle, within 450 mm (18 in.) of the outside wall.

*Exception No. 1: A recreational vehicle equipped with only a listed flexible drain system or a side-vent drain system shall be permitted to have the electrical point of entrance located on either side, provided the drain(s) for the plumbing system is (are) located on the same side.*

*Exception No. 2: A recreational vehicle shall be permitted to have the electrical point of entrance located more than*

*4.5 m (15 ft) from the rear. Where this occurs, the distance beyond the 4.5-m (15-ft) dimension shall be added to the cord's minimum length as specified in 551.46(B).*

## 551.47 Wiring Methods.

**(A) Wiring Systems.** Cables and raceways installed in accordance with Articles 320, 322, 330 through 340, 342 through 362, 386, and 388 shall be permitted in accordance with their applicable article, except as otherwise specified in this article. An equipment grounding means shall be provided in accordance with 250.118.

**(B) Conduit and Tubing.** Where rigid metal conduit or intermediate metal conduit is terminated at an enclosure with a locknut and bushing connection, two locknuts shall be provided, one inside and one outside of the enclosure. All cut ends of conduit and tubing shall be reamed or otherwise finished to remove rough edges.

**(C) Nonmetallic Boxes.** Nonmetallic boxes shall be acceptable only with nonmetallic-sheathed cable or nonmetallic raceways.

**(D) Boxes.** In walls and ceilings constructed of wood or other combustible material, boxes and fittings shall be flush with the finished surface or project therefrom.

**(E) Mounting.** Wall and ceiling boxes shall be mounted in accordance with Article 314.

*Exception No. 1: Snap-in-type boxes or boxes provided with special wall or ceiling brackets that securely fasten boxes in walls or ceilings shall be permitted.*

*Exception No. 2: A wooden plate providing a 38-mm (1½-in.) minimum width backing around the box and of a thickness of 13 mm (½ in.) or greater (actual) attached directly to the wall panel shall be considered as approved means for mounting outlet boxes.*

**(F) Raceway and Cable Continuity.** Raceways and cable sheaths shall be continuous between boxes and other enclosures.

**(G) Protected.** Metal-clad, Type AC, or nonmetallic-sheathed cables and electrical nonmetallic tubing shall be permitted to pass through the centers of the wide side of 2 by 4 wood studs. However, they shall be protected where they pass through 2 by 2 wood studs or at other wood studs or frames where the cable or tubing would be less than 32 mm (1¼ in.) from the inside or outside surface. Steel plates on each side of the cable or tubing or a steel tube, with not less than 1.35 mm (0.053 in.) wall thickness, shall be installed to protect the cable or tubing. These plates or tubes shall be securely held in place. Where nonmetallic-sheathed cables pass through punched, cut, or drilled slots or holes in metal members, the cable shall be protected by bushings or grommets securely fastened in the opening prior to installation of the cable.

**(H) Bends.** No bend shall have a radius of less than five times the cable diameter.

**(I) Cable Supports.** Where connected with cable connectors or clamps, cables shall be supported within 300 mm (12 in.) of outlet boxes, distribution panelboards, and splice boxes on appliances. Supports shall be provided every 1.4 m (4½ ft) at other places.

**(J) Nonmetallic Box Without Cable Clamps.** Nonmetallic-sheathed cables shall be supported within 200 mm (8 in.) of a nonmetallic outlet box without cable clamps. Where wiring devices with integral enclosures are employed with a loop of extra cable to permit future replacement of the device, the cable loop shall be considered as an integral portion of the device.

**(K) Physical Damage.** Where subject to physical damage, exposed nonmetallic cable shall be protected by covering boards, guard strips, raceways, or other means.

**(L) Metal Faceplates.** Metal faceplates shall be of ferrous metal not less than 0.76 mm (0.030 in.) in thickness or of nonferrous metal not less than 1.0 mm (0.040 in.) in thickness. Nonmetallic faceplates shall be listed.

**(M) Metal Faceplates Effectively Grounded.** Where metal faceplates are used, they shall be effectively grounded.

**(N) Moisture or Physical Damage.** Where outdoor or under-chassis wiring is 120 volts, nominal, or over and is exposed to moisture or physical damage, the wiring shall be protected by rigid metal conduit, by intermediate metal conduit, or by electrical metallic tubing or rigid nonmetallic conduit that is closely routed against frames and equipment enclosures or other raceway or cable identified for the application.

**(O) Component Interconnections.** Fittings and connectors that are intended to be concealed at the time of assembly shall be listed and identified for the interconnection of building components. Such fittings and connectors shall be equal to the wiring method employed in insulation, temperature rise, and fault-current withstanding and shall be capable of enduring the vibration and shock occurring in recreational vehicles.

**(P) Method of Connecting Expandable Units.** The method of connecting expandable units to the main body of the vehicle shall comply with the following as applicable:

(1) That portion of a branch circuit that is installed in an expandable unit shall be permitted to be connected to

the portion of the branch circuit in the main body of the vehicle by means of a flexible cord or attachment plug and cord listed for hard usage. The cord and its connections shall conform to all provisions of Article 400 and shall be considered as a permitted use under 400.7. Where the attachment plug and cord are located within the vehicle's interior, use of plastic thermoset or elastomer parallel cord Type SPT-3, SP-3, or SPE shall be permitted.

(2) If the receptacle provided for connection of the cord to the main circuit is located on the outside of the vehicle, it shall be protected with a ground-fault circuit interrupter for personnel and be listed for wet locations. A cord located on the outside of a vehicle shall be identified for outdoor use.

(3) Unless removable or stored within the vehicle interior, the cord assembly shall have permanent provisions for protection against corrosion and mechanical damage while the vehicle is in transit.

(4) If an attachment plug and cord is used, it shall be installed so as not to permit exposed live attachment plug pins.

**(Q) Prewiring for Air-Conditioning Installation.** Prewiring installed for the purpose of facilitating future air-conditioning installation shall conform to the applicable portions of this article and the following:

(1) An overcurrent protective device with a rating compatible with the circuit conductors shall be installed in the distribution panelboard and wiring connections completed.

(2) The load end of the circuit shall terminate in a junction box with a blank cover or a device listed for the purpose. Where a junction box with a blank cover is used, the free ends of the conductors shall be adequately capped or taped.

(3) A label conforming to 551.46(D) shall be placed on or adjacent to the junction box and shall read

AIR-CONDITIONING CIRCUIT.
THIS CONNECTION IS FOR AIR CONDITIONERS
RATED 110–125-VOLT AC, 60 HZ,
___ AMPERES MAXIMUM.
DO NOT EXCEED CIRCUIT RATING.

An ampere rating, not to exceed 80 percent of the circuit rating, shall be legibly marked in the blank space.

(4) The circuit shall serve no other purpose.

**(R) Prewiring for Generator Installation.** Prewiring installed for the purpose of facilitating future generator installation shall conform to the other applicable portions of this article and the following:

(1) Circuit conductors shall be appropriately sized in relation to the anticipated load and shall be protected by an overcurrent device in accordance with their ampacities.

(2) Where junction boxes are utilized at either of the circuit originating or terminus points, free ends of the conductors shall be adequately capped or taped.

(3) Where devices such as receptacle outlet, transfer switch, and so forth, are installed, the installation shall be complete, including circuit conductor connections. All devices shall be listed and appropriately rated.

(4) A label conforming to 551.46(D) shall be placed on the cover of each junction box containing incomplete circuitry and shall read, as appropriate, either

GENERATOR CIRCUIT. THIS CONNECTION
IS FOR GENERATORS RATED 110–125-VOLTAC,
60 HZ, _____ AMPERES MAXIMUM.

or

GENERATOR CIRCUIT. THIS CONNECTION
IS FOR GENERATORS RATED 120/240-VOLT AC,
60 HZ, _____ AMPERES MAXIMUM.

The correct ampere rating shall be legibly marked in the blank space.

**551.48 Conductors and Boxes.** The maximum number of conductors permitted in boxes shall be in accordance with 314.16.

**551.49 Grounded Conductors.** The identification of grounded conductors shall be in accordance with 200.6.

**551.50 Connection of Terminals and Splices.** Conductor splices and connections at terminals shall be in accordance with 110.14.

**551.51 Switches.** Switches shall be rated in accordance with 551.51(A) and (B).

**(A) Lighting Circuits.** For lighting circuits, switches shall be rated not less than 10 amperes, 120–125 volts and in no case less than the connected load.

**(B) Motors or Other Loads.** For motors or other loads, switches shall have ampere or horsepower ratings, or both, adequate for loads controlled. (An ac general-use snap switch shall be permitted to control a motor 2 hp or less with full-load current not over 80 percent of the switch ampere rating.)

**551.52 Receptacles.** All receptacle outlets shall be of the grounding type and installed in accordance with 406.3 and 210.21.

## 551.53 Luminaires (Lighting Fixtures).

**(A) General.** Any combustible wall or ceiling finish exposed between the edge of a luminaire (fixture) canopy, or pan and the outlet box, shall be covered with noncombustible material or a material identified for the purpose.

**(B) Shower Luminaires (Fixtures).** If a luminaire (lighting fixture) is provided over a bathtub or in a shower stall, it shall be of the enclosed and gasketed type and listed for the type of installation, and it shall be ground-fault circuit-interrupter protected.

The switch for shower luminaires (lighting fixtures) and exhaust fans, located over a tub or in a shower stall, shall be located outside the tub or shower space.

**(C) Outdoor Outlets, Luminaires (Fixtures), Air-Cooling Equipment, and So On.** Outdoor luminaires (fixtures) and other equipment shall be listed for outdoor use.

## 551.54 Grounding. (See also 551.56 on bonding of non–current-carrying metal parts.)

**(A) Power-Supply Grounding.** The grounding conductor in the supply cord or feeder shall be connected to the grounding bus or other approved grounding means in the distribution panelboard.

**(B) Distribution Panelboard.** The distribution panelboard shall have a grounding bus with sufficient terminals for all grounding conductors or other approved grounding means.

**(C) Insulated Neutral.** The grounded circuit conductor (neutral) shall be insulated from the equipment grounding conductors and from equipment enclosures and other grounded parts. The grounded (neutral) circuit terminals in the distribution panelboard and in ranges, clothes dryers, counter-mounted cooking units, and wall-mounted ovens shall be insulated from the equipment enclosure. Bonding screws, straps, or buses in the distribution panelboard or in appliances shall be removed and discarded. Connection of electric ranges and electric clothes dryers utilizing a grounded (neutral) conductor, if cord-connected, shall be made with 4-conductor cord and 3-pole, 4-wire grounding-type plug caps and receptacles.

## 551.55 Interior Equipment Grounding.

**(A) Exposed Metal Parts.** In the electrical system, all exposed metal parts, enclosures, frames, luminaire (lighting fixture) canopies, and so forth, shall be effectively bonded to the grounding terminals or enclosure of the distribution panelboard.

**(B) Equipment Grounding and Bonding Conductors.** Bare wires, insulated wire with an outer finish that is green or green with one or more yellow stripes shall be used for equipment grounding or bonding conductors only.

**(C) Grounding of Electrical Equipment.** Where grounding of electrical equipment is specified, it shall be permitted as follows:

(1) Connection of metal raceway (conduit or electrical metallic tubing), the sheath of Type MC and Type MI cable where the sheath is identified for grounding, or the armor of Type AC cable to metal enclosures.

(2) A connection between the one or more equipment grounding conductors and a metal box by means of a grounding screw, which shall be used for no other purpose, or a listed grounding device.

(3) The equipment grounding conductor in nonmetallic-sheathed cable shall be permitted to be secured under a screw threaded into the luminaire (fixture) canopy other than a mounting screw or cover screw, or attached to a listed grounding means (plate) in a nonmetallic outlet box for luminaire (fixture) mounting. [Grounding means shall also be permitted for luminaire (fixture) attachment screws.])

**(D) Grounding Connection in Nonmetallic Box.** A connection between the one or more grounding conductors brought into a nonmetallic outlet box shall be so arranged that a connection can be made to any fitting or device in that box that requires grounding.

**(E) Grounding Continuity.** Where more than one equipment grounding or bonding conductor of a branch circuit enters a box, all such conductors shall be in good electrical contact with each other, and the arrangement shall be such that the disconnection or removal of a receptacle, luminaire (fixture), or other device fed from the box will not interfere with or interrupt the grounding continuity.

**(F) Cord-Connected Appliances.** Cord-connected appliances, such as washing machines, clothes dryers, refrigerators, and the electrical system of gas ranges, and so forth, shall be grounded by means of an approved cord with equipment grounding conductor and grounding-type attachment plug.

## 551.56 Bonding of Non–Current-Carrying Metal Parts.

**(A) Required Bonding.** All exposed non–current-carrying metal parts that may become energized shall be effectively bonded to the grounding terminal or enclosure of the distribution panelboard.

**(B) Bonding Chassis.** A bonding conductor shall be connected between any distribution panelboard and an accessible terminal on the chassis. Aluminum or copper-clad aluminum conductors shall not be used for bonding if such conductors or their terminals are exposed to corrosive elements.

*Exception: Any recreational vehicle that employs a unitized metal chassis-frame construction to which the distribution panelboard is securely fastened with a bolt(s) and nut(s) or by welding or riveting shall be considered to be bonded.*

**(C) Bonding Conductor Requirements.** Grounding terminals shall be of the solderless type and listed as pressure terminal connectors recognized for the wire size used. The bonding conductor shall be solid or stranded, insulated or bare, and shall be 8 AWG copper minimum, or equal.

**(D) Metallic Roof and Exterior Bonding.** The metal roof and exterior covering shall be considered bonded where

(1) The metal panels overlap one another and are securely attached to the wood or metal frame parts by metal fasteners, and
(2) The lower panel of the metal exterior covering is secured by metal fasteners at each cross member of the chassis, or the lower panel is bonded to the chassis by a metal strap.

**(E) Gas, Water, and Waste Pipe Bonding.** The gas, water, and waste pipes shall be considered grounded if they are bonded to the chassis.

**(F) Furnace and Metal Air Duct Bonding.** Furnace and metal circulating air ducts shall be bonded.

**551.57 Appliance Accessibility and Fastening.** Every appliance shall be accessible for inspection, service, repair, and replacement without removal of permanent construction. Means shall be provided to securely fasten appliances in place when the recreational vehicle is in transit.

## VI. Factory Tests

**551.60 Factory Tests (Electrical).** Each recreational vehicle shall be subjected to the following tests.

**(A) Circuits of 120 Volts or 120/240 Volts.** Each recreational vehicle designed with a 120-volt or a 120/240-volt electrical system shall withstand the applied potential without electrical breakdown of a 1-minute, 900-volt dielectric strength test, or a 1-second, 1080-volt dielectric strength test, with all switches closed, between ungrounded and grounded conductors and the recreational vehicle ground. During the test, all switches and other controls shall be in the "on" position. Fixtures, including luminaires and permanently installed appliances shall not be required to withstand this test. The test shall be performed after branch circuits are complete prior to energizing the system and

again after all outer coverings and cabinetry have been secured.

Each recreational vehicle shall be subjected to all of the following:

(1) A continuity test to ensure that all metal parts are properly bonded
(2) Operational tests to demonstrate that all equipment is properly connected and in working order
(3) Polarity checks to determine that connections have been properly made

**(B) Low-Voltage Circuits.** An operational test of all low-voltage circuits shall be conducted to demonstrate that all equipment is connected and in electrical working order. This test shall be performed in the final stages of production after all outer coverings and cabinetry have been secured.

## VII. Recreational Vehicle Parks

**551.71 Type Receptacles Provided.** Every recreational vehicle site with electrical supply shall be equipped with at least one 20-ampere, 125-volt receptacle. A minimum of 5 percent of all recreational vehicle sites, with electrical supply, shall each be equipped with a 50-ampere, 125/250-volt receptacle conforming to the configuration as identified in Figure 551.46(C). These electrical supplies shall be permitted to include additional receptacles that have configurations in accordance with 551.81. A minimum of 70 percent of all recreational vehicle sites with electrical supply shall each be equipped with a 30-ampere, 125-volt receptacle conforming to Figure 551.46(C). This supply shall be permitted to include additional receptacle configurations conforming to 551.81. The remainder of all recreational vehicle sites with electrical supply shall be equipped with one or more of the receptacle configurations conforming to 551.81. Dedicated tent sites with a 15- or 20-ampere electrical supply shall be permitted to be excluded when determining the percentage of recreational vehicle sites with 30- or 50-ampere receptacles.

Additional receptacles shall be permitted for the connection of electrical equipment outside the recreational vehicle within the recreational vehicle park.

All 125-volt, single-phase, 15- and 20-ampere receptacles shall have listed ground-fault circuit-interrupter protection for personnel.

**551.72 Distribution System.** Receptacles rated at 50 amperes shall be supplied from a branch circuit of the voltage class and rating of the receptacle. Other recreational vehicle sites with 125-volt, 20- and 30-ampere receptacles shall be permitted to be derived from any grounded distribution system that supplies 120-volt single-phase power. The neutral conductors shall not be reduced in size below the size of

**Table 551.73 Demand Factors for Site Feeders and Service-Entrance Conductors for Park Sites**

| Number of Recreational Vehicle Sites | Demand Factor (percent) |
|---|---|
| 1 | 100 |
| 2 | 90 |
| 3 | 80 |
| 4 | 75 |
| 5 | 65 |
| 6 | 60 |
| 7 – 9 | 55 |
| 10 –12 | 50 |
| 13 –15 | 48 |
| 16 –18 | 47 |
| 19 –21 | 45 |
| 22 –24 | 43 |
| 25 –35 | 42 |
| 36 plus | 41 |

the ungrounded conductors for the site distribution. The neutral conductors shall be permitted to be reduced in size below the minimum required size of the ungrounded conductors for 240-volt, line-to-line, permanently connected loads only.

## 551.73 Calculated Load.

**(A) Basis of Calculations.** Electrical service and feeders shall be calculated on the basis of not less than 9600 volt-amperes per site equipped with 50-ampere, 120/240-volt supply facilities; 3600 volt-amperes per site equipped with both 20-ampere and 30-ampere supply facilities; 2400 volt-amperes per site equipped with only 20-ampere supply facilities; and 600 volt-amperes per site equipped with only 20-ampere supply facilities that are dedicated to tent sites. The demand factors set forth in Table 551.73 shall be the minimum allowable demand factors that shall be permitted in calculating load for service and feeders. Where the electrical supply for a recreational vehicle site has more than one receptacle, the calculated load shall only be computed for the highest rated receptacle.

**(B) Transformers and Secondary Distribution Panelboards.** For the purpose of this *Code*, where the park service exceeds 240 volts, transformers and secondary distribution panelboards shall be treated as services.

**(C) Demand Factors.** The demand factor for a given number of sites shall apply to all sites indicated. For example, 20 sites calculated at 45 percent of 3600 volt-amperes results in a permissible demand of 1620 volt-amperes per site or a total of 32,400 volt-amperes for 20 sites.

> FPN: These demand factors may be inadequate in areas of extreme hot or cold temperature with loaded circuits for heating or air conditioning.

**(D) Feeder-Circuit Capacity.** Recreational vehicle site feeder-circuit conductors shall have adequate ampacity for the loads supplied and shall be rated at not less than 30 amperes. The grounded conductors shall have the same ampacity as the ungrounded conductors.

> FPN: Due to the long circuit lengths typical in most recreational vehicle parks, feeder conductor sizes found in the ampacity tables of Article 310 may be inadequate to maintain the voltage regulation suggested in the fine print note to 210.19. Total circuit voltage drop is a sum of the voltage drops of each serial circuit segment, where the load for each segment is calculated using the load that segment sees and the demand factors of 551.73(A).

Loads for other amenities such as, but not limited to, service buildings, recreational buildings, and swimming pools shall be sized separately and then be added to the value calculated for the recreational vehicle sites where they are all supplied by one service.

**551.74 Overcurrent Protection.** Overcurrent protection shall be provided in accordance with Article 240.

**551.75 Grounding.** All electrical equipment and installations in recreational vehicle parks shall be grounded as required by Article 250.

## 551.76 Grounding — Recreational Vehicle Site Supply Equipment.

**(A) Exposed Non–Current-Carrying Metal Parts.** Exposed non–current-carrying metal parts of fixed equipment, metal boxes, cabinets, and fittings that are not electrically connected to grounded equipment shall be grounded by a continuous equipment grounding conductor run with the circuit conductors from the service equipment or from the transformer of a secondary distribution system. Equipment grounding conductors shall be sized in accordance with 250.122 and shall be permitted to be spliced by listed means.

The arrangement of equipment grounding connections shall be such that the disconnection or removal of a receptacle or other device will not interfere with, or interrupt, the grounding continuity.

**(B) Secondary Distribution System.** Each secondary distribution system shall be grounded at the transformer.

**(C) Neutral Conductor Not to Be Used as an Equipment Ground.** The neutral conductor shall not be used as an equipment ground for recreational vehicles or equipment within the recreational vehicle park.

**(D) No Connection on the Load Side.** No connection to a grounding electrode shall be made to the neutral conductor on the load side of the service disconnecting means or transformer distribution panelboard.

## 551.77 Recreational Vehicle Site Supply Equipment.

**(A) Location.** Where provided on back-in sites, the recreational vehicle site electrical supply equipment shall be located on the left (road) side of the parked vehicle, on a line that is 1.5 m to 2.1 m (5 ft to 7 ft) from the left edge (driver's side of the parked RV) of the stand and shall be located at any point on this line from the rear of the stand to 4.5 m (15 ft) forward of the rear of the stand.

For pull-through sites, the electrical supply equipment shall be permitted to be located at any point along the line that is 1.5 m to 2.1 m (5 ft to 7 ft) from the left edge (driver's side of the parked RV) from 4.9 m (16 ft) forward of the rear of the stand to the center point between the two roads that gives access to and egress from the pull-through sites.

The left edge (driver's side of the parked RV) of the stand shall be marked.

**(B) Disconnecting Means.** A disconnecting switch or circuit breaker shall be provided in the site supply equipment for disconnecting the power supply to the recreational vehicle.

**(C) Access.** All site supply equipment shall be accessible by an unobstructed entrance or passageway not less than 600 mm (2 ft) wide and 2.0 m (6 ft 6 in.) high.

**(D) Mounting Height.** Site supply equipment shall be located not less than 600 mm (2 ft) or more than 2.0 m (6 ft 6 in.) above the ground.

**(E) Working Space.** Sufficient space shall be provided and maintained about all electrical equipment to permit ready and safe operation, in accordance with 110.26.

**(F) Marking.** Where the site supply equipment contains a 125/250-volt receptacle, the equipment shall be marked as follows: "Turn disconnecting switch or circuit breaker off before inserting or removing plug. Plug must be fully inserted or removed." The marking shall be located on the equipment adjacent to the receptacle outlet.

## 551.78 Protection of Outdoor Equipment.

**(A) Wet Locations.** All switches, circuit breakers, receptacles, control equipment, and metering devices located in wet locations or outside of a building shall be rainproof equipment.

**(B) Meters.** If secondary meters are installed, meter sockets without meters installed shall be blanked off with an approved blanking plate.

## 551.79 Clearance for Overhead Conductors. Open conductors of not over 600 volts, nominal, shall have a vertical clearance of not less than 5.5 m (18 ft) and a horizontal clearance of not less than 900 mm (3 ft) in all areas subject to recreational vehicle movement. In all other areas, clearances shall conform to 225.18 and 225.19.

FPN: For clearances of conductors over 600 volts, nominal, see ANSI C2-1997, *National Electrical Safety Code.*

## 551.80 Underground Service, Feeder, Branch-Circuit, and Recreational Vehicle Site Feeder-Circuit Conductors.

**(A) General.** All direct-burial conductors, including the equipment grounding conductor if of aluminum, shall be insulated and identified for the use. All conductors shall be continuous from equipment to equipment. All splices and taps shall be made in approved junction boxes or by use of material listed and identified for the purpose.

**(B) Protection Against Physical Damage.** Direct-buried conductors and cables entering or leaving a trench shall be protected by rigid metal conduit, intermediate metal conduit, electrical metallic tubing with supplementary corrosion protection, rigid nonmetallic conduit, liquidtight flexible nonmetallic conduit, liquidtight flexible metal conduit, or other approved raceways or enclosures. Where subject to physical damage, the conductors or cables shall be protected by rigid metal conduit, intermediate metal conduit, or Schedule 80 rigid nonmetallic conduit. All such protection shall extend at least 450 mm (18 in.) into the trench from finished grade.

FPN: See 300.5 and Article 340 for conductors or Type UF cable used underground or in direct burial in earth.

## 551.81 Receptacles. A receptacle to supply electric power to a recreational vehicle shall be one of the configurations shown in Figure 551.46(C) in the following ratings.

(1) 50-ampere — 125/250-volt, 50-ampere, 3-pole, 4-wire grounding type for 120/240-volt systems
(2) 30-ampere — 125-volt, 30-ampere, 2-pole, 3-wire grounding type for 120-volt systems
(3) 20-ampere — 125-volt, 20-ampere, 2-pole, 3-wire grounding type for 120-volt systems

FPN: Complete details of these configurations can be found in ANSI/NEMA WD 6-1989, National Electrical Manufacturers Association's *Standard for Dimensions of Attachment Plugs and Receptacles*, Figures 14-50, TT, and 5-20.

# ARTICLE 552
# Park Trailers

## I. General

**552.1 Scope.** The provisions of this article cover the electrical conductors and equipment installed within or on park trailers not covered fully under Articles 550 and 551.

**552.2 Definition.** (See Articles 100, 550, and 551 for additional definitions.)

**Park Trailer.** A unit that is built on a single chassis mounted on wheels and has a gross trailer area not exceeding 37 m² (400 ft²) in the set-up mode.

**552.3 Other Articles.** Wherever the requirements of other articles of this *Code* and Article 552 differ, the requirements of Article 552 shall apply.

**552.4 General Requirements.** A park trailer as specified in 552.2 is intended for seasonal use. It is not intended as a permanent dwelling unit or for commercial uses such as banks, clinics, offices, or similar.

## II. Low-Voltage Systems

**552.10 Low-Voltage Systems.**

**(A) Low-Voltage Circuits.** Low-voltage circuits furnished and installed by the park trailer manufacturer, other than those related to braking, are subject to this *Code*. Circuits supplying lights subject to federal or state regulations shall comply with applicable government regulations and this *Code*.

**(B) Low-Voltage Wiring.**

**(1) Material.** Copper conductors shall be used for low-voltage circuits.

*Exception: A metal chassis or frame shall be permitted as the return path to the source of supply.*

**(2) Conductor Types.** Conductors shall conform to the requirements for Type GXL, HDT, SGT, SGR, or Type SXL or shall have insulation in accordance with Table 310.13 or the equivalent. Conductor sizes 6 AWG through 18 AWG or SAE shall be listed. Single-wire, low-voltage conductors shall be of the stranded type.

FPN: See SAE Standard J1128-1995 for Types GXL, HDT, and SXL and SAE Standard J1127-1995 for Types SGT and SGR.

**(3) Marking.** All insulated low-voltage conductors shall be surface marked at intervals not greater than 1.2 m (4 ft) as follows:

(1) Listed conductors shall be marked as required by the listing agency.
(2) SAE conductors shall be marked with the name or logo of the manufacturer, specification designation, and wire gauge.
(3) Other conductors shall be marked with the name or logo of the manufacturer, temperature rating, wire gauge, conductor material, and insulation thickness.

**(C) Low-Voltage Wiring Methods.**

**(1) Physical Protection.** Conductors shall be protected against physical damage and shall be secured. Where insulated conductors are clamped to the structure, the conductor insulation shall be supplemented by an additional wrap or layer of equivalent material, except that jacketed cables shall not be required to be so protected. Wiring shall be routed away from sharp edges, moving parts, or heat sources.

**(2) Splices.** Conductors shall be spliced or joined with splicing devices that provide a secure connection or by brazing, welding, or soldering with a fusible metal or alloy. Soldered splices shall first be spliced or joined to be mechanically and electrically secure without solder and then soldered. All splices, joints, and free ends of conductors shall be covered with an insulation equivalent to that on the conductors.

**(3) Separation.** Battery and other low-voltage circuits shall be physically separated by at least a 13-mm (½-in.) gap or other approved means from circuits of a different power source. Acceptable methods shall be by clamping, routing, or equivalent means that ensure permanent total separation. Where circuits of different power sources cross, the external jacket of the nonmetallic-sheathed cables shall be deemed adequate separation.

**(4) Ground Connections.** Ground connections to the chassis or frame shall be made in an accessible location and shall be mechanically secure. Ground connections shall be by means of copper conductors and copper or copper-alloy terminals of the solderless type identified for the size of wire used. The surface on which ground terminals make contact shall be cleaned and be free from oxide or paint or shall be electrically connected through the use of a cadmium, tin, or zinc-plated internal/external-toothed lockwasher or locking terminals. Ground terminal attaching screws, rivets or bolts, nuts, and lockwashers shall be cadmium, tin, or zinc-plated except rivets shall be permitted to be unanodized aluminum where attaching to aluminum structures.

The chassis-grounding terminal of the battery shall be bonded to the unit chassis with a minimum 8 AWG copper conductor. In the event the power lead from the battery exceeds 8 AWG, the bonding conductor shall be of an equal size.

**(D) Battery Installations.** Storage batteries subject to the provisions of this *Code* shall be securely attached to the unit and installed in an area vaportight to the interior and ventilated directly to the exterior of the unit. Where batteries are installed in a compartment, the compartment shall be ventilated with openings having a minimum area of 1100 mm$^2$ (1.7 in.$^2$) at both the top and at the bottom. Where compartment doors are equipped for ventilation, the openings shall be within 50 mm (2 in.) of the top and bottom. Batteries shall not be installed in a compartment containing spark- or flame-producing equipment.

**(E) Overcurrent Protection.**

**(1) Rating.** Low-voltage circuit wiring shall be protected by overcurrent protective devices rated not in excess of the ampacity of copper conductors, in accordance with Table 552.10(E)(1).

**Table 552.10(E)(1) Low-Voltage Overcurrent Protection**

| Wire Size (AWG) | Ampacity | Wire Type |
|---|---|---|
| 18 | 6 | Stranded only |
| 16 | 8 | Stranded only |
| 14 | 15 | Stranded or solid |
| 12 | 20 | Stranded or solid |
| 10 | 30 | Stranded or solid |

**(2) Type.** Circuit breakers or fuses shall be of an approved type, including automotive types. Fuseholders shall be clearly marked with maximum fuse size and shall be protected against shorting and physical damage by a cover or equivalent means.

> FPN: For further information, see ANSI/SAE J554-1987, *Standard for Electric Fuses (Cartridge Type)*; SAE J1284-1988, *Standard for Blade Type Electric Fuses*; and UL 275-1993, *Standard for Automotive Glass Tube Fuses.*

**(3) Appliances.** Higher-current-consuming dc appliances such as pumps, compressors, heater blowers, and similar motor-driven appliances shall be installed in accordance with the manufacturer's instructions.

Motors that are controlled by automatic switching or by latching-type manual switches shall be protected in accordance with 430.32(B).

**(4) Location.** The overcurrent protective device shall be installed in an accessible location on the unit within 450 mm (18 in.) of the point where the power supply connects to the unit circuits. If located outside the park trailer, the device shall be protected against weather and physical damage.

*Exception: External low-voltage supply shall be permitted to have the overcurrent protective device within 450 mm (18 in.) after entering the unit or after leaving a metal raceway.*

**(F) Switches.** Switches shall have a dc rating not less than the connected load.

**(G) Luminaires (Lighting Fixtures).** All low-voltage interior luminaires (lighting fixtures) rated more than 4 watts, employing lamps rated more than 1.2 watts, shall be listed.

### III. Combination Electrical Systems

### 552.20 Combination Electrical Systems.

**(A) General.** Unit wiring suitable for connection to a battery or other low-voltage supply source shall be permitted to be connected to a 120-volt source, provided that the entire wiring system and equipment are rated and installed in full conformity with Parts I, III, IV, and V requirements of this article covering 120-volt electrical systems. Circuits fed from ac transformers shall not supply dc appliances.

**(B) Voltage Converters (120-Volt Alternating Current to Low-Voltage Direct Current).** The 120-volt ac side of the voltage converter shall be wired in full conformity with Parts I, III, IV, and V requirements of this article for 120-volt electrical systems.

*Exception: Converters supplied as an integral part of a listed appliance shall not be subject to the above.*

All converters and transformers shall be listed for use in recreation units and designed or equipped to provide over-temperature protection. To determine the converter rating, the following formula shall be applied to the total connected load, including average battery charging rate, of all 12-volt equipment:

The first 20 amperes of load at 100 percent; plus
The second 20 amperes of load at 50 percent; plus
All load above 40 amperes at 25 percent

*Exception: A low-voltage appliance that is controlled by a momentary switch (normally open) that has no means for holding in the closed position shall not be considered as a connected load when determining the required converter rating. Momentarily energized appliances shall be limited to those used to prepare the unit for occupancy or travel.*

**(C) Bonding Voltage Converter Enclosures.** The non–current-carrying metal enclosure of the voltage converter shall be bonded to the frame of the unit with a 8 AWG copper conductor minimum. The grounding conductor for the battery and the metal enclosure shall be permitted to be the same conductor.

**(D) Dual-Voltage Fixtures Including Luminaires or Appliances.** Fixtures, including luminaires, or appliances having both 120-volt and low-voltage connections shall be listed for dual voltage.

**(E) Autotransformers.** Autotransformers shall not be used.

**(F) Receptacles and Plug Caps.** Where a park trailer is equipped with a 120-volt or 120/240-volt ac system, a low-voltage system, or both, receptacles and plug caps of the low-voltage system shall differ in configuration from those of the 120-volt or 120/240-volt system. Where a unit equipped with a battery or dc system has an external connection for low-voltage power, the connector shall have a configuration that will not accept 120-volt power.

## IV. Nominal 120-Volt or 120/240-Volt Systems

### 552.40 120-Volt or 120/240-Volt, Nominal, Systems.

**(A) General Requirements.** The electrical equipment and material of park trailers indicated for connection to a wiring system rated 120 volts, nominal, 2-wire with ground, or a wiring system rated 120/240 volts, nominal, 3-wire with ground, shall be listed and installed in accordance with the requirements of Parts I, III, IV, and V of this article.

**(B) Materials and Equipment.** Electrical materials, devices, appliances, fittings, and other equipment installed, intended for use in, or attached to the park trailer shall be listed. All products shall be used only in the manner in which they have been tested and found suitable for the intended use.

### 552.41 Receptacle Outlets Required.

**(A) Spacing.** Receptacle outlets shall be installed at wall spaces 600 mm (2 ft) wide or more so that no point along the floor line is more than 1.8 m (6 ft), measured horizontally, from an outlet in that space.

*Exception No. 1: Bath and hall areas.*

*Exception No. 2: Wall spaces occupied by kitchen cabinets, wardrobe cabinets, built-in furniture; behind doors that may open fully against a wall surface; or similar facilities.*

**(B) Location.** Receptacle outlets shall be installed as follows:

(1) Adjacent to countertops in the kitchen [at least one on each side of the sink if countertops are on each side and are 300 mm (12 in.) or over in width]

(2) Adjacent to the refrigerator and gas range space, except where a gas-fired refrigerator or cooking appliance, requiring no external electrical connection, is factory-installed

(3) Adjacent to countertop spaces of 300 mm (12 in.) or more in width that cannot be reached from a receptacle required in 552.41(B)(1) by a cord of 1.8 m (6 ft) without crossing a traffic area, cooking appliance, or sink

**(C) Ground-Fault Circuit-Interrupter Protection.** Where provided, each 125-volt, single-phase, 15- or 20-ampere receptacle outlet shall have ground-fault circuit-interrupter protection for personnel in the following locations:

(1) Adjacent to a bathroom lavatory
(2) Within 1.8 m (6 ft) of any lavatory or sink

*Exception: Receptacles installed for appliances in dedicated spaces, such as for dishwashers, disposals, refrigerators, freezers, and laundry equipment.*

(3) In the area occupied by a toilet, shower, tub, or any combination thereof

(4) On the exterior of the unit

*Exception: Receptacles that are located inside of an access panel that is installed on the exterior of the unit to supply power for an installed appliance shall not be required to have ground-fault circuit-interrupter protection.*

The receptacle outlet shall be permitted in a listed luminaire (lighting fixture). A receptacle outlet shall not be installed in a tub or combination tub–shower compartment.

**(D) Pipe Heating Cable Outlet.** Where a pipe heating cable outlet is installed, the outlet shall be as follows:

(1) Located within 600 mm (2 ft) of the cold water inlet
(2) Connected to an interior branch circuit, other than a small appliance branch circuit
(3) On a circuit where all of the outlets are on the load side of the ground-fault circuit-interrupter protection for personnel
(4) Mounted on the underside of the park trailer and shall not be considered to be the outdoor receptacle outlet required in 552.41(E)

**(E) Outdoor Receptacle Outlets.** At least one receptacle outlet shall be installed outdoors. A receptacle outlet located in a compartment accessible from the outside of the park trailer shall be considered an outdoor receptacle. Outdoor receptacle outlets shall be protected as required in 552.41(C)(4).

**(F) Receptacle Outlets Not Permitted.**

**(1) Shower or Bathtub Space.** Receptacle outlets shall not be installed in or within reach [750 mm (30 in.)] of a shower or bathtub space.

**(2) Face-Up Position.** A receptacle shall not be installed in a face-up position in any countertop.

### 552.43 Power Supply.

**(A) Feeder.** The power supply to the park trailer shall be a feeder assembly consisting of not more than one listed 30-ampere or 50-ampere park trailer power-supply cord with an integrally molded or securely attached cap, or a permanently installed feeder.

**(B) Power-Supply Cord.** If the park trailer has a power-supply cord, it shall be permanently attached to the distribution panelboard or to a junction box permanently connected to the distribution panelboard, with the free end terminating in a molded-on attachment plug cap.

Cords with adapters and pigtail ends, extension cords, and similar items shall not be attached to, or shipped with, a park trailer.

A suitable clamp or the equivalent shall be provided at the distribution panelboard knockout to afford strain relief for the cord to prevent strain from being transmitted to the terminals when the power-supply cord is handled in its intended manner.

The cord shall be a listed type with 3-wire, 120-volt or 4-wire, 120/240-volt conductors, one of which shall be identified by a continuous green color or a continuous green color with one or more yellow stripes for use as the grounding conductor.

**(C) Mast Weatherhead or Raceway.** Where the calculated load exceeds 50 amperes or where a permanent feeder is used, the supply shall be by means of one of the following:

(1) One mast weatherhead installation, installed in accordance with Article 230, containing four continuous, insulated, color-coded feeder conductors, one of which shall be an equipment grounding conductor

(2) A metal raceway, rigid nonmetallic conduit, or liquidtight flexible nonmetallic conduit from the disconnecting means in the park trailer to the underside of the park trailer, with provisions for the attachment to a suitable junction box or fitting to the raceway on the underside of the park trailer [with or without conductors as in 550.10(I)(1)]

### 552.44 Cord.

**(A) Permanently Connected.** Each power-supply assembly shall be factory supplied or factory installed and connected directly to the terminals of the distribution panelboard or conductors within a junction box and provided with means to prevent strain from being transmitted to the terminals. The ampacity of the conductors between each junction box and the terminals of each distribution panelboard shall be at least equal to the ampacity of the power-supply cord. The supply end of the assembly shall be equipped with an attachment plug of the type described in 552.44(C). Where the cord passes through the walls or floors, it shall be protected by means of conduit and bushings or equivalent. The cord assembly shall have permanent provisions for protection against corrosion and mechanical damage while the unit is in transit.

**(B) Cord Length.** The cord-exposed usable length shall be measured from the point of entrance to the park trailer or the face of the flanged surface inlet (motor-base attachment plug) to the face of the attachment plug at the supply end.

The cord-exposed usable length, measured to the point of entry on the unit exterior, shall be a minimum of 7.0 m (23 ft) where the point of entrance is at the side of the unit, or shall be a minimum 8.5 m (28 ft) where the point of entrance is at the rear of the unit. The maximum length shall not exceed 11 m (36½ ft).

Where the cord entrance into the unit is more than 900 mm (3 ft) above the ground, the minimum cord lengths above shall be increased by the vertical distance of the cord entrance heights above 900 mm (3 ft).

**(C) Attachment Plugs.**

**(1) Units with Two to Five 15- or 20-Ampere Branch Circuits.** Park trailers wired in accordance with 552.46(A) shall have an attachment plug that shall be 2-pole, 3-wire grounding-type, rated 30 amperes, 125 volts, conforming to the configuration shown in Figure 552.44(C) intended for use with units rated at 30 amperes, 125 volts.

> FPN: Complete details of this configuration can be found in ANSI/NEMA WD 6-1989, National Electrical Manufacturers Association's *Standard for Dimensions of Attachment Plugs and Receptacles*, Figure TT.

**(2) Units with 50-Ampere Power Supply Assembly.** Park trailers having a power-supply assembly rated 50 amperes as permitted by 552.43(B) shall have a 3-pole, 4-wire grounding-type attachment plug rated 50 amperes, 125/250 volts, conforming to the configuration shown in Figure 552.44(C).

> FPN: Complete details of this configuration can be found in ANSI/NEMA WD 6-1989, National Electrical Manufacturers Association *Standard for Dimensions of Attachment Plugs and Receptacles*, Figure 14-50.

**(D) Labeling at Electrical Entrance.** Each park trailer shall have permanently affixed to the exterior skin, at or near the point of entrance of the power-supply assembly, a label 75 mm × 45 mm (3 in. × 1¾ in.) minimum size, made

**Receptacles**  **Caps**

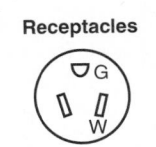

30-A, 125-V, 2-pole, 3-wire, grounding type

50-A, 125/250-V, 3-pole, 4-wire, grounding type

**Figure 552.44(C) Attachment cap and receptacle configurations.**

of etched, metal-stamped, or embossed brass, stainless steel, or anodized or alclad aluminum not less than 0.51 mm (0.020 in.) thick, or other suitable material [e.g., 0.13 mm (0.005 in.) thick plastic laminate], that reads, as appropriate, either

THIS CONNECTION IS FOR 110–125-VOLT AC, 60 HZ, 30 AMPERE SUPPLY

or

THIS CONNECTION IS FOR 120/240 VOLT AC, 3-POLE, 4-WIRE, 60 HZ, _____ AMPERE SUPPLY.

The correct ampere rating shall be marked in the blank space.

**(E) Location.** The point of entrance of a power-supply assembly shall be located within 4.5 m (15 ft) of the rear, on the left (road) side or at the rear, left of the longitudinal center of the unit, within 450 mm (18 in.) of the outside wall.

*Exception: A park trailer shall be permitted to have the electrical point of entrance located more than 4.5 m (15 ft) from the rear. Where this occurs, the distance beyond the 4.5-m (15-ft) dimension shall be added to the cord's minimum length as specified in 551.46(B).*

**552.45 Distribution Panelboard.**

**(A) Listed and Appropriately Rated.** A listed and appropriately rated distribution panelboard or other equipment specifically listed for the purpose shall be used. The grounded conductor termination bar shall be insulated from the enclosure as provided in 552.55(C). An equipment grounding terminal bar shall be attached inside the metal enclosure of the panelboard.

**(B) Location.** The distribution panelboard shall be installed in a readily accessible location. Working clearance

for the panelboard shall be not less than 600 mm (24 in.) wide and 750 mm (30 in.) deep.

*Exception: Where the panelboard cover is exposed to the inside aisle space, one of the working clearance dimensions shall be permitted to be reduced to a minimum of 550 mm (22 in.). A panelboard shall be considered exposed where the panelboard cover is within 50 mm (2 in.) of the aisle's finished surface.*

**(C) Dead-Front Type.** The distribution panelboard shall be of the dead-front type. A main disconnecting means shall be provided where fuses are used or where more than two circuit breakers are employed. A main overcurrent protective device not exceeding the power-supply assembly rating shall be provided where more than two branch circuits are employed.

**552.46 Branch Circuits.** Branch circuits shall be determined in accordance with 552.46(A) and (B).

**(A) Two to Five 15- or 20-Ampere Circuits.** Two to five 15- or 20-ampere circuits to supply lights, receptacle outlets, and fixed appliances shall be permitted. Such park trailers shall be equipped with a distribution panelboard rated at 120 volts maximum with a 30-ampere rated main power supply assembly. Not more than two 120-volt thermostatically controlled appliances (e.g., air conditioner and water heater) shall be installed in such systems unless appliance isolation switching, energy management systems, or similar methods are used.

*Exception: Additional 15- or 20-ampere circuits shall be permitted where a listed energy management system rated at 30 amperes maximum is employed within the system.*

**(B) More Than Five Circuits.** Where more than five circuits are needed, they shall be determined in accordance with 552.46(B)(1), (B)(2), and (B)(3).

**(1) Lighting.** Based on 33 volt-amperes/m$^2$ (3 VA/ft$^2$) multiplied by the outside dimensions of the park trailer (coupler excluded) divided by 120 volts to determine the number of 15- or 20-ampere lighting area circuits, for example,

$$\frac{3 \times \text{length} \times \text{width}}{120 \times 15 \text{ (or 20)}}$$

$$= \text{No. of 15 (or 20) ampere circuits}$$

The lighting circuits shall be permitted to serve built-in gas ovens with electric service only for lights, clocks or timers, or listed cord-connected garbage disposal units.

**(2) Small Appliances.** Small appliance branch circuits shall be installed in accordance with 210.11(C)(1).

**(3) General Appliances.** (including furnace, water heater, space heater, range, and central or room air conditioner, etc.) An individual branch circuit shall be permitted to supply any load for which it is rated. There shall be one or more circuits of adequate rating in accordance with (a) through (d).

> FPN No. 1:   For the laundry branch circuit, see 210.11(C)(2).

> FPN No. 2:   For central air conditioning, see Article 440.

(a)   The total rating of fixed appliances shall not exceed 50 percent of the circuit rating if lighting outlets, general-use receptacles, or both, are also supplied.

(b)   For fixed appliances with a motor(s) larger than ⅛ horsepower the total computed load shall be based on 125 percent of the largest motor plus the sum of the other loads. Where a branch circuit supplies continuous load(s) or any combination of continuous and noncontinuous loads, the branch-circuit conductor size shall be in accordance with 210.19(A).

(c)   The rating of a single cord- and plug-connected appliance supplied by other than an individual branch circuit shall not exceed 80 percent of the circuit rating.

(d)   The rating of a range branch circuit shall be based on the range demand as specified for ranges in 552.47(B)(5).

**552.47 Calculations.** The following method shall be employed in computing the supply-cord and distribution-panelboard load for each feeder assembly for each park trailer in lieu of the procedure shown in Article 220 and shall be based on a 3-wire, 120/240-volt supply with 120-volt loads balanced between the two phases of the 3-wire system.

**(A) Lighting and Small Appliance Load.** Lighting Volt-Amperes: Length times width of park trailer floor (outside dimensions) times 33 volt-amperes/m² (3 VA/ft²), for example,

$$Length \times width \times 3 = lighting\ volt\text{-}amperes$$

Small Appliance Volt-Amperes: Number of circuits times 1500 volt-amperes for each 20-ampere appliance receptacle circuit (see definition of *Appliance, Portable* with note) including 1500 volt-amperes for laundry circuit, for example,

$$No.\ of\ circuits \times 1500 = small\ appliance\ volt\text{-}amperes$$

Total: Lighting volt-amperes plus small appliance volt-amperes = total volt-amperes

First 3000 total volt-amperes at 100 percent plus remainder at 35 percent = volt-amperes to be divided by 240 volts to obtain current (amperes) per leg

**(B) Total Load for Determining Power Supply.** Total load for determining power supply is the sum of the following:

(1) Lighting and small appliance load as calculated in 552.47(A).

(2) Nameplate amperes for motors and heater loads (exhaust fans, air conditioners, electric, gas, or oil heating). Omit smaller of the heating and cooling loads, except include blower motor if used as air-conditioner evaporator motor. Where an air conditioner is not installed and a 50-ampere power-supply cord is provided, allow 15 amperes per phase for air conditioning.

(3) Twenty-five percent of current of largest motor in (2).

(4) Total of nameplate amperes for disposal, dishwasher, water heater, clothes dryer, wall-mounted oven, cooking units. Where the number of these appliances exceeds three, use 75 percent of total.

(5) Derive amperes for freestanding range (as distinguished from separate ovens and cooking units) by dividing the following values by 240 volts:

| Nameplate Rating (watts) | Use (volt-amperes) |
|---|---|
| 0–10,000 | 80 percent of rating |
| Over 10,000–12,500 | 8,000 |
| Over 12,500–13,500 | 8,400 |
| Over 13,500–14,500 | 8,800 |
| Over 14,500–15,500 | 9,200 |
| Over 15,500–16,500 | 9,600 |
| Over 16,500–17,500 | 10,000 |

(6) If outlets or circuits are provided for other than factory-installed appliances, include the anticipated load.

> FPN:   Refer to Annex D, Example D12, for an illustration of the application of this calculation.

**(C) Optional Method of Calculation for Lighting and Appliance Load.** For park trailers, the optional method for calculating lighting and appliance load shown in 220.30 shall be permitted.

**552.48 Wiring Methods.**

**(A) Wiring Systems.** Cables and raceways installed in accordance with Articles 320, 322, 330 through 340, 342 through 362, Article 386, and Article 388 shall be permitted in accordance with their applicable article, except as otherwise specified in this article. An equipment grounding means shall be provided in accordance with 250.118.

**(B) Conduit and Tubing.** Where rigid metal conduit or intermediate metal conduit is terminated at an enclosure with a locknut and bushing connection, two locknuts shall be provided, one inside and one outside of the enclosure. All cut ends of conduit and tubing shall be reamed or otherwise finished to remove rough edges.

**(C) Nonmetallic Boxes.** Nonmetallic boxes shall be acceptable only with nonmetallic-sheathed cable or nonmetallic raceways.

**(D) Boxes.** In walls and ceilings constructed of wood or other combustible material, boxes and fittings shall be flush with the finished surface or project therefrom.

**(E) Mounting.** Wall and ceiling boxes shall be mounted in accordance with Article 314.

*Exception No. 1:  Snap-in-type boxes or boxes provided with special wall or ceiling brackets that securely fasten boxes in walls or ceilings shall be permitted.*

*Exception No. 2:  A wooden plate providing a 38-mm (1½ -in.) minimum width backing around the box and of a thickness of 13 mm (½ in.) or greater (actual) attached directly to the wall panel shall be considered as approved means for mounting outlet boxes.*

**(F) Sheath Armor.** The sheath of nonmetallic-sheathed cable, metal-clad cable, and Type AC cable shall be continuous between outlet boxes and other enclosures.

**(G) Protected.** Metal-clad, Type AC, or nonmetallic-sheathed cables and electrical nonmetallic tubing shall be permitted to pass through the centers of the wide side of 2 by 4 wood studs. However, they shall be protected where they pass through 2 by 2 wood studs or at other wood studs or frames where the cable or tubing would be less than 32 mm (1¼ in.) from the inside or outside surface. Steel plates on each side of the cable or tubing, or a steel tube, with not less than 1.35 mm (0.053 in.) wall thickness, shall be installed to protect the cable or tubing. These plates or tubes shall be securely held in place. Where nonmetallic-sheathed cables pass through punched, cut, or drilled slots or holes in metal members, the cable shall be protected by bushings or grommets securely fastened in the opening prior to installation of the cable.

**(H) Cable Supports.** Where connected with cable connectors or clamps, cables shall be supported within 300 mm (12 in.) of outlet boxes, distribution panelboards, and splice boxes on appliances. Supports shall be provided every 1.4 m (4½ ft) at other places.

**(I) Nonmetallic Box Without Cable Clamps.** Nonmetallic-sheathed cables shall be supported within 200 mm (8 in.) of a nonmetallic outlet box without cable clamps.

*Exception:  Where wiring devices with integral enclosures are employed with a loop of extra cable to permit future replacement of the device, the cable loop shall be considered as an integral portion of the device.*

**(J) Physical Damage.** Where subject to physical damage, exposed nonmetallic cable shall be protected by covering boards, guard strips, raceways, or other means.

**(K) Metal Faceplates.** Metal faceplates shall be of ferrous metal not less than 0.76 mm (0.030 in.) in thickness or of nonferrous metal not less than 1.0 mm (0.040 in.) in thickness. Nonmetallic faceplates shall be listed.

**(L) Metal Faceplates Effectively Grounded.** Where metal faceplates are used, they shall be effectively grounded.

**(M) Moisture or Physical Damage.** Where outdoor or under-chassis wiring is 120 volts, nominal, or over and is exposed to moisture or physical damage, the wiring shall be protected by rigid metal conduit, by intermediate metal conduit, or by electrical metallic tubing or rigid nonmetallic conduit that is closely routed against frames and equipment enclosures or other raceway or cable identified for the application.

**(N) Component Interconnections.** Fittings and connectors that are intended to be concealed at the time of assembly shall be listed and identified for the interconnection of building components. Such fittings and connectors shall be equal to the wiring method employed in insulation, temperature rise, and fault-current withstanding, and shall be capable of enduring the vibration and shock occurring in park trailers.

**(O) Method of Connecting Expandable Units.** The method of connecting expandable units to the main body of the vehicle shall comply with the following as applicable:

(1) That portion of a branch circuit that is installed in an expandable unit shall be permitted to be connected to the branch circuit in the main body of the vehicle by means of a flexible cord or attachment plug and cord listed for hard usage. The cord and its connections shall conform to all provisions of Article 400 and shall be considered as a permitted use under 400.7.

(2) If the receptacle provided for connection of the cord to the main circuit is located on the outside of the unit, it shall be protected with a ground-fault circuit interrupter for personnel and be listed for wet locations. A cord located on the outside of a unit shall be identified for outdoor use.

(3) Unless removable or stored within the unit interior, the cord assembly shall have permanent provisions for protection against corrosion and mechanical damage while the unit is in transit.

(4) If an attachment plug and cord is used, it shall be installed so as not to permit exposed live attachment plug pins.

**(P) Prewiring for Air-Conditioning Installation.** Prewiring installed for the purpose of facilitating future air-conditioning installation shall conform to the applicable portions of this article and the following:

(1) An overcurrent protective device with a rating compatible with the circuit conductors shall be installed in the distribution panelboard and wiring connections completed.

(2) The load end of the circuit shall terminate in a junction box with a blank cover or a device listed for the purpose. Where a junction box with a blank cover is used, the free ends of the conductors shall be adequately capped or taped.

(3) A label conforming to 552.44(D) shall be placed on or adjacent to the junction box and shall read

AIR-CONDITIONING CIRCUIT.
THIS CONNECTION IS FOR AIR CONDITIONERS
RATED 110–125-VOLT AC, 60 HZ,
_____ AMPERES MAXIMUM.
DO NOT EXCEED CIRCUIT RATING.

An ampere rating, not to exceed 80 percent of the circuit rating, shall be legibly marked in the blank space.

(4) The circuit shall serve no other purpose.

**552.49 Maximum Number of Conductors in Boxes.** The maximum number of conductors permitted in boxes shall be in accordance with 314.16.

**552.50 Grounded Conductors.** The identification of grounded conductors shall be in accordance with 200.6.

**552.51 Connection of Terminals and Splices.** Conductor splices and connections at terminals shall be in accordance with 110.14.

**552.52 Switches.** Switches shall be rated as follows.

**(A) Lighting Circuits.** For lighting circuits, switches shall be rated not less than 10 amperes, 120/125 volts and in no case less than the connected load.

**(B) Motors or Other Loads.** For motors or other loads, switches shall have ampere or horsepower ratings, or both, adequate for loads controlled. (An ac general-use snap switch shall be permitted to control a motor 2 hp or less with full-load current not over 80 percent of the switch ampere rating.)

**552.53 Receptacles.** All receptacle outlets shall be of the grounding type and installed in accordance with 210.21 and 406.3.

**552.54 Luminaires (Lighting Fixtures).**

**(A) General.** Any combustible wall or ceiling finish exposed between the edge of a luminaire (fixture) canopy or pan and the outlet box shall be covered with noncombustible material or a material identified for the purpose.

**(B) Shower Luminaires (Fixtures).** If a luminaire (lighting fixture) is provided over a bathtub or in a shower stall, it shall be of the enclosed and gasketed type and listed for the type of installation, and it shall be ground-fault circuit-interrupter protected.

The switch for shower luminaires (lighting fixtures) and exhaust fans, located over a tub or in a shower stall, shall be located outside the tub or shower space.

**(C) Outdoor Outlets, Luminaires (Fixtures), Air-Cooling Equipment, and So On.** Outdoor luminaires (fixtures) and other equipment shall be listed for outdoor use.

**552.55 Grounding.** (See also 552.57 on bonding of non–current-carrying metal parts.)

**(A) Power-Supply Grounding.** The grounding conductor in the supply cord or feeder shall be connected to the grounding bus or other approved grounding means in the distribution panelboard.

**(B) Distribution Panelboard.** The distribution panelboard shall have a grounding bus with sufficient terminals for all grounding conductors or other approved grounding means.

**(C) Insulated Neutral.** The grounded circuit conductor (neutral) shall be insulated from the equipment grounding conductors and from equipment enclosures and other grounded parts. The grounded (neutral) circuit terminals in the distribution panelboard and in ranges, clothes dryers, counter-mounted cooking units, and wall-mounted ovens shall be insulated from the equipment enclosure. Bonding screws, straps, or buses in the distribution panelboard or in appliances shall be removed and discarded. Connection of electric ranges and electric clothes dryers utilizing a grounded (neutral) conductor, if cord-connected, shall be made with 4-conductor cord and 3-pole, 4-wire, grounding-type plug caps and receptacles.

**552.56 Interior Equipment Grounding.**

**(A) Exposed Metal Parts.** In the electrical system, all exposed metal parts, enclosures, frames, luminaire (lighting fixture) canopies, and so forth, shall be effectively bonded to the grounding terminals or enclosure of the distribution panelboard.

**(B) Equipment Grounding Conductors.** Bare wires, green-colored wires, or green wires with a yellow stripe(s) shall be used for equipment grounding conductors only.

**(C) Grounding of Electrical Equipment.** Where grounding of electrical equipment is specified, it shall be permitted as follows:

(1) Connection of metal raceway (conduit or electrical metallic tubing), the sheath of Type MC and Type MI cable where the sheath is identified for grounding, or the armor of Type AC cable to metal enclosures.

(2) A connection between the one or more equipment grounding conductors and a metal box by means of a grounding screw, which shall be used for no other purpose, or a listed grounding device.

(3) The equipment grounding conductor in nonmetallic-sheathed cable shall be permitted to be secured under a screw threaded into the luminaire (fixture) canopy other than a mounting screw or cover screw or attached to a listed grounding means (plate) in a nonmetallic outlet box for luminaire (fixture) mounting [grounding means shall also be permitted for luminaire (fixture) attachment screws].

**(D) Grounding Connection in Nonmetallic Box.** A connection between the one or more grounding conductors brought into a nonmetallic outlet box shall be arranged so that a connection can be made to any fitting or device in that box that requires grounding.

**(E) Grounding Continuity.** Where more than one equipment grounding conductor of a branch circuit enters a box, all such conductors shall be in good electrical contact with each other, and the arrangement shall be such that the disconnection or removal of a receptacle, fixture, including a luminaire, or other device fed from the box will not interfere with or interrupt the grounding continuity.

**(F) Cord-Connected Appliances.** Cord-connected appliances, such as washing machines, clothes dryers, refrigerators, and the electrical system of gas ranges, and so on, shall be grounded by means of an approved cord with equipment grounding conductor and grounding-type attachment plug.

### 552.57 Bonding of Non–Current-Carrying Metal Parts.

**(A) Required Bonding.** All exposed non–current-carrying metal parts that may become energized shall be effectively bonded to the grounding terminal or enclosure of the distribution panelboard.

**(B) Bonding Chassis.** A bonding conductor shall be connected between any distribution panelboard and an accessible terminal on the chassis. Aluminum or copper-clad aluminum conductors shall not be used for bonding if such conductors or their terminals are exposed to corrosive elements.

*Exception: Any park trailer that employs a unitized metal chassis-frame construction to which the distribution panelboard is securely fastened with a bolt(s) and nut(s) or by welding or riveting shall be considered to be bonded.*

**(C) Bonding Conductor Requirements.** Grounding terminals shall be of the solderless type and listed as pressure terminal connectors recognized for the wire size used. The bonding conductor shall be solid or stranded, insulated or bare, and shall be 8 AWG copper minimum or equivalent.

**(D) Metallic Roof and Exterior Bonding.** The metal roof and exterior covering shall be considered bonded where

(1) The metal panels overlap one another and are securely attached to the wood or metal frame parts by metal fasteners, and

(2) The lower panel of the metal exterior covering is secured by metal fasteners at each cross member of the chassis, or the lower panel is bonded to the chassis by a metal strap.

**(E) Gas, Water, and Waste Pipe Bonding.** The gas, water, and waste pipes shall be considered grounded if they are bonded to the chassis.

**(F) Furnace and Metal Air Duct Bonding.** Furnace and metal circulating air ducts shall be bonded.

### 552.58 Appliance Accessibility and Fastening. Every appliance shall be accessible for inspection, service, repair, and replacement without removal of permanent construction. Means shall be provided to securely fasten appliances in place when the park trailer is in transit.

### 552.59 Outdoor Outlets, Fixtures, Including Luminaires, Air-Cooling Equipment, and So On.

**(A) Listed for Outdoor Use.** Outdoor fixtures, including luminaires, and equipment shall be listed for outdoor use. Outdoor receptacle or convenience outlets shall be of a gasketed-cover type for use in wet locations.

**(B) Outside Heating Equipment, Air-Conditioning Equipment, or Both.** A park trailer provided with a branch circuit designed to energize outside heating equipment or air-conditioning equipment, or both, located outside the park trailer, other than room air conditioners, shall have such branch-circuit conductors terminate in a listed outlet box or disconnecting means located on the outside of the park trailer. A label shall be permanently affixed adjacent to the outlet box and shall contain the following information:

THIS CONNECTION IS FOR HEATING
AND/OR AIR-CONDITIONING EQUIPMENT.
THE BRANCH CIRCUIT IS RATED AT NOT
MORE THAN _____ AMPERES, AT _____ VOLTS,
60-Hz, _____ CONDUCTOR AMPACITY.
A DISCONNECTING MEANS SHALL BE
LOCATED WITHIN SIGHT OF THE EQUIPMENT.

The correct voltage and ampere rating shall be given. The tag shall not be less than 0.51 mm (0.020 in.) thick etched brass, stainless steel, anodized or alclad aluminum, or equivalent. The tag shall not be less than 75 mm × 45 mm (3 in. × 1¾ in.) minimum size.

## V. Factory Tests

**552.60 Factory Tests (Electrical).** Each park trailer shall be subjected to the following tests.

**(A) Circuits of 120 Volts or 120/240 Volts.** Each park trailer designed with a 120-volt or a 120/240-volt electrical system shall withstand the applied potential without electrical breakdown of a 1-minute, 900-volt dielectric strength test, or a 1-second, 1080-volt dielectric strength test, with all switches closed, between ungrounded and grounded conductors and the park trailer ground. During the test, all switches and other controls shall be in the on position. Fixtures, including luminaires, and permanently installed appliances shall not be required to withstand this test.

Each park trailer shall be subjected to the following:

(1) A continuity test to ensure that all metal parts are properly bonded
(2) Operational tests to demonstrate that all equipment is properly connected and in working order
(3) Polarity checks to determine that connections have been properly made
(4) Receptacles requiring GFCI protection shall be tested for correct function by the use of a GFCI testing device

**(B) Low-Voltage Circuits.** Low-voltage circuit conductors in each park trailer shall withstand the applied potential without electrical breakdown of a 1-minute, 500-volt or a 1-second, 600-volt dielectric strength test. The potential shall be applied between ungrounded and grounded conductors.

The test shall be permitted on running light circuits before the lights are installed, provided the unit's outer covering and interior cabinetry have been secured. The braking circuit shall be permitted to be tested before being connected to the brakes, provided the wiring has been completely secured.

# ARTICLE 553
## Floating Buildings

## I. General

**553.1 Scope.** This article covers wiring, services, feeders, and grounding for floating buildings.

**553.2 Definition.**

**Floating Building.** A building unit as defined in Article 100 that floats on water, is moored in a permanent location,

and has a premises wiring system served through connection by permanent wiring to an electricity supply system not located on the premises.

**553.3 Application of Other Articles.** Wiring for floating buildings shall comply with the applicable provisions of other articles of this *Code*, except as modified by this article.

## II. Services and Feeders

**553.4 Location of Service Equipment.** The service equipment for a floating building shall be located adjacent to, but not in or on, the building.

**553.5 Service Conductors.** One set of service conductors shall be permitted to serve more than one set of service equipment.

**553.6 Feeder Conductors.** Each floating building shall be supplied by a single set of feeder conductors from its service equipment.

*Exception: Where the floating building has multiple occupancy, each occupant shall be permitted to be supplied by a single set of feeder conductors extended from the occupant's service equipment to the occupant's panelboard.*

**553.7 Installation of Services and Feeders.**

**(A) Flexibility.** Flexibility of the wiring system shall be maintained between floating buildings and the supply conductors. All wiring shall be installed so that motion of the water surface and changes in the water level will not result in unsafe conditions.

**(B) Wiring Methods.** Liquidtight flexible metal conduit or liquidtight flexible nonmetallic conduit with approved fittings shall be permitted for feeders and where flexible connections are required for services. Extra-hard usage portable power cable listed for both wet locations and sunlight resistance shall be permitted for a feeder to a floating building where flexibility is required. Other raceways suitable for the location shall be permitted to be installed where flexibility is not required.

FPN: See 555.1 and 555.13.

## III. Grounding

**553.8 General Requirements.** Grounding of both electrical and nonelectrical parts in a floating building shall be through connection to a grounding bus in the building panelboard. The grounding bus shall be grounded through a green-colored insulated equipment grounding conductor run with the feeder conductors and connected to a grounding terminal in the service equipment. The grounding terminal in the service equipment shall be grounded by connection through an insulated grounding electrode conductor to a grounding electrode on shore.

**553.9 Insulated Neutral.** The grounded circuit conductor (neutral) shall be an insulated conductor identified in conformance with 200.6. The neutral conductor shall be connected to the equipment grounding terminal in the service equipment, and, except for that connection, it shall be insulated from the equipment grounding conductors, equipment enclosures, and all other grounded parts. The neutral circuit terminals in the panelboard and in ranges, clothes dryers, counter-mounted cooking units, and the like shall be insulated from the enclosures.

**553.10 Equipment Grounding.**

**(A) Electrical Systems.** All enclosures and exposed metal parts of electrical systems shall be bonded to the grounding bus.

**(B) Cord-Connected Appliances.** Where required to be grounded, cord-connected appliances shall be grounded by means of an equipment grounding conductor in the cord and a grounding-type attachment plug.

**553.11 Bonding of Non–Current-Carrying Metal Parts.** All metal parts in contact with the water, all metal piping, and all non–current-carrying metal parts that may become energized shall be bonded to the grounding bus in the panelboard.

# ARTICLE 555
## Marinas and Boatyards

**555.1 Scope.** This article covers the installation of wiring and equipment in the areas comprising fixed or floating piers, wharfs, docks, and other areas in marinas, boatyards, boat basins, boathouses, yacht clubs, boat condominiums, docking facilities associated with residential condominiums, any multiple docking facility, or similar occupancies, and facilities that are used, or intended for use, for the purpose of repair, berthing, launching, storage, or fueling of small craft and the moorage of floating buildings.

Private, noncommercial docking facilities constructed or occupied for the use of the owner or residents of the associated single-family dwelling are not covered by this article.

FPN: See NFPA 303-2000, *Fire Protection Standard for Marinas and Boatyards*, for additional information.

**555.2 Definitions.**

**Electrical Datum Plane.** The electrical datum plane is defined as follows:

(1) In land areas subject to tidal fluctuation, the electrical datum plane is a horizontal plane 606 mm (2 ft) above the highest tide level for the area occurring under normal circumstances, that is, highest high tide.

(2) In land areas not subject to tidal fluctuation, the electrical datum plane is a horizontal plane 606 mm (2 ft) above the highest water level for the area occurring under normal circumstances.

(3) The electrical datum plane for floating piers and landing stages that are (a) installed to permit rise and fall response to water level, without lateral movement, and (b) that are so equipped that they can rise to the datum plane established for (1) or (2), is a horizontal plane 762 mm (30 in.) above the water level at the floating pier or landing stage and a minimum of 305 mm (12 in.) above the level of the deck.

**Marine Power Outlet.** An enclosed assembly that can include receptacles, circuit breakers, fused switches, fuses, watt-hour meter(s), and monitoring means approved for marine use.

**555.4 Distribution System.** Yard and pier distribution systems shall not exceed 600 volts phase to phase.

**555.5 Transformers.** Transformers and enclosures shall be specifically approved for the intended location. The bottom of enclosures for transformers shall not be located below the electrical datum plane.

**555.7 Location of Service Equipment.** The service equipment for floating docks or marinas shall be located adjacent to, but not on or in, the floating structure.

**555.9 Electrical Connections.** All electrical connections shall be located at least 305 mm (12 in.) above the deck of a floating pier. All electrical connections shall be located at least 305 mm (12 in.) above the deck of a fixed pier but not below the electrical datum plane.

**555.10 Electrical Equipment Enclosures.**

**(A) Securing and Supporting.** Electrical equipment enclosures installed on piers above deck level shall be securely and substantially supported by structural members, independent of any conduit connected to them. If enclosures are not attached to mounting surfaces by means of external ears or lugs, the internal screw heads shall be sealed to prevent seepage of water through mounting holes.

**(B) Location.** Electrical equipment enclosures on piers shall be located so as not to interfere with mooring lines.

**555.11 Circuit Breakers, Switches, Panelboards, and Marine Power Outlets.** Circuit breakers and switches installed in gasketed enclosures shall be arranged to permit required manual operation without exposing the interior of the enclosure. All such enclosures shall be arranged with a weep hole to discharge condensation.

**555.12 Load Calculations for Service and Feeder Conductors.** General lighting and other loads shall be calculated in accordance with Article 220, and, in addition, the load for each service and/or feeder circuit supplying receptacles that provide shore power for boats shall be calculated using the demand factors shown in Table 555.12. These calculations shall be permitted to be modified as indicated in notes (1) and (2).

**Table 555.12 Demand Factors**

| Number of Receptacles | Sum of the Rating of the Receptacles (percent) |
|---|---|
| 1 – 4 | 100 |
| 5 – 8 | 90 |
| 9 –14 | 80 |
| 15 –30 | 70 |
| 31 –40 | 60 |
| 41 –50 | 50 |
| 51 –70 | 40 |
| 71-plus | 30 |

Notes:

1. Where shore power accommodations provide two receptacles specifically for an individual boat slip and these receptacles have different voltages (for example, one 30 ampere, 125 volt and one 50 ampere, 125/250 volt), only the receptacle with the larger kilowatt demand shall be required to be calculated.

2. If the facility being installed includes individual kilowatt-hour submeters for each slip and is being calculated using the criteria listed in Table 555.12, the total demand amperes may be multiplied by 0.9 to achieve the final demand amperes.

FPN: These demand factors may be inadequate in areas of extreme hot or cold temperatures with loaded circuits for heating, air-conditioning, or refrigerating equipment.

**555.13 Wiring Methods and Installation.**

**(A) Wiring Methods.**

**(1) General.** Wiring methods of Chapter 3 shall be permitted where identified for use in wet locations.

**(2) Portable Power Cables.** Extra-hard usage portable power cables rated not less than 167°F (75°C), 600 volts, listed for both wet locations and sunlight resistance, having an outer jacket rated to be resistant to temperature extremes, oil, gasoline, ozone, abrasion, acids, and chemicals shall be permitted as follows:

(1) As permanent wiring on the underside of piers (floating or fixed)

(2) Where flexibility is necessary as on piers composed of floating sections

**(3) Temporary Wiring.** Temporary wiring, except as permitted by Article 527, shall not be used to supply power to boats.

**(B) Installation.**

**(1) Overhead Wiring.** Overhead wiring shall be installed to avoid possible contact with masts and other parts of boats being moved in the yard.

Conductors and cables shall be routed to avoid wiring closer than 6.0 m (20 ft) from the outer edge or any portion of the yard that can be used for moving vessels or stepping or unstepping masts.

**(2) Outside Branch Circuits and Feeders.** Outside branch circuits and feeders shall comply with Article 225 except that clearances for overhead wiring in portions of the yard other than those described in 555.13(B)(1) shall not be less than 5.49 m (18 ft) above grade.

**(3) Wiring Over and Under Navigable Water.** Wiring over and under navigable water shall be subject to approval by the authority having jurisdiction.

FPN: See NFPA 303-2000, *Fire Protection Standard for Marinas and Boatyards*, for warning sign requirements.

**(4) Portable Power Cables.**

(a) Where portable power cables are permitted by 555.13(A)(2), the installation shall comply with the following:

(1) Cables shall be properly supported.

(2) Cables shall be located on the underside of the pier.

(3) Cables shall be securely fastened by nonmetallic clips to structural members other than the deck planking.

(4) Cables shall not be installed where subject to physical damage.

(5) Where cables pass through structural members, they shall be protected against chafing by a permanently installed oversized sleeve of nonmetallic material.

(b) Where portable power cables are used as permitted in 555.13(A)(2)(2), there shall be an approved junction box of corrosion-resistant construction with permanently installed terminal blocks on each pier section to which the feeder and feeder extensions are to be connected. Metal junction boxes and their covers, and metal screws and parts that are exposed externally to the boxes, shall be of corrosion-resistant materials or protected by material resistant to corrosion.

**(5) Protection.** Rigid metal or nonmetallic conduit suitable for the location shall be installed to protect wiring above decks of piers and landing stages and below the enclosure that it serves. The conduit shall be connected to the enclosure by full standard threads. The use of special fittings of nonmetallic material to provide a threaded connection into enclosures on rigid nonmetallic conduit, employing joint design as recommended by the conduit manufacturer for attachment of the fitting to the conduit shall be acceptable, provided the equipment and method of attachment are approved and the assembly meets the requirements of installation in damp or wet locations as applicable.

**555.15 Grounding.** Wiring and equipment within the scope of this article shall be grounded as specified in Article 250 and with the following additional requirements.

**(A) Equipment to Be Grounded.** The following items shall be connected to an equipment grounding conductor run with the circuit conductors in the same raceway, cable, or trench:

(1) Metal boxes, metal cabinets, and all other metal enclosures
(2) Metal frames of utilization equipment
(3) Grounding terminals of grounding-type receptacles

**(B) Type of Equipment Grounding Conductor.** The equipment grounding conductor shall be an insulated copper conductor with a continuous outer finish that is either green or green with one or more yellow stripes. The equipment grounding conductor of Type MI cable shall be permitted to be identified at terminations. For conductors larger than 6 AWG, or where multiconductor cables are used, re-identification of conductors as allowed in 250.119(A)(2) and (A)(3) or 250.119(B)(2) and (B)(3) shall be permitted.

**(C) Size of Equipment Grounding Conductor.** The insulated copper equipment grounding conductor shall be sized in accordance with 250.122 but not smaller than 12 AWG.

**(D) Branch-Circuit Equipment Grounding Conductor.** The insulated equipment grounding conductor for branch circuits shall terminate at a grounding terminal in a remote panelboard or the grounding terminal in the main service equipment.

**(E) Feeder Equipment Grounding Conductors.** Where a feeder supplies a remote panelboard, an insulated equipment grounding conductor shall extend from a grounding terminal in the service equipment to a grounding terminal in the remote panelboard.

**555.17 Disconnecting Means for Shore Power Connection(s).** Disconnecting means shall be provided to isolate each boat from its supply connection(s).

**(A) Type.** The disconnecting means shall be permitted to consist of a circuit breaker, switch, or both, and shall be properly identified as to which receptacle it controls.

**(B) Location.** The disconnecting means shall be readily accessible, located not more than 762 mm (30 in.) from the receptacle it controls, and shall be located in the supply circuit ahead of the receptacle. Circuit breakers or switches located in marine power outlets complying with this section shall be permitted as the disconnecting means.

**555.19 Receptacles.** Receptacles shall be mounted not less than 305 mm (12 in.) above the deck surface of the pier and not below the electrical datum plane on a fixed pier.

**(A) Shore Power Receptacles.**

**(1) Enclosures.** Receptacles intended to supply shore power to boats shall be housed in marine power outlets listed as marina power outlets or listed for set locations, or shall be installed in listed enclosures protected from the weather or in listed weatherproof enclosures. The integrity of the assembly shall not be affected when the receptacles are in use with any type of booted or nonbooted attachment plug/cap inserted.

**(2) Strain Relief.** Means shall be provided where necessary to reduce the strain on the plug and receptacle caused by the weight and catenary angle of the shore power cord.

**(3) Branch Circuits.** Each single receptacle that supplies shore power to boats shall be supplied from a marine power outlet or panelboard by an individual branch circuit of the voltage class and rating corresponding to the rating of the receptacle.

> FPN: Supplying receptacles at voltages other than the voltages marked on the receptacle may cause overheating or malfunctioning of connected equipment, for example, supplying single-phase, 120/240-volt, 3-wire loads from a 208Y/120-volt, 3-wire source.

**(4) Ratings.** Receptacles that provide shore power for boats shall be rated not less than 30 amperes and shall be single outlet type.

> FPN: For locking- and grounding-type receptacles for auxiliary power to boats, see NFPA 303-2000, *Fire Protection Standard for Marinas and Boatyards.*

(a) Receptacles rated not less than 30 amperes or more than 50 amperes shall be of the locking and grounding type.

> FPN: For various configurations and ratings of locking and grounding-type receptacles and caps, see ANSI/NEMA 18WD 6-1989, National Electrical Manufacturers Association's *Standard for Dimensions of Attachment Plugs and Receptacles.*

(b) Receptacles rated for 60 amperes or 100 amperes shall be of the pin and sleeve type.

FPN: For various configurations and ratings of pin and sleeve receptacles, see ANSI/UL 1686, *UL Standard for Safety Pin and Sleeve Configurations.*

**(B) Other Than Shore Power.**

**(1) Ground-Fault Circuit-Interrupter (GFCI) Protection for Personnel.** Fifteen- and 20-ampere, single-phase, 125-volt receptacles installed outdoors, in boathouses, in buildings used for storage, maintenance, or repair where portable electrical hand tools, electrical diagnostic equipment, or portable lighting equipment are to be used shall be provided with GFCI protection for personnel. Receptacles in other locations shall be protected in accordance with 210.8(B).

**(2) Marking.** Receptacles other than those supplying shore power to boats shall be permitted to be housed in marine power outlets with the receptacles that provide shore power to boats, provided they are marked to clearly indicate that they are not to be used to supply power to boats.

**555.21 Gasoline Dispensing Stations — Hazardous (Classified) Locations.** Electrical wiring and equipment located at or serving gasoline dispensing stations shall comply with Article 514 in addition to the requirements of this article.

**555.23 Marine Hoists, Railways, Cranes, and Monorails.** Motors and controls for marine hoists, railways, cranes, and monorails shall not be located below the electrical datum plane. Where it is necessary to provide electric power to a mobile crane or hoist in the yard, and a trailing cable is utilized, it shall be a listed portable power cable rated for the conditions of use and be provided with an outer jacket of distinctive color for safety.

# Chapter 6 Special Equipment

## ARTICLE 600
## Electric Signs and Outline Lighting

### I. General

**600.1 Scope.** This article covers the installation of conductors and equipment for electric signs and outline lighting as defined in Article 100.

> FPN: As defined in Article 100, electric signs and outline lighting include all products and installations utilizing neon tubing, such as signs, decorative elements, skeleton tubing, or art forms.

### 600.2 Definitions.

**Electric-Discharge Lighting.** Systems of illumination utilizing fluorescent lamps, high-intensity discharge (HID) lamps, or neon tubing.

**Neon Tubing.** Electric-discharge tubing manufactured into shapes that form letters, parts of letters, skeleton tubing, outline lighting, other decorative elements, or art forms, and filled with various inert gases.

**Sign Body.** A portion of a sign that may provide protection from the weather but is not an electrical enclosure.

**Skeleton Tubing.** Neon tubing that is itself the sign or outline lighting and not attached to an enclosure or sign body.

**600.3 Listing.** Electric signs and outline lighting — fixed, mobile, or portable — shall be listed and installed in conformance with that listing, unless otherwise approved by special permission.

**(A) Field Installed Skeleton Tubing.** Field installed skeleton tubing shall not be required to be listed where installed in conformance with this *Code*.

**(B) Outline Lighting.** Outline lighting shall not be required to be listed as a system when it consists of listed luminaires (lighting fixtures) wired in accordance with Chapter 3.

### 600.4 Markings.

**(A) Signs and Outline Lighting Systems.** Signs and outline lighting systems shall be marked with the manufacturer's name, trademark, or other means of identification; and input voltage and current rating.

**(B) With Incandescent Lamp Holders.** Signs and outline lighting systems with incandescent lamp holders shall be marked to indicate the maximum allowable wattage of lamps. The markings shall be permanently installed, in letters at least 6 mm (¼ in.) high, and shall be located where visible during relamping.

### 600.5 Branch Circuits.

**(A) Required Branch Circuit.** Each commercial building and each commercial occupancy accessible to pedestrians shall be provided with at least one outlet in an accessible location at each entrance to each tenant space for sign or outline lighting system use. The outlet(s) shall be supplied by a branch circuit rated at least 20 amperes that supplies no other load. Service hallways or corridors shall not be considered accessible to pedestrians.

**(B) Rating.** Branch circuits that supply signs shall be rated as follows.

**(1) Incandescent and Fluorescent.** Branch circuits that supply signs and outline lighting systems containing incandescent and fluorescent forms of illumination shall be rated not to exceed 20 amperes.

**(2) Neon.** Branch circuits that supply neon tubing installations shall not be rated in excess of 30 amperes.

**(C) Wiring Methods.** Wiring methods used to supply signs shall comply with 600.5(C)(1), (C)(2), and (C)(3).

**(1) Supply.** The wiring method used to supply signs and outline lighting systems shall terminate within a sign, an outline lighting system enclosure, a suitable box, or a conduit body.

**(2) Enclosures as Pull Boxes.** Signs and transformer enclosures shall be permitted to be used as pull or junction boxes for conductors supplying other adjacent signs, outline lighting systems, or floodlights that are part of a sign and shall be permitted to contain both branch and secondary circuit conductors.

**(3) Metal Poles.** Metal poles used to support signs shall be permitted to enclose supply conductors, provided the poles and conductors are installed in accordance with 410.15(B).

**600.6 Disconnects.** Each sign and outline lighting system, or feeder circuit or branch circuit supplying a sign or outline lighting system, shall be controlled by an externally operable switch or circuit breaker that will open all ungrounded conductors. Signs and outline lighting systems located within fountains shall have the disconnect located in accordance with 680.12.

*Exception No. 1: A disconnecting means shall not be required for an exit directional sign located within a building.*

*Exception No. 2: A disconnecting means shall not be required for cord-connected signs with an attachment plug.*

**(A) Location.**

**(1) Within Sight of the Sign.** The disconnecting means shall be within sight of the sign or outline lighting system that it controls. Where the disconnecting means is out of the line of sight from any section that may be energized, the disconnecting means shall be capable of being locked in the open position.

**(2) Within Sight of the Controller.** The following shall apply for signs or outline lighting systems operated by electronic or electromechanical controllers located external to the sign or outline lighting system:

(1) The disconnecting means shall be permitted to be located within sight of the controller or in the same enclosure with the controller.
(2) The disconnecting means shall disconnect the sign or outline lighting system and the controller from all ungrounded supply conductors.
(3) The disconnecting means shall be designed so that no pole can be operated independently and shall be capable of being locked in the open position.

**(B) Control Switch Rating.** Switches, flashers, and similar devices controlling transformers and electronic power supplies shall be rated for controlling inductive loads or have a current rating not less than twice the current rating of the transformer.

    FPN: See 404.14 for rating of snap switches.

**600.7 Grounding.** Signs and metal equipment of outline lighting systems shall be grounded.

**(A) Flexible Metal Conduit Length.** Listed flexible metal conduit or listed liquidtight flexible metal conduit that encloses the secondary circuit conductor from a transformer or power supply for use with electric discharge tubing shall be permitted as a bonding means if the total accumulative length of the conduit in the secondary circuit does not exceed 30 m (100 ft).

**(B) Small Metal Parts.** Small metal parts not exceeding 50 mm (2 in.) in any dimension, not likely to be energized, and spaced at least 19 mm (¾ in.) from neon tubing shall not require bonding.

**(C) Nonmetallic Conduit.** Where listed nonmetallic conduit is used to enclose the secondary circuit conductor from a transformer or power supply and a bonding conductor is required, the bonding conductor shall be installed separate and remote from the nonmetallic conduit and be spaced at least 38 mm (1½ in.) from the conduit when the circuit is operated at 100 Hz or less or 45 mm (1¾ in.) when the circuit is operated at over 100 Hz.

**(D) Bonding Conductors.** Bonding conductors shall be copper and not smaller than 14 AWG.

**(E) Metal Building Parts.** Metal parts of a building shall not be permitted as a secondary return conductor or an equipment grounding conductor.

**(F) Signs in Fountains.** Signs or outline lighting installed inside a fountain shall have all metal parts and equipment grounding conductors bonded to the equipment grounding conductor for the fountain recirculating system. The bonding connection shall be as near as practicable to the fountain and shall be permitted to be made to metal piping systems that are bonded in accordance with 680.53.

    FPN: Refer to 600.32(J) for restrictions on length of high-voltage secondary conductors.

**600.8 Enclosures.** Live parts other than lamps and neon tubing shall be enclosed.

*Exception: A transformer or electronic power supply provided with an integral enclosure, including a primary and secondary circuit splice enclosure, shall not be required to be provided with an additional enclosure.*

**(A) Strength.** Enclosures shall have ample structural strength and rigidity.

**(B) Material.** Sign and outline lighting system enclosures shall be constructed of metal or shall be listed.

**(C) Minimum Thickness of Enclosure Metal.** Sheet copper or aluminum shall be at least 0.51 mm (0.020 in.) thick. Sheet steel shall be at least 0.41 mm (0.016 in.) thick.

**(D) Protection of Metal.** Metal parts of equipment shall be protected from corrosion.

**600.9 Location.**

**(A) Vehicles.** Sign or outline lighting system equipment shall be at least 4.3 m (14 ft) above areas accessible to vehicles unless protected from physical damage.

**(B) Pedestrians.** Neon tubing, other than dry-location portable signs, accessible to pedestrians shall be protected from physical damage.

**(C) Adjacent to Combustible Materials.** Signs and outline lighting systems shall be installed so that adjacent combustible materials are not subjected to temperatures in excess of 90°C (194°F).

The spacing between wood or other combustible materials and an incandescent or HID lamp or lampholder shall not be less than 50 mm (2 in.).

**(D) Wet Location.** Signs and outline lighting system equipment for wet location use, other than listed watertight type, shall be weatherproof and have drain holes, as necessary, in accordance with the following:

(1) Drain holes shall not be larger than 13 mm (½ in.) or smaller than 6 mm (¼ in.).
(2) Every low point or isolated section of the equipment shall have at least one drain hole.
(3) Drain holes shall be positioned such that there will be no external obstructions.

### 600.10 Portable or Mobile Signs.

**(A) Support.** Portable or mobile signs shall be adequately supported and readily movable without the use of tools.

**(B) Attachment Plug.** An attachment plug shall be provided for each portable or mobile sign.

**(C) Wet or Damp Location.** Portable or mobile signs in wet or damp locations shall comply with 600.10(C)(1) and (C)(2).

**(1) Cords.** All cords shall be junior hard service or hard service types as designated in Table 400.4 and have an equipment grounding conductor.

**(2) Ground-Fault Circuit Interrupter.** Portable or mobile signs shall be provided with factory-installed ground-fault circuit-interrupter protection for personnel. The ground-fault circuit interrupter shall be an integral part of the attachment plug or shall be located in the power-supply cord within 300 mm (12 in.) of the attachment plug.

**(D) Dry Location.** Portable or mobile signs in dry locations shall meet the following:

(1) Cords shall be SP-2, SPE-2, SPT-2, or heavier, as designated in Table 400.4.
(2) The cord shall not exceed 4.5 m (15 ft) in length.

### 600.21 Ballasts, Transformers, and Electronic Power Supplies.

**(A) Accessibility.** Ballasts, transformers, and electronic power supplies shall be located where accessible and shall be securely fastened in place.

**(B) Location.** Ballasts, transformers, and electronic power supplies shall be installed as near to the lamps or neon tubing as practicable to keep the secondary conductors as short as possible.

**(C) Wet Location.** Ballasts, transformers, and electronic power supplies used in wet locations shall be of the weatherproof type or be of the outdoor type and protected from the weather by placement in a sign body or separate enclosure.

**(D) Working Space.** A working space at least 900 mm (3 ft) high, 900 mm (3 ft) wide, by 900 mm (3 ft) deep shall be provided at each ballast, transformer, and electronic power supply or its enclosure where not installed in a sign.

**(E) Attic and Soffit Locations.** Ballasts, transformers, and electronic power supplies shall be permitted to be located in attics and soffits, provided there is an access door at least 900 mm by 600 mm (3 ft by 2 ft) and a passageway of at least 900 mm (3 ft) high by 600 mm (2 ft) wide with a suitable permanent walkway at least 300 mm (12 in.) wide extending from the point of entry to each component.

**(F) Suspended Ceilings.** Ballasts, transformers, and electronic power supplies shall be permitted to be located above suspended ceilings, provided their enclosures are securely fastened in place and not dependent on the suspended ceiling grid for support. Ballasts, transformers, and electronic power supplies installed in suspended ceilings shall not be connected to the branch circuit by flexible cord.

### 600.22 Ballasts.

**(A) Type.** Ballasts shall be identified for the use and shall be listed.

**(B) Thermal Protection.** Ballasts shall be thermally protected.

### 600.23 Transformers and Electronic Power Supplies.

**(A) Type.** Transformers and electronic power supplies shall be identified for the use and shall be listed.

**(B) Secondary-Circuit Ground-Fault Protection.** Transformers and electronic power supplies other than the following shall have secondary-circuit ground-fault protection:

(1) Transformers with isolated ungrounded secondaries and with a maximum open circuit voltage of 7500 volts or less
(2) Transformers with integral porcelain or glass secondary housing for the neon tubing and requiring no field wiring of the secondary circuit

**(C) Voltage.** Secondary-circuit voltage shall not exceed 15,000 volts, nominal, under any load condition. The voltage to ground of any output terminals of the secondary circuit shall not exceed 7500 volts, under any load condition.

**(D) Rating.** Transformers and electronic power supplies shall have a secondary-circuit current rating of not more than 300 mA.

**(E) Secondary Connections.** Secondary circuit outputs shall not be connected in parallel or in series.

**(F) Marking.** A transformer or power supply shall be marked to indicate that it has secondary-circuit ground-fault protection.

## II. Field-Installed Skeleton Tubing

**600.30 Applicability.** Part II of this article shall apply only to field-installed skeleton tubing. These requirements are in addition to the requirements of Part I.

**600.31 Neon Secondary-Circuit Conductors, 1000 Volts or Less, Nominal.**

**(A) Wiring Method.** Conductors shall be installed using any wiring method included in Chapter 3 suitable for the conditions.

**(B) Insulation and Size.** Conductors shall be insulated, listed for the purpose, and not smaller than 18 AWG.

**(C) Number of Conductors in Raceway.** The number of conductors in a raceway shall be in accordance with Table 1 of Chapter 9.

**(D) Installation.** Conductors shall be installed so they are not subject to physical damage.

**(E) Protection of Leads.** Bushings shall be used to protect wires passing through an opening in metal.

**600.32 Neon Secondary Circuit Conductors, Over 1000 Volts, Nominal.**

**(A) Wiring Methods.**

**(1) Installation.** Conductors shall be installed on insulators, in rigid metal conduit, intermediate metal conduit, rigid nonmetallic conduit, liquidtight flexible nonmetallic conduit, flexible metal conduit, liquidtight flexible metal conduit, electrical metallic tubing, metal enclosures, or other equipment listed for the purpose and shall be installed in accordance with the requirements of Chapter 3.

**(2) Number of Conductors.** Conduit or tubing shall contain only one conductor.

**(3) Size.** Conduit or tubing shall be a minimum of metric designator 16 (trade size ½).

**(4) Spacing from Ground.** Other than at the location of connection to a metal enclosure or sign body, nonmetallic conduit or flexible nonmetallic conduit shall be spaced no less than 38 mm (1½ in.) from grounded or bonded parts when the conduit contains a conductor operating at 100 Hz or less and shall be spaced no less than 45 mm (1¾ in.) from grounded or bonded parts when the conduit contains a conductor operating at more than 100 Hz.

**(5) Metal Building Parts.** Metal parts of a building shall not be permitted as a secondary return conductor or an equipment grounding conductor.

**(B) Insulation and Size.** Conductors shall be insulated, listed as Gas Tube Sign and Ignition Cable Type GTO, rated for 5, 10, or 15 kV, not smaller than 18 AWG, and have a minimum temperature rating of 105°C (221°F).

**(C) Installation.** Conductors shall be installed so they are not subject to physical damage.

**(D) Bends in Conductors.** Sharp bends in insulated conductors shall be avoided.

**(E) Spacing.** Secondary conductors shall be separated from each other and from all objects other than insulators or neon tubing by a spacing of not less than 38 mm (1½ in.). GTO cable installed in metal conduit or tubing requires no spacing between the cable insulation and the conduit or tubing.

**(F) Insulators and Bushings.** Insulators and bushings for conductors shall be listed for the purpose.

**(G) Conductors in Raceways.**

**(1) Damp or Wet Locations.** In damp or wet locations, the insulation on all conductors shall extend not less than 100 mm (4 in.) beyond the metal conduit or tubing.

**(2) Dry Locations.** In dry locations, the insulation on all conductors shall extend not less than 65 mm (2½ in.) beyond the metal conduit or tubing.

**(H) Between Neon Tubing and Midpoint Return.** Conductors shall be permitted to run between the ends of neon tubing or to the secondary circuit midpoint return of transformers or electronic power supplies listed for the purpose and provided with terminals or leads at the midpoint.

**(I) Dwelling Occupancies.** Equipment having an open circuit voltage exceeding 1000 volts shall not be installed in or on dwelling occupancies.

**(J) Length of Secondary Circuit Conductors.**

**(1) Secondary Conductor to the First Electrode.** The length of secondary circuit conductors from a high-voltage terminal or lead of a transformer or electronic power supply to the first neon tube electrode shall not exceed the following:

(1) 6 m (20 ft) where installed in metal conduit or tubing
(2) 15 m (50 ft) where installed in nonmetallic conduit

**(2) Other Secondary Circuit Conductors.** All other sections of secondary circuit conductor in a neon tube circuit shall be as short as practicable.

### 600.41 Neon Tubing.

**(A) Design.** The length and design of the tubing shall not cause a continuous overcurrent beyond the design loading of the transformer or electronic power supply.

**(B) Support.** Tubing shall be supported by listed tube supports.

**(C) Spacing.** A spacing of not less than 6 mm (¼ in.) shall be maintained between the tubing and the nearest surface, other than its support.

### 600.42 Electrode Connections.

**(A) Accessibility.** Terminals of the electrode shall not be accessible to unqualified persons.

**(B) Electrode Connections.** Connections shall be made by use of a connection device, twisting of the wires together, or use of an electrode receptacle. Connections shall be electrically and mechanically secure and shall be in an enclosure listed for the purpose.

**(C) Support.** The neon tubing and conductor shall be supported not more than 150 mm (6 in.) from the electrode connection.

**(D) Receptacles.** Electrode receptacles shall be listed for the purpose.

**(E) Bushings.** Where electrodes penetrate an enclosure, bushings listed for the purpose shall be used unless receptacles are provided.

**(F) Wet Locations.** A listed cap shall be used to close the opening between neon tubing and a receptacle where the receptacle penetrates a building. Where a bushing or neon tubing penetrates a building, the opening between neon tubing and the bushing shall be sealed.

**(G) Electrode Enclosures.** Electrode enclosures shall be listed for the purpose.

# ARTICLE 604
# Manufactured Wiring Systems

**604.1 Scope.** The provisions of this article apply to field-installed wiring using off-site manufactured subassemblies for branch circuits, remote-control circuits, signaling circuits, and communications circuits in accessible areas.

**604.2 Definition.**

**Manufactured Wiring System.** A system containing component parts that are assembled in the process of manufacture and cannot be inspected at the building site without damage or destruction to the assembly.

**604.3 Other Articles.** Except as modified by the requirements of this article, all other applicable articles of this *Code* shall apply.

**604.4 Uses Permitted.** The manufactured wiring systems shall be permitted in accessible and dry locations and in plenums and spaces used for environmental air, where listed for this application and installed in accordance with 300.22.

*Exception No. 1: In concealed spaces, one end of tapped cable shall be permitted to extend into hollow walls for direct termination at switch and outlet points.*

*Exception No. 2: For use in outdoor locations where listed for the purpose.*

**604.5 Uses Not Permitted.** Manufactured wiring system types shall not be permitted where limited by the applicable article in Chapter 3 for the wiring method used in its construction.

**604.6 Construction.**

**(A) Types.**

**(1) Cables.** Cable shall be listed armored cable or metal-clad cable containing nominal 600-volt 10 or 12 AWG copper-insulated conductors with a bare or insulated copper equipment grounding conductor equivalent in size to the ungrounded conductor.

Other cables as listed in 725.61, 800.50, 820.50, and 830.5 shall be permitted in manufactured wiring systems for wiring of equipment within the scope of their respective articles.

**(2) Conduits.** Conduit shall be listed flexible metal conduit or listed liquidtight flexible conduit containing nominal 600-volt 10 or 12 AWG copper-insulated conductors with a bare or insulated copper equipment grounding conductor equivalent in size to the ungrounded conductor.

*Exception No. 1 to (1) and (2): A luminaire (fixture) tap, maximum 1.8 m (6 ft) long, intended for connection to a single luminaire (fixture) shall be permitted to contain conductors smaller than 12 AWG but not smaller than 18 AWG.*

*Exception No. 2 to (1) and (2): Conductors smaller than 12 AWG shall be permitted for remote-control, signaling, or communications circuits. The assembly shall be listed for the purpose.*

**(3) Flexible Cord.** Flexible cord suitable for hard usage, with minimum 12 AWG conductors, shall be permitted as part of a listed factory-made assembly not exceeding 1.8 m (6 ft) in length when making a transition between components of a manufactured wiring system and utilization equipment not permanently secured to the building structure. The cord shall be visible for its entire length and shall not be subject to strain or physical damage.

**(B) Marking.** Each section shall be marked to identify the type of cable, flexible cord, or conduit.

**(C) Receptacles and Connectors.** Receptacles and connectors shall be of the locking type, uniquely polarized and identified for the purpose, and shall be part of a listed assembly for the appropriate system.

**(D) Other Component Parts.** Other component parts shall be listed for the appropriate system.

**(E) Support.** Manufactured wiring systems shall be supported in accordance with the applicable cable or conduit article for the cable or conduit type employed.

**604.7 Unused Outlets.** All unused outlets shall be capped to effectively close the connector openings.

## ARTICLE 605
## Office Furnishings (Consisting of Lighting Accessories and Wired Partitions)

**605.1 Scope.** This article covers electrical equipment, lighting accessories, and wiring systems used to connect, or contained within, or installed on relocatable wired partitions.

**605.2 General.** Wiring systems shall be identified as suitable for providing power for lighting accessories and appliances in wired partitions. These partitions shall not extend from floor to ceiling.

*Exception:   Where permitted by the authority having jurisdiction, these relocatable wired partitions shall be permitted to extend to the ceiling but shall not penetrate the ceiling.*

**(A) Use.** These assemblies shall be installed and used only as provided for by this article.

**(B) Other Articles.** Except as modified by the requirements of this article, all other articles of this *Code* shall apply.

**(C) Hazardous (Classified) Locations.** Where used in hazardous (classified) locations, these assemblies shall conform with Articles 500 through 517 in addition to this article.

**605.3 Wireways.** All conductors and connections shall be contained within wiring channels of metal or other material identified as suitable for the conditions of use. Wiring channels shall be free of projections or other conditions that may damage conductor insulation.

**605.4 Partition Interconnections.** The electrical connection between partitions shall be a flexible assembly identified for use with wired partitions or shall be permitted to be installed using flexible cord, provided all the following conditions are met:

(1) The cord is extra-hard usage type with 12 AWG or larger conductors, with an insulated grounding conductor.
(2) The partitions are mechanically contiguous.
(3) The cord is not longer than necessary for maximum positioning of the partitions but is in no case to exceed 600 mm (2 ft).
(4) The cord is terminated at an attachment plug and cord connector with strain relief.

**605.5 Lighting Accessories.** Lighting equipment listed and identified for use with wired partitions shall comply with 605.5(A), (B), and (C).

**(A) Support.** A means for secure attachment or support shall be provided.

**(B) Connection.** Where cord and plug connection is provided, the cord length shall be suitable for the intended application but shall not exceed 2.7 m (9 ft) in length. The cord shall not be smaller than 18 AWG, shall contain an equipment grounding conductor, and shall be of the hard usage type. Connection by other means shall be identified as suitable for the condition of use.

**(C) Receptacle Outlet.** Convenience receptacles shall not be permitted in lighting accessories.

**605.6 Fixed-Type Partitions.** Wired partitions that are fixed (secured to building surfaces) shall be permanently connected to the building electrical system by one of the wiring methods of Chapter 3.

**605.7 Freestanding-Type Partitions.** Partitions of the freestanding type (not fixed) shall be permitted to be permanently connected to the building electrical system by one of the wiring methods of Chapter 3.

**605.8 Freestanding-Type Partitions, Cord-and-Plug-Connected.** Individual partitions of the freestanding type, or groups of individual partitions that are electrically connected, are mechanically contiguous, and do not exceed 9.0 m (30 ft) when assembled, shall be permitted to be connected to the building electrical system by a single flexible cord and plug, provided all of the conditions of 605.8(A) through (D) are met.

**(A) Flexible Power-Supply Cord.** The flexible power-supply cord shall be extra-hard usage type with 12 AWG or larger conductors with an insulated equipment grounding conductor and not exceeding 600 mm (2 ft) in length.

**(B) Receptacle Supplying Power.** The receptacle(s) supplying power shall be on a separate circuit serving only panels and no other loads and shall be located not more than 300 mm (12 in.) from the partition that is connected to it.

**(C) Receptacle Outlets, Maximum.** Individual partitions or groups of interconnected individual partitions shall not contain more than thirteen 15-ampere, 125-volt receptacle outlets.

**(D) Multiwire Circuits, Not Permitted.** Individual partitions or groups of interconnected individual partitions shall not contain multiwire circuits.

FPN: See 210.4 for circuits supplying partitions in 605.6 and 605.7.

## ARTICLE 610
## Cranes and Hoists

### I. General

**610.1 Scope.** This article covers the installation of electrical equipment and wiring used in connection with cranes, monorail hoists, hoists, and all runways.

FPN: For further information, see ANSI B-30, *Safety Code for Cranes, Derricks, Hoists, Jacks, and Slings.*

**610.2 Special Requirements for Particular Locations.**

**(A) Hazardous (Classified) Locations.** All equipment that operates in a hazardous (classified) location shall conform to Article 500.

**(1) Class I Locations.** Equipment used in locations that are hazardous because of the presence of flammable gases or vapors shall conform to Article 501.

**(2) Class II Locations.** Equipment used in locations that are hazardous because of combustible dust shall conform to Article 502.

**(3) Class III Locations.** Equipment used in locations that are hazardous because of the presence of easily ignitible fibers or flyings shall conform to Article 503.

**(B) Combustible Materials.** Where a crane, hoist, or monorail hoist operates over readily combustible material, the resistors shall be located as permitted in the following:

(1) A well-ventilated cabinet composed of noncombustible material constructed so that it does not emit flames or molten metal
(2) A cage or cab constructed of noncombustible material that encloses the sides of the cage or cab from the floor to a point at least 150 mm (6 in.) above the top of the resistors

**(C) Electrolytic Cell Lines.** See 668.32.

### II. Wiring

**610.11 Wiring Method.** Conductors shall be enclosed in raceways or be Type AC cable with insulated grounding conductor, Type MC cable, or Type MI cable unless otherwise permitted or required in 610.11(A) through (E).

**(A) Contact Conductor.** Contact conductors are not required to be enclosed in raceways.

**(B) Open Conductors.** Short lengths of open conductors at resistors, collectors, and other equipment are not required to be enclosed in raceways.

**(C) Flexible Connections to Motors and Similar Equipment.** Where flexible connections are necessary, flexible stranded conductors shall be used. Conductors shall be in flexible metal conduit, liquidtight flexible metal conduit, liquidtight flexible nonmetallic conduit, multiconductor cable, or an approved nonmetallic flexible raceway.

**(D) Pushbutton Stations Multiconductor Cable.** Where multiconductor cable is used with a suspended pushbutton

**Table 610.14(A). Ampacities of Insulated Copper Conductors Used with Short-Time Rated Crane and Hoist Motors. Based on Ambient Temperature of 30°C (86°F). Up to Four Conductors in Raceway or Cable.[1] Up to 3 ac[2] or 4 dc[1] Conductors in Raceway or Cable**

| Maximum Operating Temperature | 75°C (167°F) | | 90°C (194°F) | | 125°C (257°F) | | Maximum Operating Temperature |
|---|---|---|---|---|---|---|---|
| Size (AWG or kcmil) | Types MTW, RHW, THW, THWN, XHHW, USE, ZW | | Types TA, TBS, SA, SIS, PFA, FEP, FEPB, RHH, THHN, XHHW, Z, ZW | | Types FEP, FEPB, PFA, PFAH, SA, TFE, Z, ZW | | Size (AWG or kcmil) |
| | 60 Min | 30 Min | 60 Min | 30 Min | 60 Min | 30 Min | |
| 16 | 10 | 12 | — | — | — | — | 16 |
| 14 | 25 | 26 | 31 | 32 | 38 | 40 | 14 |
| 12 | 30 | 33 | 36 | 40 | 45 | 50 | 12 |
| 10 | 40 | 43 | 49 | 52 | 60 | 65 | 10 |
| 8 | 55 | 60 | 63 | 69 | 73 | 80 | 8 |
| 6 | 76 | 86 | 83 | 94 | 101 | 119 | 6 |
| 5 | 85 | 95 | 95 | 106 | 115 | 134 | 5 |
| 4 | 100 | 117 | 111 | 130 | 133 | 157 | 4 |
| 3 | 120 | 141 | 131 | 153 | 153 | 183 | 3 |
| 2 | 137 | 160 | 148 | 173 | 178 | 214 | 2 |
| 1 | 143 | 175 | 158 | 192 | 210 | 253 | 1 |
| 1/0 | 190 | 233 | 211 | 259 | 253 | 304 | 1/0 |
| 2/0 | 222 | 267 | 245 | 294 | 303 | 369 | 2/0 |
| 3/0 | 280 | 341 | 305 | 372 | 370 | 452 | 3/0 |
| 4/0 | 300 | 369 | 319 | 399 | 451 | 555 | 4/0 |
| 250 | 364 | 420 | 400 | 461 | 510 | 635 | 250 |
| 300 | 455 | 582 | 497 | 636 | 587 | 737 | 300 |
| 350 | 486 | 646 | 542 | 716 | 663 | 837 | 350 |
| 400 | 538 | 688 | 593 | 760 | 742 | 941 | 400 |
| 450 | 600 | 765 | 660 | 836 | 818 | 1042 | 450 |
| 500 | 660 | 847 | 726 | 914 | 896 | 1143 | 500 |

### AMPACITY CORRECTION FACTORS

| Ambient Temperature (°C) | For ambient temperatures other than 30°C (86°F), multiply the ampacities shown above by the appropriate factor shown below. | | | | | | Ambient Temperature (°F) |
|---|---|---|---|---|---|---|---|
| 21–25 | 1.05 | 1.05 | 1.04 | 1.04 | 1.02 | 1.02 | 70–77 |
| 26–30 | 1.00 | 1.00 | 1.00 | 1.00 | 1.00 | 1.00 | 79–86 |
| 31–35 | 0.94 | 0.94 | 0.96 | 0.96 | 0.97 | 0.97 | 88–95 |
| 36–40 | 0.88 | 0.88 | 0.91 | 0.91 | 0.95 | 0.95 | 97–104 |
| 41–45 | 0.82 | 0.82 | 0.87 | 0.87 | 0.92 | 0.92 | 106–113 |
| 46–50 | 0.75 | 0.75 | 0.82 | 0.82 | 0.89 | 0.89 | 115–122 |
| 51–55 | 0.67 | 0.67 | 0.76 | 0.76 | 0.86 | 0.86 | 124–131 |
| 56–60 | 0.58 | 0.58 | 0.71 | 0.71 | 0.83 | 0.83 | 133–140 |
| 61–70 | 0.33 | 0.33 | 0.58 | 0.58 | 0.76 | 0.76 | 142–158 |
| 71–80 | — | — | 0.41 | 0.41 | 0.69 | 0.69 | 160–176 |
| 81–90 | — | — | — | — | 0.61 | 0.61 | 177–194 |
| 91–100 | — | — | — | — | 0.51 | 0.51 | 195–212 |
| 101–120 | — | — | — | — | 0.40 | 0.40 | 213–248 |

Note: Other insulations shown in Table 310.13 and approved for the temperature and location shall be permitted to be substituted for those shown in Table 610.14(A). The allowable ampacities of conductors used with 15-minute motors shall be the 30-minute ratings increased by 12 percent.

[1] For 5 to 8 simultaneously energized power conductors in raceway or cable, the ampacity of each power conductor shall be reduced to a value of 80 percent of that shown in the table.

[2] For 4 to 6 simultaneously energized 125°C (257°F) ac power conductors in raceway or cable, the ampacity of each power conductor shall be reduced to a value of 80 percent of that shown in the table.

station, the station shall be supported in some satisfactory manner that protects the electric conductors against strain.

**(E) Flexibility to Moving Parts.** Where flexibility is required for power or control to moving parts, a cord suitable for the purpose shall be permitted provided the following apply:

(1) Suitable strain relief and protection from physical damage is provided.
(2) In Class I, Division 2 locations, the cord is approved for extra-hard usage.

**610.12 Raceway or Cable Terminal Fittings.** Conductors leaving raceways or cables shall comply with either of 610.12(A) or (B).

**(A) Separately Bushed Hole.** A box or terminal fitting that has a separately bushed hole for each conductor shall be used wherever a change is made from a raceway or cable to open wiring. A fitting used for this purpose shall not contain taps or splices and shall not be used at luminaire (fixture) outlets.

**(B) Bushing in Lieu of a Box.** A bushing shall be permitted to be used in lieu of a box at the end of a rigid metal conduit, intermediate metal conduit, or electrical metallic tubing where the raceway terminates at unenclosed controls or similar equipment, including contact conductors, collectors, resistors, brakes, power-circuit limit switches, and dc split-frame motors.

**610.13 Types of Conductors.** Conductors shall comply with Table 310.13 unless otherwise permitted in 610.13(A) through (D).

**(A) Exposed to External Heat or Connected to Resistors.** A conductor(s) exposed to external heat or connected to resistors shall have a flame-resistant outer covering or be covered with flame-resistant tape individually or as a group.

**(B) Contact Conductors.** Contact conductors along runways, crane bridges, and monorails shall be permitted to be bare and shall be copper, aluminum, steel, or other alloys or combinations thereof in the form of hard drawn wire, tees, angles, tee rails, or other stiff shapes.

**(C) Flexibility.** Where flexibility is required, flexible cord or cable shall be permitted to be used and, where necessary, cable reels or take-up devices shall be used.

**(D) Class 1, Class 2, and Class 3 Circuits.** Conductors for Class 1, Class 2, and Class 3 remote-control, signaling, and power-limited circuits, installed in accordance with Article 725, shall be permitted.

**610.14 Rating and Size of Conductors.**

**(A) Ampacity.** The allowable ampacities of conductors shall be as shown in Table 610.14(A).

FPN:  For the ampacities of conductors between controllers and resistors, see 430.23.

**(B) Secondary Resistor Conductors.** Where the secondary resistor is separate from the controller, the minimum size of the conductors between controller and resistor shall be calculated by multiplying the motor secondary current by the appropriate factor from Table 610.14(B) and selecting a wire from Table 610.14(A).

**Table 610.14(B) Secondary Conductor Rating Factors**

| Time in Seconds | | Ampacity of Wire in Percent of Full-Load Secondary Current |
|---|---|---|
| **On** | **Off** | |
| 5 | 75 | 35 |
| 10 | 70 | 45 |
| 15 | 75 | 55 |
| 15 | 45 | 65 |
| 15 | 30 | 75 |
| 15 | 15 | 85 |
| Continuous Duty | | 110 |

**(C) Minimum Size.** Conductors external to motors and controls shall not be smaller than 16 AWG unless otherwise permitted in (1) and (2).

(1) 18 AWG wire in multiconductor cord shall be permitted for control circuits at not over 7 amperes.
(2) Wires not smaller than 20 AWG shall be permitted for electronic circuits.

**(D) Contact Conductors.** Contact wires shall have an ampacity not less than that required by Table 610.14(A). for 75°C (167°F) wire, and in no case shall they be smaller than as shown in Table 610.14(D).

**Table 610.14(D) Contact Conductor Supports**

| Distance Between End Strain Insulators or Clamp-Type Intermediate Supports | Size of Wire (AWG) |
|---|---|
| Less than 9.0 m (30 ft) | 6 |
| 9.0 m–18 m (30 ft–60 ft) | 4 |
| Over 18 m (60 ft) | 2 |

**(E) Calculation of Motor Load.**

**(1) Single Motor.** For one motor, 100 percent of motor nameplate full-load ampere rating shall be used.

**(2) Multiple Motors on Single Crane or Hoist.** For multiple motors on a single crane or hoist, the minimum ampacity of the power supply conductors shall be the nameplate full-load ampere rating of the largest motor or group of motors for any single crane motion, plus 50 percent of the nameplate full-load ampere rating of the next largest motor or group of motors, using that column of Table 610.14(A) that applies to the longest time-rated motor.

**(3) Multiple Cranes or Hoists on a Common Conductor System.** For multiple cranes, hoists, or both, supplied by a common conductor system, compute the motor minimum ampacity for each crane as defined in 610.14(E), add them together, and multiply the sum by the appropriate demand factor from Table 610.14(E).

**Table 610.14(E) Demand Factors**

| Number of Cranes or Hoists | Demand Factor |
|:---:|:---:|
| 2 | 0.95 |
| 3 | 0.91 |
| 4 | 0.87 |
| 5 | 0.84 |
| 6 | 0.81 |
| 7 | 0.78 |

**(F) Other Loads.** Additional loads, such as heating, lighting, and air conditioning, shall be provided for by application of the appropriate sections of this *Code*.

**(G) Nameplate.** Each crane, monorail, or hoist shall be provided with a visible nameplate marked with the manufacturer's name, the rating in volts, frequency, number of phases, and circuit amperes as calculated in 610.14(E) and (F).

**610.15 Common Return.** Where a crane or hoist is operated by more than one motor, a common-return conductor of proper ampacity shall be permitted.

## III. Contact Conductors

**610.21 Installation of Contact Conductors.** Contact conductors shall comply with 610.21(A) through (H).

**(A) Locating or Guarding Contact Conductors.** Runway contact conductors shall be guarded, and bridge contact conductors shall be located or guarded in such a manner that persons cannot inadvertently touch energized current-carrying parts.

**(B) Contact Wires.** Wires that are used as contact conductors shall be secured at the ends by means of approved strain insulators and shall be mounted on approved insula-tors so that the extreme limit of displacement of the wire does not bring the latter within less than 38 mm (1½ in.) from the surface wired over.

**(C) Supports Along Runways.** Main contact conductors carried along runways shall be supported on insulating supports placed at intervals not exceeding 6.0 m (20 ft) unless otherwise permitted in 610.21(F).

Such conductors shall be separated not less than 150 mm (6 in.), other than for monorail hoists where a spacing of not less than 75 mm (3 in.) shall be permitted. Where necessary, intervals between insulating supports shall be permitted to be increased up to 12 m (40 ft), the separation between conductors being increased proportionately.

**(D) Supports on Bridges.** Bridge wire contact conductors shall be kept at least 65 mm (2½ in.) apart, and, where the span exceeds 25 m (80 ft), insulating saddles shall be placed at intervals not exceeding 15 m (50 ft).

**(E) Supports for Rigid Conductors.** Conductors along runways and crane bridges, that are of the rigid type specified in 610.13(B) and not contained within an approved enclosed assembly, shall be carried on insulating supports spaced at intervals of not more than 80 times the vertical dimension of the conductor, but in no case greater than 4.5 m (15 ft), and spaced apart sufficiently to give a clear electrical separation of conductors or adjacent collectors of not less than 25 mm (1 in.).

**(F) Track as Circuit Conductor.** Monorail, tram rail, or crane runway tracks shall be permitted as a conductor of current for one phase of a 3-phase, ac system furnishing power to the carrier, crane, or trolley, provided all of the following conditions are met:

(1) The conductors supplying the other two phases of the power supply are insulated.
(2) The power for all phases is obtained from an insulating transformer.
(3) The voltage does not exceed 300 volts.
(4) The rail serving as a conductor is effectively grounded at the transformer and also shall be permitted to be grounded by the fittings used for the suspension or attachment of the rail to a building or structure.

**(G) Electrical Continuity of Contact Conductors.** All sections of contact conductors shall be mechanically joined to provide a continuous electrical connection.

**(H) Not to Supply Other Equipment.** Contact conductors shall not be used as feeders for any equipment other than the crane(s) or hoist(s) that they are primarily designed to serve.

**610.22 Collectors.** Collectors shall be designed so as to reduce to a minimum sparking between them and the contact conductor; and, where operated in rooms used for the storage of easily ignitible combustible fibers and materials, they shall comply with 503.13.

## IV. Disconnecting Means

**610.31 Runway Conductor Disconnecting Means.** A disconnecting means that has a continuous ampere rating not less than that computed in 610.14(E) and (F) shall be provided between the runway contact conductors and the power supply. Such disconnecting means shall consist of a motor-circuit switch, circuit breaker, or molded case switch. This disconnecting means shall be as follows:

(1) Readily accessible and operable from the ground or floor level
(2) Capable of being locked in the open position
(3) Open all ungrounded conductors simultaneously
(4) Placed within view of the runway contact conductors

**610.32 Disconnecting Means for Cranes and Monorail Hoists.** A motor-circuit switch, molded-case switch, or circuit breaker shall be provided in the leads from the runway contact conductors or other power supply on all cranes and monorail hoists. The disconnecting means shall be capable of being locked in the open position.

Where a monorail hoist or hand-propelled crane bridge installation meets all of the following, the disconnecting means shall be permitted to be omitted:

(1) The unit is controlled from the ground or floor level.
(2) The unit is within view of the power supply disconnecting means.
(3) No fixed work platform has been provided for servicing the unit.

Where the disconnecting means is not readily accessible from the crane or monorail hoist operating station, means shall be provided at the operating station to open the power circuit to all motors of the crane or monorail hoist.

**610.33 Rating of Disconnecting Means.** The continuous ampere rating of the switch or circuit breaker required by 610.32 shall not be less than 50 percent of the combined short-time ampere rating of the motors or less than 75 percent of the sum of the short-time ampere rating of the motors required for any single motion.

## V. Overcurrent Protection

**610.41 Feeders, Runway Conductors.**

**(A) Single Feeder.** The runway supply conductors and main contact conductors of a crane or monorail shall be protected by an overcurrent device(s) that shall not be greater than the largest rating or setting of any branch-

circuit protective device plus the sum of the nameplate ratings of all the other loads with application of the demand factors from Table 610.14(E).

**(B) More Than One Feeder Circuit.** Where more than one feeder circuit is installed to supply runway conductors, each feeder circuit shall be sized and protected in compliance with 610.41(A).

**610.42 Branch-Circuit Short-Circuit and Ground-Fault Protection.** Branch circuits shall be protected in accordance with 610.42(A). Branch-circuit taps, where made, shall comply with 610.42(B).

**(A) Fuse or Circuit Breaker Rating.** Crane, hoist, and monorail hoist motor branch circuits shall be protected by fuses or inverse-time circuit breakers that have a rating in accordance with Table 430.52. Where two or more motors operate a single motion, the sum of their nameplate current ratings shall be considered as that of a single motor.

**(B) Taps.**

**(1) Multiple Motors.** Where two or more motors are connected to the same branch circuit, each tap conductor to an individual motor shall have an ampacity not less than one-third that of the branch circuit. Each motor shall be protected from overload according to 610.43.

**(2) Control Circuits.** Where taps to control circuits originate on the load side of a branch-circuit protective device, each tap and piece of equipment shall be protected in accordance with 430.72.

**(3) Brake Coils.** Taps without separate overcurrent protection shall be permitted to brake coils.

**610.43 Overload Protection.**

**(A) Motor and Branch-Circuit Overload Protection.** Each motor, motor controller, and branch-circuit conductor shall be protected from overload by one of the following means:

(1) A single motor shall be considered as protected where the branch-circuit overcurrent device meets the rating requirements of 610.42.
(2) Overload relay elements in each ungrounded circuit conductor, with all relay elements protected from short circuit by the branch-circuit protection.
(3) Thermal sensing devices, sensitive to motor temperature or to temperature and current, that are thermally in contact with the motor winding(s). A hoist or trolley shall be considered to be protected if the sensing device is connected in the hoist's upper limit switch circuit so as to prevent further hoisting during an overload condition of either motor.

**(B) Manually Controlled Motor.** If the motor is manually controlled, with spring return controls, the overload protective device shall not be required to protect the motor against stalled rotor conditions.

**(C) Multimotor.** Where two or more motors drive a single trolley, truck, or bridge and are controlled as a unit and protected by a single set of overload devices with a rating equal to the sum of their rated full-load currents, a hoist or trolley shall be considered to be protected if the sensing device is connected in the hoist's upper limit switch circuit so as to prevent further hoisting during an overtemperature condition of either motor.

**(D) Hoists and Monorail Hoists.** Hoists and monorail hoists and their trolleys that are not used as part of an overhead traveling crane shall not require individual motor overload protection, provided the largest motor does not exceed 7½ hp and all motors are under manual control of the operator.

## VI. Control

**610.51 Separate Controllers.** Each motor shall be provided with an individual controller unless otherwise permitted in 610.51(A) or (B).

**(A) Motions with More Than One Motor.** Where two or more motors drive a single hoist, carriage, truck, or bridge, they shall be permitted to be controlled by a single controller.

**(B) Multiple Motion Controller.** One controller shall be permitted to be switched between motors, under the following conditions:

(1) The controller has a horsepower rating that is not lower than the horsepower rating of the largest motor.
(2) Only one motor is operated at one time.

**610.53 Overcurrent Protection.** Conductors of control circuits shall be protected against overcurrent. Control circuits shall be considered as protected by overcurrent devices that are rated or set at not more than 300 percent of the ampacity of the control conductors, unless otherwise permitted in 610.53(A) or (B).

**(A) Taps to Control Transformers.** Taps to control transformers shall be considered as protected where the secondary circuit is protected by a device rated or set at not more than 200 percent of the rated secondary current of the transformer and not more than 200 percent of the ampacity of the control circuit conductors.

**(B) Continuity of Power.** Where the opening of the control circuit would create a hazard, as for example, the con-

trol circuit of a hot metal crane, the control circuit conductors shall be considered as being properly protected by the branch-circuit overcurrent devices.

**610.55 Limit Switch.** A limit switch or other device shall be provided to prevent the load block from passing the safe upper limit of travel of all hoisting mechanisms.

**610.57 Clearance.** The dimension of the working space in the direction of access to live parts that are likely to require examination, adjustment, servicing, or maintenance while energized shall be a minimum of 750 mm (2½ ft). Where controls are enclosed in cabinets, the door(s) shall either open at least 90 degrees or be removable.

## VII. Grounding

**610.61 Grounding.** All exposed non–current-carrying metal parts of cranes, monorail hoists, hoists, and accessories, including pendant controls, shall be metallically joined together into a continuous electrical conductor so that the entire crane or hoist will be grounded in accordance with Article 250. Moving parts, other than removable accessories or attachments, that have metal-to-metal bearing surfaces shall be considered to be electrically connected to each other through the bearing surfaces for grounding purposes. The trolley frame and bridge frame shall be considered as electrically grounded through the bridge and trolley wheels and their respective tracks unless local conditions, such as paint or other insulating material, prevent reliable metal-to-metal contact. In this case, a separate bonding conductor shall be provided.

## ARTICLE 620
## Elevators, Dumbwaiters, Escalators, Moving Walks, Wheelchair Lifts, and Stairway Chair Lifts

### I. General

**620.1 Scope.** This article covers the installation of electrical equipment and wiring used in connection with elevators, dumbwaiters, escalators, moving walks, wheelchair lifts, and stairway chair lifts.

> FPN No. 1:  For further information, see ASME/ANSI A17.1-1996, *Safety Code for Elevators and Escalators.*

> FPN No. 2:  For further information, see ASME/ANSI A17.5-1996 (CSA B44.1-1996), *Elevator and Escalator Electrical Equipment Certification Standard.*

## 620.2 Definitions.

**Control System.** The overall system governing the starting, stopping, direction of motion, acceleration, speed, and retardation of the moving member.

**Controller, Motion.** The electric device(s) for that part of the control system that governs the acceleration, speed, retardation, and stopping of the moving member.

**Controller, Motor.** The operative units of the control system comprised of the starter device(s) and power conversion equipment used to drive an electric motor, or the pumping unit used to power hydraulic control equipment.

**Controller, Operation.** The electric device(s) for that part of the control system that initiates the starting, stopping, and direction of motion in response to a signal from an operating device.

**Operating Device.** The car switch, push buttons, key or toggle switch(s), or other devices used to activate the operation controller.

**Signal Equipment.** Includes audible and visual equipment such as chimes, gongs, lights, and displays that convey information to the user.

> FPN No. 1: The motor controller, motion controller, and operation controller may be located in a single enclosure or a combination of enclosures.
>
> FPN No. 2: FPN Figure 620.2 is for information only.

## 620.3 Voltage Limitations.
The supply voltage shall not exceed 300 volts between conductors unless otherwise permitted in 620.3(A) through (C).

**(A) Power Circuits.** Branch circuits to door operator controllers and door motors and branch circuits and feeders to motor controllers, driving machine motors, machine brakes, and motor-generator sets shall not have a circuit voltage in excess of 600 volts. Internal voltages of power conversion and functionally associated equipment, including the interconnecting wiring, shall be permitted to have higher voltages, provided that all such equipment and wiring shall be listed for the higher voltages. Where the voltage exceeds 600 volts, warning labels or signs that read "DANGER — HIGH VOLTAGE" shall be attached to the equipment and shall be plainly visible.

**(B) Lighting Circuits.** Lighting circuits shall comply with the requirements of Article 410.

**(C) Heating and Air-Conditioning Circuits.** Branch circuits for heating and air-conditioning equipment located on the elevator car shall not have a circuit voltage in excess of 600 volts.

## 620.4 Live Parts Enclosed.
All live parts of electrical apparatus in the hoistways, at the landings, in or on the cars of elevators and dumbwaiters, in the wellways or the landings of escalators or moving walks, or in the runways and ma-

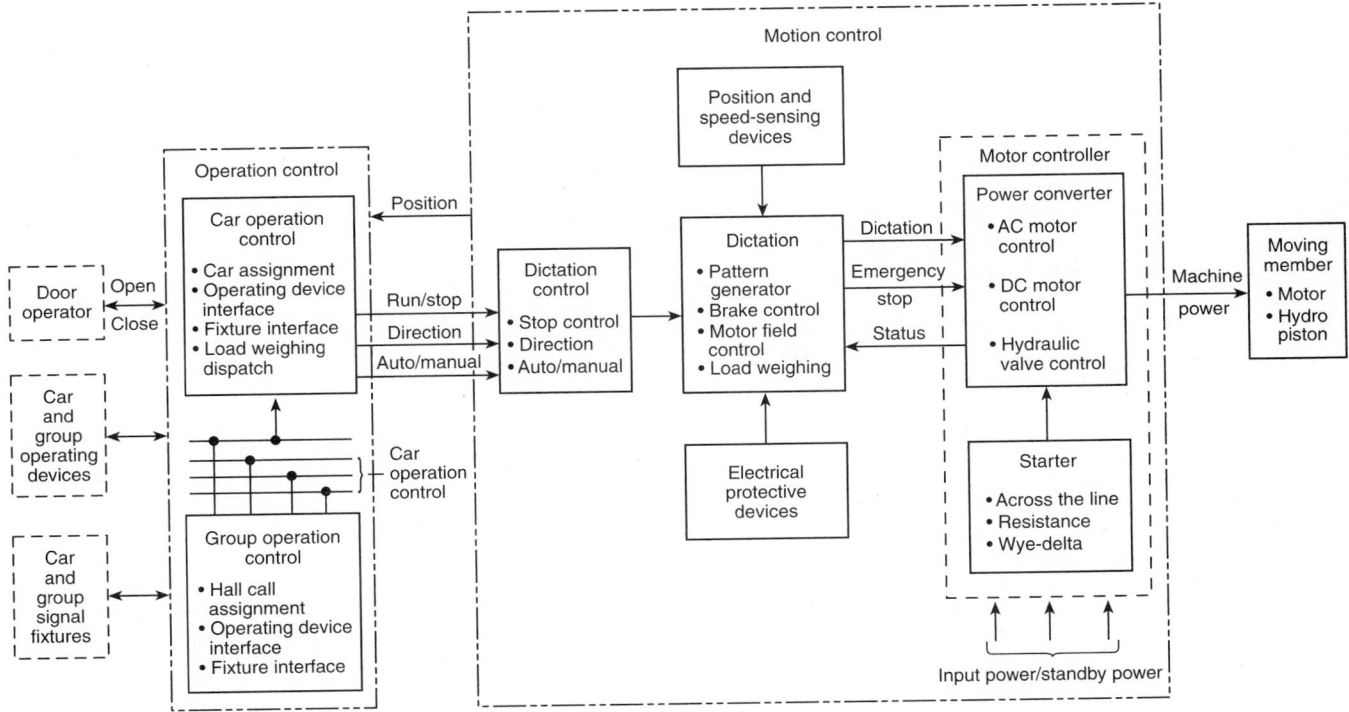

**FPN Figure 620.2  Control system.**

chinery spaces of wheelchair lifts and stairway chair lifts shall be enclosed to protect against accidental contact.

FPN: See 110.27 for guarding of live parts (600 volts, nominal, or less).

**620.5 Working Clearances.** Working space shall be provided about controllers, disconnecting means, and other electrical equipment. The minimum working space shall not be less than that specified in 110.26(A).

Where conditions of maintenance and supervision ensure that only qualified persons examine, adjust, service, and maintain the equipment, the clearance requirements of 110.26(A) shall be waived as permitted in 620.5(A) through (D).

**(A) Flexible Connections to Equipment.** Electrical equipment in (1) through (4) shall be permitted to be provided with flexible leads to all external connections so that it can be repositioned to meet the clear working space requirements of 110.26(A).

(1) Controllers and disconnecting means for dumbwaiters, escalators, moving walks, wheelchair lifts, and stairway chair lifts installed in the same space with the driving machine
(2) Controllers and disconnecting means for elevators installed in the hoistway or on the car
(3) Controllers for door operators
(4) Other electrical equipment installed in the hoistway or on the car

**(B) Guards.** Live parts of the electrical equipment are suitably guarded, isolated, or insulated, and the equipment can be examined, adjusted, serviced, or maintained while energized without removal of this protection.

FPN: See definition of *Exposed* in Article 100.

**(C) Examination, Adjusting, and Servicing.** Electrical equipment is not required to be examined, adjusted, serviced, or maintained while energized.

**(D) Low Voltage.** Uninsulated parts are at a voltage not greater than 30 volts rms, 42 volts peak, or 60 volts dc.

## II. Conductors

**620.11 Insulation of Conductors.** The insulation of conductors shall comply with 620.11(A) through (D).

FPN: One method of determining that conductors are flame retardant is by testing the conductors to the VW-1 (Vertical-Wire) Flame Test in ANSI/UL 1581-1991, *Reference Standard for Electrical Wires, Cables, and Flexible Cords.*

**(A) Hoistway Door Interlock Wiring.** The conductors to the hoistway door interlocks from the hoistway riser shall be flame retardant and suitable for a temperature of not less than 200°C (392°F). Conductors shall be Type SF or equivalent.

**(B) Traveling Cables.** Traveling cables used as flexible connections between the elevator or dumbwaiter car or counterweight and the raceway shall be of the types of elevator cable listed in Table 400.4 or other approved types.

**(C) Other Wiring.** All conductors in raceways shall have flame-retardant insulation.

Conductors shall be Type MTW, TF, TFF, TFN, TFFN, THHN, THW, THWN, TW, XHHW, hoistway cable, or any other conductor with insulation designated as flame retardant. Shielded conductors shall be permitted if such conductors are insulated for the maximum nominal circuit voltage applied to any conductor within the cable or raceway system.

**(D) Insulation.** All conductors shall have an insulation voltage rating equal to at least the maximum nominal circuit voltage applied to any conductor within the enclosure, cable, or raceway. Insulations and outer coverings that are marked for limited smoke and are so listed shall be permitted.

**620.12 Minimum Size of Conductors.** The minimum size of conductors, other than conductors that form an integral part of control equipment, shall be in accordance with 620.12(A) and (B).

**(A) Traveling Cables.**

**(1) Lighting Circuits.** For lighting circuits: 14 AWG copper; 20 AWG copper or larger conductors shall be permitted in parallel, provided the ampacity is equivalent to at least that of 14 AWG copper.

**(2) Other Circuits.** For other circuits, 20 AWG copper.

**(B) Other Wiring.** 24 AWG copper; smaller size listed conductors shall be permitted.

**620.13 Feeder and Branch-Circuit Conductors.** Conductors shall have an ampacity in accordance with 620.13(A) through (D). With generator field control, the conductor ampacity shall be based on the nameplate current rating of the driving motor of the motor-generator set that supplies power to the elevator motor.

FPN No. 1: The heating of conductors depends on root-mean-square current values, which, with generator field control, are reflected by the nameplate current rating of the motor-generator driving motor rather than by the rating of the elevator motor, which represents actual but short-time and intermittent full-load current values.

FPN No. 2: See Figure 620.13.

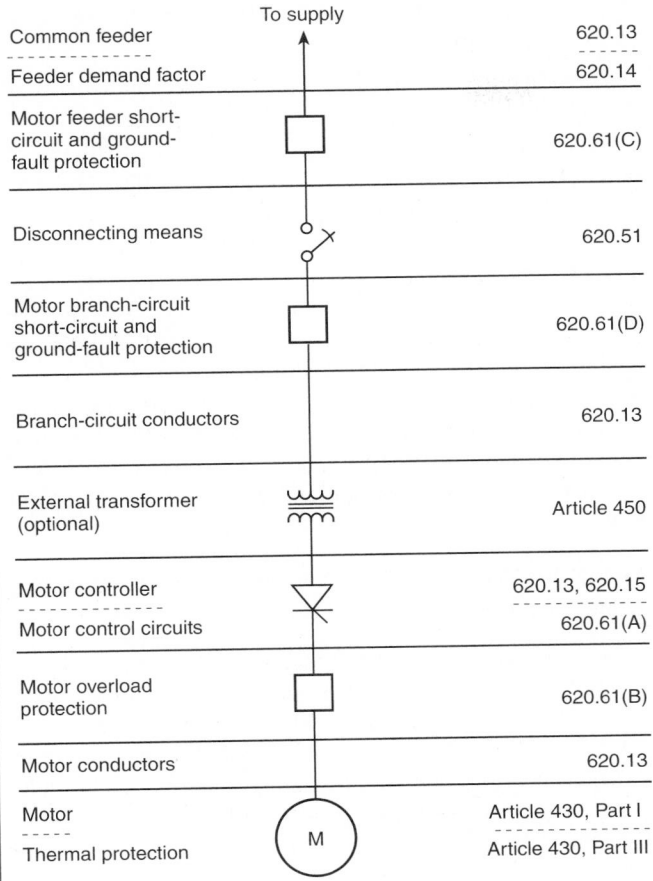

| | To supply | |
|---|---|---|
| Common feeder | | 620.13 |
| Feeder demand factor | | 620.14 |
| Motor feeder short-circuit and ground-fault protection | | 620.61(C) |
| Disconnecting means | | 620.51 |
| Motor branch-circuit short-circuit and ground-fault protection | | 620.61(D) |
| Branch-circuit conductors | | 620.13 |
| External transformer (optional) | | Article 450 |
| Motor controller | | 620.13, 620.15 |
| Motor control circuits | | 620.61(A) |
| Motor overload protection | | 620.61(B) |
| Motor conductors | | 620.13 |
| Motor | | Article 430, Part I |
| Thermal protection | | Article 430, Part III |

**Figure 620.13 Single-line diagram.**

**(A) Conductors Supplying Single Motor.** Conductors supplying a single motor shall have an ampacity not less than the percentage of motor nameplate current determined from 430.22(A) and (E).

FPN: Elevator motor currents, or those of similar functions, may exceed the nameplate value, but since they are inherently intermittent duty and the heating of the motor and conductors is dependent on the root-mean-square (rms) current value, conductors are sized for duty cycle service as shown in Table 430.22(E).

**(B) Conductors Supplying a Single Motor Controller.** Conductors supplying a single motor controller shall have an ampacity not less than the motor controller nameplate current rating, plus all other connected loads.

FPN: Motor controller nameplate current rating may be derived based on the rms value of the motor current using an intermittent duty cycle and other control system loads, if applicable.

**(C) Conductors Supplying a Single Power Transformer.** Conductors supplying a single power transformer shall have an ampacity not less than the nameplate current rating of the power transformer plus all other connected loads.

FPN No. 1: The nameplate current rating of a power transformer supplying a motor controller reflects the nameplate current rating of the motor controller at line voltage (transformer primary).

FPN No. 2: See Annex D, Example No. D10.

**(D) Conductors Supplying More Than One Motor, Motor Controller, or Power Transformer.** Conductors supplying more than one motor, motor controller, or power transformer shall have an ampacity not less than the sum of the nameplate current ratings of the equipment plus all other connected loads. The ampere ratings of motors to be used in the summation shall be determined from Table 430.22(E), and 430.24 and 430.24, Exception No. 1.

FPN: See Annex D, Example Nos. D9 and D10.

**620.14 Feeder Demand Factor.** Feeder conductors of less ampacity than required by 620.13 shall be permitted subject to the requirements of Table 620.14.

**Table 620.14 Feeder Demand Factors for Elevators**

| Number of Elevators on a Single Feeder | Demand Factor |
|---|---|
| 1 | 1.00 |
| 2 | 0.95 |
| 3 | 0.90 |
| 4 | 0.85 |
| 5 | 0.82 |
| 6 | 0.79 |
| 7 | 0.77 |
| 8 | 0.75 |
| 9 | 0.73 |
| 10 or more | 0.72 |

FPN: Demand factors are based on 50 percent duty cycle (i.e., half time on and half time off).

**620.15 Motor Controller Rating.** The motor controller rating shall comply with 430.83. The rating shall be permitted to be less than the nominal rating of the elevator motor, when the controller inherently limits the available power to the motor and is marked as power limited.

FPN: For controller markings, see 430.8.

## III. Wiring

**620.21 Wiring Methods.** Conductors and optical fibers located in hoistways, in escalator and moving walk wellways, in wheelchair lifts, stairway chair lift runways, machinery spaces, control spaces, in or on cars, in machine rooms and control rooms, not including the traveling cables connecting the car or counterweight and hoistway wiring, shall be installed in rigid metal conduit, intermediate metal conduit, electrical metallic tubing, rigid nonmetallic conduit, or wireways, or shall be Type MC, MI, or AC cable unless otherwise permitted in 620.21(A) through (C).

**(A) Elevators.**

**(1) Hoistways.**

(a)   Flexible metal conduit, liquidtight flexible metal conduit, or liquidtight flexible nonmetallic conduit shall be permitted in hoistways between risers and limit switches, interlocks, operating buttons, and similar devices.

(b)   Cables used in Class 2 power-limited circuits shall be permitted to be installed between risers and signal equipment and operating devices, provided the cables are supported and protected from physical damage and are of a jacketed and flame-retardant type.

(c)   Flexible cords and cables that are components of listed equipment and used in circuits operating at 30 volts rms or less or 42 volts dc or less shall be permitted in lengths not to exceed 1.8 m (6 ft), provided the cords and cables are supported and protected from physical damage and are of a jacketed and flame-retardant type.

(d)   Flexible metal conduit, liquidtight flexible metal conduit, liquidtight flexible nonmetallic conduit or flexible cords and cables, or conductors grouped together and taped or corded that are part of listed equipment, a driving machine, or a driving machine brake shall be permitted in the hoistway, in lengths not to exceed 1.8 m (6 ft), without being installed in a raceway and where located to be protected from physical damage and are of a flame-retardant type.

**(2) Cars.**

(a)   Flexible metal conduit, liquidtight flexible metal conduit, or liquidtight flexible nonmetallic conduit of metric designator 12 (trade size ⅜), or larger, not exceeding 1.8 m (6 ft) in length, shall be permitted on cars where located so as to be free from oil and if securely fastened in place.

*Exception: Liquidtight flexible nonmetallic conduit of metric designator 12 (trade size ⅜), or larger, as defined by 356.2, shall be permitted in lengths in excess of 1.8 m (6 ft).*

(b)   Hard-service cords and junior hard-service cords that conform to the requirements of Article 400 (Table 400.4) shall be permitted as flexible connections between the fixed wiring on the car and devices on the car doors or gates. Hard-service cords only shall be permitted as flexible connections for the top-of-car operating device or the car-top work light. Devices or luminaires (fixtures) shall be grounded by means of an equipment grounding conductor run with the circuit conductors. Cables with smaller conductors and other types and thicknesses of insulation and jackets shall be permitted as flexible connections between

the fixed wiring on the car and devices on the car doors or gates, if listed for this use.

(c)   Flexible cords and cables that are components of listed equipment and used in circuits operating at 30 volts rms or less or 42 volts dc or less shall be permitted in lengths not to exceed 1.8 m (6 ft), provided the cords and cables are supported and protected from physical damage and are of a jacketed and flame-retardant type.

(d)   Flexible metal conduit, liquidtight flexible metal conduit, liquidtight flexible nonmetallic conduit or flexible cords and cables, or conductors grouped together and taped or corded that are part of listed equipment, a driving machine, or a driving machine brake shall be permitted on the car assembly, in lengths not to exceed 1.8 m (6 ft) without being installed in a raceway and where located to be protected from physical damage and are of a flame-retardant type.

**(3) Within Machine Rooms, Control Rooms, and Machinery Spaces and Control Spaces.**

(a)   Flexible metal conduit, liquidtight flexible metal conduit, or liquidtight flexible nonmetallic conduit of metric designator 12 (trade size ⅜), or larger, not exceeding 1.8 m (6 ft) in length, shall be permitted between control panels and machine motors, machine brakes, motor-generator sets, disconnecting means, and pumping unit motors and valves.

*Exception: Liquidtight flexible nonmetallic conduit metric designator 12 (trade size ⅜) or larger, as defined in 356.2(2), shall be permitted to be installed in lengths in excess of 1.8 m (6 ft).*

(b)   Where motor-generators, machine motors, or pumping unit motors and valves are located adjacent to or underneath control equipment and are provided with extra-length terminal leads not exceeding 1.8 m (6 ft) in length, such leads shall be permitted to be extended to connect directly to controller terminal studs without regard to the carrying-capacity requirements of Articles 430 and 445. Auxiliary gutters shall be permitted in machine and control rooms between controllers, starters, and similar apparatus.

(c)   Flexible cords and cables that are components of listed equipment and used in circuits operating at 30 volts rms or less or 42 volts dc or less shall be permitted in lengths not to exceed 1.8 m (6 ft), provided the cords and cables are supported and protected from physical damage and are of a jacketed and flame-retardant type.

(d)   On existing or listed equipment, conductors shall also be permitted to be grouped together and taped or corded without being installed in a raceway. Such cable groups shall be supported at intervals not over 900 mm (3 ft) and located so as to be protected from physical damage.

**(4) Counterweight.** Flexible metal conduit, liquidtight flexible metal conduit, liquidtight flexible nonmetallic conduit or flexible cords and cables, or conductors grouped together and taped or corded that are part of listed equipment, a driving machine, or a driving machine brake shall be permitted on the counterweight assembly, in lengths not to exceed 1.8 m (6 ft) without being installed in a raceway and where located to be protected from physical damage and are of a flame-retardant type.

**(B) Escalators.**

**(1) Wiring Methods.** Flexible metal conduit, liquidtight flexible metal conduit, or liquidtight flexible nonmetallic conduit shall be permitted in escalator and moving walk wellways. Flexible metal conduit or liquidtight flexible conduit of metric designator 12 (trade size ⅜) shall be permitted in lengths not in excess of 1.8 m (6 ft).

*Exception: Metric designator 12 (trade size ⅜), nominal, or larger liquidtight flexible nonmetallic conduit, as defined in 356.2(2), shall be permitted to be installed in lengths in excess of 1.8 m (6 ft).*

**(2) Class 2 Circuit Cables.** Cables used in Class 2 power-limited circuits shall be permitted to be installed within escalators and moving walkways, provided the cables are supported and protected from physical damage and are of a jacketed and flame-retardant type.

**(3) Flexible Cords.** Hard-service cords that conform to the requirements of Article 400 (Table 400.4) shall be permitted as flexible connections on escalators and moving walk control panels and disconnecting means where the entire control panel and disconnecting means are arranged for removal from machine spaces as permitted in 620.5.

**(C) Wheelchair Lifts and Stairway Chair Lift Raceways.**

**(1) Wiring Methods.** Flexible metal conduit or liquidtight flexible metal conduit shall be permitted in wheelchair lifts and stairway chair lift runways and machinery spaces. Flexible metal conduit or liquidtight flexible conduit of metric designator 12 (trade size ⅜) shall be permitted in lengths not in excess of 1.8 m (6 ft).

*Exception: Metric designator 12 (trade size ⅜) or larger liquidtight flexible nonmetallic conduit, as defined in 356.2(2), shall be permitted to be installed in lengths in excess of 1.8 m (6 ft).*

**(2) Class 2 Circuit Cables.** Cables used in Class 2 power-limited circuits shall be permitted to be installed within wheelchair lifts and stairway chair lift runways and machinery spaces, provided the cables are supported and pro-

tected from physical damage and are of a jacketed and flame-retardant type.

**620.22 Branch Circuits for Car Lighting, Receptacle(s), Ventilation, Heating, and Air Conditioning.**

**(A) Car Light Source.** A separate branch circuit shall supply the car lights, receptacle(s), auxiliary lighting power source, and ventilation on each elevator car. The overcurrent device protecting the branch circuit shall be located in the elevator machine room or control room/machinery space or control space.

**(B) Air-Conditioning and Heating Source.** A dedicated branch circuit shall supply the air-conditioning and heating units on each elevator car. The overcurrent device protecting the branch circuit shall be located in the elevator machine room or control room/machinery space or control space.

**620.23 Branch Circuits for Machine Room or Control Room/Machinery Space or Control Space Lighting and Receptacle(s).**

**(A) Separate Branch Circuit.** A separate branch circuit shall supply the machine room or control room/machinery space or control space lighting and receptacle(s).

Required lighting shall not be connected to the load side of a ground-fault circuit interrupter.

**(B) Lighting Switch.** The machine room or control room/machinery space or control space lighting switch shall be located at the point of entry.

**(C) Duplex Receptacle.** At least one 125-volt, single-phase, duplex receptacle shall be provided in each machine room or control room and machinery space or control space.

FPN: See ANSI/ASME A17.1-1996, *Safety Code for Elevators and Escalators*, for illumination levels.

**620.24 Branch Circuit for Hoistway Pit Lighting and Receptacle(s).**

**(A) Separate Branch Circuit.** A separate branch circuit shall supply the hoistway pit lighting and receptacle(s).

Required lighting shall not be connected to the load side of a ground-fault circuit interrupter.

**(B) Lighting Switch.** The lighting switch shall be located so as to be readily accessible from the pit access door.

**(C) Duplex Receptacle.** At least one 125-volt, single-phase, duplex receptacle shall be provided in the hoistway pit.

FPN: See ANSI/ASME A17.1-1996, *Safety Code for Elevators and Escalators*, for illumination levels.

**620.25 Branch Circuits for Other Utilization Equipment.**

**(A) Additional Branch Circuits.** Additional branch circuit(s) shall supply utilization equipment not identified in 620.22, 620.23, and 620.24. Other utilization equipment shall be restricted to that equipment identified in 620.1.

**(B) Overcurrent Devices.** The overcurrent devices protecting the branch circuit(s) shall be located in the elevator machinery room or control room/machinery space or control space.

## IV. Installation of Conductors

**620.32 Metal Wireways and Nonmetallic Wireways.** The sum of the cross-sectional area of the individual conductors in a wireway shall not be more than 50 percent of the interior cross-sectional area of the wireway.

Vertical runs of wireways shall be securely supported at intervals not exceeding 4.5 m (15 ft) and shall have not more than one joint between supports. Adjoining wireway sections shall be securely fastened together to provide a rigid joint.

**620.33 Number of Conductors in Raceways.** The sum of the cross-sectional area of the individual conductors in raceways shall not exceed 40 percent of the interior cross-sectional area of the raceway, except as permitted in 620.32 for wireways.

**620.34 Supports.** Supports for cables or raceways in a hoistway or in an escalator or moving walk wellway or wheelchair lift and stairway chair lift runway shall be securely fastened to the guide rail; escalator or moving walk truss; or to the hoistway, wellway, or runway construction.

**620.35 Auxiliary Gutters.** Auxiliary gutters shall not be subject to the restrictions of 366.3 as to length or of 366.6 as to number of conductors.

**620.36 Different Systems in One Raceway or Traveling Cable.** Optical fiber cables and conductors for operating devices, operation and motion control, power, signaling, fire alarm, lighting, heating, and air-conditioning circuits of 600 volts or less shall be permitted to be run in the same traveling cable or raceway system if all conductors are insulated for the maximum voltage applied to any conductor within the cables or raceway system and if all live parts of the equipment are insulated from ground for this maximum voltage. Such a traveling cable or raceway shall also be permitted to include shielded conductors and/or one or more coaxial cables, if such conductors are insulated for the maximum voltage applied to any conductor within the cable or raceway system. Conductors shall be permitted to

be covered with suitable shielding for telephone, audio, video, or higher frequency communications circuits.

**620.37 Wiring in Hoistways, Machine Rooms, Control Rooms, Machinery Spaces, and Control Spaces.**

**(A) Uses Permitted.** Only such electric wiring, raceways, and cables used directly in connection with the elevator or dumbwaiter, including wiring for signals, for communication with the car, for lighting, heating, air conditioning, and ventilating the elevator car, for fire detecting systems, for pit sump pumps, and for heating, lighting, and ventilating the hoistway, shall be permitted inside the hoistway, machine rooms, control rooms, machinery spaces, and control spaces.

**(B) Lightning Protection.** Bonding of elevator rails (car and/or counterweight) to a lightning protection system grounding down conductor(s) shall be permitted. The lightning protection system grounding down conductor(s) shall not be located within the hoistway. Elevator rails or other hoistway equipment shall not be used as the grounding down conductor for lightning protection systems.

> FPN: See 250.106 for bonding requirements. For further information, see NFPA 780-1997, *Standard for the Installation of Lightning Protection Systems.*

**(C) Main Feeders.** Main feeders for supplying power to elevators and dumbwaiters shall be installed outside the hoistway unless as follows:

(1) By special permission, feeders for elevators shall be permitted within an existing hoistway if no conductors are spliced within the hoistway.
(2) Feeders shall be permitted inside the hoistway for elevators with driving machine motors located in the hoistway or on the car or counterweight.

**620.38 Electrical Equipment in Garages and Similar Occupancies.** Electrical equipment and wiring used for elevators, dumbwaiters, escalators, moving walks, and wheelchair lifts and stairway chair lifts in garages shall comply with the requirements of Article 511.

> FPN: Garages used for parking or storage and where no repair work is done in accordance with 511.3 are not classified.

## V. Traveling Cables

**620.41 Suspension of Traveling Cables.** Traveling cables shall be suspended at the car and hoistways' ends, or counterweight end where applicable, so as to reduce the strain on the individual copper conductors to a minimum.

Traveling cables shall be supported by one of the following means:

(1) By its steel supporting member(s)
(2) By looping the cables around supports for unsupported lengths less than 30 m (100 ft)
(3) By suspending from the supports by a means that automatically tightens around the cable when tension is increased for unsupported lengths up to 60 m (200 ft)

FPN: Unsupported length for the hoistway suspension means is that length of cable as measured from the point of suspension in the hoistway to the bottom of the loop, with the elevator car located at the bottom landing. Unsupported length for the car suspension means is that length of cable as measured from the point of suspension on the car to the bottom of the loop, with the elevator car located at the top landing.

**620.42 Hazardous (Classified) Locations.** In hazardous (classified) locations, traveling cables shall be of a type approved for hazardous (classified) locations and shall comply with 501.11, 502.12, or 503.10, as applicable.

**620.43 Location of and Protection for Cables.** Traveling cable supports shall be located so as to reduce to a minimum the possibility of damage due to the cables coming in contact with the hoistway construction or equipment in the hoistway. Where necessary, suitable guards shall be provided to protect the cables against damage.

**620.44 Installation of Traveling Cables.** Traveling cable shall be permitted to be run without the use of a raceway for a distance not exceeding 1.8 m (6 ft) in length as measured from the first point of support on the elevator car or hoistway wall, or counterweight where applicable, provided the conductors are grouped together and taped or corded, or in the original sheath.

Traveling cables shall be permitted to be continued to elevator controller enclosures and to elevator car and machine room, control room, machinery space, and control space connections, as fixed wiring, provided they are suitably supported and protected from physical damage.

## VI. Disconnecting Means and Control

**620.51 Disconnecting Means.** A single means for disconnecting all ungrounded main power supply conductors for each unit shall be provided and be designed so that no pole can be operated independently. Where multiple driving machines are connected to a single elevator, escalator, moving walk, or pumping unit, there shall be one disconnecting means to disconnect the motor(s) and control valve operating magnets.

The disconnecting means for the main power supply conductors shall not disconnect the branch circuit required in 620.22, 620.23, and 620.24.

**(A) Type.** The disconnecting means shall be an enclosed externally operable fused motor circuit switch or circuit breaker capable of being locked in the open position. The disconnecting means shall be a listed device.

FPN: For additional information, see ASME/ANSI A17.1-1996, *Safety Code for Elevators and Escalators.*

*Exception: Where an individual branch circuit supplies a wheelchair lift, the disconnecting means required by 620.51(C)(4) shall be permitted to comply with 430.109(C). This disconnecting means shall be listed and shall be capable of being locked in the open position.*

**(B) Operation.** No provision shall be made to open or close this disconnecting means from any other part of the premises. If sprinklers are installed in hoistways, machine rooms, control rooms, machinery spaces, or control spaces, the disconnecting means shall be permitted to automatically open the power supply to the affected elevator(s) prior to the application of water. No provision shall be made to automatically close this disconnecting means. Power shall only be restored by manual means.

FPN: To reduce hazards associated with water on live elevator electrical equipment.

**(C) Location.** The disconnecting means shall be located where it is readily accessible to qualified persons.

**(1) On Elevators Without Generator Field Control.** On elevators without generator field control, the disconnecting means shall be located within sight of the motor controller. Driving machines or motion and operation controllers not within sight of the disconnecting means shall be provided with a manually operated switch installed in the control circuit to prevent starting. The manually operated switch(es) shall be installed adjacent to this equipment.

Where the driving machine of an electric elevator or the hydraulic machine of a hydraulic elevator is located in a remote machine room or remote machinery space, a single means for disconnecting all ungrounded main power supply conductors shall be provided and be capable of being locked in the open position.

**(2) On Elevators with Generator Field Control.** On elevators with generator field control, the disconnecting means shall be located within sight of the motor controller for the driving motor of the motor-generator set. Driving machines, motor-generator sets, or motion and operation controllers not within sight of the disconnecting means shall be provided with a manually operated switch installed in the control circuit to prevent starting. The manually operated switch(es) shall be installed adjacent to this equipment.

Where the driving machine or the motor-generator set is located in a remote machine room or remote machinery space, a single means for disconnecting all ungrounded main power supply conductors shall be provided and be capable of being locked in the open position.

**(3) On Escalators and Moving Walks.** On escalators and moving walks, the disconnecting means shall be installed in the space where the controller is located.

**(4) On Wheelchair Lifts and Stairway Chair Lifts.** On wheelchair lifts and stairway chair lifts, the disconnecting means shall be located within sight of the motor controller.

**(D) Identification and Signs.** Where there is more than one driving machine in a machine room, the disconnecting means shall be numbered to correspond to the identifying number of the driving machine that they control.

The disconnecting means shall be provided with a sign to identify the location of the supply side overcurrent protective device.

### 620.52 Power from More Than One Source.

**(A) Single-Car and Multicar Installations.** On single-car and multicar installations, equipment receiving electrical power from more than one source shall be provided with a disconnecting means for each source of electrical power. The disconnecting means shall be within sight of the equipment served.

**(B) Warning Sign for Multiple Disconnecting Means.** Where multiple disconnecting means are used and parts of the controllers remain energized from a source other than the one disconnected, a warning sign shall be mounted on or next to the disconnecting means. The sign shall be clearly legible and shall read

WARNING
PARTS OF THE CONTROLLER ARE NOT
DE-ENERGIZED BY THIS SWITCH.

**(C) Interconnection Multicar Controllers.** Where interconnections between controllers are necessary for the operation of the system on multicar installations that remain energized from a source other than the one disconnected, a warning sign in accordance with 620.52(B) shall be mounted on or next to the disconnecting means.

### 620.53 Car Light, Receptacle(s), and Ventilation Disconnecting Means. Elevators shall have a single means for disconnecting all ungrounded car light, receptacle(s), and ventilation power-supply conductors for that elevator car.

The disconnecting means shall be an enclosed externally operable fused motor circuit switch or circuit breaker capable of being locked in the open position and shall be located in the machine room or control room for that elevator car. Where there is no machine room or control room, the disconnecting means shall be located in the same space as the disconnecting means required by 620.51.

Disconnecting means shall be numbered to correspond to the identifying number of the elevator car whose light source they control.

The disconnecting means shall be provided with a sign to identify the location of the supply side overcurrent protective device.

### 620.54 Heating and Air-Conditioning Disconnecting Means. Elevators shall have a single means for disconnecting all ungrounded car heating and air-conditioning power-supply conductors for that elevator car.

The disconnecting means shall be an enclosed externally operable fused motor circuit switch or circuit breaker capable of being locked in the open position and shall be located in the machine room or control room for that elevator car. Where there is no machine room or control room, the disconnecting means shall be located in the same space as the disconnecting means required by 620.51.

Where there is equipment for more than one elevator car in the machine room, the disconnecting means shall be numbered to correspond to the identifying number of the elevator car whose heating and air-conditioning source they control.

The disconnecting means shall be provided with a sign to identify the location of the supply side overcurrent protective device.

### 620.55 Utilization Equipment Disconnecting Means. Each branch circuit for other utilization equipment shall have a single means for disconnecting all ungrounded conductors. The disconnecting means shall be capable of being locked in the open position and shall be located in the machine room or control room/machine space or control space. Where there is more than one branch circuit for other utilization equipment, the disconnecting means shall be numbered to correspond to the identifying number of the equipment served. The disconnecting means shall be provided with a sign to identify the location of the supply side overcurrent protective device.

## VII. Overcurrent Protection

### 620.61 Overcurrent Protection. Overcurrent protection shall be provided in accordance with 620.61(A) through (D).

**(A) Operating Devices and Control and Signaling Circuits.** Operating devices and control and signaling circuits shall be protected against overcurrent in accordance with the requirements of 725.23 and 725.24.

Class 2 power-limited circuits shall be protected against overcurrent in accordance with the requirements of Chapter 9, Notes to Tables 11(A) and 11(B).

**(B) Overload Protection for Motors.**

**(1) Duty Rating on Elevator, Dumbwaiter, and Motor-Generator Sets Driving Motors.** Duty on elevator and dumbwaiter driving machine motors and driving motors of motor-generators used with generator field control shall be rated as intermittent. Such motors shall be protected against overload in accordance with 430.33.

**(2) Duty Rating on Escalator Motors.** Duty on escalator and moving walk driving machine motors shall be rated as continuous. Such motors shall be protected against overload in accordance with 430.32.

**(3) Overload Protection.** Escalator and moving walk driving machine motors and driving motors of motor-generator sets shall be protected against running overload as provided in Table 430.37.

**(4) Duty Rating and Overload Protection on Wheelchair and Stairway Chair Lift Motors.** Duty on wheelchair lift and stairway chair lift driving machine motors shall be rated as intermittent. Such motors shall be protected against overload in accordance with 430.33.

FPN: For further information, see 430.44 for orderly shutdown.

**(C) Motor Feeder Short-Circuit and Ground-Fault Protection.** Motor feeder short-circuit and ground-fault protection shall be as required in Article 430, Part V.

**(D) Motor Branch-Circuit Short-Circuit and Ground-Fault Protection.** Motor branch-circuit short-circuit and ground-fault protection shall be as required in Article 430, Part IV.

**620.62 Selective Coordination.** Where more than one driving machine disconnecting means is supplied by a single feeder, the overcurrent protective devices in each disconnecting means shall be selectively coordinated with any other supply side overcurrent protective devices.

## VIII. Machine Rooms, Control Rooms, Machinery Spaces, and Control Spaces

**620.71 Guarding Equipment.** Elevator, dumbwaiter, escalator, and moving walk driving machines; motor-generator sets; motor controllers; and disconnecting means shall be installed in a room or space set aside for that purpose unless otherwise permitted in 620.71(A) or (B). The room or space shall be secured against unauthorized access.

**(A) Motor Controllers.** Motor controllers shall be permitted outside the spaces herein specified, provided they are in enclosures with doors or removable panels that are capable of being locked in the closed position and the disconnecting means is located adjacent to or is an integral part of the motor controller. Motor controller enclosures for escalator or moving walks shall be permitted in the balustrade on the side located away from the moving steps or moving treadway. If the disconnecting means is an integral part of the motor controller, it shall be operable without opening the enclosure.

**(B) Driving Machines.** Elevators with driving machines located on the car, on the counterweight, or in the hoistway, and driving machines for dumbwaiters, wheelchair lifts, and stairway lifts shall be permitted outside the spaces herein specified.

## IX. Grounding

**620.81 Metal Raceways Attached to Cars.** Metal raceways, Type MC cable, Type MI cable, or Type AC cable attached to elevator cars shall be bonded to grounded metal parts of the car that they contact.

**620.82 Electric Elevators.** For electric elevators, the frames of all motors, elevator machines, controllers, and the metal enclosures for all electrical equipment in or on the car or in the hoistway shall be grounded in accordance with Article 250.

**620.83 Nonelectric Elevators.** For elevators other than electric having any electric conductors attached to the car, the metal frame of the car, where normally accessible to persons, shall be grounded in accordance with Article 250.

**620.84 Escalators, Moving Walks, Wheelchair Lifts, and Stairway Chair Lifts.** Escalators, moving walks, wheelchair lifts, and stairway chair lifts shall comply with Article 250.

**620.85 Ground-Fault Circuit-Interrupter Protection for Personnel.** Each 125-volt, single-phase, 15- and 20-ampere receptacle installed in pits, in hoistways, on elevator car tops, and in escalator and moving walk wellways shall be of the ground-fault circuit-interrupter type.

All 125-volt, single-phase, 15- and 20-ampere receptacles installed in machine rooms and machinery spaces shall have ground-fault circuit-interrupter protection for personnel.

A single receptacle supplying a permanently installed sump pump shall not require ground-fault circuit-interrupter protection.

## X. Emergency and Standby Power Systems

**620.91 Emergency and Standby Power Systems.** An elevator(s) shall be permitted to be powered by an emergency or standby power system.

FPN: See ASME/ANSI A17.1-1996, Rule 211.2, and CAN/CSA-B44-1994, Clause 3.12.13, for additional information.

**(A) Regenerative Power.** For elevator systems that regenerate power back into the power source that is unable to absorb the regenerative power under overhauling elevator load conditions, a means shall be provided to absorb this power.

**(B) Other Building Loads.** Other building loads, such as power and lighting, shall be permitted as the energy absorption means required in 620.91(A), provided that such loads are automatically connected to the emergency or standby power system operating the elevators and are large enough to absorb the elevator regenerative power.

**(C) Disconnecting Means.** The disconnecting means required by 620.51 shall disconnect the elevator from both the emergency or standby power system and the normal power system.

Where an additional power source is connected to the load side of the disconnecting means, the disconnecting means required in 620.51 shall be provided with an auxiliary contact that is positively opened mechanically, and the opening shall not be solely dependent on springs. This contact shall cause the additional power source to be disconnected from its load when the disconnecting means is in the open position.

# ARTICLE 625
# Electric Vehicle Charging System

## I. General

**625.1 Scope.** The provisions of this article cover the electrical conductors and equipment external to an electric vehicle that connect an electric vehicle to a supply of electricity by conductive or inductive means, and the installation of equipment and devices related to electric vehicle charging.

> FPN: For industrial trucks, see NFPA 505-1999, *Fire Safety Standard for Powered Industrial Trucks Including Type Designations, Areas of Use, Conversions, Maintenance, and Operation.*

**625.2 Definitions.**

**Electric Vehicle.** An automotive-type vehicle for highway use, such as passenger automobiles, buses, trucks, vans, and the like, primarily powered by an electric motor that draws current from a rechargeable storage battery, fuel cell, photovoltaic array, or other source of electric current. For the purpose of this article, electric motorcycles and similar type vehicles and off-road self-propelled electric vehicles, such as industrial trucks, hoists, lifts, transports, golf carts,

airline ground support equipment, tractors, boats, and the like, are not included.

**Electric Vehicle Connector.** A device that, by insertion into an electric vehicle inlet, establishes an electrical connection to the electric vehicle for the purpose of charging and information exchange. This device is part of the electric vehicle coupler.

**Electric Vehicle Coupler.** A mating electric vehicle inlet and electric vehicle connector set.

**Electric Vehicle Inlet.** The device on the electric vehicle into which the electric vehicle connector is inserted for charging and information exchange. This device is part of the electric vehicle coupler. For the purposes of this *Code,* the electric vehicle inlet is considered to be part of the electric vehicle and not part of the electric vehicle supply equipment.

**Electric Vehicle Nonvented Storage Battery.** A hermetically sealed battery comprised of one or more rechargeable electrochemical cells that has no provision for release of excessive gas pressure, or for the addition of water or electrolyte, or for external measurements of electrolyte specific gravity.

**Electric Vehicle Supply Equipment.** The conductors, including the ungrounded, grounded, and equipment grounding conductors and the electric vehicle connectors, attachment plugs, and all other fittings, devices, power outlets, or apparatus installed specifically for the purpose of delivering energy from the premises wiring to the electric vehicle.

**Personnel Protection System.** A system of personnel protection devices and constructional features that when used together provide protection against electric shock of personnel.

**625.3 Other Articles.** Wherever the requirements of other articles of this *Code* and Article 625 differ, the requirements of Article 625 shall apply.

**625.4 Voltages.** Unless other voltages are specified, the nominal ac system voltages of 120, 120/240, 208Y/120, 240, 480Y/277, 480, 600Y/347, and 600 volts shall be used to supply equipment covered by this article.

**625.5 Listed or Labeled.** All electrical materials, devices, fittings, and associated equipment shall be listed or labeled.

## II. Wiring Methods

**625.9 Electric Vehicle Coupler.** The electric vehicle coupler shall comply with 625.9(A) through (F).

**(A) Polarization.** The electric vehicle coupler shall be polarized unless part of a system identified and listed as suitable for the purpose.

**(B) Noninterchangeability.** The electric vehicle coupler shall have a configuration that is noninterchangeable with wiring devices in other electrical systems. Nongrounding-type electric vehicle couplers shall not be interchangeable with grounding-type electric vehicle couplers.

**(C) Construction and Installation.** The electric vehicle coupler shall be constructed and installed so as to guard against inadvertent contact by persons with parts made live from the electric vehicle supply equipment or the electric vehicle battery.

**(D) Unintentional Disconnection.** The electric vehicle coupler shall be provided with a positive means to prevent unintentional disconnection.

**(E) Grounding Pole.** The electric vehicle coupler shall be provided with a grounding pole, unless part of a system identified and listed as suitable for the purpose in accordance with Article 250.

**(F) Grounding Pole Requirements.** If a grounding pole is provided, the electric vehicle coupler shall be designed so that the grounding pole connection is the first to make and the last to break contact.

## III. Equipment Construction

**625.13 Electric Vehicle Supply Equipment.** Electric vehicle supply equipment rated at 125 volts, single phase, 15 or 20 amperes or part of a system identified and listed as suitable for the purpose and meeting the requirements of 625.18, 625.19, and 625.29 shall be permitted to be cord-and-plug connected. All other electric vehicle supply equipment shall be permanently connected and fastened in place. This equipment shall have no exposed live parts.

**625.14 Rating.** Electric vehicle supply equipment shall have sufficient rating to supply the load served. For the purposes of this article, electric vehicle charging loads shall be considered to be continuous loads.

**625.15 Markings.** The electric vehicle supply equipment shall comply with 625.15(A) through (C).

**(A) General.** All electric vehicle supply equipment shall be marked by the manufacturer as follows:

FOR USE WITH ELECTRIC VEHICLES.

**(B) Ventilation Not Required.** Where marking is required by 625.29(C), the electric vehicle supply equipment shall be clearly marked by the manufacturer as follows:

VENTILATION NOT REQUIRED.

The marking shall be located so as to be clearly visible after installation.

**(C) Ventilation Required.** Where marking is required by 625.29(D), the electric vehicle supply equipment shall be clearly marked by the manufacturer "Ventilation Required." The marking shall be located so as to be clearly visible after installation.

**625.16 Means of Coupling.** The means of coupling to the electric vehicle shall be either conductive or inductive. Attachment plugs, electric vehicle connectors, and electric vehicle inlets shall be listed or labeled for the purpose.

**625.17 Cable.** The electric vehicle supply equipment cable shall be Type EV, EVJ, EVE, EVJE, EVT, or EVJT flexible cable as specified in Article 400 and Table 400.4. Ampacities shall be as specified in Table 400.5(A) for 10 AWG and smaller and Table 400.5(B) for 8 AWG and larger. The overall length of the cable shall not exceed 7.5 m (25 ft) unless equipped with a cable management system that is listed as suitable for the purpose. Other cable types and assemblies listed as being suitable for the purpose, including optional hybrid communications, signal, and optical fiber cables, shall be permitted.

**625.18 Interlock.** Electric vehicle supply equipment shall be provided with an interlock that de-energizes the electric vehicle connector and its cable whenever the electric connector is uncoupled from the electric vehicle. An interlock shall not be required for portable cord-and-plug-connected electric vehicle supply equipment intended for connection to receptacle outlets rated at 125 volts, single phase, 15 and 20 amperes.

**625.19 Automatic De-Energization of Cable.** The electric vehicle supply equipment or the cable-connector combination of the equipment shall be provided with an automatic means to de-energize the cable conductors and electric vehicle connector upon exposure to strain that could result in either cable rupture or separation of the cable from the electric connector and exposure of live parts. Automatic means to de-energize the cable conductors and electric vehicle connector shall not be required for portable cord-and-plug-connected electric vehicle supply equipment intended for connection to receptacle outlets rated at 125 volts, single phase, 15 and 20 amperes.

## IV. Control and Protection

**625.21 Overcurrent Protection.** Overcurrent protection for feeders and branch circuits supplying electric vehicle supply equipment shall be sized for continuous duty and shall have a rating of not less than 125 percent of the maximum load of the electric vehicle supply equipment. Where noncontinuous loads are supplied from the same feeder or branch circuit, the overcurrent device shall have a rating of not less than the sum of the noncontinuous loads plus 125 percent of the continuous loads.

**625.22 Personnel Protection System.** The electric vehicle supply equipment shall have a listed system of protection against electric shock of personnel. The personnel protection system shall be composed of listed personnel protection devices and constructional features. Where cord-and-plug-connected electric vehicle supply equipment is used, the interrupting device of a listed personnel protection system shall be provided and shall be an integral part of the attachment plug or shall be located in the power supply cable not more than 300 mm (12 in.) from the attachment plug.

**625.23 Disconnecting Means.** For electric vehicle supply equipment rated more than 60 amperes or more than 150 volts to ground, the disconnecting means shall be provided and installed in a readily accessible location. The disconnecting means shall be capable of being locked in the open position.

**625.25 Loss of Primary Source.** Means shall be provided such that, upon loss of voltage from the utility or other electric system(s), energy cannot be backfed through the electric vehicle supply equipment to the premises wiring system. The electric vehicle shall not be permitted to serve as a standby power supply.

## V. Electric Vehicle Supply Equipment Locations

**625.28 Hazardous (Classified) Locations.** Where electric vehicle supply equipment or wiring is installed in a hazardous (classified) location, the requirements of Articles 500 through 516 shall apply.

**625.29 Indoor Sites.** Indoor sites shall include, but not be limited to, integral, attached, and detached residential garages; enclosed and underground parking structures; repair and nonrepair commercial garages; and agricultural buildings.

**(A) Location.** The electric vehicle supply equipment shall be located to permit direct connection to the electric vehicle.

**(B) Height.** Unless specifically listed for the purpose and location, the coupling means of the electric vehicle supply equipment shall be stored or located at a height of not less than 450 mm (18 in.) and not more than 1.2 m (4 ft) above the floor level.

**(C) Ventilation Not Required.** Where electric vehicle nonvented storage batteries are used or where the electric vehicle supply equipment is listed or labeled as suitable for charging electric vehicles indoors without ventilation and marked in accordance with 625.15(B), mechanical ventilation shall not be required.

**(D) Ventilation Required.** Where the electric vehicle supply equipment is listed or labeled as suitable for charging electric vehicles that require ventilation for indoor charging and marked in accordance with 625.15(C), mechanical ventilation, such as a fan, shall be provided. The ventilation shall include both supply and exhaust equipment and shall be permanently installed and located to intake from, and vent directly to, the outdoors. Positive pressure ventilation systems shall be permitted only in buildings or areas that have been specifically designed and approved for that application. Mechanical ventilation requirements shall be determined by one of the methods specified in 625.29(D)(1) through (D)(4).

**(1) Table Values.** For supply voltages and currents specified in Table 625.29(D)(1) or Table 625.29(D)(2), the minimum ventilation requirements shall be as specified in Table 625.29(D)(1) or Table 625.29(D)(2) for each of the total number of electric vehicles that can be charged at one time.

**(2) Other Values.** For supply voltages and currents other than specified in Table 625.29(D)(1) or Table 625.29(D)(2), the minimum ventilation requirements shall be calculated by means of the following general formulas as applicable.

(1) Single phase:
Ventilation $_{single\ phase}$ in cubic meters per minute ($m^3$/min) =

$$\frac{(\text{volts})\,(\text{amperes})}{1718}$$

Ventilation $_{single\ phase}$ in cubic feet per minute (cfm) =

$$\frac{(\text{volts})\,(\text{amperes})}{48.7}$$

(2) Three phase:
Ventilation $_{three\ phase}$ in cubic meters per minute ($m^3$/min) =

$$\frac{1.732\,(\text{volts})\,(\text{amperes})}{1718}$$

Ventilation$_{three\ phase}$ in cubic feet per minute (cfm) =

$$\frac{1.732\,(\text{volts})\,(\text{amperes})}{48.7}$$

**(3) Engineered Systems.** For an electric vehicle supply equipment ventilation system designed by a person qualified to perform such calculations as an integral part of a

**Table 625.29(D)(1) Minimum Ventilation Required in Cubic Meters per Minute (m³/min) for Each of the Total Number of Electric Vehicles That Can Be Charged at One Time**

| Branch-Circuit Ampere Rating | Branch-Circuit Voltage | | | | | | |
| | Single Phase | | | 3 Phase | | | |
| | 120 V | 208 V | 240 V or 120/240 V | 208 V or 208 Y/120 V | 240V | 480 V or 480 Y/277 V | 600 V or 600 Y/347 V |
|---|---|---|---|---|---|---|---|
| 15 | 1.1 | 1.8 | 2.1 | — | — | — | — |
| 20 | 1.4 | 2.4 | 2.8 | 4.2 | 4.8 | 9.7 | 12 |
| 30 | 2.1 | 3.6 | 4.2 | 6.3 | 7.2 | 15 | 18 |
| 40 | 2.8 | 4.8 | 5.6 | 8.4 | 9.7 | 19 | 24 |
| 50 | 3.5 | 6.1 | 7.0 | 10 | 12 | 24 | 30 |
| 60 | 4.2 | 7.3 | 8.4 | 13 | 15 | 29 | 36 |
| 100 | 7.0 | 12 | 14 | 21 | 24 | 48 | 60 |
| 150 | — | — | — | 31 | 36 | 73 | 91 |
| 200 | — | — | — | 42 | 48 | 97 | 120 |
| 250 | — | — | — | 52 | 60 | 120 | 150 |
| 300 | — | — | — | 63 | 73 | 145 | 180 |
| 350 | — | — | — | 73 | 85 | 170 | 210 |
| 400 | — | — | — | 84 | 97 | 195 | 240 |

building's total ventilation system, the minimum ventilation requirements shall be permitted to be determined per calculations specified in the engineering study.

**(4) Supply Circuits.** The supply circuit to the mechanical ventilation equipment shall be electrically interlocked with the electric vehicle supply equipment and shall remain energized during the entire electric vehicle charging cycle. Electric vehicle supply equipment shall be marked in accordance with 625.15. Electric vehicle supply equipment receptacles rated at 125 volts, single phase, 15 and 20 amperes shall be marked in accordance with 625.15(C) and shall be switched and the mechanical ventilation system shall be electrically interlocked through the switch supply power to the receptacle.

**625.30 Outdoor Sites.** Outdoor sites shall include but not be limited to residential carports and driveways, curbside, open parking structures, parking lots, and commercial charging facilities.

**(A) Location.** The electric vehicle supply equipment shall be located to permit direct connection to the electric vehicle.

**Table 625.29(D)(2) Minimum Ventilation Required in Cubic Feet per Minute (cfm) for Each of the Total Number of Electric Vehicles That Can Be Charged at One Time**

| Branch-Circuit Ampere Rating | Branch-CircuitVoltage | | | | | | |
| | Single Phase | | | 3 Phase | | | |
| | 120 V | 208 V | 240 V or 120/240 V | 208 V or 208 Y/120 V | 240 V | 480 V or 480 Y/277 V | 600 V or 600 Y/347 V |
|---|---|---|---|---|---|---|---|
| 15 | 37 | 64 | 74 | — | — | — | — |
| 20 | 49 | 85 | 99 | 148 | 171 | 342 | 427 |
| 30 | 74 | 128 | 148 | 222 | 256 | 512 | 641 |
| 40 | 99 | 171 | 197 | 296 | 342 | 683 | 854 |
| 50 | 123 | 214 | 246 | 370 | 427 | 1025 | 1281 |
| 60 | 148 | 256 | 296 | 444 | 512 | 1708 | 1281 |
| 100 | 246 | 427 | 493 | 740 | 854 | 1708 | 2135 |
| 150 | — | — | — | 1110 | 1281 | 2562 | 3203 |
| 200 | — | — | — | 1480 | 1708 | 3416 | 4270 |
| 250 | — | — | — | 1850 | 2135 | 4270 | 5338 |
| 300 | — | — | — | 2221 | 2562 | 5125 | 6406 |
| 350 | — | — | — | 2591 | 2989 | 5979 | 7473 |
| 400 | — | — | — | 2961 | 3416 | 6832 | 8541 |

**(B) Height.** Unless specifically listed for the purpose and location, the coupling means of electric vehicle supply equipment shall be stored or located at a height of not less than 600 mm (24 in.) and not more than 1.2 m (4 ft) above the parking surface.

# ARTICLE 630
# Electric Welders

## I. General

**630.1 Scope.** This article covers electric arc welding, resistance welding apparatus, and other similar welding equipment that is connected to an electric supply system.

## II. Arc Welders

**630.11 Ampacity of Supply Conductors.** The ampacity of conductors for arc welders shall be in accordance with 630.11(A) and (B).

**(A) Individual Welders.** The ampacity of the supply conductors shall not be less than the $I_{1eff}$ value on the rating plate. Alternatively, if the $I_{1eff}$ is not given, the ampacity of the supply conductors shall not be less than the current value determined by multiplying the rated primary current in amperes given on the welder rating plate and the factor shown in Table 630.11(A) based on the duty cycle of the welder.

**Table 630.11(A) Duty Cycle Multiplication Factors for Arc Welders**

| Duty Cycle | Multiplier for Arc Welders | |
| --- | --- | --- |
| | Nonmotor Generator | Motor Generator |
| 100 | 1.00 | 1.00 |
| 90 | 0.95 | 0.96 |
| 80 | 0.89 | 0.91 |
| 70 | 0.84 | 0.86 |
| 60 | 0.78 | 0.81 |
| 50 | 0.71 | 0.75 |
| 40 | 0.63 | 0.69 |
| 30 | 0.55 | 0.62 |
| 20 or less | 0.45 | 0.55 |

**(B) Group of Welders.** Conductor ampacity shall be based on the individual currents determined in 630.11(A) as the sum of 100 percent of the two largest welders, plus 85 percent of the third largest welder, plus 70 percent of the fourth largest welder, plus 60 percent of all remaining welders.

*Exception: Percentage values lower than those given in (B) shall be permitted in cases where the work is such that a high-operating duty cycle for individual welders is impossible.*

FPN: Duty cycle considers welder loading based on the use to be made of each welder and the number of welders supplied by the conductors that will be in use at the same time. The load value used for each welder considers both the magnitude and the duration of the load while the welder is in use.

**630.12 Overcurrent Protection.** Overcurrent protection for arc welders shall be as provided in 630.12(A) and (B). Where the values as determined by this section do not correspond with the standard ampere ratings provided in 240.6 or the rating or setting specified results in unnecessary opening of the overcurrent device, the next higher standard rating or setting shall be permitted.

**(A) For Welders.** Each welder shall have overcurrent protection rated or set at not more than 200 percent of $I_{1max}$. Alternatively, if the $I_{1max}$ is not given, the overcurrent protection shall be rated or set at not more than 200 percent of the rated primary current of the welder.

An overcurrent device shall not be required for a welder that has supply conductors protected by an overcurrent device rated or set at not more than 200 percent of $I_{1max}$ or the rated primary current of the welder.

If the supply conductors for a welder are protected by an overcurrent device rated or set at not more than 200 percent of $I_{1max}$ or rated primary current of the welder, a separate overcurrent device shall not be required.

**(B) For Conductors.** Conductors that supply one or more welders shall be protected by an overcurrent device rated or set at not more than 200 percent of the conductor ampacity.

FPN: $I_{1max}$ is the maximum value of the rated supply current at maximum rated output. $I_{1eff}$ is the maximum value of the effective supply current, calculated from the rated supply current ($I_1$), the corresponding duty cycle (duty factor) ($X$), and the supply current at no-load ($I_0$) by the following formula:

$$I_{1eff} = \sqrt{I_1^2 X + I_0^2 (1 - X)}$$

**630.13 Disconnecting Means.** A disconnecting means shall be provided in the supply circuit for each arc welder that is not equipped with a disconnect mounted as an integral part of the welder.

The disconnecting means shall be a switch or circuit breaker, and its rating shall not be less than that necessary to accommodate overcurrent protection as specified under 630.12.

**630.14 Marking.** A rating plate shall be provided for arc welders giving the following information:

(1) Name of manufacturer
(2) Frequency
(3) Number of phases
(4) Primary voltage
(5) $I_{1max}$ and $I_{1eff}$, or rated primary current
(6) Maximum open-circuit voltage
(7) Rated secondary current and
(8) Basis of rating, such as the duty cycle

**630.15 Grounding of Welder Secondary Circuit.** The secondary circuit conductors of an arc welder, consisting of the electrode conductor and the work conductor, shall not be considered as premises wiring for the purpose of applying Article 250.

FPN: Connecting welder secondary circuits to grounded objects can create parallel paths and can cause objectionable current over equipment grounding conductors.

## III. Resistance Welders

**630.31 Ampacity of Supply Conductors.** The ampacity of the supply conductors for resistance welders necessary to limit the voltage drop to a value permissible for the satisfactory performance of the welder is usually greater than that required to prevent overheating as described in 630.31(A) and (B).

**(A) Individual Welders.** The rated ampacity for conductors for individual welders shall comply with the following:

(1) The ampacity of the supply conductors for a welder that may be operated at different times at different values of primary current or duty cycle shall not be less than 70 percent of the rated primary current for seam and automatically fed welders and 50 percent of the rated primary current for manually operated nonautomatic welders.
(2) The ampacity of the supply conductors for a welder wired for a specific operation for which the actual primary current and duty cycle are known and remain unchanged shall not be less than the product of the actual primary current and the multiplier specified in Table 630.31(A) for the duty cycle at which the welder will be operated.

**(B) Groups of Welders.** The ampacity of conductors that supply two or more welders shall not be less than the sum of the value obtained in accordance with 630.31(A) for the largest welder supplied and 60 percent of the values obtained for all the other welders supplied.

**Table 630.31(A)(2) Duty Cycle Multiplication Factors for Resistance Welders**

| Duty Cycle (percent) | Multiplier |
|---|---|
| 50 | 0.71 |
| 40 | 0.63 |
| 30 | 0.55 |
| 25 | 0.50 |
| 20 | 0.45 |
| 15 | 0.39 |
| 10 | 0.32 |
| 7.5 | 0.27 |
| 5 or less | 0.22 |

FPN: **Explanation of Terms**

(1) The *rated primary current* is the rated kilovolt-amperes (kVA) multiplied by 1000 and divided by the rated primary voltage, using values given on the nameplate.
(2) The *actual primary current* is the current drawn from the supply circuit during each welder operation at the particular heat tap and control setting used.
(3) The *duty cycle* is the percentage of the time during which the welder is loaded. For instance, a spot welder supplied by a 60-Hz system (216,000 cycles per hour) making four hundred 15-cycle welds per hour would have a duty cycle of 2.8 percent (400 multiplied by 15, divided by 216,000, multiplied by 100). A seam welder operating 2 cycles "on" and 2 cycles "off" would have a duty cycle of 50 percent.

**630.32 Overcurrent Protection.** Overcurrent protection for resistance welders shall be as provided in 630.32(A) and (B). Where the values as determined by this section do not correspond with the standard ampere ratings provided in 240.6 or the rating or setting specified results in unnecessary opening of the overcurrent device, a higher rating or setting that does not exceed the next higher standard ampere rating shall be permitted.

**(A) For Welders.** Each welder shall have an overcurrent device rated or set at not more than 300 percent of the rated primary current of the welder. If the supply conductors for a welder are protected by an overcurrent device rated or set at not more than 200 percent of the rated primary current of the welder, a separate overcurrent device shall not be required.

**(B) For Conductors.** Conductors that supply one or more welders shall be protected by an overcurrent device rated or set at not more than 300 percent of the conductor rating.

**630.33 Disconnecting Means.** A switch or circuit breaker shall be provided by which each resistance welder and its control equipment can be disconnected from the supply circuit. The ampere rating of this disconnecting means shall not be less than the supply conductor ampacity determined in accordance with 630.31. The supply circuit switch shall be permitted as the welder disconnecting means where the circuit supplies only one welder.

**630.34 Marking.** A nameplate shall be provided for each resistance welder, giving the following information:

(1) Name of manufacturer
(2) Frequency
(3) Primary voltage
(4) Rated kilovolt-amperes (kVA) at 50 percent duty cycle
(5) Maximum and minimum open-circuit secondary voltage
(6) Short-circuit secondary current at maximum secondary voltage
(7) Specified throat and gap setting

## IV. Welding Cable

**630.41 Conductors.** Insulation of conductors intended for use in the secondary circuit of electric welders shall be flame retardant.

**630.42 Installation.** Cables shall be permitted to be installed in a dedicated cable tray as provided in 630.42(A), (B), and (C).

**(A) Cable Support.** The cable tray shall provide support at not greater than 150-mm (6-in.) intervals.

**(B) Spread of Fire and Products of Combustion.** The installation shall comply with 300.21.

**(C) Signs.** A permanent sign shall be attached to the cable tray at intervals not greater than 6.0 m (20 ft). The sign shall read as follows:

CABLE TRAY
FOR WELDING CABLES ONLY

# ARTICLE 640
## Audio Signal Processing, Amplification, and Reproduction Equipment

## I. General

**640.1 Scope.** This article covers equipment and wiring for audio signal generation, recording, processing, amplification and reproduction; distribution of sound; public address; speech input systems; temporary audio system installations; and electronic organs or other electronic musical instruments. This also includes audio systems subject to Article 517, Part VI, and Articles 518, 520, 525, and 530.

FPN No. 1:  Examples of permanently installed distributed audio system locations include, but are not limited to, restaurant, hotel, business office, commercial and retail sales environments, churches, and schools. Both portable and permanently installed equipment locations include, but are not limited to, residences, auditoriums, theaters, stadiums, and movie and television studios. Temporary installations include, but are not limited to, auditoriums, theaters, stadiums (which use both temporary and permanently installed systems), and outdoor events such as fairs, festivals, circuses, public events, and concerts.

FPN No. 2:  Fire and burglary alarm signaling devices are specifically not encompassed by this article.

**640.2 Definitions.** For purposes of this article, the following definitions apply.

**Abandoned Audio Distribution Cable.** Installed audio distribution cable that is not terminated at equipment and not identified for future use with a tag.

**Audio Amplifier or Pre-Amplifier.** Electronic equipment that increases the current or voltage, or both, potential of an audio signal intended for use by another piece of audio equipment. *Amplifier* is the term used to denote an audio amplifier within this article.

**Audio Autotransformer.** A transformer with a single winding and multiple taps intended for use with an amplifier loudspeaker signal output.

**Audio Signal Processing Equipment.** Electrically operated equipment that produces or processes, or both, electronic signals that, when appropriately amplified and reproduced by a loudspeaker, produce an acoustic signal within the range of normal human hearing (typically 20-20 kHz). Within this article, the terms *equipment* and *audio equipment* are assumed to be equivalent to audio signal processing equipment.

FPN:  This equipment includes, but is not limited to, loudspeakers; headphones; pre-amplifiers; microphones and their power supplies; mixers; MIDI (musical instrument digital interface) equipment or other digital control systems; equalizers, compressors, and other audio signal processing equipment; audio media recording and playback equipment, including turntables, tape decks and disk players (audio and multimedia), synthesizers, tone generators, and electronic organs. Electronic organs and synthesizers may have integral or separate amplification and loudspeakers. With the exception of amplifier outputs, virtually all such equipment is used to process signals (utilizing analog or digital techniques) that have nonhazardous levels of voltage or current potential.

**Audio System.** Within this article, the totality of all equipment and interconnecting wiring used to fabricate a fully functional audio signal processing, amplification, and reproduction system.

**Audio Transformer.** A transformer with two or more electrically isolated windings and multiple taps intended for use with an amplifier loudspeaker signal output.

**Equipment Rack.** A framework for the support, enclosure, or both, of equipment. May be portable or stationary. See ANSI/EIA/310-D-1992, *Cabinets, Racks, Panels and Associated Equipment.*

**Loudspeaker.** Equipment that converts an ac electric signal into an acoustic signal. The term *speaker* is commonly used to mean *loudspeaker.*

**Maximum Output Power.** The maximum output power delivered by an amplifier into its rated load as determined under specified test conditions. This may exceed the manufacturer's rated output power for the same amplifier.

**Mixer.** Equipment used to combine and level match a multiplicity of electronic signals, such as from microphones, electronic instruments, and recorded audio.

**Mixer–Amplifier.** Equipment that combines the functions of a mixer and amplifier within a single enclosure.

**Portable Equipment.** Equipment fed with portable cords or cables intended to be moved from one place to another.

**Powered Loudspeaker.** Equipment that consists of a loudspeaker and amplifier within the same enclosure. Other signal processing may also be included.

**Rated Load Impedance.** The amplifier manufacturer's stated or marked speaker impedance into which an amplifier will deliver its rated output power. $2\Omega$, $4\Omega$, and $8\Omega$ are typical ratings.

**Rated Output Power.** The amplifier manufacturer's stated or marked output power capability into its rated load.

**Rated Output Voltage.** For audio amplifiers of the constant-voltage type, the nominal output voltage when the amplifier is delivering full rated power. Rated output voltage is used for determining approximate acoustic output in distributed speaker systems that typically employ impedance matching transformers. Typical ratings are 25 volts, 70.7 volts, and 100 volts.

**Technical Power System.** An electrical distribution system with grounding in accordance with 250.146(D), where the equipment grounding conductor is isolated from the premises grounded conductor except at a single grounded termination point within a branch-circuit panelboard, at the originating (main breaker) branch-circuit panelboard, or at the premises grounding electrode.

**Temporary Equipment.** Portable wiring and equipment intended for use with events of a transient or temporary nature where all equipment is presumed to be removed at the conclusion of the event.

**640.3 Locations and Other Articles.** Circuits and equipment shall comply with 640.3(A) through (L), as applicable.

**(A) Spread of Fire or Products of Combustion.** The accessible portion of abandoned audio distribution cables shall not be permitted to remain. See 300.21.

**(B) Ducts, Plenums, and Other Air-Handling Spaces.** See 300.22 for circuits and equipment installed in ducts or plenums or other space used for environmental air.

FPN: NFPA 90A-1999, *Standard for the Installation of Air Conditioning and Ventilation Systems,* 2-3.10.2(a), Exception No. 3, permits loudspeakers, loudspeaker assemblies, and their accessories listed in accordance with UL 2043-1996, *Fire Test for Heat and Visible Smoke Release for Discrete Products and Their Accessories Installed in Air-Handling Spaces,* to be installed in other spaces used for environmental air (ceiling cavity plenums).

**(C) Cable Trays.** Cable trays shall be used in accordance with Article 392.

FPN: See 725.61(C) for the use of Class 2, Class 3, and Type PLTC cable in cable trays.

**(D) Hazardous (Classified) Locations.** Equipment used in hazardous (classified) locations shall comply with the applicable requirements of Chapter 5.

**(E) Places of Assembly.** Equipment used in places of assembly shall comply with Article 518.

**(F) Theaters, Audience Areas of Motion Picture and Television Studios, and Similar Locations.** Equipment used in theaters, audience areas of motion picture and television studios, and similar locations shall comply with Article 520.

**(G) Carnivals, Circuses, Fairs, and Similar Events.** Equipment used in carnivals, circuses, fairs, and similar events shall comply with Article 525.

**(H) Motion Picture and Television Studios.** Equipment used in motion picture and television studios shall comply with Article 530.

**(I) Swimming Pools, Fountains, and Similar Locations.** Audio equipment used in or near swimming pools, fountains, and similar locations shall comply with Article 680.

**(J) Combination Systems.** Where the authority having jurisdiction permits audio systems for paging or music, or both, to be combined with fire alarm systems, the wiring shall comply with Article 760.

FPN: For installation requirements for such combination systems, refer to NFPA 72-1999, *National Fire Alarm Code®,* and NFPA *101®*-2000, *Life Safety Code®.*

**(K) Antennas.** Equipment used in audio systems that contain an audio or video tuner and an antenna input shall comply with Article 810. Wiring other than antenna wiring that connects such equipment to other audio equipment shall comply with this article.

**(L) Generators.** Generators shall be installed in accordance with 445.2 through 445.10. Grounding of portable and vehicle-mounted generators shall be in accordance with 250.34.

**640.4 Protection of Electrical Equipment.** Amplifiers, loudspeakers, and other equipment shall be so located or protected so as to guard against environmental exposure or physical damage, such as might result in fire, shock, or personal hazard.

**640.5 Access to Electrical Equipment Behind Panels Designed to Allow Access.** Access to equipment shall not be denied by an accumulation of wires and cables that prevents removal of panels, including suspended ceiling panels.

**640.6 Mechanical Execution of Work.** Equipment and cabling shall be installed in a neat and workmanlike manner. Cables installed exposed on the surface of ceiling and sidewalls shall be supported by the structural components of the building structure in such a manner that the cable will not be damaged by normal building use. Such cables shall be attached to structural components by straps, staples, hangers, or similar fittings designed and installed so as not to damage the cable. The installation shall also conform with 300.4(D).

**640.7 Grounding.**

**(A) General.** Wireways and auxiliary gutters shall be grounded and bonded in accordance with the requirements of Article 250. Where the wireway or auxiliary gutter does not contain power-supply wires, the equipment grounding conductor shall not be required to be larger than 14 AWG copper or its equivalent. Where the wireway or auxiliary gutter contains power-supply wires, the equipment grounding conductor shall not be smaller than specified in 250.122.

**(B) Separately Derived Systems with 60 Volts to Ground.** Grounding of separately derived systems with 60 volts to ground shall be in accordance with 647.6.

**(C) Isolated Ground Receptacles.** Isolated grounding-type receptacles shall be permitted as described in 250.146(D), and for the implementation of other technical power systems in compliance with Article 250. For separately derived systems with 60 volts to ground, the branch-circuit equipment grounding conductor shall be terminated as required in 647.6(B).

FPN: See 406.2(D) for grounding-type receptacles and required identification.

**640.8 Grouping of Conductors.** Insulated conductors of different systems grouped or bundled so as to be in close physical contact with each other in the same raceway or other enclosure, or in portable cords or cables, shall comply with 300.3(C)(1).

**640.9 Wiring Methods.**

**(A) Wiring to and Between Audio Equipment.**

**(1) Power Wiring.** Wiring and equipment from source of power to and between devices connected to the premises wiring systems shall comply with the requirements of Chapters 1 through 4, except as modified by this article.

**(2) Separately Derived Power Systems.** Separately derived systems shall comply with the applicable articles of this *Code*, except as modified by this article. Separately derived systems with 60 volts to ground shall be permitted for use in audio system installations as specified in Article 647.

**(3) Other Wiring.** All wiring not connected to the premises wiring system or to a wiring system separately derived from the premises wiring system shall comply with Article 725.

**(B) Auxiliary Power Supply Wiring.** Equipment that has a separate input for an auxiliary power supply shall be wired in compliance with Article 725. Battery installation shall be in accordance with Article 480.

FPN No. 1: This section does not apply to the use of uninterruptible power supply (*ups*) equipment, or other sources of supply, that are intended to act as a direct replacement for the primary circuit power source and are connected to the primary circuit input.

FPN No. 2: Refer to NFPA 72-1999, *National Fire Alarm Code*, where equipment is used for a fire alarm system.

**(C) Output Wiring and Listing of Amplifiers.** Amplifiers with output circuits carrying audio program signals shall be permitted to employ Class 1, Class 2, or Class 3 wiring where the amplifier is listed and marked for use with the specific class of wiring method. Such listing shall ensure the energy output is equivalent to the shock and fire risk of the same class as stated in Article 725. Overcurrent protection shall be provided and shall be permitted to be inherent in the amplifier.

Audio amplifier output circuits wired using Class 1 wiring methods shall be considered equivalent to Class 1 circuits and be installed in accordance with 725.25, where applicable.

Audio amplifier output circuits wired using Class 2 or Class 3 wiring methods shall be considered equivalent to Class 2 or Class 3 circuits respectively. They shall use conductors insulated at not less than the requirements of 725.71, and shall be installed in accordance with 725.54 and 725.61.

FPN No. 1: ANSI/UL 1711-1994, *Amplifiers for Fire Protective Signaling Systems*, contains requirements for the listing of amplifiers used for fire alarm systems in compliance with NFPA 72-1999, *National Fire Alarm Code*.

FPN No. 2: Examples of requirements for listing amplifiers used in residential, commercial, and professional use are found in ANSI/UL 813-1996, *Commercial Audio Equipment*, ANSI/UL 1419-1997, *Professional Video and Audio Equipment*, ANSI/UL 1492-1996, *Audio-Video Products and Accessories*, or ANSI/UL 6500-1996, *Audio/Video and Musical Instrument Apparatus for Household, Commercial, and Similar Use*.

**(D) Use of Audio Transformers and Autotransformers.** Audio transformers and autotransformers shall only be used for audio signals in a manner so as not to exceed the manufacturer's stated input or output voltage, impedance, or power limitations. The input or output wires of an audio transformer or autotransformer shall be allowed to connect directly to the amplifier or loudspeaker terminals. No electrical terminal or lead shall be required to be grounded or bonded.

**640.10 Audio Systems Near Bodies of Water.** Audio systems near bodies of water, either natural or artificial, shall be subject to the restrictions specified in 640.10(A) and (B).

*Exception: This section does not include audio systems intended for use on boats, yachts, or other forms of land or water transportation used near bodies of water, whether or not supplied by branch-circuit power.*

FPN: See 680.27(A) for installation of underwater audio equipment.

**(A) Equipment Supplied by Branch-Circuit Power.** Audio system equipment supplied by branch-circuit power shall not be placed laterally within 1.5 m (5 ft) of the inside wall of a pool, spa, hot tub, or fountain, or within 1.5 m (5 ft) of the prevailing or tidal high water mark. The equipment shall be provided with branch-circuit power protected by a ground-fault circuit interrupter where required by other articles.

**(B) Equipment Not Supplied by Branch-Circuit Power.** Audio system equipment powered by a listed Class 2 power supply or by the output of an amplifier listed as permitting the use of Class 2 wiring shall only be restricted in placement by the manufacturer's recommendations.

FPN: Placement of the power supply or amplifier, if supplied by branch-circuit power, is still subject to 640.10(A).

## II. Permanent Audio System Installations

**640.21 Use of Flexible Cords and Cables.**

**(A) Between Equipment and Branch-Circuit Power.** Power supply cords for audio equipment shall be suitable for the use and shall be permitted to be used where the interchange, maintenance, or repair of such equipment is facilitated through the use of a power supply cord.

**(B) Between Loudspeakers and Amplifiers or Between Loudspeakers.** Cables used to connect loudspeakers to each other or to an amplifier shall comply with Article 725. Other listed cable types and assemblies, including optional hybrid communications, signal, and optical fiber cables, shall be permitted.

**(C) Between Equipment.** Cables used for the distribution of audio signals between equipment shall comply with Article 725. Other listed cable types and assemblies, including optional hybrid communications, signal, and optical fiber cables, shall be permitted. Other cable types and assemblies specified by the equipment manufacturer as acceptable for the use shall be permitted in accordance with 110.3(B).

**(D) Between Equipment and Power Supplies Other Than Branch-Circuit Power.** The following power supplies, other than branch-circuit power supplies, shall be installed and wired between equipment in accordance with the requirements of this *Code* for the voltage and power delivered:

(1) Storage batteries
(2) Transformers
(3) Transformer rectifiers
(4) Other ac or dc power supplies

FPN: For some equipment, these sources such as in items (1) and (2) serve as the only source of power. These could, in turn, be supplied with intermittent or continuous branch-circuit power.

**(E) Between Equipment Racks and Premises Wiring System.** Flexible cords and cables shall be permitted for the electrical connection of permanently installed equipment racks to the premises wiring system to facilitate access to equipment or for the purpose of isolating the technical power system of the rack from the premises ground. Connection shall be made either using approved plugs and receptacles or by direct connection within an approved enclosure. Flexible cords and cables shall not be subjected to physical manipulation or abuse while the rack is in use.

**640.22 Wiring of Equipment Racks and Enclosures.** Metal equipment racks and enclosures shall be grounded. Bonding shall not be required if the rack is connected to a technical power ground.

Equipment racks shall be wired in a neat and workmanlike manner. Wires, cables, structural components, or other equipment shall not be placed in such a manner as to prevent reasonable access to equipment power switches and resettable or replaceable circuit overcurrent protection devices.

Supply cords or cables, if used, shall terminate within the equipment rack enclosure in an identified connector assembly. The supply cords or cable (and connector assembly, if used) shall have sufficient ampacity to carry the total load connected to the equipment rack and shall be protected by overcurrent devices.

## 640.23 Conduit or Tubing.

**(A) Number of Conductors.** The number of conductors permitted in a single conduit or tubing shall not exceed the percentage fill specified in Table 1, Chapter 9.

**(B) Nonmetallic Conduit or Tubing and Insulating Bushings.** The use of nonmetallic conduit or tubing and insulating bushings shall be permitted where a technical power system is employed and shall comply with applicable articles.

## 640.24 Wireways, Gutters, and Auxiliary Gutters.
The use of metallic and nonmetallic wireways, gutters, and auxiliary gutters shall be permitted for use with audio signal conductors and shall comply with applicable articles with respect to permitted locations, construction, and fill.

## 640.25 Loudspeaker Installation in Fire Resistance-Rated Partitions, Walls, and Ceilings.
Loudspeakers installed in a fire resistance-rated partition, wall, or ceiling shall be listed for the purpose or installed in an enclosure or recess that maintains the fire resistance rating.

FPN:   Fire-rated construction is the fire-resistive classification used in building codes. One method of determining fire rating is testing in accordance with NFPA 251-1999, *Standard Methods of Tests of Fire Endurance of Building Construction and Materials.*

## III. Portable and Temporary Audio System Installations

## 640.41 Multipole Branch-Circuit Cable Connectors.
Multipole branch-circuit cable connectors, male and female, for power supply cords and cables shall be constructed so that tension on the cord or cable is not transmitted to the connections. The female half shall be attached to the load end of the power supply cord or cable. The connector shall be rated in amperes and designed so that differently rated devices cannot be connected together. Alternating-current multipole connectors shall be polarized and comply with 406.6(A) and (B) and 406.9. Alternating-current or direct-current multipole connectors utilized for connection between loudspeakers and amplifiers, or between loudspeakers, shall not be compatible with nonlocking 15- or 20-ampere rated connectors intended for branch-circuit power or with connectors rated 250 volts or greater and of either the locking or nonlocking type. Signal cabling not intended for such loudspeaker and amplifier interconnection shall not be permitted to be compatible with multipole branch-circuit cable connectors of any accepted configuration.

FPN:   See 400.10 for pull at terminals.

## 640.42 Use of Flexible Cords and Cables.

**(A) Between Equipment and Branch-Circuit Power.** Power supply cords for audio equipment shall be listed and shall be permitted to be used where the interchange, maintenance, or repair of such equipment is facilitated through the use of a power supply cord.

**(B) Between Loudspeakers and Amplifiers, or Between Loudspeakers.** Flexible cords and cables used to connect loudspeakers to each other or to an amplifier shall comply with Article 400 and Article 725, respectively. Cords and cables listed for portable use, either hard or extra-hard usage as defined by Article 400, shall also be permitted. Other listed cable types and assemblies, including optional hybrid communications, signal, and optical fiber cables, shall be permitted.

**(C) Between Equipment and/or Between Equipment Racks.** Flexible cords and cables used for the distribution of audio signals between equipment shall comply with Article 400 and Article 725, respectively. Cords and cables listed for portable use, either hard or extra-hard service as defined by Article 400, shall also be permitted. Other listed cable types and assemblies, including optional hybrid communications, signal, and optical fiber cables, shall be permitted.

**(D) Between Equipment, Equipment Racks, and Power Supplies Other Than Branch-Circuit Power.** Wiring between the following power supplies, other than branch-circuit power supplies, shall be installed, connected, or wired in accordance with the requirements of this *Code* for the voltage and power required:

(1) Storage batteries
(2) Transformers
(3) Transformer rectifiers
(4) Other ac or dc power supplies

**(E) Between Equipment Racks and Branch-Circuit Power.** The supply to a portable equipment rack shall be by means of listed extra-hard usage cords or cables, as defined in Table 400.4. For outdoor portable or temporary use, the cords or cables shall be further listed as being suitable for wet locations and sunlight resistant. Sections 520.5, 520.10, and 525.3 shall apply as appropriate when the following conditions exist:

(1) Where equipment racks include audio and lighting and/or power equipment
(2) When using or constructing cable extensions, adapters, and breakout assemblies

**640.43 Wiring of Equipment Racks.** Equipment racks fabricated of metal shall be grounded. Nonmetallic racks with covers (if provided) removed shall not allow access to Class 1, Class 3, or primary circuit power without the removal of covers over terminals or the use of tools.

Equipment racks shall be wired in a neat and workmanlike manner. Wires, cables, structural components, or other equipment shall not be placed in such a manner as to prevent reasonable access to equipment power switches and resettable or replaceable circuit overcurrent protection devices.

Wiring that exits the equipment rack for connection to other equipment or to a power supply shall be relieved of strain or otherwise suitably terminated such that a pull on the flexible cord or cable shall not increase the risk of damage to the cable or connected equipment such as to cause an unreasonable risk of fire or electric shock.

**640.44 Environmental Protection of Equipment.** Temporary outdoor, unsheltered placement or use of portable equipment not listed for the purpose shall be permitted only where appropriate protection of such equipment from adverse weather conditions is provided to prevent risk of fire or electrical shock. Where the system is intended to remain operable during adverse weather, arrangements shall be made for maintaining operation and ventilation of heat dissipating equipment.

**640.45 Protection of Wiring.** Where accessible to the public, flexible cords and cables laid or run on the ground or on the floor shall be covered with approved nonconductive mats. Cables and mats shall be arranged so as not to present a tripping hazard.

**640.46 Equipment Access.** Equipment likely to present a risk of fire, electrical shock, or physical injury to the public shall be protected by barriers or supervised by qualified personnel so as to prevent public access.

# ARTICLE 645
# Information Technology Equipment

**645.1 Scope.** This article covers equipment, power-supply wiring, equipment interconnecting wiring, and grounding of information technology equipment and systems, including terminal units, in an information technology equipment room.

FPN: For further information, see NFPA 75-1999, *Standard for the Protection of Electronic Computer/Data Processing Equipment.*

**645.2 Special Requirements for Information Technology Equipment Room.** This article shall apply, provided all the following conditions are met:

(1) Disconnecting means complying with 645.10 are provided.
(2) A separate heating/ventilating/air-conditioning (HVAC) system is provided that is dedicated for information technology equipment use and is separated from other areas of occupancy. Any HVAC system that serves other occupancies shall be permitted to also serve the information technology equipment room if fire/smoke dampers are provided at the point of penetration of the room boundary. Such dampers shall operate on activation of smoke detectors and also by operation of the disconnecting means required by 645.10.

FPN: For further information, see NFPA 75-1999, *Standard for the Protection of Electronic Computer/Data Processing Equipment*, Chapter 8, 8-1, 8-1.1, 8-1.2, and 8-1.3.

(3) Listed information technology equipment is installed.
(4) The room is occupied only by those personnel needed for the maintenance and functional operation of the installed information technology equipment.
(5) The room is separated from other occupancies by fire-resistant-rated walls, floors, and ceilings with protected openings.

FPN: For further information on room construction requirements, see NFPA 75-1999, *Standard for the Protection of Electronic Computer/Data Processing Equipment*, Chapter 3.

**645.5 Supply Circuits and Interconnecting Cables.**

**(A) Branch-Circuit Conductors.** The branch-circuit conductors supplying one or more units of a data processing system shall have an ampacity not less than 125 percent of the total connected load.

**(B) Cord-and-Plug Connections.** The data processing system shall be permitted to be connected to a branch circuit by any of the following means listed for the purpose:

(1) Flexible cord and attachment plug cap not to exceed 4.5 m (15 ft).
(2) Cord set assembly. Where run on the surface of the floor, they shall be protected against physical damage.

**(C) Interconnecting Cables.** Separate data processing units shall be permitted to be interconnected by means of cables and cable assemblies listed for the purpose. Where run on the surface of the floor, they shall be protected against physical damage.

**(D) Under Raised Floors.** Power cables, communications cables, connecting cables, interconnecting cables, and receptacles associated with the information technology equipment shall be permitted under a raised floor, provided the following conditions are met.

(1) The raised floor is of suitable construction, and the area under the floor is accessible.

(2) The branch-circuit supply conductors to receptacles or field-wired equipment are in rigid metal conduit, rigid nonmetallic conduit, intermediate metal conduit, electrical metallic tubing, electrical nonmetallic tubing, metal wireway, nonmetallic wireway, surface metal raceway with metal cover, nonmetallic surface raceway, flexible metal conduit, liquidtight flexible metal conduit, or liquidtight flexible nonmetallic conduit, Type MI cable, Type MC cable, or Type AC cable. These supply conductors shall be installed in accordance with the requirements of 300.11.

(3) Ventilation in the underfloor area is used for the information equipment room only. The ventilation system shall be so arranged, with approved smoke detection devices, that upon the detection of fire or products of combustion in the underfloor space the circulation of air will cease.

(4) Openings in raised floors for cables protect cables against abrasions and minimize the entrance of debris beneath the floor.

(5) Cables, other than those covered in (2) and those complying with (a), (b), and (c), shall be listed as Type DP cable having adequate fire-resistant characteristics suitable for use under raised floors of an information technology equipment room.

   (a) Interconnecting cables enclosed in a raceway.

   (b) Interconnecting cables listed with equipment manufactured prior to July 1, 1994, being installed with that equipment.

   (c) Cable type designations Type TC (Article 336); Types CL2, CL3, and PLTC (Article 725); Type ITC (Article 727); Types NPLF and FPL (Article 760); Types OFC and OFN (Article 770); Types CM and MP (Article 800); and Type CATV (Article 820). These designations shall be permitted to have an additional letter P or R or G. Green, with one or more yellow stripes, insulated single conductor cables, 4 AWG and larger, marked "for use in cable trays" or "for CT use" shall be permitted for equipment grounding.

FPN: One method of defining fire resistance is by establishing that the cables do not spread fire to the top of the tray in the "Vertical Tray Flame Test" referenced in

ANSI/UL 1581-1991, *Standard for Electrical Wires, Cables, and Flexible Cords*. Another method of defining fire resistance is for the damage (char length) not to exceed 1.5 m (4 ft 11 in.) when performing the CSA "Vertical Flame Test — Cables in Cable Trays," as described in CSA C22.2 No. 0.3-M-1985, *Test Methods for Electrical Wires and Cables*.

(6) Abandoned cables shall not be permitted to remain unless contained in metal raceways.

**(E) Securing in Place.** Power cables; communications cables; connecting cables; interconnecting cables; and associated boxes, connectors, plugs, and receptacles that are listed as part of, or for, information technology equipment shall not be required to be secured in place.

**645.6 Cables Not in Information Technology Equipment Room.** Cables extending beyond the information technology equipment room shall be subject to the applicable requirements of this *Code*.

FPN: For signaling circuits, refer to Article 725; for fiber optic circuits, refer to Article 770; and for communications circuits, refer to Article 800. For fire alarm systems, refer to Article 760.

**645.7 Penetrations.** Penetrations of the fire-resistant room boundary shall be in accordance with 300.21.

**645.10 Disconnecting Means.** A means shall be provided to disconnect power to all electronic equipment in the information technology equipment room. There shall also be a similar means to disconnect the power to all dedicated HVAC systems serving the room and cause all required fire/smoke dampers to close. The control for these disconnecting means shall be grouped and identified and shall be readily accessible at the principal exit doors. A single means to control both the electronic equipment and HVAC systems shall be permitted. Where a pushbutton is used as a means to disconnect power, pushing the button in shall disconnect the power.

*Exception: Installations qualifying under the provisions of Article 685.*

**645.11 Uninterruptible Power Supplies (UPS).** Unless otherwise permitted in (1) or (2), UPS systems installed within the information technology room, and their supply and output circuits, shall comply with 645.10. The disconnecting means shall also disconnect the battery from its load.

(1) Installations qualifying under the provisions of Article 685

(2) Power sources capable of supplying 750 volt-amperes or less derived either from UPS equipment or from battery circuits integral to electronic equipment

**645.15 Grounding.** All exposed non–current-carrying metal parts of an information technology system shall be grounded in accordance with Article 250 or shall be double insulated. Power systems derived within listed information technology equipment that supply information technology systems through receptacles or cable assemblies supplied as part of this equipment shall not be considered separately derived for the purpose of applying 250.20(D). Where signal reference structures are installed, they shall be bonded to the equipment grounding system provided for the information technology equipment.

FPN No. 1: The bonding and grounding requirements in the product standards governing this listed equipment ensure that it complies with Article 250.

FPN No. 2: Where isolated grounding-type receptacles are used, see 250.146(D) and 406.2(D).

**645.16 Marking.** Each unit of an information technology system supplied by a branch circuit shall be provided with a manufacturer's nameplate, which shall also include the input power requirements for voltage, frequency, and maximum rated load in amperes.

# ARTICLE 647
# Sensitive Electronic Equipment

**647.1 Scope.** This article covers the installation and wiring of separately derived systems operating at 120 volts line-to-line and 60 volts to ground for sensitive electronic equipment.

**647.3 General.** Use of a separately derived 120-volt single-phase 3-wire system with 60 volts on each of two ungrounded conductors to a grounded neutral conductor shall be permitted for the purpose of reducing objectionable noise in senstive electronic equipment locations provided that the following conditions apply:

(1) The system is installed only in commercial or industrial occupancies.
(2) The system's use is restricted to areas under close supervision by qualified personnel.
(3) All of the requirements in 647.4 through 647.8 are met.

**647.4 Wiring Methods.**

**(A) Panelboards and Overcurrent Protection.** Use of standard single-phase panelboards and distribution equipment with a higher voltage rating shall be permitted. The system shall be clearly marked on the face of the panel or on the inside of the panel doors. Common-trip two-pole circuit breakers that are identified for operation at the system voltage shall be provided for both ungrounded conductors in all feeders and branch circuits.

**(B) Junction Boxes.** All junction box covers shall be clearly marked to indicate the distribution panel and the system voltage.

**(C) Color Coding.** All feeders and branch-circuit conductors installed under this section shall be identified as to system at all splices and terminations by color, marking, tagging, or equally effective means. The means of identification shall be posted at each branch-circuit panelboard and at the disconnecting means for the building.

**(D) Voltage Drop.** The voltage drop on any branch circuit shall not exceed 1.5 percent. The combined voltage drop of feeder and branch-circuit conductors shall not exceed 2.5 percent.

**(1) Fixed Equipment.** The voltage drop on branch circuits supplying equipment connected using wiring methods in Chapter 3 shall not exceed 1.5 percent. The combined voltage drop of feeder and branch-circuit conductors shall not exceed 2.5 percent.

**(2) Cord-Connected Equipment.** The voltage drop on branch circuits supplying receptacles shall not exceed 1 percent. For the purposes of making this calculation, the load connected to the receptacle outlet shall be considered to be 50 percent of the branch-circuit rating. The combined voltage drop of feeder and branch-circuit conductors shall not exceed 2.0 percent.

FPN: The purpose of this provision is to limit voltage drop to 1.5 percent where portable cords may be used as a means of connecting equipment.

**647.5 Three-Phase Systems.** Where 3-phase power is supplied, a separately derived 6-phase "wye" system with 60 volts to ground installed under this article shall be configured as three separately derived 120-volt single-phase systems having a combined total of no more than six main disconnects.

**647.6 Grounding.**

**(A) General.** The system shall be grounded as provided in 250.30 as a separately derived single-phase 3-wire system.

**(B) Grounding Conductors Required.** Permanently wired utilization equipment and receptacles shall be grounded by means of an equipment grounding conductor run with the circuit conductors to an equipment grounding bus prominently marked "Technical Equipment Ground" in the originating branch-circuit panelboard. The grounding bus shall be connected to the grounded conductor on the line side of the separately derived system's disconnecting means. The grounding conductor shall not be smaller than that specified in Table 250.122 and run with the feeder conductors. The technical equipment grounding bus need not be bonded to the panelboard enclosure. Other grounding methods authorized elsewhere in this *Code* shall be permitted where the impedance of the grounding return path does not exceed the impedance of equipment grounding conductors sized and installed in accordance with this article.

FPN No. 1:  See 250.122 for equipment grounding conductor sizing requirements where circuit conductors are adjusted in size to compensate for voltage drop.

FPN No. 2:  These requirements limit the impedance of the ground fault path where only 60 volts apply to a fault condition instead of the usual 120 volts.

### 647.7 Receptacles.

**(A) General.** Where receptacles are used as a means of connecting equipment, the following conditions shall be met:

(1) All 15- and 20-ampere receptacles shall be GFCI protected.

(2) All outlet strips, adapters, receptacle covers, and faceplates shall be marked with the following words or equivalent:

> WARNING — TECHNICAL POWER
> Do not connect to lighting equipment.
> For electronic equipment use only.
> 60/120 V. 1φac
> GFCI protected

(3) A 125-volt, single-phase, 15- or 20-ampere-rated receptacle outlet having one of its current-carrying poles connected to a grounded circuit conductor shall be located within 1.8 m (6 ft) of all permanently installed 15- or 20-ampere-rated 60/120-volt technical power-system receptacles.

(4) All 125-volt receptacles used for 60/120-volt technical power shall have a unique configuration and be identified for use with this class of system. All 125-volt, single-phase, 15- or 20-ampere-rated receptacle outlets and attachment plugs that are identified for use with grounded circuit conductors shall be permitted in machine rooms, control rooms, equipment rooms, equipment racks, and other similar locations that are restricted to use by qualified personnel.

**(B) Isolated Ground Receptacles.** Isolated ground receptacles shall be permitted as described in 250.146(D); however, the branch circuit equipment grounding conductor shall be terminated as required in 647.6(B).

### 647.8 Lighting Equipment.
Lighting equipment installed under this article for the purpose of reducing electrical noise originating from lighting equipment shall meet the conditions of 647.8(A) through (C).

**(A) Disconnecting Means.** All luminaires (lighting fixtures) connected to separately derived systems operating at 60 volts to ground and associated control equipment, if provided, shall have a disconnecting means that simultaneously opens all ungrounded conductors. The disconnecting means shall be located within sight of the luminaire (lighting fixture) or be capable of being locked in the open position.

**(B) Luminaires (Lighting Fixtures).** All luminaires (lighting fixtures) shall be permanently installed and listed for connection to a separately derived system at 120 volts line-to-line and 60 volts to ground.

**(C) Screw-shell.** Luminaires installed under this section shall not have an exposed lamp screw-shell.

## ARTICLE 650
## Pipe Organs

**650.1 Scope.** This article covers those electrical circuits and parts of electrically operated pipe organs that are employed for the control of the sounding apparatus and keyboards.

**650.2 Other Articles.** Electronic organs shall comply with the appropriate provisions of Article 640.

**650.3 Source of Energy.** The source of power shall be a transformer-type rectifier, the dc potential of which shall not exceed 30 volts dc.

**650.4 Grounding.** The rectifier shall be grounded according to the provisions in 250.112(B).

**650.5 Conductors.** Conductors shall comply with 650.5(A) through (D).

**(A) Size.** Conductors shall be not less than 28 AWG for electronic signal circuits and not less than 26 AWG for electromagnetic valve supply and the like. A main common-return conductor in the electromagnetic supply shall not be less than 14 AWG.

**(B) Insulation.** Conductors shall have thermoplastic or thermosetting insulation.

**(C) Conductors to Be Cabled.** Except for the common-return conductor and conductors inside the organ proper, the organ sections and the organ console conductors shall be cabled. The common-return conductors shall be permitted under an additional covering enclosing both cable and return conductor, or they shall be permitted as a separate conductor and shall be permitted to be in contact with the cable.

**(D) Cable Covering.** Each cable shall be provided with an outer covering, either overall or on each of any subassemblies of grouped conductors. Tape shall be permitted in place of a covering. Where not installed in metal raceway, the covering shall be resistant to flame spread or the cable or each cable subassembly shall be covered with a closely wound listed fireproof tape.

> FPN: One method of determining that cable is resistant to flame spread is by testing the cable to the VW-1 (vertical-wire) flame test in the ANSI/UL 1581-1991, *Reference Standard for Electrical Wires, Cables and Flexible Cords.*

**650.6 Installation of Conductors.** Cables shall be securely fastened in place and shall be permitted to be attached directly to the organ structure without insulating supports. Cables shall not be placed in contact with other conductors.

**650.7 Overcurrent Protection.** Circuits shall be so arranged that 26 AWG and 28 AWG conductors shall be protected by an overcurrent device rated at not more than 6 amperes. Other conductor sizes shall be protected in accordance with their ampacity. A common return conductor shall not require overcurrent protection.

# ARTICLE 660
# X-Ray Equipment

## I. General

**660.1 Scope.** This article covers all X-ray equipment operating at any frequency or voltage for industrial or other nonmedical or nondental use.

> FPN: See Article 517, Part V, for X-ray installations in health care facilities.

Nothing in this article shall be construed as specifying safeguards against the useful beam or stray X-ray radiation.

> FPN No. 1: Radiation safety and performance requirements of several classes of X-ray equipment are regulated under Public Law 90-602 and are enforced by the Department of Health and Human Services.

> FPN No. 2: In addition, information on radiation protection by the National Council on Radiation Protection and Measurements is published as *Reports of the National Council on Radiation Protection and Measurement.* These

reports can be obtained from NCRP Publications, 7910 Woodmont Ave., Suite 1016, Bethesda, MD 20814.

**660.2 Definitions.**

**Long-Time Rating.** A rating based on an operating interval of 5 minutes or longer.

**Mobile.** X-ray equipment mounted on a permanent base with wheels and/or casters for moving while completely assembled.

**Momentary Rating.** A rating based on an operating interval that does not exceed 5 seconds.

**Portable.** X-ray equipment designed to be hand-carried.

**Transportable.** X-ray equipment that is to be installed in a vehicle or that may be readily disassembled for transport in a vehicle.

**660.3 Hazardous (Classified) Locations.** Unless approved for the location, X-ray and related equipment shall not be installed or operated in hazardous (classified) locations.

> FPN: See Article 517, Part IV.

**660.4 Connection to Supply Circuit.**

**(A) Fixed and Stationary Equipment.** Fixed and stationary X-ray equipment shall be connected to the power supply by means of a wiring method meeting the general requirements of this *Code.* Equipment properly supplied by a branch circuit rated at not over 30 amperes shall be permitted to be supplied through a suitable attachment plug cap and hard-service cable or cord.

**(B) Portable, Mobile, and Transportable Equipment.** Individual branch circuits shall not be required for portable, mobile, and transportable X-ray equipment requiring a capacity of not over 60 amperes. Portable and mobile types of X-ray equipment of any capacity shall be supplied through a suitable hard-service cable or cord. Transportable X-ray equipment of any capacity shall be permitted to be connected to its power supply by suitable connections and hard-service cable or cord.

**(C) Over 600 Volts, Nominal.** Circuits and equipment operated at more than 600 volts, nominal, shall comply with Article 490.

**660.5 Disconnecting Means.** A disconnecting means of adequate capacity for at least 50 percent of the input required for the momentary rating or 100 percent of the input required for the long-time rating of the X-ray equipment, whichever is greater, shall be provided in the supply circuit. The disconnecting means shall be operable from a location readily accessible from the X-ray control. For equipment connected to a 120-volt, nominal, branch circuit of 30 amperes or less, a grounding-type attachment plug cap and receptacle of proper rating shall be permitted to serve as a disconnecting means.

### 660.6 Rating of Supply Conductors and Overcurrent Protection.

**(A) Branch-Circuit Conductors.** The ampacity of supply branch-circuit conductors and the overcurrent protective devices shall not be less than 50 percent of the momentary rating or 100 percent of the long-time rating, whichever is greater.

**(B) Feeder Conductors.** The rated ampacity of conductors and overcurrent devices of a feeder for two or more branch circuits supplying X-ray units shall not be less than 100 percent of the momentary demand rating [as determined by 660.6(A)] of the two largest X-ray apparatus plus 20 percent of the momentary ratings of other X-ray apparatus.

> FPN: The minimum conductor size for branch and feeder circuits is also governed by voltage regulation requirements. For a specific installation, the manufacturer usually specifies minimum distribution transformer and conductor sizes, rating of disconnect means, and overcurrent protection.

### 660.7 Wiring Terminals.
X-ray equipment not provided with a permanently attached cord or cord set shall be provided with suitable wiring terminals or leads for the connection of power-supply conductors of the size required by the rating of the branch circuit for the equipment.

### 660.8 Number of Conductors in Raceway.
The number of control circuit conductors installed in a raceway shall be determined in accordance with 300.17.

### 660.9 Minimum Size of Conductors.
Size 18 AWG or 16 AWG fixture wires, as specified in 725.27, and flexible cords shall be permitted for the control and operating circuits of X-ray and auxiliary equipment where protected by not larger than 20-ampere overcurrent devices.

### 660.10 Equipment Installations.
All equipment for new X-ray installations and all used or reconditioned X-ray equipment moved to and reinstalled at a new location shall be of an approved type.

## II. Control

### 660.20 Fixed and Stationary Equipment.

**(A) Separate Control Device.** A separate control device, in addition to the disconnecting means, shall be incorpo-

rated in the X-ray control supply or in the primary circuit to the high-voltage transformer. This device shall be a part of the X-ray equipment but shall be permitted in a separate enclosure immediately adjacent to the X-ray control unit.

**(B) Protective Device.** A protective device, which shall be permitted to be incorporated into the separate control device, shall be provided to control the load resulting from failures in the high-voltage circuit.

### 660.21 Portable and Mobile Equipment.
Portable and mobile equipment shall comply with 660.20, but the manually controlled device shall be located in or on the equipment.

### 660.23 Industrial and Commercial Laboratory Equipment.

**(A) Radiographic and Fluoroscopic Types.** All radiographic- and fluoroscopic-type equipment shall be effectively enclosed or shall have interlocks that de-energize the equipment automatically to prevent ready access to live current-carrying parts.

**(B) Diffraction and Irradiation Types.** Diffraction- and irradiation-type equipment or installations not effectively enclosed or provided with interlocks to prevent access to live current-carrying parts during operation shall be provided with a positive means to indicate when they are energized. The indicator shall be a pilot light, readable meter deflection, or equivalent means.

### 660.24 Independent Control.
Where more than one piece of equipment is operated from the same high-voltage circuit, each piece or each group of equipment as a unit shall be provided with a high-voltage switch or equivalent disconnecting means. This disconnecting means shall be constructed, enclosed, or located so as to avoid contact by persons with its live parts.

## III. Transformers and Capacitors

### 660.35 General.
Transformers and capacitors that are part of an X-ray equipment shall not be required to comply with Articles 450 and 460.

### 660.36 Capacitors.
Capacitors shall be mounted within enclosures of insulating material or grounded metal.

## IV. Guarding and Grounding

### 660.47 General.

**(A) High-Voltage Parts.** All high-voltage parts, including X-ray tubes, shall be mounted within grounded enclosures. Air, oil, gas, or other suitable insulating media shall be used to insulate the high voltage from the grounded enclosure. The connection from the high-voltage equipment to X-ray tubes and other high-voltage components shall be made with high-voltage shielded cables.

**(B) Low-Voltage Cables.** Low-voltage cables connecting to oil-filled units that are not completely sealed, such as transformers, condensers, oil coolers, and high-voltage switches, shall have insulation of the oil-resistant type.

**660.48 Grounding.** Non–current-carrying metal parts of X-ray and associated equipment (controls, tables, X-ray tube supports, transformer tanks, shielded cables, X-ray tube heads, and so forth) shall be grounded in the manner specified in Article 250. Portable and mobile equipment shall be provided with an approved grounding-type attachment plug cap.

*Exception: Battery-operated equipment.*

# ARTICLE 665
# Induction and Dielectric Heating Equipment

## I. General

**665.1 Scope.** This article covers the construction and installation of dielectric heating, induction heating, induction melting, and induction welding equipment and accessories for industrial and scientific applications. Medical or dental applications, appliances, or line frequency pipeline and vessel heating are not covered in this article.

FPN: See Article 427, Part V, for line frequency induction heating of pipelines and vessels.

## 665.2 Definitions.

**Converting Device.** That part of the heating equipment that converts input mechanical or electrical energy to the voltage, current, and frequency suitable for the heating applicator. A converting device shall consist of equipment using mains frequency, all static multipliers, oscillator-type units using vacuum tubes, inverters using solid state devices, or motor generator equipment.

**Dielectric Heating.** Heating of a nominally insulating material due to its own dielectric losses when the material is placed in a varying electric field.

**Heating Equipment.** As used in this article, any equipment that is used for heating purposes and whose heat is generated by induction or dielectric methods.

**Heating Equipment Applicator.** The heating equipment applicator is the device used to transfer energy between the output circuit and the object or mass to be heated.

**Induction Heating, Melting, and Welding.** The heating, melting, or welding of a nominally conductive material due to its own $I^2R$ losses when the material is placed in a varying electromagnetic field.

**665.3 Other Articles.** Unless specifically amended by this article, wiring from the source of power to the heating equipment shall comply with Chapters 1 through 4.

**665.4 Hazardous (Classified) Locations.** Heating equipment shall not be installed in hazardous (classified) locations as defined in Article 500 unless the equipment and wiring are designed and approved for the hazardous (classified) locations.

**665.5 Output Circuit.** The output circuit shall include all output components external to the converting device, including contactors, switches, bus bars, and other conductors. The current flow from the output circuit to ground under operating and ground-fault conditions shall be limited to a value that does not cause 50 volts or more to ground to appear on any accessible part of the heating equipment and its load. The output circuit shall be permitted to be isolated from ground.

**665.7 Remote Control.**

**(A) Multiple Control Points.** Where multiple control points are used for applicator energization, a means shall be provided and interlocked so that the applicator can be energized from only one control point at a time. A means for de-energizing the applicator shall be provided at each control point.

**(B) Foot Switches.** Switches operated by foot pressure shall be provided with a shield over the contact button to avoid accidental closing of a foot switch.

**665.10 Ampacity of Supply Conductors.** The ampacity of supply conductors shall be determined by 665.10(A) or (B).

**(A) Nameplate Rating.** The ampacity of conductors supplying one or more pieces of equipment shall not be less than the sum of the nameplate ratings for the largest group of machines capable of simultaneous operation, plus 100 percent of the standby currents of the remaining machines. Where standby currents are not given on the nameplate, the nameplate rating shall be used as the standby current.

**(B) Motor-Generator Equipment.** The ampacity of supply conductors for motor generator equipment shall be determined in accordance with Article 430, Part II.

**665.11 Overcurrent Protection.** Overcurrent protection for the heating equipment shall be provided as specified in Article 240. This overcurrent protection shall be permitted to be provided separately or as a part of the equipment.

**665.12 Disconnecting Means.** A readily accessible disconnecting means shall be provided to disconnect each heating equipment from its supply circuit. The disconnecting means shall be located within sight from the controller or be capable of being locked in the open position. The rating of this disconnecting means shall not be less than the nameplate rating of the heating equipment. Motor-generator equipment shall comply with Article 430, Part IX. The supply circuit disconnecting means shall be permitted to serve as the heating equipment disconnecting means where only one heating equipment is supplied.

## II. Guarding, Grounding, and Labeling

**665.19 Component Interconnection.** The interconnection components required for a complete heating equipment installation shall be guarded.

**665.20 Enclosures.** The converting device (excluding the component interconnections) shall be completely contained within an enclosure(s) of noncombustible material.

**665.21 Control Panels.** All control panels shall be of dead-front construction.

**665.22 Access to Internal Equipment.** Access doors or detachable access panels shall be employed for internal access to heating equipment. Access doors to internal compartments containing equipment employing voltages from 150 volts to 1000 volts ac or dc shall be capable of being locked closed or shall be interlocked to prevent the supply circuit from being energized while the door(s) is open. Access doors to internal compartments containing equipment employing voltages exceeding 1000 volts ac or dc shall be provided with a disconnecting means equipped with mechanical lockouts to prevent access while the heating equipment is energized, or the access doors shall be capable of being locked closed and interlocked to prevent the supply circuit from being energized while the door(s) is open. Detachable panels not normally used for access to such parts shall be fastened in a manner that will make them inconvenient to remove.

**665.23 Warning Labels or Signs.** Warning labels or signs that read "DANGER – HIGH VOLTAGE – KEEP OUT" shall be attached to the equipment and shall be plainly visible where persons might come in contact with energized parts when doors are open or closed or when panels are removed from compartments containing over 150 volts ac or dc.

**665.24 Capacitors.** The time and means of discharge shall be in accordance with 460.6 for capacitors rated 600 volts, nominal, and under. The time and means of discharge shall be in accordance with 460.28 for capacitors rated over 600 volts, nominal. Capacitor internal pressure switches connected to a circuit-interruptor device shall be permitted for capacitor overcurrent protection.

**665.25 Dielectric Heating Applicator Shielding.** Protective cages or adequate shielding shall be used to guard dielectric heating applicators. Interlock switches shall be used on all hinged access doors, sliding panels, or other easy means of access to the applicator. All interlock switches shall be connected in such a manner as to remove all power from the applicator when any one of the access doors or panels is open.

**665.26 Grounding and Bonding.** Grounding or inter-unit bonding, or both, shall be used wherever required for circuit operation, for limiting to a safe value radio frequency voltages between all exposed non–current-carrying parts of the equipment and earth ground, between all equipment parts and surrounding objects, and between such objects and earth ground. Such grounding and bonding shall be installed in accordance with Article 250, Parts II and V.

> FPN: Under certain conditions, contact between the object being heated and the applicator results in an unsafe condition, such as eruption of heated materials. This unsafe condition may be prevented by grounding of the object being heated and ground detection.

**665.27 Marking.** Each heating equipment shall be provided with a nameplate giving the manufacturer's name and model identification and the following input data: line volts, frequency, number of phases, maximum current, full-load kilovolt-amperes (kVA), and full-load power factor. Additional data shall be permitted.

## ARTICLE 668
## Electrolytic Cells

**668.1 Scope.** The provisions of this article apply to the installation of the electrical components and accessory equipment of electrolytic cells, electrolytic cell lines, and process power supply for the production of aluminum, cadmium, chlorine, copper, fluorine, hydrogen peroxide, magnesium, sodium, sodium chlorate, and zinc.

Not covered by this article are cells used as a source of electric energy and for electroplating processes and cells used for the production of hydrogen.

> FPN No. 1: In general, any cell line or group of cell lines operated as a unit for the production of a particular metal, gas, or chemical compound may differ from any other cell lines producing the same product because of variations in the particular raw materials used, output capacity, use of proprietary methods or process practices, or other modifying factors to the extent that detailed *Code* requirements become overly restrictive and do not accomplish the stated purpose of this *Code*.

> FPN No. 2: For further information, see IEEE 463-1993, *Standard for Electrical Safety Practices in Electrolytic Cell Line Working Zones*.

### 668.2 Definitions.

**Cell Line.** An assembly of electrically interconnected electrolytic cells supplied by a source of direct-current power.

**Cell Line Attachments and Auxiliary Equipment.** As applied to this article, a term that includes, but is not limited to, auxiliary tanks; process piping; ductwork; structural supports; exposed cell line conductors; conduits and other raceways; pumps, positioning equipment, and cell cutout or bypass electrical devices. Auxiliary equipment includes tools, welding machines, crucibles, and other portable equipment used for operation and maintenance within the electrolytic cell line working zone.

In the cell line working zone, auxiliary equipment includes the exposed conductive surfaces of ungrounded cranes and crane-mounted cell-servicing equipment.

**Electrically Connected.** A connection capable of carrying current as distinguished from connection through electromagnetic induction.

**Electrolytic Cell.** A tank or vat in which electrochemical reactions are caused by applying electrical energy for the purpose of refining or producing usable materials.

**Electrolytic Cell Line Working Zone.** The space envelope wherein operation or maintenance is normally performed on or in the vicinity of exposed energized surfaces of electrolytic cell lines or their attachments.

### 668.3 Other Articles.

**(A) Lighting, Ventilating, Material Handling.** Chapters 1 through 4 shall apply to services, feeders, branch circuits, and apparatus for supplying lighting, ventilating, material handling, and the like that are outside the electrolytic cell line working zone.

**(B) Systems Not Electrically Connected.** Those elements of a cell line power-supply system that are not electrically connected to the cell supply system, such as the primary winding of a two-winding transformer, the motor of a motor-generator set, feeders, branch circuits, disconnecting means, motor controllers, and overload protective equipment, shall be required to comply with all applicable provisions of this *Code*.

**(C) Electrolytic Cell Lines.** Electrolytic cell lines shall comply with the provisions of Chapters 1, 2, 3, and 4 except as amended in 668.3(C)(1), (C)(2), (C)(3), or (C)(4).

**(1) Conductors.** The electrolytic cell line conductors shall not be required to comply with the provisions of Articles 110, 210, 215, 220, and 225. See 668.11.

**(2) Overcurrent Protection.** Overcurrent protection of electrolytic cell dc process power circuits shall not be required to comply with the requirements of Article 240.

**(3) Grounding.** Equipment located or used within the electrolytic cell line working zone or associated with the cell line dc power circuits shall not be required to comply with the provisions of Article 250.

**(4) Working Zone.** The electrolytic cells, cell line attachments, and the wiring of auxiliary equipments and devices within the cell line working zone shall not be required to comply with the provisions of Articles 110, 210, 215, 220, and 225. See 668.30.

> FPN: See 668.15 for equipment, apparatus, and structural component grounding.

### 668.10 Cell Line Working Zone.

**(A) Area Covered.** The space envelope of the cell line working zone shall encompass spaces that meet any of the following conditions:

(1) Is within 2.5 m (96 in.) above energized surfaces of electrolytic cell lines or their energized attachments.
(2) Is below energized surfaces of electrolytic cell lines or their energized attachments, provided the headroom in the space beneath is less than 2.5 m (96 in.).
(3) Is within 1.0 m (42 in.) horizontally from energized surfaces of electrolytic cell lines or their energized attachments or from the space envelope described in 668.10(A)(1) or (A)(2).

**(B) Area Not Covered.** The cell line working zone shall not be required to extend through or beyond walls, floors, roofs, partitions, barriers, or the like.

### 668.11 Direct-Current Cell Line Process Power Supply.

**(A) Not Grounded.** The dc cell line process power-supply conductors shall not be required to be grounded.

**(B) Metal Enclosures Grounded.** All metal enclosures of dc cell line process power-supply apparatus operating at a power-supply potential between terminals of over 50 volts shall be grounded as follows:

(1) Through protective relaying equipment, or
(2) By a minimum 2/0 AWG copper grounding conductor or a conductor of equal or greater conductance

**(C) Grounding Requirements.** The grounding connections required by 668.11(B) shall be installed in accordance with 250.8, 250.10, 250.12, 250.68, and 250.70.

### 668.12 Cell Line Conductors.

**(A) Insulation and Material.** Cell line conductors shall be either bare, covered, or insulated and of copper, aluminum, copper-clad aluminum, steel, or other suitable material.

**(B) Size.** Cell line conductors shall be of such cross-sectional area that the temperature rise under maximum load conditions and at maximum ambient shall not exceed the safe operating temperature of the conductor insulation or the material of the conductor supports.

**(C) Connections.** Cell line conductors shall be joined by bolted, welded, clamped, or compression connectors.

### 668.13 Disconnecting Means.

**(A) More Than One Process Power Supply.** Where more than one dc cell line process power supply serves the same cell line, a disconnecting means shall be provided on the cell line circuit side of each power supply to disconnect it from the cell line circuit.

**(B) Removable Links or Conductors.** Removable links or removable conductors shall be permitted to be used as the disconnecting means.

### 668.14 Shunting Means.

**(A) Partial or Total Shunting.** Partial or total shunting of cell line circuit current around one or more cells shall be permitted.

**(B) Shunting One or More Cells.** The conductors, switches, or combination of conductors and switches used for shunting one or more cells shall comply with the applicable requirements of 668.12.

### 668.15 Grounding.
For equipment, apparatus, and structural components that are required to be grounded by provisions of Article 668, the provisions of Article 250 shall apply, except a water pipe electrode shall not be required to be used. Any electrode or combination of electrodes described in 250.52 shall be permitted.

### 668.20 Portable Electrical Equipment.

**(A) Portable Electrical Equipment Not to Be Grounded.** The frames and enclosures of portable electrical equipment used within the cell line working zone shall not be grounded.

*Exception No. 1: Where the cell line voltage does not exceed 200 volts dc, these frames and enclosures shall be permitted to be grounded.*

*Exception No. 2: These frames and enclosures shall be permitted to be grounded where guarded.*

**(B) Isolating Transformers.** Electrically powered, hand-held, cord-connected portable equipment with ungrounded frames or enclosures used within the cell line working zone shall be connected to receptacle circuits that have only ungrounded conductors such as a branch circuit supplied by an isolating transformer with an ungrounded secondary.

**(C) Marking.** Ungrounded portable electrical equipment shall be distinctively marked and shall employ plugs and receptacles of a configuration that prevents connection of this equipment to grounding receptacles and that prevents inadvertent interchange of ungrounded and grounded portable electrical equipments.

### 668.21 Power Supply Circuits and Receptacles for Portable Electrical Equipment.

**(A) Isolated Circuits.** Circuits supplying power to ungrounded receptacles for hand-held, cord-connected equipments shall be electrically isolated from any distribution system supplying areas other than the cell line working zone and shall be ungrounded. Power for these circuits shall be supplied through isolating transformers. Primaries of such transformers shall operate at not more than 600 volts between conductors and shall be provided with proper overcurrent protection. The secondary voltage of such transformers shall not exceed 300 volts between conductors, and all circuits supplied from such secondaries shall be ungrounded and shall have an approved overcurrent device of proper rating in each conductor.

**(B) Noninterchangeability.** Receptacles and their mating plugs for ungrounded equipment shall not have provision for a grounding conductor and shall be of a configuration that prevents their use for equipment required to be grounded.

**(C) Marking.** Receptacles on circuits supplied by an isolating transformer with an ungrounded secondary shall be a distinctive configuration, distinctively marked, and shall not be used in any other location in the plant.

### 668.30 Fixed and Portable Electrical Equipment.

**(A) Electrical Equipment Not Required to Be Grounded.** Alternating-current systems supplying fixed and portable electrical equipments within the cell line working zone shall not be required to be grounded.

**(B) Exposed Conductive Surfaces Not Required to Be Grounded.** Exposed conductive surfaces, such as electrical equipment housings, cabinets, boxes, motors, raceways, and the like, that are within the cell line working zone shall not be required to be grounded.

**(C) Wiring Methods.** Auxiliary electrical equipment such as motors, transducers, sensors, control devices, and alarms, mounted on an electrolytic cell or other energized surface, shall be connected to premises wiring systems by any of the following means:

(1) Multiconductor hard usage cord.
(2) Wire or cable in suitable raceways or metal or nonmetallic cable trays. If metal conduit, cable tray, armored cable, or similar metallic systems are used, they shall be installed with insulating breaks such that they do not cause a potentially hazardous electrical condition.

**(D) Circuit Protection.** Circuit protection shall not be required for control and instrumentation that are totally within the cell line working zone.

**(E) Bonding.** Bonding of fixed electrical equipment to the energized conductive surfaces of the cell line, its attachments, or auxiliaries shall be permitted. Where fixed electrical equipment is mounted on an energized conductive surface, it shall be bonded to that surface.

**668.31 Auxiliary Nonelectric Connections.** Auxiliary nonelectric connections, such as air hoses, water hoses, and the like, to an electrolytic cell, its attachments, or auxiliary equipments shall not have continuous conductive reinforcing wire, armor, braids, and the like. Hoses shall be of a nonconductive material.

**668.32 Cranes and Hoists.**

**(A) Conductive Surfaces to Be Insulated from Ground.** The conductive surfaces of cranes and hoists that enter the cell line working zone shall not be required to be grounded. The portion of an overhead crane or hoist that contacts an energized electrolytic cell or energized attachments shall be insulated from ground.

**(B) Hazardous Electrical Conditions.** Remote crane or hoist controls that could introduce hazardous electrical conditions into the cell line working zone shall employ one or more of the following systems:

(1) Isolated and ungrounded control circuit in accordance with 668.21(A)
(2) Nonconductive rope operator

(3) Pendant pushbutton with nonconductive supporting means and having nonconductive surfaces or ungrounded exposed conductive surfaces
(4) Radio

**668.40 Enclosures.** General-purpose electrical equipment enclosures shall be permitted where a natural draft ventilation system prevents the accumulation of gases.

## ARTICLE 669
## Electroplating

**669.1 Scope.** The provisions of this article apply to the installation of the electrical components and accessory equipment that supply the power and controls for electroplating, anodizing, electropolishing, and electrostripping. For purposes of this article, the term *electroplating* shall be used to identify any or all of these processes.

**669.2 Other Articles.** Except as modified by this article, wiring and equipment used for electroplating processes shall comply with the applicable requirements of Chapters 1 through 4.

**669.3 General.** Equipment for use in electroplating processes shall be identified for such service.

**669.5 Branch-Circuit Conductors.** Branch-circuit conductors supplying one or more units of equipment shall have an ampacity of not less than 125 percent of the total connected load. The ampacities for busbars shall be in accordance with 366.7.

**669.6 Wiring Methods.** Conductors connecting the electrolyte tank equipment to the conversion equipment shall be in accordance with 669.6(A) and (B).

**(A) Systems Not Exceeding 50 Volts Direct Current.** Insulated conductors shall be permitted to be run without insulated support, provided they are protected from physical damage. Bare copper or aluminum conductors shall be permitted where supported on insulators.

**(B) Systems Exceeding 50 Volts Direct Current.** Insulated conductors shall be permitted to be run on insulated supports, provided they are protected from physical damage. Bare copper or aluminum conductors shall be permitted where supported on insulators and guarded against accidental contact up to the point of termination in accordance with 110.27.

**669.7 Warning Signs.** Warning signs shall be posted to indicate the presence of bare conductors.

**669.8 Disconnecting Means.**

**(A) More Than One Power Supply.** Where more than one power supply serves the same dc system, a disconnecting means shall be provided on the dc side of each power supply.

**(B) Removable Links or Conductors.** Removable links or removable conductors shall be permitted to be used as the disconnecting means.

**669.9 Overcurrent Protection.** Direct-current conductors shall be protected from overcurrent by one or more of the following:

(1) Fuses or circuit breakers
(2) A current-sensing device that operates a disconnecting means
(3) Other approved means

# ARTICLE 670
# Industrial Machinery

**670.1 Scope.** This article covers the definition of, the nameplate data for, and the size and overcurrent protection of supply conductors to industrial machinery.

FPN: For further information, see NFPA 79-1997, *Electrical Standard for Industrial Machinery.*

**670.2 Definitions.**

**Industrial Machinery (Machine).** A power-driven machine (or a group of machines working together in a coordinated manner), not portable by hand while working, that is used to process material by cutting; forming; pressure; electrical, thermal, or optical techniques; lamination; or a combination of these processes. It can include associated equipment used to transfer material or tooling, including fixtures, to assemble/disassemble, to inspect or test, or to package. [The associated electrical equipment, including the logic controller(s) and associated software or logic together with the machine actuators and sensors, are considered as part of the industrial machine.]

**Industrial Manufacturing System.** A systematic array of one or more industrial machines that is not portable by hand and includes any associated material handling, manipulating, gauging, measuring, or inspection equipment.

**670.3 Machine Nameplate Data.**

**(A) Permanent Nameplate.** A permanent nameplate that lists supply voltage, phase, frequency, full-load current, the maximum ampere rating of the short-circuit and ground-fault protective device, ampere rating of largest motor or load, short-circuit interrupting rating of the machine overcurrent-protective device, if furnished, and diagram number shall be attached to the control equipment enclosure or machine where plainly visible after installation.

The full-load current shown on the nameplate shall not be less than the sum of the full-load currents required for all motors and other equipment that may be in operation at the same time under normal conditions of use. Where unusual type loads, duty cycles, and so forth require oversized conductors or permit reduced-size conductors, the required capacity shall be included in the marked "full-load current." Where more than one incoming supply circuit is to be provided, the nameplate shall state the above information for each circuit.

**(B) Overcurrent Protection.** Where overcurrent protection is provided in accordance with 670.4(B), the machine shall be marked "overcurrent protection provided at machine supply terminals."

**670.4 Supply Conductors and Overcurrent Protection.**

**(A) Size.** The size of the supply conductor shall be such as to have an ampacity not less than 125 percent of the full-load current rating of all resistance heating loads plus 125 percent of the full-load current rating of the highest rated motor plus the sum of the full-load current ratings of all other connected motors and apparatus based on their duty cycle that may be in operation at the same time.

FPN: See the 0–2000-volt ampacity tables of Article 310 for ampacity of conductors rated 600 volts and below.

**(B) Overcurrent Protection.** A machine shall be considered as an individual unit and therefore shall be provided with a disconnecting means. The disconnecting means shall be permitted to be supplied by branch circuits protected by either fuses or circuit breakers. The disconnecting means shall not be required to incorporate overcurrent protection. Where furnished as part of the machine, overcurrent protection shall consist of a single circuit breaker or set of fuses, the machine shall bear the marking required in 670.3, and the supply conductors shall be considered either as feeders or taps as covered by 240.21.

The rating or setting of the overcurrent protective device for the circuit supplying the machine shall not be greater than the sum of the largest rating or setting of the branch-circuit short-circuit and ground-fault protective device provided with the machine, plus 125 percent of the full-load current rating of all resistance heating loads, plus the sum of the full-load currents of all other motors and apparatus that could be in operation at the same time.

*Exception: Where one or more instantaneous trip circuit breakers or motor short-circuit protectors are used for motor branch-circuit short-circuit and ground-fault protection as permitted by 430.52(C), the procedure specified above for determining the maximum rating of the protective device for the circuit supplying the machine shall apply with the following provision: For the purpose of the calculation, each instantaneous trip circuit breaker or motor short-circuit protector shall be assumed to have a rating not exceeding the maximum percentage of motor full-load current permitted by Table 430.52 for the type of machine supply circuit protective device employed.*

Where no branch-circuit short-circuit and ground-fault protective device is provided with the machine, the rating or setting of the overcurrent protective device shall be based on 430.52 and 430.53, as applicable.

**670.5 Clearance.** Where the conditions of maintenance and supervision ensure that only qualified persons service the installation, the dimensions of the working space in the direction of access to live parts operating at not over 150 volts line-to-line or line-to-ground that are likely to require examination, adjustment, servicing, or maintenance while energized shall be a minimum of 750 mm (2½ ft). Where controls are enclosed in cabinets, the door(s) shall open at least 90 degrees or be removable.

*Exception: Where the enclosure requires a tool to open, and where only diagnostic and troubleshooting testing is involved on live parts, the clearances shall be permitted to be less than 750 mm (2½ ft).*

# ARTICLE 675
# Electrically Driven or Controlled Irrigation Machines

## I. General

**675.1 Scope.** The provisions of this article apply to electrically driven or controlled irrigation machines, and to the branch circuits and controllers for such equipment.

**675.2 Definitions.**

**Center Pivot Irrigation Machines.** A multimotored irrigation machine that revolves around a central pivot and employs alignment switches or similar devices to control individual motors.

**Collector Rings.** An assembly of slip rings for transferring electrical energy from a stationary to a rotating member.

**Irrigation Machines.** An electrically driven or controlled machine, with one or more motors, not hand portable, and used primarily to transport and distribute water for agricultural purposes.

**675.3 Other Articles.** These provisions are in addition to, or amendatory of, the provisions of Article 430 and other articles in this *Code* that apply except as modified in this article.

**675.4 Irrigation Cable.**

**(A) Construction.** The cable used to interconnect enclosures on the structure of an irrigation machine shall be an assembly of stranded, insulated conductors with nonhygroscopic and nonwicking filler in a core of moisture- and flame-resistant nonmetallic material overlaid with a metallic covering and jacketed with a moisture-, corrosion-, and sunlight-resistant nonmetallic material.

The conductor insulation shall be of a type listed in Table 310.13 for an operating temperature of 75°C (167°F) and for use in wet locations. The core insulating material thickness shall not be less than 0.76 mm (30 mils), and the metallic overlay thickness shall not be less than 0.20 mm (8 mils). The jacketing material thickness shall not be less than 1.27 mm (50 mils).

A composite of power, control, and grounding conductors in the cable shall be permitted.

**(B) Alternate Wiring Methods.** Other cables listed for the purpose.

**(C) Supports.** Irrigation cable shall be secured by straps, hangers, or similar fittings identified for the purpose and installed as not to damage the cable. Cable shall be supported at intervals not exceeding 1.2 m (4 ft).

**(D) Fittings.** Fittings shall be used at all points where irrigation cable terminates. The fittings shall be designed for use with the cable and shall be suitable for the conditions of service.

**675.5 More Than Three Conductors in a Raceway or Cable.** The signal and control conductors of a raceway or cable shall not be counted for the purpose of derating the conductors as required in 310.15(B)(2)(a).

**675.6 Marking on Main Control Panel.** The main control panel shall be provided with a nameplate that shall give the following information:

(1) The manufacturer's name, the rated voltage, the phase, and the frequency

(2) The current rating of the machine
(3) The rating of the main disconnecting means and size of overcurrent protection required

**675.7 Equivalent Current Ratings.** Where intermittent duty is not involved, the provisions of Article 430 shall be used for determining ratings for controllers, disconnecting means, conductors, and the like. Where irrigation machines have inherent intermittent duty, the determinations of equivalent current ratings in 675.7(A) and (B) shall be used.

**(A) Continuous-Current Rating.** The equivalent continuous-current rating for the selection of branch-circuit conductors and overcurrent protection shall be equal to 125 percent of the motor nameplate full-load current rating of the largest motor plus a quantity equal to the sum of each of the motor nameplate full-load current ratings of all remaining motors on the circuit multiplied by the maximum percent duty cycle at which they can continuously operate.

**(B) Locked-Rotor Current.** The equivalent locked-rotor current rating shall be equal to the numerical sum of the locked-rotor current of the two largest motors plus 100 percent of the sum of the motor nameplate full-load current ratings of all the remaining motors on the circuit.

**675.8 Disconnecting Means.**

**(A) Main Controller.** A controller that is used to start and stop the complete machine shall meet all of the following requirements:

(1) An equivalent continuous current rating not less than specified in 675.7(A) or 675.22(A)
(2) A horsepower rating not less than the value from Tables 430.151(A) and (B), based on the equivalent locked-rotor current specified in 675.7(B) or 675.22(B)

**(B) Main Disconnecting Means.** The main disconnecting means for the machine shall provide overcurrent protection, shall be at the point of connection of electrical power to the machine or shall be visible and not more than 15 m (50 ft) from the machine, and shall be readily accessible and capable of being locked in the open position. This disconnecting means shall have a horsepower and current rating not less than required for the main controller.

*Exception: Circuit breakers without marked horsepower ratings shall be permitted in accordance with 430.109.*

**(C) Disconnecting Means for Individual Motors and Controllers.** A disconnecting means shall be provided to simultaneously disconnect all ungrounded conductors for each motor and controller and shall be located as required by Article 430, Part IX. The disconnecting means shall not be required to be readily accessible.

**675.9 Branch-Circuit Conductors.** The branch-circuit conductors shall have an ampacity not less than specified in 675.7(A) or 675.22(A).

**675.10 Several Motors on One Branch Circuit.**

**(A) Protection Required.** Several motors, each not exceeding 2 hp rating, shall be permitted to be used on an irrigation machine circuit protected at not more than 30 amperes at 600 volts, nominal, or less, provided all of the following conditions are met:

(1) The full-load rating of any motor in the circuit shall not exceed 6 amperes.
(2) Each motor in the circuit shall have individual overload protection in accordance with 430.32.
(3) Taps to individual motors shall not be smaller than 14 AWG copper and not more than 7.5 m (25 ft) in length.

**(B) Individual Protection Not Required.** Individual branch-circuit short-circuit protection for motors and motor controllers shall not be required where the requirements of 675.10(A) are met.

**675.11 Collector Rings.**

**(A) Transmitting Current for Power Purposes.** Collector rings shall have a current rating not less than 125 percent of the full-load current of the largest device served plus the full-load current of all other devices served, or as determined from 675.7(A) or 675.22(A).

**(B) Control and Signal Purposes.** Collector rings for control and signal purposes shall have a current rating not less than 125 percent of the full-load current of the largest device served plus the full-load current of all other devices served.

**(C) Grounding.** The collector ring used for grounding shall have a current rating of not less than that sized in accordance with 675.11(A).

**(D) Protection.** Collector rings shall be protected from the expected environment and from accidental contact by means of a suitable enclosure.

**675.12 Grounding.** The following equipment shall be grounded:

(1) All electrical equipment on the irrigation machine
(2) All electrical equipment associated with the irrigation machine
(3) Metal junction boxes and enclosures
(4) Control panels or control equipment that supply or control electrical equipment to the irrigation machine

*Exception: Grounding shall not be required on machines where all of the following provisions are met:*

*(a) The machine is electrically controlled but not electrically driven.*

*(b) The control voltage is 30 volts or less.*

*(c) The control or signal circuits are current limited as specified in Chapter 9, Tables 11(A) and 11(B).*

**675.13 Methods of Grounding.** Machines that require grounding shall have a non–current-carrying equipment grounding conductor provided as an integral part of each cord, cable, or raceway. This grounding conductor shall be sized not less than the largest supply conductor in each cord, cable, or raceway. Feeder circuits supplying power to irrigation machines shall have an equipment grounding conductor sized according to Table 250.122.

**675.14 Bonding.** Where electrical grounding is required on an irrigation machine, the metallic structure of the machine, metallic conduit, or metallic sheath of cable shall be bonded to the grounding conductor. Metal-to-metal contact with a part that is bonded to the grounding conductor and the non–current-carrying parts of the machine shall be considered as an acceptable bonding path.

**675.15 Lightning Protection.** If an irrigation machine has a stationary point, a grounding electrode system in accordance with Article 250, Part III, shall be connected to the machine at the stationary point for lightning protection.

**675.16 Energy from More Than One Source.** Equipment within an enclosure receiving electrical energy from more than one source shall not be required to have a disconnecting means for the additional source, provided that its voltage is 30 volts or less and it meets the requirements of Part III of Article 725.

**675.17 Connectors.** External plugs and connectors on the equipment shall be of the weatherproof type.

Unless provided solely for the connection of circuits meeting the requirements of Part III of Article 725, external plugs and connectors shall be constructed as specified in 250.124(A).

## II. Center Pivot Irrigation Machines

**675.21 General.** The provisions of Part II are intended to cover additional special requirements that are peculiar to center pivot irrigation machines. See 675.2 for definition of *Center Pivot Irrigation Machines.*

**675.22 Equivalent Current Ratings.** In order to establish ratings of controllers, disconnecting means, conductors, and the like, for the inherent intermittent duty of center pivot irrigation machines, the determinations in 675.22(A) and (B) shall be used.

**(A) Continuous-Current Rating.** The equivalent continuous-current rating for the selection of branch-circuit conductors and branch-circuit devices shall be equal to 125 percent of the motor nameplate full-load current rating of the largest motor plus 60 percent of the sum of the motor nameplate full-load current ratings of all remaining motors on the circuit.

**(B) Locked-Rotor Current.** The equivalent locked-rotor current rating shall be equal to the numerical sum of two times the locked-rotor current of the largest motor plus 80 percent of the sum of the motor nameplate full-load current ratings of all the remaining motors on the circuit.

## ARTICLE 680
## Swimming Pools, Fountains, and Similar Installations

## I. General

**680.1 Scope.** The provisions of this article apply to the construction and installation of electrical wiring for and equipment in or adjacent to all swimming, wading, therapeutic, and decorative pools, fountains, hot tubs, spas, and hydromassage bathtubs, whether permanently installed or storable, and to metallic auxiliary equipment, such as pumps, filters, and similar equipment. The term *body of water* used throughout Part I applies to all bodies of water covered in this scope unless otherwise amended.

**680.2 Definitions.**

**Cord-and-Plug-Connected Lighting Assembly.** A lighting assembly consisting of a luminaire (lighting fixture) intended for installation in the wall of a spa, hot tub, or storable pool, and a cord-and-plug-connected transformer.

**Dry-Niche Luminaire (Lighting Fixture).** A luminaire (lighting fixture) intended for installation in the wall of a pool or fountain in a niche that is sealed against the entry of pool water.

**Equipment, Fixed.** Equipment that is fastened or otherwise secured at a specific location.

**Equipment, Portable.** Equipment that is actually moved or can easily be moved from one place to another in normal use.

**Equipment, Stationary.** Equipment that is not easily moved from one place to another in normal use.

**Forming Shell.** A structure designed to support a wet-niche luminaire (lighting fixture) assembly and intended for mounting in a pool or fountain structure.

**Fountain.** Fountains, ornamental pools, display pools, and reflection pools. The definition does not include drinking fountains.

**Hydromassage Bathtub.** A permanently installed bathtub equipped with a recirculating piping system, pump, and associated equipment. It is designed so it can accept, circulate, and discharge water upon each use.

**Maximum Water Level.** The highest level that water can reach before it spills out.

**No-Niche Luminaire (Lighting Fixture).** A luminaire (lighting fixture) intended for installation above or below the water without a niche.

**Packaged Spa or Hot Tub Equipment Assembly.** A factory-fabricated unit consisting of water-circulating, heating, and control equipment mounted on a common base, intended to operate a spa or hot tub. Equipment can include pumps, air blowers, heaters, lights, controls, sanitizer generators, and so forth.

**Packaged Therapeutic Tub or Hydrotherapeutic Tank Equipment Assembly.** A factory-fabricated unit consisting of water-circulating, heating, and control equipment mounted on a common base, intended to operate a therapeutic tub or hydrotherapeutic tank. Equipment can include pumps, air blowers, heaters, lights, controls, sanitizer generators, and so forth.

**Permanently Installed Decorative Fountains and Reflection Pools.** Those that are constructed in the ground, on the ground, or in a building in such a manner that the fountain cannot be readily disassembled for storage, whether or not served by electrical circuits of any nature. These units are primarily constructed for their aesthetic value and are not intended for swimming or wading.

**Permanently Installed Swimming, Wading, and Therapeutic Pools.** Those that are constructed in the ground or partially in the ground, and all others capable of holding water in a depth greater than 1.0 m (42 in.), and all pools installed inside of a building, regardless of water depth, whether or not served by electrical circuits of any nature.

**Pool.** Manufactured or field-constructed equipment designed to contain water on a permanent or semipermanent basis and used for swimming, wading, or other purposes.

**Pool Cover, Electrically Operated.** Motor-driven equipment designed to cover and uncover the water surface of a pool by means of a flexible sheet or rigid frame.

**Self-Contained Spa or Hot Tub.** Factory-fabricated unit consisting of a spa or hot tub vessel with all water-circulating, heating, and control equipment integral to the unit. Equipment can include pumps, air blowers, heaters, lights, controls, sanitizer generators, and so forth.

**Self-Contained Therapeutic Tubs or Hydrotherapeutic Tanks.** A factory-fabricated unit consisting of a therapeutic tub or hydrotherapeutic tank with all water-circulating, heating, and control equipment integral to the unit. Equipment may include pumps, air blowers, heaters, light controls, sanitizer generators, and so forth.

**Spa or Hot Tub.** A hydromassage pool, or tub for recreational or therapeutic use, not located in health care facilities, designed for immersion of users, and usually having a filter, heater, and motor-driven blower. It may be installed indoors or outdoors, on the ground or supporting structure, or in the ground or supporting structure. Generally, a spa or hot tub is not designed or intended to have its contents drained or discharged after each use.

**Storable Swimming or Wading Pool.** Those that are constructed on or above the ground and are capable of holding water to a maximum depth of 1.0 m (42 in.), or a pool with nonmetallic, molded polymeric walls or inflatable fabric walls regardless of dimension.

**Through-Wall Lighting Assembly.** A lighting assembly intended for installation above grade, on or through the wall of a pool, consisting of two interconnected groups of components separated by the pool wall.

**Wet-Niche Luminaire (Lighting Fixture).** A luminaire (lighting fixture) intended for installation in a forming shell mounted in a pool or fountain structure where the luminaire (fixture) will be completely surrounded by water.

**680.3 Other Articles.** Except as modified by this article, wiring and equipment in or adjacent to pools and fountains shall comply with other applicable provisions of this *Code*, including those provisions identified in Table 680.3.

**680.4 Approval of Equipment.** All electrical equipment installed in the water, walls, or decks of pools, fountains, and similar installations shall comply with the provisions of this article.

**680.5 Ground-Fault Circuit Interrupters.** Ground-fault circuit interrupters (GFCIs) shall be self-contained units, circuit-breaker or receptacle types, or other listed types.

**Table 680.3 Other Articles**

| Topic | Section or Article |
|---|---|
| Wiring | Chapters 1–4 |
|   Junction box support | 314.23 |
|   Rigid nonmetallic conduit | 352.12 |
| Audio Equipment | Article 640, Parts I and II |
|   Adjacent to pools and fountains | 640.10 |
|   Underwater speakers* | |

*Underwater loudspeakers shall be installed in accordance with 680.27(A).

**680.6 Grounding.** Electrical equipment shall be grounded in accordance with Parts V, VI, and VII of Article 250 and connected by wiring methods of Chapter 3, except as modified by this article. The following equipment shall be grounded:

(1) Through-wall lighting assemblies and underwater luminaires (lighting fixtures), other than those low-voltage systems listed for the application without a grounding conductor

(2) All electrical equipment located within 1.5 m (5 ft) of the inside wall of the specified body of water

(3) All electrical equipment associated with the recirculating system of the specified body of water

(4) Junction boxes

(5) Transformer enclosures

(6) Ground-fault circuit interrupters

(7) Panelboards that are not part of the service equipment and that supply any electrical equipment associated with the specified body of water

**680.7 Cord-and-Plug-Connected Equipment.** Fixed or stationary equipment other than an underwater luminaire (lighting fixture) for a permanently installed pool shall be permitted to be connected with a flexible cord to facilitate the removal or disconnection for maintenance or repair.

**(A) Length.** For other than storable pools, the flexible cord shall not exceed 900 mm (3 ft) in length.

**(B) Equipment Grounding.** The flexible cord shall have a copper equipment grounding conductor sized in accordance

with 250.122 but not smaller than 12 AWG. The cord shall terminate in a grounding-type attachment plug.

**(C) Construction.** The equipment grounding conductors shall be connected to a fixed metal part of the assembly. The removable part shall be mounted on or bonded to the fixed metal part.

**680.8 Overhead Conductor Clearances.**

**(A) Power.** With respect to service drop conductors and open overhead wiring, swimming pool and similar installations shall comply with the minimum clearances given in Table 680.8 and illustrated in Figure 680.8.

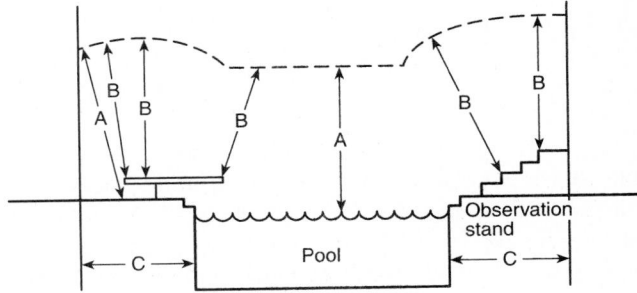

**Figure 680.8 Clearances from pool structures.**

FPN: Open overhead wiring as used in this article typically refers to conductor(s) not in an enclosed raceway.

**(B) Communications Systems.** Communication, radio, and television coaxial cables within the scope of Articles

**Table 680.8 Overhead Conductor Clearances**

| | Clearance Parameters | Insulated Cables, 0–750 Volts to Ground, Supported on and Cabled Together with an Effectively Grounded Bare Messenger or Effectively Grounded Neutral Conductor | | All Other Conductors Voltage to Ground | | | |
|---|---|---|---|---|---|---|---|
| | | | | 0 through 15 kV | | Over 15 through 50 kV | |
| | | m | ft | m | ft | m | ft |
| A. | Clearance in any direction to the water level, edge of water surface, base of diving platform, or permanently anchored raft | 6.9 | 22.5 | 7.5 | 25 | 8.0 | 27 |
| B. | Clearance in any direction to the observation stand, tower, or diving platform | 4.4 | 14.5 | 5.2 | 17 | 5.5 | 18 |
| C. | Horizontal limit of clearance measured from inside wall of the pool | This limit shall extend to the outer edge of the structures listed in A and B of this table but not less than 3 m (10 ft). | | | | | |

800 through 820 shall be permitted at a height of not less than 3.0 m (10 ft) above swimming and wading pools, diving structures, and observation stands, towers, or platforms.

**(C) Network-Powered Broadband Communications Systems.** The minimum clearances for overhead network-powered broadband communications systems conductors from pools or fountains shall comply with the provisions in Table 680.8 for conductors operating at 0 to 750 volts to ground.

**680.9 Electric Pool Water Heaters.** All electric pool water heaters shall have the heating elements subdivided into loads not exceeding 48 amperes and protected at not over 60 amperes. The ampacity of the branch-circuit conductors and the rating or setting of overcurrent protective devices shall not be less than 125 percent of the total nameplate-rated load.

**680.10 Underground Wiring Location.** Underground wiring shall not be permitted under the pool or within the area extending 1.5 m (5 ft) horizontally from the inside wall of the pool unless this wiring is necessary to supply pool equipment permitted by this article. Where space limitations prevent wiring from being routed a distance 1.5 m (5 ft) or more from the pool, such wiring shall be permitted where installed in rigid metal conduit, intermediate metal conduit, or a nonmetallic raceway system. All metal conduit shall be corrosion resistant and suitable for the location. The minimum burial depth shall be as given in Table 680.10.

**Table 680.10 Minimum Burial Depths**

| Wiring Method | Minimum Burial | |
|---|---|---|
| | mm | in. |
| Rigid metal conduit | 150 | 6 |
| Intermediate metal conduit | 150 | 6 |
| Nonmetallic raceways listed for direct burial without concrete encasement | 450 | 18 |
| Other approved raceways* | 450 | 18 |

*Raceways approved for burial only where concrete encased shall require a concrete envelope not less than 50 mm (2 in.) thick.

**680.11 Equipment Rooms and Pits.** Electric equipment shall not be installed in rooms or pits that do not have drainage that adequately prevents water accumulation during normal operation or filter maintenance.

**680.12 Maintenance Disconnecting Means.** One or more means to disconnect all ungrounded conductors shall be provided for all utilization equipment other than lighting. Each means shall be accessible and within sight from its equipment.

## II. Permanently Installed Pools

**680.20 General.** Electrical installations at permanently installed pools shall comply with the provisions of Part I and Part II of this article.

**680.21 Motors.**

**(A) Wiring Methods.**

**(1) General.** The branch circuits for pool-associated motors shall be installed in rigid metal conduit, intermediate metal conduit, rigid nonmetallic conduit, or Type MC cable listed for the location. Other wiring methods and materials shall be permitted in specific locations or applications as covered in this section. Any wiring method employed shall contain a copper equipment grounding conductor sized in accordance with 250.122 but not smaller than 12 AWG.

**(2) On or Within Buildings.** Where installed on or within buildings, electrical metallic tubing shall be permitted.

**(3) Flexible Connections.** Where necessary to employ flexible connections at or adjacent to the motor, liquidtight flexible metal or nonmetallic conduit with approved fittings shall be permitted.

**(4) One-Family Dwellings.** In the interior of one-family dwellings, or in the interior of accessory buildings associated with a one-family dwelling, any of the wiring methods recognized in Chapter 3 of this *Code* shall be permitted that comply with the provisions of this paragraph. Where run in a raceway, the equipment grounding conductor shall be insulated. Where run in a cable assembly, the equipment grounding conductor shall be permitted to be uninsulated, but it shall be enclosed within the outer sheath of the cable assembly.

**(5) Cord-and-Plug Connections.** Pool-associated motors shall be permitted to employ cord-and-plug connections. The flexible cord shall not exceed 900 mm (3 ft) in length. The flexible cord shall include an equipment grounding conductor sized in accordance with 250.122 and shall terminate in a grounding-type attachment plug.

**(B) Double Insulated Pool Pumps.** A listed cord-and-plug-connected pool pump incorporating an approved system of double insulation that provides a means for grounding only the internal and nonaccessible, non–current-carrying metal parts of the pump shall be connected to any wiring method recognized in Chapter 3 that is suitable for the location.

## 680.22 Area Lighting, Receptacles, and Equipment.

### (A) Receptacles.

**(1) Circulation and Sanitation System, Location.** Receptacles that provide power for water-pump motors or for other loads directly related to the circulation and sanitation system shall be located at least 3.0 m (10 ft) from the inside walls of the pool, or not less than 1.5 m (5 ft) from the inside walls of the pool if they meet all of the following conditions:

(1) Consist of single receptacles
(2) Employ a locking configuration
(3) Are of the grounding type
(4) Have GFCI protection

**(2) Other Receptacles, Location.** Other receptacles shall be not less than 3.0 m (10 ft) from the inside walls of a pool.

**(3) Dwelling Unit(s).** Where a permanently installed pool is installed at a dwelling unit(s), no fewer than one 125-volt 15- or 20-ampere receptacle on a general-purpose branch circuit shall be located not less than 3.0 m (10 ft) from and not more than 6.0 m (20 ft) from the inside wall of the pool. This receptacle shall be located not more than 2.0 m (6 ft 6 in.) above the floor, platform, or grade level serving the pool.

**(4) Restricted Space.** Where a pool is within 3.0 m (10 ft) of a dwelling and the dimensions of the lot preclude meeting the required clearances, not more than one receptacle outlet shall be permitted if not less than 1.5 m (5 ft) measured horizontally from the inside wall of the pool.

**(5) GFCI Protection.** All 125-volt receptacles located within 6.0 m (20 ft) of the inside walls of a pool or fountain shall be protected by a ground-fault circuit interrupter. Receptacles that supply pool pump motors and that are rated 15 or 20 amperes, 120 volt through 240 volts, single phase, shall be provided with GFCI protection.

**(6) Measurements.** In determining the dimensions in this section addressing receptacle spacings, the distance to be measured shall be the shortest path the supply cord of an appliance connected to the receptacle would follow without piercing a floor, wall, ceiling, doorway with hinged or sliding door, window opening, or other effective permanent barrier.

### (B) Luminaires (Lighting Fixtures), Lighting Outlets, and Ceiling-Suspended (Paddle) Fans.

**(1) New Outdoor Installation Clearances.** In outdoor pool areas, luminaires (lighting fixtures), lighting outlets, and ceiling-suspended (paddle) fans installed above the pool or the area extending 1.5 m (5 ft) horizontally from the inside walls of the pool shall be installed at a height not less than 3.7 m (12 ft) above the maximum water level of the pool.

**(2) Indoor Clearances.** For installations in indoor pool areas, the clearances shall be the same as for outdoor areas unless modified as provided in this paragraph. If the branch circuit supplying the equipment is protected by a ground-fault circuit interrupter, the following equipment shall be permitted at a height not less than 2.3 m (7 ft 6 in.) above the maximum pool water level:

(1) Totally enclosed luminaires (fixtures)
(2) Ceiling-suspended (paddle) fans identified for use beneath ceiling structures such as provided on porches or patios

**(3) Existing Installations.** Existing luminaires (lighting fixtures) and lighting outlets located less than 1.5 m (5 ft) measured horizontally from the inside walls of a pool shall be not less than 1.5 m (5 ft) above the surface of the maximum water level, shall be rigidly attached to the existing structure, and shall be protected by a ground-fault circuit interrupter.

**(4) GFCI Protection in Adjacent Areas.** Luminaires (lighting fixtures), lighting outlets, and ceiling-suspended (paddle) fans installed in the area extending between 1.5 m (5 ft) and 3.0 m (10 ft) horizontally from the inside walls of a pool shall be protected by a ground-fault circuit interrupter unless installed not less than 1.5 m (5 ft) above the maximum water level and rigidly attached to the structure adjacent to or enclosing the pool.

**(5) Cord-and-Plug-Connected Luminaires (Lighting Fixtures).** Cord-and-plug-connected luminaires (lighting fixtures) shall comply with the requirements of 680.7 where installed within 4.9 m (16 ft) of any point on the water surface, measured radially.

### (C) Switching Devices.
Switching devices shall be located at least 1.5 m (5 ft) horizontally from the inside walls of a pool unless separated from the pool by a solid fence, wall, or other permanent barrier. Alternatively, a switch that is listed as being acceptable for use within 1.5 m (5 ft) shall be permitted.

**680.23 Underwater Luminaires (Lighting Fixtures).** This section covers all luminaires (lighting fixtures) installed below the normal water level of the pool.

### (A) General.

**(1) Luminaire (Fixture) Design, Normal Operation.** The design of an underwater luminaire (lighting fixture) supplied from a branch circuit either directly or by way of a transformer meeting the requirements of this section shall be such that, where the luminaire (fixture) is properly installed without a ground-fault circuit interrupter, there is no shock hazard with any likely combination of fault conditions during normal use (not relamping).

**(2) Transformers.** Transformers used for the supply of underwater luminaires (fixtures), together with the transformer enclosure, shall be listed for the purpose. The transformer shall be an isolated winding type with an ungrounded secondary that has a grounded metal barrier between the primary and secondary windings.

**(3) GFCI Protection, Relamping.** A ground-fault circuit interrupter shall be installed in the branch circuit supplying luminaires (fixtures) operating at more than 15 volts, so that there is no shock hazard during relamping. The installation of the ground-fault circuit interrupter shall be such that there is no shock hazard with any likely fault-condition combination that involves a person in a conductive path from any ungrounded part of the branch circuit or the luminaire (fixture) to ground.

**(4) Voltage Limitation.** No luminaires (lighting fixtures) shall be installed for operation on supply circuits over 150 volts between conductors.

**(5) Location, Wall-Mounted Luminaires (Fixtures).** Luminaires (lighting fixtures) mounted in walls shall be installed with the top of the luminaire (fixture) lens not less than 450 mm (18 in.) below the normal water level of the pool, unless the luminaire (lighting fixture) is listed and identified for use at lesser depths. No luminaire (fixture) shall be installed less than 100 mm (4 in.) below the normal water level of the pool.

**(6) Bottom-Mounted Luminaires (Fixtures).** A luminaire (lighting fixture) facing upward shall have the lens adequately guarded to prevent contact by any person.

**(7) Dependence on Submersion.** Luminaires (fixtures) that depend on submersion for safe operation shall be inherently protected against the hazards of overheating when not submerged.

**(8) Compliance.** Compliance with these requirements shall be obtained by the use of a listed underwater luminaire (lighting fixture) and by installation of a listed ground-fault circuit interrupter in the branch circuit or a listed transformer for luminaires (fixtures) operating at not more than 15 volts.

**(B) Wet-Niche Luminaires (Fixtures).**

**(1) Forming Shells.** Forming shells shall be installed for the mounting of all wet-niche underwater luminaires (fixtures) and shall be equipped with provisions for conduit entries. Metal parts of the luminaire (fixture) and forming shell in contact with the pool water shall be of brass or other approved corrosion-resistant metal. All forming shells used with nonmetallic conduit systems, other than those that are part of a listed low-voltage lighting system not requiring grounding, shall include provisions for terminating an 8 AWG copper conductor.

**(2) Wiring Extending Directly to the Forming Shell.** Conduit shall be installed from the forming shell to a suitable junction box or other enclosure located as provided in 680.24. Conduit shall be rigid metal, intermediate metal, liquidtight flexible nonmetallic, or rigid nonmetallic.

(a) Metal Conduit. Metal conduit shall be approved and shall be of brass or other approved corrosion-resistant metal.

(b) Nonmetallic Conduit. Where a nonmetallic conduit is used, an 8 AWG insulated solid or stranded copper equipment grounding conductor shall be installed in this conduit unless a listed low-voltage lighting system not requiring grounding is used. The equipment grounding conductor shall be terminated in the forming shell, junction box or transformer enclosure, or ground-fault circuit-interrupter enclosure. The termination of the 8 AWG equipment grounding conductor in the forming shell shall be covered with, or encapsulated in, a listed potting compound to protect the connection from the possible deteriorating effect of pool water.

**(3) Equipment Grounding Provisions for Cords.** Wet-niche luminaires (lighting fixtures) that are supplied by a flexible cord or cable shall have all exposed non–current-carrying metal parts grounded by an insulated copper equipment grounding conductor that is an integral part of the cord or cable. This grounding conductor shall be connected to a grounding terminal in the supply junction box, transformer enclosure, or other enclosure. The grounding conductor shall not be smaller than the supply conductors and not smaller than 16 AWG.

**(4) Luminaire (Fixture) Grounding Terminations.** The end of the flexible-cord jacket and the flexible-cord conductor terminations within a luminaire (fixture) shall be covered with, or encapsulated in, a suitable potting compound to prevent the entry of water into the luminaire (fixture) through the cord or its conductors. In addition, the grounding connection within a luminaire (fixture) shall be similarly treated to protect such connection from the deteriorating effect of pool water in the event of water entry into the luminaire (fixture).

**(5) Luminaire (Fixture) Bonding.** The luminaire (fixture) shall be bonded to and secured to the forming shell by a positive locking device that ensures a low-resistance contact and requires a tool to remove the luminaire (fixture) from the forming shell. Bonding shall not be required for luminaires (fixtures) that are listed for the application and have no non–current-carrying metal parts.

**(C) Dry-Niche Luminaires (Fixtures).**

**(1) Construction.** A dry-niche luminaire (lighting fixture) shall be provided with a provision for drainage of water and a means for accommodating one equipment grounding conductor for each conduit entry.

**(2) Junction Box.** A junction box shall not be required but, if used, shall not be required to be elevated or located as specified in 680.24(A)(2) if the luminaire (fixture) is specifically identified for the purpose.

**(D) No-Niche Luminaires (Fixtures).** A no-niche luminaire (fixture) shall meet the construction requirements of 680.23(B)(3) and be installed in accordance with the requirements of 680.23(B). Where connection to a forming shell is specified, the connection shall be to the mounting bracket.

**(E) Through-Wall Lighting Assembly.** A through-wall lighting assembly shall be equipped with a threaded entry or hub, or a nonmetallic hub listed for the purpose, for the purpose of accommodating the termination of the supply conduit. A through-wall lighting assembly shall meet the construction requirements of 680.23(B)(3) and be installed in accordance with the requirements of 680.23. Where connection to a forming shell is specified, the connection shall be to the conduit termination point.

**(F) Branch-Circuit Wiring.**

**(1) Wiring Methods.** Branch-circuit wiring on the supply side of enclosures and junction boxes connected to conduits run to wet-niche and no-niche luminaires (fixtures), and the field wiring compartments of dry-niche luminaires (fixtures), shall be installed using rigid metal conduit, intermediate metal conduit, liquidtight flexible nonmetallic conduit, or rigid nonmetallic conduit. Where installed on buildings, electrical metallic tubing shall be permitted, and where installed within buildings, electrical nonmetallic tubing or electrical metallic tubing shall be permitted.

*Exception: Where connecting to transformers for pool lights, liquidtight flexible metal conduit or liquidtight flexible nonmetallic conduit shall be permitted. The length shall not exceed 1.8 m (6 ft) for any one length or exceed 3.0 m (10 ft) in total length used. Liquidtight flexible nonmetallic conduit, Type B (LFNC-B), shall be permitted in lengths longer than 1.8 m (6 ft).*

**(2) Equipment Grounding.** Through-wall lighting assemblies, wet-niche, dry-niche, or no-niche luminaires (lighting fixtures) shall be connected to an insulated copper equipment grounding conductor installed with the circuit con-

ductors. The equipment grounding conductor shall be installed without joint or splice except as permitted in (a) and (b). The equipment grounding conductor shall be sized in accordance with Table 250.122 but shall not be smaller than 12 AWG.

*Exception: An equipment grounding conductor between the wiring chamber of the secondary winding of a transformer and a junction box shall be sized in accordance with the overcurrent device in this circuit.*

(a) If more than one underwater luminaire (lighting fixture) is supplied by the same branch circuit, the equipment grounding conductor, installed between the junction boxes, transformer enclosures, or other enclosures in the supply circuit to wet-niche luminaires (fixtures), or between the field-wiring compartments of dry-niche luminaires (fixtures), shall be permitted to be terminated on grounding terminals.

(b) If the underwater luminaire (lighting fixture) is supplied from a transformer, ground-fault circuit interrupter, clock-operated switch, or a manual snap switch that is located between the panelboard and a junction box connected to the conduit that extends directly to the underwater luminaire (lighting fixture), the equipment grounding conductor shall be permitted to terminate on grounding terminals on the transformer, ground-fault circuit interrupter, clock-operated switch enclosure, or an outlet box used to enclose a snap switch.

**(3) Conductors.** Conductors on the load side of a ground-fault circuit interrupter or of a transformer, used to comply with the provisions of 680.23(A)(8), shall not occupy raceways, boxes, or enclosures containing other conductors unless one of the following conditions applies:

(1) The other conductors are protected by ground-fault circuit interrupters.
(2) The other conductors are grounding conductors.
(3) The other conductors are supply conductors to a feed-through type ground-fault circuit interrupter.
(4) Ground-fault circuit interrupters shall be permitted in a panelboard that contains circuits protected by other than ground-fault circuit interrupters.

**680.24 Junction Boxes and Enclosures for Transformers or Ground-Fault Circuit Interrupters.**

**(A) Junction Boxes.** A junction box connected to a conduit that extends directly to a forming shell or mounting bracket of a no-niche luminaire (fixture) shall meet the requirements of this section.

**(1) Construction.** The junction box shall be listed and labeled for the purpose and shall comply with the following conditions:

(1) Be equipped with threaded entries or hubs or a nonmetallic hub listed for the purpose

(2) Be comprised of copper, brass, suitable plastic, or other approved corrosion-resistant material

(3) Be provided with electrical continuity between every connected metal conduit and the grounding terminals by means of copper, brass, or other approved corrosion-resistant metal that is integral with the box

**(2) Installation.** Where the luminaire (fixture) operates over 15 volts, the junction box location shall comply with (a) and (b). Where the luminaire (fixture) operates at less than 15 volts, the junction box location shall be permitted to comply with (c).

(a) Vertical Spacing. The junction box shall be located not less than 100 mm (4 in.), measured from the inside of the bottom of the box, above the ground level, or pool deck, or not less than 200 mm (8 in.) above the maximum pool water level, whichever provides the greater elevation.

(b) Horizontal Spacing. The junction box shall be located not less than 1.2 m (4 ft) from the inside wall of the pool, unless separated from the pool by a solid fence, wall, or other permanent barrier.

(c) Flush Deck Box. If used on a lighting system operating at 15 volts or less, a flush deck box shall be permitted if both of the following apply:

(1) An approved potting compound is used to fill the box to prevent the entrance of moisture.

(2) The flush deck box is located not less than 1.2 m (4 ft) from the inside wall of the pool.

**(B) Other Enclosures.** An enclosure for a transformer, ground-fault circuit interrupter, or a similar device connected to a conduit that extends directly to a forming shell or mounting bracket of a no-niche luminaire (fixture) shall meet the requirements of this section.

**(1) Construction.** The enclosure shall be listed and labeled for the purpose and meet the following requirements:

(1) Equipped with threaded entries or hubs or a nonmetallic hub listed for the purpose

(2) Comprised of copper, brass, suitable plastic, or other approved corrosion-resistant material

(3) Provided with an approved seal, such as duct seal at the conduit connection, that prevents circulation of air between the conduit and the enclosures

(4) Provided with electrical continuity between every connected metal conduit and the grounding terminals by means of copper, brass, or other approved corrosion-resistant metal that is integral with the box

**(2) Installation.**

(a) Vertical Spacing. The enclosure shall be located not less than 100 mm (4 in.), measured from the inside of

the bottom of the box, above the ground level, or pool deck, or not less than 200 mm (8 in.) above the maximum pool water level, whichever provides the greater elevation.

(b) Horizontal Spacing. The enclosure shall be located not less than 1.2 m (4 ft) from the inside wall of the pool, unless separated from the pool by a solid fence, wall, or other permanent barrier.

**(C) Protection.** Junction boxes and enclosures mounted above the grade of the finished walkway around the pool shall not be located in the walkway unless afforded additional protection, such as by location under diving boards, adjacent to fixed structures, and the like.

**(D) Grounding Terminals.** Junction boxes, transformer enclosures, and ground-fault circuit-interrupter enclosures connected to a conduit that extends directly to a forming shell or mounting bracket of a no-niche luminaire (fixture) shall be provided with a number of grounding terminals that shall be no fewer than one more than the number of conduit entries.

**(E) Strain Relief.** The termination of a flexible cord of an underwater luminaire (lighting fixture) within a junction box, transformer enclosure, ground-fault circuit interrupter, or other enclosure shall be provided with a strain relief.

**(F) Grounding.** The junction box, transformer enclosure, or other enclosure in the supply circuit to a wet-niche or no-niche luminaire (lighting fixture) and the field-wiring chamber of a dry-niche luminaire (lighting fixture) shall be grounded to the equipment grounding terminal of the panelboard. This terminal shall be directly connected to the panelboard enclosure.

**680.25 Feeders.** These provisions shall apply to any feeder on the supply side of panelboards supplying branch circuits for pool equipment covered in Part II of this article and on the load side of the service equipment or the source of a separately derived system.

**(A) Wiring Methods.** Feeders shall be installed in rigid metal conduit, intermediate metal conduit, liquidtight flexible nonmetallic conduit, or rigid nonmetallic conduit. Electrical metallic tubing shall be permitted where installed on or within a building, and electrical nonmetallic tubing shall be permitted where installed within a building.

*Exception: An existing feeder between an existing remote panelboard and service equipment shall be permitted to run in flexible metal conduit or an approved cable assembly that includes an equipment grounding conductor within its outer sheath. The equipment grounding conductor shall comply with 250.24(A)(5).*

**(B) Grounding.** An equipment grounding conductor shall be installed with the feeder conductors between the grounding terminal of the pool equipment panelboard and the grounding terminal of the applicable service equipment or source of a separately derived system. For other than (1) existing feeders covered in 680.25(A), Exception or (2) feeders to separate buildings that do not utilize an insulated equipment grounding conductor in accordance with 680.25(B)(2), this equipment grounding conductor shall be insulated.

**(1) Size.** This conductor shall be sized in accordance with 250.122 but not smaller than 12 AWG. On separately derived systems, this conductor shall be sized in accordance with Table 250.66 but not smaller than 8 AWG.

**(2) Separate Buildings.** A feeder to a separate building shall be permitted to supply swimming pool equipment branch circuits, or feeders supplying swimming pool equipment branch circuits, if the grounding arrangements in the separate building meet the requirements in 250.32. Where installed, a separate equipment grounding conductor shall be an insulated conductor.

### 680.26 Bonding.

**(A) Performance.** The bonding required by this section shall be installed to eliminate voltage gradients in the pool area as prescribed.

> FPN: This section does not require that the 8 AWG or larger solid copper bonding conductor be extended or attached to any remote panelboard, service equipment, or any electrode.

**(B) Bonded Parts.** The parts specified in 680.26(B)(1) through (B)(5) shall be bonded together.

**(1) Metallic Structural Components.** All metallic parts of the pool structure, including the reinforcing metal of the pool shell, coping stones, and deck, shall be bonded. The usual steel tie wires shall be considered suitable for bonding the reinforcing steel together, and welding or special clamping shall not be required. These tie wires shall be made tight. If reinforcing steel is effectively insulated by an encapsulating nonconductive compound at the time of manufacture and installation, it shall not be required to be bonded. Where reinforcing steel is encapsulated with a nonconductive compound, provisions shall be made for an alternate means to eliminate voltage gradients that would otherwise be provided by unencapsulated, bonded reinforcing steel.

**(2) Underwater Lighting.** All forming shells and mounting brackets of no-niche luminaires (fixtures) shall be bonded unless a listed low-voltage lighting system with nonmetallic forming shells not requiring bonding is used.

**(3) Metal Fittings.** All metal fittings within or attached to the pool structure shall be bonded. Isolated parts that are

not over 100 mm (4 in.) in any dimension and do not penetrate into the pool structure more than 25 mm (1 in.) shall not require bonding.

**(4) Electrical Equipment.** Metal parts of electrical equipment associated with the pool water circulating system, including pump motors and metal parts of equipment associated with pool covers, including electric motors, shall be bonded. Metal parts of listed equipment incorporating an approved system of double insulation and providing a means for grounding internal nonaccessible, non–current-carrying metal parts shall not be bonded.

Where a double-insulated water-pump motor is installed under the provisions of this rule, a solid 8 AWG copper conductor that is of sufficient length to make a bonding connection to a replacement motor shall be extended from the bonding grid to an accessible point in the motor vicinity. Where there is no connection between the swimming pool bonding grid and the equipment grounding system for the premises, this bonding conductor shall be connected to the equipment grounding conductor of the motor circuit.

**(5) Metal Wiring Methods and Equipment.** Metal-sheathed cables and raceways, metal piping, and all fixed metal parts except those separated from the pool by a permanent barrier shall be bonded that are within the following distances of the pool:

(1) Within 1.5 m (5 ft) horizontally of the inside walls of the pool
(2) Within 3.7 m (12 ft) measured vertically above the maximum water level of the pool, or any observation stands, towers, or platforms, or any diving structures

**(C) Common Bonding Grid.** The parts specified in 680.26(B) shall be connected to a common bonding grid with a solid copper conductor, insulated, covered, or bare, not smaller than 8 AWG. Connection shall be made by exothermic welding or by pressure connectors or clamps that are labeled as being suitable for the purpose and are of stainless steel, brass, copper, or copper alloy. The common bonding grid shall be permitted to be any of the following:

(1) The structural reinforcing steel of a concrete pool where the reinforcing rods are bonded together by the usual steel tie wires or the equivalent
(2) The wall of a bolted or welded metal pool
(3) A solid copper conductor, insulated, covered, or bare, not smaller than 8 AWG
(4) Rigid metal conduit or intermediate metal conduit of brass or other identified corrosion-resistant metal conduit

**(D) Connections.** Where structural reinforcing steel or the walls of bolted or welded metal pool structures are used as a common bonding grid for nonelectrical parts, the connections shall be made in accordance with 250.8.

**(E) Pool Water Heaters.** For pool water heaters rated at more than 50 amperes that have specific instructions regarding bonding and grounding, only those parts designated to be bonded shall be bonded, and only those parts designated to be grounded shall be grounded.

**680.27 Specialized Pool Equipment.**

**(A) Underwater Audio Equipment.** All underwater audio equipment shall be identified for the purpose.

**(1) Speakers.** Each speaker shall be mounted in an approved metal forming shell, the front of which is enclosed by a captive metal screen, or equivalent, that is bonded to and secured to the forming shell by a positive locking device that ensures a low-resistance contact and requires a tool to open for installation or servicing of the speaker. The forming shell shall be installed in a recess in the wall or floor of the pool.

**(2) Wiring Methods.** Rigid metal conduit or intermediate metal conduit of brass or other identified corrosion-resistant metal, liquidtight flexible nonmetallic conduit (LFNC-B), or rigid nonmetallic conduit shall extend from the forming shell to a listed junction box or other enclosure as provided in 680.24. Where rigid nonmetallic conduit or liquidtight flexible nonmetallic conduit is used, an 8 AWG insulated solid or stranded copper equipment grounding conductor shall be installed in this conduit. The equipment grounding conductor shall be terminated in the forming shell and the junction box. The termination of the 8 AWG equipment grounding conductor in the forming shell shall be covered with, or encapsulated in, a listed potting compound to protect such connection from the possible deteriorating effect of pool water.

**(3) Forming Shell and Metal Screen.** The forming shell and metal screen shall be of brass or other approved corrosion-resistant metal. All forming shells shall include provisions for terminating an 8 AWG copper conductor.

**(B) Electrically Operated Pool Covers.**

**(1) Motors and Controllers.** The electric motors, controllers, and wiring shall be located not less than 1.5 m (5 ft) from the inside wall of the pool unless separated from the pool by a wall, cover, or other permanent barrier. Electric motors installed below grade level shall be of the totally enclosed type. The device that controls the operation of the motor for an electrically operated pool cover shall be located so that the operator has full view of the pool.

FPN No. 1:  For cabinets installed in damp and wet locations, see 312.2(A).

FPN No. 2:  For switches or circuit breakers installed in wet locations, see 404.4.

FPN No. 3:  For protection against liquids, see 430.11.

**(2) Protection.** The electric motor and controller shall be connected to a circuit protected by a ground-fault circuit interrupter.

**(C) Deck Area Heating.** These provisions of this section shall apply to all pool deck areas, including a covered pool, where electrically operated comfort heating units are installed within 6.0 m (20 ft) of the inside wall of the pool.

**(1) Unit Heaters.** Unit heaters shall be rigidly mounted to the structure and shall be of the totally enclosed or guarded types. Unit heaters shall not be mounted over the pool or within the area extending 1.5 m (5 ft) horizontally from the inside walls of a pool.

**(2) Permanently Wired Radiant Heaters.** Radiant electric heaters shall be suitably guarded and securely fastened to their mounting device(s). Heaters shall not be installed over a pool or within the area extending 1.5 m (5 ft) horizontally from the inside walls of the pool and shall be mounted at least 3.7 m (12 ft) vertically above the pool deck unless otherwise approved.

**(3) Radiant Heating Cables Not Permitted.** Radiant heating cables embedded in or below the deck shall not be permitted.

## III. Storable Pools

**680.30 General.** Electrical installations at storable pools shall comply with the provisions of Part I and Part III of this article.

**680.31 Pumps.** A cord-connected pool filter pump shall incorporate an approved system of double insulation or its equivalent and shall be provided with means for grounding only the internal and nonaccessible non–current-carrying metal parts of the appliance.

The means for grounding shall be an equipment grounding conductor run with the power-supply conductors in the flexible cord that is properly terminated in a grounding-type attachment plug having a fixed grounding contact member.

**680.32 Ground-Fault Circuit Interrupters Required.** All electrical equipment, including power-supply cords, used with storable pools shall be protected by ground-fault circuit interrupters.

FPN:  For flexible cord usage, see 400.4.

**680.33 Luminaires (Lighting Fixtures).** An underwater luminaire (lighting fixture), if installed, shall be installed in or on the wall of the storable pool. It shall comply with one of the following two provisions.

**(A) 15 Volts or Less.** A luminaire (lighting fixture) shall be part of a cord-and-plug-connected lighting assembly. This assembly shall be listed as an assembly for the purpose and have the following construction features:

(1) No exposed metal parts
(2) A luminaire (fixture) lamp that operates at 15 volts or less
(3) An impact-resistant polymeric lens, luminaire (fixture) body, and transformer enclosure
(4) A transformer meeting the requirements of 680.23(A)(2) with a primary rating not over 150 volts

**(B) Over 15 Volts But Not Over 150 Volts.** A lighting assembly without a transformer and with the luminaire (fixture) lamp(s) operating at not over 150 volts shall be permitted to be cord-and-plug connected where the assembly is listed as an assembly for the purpose. The installation shall comply with 680.23(A)(5), and the assembly shall have the following construction features:

(1) No exposed metal parts
(2) An impact-resistant polymeric lens and luminaire (fixture) body
(3) A ground-fault circuit interrupter with open neutral protection as an integral part of the assembly
(4) The luminaire (fixture) lamp permanently connected to the ground-fault circuit interrupter with open-neutral protection
(5) Compliance with the requirements of 680.23(A)

## IV. Spas and Hot Tubs

**680.40 General.** Electrical installations at spas and hot tubs shall comply with the provisions of Part I and Part IV of this article.

**680.41 Emergency Switch for Spas and Hot Tubs.** A clearly labeled emergency shutoff or control switch for the purpose of stopping the motor(s) that provide power to the recirculation system and jet system shall be installed at a point readily accessible to the users and not less than 1.5 m (5 ft) away, adjacent to, and within sight of the spa or hot tub. This requirement shall not apply to single-family dwellings.

**680.42 Outdoor Installations.** A spa or hot tub installed outdoors shall comply with the provisions of Parts I and II of this article, except as permitted in 680.42(A) and (B), that would otherwise apply to pools installed outdoors.

**(A) Flexible Connections.** Listed packaged spa or hot tub equipment assemblies or self-contained spas or hot tubs utilizing a factory-installed or assembled control panel or panelboard shall be permitted to use flexible connections as covered in 680.42(A)(1) and (A)(2).

**(1) Flexible Conduit.** Liquidtight flexible metal conduit or liquidtight flexible nonmetallic conduit shall be permitted in lengths of not more than 1.8 m (6 ft).

**(2) Cord-and-Plug Connections.** Cord-and-plug connections with a cord not longer than 4.6 m (15 ft) shall be permitted where protected by a ground-fault circuit interrupter.

**(B) Bonding.** Bonding by metal-to-metal mounting on a common frame or base shall be permitted. The metal bands or hoops used to secure wooden staves shall not be required to be bonded as required in 680.26.

**(C) Interior Wiring to Outdoor Installations.** In the interior of a one-family dwelling or in the interior of another building or structure associated with a one-family dwelling, any of the wiring methods recognized in Chapter 3 of this *Code* that contain a copper equipment grounding conductor that is insulated or enclosed within the outer sheath of the wiring method and not smaller than 12 AWG shall be permitted to be used for the connection to motor, heating, and control loads that are part of a self-contained spa or hot tub, or a packaged spa or hot tub equipment assembly. Wiring to an underwater light shall comply with 680.23 or 680.33.

**680.43 Indoor Installations.** A spa or hot tub installed indoors shall comply with the provisions of Parts I and II of this article except as modified by this section, and shall be connected by the wiring methods of Chapter 3.

*Exception: Listed spa and hot tub packaged units rated 20 amperes or less shall be permitted to be cord-and-plug connected to facilitate the removal or disconnection of the unit for maintenance and repair.*

**(A) Receptacles.** At least one 125-volt, 15- or 20-ampere receptacle on a general-purpose branch circuit shall be located not less than of 1.5 m (5 ft) from and not exceeding 3.0 m (10 ft) from the inside wall of the spa or hot tub.

**(1) Location.** Receptacles shall be located at least 1.5 m (5 ft) measured horizontally from the inside walls of the spa or hot tub.

**(2) Protection, General.** Receptacles rated 125 volts and 30 amperes or less and located within 3.0 m (10 ft) of the inside walls of a spa or hot tub shall be protected by a ground-fault circuit interrupter.

**(3) Protection, Spa or Hot Tub Supply Receptacle.** Receptacles that provide power for a spa or hot tub shall be ground-fault circuit-interrupter protected.

**(4) Measurements.** In determining the dimensions in this section addressing receptacle spacings, the distance to be measured shall be the shortest path the supply cord of an appliance connected to the receptacle would follow without piercing a floor, wall, ceiling, doorway with hinged or sliding door, window opening, or other effective permanent barrier.

**(B) Installation of Luminaires (Lighting Fixtures), Lighting Outlets, and Ceiling-Suspended (Paddle) Fans.**

**(1) Elevation.** Luminaires (lighting fixtures), except as covered in 680.43(B)(2), lighting outlets, and ceiling-suspended (paddle) fans located over the spa or hot tub or within 1.5 m (5 ft) from the inside walls of the spa or hot tub shall comply with the clearances specified in (a), (b), and (c) above the maximum water level.

(a) Without GFCI. Where no GFCI protection is provided, the mounting height shall be not less than 3.7 m (12 ft).

(b) With GFCI. Where GFCI protection is provided, the mounting height shall be permitted to be not less than 2.3 m (7 ft 6 in.).

(c) Below 2.3 m (7 ft 6 in.). Luminaires (lighting fixtures) meeting the requirements of item (1) or (2) and protected by a ground-fault circuit interrupter shall be permitted to be installed less than 2.3 m (7 ft 6 in.) over a spa or hot tub.

(1) Recessed luminaires (fixtures) with a glass or plastic lens, nonmetallic or electrically isolated metal trim, and suitable for use in damp locations

(2) Surface-mounted luminaires (fixtures) with a glass or plastic globe, a nonmetallic body, or a metallic body isolated from contact, and suitable for use in damp locations

**(2) Underwater Applications.** Underwater luminaires (lighting fixtures) shall comply with the provisions of 680.23 or 680.33.

**(C) Wall Switches.** Switches shall be located at least 1.5 m (5 ft), measured horizontally, from the inside walls of the spa or hot tub.

**(D) Bonding.** The following parts shall be bonded together:

(1) All metal fittings within or attached to the spa or hot tub structure

(2) Metal parts of electrical equipment associated with the spa or hot tub water circulating system, including pump motors

(3) Metal conduit and metal piping that are within 1.5 m (5 ft) of the inside walls of the spa or hot tub and that are not separated from the spa or hot tub by a permanent barrier

(4) All metal surfaces that are within 1.5 m (5 ft) of the inside walls of the spa or hot tub and that are not separated from the spa or hot tub area by a permanent barrier

*Exception: Small conductive surfaces not likely to become energized, such as air and water jets and drain fittings, where not connected to metallic piping, towel bars, mirror frames, and similar nonelectrical equipment, shall not be required to be bonded.*

(5) Electrical devices and controls that are not associated with the spas or hot tubs and that are located not less than 1.5 m (5 ft) from such units; otherwise they shall be bonded to the spa or hot tub system

**(E) Methods of Bonding.** All metal parts associated with the spa or hot tub shall be bonded by any of the following methods:

(1) The interconnection of threaded metal piping and fittings

(2) Metal-to-metal mounting on a common frame or base

(3) The provisions of a copper bonding jumper, insulated, covered, or bare, not smaller than 8 AWG solid

**(F) Grounding.** The following equipment shall be grounded:

(1) All electric equipment located within 1.5 m (5 ft) of the inside wall of the spa or hot tub

(2) All electric equipment associated with the circulating system of the spa or hot tub

**(G) Underwater Audio Equipment.** Underwater audio equipment shall comply with the provisions of Part II of this article.

**680.44 Protection.** Except as otherwise provided in this section, the outlet(s) that supplies a self-contained spa or hot tub, a packaged spa or hot tub equipment assembly, or a field-assembled spa or hot tub shall be protected by a ground-fault circuit interrupter.

**(A) Listed Units.** If so marked, a listed self-contained unit or listed packaged equipment assembly that includes integral ground-fault circuit-interrupter protection for all electrical parts within the unit or assembly (pumps, air blowers, heaters, lights, controls, sanitizer generators, wiring, and so forth) shall be permitted without additional GFCI protection.

**(B) Other Units.** A field assembled spa or hot tub rated 3 phase or rated over 250 volts or with a heater load of more than 50 amperes shall not require the supply to be protected by a ground-fault circuit interrupter.

**(C) Combination Pool and Spa or Hot Tub.** A combination pool/hot tub or spa assembly commonly bonded need not be protected by a ground-fault circuit interrupter.

> FPN: See 680.2 for definitions of *self-contained spa or hot tub* and for *packaged spa or hot tub equipment assembly.*

## V. Fountains

**680.50 General.** The provisions of Part I and Part V of this article shall apply to all permanently installed fountains as defined in 680.2. Fountains that have water common to a pool shall additionally comply with the requirements in Part II of this article. Part V does not cover self-contained, portable fountains not larger than 1.5 m (5 ft) in any dimension. Portable fountains shall comply with Parts II and III of Article 422.

**680.51 Luminaires (Lighting Fixtures), Submersible Pumps, and Other Submersible Equipment.**

**(A) Ground-Fault Circuit Interrupter.** Fountain equipment, unless listed for operation at 15 volts or less and supplied by a transformer that complies with 680.23(A)(2), shall be protected by a ground-fault circuit interrupter.

**(B) Operating Voltage.** No luminaires (lighting fixtures) shall be installed for operation on supply circuits over 150 volts between conductors. Submersible pumps and other submersible equipment shall operate at 300 volts or less between conductors.

**(C) Luminaire (Lighting Fixture) Lenses.** Luminaires (lighting fixtures) shall be installed with the top of the luminaire (fixture) lens below the normal water level of the fountain unless listed for above-water locations. A luminaire (lighting fixture) facing upward shall have the lens adequately guarded to prevent contact by any person.

**(D) Overheating Protection.** Electrical equipment that depends on submersion for safe operation shall be protected against overheating by a low-water cutoff or other approved means when not submerged.

**(E) Wiring.** Equipment shall be equipped with provisions for threaded conduit entries or be provided with a suitable flexible cord. The maximum length of exposed cord in the fountain shall be limited to 3.0 m (10 ft). Cords extending beyond the fountain perimeter shall be enclosed in approved wiring enclosures. Metal parts of equipment in contact with water shall be of brass or other approved corrosion-resistant metal.

**(F) Servicing.** All equipment shall be removable from the water for relamping or normal maintenance. Luminaires (fixtures) shall not be permanently embedded into the fountain structure such that the water level must be reduced or the fountain drained for relamping, maintenance, or inspection.

**(G) Stability.** Equipment shall be inherently stable or be securely fastened in place.

**680.52 Junction Boxes and Other Enclosures.**

**(A) General.** Junction boxes and other enclosures used for other than underwater installation shall comply with 680.24.

**(B) Underwater Junction Boxes and Other Underwater Enclosures.** Junction boxes and other underwater enclosures shall meet the requirements of 680.52(B)(1) and (B)(2).

**(1) Construction.**

(a) Underwater enclosures shall be equipped with provisions for threaded conduit entries or compression glands or seals for cord entry.

(b) Underwater enclosures shall be submersible, and made of copper, brass, or other approved corrosion-resistant material.

**(2) Installation.** Underwater enclosure installations shall comply with (a) and (b).

(a) Underwater enclosures shall be filled with an approved potting compound to prevent the entry of moisture.

(b) Underwater enclosures shall be firmly attached to the supports or directly to the fountain surface and bonded as required. Where the junction box is supported only by the conduit, the conduit shall be of copper, brass, or other approved corrosion-resistant metal. Where the box is fed by nonmetallic conduit, it shall have additional supports and fasteners of copper, brass, or other approved corrosion-resistant material.

> FPN: See 314.23 for support of enclosures.

**680.53 Bonding.** All metal piping systems associated with the fountain shall be bonded to the equipment grounding conductor of the branch circuit supplying the fountain.

> FPN: See 250.122 for sizing of these conductors.

**680.54 Grounding.** The following equipment shall be grounded:

(1) All electrical equipment located within the fountain or within 1.5 m (5 ft) of the inside wall of the fountain

(2) All electrical equipment associated with the recirculating system of the fountain

(3) Panelboards that are not part of the service equipment and that supply any electrical equipment associated with the fountain

## 680.55 Methods of Grounding.

**(A) Applied Provisions.** The provisions of 680.21(A), 680.23(B)(3), 680.23(F)(1) and (2), 680.24(F), and 680.25 shall apply.

**(B) Supplied by a Flexible Cord.** Electrical equipment that is supplied by a flexible cord shall have all exposed non–current-carrying metal parts grounded by an insulated copper equipment grounding conductor that is an integral part of this cord. The grounding conductor shall be connected to a grounding terminal in the supply junction box, transformer enclosure, or other enclosure.

## 680.56 Cord-and-Plug-Connected Equipment.

**(A) Ground-Fault Circuit Interrupter.** All electrical equipment, including power-supply cords, shall be protected by ground-fault circuit interrupters.

**(B) Cord Type.** Flexible cord immersed in or exposed to water shall be of a type for extra-hard usage, as designated in Table 400.4 and shall be a listed type with a "W" suffix.

**(C) Sealing.** The end of the flexible cord jacket and the flexible cord conductor termination within equipment shall be covered with, or encapsulated in, a suitable potting compound to prevent the entry of water into the equipment through the cord or its conductors. In addition, the ground connection within equipment shall be similarly treated to protect such connections from the deteriorating effect of water that may enter into the equipment.

**(D) Terminations.** Connections with flexible cord shall be permanent, except that grounding-type attachment plugs and receptacles shall be permitted to facilitate removal or disconnection for maintenance, repair, or storage of fixed or stationary equipment not located in any water-containing part of a fountain.

## 680.57 Signs.

**(A) General.** This section covers electric signs installed within or adjacent to fountains.

**(B) Ground-Fault Circuit-Interrupter Protection for Personnel.** All circuits supplying the sign shall have ground-fault circuit-interrupter protection for personnel.

**(C) Location.**

**(1) Fixed or Stationary.** A fixed or stationary electric sign installed within a fountain shall be not less than 1.5 m (5 ft) inside the fountain measured from the outside edges of the fountain.

**(2) Portable.** A portable electric sign shall not be placed within a pool or fountain or within 1.5 m (5 ft) measured horizontally from the inside walls of the fountain.

**(D) Disconnect.** A sign shall have a local disconnecting means in accordance with 600.6 and 680.12.

**(E) Bonding and Grounding.** A sign shall be grounded and bonded in accordance with 600.7.

## VI. Pools and Tubs for Therapeutic Use

**680.60 General.** The provisions of Part I and Part VI of this article shall apply to pools and tubs for therapeutic use in health care facilities, gymnasiums, athletic training rooms, and similar areas. Portable therapeutic appliances shall comply with Parts II and III of Article 422.

    FPN: See 517.2 for definition of health care facilities.

**680.61 Permanently Installed Therapeutic Pools.** Therapeutic pools that are constructed in the ground, on the ground, or in a building in such a manner that the pool cannot be readily disassembled shall comply with Parts I and II of this article.

*Exception: The limitations of 680.22(B)(1) through (B)(4) shall not apply where all luminaires (lighting fixtures) are of the totally enclosed type.*

**680.62 Therapeutic Tubs (Hydrotherapeutic Tanks).** Therapeutic tubs, used for the submersion and treatment of patients, that are not easily moved from one place to another in normal use or that are fastened or otherwise secured at a specific location, including associated piping systems, shall conform to this part.

**(A) Protection.** Except as otherwise provided in this section, the outlet(s) that supplies a self-contained therapeutic tub or hydrotherapeutic tank, a packaged therapeutic tub or hydrotherapeutic tank, or a field-assembled therapeutic tub or hydrotherapeutic tank shall be protected by a ground-fault circuit interrupter.

**(1) Listed Units.** If so marked, a listed self-contained unit or listed packaged equipment assembly that includes integral ground-fault circuit-interrupter protection for all electrical parts within the unit or assembly (pumps, air blowers, heaters, lights, controls, sanitizer generators, wiring, and so forth) shall be permitted without additional GFCI protection.

**(2) Other Units.** A therapeutic tub or hydrotherapeutic tank rated 3 phase or rated over 250 volts or with a heater load of more than 50 amperes shall not require the supply to be protected by a ground-fault circuit interrupter.

**(B) Bonding.** The following parts shall be bonded together:

(1) All metal fittings within or attached to the tub structure
(2) Metal parts of electrical equipment associated with the tub water circulating system, including pump motors
(3) Metal-sheathed cables and raceways and metal piping that are within 1.5 m (5 ft) of the inside walls of the tub and not separated from the tub by a permanent barrier
(4) All metal surfaces that are within 1.5 m (5 ft) of the inside walls of the tub and not separated from the tub area by a permanent barrier
(5) Electrical devices and controls that are not associated with the therapeutic tubs and that are not located a minimum of 1.5 m (5 ft) from such units

**(C) Methods of Bonding.** All metal parts required to be bonded by this section shall be bonded by any of the following methods:

(1) The interconnection of threaded metal piping and fittings
(2) Metal-to-metal mounting on a common frame or base
(3) Connections by suitable metal clamps
(4) By the provisions of a solid copper bonding jumper, insulated, covered, or bare, not smaller than 8 AWG

**(D) Grounding.**

**(1) Fixed or Stationary Equipment.** The equipment specified in (a) and (b) shall be grounded.

(a) Location. All electrical equipment located within 1.5 m (5 ft) of the inside wall of the tub shall be grounded.
(b) Circulation System. All electrical equipment associated with the circulating system of the tub shall be grounded.

**(2) Portable Equipment.** Portable therapeutic appliances shall meet the grounding requirements in 250.114.

**(E) Receptacles.** All receptacles within 1.5 m (5 ft) of a therapeutic tub shall be protected by a ground-fault circuit interrupter.

**(F) Luminaires (Lighting Fixtures).** All luminaires (lighting fixtures) used in therapeutic tub areas shall be of the totally enclosed type.

## VII. Hydromassage Bathtubs

**680.70 General.** Hydromassage bathtubs as defined in 680.2 shall comply with Part VII of this article. They shall not be required to comply with other parts of this article.

**680.71 Protection.** Hydromassage bathtubs and their associated electrical components shall be protected by a ground-fault circuit interrupter. All 125-volt, single-phase receptacles not exceeding 30 amperes and located within 1.5 m (5 ft) measured horizontally of the inside walls of a hydromassage tub shall be protected by a ground-fault circuit interrupter(s).

**680.72 Other Electrical Equipment.** Luminaires (lighting fixtures), switches, receptacles, and other electrical equipment located in the same room, and not directly associated with a hydromassage bathtub, shall be installed in accordance with the requirements of Chapters 1 through 4 in this *Code* covering the installation of that equipment in bathrooms.

**680.73 Accessibility.** Hydromassage bathtub electrical equipment shall be accessible without damaging the building structure or building finish.

**680.74 Bonding.** All metal piping systems, metal parts of electrical equipment, and pump motors associated with the hydromassage tub shall be bonded together using a copper bonding jumper, insulated, covered, or bare, not smaller than 8 AWG solid. Metal parts of listed equipment incorporating an approved system of double insulation and providing a means for grounding internal nonaccessible, non–current-carrying metal parts shall not be bonded.

## ARTICLE 685
## Integrated Electrical Systems

### I. General

**685.1 Scope.** This article covers integrated electrical systems, other than unit equipment, in which orderly shutdown is necessary to ensure safe operation. An *integrated electrical system* as used in this article is a unitized segment of an industrial wiring system where all of the following conditions are met:

(1) An orderly shutdown is required to minimize personnel hazard and equipment damage.
(2) The conditions of maintenance and supervision ensure that qualified persons service the system.
(3) Effective safeguards, acceptable to the authority having jurisdiction, are established and maintained.

**685.2 Application of Other Articles.** The articles/sections in Table 685.2 apply to particular cases of installation of conductors and equipment, where there are orderly shutdown requirements that are in addition to those of this article or are modifications of them.

**Table 685.2 Application of Other Articles**

| Conductor/Equipment | Section |
|---|---|
| More than one building or other structure | 225, Part II |
| Ground-fault protection of equipment | 230.95, Exception No. 1 |
| Protection of conductors | 240.4 |
| Electrical system coordination | 240.12 |
| Ground-fault protection of equipment | 240.13(1) |
| Grounding ac systems of 50 volts to 1000 volts | 250.21 |
| Equipment protection | 427.22 |
| Orderly shutdown | 430.44 |
| Disconnection | 430.74, Exception Nos. 1 and 2 |
| Disconnecting means in sight from controller | 430.102(A), Exception No. 2 |
| Energy from more than one source | 430.113, Exception Nos. 1 and 2 |
| Disconnecting means | 645.10, Exception |
| Uninterruptible power supplies (UPS) | 645.11(1) |
| Point of connection | 705.12(A) |

## II. Orderly Shutdown

**685.10 Location of Overcurrent Devices in or on Premises.** Location of overcurrent devices that are critical to integrated electrical systems shall be permitted to be accessible, with mounting heights permitted to ensure security from operation by nonqualified personnel.

**685.12 Direct-Current System Grounding.** Two-wire dc circuits shall be permitted to be ungrounded.

**685.14 Ungrounded Control Circuits.** Where operational continuity is required, control circuits of 150 volts or less from separately derived systems shall be permitted to be ungrounded.

# ARTICLE 690
# Solar Photovoltaic Systems

## I. General

**690.1 Scope.** The provisions of this article apply to solar photovoltaic electrical energy systems, including the array circuit(s), inverter(s), and controller(s) for such systems.

[See Figures 690.1(A) and (B).] Solar photovoltaic systems covered by this article may be interactive with other electrical power production sources or stand-alone, with or without electrical energy storage such as batteries. These systems may have ac or dc output for utilization.

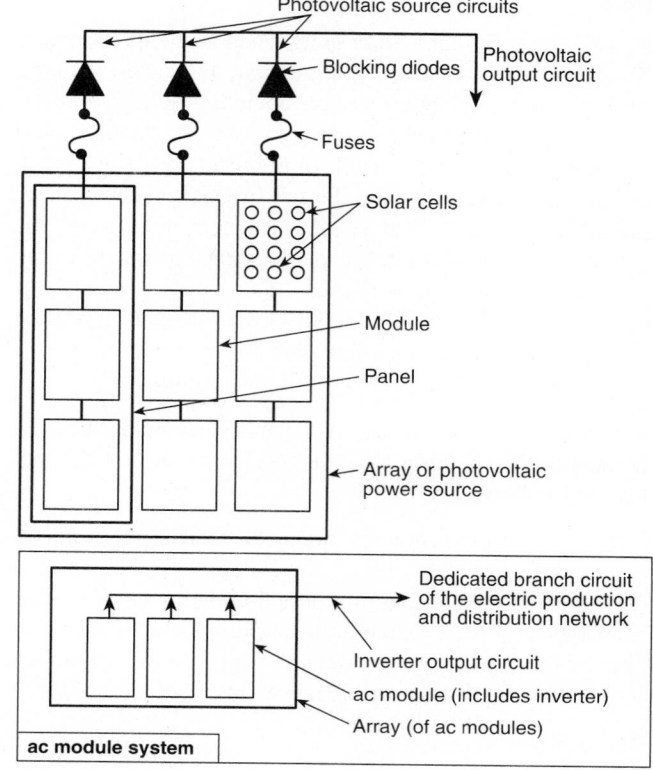

**Notes:**
1. These diagrams are intended to be a means of identification for photovoltaic system components, circuits, and connections.
2. Disconnecting means required by Article 690, Part III are not shown.
3. System grounding and equipment grounding are not shown. See Article 690, Part V.

**Figure 690.1(A) Identification of solar photovoltaic system components.**

## 690.2 Definitions.

**Alternating-Current Module (Alternating-Current Photovoltaic Module).** A complete, environmentally protected unit consisting of solar cells, optics, inverter, and other components, exclusive of tracker, designed to generate ac power when exposed to sunlight.

**Array.** A mechanically integrated assembly of modules or panels with a support structure and foundation, tracker, and other components, as required, to form a direct-current power-producing unit.

**Bipolar Photovoltaic Array.** A photovoltaic array that has two outputs, each having opposite polarity to a common reference point or center tap.

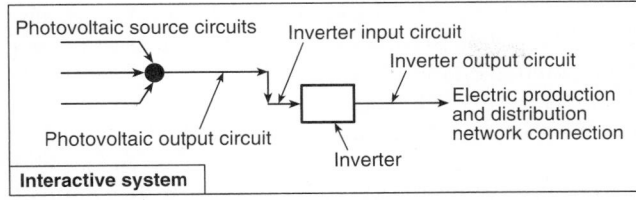

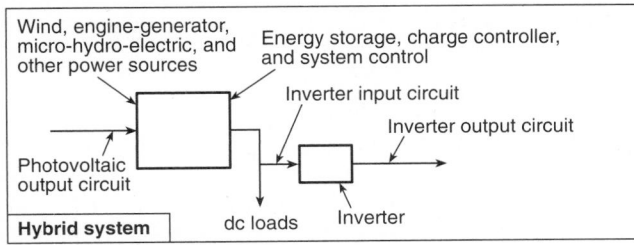

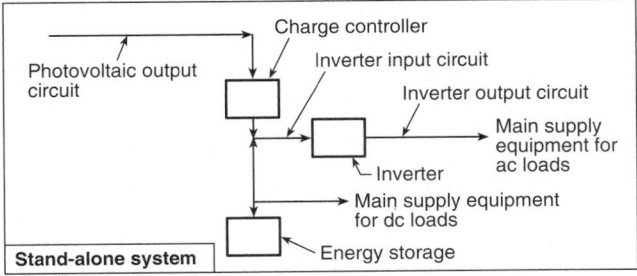

Notes:
1. These diagrams are intended to be a means of identification for photovoltaic system components, circuits, and connections.
2. Disconnecting means and overcurrent protection required by Article 690 are not shown.
3. System grounding and equipment grounding are not shown. See Article 690, Part V.
4. Custom designs occur in each configuration and some components are optional.

**Figure 690.1(B) Identification of solar photovoltaic system components in common system configurations.**

**Blocking Diode.** A diode used to block reverse flow of current into a photovoltaic source circuit.

**Charge Controller.** Equipment that controls dc voltage or dc current, or both, used to charge a battery.

**Diversion Charge Controller.** Equipment that regulates the charging process of a battery by diverting power from energy storage to direct-current or alternating-current loads or to an interconnected utility service.

**Electrical Production and Distribution Network.** A power production, distribution, and utilization system, such as a utility system and connected loads, that is external to and not controlled by the photovoltaic power system.

**Hybrid System.** A system comprised of multiple power sources. These power sources may include photovoltaic, wind, micro-hydro generators, engine-driven generators, and others, but do not include electrical production and distribution network systems. Energy storage systems, such

as batteries, do not constitute a power source for the purpose of this definition.

**Interactive System.** A solar photovoltaic system that operates in parallel with and may deliver power to an electrical production and distribution network. For the purpose of this definition, an energy storage subsystem of a solar photovoltaic system, such as a battery, is not another electrical production source.

**Inverter.** Equipment that is used to change voltage level or waveform, or both, of electrical energy. Commonly, an inverter [also known as a power conditioning unit (PCU) or power conversion system (PCS)] is a device that changes dc input to an ac output. Inverters may also function as battery chargers that use alternating current from another source and convert it into direct current for charging batteries.

**Inverter Input Circuit.** Conductors between the inverter and the battery in stand-alone systems or the conductors between the inverter and the photovoltaic output circuits for electrical production and distribution network.

**Inverter Output Circuit.** Conductors between the inverter and an ac load center for stand-alone systems or the conductors between the inverter and the service equipment or another electric power production source, such as a utility, for electrical production and distribution network.

**Module.** A complete, environmentally protected unit consisting of solar cells, optics, and other components, exclusive of tracker, designed to generate dc power when exposed to sunlight.

**Panel.** A collection of modules mechanically fastened together, wired, and designed to provide a field-installable unit.

**Photovoltaic Output Circuit.** Circuit conductors between the photovoltaic source circuit(s) and the inverter or dc utilization equipment.

**Photovoltaic Power Source.** An array or aggregate of arrays that generates dc power at system voltage and current.

**Photovoltaic Source Circuit.** Circuits between modules and from modules to the common connection point(s) of the dc system.

**Photovoltaic Systems Voltage.** The direct current (dc) voltage of any photovoltaic source or photovoltaic output circuit. For bipolar or multiwire installations, the photovoltaic systems voltage is the highest voltage between any two dc conductors.

**Solar Cell.** The basic photovoltaic device that generates electricity when exposed to light.

**Solar Photovoltaic System.** The total components and subsystems that, in combination, convert solar energy into electrical energy suitable for connection to a utilization load.

**Stand-Alone System.** A solar photovoltaic system that supplies power independently of an electrical production and distribution network.

**690.3 Other Articles.** Wherever the requirements of other articles of this *Code* and Article 690 differ, the requirements of Article 690 shall apply and, if the system is operated in parallel with a primary source(s) of electricity, the requirements in 705.14, 705.16, 705.32, and 705.43 shall apply.

**690.4 Installation.**

**(A) Solar Photovoltaic System.** A solar photovoltaic system shall be permitted to supply a building or other structure in addition to any service(s) of another electricity supply system(s).

**(B) Conductors of Different Systems.** Photovoltaic source circuits and photovoltaic output circuits shall not be contained in the same raceway, cable tray, cable, outlet box, junction box, or similar fitting as feeders or branch circuits of other systems, unless the conductors of the different systems are separated by a partition or are connected together.

**(C) Module Connection Arrangement.** The connections to a module or panel shall be arranged so that removal of a module or panel from a photovoltaic source circuit does not interrupt a grounded conductor to another photovoltaic source circuit. Sets of modules interconnected as systems rated at 50 volts or less, with or without blocking diodes, and having a single overcurrent device shall be considered as a single-source circuit. Supplementary overcurrent devices used for the exclusive protection of the photovoltaic modules are not considered as overcurrent devices for the purpose of this section.

**(D) Equipment.** Inverters or motor generators shall be identified for use in solar photovoltaic systems.

**690.5 Ground-Fault Protection.** Roof-mounted dc photovoltaic arrays located on dwellings shall be provided with dc ground-fault protection to reduce fire hazards.

**(A) Ground-Fault Detection and Interruption.** The ground-fault protection device or system shall be capable of detecting a ground fault, interrupting the flow of fault current, and providing an indication of the fault.

**(B) Disconnection of Conductors.** The ungrounded conductors of the faulted source circuit shall be automatically disconnected. If the grounded conductors of the faulted source circuit are disconnected to comply with the requirements of 690.5(A), all conductors of the faulted source circuit shall be opened automatically and simultaneously.

Opening the grounded conductor of the array or opening the faulted sections of the array shall be permitted to interrupt the ground-fault current path.

**(C) Labels and Markings.** Labels and markings shall be applied near the ground-fault indicator at a visible location, stating that, if a ground fault is indicated, the normally grounded conductors may be energized and ungrounded.

**690.6 Alternating-Current Modules.**

**(A) Photovoltaic Source Circuits.** The requirements of Article 690 pertaining to photovoltaic source circuits shall not apply to ac modules. The photovoltaic source circuit, conductors, and inverters shall be considered as internal wiring of an ac module.

**(B) Inverter Output Circuit.** The output of an ac module shall be considered an inverter output circuit.

**(C) Disconnecting Means.** A single disconnecting means, in accordance with 690.15 and 690.17, shall be permitted for the combined ac output of one or more ac modules. Additionally, each ac module in a multiple ac-module system shall be provided with a connector, bolted, or terminal-type disconnecting means.

**(D) Ground-Fault Detection.** Alternating-current-module systems shall be permitted to use a single detection device to detect only ac ground faults and to disable the array by removing ac power to the ac module(s).

**(E) Overcurrent Protection.** The output circuits of ac modules shall be permitted to have overcurrent protection and conductor sizing in accordance with 240.5(B)(2).

**II. Circuit Requirements**

**690.7 Maximum Voltage.**

**(A) Maximum Photovoltaic System Voltage.** In a dc photovoltaic source circuit or output circuit, the maximum photovoltaic system voltage for that circuit shall be computed as the sum of the rated open-circuit voltage of the series-connected photovoltaic modules corrected for the lowest expected ambient temperature. For crystalline and multicrystalline silicon modules, the rated open-circuit voltage shall be multiplied by the correction factor provided in Table 690.7. This voltage shall be used to determine the voltage rating of cables, disconnects, overcurrent devices, and other equipment. Where the lowest expected ambient temperature is below −40°C (−40°F), or where other than crystalline or multicrystalline silicon photovoltaic modules are used, the system voltage adjustment shall be made in accordance with the manufacturer's instructions.

**Table 690.7 Voltage Correction Factors for Crystalline and Multicrystalline Silicon Modules**

| Ambient Temperature (°C) | Correction Factors for Ambient Temperatures Below 25°C (77°F) (Multiply the rated open-circuit voltage by the appropriate correction factor shown below.) | Ambient Temperature (°F) |
| --- | --- | --- |
| 25 to 10 | 1.06 | 77 to 50 |
| 9 to 0 | 1.10 | 49 to 32 |
| −1 to −10 | 1.13 | 31 to 14 |
| −11 to −20 | 1.17 | 13 to −4 |
| −21 to −40 | 1.25 | −5 to −40 |

**(B) Direct-Current Utilization Circuits.** The voltage of dc utilization circuits shall conform with 210.6.

**(C) Photovoltaic Source and Output Circuits.** In one- and two-family dwellings, photovoltaic source circuits and photovoltaic output circuits that do not include lampholders, fixtures, or receptacles shall be permitted to have a maximum system voltage up to 600 volts. Other installations with a maximum system voltage over 600 volts shall comply with Article 690, Part I.

**(D) Circuits Over 150 Volts to Ground.** In one- and two-family dwellings, live parts in photovoltaic source circuits and photovoltaic output circuits over 150 volts to ground shall not be accessible to other than qualified persons while energized.

FPN: See 110.27 for guarding of live parts, and 210.6 for voltage to ground and between conductors.

**(E) Bipolar Source and Output Circuits.** For 2-wire circuits connected to bipolar systems, the maximum system voltage shall be the highest voltage between the conductors of the 2-wire circuit if all of the following conditions apply:

(1) One conductor of each circuit is solidly grounded.
(2) Each circuit is connected to a separate subarray.
(3) The equipment is clearly marked with a label as follows: Warning—Bipolar Photovoltaic Array. Disconnection of neutral or grounded conductors may result in overvoltage on array or inverter.

**690.8 Circuit Sizing and Current.**

**(A) Computation of Maximum Circuit Current.** The maximum current for the specific circuit shall be computed in accordance with 690.8(A)(1) through (A)(4).

**(1) Photovoltaic Source Circuit Currents.** The maximum current shall be the sum of parallel module rated short-circuit currents multiplied by 125 percent.

**(2) Photovoltaic Output Circuit Currents.** The maximum current shall be the sum of parallel source circuit maximum currents as calculated in 690.8(A)(1).

**(3) Inverter Output Circuit Current.** The maximum current shall be the inverter continuous output current rating.

**(4) Stand-Alone Inverter Input Circuit Current.** The maximum current shall be the stand-alone continuous inverter input current rating when the inverter is producing rated power at the lowest input voltage.

**(B) Ampacity and Overcurrent Device Ratings.** Photovoltaic system currents shall be considered continuous.

**(1) Sizing of Conductors and Overcurrent Devices.** The circuit conductors and overcurrent devices shall be sized to carry not less than 125 percent of the maximum currents as computed in 690.8(A). The rating or setting of overcurrent devices shall be permitted in accordance with 240.4(B) and (C).

*Exception: Circuits containing an assembly, together with its overcurrent device(s), that is listed for continuous operation at 100 percent of its rating shall be permitted to be utilized at 100 percent of its rating.*

**(2) Internal Current Limitation.** Overcurrent protection for photovoltaic output circuits with devices that internally limit the current from the photovoltaic output circuit shall be permitted to be rated at less than the value computed in 690.8(B)(1). This reduced rating shall be at least 125 percent of the limited current value. Photovoltaic output circuit conductors shall be sized in accordance with 690.8(B)(1).

*Exception: An overcurrent device in an assembly listed for continuous operation at 100 percent of its rating shall be permitted to be utilized at 100 percent of its rating.*

**(C) Systems with Multiple Direct-Current Voltages.** For a photovoltaic power source that has multiple output circuit voltages and employs a common-return conductor, the ampacity of the common-return conductor shall not be less than the sum of the ampere ratings of the overcurrent devices of the individual output circuits.

**(D) Sizing of Module Interconnection Conductors.** Where a single overcurrent device is used to protect a set of two or more parallel-connected module circuits, the ampacity of each of the module interconnection conductors shall not be less than the sum of the rating of the single fuse plus 125 percent of the short-circuit current from the other parallel-connected modules.

**690.9 Overcurrent Protection.**

**(A) Circuits and Equipment.** Photovoltaic source circuit, photovoltaic output circuit, inverter output circuit, and storage battery circuit conductors and equipment shall be protected in accordance with the requirements of Article 240. Circuits connected to more than one electrical source shall have overcurrent devices located so as to provide overcurrent protection from all sources.

*Exception: An overcurrent device shall not be required for circuit conductors sized in accordance with 690.8(B) and located where one of the following apply:*

(a) *There are no external sources such as parallel-connected source circuits, batteries, or backfeed from inverters.*

(b) *The short-circuit currents from all sources do not exceed the ampacity of the conductors.*

> FPN: Possible backfeed of current from any source of supply, including a supply through an inverter into the photovoltaic output circuit and photovoltaic source circuits, is a consideration in determining whether adequate overcurrent protection from all sources is provided for conductors and modules.

**(B) Power Transformers.** Overcurrent protection for a transformer with a source(s) on each side shall be provided in accordance with 450.3 by considering first one side of the transformer, then the other side of the transformer, as the primary.

*Exception: A power transformer with a current rating on the side connected toward the photovoltaic power source not less than the short-circuit output current rating of the inverter shall be permitted without overcurrent protection from that source.*

**(C) Photovoltaic Source Circuits.** Branch-circuit or supplementary-type overcurrent devices shall be permitted to provide overcurrent protection in photovoltaic source circuits. The overcurrent devices shall be accessible but shall not be required to be readily accessible.

Standard values of supplementary overcurrent devices allowed by this section shall be in one ampere size increments starting at one ampere up to and including 15 amperes. Higher standard values above 15 amperes for supplementary overcurrent devices shall be based on the standard sizes provided in 240.6(A).

**(D) Direct-Current Rating.** Overcurrent devices, either fuses or circuit breakers, used in any dc portion of a photovoltaic power system shall be listed for use in dc circuits and shall have the appropriate voltage, current, and interrupt ratings.

**(E) Series Overcurrent Protection.** In series-connected strings of two or more modules, a single overcurrent protection device shall be permitted.

**690.10 Stand-Alone Systems.** The premises wiring system shall be adequate to meet the requirements of this *Code*

for a similar installation connected to a service. The wiring on the supply side of the building or structure disconnecting means shall comply with this *Code* except as modified by 690.10(A), (B), and (C).

**(A) Inverter Output.** The ac inverter output from a stand-alone system shall be permitted to supply ac power to the building or structure disconnecting means at current levels below the rating of that disconnecting means.

**(B) Sizing and Protection.** The circuit conductors between the inverter output and the building or structure disconnecting means shall be sized based on the output rating of the inverter. These conductors shall be protected from overcurrents in accordance with Article 240. The overcurrent protection shall be located at the output of the inverter.

**(C) Single 120-Volt Supply.** The inverter output of a stand-alone solar photovoltaic system shall be permitted to supply 120 volts to single-phase, 3-wire, 120/240-volt service equipment or distribution panels where there are no 240-volt outlets and where there are no multiwire branch circuits. In all installations, the rating of the overcurrent device connected to the output of the inverter shall be less than the rating of the neutral bus in the service equipment. This equipment shall be marked with the following words or equivalent:

WARNING
SINGLE 120-VOLT SUPPLY, DO NOT CONNECT
MULTIWIRE BRANCH CIRCUITS!

### III. Disconnecting Means

**690.13 All Conductors.** Means shall be provided to disconnect all current-carrying conductors of a photovoltaic power source from all other conductors in a building or other structure. Where a circuit grounding connection is not designed to be automatically interrupted as part of the ground-fault protection system required by 690.5, a switch or circuit breaker used as a disconnecting means shall not have a pole in the grounded conductor.

> FPN: The grounded conductor may have a bolted or terminal disconnecting means to allow maintenance or troubleshooting by qualified personnel.

**690.14 Additional Provisions.** Photovoltaic disconnecting means shall comply with 690.14(A) through (C).

**(A) Disconnecting Means.** The disconnecting means shall not be required to be suitable as service equipment and shall be rated in accordance with 690.17.

**(B) Equipment.** Equipment such as photovoltaic source circuit isolating switches, overcurrent devices, and blocking diodes shall be permitted on the photovoltaic side of the photovoltaic disconnecting means.

**(C) Requirements for Disconnecting Means.** Means shall be provided to disconnect all conductors in a building or other structure from the photovoltaic system conductors.

**(1) Location.** The photovoltaic disconnecting means shall be installed at a readily accessible location either outside of a building or structure or inside nearest the point of entrance of the system conductors.

The photovoltaic system disconnecting means shall not be installed in bathrooms.

**(2) Marking.** Each photovoltaic system disconnecting means shall be permanently marked to identify it as a photovoltaic system disconnect.

**(3) Suitable for Use.** Each photovoltaic system disconnecting means shall be suitable for the prevailing conditions. Equipment installed in hazardous (classified) locations shall comply with the requirements of Articles 500 through 517.

**(4) Maximum Number of Disconnects.** The photovoltaic system disconnecting means shall consist of not more than six switches or six circuit breakers mounted in a single enclosure, in a group of separate enclosures, or in or on a switchboard.

**(5) Grouping.** The photovoltaic system disconnecting means shall be grouped with other disconnecting means for the system to comply with 690.14(C)(4). A photovoltaic disconnecting means shall not be required at the photovoltaic module or array location.

**690.15 Disconnection of Photovoltaic Equipment.** Means shall be provided to disconnect equipment, such as inverters, batteries, charge controllers, and the like, from all ungrounded conductors of all sources. If the equipment is energized from more than one source, the disconnecting means shall be grouped and identified.

A single disconnecting means in accordance with 690.17 shall be permitted for the combined ac output of one or more inverters or ac modules in an interactive system.

**690.16 Fuses.** Disconnecting means shall be provided to disconnect a fuse from all sources of supply if the fuse is energized from both directions and is accessible to other than qualified persons. Such a fuse in a photovoltaic source circuit shall be capable of being disconnected independently of fuses in other photovoltaic source circuits.

**690.17 Switch or Circuit Breaker.** The disconnecting means for ungrounded conductors shall consist of a manu-

ally operable switch(es) or circuit breaker(s) complying with all of the following requirements:

(1) Located where readily accessible
(2) Externally operable without exposing the operator to contact with live parts
(3) Plainly indicating whether in the open or closed position
(4) Having an interrupting rating sufficient for the nominal circuit voltage and the current that is available at the line terminals of the equipment

Where all terminals of the disconnecting means may be energized in the open position, a warning sign shall be mounted on or adjacent to the disconnecting means. The sign shall be clearly legible and have the following words or equivalent:

WARNING.
ELECTRIC SHOCK HAZARD.
DO NOT TOUCH TERMINALS.
TERMINALS ON BOTH THE LINE AND
LOAD SIDES MAY BE ENERGIZED
IN THE OPEN POSITION.

*Exception No. 1: A disconnecting means located on the dc side shall be permitted to have an interrupting rating less than the current-carrying rating where the system is designed so that the dc switch cannot be opened under load.*

*Exception No. 2: A connector shall be permitted to be used as an ac or a dc disconnecting means provided that it complies with the requirements of 690.33 and is listed and identified for the use.*

**690.18 Installation and Service of an Array.** Open circuiting, short circuiting, or opaque covering shall be used to disable an array or portions of an array for installation and service.

> FPN: Photovoltaic modules are energized while exposed to light. Installation, replacement, or servicing of array components while a module(s) is irradiated may expose persons to electric shock.

## IV. Wiring Methods

### 690.31 Methods Permitted.

**(A) Wiring Systems.** All raceway and cable wiring methods included in this *Code* and other wiring systems and fittings specifically intended and identified for use on photovoltaic arrays shall be permitted. Where wiring devices with integral enclosures are used, sufficient length of cable shall be provided to facilitate replacement.

**(B) Single Conductor Cable.** Types SE, UF, USE, and USE-2 single-conductor cable shall be permitted in photovoltaic source circuits where installed in the same manner as a Type UF multiconductor cable in accordance with Article 339. Where exposed to sunlight, Type UF cable identified as sunlight-resistant shall be used.

**(C) Flexible Cords and Cables.** Flexible cords and cables, where used to connect the moving parts of tracking PV modules, shall comply with Article 400 and shall be of a type identified as a hard service cord or portable power cable; they shall be suitable for extra-hard usage, listed for outdoor use, water resistant, and sunlight resistant. Allowable ampacities shall be in accordance with 400.5. For ambient temperatures exceeding 30°C (86°F), the ampacities shall be derated by the appropriate factors given in Table 690.31(C).

**(D) Small-Conductor Cables.** Single-conductor cables listed for outdoor use that are sunlight resistant and moisture resistant in sizes 16 AWG and 18 AWG shall be permitted for module interconnections where such cables meet the ampacity requirements of 690.8. Section 310.15 shall be used to determine the cable ampacity and temperature derating factors.

**690.32 Component Interconnections.** Fittings and connectors that are intended to be concealed at the time of on-site assembly, where listed for such use, shall be permitted for on-site interconnection of modules or other array components. Such fittings and connectors shall be equal to the wiring method employed in insulation, temperature rise, and fault-current withstand, and shall be capable of resisting the effects of the environment in which they are used.

**690.33 Connectors.** The connectors permitted by Article 690 shall comply with 690.33(A) through (E).

**(A) Configuration.** The connectors shall be polarized and shall have a configuration that is noninterchangeable with receptacles in other electrical systems on the premises.

**(B) Guarding.** The connectors shall be constructed and installed so as to guard against inadvertent contact with live parts by persons.

**(C) Type.** The connectors shall be of the latching or locking type.

**(D) Grounding Member.** The grounding member shall be the first to make and the last to break contact with the mating connector.

**(E) Interruption of Circuit.** The connectors shall be capable of interrupting the circuit current without hazard to the operator.

**690.34 Access to Boxes.** Junction, pull, and outlet boxes located behind modules or panels shall be installed so that the wiring contained in them can be rendered accessible directly or by displacement of a module(s) or panel(s) secured by removable fasteners and connected by a flexible wiring system.

## V. Grounding

**690.41 System Grounding.** For a photovoltaic power source, one conductor of a two-wire system with a system voltage over 50 volts and the reference (center tap) conductor of a bipolar system shall be solidly grounded or shall use other methods that accomplish equivalent system protection in accordance with 250.4(A) and that utilize equipment listed and identified for the use.

**690.42 Point of System Grounding Connection.** The dc circuit grounding connection shall be made at any single point on the photovoltaic output circuit.

> FPN: Locating the grounding connection point as close as practicable to the photovoltaic source better protects the system from voltage surges due to lightning.

**690.43 Equipment Grounding.** Exposed non–current-carrying metal parts of module frames, equipment, and conductor enclosures shall be grounded in accordance with 250.134 or 250.136(A) regardless of voltage.

**Table 690.31(C) Correction Factors**

| Ambient Temperature (°C) | Temperature Rating of Conductor | | | | Ambient Temperature (°F) |
| | 60°C (140°F) | 75°C (167°F) | 90°C (194°F) | 105°C (221°F) | |
|---|---|---|---|---|---|
| 30 | 1.00 | 1.00 | 1.00 | 1.00 | 86 |
| 31–35 | 0.91 | 0.94 | 0.96 | 0.97 | 87–95 |
| 36–40 | 0.82 | 0.88 | 0.91 | 0.93 | 96–104 |
| 41–45 | 0.71 | 0.82 | 0.87 | 0.89 | 105–113 |
| 46–50 | 0.58 | 0.75 | 0.82 | 0.86 | 114–122 |
| 51–55 | 0.41 | 0.67 | 0.76 | 0.82 | 123–131 |
| 56–60 | — | 0.58 | 0.71 | 0.77 | 132–140 |
| 61–70 | — | 0.33 | 0.58 | 0.68 | 141–158 |
| 71–80 | — | — | 0.41 | 0.58 | 159–176 |

**690.45 Size of Equipment Grounding Conductor.** Where not protected by the ground-fault protection equipment required by 690.5, the equipment-grounding conductor for photovoltaic source and photovoltaic output circuits shall be sized for 125 percent of the photovoltaic-originated short-circuit currents in that circuit. Where protected by the ground-fault protection equipment required by 690.5, the equipment-grounding conductors for photovoltaic source and photovoltaic output circuits shall be sized in accordance with 250.122.

**690.47 Grounding Electrode System.**

**(A) Alternating-Current Systems.** If installing an ac system, a grounding electrode system shall be provided in accordance with 250.50 through 250.60. The grounding electrode conductor shall be installed in accordance with 250.64.

**(B) Direct-Current Systems.** If installing a dc system, a grounding electrode system shall be provided in accordance with 250.166 for grounded systems or 250.169 for ungrounded systems. The grounding electrode conductor shall be installed in accordance with 250.64.

**VI. Marking**

**690.51 Modules.** Modules shall be marked with identification of terminals or leads as to polarity, maximum overcurrent device rating for module protection, and with the following ratings:

(1) Open-circuit voltage
(2) Operating voltage
(3) Maximum permissible system voltage
(4) Operating current
(5) Short-circuit current
(6) Maximum power

**690.52 Alternating-Current Photovoltaic Modules.** Alternating-current modules shall be marked with identification of terminals or leads and with identification of the following ratings:

(1) Nominal operating ac voltage
(2) Nominal operating ac frequency
(3) Maximum ac power
(4) Maximum ac current
(5) Maximum overcurrent device rating for ac module protection

**690.53 Photovoltaic Power Source.** A marking, specifying the photovoltaic power source rated as follows, shall be provided by the installer at the site at an accessible location at the disconnecting means for the photovoltaic power source:

(1) Operating current
(2) Operating voltage
(3) Maximum system voltage
(4) Short-circuit current

FPN: Reflecting systems used for irradiance enhancement may result in increased levels of output current and power.

**690.54 Interactive System Point of Interconnection.** All interactive system(s) points of interconnection with other sources shall be marked at an accessible location at the disconnecting means as a power source with the maximum ac output operating current and the operating ac voltage.

**690.55 Photovoltaic Power Systems Employing Energy Storage.** Photovoltaic power systems employing energy storage shall also be marked with the maximum operating voltage, including any equalization voltage and the polarity of the grounded circuit conductor.

**690.56 Identification of Power Sources.**

**(A) Facilities with Stand-Alone Systems.** Any structure or building with a photovoltaic power system that is not connected to a utility service source and is a stand-alone system shall have a permanent plaque or directory installed on the exterior of the building or structure at a readily visible location acceptable to the authority having jurisdiction. The plaque or directory shall indicate the location of system disconnecting means and that the structure contains a stand-alone electrical power system.

**(B) Facilities with Utility Services and PV Systems.** Buildings or structures with both utility service and a photovoltaic system shall have a permanent plaque or directory providing the location of the service disconnecting means and the photovoltaic system disconnecting means, if not located at the same location.

**VII. Connection to Other Sources**

**690.60 Identified Interactive Equipment.** Only inverters and ac modules listed and identified as interactive shall be permitted in interactive systems.

**690.61 Loss of Interactive System Power.** An inverter or an ac module in an interactive solar photovoltaic system shall automatically de-energize its output to the connected electrical production and distribution network upon loss of voltage in that system and shall remain in that state until the electrical production and distribution network voltage has been restored.

A normally interactive solar photovoltaic system shall be permitted to operate as a stand-alone system to supply loads that have been disconnected from electrical production and distribution network sources.

**690.62 Ampacity of Neutral Conductor.** If a single-phase, 2-wire inverter output is connected to the neutral and one ungrounded conductor (only) of a 3-wire system or of a 3-phase, 4-wire wye-connected system, the maximum load connected between the neutral and any one ungrounded conductor plus the inverter output rating shall not exceed the ampacity of the neutral conductor.

**690.63 Unbalanced Interconnections.**

**(A) Single Phase.** Single-phase inverters for photovoltaic systems and ac modules in interactive solar photovoltaic systems shall not be connected to 3-phase power systems unless the interconnected system is designed so that significant unbalanced voltages cannot result.

**(B) Three Phase.** Three-phase inverters and 3-phase ac modules in interactive systems shall have all phases automatically de-energized upon loss of, or unbalanced, voltage in one or more phases unless the interconnected system is designed so that significant unbalanced voltages will not result.

**690.64 Point of Connection.** The output of a photovoltaic power source shall be connected as specified in 690.64(A) or (B).

**(A) Supply Side.** A photovoltaic power source shall be permitted to be connected to the supply side of the service disconnecting means as permitted in 230.82(5).

**(B) Load Side.** A photovoltaic power source shall be permitted to be connected to the load side of the service disconnecting means of the other source(s) at any distribution equipment on the premises, provided that all of the following conditions are met:

(1) Each source interconnection shall be made at a dedicated circuit breaker or fusible disconnecting means.
(2) The sum of the ampere ratings of overcurrent devices in circuits supplying power to a busbar or conductor shall not exceed the rating of the busbar or conductor.

*Exception: For a dwelling unit, the sum of the ampere ratings of the overcurrent devices shall not exceed 120 percent of the rating of the busbar or conductor.*

(3) The interconnection point shall be on the line side of all ground-fault protection equipment.

*Exception: Connection shall be permitted to be made to the load side of ground-fault protection, provided that there is ground-fault protection for equipment from all ground-fault current sources.*

(4) Equipment containing overcurrent devices in circuits supplying power to a busbar or conductor shall be marked to indicate the presence of all sources.

*Exception: Equipment with power supplied from a single point of connection.*

(5) Equipment such as circuit breakers, if backfed, shall be identified for such operation.

## VIII. Storage Batteries

**690.71 Installation.**

**(A) General.** Storage batteries in a solar photovoltaic system shall be installed in accordance with the provisions of Article 480. The interconnected battery cells shall be considered grounded where the photovoltaic power source is installed in accordance with 690.41.

**(B) Dwellings.**

**(1) Operating Voltage.** Storage batteries for dwellings shall have the cells connected so as to operate at less than 50 volts, nominal.

*Exception: Where live parts are not accessible during routine battery maintenance, a battery system voltage in accordance with 690.7 shall be permitted.*

**(2) Guarding of Live Parts.** Live parts of battery systems for dwellings shall be guarded to prevent accidental contact by persons or objects, regardless of voltage or battery type.

> FPN: Batteries in solar photovoltaic systems are subject to extensive charge–discharge cycles and typically require frequent maintenance, such as checking electrolyte and cleaning connections.

**(C) Current Limiting.** A listed, current-limiting, overcurrent device shall be installed in each circuit adjacent to the batteries where the available short-circuit current from a battery or battery bank exceeds the interrupting or withstand ratings of other equipment in that circuit. The installation of current-limiting fuses shall comply with 690.16.

**(D) Battery Nonconductive Cases and Conductive Racks.** Flooded, vented, lead-acid batteries with more than twenty-four 2-volt cells connected in series (48 volts, nominal) shall not use conductive cases or shall not be installed in conductive cases. Conductive racks used to support the nonconductive cases shall be permitted where no rack material is located within 150 mm (6 in.) of the tops of the nonconductive cases.

This requirement shall not apply to any type of valve-regulated lead-acid (VRLA) battery or any other types of sealed batteries that may require steel cases for proper operation.

**(E) Disconnection of Series Battery Circuits.** Battery circuits subject to field servicing, where more than twenty-four 2-volt cells are connected in series (48 volts, nominal), shall have provisions to disconnect the series-connected strings into segments of 24 cells or less for maintenance by qualified persons. Non–load-break bolted or plug-in disconnects shall be permitted.

**(F) Battery Maintenance Disconnecting Means.** Battery installations, where there are more than twenty-four 2-volt cells connected in series (48 volts, nominal), shall have a disconnecting means, accessible only to qualified persons, that disconnects the grounded circuit conductor(s) in the battery electrical system for maintenance. This disconnecting means shall not disconnect the grounded circuit conductor(s) for the remainder of the photovoltaic electrical system. A non–load-break-rated switch shall be permitted to be used as the disconnecting means.

**(G) Battery Systems of More Than 48 Volts.** On photovoltaic systems where the battery system consists of more than twenty-four 2-volt cells connected in series (more than 48 volts, nominal), the battery system shall be permitted to operate with ungrounded conductors, provided the conditions in 690.71(G)(1) through (G)(4) are met:

(1) The photovoltaic array source and output circuits shall comply with 690.41.
(2) The dc and ac load circuits shall be solidly grounded.
(3) All main ungrounded battery input/output circuit conductors shall be provided with switched disconnects and overcurrent protection.
(4) A ground-fault detector and indicator shall be installed to monitor for ground faults in the battery bank.

**690.72 Charge Control.**

**(A) General.** Equipment shall be provided to control the charging process of the battery. Charge control shall not be required where the design of the photovoltaic source circuit is matched to the voltage rating and charge current requirements of the interconnected battery cells and the maximum charging current multiplied by 1 hour is less than 3 percent of the rated battery capacity expressed in ampere-hours or as recommended by the battery manufacturer.

All adjusting means for control of the charging process shall be accessible only to qualified persons.

> FPN: Certain battery types such as valve-regulated lead acid or nickel cadmium can experience thermal failure when overcharged.

**(B) Diversion Charge Controller.**

**(1) Sole Means of Regulating Charging.** A photovoltaic power system employing a diversion charge controller as the sole means of regulating the charging of a battery shall

be equipped with a second independent means to prevent overcharging of the battery.

**(2) Circuits with Direct Current Diversion Charge Controller and Diversion Load.** Circuits containing a dc diversion charge controller and a dc diversion load shall comply with the following:

(1) The current rating of the diversion load shall be rated at least 150 percent of the current rating of the diversion charge controller.
(2) The conductor ampacity and the rating of the overcurrent device for this circuit shall be at least 150 percent of the maximum current rating of the diversion charge controller.

**(3) Circuits Using Inverters.** The requirements in 690.72(B)(2) shall not apply to ac or dc circuits using inverters to control the battery charging process by feeding power into the utility system. These circuits, used in several modes, shall be sized and protected as required in 690.8.

**690.74 Battery Interconnections.** Flexible cables, as identified in Article 400, in sizes 2/0 AWG and larger shall be permitted within the battery enclosure from battery terminals to a nearby junction box where they shall be connected to an approved wiring method. Flexible battery cables shall also be permitted between batteries and cells within the battery enclosure. Such cables shall be listed for hard service use and identified as moisture resistant.

**IX. Systems Over 600 Volts**

**690.80 General.** Solar photovoltaic systems with a maximum system voltage over 600 volts dc shall comply with Article 490 and other requirements applicable to installations rated over 600 volts.

**690.85 Definitions.** For the purposes of Part IX of this article, the voltages used to determine cable and equipment ratings are as follows.

**Battery Circuits.** In battery circuits, the highest voltage experienced under charging or equalizing conditions.

**Photovoltaic Circuits.** In dc photovoltaic source circuits and photovoltaic output circuits, the maximum system voltage.

## ARTICLE 692
## Fuel Cell Systems

**I. General**

**692.1 Scope.** This article identifies the requirements for the installation of fuel cell power systems, which may be stand-alone or interactive with other electrical power production sources and may be with or without electrical energy storage such as batteries.

## 692.2 Definitions.

**Fuel Cell.** An electrochemical system that consumes fuel to produce an electrical current. The main chemical reaction used in a fuel cell for producing electrical power is not combustion. There may, however, be sources of combustion used within the overall fuel cell system such as reformers/fuel processors.

**Fuel Cell System.** The complete aggregate of equipment used to convert chemical fuel into usable electricity. A fuel cell system typically consists of a reformer, stack, power inverter, and auxiliary equipment.

**Interactive System.** A fuel cell system that operates in parallel with and may deliver power to an electrical production and distribution network. For the purpose of this definition, an energy storage subsystem of a fuel cell system, such as a battery, is not another electrical production source.

**Maximum System Voltage.** The highest fuel cell inverter output voltage between any ungrounded conductors present at accessible output terminals.

**Output Circuit.** The conductors used to connect the fuel cell system to its electrical point of delivery. In the case of sites that have series- or parallel-connected multiple units, the term *output circuit* also refers to the conductors used to electrically interconnect the fuel cell system(s).

**Point of Common Coupling.** The point at which the power production and distribution network and the customer interface occurs in an interactive system. Typically, this is the load side of the power network meter.

**Stand-Alone System.** A fuel cell system that supplies power independently of an electrical production and distribution network.

## 692.3 Other Articles.
Wherever the requirements of other articles of this *Code* and Article 692 differ, the requirements of Article 692 shall apply, and, if the system is operated in parallel with a primary source(s) of electricity, the requirements in 705.14, 705.16, 705.32, and 705.43 shall apply.

## 692.4 Installation.

**(A) Fuel Cell System.** A fuel cell system shall be permitted to supply a building or other structure in addition to any service(s) of another electricity supply system(s).

**(B) Identification.** A permanent plaque or directory, denoting all electrical power sources on or in the premises, shall be installed at each service equipment location.

## 692.6 Listing Requirement.
The fuel cell system shall be evaluated and listed for its intended application prior to installation.

## II. Circuit Requirements

## 692.8 Circuit Sizing and Current.

**(A) Nameplate Rated Circuit Current.** The nameplate(s) rated circuit current shall be the rated current indicated on the fuel cell nameplate(s).

**(B) Conductor Ampacity and Overcurrent Device Ratings.** The ampacity of the feeder circuit conductors from the fuel cell system(s) to the premises wiring system shall not be less than the greater of (1) nameplate(s) rated circuit current or (2) the rating of the fuel cell system(s) overcurrent protective device(s).

**(C) Ampacity of Grounded or Neutral Conductor.** If interactive single-phase, 2-wire fuel cell output(s) is connected to the grounded or neutral conductor and a single ungrounded conductor of a 3-wire system or of a 3-phase, 4-wire wye-connected system, the maximum unbalanced neutral load current plus the fuel cell system(s) output rating shall not exceed the ampacity of the grounded or neutral conductor.

## 692.9 Overcurrent Protection.

**(A) Circuits and Equipment.** If the fuel cell system is provided with overcurrent protection sufficient to protect the circuit conductors that supply the load, additional circuit overcurrent devices shall not be required. Equipment and conductors connected to more than one electrical source shall be protected.

**(B) Accessibility.** Overcurrent devices shall be readily accessible.

## 692.10 Stand-Alone Systems.
The premises wiring system shall meet the requirements of this *Code* except as modified by 692.10(A), (B), and (C).

**(A) Fuel Cell System Output.** The fuel cell system output from a stand-alone system shall be permitted to supply ac power to the building or structure disconnecting means at current levels below the rating of that disconnecting means.

**(B) Sizing and Protection.** The circuit conductors between the fuel cell system(s) output and the building or structure disconnecting means shall be sized based on the output rating of the fuel cell system(s). These conductors shall be protected from overcurrents in accordance with 240.4. The overcurrent protection shall be located at the output of the fuel cell system(s).

**(C) Single 120-Volt Nominal Supply.** The inverter output of a stand-alone fuel cell system shall be permitted to supply 120 volts, nominal, to single-phase, 3-wire 120/240-volt service equipment or distribution panels where there are no 240-volt loads and where there are no multiwire branch circuits. In all installations, the rating of the overcurrent device connected to the output of the fuel cell system(s) shall be less than the rating of the service equipment. This equipment shall be marked as follows:

WARNING
SINGLE 120-VOLT SUPPLY.
DO NOT CONNECT MULTIWIRE
BRANCH CIRCUITS!

### III. Disconnecting Means

**692.13 All Conductors.** Means shall be provided to disconnect all current-carrying conductors of a fuel cell system power source from all other conductors in a building or other structure.

**692.14 Provisions.** The provisions of 225.31 and 225.33 through 225.40 shall apply to the fuel cell source disconnecting means. The disconnecting means shall not be required to be suitable as service equipment and shall be rated in accordance with 692.17.

**692.17 Switch or Circuit Breaker.** The disconnecting means for ungrounded conductors shall consist of readily accessible, manually operable switch(es) or circuit breaker(s).

Where all terminals of the disconnecting means may be energized in the open position, a warning sign shall be mounted on or adjacent to the disconnecting means. The sign shall be clearly legible and shall have the following words or equivalent:

DANGER
ELECTRIC SHOCK HAZARD.
DO NOT TOUCH TERMINALS.
TERMINALS ON BOTH THE LINE AND
LOAD SIDES MAY BE ENERGIZED
IN THE OPEN POSITION.

### IV. Wiring Methods

**692.31 Wiring Systems.** All raceway and cable wiring methods included in Chapter 3 of this *Code* and other wiring systems and fittings specifically intended and identified for use with fuel cell systems shall be permitted. Where wiring devices with integral enclosures are used, sufficient length of cable shall be provided to facilitate replacement.

### V. Grounding

**692.41 System Grounding.** For a fuel cell system output circuit, one conductor of a 2-wire system rated over 50 volts and a neutral conductor of a 3-wire system shall be solidly grounded by either 692.41(A) or (B).

**(A) Stand-Alone Systems.** Grounding and bonding shall be in accordance with 250.30.

**(B) Other Than Stand-Alone Systems.**

**(1) Two-Wire Systems.** One conductor shall be terminated at the grounded circuit conductor terminal of the premises wiring system.

**(2) Three-Wire Systems.** The neutral conductor shall be terminated at the grounded circuit conductor terminal of the premises wiring system.

**692.44 Equipment Grounding Conductor.** A separate equipment grounding conductor shall be installed.

**692.45 Size of Equipment Grounding Conductor.** The equipment grounding conductor shall be sized in accordance with 250.122.

**692.47 Grounding Electrode System.** Any supplementary grounding electrode(s) required by the manufacturer shall be connected to the equipment grounding conductor specified in 250.118.

### VI. Marking

**692.53 Fuel Cell Power Sources.** A marking specifying the fuel cell system, output voltage, output power rating, and continuous output current rating shall be provided at the disconnecting means for the fuel cell power source at an accessible location on the site.

**692.54 Fuel Shut-Off.** The location of the manual fuel shut-off valve shall be at the location of the primary disconnecting means of the building or circuits supplied.

**692.56 Stored Energy.** A fuel cell system that stores electrical energy shall require the following warning sign, or equivalent, at the location of the service disconnecting means of the premises:

WARNING
FUEL CELL POWER SYSTEM CONTAINS
ELECTRICAL ENERGY STORAGE DEVICES.

### VII. Connection to Other Circuits

**692.59 Transfer Switch.** A transfer switch shall be required in non–grid-interactive systems that use utility grid backup. The transfer switch shall maintain isolation between the electrical production and distribution network and the fuel cell system. The transfer switch shall be permitted to be located externally or internally to the fuel cell system unit. When the utility service conductors of the structure are connected to the transfer switch, the switch shall comply with Article 230, Part V.

**692.60 Identified Interactive Equipment.** Only fuel cell systems listed and identified as interactive shall be permitted in interactive systems.

**692.61 Output Characteristics.** The output of a fuel cell system operating in parallel with an electric supply system shall be compatible with the voltage, wave shape, and frequency of the system to which it is connected.

> FPN: The term *compatible* does not necessarily mean matching the primary source wave shape.

**692.62 Loss of Interactive System Power.** The fuel cell system shall be provided with a means of detecting when the electrical production and distribution network has become de-energized and shall not feed the electrical production and distribution network side of the point of common coupling during this condition. The fuel cell system shall remain in that state until the electrical production and distribution network voltage has been restored.

A normally interactive fuel cell system shall be permitted to operate as a stand-alone system to supply loads that have been disconnected from electrical production and distribution network sources.

**692.64 Unbalanced Interconnections.**

**(A) Single Phase.** Single-phase interactive fuel cell systems shall not be connected to a 3-phase power system unless the interactive system is designed so that significant unbalanced voltages cannot result.

**(B) Three Phase.** Three-phase interactive fuel cell systems shall have all phases automatically de-energized upon loss of voltage, or unbalance of voltage in one or more phases, unless the interactive system is designed so that significant unbalanced voltages will not result.

**692.65 Point of Connection.** The output of a fuel cell system power source shall be connected as specified in 692.65(A) or (B).

**(A) Supply Side.** A fuel cell system power source shall be permitted to be connected to the supply side of the service disconnecting means as permitted in 230.82(5).

**(B) Load Side.** A fuel cell system power source shall be permitted to be connected to the load side of the service

disconnecting means of the other source(s) at any distribution equipment on the premises, provided that all of the following conditions are met:

(1) Each source interconnection shall be made at a dedicated circuit breaker or fusible disconnecting means.
(2) The sum of the ampere ratings of overcurrent devices in circuits supplying power to a busbar or conductor shall not exceed the rating of the busbar or conductor.

*Exception: For a dwelling unit, the sum of the ampere ratings of the overcurrent devices shall not exceed 120 percent of the rating of the busbar or conductor.*

(3) The interconnection point shall be on the line side of all ground-fault protection equipment.
(4) Equipment containing overcurrent devices in circuits supplying power to a busbar or conductor shall be marked to indicate the presence of all sources.
(5) Equipment such as circuit breakers, if backfed, shall be identified for such operation.
(6) The circuit breaker on the dedicated output of a utility-interactive inverter shall be positioned in the distribution equipment at the opposite (load) end from the input feeder connection or main circuit location. A permanent warning label shall be applied to the distribution equipment with the following, or equivalent:

WARNING
FUEL CELL POWER SYSTEM OUTPUT.
DO NOT RELOCATE THIS CIRCUIT BREAKER.

## VIII. Outputs Over 600 Volts

**692.80 General.** Fuel cell systems with a maximum output voltage over 600 volts ac shall comply with the requirements of other articles applicable to such installations.

## ARTICLE 695
## Fire Pumps

> FPN: Rules that are followed by a reference in brackets contain text that has been extracted from NFPA 20-1999, *Standard for the Installation of Stationary Pumps for Fire Protection.* Only editorial changes were made to the extracted text to make it consistent with this *Code.*

**695.1 Scope.**

**(A) Covered.** This article covers the installation of the following:

(1) Electric power sources and interconnecting circuits

(2) Switching and control equipment dedicated to fire pump drivers

**(B) Not Covered.** This article does not cover the following:

(1) The performance, maintenance, and acceptance testing of the fire pump system, and the internal wiring of the components of the system
(2) Pressure maintenance (jockey or makeup) pumps

> FPN: See NFPA 20-1999, *Standard for the Installation of Stationary Pumps for Fire Protection,* for further information.

### 695.2 Definitions.

**Fault Tolerant External Control Conductors.** Those control conductors entering and/or leaving the fire pump controller enclosure, which if broken, disconnected, or shorted will not prevent the controller from starting the fire pump from all other internal or external means and may cause the controller to start the pump under these conditions.

**On-Site Power Production Facility.** The normal supply of electric power for the site that is expected to be constantly producing power.

**On-Site Standby Generator.** A facility producing electric power on site as the alternate supply of electric power. It differs from an on-site power production facility in that it is not constantly producing power.

### 695.3 Power Source(s) for Electric Motor-Driven Fire Pumps. Electric motor-driven fire pumps shall have a reliable source of power.

**(A) Individual Sources.** Where reliable, and where capable of carrying indefinitely the sum of the locked-rotor current of the fire pump motor(s) and the pressure maintenance pump motor(s) and the full-load current of the associated fire pump accessory equipment when connected to this power supply, the power source for an electric motor-driven fire pump shall be one or more of the following.

**(1) Electric Utility Service Connection.** A fire pump shall be permitted to be supplied by a separate service, or by a tap located ahead of and not within the same cabinet, enclosure, or vertical switchboard section as the service disconnecting means. The connection shall be located and arranged so as to minimize the possibility of damage by fire from within the premises and from exposing hazards. A tap ahead of the service disconnecting means shall comply with 230.82(4). The service equipment shall comply with the labeling requirements in 230.2 and the location requirements in 230.72(B).

**(2) On-Site Power Production Facility.** A fire pump shall be permitted to be supplied by an on-site power production facility. The source facility shall be located and protected to minimize the possibility of damage by fire. [NFPA 20, 6.2.1, 6.2.2, 6.2.4.4]

**(B) Multiple Sources.** Where reliable power cannot be obtained from a source described in 695.3(A), power shall be supplied from an approved combination of two or more of either of such sources, or from an approved combination of feeders constituting two or more power sources as covered in 695.3(B)(2), or from an approved combination of one or more of such power sources in combination with an on-site standby generator complying with 695.3(B)(1) and (B)(3).

**(1) Generator Capacity.** An on-site generator(s) used to comply with this section shall be of sufficient capacity to allow normal starting and running of the motor(s) driving the fire pump(s) while supplying all other simultaneously operated load. Automatic shedding of one or more optional standby loads in order to comply with this capacity requirement shall be permitted. A tap ahead of the on-site generator disconnecting means shall not be required. The requirements of 430.113 shall not apply.

**(2) Feeder Sources.** This section applies to multibuilding campus-style complexes with fire pumps at one or more buildings. Where sources in 695.3(A) are not practicable, and with the approval of the authority having jurisdiction, two or more feeder sources shall be permitted as one power source or as more than one power source where such feeders are connected to or derived from separate utility services. The connection(s), overcurrent protective device(s), and disconnecting means for such feeders shall meet the requirements of 695.4(B).

**(3) Arrangement.** The power sources shall be arranged so that a fire at one source will not cause an interruption at the other source. [NFPA 20, 6.2.3, 6.2.4.1, 6.2.4.3, 6.6.1]

### 695.4 Continuity of Power. Circuits that supply electric motor-driven fire pumps shall be supervised from inadvertent disconnection as covered in 695.4(A) or (B).

**(A) Direct Connection.** The supply conductors shall directly connect the power source to either a listed fire pump controller or listed combination fire pump controller and power transfer switch.

**(B) Supervised Connection.** A single disconnecting means and associated overcurrent protective device(s) shall be permitted to be installed between a remote power source and one of the following:

(1) A listed fire pump controller
(2) A listed fire pump power transfer switch
(3) A listed combination fire pump controller and power transfer switch

For systems installed under the provisions of 695.3(B)(2) only, such additional disconnecting means and associated overcurrent protective device(s) shall be permitted as required to comply with other provisions of this *Code*. Overcurrent protective devices between an on-site standby generator and a fire pump controller shall be selected and sized according to 430.62 to provide short-circuit protection only. All disconnecting devices and overcurrent protective devices that are unique to the fire pump loads shall comply with 695.4(B)(1) through (B)(4).

**(1) Overcurrent Device Selection.** The overcurrent protective device(s) shall be selected or set to carry indefinitely the sum of the locked-rotor current of the fire pump motor(s) and the pressure maintenance pump motor(s) and the full-load current of the associated fire pump accessory equipment when connected to this power supply.

**(2) Disconnecting Means.** The disconnecting means shall comply with all of the following:

(1) Be identified as suitable for use as service equipment
(2) Be lockable in the closed position
(3) Be located sufficiently remote from other building or other fire pump source disconnecting means such that inadvertent contemporaneous operation would be unlikely

**(3) Disconnect Marking.** The disconnecting means shall be marked "Fire Pump Disconnecting Means." The letters shall be at least 25 mm (1 in.) in height, and they shall be visible without opening enclosure doors or covers.

**(4) Controller Marking.** A placard shall be placed adjacent to the fire pump controller, stating the location of this disconnecting means and the location of the key (if the disconnecting means is locked).

**(5) Supervision.** The disconnecting means shall be supervised in the closed position by one of the following methods:

(1) Central station, proprietary, or remote station signal device
(2) Local signaling service that causes the sounding of an audible signal at a constantly attended point
(3) Locking the disconnecting means in the closed position
(4) Sealing of disconnecting means and approved weekly recorded inspections when the disconnecting means are located within fenced enclosures or in buildings under the control of the owner [NFPA 20, 6.3.2.2.1, 6.3.2.2.2, 6.3.2.2.3]

**695.5 Transformers.** Where the service or system voltage is different from the utilization voltage of the fire pump motor, transformer(s) protected by disconnecting means and overcurrent protective devices shall be permitted to be installed between the system supply and the fire pump controller in accordance with 695.5(A) and (B), or (C). Only transformers covered in 695.5(C) shall be permitted to supply loads not directly associated with the fire pump system.

**(A) Size.** Where a transformer supplies an electric motor-driven fire pump, it shall be rated at a minimum of 125 percent of the sum of the fire pump motor(s) and pressure maintenance pump(s) motor loads, and 100 percent of the associated fire pump accessory equipment supplied by the transformer.

**(B) Overcurrent Protection.** The primary overcurrent protective device(s) shall be selected or set to carry indefinitely the sum of the locked-rotor current of the fire pump motor(s) and the pressure maintenance pump motor(s) and the full-load current of the associated fire pump accessory equipment when connected to this power supply. Secondary overcurrent protection shall not be permitted.

**(C) Feeder Source.** Where a feeder source is provided in accordance with 695.3(B)(2), transformers supplying the fire pump system shall be permitted to supply other loads. All other loads shall be calculated in accordance with Article 220, including demand factors as applicable.

**(1) Size.** Transformers shall be rated at a minimum of 125 percent of the sum of the fire pump motor(s) and pressure maintenance pump(s) motor loads, and 100 percent of the remaining load supplied by the transformer.

**(2) Overcurrent Protection.** The transformer size, the feeder size, and the overcurrent protective device(s) shall be coordinated such that overcurrent protection is provided for the transformer in accordance with 450.3 and for the feeder in accordance with 215.3, and such that the overcurrent protective device(s) is selected or set to carry indefinitely the sum of the locked-rotor current of the fire pump motor(s), the pressure maintenance pump motor(s), the full-load current of the associated fire pump accessory equipment, and 100 percent of the remaining loads supplied by the transformer.

**695.6 Power Wiring.** Power circuits and wiring methods shall comply with the requirements in 695.6(A) through (G), and as permitted in 230.90(A), Exception No. 4; 230.94, Exception No. 4; 230.95, Exception No. 2; 240.13; 230.208; 240.4(A); and 430.31.

**(A) Service Conductors.** Supply conductors shall be physically routed outside a building(s) and shall be installed as service entrance conductors in accordance with Article 230. Where supply conductors cannot be physically routed outside buildings, they shall be permitted to be routed through buildings where installed in accordance with 230.6(1) or (2). Where a fire pump is wired under the provisions of 695.3(B)(2), this requirement shall apply to all supply conductors on the load side of the service disconnecting means that constitute the normal source of supply to that fire pump.

*Exception: Where there are multiple sources of supply with means for automatic connection from one source to the other, the requirement shall only apply to those conductors on the load side of that point of automatic connection between sources.*

**(B) Circuit Conductors.** Fire pump supply conductors on the load side of the final disconnecting means and overcurrent device(s) permitted by 695.4(B) shall be kept entirely independent of all other wiring. They shall only supply loads that are directly associated with the fire pump system, and they shall be protected to resist potential damage by fire, structural failure, or operational accident. They shall be permitted to be routed through a building(s) using one of the following methods:

(1) Be encased in a minimum 50 mm (2 in.) of concrete
(2) Be within an enclosed construction dedicated to the fire pump circuit(s) and having a minimum of a 1-hour fire resistive rating
(3) Be a listed electrical circuit protective system with a minimum 1-hour fire rating

*Exception: The supply conductors located in the electrical equipment room where they originate and in the fire pump room shall not be required to have the minimum 1-hour fire separation or fire resistance rating, unless otherwise required by 700.9(D) of this Code.*

**(C) Conductor Size.**

**(1) Fire Pump Motors and Other Equipment.** Conductors supplying a fire pump motor(s), pressure maintenance pumps, and associated fire pump accessory equipment shall have a rating not less than 125 percent of the sum of the fire pump motor(s) and pressure maintenance motor(s) full-load current(s), and 100 percent of the associated fire pump accessory equipment.

**(2) Fire Pump Motors Only.** Conductors supplying only a fire pump motor(s) shall have a rating not less than 125 percent of the fire pump motor(s) full-load current(s).

**(D) Overload Protection.** Power circuits shall not have automatic protection against overloads. Except as provided in 695.5(C)(2), branch-circuit and feeder conductors shall be protected against short circuit only. Where a tap is made to supply a fire pump, and the tap wiring is run in accordance with 230.6, the applicable distance and size restrictions in 240.21 shall not apply.

*Exception No. 1: Conductors between storage batteries and the engine shall not require overcurrent protection or disconnecting means.*

*Exception No. 2: For on-site standby generator(s) that produce continuous currents in excess of 225 percent of the full-load amperes of the fire pump motor, the conductors between the on-site generator(s) and the combination fire pump transfer switch controller or separately mounted transfer switch shall be installed in accordance with 695.6(B) or protected in accordance with 430.52.*

*The protection provided shall be in accordance with the short-circuit current rating of the combination fire pump transfer switch controller or separately mounted transfer switch.*

**(E) Pump Wiring.** All wiring from the controllers to the pump motors shall be in rigid metal conduit, intermediate metal conduit, liquidtight flexible metal conduit, or liquidtight flexible nonmetallic conduit Type LFNC-B, or Type MI cable.

**(F) Junction Points.** Where wire connectors are used in the fire pump circuit, the connectors shall be listed. A fire pump controller or fire pump power transfer switch, where provided, shall not be used as a junction box to supply other equipment, including a pressure maintenance (jockey) pump(s). A fire pump controller and fire pump power transfer switch, where provided, shall not serve any load other than the fire pump for which it is intended.

**(G) Mechanical Protection.** All wiring from engine controllers and batteries shall be protected against physical damage and shall be installed in accordance with the controller and engine manufacturer's instructions.

**695.7 Voltage Drop.** The voltage at the controller line terminals shall not drop more than 15 percent below normal (controller-rated voltage) under motor starting conditions. The voltage at the motor terminals shall not drop more than 5 percent below the voltage rating of the motor when the motor is operating at 115 percent of the full-load current rating of the motor.

*Exception: This limitation shall not apply for emergency run mechanical starting. [NFPA 20, 6.4]*

**695.10 Listed Equipment.** Diesel engine fire pump controllers, electric fire pump controllers, electric motors, fire pump power transfer switches, foam pump controllers, and limited service controllers shall be listed for fire pump service. [NFPA 20, 6.5.1.1, 7.1.2.1, 9.1.1.1]

**695.12 Equipment Location.**

**(A) Controllers and Transfer Switches.** Electric motor-driven fire pump controllers and power transfer switches shall be located as close as practicable to the motors that they control and shall be within sight of the motors.

**(B) Engine-Drive Controllers.** Engine-drive fire pump controllers shall be located as close as is practical to the engines that they control and shall be within sight of the engines.

**(C) Storage Batteries.** Storage batteries for diesel engine drives shall be rack supported above the floor, secured against displacement, and located where they will not be subject to physical damage, flooding with water, excessive temperature, or excessive vibration.

**(D) Energized Equipment.** All energized equipment parts shall be located at least 300 mm (12 in.) above the floor level.

**(E) Protection Against Pump Water.** Fire pump controllers and power transfer switches shall be located or protected so that they will not be damaged by water escaping from pumps or pump connections.

**(F) Mounting.** All fire pump control equipment shall be mounted in a substantial manner on noncombustible supporting structures.

### 695.14 Control Wiring.

**(A) Control Circuit Failures.** External control circuits that extend outside the fire pump room shall be arranged so that failure of any external circuit (open or short circuit) shall not prevent the operation of a pump(s) from all other internal or external means. Breakage, disconnecting, shorting of the wires, or loss of power to these circuits could cause continuous running of the fire pump but shall not prevent the controller(s) from starting the fire pump(s) due to causes other than these external control circuits. All control conductors within the fire pump room that are not fault tolerant shall be protected against physical damage. [NFPA 20, 7.5.2.5]

**(B) Sensor Functioning.** No undervoltage, phase-loss, frequency-sensitive, or other sensor(s) shall be installed that automatically or manually prohibit actuation of the motor contactor. [NFPA 20, 7.4.5.6]

*Exception: A phase loss sensor(s) shall be permitted only as a part of a listed fire pump controller.*

**(C) Remote Device(s).** No remote device(s) shall be installed that will prevent automatic operation of the transfer switch. [NFPA 20, 7.8.1.3]

**(D) Engine-Drive Control Wiring.** All wiring between the controller and the diesel engine shall be stranded and sized to continuously carry the charging or control currents as required by the controller manufacturer. Such wiring shall be protected against physical damage. Controller manufacturer's specifications for distance and wire size shall be followed. [NFPA 20, 9.3.5.1]

**(E) Electric Fire Pump Control Wiring Methods.** All electric motor-driven fire pump control wiring shall be in rigid metal conduit, intermediate metal conduit, liquidtight flexible metal conduit, liquidtight flexible nonmetallic conduit Type B (LFNC-B), or Type MI cable.

**(F) Generator Control Wiring Methods.** Control conductors installed between the fire pump power transfer switch and the standby generator supplying the fire pump during normal power loss shall be kept entirely independent of all other wiring. They shall be protected to resist potential damage by fire or structural failure. They shall be permitted to be routed through a building(s) encased in 50 mm (2 in.) of concrete or within enclosed construction dedicated to the fire pump circuits and having a minimum 1-hour fire resistance rating, or circuit protective systems with a minimum of 1-hour fire resistance. The installation shall comply with any restrictions provided in the listing of the electrical circuit protective system used.

# Chapter 7 Special Conditions

## ARTICLE 700
## Emergency Systems

### I. General

**700.1 Scope.** The provisions of this article apply to the electrical safety of the installation, operation, and maintenance of emergency systems consisting of circuits and equipment intended to supply, distribute, and control electricity for illumination, power, or both, to required facilities when the normal electrical supply or system is interrupted.

Emergency systems are those systems legally required and classed as emergency by municipal, state, federal, or other codes, or by any governmental agency having jurisdiction. These systems are intended to automatically supply illumination, power, or both, to designated areas and equipment in the event of failure of the normal supply or in the event of accident to elements of a system intended to supply, distribute, and control power and illumination essential for safety to human life.

FPN No. 1: For further information regarding wiring and installation of emergency systems in health care facilities, see Article 517.

FPN No. 2: For further information regarding performance and maintenance of emergency systems in health care facilities, see NFPA 99-1999, *Standard for Health Care Facilities.*

FPN No. 3: Emergency systems are generally installed in places of assembly where artificial illumination is required for safe exiting and for panic control in buildings subject to occupancy by large numbers of persons, such as hotels, theaters, sports arenas, health care facilities, and similar institutions. Emergency systems may also provide power for such functions as ventilation where essential to maintain life, fire detection and alarm systems, elevators, fire pumps, public safety communications systems, industrial processes where current interruption would produce serious life safety or health hazards, and similar functions.

FPN No. 4: For specification of locations where emergency lighting is considered essential to life safety, see NFPA *101*®-2000, *Life Safety Code*®.

FPN No. 5: For further information regarding performance of emergency and standby power systems, see NFPA 110-1999, *Standard for Emergency and Standby Power Systems.*

**700.2 Application of Other Articles.** Except as modified by this article, all applicable articles of this *Code* shall apply.

**700.3 Equipment Approval.** All equipment shall be approved for use on emergency systems.

**700.4 Tests and Maintenance.**

**(A) Conduct or Witness Test.** The authority having jurisdiction shall conduct or witness a test of the complete system upon installation and periodically afterward.

**(B) Tested Periodically.** Systems shall be tested periodically on a schedule acceptable to the authority having jurisdiction to ensure the systems are maintained in proper operating condition.

**(C) Battery Systems Maintenance.** Where battery systems or unit equipments are involved, including batteries used for starting, control, or ignition in auxiliary engines, the authority having jurisdiction shall require periodic maintenance.

**(D) Written Record.** A written record shall be kept of such tests and maintenance.

**(E) Testing Under Load.** Means for testing all emergency lighting and power systems during maximum anticipated load conditions shall be provided.

FPN: For testing and maintenance procedures of emergency power supply systems (EPSSs), see NFPA 110-1999, *Standard for Emergency and Standby Power Systems.*

**700.5 Capacity.**

**(A) Capacity and Rating.** An emergency system shall have adequate capacity and rating for all loads to be operated simultaneously. The emergency system equipment shall be suitable for the maximum available fault current at its terminals.

**(B) Selective Load Pickup, Load Shedding, and Peak Load Shaving.** The alternate power source shall be permitted to supply emergency, legally required standby, and optional standby system loads where automatic selective load pickup and load shedding is provided as needed to ensure adequate power to (1) the emergency circuits, (2) the legally required standby circuits, and (3) the optional standby circuits, in that order of priority. The alternate power source shall be permitted to be used for peak load shaving, provided the above conditions are met.

Peak load-shaving operation shall be permitted for satisfying the test requirement of 700.4(B), provided all other conditions of 700.4 are met.

A portable or temporary alternate source shall be available whenever the emergency generator is out of service for major maintenance or repair.

**700.6 Transfer Equipment.**

**(A) General.** Transfer equipment, including automatic transfer switches, shall be automatic, identified for emergency use, and approved by the authority having jurisdiction. Transfer equipment shall be designed and installed to prevent the inadvertent interconnection of normal and emergency sources of supply in any operation of the transfer equipment. Transfer equipment and electric power production systems installed to permit operation in parallel with the normal source shall meet the requirements of Article 705.

**(B) Bypass Isolation Switches.** Means shall be permitted to bypass and isolate the transfer equipment. Where bypass isolation switches are used, inadvertent parallel operation shall be avoided.

**(C) Automatic Transfer Switches.** Automatic transfer switches shall be electrically operated and mechanically held.

**(D) Use.** Transfer equipment shall supply only emergency loads.

**700.7 Signals.** Audible and visual signal devices shall be provided, where practicable, for the purpose described in 700.7(A) through (D).

**(A) Derangement.** To indicate derangement of the emergency source.

**(B) Carrying Load.** To indicate that the battery is carrying load.

**(C) Not Functioning.** To indicate that the battery charger is not functioning.

**(D) Ground Fault.** To indicate a ground fault in solidly grounded wye emergency systems of more than 150 volts to ground and circuit-protective devices rated 1000 amperes or more. The sensor for the ground-fault signal devices shall be located at, or ahead of, the main system disconnecting means for the emergency source, and the maximum setting of the signal devices shall be for a ground-fault current of 1200 amperes. Instructions on the course of action to be taken in event of indicated ground fault shall be located at or near the sensor location.

> FPN: For signals for generator sets, see NFPA 110-1999, *Standard for Emergency and Standby Power Systems.*

**700.8 Signs.**

**(A) Emergency Sources.** A sign shall be placed at the service entrance equipment indicating type and location of on-site emergency power sources.

*Exception: A sign shall not be required for individual unit equipment as specified in 700.12(E).*

**(B) Grounding.** Where the grounded circuit conductor connected to the emergency source is connected to a grounding electrode conductor at a location remote from the emergency source, there shall be a sign at the grounding location that shall identify all emergency and normal sources connected at that location.

## II. Circuit Wiring

**700.9 Wiring, Emergency System.**

**(A) Identification.** All boxes and enclosures (including transfer switches, generators, and power panels) for emergency circuits shall be permanently marked so they will be readily identified as a component of an emergency circuit or system.

**(B) Wiring.** Wiring of two or more emergency circuits supplied from the same source shall be permitted in the same raceway, cable, box, or cabinet. Wiring from an emergency source or emergency source distribution overcurrent protection to emergency loads shall be kept entirely independent of all other wiring and equipment, unless otherwise permitted in (1) through (4):

(1) Wiring from the normal power source located in transfer equipment enclosures
(2) Wiring supplied from two sources in exit or emergency luminaires (lighting fixtures)
(3) Wiring from two sources in a common junction box, attached to exit or emergency luminaires (lighting fixtures)
(4) Wiring within a common junction box attached to unit equipment, containing only the branch circuit supplying the unit equipment and the emergency circuit supplied by the unit equipment

**(C) Wiring Design and Location.** Emergency wiring circuits shall be designed and located so as to minimize the hazards that might cause failure due to flooding, fire, icing, vandalism, and other adverse conditions.

**(D) Fire Protection.** Emergency systems shall meet the following additional requirements in assembly occupancies for not less than 1000 persons or in buildings above 23 m (75 ft) in height with any of the following occupancy classes: assembly, educational, residential, detention and correctional, business, and mercantile.

**(1) Feeder-Circuit Wiring.** Feeder-circuit wiring shall meet one of the following conditions:

(1) Be installed with buildings that are fully protected by an approved automatic fire suppression system
(2) Be a listed electrical circuit protective system with a minimum 1-hour fire rating

(3) Be protected by a listed thermal barrier system for electrical system components

(4) Be protected by a fire-rated assembly listed to achieve a minimum fire rating of 1 hour

(5) Be embedded in not less than 50 mm (2 in.) of concrete

(6) Be a cable listed to maintain circuit integrity for not less than 1 hour when installed in accordance with the listing requirements

**(2) Feeder-Circuit Equipment.** Equipment for feeder circuits (including transfer switches, transformers, and panelboards) shall be located either in spaces fully protected by approved automatic fire suppression systems (including sprinklers, carbon dioxide systems) or in spaces with a 1-hour fire resistance rating.

FPN: For the definition of *occupancy class*, see 4.1 of NFPA *101-2000*, *Life Safety Code*.

## III. Sources of Power

**700.12 General Requirements.** Current supply shall be such that, in the event of failure of the normal supply to, or within, the building or group of buildings concerned, emergency lighting, emergency power, or both shall be available within the time required for the application but not to exceed 10 seconds. The supply system for emergency purposes, in addition to the normal services to the building and meeting the general requirements of this section, shall be one or more of the types of systems described in 700.12(A) through (D). Unit equipment in accordance with 700.12(E) shall satisfy the applicable requirements of this article.

In selecting an emergency source of power, consideration shall be given to the occupancy and the type of service to be rendered, whether of minimum duration, as for evacuation of a theater, or longer duration, as for supplying emergency power and lighting due to an indefinite period of current failure from trouble either inside or outside the building.

Equipment shall be designed and located to minimize the hazards that might cause complete failure due to flooding, fires, icing, and vandalism.

Equipment for sources of power as described in 700.12(A) through (D) where located within assembly occupancies for greater than 1000 persons or in buildings above 23 m (75 ft) in height with any of the following occupancy classes — assembly, educational, residential, detention and correctional, business, and mercantile — shall be installed either in spaces fully protected by approved automatic fire suppression systems (sprinklers, carbon dioxide systems, and so forth), or in spaces with a 1-hour fire rating.

FPN No. 1: For the definition of occupancy class, see 4.1 of NFPA *101-2000*, *Life Safety Code*.

FPN No. 2: Assignment of degree of reliability of the recognized emergency supply system depends on the careful evaluation of the variables at each particular installation.

**(A) Storage Battery.** Storage batteries used as a source of power for emergency systems shall be of suitable rating and capacity to supply and maintain the total load for a period of 1½ hours minimum, without the voltage applied to the load falling below 87½ percent of normal.

Batteries, whether of the acid or alkali type, shall be designed and constructed to meet the requirements of emergency service and shall be compatible with the charger for that particular installation.

For a sealed battery, the container shall not be required to be transparent. However, for the lead acid battery that requires water additions, transparent or translucent jars shall be furnished. Automotive-type batteries shall not be used.

An automatic battery charging means shall be provided.

**(B) Generator Set.**

**(1) Prime Mover-Driven.** For a generator set driven by a prime mover acceptable to the authority having jurisdiction and sized in accordance with 700.5, means shall be provided for automatically starting the prime mover on failure of the normal service and for automatic transfer and operation of all required electrical circuits. A time-delay feature permitting a 15-minute setting shall be provided to avoid retransfer in case of short-time reestablishment of the normal source.

**(2) Internal Combustion as Prime Movers.** Where internal combustion engines are used as the prime mover, an on-site fuel supply shall be provided with an on-premise fuel supply sufficient for not less than 2 hours' full-demand operation of the system. Where power is needed for the operation of the fuel transfer pumps to deliver fuel to a generator set day tank, this pump shall be connected to the emergency power system.

**(3) Dual Supplies.** Prime movers shall not be solely dependent on a public utility gas system for their fuel supply or municipal water supply for their cooling systems. Means shall be provided for automatically transferring from one fuel supply to another where dual fuel supplies are used.

*Exception: Where acceptable to the authority having jurisdiction, the use of other than on-site fuels shall be permitted where there is a low probability of a simultaneous failure of both the off-site fuel delivery system and power from the outside electrical utility company.*

**(4) Battery Power and Dampers.** Where a storage battery is used for control or signal power or as the means of starting the prime mover, it shall be suitable for the purpose and shall be equipped with an automatic charging means independent of the generator set. Where the battery charger is required for the operation of the generator set, it shall be connected to the emergency system. Where power is required for the operation of dampers used to ventilate the generator set, the dampers shall be connected to the emergency system.

**(5) Auxiliary Power Supply.** Generator sets that require more than 10 seconds to develop power shall be permitted if an auxiliary power supply energizes the emergency system until the generator can pick up the load.

**(6) Outdoor Generator Sets.** Where an outdoor housed generator set is equipped with a readily accessible disconnecting means located within sight of the building or structure supplied, an additional disconnecting means shall not be required where ungrounded conductors pass through the building or structure.

**(C) Uninterruptible Power Supplies.** Uninterruptible power supplies used to provide power for emergency systems shall comply with the applicable provisions of 700.12(A) and (B).

**(D) Separate Service.** Where acceptable to the authority having jurisdiction as suitable for use as an emergency source, a second service shall be permitted. This service shall be in accordance with Article 230, with separate service drop or lateral, widely separated electrically and physically from the normal service to minimize the possibility of simultaneous interruption of supply.

**(E) Unit Equipment.** Individual unit equipment for emergency illumination shall consist of the following:

(1) A rechargeable battery
(2) A battery charging means
(3) Provisions for one or more lamps mounted on the equipment, or shall be permitted to have terminals for remote lamps, or both
(4) A relaying device arranged to energize the lamps automatically upon failure of the supply to the unit equipment

The batteries shall be of suitable rating and capacity to supply and maintain at not less than 87½ percent of the nominal battery voltage for the total lamp load associated with the unit for a period of at least 1½ hours, or the unit equipment shall supply and maintain not less than 60 percent of the initial emergency illumination for a period of at least 1½ hours. Storage batteries, whether of the acid or alkali type, shall be designed and constructed to meet the requirements of emergency service.

Unit equipment shall be permanently fixed in place (i.e., not portable) and shall have all wiring to each unit installed in accordance with the requirements of any of the wiring methods in Chapter 3. Flexible cord-and-plug connection shall be permitted, provided that the cord does not exceed 900 mm (3 ft) in length. The branch circuit feeding the unit equipment shall be the same branch circuit as that serving the normal lighting in the area and connected ahead of any local switches. The branch circuit that feeds unit equipment shall be clearly identified at the distribution panel. Emergency luminaires (illumination fixtures) that obtain power from a unit equipment and are not part of the unit equipment shall be wired to the unit equipment as required by 700.9 and by one of the wiring methods of Chapter 3.

*Exception: In a separate and uninterrupted area supplied by a minimum of three normal lighting circuits, a separate branch circuit for unit equipment shall be permitted if it originates from the same panelboard as that of the normal lighting circuits and is provided with a lock-on feature.*

## IV. Emergency System Circuits for Lighting and Power

**700.15 Loads on Emergency Branch Circuits.** No appliances and no lamps, other than those specified as required for emergency use, shall be supplied by emergency lighting circuits.

**700.16 Emergency Illumination.** Emergency illumination shall include all required means of egress lighting, illuminated exit signs, and all other lights specified as necessary to provide required illumination.

Emergency lighting systems shall be designed and installed so that the failure of any individual lighting element, such as the burning out of a light bulb, cannot leave in total darkness any space that requires emergency illumination.

Where high-intensity discharge lighting such as high- and low-pressure sodium, mercury vapor, and metal halide is used as the sole source of normal illumination, the emergency lighting system shall be required to operate until normal illumination has been restored.

*Exception: Alternative means that ensure emergency lighting illumination level is maintained shall be permitted.*

**700.17 Circuits for Emergency Lighting.** Branch circuits that supply emergency lighting shall be installed to provide service from a source complying with 700.12 when the normal supply for lighting is interrupted. Such installations shall provide either one of the following:

(1) An emergency lighting supply, independent of the general lighting supply, with provisions for automatically transferring the emergency lights upon the event of failure of the general lighting system supply

(2) Two or more separate and complete systems with independent power supply, each system providing sufficient current for emergency lighting purposes

Unless both systems are used for regular lighting purposes and are both kept lighted, means shall be provided for automatically energizing either system upon failure of the other. Either or both systems shall be permitted to be a part of the general lighting system of the protected occupancy if circuits supplying lights for emergency illumination are installed in accordance with other sections of this article.

**700.18 Circuits for Emergency Power.** For branch circuits that supply equipment classed as emergency, there shall be an emergency supply source to which the load will be transferred automatically upon the failure of the normal supply.

## V. Control — Emergency Lighting Circuits

**700.20 Switch Requirements.** The switch or switches installed in emergency lighting circuits shall be arranged so that only authorized persons have control of emergency lighting.

*Exception No. 1: Where two or more single-throw switches are connected in parallel to control a single circuit, at least one of these switches shall be accessible only to authorized persons.*

*Exception No. 2: Additional switches that act only to put emergency lights into operation but not disconnect them shall be permissible.*

Switches connected in series or 3- and 4-way switches shall not be used.

**700.21 Switch Location.** All manual switches for controlling emergency circuits shall be in locations convenient to authorized persons responsible for their actuation. In places of assembly, such as theaters, a switch for controlling emergency lighting systems shall be located in the lobby or at a place conveniently accessible thereto.

In no case shall a control switch for emergency lighting in a theater, motion-picture theater, or place of assembly be placed in a motion-picture projection booth or on a stage or platform.

*Exception: Where multiple switches are provided, one such switch shall be permitted in such locations where arranged so that it can energize the circuit only but cannot de-energize the circuit.*

**700.22 Exterior Lights.** Those lights on the exterior of a building that are not required for illumination when there is sufficient daylight shall be permitted to be controlled by an automatic light-actuated device.

## VI. Overcurrent Protection

**700.25 Accessibility.** The branch-circuit overcurrent devices in emergency circuits shall be accessible to authorized persons only.

FPN: Fuses and circuit breakers for emergency circuit overcurrent protection, where coordinated to ensure selective clearing of fault currents, increase overall reliability of the system.

**700.26 Ground-Fault Protection of Equipment.** The alternate source for emergency systems shall not be required to have ground-fault protection of equipment with automatic disconnecting means. Ground-fault indication of the emergency source shall be provided per 700.7(D).

# ARTICLE 701
# Legally Required Standby Systems

## I. General

**701.1 Scope.** The provisions of this article apply to the electrical safety of the installation, operation, and maintenance of legally required standby systems consisting of circuits and equipment intended to supply, distribute, and control electricity to required facilities for illumination or power, or both, when the normal electrical supply or system is interrupted.

The systems covered by this article consist only of those that are permanently installed in their entirety, including the power source.

FPN No. 1: For additional information, see NFPA 99-1999, *Standard for Health Care Facilities.*

FPN No. 2: For further information regarding performance of emergency and standby power systems, see NFPA 110-1999, *Standard for Emergency and Standby Power Systems.*

FPN No. 3: For further information, see ANSI/IEEE 446-1995, *Recommended Practice for Emergency and Standby Power Systems for Industrial and Commercial Applications.*

## 701.2 Definition.

**Legally Required Standby Systems.** Those systems required and so classed as legally required standby by municipal, state, federal, or other codes or by any governmental agency having jurisdiction. These systems are intended to automatically supply power to selected loads (other than those classed as emergency systems) in the event of failure of the normal source.

> FPN: Legally required standby systems are typically installed to serve loads, such as heating and refrigeration systems, communications systems, ventilation and smoke removal systems, sewage disposal, lighting systems, and industrial processes, that, when stopped during any interruption of the normal electrical supply, could create hazards or hamper rescue or fire-fighting operations.

**701.3 Application of Other Articles.** Except as modified by this article, all applicable articles of this *Code* shall apply.

**701.4 Equipment Approval.** All equipment shall be approved for the intended use.

**701.5 Tests and Maintenance for Legally Required Standby Systems.**

**(A) Conduct or Witness Test.** The authority having jurisdiction shall conduct or witness a test of the complete system upon installation.

**(B) Tested Periodically.** Systems shall be tested periodically on a schedule and in a manner acceptable to the authority having jurisdiction to ensure the systems are maintained in proper operating condition.

**(C) Battery Systems Maintenance.** Where batteries are used for control, starting, or ignition of prime movers, the authority having jurisdiction shall require periodic maintenance.

**(D) Written Record.** A written record shall be kept on such tests and maintenance.

**(E) Testing Under Load.** Means for testing legally required standby systems under load shall be provided.

> FPN: For testing and maintenance procedures of emergency power supply systems (EPSSs), see NFPA 110-1999, *Standard for Emergency and Standby Power Systems.*

**701.6 Capacity and Rating.** A legally required standby system shall have adequate capacity and rating for the supply of all equipment intended to be operated at one time. Legally required standby system equipment shall be suitable for the maximum available fault current at its terminals.

The alternate power source shall be permitted to supply legally required standby and optional standby system loads where automatic selective load pickup and load shedding is provided as needed to ensure adequate power to the legally required standby circuits.

**701.7 Transfer Equipment.**

**(A) General.** Transfer equipment, including automatic transfer switches, shall be automatic and identified for standby use and approved by the authority having jurisdiction. Transfer equipment shall be designed and installed to prevent the inadvertent interconnection of normal and alternate sources of supply in any operation of the transfer equipment. Transfer equipment and electric power production systems installed to permit operation in parallel with the normal source shall meet the requirements of Article 705.

**(B) Bypass Isolation Switches.** Means to bypass and isolate the transfer switch equipment shall be permitted. Where bypass isolation switches are used, inadvertent parallel operation shall be avoided.

**(C) Automatic Transfer Switches.** Automatic transfer switches shall be electrically operated and mechanically held.

**701.8 Signals.** Audible and visual signal devices shall be provided, where practicable, for the purposes described in 701.8(A), (B), and (C).

**(A) Derangement.** To indicate derangement of the standby source.

**(B) Carrying Load.** To indicate that the standby source is carrying load.

**(C) Not Functioning.** To indicate that the battery charger is not functioning.

> FPN: For signals for generator sets, see NFPA 110-1999, *Standard for Emergency and Standby Power Systems.*

**701.9 Signs.**

**(A) Mandated Standby.** A sign shall be placed at the service entrance indicating type and location of on-site legally required standby power sources.

*Exception: A sign shall not be required for individual unit equipment as specified in 701.11(F).*

**(B) Grounding.** Where the grounded circuit conductor connected to the legally required standby power source is connected to a grounding electrode conductor at a location remote from the legally required standby power source, there shall be a sign at the grounding location that shall identify all legally required standby power and normal sources connected at that location.

## II. Circuit Wiring

**701.10 Wiring Legally Required Standby Systems.** The legally required standby system wiring shall be permitted to occupy the same raceways, cables, boxes, and cabinets with other general wiring.

## III. Sources of Power

**701.11 Legally Required Standby Systems.** Current supply shall be such that, in the event of failure of the normal supply to, or within, the building or group of buildings concerned, legally required standby power will be available within the time required for the application but not to exceed 60 seconds. The supply system for legally required standby purposes, in addition to the normal services to the building, shall be permitted to comprise one or more of the types of systems described in 701.11(A) through (E). Unit equipment in accordance with 701.11(F) shall satisfy the applicable requirements of this article.

In selecting a legally required standby source of power, consideration shall be given to the type of service to be rendered, whether of short-time duration or long duration.

Consideration shall be given to the location or design, or both, of all equipment to minimize the hazards that might cause complete failure due to floods, fires, icing, and vandalism.

> FPN: Assignment of degree of reliability of the recognized legally required standby supply system depends on the careful evaluation of the variables at each particular installation.

**(A) Storage Battery.** A storage battery shall be of suitable rating and capacity to supply and maintain at not less than 87½ percent of system voltage the total load of the circuits supplying legally required standby power for a period of at least 1½ hours.

Batteries, whether of the acid or alkali type, shall be designed and constructed to meet the service requirements of emergency service and shall be compatible with the charger for that particular installation.

For a sealed battery, the container shall not be required to be transparent. However, for the lead acid battery that requires water additions, transparent or translucent jars shall be furnished. Automotive-type batteries shall not be used.

An automatic battery charging means shall be provided.

**(B) Generator Set.**

**(1) Prime Mover-Driven.** For a generator set driven by a prime mover acceptable to the authority having jurisdiction and sized in accordance with 701.6, means shall be provided for automatically starting the prime mover upon failure of the normal service and for automatic transfer and operation of all required electrical circuits. A time-delay feature permitting a 15-minute setting shall be provided to avoid retransfer in case of short-time re-establishment of the normal source.

**(2) Internal Combustion Engines as Prime Mover.** Where internal combustion engines are used as the prime mover, an on-site fuel supply shall be provided with an on-premise fuel supply sufficient for not less than 2 hours' full-demand operation of the system.

**(3) Dual Fuel Supplies.** Prime movers shall not be solely dependent on a public utility gas system for their fuel supply or municipal water supply for their cooling systems. Means shall be provided for automatically transferring one fuel supply to another where dual fuel supplies are used.

*Exception: Where acceptable to the authority having jurisdiction, the use of other than on-site fuels shall be permitted where there is a low probability of a simultaneous failure of both the off-site fuel delivery system and power from the outside electrical utility company.*

**(4) Battery Power.** Where a storage battery is used for control or signal power, or as the means of starting the prime mover, it shall be suitable for the purpose and shall be equipped with an automatic charging means independent of the generator set.

**(5) Outdoor Generator Sets.** Where an outdoor housed generator set is equipped with a readily accessible disconnecting means located within sight of the building or structure supplied, an additional disconnecting means shall not be required where ungrounded conductors pass through the building or structure.

**(C) Uninterruptible Power Supplies.** Uninterruptible power supplies used to provide power for legally required standby systems shall comply with the applicable provisions of 701.11(A) and (B).

**(D) Separate Service.** Where acceptable to the authority having jurisdiction, a second service shall be permitted. This service shall be in accordance with Article 230, with separate service drop or lateral widely separated electrically and physically from the normal service to minimize the possibility of simultaneous interruption of supply.

**(E) Connection Ahead of Service Disconnecting Means.** Where acceptable to the authority having jurisdiction, connections located ahead of and not within the same cabinet, enclosure, or vertical switchboard section as the service disconnecting means shall be permitted. The legally required standby service shall be sufficiently separated from the normal main service disconnecting means to prevent simultaneous interruption of supply through an occurrence within the building or groups of buildings served.

> FPN: See 230.82 for equipment permitted on the supply side of a service disconnecting means.

**(F) Unit Equipment.** Individual unit equipment for legally required standby illumination shall consist of the following:

(1) A rechargeable battery
(2) A battery charging means
(3) Provisions for one or more lamps mounted on the equipment and shall be permitted to have terminals for remote lamps
(4) A relaying device arranged to energize the lamps automatically upon failure of the supply to the unit equipment

The batteries shall be of suitable rating and capacity to supply and maintain at not less than 87½ percent of the nominal battery voltage for the total lamp load associated with the unit for a period of at least 1½ hours, or the unit equipment shall supply and maintain not less than 60 percent of the initial legally required standby illumination for a period of at least 1½ hours. Storage batteries, whether of the acid or alkali type, shall be designed and constructed to meet the requirements of emergency service.

Unit equipment shall be permanently fixed in place (i.e., not portable) and shall have all wiring to each unit installed in accordance with the requirements of any of the wiring methods in Chapter 3. Flexible cord-and-plug connection shall be permitted, provided that the cord does not exceed 900 mm (3 ft) in length. The branch circuit feeding the unit equipment shall be the same branch circuit as that serving the normal lighting in the area and connected ahead of any local switches. Legally required standby luminaires (illumination fixtures) that obtain power from a unit equipment and are not part of the unit equipment shall be wired to the unit equipment by one of the wiring methods of Chapter 3.

*Exception: In a separate and uninterrupted area supplied by a minimum of three normal lighting circuits, a separate branch circuit for unit equipment shall be permitted if it originates from the same panelboard as that of the normal lighting circuits and is provided with a lock-on feature.*

## IV. Overcurrent Protection

**701.15 Accessibility.** The branch-circuit overcurrent devices in legally required standby circuits shall be accessible to authorized persons only.

**701.17 Ground-Fault Protection of Equipment.** The alternate source for legally required standby systems shall not be required to have ground-fault protection of equipment.

## ARTICLE 702
## Optional Standby Systems

### I. General

**702.1 Scope.** The provisions of this article apply to the installation and operation of optional standby systems.

The systems covered by this article consist of those that are permanently installed in their entirety, including prime movers, and those that are arranged for a connection to a premises wiring system from a portable alternate power supply.

**702.2 Definition.**

**Optional Standby Systems.** Those systems intended to protect public or private facilities or property where life safety does not depend on the performance of the system. Optional standby systems are intended to supply on-site generated power to selected loads either automatically or manually.

FPN: Optional standby systems are typically installed to provide an alternate source of electric power for such facilities as industrial and commercial buildings, farms, and residences and to serve loads such as heating and refrigeration systems, data processing and communications systems, and industrial processes that, when stopped during any power outage, could cause discomfort, serious interruption of the process, damage to the product or process, or the like.

**702.3 Application of Other Articles.** Except as modified by this article, all applicable articles of this *Code* shall apply.

**702.4 Equipment Approval.** All equipment shall be approved for the intended use.

**702.5 Capacity and Rating.** An optional standby system shall have adequate capacity and rating for the supply of all equipment intended to be operated at one time. Optional standby system equipment shall be suitable for the maximum available fault current at its terminals. The user of the optional standby system shall be permitted to select the load connected to the system.

**702.6 Transfer Equipment.** Transfer equipment shall be suitable for the intended use and designed and installed so as to prevent the inadvertent interconnection of normal and alternate sources of supply in any operation of the transfer equipment. Transfer equipment and electric power production systems installed to permit operation in parallel with the normal source shall meet the requirements of Article 705.

Transfer equipment, located on the load side of branch circuit protection, shall be permitted to contain supplementary overcurrent protection having an interrupting rating sufficient for the available fault current that the generator can deliver. The supplementary overcurrent protection devices shall be part of a listed transfer equipment.

Transfer equipment shall be required for all standby systems subject to the provisions of this article and for which an electric-utility supply is either the normal or standby source.

**702.7 Signals.** Audible and visual signal devices shall be provided, where practicable, for the following purposes.

**(1) Derangement.** To indicate derangement of the optional standby source.

**(2) Carrying Load.** To indicate that the optional standby source is carrying load.

**702.8 Signs.**

**(A) Standby.** A sign shall be placed at the service-entrance equipment that indicates the type and location of on-site optional standby power sources. A sign shall not be required for individual unit equipment for standby illumination.

**(B) Grounding.** Where the grounded circuit conductor connected to the optional standby power source is connected to a grounding electrode conductor at a location remote from the optional standby power source, there shall be a sign at the grounding location that shall identify all optional standby power and normal sources connected at that location.

## II. Circuit Wiring

**702.9 Wiring Optional Standby Systems.** The optional standby system wiring shall be permitted to occupy the same raceways, cables, boxes, and cabinets with other general wiring.

## III. Grounding

**702.10 Portable Generator Grounding.**

**(A) Separately Derived System.** Where a portable optional standby source is used as a separately derived system, it shall be grounded to a grounding electrode in accordance with 250.30.

**(B) Nonseparately Derived System.** Where a portable optional standby source is used as a nonseparately derived system, the equipment grounding conductor shall be bonded to the system grounding electrode.

## ARTICLE 705
## Interconnected Electric Power Production Sources

**705.1 Scope.** This article covers installation of one or more electric power production sources operating in parallel with a primary source(s) of electricity.

FPN: Examples of the types of primary sources are a utility supply, on-site electric power source(s), or other sources.

**705.2 Definition.** For purposes of this article, the following definition applies.

**Interactive System.** An electric power production system that is operating in parallel with and capable of delivering energy to an electric primary source supply system.

**705.3 Other Articles.** Interconnected electric power production sources shall comply with this article and also the applicable requirements of the articles in Table 705.3.

**Table 705.3 Other Articles**

| Equipment/System | Article |
|---|---|
| Generators | 445 |
| Emergency systems | 700 |
| Legally required standby systems | 701 |
| Optional standby systems | 702 |

*Exception No. 1: Installation of solar photovoltaic systems operated as interconnected power sources shall be in accordance with Article 690.*

*Exception No. 2: Installation of fuel cell systems operated as interconnected power sources shall be in accordance with Article 692.*

**705.10 Directory.** A permanent plaque or directory, denoting all electrical power sources on or in the premises, shall be installed at each service equipment location and at locations of all electric power production sources capable of being interconnected.

*Exception: Installations with large numbers of power production sources shall be permitted to be designated by groups.*

**705.12 Point of Connection.** The outputs of electric power production systems shall be interconnected at the premises service disconnecting means.

**(A) Integrated Electric System.** The outputs shall be permitted to be interconnected at a point or points elsewhere on the premises where the system qualifies as an integrated electric system and incorporates protective equipment in accordance with all applicable sections of Article 685.

**(B) General.** The outputs shall be permitted to be interconnected at a point or points elsewhere on the premises where all of the following conditions are met:

(1) The aggregate of nonutility sources of electricity has a capacity in excess of 100 kW, or the service is above 1000 volts.

(2) The conditions of maintenance and supervision ensure that qualified persons service and operate the system.

(3) Safeguards and protective equipment are established and maintained.

**705.14 Output Characteristics.** The output of a generator or other electric power production source operating in parallel with an electric supply system shall be compatible with the voltage, wave shape, and frequency of the system to which it is connected.

> FPN: The term *compatible* does not necessarily mean matching the primary source wave shape.

**705.16 Interrupting and Short-Circuit Current Rating.** Consideration shall be given to the contribution of fault currents from all interconnected power sources for the interrupting and short-circuit current ratings of equipment on interactive systems.

**705.20 Disconnecting Means, Sources.** Means shall be provided to disconnect all ungrounded conductors of an electric power production source(s) from all other conductors.

**705.21 Disconnecting Means, Equipment.** Means shall be provided to disconnect equipment, such as inverters or transformers associated with a power production source from all ungrounded conductors of all sources of supply. Equipment intended to be operated and maintained as an integral part of a power production source exceeding 1000 volts shall not be required to have a disconnecting means.

**705.22 Disconnect Device.** The disconnecting means for ungrounded conductors shall consist of a manually or power operable switch(es) or circuit breaker(s) with the following features:

(1) Located where accessible

(2) Externally operable without exposing the operator to contact with live parts and, if power operable, of a type that can be opened by hand in the event of a power supply failure

(3) Plainly indicating whether in the open or closed position

(4) Having ratings not less than the load to be carried and the fault current to be interrupted

For disconnect equipment energized from both sides, a marking shall be provided to indicate that all contacts of the disconnect equipment may be energized.

> FPN No. 1: In parallel generation systems, some equipment, including knife blade switches and fuses, is likely to be energized from both directions. See 240.40.

> FPN No. 2: Interconnection to an off-premises primary source could require a visibly verifiable disconnecting device.

**705.30 Overcurrent Protection.** Conductors shall be protected in accordance with Article 240. Equipment and conductors connected to more than one electrical source shall have a sufficient number of overcurrent devices located so as to provide protection from all sources.

**(A) Generators.** Generators shall be protected in accordance with 445.12.

**(B) Solar Photovoltaic Systems.** Solar photovoltaic systems shall be protected in accordance with Article 690.

**(C) Transformers.** Overcurrent protection for a transformer with a source(s) on each side shall be provided in accordance with 450.3 by considering first one side of the transformer, then the other side of the transformer, as the primary.

**(D) Fuel Cell Systems.** Fuel cell systems shall be protected in accordance with Article 692.

**705.32 Ground-Fault Protection.** Where ground-fault protection is used, the output of an interactive system shall be connected to the supply side of the ground-fault protection.

*Exception: Connection shall be permitted to be made to the load side of ground-fault protection, provided that there is ground-fault protection for equipment from all ground-fault current sources.*

**705.40 Loss of Primary Source.** Upon loss of primary source, an electric power production source shall be automatically disconnected from all ungrounded conductors of the primary source and shall not be reconnected until the primary source is restored.

> FPN No. 1: Risks to personnel and equipment associated with the primary source could occur if an interactive electric power production source can operate as an island. Special detection methods can be required to determine that a primary source supply system outage has occurred and whether there should be automatic disconnection. When the

primary source supply system is restored, special detection methods can be required to limit exposure of power production sources to out-of-phase reconnection.

FPN No. 2: Induction-generating equipment on systems with significant capacitance can become self-excited upon loss of primary source and experience severe overvoltage as a result.

**705.42 Unbalanced Interconnections.** A 3-phase electric power production source shall be automatically disconnected from all ungrounded conductors of the interconnected systems when one of the phases of that source opens. This requirement shall not be applicable to an electric power production source providing power for an emergency or legally required standby system.

**705.43 Synchronous Generators.** Synchronous generators in a parallel system shall be provided with the necessary equipment to establish and maintain a synchronous condition.

**705.50 Grounding.** Interconnected electric power production sources shall be grounded in accordance with Article 250.

*Exception: For direct-current systems connected through an inverter directly to a grounded service, other methods that accomplish equivalent system protection and that utilize equipment listed and identified for the use shall be permitted.*

ance receptacle shall not be smaller than 10 AWG copper or equivalent.

**720.5 Lampholders.** Standard lampholders that have a rating of not less than 660 watts shall be used.

**720.6 Receptacle Rating.** Receptacles shall have a rating of not less than 15 amperes.

**720.7 Receptacles Required.** Receptacles of not less than 20-ampere rating shall be provided in kitchens, laundries, and other locations where portable appliances are likely to be used.

**720.8 Overcurrent Protection.** Overcurrent protection shall comply with Article 240.

**720.9 Batteries.** Installations of storage batteries shall comply with Article 480.

**720.10 Grounding.** Grounding shall be as provided in Article 250.

**720.11 Mechanical Execution of Work.** Circuits operating at less than 50 volts shall be installed in a neat and workmanlike manner. Cables shall be supported by the building structure in such a manner that the cable will not be damaged by normal building use.

# ARTICLE 720
## Circuits and Equipment Operating at Less Than 50 Volts

**720.1 Scope.** This article covers installations operating at less than 50 volts, direct current or alternating current.

**720.2 Other Articles.** Installations operating at less than 50 volts, direct current or alternating current as covered in Articles 411, 551, 650, 669, 690, 725, and 760 shall not be required to comply with this article.

**720.3 Hazardous (Classified) Locations.** Installations coming within the scope of this article and installed in hazardous (classified) locations shall also comply with the appropriate provisions of Articles 500 through 517.

**720.4 Conductors.** Conductors shall not be smaller than 12 AWG copper or equivalent. Conductors for appliance branch circuits supplying more than one appliance or appli-

# ARTICLE 725
## Class 1, Class 2, and Class 3 Remote-Control, Signaling, and Power-Limited Circuits

### I. General

**725.1 Scope.** This article covers remote-control, signaling, and power-limited circuits that are not an integral part of a device or appliance.

FPN: The circuits described herein are characterized by usage and electrical power limitations that differentiate them from electric light and power circuits; therefore, alternative requirements to those of Chapters 1 through 4 are given with regard to minimum wire sizes, derating factors, overcurrent protection, insulation requirements, and wiring methods and materials.

**725.2 Definitions.** For purposes of this article, the following definitions apply.

**Abandoned Class 2, Class 3, and PLTC Cable.** Installed Class 2, Class 3, and PLTC cable that is not terminated at equipment and not identified for future use with a tag.

**Class 1 Circuit.** The portion of the wiring system between the load side of the overcurrent device or power-limited supply and the connected equipment. The voltage and power limitations of the source are in accordance with 725.21.

**Class 2 Circuit.** The portion of the wiring system between the load side of a Class 2 power source and the connected equipment. Due to its power limitations, a Class 2 circuit considers safety from a fire initiation standpoint and provides acceptable protection from electric shock.

**Class 3 Circuit.** The portion of the wiring system between the load side of a Class 3 power source and the connected equipment. Due to its power limitations, a Class 3 circuit considers safety from a fire initiation standpoint. Since higher levels of voltage and current than Class 2 are permitted, additional safeguards are specified to provide protection from an electric shock hazard that could be encountered.

**725.3 Locations and Other Articles.** Circuits and equipment shall comply with the articles or sections listed in 725.3(A) through (F). Only those sections of Article 300 referenced in this article shall apply to Class 1, Class 2, and Class 3 circuits.

**(A) Number and Size of Conductors in Raceway.** Section 300.17.

**(B) Spread of Fire or Products of Combustion.** Section 300.21. The accessible portion of abandoned Class 2, Class 3, and PLTC cables shall not be permitted to remain.

**(C) Ducts, Plenums, and Other Air-Handling Spaces.** Section 300.22 for Class 1, Class 2, and Class 3 circuits installed in ducts, plenums, or other space used for environmental air. Type CL2P or CL3P cables shall be permitted for Class 2 and Class 3 circuits.

**(D) Hazardous (Classified) Locations.** Articles 500 through 516 and Article 517, Part IV, where installed in hazardous (classified) locations.

**(E) Cable Trays.** Article 392, where installed in cable tray.

**(F) Motor Control Circuits.** Article 430, Part VI, where tapped from the load side of the motor branch-circuit protective device(s) as specified in 430.72(A).

**725.5 Access to Electrical Equipment Behind Panels Designed to Allow Access.** Access to electrical equipment shall not be denied by an accumulation of wires and cables that prevents removal of panels, including suspended ceiling panels.

**725.6 Mechanical Execution of Work.** Class 1, Class 2, and Class 3 circuits shall be installed in a neat and workmanlike manner. Cables and conductors installed exposed on the outer surface of ceiling and sidewalls shall be supported by structural components of the building in such a manner that the cable or conductors will not be damaged by normal building use. Such cables shall be attached to structural components by straps, staples, hangers, or similar fittings designed and installed so as not to damage the cable. The installation shall also conform with 300.4(D).

**725.8 Safety-Control Equipment.**

**(A) Remote-Control Circuits.** Remote-control circuits for safety-control equipment shall be classified as Class 1 if the failure of the equipment to operate introduces a direct fire or life hazard. Room thermostats, water temperature regulating devices, and similar controls used in conjunction with electrically controlled household heating and air conditioning shall not be considered safety-control equipment.

**(B) Physical Protection.** Where damage to remote-control circuits of safety control equipment would introduce a hazard, as covered in 725.8(A), all conductors of such remote-control circuits shall be installed in rigid metal conduit, intermediate metal conduit, rigid nonmetallic conduit, electrical metallic tubing, Type MI cable, Type MC cable, or be otherwise suitably protected from physical damage.

**725.9 Class 1, Class 2, and Class 3 Circuit Grounding.** Class 1, Class 2, and Class 3 circuits and equipment shall be grounded in accordance with Article 250.

**725.10 Class 1, Class 2, and Class 3 Circuit Identification.** Class 1, Class 2, and Class 3 circuits shall be identified at terminal and junction locations, in a manner that prevents unintentional interference with other circuits during testing and servicing.

**725.15 Class 1, Class 2, and Class 3 Circuit Requirements.** A remote-control, signaling, or power-limited circuit shall comply with the following parts of this article:

(1) Class 1 Circuits, Parts I and II
(2) Class 2 and Class 3 Circuits, Parts I and III

**II. Class 1 Circuits**

**725.21 Class 1 Circuit Classifications and Power Source Requirements.** Class 1 circuits shall be classified as either Class 1 power-limited circuits where they comply with the power limitations of 725.21(A) or as Class 1 remote-control and signaling circuits where they are used for remote control or signaling purposes and comply with the power limitations of 725.21(B).

**(A) Class 1 Power-Limited Circuits.** These circuits shall be supplied from a source that has a rated output of not more than 30 volts and 1000 volt-amperes.

**(1) Class 1 Transformers.** Transformers used to supply power-limited Class 1 circuits shall comply with Article 450.

**(2) Other Class 1 Power Sources.** Power sources other than transformers shall be protected by overcurrent devices rated at not more than 167 percent of the volt-ampere rating of the source divided by the rated voltage. The overcurrent devices shall not be interchangeable with overcurrent devices of higher ratings. The overcurrent device shall be permitted to be an integral part of the power supply.

To comply with the 1000 volt-ampere limitation of 725.21(A), the maximum output ($VA_{max}$) of power sources other than transformers shall be limited to 2500 volt-amperes, and the product of the maximum current ($I_{max}$) and maximum voltage ($V_{max}$) shall not exceed 10,000 volt-amperes. These ratings shall be determined with any overcurrent-protective device bypassed.

$VA_{max}$ is the maximum volt-ampere output after one minute of operation regardless of load and with overcurrent protection bypassed, if used. Current-limiting impedance shall not be bypassed when determining $VA_{max}$.

$I_{max}$ is the maximum output current under any noncapacitive load, including short circuit, and with overcurrent protection bypassed, if used. Current-limiting impedance should not be bypassed when determining $I_{max}$. Where a current-limiting impedance, listed for the purpose or as part of a listed product, is used in combination with a stored energy source, for example, storage battery, to limit the output current, $I_{max}$ limits apply after 5 seconds.

$V_{max}$ is the maximum output voltage regardless of load with rated input applied.

**(B) Class 1 Remote-Control and Signaling Circuits.** These circuits shall not exceed 600 volts. The power output of the source shall not be required to be limited.

**725.23 Class 1 Circuit Overcurrent Protection.** Overcurrent protection for conductors 14 AWG and larger shall be provided in accordance with the conductor ampacity, without applying the derating factors of 310.15 to the ampacity calculation. Overcurrent protection shall not exceed 7 amperes for 18 AWG conductors and 10 amperes for 16 AWG.

*Exception: Where other articles of this Code permit or require other overcurrent protection.*

FPN: For example, see 430.72 for motors, 610.53 for cranes and hoists, and 517.74(B) and 660.9 for X-ray equipment.

**725.24 Class 1 Circuit Overcurrent Device Location.** Overcurrent devices shall be located as specified in 725.24(A) through (E).

**(A) Point of Supply.** Overcurrent devices shall be located at the point where the conductor to be protected receives its supply.

**(B) Feeder Taps.** Class 1 circuit conductors shall be permitted to be tapped, without overcurrent protection at the tap, where the overcurrent device protecting the circuit conductor is sized to protect the tap conductor.

**(C) Transformer Taps.** Class 1 circuit conductors 14 AWG and larger that are tapped from the load side of the overcurrent-protective device(s) of a controlled light and power circuit shall require only short-circuit and ground-fault protection and shall be permitted to be protected by the branch-circuit overcurrent protective device(s) where the rating of the protective device(s) is not more than 300 percent of the ampacity of the Class 1 circuit conductor.

**(D) Primary Side of Transformer.** Class 1 circuit conductors supplied by the secondary of a single-phase transformer having only a 2-wire (single-voltage) secondary shall be permitted to be protected by overcurrent protection provided on the primary side of the transformer, provided this protection is in accordance with 450.3 and does not exceed the value determined by multiplying the secondary conductor ampacity by the secondary-to-primary transformer voltage ratio. Transformer secondary conductors other than 2 wire shall not be considered to be protected by the primary overcurrent protection.

**(E) Input Side of Electronic Power Source.** Class 1 circuit conductors supplied by the output of a single-phase, listed electronic power source, other than a transformer, having only a 2-wire (single voltage) output for connection to Class 1 circuits shall be permitted to be protected by overcurrent protection provided on the input side of the electronic power source, provided this protection does not exceed the value determined by multiplying the Class 1 circuit conductor ampacity by the output-to-input voltage ratio. Electronic power source outputs, other than 2 wire (single voltage), shall not be considered to be protected by the primary overcurrent protection.

**725.25 Class 1 Circuit Wiring Methods.** Installations of Class 1 circuits shall be in accordance with Article 300 and the other appropriate articles in Chapter 3.

*Exception No. 1: The provisions of 725.26 through 725.28 shall be permitted to apply in installations of Class 1 circuits.*

*Exception No. 2: Methods permitted or required by other articles of this Code shall apply to installations of Class 1 circuits.*

**725.26 Conductors of Different Circuits in the Same Cable, Cable Tray, Enclosure, or Raceway.** Class 1 circuits shall be permitted to be installed with other circuits as specified in 725.26(A) and (B).

**(A) Two or More Class 1 Circuits.** Class 1 circuits shall be permitted to occupy the same cable, cable tray, enclosure, or raceway without regard to whether the individual circuits are alternating current or direct current, provided all conductors are insulated for the maximum voltage of any conductor in the cable, cable tray, enclosure, or raceway.

**(B) Class 1 Circuits with Power Supply Circuits.** Class 1 circuits shall be permitted to be installed with power supply conductors as specified in 725.26(B)(1) through (B)(4).

**(1) In a Cable, Enclosure, or Raceway.** Class 1 circuits and power supply circuits shall be permitted to occupy the same cable, enclosure, or raceway only where the equipment powered is functionally associated.

**(2) In Factory- or Field-Assembled Control Centers.** Class 1 circuits and power supply circuits shall be permitted to be installed in factory- or field-assembled control centers.

**(3) In a Manhole.** Class 1 circuits and power supply circuits shall be permitted to be installed as underground conductors in a manhole in accordance with one of the following:

(1) The power-supply or Class 1 circuit conductors are in a metal-enclosed cable or Type UF cable.
(2) The conductors are permanently separated from the power-supply conductors by a continuous firmly fixed nonconductor, such as flexible tubing, in addition to the insulation on the wire.
(3) The conductors are permanently and effectively separated from the power supply conductors and securely fastened to racks, insulators, or other approved supports.

**(4)** In cable trays, where the Class 1 circuit conductors and power-supply conductors not functionally associated with them are separated by a solid fixed barrier of a material compatible with the cable tray, or where the power-supply or Class 1 circuit conductors are in a metal-enclosed cable.

**725.27 Class 1 Circuit Conductors.**

**(A) Sizes and Use.** Conductors of sizes 18 AWG and 16 AWG shall be permitted to be used, provided they supply loads that do not exceed the ampacities given in 402.5 and are installed in a raceway, an approved enclosure, or a listed cable. Conductors larger than 16 AWG shall not supply loads greater than the ampacities given in 310.15. Flexible cords shall comply with Article 400.

**(B) Insulation.** Insulation on conductors shall be suitable for 600 volts. Conductors larger than 16 AWG shall comply with Article 310. Conductors in sizes 18 AWG and 16 AWG shall be Type FFH-2, KF-2, KFF-2, PAF, PAFF, PF, PFF, PGF, PGFF, PTF, PTFF, RFH-2, RFHH-2, RFHH-3, SF-2, SFF-2, TF, TFF, TFFN, TFN, ZF, or ZFF. Conductors with other types and thicknesses of insulation shall be permitted if listed for Class 1 circuit use.

**725.28 Number of Conductors in Cable Trays and Raceway, and Derating.**

**(A) Class 1 Circuit Conductors.** Where only Class 1 circuit conductors are in a raceway, the number of conductors shall be determined in accordance with 300.17. The derating factors given in 310.15(B)(2)(a) shall apply only if such conductors carry continuous loads in excess of 10 percent of the ampacity of each conductor.

**(B) Power-Supply Conductors and Class 1 Circuit Conductors.** Where power-supply conductors and Class 1 circuit conductors are permitted in a raceway in accordance with 725.26, the number of conductors shall be determined in accordance with 300.17. The derating factors given in 310.15(B)(2)(a) shall apply as follows:

(1) To all conductors where the Class 1 circuit conductors carry continuous loads in excess of 10 percent of the ampacity of each conductor and where the total number of conductors is more than three
(2) To the power-supply conductors only, where the Class 1 circuit conductors do not carry continuous loads in excess of 10 percent of the ampacity of each conductor and where the number of power-supply conductors is more than three

**(C) Class 1 Circuit Conductors in Cable Trays.** Where Class 1 circuit conductors are installed in cable trays, they shall comply with the provisions of 392.9 through 392.11.

**725.29 Circuits Extending Beyond One Building.** Class 1 circuits that extend aerially beyond one building shall also meet the requirements of Article 225.

**III. Class 2 and Class 3 Circuits**

**725.41 Power Sources for Class 2 and Class 3 Circuits.**

**(A) Power Source.** The power source for a Class 2 or a Class 3 circuit shall be as specified in 725.41(A)(1), (2), (3), (4), or (5):

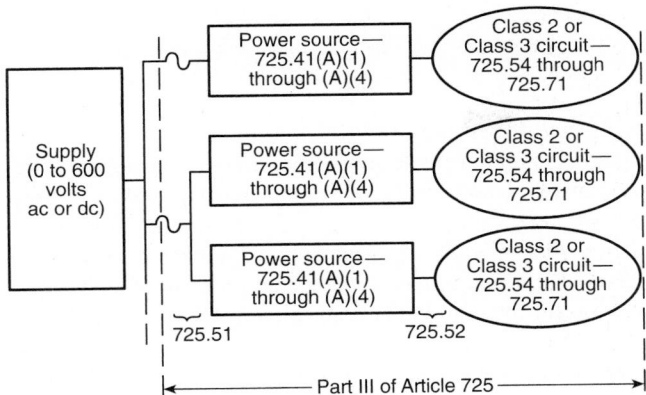

**Figure 725.41 Class 2 and Class 3 circuits.**

FPN No. 1:  Figure 725.41 illustrates the relationships between Class 2 or Class 3 power sources, their supply, and the Class 2 or Class 3 circuits.

FPN No. 2:  Tables 11(A) and 11(B) in Chapter 9 provide the requirements for listed Class 2 and Class 3 power sources.

(1)  A listed Class 2 or Class 3 transformer
(2)  A listed Class 2 or Class 3 power supply
(3)  Other listed equipment marked to identify the Class 2 or Class 3 power source

*Exception:  Thermocouples shall not require listing as a Class 2 power source.*

FPN:  Examples of other listed equipment are as follows:

(1)  A circuit card listed for use as a Class 2 or Class 3 power source where used as part of a listed assembly
(2)  A current-limiting impedance, listed for the purpose, or part of a listed product, used in conjunction with a non–power-limited transformer or a stored energy source, for example, storage battery, to limit the output current
(3)  A thermocouple

(4)  Listed information technology (computer) equipment limited power circuits.

FPN:  One way to determine applicable requirements for listing of information technology (computer) equipment is to refer to UL 1950-1995, *Standard for Safety of Information Technology Equipment, Including Electrical Business Equipment.* Typically such circuits are used to interconnect information technology equipment for the purpose of exchanging information (data).

(5)  A dry cell battery shall be considered an inherently limited Class 2 power source, provided the voltage is 30 volts or less and the capacity is equal to or less than that available from series connected No. 6 carbon zinc cells.

**(B)  Interconnection of Power Sources.** Class 2 or Class 3 power sources shall not have the output connections paralleled or otherwise interconnected unless listed for such interconnection.

**725.42 Circuit Marking.** The equipment shall be durably marked where plainly visible to indicate each circuit that is a Class 2 or Class 3 circuit.

**725.51 Wiring Methods on Supply Side of the Class 2 or Class 3 Power Source.** Conductors and equipment on the supply side of the power source shall be installed in accordance with the appropriate requirements of Chapters 1 through 4. Transformers or other devices supplied from electric light or power circuits shall be protected by an overcurrent device rated not over 20 amperes.

*Exception:  The input leads of a transformer or other power source supplying Class 2 and Class 3 circuits shall be permitted to be smaller than 14 AWG, but not smaller than 18 AWG if they are not over 12 in. (305 mm) long and if they have insulation that complies with 725.27(B).*

**725.52 Wiring Methods and Materials on Load Side of the Class 2 or Class 3 Power Source.** Class 2 and Class 3 circuits on the load side of the power source shall be permitted to be installed using wiring methods and materials in accordance with either 725.52(A) or (B).

**(A)  Class 1 Wiring Methods and Materials.** Installation shall be in accordance with 725.25.

*Exception No. 1:  The derating factors that are given in 310.15(B)(2)(a) shall not apply.*

*Exception No. 2:  Class 2 and Class 3 circuits shall be permitted to be reclassified and installed as Class 1 circuits if the Class 2 and Class 3 markings required in 725.42 are eliminated and the entire circuit is installed using the wiring methods and materials in accordance with Part II, Class 1 circuits.*

FPN:  Class 2 and Class 3 circuits reclassified and installed as Class 1 circuits are no longer Class 2 or Class 3 circuits, regardless of the continued connection to a Class 2 or Class 3 power source.

**(B)  Class 2 and Class 3 Wiring Methods.** Conductors on the load side of the power source shall be insulated at not less than the requirements of 725.71 and shall be installed in accordance with 725.54 and 725.61.

*Exception No. 1:  As provided for in 620.21 for elevators and similar equipment.*

*Exception No. 2:  Other wiring methods and materials installed in accordance with the requirements of 725.3 shall be permitted to extend or replace the conductors and cables described in 725.71 and permitted by 725.52(B).*

**725.54 Installation of Conductors and Equipment in Cables, Compartments, Cable Trays, Enclosures, Manholes, Outlet Boxes, Device Boxes, and Raceways for Class 2 and Class 3 Circuits.** Conductors and equipment for Class 2 and Class 3 circuits shall be installed in accordance with 725.55 through 725.58.

**725.55 Separation from Electric Light, Power, Class 1, Non–Power-Limited Fire Alarm Circuit Conductors, and Medium Power Network-Powered Broadband Communications Cables.**

**(A) General.** Cables and conductors of Class 2 and Class 3 circuits shall not be placed in any cable, cable tray, compartment, enclosure, manhole, outlet box, device box, raceway, or similar fitting with conductors of electric light, power, Class 1, non–power-limited fire alarm circuits, and medium power network-powered broadband communications circuits unless permitted by 725.55(B) through (J).

**(B) Separated by Barriers.** Class 2 and Class 3 circuits shall be permitted to be installed together with Class 1, non–power-limited fire alarm and medium power network-powered broadband communications circuits where they are separated by a barrier.

**(C) Raceways Within Enclosures.** In enclosures, Class 2 and Class 3 circuits shall be permitted to be installed in a raceway to separate them from Class 1, non–power-limited fire alarm and medium power network-powered broadband communications circuits.

**(D) Associated Systems Within Enclosures.** Class 2 and Class 3 circuit conductors in compartments, enclosures, device boxes, outlet boxes, or similar fittings shall be permitted to be installed with electric light, power, Class 1, non–power-limited fire alarm, and medium power network-powered broadband communications circuits where they are introduced solely to connect the equipment connected to Class 2 and Class 3 circuits, and where (1) or (2) applies:

(1) The electric light, power, Class 1, non–power-limited fire alarm, and medium power network-powered broadband communications circuit conductors are routed to maintain a minimum of 6 mm (0.25 in.) separation from the conductors and cables of Class 2 and Class 3 circuits.

(2) The circuit conductors operate at 150 volts or less to ground and also comply with one of the following:

 a. The Class 2 and Class 3 circuits are installed using Type CL3, CL3R, or CL3P or permitted substitute cables, provided these Class 3 cable conductors extending beyond the jacket are separated by a minimum of 6 mm (0.25 in.) or by a nonconductive sleeve or nonconductive barrier from all other conductors.

 b. The Class 2 and Class 3 circuit conductors are installed as a Class 1 circuit in accordance with 725.21.

**(E) Enclosures with Single Opening.** Class 2 and Class 3 circuit conductors entering compartments, enclosures, device boxes, outlet boxes, or similar fittings shall be permitted to be installed with Class 1, non–power-limited fire alarm and medium power network-powered broadband communications circuits where they are introduced solely to connect the equipment connected to Class 2 and Class 3 circuits. Where Class 2 and Class 3 circuit conductors must enter an enclosure that is provided with a single opening, they shall be permitted to enter through a single fitting (such as a tee), provided the conductors are separated from the conductors of the other circuits by a continuous and firmly fixed nonconductor, such as flexible tubing.

**(F) Manholes.** Underground Class 2 and Class 3 circuit conductors in a manhole shall be permitted to be installed with Class 1, non–power-limited fire alarm and medium power network-powered broadband communications circuits where one of the following conditions is met:

(1) The electric light, power, Class 1, non–power-limited fire alarm and medium power network-powered broadband communications circuit conductors are in a metal-enclosed cable or Type UF cable.

(2) The Class 2 and Class 3 circuit conductors are permanently and effectively separated from the conductors of other circuits by a continuous and firmly fixed nonconductor, such as flexible tubing, in addition to the insulation or covering on the wire.

(3) The Class 2 and Class 3 circuit conductors are permanently and effectively separated from conductors of the other circuits and securely fastened to racks, insulators, or other approved supports.

**(G) Article 780.** Class 2 and Class 3 conductors as permitted by 780.6(A) shall be permitted to be installed in accordance with Article 780.

**(H) Cable Trays.** Class 2 and Class 3 circuit conductors shall be permitted to be installed in cable trays, where the conductors of the electric light, Class 1, and non–power-limited fire alarm circuits are separated by a solid fixed barrier of a material compatible with the cable tray or where the Class 2 or Class 3 circuits are installed in Type MC cable.

**(I) In Hoistways.** In hoistways, Class 2 or Class 3 circuit conductors shall be installed in rigid metal conduit, rigid nonmetallic conduit, intermediate metal conduit, liquidtight flexible nonmetallic conduit, or electrical metallic tubing. For elevators or similar equipment, these conductors shall be permitted to be installed as provided in 620.21.

**(J) Other Applications.** For other applications, conductors of Class 2 and Class 3 circuits shall be separated by at least 50 mm (2 in.) from conductors of any electric light, power, Class 1 non–power-limited fire alarm or medium power network-powered broadband communications circuits unless one of the following conditions is met:

(1) Either (a) all of the electric light, power, Class 1, non–power-limited fire alarm and medium power network-powered broadband communications circuit conductors or (b) all of the Class 2 and Class 3 circuit conductors are in a raceway or in metal-sheathed, metal-clad, non–metallic-sheathed, or Type UF cables.

(2) All of the electric light, power, Class 1 non–power-limited fire alarm, and medium power network-powered broadband communications circuit conductors are permanently separated from all of the Class 2 and Class 3 circuit conductors by a continuous and firmly fixed nonconductor, such as porcelain tubes or flexible tubing, in addition to the insulation on the conductors.

**725.56 Installation of Conductors of Different Circuits in the Same Cable, Enclosure, or Raceway.**

**(A) Two or More Class 2 Circuits.** Conductors of two or more Class 2 circuits shall be permitted within the same cable, enclosure, or raceway.

**(B) Two or More Class 3 Circuits.** Conductors of two or more Class 3 circuits shall be permitted within the same cable, enclosure, or raceway.

**(C) Class 2 Circuits with Class 3 Circuits.** Conductors of one or more Class 2 circuits shall be permitted within the same cable, enclosure, or raceway with conductors of Class 3 circuits, provided that the insulation of the Class 2 circuit conductors in the cable, enclosure, or raceway is at least that required for Class 3 circuits.

**(D) Class 2 and Class 3 Circuits with Communications Circuits.**

**(1) Classified as Communications Circuits.** Class 2 and Class 3 circuit conductors shall be permitted in the same cable with communications circuits, in which case the Class 2 and Class 3 circuits shall be classified as communications circuits and shall be installed in accordance with the requirements of Article 800. The cables shall be listed as communications cables or multipurpose cables.

**(2) Composite Cables.** Cables constructed of individually listed Class 2, Class 3, and communications cables under a common jacket shall be permitted to be classified as communications cables. The fire resistance rating of the composite cable shall be determined by the performance of the composite cable.

**(E) Class 2 or Class 3 Cables with Other Circuit Cables.** Jacketed cables of Class 2 or Class 3 circuits shall be permitted in the same enclosure or raceway with jacketed cables of any of the following:

(1) Power-limited fire alarm systems in compliance with Article 760

(2) Nonconductive and conductive optical fiber cables in compliance with Article 770

(3) Communications circuits in compliance with Article 800

(4) Community antenna television and radio distribution systems in compliance with Article 820

(5) Low-power, network-powered broadband communications in compliance with Article 830

**725.57 Installation of Circuit Conductors Extending Beyond One Building.** Where Class 2 or Class 3 circuit conductors extend beyond one building and are run so as to be subject to accidental contact with electric light or power conductors operating over 300 volts to ground, or are exposed to lightning on interbuilding circuits on the same premises, the requirements of the following shall also apply:

(1) Sections 800.10, 800.12, 800.13, 800.31, 800.32, 800.33, and 800.40 for other than coaxial conductors

(2) Sections 820.10, 820.33, and 820.40 for coaxial conductors

**725.58 Support of Conductors.** Class 2 or Class 3 circuit conductors shall not be strapped, taped, or attached by any means to the exterior of any conduit or other raceway as a means of support. These conductors shall be permitted to be installed as permitted by 300.11(B)(2).

**725.61 Applications of Listed Class 2, Class 3, and PLTC Cables.** Class 2, Class 3, and PLTC cables shall comply with any of the requirements described in 725.61(A) through (F).

**(A) Plenum.** Cables installed in ducts, plenums, and other spaces used for environmental air shall be Type CL2P or CL3P. Abandoned cables shall not be permitted to remain. Listed wires and cables installed in compliance with 300.22 shall be permitted.

**(B) Riser.** Cables installed in risers shall be as described in any of (1), (2), or (3):

(1) Cables installed in vertical runs and penetrating more than one floor, or cables installed in vertical runs in a shaft, shall be Type CL2R or CL3R. Floor penetrations requiring Type CL2R or CL3R shall contain only cables suitable for riser or plenum use. Abandoned cables shall not be permitted to remain.

**Table 725.61 Cable Uses and Permitted Substitutions**

| Cable Type | Use | References | Permitted Substitutions |
|---|---|---|---|
| CL3P | Class 3 plenum cable | 725.61(A) | CMP |
| CL2P | Class 2 plenum cable | 725.61(A) | CMP, CL3P |
| CL3R | Class 3 riser cable | 725.61(B) | CMP, CL3P, CMR |
| CL2R | Class 2 riser cable | 725.61(B) | CMP, CL3P, CL2P, CMR, CL3R |
| PLTC | Power-limited tray cable | 725.61(C) and (D) | |
| CL3 | Class 3 cable | 725.61(B), (E), and (F) | CMP, CL3P, CMR, CL3R, CMG, CM, PLTC |
| CL2 | Class 2 cable | 725.61(B), (E), and (F) | CMP, CL3P, CL2P, CMR, CL3R, CL2R, CMG, CM, PLTC, CL3 |
| CL3X | Class 3 cable, limited use | 725.61(B) and (E) | CMP, CL3P, CMR, CL3R, CMG, CM, PLTC, CL3, CMX |
| CL2X | Class 2 cable, limited use | 725.61(B) and (E) | CMP, CL3P, CL2P, CMR, CL3R, CL2R, CMG, CM, PLTC, CL3, CL2, CMX, CL3X |

(2) Other cables as covered in Table 725.61 and other listed wiring methods as covered in Chapter 3 shall be installed in metal raceways or located in a fireproof shaft having firestops at each floor.

(3) Type CL2, CL3, CL2X, and CL3X cables shall be permitted in one- and two-family dwellings.

FPN: See 300.21 for firestop requirements for floor penetrations.

**(C) Cable Trays.** Cables installed in cable trays outdoors shall be Type PLTC. Cables installed in cable trays indoors shall be Types PLTC, CL3P, CL3R, CL3, CL2P, CL2R, and CL2.

FPN: See 800.52(D) for cables permitted in cable trays.

**(D) Hazardous (Classified) Locations.** Cables installed in hazardous locations shall be as described in 725.61(D)(1) through (D)(4).

**(1) Type PLTC.** Cables installed in hazardous (classified) locations shall be Type PLTC. Where the use of Type PLTC cable is permitted by 501.4(B), 502.4(B), and 504.20, the cable shall be installed in cable trays, in raceways supported by messenger wire, or otherwise adequately supported and mechanically protected by angles, struts, channels, or other mechanical means. The cable shall be permitted to be directly buried where the cable is listed for this use.

**(2) Nonincendive Field Wiring.** Wiring for Class 2 circuits as permitted by 501.4(B)(3) shall be permitted.

**(3) Thermocouple Circuits.** Conductors in Type PLTC cables used for Class 2 thermocouple circuits shall be permitted to be any of the materials used for thermocouple extension wire.

**(4) In Industrial Establishments.** In industrial establishments where the conditions of maintenance and supervision ensure that only qualified persons service the installation, and where the cable is not subject to physical damage, Type PLTC cable that complies with the crush and impact requirements of Type MC cable and is identified for such use shall be permitted as open wiring between cable tray and utilization equipment in lengths not to exceed 15 m (50 ft). The cable shall be supported and protected against physical damage using mechanical protection such as dedicated struts, angles, or channels. The cable shall be supported and secured at intervals not exceeding 1.75 m (6 ft).

**(E) Other Wiring Within Buildings.** Cables installed in building locations other than those covered in 725.61(A) through (D) shall be as described in any of (1) through (6). Abandoned cables in hollow spaces shall not be permitted to remain.

(1) Type CL2 or CL3 shall be permitted.
(2) Type CL2X or CL3X shall be permitted to be installed in a raceway or in accordance with other wiring methods covered in Chapter 3.
(3) Cables shall be permitted to be installed in nonconcealed spaces where the exposed length of cable does not exceed 3 m (10 ft).
(4) Listed Type CL2X cables less than 6 mm (0.25 in.) in diameter and listed Type CL3X cables less than 6 mm (0.25 in.) in diameter shall be permitted to be installed in one- and two-family dwellings.
(5) Listed Type CL2X cables less than 6 mm (0.25 in.) in diameter and listed Type CL3X cables less than 6 mm (0.25 in.) in diameter shall be permitted to be installed in nonconcealed spaces in multifamily dwellings.
(6) Type CMUC undercarpet communications wires and cables shall be permitted to be installed under carpet.

**(F) Cross-Connect Arrays.** Type CL2 or CL3 conductors or cables shall be used for cross-connect arrays.

**(G) Class 2 and Class 3 Cable Uses and Permitted Substitutions.** The uses and permitted substitutions for Class 2 and Class 3 cables listed in Table 725.61 shall be considered suitable for the purpose and shall be permitted.

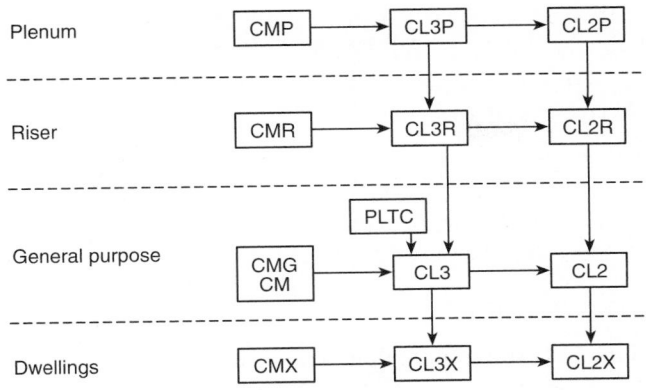

Type CM—Communications wires and cables
Type CL2 and CL3—Class 2 and Class 3 remote-control, signaling, and power-limited cables
Type PLTC—Power-limited tray cable

A ▶ B    Cable A shall be permitted to be used in place of cable B.

**Figure 725.61 Cable substitution hierarchy**

> FPN: For information on Types CMP, CMR, CM, and CMX cables, see 800.51.

### 725.71 Listing and Marking of Class 2, Class 3, and Type PLTC Cables.
Class 2, Class 3, and Type PLTC cables installed as wiring within buildings shall be listed as being resistant to the spread of fire and other criteria in accordance with 725.71(A) through (G) and shall be marked in accordance with 725.71(H).

**(A) Types CL2P and CL3P.** Types CL2P and CL3P plenum cables shall be listed as being suitable for use in ducts, plenums, and other space used for environmental air and shall also be listed as having adequate fire-resistant and low smoke-producing characteristics.

> FPN: One method of defining *low smoke-producing cable* is by establishing an acceptable value of the smoke produced when tested in accordance with NFPA 262-1999, *Standard Method of Test for Flame Travel and Smoke of Wires and Cables for Use in Air-Handling Spaces*, to a maximum peak optical density of 0.5 and a maximum average optical density of 0.15. Similarly, one method of defining fire-resistant cables is by establishing a maximum allowable flame travel distance of 1.52 m (5 ft) when tested in accordance with the same test.

**(B) Types CL2R and CL3R.** Types CL2R and CL3R riser cables shall be listed as being suitable for use in a vertical run in a shaft or from floor to floor and shall also be listed as having fire-resistant characteristics capable of preventing the carrying of fire from floor to floor.

> FPN: One method of defining fire-resistant characteristics capable of preventing the carrying of fire from floor to floor is that the cables pass the requirements of ANSI/UL 1666-

1997, *Test for Flame Propagation Height of Electrical and Optical-Fiber Cable Installed Vertically in Shafts.*

**(C) Types CL2 and CL3.** Types CL2 and CL3 cables shall be listed as being suitable for general-purpose use, with the exception of risers, ducts, plenums, and other space used for environmental air and shall also be listed as being resistant to the spread of fire.

> FPN: One method of defining *resistant to the spread of fire* is that the cables do not spread fire to the top of the tray in the vertical tray flame test in ANSI/UL 1581-1991, *Reference Standard for Electrical Wires, Cables and Flexible Cords.*
>
> Another method of defining *resistant to the spread of fire* is for the damage (char length) not to exceed 1.5 m (4 ft 11 in.) when performing the CSA vertical flame test for cables in cable trays, as described in CSA C22.2 No. 0.3-M-1985, *Test Methods for Electrical Wires and Cables.*

**(D) Types CL2X and CL3X.** Types CL2X and CL3X limited-use cables shall be listed as being suitable for use in dwellings and for use in raceway and shall also be listed as being resistant to flame spread.

> FPN: One method of determining that cable is resistant to flame spread is by testing the cable to the VW-1 (vertical-wire) flame test in ANSI/UL 1581-1991, *Reference Standard for Electrical Wires, Cables and Flexible Cords.*

**(E) Type PLTC.** Type PLTC nonmetallic-sheathed, power-limited tray cable shall be listed as being suitable for cable trays and shall consist of a factory assembly of two or more insulated conductors under a nonmetallic jacket. The insulated conductors shall be 22 AWG through 12 AWG. The conductor material shall be copper (solid or stranded). Insulation on conductors shall be suitable for 300 volts. The cable core shall be either (1) two or more parallel conductors, (2) one or more group assemblies of twisted or parallel conductors, or (3) a combination thereof. A metallic shield or a metallized foil shield with drain wire(s) shall be permitted to be applied either over the cable core, over groups of conductors, or both. The cable shall be listed as being resistant to the spread of fire. The outer jacket shall be a sunlight- and moisture-resistant nonmetallic material.

*Exception No. 1: Where a smooth metallic sheath, continuous corrugated metallic sheath, or interlocking tape armor is applied over the nonmetallic jacket, an overall nonmetallic jacket shall not be required. On metallic-sheathed cable without an overall nonmetallic jacket, the information required in 310.11 shall be located on the nonmetallic jacket under the sheath.*

*Exception No. 2: Conductors in PLTC cables used for Class 2 thermocouple circuits shall be permitted to be any of the materials used for thermocouple extension wire.*

> FPN: One method of defining *resistant to the spread of fire* is that the cables do not spread fire to the top of the tray in the vertical tray flame test in ANSI/UL 1581-1991, *Ref-*

*erence Standard for Electrical Wires, Cables and Flexible Cords.*

Another method of defining *resistant to the spread of fire* is for the damage (char length) not to exceed 1.5 m (4 ft 11 in.) when performing the CSA vertical flame test for cables in cable trays, as described in CSA C22.2 No. 0.3-M-1985, *Test Methods for Electrical Wires and Cables.*

**(F) Class 2 and Class 3 Cable Voltage Ratings.** Class 2 cables shall have a voltage rating of not less than 150 volts. Class 3 cables shall have a voltage rating of not less than 300 volts.

**(G) Class 3 Single Conductors.** Class 3 single conductors used as other wiring within buildings shall not be smaller than 18 AWG and shall be Type CL3. Conductor types described in 725.27(B) that are also listed as Type CL3 shall be permitted.

> FPN: One method of defining *resistant to the spread of fire* is that the cables do not spread fire to the top of the tray in the vertical tray flame test in ANSI/UL 1581-1991, *Reference Standard for Electrical Wires, Cables and Flexible Cords.*
>
> Another method of defining *resistant to the spread of fire* is for the damage (char length) not to exceed 1.5 m (4 ft 11 in.) when performing the CSA vertical flame test for cables in cable trays as described in CSA C22.2 No. 0.3-M-1985, *Test Methods for Electrical Wires and Cables.*

**(H) Marking.** Cables shall be marked in accordance with 310.11(A)(2), (3), (4), and (5) and Table 725.71. Voltage ratings shall not be marked on the cables.

> FPN: Voltage markings on cables may be misinterpreted to suggest that the cables may be suitable for Class 1 electric light and power applications.

*Exception: Voltage markings shall be permitted where the cable has multiple listings and a voltage marking is required for one or more of the listings.*

> FPN: Class 2 and Class 3 cable types are listed in descending order of fire resistance rating, and Class 3 cables are listed above Class 2 cables, because Class 3 cables can substitute for Class 2 cables.

**Table 725.71 Cable Markings**

| Cable Marking | Type | Listing References |
|---|---|---|
| CL3P | Class 3 plenum cable | 725.71(A), (F), and (H) |
| CL2P | Class 2 plenum cable | 725.71(A) and (H) |
| CL3R | Class 3 riser cable | 725.71(B), (F), and (H) |
| CL2R | Class 2 riser cable | 725.71(B) and (H) |
| PLTC | Power-limited tray cable | 725.71(E) and (H) |
| CL3 | Class 3 cable | 725.71(C), (F), and (H) |
| CL2 | Class 2 cable | 725.71(C), (F), and (H) |
| CL3X | Class 3 cable, limited use | 725.71(D), (F), and (H) |
| CL2X | Class 2 cable, limited use | 725.71(D), (F), and (H) |

# ARTICLE 727
# Instrumentation Tray Cable: Type ITC

**727.1 Scope.** This article covers the use, installation, and construction specifications of instrumentation tray cable for application to instrumentation and control circuits operating at 150 volts or less and 5 amperes or less.

**727.2 Definition.**

**Type ITC Instrumentation Tray Cable.** A factory assembly of two or more insulated conductors, with or without a grounding conductor(s), enclosed in a nonmetallic sheath.

**727.3 Other Articles.** In addition to the provisions of this article, installation of Type ITC cable shall comply with other applicable articles of this *Code*, such as Articles 240, 250, 300, and 318.

**727.4 Uses Permitted.** Type ITC cable shall be permitted to be used as follows in industrial establishments where the conditions of maintenance and supervision ensure that only qualified persons service the installation:

(1) In cable trays.
(2) In raceways.
(3) In hazardous locations as permitted in 501.4, 502.4, 503.3, 504.20, 504.30, 504.80, and 505.15.
(4) As open wiring where enclosed in a smooth metallic sheath, continuous corrugated metallic sheath, or interlocking tape armor applied over the nonmetallic sheath in accordance with 727.6. The cable shall be supported and secured at intervals not exceeding 1.8 m (6 ft).
(5) As open wiring without a metallic sheath or armor between cable tray and equipment in lengths not to exceed 15 m (50 ft), where the cable is supported and protected against physical damage using mechanical protection, such as struts, angles, or channels. The cable shall be supported and secured at intervals not exceeding 1.8 m (6 ft).
(6) As open wiring between cable tray and equipment in lengths not to exceed 15 m (50 ft), where the cable complies with the crush and impact requirements of Type MC cable and is identified for such use. The cable shall be supported and secured at intervals not exceeding 1.8 m (6 ft).
(7) As aerial cable on a messenger.
(8) Direct buried where identified for the use.
(9) Under raised floors in rooms containing industrial process control equipment and rack rooms where arranged to prevent damage to the cable.
(10) Under raised floors in information technology equipment rooms in accordance with 645.5(D)(5)(c).

**727.5 Uses Not Permitted.** Type ITC cable shall not be installed on circuits operating at more than 150 volts or more than 5 amperes.

Installation of Type ITC cable with other cables shall be subject to the stated provisions of the specific articles for the other cables. Where the governing articles do not contain stated provisions for installation with Type ITC cable, the installation of Type ITC cable with the other cables shall not be permitted.

Type ITC cable shall not be installed with power, lighting, Class 1, or non–power-limited circuits.

*Exception No. 1: Where terminated within equipment or junction boxes and separations are maintained by insulating barriers or other means.*

*Exception No. 2: Where a metallic sheath or armor is applied over the nonmetallic sheath of the Type ITC cable.*

**727.6 Construction.** The insulated conductors of Type ITC cable shall be in sizes 22 AWG through 12 AWG. The conductor material shall be copper or thermocouple alloy. Insulation on the conductors shall be rated for 300 volts. Shielding shall be permitted.

The cable shall be listed as being resistant to the spread of fire. The outer jacket shall be sunlight and moisture resistant.

Where a smooth metallic sheath, continuous corrugated metallic sheath, or interlocking tape armor is applied over the nonmetallic sheath, an overall nonmetallic jacket shall not be required.

**727.7 Marking.** The cable shall be marked in accordance with 310.11(A)(2), (3), (4), and (5). Voltage ratings shall not be marked on the cable.

**727.8 Allowable Ampacity.** The allowable ampacity of the conductors shall be 5 amperes, except for 22 AWG conductors that shall have an allowable ampacity of 3 amperes.

**727.9 Overcurrent Protection.** Overcurrent protection shall not exceed 5 amperes for 20 AWG and larger conductors, and 3 amperes for 22 AWG conductors.

**727.10 Bends.** Bends in Type ITC cables shall be made so as not to damage the cable.

# ARTICLE 760
# Fire Alarm Systems

## I. General

**760.1 Scope.** This article covers the installation of wiring and equipment of fire alarm systems including all circuits controlled and powered by the fire alarm system.

FPN No. 1: Fire alarm systems include fire detection and alarm notification, guard's tour, sprinkler waterflow, and sprinkler supervisory systems. Circuits controlled and powered by the fire alarm system include circuits for the control of building systems safety functions, elevator capture, elevator shutdown, door release, smoke doors and damper control, fire doors and damper control and fan shutdown, but only where these circuits are powered by and controlled by the fire alarm system. For further information on the installation and monitoring for integrity requirements for fire alarm systems, refer to the NFPA 72-1999, *National Fire Alarm Code®*.

FPN No. 2: Class 1, 2, and 3 circuits are defined in Article 725.

**760.2 Definitions.** For purposes of this article, the following definitions apply.

**Abandoned Fire Alarm Cable.** Installed fire alarm cable that is not terminated at equipment other than a connector and not identified for future use with a tag.

**Fire Alarm Circuit.** The portion of the wiring system between the load side of the overcurrent device or the power-limited supply and the connected equipment of all circuits powered and controlled by the fire alarm system. Fire alarm circuits are classified as either non–power-limited or power-limited.

**Fire Alarm Circuit Integrity (CI) Cable.** Cable used in fire alarm systems to ensure continued operation of critical circuits during a specified time under fire conditions.

**Non–Power-Limited Fire Alarm Circuit (NPLFA).** A fire alarm circuit powered by a source that complies with 760.21 and 760.23.

**Power-Limited Fire Alarm Circuit (PLFA).** A fire alarm circuit powered by a source that complies with 760.41.

**760.3 Locations and Other Articles.** Circuits and equipment shall comply with 760.3(A) through (F). Only those sections of Article 300 referenced in this article shall apply to fire alarm systems.

**(A) Spread of Fire or Products of Combustion.** Section 300.21. The accessible portion of abandoned fire alarm cables shall not be permitted to remain.

**(B) Ducts, Plenums, and Other Air-Handling Spaces.** Section 300.22, where installed in ducts or plenums or other spaces used for environmental air.

*Exception: As permitted in 760.30(B)(1) and (2) and 760.61(A).*

**(C) Hazardous (Classified) Locations.** Articles 500 through 516 and Article 517, Part IV, where installed in hazardous (classified) locations.

**(D) Corrosive, Damp, or Wet Locations.** Sections 110.11, 300.6, and 310.9 where installed in corrosive, damp, or wet locations.

**(E) Building Control Circuits.** Article 725 where building control circuits (e.g., elevator capture, fan shutdown) are associated with the fire alarm system.

**(F) Optical Fiber Cables.** Where optical fiber cables are utilized for fire alarm circuits, the cables shall be installed in accordance with Article 770.

**760.5 Access to Electrical Equipment Behind Panels Designed to Allow Access.** Access to electrical equipment shall not be denied by an accumulation of conductors and cables that prevents removal of panels, including suspended ceiling panels.

**760.6 Mechanical Execution of Work.** Fire alarm circuits shall be installed in a neat and workmanlike manner. Cables and conductors installed exposed on the surface of ceiling and sidewalls shall be supported by structural components of the building in such a manner that the cable or conductors will not be damaged by normal building use. Such cables shall be attached to structural components by straps, staples, hangers, or similar fittings designed and installed so as not to damage the cable. The installation shall also conform with 300.4(D).

**760.7 Fire Alarm Circuits Extending Beyond One Building.** Power-limited fire alarm circuits that extend beyond one building and run outdoors either shall meet the installation requirements of Parts II, III, and IV of Article 800 or shall meet the installation requirements of Part I of Article 300. Non–power-limited fire alarm circuits that extend beyond one building and run outdoors shall meet the installation requirements of Part I of Article 300 and the applicable sections of Part I of Article 225.

**760.9 Fire Alarm Circuit and Equipment Grounding.** Fire alarm circuits and equipment shall be grounded in accordance with Article 250.

**760.10 Fire Alarm Circuit Identification.** Fire alarm circuits shall be identified at terminal and junction locations, in a manner that will prevent unintentional interference with the signaling circuit during testing and servicing.

**760.15 Fire Alarm Circuit Requirements.** Fire alarm circuits shall comply with the following parts of this article.

**(A) Non–Power–Limited Fire Alarm (NPLFA) Circuits.** See Parts I and II.

**(B) Power-Limited Fire Alarm (PLFA) Circuits.** See Parts I and III.

## II. Non–Power-Limited Fire Alarm (NPLFA) Circuits

**760.21 NPLFA Circuit Power Source Requirements.** The power source of non–power-limited fire alarm circuits shall comply with Chapters 1 through 4, and the output voltage shall not be more than 600 volts, nominal. These circuits shall not be supplied through ground-fault circuit interrupters.

> FPN: See 210.8(A)(5), Exception No. 3 for receptacles in dwelling-unit unfinished basements that supply power for fire alarm systems.

**760.23 NPLFA Circuit Overcurrent Protection.** Overcurrent protection for conductors 14 AWG and larger shall be provided in accordance with the conductor ampacity without applying the derating factors of 310.15 to the ampacity calculation. Overcurrent protection shall not exceed 7 amperes for 18 AWG conductors and 10 amperes for 16 AWG conductors.

*Exception: Where other articles of this Code permit or require other overcurrent protection.*

**760.24 NPLFA Circuit Overcurrent Device Location.** Overcurrent devices shall be located at the point where the conductor to be protected receives its supply.

*Exception No. 1: Where the overcurrent device protecting the larger conductor also protects the smaller conductor.*

*Exception No. 2: Transformer secondary conductors. Non–power-limited fire alarm circuit conductors supplied by the secondary of a single-phase transformer that has only a 2-wire (single-voltage) secondary shall be permitted to be protected by overcurrent protection provided by the primary (supply) side of the transformer, provided the protection is in accordance with 450.3 and does not exceed the value determined by multiplying the secondary conductor ampacity by the secondary-to-primary transformer voltage ratio. Transformer secondary conductors other than 2-wire shall not be considered to be protected by the primary overcurrent protection.*

*Exception No. 3: Electronic power source output conductors. Non–power-limited circuit conductors supplied by the output of a single-phase, listed electronic power source, other than a transformer, having only a 2-wire (single-voltage) output for connection to non–power-limited circuits shall be permitted to be protected by overcurrent protection provided on the input side of the electronic power source, provided this protection does not exceed the value determined by multiplying the non–power-limited circuit conductor ampacity by the output-to-input voltage ratio.*

*Electronic power source outputs, other than 2-wire (single voltage), connected to non–power-limited circuits shall not be considered to be protected by overcurrent protection on the input of the electronic power source.*

> FPN: A single-phase, listed electronic power supply whose output supplies a 2-wire (single-voltage) circuit is an example of a non–power-limited power source that meets the requirements of 760.21.

**760.25 NPLFA Circuit Wiring Methods.** Installation of non–power-limited fire alarm circuits shall be in accordance with 110.3(B), 300.11(A), 300.15, 300.17, and other appropriate articles of Chapter 3.

*Exception No. 1: As provided in 760.26 through 760.30.*

*Exception No. 2: Where other articles of this Code require other methods.*

**760.26 Conductors of Different Circuits in Same Cable, Enclosure, or Raceway.**

**(A) Class 1 with NPLFA Circuits.** Class 1 and non–power-limited fire alarm circuits shall be permitted to occupy the same cable, enclosure, or raceway without regard to whether the individual circuits are alternating current or direct current, provided all conductors are insulated for the maximum voltage of any conductor in the enclosure or raceway.

**(B) Fire Alarm with Power-Supply Circuits.** Power-supply and fire alarm circuit conductors shall be permitted in the same cable, enclosure, or raceway only where connected to the same equipment.

**760.27 NPLFA Circuit Conductors.**

**(A) Sizes and Use.** Only copper conductors shall be permitted to be used for fire alarm systems. Size 18 AWG and 16 AWG conductors shall be permitted to be used, provided they supply loads that do not exceed the ampacities given in Table 402.5 and are installed in a raceway, an approved enclosure, or a listed cable. Conductors larger than 16 AWG shall not supply loads greater than the ampacities given in 310.15, as applicable.

**(B) Insulation.** Insulation on conductors shall be suitable for 600 volts. Conductors larger than 16 AWG shall comply with Article 310. Conductors 18 AWG and 16 AWG shall be Type KF-2, KFF-2, PAFF, PTFF, PF, PFF, PGF, PGFF, RFH-2, RFHH-2, RFHH-3, SF-2, SFF-2, TF, TFF, TFN, TFFN, ZF, or ZFF. Conductors with other types and thickness of insulation shall be permitted if listed for non–power-limited fire alarm circuit use.

> FPN: For application provisions, see Table 402.3.

**(C) Conductor Materials.** Conductors shall be solid or stranded copper.

*Exception to (B) and (C): Wire Types PAF and PTF shall be permitted only for high-temperature applications between 90°C (194°F) and 250°C (482°F).*

**760.28 Number of Conductors in Cable Trays and Raceways, and Derating.**

**(A) NPLFA Circuits and Class 1 Circuits.** Where only non–power-limited fire alarm circuit and Class 1 circuit conductors are in a raceway, the number of conductors shall be determined in accordance with 300.17. The derating factors given in 310.15(B)(2)(a) shall apply if such conductors carry continuous load in excess of 10 percent of the ampacity of each conductor.

**(B) Power-Supply Conductors and Fire Alarm Circuit Conductors.** Where power-supply conductors and fire alarm circuit conductors are permitted in a raceway in accordance with 760.26, the number of conductors shall be determined in accordance with 300.17. The derating factors given in 310.15(B)(2)(a) shall apply as follows:

(1) To all conductors where the fire alarm circuit conductors carry continuous loads in excess of 10 percent of the ampacity of each conductor and where the total number of conductors is more than three

(2) To the power-supply conductors only, where the fire alarm circuit conductors do not carry continuous loads in excess of 10 percent of the ampacity of each conductor and where the number of power-supply conductors is more than three

**(C) Cable Trays.** Where fire alarm circuit conductors are installed in cable trays, they shall comply with 392.9 through 392.11.

**760.30 Multiconductor NPLFA Cables.** Multiconductor non–power-limited fire alarm cables that meet the requirements of 760.31 shall be permitted to be used on fire alarm circuits operating at 150 volts or less and shall be installed in accordance with 760.30(A) and (B).

**(A) NPLFA Wiring Method.** Multiconductor non–power-limited fire alarm circuit cables shall be installed in accordance with 760.30(A)(1), (A)(2), and (A)(3).

**(1) Exposed or Fished in Concealed Spaces.** In raceway or exposed on surface of ceiling and sidewalls or fished in concealed spaces. Cable splices or terminations shall be made in listed fittings, boxes, enclosures, fire alarm devices, or utilization equipment. Where installed exposed, cables shall be adequately supported and installed in such a way that maximum protection against physical damage is afforded by building construction such as baseboards, door frames, ledges, and so forth. Where located within 2.1 m (7 ft) of the floor, cables shall be securely fastened in an approved manner at intervals of not more than 450 mm (18 in.).

**(2) Passing Through a Floor or Wall.** In metal raceway or rigid nonmetallic conduit where passing through a floor or wall to a height of 2.1 m (7 ft) above the floor unless adequate protection can be afforded by building construction such as detailed in 760.30(A)(1) or unless an equivalent solid guard is provided.

**(3) In Hoistways.** In rigid metal conduit, rigid nonmetallic conduit, intermediate metal conduit, liquidtight flexible nonmetallic tubing, or electrical metallic tubing where installed in hoistways.

*Exception: As provided for in 620.21 for elevators and similar equipment.*

**(B) Applications of Listed NPLFA Cables.** The use of non–power-limited fire alarm circuit cables shall comply with 760.30(B)(1) through (B)(4).

**(1) Ducts and Plenums.** Multiconductor non–power-limited fire alarm circuit cables, Types NPLFP, NPLFR, and NPLF, shall not be installed exposed in ducts or plenums.

    FPN: See 300.22(B).

**(2) Other Spaces Used for Environmental Air.** Cables installed in other spaces used for environmental air shall be Type NPLFP.

*Exception No. 1: Types NPLFR and NPLF cables installed in compliance with 300.22(C).*

*Exception No. 2: Other wiring methods in accordance with 300.22(C) and conductors in compliance with 760.27(C).*

**(3) Riser.** Cables installed in vertical runs and penetrating more than one floor or cables installed in vertical runs in a shaft shall be Type NPLFR. Floor penetrations requiring Type NPLFR shall contain only cables suitable for riser or plenum use.

*Exception No. 1: Type NPLF or other cables that are specified in Chapter 3 and are in compliance with 760.27(C) and encased in metal raceway.*

*Exception No. 2: Type NPLF cables located in a fireproof shaft having firestops at each floor.*

    FPN: See 300.21 for firestop requirements for floor penetrations.

**(4) Other Wiring Within Buildings.** Cables installed in building locations other than the locations covered in 760.30(B)(1), (B)(2), and (B)(3) shall be Type NPLF.

*Exception No. 1: Chapter 3 wiring methods with conductors in compliance with 760.27(C).*

*Exception No. 2: Type NPLFP or Type NPLFR cables shall be permitted.*

**760.31 Listing and Marking of NPLFA Cables.**
Non–power-limited fire alarm cables installed as wiring within buildings shall be listed in accordance with 760.31(A) and (B) and as being resistant to the spread of fire in accordance with 760.31(C) through (F), and shall be marked in accordance with 760.31(G).

**(A) NPLFA Conductor Materials.** Conductors shall be 18 AWG or larger solid or stranded copper.

**(B) Insulated Conductors.** Insulated conductors shall be suitable for 600 volts. Insulated conductors 14 AWG and larger shall be one of the types listed in Table 310.13 or one that is identified for this use. Insulated conductors 18 AWG and 16 AWG shall be in accordance with 760.27.

**(C) Type NPLFP.** Type NPLFP non–power-limited fire alarm cable for use in other space used for environmental air shall be listed as being suitable for use in other space used for environmental air as described in 300.22(C) and shall also be listed as having adequate fire-resistant and low smoke-producing characteristics.

    FPN: One method of defining low smoke-producing cable is by establishing an acceptable value of the smoke produced when tested in accordance with NFPA 262-1999, *Standard Method of Test for Flame Travel and Smoke of Wires and Cables for Use in Air-Handling Spaces*, to a maximum peak optical density of 0.5 and a maximum average optical density of 0.15. Similarly, one method of defining fire-resistant cables is by establishing a maximum allowable flame travel distance of 1.52 m (5 ft) when tested in accordance with the same test.

**(D) Type NPLFR.** Type NPLFR non–power-limited fire alarm riser cable shall be listed as being suitable for use in a vertical run in a shaft or from floor to floor and shall also be listed as having fire-resistant characteristics capable of preventing the carrying of fire from floor to floor.

    FPN: One method of defining fire-resistant characteristics capable of preventing the carrying of fire from floor to floor is that the cables pass ANSI/UL 1666-1997, *Test for Flame Propagation Height of Electrical and Optical-Fiber Cables Installed Vertically in Shafts.*

**(E) Type NPLF.** Type NPLF non–power-limited fire alarm cable shall be listed as being suitable for general-purpose fire alarm use, with the exception of risers, ducts, plenums, and other space used for environmental air, and shall also be listed as being resistant to the spread of fire.

    FPN No. 1: One method of defining *resistant to the spread of fire* is that the cables do not spread fire to the top of the tray in the vertical-tray flame test in ANSI/UL 1581-1991,

*Reference Standard for Electrical Wires, Cables and Flexible Cords.*

FPN No. 2: Another method of defining *resistant to the spread of fire* is for the damage (char length) not to exceed 1.5 m (4 ft 11 in.) when performing the CSA vertical flame test for cables in cable trays, as described in CSA C22.2 No. 0.3-M-1985, *Test Methods for Electrical Wires and Cables.*

**(F) Fire Alarm Circuit Integrity (CI) Cable.** Cables suitable for use in fire alarm systems to ensure survivability of critical circuits during a specified time under fire conditions shall be listed as circuit integrity (CI) cable. Cables identified in 760.31(C), (D), and (E) that meet the requirements for circuit integrity shall have the additional classification using the suffix "CI" (for example, NPLFP-CI, NPLFR-CI, and NPLF-CI).

FPN No. 1: This cable may be used for fire alarm circuits to comply with the survivability requirements of NFPA 72-1999, *National Fire Alarm Code*®, 3-4.2.2.2, 3-8.4.1.1.4, and 3-8.4.1.3.3.3(3), that the cable maintain its electrical function during fire conditions for a defined period of time.

FPN No. 2: One method of defining circuit integrity (CI) cable is by establishing a minimum 2-hour fire resistance rating for the cable when tested in accordance with UL 2196-1995, *Standard for Tests of Fire Resistive Cables.*

**(G) NPLFA Cable Markings.** Multiconductor non–power-limited fire alarm cables shall be marked in accordance with Table 760.31(G). Non–power-limited fire alarm circuit cables shall be permitted to be marked with a maximum usage voltage rating of 150 volts. Cables that are listed for circuit integrity shall be identified with the suffix "CI" as defined in 760.31(F).

**Table 760.31(G) NPLFA Cable Markings**

| Cable Marking | Type | Reference |
|---|---|---|
| NPLFP | Non–power-limited fire alarm circuit cable for use in other space used for environmental air | 760.31(C) and (G) |
| NPLFR | Non–power-limited fire alarm circuit riser cable | 760.31(D) and (G) |
| NPLF | Non–power-limited fire alarm circuit cable | 760.31(E) and (G) |

Note: Cables identified in 760.31(C), (D), and (E) and meeting the requirements for circuit integrity shall have the additional classification using the suffix "CI" (for example, NPLFP-CI, NPLFR-CI, and NPLF-CI).

FPN: Cable types are listed in descending order of fire resistance rating.

## III. Power-Limited Fire Alarm (PLFA) Circuits

**760.41 Power Sources for PLFA Circuits.** The power source for a power-limited fire alarm circuit shall be as specified in 760.41(A), (B), or (C). These circuits shall not be supplied through ground-fault circuit interrupters.

FPN No. 1: Tables 12(A) and 12(B) in Chapter 9 provide the listing requirements for power-limited fire alarm circuit sources.

FPN No. 2: See 210.8(A)(5), Exception No. 3 for receptacles in dwelling-unit unfinished basements that supply power for fire alarm systems.

**(A) Transformers.** A listed PLFA or Class 3 transformer.

**(B) Power Supplies.** A listed PLFA or Class 3 power supply.

**(C) Listed Equipment.** Listed equipment marked to identify the PLFA power source.

FPN: Examples of listed equipment are a fire alarm control panel with integral power source; a circuit card listed for use as a PLFA source, where used as part of a listed assembly; a current-limiting impedance, listed for the purpose or part of a listed product, used in conjunction with a non–power-limited transformer or a stored energy source, for example, storage battery, to limit the output current.

**760.42 Circuit Marking.** The equipment shall be durably marked where plainly visible to indicate each circuit that is a power-limited fire alarm circuit.

FPN: See 760.52(A), Exception No. 3 where a power-limited circuit is to be reclassified as a non–power-limited circuit.

**760.51 Wiring Methods on Supply Side of the PLFA Power Source.** Conductors and equipment on the supply side of the power source shall be installed in accordance with the appropriate requirements of Part II and Chapters 1 through 4. Transformers or other devices supplied from power-supply conductors shall be protected by an overcurrent device rated not over 20 amperes.

*Exception: The input leads of a transformer or other power source supplying power-limited fire alarm circuits shall be permitted to be smaller than 14 AWG, but not smaller than 18 AWG, if they are not over 300 mm (12 in.) long and if they have insulation that complies with 760.27(B).*

**760.52 Wiring Methods and Materials on Load Side of the PLFA Power Source.** Fire alarm circuits on the load side of the power source shall be permitted to be installed using wiring methods and materials in accordance with either 760.52(A) or (B).

**(A) NPLFA Wiring Methods and Materials.** Installation shall be in accordance with 760.25, and conductors shall be solid or stranded copper.

*Exception No. 1: The derating factors given in 310.15(B)(2)(a) shall not apply.*

*Exception No. 2: Conductors and multiconductor cables described in and installed in accordance with 760.27 and 760.30 shall be permitted.*

*Exception No. 3: Power-limited circuits shall be permitted to be reclassified and installed as non–power-limited circuits if the power-limited fire alarm circuit markings required by 760.42 are eliminated and the entire circuit is installed using the wiring methods and materials in accordance with Part II, Non–Power-Limited Fire Alarm Circuits.*

> FPN: Power-limited circuits reclassified and installed as non–power-limited circuits are no longer power-limited circuits, regardless of the continued connection to a power-limited source.

**(B) PLFA Wiring Methods and Materials.** Power-limited fire alarm conductors and cables described in 760.71 shall be installed as detailed in 760.52(B)(1), (2), or (3) of this section. Devices shall be installed in accordance with 110.3(B), 300.11(A), and 300.15.

**(1) Exposed or Fished in Concealed Spaces.** In raceway or exposed on the surface of ceiling and sidewalls or fished in concealed spaces. Cable splices or terminations shall be made in listed fittings, boxes, enclosures, fire alarm devices, or utilization equipment. Where installed exposed, cables shall be adequately supported and installed in such a way that maximum protection against physical damage is afforded by building construction such as baseboards, door frames, ledges, and so forth. Where located within 2.1 m (7 ft) of the floor, cables shall be securely fastened in an approved manner at intervals of not more than 450 mm (18 in.).

**(2) Passing Through a Floor or Wall.** In metal raceways or rigid nonmetallic conduit where passing through a floor or wall to a height of 2.1 m (7 ft) above the floor, unless adequate protection can be afforded by building construction such as detailed in 760.52(B)(1) or unless an equivalent solid guard is provided.

**(3) In Hoistways.** In rigid metal conduit, rigid nonmetallic conduit, intermediate metal conduit, or electrical metallic tubing where installed in hoistways.

*Exception No. 1: As provided for in 620.21 for elevators and similar equipment.*

*Exception No. 2: Other wiring methods and materials installed in accordance with the requirements of 760.3 shall be permitted to extend or replace the conductors and cables described in 760.71 and permitted by 760.52(B).*

**760.54 Installation of Conductors and Equipment in Cables, Compartments, Cable Trays, Enclosures, Manholes, Outlet Boxes, Device Boxes, and Raceways for**

**Power-Limited Circuits.** Conductors and equipment for power-limited fire alarm circuits shall be installed in accordance with 760.55 through 760.58.

**760.55 Separation from Electric Light, Power, Class 1, NPLFA, and Medium Power Network-Powered Broadband Communications Circuit Conductors.**

**(A) General.** Power-limited fire alarm circuit cables and conductors shall not be placed in any cable, cable tray, compartment, enclosure, manhole, outlet box, device box, raceway, or similar fitting with conductors of electric light, power, Class 1, non–power-limited fire alarm circuits, and medium power network-powered broadband communications circuits unless permitted by 760.55(B) through (G).

**(B) Separated by Barriers.** Power-limited fire alarm circuit cables shall be permitted to be installed together with Class 1, non–power-limited fire alarm, and medium power network-powered broadband communications circuits where they are separated by a barrier.

**(C) Raceways Within Enclosures.** In enclosures, power-limited fire alarm circuits shall be permitted to be installed in a raceway within the enclosure to separate them from Class 1, non–power-limited fire alarm, and medium power network-powered broadband communications circuits.

**(D) Associated Systems Within Enclosures.** Power-limited fire alarm conductors in compartments, enclosures, device boxes, outlet boxes, or similar fittings shall be permitted to be installed with electric light, power, Class 1, non–power-limited fire alarm, and medium power network-powered broadband communications circuits where they are introduced solely to connect the equipment connected to power-limited fire alarm circuits, and comply with either of the following conditions:

(1) The electric light, power, Class 1, non–power-limited fire alarm, and medium power network-powered broadband communications circuit conductors are routed to maintain a minimum of 6 mm (0.25 in.) separation from the conductors and cables of power-limited fire alarm circuits.

(2) The circuit conductors operate at 150 volts or less to ground and also comply with one of the following:

    a. The fire alarm power-limited circuits are installed using Type FPL, FPLR, FPLP, or permitted substitute cables, provided these power-limited cable conductors extending beyond the jacket are separated by a minimum of 6 mm (0.25 in.) or by a nonconductive sleeve or nonconductive barrier from all other conductors.

    b. The power-limited fire alarm circuit conductors are installed as non–power-limited circuits in accordance with 760.25.

**(E) Enclosures with Single Opening.** Power-limited fire alarm circuit conductors entering compartments, enclosures, device boxes, outlet boxes, or similar fittings shall be permitted to be installed with electric light, power, Class 1 non–power-limited fire alarm, and medium power network-powered broadband communications circuits where they are introduced solely to connect the equipment connected to power-limited fire alarm circuits or to other circuits controlled by the fire alarm system to which the other conductors in the enclosure are connected. Where power-limited fire alarm circuit conductors must enter an enclosure that is provided with a single opening, they shall be permitted to enter through a single fitting (such as a tee), provided the conductors are separated from the conductors of the other circuits by a continuous and firmly fixed nonconductor, such as flexible tubing.

**(F) In Hoistways.** In hoistways, power-limited fire alarm circuit conductors shall be installed in rigid metal conduit, rigid nonmetallic conduit, intermediate metal conduit, liquidtight flexible nonmetallic conduit, or electrical metallic tubing. For elevators or similar equipment, these conductors shall be permitted to be installed as provided in 620.21.

**(G) Other Applications.** For other applications, power-limited fire alarm circuit conductors shall be separated by at least 50 mm (2 in.) from conductors of any electric light, power, Class 1, non–power-limited fire alarm, or medium power network-powered broadband communications circuits unless one of the following conditions is met:

(1) Either (a) all of the electric light, power, Class 1, non–power-limited fire alarm, and medium power network-powered broadband communications circuit conductors or (b) all of the power-limited fire alarm circuit conductors are in a raceway or in metal-sheathed, metal-clad, nonmetallic-sheathed, or Type UF cables.

(2) All of the electric light, power, Class 1 non–power-limited fire alarm, and medium power network-powered broadband communications circuit conductors are permanently separated from all of the power-limited fire alarm circuit conductors by a continuous and firmly fixed nonconductor, such as porcelain tubes or flexible tubing, in addition to the insulation on the conductors.

**760.56 Installation of Conductors of Different PLFA Circuits, Class 2, Class 3, and Communications Circuits in the Same Cable, Enclosure, or Raceway.**

**(A) Two or More PLFA Circuits.** Cable and conductors of two or more power-limited fire alarm circuits, communications circuits, or Class 3 circuits shall be permitted within the same cable, enclosure, or raceway.

**(B) Class 2 Circuits with PLFA Circuits.** Conductors of one or more Class 2 circuits shall be permitted within the same cable, enclosure, or raceway with conductors of power-limited fire alarm circuits, provided that the insulation of the Class 2 circuit conductors in the cable, enclosure, or raceway is at least that required by the power-limited fire alarm circuits.

**(C) Low-Power Network-Powered Broadband Communications Cables and PLFA Cables.** Low-power network-powered broadband communications circuits shall be permitted in the same enclosure or raceway with PLFA cables.

**760.57 Support of Conductors.** Power-limited fire alarm circuit conductors shall not be strapped, taped, or attached by any means to the exterior of any conduit or other raceway as a means of support.

**760.58 Conductor Size.** Conductors of 26 AWG shall be permitted only where spliced with a connector listed as suitable for 26 AWG to 24 AWG or larger conductors that are terminated on equipment or where the 26 AWG conductors are terminated on equipment listed as suitable for 26 AWG conductors. Single conductors shall not be smaller than 18 AWG.

**760.59 Current-Carrying Continuous Line-Type Fire Detectors.**

**(A) Application.** Listed continuous line-type fire detectors, including insulated copper tubing of pneumatically operated detectors, employed for both detection and carrying signaling currents shall be permitted to be used in power-limited circuits.

**(B) Installation.** Continuous line-type fire detectors shall be installed in accordance with 760.42 through 760.52 and 760.54.

**760.61 Applications of Listed PLFA Cables.** PLFA cables shall comply with the requirements described in either 760.61(A), (B), or (C) or where cable substitutions are made as shown in 760.61(D).

**(A) Plenum.** Cables installed in ducts, plenums, and other spaces used for environmental air shall be Type FPLP. Abandoned cables shall not be permitted to remain. Types FPLP, FPLR, and FPL cables installed in compliance with 300.22 shall be permitted.

**(B) Riser.** Cables installed in risers shall be as described in either (1), (2), or (3):

(1) Cables installed in vertical runs and penetrating more than one floor, or cables installed in vertical runs in a shaft, shall be Type FPLR. Floor penetrations requiring Type FPLR shall contain only cables suitable for riser

**Table 760.61 Cable Uses and Permitted Substitutions**

| Cable Type | Use | References | Permitted Substitutions | |
|---|---|---|---|---|
| | | | Multiconductor | Coaxial |
| FPLP | Power-limited fire alarm plenum cable | 760.61(A) | CMP | MPP |
| FPLR | Power-limited fire alarm riser cable | 760.61(B) | CMP, FPLP, CMR | MPP, MPR |
| FPL | Power-limited fire alarm cable | 760.61(C) | CMP, FPLP, CMR, FPLR, CMG, CM | MPP, MPR, MPG, MP |

or plenum use. Abandoned cables shall not be permitted to remain.

(2) Other cables shall be installed in metal raceways or located in a fireproof shaft having firestops at each floor.

(3) Type FPL cable shall be permitted in one- and two-family dwellings.

FPN: See 300.21 for firestop requirements for floor penetrations.

**(C) Other Wiring Within Buildings.** Cables installed in building locations other than those covered in 760.61(A) or (B) shall be as described in either (1), (2), (3), or (4).

(1) Type FPL shall be permitted.

(2) Cables shall be permitted to be installed in raceways.

(3) Cables specified in Chapter 3 and meeting the requirements of 760.71(A) and (B) shall be permitted to be installed in nonconcealed spaces where the exposed length of cable does not exceed 3 m (10 ft).

(4) A portable fire alarm system provided to protect a stage or set when not in use shall be permitted to use wiring methods in accordance with 530.12.

**(D) Fire Alarm Cable Uses and Permitted Substitutions.** The uses and permitted substitutions for fire alarm cables listed in Table 760.61 shall be considered suitable for the purpose and shall be permitted.

FPN: For information on multipurpose cables (Types MPP, MPR, MPG, MP) and communications cables (Types CMP, CMR, CMG, CM), see 800.51.

**760.71 Listing and Marking of PLFA Cables and Insulated Continuous Line-Type Fire Detectors.** Type FPL cables installed as wiring within buildings shall be listed as being resistant to the spread of fire and other criteria in accordance with 760.71(A) through (H) and shall be marked in accordance with 760.71(I). Insulated continuous line-type fire detectors shall be listed in accordance with 760.71(J).

**(A) Conductor Materials.** Conductors shall be solid or stranded copper.

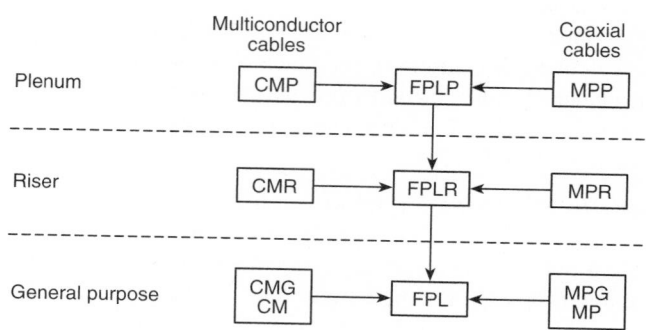

Type CM—Communications wires and cables
Type FPL—Power-limited fire alarm cables
Type MP—Multipurpose cables (coaxial cables only)

[A]→[B]   Cable A shall be permitted to be used in place of cable B.
26 AWG minimum

**Figure 760.61 Cable substitution hierarchy.**

**(B) Conductor Size.** The size of conductors in a multiconductor cable shall not be smaller than 26 AWG. Single conductors shall not be smaller than 18 AWG.

**(C) Ratings.** The cable shall have a voltage rating of not less than 300 volts.

**(D) Type FPLP.** Type FPLP power-limited fire alarm plenum cable shall be listed as being suitable for use in ducts, plenums, and other space used for environmental air and shall also be listed as having adequate fire-resistant and low smoke-producing characteristics.

FPN: One method of defining low smoke-producing cable is by establishing an acceptable value of the smoke produced when tested in accordance with NFPA 262-1999, *Standard Method of Test for Flame Travel and Smoke of Wires and Cables for Use in Air-Handling Spaces*, to a maximum peak optical density of 0.5 and a maximum average optical density of 0.15. Similarly, one method of defining fire-resistant cables is by establishing maximum allowable flame travel distance of 1.52 m (5 ft) when tested in accordance with the same test.

**(E) Type FPLR.** Type FPLR power-limited fire alarm riser cable shall be listed as being suitable for use in a vertical run in a shaft or from floor to floor and shall also be listed as having fire-resistant characteristics capable of preventing the carrying of fire from floor to floor.

> FPN: One method of defining fire-resistant characteristics capable of preventing the carrying of fire from floor to floor is that the cables pass the requirements of ANSI/UL 1666-1997, *Standard Test for Flame Propagation Height of Electrical and Optical-Fiber Cable Installed Vertically in Shafts.*

**(F) Type FPL.** Type FPL power-limited fire alarm cable shall be listed as being suitable for general-purpose fire alarm use, with the exception of risers, ducts, plenums, and other spaces used for environmental air and shall also be listed as being resistant to the spread of fire.

> FPN: One method of defining *resistant to the spread of fire* is that the cables do not spread fire to the top of the tray in the vertical-tray flame test in ANSI/UL 1581-1991, *Reference Standard for Electrical Wires, Cables and Flexible Cords.* Another method of defining *resistant to the spread of fire* is for the damage (char length) not to exceed 1.5 m (4 ft 11 in.) when performing the CSA vertical flame test — cables in cable trays, as described in CSA C22.2 No. 0.3-M-1985, *Test Methods for Electrical Wires and Cables.*

**(G) Fire Alarm Circuit Integrity (CI) Cable.** Cables suitable for use in fire alarm systems to ensure survivability of critical circuits during a specified time under fire conditions shall be listed as circuit integrity (CI) cable. Cables identified in 760.71(D), (E), and (F) that meet the requirements for circuit integrity shall have the additional classification using the suffix "CI" (for example, FPLP-CI, FPLR-CI, and FPL-CI).

> FPN No. 1: This cable is used for fire alarm circuits as one method of complying with the survivability requirements of NFPA 72-1999, *National Fire Alarm Code*, 3-4.2.2.2, 3-8.4.1.1.4, and 3-8.4.1.3.3.3(3), that the cable maintain its electrical function during fire conditions for a defined period of time.

> FPN No. 2: One method of defining circuit integrity (CI) cable is by establishing a minimum 2-hour fire resistance rating for the cable when tested in accordance with UL 2196-1995, *Standard for Tests of Fire Resistive Cables.*

**(H) Coaxial Cables.** Coaxial cables shall be permitted to use 30 percent conductivity copper-covered steel center conductor wire and shall be listed as Type FPLP, FPLR, or FPL cable.

**(I) Cable Marking.** The cable shall be marked in accordance with Table 760.71(I). The voltage rating shall not be marked on the cable. Cables that are listed for circuit integrity shall be identified with the suffix CI as defined in 760.71(G).

> FPN: Voltage ratings on cables may be misinterpreted to suggest that the cables may be suitable for Class 1, electric light, and power applications.

*Exception: Voltage markings shall be permitted where the cable has multiple listings and voltage marking is required for one or more of the listings.*

**Table 760.71(I) Cable Markings**

| Cable Marking | Type | Listing References |
|---|---|---|
| FPLP | Power-limited fire alarm plenum cable | 760.71(D) and (I) |
| FPLR | Power-limited fire alarm riser cable | 760.71(E) and (I) |
| FPL | Power-limited fire alarm cable | 760.71(F) and (I) |

Note: Cables identified in (D), (E), and (F) meeting the requirements for circuit integrity shall have the additional classification using the suffix "CI" (for example, FPLP-CI, FPLR-CI, and FPL-CI).

> FPN: Cable types are listed in descending order of fire-resistance rating.

**(J) Insulated Continuous Line-Type Fire Detectors.** Insulated continuous line-type fire detectors shall be rated in accordance with 760.71(C), listed as being resistant to the spread of fire in accordance with 760.71(D) through (F), marked in accordance with 760.71(I), and the jacket compound shall have a high degree of abrasion resistance.

# ARTICLE 770
## Optical Fiber Cables and Raceways

### I. General

**770.1 Scope.** The provisions of this article apply to the installation of optical fiber cables and raceways. This article does not cover the construction of optical fiber cables and raceways.

### 770.2 Definitions.

**Abandoned Optical Fiber Cable.** Installed optical fiber cable that is not terminated at equipment other than a connector and not identified for future use with a tag.

**Exposed.** The circuit is in such a position that, in case of failure of supports and insulation, contact with another circuit may result.

> FPN: See Article 100 for two other definitions of *Exposed.*

**Optical Fiber Raceway.** A raceway designed for enclosing and routing listed optical fiber cables.

**Point of Entrance.** The point at which the wire or cable emerges from an external wall, from a concrete floor slab, or from a rigid metal conduit or an intermediate metal conduit grounded to an electrode in accordance with 800.40(B).

**770.3 Locations and Other Articles.** Circuits and equipment shall comply with 770.3(A) and (B). Only those sections of Article 300 referenced in this article shall apply to optical fiber cables and raceways.

**(A) Spread of Fire or Products of Combustion.** The requirements of 300.21 for electrical installations shall also apply to installations of optical fiber cables and raceways. The accessible portion of abandoned optical fiber cables shall not be permitted to remain.

**(B) Ducts, Plenums, and Other Air-Handling Spaces.** The requirements of 300.22 for electric wiring shall also apply to installations of optical fiber cables and raceways where they are installed in ducts or plenums or other space used for environmental air.

*Exception: As permitted in 770.53(A).*

**770.4 Optical Fiber Cables.** Optical fiber cables transmit light for control, signaling, and communications through an optical fiber.

**770.5 Types.** Optical fiber cables can be grouped into three types.

**(A) Nonconductive.** These cables contain no metallic members and no other electrically conductive materials.

**(B) Conductive.** These cables contain non–current-carrying conductive members such as metallic strength members, metallic vapor barriers, and metallic armor or sheath.

**(C) Composite.** These cables contain optical fibers and current-carrying electrical conductors, and shall be permitted to contain non–current-carrying conductive members such as metallic strength members and metallic vapor barriers. Composite optical fiber cables shall be classified as electrical cables in accordance with the type of electrical conductors.

**770.6 Raceways for Optical Fiber Cables.** The raceway shall be of a type permitted in Chapter 3 and installed in accordance with Chapter 3.

*Exception: Listed nonmetallic optical fiber raceway identified as general-purpose, riser, or plenum optical fiber raceway in accordance with 770.51 and installed in accordance with 362.24 through 362.56, where the requirements applicable to electrical nonmetallic tubing shall apply. Unlisted underground or outside plant construction plastic innerduct shall be terminated at the point of entrance.*

FPN: For information on listing requirements for optical fiber raceways, see UL 2024, *Standard for Optical Fiber Raceways*.

Where optical fiber cables are installed within the raceway without current-carrying conductors, the raceway fill tables of Chapter 3 and Chapter 9 shall not apply.

Where nonconductive optical fiber cables are installed with electric conductors in a raceway, the raceway fill tables of Chapter 3 and Chapter 9 shall apply.

**770.7 Access to Electrical Equipment Behind Panels Designed to Allow Access.** Access to electrical equipment shall not be denied by an accumulation of cables that prevents removal of panels, including suspended ceiling panels.

**770.8 Mechanical Execution of Work.** Optical fiber cables shall be installed in a neat and workmanlike manner. Cables installed exposed on the surface of ceiling and sidewalls shall be supported by the structural components of the building structure in such a manner that the cable will not be damaged by normal building use. Such cables shall be attached to structural components by straps, staples, hangers, or similar fittings designed and installed so as not to damage the cable. The installation shall also conform with 300.4(D).

## II. Protection

**770.33 Grounding of Entrance Cables.** Where exposed to contact with electric light or power conductors, the non–current-carrying metallic members of optical fiber cables entering buildings shall be grounded as close to the point of entrance as practicable or shall be interrupted as close to the point of entrance as practicable by an insulating joint or equivalent device.

## III. Cables Within Buildings

**770.49 Fire Resistance of Optical Fiber Cables.** Optical fiber cables installed as wiring within buildings shall be listed as being resistant to the spread of fire in accordance with 770.50 and 770.51.

**770.50 Listing, Marking, and Installation of Optical Fiber Cables.** Optical fiber cables in a building shall be listed as being suitable for the purpose, and cables shall be marked in accordance with Table 770.50.

*Exception No. 1: Optical fiber cables shall not be required to be listed and marked where the length of the cable within the building, measured from its point of entrance, does not exceed 15 m (50 ft) and the cable enters the building from the outside and is terminated in an enclosure.*

FPN: Splice cases or terminal boxes, both metallic and plastic types, are typically used as enclosures for splicing or terminating optical fiber cables.

*Exception No. 2: Conductive optical fiber cable shall not be required to be listed and marked where the cable enters the building from the outside and is run in rigid metal conduit or intermediate metal conduit and such conduits are grounded to an electrode in accordance with 800.40(B).*

*Exception No. 3: Nonconductive optical fiber cables shall not be required to be listed and marked where the cable enters the building from the outside and is run in raceway installed in compliance with Chapter 3.*

**Table 770.50 Cable Markings**

| Cable Marking | Type | Reference |
|---|---|---|
| OFNP | Nonconductive optical fiber plenum cable | 770.51(A) and 770.53(A) |
| OFCP | Conductive optical fiber plenum cable | 770.51(A) and 770.53(A) |
| OFNR | Nonconductive optical fiber riser cable | 770.51(B) and 770.53(B) |
| OFCR | Conductive optical fiber riser cable | 770.51(B) and 770.53(B) |
| OFNG | Nonconductive optical fiber general-purpose cable | 770.51(C) and 770.53(C) |
| OFCG | Conductive optical fiber general-purpose cable | 770.51(C) and 770.53(C) |
| OFN | Nonconductive optical fiber general-purpose cable | 770.51(D) and 770.53(C) |
| OFC | Conductive optical fiber general-purpose cable | 770.51(D) and 770.53(C) |

FPN No. 1: Cable types are listed in descending order of fire resistance rating. Within each fire resistance rating, nonconductive cable is listed first, since it may substitute for the conductive cable.

FPN No. 2: See the referenced sections for requirements and permitted uses.

**770.51 Listing Requirements for Optical Fiber Cables and Raceways.** Optical fiber cables shall be listed in accordance with 770.51(A) through (D), and optical fiber raceways shall be listed in accordance with 770.51(E) through (G).

**(A) Types OFNP and OFCP.** Types OFNP and OFCP nonconductive and conductive optical fiber plenum cables shall be listed as being suitable for use in ducts, plenums, and other space used for environmental air and shall also be listed as having adequate fire-resistant and low smoke-producing characteristics.

FPN: One method of defining low smoke-producing cables is by establishing an acceptable value of the smoke produced when tested in accordance with NFPA 262-1999, *Standard Method of Test for Flame Travel and Smoke of Wires and Cables for Use in Air-Handling Spaces*, to a maximum peak optical density of 0.5 and a maximum av-

erage optical density of 0.15. Similarly, one method of defining fire-resistant cables is by defining maximum allowable flame travel distance of 1.52 m (5 ft) when tested in accordance with the same test.

**(B) Types OFNR and OFCR.** Types OFNR and OFCR nonconductive and conductive optical fiber riser cables shall be listed as being suitable for use in a vertical run in a shaft or from floor to floor and shall also be listed as having fire-resistant characteristics capable of preventing the carrying of fire from floor to floor.

FPN: One method of defining fire-resistant characteristics capable of preventing the carrying of fire from floor to floor is that the cables pass the requirements of ANSI/UL 1666-1997, *Standard Test for Flame Propagation Height of Electrical and Optical-Fiber Cable Installed Vertically in Shafts*.

**(C) Types OFNG and OFCG.** Types OFNG and OFCG nonconductive and conductive general-purpose optical fiber cables shall be listed as being suitable for general-purpose use, with the exception of risers and plenums, and shall also be listed as being resistant to the spread of fire.

FPN: One method of defining *resistance to the spread of fire* is for the damage (char length) not to exceed 1.5 m (4 ft 11 in.) when performing the vertical flame test for cables in cable trays, as described in CSA C22.2 No. 0.3-M-1985, *Test Methods for Electrical Wires and Cables*.

**(D) Types OFN and OFC.** Types OFN and OFC nonconductive and conductive optical fiber cables shall be listed as being suitable for general-purpose use, with the exception of risers, plenums, and other spaces used for environmental air, and shall also be listed as being resistant to the spread of fire.

FPN: One method of defining *resistant to the spread of fire* is that the cables do not spread fire to the top of the tray in the vertical-tray flame test in ANSI/UL 1581-1991, *Reference Standard for Electrical Wires, Cables, and Flexible Cords*.

Another method of defining *resistant to the spread of fire* is for the damage (char length) not to exceed 1.5 m (4 ft 11 in.) when performing the vertical flame test for cables in cable trays, as described in CSA C22.2 No. 0.3-M-1985, *Test Methods for Electrical Wires and Cables*.

**(E) Plenum Optical Fiber Raceway.** Plenum optical fiber raceways shall be listed as having adequate fire-resistant and low smoke-producing characteristics.

**(F) Riser Optical Fiber Raceway.** Riser optical fiber raceways shall be listed as having fire-resistant characteristics capable of preventing the carrying of fire from floor to floor.

**(G) General-Purpose Optical Fiber Cable Raceway.** General-purpose optical fiber cable raceway shall be listed as being resistant to the spread of fire.

## 770.52 Installation of Optical Fibers and Electrical Conductors.

**(A) With Conductors for Electric Light, Power, Class 1, Non–Power-Limited Fire Alarm, or Medium Power Network-Powered Broadband Communications Circuits.** Optical fibers shall be permitted within the same composite cable for electric light, power, Class 1, non–power-limited fire alarm, or medium power network-powered broadband communications circuits operating at 600 volts or less only where the functions of the optical fibers and the electrical conductors are associated.

Nonconductive optical fiber cables shall be permitted to occupy the same cable tray or raceway with conductors for electric light, power, Class 1, non–power-limited fire alarm, or medium power network-powered broadband communications circuits operating at 600 volts or less. Conductive optical fiber cables shall not be permitted to occupy the same cable tray or raceway with conductors for electric light, power, Class 1, non–power-limited fire alarm, or medium power network-powered broadband communications circuits.

Composite optical fiber cables containing only current-carrying conductors for electric light, power, Class 1 circuits rated 600 volts or less shall be permitted to occupy the same cabinet, cable tray, outlet box, panel, raceway, or other termination enclosure with conductors for electric light, power, or Class 1 circuits operating at 600 volts or less.

Nonconductive optical fiber cables shall not be permitted to occupy the same cabinet, outlet box, panel, or similar enclosure housing the electrical terminations of an electric light, power, Class 1, non–power-limited fire alarm, or medium power network-powered broadband communications circuit.

*Exception No. 1: Occupancy of the same cabinet, outlet box, panel, or similar enclosure shall be permitted where nonconductive optical fiber cable is functionally associated with the electric light, power, Class 1, non–power-limited fire alarm, or medium power network-powered broadband communications circuit.*

*Exception No. 2: Occupancy of the same cabinet, outlet box, panel, or similar enclosure shall be permitted where nonconductive optical fiber cables are installed in factory- or field-assembled control centers.*

*Exception No. 3: In industrial establishments only, where conditions of maintenance and supervision ensure that only qualified persons service the installation, nonconductive optical fiber cables shall be permitted with circuits exceeding 600 volts.*

*Exception No. 4: In industrial establishments only, where conditions of maintenance and supervision ensure that only qualified persons service the installation, composite optical fiber cables shall be permitted to contain current-carrying conductors operating over 600 volts.*

Installations in raceway shall comply with 300.17.

**(B) With Other Conductors.** Optical fibers shall be permitted in the same cable, and conductive and nonconductive optical fiber cables shall be permitted in the same cable tray, enclosure, or raceway with conductors of any of the following:

(1) Class 2 and Class 3 remote-control, signaling, and power-limited circuits in compliance with Article 725
(2) Power-limited fire alarm systems in compliance with Article 760
(3) Communications circuits in compliance with Article 800
(4) Community antenna television and radio distribution systems in compliance with Article 820
(5) Low-power network-powered broadband communications circuits in compliance with Article 830

**(C) Grounding.** Non–current-carrying conductive members of optical fiber cables shall be grounded in accordance with Article 250.

## 770.53 Applications of Listed Optical Fiber Cables and Raceways.
Nonconductive and conductive optical fiber cables shall comply with any of the requirements given in 770.53(A) through (E) or where cable substitutions are made as shown in 770.53(F).

**(A) Plenum.** Cables installed in ducts, plenums, and other spaces used for environmental air shall be Type OFNP or OFCP. Abandoned cables shall not be permitted to remain. Types OFNR, OFCR, OFNG, OFN, OFCG, and OFC cables installed in compliance with 300.22 shall be permitted. Listed plenum optical fiber raceways shall be permitted to be installed in ducts and plenums as described in 300.22(B) and in other spaces used for environmental air as described in 300.22(C). Only types OFNP and OFCP cables shall be permitted to be installed in these raceways.

**(B) Riser.** Cables installed in risers shall be as described in any of the following:

(1) Cables installed in vertical runs and penetrating more than one floor, or cables installed in vertical runs in a shaft, shall be Type OFNR or OFCR. Floor penetrations requiring Type OFNR or OFCR shall contain only cables suitable for riser or plenum use. Abandoned cables shall not be permitted to remain. Listed riser optical fiber raceways shall be permitted to be installed in vertical riser runs in a shaft from floor to floor. Only Types OFNP, OFCP, OFNR and OFCR cables shall be permitted to be installed in these raceways.

(2) Types OFNG, OFN, OFCG, and OFC cables shall be permitted to be encased in a metal raceway or located in a fireproof shaft having firestops at each floor.

(3) Types OFNG, OFN, OFCG, and OFC cables shall be permitted in one- and two-family dwellings.

FPN: See 300.21 for firestop requirements for floor penetrations.

**(C) Other Wiring Within Buildings.** Cables installed in building locations other than the locations covered in 770.53(A) and (B) shall be Type OFNG, OFN, OFCG, or OFC. Such cables shall be permitted to be installed in listed general-purpose optical fiber raceways.

**(D) Hazardous (Classified) Locations.** Cables installed in hazardous (classified) locations shall be any type indicated in Table 770.53.

**(E) Cable Trays.** Optical fiber cables of the types listed in Table 770.50 shall be permitted to be installed in cable trays.

FPN: It is not the intent to require that these optical fiber cables be listed specifically for use in cable trays.

**(F) Cable Substitutions.** The substitutions for optical fiber cables listed in Table 770.53 shall be permitted.

**Table 770.53 Cable Substitutions**

| Cable Type | Permitted Substitutions |
|---|---|
| OFNP | None |
| OFCP | OFNP |
| OFNR | OFNP |
| OFCR | OFNP, OFCP, OFNR |
| OFNG, OFN | OFNP, OFNR |
| OFCG, OFC | OFNP, OFCP, OFNR, OFCR, OFNG, OFN |

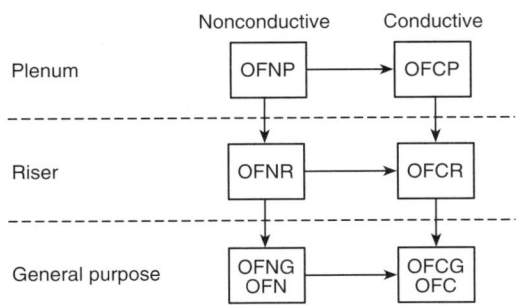

A →B  Cable A may be used in place of cable B.

**Figure 770.53 Cable substitution hierarchy.**

# ARTICLE 780
## Closed-Loop and Programmed Power Distribution

**780.1 Scope.** The provisions of this article apply to premise power distribution systems jointly controlled by a signaling between the energy controlling equipment and utilization equipment.

**780.2 General.**

**(A) Other Articles.** Except as modified by the requirements of this article, all other applicable articles of this *Code* shall apply.

**(B) Component Parts.** All equipment and conductors shall be listed and identified.

**780.3 Control.** The control equipment and all power switching devices operated by the control equipment shall be listed and identified. The system shall operate in accordance with 780.2(A) through (D).

**(A) Characteristic Electrical Identification Required.** Outlets shall not be energized unless the utilization equipment first exhibits a characteristic electrical identification.

**(B) Conditions for De-Energization.** Outlets shall be de-energized when any of the following conditions occur:

(1) A nominal-operation acknowledgment signal is not being received from the utilization equipment connected to the outlet.
(2) A ground-fault condition exists.
(3) An overcurrent condition exists.

**(C) Additional Conditions for De-Energization When an Alternate Source of Power Is Used.** In addition to the requirements in 780.2(B), outlets shall be de-energized when any of the following conditions occur:

(1) The grounded conductor is not properly grounded.
(2) Any ungrounded conductor is not at nominal voltage.

**(D) Controller Malfunction.** In the event of a controller malfunction, all associated outlets shall be de-energized.

**780.5 Power Limitation in Signaling Circuits.** For signaling circuits not exceeding 24 volts, the current required shall not exceed 1 ampere where protected by an overcurrent device or an inherently limited power source.

**780.6 Cables and Conductors.**

**(A) Hybrid Cable.** Listed hybrid cable consisting of power, communications, and signaling conductors shall be permitted under a common jacket. The jacket shall be applied so as to separate the power conductors from the communications and signaling conductors. An optional outer jacket shall be permitted to be applied. The individual conductors of a hybrid cable shall conform to the *Code* provisions applicable to their current, voltage, and insulation rating. The signaling conductors shall not be smaller than 24 AWG copper.

**(B) Cables and Conductors in the Same Cabinet, Panel, or Box.** The power, communications, and signaling conductors of listed hybrid cable are permitted to occupy the same cabinet, panel, or outlet box (or similar enclosure housing the electrical terminations of electric light or power circuits) only if connectors specifically listed for hybrid cable are employed.

**780.7 Noninterchangeability.** Receptacles, cord connectors, and attachment plugs used on closed-loop power distribution systems shall be constructed so that they are not interchangeable with other receptacles, cord connectors, and attachment plugs.

# Chapter 8  Communications Systems

## ARTICLE 800
## Communications Circuits

## I. General

**800.1 Scope.** This article covers telephone, telegraph (except radio), outside wiring for fire alarm and burglar alarm, and similar central station systems; and telephone systems not connected to a central station system but using similar types of equipment, methods of installation, and maintenance.

> FPN No. 1:  For further information for fire alarm, guard tour, sprinkler waterflow, and sprinkler supervisory systems, see Article 760.
>
> FPN No. 2:  For installation requirements of optical fiber cables, see Article 770.
>
> FPN No. 3:  For installation requirements for network-powered broadband communications circuits, see Article 830.

**800.2 Definitions.** See Article 100. For purposes of this article, the following additional definitions apply.

**Abandoned Communications Cable.** Installed communications cable that is not terminated at both ends at a connector or other equipment and not identified for future use with a tag.

**Block.** A square or portion of a city, town, or village enclosed by streets and including the alleys so enclosed, but not any street.

**Cable.** A factory assembly of two or more conductors having an overall covering.

**Cable Sheath.** A covering over the conductor assembly that may include one or more metallic members, strength members, or jackets.

**Exposed.** A circuit that is in such a position that, in case of failure of supports and insulation, contact with another circuit may result.

> FPN:  See Article 100 for two other definitions of *Exposed.*

**Point of Entrance.** Within a building, the point at which the wire or cable emerges from an external wall, from a concrete floor slab, or from a rigid metal conduit or an intermediate metal conduit grounded to an electrode in accordance with 800.40(B).

**Premises.** The land and buildings of a user located on the user side of the utility-user network point of demarcation.

**Wire.** A factory assembly of one or more insulated conductors without an overall covering.

**800.3 Hybrid Power and Communications Cables.** The provisions of 780.6 shall apply for listed hybrid power and communications cables in closed-loop and programmed power distribution.

> FPN:  See 800.51(I) for hybrid power and communications cable in other applications.

**800.4 Equipment.** Equipment intended to be electrically connected to a telecommunications network shall be listed for the purpose. Installation of equipment shall also comply with 110.3(B).

> FPN:  One way to determine applicable requirements is to refer to UL 1950-1993, *Standard for Safety of Information Technology Equipment, Including Electrical Business Equipment*, third edition; UL 1459-1995, *Standard for Safety, Telephone Equipment*, third edition; or UL 1863-1995, *Standard for Safety, Communications Circuit Accessories*, second edition. For information on listing requirements for communications raceways, see UL 2024-1995, *Standard for Optical Fiber Raceways.*

*Exception: This listing requirement shall not apply to test equipment that is intended for temporary connection to a telecommunications network by qualified persons during the course of installation, maintenance, or repair of telecommunications equipment or systems.*

**800.5 Access to Electrical Equipment Behind Panels Designed to Allow Access.** Access to electrical equipment shall not be denied by an accumulation of wires and cables that prevents removal of panels, including suspended ceiling panels.

**800.6 Mechanical Execution of Work.** Communications circuits and equipment shall be installed in a neat and workmanlike manner. Cables installed exposed on the outer surface of ceiling and sidewalls shall be supported by the structural components of the building structure in such a manner that the cable is not be damaged by normal building use. Such cables shall be attached to structural components by straps, staples, hangers, or similar fittings designed and installed so as not to damage the cable. The installation shall also conform with 300.4(D).

**800.8 Hazardous (Classified) Locations.** Communications circuits and equipment installed in a location that is classified in accordance with Article 500 shall comply with the applicable requirements of Chapter 5.

## II. Conductors Outside and Entering Buildings

**800.10 Overhead Communications Wires and Cables.** Overhead communications wires and cables entering buildings shall comply with 800.10(A) and (B).

**(A) On Poles and In-Span.** Where communications wires and cables and electric light or power conductors are supported by the same pole or run parallel to each other in-span, the conditions described in 800.10(A)(1) through (A)(4) shall be met.

**(1) Relative Location.** Where practicable, the communications wires and cables shall be located below the electric light or power conductors.

**(2) Attachment to Crossarms.** Communications wires and cables shall not be attached to a cross-arm that carries electric light or power conductors.

**(3) Climbing Space.** The climbing space through communications wires and cables shall comply with the requirements of 225.14(D).

**(4) Clearance.** Supply service drops of 0–750 volts running above and parallel to communications service drops shall have a minimum separation of 300 mm (12 in.) at any point in the span, including the point of and at their attachment to the building, provided the nongrounded conductors are insulated and that a clearance of not less than 1.0 m (40 in.) is maintained between the two services at the pole.

**(B) Above Roofs.** Communications wires and cables shall have a vertical clearance of not less than 2.5 m (8 ft) from all points of roofs above which they pass.

*Exception No. 1: Auxiliary buildings, such as garages and the like.*

*Exception No. 2: A reduction in clearance above only the overhanging portion of the roof to not less than 450 mm (18 in.) shall be permitted if (a) not more than 1.2 m (4 ft) of communications service-drop conductors pass above the roof overhang and (b) they are terminated at a through- or above-the-roof raceway or approved support.*

*Exception No. 3: Where the roof has a slope of not less than 100 mm in 300 mm (4 in. in 12 in.), a reduction in clearance to not less than 900 mm (3 ft) shall be permitted.*

> FPN: For additional information regarding overhead wires and cables, see ANSI C2-1997, *National Electric Safety Code*, Part 2 Safety Rules For Overhead Lines.

**800.11 Underground Circuits Entering Buildings.** Underground communications wires and cables entering buildings shall comply with 800.11(A) and (B).

**(A) With Electric Light or Power Conductors.** Underground communications wires and cables in a raceway, handhole, or manhole containing electric light, power, Class 1, or non–power-limited fire alarm circuit conductors shall be in a section separated from such conductors by means of brick, concrete, or tile partitions or by means of a suitable barrier.

**(B) Underground Block Distribution.** Where the entire street circuit is run underground and the circuit within the block is placed so as to be free from the likelihood of accidental contact with electric light or power circuits of over 300 volts to ground, the insulation requirements of 800.12(A) and (C) shall not apply, insulating supports shall not be required for the conductors, and bushings shall not be required where the conductors enter the building.

**800.12 Circuits Requiring Primary Protectors.** Circuits that require primary protectors as provided in 800.30 shall comply with 800.12(A), (B), and (C).

**(A) Insulation, Wires, and Cables.** Communications wires and cables without a metallic shield, running from the last outdoor support to the primary protector, shall be listed as being suitable for the purpose and shall have current-carrying capacity as specified in 800.30(A)(1)(b) or 800.30(A)(1)(c).

**(B) On Buildings.** Communications wires and cables in accordance with 800.12(A) shall be separated at least 100 mm (4 in.) from electric light or power conductors not in a raceway or cable or be permanently separated from conductors of the other system by a continuous and firmly fixed nonconductor in addition to the insulation on the wires, such as porcelain tubes or flexible tubing. Communications wires and cables in accordance with 800.12(A) exposed to accidental contact with electric light and power conductors operating at over 300 volts to ground and attached to buildings shall be separated from woodwork by being supported on glass, porcelain, or other insulating material.

*Exception: Separation from woodwork shall not be required where fuses are omitted as provided for in 800.30(A)(1), or where conductors are used to extend circuits to a building from a cable having a grounded metal sheath.*

**(C) Entering Buildings.** Where a primary protector is installed inside the building, the communications wires and cables shall enter the building either through a noncombustible, nonabsorbent insulating bushing or through a metal raceway. The insulating bushing shall not be required where the entering communications wires and cables (1) are in metal-sheathed cable, (2) pass through masonry, (3) meet the requirements of 800.12(A) and fuses are omitted as provided in 800.30(A)(1), or (4) meet the requirements of 800.12(A) and are used to extend circuits to a building from a cable having a grounded metallic sheath. Raceways or bushings shall slope upward from the outside or, where this cannot be done, drip loops shall be formed in the communications wires and cables immediately before they enter the building.

Raceways shall be equipped with an approved service head. More than one communications wire and cable shall be permitted to enter through a single raceway or bushing. Conduits or other metal raceways located ahead of the primary protector shall be grounded.

**800.13 Lightning Conductors.** Where practicable, a separation of at least 1.8 m (6 ft) shall be maintained between communications wires and cables on buildings and lightning conductors.

## III. Protection

### 800.30 Protective Devices.

**(A) Application.** A listed primary protector shall be provided on each circuit run partly or entirely in aerial wire or aerial cable not confined within a block. Also, a listed primary protector shall be provided on each circuit, aerial or underground, located within the block containing the building served so as to be exposed to accidental contact with electric light or power conductors operating at over 300 volts to ground. In addition, where there exists a lightning exposure, each interbuilding circuit on a premises shall be protected by a listed primary protector at each end of the interbuilding circuit. Installation of primary protectors shall also comply with 110.3(B).

FPN No. 1: On a circuit not exposed to accidental contact with power conductors, providing a listed primary protector in accordance with this article helps protect against other hazards, such as lightning and above-normal voltages induced by fault currents on power circuits in proximity to the communications circuit.

FPN No. 2: Interbuilding circuits are considered to have a lightning exposure unless one or more of the following conditions exist:

(1) Circuits in large metropolitan areas where buildings are close together and sufficiently high to intercept lightning.
(2) Interbuilding cable runs of 42 m (140 ft) or less, directly buried or in underground conduit, where a continuous metallic cable shield or a continuous metallic conduit containing the cable is bonded to each building grounding electrode system.
(3) Areas having an average of five or fewer thunderstorm days per year and earth resistivity of less than 100 ohm-meters. Such areas are found along the Pacific coast.

**(1) Fuseless Primary Protectors.** Fuseless-type primary protectors shall be permitted under any of the conditions given in (a) through (e).

(a) Where conductors enter a building through a cable with grounded metallic sheath member(s) and if the conductors in the cable safely fuse on all currents greater than the current-carrying capacity of the primary protector and of the primary protector grounding conductor

(b) Where insulated conductors in accordance with 800.12(A) are used to extend circuits to a building from a cable with an effectively grounded metallic sheath member(s) and if the conductors in the cable or cable stub, or the connections between the insulated conductors and the exposed plant, safely fuse on all currents greater than the current-carrying capacity of the primary protector, or the associated insulated conductors and of the primary protector grounding conductor

(c) Where insulated conductors in accordance with 800.12(A) or (B) are used to extend circuits to a building from other than a cable with a metallic sheath member(s) if (1) the primary protector is listed for this purpose, and (2) the connections of the insulated conductors to the exposed plant or the conductors of the exposed plant safely fuse on all currents greater than the current-carrying capacity of the primary protector, or the associated insulated conductors and of the primary protector grounding conductor

(d) Where insulated conductors in accordance with 800.12(A) are used to extend circuits aerially to a building from an unexposed buried or underground circuit

(e) Where insulated conductors in accordance with 800.12(A) are used to extend circuits to a building from cable with an effectively grounded metallic sheath member(s) and if (1) the combination of the primary protector and insulated conductors is listed for this purpose, and (2) the insulated conductors safely fuse on all currents greater than the current-carrying capacity of the primary protector and of the primary protector grounding conductor

**(2) Fused Primary Protectors.** Where the requirements listed under 800.30(A)(1)(a) through (1)(e) are not met, fused-type primary protectors shall be used. Fused-type primary protectors shall consist of an arrester connected between each line conductor and ground, a fuse in series with each line conductor, and an appropriate mounting arrangement. Primary protector terminals shall be marked to indicate line, instrument, and ground, as applicable.

**(B) Location.** The primary protector shall be located in, on, or immediately adjacent to the structure or building served and as close as practicable to the point of entrance.

FPN: See 800.2 for the definition of *point of entrance*.

For purposes of this section, primary protectors located at mobile home service equipment located in sight from and not more than 9.0 m (30 ft) from the exterior wall of the mobile home it serves, or at a mobile home disconnecting means grounded in accordance with 250.32 and located in sight from and not more than 9.0 m (30 ft) from the exterior wall of the mobile home it serves, shall be considered to meet the requirements of this section.

FPN: Selecting a primary protector location to achieve the shortest practicable primary protector grounding conductor helps limit potential differences between communications circuits and other metallic systems.

**(C) Hazardous (Classified) Locations.** The primary protector shall not be located in any hazardous (classified) location as defined in Article 500 or in the vicinity of easily ignitible material.

*Exception: As permitted in 501.14, 502.14, and 503.12.*

**800.31 Primary Protector Requirements.** The primary protector shall consist of an arrester connected between each line conductor and ground in an appropriate mounting. Primary protector terminals shall be marked to indicate line and ground as applicable.

FPN: One way to determine applicable requirements for a listed primary protector is to refer to ANSI/UL 497-1995, *Standard for Protectors for Paired Conductor Communications Circuits.*

**800.32 Secondary Protector Requirements.** Where a secondary protector is installed in series with the indoor communications wire and cable between the primary protector and the equipment, it shall be listed for the purpose. The secondary protector shall provide means to safely limit currents to less than the current-carrying capacity of listed indoor communications wire and cable, listed telephone set line cords, and listed communications terminal equipment having ports for external wire line communications circuits. Any overvoltage protection, arresters, or grounding connection shall be connected on the equipment terminals side of the secondary protector current-limiting means.

FPN No. 1: One way to determine applicable requirements for a listed secondary protector is to refer to UL 497A-1996, *Standard for Secondary Protectors for Communications Circuits.*

FPN No. 2: Secondary protectors on exposed circuits are not intended for use without primary protectors.

**800.33 Cable Grounding.** The metallic sheath of communications cables entering buildings shall be grounded as close as practicable to the point of entrance or shall be interrupted as close to the point of entrance as practicable by an insulating joint or equivalent device.

FPN: See 800.2 for the definition of *point of entrance.*

## IV. Grounding Methods

**800.40 Cable and Primary Protector Grounding.** The metallic member(s) of the cable sheath, where required to

be grounded by 800.33, and primary protectors shall be grounded as specified in 800.40(A) through (D).

**(A) Grounding Conductor.**

**(1) Insulation.** The grounding conductor shall be insulated and shall be listed as suitable for the purpose.

**(2) Material.** The grounding conductor shall be copper or other corrosion-resistant conductive material, stranded or solid.

**(3) Size.** The grounding conductor shall not be smaller than 14 AWG.

**(4) Length.** The primary protector grounding conductor shall be as short as practicable. In one- and two-family dwellings, the primary protector grounding conductor shall be as short as practicable, not to exceed 6.0 m (20 ft) in length.

*Exception: In one- and two-family dwellings where it is not practicable to achieve an overall maximum primary protector grounding conductor length of 6.0 m (20 ft), a separate communications ground rod meeting the minimum dimensional criteria of 800.40(B)(2)(2) shall be driven, the primary protector shall be grounded to the communications ground rod in accordance with 800.40(C), and the communications ground rod bonded to the power grounding electrode system in accordance with 800.40(D).*

**(5) Run in Straight Line.** The grounding conductor shall be run to the grounding electrode in as straight a line as practicable.

**(6) Physical Damage.** Where necessary, the grounding conductor shall be guarded from physical damage. Where the grounding conductor is run in a metal raceway, both ends of the raceway shall be bonded to the grounding conductor or the same terminal or electrode to which the grounding conductor is connected.

**(B) Electrode.** The grounding conductor shall be connected in accordance with 800.40(B)(1) and (B)(2).

**(1) In Buildings or Structures with Grounding Means.** To the nearest accessible location on the following:

(1) The building or structure grounding electrode system as covered in 250.50
(2) The grounded interior metal water piping system, within 1.5 m (5 ft) from its point of entrance to the building, as covered in 250.52
(3) The power service accessible means external to enclosures as covered in 250.94
(4) The metallic power service raceway
(5) The service equipment enclosure
(6) The grounding electrode conductor or the grounding electrode conductor metal enclosure

(7) The grounding conductor or the grounding electrode of a building or structure disconnecting means that is grounded to an electrode as covered in 250.32.

For purposes of this section, the mobile home service equipment or the mobile home disconnecting means, as described in 800.30(B), shall be considered accessible.

**(2) In Buildings or Structures Without Grounding Means.** If the building or structure served has no grounding means, as described in 800.40(B)(1):

(1) To any one of the individual electrodes described in 250.52(A)(1), (2), (3), (4); or

(2) If the building or structure served has no grounding means, as described in 800.40(B)(1) or (B)(2)(1), to an effectively grounded metal structure or to a ground rod or pipe not less than 1.5 m (5 ft) in length and 12.7 mm (½ in.) in diameter, driven, where practicable, into permanently damp earth and separated from lightning conductors as covered in 800.13 and at least 1.8 m (6 ft) from electrodes of other systems. Steam or hot water pipes or air terminal conductors (lightning-rod conductors) shall not be employed as electrodes for protectors.

**(C) Electrode Connection.** Connections to grounding electrodes shall comply with 250.70. Connectors, clamps, fittings, or lugs used to attach grounding conductors and bonding jumpers to grounding electrodes or to each other that are to be concrete-encased or buried in the earth shall be suitable for its application.

**(D) Bonding of Electrodes.** A bonding jumper not smaller than 6 AWG copper or equivalent shall be connected between the communications grounding electrode and power grounding electrode system at the building or structure served where separate electrodes are used. Bonding together of all separate electrodes shall be permitted.

*Exception: At mobile homes as covered in 800.41.*

FPN No. 1: See 250.60 for use of air terminals (lightning rods).

FPN No. 2: Bonding together of all separate electrodes limits potential differences between them and between their associated wiring systems.

**800.41 Primary Protector Grounding and Bonding at Mobile Homes.**

**(A) Grounding.** Where there is no mobile home service equipment located in sight from, and not more than 9.0 m (30 ft) from, the exterior wall of the mobile home it serves, or there is no mobile home disconnecting means grounded in accordance with 250.32 and located within sight from, and not more than 9.0 m (30 ft) from, the exterior wall of the mobile home it serves, the primary protector ground shall be in accordance with 800.40(B)(2).

**(B) Bonding.** The primary protector grounding terminal or grounding electrode shall be bonded to the metal frame or available grounding terminal of the mobile home with a copper grounding conductor not smaller than 12 AWG under any of the following conditions:

(1) Where there is no mobile home service equipment or disconnecting means as in 800.41(A)

(2) Where the mobile home is supplied by cord and plug

### V. Communications Wires and Cables Within Buildings

**800.48 Raceways for Communications Wires and Cables.** Where communications wire and cables are installed in a raceway, the raceway shall be either of a type permitted in Chapter 3 and installed in accordance with Chapter 3 or a listed nonmetallic raceway complying with 800.51(J), (K), or (L), as applicable, and installed in accordance with 362.24 through 362.56, where the requirements applicable to electrical nonmetallic tubing apply.

*Exception: Conduit fill restrictions shall not apply.*

**800.49 Fire Resistance of Communications Wires and Cables.** Communications wires and cables installed as wiring within a building shall be listed as being resistant to the spread of fire in accordance with 800.50 and 800.51.

**800.50 Listing, Marking, and Installation of Communications Wires and Cables.** Communications wires and cables installed as wiring within buildings shall be listed as being suitable for the purpose and installed in accordance with 800.52. Communications cables and undercarpet communications wires shall be marked in accordance with Table 800.50. The cable voltage rating shall not be marked on the cable or on the undercarpet communications wire.

FPN: Voltage markings on cables may be misinterpreted to suggest that the cables may be suitable for Class 1, electric light, and power applications.

*Exception No. 1: Voltage markings shall be permitted where the cable has multiple listings and voltage marking is required for one or more of the listings.*

*Exception No. 2: Listing and marking shall not be required where the cable enters the building from the outside and is continuously enclosed in a rigid metal conduit system or an intermediate metal conduit system and such conduit systems are grounded to an electrode in accordance with 800.40(B).*

*Exception No. 3: Listing and marking shall not be required where the length of the cable within the building, measured from its point of entrance, does not exceed 15 m (50 ft) and the cable enters the building from the outside*

**Table 800.50 Cable Markings**

| Cable Marking | Type | Reference |
|---|---|---|
| MPP | Multipurpose plenum cable | 800.51(G) and 800.53(A) |
| CMP | Communications plenum cable | 800.51(A) and 800.53(A) |
| MPR | Multipurpose riser cable | 800.51(G) and 800.53(B) |
| CMR | Communications riser cable | 800.51(B) and 800.53(B) |
| MPG | Multipurpose general-purpose cable | 800.51(G) and 800.53(D) and (E)(1) |
| CMG | Communications general-purpose cable | 800.51(C) and 800.53(D) and (E)(1) |
| MP | Multipurpose general-purpose cable | 800.51(G) and 800.53(D) and (E)(1) |
| CM | Communications general-purpose cable | 800.51(D) and 800.53(D) and (E)(1) |
| CMX | Communications cable, limited use | 800.51(E) and 800.53(C), (D), and (E) |
| CMUC | Undercarpet communications wire and cable | 800.51(F) and 800.53(F)(6) |

*and is terminated in an enclosure or on a listed primary protector.*

FPN No. 1: Splice cases or terminal boxes, both metallic and plastic types, are typically used as enclosures for splicing or terminating telephone cables.

FPN No. 2: This exception limits the length of unlisted outside plant cable to 15 m (50 ft), while 800.30(B) requires that the primary protector be located as close as practicable to the point at which the cable enters the building. Therefore, in installations requiring a primary protector, the outside plant cable may not be permitted to extend 15 m (50 ft) into the building if it is practicable to place the primary protector closer than 15 m (50 ft) to the entrance point.

*Exception No. 4: Multipurpose cables shall be considered as being suitable for the purpose and shall be permitted to substitute for communications cables as provided for in 800.53(G).*

FPN No. 1: Cable types are listed in descending order of fire resistance rating, and multipurpose cables are listed above communications cables because multipurpose cables may substitute for communications cables.

FPN No. 2: See the referenced sections for permitted uses.

## 800.51 Listing Requirements for Communications Wires and Cables and Communications Raceways. Communi-

cations wires and cables shall have a voltage rating of not less than 300 volts and shall be listed in accordance with 800.51(A) through (J), and communications raceways shall be listed in accordance with 800.51(K) through (L). Conductors in communications cables, other than in a coaxial cable, shall be copper.

FPN: See 800.4 for listing requirement for equipment.

**(A) Type CMP.** Type CMP communications plenum cable shall be listed as being suitable for use in ducts, plenums, and other spaces used for environmental air and shall also be listed as having adequate fire-resistant and low smoke-producing characteristics.

FPN: One method of defining low smoke-producing cables is by establishing an acceptable value of the smoke produced when tested in accordance with NFPA 262-1999, *Standard Method of Test for Flame Travel and Smoke of Wire and Cables for Use in Air-Handling Spaces*, to a maximum peak optical density of 0.5 and a maximum average optical density of 0.15. Similarly, one method of defining fire-resistant cables is by establishing a maximum allowable flame travel distance of 1.52 m (5 ft) when tested in accordance with the same test.

**(B) Type CMR.** Type CMR communications riser cable shall be listed as being suitable for use in a vertical run in a shaft or from floor to floor and shall also be listed as having fire-resistant characteristics capable of preventing the carrying of fire from floor to floor.

FPN: One method of defining fire-resistant characteristics capable of preventing the carrying of fire from floor to floor is that the cables pass the requirements of ANSI/UL 1666-1997, *Standard Test for Flame Propagation Height of Electrical and Optical-Fiber Cable Installed Vertically in Shafts*.

**(C) Type CMG.** Type CMG general-purpose communications cable shall be listed as being suitable for general-purpose communications use, with the exception of risers and plenums, and shall also be listed as being resistant to the spread of fire.

FPN: One method of defining *resistant to the spread of fire* is for the damage (char length) not to exceed 1.5 m (4 ft 11 in.) when performing the vertical flame test for cables in cable trays, as described in CSA C22.2 No. 0.3-M 1985, *Test Methods for Electrical Wires and Cables*.

**(D) Type CM.** Type CM communications cable shall be listed as being suitable for general-purpose communications use, with the exception of risers and plenums, and shall also be listed as being resistant to the spread of fire.

FPN: One method of defining *resistant to the spread of fire* is that the cables do not spread fire to the top of the tray in the vertical-tray flame test in ANSI/UL 1581-1991, *Reference Standard for Electrical Wires, Cables and Flexible Cords*. Another method of defining *resistant to the spread of fire* is for the damage (char length) not to exceed 1.5 m (4 ft 11 in.) when performing the vertical flame test for cables in cable trays, as described in CSA C22.2 No. 0.3-M-1985, *Test Methods for Electrical Wires and Cables.*

**(E) Type CMX.** Type CMX limited-use communications cable shall be listed as being suitable for use in dwellings and for use in raceway and shall also be listed as being resistant to flame spread.

> FPN: One method of determining that cable is resistant to flame spread is by testing the cable to the VW-1 (vertical-wire) flame test in ANSI/UL 1581-1991, *Reference Standard for Electrical Wires, Cables and Flexible Cords.*

**(F) Type CMUC Undercarpet Wire and Cable.** Type CMUC undercarpet communications wire and cable shall be listed as being suitable for undercarpet use and shall also be listed as being resistant to flame spread.

> FPN: One method of determining that cable is resistant to flame spread is by testing the cable to the VW-1 (vertical-wire) flame test in ANSI/UL 1581-1991, *Reference Standard for Electrical Wires, Cables and Flexible Cords.*

**(G) Multipurpose (MP) Cables.** Until July 1, 2003, cables that meet the requirements for Types CMP, CMR, CMG, and CM and also satisfy the requirements of 760.71(B) for multiconductor cables and 760.71(H) for co-axial cables shall be permitted to be listed and marked as multipurpose cable Types MPP, MPR, MPG, and MP, respectively.

**(H) Communications Wires.** Communications wires, such as distributing frame wire and jumper wire, shall be listed as being resistant to the spread of fire.

> FPN: One method of defining *resistant to the spread of fire* is that the cables do not spread fire to the top of the tray in the vertical-tray flame test in ANSI/UL 1581-1991, *Reference Standard for Electrical Wires, Cables and Flexible Cords.* Another method of defining *resistant to the spread of fire* is for the damage (char length) not to exceed 1.5 m (4 ft 11 in.) when performing the vertical flame test for cables in cable trays, as described in CSA C22.2 No. 0.3-M-1985, *Test Methods for Electrical Wires and Cables.*

**(I) Hybrid Power and Communications Cable.** Listed hybrid power and communications cable shall be permitted where the power cable is a listed Type NM or NM-B conforming to the provisions of Article 334, and the communications cable is a listed Type CM, the jackets on the listed NM or NM-B and listed CM cables are rated for 600 volts minimum, and the hybrid cable is listed as being resistant to the spread of fire.

> FPN: One method of defining *resistant to the spread of fire* is that the cables do not spread fire to the top of the tray in the vertical-tray flame test in ANSI/UL 1581-1991, *Reference Standard for Electrical Wires, Cables and Flexible Cords.* Another method of defining *resistant to the spread of fire* is for the damage (char length) not to exceed 1.5 m (4 ft 11 in.) when performing the vertical flame test for cables in cable trays, as described in CSA C22.2 No. 0.3-M-1985, *Test Methods for Electrical Wires and Cables.*

**(J) Plenum Communications Raceways.** Plenum communications raceways listed as plenum optical fiber raceways shall be permitted for use in ducts, plenums, and other spaces used for environmental air and shall also be listed as having adequate fire-resistant and low smoke-producing characteristics.

**(K) Riser Communications Raceway.** Riser communications raceways shall be listed as having adequate fire-resistant characteristics capable of preventing the carrying of fire from floor to floor.

**(L) General-Purpose Communications Raceway.** General-purpose communications raceways shall be listed as being resistant to the spread of fire.

**800.52 Installation of Communications Wires, Cables, and Equipment.** Communications wires and cables from the protector to the equipment or, where no protector is required, communications wires and cables attached to the outside or inside of the building shall comply with 800.52(A) through (E).

**(A) Separation from Other Conductors.**

**(1) In Raceways, Boxes, and Cables.**

(a) Other Power-Limited Circuits. Communications cables shall be permitted in the same raceway or enclosure with cables of any of the following:

(1) Class 2 and Class 3 remote-control, signaling, and power-limited circuits in compliance with Article 725
(2) Power-limited fire alarm systems in compliance with Article 760
(3) Nonconductive and conductive optical fiber cables in compliance with Article 770
(4) Community antenna television and radio distribution systems in compliance with Article 820
(5) Low-power network-powered broadband communications circuits in compliance with Article 830

(b) Class 2 and Class 3 Circuits. Class 1 circuits shall not be run in the same cable with communications circuits. Class 2 and Class 3 circuit conductors shall be permitted in the same cable with communications circuits, in which case the Class 2 and Class 3 circuits shall be classified as communications circuits and shall meet the requirements of this article. The cables shall be listed as communications cables or multipurpose cables.

*Exception: Cables constructed of individually listed Class 2, Class 3, and communications cables under a common jacket shall not be required to be classified as communications cable. The fire-resistance rating of the composite cable shall be determined by the performance of the composite cable.*

**(c)  Electric Light, Power, Class 1, Non–Power-Limited Fire Alarm, and Medium Power Network-Powered Broadband Communications Circuits in Raceways, Compartments, and Boxes.** Communications conductors shall not be placed in any raceway, compartment, outlet box, junction box, or similar fitting with conductors of electric light, power, Class 1, non–power-limited fire alarm or medium power network-powered broadband communications circuits.

*Exception No. 1: Where all of the conductors of electric light, power, Class 1, non–power-limited fire alarm, and medium power network-powered broadband communications circuits are separated from all of the conductors of communications circuits by a barrier.*

*Exception No. 2: Power conductors in outlet boxes, junction boxes, or similar fittings or compartments where such conductors are introduced solely for power supply to communications equipment. The power circuit conductors shall be routed within the enclosure to maintain a minimum of 6 mm (0.25 in.) separation from the communications circuit conductors.*

*Exception No. 3:  As permitted by 620.36.*

**(2)  Other Applications.** Communications wires and cables shall be separated at least 50 mm (2 in.) from conductors of any electric light, power, Class 1, non–power-limited fire alarm, or medium power network-powered broadband communications circuits.

*Exception No. 1:  Where either (1) all of the conductors of the electric light, power, Class 1, non–power-limited fire alarm, and medium power network-powered broadband communications circuits are in a raceway or in metal-sheathed, metal-clad, nonmetallic-sheathed, Type AC, or Type UF cables, or (2) all of the conductors of communications circuits are encased in raceway.*

*Exception No. 2:  Where the communications wires and cables are permanently separated from the conductors of electric light, power, Class 1, non–power-limited fire alarm, and medium power network-powered broadband communications circuits by a continuous and firmly fixed nonconductor, such as porcelain tubes or flexible tubing, in addition to the insulation on the wire.*

**(B)  Spread of Fire or Products of Combustion.** Installations in hollow spaces, vertical shafts, and ventilation or air-handling ducts shall be made so that the possible spread of fire or products of combustion is not substantially increased. Openings around penetrations through fire resistance-rated walls, partitions, floors, or ceilings shall be firestopped using approved methods to maintain the fire resistance rating.

The accessible portion of abandoned communications cables shall not be permitted to remain.

FPN:  Directories of electrical construction materials published by qualified testing laboratories contain many listing installation restrictions necessary to maintain the fire-resistive rating of assemblies where penetrations or openings are made.

**(C)  Equipment in Other Space Used for Environmental Air.** Section 300.22(C) shall apply.

**(D)  Cable Trays.** Types MPP, MPR, MPG, and MP multipurpose cables and Types CMP, CMR, CMG, and CM communications cables shall be permitted to be installed in cable trays. Communications raceways, as described in 800.51, shall be permitted to be installed in cable trays.

**(E)  Support of Conductors.** Raceways shall be used for their intended purpose. Communications cables or wires shall not be strapped, taped, or attached by any means to the exterior of any conduit or raceway as a means of support.

*Exception:  Overhead (aerial) spans of communications cables or wires shall be permitted to be attached to the exterior of a raceway-type mast intended for the attachment and support of such conductors.*

**800.53  Applications of Listed Communications Wires and Cables and Communications Raceways.** Communications wires and cables shall comply with the requirements of 800.53(A) through (F) or where cable substitutions are made in accordance with 800.53(G).

**(A)  Plenum.** Cables installed in ducts, plenums, and other spaces used for environmental air shall be Type CMP. Abandoned cables shall not be permitted to remain. Types CMP, CMR, CMG, CM, and CMX and communications wire installed in compliance with 300.22 shall be permitted. Listed plenum communications raceways shall be permitted to be installed in ducts and plenums as described in 300.22(B) and in other spaces used for environmental air as described in 300.22(C). Only Type CMP cable shall be permitted to be installed in these raceways.

**(B)  Riser.** Cables installed in risers shall comply with 800.53(B)(1), (B)(2), or (B)(3).

**(1)  Cables in Vertical Runs.** Cables installed in vertical runs and penetrating more than one floor, or cables installed in vertical runs in a shaft, shall be Type CMR. Floor penetrations requiring Type CMR shall contain only cables suitable for riser or plenum use. Abandoned cables shall not be permitted to remain. Listed riser communications raceways shall be permitted to be installed in vertical riser runs in a shaft from floor to floor. Only Type CMR and CMP cables shall be permitted to be installed in these raceways.

**(2)  Metal Raceways or Fireproof Shafts.** Listed communications cables shall be encased in a metal raceway or located in a fireproof shaft having firestops at each floor.

**(3) One- and Two-Family Dwellings.** Type CM and CMX cable shall be permitted in one- and two-family dwellings.

FPN: See 800.52(B) for firestop requirements for floor penetrations.

**(C) Distributing Frames and Cross-Connect Arrays.** Listed communications wire and Types CMP, CMR, CMG, and CM communications cables shall be used in distributing frames and cross-connect arrays.

**(D) Cable Trays.** Types MPP, MPR, MPG, and MP multipurpose cables and Types CMP, CMR, CMG, and CM communications cables shall be permitted to be installed in cable trays.

**(E) Other Wiring Within Buildings.** Cables installed in building locations other than the locations covered in 800.53(A) through (D) shall be in accordance with 800.53(E)(1) through (E)(6).

**(1) General.** Cables shall be Type CMG or Type CM. Listed communications general-purpose raceways shall be permitted. Only Types CMG, CM, CMR, or CMP cables shall be permitted to be installed in general-purpose communications raceways.

**(2) In Raceways.** Listed communications wires that are enclosed in a raceway of a type included in Chapter 3 shall be permitted.

**(3) Nonconcealed Spaces.** Type CMX communications cable shall be permitted to be installed in nonconcealed spaces where the exposed length of cable does not exceed 3 m (10 ft).

**(4) One- and Two-Family Dwellings.** Type CMX communications cable less than 6 mm (0.25 in.) in diameter shall be permitted to be installed in one- and two-family dwellings.

**(5) Multi-Family Dwellings.** Type CMX communications cable less than 6 mm (0.25 in.) in diameter shall be permitted to be installed in nonconcealed spaces in multi-family dwellings.

**(6) Under Carpets.** Type CMUC undercarpet communications wires and cables shall be permitted to be installed under carpet.

**(F) Hybrid Power and Communications Cable.** Hybrid power and communications cable listed in accordance with 800.51(J) shall be permitted to be installed in one- and two-family dwellings.

**(G) Cable Substitutions.** The uses and permitted substitutions for communications cables listed in Table 800.53 shall

**Table 800.53 Cable Uses and Permitted Substitutions**

| Cable Type | Use | References | Permitted Substitutions |
|---|---|---|---|
| CMP | Communications plenum cable | 800.53(A) | MPP |
| CMR | Communications riser cable | 800.53(B) | MPP, CMP, MPR |
| CMG, CM | Communications general-purpose cable | 800.53(E)(1) | MPP, CMP, MPR, CMR, MPG, MP |
| CMX | Communications cable, limited use | 800.53(E) | MPP, CMP, MPR, CMR, MPG, MP, CMG, CM |

Note: See Figure 800.53, Cable substitution hierarchy.

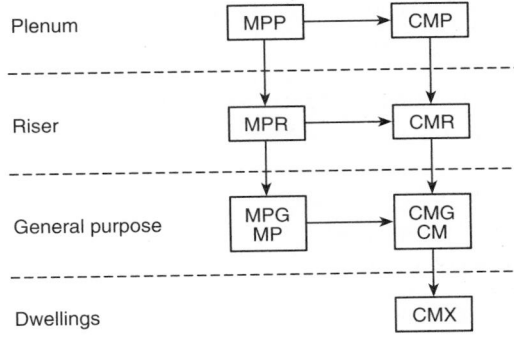

Type CM—Communications cables
Type MP—Multipurpose cable

[A]→[B]  Cable A shall be permitted to be used in place of cable B.

**Figure 800.53 Cable substitution hierarchy.**

be considered suitable for the purpose and shall be permitted.

FPN: For information on Types CMP, CMR, CMG, CM, and CMX cables, see 800.51.

# ARTICLE 810
# Radio and Television Equipment

## I. General

**810.1 Scope.** This article covers antenna systems for radio and television receiving equipment, amateur radio transmitting and receiving equipment, and certain features of transmitter safety. This article covers antennas such as multi-element, vertical rod, and dish, and also covers the wiring and cabling that connects them to equipment. This article does not cover equipment and antennas used for coupling carrier current to power line conductors.

**810.2 Definitions.** For definitions applicable to this article, see Article 100.

**810.3 Other Articles.** Wiring from the source of power to and between devices connected to the interior wiring system shall comply with Chapters 1 through 4 other than as modified by Parts I and II of Article 640. Wiring for audio signal processing, amplification, and reproduction equipment shall comply with Article 640. Coaxial cables that connect antennas to equipment shall comply with Article 820.

**810.4 Community Television Antenna.** The antenna shall comply with this article. The distribution system shall comply with Article 820.

**810.5 Radio Noise Suppressors.** Radio interference eliminators, interference capacitors, or noise suppressors connected to power-supply leads shall be of a listed type. They shall not be exposed to physical damage.

## II. Receiving Equipment — Antenna Systems

**810.11 Material.** Antennas and lead-in conductors shall be of hard-drawn copper, bronze, aluminum alloy, copper-clad steel, or other high-strength, corrosion-resistant material.

*Exception: Soft-drawn or medium-drawn copper shall be permitted for lead-in conductors where the maximum span between points of support is less than 11 m (35 ft).*

**810.12 Supports.** Outdoor antennas and lead-in conductors shall be securely supported. The antennas or lead-in conductors shall not be attached to the electric service mast. They shall not be attached to poles or similar structures carrying open electric light or power wires or trolley wires of over 250 volts between conductors. Insulators supporting the antenna conductors shall have sufficient mechanical strength to safely support the conductors. Lead-in conductors shall be securely attached to the antennas.

**810.13 Avoidance of Contacts with Conductors of Other Systems.** Outdoor antennas and lead-in conductors from an antenna to a building shall not cross over open conductors of electric light or power circuits and shall be kept well away from all such circuits so as to avoid the possibility of accidental contact. Where proximity to open electric light or power service conductors of less than 250 volts between conductors cannot be avoided, the installation shall be such as to provide a clearance of at least 600 mm (2 ft).

Where practicable, antenna conductors shall be installed so as not to cross under open electric light or power conductors.

**810.14 Splices.** Splices and joints in antenna spans shall be made mechanically secure with approved splicing devices or by such other means as will not appreciably weaken the conductors.

**810.15 Grounding.** Masts and metal structures supporting antennas shall be grounded in accordance with 810.21.

**810.16 Size of Wire-Strung Antenna — Receiving Station.**

**(A) Size of Antenna Conductors.** Outdoor antenna conductors for receiving stations shall be of a size not less than given in Table 810.16(A).

**Table 810.16(A) Size of Receiving Station Outdoor Antenna Conductors**

| | Minimum Size of Conductors (AWG) Where Maximum Open Span Length Is | | |
|---|---|---|---|
| **Material** | **Less Than 11 m (35 ft)** | **11 m to 45 m (35 ft to 150 ft)** | **Over 45 m (150 ft)** |
| Aluminum alloy, hard-drawn copper | 19 | 14 | 12 |
| Copper-clad steel, bronze, or other high-strength material | 20 | 17 | 14 |

**(B) Self-Supporting Antennas.** Outdoor antennas, such as vertical rods, dishes, or dipole structures, shall be of corrosion-resistant materials and of strength suitable to withstand ice and wind loading conditions and shall be located well away from overhead conductors of electric light and power circuits of over 150 volts to ground, so as to avoid the possibility of the antenna or structure falling into or making accidental contact with such circuits.

**810.17 Size of Lead-in — Receiving Station.** Lead-in conductors from outside antennas for receiving stations shall, for various maximum open span lengths, be of such size as to have a tensile strength at least as great as that of the conductors for antennas as specified in 810.16. Where the lead-in consists of two or more conductors that are twisted together, are enclosed in the same covering, or are concentric, the conductor size shall, for various maximum open span lengths, be such that the tensile strength of the combination is at least as great as that of the conductors for antennas as specified in 810.16.

## 810.18 Clearances — Receiving Stations.

**(A) Outside of Buildings.** Lead-in conductors attached to buildings shall be installed so that they cannot swing closer than 600 mm (2 ft) to the conductors of circuits of 250 volts or less between conductors, or 3.0 m (10 ft) to the conductors of circuits of over 250 volts between conductors, except that in the case of circuits not over 150 volts between conductors, where all conductors involved are supported so as to ensure permanent separation, the clearance shall be permitted to be reduced but shall not be less than 100 mm (4 in.). The clearance between lead-in conductors and any conductor forming a part of a lightning rod system shall not be less than 1.8 m (6 ft) unless the bonding referred to in 250.60 is accomplished. Underground conductors shall be separated at least 300 mm (12 in.) from conductors of any light or power circuits or Class 1 circuits.

*Exception: Where the electric light or power conductors, Class 1 conductors, or lead-in conductors are installed in raceways or metal cable armor.*

**(B) Antennas and Lead-ins — Indoors.** Indoor antennas and indoor lead-ins shall not be run nearer than 50 mm (2 in.) to conductors of other wiring systems in the premises.

*Exception No. 1: Where such other conductors are in metal raceways or cable armor.*

*Exception No. 2: Where permanently separated from such other conductors by a continuous and firmly fixed nonconductor, such as porcelain tubes or flexible tubing.*

**(C) In Boxes or Other Enclosures.** Indoor antennas and indoor lead-ins shall be permitted to occupy the same box or enclosure with conductors of other wiring systems where separated from such other conductors by an effective permanently installed barrier.

## 810.19 Electric Supply Circuits Used in Lieu of Antenna — Receiving Stations.
Where an electric supply circuit is used in lieu of an antenna, the device by which the radio receiving set is connected to the supply circuit shall be listed.

## 810.20 Antenna Discharge Units — Receiving Stations.

**(A) Where Required.** Each conductor of a lead-in from an outdoor antenna shall be provided with a listed antenna discharge unit.

*Exception: Where the lead-in conductors are enclosed in a continuous metallic shield that either is permanently and effectively grounded or is protected by an antenna discharge unit.*

**(B) Location.** Antenna discharge units shall be located outside the building or inside the building between the point of entrance of the lead-in and the radio set or transformers and as near as practicable to the entrance of the conductors to the building. The antenna discharge unit shall not be located near combustible material or in a hazardous (classified) location as defined in Article 500.

**(C) Grounding.** The antenna discharge unit shall be grounded in accordance with 810.21.

## 810.21 Grounding Conductors — Receiving Stations.
Grounding conductors shall comply with 810.21(A) through (J).

**(A) Material.** The grounding conductor shall be of copper, aluminum, copper-clad steel, bronze, or similar corrosion-resistant material. Aluminum or copper-clad aluminum grounding conductors shall not be used where in direct contact with masonry or the earth or where subject to corrosive conditions. Where used outside, aluminum or copper-clad aluminum shall not be installed within 450 mm (18 in.) of the earth.

**(B) Insulation.** Insulation on grounding conductors shall not be required.

**(C) Supports.** The grounding conductors shall be securely fastened in place and shall be permitted to be directly attached to the surface wired over without the use of insulating supports.

*Exception: Where proper support cannot be provided, the size of the grounding conductors shall be increased proportionately.*

**(D) Mechanical Protection.** The grounding conductor shall be protected where exposed to physical damage, or the size of the grounding conductors shall be increased proportionately to compensate for the lack of protection. Where the grounding conductor is run in a metal raceway, both ends of the raceway shall be bonded to the grounding conductor or to the same terminal or electrode to which the grounding conductor is connected.

**(E) Run in Straight Line.** The grounding conductor for an antenna mast or antenna discharge unit shall be run in as straight a line as practicable from the mast or discharge unit to the grounding electrode.

**(F) Electrode.** The grounding conductor shall be connected as follows:

(1) To the nearest accessible location on the following:

  a. The building or structure grounding electrode system as covered in 250.50

b. The grounded interior metal water piping systems, within 1.52 m (5 ft) from its point of entrance to the building, as covered in 250.52

c. The power service accessible means external to the building, as covered in 250.94

d. The metallic power service raceway

e. The service equipment enclosure, or

f. The grounding electrode conductor or the grounding electrode conductor metal enclosures; or

(2) If the building or structure served has no grounding means, as described in 810.21(F)(1), to any one of the individual electrodes described in 250.52; or

(3) If the building or structure served has no grounding means, as described in 810.21(F)(1) or (F)(2), to an effectively grounded metal structure or to any of the individual electrodes described in 250.52.

**(G) Inside or Outside Building.** The grounding conductor shall be permitted to be run either inside or outside the building.

**(H) Size.** The grounding conductor shall not be smaller than 10 AWG copper, 8 AWG aluminum, or 17 AWG copper-clad steel or bronze.

**(I) Common Ground.** A single grounding conductor shall be permitted for both protective and operating purposes.

**(J) Bonding of Electrodes.** A bonding jumper not smaller than 6 AWG copper or equivalent shall be connected between the radio and television equipment grounding electrode and the power grounding electrode system at the building or structure served where separate electrodes are used.

## III. Amateur Transmitting and Receiving Stations — Antenna Systems

**810.51 Other Sections.** In addition to complying with Part III, antenna systems for amateur transmitting and receiving stations shall also comply with 810.11 through 810.15.

**810.52 Size of Antenna.** Antenna conductors for transmitting and receiving stations shall be of a size not less than given in Table 810.52.

**810.53 Size of Lead-In Conductors.** Lead-in conductors for transmitting stations shall, for various maximum span lengths, be of a size at least as great as that of conductors for antennas as specified in 810.52.

**810.54 Clearance on Building.** Antenna conductors for transmitting stations, attached to buildings, shall be firmly mounted at least 75 mm (3 in.) clear of the surface of the building on nonabsorbent insulating supports, such as

**Table 810.52 Size of Amateur Station Outdoor Antenna Conductors**

| Material | Minimum Size of Conductors (AWG) Where Maximum Open Span Length Is | |
| --- | --- | --- |
| | Less Than 45 m (150 ft) | Over 45 m (150 ft) |
| Hard-drawn copper | 14 | 10 |
| Copper-clad steel, bronze, or other high-strength material | 14 | 12 |

treated pins or brackets equipped with insulators having not less than 75-mm (3-in.) creepage and airgap distances. Lead-in conductors attached to buildings shall also comply with these requirements.

*Exception: Where the lead-in conductors are enclosed in a continuous metallic shield that is permanently and effectively grounded, they shall not be required to comply with these requirements. Where grounded, the metallic shield shall also be permitted to be used as a conductor.*

**810.55 Entrance to Building.** Except where protected with a continuous metallic shield that is permanently and effectively grounded, lead-in conductors for transmitting stations shall enter buildings by one of the following methods:

(1) Through a rigid, noncombustible, nonabsorbent insulating tube or bushing

(2) Through an opening provided for the purpose in which the entrance conductors are firmly secured so as to provide a clearance of at least 50 mm (2 in.)

(3) Through a drilled window pane

**810.56 Protection Against Accidental Contact.** Lead-in conductors to radio transmitters shall be located or installed so as to make accidental contact with them difficult.

**810.57 Antenna Discharge Units — Transmitting Stations.** Each conductor of a lead-in for outdoor antennas shall be provided with an antenna discharge unit or other suitable means that drain static charges from the antenna system.

*Exception No. 1: Where protected by a continuous metallic shield that is permanently and effectively grounded.*

*Exception No. 2: Where the antenna is permanently and effectively grounded.*

**810.58 Grounding Conductors — Amateur Transmitting and Receiving Stations.** Grounding conductors shall comply with 810.58(A) through (C).

**(A) Other Sections.** All grounding conductors for amateur transmitting and receiving stations shall comply with 810.21(A) through (J).

**(B) Size of Protective Grounding Conductor.** The protective grounding conductor for transmitting stations shall be as large as the lead-in but not smaller than 10 AWG copper, bronze, or copper-clad steel.

**(C) Size of Operating Grounding Conductor.** The operating grounding conductor for transmitting stations shall not be less than 14 AWG copper or its equivalent.

## IV. Interior Installation — Transmitting Stations

**810.70 Clearance from Other Conductors.** All conductors inside the building shall be separated at least 100 mm (4 in.) from the conductors of any electric light, power, or signaling circuit.

*Exception No. 1: As provided in Article 640.*

*Exception No. 2: Where separated from other conductors by raceway or some firmly fixed nonconductor, such as porcelain tubes or flexible tubing.*

**810.71 General.** Transmitters shall comply with 810.71(A) through (C).

**(A) Enclosing.** The transmitter shall be enclosed in a metal frame or grille, or separated from the operating space by a barrier or other equivalent means, all metallic parts of which are effectively connected to ground.

**(B) Grounding of Controls.** All external metal handles and controls accessible to the operating personnel shall be effectively grounded.

**(C) Interlocks on Doors.** All access doors shall be provided with interlocks that disconnect all voltages of over 350 volts between conductors when any access door is opened.

## ARTICLE 820
## Community Antenna Television and Radio Distribution Systems

## I. General

**820.1 Scope.** This article covers coaxial cable distribution of radio frequency signals typically employed in community antenna television (CATV) systems.

**820.2 Definitions.** See Article 100. For the purposes of this article, the following additional definitions apply.

**Abandoned Coaxial Cable.** Installed coaxial cable that is not terminated at equipment other than a coaxial connector and not identified for future use with a tag.

**Exposed.** An exposed cable is one that is in such a position that, in case of failure of supports and insulation, contact with another circuit could result.

FPN: See Article 100 for two other definitions of *exposed*.

**Point of Entrance.** The point within a building at which the cable emerges from an external wall, from a concrete floor slab, or from a rigid metal conduit or an intermediate metal conduit grounded to an electrode in accordance with 820.40(B).

**Premises.** The land and buildings of a user located on the user side of utility-user network point of demarcation.

**820.3 Locations and Other Articles.** Circuits and equipment shall comply with 820.3(A) through (G).

**(A) Spread of Fire or Products of Combustion.** Section 300.21 shall apply. The accessible portion of abandoned coaxial cables shall not be permitted to remain.

**(B) Ducts, Plenums, and Other Air-Handling Spaces.** Section 300.22, where installed in ducts or plenums or other spaces used for environmental air, shall apply.

*Exception: As permitted in 820.53(A).*

**(C) Installation and Use.** Section 110.3 shall apply.

**(D) Installations of Conductive and Nonconductive Optical Fiber Cables.** Article 770 shall apply.

**(E) Communications Circuits.** Article 800 shall apply.

**(F) Network-Powered Broadband Communications Systems.** Article 830 shall apply.

**(G) Alternate Wiring Methods.** The wiring methods of Article 830 shall be permitted to substitute for the wiring methods of Article 820.

FPN: Use of Article 830 wiring methods will facilitate the upgrading of Article 820 installations to network-powered broadband applications.

**820.4 Energy Limitations.** The coaxial cable shall be permitted to deliver low-energy power to equipment that is directly associated with the radio frequency distribution system if the voltage is not over 60 volts and if the current supply is from a transformer or other device that has energy-limiting characteristics.

**820.5 Access to Electrical Equipment Behind Panels Designed to Allow Access.** Access to electrical equipment shall not be denied by an accumulation of wires and cables that prevents removal of panels, including suspended ceiling panels.

**820.6 Mechanical Execution of Work.** Community antenna television and radio distribution systems shall be installed in a neat and workmanlike manner. Cables installed exposed on the surface of ceiling and sidewalls shall be supported by the structural components of the building structure in such a manner that the cable is not damaged by normal building use. Such cables shall be attached to structural components by straps, staples, hangers, or similar fittings designed and installed so as not to damage the cable. The installation shall also conform with 300.4(D).

## II. Cables Outside and Entering Buildings

**820.10 Outside Cables.** Coaxial cables, prior to the point of grounding, as defined in 820.33, shall comply with 820.10(A) through (F).

**(A) On Poles.** Where practicable, conductors on poles shall be located below the electric light, power, Class 1, or non–power-limited fire alarm circuit conductors and shall not be attached to a crossarm that carries electric light or power conductors.

**(B) Lead-in Clearance.** Lead-in or aerial-drop cables from a pole or other support, including the point of initial attachment to a building or structure, shall be kept away from electric light, power, Class 1, or non–power-limited fire alarm circuit conductors so as to avoid the possibility of accidental contact.

*Exception: Where proximity to electric light, power, Class 1, or non–power-limited fire alarm circuit service conductors cannot be avoided, the installation shall be such as to provide clearances of not less than 300 mm (12 in.) from light, power, Class 1, or non–power-limited fire alarm circuit service drops. The clearance requirement shall apply at all points along the drop, and it shall increase to 1.02 m (40 in.) at the pole.*

**(C) On Masts.** Aerial cable shall be permitted to be attached to an above-the-roof raceway mast that does not enclose or support conductors of electric light or power circuits.

**(D) Above Roofs.** Cables shall have a vertical clearance of not less than 2.5 m (8 ft) from all points of roofs above which they pass.

*Exception No. 1: Auxiliary buildings such as garages and the like.*

*Exception No. 2: A reduction in clearance above only the overhanging portion of the roof to not less than 450 mm (18 in.) shall be permitted if (1) not more than 1.2 m (4 ft) of communications service drop conductors pass above the roof overhang, and (2) they are terminated at a raceway mast or other approved support.*

*Exception No. 3: Where the roof has a slope of not less than 100 mm (4 in.) in 300 mm (12 in.), a reduction in clearance to not less than 900 mm (3 ft) shall be permitted.*

**(E) Between Buildings.** Cables extending between buildings and also the supports or attachment fixtures shall be acceptable for the purpose and shall have sufficient strength to withstand the loads to which they may be subjected.

*Exception: Where a cable does not have sufficient strength to be self-supporting, it shall be attached to a supporting messenger cable that, together with the attachment fixtures or supports, shall be acceptable for the purpose and shall have sufficient strength to withstand the loads to which they may be subjected.*

**(F) On Buildings.** Where attached to buildings, cables shall be securely fastened in such a manner that they will be separated from other conductors in accordance with 820.10(F)(1), (F)(2), and (F)(3).

**(1) Electric Light or Power.** The coaxial cable shall have a separation of at least 100 mm (4 in.) from electric light, power, Class 1, or non–power-limited fire alarm circuit conductors not in raceway or cable or be permanently separated from conductors of the other system by a continuous and firmly fixed nonconductor in addition to the insulation on the wires.

**(2) Other Communications Systems.** Coaxial cable shall be installed so that there will be no unnecessary interference in the maintenance of the separate systems. In no case shall the conductors, cables, messenger strand, or equipment of one system cause abrasion to the conductors, cable, messenger strand, or equipment of any other system.

**(3) Lightning Conductors.** Where practicable, a separation of at least 1.8 m (6 ft) shall be maintained between any coaxial cable and lightning conductors.

FPN: For additional information regarding overhead wires and cables, see ANSI C2-1997, *National Electric Safety Code*, Part 2, Safety Rules for Overhead Lines.

**820.11 Entering Buildings.**

**(A) Underground Systems.** Underground coaxial cables in a duct, pedestal, handhole, or manhole that contains electric light or power conductors or Class 1 circuits shall be in a section permanently separated from such conductors by means of a suitable barrier.

**(B) Direct-Buried Cables and Raceways.** Direct-buried coaxial cable shall be separated at least 300 mm (12 in.) from conductors of any light or power or Class 1 circuit.

*Exception No. 1: Where electric service conductors or coaxial cables are installed in raceways or have metal cable armor.*

*Exception No. 2: Where electric light or power branch-circuit or feeder conductors or Class 1 circuit conductors are installed in a raceway or in metal-sheathed, metal-clad, or Type UF or Type USE cables; or the coaxial cables have metal cable armor or are installed in a raceway.*

## III. Protection

**820.33 Grounding of Outer Conductive Shield of a Co-axial Cable.** The outer conductive shield of the coaxial cable shall be grounded at the building premises as close to the point of cable entrance or attachment as practicable.

For purposes of this section, grounding located at mobile home service equipment located in sight from, and not more than 9.0 m (30 ft) from, the exterior wall of the mobile home it serves, or at a mobile home disconnecting means grounded in accordance with 250.32 and located in sight from and not more than 9.0 m (30 ft) from the exterior wall of the mobile home it serves, shall be considered to meet the requirements of this section.

> FPN: Selecting a grounding location to achieve the shortest practicable grounding conductor helps limit potential differences between CATV and other metallic systems.

**(A) Shield Grounding.** Where the outer conductive shield of a coaxial cable is grounded, no other protective devices shall be required.

**(B) Shield Protection Devices.** Grounding of a coaxial drop cable shield by means of a protective device that does not interrupt the grounding system within the premises shall be permitted.

## IV. Grounding Methods

**820.40 Cable Grounding.** Where required by 820.33, the shield of the coaxial cable shall be grounded as specified in 820.40(A) through (D).

**(A) Grounding Conductor.**

**(1) Insulation.** The grounding conductor shall be insulated and shall be listed as suitable for the purpose.

**(2) Material.** The grounding conductor shall be copper or other corrosion-resistant conductive material, stranded or solid.

**(3) Size.** The grounding conductor shall not be smaller than 14 AWG. It shall have a current-carrying capacity approximately equal to that of the outer conductor of the coaxial cable. The grounding conductor shall not be required to exceed 6 AWG.

**(4) Length.** The grounding conductor shall be as short as practicable. In one- and two-family dwellings, the grounding conductor shall be as short as practicable, not to exceed 6.0 m (20 ft) in length.

*Exception: In one- and two-family dwellings where it is not practicable to achieve an overall maximum grounding conductor length of 6.0 m (20 ft), a separate ground as specified in 250.52(A)(5), (6), or (7) shall be used, the grounding conductor shall be grounded to the separate ground in accordance with 250.70, and the separate ground bonded to the power grounding electrode system in accordance with 820.40(D).*

**(5) Run in Straight Line.** The grounding conductor shall be run to the grounding electrode in as straight a line as practicable.

**(6) Physical Protection.** Where subject to physical damage, the grounding conductor shall be adequately protected. Where the grounding conductor is run in a metal raceway, both ends of the raceway shall be bonded to the grounding conductor or the same terminal or electrode to which the grounding conductor is connected.

**(B) Electrode.** The grounding conductor shall be connected in accordance with 820.40(B)(1) and (B)(2).

**(1) In Buildings or Structures with Grounding Means.** To the nearest accessible location on the following:

(1) The building or structure grounding electrode system as covered in 250.50;
(2) The grounded interior metal water piping system, within 1.52 m (5 ft) from its point of entrance to the building, as covered in 250.52;
(3) The power service accessible means external to enclosures as covered in 250.94;
(4) The metallic power service raceway;
(5) The service equipment enclosure;
(6) The grounding electrode conductor or the grounding electrode conductor metal enclosure; or
(7) The grounding conductor or the grounding electrode of a building or structure disconnecting means that is grounded to an electrode as covered in 250.32.

**(2) In Buildings or Structures Without Grounding Means.** If the building or structure served has no grounding means, as described in 820.40(B)(1):

(1) To any one of the individual electrodes described in 250.52(A)(1), (2), (3), (4); or,

(2) If the building or structure served has no grounding means, as described in 820.40(B)(1) or (B)(2)(1), to an effectively grounded metal structure or to any one of the individual electrodes described in 250.52(A)(5), (6), and (7).

**(C) Electrode Connection.** Connections to grounding electrodes shall comply with 250.70.

**(D) Bonding of Electrodes.** A bonding jumper not smaller than 6 AWG copper or equivalent shall be connected between the antenna systems grounding electrode and the power grounding electrode system at the building or structure served where separate electrodes are used.

*Exception: At mobile homes as covered in 820.42.*

FPN No. 1: See 250.60 for use of air terminals (lightning rods).

FPN No. 2: Bonding together of all separate electrodes limits potential differences between them and between their associated wiring systems.

**820.41 Equipment Grounding.** Unpowered equipment and enclosures or equipment powered by the coaxial cable shall be considered grounded where connected to the metallic cable shield.

**820.42 Bonding and Grounding at Mobile Homes.**

**(A) Grounding.** Where there is no mobile home service equipment located in sight from, and not more than 9.0 m (30 ft) from, the exterior wall of the mobile home it serves or there is no mobile home disconnecting means grounded in accordance with 250.32 and located within sight from, and not more than 9.0 m (30 ft) from, the exterior wall of the mobile home it serves, the coaxial cable shield ground, or surge arrester ground, shall be in accordance with 820.40(B)(2).

**(B) Bonding.** The coaxial cable shield grounding terminal, surge arrester grounding terminal, or grounding electrode shall be bonded to the metal frame or available grounding terminal of the mobile home with a copper grounding conductor not smaller than 12 AWG under any of the following conditions:

(1) Where there is no mobile home service equipment or disconnecting means as in 820.42(A)
(2) Where the mobile home is supplied by cord and plug

## V. Cables Within Buildings

**820.49 Fire Resistance of CATV Cables.** Coaxial cables installed as wiring within buildings shall be listed as being resistant to the spread of fire in accordance with 820.50 and 820.51.

**820.50 Listing, Marking, and Installation of Coaxial Cables.** Coaxial cables in a building shall be listed as being suitable for the purpose, and cables shall be marked in accordance with Table 820.50. The cable voltage rating shall not be marked on the cable.

FPN: Voltage markings on cables could be misinterpreted to suggest that the cables may be suitable for Class 1, electric light, and power applications.

*Exception No. 1: Voltage markings shall be permitted where the cable has multiple listings and voltage marking is required for one or more of the listings.*

*Exception No. 2: Listing and marking shall not be required where the cable enters the building from the outside and is run in rigid metal conduit or intermediate metal conduit, and such conduits are grounded to an electrode in accordance with 820.40(B).*

*Exception No. 3: Listing and marking shall not be required where the length of the cable within the building, measured from its point of entrance, does not exceed 15 m (50 ft) and the cable enters the building from the outside and is terminated at a grounding block.*

**Table 820.50 Cable Markings**

| Cable Marking | Type | Reference |
|---|---|---|
| CATVP | CATV plenum cable | 820.51(A) and 820.53(A) |
| CATVR | CATV riser cable | 820.51(B) and 820.53(B) |
| CATV | CATV cable | 820.51(C) and 820.53(C) |
| CATVX | CATV cable, limited use | 820.51(D) and 820.53(C) |

FPN No. 1: Cable types are listed in descending order of fire-resistance rating.

FPN No. 2: See the referenced sections for listing requirements and permitted uses.

**820.51 Additional Listing Requirements.** Cables shall be listed in accordance with 820.51(A) through (D).

**(A) Type CATVP.** Type CATVP community antenna television plenum cable shall be listed as being suitable for use in ducts, plenums, and other spaces used for environmental air and shall also be listed as having adequate fire-resistant and low smoke-producing characteristics.

FPN: One method of defining low smoke-producing cables is by establishing an acceptable value of the smoke produced when tested in accordance with NFPA 262-1999, *Standard Method for Test for Flame Travel and Smoke of Wire and Cables for Use in Air-Handling Spaces*, to a maximum peak optical density of 0.5 and a maximum average optical density of 0.15. Similarly, one method of defining fire-resistant cables is by establishing maximum allowable flame travel distance of 1.52 m (5 ft) when tested in accordance with the same test.

**(B) Type CATVR.** Type CATVR community antenna television riser cable shall be listed as being suitable for use in a vertical run in a shaft or from floor to floor and shall also be listed as having fire-resistant characteristics capable of preventing the carrying of fire from floor to floor.

FPN: One method of defining fire-resistant characteristics capable of preventing the carrying of fire from floor to floor is that the cables pass the requirements of ANSI/UL 1666-1997, *Standard Test for Flame Propagation Height of Electrical and Optical-Fiber Cable Installed Vertically in Shafts.*

**(C) Type CATV.** Type CATV community antenna television cable shall be listed as being suitable for general-purpose CATV use, with the exception of risers and plenums, and shall also be listed as being resistant to the spread of fire.

FPN: One method of defining *resistant to the spread of fire* is that the cables do not spread fire to the top of the tray in the vertical-tray flame test in ANSI/UL 1581-1991, *Reference Standard for Electrical Wires, Cables and Flexible Cords.*

Another method of defining *resistant to the spread of fire* is for the damage (char length) not to exceed 1.5 m (4 ft 11 in.) when performing the vertical flame test for cables in cable trays, as described in CSA C22.2 No. 0.3-M-1985, *Test Methods for Electrical Wires and Cables.*

**(D) Type CATVX.** Type CATVX limited-use community antenna television cable shall be listed as being suitable for use in dwellings and for use in raceway and shall also be listed as being resistant to flame spread.

FPN: One method of determining that cable is resistant to flame spread is by testing the cable to the VW-1 (vertical-wire) flame test in ANSI/UL 1581-1991, *Reference Standard for Electrical Wires, Cables and Flexible Cords.*

**820.52 Installation of Cables and Equipment.** Beyond the point of grounding, as defined in 820.33, the cable installation shall comply with 820.33(A) through (D).

**(A) Separation from Other Conductors.**

**(1) In Raceways and Boxes.**

(a) Other Circuits. Coaxial cables shall be permitted in the same raceway or enclosure with jacketed cables of any of the following:

(1) Class 2 and Class 3 remote-control, signaling, and power-limited circuits in compliance with Article 725
(2) Power-limited fire alarm systems in compliance with Article 760
(3) Nonconductive and conductive optical fiber cables in compliance with Article 770

(4) Communications circuits in compliance with Article 800
(5) Low-power network-powered broadband communications circuits in compliance with Article 830

(b) Electric Light, Power, Class 1, Non–Power-Limited Fire Alarm, and Medium Power Network-Powered Broadband Communications Circuits. Coaxial cable shall not be placed in any raceway, compartment, outlet box, junction box, or other enclosures with conductors of electric light, power, Class 1, non–power-limited fire alarm, or medium power network-powered broadband communications circuits.

*Exception No. 1: Where all of the conductors of electric light, power, Class 1, non–power-limited fire alarm, and medium power network-powered broadband communications circuits are separated from all of the coaxial cables by a barrier.*

*Exception No. 2: Power circuit conductors in outlet boxes, junction boxes, or similar fittings or compartments where such conductors are introduced solely for power supply to the coaxial cable system distribution equipment. The power circuit conductors shall be routed within the enclosure to maintain a minimum 6-mm (0.25-in.) separation from coaxial cables.*

**(2) Other Applications.** Coaxial cable shall be separated at least 50 mm (2 in.) from conductors of any electric light, power, Class 1, non–power-limited fire alarm, or medium power network-powered broadband communications circuits.

*Exception No. 1: Where either (1) all of the conductors of electric light, power, Class 1, non–power-limited fire alarm, and medium power network-powered broadband communications and circuits are in a raceway, or in metal-sheathed, metal-clad, nonmetallic-sheathed Type AC or Type UF cables, or (2) all of the coaxial cables are encased in raceway.*

*Exception No. 2: Where the coaxial cables are permanently separated from the conductors of electric light, power, Class 1, non–power-limited fire alarm, and medium power network-powered broadband communications circuits by a continuous and firmly fixed nonconductor, such as porcelain tubes or flexible tubing, in addition to the insulation on the wire.*

**(B) Equipment in Other Space Used for Environmental Air.** Section 300.22(C) shall apply.

**(C) Hybrid Power and Coaxial Cabling.** The provisions of 780.6 shall apply for listed hybrid power and coaxial cabling in closed-loop and programmed power distribution.

**(D) Support of Cables.** Raceways shall be used for their intended purpose. Coaxial cables shall not be strapped, taped, or attached by any means to the exterior of any conduit or raceway as a means of support.

*Exception: Overhead (aerial) spans of coaxial cables shall be permitted to be attached to the exterior of a raceway-type mast intended for the attachment and support of such cables.*

**820.53 Applications of Listed CATV Cables.** CATV cables shall comply with the requirements of 820.53(A) through (D) or where cable substitutions are made as shown in Table 820.53.

**Table 820.53 Coaxial Cable Uses and Permitted Substitutions**

| Cable Type | Use | References | Permitted Substitutions |
|---|---|---|---|
| CATVP | Coaxial plenum cable | 820.53(A) | CMP |
| CATVR | Coaxial riser cable | 820.53(B) | CATVP, CMP, CMR |
| CATV | Coaxial general-purpose cable | 820.53(D) | CATVP, CMP, CATVR, CMR, CMG, CM |
| CATVX | Coaxial cable, limited use | 820.53(D) | CATVP, CMP, CATVR, CMR, CATV, CMG, CM |

Note: See Figure 820.53, Cable substitution hierarchy.

FPN: The substitute cables in Table 820.53 are only coaxial-type cables.

**(A) Plenum.** Cables installed in ducts, plenums, and other spaces used for environmental air shall be Type CATVP. Abandoned cables shall not be permitted to remain. Types CATVP, CATVR, CATV, and CATVX cables installed in compliance with 300.22 shall be permitted.

**(B) Riser.** Cables installed in risers shall comply with any of the requirements of 820.53(B)(1) through (B)(3).

**(1) Cables in Vertical Runs.** Cables installed in vertical runs and penetrating more than one floor, or cables installed in vertical runs in a shaft, shall be Type CATVR. Floor penetrations requiring Type CATVR shall contain only cables suitable for riser or plenum use. Abandoned cables shall not be permitted to remain.

**(2) Metal Raceways or Fireproof Shafts.** Types CATV and CATVX cables shall be permitted to be encased in a metal raceway or located in a fireproof shaft having firestops at each floor.

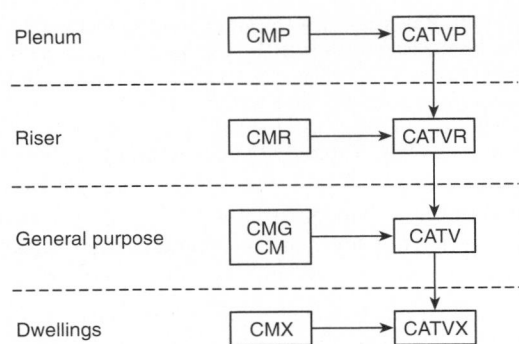

Type CATV—Community antenna television cables
Type CM—Communications cables

A ➞ B    Coaxial cable A shall be permitted to be used in place of coaxial cable B.

**Figure 820.53 Cable substitution hierarchy.**

**(3) One- and Two-Family Dwellings.** Types CATV and CATVX cables shall be permitted in one- and two-family dwellings.

FPN: See 820.53(A) for the firestop requirements for floor penetrations.

**(C) Cable Trays.** Cables installed in cable trays shall be Types CATVP, CATVR, and CATV.

**(D) Other Wiring Within Buildings.** Cables installed in building locations other than the locations covered in 820.53(A) and (B) shall be with any of the requirements in 820.53(D)(1) through (5). Abandoned cables in hollow spaces shall not be permitted to remain.

**(1) General.** Type CATV shall be permitted.

**(2) In Raceways.** Type CATVX shall be permitted to be installed in a raceway.

**(3) Nonconcealed Spaces.** Type CATVX shall be permitted to be installed in nonconcealed spaces where the exposed length of cable does not exceed 3 m (10 ft).

**(4) One- and Two-Family Dwellings.** Type CATVX cables less than 10 mm (0.375 in.) in diameter shall be permitted to be installed in one- and two-family dwellings.

**(5) Multifamily Dwellings.** Type CATVX cables less than 10 mm (0.375 in.) in diameter shall be permitted to be installed in multifamily dwellings.

## ARTICLE 830
## Network-Powered Broadband Communications Systems

### I. General

**830.1 Scope.** This article covers network-powered broadband communications systems that provide any combination of voice, audio, video, data, and interactive services through a network interface unit.

> FPN No. 1: A typical basic system configuration includes a cable supplying power and broadband signal to a network interface unit that converts the broadband signal to the component signals. Typical cables are coaxial cable with both broadband signal and power on the center conductor, composite metallic cable with a coaxial member for the broadband signal and a twisted pair for power, and composite optical fiber cable with a pair of conductors for power. Larger systems may also include network components such as amplifiers that require network power.

> FPN No. 2: See 90.2(B)(4) for installations of broadband communications systems that are not covered.

**830.2 Definitions.** See Article 100. For purposes of this article, the following additional definitions apply.

**Abandoned Network-Powered Broadband Communications Cable.** Installed network-powered broadband communications cable that is not terminated at equipment other than a connector and not identified for future use with a tag.

**Block.** A square or portion of a city, town, or village enclosed by streets, including the alleys so enclosed but not any street.

**Exposed to Accidental Contact with Electrical Light or Power Conductors.** A circuit in such a position that, in case of failure of supports or insulation, contact with another circuit may result.

**Fault Protection Device.** An electronic device that is intended for the protection of personnel and functions under fault conditions, such as network-powered broadband communications cable short or open circuit, to limit the current or voltage, or both, for a low-power network-powered broadband communications circuit and provide acceptable protection from electric shock.

**Network Interface Unit (NIU).** A device that converts a broadband signal into component voice, audio, video, data, and interactive services signals. The NIU provides isolation between the network power and the premises signal circuits. The NIU may also contain primary and secondary protectors.

**Network-Powered Broadband Communications Circuit.** The circuit extending from the communications utility's serving terminal or tap up to and including the NIU.

> FPN: A typical single-family network-powered communications circuit consists of a communications drop or communications service cable and an NIU and includes the communications utility's serving terminal or tap where it is not under the exclusive control of the communications utility.

**Point of Entrance.** The point within a building at which the cable emerges from an external wall, from a concrete floor slab, or from a rigid metal conduit or an intermediate metal conduit grounded to an electrode in accordance with 830.40(B).

**Premises Wiring.** The circuits located on the user side of the network interface unit.

**830.3 Locations and Other Articles.** Circuits and equipment shall comply with 830.3(A) through (D).

**(A) Spread of Fire or Products of Combustion.** Section 300.21 shall apply. The accessible portion of abandoned network-powered broadband communications cables shall not be permitted to remain.

**(B) Ducts, Plenums, and Other Air-Handling Spaces.** Section 300.22 shall apply, where installed in ducts or plenums or other spaces used for environmental air.

*Exception: As permitted in 830.55(B).*

**(C) Installation and Use.** Section 110.3(B) shall apply.

**(D) Output Circuits.** As appropriate for the services provided, the output circuits derived from the network interface unit shall comply with the requirements of the following:

(1) Installations of communications circuits — Article 800
(2) Installations of community antenna television and radio distribution circuits — Article 820

*Exception: 830.30(B)(3) shall apply where protection is provided in the output of the NIU.*

(3) Installations of optical fiber cables — Article 770
(4) Installations of Class 2 and Class 3 circuits — Article 725
(5) Installations of power-limited fire alarm circuits — Article 760

**830.4 Power Limitations.** Network-powered broadband communications systems shall be classified as having low or medium power sources as defined in Table 830.4.

**830.5 Network-Powered Broadband Communications Equipment and Cables.** Network-powered broadband communications equipment and cables shall be listed as suitable for the purpose.

**Table 830.4 Limitations for Network-Powered Broadband Communications Systems**

| Network Power Source | Low | Medium |
|---|---|---|
| Circuit voltage, $V_{max}$ (volts)[1] | 0–100 | 0–150 |
| Power limitation, $VA_{max}$(volt-amperes)[1] | 250 | 250 |
| Current limitation, $I_{max}$ (amperes)[1] | $1000/V_{max}$ | $1000/V_{max}$ |
| Maximum power rating (volt-amperes) | 100 | 100 |
| Maximum voltage rating (volts) | 100 | 150 |
| Maximum overcurrent protection (amperes)[2] | $100/V_{max}$ | NA |

[1]$V_{max}$, $I_{max}$, and $VA_{max}$ are determined with the current-limiting impedance in the circuit (not bypassed) as follows:

$V_{max}$—Maximum system voltage regardless of load with rated input applied.

$I_{max}$—Maximum system current under any noncapacitive load, including short circuit, and with overcurrent protection bypassed if used. $I_{max}$ limits apply after 1 minute of operation.

$VA_{max}$—Maximum volt-ampere output after 1 minute of operation regardless of load and overcurrent protection bypassed if used.

[2]Overcurrent protection is not required where the current-limiting device provides equivalent current limitation and the current-limiting device does not reset until power or the load is removed.

*Exception No. 1:  This listing requirement shall not apply to community antenna television and radio distribution system coaxial cables that were installed prior to January 1, 2000, in accordance with Article 820 and are used for low-power network-powered broadband communications circuits. See 830.9.*

*Exception No. 2:  Substitute cables for network-powered broadband communications cables shall be permitted as shown in Table 830.58.*

**(A) Listing and Marking.** Listing and marking of network-powered broadband communications cables shall comply with 830.5(A)(1) or (A)(2).

**(1) Type BMU, Type BM, and Type BMR Cables.** Network-powered broadband communications medium power underground cable, Type BMU; network-powered broadband communications medium power cable, Type BM; and network-powered broadband communications medium power riser cable, Type BMR, shall be factory-assembled cables consisting of a jacketed coaxial cable, a jacketed combination of coaxial cable and multiple individual conductors, or a jacketed combination of an optical fiber cable and multiple individual conductors. The insulation for the individual conductors shall be rated for 300 volts minimum. Cables intended for outdoor use shall be listed as suitable for the application. Cables shall be marked in accordance with 310.11. Type BMU cables shall be jack-eted and listed as being suitable for outdoor underground use. Type BM cables shall be listed as being general-purpose use, with the exception of risers and plenums, and shall also be listed as being resistant to the spread of fire. Type BMR cables shall be listed as being suitable for use in a vertical run in a shaft or from floor to floor and shall also be listed as having fire-resistant characteristics capable of preventing the carrying of fire from floor to floor.

FPN No. 1:  One method of defining *resistant to spread of fire* is that the cables do not spread fire to the top of the tray in the vertical tray flame test in ANSI/UL 1581-1991, *Reference Standard for Electrical Wires, Cables and Flexible Cords*. Another method of defining *resistant to the spread of fire* is for the damage (char length) not to exceed 1.5 m (4 ft 11 in.) when performing the CSA vertical flame test for cables in cable trays, as described in CSA C22.2 No. 0.3-M-1985, *Test Methods for Electrical Wires and Cables*.

FPN No. 2:  One method of defining fire-resistant characteristics capable of preventing the carrying of fire from floor to floor is that the cables pass the requirements of ANSI/UL 1666-1997, *Standard Test for Flame Propagation Height of Electrical and Optical-Fiber Cable Installed Vertically in Shafts.*

**(2) Type BLU, Type BLX, and Type BLP Cables.** Network-powered broadband communications low-power underground cable, Type BLU; limited use network-powered broadband communications low-power cable, Type BLX; and network-powered broadband communications low-power plenum cable, Type BLP, shall be factory assembled cables consisting of a jacketed coaxial cable, a jacketed combination of coaxial cable and multiple individual conductors, or a jacketed combination of an optical fiber cable and multiple individual conductors. The insulation for the individual conductors shall be rated for 300 volts minimum. Cables intended for outdoor use shall be listed as suitable for the application. Cables shall be marked in accordance with 310.11. Type BLU cables shall be jack-eted and listed as being suitable for outdoor underground use. Type BLX limited-use cables shall be listed as being suitable for use outside, for use in dwellings, and for use in raceways and shall also be listed as being resistant to flame spread. Type BLP cables shall be listed as being suitable for use in ducts, plenums, and other spaces for environmental air and shall also be listed as having adequate fire-resistant and low smoke-producing characteristics.

FPN No. 1:  One method of determining that cable is resistant to flame spread is by testing the cable to VW-1 (vertical-wire) flame test in ANSI/UL 1581-1991, *Reference Standard for Electrical Wires, Cables and Flexible Cords.*

FPN No. 2:  One method of defining low smoke-producing cable is by establishing an acceptable value of the smoke produced when tested in accordance with NFPA 262-1999, *Standard Method of Test for Flame Travel and Smoke of*

*Wires and Cables for Use in Air-Handling Spaces*, to a maximum peak optical density of 0.5 and a maximum average optical density of 0.15. Similarly, one method of defining fire-resistant cables is by establishing maximum allowable flame travel distance of 1.52 m (5 ft) when tested in accordance with the same test.

**830.6 Access to Electrical Equipment Behind Panels Designed to Allow Access.** Access to electrical equipment shall not be denied by an accumulation of wires and cables that prevents removal of panels. including suspended ceiling panels.

**830.7 Mechanical Execution of Work.** Network-powered broadband communications circuits and equipment shall be installed in a neat and workmanlike manner. Cables installed exposed on the surface of ceiling and sidewalls shall be supported by the structural components of the building structure in such a manner that the cable is not damaged by normal building use. Such cables shall be attached to structural components by straps, staples, hangers, or similar fittings designed and installed so as not to damage the cable. The installation shall also conform with 300.4(D).

**830.9 Hazardous (Classified) Locations.** Network-powered broadband communications circuits and equipment installed in a location that is classified in accordance with Article 500 shall comply with the applicable requirements of Chapter 5.

## II. Cables Outside and Entering Buildings

**830.10 Entrance Cables.** Cables installed outdoors shall be listed as suitable for the application. In addition, network-powered broadband communications cables located outside and entering buildings shall comply with 830.10(A) and (B).

**(A) Medium Power Circuits.** Medium power network-powered broadband communications circuits located outside and entering buildings shall be installed using Type BMU, Type BM, or Type BMR network-powered broadband communications medium power cables.

**(B) Low-Power Circuits.** Low-power network-powered broadband communications circuits located outside and entering buildings shall be installed using Type BLU or Type BLX low-power network-powered broadband communications cables. Cables shown in Table 830.58 shall be permitted to substitute.

*Exception: Outdoor community antenna television and radio distribution system coaxial cables installed prior to January 1, 2000, and installed in accordance with Article 820, shall be permitted for low-power-type, network-powered broadband communications circuits.*

**830.11 Aerial Cables.** Aerial powered broadband communications cables shall comply with 830.11(A) through (I).

FPN: For additional information regarding overhead wires and cables, see ANSI C2-1997, *National Electric Safety Code*, Part 2, Safety Rules For Overhead Lines.

**(A) On Poles.** Where practicable, network-powered broadband communications cables on poles shall be located below the electric light, power, Class 1, or non–power-limited fire alarm circuit conductors and shall not be attached to a cross-arm that carries electric light or power conductors.

**(B) Climbing Space.** The climbing space through network-powered broadband communications cables shall comply with the requirements of 225.14(D).

**(C) Lead-in Clearance.** Lead-in or aerial-drop network-powered broadband communications cables from a pole or other support, including the point of initial attachment to a building or structure, shall be kept away from electric light, power, Class 1, or non–power-limited fire alarm circuit conductors so as to avoid the possibility of accidental contact.

*Exception: ·Where proximity to electric light, power, Class 1, or non–power-limited fire alarm circuit service conductors cannot be avoided, the installation shall be such as to provide clearances of not less than 300 mm (12 in.) from light, power, Class 1, or non–power-limited fire alarm circuit service drops. The clearance requirement shall apply to all points along the drop, and it shall increase to 1.02 m (40 in.) at the pole.*

**(D) Clearance from Ground.** Overhead spans of network-powered broadband communication cables shall conform to not less than the following:

(1) 2.9 m (9.5 ft) — above finished grade, sidewalks, or from any platform or projection from which they might be reached and accessible to pedestrians only

(2) 3.5 m (11.5 ft) — over residential property and driveways, and those commercial areas not subject to truck traffic

(3) 4.7 m (15.5 ft) — over public streets, alleys, roads, parking areas subject to truck traffic, driveways on other than residential property, and other land traversed by vehicles such as cultivated, grazing, forest, and orchard

FPN: These clearances have been specifically chosen to correlate with ANSI C2-1997, *National Electrical Safety Code*, Table 232-1, which provides for clearances of wires, conductors, and cables above ground and roadways, rather than using the clearances referenced in 225.18. Because Article 800 and Article 820 have had no required clearances, the communications industry has used the clearances from the NESC for their installed cable plant.

**(E) Over Pools.** Clearance of network-powered broadband communications cable in any direction from the water level, edge of pool, base of diving platform, or anchored raft shall comply with those clearances in 680.8.

**(F) Above Roofs.** Network-powered broadband communications cables shall have a vertical clearance of not less than 2.5 m (8 ft) from all points of roofs above which they pass.

*Exception No. 1: Auxiliary buildings such as garages and the like.*

*Exception No. 2: A reduction in clearance above only the overhanging portion of the roof to not less than 450 mm (18 in.) shall be permitted if (1) not more than 1.2 m (4 ft) of the broadband communications drop cables pass above the roof overhang, and (2) they are terminated at a through-the-roof raceway or support.*

*Exception No. 3: Where the roof has a slope of not less than 100 mm (4 in.) in 300 mm (12 in.), a reduction in clearance to not less than 900 mm (3 ft) shall be permitted.*

**(G) Final Spans.** Final spans of network-powered broadband communications cables without an outer jacket shall be permitted to be attached to the building, but they shall be kept not less than 900 mm (3 ft) from windows that are designed to be opened, doors, porches, balconies, ladders, stairs, fire escapes, or similar locations.

*Exception: Conductors run above the top level of a window shall be permitted to be less than the 900-mm (3-ft) requirement above.*

Overhead network-powered broadband communications cables shall not be installed beneath openings through which materials may be moved, such as openings in farm and commercial buildings, and shall not be installed where they will obstruct entrance to these building openings.

**(H) Between Buildings.** Network-powered broadband communications cables extending between buildings and also the supports or attachment fixtures shall be acceptable for the purpose and shall have sufficient strength to withstand the loads to which they may be subjected.

*Exception: Where a network-powered broadband communications cable does not have sufficient strength to be self-supporting, it shall be attached to a supporting messenger cable that, together with the attachment fixtures or supports, shall be acceptable for the purpose and shall have sufficient strength to withstand the loads to which they may be subjected.*

**(I) On Buildings.** Where attached to buildings, network-powered broadband communications cables shall be securely fastened in such a manner that they are separated from other conductors in accordance with 830.11(I)(1) through (I)(4).

**(1) Electric Light or Power.** The network-powered broadband communications cable shall have a separation of at least 100 mm (4 in.) from electric light, power, Class 1, or non–power-limited fire alarm circuit conductors not in raceway or cable, or be permanently separated from conductors of the other system by a continuous and firmly fixed nonconductor in addition to the insulation on the wires.

**(2) Other Communications Systems.** Network-powered broadband communications cables shall be installed so that there will be no unnecessary interference in the maintenance of the separate systems. In no case shall the conductors, cables, messenger strand, or equipment of one system cause abrasion to the conductors, cables, messenger strand, or equipment of any other system.

**(3) Lightning Conductors.** Where practicable, a separation of at least 1.8 m (6 ft) shall be maintained between any network-powered broadband communications cable and lightning conductors.

**(4) Protection from Damage.** Network-powered broadband communications cables attached to buildings and located within 2.5 m (8 ft) of finished grade shall be protected by enclosures, raceways, or other approved means.

*Exception: A low-power network-powered broadband communications circuit that is equipped with a listed fault protection device, appropriate to the network-powered broadband communications cable used, and located on the network side of the network-powered broadband communications cable being protected.*

**830.12 Underground Circuits Entering Buildings.**

**(A) Underground Systems.** Underground network-powered broadband communications cables in a duct, pedestal, handhole, or manhole that contains electric light, power conductors, non–power-limited fire alarm circuit conductors, or Class 1 circuits shall be in a section permanently separated from such conductors by means of a suitable barrier.

**(B) Direct-Buried Cables and Raceways.** Direct-buried network-powered broadband communications cables shall be separated at least 300 mm (12 in.) from conductors of any light, power, non–power-limited fire alarm circuit conductors or Class 1 circuit.

*Exception No. 1: Where electric service conductors or network-powered broadband communications cables are installed in raceways or have metal cable armor.*

*Exception No. 2: Where electric light or power branch-circuit or feeder conductors, non–power-limited fire alarm* circuit conductors, or Class 1 circuit conductors are installed in a raceway or in metal-sheathed, metal-clad, or Type UF or Type USE cables; or the network-powered broadband communications cables have metal cable armor or are installed in a raceway.

**(C) Mechanical Protection.** Direct-buried cable, conduit, or other raceways shall be installed to meet the minimum cover requirements of Table 830.12. In addition, direct-buried cables emerging from the ground shall be protected by enclosures, raceways, or other approved means extending from the minimum cover distance required by Table 830.12 below grade to a point at least 2.5 m (8 ft) above finished grade. In no case shall the protection be required to exceed 450 mm (18 in.) below finished grade. Type BMU and BLU direct-buried cables emerging from the ground

shall be installed in rigid metal conduit, intermediate metal conduit, rigid nonmetallic conduit, or other approved means extending from the minimum cover distance required by Table 830.12 below grade to the point of entrance.

*Exception: A low-power network-powered broadband communications circuit that is equipped with a listed fault protection device, appropriate to the network-powered broadband communications cable used, and located on the network side of the network-powered broadband communications cable being protected.*

**(D) Pools.** Cables located under the pool or within the area extending 1.5 m (5 ft) horizontally from the inside wall of the pool shall meet those clearances and requirements specified in 680.10.

**Table 830.12 Network-Powered Broadband Communications Systems Minimum Cover Requirements**
(*Cover* is the shortest distance measured between a point on the top surface of any direct-buried cable, conduit, or other raceway and the top surface of finished grade, concrete, or similar cover.)

| Location of Wiring Method or Circuit | Direct Burial Cables | | Rigid Metal Conduit or Intermediate Metal Conduit | | Nonmetallic Raceways Listed for Direct Burial; Without Concrete Encasement or Other Approved Raceways | |
|---|---|---|---|---|---|---|
| | mm | in. | mm | in. | mm | in. |
| All locations not specified below | 450 | 18 | 150 | 6 | 300 | 12 |
| In trench below 50-mm (2-in.) thick concrete or equivalent | 300 | 12 | 150 | 6 | 150 | 6 |
| Under a building (in raceway only) | 0 | 0 | 0 | 0 | 0 | 0 |
| Under minimum of 100-mm (4-in.) thick concrete exterior slab with no vehicular traffic and the slab extending not less than 150 mm (6 in.) beyond the underground installation | 300 | 12 | 100 | 4 | 100 | 4 |
| One- and two-family dwelling driveways and outdoor parking areas and used only for dwelling-related purposes | 300 | 12 | 300 | 12 | 300 | 12 |

Notes:
1. Raceways approved for burial only where concrete encased shall require a concrete envelope not less than 50 mm (2 in.) thick.
2. Lesser depths shall be permitted where cables rise for terminations or splices or where access is otherwise required.
3. Where solid rock is encountered, all wiring shall be installed in metal or nonmetallic raceway permitted for direct burial. The raceways shall be covered by a minimum of 50 mm (2 in.) of concrete extending down to rock.
4. Low-power network-powered broadband communications circuits using directly buried community antenna television and radio distribution system coaxial cables that were installed outside and entering buildings prior to January 1, 2000, in accordance with Article 820 shall be permitted where buried to a minimum depth of 300 mm (12 in.).

## III. Protection

### 830.30 Primary Electrical Protection.

**(A) Application.** Primary electrical protection shall be provided on all network-powered broadband communications conductors that are neither grounded nor interrupted and are run partly or entirely in aerial cable not confined within a block. Also, primary electrical protection shall be provided on all aerial or underground network-powered broadband communications conductors that are neither grounded nor interrupted and are located within the block containing the building served so as to be exposed to lightning or accidental contact with electric light or power conductors operating at over 300 volts to ground.

*Exception: Where electrical protection is provided on the derived circuit(s) (output side of the NIU) in accordance with 830.30(B)(3).*

> FPN No. 1: On network-powered broadband communications conductors not exposed to lightning or accidental contact with power conductors, providing primary electrical protection in accordance with this article helps protect against other hazards, such as ground potential rise caused by power fault currents, and above-normal voltages induced by fault currents on power circuits in proximity to the network-powered broadband communications conductors.

> FPN No. 2: Network-powered broadband communications circuits are considered to have a lightning exposure unless one or more of the following conditions exist:

> (1) Circuits in large metropolitan areas where buildings are close together and sufficiently high to intercept lightning.
> (2) Areas having an average of five or fewer thunderstorm days per year and earth resistivity of less than 100 ohm-meters. Such areas are found along the Pacific coast.

**(1) Fuseless Primary Protectors.** Fuseless-type primary protectors shall be permitted where power fault currents on all protected conductors in the cable are safely limited to a value no greater than the current-carrying capacity of the primary protector and of the primary protector grounding conductor.

**(2) Fused Primary Protectors.** Where the requirements listed in 830.30(A)(1) are not met, fused-type primary protectors shall be used. Fused-type primary protectors shall consist of an arrester connected between each conductor to be protected and ground, a fuse in series with each conductor to be protected, and an appropriate mounting arrangement. Fused primary protector terminals shall be marked to indicate line, instrument, and ground, as applicable.

**(B) Location.** The location of the primary protector, where required, shall comply with (1), (2), or (3):

(1) A listed primary protector shall be applied on each network-powered broadband communications cable external to and on the network side of the network interface unit.
(2) The primary protection function shall be an integral part of and contained in the network interface unit. The network interface unit shall be listed for the purpose and shall have an external marking indicating that it contains primary electrical protection.
(3) The primary protector(s) shall be provided on the derived circuit(s) (output side of the NIU), and the combination of the NIU and the protector(s) shall be listed for the purpose.

A primary protector, whether provided integrally or external to the network interface unit, shall be located as close as practicable to the point of entrance.

For purposes of this section, a network interface unit and any externally provided primary protectors located at mobile home service equipment located in sight from and not more than 9.0 m (30 ft) from the exterior wall of the mobile home it serves, or at a mobile home disconnecting means grounded in accordance with 250.32 and located in sight from and not more than 9.0 m (30 ft) from the exterior wall of the mobile home it serves, shall be considered to meet the requirements of this section.

> FPN: Selecting a network interface unit and primary protector location to achieve the shortest practicable primary protector grounding conductor helps limit potential differences between communications circuits and other metallic systems.

**(C) Hazardous (Classified) Locations.** The primary protector or equipment providing the primary protection function shall not be located in any hazardous (classified) location as defined in Article 500 or in the vicinity of easily ignitible material.

*Exception: As permitted in 501.14, 502.14, and 503.12.*

### 830.33 Grounding or Interruption of Metallic Members of Network-Powered Broadband Communications Cables. The shields of network-powered broadband communications cables used for communications or powering shall be grounded at the building as close as practicable to the point of entrance or attachment of the NIU. Metallic cable members not used for communications or powering shall be grounded or interrupted by an insulating joint or equivalent device as close as practicable to the point of entrance or attachment of the NIU.

For purposes of this section, grounding or interruption of network-powered broadband communications cable metallic members installed at mobile home service equipment located in sight from and no more than 9.0 m (30 ft) from the exterior wall of the mobile home it serves, or at a mobile home disconnecting means grounded in accordance with 250.32 and located in sight from and not more than 9.0 m (30 ft) from the exterior wall of the mobile home it serves, shall be considered to meet the requirements of this section.

> FPN: Selecting a grounding location to achieve the shortest practicable grounding conductor helps limit potential differences between the network-powered broadband communications circuits and other metallic systems.

## IV. Grounding Methods

**830.40 Cable, Network Interface Unit, and Primary Protector Grounding.** Network interface units containing protectors, NIUs with metallic enclosures, primary protectors, and the metallic members of the network-powered broadband communications cable that are intended to be grounded shall be grounded as specified in 830.40(A) through (D).

**(A) Grounding Conductor.**

**(1) Insulation.** The grounding conductor shall be insulated and shall be listed as suitable for the purpose.

**(2) Material.** The grounding conductor shall be copper or other corrosion-resistant conductive material, stranded or solid.

**(3) Size.** The grounding conductor shall not be smaller than 14 AWG and shall have a current-carrying capacity approximately equal to that of the grounded metallic member(s) and protected conductor(s) of the network-powered broadband communications cable. The grounding conductor shall not be required to exceed 6 AWG.

**(4) Length.** The grounding conductor shall be as short as practicable. In one-family and multifamily dwellings, the grounding conductor shall be as short as permissible, not to exceed 6.0 m (20 ft) in length.

*Exception: In one- and two-family dwellings where it is not practicable to achieve an overall maximum grounding conductor length of 6.0 m (20 ft), a separate communications ground rod meeting the minimum dimensional criteria of 830.40(B)(2)(2) shall be driven, and the grounding conductor connected to the communications ground rod in accordance with 830.40(C). The communications ground rod shall be bonded to the power grounding electrode system in accordance with 830.40(D).*

**(5) Run in Straight Line.** The grounding conductor shall be run to the grounding electrode in as straight a line as practicable.

**(6) Physical Protection.** Where subject to physical damage, the grounding conductor shall be adequately protected.

Where the grounding conductor is run in a metal raceway, both ends of the raceway shall be bonded to the grounding conductor or the same terminal or electrode to which the grounding conductor is connected.

**(B) Electrode.** The grounding conductor shall be connected as follows.

**(1) In Buildings or Structures with Grounding Means.** To the nearest accessible location on the following:

(1) The building or structure grounding electrode system as covered in 250.50;
(2) The grounded interior metal water piping system, within 1.5 m (5 ft) from its point of entrance to the building, as covered in 250.52;
(3) The power service accessible means external to enclosures as covered in 250.94;
(4) The metallic power service raceway;
(5) The service equipment enclosure;
(6) The grounding electrode conductor or the grounding electrode metal enclosure; or
(7) The grounding conductor or the grounding electrode of a building or structure disconnecting means that is grounded to an electrode as covered in 250.32.

For purposes of this section, the mobile home service equipment or the mobile home disconnecting means, as described in 830.33, shall be considered accessible.

**(2) In Buildings or Structures Without Grounding Means.** If the building or structure served has no grounding means, as described in (B)(1):

(1) To any one of the individual electrodes described in 250.52(A)(1), (2), (3), (4); or
(2) If the building or structure served has no grounding means, as described in 830.40(B)(1) or (B)(2)(1), to an effectively grounded metal structure or to a ground rod or pipe not less than 1.5 m (5 ft) in length and 12.7 mm (½ in.) in diameter, driven, where practicable, into permanently damp earth and separated from lightning conductors as covered in 800.13 and at least 1.8 m (6 ft) from electrodes of other systems. Steam or hot water pipes or lightning-rod conductors shall not be employed as electrodes for protectors, NIUs with integral protection, grounded metallic members, NIUs with metallic enclosures, and other equipment.

**(C) Electrode Connection.** Connections to grounding electrodes shall comply with 250.70. Connectors, clamps, fittings, or lugs used to attach grounding conductors and bonding jumpers to grounding electrodes or to each other that are to be concrete encased or buried in the earth shall be suitable for its application.

**(D) Bonding of Electrodes.** A bonding jumper not smaller than 6 AWG copper or equivalent shall be connected between the network-powered broadband communications system grounding electrode and the power grounding electrode system at the building or structure served where separate electrodes are used.

*Exception:  At mobile homes as covered in 830.42.*

> FPN No. 1:  See 250.60 for use of lightning rods.

> FPN No. 2:  Bonding together of all separate electrodes limits potential differences between them and between their associated wiring systems.

## 830.42 Bonding and Grounding at Mobile Homes.

**(A) Grounding.** Where there is no mobile home service equipment located in sight from and not more than 9.0 m (30 ft) from the exterior wall of the mobile home it serves, or there is no mobile home disconnecting means grounded in accordance with 250.32 and located within sight from and not more than 9.0 m (30 ft) from the exterior wall of the mobile home it serves, the network-powered broadband communications cable, network interface unit, and primary protector ground shall be installed in accordance with 830.40(B)(2).

**(B) Bonding.** The network-powered broadband communications cable grounding terminal, network interface unit grounding terminal, if present, and primary protector grounding terminal shall be bonded together with a copper bonding conductor not smaller than 12 AWG. The network-powered broadband communications cable grounding terminal, network interface unit grounding terminal, primary protector grounding terminal, or the grounding electrode shall be bonded to the metal frame or available grounding terminal of the mobile home with a copper bonding conductor not smaller than 12 AWG under any of the following conditions:

(1) Where there is no mobile home service equipment or disconnecting means as in (A)
(2) Where the mobile home is supplied by cord and plug

## V. Wiring Methods Within Buildings

## 830.54 Medium Power Network-Powered Broadband Communications System Wiring Methods. Medium power network-powered broadband communications systems shall be installed within buildings using listed Type BM or Type BMR, network-powered broadband communications medium power cables.

**(A) Ducts, Plenums, and Other Air-Handling Spaces.** Section 300.22 shall apply.

**(B) Riser.** Cables installed in vertical runs and penetrating more than one floor, or cables installed in vertical runs in a shaft, shall be Type BMR. Floor penetrations requiring Type BMR shall contain only cables suitable for riser or plenum use.

*Exception No. 1:  Type BM cables encased in metal raceway or located in a fireproof shaft that has firestops at each floor.*

*Exception No. 2:  Type BM cables in one- and two-family dwellings.*

**(C) Other Wiring.** Cables installed in locations other than the locations covered in 830.54(A) and (B) shall be Type BM.

*Exception:  Type BMU cable where the cable enters the building from the outside and is run in rigid metal conduit or intermediate metal conduit, and such conduits are grounded to an electrode in accordance with 830.40(B).*

## 830.55 Low-Power Network-Powered Broadband Communications System Wiring Methods. Low-power network-powered broadband communications systems shall comply with any of the requirements of 830.55(A) through (D).

**(A) In Buildings.** Low-power network-powered broadband communications systems shall be installed within buildings using listed Type BLX or Type BLP network-powered broadband communications low power cables.

**(B) Ducts, Plenums, and Other Air-Handling Spaces.** Cables installed in ducts, plenums, and other spaces used for environmental air shall be Type BLP. Abandoned cables shall not be permitted to remain. Type BLX cable installed in compliance with 300.22 shall be permitted.

**(C) Riser.** Cables installed in risers shall comply with any of the requirements in 830.55(C)(1), (C)(2), or (C)(3).

**(1) Cables in Vertical Runs.** Cables installed in vertical runs and penetrating more than one floor, or cables installed in vertical runs in a shaft, shall be Type BLP or BMR. Floor penetrations requiring Type BMR shall contain only cables suitable for riser or plenum use. Abandoned cables shall not be permitted to remain.

**(2) Metal Raceways or Fireproof Shafts.** Type BLX cables shall be permitted to be encased in a metal raceway or located in a fireproof shaft having firestops at each floor.

**(3) One- and Two-Family Dwellings.** Type BLX cables less than 10 mm (0.375 in.) in diameter shall be permitted in one- and two-family dwellings.

**(D) Other Wiring.** Cables installed in locations other than the locations covered in 830.55(A), (B), and (C) shall comply with the requirements of 830.55(D)(1) through (D)(5).

**(1) General.** Type BLP or BM shall be permitted.

**(2) In Raceways.** Type BLX shall be permitted to be installed in a raceway.

**(3) Type BLU Cable.** Type BLU cable entering the building from outside shall be permitted to be run in rigid metal conduit or intermediate metal conduit. Such conduits shall be grounded to an electrode in accordance with 830.40(B).

**(4) One- and Two-Family Dwellings.** Type BLX cable less than 10 mm (0.375 in.) in diameter shall be permitted to be installed in one- and two-family dwellings.

**(5) Type BLX Cable.** Type BLX cable entering the building from outside and terminated at a grounding block or a primary protection location shall be permitted to be installed, provided that the length of cable within the building does not exceed 15 m (50 ft).

> FPN: This provision limits the length of Type BLX cable to 15 m (50 ft), while 830.30(B) requires that the primary protector, or NIU with integral protection, be located as close as practicable to the point at which the cable enters the building. Therefore, in installations requiring a primary protector, or NIU with integral protection, Type BLX cable may not be permitted to extend 15 m (50 ft) into the building if it is practicable to place the primary protector closer than 15 m (50 ft) to the entrance point.

**830.56 Protection Against Physical Damage.** Section 300.4 shall apply.

**830.57 Bends.** Bends in network broadband cable shall be made so as not to damage the cable.

**830.58 Installation of Network-Powered Broadband Communications Cables and Equipment.** Cable and equipment installations within buildings shall comply with 830.58(A) through (E), as applicable.

**(A) Separation of Conductors.**

**(1) In Raceways and Enclosures.**

(a) Low and Medium Power Network-Powered Broadband Communications Circuit Cables. Low and medium power network-powered broadband communications cables shall be permitted in the same raceway or enclosure.

(b) Low Power Network-Powered Broadband Communications Circuit Cables. Low Power network-powered broadband communications cables shall be permitted in the same raceway or enclosure with jacketed cables of any of the following circuits:

(1) Class 2 and Class 3 remote-control, signaling, and power-limited circuits in compliance with Article 725
(2) Power-limited fire alarm systems in compliance with Article 760
(3) Communications circuits in compliance with Article 800

(4) Nonconductive and conductive optical fiber cables in compliance with Article 770
(5) Community antenna television and radio distribution systems in compliance with Article 820

(c) Medium Power Network-Powered Broadband Communications Circuit Cables. Medium power network-powered broadband communications cables shall not be permitted in the same raceway or enclosure with conductors of any of the following circuits:

(1) Class 2 and Class 3 remote-control, signaling, and power-limited circuits in compliance with Article 725
(2) Power-limited fire alarm systems in compliance with Article 760
(3) Communications circuits in compliance with Article 800
(4) Conductive optical fiber cables in compliance with Article 770
(5) Community antenna television and radio distribution systems in compliance with Article 820

(d) Electric Light, Power, Class 1, Non–Powered Broadband Communications Circuit Cables. Network-powered broadband communications cable shall not be placed in any raceway, compartment, outlet box, junction box, or similar fittings with conductors of electric light, power, Class 1, or non–power-limited fire alarm circuit cables.

*Exception No. 1: Where all of the conductors of electric light, power, Class 1, non–power-limited fire alarm circuits are separated from all of the network-powered broadband communications cables by a barrier.*

*Exception No. 2: Power circuit conductors in outlet boxes, junction boxes, or similar fittings or compartments where such conductors are introduced solely for power supply to the network-powered broadband communications system distribution equipment. The power circuit conductors shall be routed within the enclosure to maintain a minimum 6-mm (0.25-in.) separation from network-powered broadband communications cables.*

**(2) Other Applications.** Network-powered broadband communications cable shall be separated at least 50 mm (2 in.) from conductors of any electric light, power, Class 1, and non–power-limited fire alarm circuits.

*Exception No. 1: Where either (1) all of the conductors of electric light, power, Class 1, and non–power-limited fire alarm circuits are in a raceway, or in metal-sheathed, metal-clad, nonmetallic-sheathed, Type AC, or Type UF cables, or (2) all of the network-powered broadband communications cables are encased in raceway.*

*Exception No. 2: Where the network-powered broadband communications cables are permanently separated from the conductors of electric light, power, Class 1, and non–*

*power-limited fire alarm circuits by a continuous and firmly fixed nonconductor, such as porcelain tubes or flexible tubing, in addition to the insulation on the wire.*

**(B) Spread of Fire or Products of Combustion.** Installations in hollow spaces, vertical shafts, and ventilation or air-handling ducts shall be made so that the possible spread of fire or products of combustion will not be substantially increased. Openings around penetrations through fire-resistance–rated walls, partitions, floors, or ceilings shall be firestopped using approved methods to maintain the fire resistance rating.

**(C) Equipment in Other Space Used for Environmental Air.** Section 300.22(C) shall apply.

**(D) Support of Conductors.** Raceways shall be used for their intended purpose. Network-powered broadband com-

munications cables shall not be strapped, taped, or attached by any means to the exterior of any conduit or raceway as a means of support.

**(E) Cable Substitutions.** The substitutions for network-powered broadband cables listed in Table 830.58 shall be permitted. All cables in Table 830.58, other than network-powered broadband cables, shall be coaxial cables.

**Table 830.58 Cable Substitutions**

| Cable Type | Permitted Cable Substitutions |
|---|---|
| BM | BMR |
| BLP | CMP, CL3P |
| BLX | CMP, CL3P, CMR, CL3R, CMG, CM, CL3, CMX, CL3X, BMR, BM, BLP |

# Chapter 9 Tables

**Table 1 Percent of Cross Section of Conduit and Tubing for Conductors**

| Number of Conductors | All Conductor Types |
|:---:|:---:|
| 1 | 53 |
| 2 | 31 |
| Over 2 | 40 |

FPN No. 1:  Table 1 is based on common conditions of proper cabling and alignment of conductors where the length of the pull and the number of bends are within reasonable limits. It should be recognized that, for certain conditions, a larger size conduit or a lesser conduit fill should be considered.

FPN No. 2:  When pulling three conductors or cables into a raceway, if the ratio of the raceway (inside diameter) to the conductor or cable (outside diameter) is between 2.8 and 3.2, jamming can occur. While jamming can occur when pulling four or more conductors or cables into a raceway, the probability is very low.

**Notes to Tables**

(1) See Annex C for the maximum number of conductors and fixture wires, all of the same size (total cross-sectional area including insulation) permitted in trade sizes of the applicable conduit or tubing.

(2) Table 1 applies only to complete conduit or tubing systems and is not intended to apply to sections of conduit or tubing used to protect exposed wiring from physical damage.

(3) Equipment grounding or bonding conductors, where installed, shall be included when calculating conduit or tubing fill. The actual dimensions of the equipment grounding or bonding conductor (insulated or bare) shall be used in the calculation.

(4) Where conduit or tubing nipples having a maximum length not to exceed 600 mm (24 in.) are installed between boxes, cabinets, and similar enclosures, the nipples shall be permitted to be filled to 60 percent of their total cross-sectional area, and 310.15(B)(2)(a) adjustment factors need not apply to this condition.

(5) For conductors not included in Chapter 9, such as multiconductor cables, the actual dimensions shall be used.

(6) For combinations of conductors of different sizes, use Table 5 and Table 5A for dimensions of conductors and Table 4 for the applicable conduit or tubing dimensions.

(7) When calculating the maximum number of conductors permitted in a conduit or tubing, all of the same size (total cross-sectional area including insulation), the next higher whole number shall be used to determine the maximum number of conductors permitted when the calculation results in a decimal of 0.8 or larger.

(8) Where bare conductors are permitted by other sections of this *Code*, the dimensions for bare conductors in Table 8 shall be permitted.

(9) A multiconductor cable of two or more conductors shall be treated as a single conductor for calculating percentage conduit fill area. For cables that have elliptical cross sections, the cross-sectional area calculation shall be based on using the major diameter of the ellipse as a circle diameter.

**Table 4 Dimensions and Percent Area of Conduit and Tubing (Areas of Conduit or Tubing for the Combinations of Wires Permitted in Table 1, Chapter 9)**

| | | Nominal Internal Diameter | | Total Area 100% | | 2 Wires 31% | | Over 2 Wires 40% | | 1 Wire 53% | | 60% | |
|---|---|---|---|---|---|---|---|---|---|---|---|---|---|
| **Metric Designator** | **Trade Size** | **mm** | **in.** | **mm²** | **in.²** | **mm²** | **in.²** | **mm²** | **in.²** | **mm²** | **in.²** | **mm²** | **in.²** |
| 16 | ½ | 15.8 | 0.622 | 196 | 0.304 | 61 | 0.094 | 78 | 0.122 | 104 | 0.161 | 118 | 0.182 |
| 21 | ¾ | 20.9 | 0.824 | 343 | 0.533 | 106 | 0.165 | 137 | 0.213 | 182 | 0.283 | 206 | 0.320 |
| 27 | 1 | 26.6 | 1.049 | 556 | 0.864 | 172 | 0.268 | 222 | 0.346 | 295 | 0.458 | 333 | 0.519 |
| 35 | 1¼ | 35.1 | 1.380 | 968 | 1.496 | 300 | 0.464 | 387 | 0.598 | 513 | 0.793 | 581 | 0.897 |
| 41 | 1½ | 40.9 | 1.610 | 1314 | 2.036 | 407 | 0.631 | 526 | 0.814 | 696 | 1.079 | 788 | 1.221 |
| 53 | 2 | 52.5 | 2.067 | 2165 | 3.356 | 671 | 1.040 | 866 | 1.342 | 1147 | 1.778 | 1299 | 2.013 |
| 63 | 2½ | 69.4 | 2.731 | 3783 | 5.858 | 1173 | 1.816 | 1513 | 2.343 | 2005 | 3.105 | 2270 | 3.515 |
| 78 | 3 | 85.2 | 3.356 | 5701 | 8.846 | 1767 | 2.742 | 2280 | 3.538 | 3022 | 4.688 | 3421 | 5.307 |
| 91 | 3½ | 97.4 | 3.834 | 7451 | 11.545 | 2310 | 3.579 | 2980 | 4.618 | 3949 | 6.119 | 4471 | 6.927 |
| 103 | 4 | 110.1 | 4.334 | 9521 | 14.753 | 2951 | 4.573 | 3808 | 5.901 | 5046 | 7.819 | 5712 | 8.852 |

*Table header within table:* **Article 358 — Electrical Metallic Tubing (EMT)**

**Table 4** *Continued*

### Article 362 — Electrical Nonmetallic Tubing (ENT)

| Metric Designator | Trade Size | Nominal Internal Diameter | | Total Area 100% | | 2 Wires 31% | | Over 2 Wires 40% | | 1 Wire 53% | | 60% | |
|---|---|---|---|---|---|---|---|---|---|---|---|---|---|
| | | mm | in. | mm² | in.² | mm² | in.² | mm² | in.² | mm² | in.² | mm² | in.² |
| 16 | ½ | 14.2 | 0.560 | 158 | 0.246 | 49 | 0.076 | 63 | 0.099 | 84 | 0.131 | 95 | 0.148 |
| 21 | ¾ | 19.3 | 0.760 | 293 | 0.454 | 91 | 0.141 | 117 | 0.181 | 155 | 0.240 | 176 | 0.272 |
| 27 | 1 | 25.4 | 1.000 | 507 | 0.785 | 157 | 0.243 | 203 | 0.314 | 269 | 0.416 | 304 | 0.471 |
| 35 | 1¼ | 34.0 | 1.340 | 908 | 1.410 | 281 | 0.437 | 363 | 0.564 | 481 | 0.747 | 545 | 0.846 |
| 41 | 1½ | 39.9 | 1.570 | 1250 | 1.936 | 388 | 0.600 | 500 | 0.774 | 663 | 1.026 | 750 | 1.162 |
| 53 | 2 | 51.3 | 2.020 | 2067 | 3.205 | 641 | 0.993 | 827 | 1.282 | 1095 | 1.699 | 1240 | 1.923 |
| 63 | 2½ | — | — | — | — | — | — | — | — | — | — | — | — |
| 78 | 3 | — | — | — | — | — | — | — | — | — | — | — | — |
| 91 | 3½ | — | — | — | — | — | — | — | — | — | — | — | — |

### Article 348 — Flexible Metal Conduit (FMC)

| Metric Designator | Trade Size | Nominal Internal Diameter | | Total Area 100% | | 2 Wires 31% | | Over 2 Wires 40% | | 1 Wire 53% | | 60% | |
|---|---|---|---|---|---|---|---|---|---|---|---|---|---|
| | | mm | in. | mm² | in.² | mm² | in.² | mm² | in.² | mm² | in.² | mm² | in.² |
| 12 | ⅜ | 9.7 | 0.384 | 74 | 0.116 | 23 | 0.036 | 30 | 0.046 | 39 | 0.061 | 44 | 0.069 |
| 16 | ½ | 16.1 | 0.635 | 204 | 0.317 | 63 | 0.098 | 81 | 0.127 | 108 | 0.168 | 122 | 0.190 |
| 21 | ¾ | 20.9 | 0.824 | 343 | 0.533 | 106 | 0.165 | 137 | 0.213 | 182 | 0.283 | 206 | 0.320 |
| 27 | 1 | 25.9 | 1.020 | 527 | 0.817 | 163 | 0.253 | 211 | 0.327 | 279 | 0.433 | 316 | 0.490 |
| 35 | 1¼ | 32.4 | 1.275 | 824 | 1.277 | 256 | 0.396 | 330 | 0.511 | 437 | 0.677 | 495 | 0.766 |
| 41 | 1½ | 39.1 | 1.538 | 1201 | 1.858 | 372 | 0.576 | 480 | 0.743 | 636 | 0.985 | 720 | 1.115 |
| 53 | 2 | 51.8 | 2.040 | 2107 | 3.269 | 653 | 1.013 | 843 | 1.307 | 1117 | 1.732 | 1264 | 1.961 |
| 63 | 2½ | 63.5 | 2.500 | 3167 | 4.909 | 982 | 1.522 | 1267 | 1.963 | 1678 | 2.602 | 1900 | 2.945 |
| 78 | 3 | 76.2 | 3.000 | 4560 | 7.069 | 1414 | 2.191 | 1824 | 2.827 | 2417 | 3.746 | 2736 | 4.241 |
| 91 | 3½ | 88.9 | 3.500 | 6207 | 9.621 | 1924 | 2.983 | 2483 | 3.848 | 3290 | 5.099 | 3724 | 5.773 |
| 103 | 4 | 101.6 | 4.000 | 8107 | 12.566 | 2513 | 3.896 | 3243 | 5.027 | 4297 | 6.660 | 4864 | 7.540 |

### Article 342 — Intermediate Metal Conduit (IMC)

| Metric Designator | Trade Size | Nominal Internal Diameter | | Total Area 100% | | 2 Wires 31% | | Over 2 Wires 40% | | 1 Wire 53% | | 60% | |
|---|---|---|---|---|---|---|---|---|---|---|---|---|---|
| | | mm | in. | mm² | in.² | mm² | in.² | mm² | in.² | mm² | in.² | mm² | in.² |
| 12 | ⅜ | — | — | — | — | — | — | — | — | — | — | — | — |
| 16 | ½ | 16.8 | 0.660 | 222 | 0.342 | 69 | 0.106 | 89 | 0.137 | 117 | 0.181 | 133 | 0.205 |
| 21 | ¾ | 21.9 | 0.864 | 377 | 0.586 | 117 | 0.182 | 151 | 0.235 | 200 | 0.311 | 226 | 0.352 |
| 27 | 1 | 28.1 | 1.105 | 620 | 0.959 | 192 | 0.297 | 248 | 0.384 | 329 | 0.508 | 372 | 0.575 |
| 35 | 1¼ | 36.8 | 1.448 | 1064 | 1.647 | 330 | 0.510 | 425 | 0.659 | 564 | 0.873 | 638 | 0.988 |
| 41 | 1½ | 42.7 | 1.683 | 1432 | 2.225 | 444 | 0.690 | 573 | 0.890 | 759 | 1.179 | 859 | 1.335 |
| 53 | 2 | 54.6 | 2.150 | 2341 | 3.630 | 726 | 1.125 | 937 | 1.452 | 1241 | 1.924 | 1405 | 2.178 |
| 63 | 2½ | 64.9 | 2.557 | 3308 | 5.135 | 1026 | 1.592 | 1323 | 2.054 | 1753 | 2.722 | 1985 | 3.081 |
| 78 | 3 | 80.7 | 3.176 | 5115 | 7.922 | 1586 | 2.456 | 2046 | 3.169 | 2711 | 4.199 | 3069 | 4.753 |
| 91 | 3½ | 93.2 | 3.671 | 6822 | 10.584 | 2115 | 3.281 | 2729 | 4.234 | 3616 | 5.610 | 4093 | 6.351 |
| 103 | 4 | 105.4 | 4.166 | 8725 | 13.631 | 2705 | 4.226 | 3490 | 5.452 | 4624 | 7.224 | 5235 | 8.179 |

**Table 4** *Continued*

### Article 356— Liquidtight Flexible Nonmetallic Conduit (LFNC-B*)

| Metric Designator | Trade Size | Nominal Internal Diameter | | Total Area 100% | | 2 Wires 31% | | Over 2 Wires 40% | | 1 Wire 53% | | 60% | |
|---|---|---|---|---|---|---|---|---|---|---|---|---|---|
| | | mm | in. | mm² | in.² | mm² | in.² | mm² | in.² | mm² | in.² | mm² | in.² |
| 12 | ⅜ | 12.5 | 0.494 | 123 | 0.192 | 38 | 0.059 | 49 | 0.077 | 65 | 0.102 | 74 | 0.115 |
| 16 | ½ | 16.1 | 0.632 | 204 | 0.314 | 63 | 0.097 | 81 | 0.125 | 108 | 0.166 | 122 | 0.188 |
| 21 | ¾ | 21.1 | 0.830 | 350 | 0.541 | 108 | 0.168 | 140 | 0.216 | 185 | 0.287 | 210 | 0.325 |
| 27 | 1 | 26.8 | 1.054 | 564 | 0.873 | 175 | 0.270 | 226 | 0.349 | 299 | 0.462 | 338 | 0.524 |
| 35 | 1¼ | 35.4 | 1.395 | 984 | 1.528 | 305 | 0.474 | 394 | 0.611 | 522 | 0.810 | 591 | 0.917 |
| 41 | 1½ | 40.3 | 1.588 | 1276 | 1.981 | 395 | 0.614 | 510 | 0.792 | 676 | 1.050 | 765 | 1.188 |
| 53 | 2 | 51.6 | 2.033 | 2091 | 3.246 | 648 | 1.006 | 836 | 1.298 | 1108 | 1.720 | 1255 | 1.948 |

*Corresponds to 356.2(2)

### Article 356 — Liquidtight Flexible Nonmetallic Conduit (LFNC-A*)

| Metric Designator | Trade Size | Nominal Internal Diameter | | Total Area 100% | | 2 Wires 31% | | Over 2 Wires 40% | | 1 Wire 53% | | 60% | |
|---|---|---|---|---|---|---|---|---|---|---|---|---|---|
| | | mm | in. | mm² | in.² | mm² | in.² | mm² | in.² | mm² | in.² | mm² | in.² |
| 12 | ⅜ | 12.6 | 0.495 | 125 | 0.192 | 39 | 0.060 | 50 | 0.077 | 66 | 0.102 | 75 | 0.115 |
| 16 | ½ | 16.0 | 0.630 | 201 | 0.312 | 62 | 0.097 | 80 | 0.125 | 107 | 0.165 | 121 | 0.187 |
| 21 | ¾ | 21.0 | 0.825 | 346 | 0.535 | 107 | 0.166 | 139 | 0.214 | 184 | 0.283 | 208 | 0.321 |
| 27 | 1 | 26.5 | 1.043 | 552 | 0.854 | 171 | 0.265 | 221 | 0.342 | 292 | 0.453 | 331 | 0.513 |
| 35 | 1¼ | 35.1 | 1.383 | 968 | 1.502 | 300 | 0.466 | 387 | 0.601 | 513 | 0.796 | 581 | 0.901 |
| 41 | 1½ | 40.7 | 1.603 | 1301 | 2.018 | 403 | 0.626 | 520 | 0.807 | 690 | 1.070 | 781 | 1.211 |
| 53 | 2 | 52.4 | 2.063 | 2157 | 3.343 | 669 | 1.036 | 863 | 1.337 | 1143 | 1.772 | 1294 | 2.006 |

*Corresponds to 356.2(1)

### Article 350 — Liquidtight Flexible Metal Conduit (LFMC)

| Metric Designator | Trade Size | Nominal Internal Diameter | | Total Area 100% | | 2 Wires 31% | | Over 2 Wires 40% | | 1 Wire 53% | | 60% | |
|---|---|---|---|---|---|---|---|---|---|---|---|---|---|
| | | mm | in. | mm² | in.² | mm² | in.² | mm² | in.² | mm² | in.² | mm² | in.² |
| 12 | ⅜ | 12.5 | 0.494 | 123 | 0.192 | 38 | 0.059 | 49 | 0.077 | 65 | 0.102 | 74 | 0.115 |
| 16 | ½ | 16.1 | 0.632 | 204 | 0.314 | 63 | 0.097 | 81 | 0.125 | 108 | 0.166 | 122 | 0.188 |
| 21 | ¾ | 21.1 | 0.830 | 350 | 0.541 | 108 | 0.168 | 140 | 0.216 | 185 | 0.287 | 210 | 0.325 |
| 27 | 1 | 26.8 | 1.054 | 564 | 0.873 | 175 | 0.270 | 226 | 0.349 | 299 | 0.462 | 338 | 0.524 |
| 35 | 1¼ | 35.4 | 1.395 | 984 | 1.528 | 305 | 0.474 | 394 | 0.611 | 522 | 0.810 | 591 | 0.917 |
| 41 | 1½ | 40.3 | 1.588 | 1276 | 1.981 | 395 | 0.614 | 510 | 0.792 | 676 | 1.050 | 765 | 1.188 |
| 53 | 2 | 51.6 | 2.033 | 2091 | 3.246 | 648 | 1.006 | 836 | 1.298 | 1108 | 1.720 | 1255 | 1.948 |
| 63 | 2½ | 63.3 | 2.493 | 3147 | 4.881 | 976 | 1.513 | 1259 | 1.953 | 1668 | 2.587 | 1888 | 2.929 |
| 78 | 3 | 78.4 | 3.085 | 4827 | 7.475 | 1497 | 2.317 | 1931 | 2.990 | 2559 | 3.962 | 2896 | 4.485 |
| 91 | 3½ | 89.4 | 3.520 | 6277 | 9.731 | 1946 | 3.017 | 2511 | 3.893 | 3327 | 5.158 | 3766 | 5.839 |
| 103 | 4 | 102.1 | 4.020 | 8187 | 12.692 | 2538 | 3.935 | 3275 | 5.077 | 4339 | 6.727 | 4912 | 7.615 |
| 129 | 5 | — | — | — | — | — | — | — | — | — | — | — | — |
| 155 | 6 | — | — | — | — | — | — | — | — | — | — | — | — |

**Table 4** *Continued*

### Article 344 — Rigid Metal Conduit (RMC)

| Metric Designator | Trade Size | Nominal Internal Diameter | | Total Area 100% | | 2 Wires 31% | | Over 2 Wires 40% | | 1 Wire 53% | | 60% | |
|---|---|---|---|---|---|---|---|---|---|---|---|---|---|
| | | mm | in. | mm² | in.² | mm² | in.² | mm² | in.² | mm² | in.² | mm² | in.² |
| 12 | ⅜ | — | — | — | — | — | — | — | — | — | — | — | — |
| 16 | ½ | 16.1 | 0.632 | 204 | 0.314 | 63 | 0.097 | 81 | 0.125 | 108 | 0.166 | 122 | 0.188 |
| 21 | ¾ | 21.2 | 0.836 | 353 | 0.549 | 109 | 0.170 | 141 | 0.220 | 187 | 0.291 | 212 | 0.329 |
| 27 | 1 | 27.0 | 1.063 | 573 | 0.887 | 177 | 0.275 | 229 | 0.355 | 303 | 0.470 | 344 | 0.532 |
| 35 | 1¼ | 35.4 | 1.394 | 984 | 1.526 | 305 | 0.473 | 394 | 0.610 | 522 | 0.809 | 591 | 0.916 |
| 41 | 1½ | 41.2 | 1.624 | 1333 | 2.071 | 413 | 0.642 | 533 | 0.829 | 707 | 1.098 | 800 | 1.243 |
| 53 | 2 | 52.9 | 2.083 | 2198 | 3.408 | 681 | 1.056 | 879 | 1.363 | 1165 | 1.806 | 1319 | 2.045 |
| 63 | 2½ | 63.2 | 2.489 | 3137 | 4.866 | 972 | 1.508 | 1255 | 1.946 | 1663 | 2.579 | 1882 | 2.919 |
| 78 | 3 | 78.5 | 3.090 | 4840 | 7.499 | 1500 | 2.325 | 1936 | 3.000 | 2565 | 3.974 | 2904 | 4.499 |
| 91 | 3½ | 90.7 | 3.570 | 6461 | 10.010 | 2003 | 3.103 | 2584 | 4.004 | 3424 | 5.305 | 3877 | 6.006 |
| 103 | 4 | 102.9 | 4.050 | 8316 | 12.882 | 2578 | 3.994 | 3326 | 5.153 | 4408 | 6.828 | 4990 | 7.729 |
| 129 | 5 | 128.9 | 5.073 | 13050 | 20.212 | 4045 | 6.266 | 5220 | 8.085 | 6916 | 10.713 | 7830 | 12.127 |
| 155 | 6 | 154.8 | 6.093 | 18821 | 29.158 | 5834 | 9.039 | 7528 | 11.663 | 9975 | 15.454 | 11292 | 17.495 |

### Article 352 — Rigid PVC Conduit (RNC), Schedule 80

| Metric Designator | Trade Size | Nominal Internal Diameter | | Total Area 100% | | 2 Wires 31% | | Over 2 Wires 40% | | 1 Wire 53% | | 60% | |
|---|---|---|---|---|---|---|---|---|---|---|---|---|---|
| | | mm | in. | mm² | in.² | mm² | in.² | mm² | in.² | mm² | in.² | mm² | in.² |
| 12 | ⅜ | — | — | — | — | — | — | — | — | — | — | — | — |
| 16 | ½ | 13.4 | 0.526 | 141 | 0.217 | 44 | 0.067 | 56 | 0.087 | 75 | 0.115 | 85 | 0.130 |
| 21 | ¾ | 18.3 | 0.722 | 263 | 0.409 | 82 | 0.127 | 105 | 0.164 | 139 | 0.217 | 158 | 0.246 |
| 27 | 1 | 23.8 | 0.936 | 445 | 0.688 | 138 | 0.213 | 178 | 0.275 | 236 | 0.365 | 267 | 0.413 |
| 35 | 1¼ | 31.9 | 1.255 | 799 | 1.237 | 248 | 0.383 | 320 | 0.495 | 424 | 0.656 | 480 | 0.742 |
| 41 | 1½ | 37.5 | 1.476 | 1104 | 1.711 | 342 | 0.530 | 442 | 0.684 | 585 | 0.907 | 663 | 1.027 |
| 53 | 2 | 48.6 | 1.913 | 1855 | 2.874 | 575 | 0.891 | 742 | 1.150 | 983 | 1.523 | 1113 | 1.725 |
| 63 | 2½ | 58.2 | 2.290 | 2660 | 4.119 | 825 | 1.277 | 1064 | 1.647 | 1410 | 2.183 | 1596 | 2.471 |
| 78 | 3 | 72.7 | 2.864 | 4151 | 6.442 | 1287 | 1.997 | 1660 | 2.577 | 2200 | 3.414 | 2491 | 3.865 |
| 91 | 3½ | 84.5 | 3.326 | 5608 | 8.688 | 1738 | 2.693 | 2243 | 3.475 | 2972 | 4.605 | 3365 | 5.213 |
| 103 | 4 | 96.2 | 3.786 | 7268 | 11.258 | 2253 | 3.490 | 2907 | 4.503 | 3852 | 5.967 | 4361 | 6.755 |
| 129 | 5 | 121.1 | 4.768 | 11518 | 17.855 | 3571 | 5.535 | 4607 | 7.142 | 6105 | 9.463 | 6911 | 10.713 |
| 155 | 6 | 145.0 | 5.709 | 16513 | 25.598 | 5119 | 7.935 | 6605 | 10.239 | 8752 | 13.567 | 9908 | 15.359 |

### Article 352 — Rigid PVC Conduit (RNC), Schedule 40, and HDPE Conduit

| Metric Designator | Trade Size | Nominal Internal Diameter | | Total Area 100% | | 2 Wires 31% | | Over 2 Wires 40% | | 1 Wire 53% | | 60% | |
|---|---|---|---|---|---|---|---|---|---|---|---|---|---|
| | | mm | in. | mm² | in.² | mm² | in.² | mm² | in.² | mm² | in.² | mm² | in.² |
| 12 | ⅜ | — | — | — | — | — | — | — | — | — | — | — | — |
| 16 | ½ | 15.3 | 0.602 | 184 | 0.285 | 57 | 0.088 | 74 | 0.114 | 97 | 0.151 | 110 | 0.171 |
| 21 | ¾ | 20.4 | 0.804 | 327 | 0.508 | 101 | 0.157 | 131 | 0.203 | 173 | 0.269 | 196 | 0.305 |
| 27 | 1 | 26.1 | 1.029 | 535 | 0.832 | 166 | 0.258 | 214 | 0.333 | 284 | 0.441 | 321 | 0.499 |
| 35 | 1¼ | 34.5 | 1.360 | 935 | 1.453 | 290 | 0.450 | 374 | 0.581 | 495 | 0.770 | 561 | 0.872 |
| 41 | 1½ | 40.4 | 1.590 | 1282 | 1.986 | 397 | 0.616 | 513 | 0.794 | 679 | 1.052 | 769 | 1.191 |
| 53 | 2 | 52.0 | 2.047 | 2124 | 3.291 | 658 | 1.020 | 849 | 1.316 | 1126 | 1.744 | 1274 | 1.975 |

**Table 4** *Continued*

### Article 352 — Rigid PVC Conduit (RNC), Schedule 40, and HDPE Conduit

| Metric Designator | Trade Size | Nominal Internal Diameter | | Total Area 100% | | 2 Wires 31% | | Over 2 Wires 40% | | 1 Wire 53% | | 60% | |
|---|---|---|---|---|---|---|---|---|---|---|---|---|---|
| | | mm | in. | mm² | in.² | mm² | in.² | mm² | in.² | mm² | in.² | mm² | in.² |
| 63 | 2½ | 62.1 | 2.445 | 3029 | 4.695 | 939 | 1.455 | 1212 | 1.878 | 1605 | 2.488 | 1817 | 2.817 |
| 78 | 3 | 77.3 | 3.042 | 4693 | 7.268 | 1455 | 2.253 | 1877 | 2.907 | 2487 | 3.852 | 2816 | 4.361 |
| 91 | 3½ | 89.4 | 3.521 | 6277 | 9.737 | 1946 | 3.018 | 2511 | 3.895 | 3327 | 5.161 | 3766 | 5.842 |
| 103 | 4 | 101.5 | 3.998 | 8091 | 12.554 | 2508 | 3.892 | 3237 | 5.022 | 4288 | 6.654 | 4855 | 7.532 |
| 129 | 5 | 127.4 | 5.016 | 12748 | 19.761 | 3952 | 6.126 | 5099 | 7.904 | 6756 | 10.473 | 7649 | 11.856 |
| 155 | 6 | 153.2 | 6.031 | 18433 | 28.567 | 5714 | 8.856 | 7373 | 11.427 | 9770 | 15.141 | 11060 | 17.140 |

### Article 352 — Type A, Rigid PVC Conduit (RNC)

| Metric Designator | Trade Size | Nominal Internal Diameter | | Total Area 100% | | 2 Wires 31% | | Over 2 Wires 40% | | 1 Wire 53% | | 60% | |
|---|---|---|---|---|---|---|---|---|---|---|---|---|---|
| | | mm | in. | mm² | in.² | mm² | in.² | mm² | in.² | mm² | in.² | mm² | in.² |
| 16 | ½ | 17.8 | 0.700 | 249 | 0.385 | 77 | 0.119 | 100 | 0.154 | 132 | 0.204 | 149 | 0.231 |
| 21 | ¾ | 23.1 | 0.910 | 419 | 0.650 | 130 | 0.202 | 168 | 0.260 | 222 | 0.345 | 251 | 0.390 |
| 27 | 1 | 29.8 | 1.175 | 697 | 1.084 | 216 | 0.336 | 279 | 0.434 | 370 | 0.575 | 418 | 0.651 |
| 35 | 1¼ | 38.1 | 1.500 | 1140 | 1.767 | 353 | 0.548 | 456 | 0.707 | 604 | 0.937 | 684 | 1.060 |
| 41 | 1½ | 43.7 | 1.720 | 1500 | 2.324 | 465 | 0.720 | 600 | 0.929 | 795 | 1.231 | 900 | 1.394 |
| 53 | 2 | 54.7 | 2.155 | 2350 | 3.647 | 728 | 1.131 | 940 | 1.459 | 1245 | 1.933 | 1410 | 2.188 |
| 63 | 2½ | 66.9 | 2.635 | 3515 | 5.453 | 1090 | 1.690 | 1406 | 2.181 | 1863 | 2.890 | 2109 | 3.272 |
| 78 | 3 | 82.0 | 3.230 | 5281 | 8.194 | 1637 | 2.540 | 2112 | 3.278 | 2799 | 4.343 | 3169 | 4.916 |
| 91 | 3½ | 93.7 | 3.690 | 6896 | 10.694 | 2138 | 3.315 | 2758 | 4.278 | 3655 | 5.668 | 4137 | 6.416 |
| 103 | 4 | 106.2 | 4.180 | 8858 | 13.723 | 2746 | 4.254 | 3543 | 5.489 | 4695 | 7.273 | 5315 | 8.234 |
| 129 | 5 | — | — | — | — | — | — | — | — | — | — | — | — |
| 155 | 6 | — | — | — | — | — | — | — | — | — | — | — | — |

### Article 352 — Type EB, PVC Conduit (RNC)

| Metric Designator | Trade Size | Nominal Internal Diameter | | Total Area 100% | | 2 Wires 31% | | Over 2 Wires 40% | | 1 Wire 53% | | 60% | |
|---|---|---|---|---|---|---|---|---|---|---|---|---|---|
| | | mm | in. | mm² | in.² | mm² | in.² | mm² | in.² | mm² | in.² | mm² | in.² |
| 16 | ½ | — | — | — | — | — | — | — | — | — | — | — | — |
| 21 | ¾ | — | — | — | — | — | — | — | — | — | — | — | — |
| 27 | 1 | — | — | — | — | — | — | — | — | — | — | — | — |
| 35 | 1¼ | — | — | — | — | — | — | — | — | — | — | — | — |
| 41 | 1½ | — | — | — | — | — | — | — | — | — | — | — | — |
| 53 | 2 | 56.4 | 2.221 | 2498 | 3.874 | 774 | 1.201 | 999 | 1.550 | 1324 | 2.053 | 1499 | 2.325 |
| 63 | 2½ | — | — | — | — | — | — | — | — | 2979 | 4.616 | 3373 | 5.226 |
| 78 | 3 | 84.6 | 3.330 | 5621 | 8.709 | 1743 | 2.700 | 2248 | 3.484 | 3884 | 6.023 | 4397 | 6.819 |
| 91 | 3½ | 96.6 | 3.804 | 7329 | 11.365 | 2272 | 3.523 | 2932 | 4.546 | 4937 | 7.657 | 5589 | 8.669 |
| 103 | 4 | 108.9 | 4.289 | 9314 | 14.448 | 2887 | 4.479 | 3726 | 5.779 | 7586 | 11.763 | 8588 | 13.317 |
| 129 | 5 | 135.0 | 5.316 | 14314 | 22.195 | 4437 | 6.881 | 5726 | 8.878 | 10776 | 16.711 | 12200 | 18.918 |
| 155 | 6 | 160.9 | 6.336 | 20333 | 31.530 | 6303 | 9.774 | 8133 | 12.612 | | | | |

## Table 5  Dimensions of Insulated Conductors and Fixture Wires

| Type | Size (AWG or kcmil) | Approximate Diameter | | Approximate Area | |
|---|---|---|---|---|---|
| | | mm | in. | mm² | in.² |
| **Type: FFH-2, RFH-1, RFH-2, RHH\*, RHW\*, RHW-2\*, RHH, RHW, RHW-2, SF-1, SF-2, SFF-1, SFF-2, TF, TFF, THHW, THW, THW-2, TW, XF, XFF** | | | | | |
| RFH-2, | 18 | 3.454 | 0.136 | 9.355 | 0.0145 |
| FFH-2 | 16 | 3.759 | 0.148 | 11.10 | 0.0172 |
| RHW-2, RHH, | 14 | 4.902 | 0.193 | 18.90 | 0.0293 |
| RHW | 12 | 5.385 | 0.212 | 22.77 | 0.0353 |
| | 10 | 5.994 | 0.236 | 28.19 | 0.0437 |
| | 8 | 8.280 | 0.326 | 53.87 | 0.0835 |
| | 6 | 9.246 | 0.364 | 67.16 | 0.1041 |
| | 4 | 10.46 | 0.412 | 86.00 | 0.1333 |
| | 3 | 11.18 | 0.440 | 98.13 | 0.1521 |
| | 2 | 11.99 | 0.472 | 112.9 | 0.1750 |
| | 1 | 14.78 | 0.582 | 171.6 | 0.2660 |
| | 1/0 | 15.80 | 0.622 | 196.1 | 0.3039 |
| | 2/0 | 16.97 | 0.668 | 226.1 | 0.3505 |
| | 3/0 | 18.29 | 0.720 | 262.7 | 0.4072 |
| | 4/0 | 19.76 | 0.778 | 306.7 | 0.4754 |
| | 250 | 22.73 | 0.895 | 405.9 | 0.6291 |
| | 300 | 24.13 | 0.950 | 457.3 | 0.7088 |
| | 350 | 25.43 | 1.001 | 507.7 | 0.7870 |
| | 400 | 26.62 | 1.048 | 556.5 | 0.8626 |
| | 500 | 28.78 | 1.133 | 650.5 | 1.0082 |
| | 600 | 31.57 | 1.243 | 782.9 | 1.2135 |
| | 700 | 33.38 | 1.314 | 874.9 | 1.3561 |
| | 750 | 34.24 | 1.348 | 920.8 | 1.4272 |
| | 800 | 35.05 | 1.380 | 965.0 | 1.4957 |
| | 900 | 36.68 | 1.444 | 1057 | 1.6377 |
| | 1000 | 38.15 | 1.502 | 1143 | 1.7719 |
| | 1250 | 43.92 | 1.729 | 1515 | 2.3479 |
| | 1500 | 47.04 | 1.852 | 1738 | 2.6938 |
| | 1750 | 49.94 | 1.966 | 1959 | 3.0357 |
| | 2000 | 52.63 | 2.072 | 2175 | 3.3719 |
| SF-2, SFF-2 | 18 | 3.073 | 0.121 | 7.419 | 0.0115 |
| | 16 | 3.378 | 0.133 | 8.968 | 0.0139 |
| | 14 | 3.759 | 0.148 | 11.10 | 0.0172 |
| SF-1, SFF-1 | 18 | 2.311 | 0.091 | 4.194 | 0.0065 |
| RFH-1, XF, XFF | 18 | 2.692 | 0.106 | 5.161 | 0.0080 |
| TF, TFF, XF, XFF | 16 | 2.997 | 0.118 | 7.032 | 0.0109 |
| TW, XF, XFF, THHW, THW, THW-2 | 14 | 3.378 | 0.133 | 8.968 | 0.0139 |
| TW, THHW, THW, THW-2 | 12 | 3.861 | 0.152 | 11.68 | 0.0181 |
| | 10 | 4.470 | 0.176 | 15.68 | 0.0243 |
| | 8 | 5.994 | 0.236 | 28.19 | 0.0437 |
| RHH\*, RHW\*, RHW-2\* | 14 | 4.140 | 0.163 | 13.48 | 0.0209 |
| RHH\*, RHW\*, RHW-2\*, XF, XFF | 12 | 4.623 | 0.182 | 16.77 | 0.0260 |

### Table 5  *Continued*

| Type | Size (AWG or kcmil) | Approximate Diameter | | Approximate Area | |
|---|---|---|---|---|---|
| | | mm | in. | mm² | in.² |
| **Type: RHH\*, RHW\*, RHW-2\*, THHN, THHW, THW, THW-2, TFN, TFFN, THWN, THWN-2, XF, XFF** | | | | | |
| RHH\*, RHW\*, RHW-2\*, XF, XFF | 10 | 5.232 | 0.206 | 21.48 | 0.0333 |
| RHH\*, RHW\*, RHW-2\* | 8 | 6.756 | 0.266 | 35.87 | 0.0556 |
| TW, THW, THHW, THW-2, RHH\*, RHW\*, RHW-2\* | 6 | 7.722 | 0.304 | 46.84 | 0.0726 |
| | 4 | 8.941 | 0.352 | 62.77 | 0.0973 |
| | 3 | 9.652 | 0.380 | 73.16 | 0.1134 |
| | 2 | 10.46 | 0.412 | 86.00 | 0.1333 |
| | 1 | 12.50 | 0.492 | 122.6 | 0.1901 |
| | 1/0 | 13.51 | 0.532 | 143.4 | 0.2223 |
| | 2/0 | 14.68 | 0.578 | 169.3 | 0.2624 |
| | 3/0 | 16.00 | 0.630 | 201.1 | 0.3117 |
| | 4/0 | 17.48 | 0.688 | 239.9 | 0.3718 |
| | 250 | 19.43 | 0.765 | 296.5 | 0.4596 |
| | 300 | 20.83 | 0.820 | 340.7 | 0.5281 |
| | 350 | 22.12 | 0.871 | 384.4 | 0.5958 |
| | 400 | 23.32 | 0.918 | 427.0 | 0.6619 |
| | 500 | 25.48 | 1.003 | 509.7 | 0.7901 |
| | 600 | 28.27 | 1.113 | 627.7 | 0.9729 |
| | 700 | 30.07 | 1.184 | 710.3 | 1.1010 |
| | 750 | 30.94 | 1.218 | 751.7 | 1.1652 |
| | 800 | 31.75 | 1.250 | 791.7 | 1.2272 |
| | 900 | 33.38 | 1.314 | 874.9 | 1.3561 |
| | 1000 | 34.85 | 1.372 | 953.8 | 1.4784 |
| | 1250 | 39.09 | 1.539 | 1200 | 1.8602 |
| | 1500 | 42.21 | 1.662 | 1400 | 2.1695 |
| | 1750 | 45.11 | 1.776 | 1598 | 2.4773 |
| | 2000 | 47.80 | 1.882 | 1795 | 2.7818 |
| TFN, TFFN | 18 | 2.134 | 0.084 | 3.548 | 0.0055 |
| | 16 | 2.438 | 0.096 | 4.645 | 0.0072 |
| THHN, THWN, THWN-2 | 14 | 2.819 | 0.111 | 6.258 | 0.0097 |
| | 12 | 3.302 | 0.130 | 8.581 | 0.0133 |
| | 10 | 4.166 | 0.164 | 13.61 | 0.0211 |
| | 8 | 5.486 | 0.216 | 23.61 | 0.0366 |
| | 6 | 6.452 | 0.254 | 32.71 | 0.0507 |
| | 4 | 8.230 | 0.324 | 53.16 | 0.0824 |
| | 3 | 8.941 | 0.352 | 62.77 | 0.0973 |
| | 2 | 9.754 | 0.384 | 74.71 | 0.1158 |
| | 1 | 11.33 | 0.446 | 100.8 | 0.1562 |
| | 1/0 | 12.34 | 0.486 | 119.7 | 0.1855 |
| | 2/0 | 13.51 | 0.532 | 143.4 | 0.2223 |
| | 3/0 | 14.83 | 0.584 | 172.8 | 0.2679 |
| | 4/0 | 16.31 | 0.642 | 208.8 | 0.3237 |
| | 250 | 18.06 | 0.711 | 256.1 | 0.3970 |
| | 300 | 19.46 | 0.766 | 297.3 | 0.4608 |

**Table 5** *Continued*

| Type | Size (AWG or kcmil) | Approximate Diameter mm | Approximate Diameter in. | Approximate Area mm² | Approximate Area in.² |
|---|---|---|---|---|---|
| **Type: FEP, FEPB, PAF, PAFF, PF, PFA, PFAH, PFF, PGF, PGFF, PTF, PTFF, TFE, THHN, THWN, THWN-2, Z, ZF, ZFF** | | | | | |
| THHN, | 350 | 20.75 | 0.817 | 338.2 | 0.5242 |
| THWN, | 400 | 21.95 | 0.864 | 378.3 | 0.5863 |
| THWN-2 | 500 | 24.10 | 0.949 | 456.3 | 0.7073 |
| | 600 | 26.70 | 1.051 | 559.7 | 0.8676 |
| | 700 | 28.50 | 1.12 2 | 637.9 | 0.9887 |
| | 750 | 29.36 | 1.156 | 677.2 | 1.0496 |
| | 800 | 30.18 | 1.188 | 715.2 | 1.1085 |
| | 900 | 31.80 | 1.252 | 794.3 | 1.2311 |
| | 1000 | 33.27 | 1.310 | 869.5 | 1.3478 |
| PF, PGFF, PGF, | 18 | 2.184 | 0.086 | 3.742 | 0.0058 |
| PFF, PTF, PAF, PTFF, PAFF | 16 | 2.489 | 0.098 | 4.839 | 0.0075 |
| PF, PGFF, PGF, PFF, PTF, PAF, PTFF, PAFF, TFE, FEP, PFA, FEPB, PFAH | 14 | 2.870 | 0.113 | 6.452 | 0.0100 |
| TFE, FEP, | 12 | 3.353 | 0.132 | 8.839 | 0.0137 |
| PFA, FEPB, | 10 | 3.962 | 0.156 | 12.32 | 0.0191 |
| PFAH | 8 | 5.232 | 0.206 | 21.48 | 0.0333 |
| | 6 | 6.198 | 0.244 | 30.19 | 0.0468 |
| | 4 | 7.417 | 0.292 | 43.23 | 0.0670 |
| | 3 | 8.128 | 0.320 | 51.87 | 0.0804 |
| | 2 | 8.941 | 0.352 | 62.77 | 0.0973 |
| TFE, PFAH | 1 | 10.72 | 0.422 | 90.26 | 0.1399 |
| TFE, PFA | 1/0 | 11.73 | 0.462 | 108.1 | 0.1676 |
| PFAH, Z | 2/0 | 12.90 | 0.508 | 130.8 | 0.2027 |
| | 3/0 | 14.22 | 0.560 | 158.9 | 0.2463 |
| | 4/0 | 15.70 | 0.618 | 193.5 | 0.3000 |
| ZF, ZFF | 18 | 1.930 | 0.076 | 2.903 | 0.0045 |
| | 16 | 2.235 | 0.088 | 3.935 | 0.0061 |
| Z, ZF, ZFF | 14 | 2.616 | 0.103 | 5.355 | 0.0083 |
| Z | 12 | 3.099 | 0.122 | 7.548 | 0.0117 |
| | 10 | 3.962 | 0.156 | 12.32 | 0.0191 |
| | 8 | 4.978 | 0.196 | 19.48 | 0.0302 |
| | 6 | 5.944 | 0.234 | 27.74 | 0.0430 |
| | 4 | 7.163 | 0.282 | 40.32 | 0.0625 |
| | 3 | 8.382 | 0.330 | 55.16 | 0.0855 |
| | 2 | 9.195 | 0.362 | 66.39 | 0.1029 |
| | 1 | 10.21 | 0.402 | 81.87 | 0.1269 |

**Table 5** *Continued*

| Type | Size (AWG or kcmil) | Approximate Diameter mm | Approximate Diameter in. | Approximate Area mm² | Approximate Area in.² |
|---|---|---|---|---|---|
| **Type: KF-1, KF-2, KFF-1, KFF-2, XHH, XHHW, XHHW-2, ZW** | | | | | |
| XHHW, ZW, | 14 | 3.378 | 0.133 | 8.968 | 0.0139 |
| XHHW-2, | 12 | 3.861 | 0.152 | 11.68 | 0.0181 |
| XHH | 10 | 4.470 | 0.176 | 15.68 | 0.0243 |
| | 8 | 5.994 | 0.236 | 28.19 | 0.0437 |
| | 6 | 6.960 | 0.274 | 38.06 | 0.0590 |
| | 4 | 8.179 | 0.322 | 52.52 | 0.0814 |
| | 3 | 8.890 | 0.350 | 62.06 | 0.0962 |
| | 2 | 9.703 | 0.382 | 73.94 | 0.1146 |
| XHHW, | 1 | 11.23 | 0.442 | 98.97 | 0.1534 |
| XHHW-2, | 1/0 | 12.24 | 0.482 | 117.7 | 0.1825 |
| XHH | 2/0 | 13.41 | 0.528 | 141.3 | 0.2190 |
| | 3/0 | 14.73 | 0.58 | 170.5 | 0.2642 |
| | 4/0 | 16.21 | 0.638 | 206.3 | 0.3197 |
| | 250 | 17.91 | 0.705 | 251.9 | 0.3904 |
| | 300 | 19.30 | 0.76 | 292.6 | 0.4536 |
| | 350 | 20.60 | 0.811 | 333.3 | 0.5166 |
| | 400 | 21.79 | 0.858 | 373.0 | 0.5782 |
| | 500 | 23.95 | 0.943 | 450.6 | 0.6984 |
| | 600 | 26.75 | 1.053 | 561.9 | 0.8709 |
| | 700 | 28.55 | 1.124 | 640.2 | 0.9923 |
| | 750 | 29.41 | 1.158 | 679.5 | 1.0532 |
| | 800 | 30.23 | 1.190 | 717.5 | 1.1122 |
| | 900 | 31.85 | 1.254 | 796.8 | 1.2351 |
| | 1000 | 33.32 | 1.312 | 872.2 | 1.3519 |
| | 1250 | 37.57 | 1.479 | 1108 | 1.7180 |
| | 1500 | 40.69 | 1.602 | 1300 | 2.0157 |
| | 1750 | 43.59 | 1.716 | 1492 | 2.3127 |
| | 2000 | 46.28 | 1.822 | 1682 | 2.6073 |
| KF-2, | 18 | 1.600 | 0.063 | 2.000 | 0.0031 |
| KFF-2 | 16 | 1.905 | 0.075 | 2.839 | 0.0044 |
| | 14 | 2.286 | 0.090 | 4.129 | 0.0064 |
| | 12 | 2.769 | 0.109 | 6.000 | 0.0093 |
| | 10 | 3.378 | 0.133 | 8.968 | 0.0139 |
| KF-1, | 18 | 1.448 | 0.057 | 1.677 | 0.0026 |
| KFF-1 | 16 | 1.753 | 0.069 | 2.387 | 0.0037 |
| | 14 | 2.134 | 0.084 | 3.548 | 0.0055 |
| | 12 | 2.616 | 0.103 | 5.355 | 0.0083 |
| | 10 | 3.226 | 0.127 | 8.194 | 0.0127 |

*Types RHH, RHW, and RHW-2 without outer covering.

## Table 5A  Compact Aluminum Building Wire Nominal Dimensions* and Areas

| Size (AWG or kcmil) | Number of Strands | Bare Conductor Diameter | | Types THW and THHW Approximate Diameter | | Approximate Area | | Type THHN Approximate Diameter | | Approximate Area | | Type XHHW Approximate Diameter | | Approximate Area | | Size (AWG or kcmil) |
|---|---|---|---|---|---|---|---|---|---|---|---|---|---|---|---|---|
| | | mm | in. | mm | in. | mm$^2$ | in.$^2$ | mm | in. | mm$^2$ | in.$^2$ | mm | in. | mm$^2$ | in.$^2$ | |
| 8 | 7 | 3.404 | 0.134 | 6.477 | 0.255 | 32.90 | 0.0510 | — | — | — | — | 5.690 | 0.224 | 25.42 | 0.0394 | 8 |
| 6 | 7 | 4.293 | 0.169 | 7.366 | 0.290 | 42.58 | 0.0660 | 6.096 | 0.240 | 29.16 | 0.0452 | 6.604 | 0.260 | 34.19 | 0.0530 | 6 |
| 4 | 7 | 5.410 | 0.213 | 8.509 | 0.335 | 56.84 | 0.0881 | 7.747 | 0.305 | 47.10 | 0.0730 | 7.747 | 0.305 | 47.10 | 0.0730 | 4 |
| 2 | 7 | 6.807 | 0.268 | 9.906 | 0.390 | 77.03 | 0.1194 | 9.144 | 0.360 | 65.61 | 0.1017 | 9.144 | 0.360 | 65.61 | 0.1017 | 2 |
| 1 | 19 | 7.595 | 0.299 | 11.81 | 0.465 | 109.5 | 0.1698 | 10.54 | 0.415 | 87.23 | 0.1352 | 10.54 | 0.415 | 87.23 | 0.1352 | 1 |
| 1/0 | 19 | 8.534 | 0.336 | 12.70 | 0.500 | 126.6 | 0.1963 | 11.43 | 0.450 | 102.6 | 0.1590 | 11.43 | 0.450 | 102.6 | 0.1590 | 1/0 |
| 2/0 | 19 | 9.550 | 0.376 | 13.84 | 0.545 | 150.5 | 0.2332 | 12.57 | 0.495 | 124.1 | 0.1924 | 12.45 | 0.490 | 121.6 | 0.1885 | 2/0 |
| 3/0 | 19 | 10.74 | 0.423 | 14.99 | 0.590 | 176.3 | 0.2733 | 13.72 | 0.540 | 147.7 | 0.2290 | 13.72 | 0.540 | 147.7 | 0.2290 | 3/0 |
| 4/0 | 19 | 12.07 | 0.475 | 16.38 | 0.645 | 210.8 | 0.3267 | 15.11 | 0.595 | 179.4 | 0.2780 | 14.99 | 0.590 | 176.3 | 0.2733 | 4/0 |
| 250 | 37 | 13.21 | 0.520 | 18.42 | 0.725 | 266.3 | 0.4128 | 17.02 | 0.670 | 227.4 | 0.3525 | 16.76 | 0.660 | 220.7 | 0.3421 | 250 |
| 300 | 37 | 14.48 | 0.570 | 19.69 | 0.775 | 304.3 | 0.4717 | 18.29 | 0.720 | 262.6 | 0.4071 | 18.16 | 0.715 | 259.0 | 0.4015 | 300 |
| 350 | 37 | 15.65 | 0.616 | 20.83 | 0.820 | 340.7 | 0.5281 | 19.56 | 0.770 | 300.4 | 0.4656 | 19.30 | 0.760 | 292.6 | 0.4536 | 350 |
| 400 | 37 | 16.74 | 0.659 | 21.97 | 0.865 | 379.1 | 0.5876 | 20.70 | 0.815 | 336.5 | 0.5216 | 20.32 | 0.800 | 324.3 | 0.5026 | 400 |
| 500 | 37 | 18.69 | 0.736 | 23.88 | 0.940 | 447.7 | 0.6939 | 22.48 | 0.885 | 396.8 | 0.6151 | 22.35 | 0.880 | 392.4 | 0.6082 | 500 |
| 600 | 61 | 20.65 | 0.813 | 26.67 | 1.050 | 558.6 | 0.8659 | 25.02 | 0.985 | 491.6 | 0.7620 | 24.89 | 0.980 | 486.6 | 0.7542 | 600 |
| 700 | 61 | 22.28 | 0.877 | 28.19 | 1.110 | 624.3 | 0.9676 | 26.67 | 1.050 | 558.6 | 0.8659 | 26.67 | 1.050 | 558.6 | 0.8659 | 700 |
| 750 | 61 | 23.06 | 0.908 | 29.21 | 1.150 | 670.1 | 1.0386 | 27.31 | 1.075 | 585.5 | 0.9076 | 27.69 | 1.090 | 602.0 | 0.9331 | 750 |
| 1000 | 61 | 26.92 | 1.060 | 32.64 | 1.285 | 836.6 | 1.2968 | 31.88 | 1.255 | 798.1 | 1.2370 | 31.24 | 1.230 | 766.6 | 1.1882 | 1000 |

*Dimensions are from industry sources.

## Table 8  Conductor Properties

| Size (AWG or kcmil) | Area mm² | Area Circular mils | Stranding Quantity | Stranding Diameter mm | Stranding Diameter in. | Overall Diameter mm | Overall Diameter in. | Overall Area mm² | Overall Area in.² | Copper Uncoated ohm/km | Copper Uncoated ohm/kFT | Copper Coated ohm/km | Copper Coated ohm/kFT | Aluminum ohm/km | Aluminum ohm/kFT |
|---|---|---|---|---|---|---|---|---|---|---|---|---|---|---|---|
| 18 | 0.823 | 1620 | 1 | — | — | 1.02 | 0.040 | 0.823 | 0.001 | 25.5 | 7.77 | 26.5 | 8.08 | 42.0 | 12.8 |
| 18 | 0.823 | 1620 | 7 | 0.39 | 0.015 | 1.16 | 0.046 | 1.06 | 0.002 | 26.1 | 7.95 | 27.7 | 8.45 | 42.8 | 13.1 |
| 16 | 1.31 | 2580 | 1 | — | — | 1.29 | 0.051 | 1.31 | 0.002 | 16.0 | 4.89 | 16.7 | 5.08 | 26.4 | 8.05 |
| 16 | 1.31 | 2580 | 7 | 0.49 | 0.019 | 1.46 | 0.058 | 1.68 | 0.003 | 16.4 | 4.99 | 17.3 | 5.29 | 26.9 | 8.21 |
| 14 | 2.08 | 4110 | 1 | — | — | 1.63 | 0.064 | 2.08 | 0.003 | 10.1 | 3.07 | 10.4 | 3.19 | 16.6 | 5.06 |
| 14 | 2.08 | 4110 | 7 | 0.62 | 0.024 | 1.85 | 0.073 | 2.68 | 0.004 | 10.3 | 3.14 | 10.7 | 3.26 | 16.9 | 5.17 |
| 12 | 3.31 | 6530 | 1 | — | — | 2.05 | 0.081 | 3.31 | 0.005 | 6.34 | 1.93 | 6.57 | 2.01 | 10.45 | 3.18 |
| 12 | 3.31 | 6530 | 7 | 0.78 | 0.030 | 2.32 | 0.092 | 4.25 | 0.006 | 6.50 | 1.98 | 6.73 | 2.05 | 10.69 | 3.25 |
| 10 | 5.261 | 10380 | 1 | — | — | 2.588 | 0.102 | 5.26 | 0.008 | 3.984 | 1.21 | 4.148 | 1.26 | 6.561 | 2.00 |
| 10 | 5.261 | 10380 | 7 | 0.98 | 0.038 | 2.95 | 0.116 | 6.76 | 0.011 | 4.070 | 1.24 | 4.226 | 1.29 | 6.679 | 2.04 |
| 8 | 8.367 | 16510 | 1 | — | — | 3.264 | 0.128 | 8.37 | 0.013 | 2.506 | 0.764 | 2.579 | 0.786 | 4.125 | 1.26 |
| 8 | 8.367 | 16510 | 7 | 1.23 | 0.049 | 3.71 | 0.146 | 10.76 | 0.017 | 2.551 | 0.778 | 2.653 | 0.809 | 4.204 | 1.28 |
| 6 | 13.30 | 26240 | 7 | 1.56 | 0.061 | 4.67 | 0.184 | 17.09 | 0.027 | 1.608 | 0.491 | 1.671 | 0.510 | 2.652 | 0.808 |
| 4 | 21.15 | 41740 | 7 | 1.96 | 0.077 | 5.89 | 0.232 | 27.19 | 0.042 | 1.010 | 0.308 | 1.053 | 0.321 | 1.666 | 0.508 |
| 3 | 26.67 | 52620 | 7 | 2.20 | 0.087 | 6.60 | 0.260 | 34.28 | 0.053 | 0.802 | 0.245 | 0.833 | 0.254 | 1.320 | 0.403 |
| 2 | 33.62 | 66360 | 7 | 2.47 | 0.097 | 7.42 | 0.292 | 43.23 | 0.067 | 0.634 | 0.194 | 0.661 | 0.201 | 1.045 | 0.319 |
| 1 | 42.41 | 83690 | 19 | 1.69 | 0.066 | 8.43 | 0.332 | 55.80 | 0.087 | 0.505 | 0.154 | 0.524 | 0.160 | 0.829 | 0.253 |
| 1/0 | 53.49 | 105600 | 19 | 1.89 | 0.074 | 9.45 | 0.372 | 70.41 | 0.109 | 0.399 | 0.122 | 0.415 | 0.127 | 0.660 | 0.201 |
| 2/0 | 67.43 | 133100 | 19 | 2.13 | 0.084 | 10.62 | 0.418 | 88.74 | 0.137 | 0.3170 | 0.0967 | 0.329 | 0.101 | 0.523 | 0.159 |
| 3/0 | 85.01 | 167800 | 19 | 2.39 | 0.094 | 11.94 | 0.470 | 111.9 | 0.173 | 0.2512 | 0.0766 | 0.2610 | 0.0797 | 0.413 | 0.126 |
| 4/0 | 107.2 | 211600 | 19 | 2.68 | 0.106 | 13.41 | 0.528 | 141.1 | 0.219 | 0.1996 | 0.0608 | 0.2050 | 0.0626 | 0.328 | 0.100 |
| 250 | | — | 37 | 2.09 | 0.082 | 14.61 | 0.575 | 168 | 0.260 | 0.1687 | 0.0515 | 0.1753 | 0.0535 | 0.2778 | 0.0847 |
| 300 | | — | 37 | 2.29 | 0.090 | 16.00 | 0.630 | 201 | 0.312 | 0.1409 | 0.0429 | 0.1463 | 0.0446 | 0.2318 | 0.0707 |
| 350 | | — | 37 | 2.47 | 0.097 | 17.30 | 0.681 | 235 | 0.364 | 0.1205 | 0.0367 | 0.1252 | 0.0382 | 0.1984 | 0.0605 |
| 400 | | — | 37 | 2.64 | 0.104 | 18.49 | 0.728 | 268 | 0.416 | 0.1053 | 0.0321 | 0.1084 | 0.0331 | 0.1737 | 0.0529 |
| 500 | | — | 37 | 2.95 | 0.116 | 20.65 | 0.813 | 336 | 0.519 | 0.0845 | 0.0258 | 0.0869 | 0.0265 | 0.1391 | 0.0424 |
| 600 | | — | 61 | 2.52 | 0.099 | 22.68 | 0.893 | 404 | 0.626 | 0.0704 | 0.0214 | 0.0732 | 0.0223 | 0.1159 | 0.0353 |
| 700 | | — | 61 | 2.72 | 0.107 | 24.49 | 0.964 | 471 | 0.730 | 0.0603 | 0.0184 | 0.0622 | 0.0189 | 0.0994 | 0.0303 |
| 750 | | — | 61 | 2.82 | 0.111 | 25.35 | 0.998 | 505 | 0.782 | 0.0563 | 0.0171 | 0.0579 | 0.0176 | 0.0927 | 0.0282 |
| 800 | | — | 61 | 2.91 | 0.114 | 26.16 | 1.030 | 538 | 0.834 | 0.0528 | 0.0161 | 0.0544 | 0.0166 | 0.0868 | 0.0265 |
| 900 | | — | 61 | 3.09 | 0.122 | 27.79 | 1.094 | 606 | 0.940 | 0.0470 | 0.0143 | 0.0481 | 0.0147 | 0.0770 | 0.0235 |
| 1000 | | — | 61 | 3.25 | 0.128 | 29.26 | 1.152 | 673 | 1.042 | 0.0423 | 0.0129 | 0.0434 | 0.0132 | 0.0695 | 0.0212 |
| 1250 | | — | 91 | 2.98 | 0.117 | 32.74 | 1.289 | 842 | 1.305 | 0.0338 | 0.0103 | 0.0347 | 0.0106 | 0.0554 | 0.0169 |
| 1500 | | — | 91 | 3.26 | 0.128 | 35.86 | 1.412 | 1011 | 1.566 | 0.02814 | 0.00858 | 0.02814 | 0.00883 | 0.0464 | 0.0141 |
| 1750 | | — | 127 | 2.98 | 0.117 | 38.76 | 1.526 | 1180 | 1.829 | 0.02410 | 0.00735 | 0.02410 | 0.00756 | 0.0397 | 0.0121 |
| 2000 | | — | 127 | 3.19 | 0.126 | 41.45 | 1.632 | 1349 | 2.092 | 0.02109 | 0.00643 | 0.02109 | 0.00662 | 0.0348 | 0.0106 |

Notes:

1. These resistance values are valid **only** for the parameters as given. Using conductors having coated strands, different stranding type, and, especially, other temperatures changes the resistance.

2. Formula for temperature change: $R_2 = R_1 [1 + \alpha (T_2 - 75)]$ where $\alpha_{cu} = 0.00323$, $\alpha_{AL} = 0.00330$ at 75°C.

3. Conductors with compact and compressed stranding have about 9 percent and 3 percent, respectively, smaller bare conductor diameters than those shown. See Table 5A for actual compact cable dimensions.

4. The IACS conductivities used: bare copper = 100%, aluminum = 61%.

5. Class B stranding is listed as well as solid for some sizes. Its overall diameter and area is that of its circumscribing circle.

FPN:  The construction information is per NEMA WC8-1992 or ANSI/UL 1581-1998. The resistance is calculated per National Bureau of Standards Handbook 100, dated 1966, and Handbook 109, dated 1972.

## Table 9 Alternating-Current Resistance and Reactance for 600-Volt Cables, 3-Phase, 60 Hz, 75°C (167°F) — Three Single Conductors in Conduit

| | | | | | | Ohms to Neutral per Kilometer / Ohms to Neutral per 1000 Feet | | | | | | | | | | |
|---|---|---|---|---|---|---|---|---|---|---|---|---|---|---|---|---|
| | $X_L$ (Reactance) for All Wires | | Alternating-Current Resistance for Uncoated Copper Wires | | | Alternating-Current Resistance for Aluminum Wires | | | Effective $Z$ at 0.85 $PF$ for Uncoated Copper Wires | | | Effective $Z$ at 0.85 $PF$ for Aluminum Wires | | | |
| Size (AWG or kcmil) | PVC, Aluminum Conduits | Steel Conduit | PVC Conduit | Aluminum Conduit | Steel Conduit | PVC Conduit | Aluminum Conduit | Steel Conduit | PVC Conduit | Aluminum Conduit | Steel Conduit | PVC Conduit | Aluminum Conduit | Steel Conduit | Size (AWG or kcmil) |
| 14 | 0.190 / 0.058 | 0.240 / 0.073 | 10.2 / 3.1 | 10.2 / 3.1 | 10.2 / 3.1 | — / — | — / — | — / — | 8.9 / 2.7 | 8.9 / 2.7 | 8.9 / 2.7 | — / — | — / — | — / — | 14 |
| 12 | 0.177 / 0.054 | 0.223 / 0.068 | 6.6 / 2.0 | 6.6 / 2.0 | 6.6 / 2.0 | 10.5 / 3.2 | 10.5 / 3.2 | 10.5 / 3.2 | 5.6 / 1.7 | 5.6 / 1.7 | 5.6 / 1.7 | 9.2 / 2.8 | 9.2 / 2.8 | 9.2 / 2.8 | 12 |
| 10 | 0.164 / 0.050 | 0.207 / 0.063 | 3.9 / 1.2 | 3.9 / 1.2 | 3.9 / 1.2 | 6.6 / 2.0 | 6.6 / 2.0 | 6.6 / 2.0 | 3.6 / 1.1 | 3.6 / 1.1 | 3.6 / 1.1 | 5.9 / 1.8 | 5.9 / 1.8 | 5.9 / 1.8 | 10 |
| 8 | 0.171 / 0.052 | 0.213 / 0.065 | 2.56 / 0.78 | 2.56 / 0.78 | 2.56 / 0.78 | 4.3 / 1.3 | 4.3 / 1.3 | 4.3 / 1.3 | 2.26 / 0.69 | 2.26 / 0.69 | 2.30 / 0.70 | 3.6 / 1.1 | 3.6 / 1.1 | 3.6 / 1.1 | 8 |
| 6 | 0.167 / 0.051 | 0.210 / 0.064 | 1.61 / 0.49 | 1.61 / 0.49 | 1.61 / 0.49 | 2.66 / 0.81 | 2.66 / 0.81 | 2.66 / 0.81 | 1.44 / 0.44 | 1.48 / 0.45 | 1.48 / 0.45 | 2.33 / 0.71 | 2.36 / 0.72 | 2.36 / 0.72 | 6 |
| 4 | 0.157 / 0.048 | 0.197 / 0.060 | 1.02 / 0.31 | 1.02 / 0.31 | 1.02 / 0.31 | 1.67 / 0.51 | 1.67 / 0.51 | 1.67 / 0.51 | 0.95 / 0.29 | 0.95 / 0.29 | 0.98 / 0.30 | 1.51 / 0.46 | 1.51 / 0.46 | 1.51 / 0.46 | 4 |
| 3 | 0.154 / 0.047 | 0.194 / 0.059 | 0.82 / 0.25 | 0.82 / 0.25 | 0.82 / 0.25 | 1.31 / 0.40 | 1.35 / 0.41 | 1.31 / 0.40 | 0.75 / 0.23 | 0.79 / 0.24 | 0.79 / 0.24 | 1.21 / 0.37 | 1.21 / 0.37 | 1.21 / 0.37 | 3 |
| 2 | 0.148 / 0.045 | 0.187 / 0.057 | 0.62 / 0.19 | 0.66 / 0.20 | 0.66 / 0.20 | 1.05 / 0.32 | 1.05 / 0.32 | 1.05 / 0.32 | 0.62 / 0.19 | 0.62 / 0.19 | 0.66 / 0.20 | 0.98 / 0.30 | 0.98 / 0.30 | 0.98 / 0.30 | 2 |
| 1 | 0.151 / 0.046 | 0.187 / 0.057 | 0.49 / 0.15 | 0.52 / 0.16 | 0.52 / 0.16 | 0.82 / 0.25 | 0.85 / 0.26 | 0.82 / 0.25 | 0.52 / 0.16 | 0.52 / 0.16 | 0.52 / 0.16 | 0.79 / 0.24 | 0.79 / 0.24 | 0.82 / 0.25 | 1 |
| 1/0 | 0.144 / 0.044 | 0.180 / 0.055 | 0.39 / 0.12 | 0.43 / 0.13 | 0.39 / 0.12 | 0.66 / 0.20 | 0.69 / 0.21 | 0.66 / 0.20 | 0.43 / 0.13 | 0.43 / 0.13 | 0.43 / 0.13 | 0.62 / 0.19 | 0.66 / 0.20 | 0.66 / 0.20 | 1/0 |
| 2/0 | 0.141 / 0.043 | 0.177 / 0.054 | 0.33 / 0.10 | 0.33 / 0.10 | 0.33 / 0.10 | 0.52 / 0.16 | 0.52 / 0.16 | 0.52 / 0.16 | 0.36 / 0.11 | 0.36 / 0.11 | 0.36 / 0.11 | 0.52 / 0.16 | 0.52 / 0.16 | 0.52 / 0.16 | 2/0 |
| 3/0 | 0.138 / 0.042 | 0.171 / 0.052 | 0.253 / 0.077 | 0.269 / 0.082 | 0.259 / 0.079 | 0.43 / 0.13 | 0.43 / 0.13 | 0.43 / 0.13 | 0.289 / 0.088 | 0.302 / 0.092 | 0.308 / 0.094 | 0.43 / 0.13 | 0.43 / 0.13 | 0.46 / 0.14 | 3/0 |
| 4/0 | 0.135 / 0.041 | 0.167 / 0.051 | 0.203 / 0.062 | 0.220 / 0.067 | 0.207 / 0.063 | 0.33 / 0.10 | 0.36 / 0.11 | 0.33 / 0.10 | 0.243 / 0.074 | 0.256 / 0.078 | 0.262 / 0.080 | 0.36 / 0.11 | 0.36 / 0.11 | 0.36 / 0.11 | 4/0 |
| 250 | 0.135 / 0.041 | 0.171 / 0.052 | 0.171 / 0.052 | 0.187 / 0.057 | 0.177 / 0.054 | 0.279 / 0.085 | 0.295 / 0.090 | 0.282 / 0.086 | 0.217 / 0.066 | 0.230 / 0.070 | 0.240 / 0.073 | 0.308 / 0.094 | 0.322 / 0.098 | 0.33 / 0.10 | 250 |
| 300 | 0.135 / 0.041 | 0.167 / 0.051 | 0.144 / 0.044 | 0.161 / 0.049 | 0.148 / 0.045 | 0.233 / 0.071 | 0.249 / 0.076 | 0.236 / 0.072 | 0.194 / 0.059 | 0.207 / 0.063 | 0.213 / 0.065 | 0.269 / 0.082 | 0.282 / 0.086 | 0.289 / 0.088 | 300 |
| 350 | 0.131 / 0.040 | 0.164 / 0.050 | 0.125 / 0.038 | 0.141 / 0.043 | 0.128 / 0.039 | 0.200 / 0.061 | 0.217 / 0.066 | 0.207 / 0.063 | 0.174 / 0.053 | 0.190 / 0.058 | 0.197 / 0.060 | 0.240 / 0.073 | 0.253 / 0.077 | 0.262 / 0.080 | 350 |
| 400 | 0.131 / 0.040 | 0.161 / 0.049 | 0.108 / 0.033 | 0.125 / 0.038 | 0.115 / 0.035 | 0.177 / 0.054 | 0.194 / 0.059 | 0.180 / 0.055 | 0.161 / 0.049 | 0.174 / 0.053 | 0.184 / 0.056 | 0.217 / 0.066 | 0.233 / 0.071 | 0.240 / 0.073 | 400 |
| 500 | 0.128 / 0.039 | 0.157 / 0.048 | 0.089 / 0.027 | 0.105 / 0.032 | 0.095 / 0.029 | 0.141 / 0.043 | 0.157 / 0.048 | 0.148 / 0.045 | 0.141 / 0.043 | 0.157 / 0.048 | 0.164 / 0.050 | 0.187 / 0.057 | 0.200 / 0.061 | 0.210 / 0.064 | 500 |
| 600 | 0.128 / 0.039 | 0.157 / 0.048 | 0.075 / 0.023 | 0.092 / 0.028 | 0.082 / 0.025 | 0.118 / 0.036 | 0.135 / 0.041 | 0.125 / 0.038 | 0.131 / 0.040 | 0.144 / 0.044 | 0.154 / 0.047 | 0.167 / 0.051 | 0.180 / 0.055 | 0.190 / 0.058 | 600 |

Table 9    *Continued*

| | Ohms to Neutral per Kilometer | | | | | | | | | | | | | | | |
| | Ohms to Neutral per 1000 Feet | | | | | | | | | | | | | | | |
| Size (AWG or kcmil) | $X_L$ (Reactance) for All Wires | | Alternating-Current Resistance for Uncoated Copper Wires | | | Alternating-Current Resistance for Aluminum Wires | | | Effective $Z$ at 0.85 *PF* for Uncoated Copper Wires | | | Effective $Z$ at 0.85 *PF* for Aluminum Wires | | | Size (AWG or kcmil) |
| | PVC, Aluminum Conduits | Steel Conduit | PVC Conduit | Aluminum Conduit | Steel Conduit | PVC Conduit | Aluminum Conduit | Steel Conduit | PVC Conduit | Aluminum Conduit | Steel Conduit | PVC Conduit | Aluminum Conduit | Steel Conduit | |
| 750 | 0.125 0.038 | 0.157 0.048 | 0.062 0.019 | 0.079 0.024 | 0.069 0.021 | 0.095 0.029 | 0.112 0.034 | 0.102 0.031 | 0.118 0.036 | 0.131 0.040 | 0.141 0.043 | 0.148 0.045 | 0.161 0.049 | 0.171 0.052 | 750 |
| 1000 | 0.121 0.037 | 0.151 0.046 | 0.049 0.015 | 0.062 0.019 | 0.059 0.018 | 0.075 0.023 | 0.089 0.027 | 0.082 0.025 | 0.105 0.032 | 0.118 0.036 | 0.131 0.040 | 0.128 0.039 | 0.138 0.042 | 0.151 0.046 | 1000 |

Notes:
1. These values are based on the following constants: UL-Type RHH wires with Class B stranding, in cradled configuration. Wire conductivities are 100 percent IACS copper and 61 percent IACS aluminum, and aluminum conduit is 45 percent IACS. Capacitive reactance is ignored, since it is negligible at these voltages. These resistance values are valid only at 75°C (167°F) and for the parameters as given, but are representative for 600-volt wire types operating at 60 Hz.
2. *Effective Z* is defined as $R \cos(\theta) + X \sin(\theta)$, where $\theta$ is the power factor angle of the circuit. Multiplying current by effective impedance gives a good approximation for line-to-neutral voltage drop. Effective impedance values shown in this table are valid only at 0.85 power factor. For another circuit power factor (*PF*), effective impedance (*Ze*) can be calculated from $R$ and $X_L$ values given in this table as follows: $Ze = R \times PF + X_L \sin[\arccos(PF)]$.

### Tables 11(A) and 11(B)

For listing purposes, Table 11(A) and Table 11(B) provide the required power source limitations for Class 2 and Class 3 power sources. Table 11(A) applies for alternating-current sources, and Table 11(B) applies for direct-current sources.

The power for Class 2 and Class 3 circuits shall be either (1) inherently limited, requiring no overcurrent protection, or (2) not inherently limited, requiring a combination of power source and overcurrent protection. Power sources designed for interconnection shall be listed for the purpose.

As part of the listing, the Class 2 or Class 3 power source shall be durably marked where plainly visible to indicate the class of supply and its electrical rating. A Class 2 power source not suitable for wet location use shall be so marked.

*Exception:   Limited power circuits used by listed information technology equipment.*

Overcurrent devices, where required, shall be located at the point where the conductor to be protected receives its supply and shall not be interchangeable with devices of higher ratings. The overcurrent device shall be permitted as an integral part of the power source.

**Table 11(A) Class 2 and Class 3 Alternating-Current Power Source Limitations**

| Power Source | | Inherently Limited Power Source (Overcurrent Protection Not Required) | | | | Not Inherently Limited Power Source (Overcurrent Protection Required) | | | |
|---|---|---|---|---|---|---|---|---|---|
| | | Class 2 | | | Class 3 | Class 2 | | Class 3 | |
| Source voltage $V_{max}$ (volts) (see Note 1) | | 0 through 20* | Over 20 and through 30* | Over 30 and through 150 | Over 30 and through 100 | 0 through 20* | Over 20 and through 30* | Over 30 and through 100 | Over 100 and through 150 |
| Power limitations $VA_{max}$ (volt-amperes) (see Note 1) | | — | — | — | — | 250 (see Note 3) | 250 | 250 | N.A. |
| Current limitations $I_{max}$ (amperes) (see Note 1) | | 8.0 | 8.0 | 0.005 | $150/V_{max}$ | $1000/V_{max}$ | $1000/V_{max}$ | $1000/V_{max}$ | 1.0 |
| Maximum overcurrent protection (amperes) | | — | — | — | — | 5.0 | $100/V_{max}$ | $100/V_{max}$ | 1.0 |
| Power source maximum nameplate rating | VA (volt-amperes) | $5.0 \times V_{max}$ | 100 | $0.005 \times V_{max}$ | 100 | $5.0 \times V_{max}$ | 100 | 100 | 100 |
| | Current (amperes) | 5.0 | $100/V_{max}$ | 0.005 | $100/V_{max}$ | 5.0 | $100/V_{max}$ | $100/V_{max}$ | $100/V_{max}$ |

*Voltage ranges shown are for sinusoidal ac in indoor locations or where wet contact is not likely to occur. For nonsinusoidal or wet contact conditions, see Note 2.

**Table 11(B) Class 2 and Class 3 Direct-Current Power Source Limitations**

| Power Source | | Inherently Limited Power Source (Overcurrent Protection Not Required) | | | | | Not Inherently Limited Power Source (Overcurrent Protection Required) | | | |
|---|---|---|---|---|---|---|---|---|---|---|
| | | Class 2 | | | | Class 3 | Class 2 | | Class 3 | |
| Source voltage $V_{max}$ (volts) (see Note 1) | | 0 through 20* | Over 20 and through 30* | Over 30 and through 60* | Over 60 and through 150 | Over 60 and through 100 | 0 through 20* | Over 20 and through 60* | Over 60 and through 100 | Over 100 and through 150 |
| Power limitations $VA_{max}$ (volt-amperes) (see Note 1) | | — | — | — | — | — | 250 (see Note 3) | 250 | 250 | N.A. |
| Current limitations $I_{max}$ (amperes) (see Note 1) | | 8.0 | 8.0 | $150/V_{max}$ | 0.005 | $150/V_{max}$ | $1000/V_{max}$ | $1000/V_{max}$ | $1000/V_{max}$ | 1.0 |
| Maximum overcurrent protection (amperes) | | — | — | — | — | — | 5.0 | $100/V_{max}$ | $100/V_{max}$ | 1.0 |
| Power source maximum nameplate rating | VA (volt-amperes) | $5.0 \times V_{max}$ | 100 | 100 | $0.005 \times V_{max}$ | 100 | $5.0 \times V_{max}$ | 100 | 100 | 100 |
| | Current (amperes) | 5.0 | $100/V_{max}$ | $100/V_{max}$ | 0.005 | $100/V_{max}$ | 5.0 | $100/V_{max}$ | $100/V_{max}$ | $100/V_{max}$ |

*Voltage ranges shown are for continuous dc in indoor locations or where wet contact is not likely to occur. For interrupted dc or wet contact conditions, see Note 4.

**Notes for Tables 11(A) and 11(B)**

1. $V_{max}$, $I_{max}$, and $VA_{max}$ are determined with the current-limiting impedance in the circuit (not bypassed) as follows:

$V_{max}$: Maximum output voltage regardless of load with rated input applied.

$I_{max}$: Maximum output current under any noncapacitive load, including short circuit, and with overcurrent protection bypassed if used. Where a transformer limits the output current, $I_{max}$ limits apply after 1 minute of operation. Where a current-limiting impedance, listed for the purpose, or as part of a listed product, is used in combination with a nonpower-limited transformer or a stored energy source, e.g., storage battery, to limit the output current, $I_{max}$ limits apply after 5 seconds.

$VA_{max}$: Maximum volt-ampere output after 1 minute of operation regardless of load and overcurrent protection bypassed if used.

2. For nonsinusoidal ac, $V_{max}$ shall not be greater than 42.4 volts peak. Where wet contact (immersion not included) is likely to occur, Class 3 wiring methods shall be used or $V_{max}$ shall not be greater than 15 volts for sinusoidal ac and 21.2 volts peak for nonsinusoidal ac.

3. If the power source is a transformer, $VA_{max}$ is 350 or less when $V_{max}$ is 15 or less.

4. For dc interrupted at a rate of 10 to 200 Hz, $V_{max}$ shall not be greater than 24.8 volts peak. Where wet contact (immersion not included) is likely to occur, Class 3 wiring methods shall be used, or $V_{max}$ shall not be greater than 30 volts for continuous dc; 12.4 volts peak for dc that is interrupted at a rate of 10 to 200 Hz.

### Tables 12(A) and 12(B)

For listing purposes, Tables 12(A) and 12(B) provide the required power source limitations for power-limited fire alarm sources. Table 12(A) applies for alternating-current sources, and Table 12(B) applies for direct-current sources.

The power for power-limited fire alarm circuits shall be either (1) inherently limited, requiring no overcurrent protection, or (2) not inherently limited, requiring the power to be limited by a combination of power source and overcurrent protection.

As part of the listing, the PLFA power source shall be durably marked where plainly visible to indicate that it is a power-limited fire alarm power source. The overcurrent device, where required, shall be located at the point where the conductor to be protected receives its supply and shall not be interchangeable with devices of higher ratings. The overcurrent device shall be permitted as an integral part of the power source.

#### Table 12(A)  PLFA Alternating-Current Power Source Limitations

| Power Source | | Inherently Limited Power Source (Overcurrent Protection Not Required) | | | Not Inherently Limited Power Source (Overcurrent Protection Required) | | |
|---|---|---|---|---|---|---|---|
| Circuit voltage $V_{max}$ (volts) (see Note 1) | | 0 through 20 | Over 20 and through 30 | Over 30 and through 100 | 0 through 20 | Over 20 and through 100 | Over 100 and through 150 |
| Power limitations $VA_{max}$ (volt-amperes) (see Note 1) | | — | — | — | 250 (see Note 2) | 250 | N.A. |
| Current limitations $I_{max}$ (amperes) (see Note 1) | | 8.0 | 8.0 | $150/V_{max}$ | $1000/V_{max}$ | $1000/V_{max}$ | 1.0 |
| Maximum overcurrent protection (amperes) | | — | — | — | 5.0 | $100/V_{max}$ | 1.0 |
| Power source maximum nameplate ratings | VA (volt-amperes) | $5.0 \times V_{max}$ | 100 | 100 | $5.0 \times V_{max}$ | 100 | 100 |
| | Current (amperes) | 5.0 | $100/V_{max}$ | $100/V_{max}$ | 5.0 | $100/V_{max}$ | $100/V_{max}$ |

#### Table 12(B)  PLFA Direct-Current Power Source Limitations

| Power Source | | Inherently Limited Power Source (Overcurrent Protection Not Required) | | | | Not Inherently Limited Power Source (Overcurrent Protection Required) | | |
|---|---|---|---|---|---|---|---|---|
| Circuit voltage $V_{max}$ (volts) (see Note 1) | | 0 through 20 | Over 20 and through 30 | Over 30 and through 100 | Over 100 and through 250 | 0 through 20 | Over 20 and through 100 | Over 100 and through 150 |
| Power limitations $VA_{max}$ (volt-amperes) (see Note 1) | | — | — | — | — | 250 (see Note 2) | 250 | N.A. |
| Current limitations $I_{max}$ (amperes) (see Note 1) | | 8.0 | 8.0 | $150/V_{max}$ | 0.030 | $1000/V_{max}$ | $1000/V_{max}$ | 1.0 |
| Maximum overcurrent protection (amperes) | | — | — | — | — | 5.0 | $100/V_{max}$ | 1.0 |
| Power source maximum nameplate ratings | VA (volt-amperes) | $5.0 \times V_{max}$ | 100 | 100 | $0.030 \times V_{max}$ | $5.0 \times V_{max}$ | 100 | 100 |
| | Current (amperes) | 5.0 | $100/V_{max}$ | $100/V_{max}$ | 0.030 | 5.0 | $100/V_{max}$ | $100/V_{max}$ |

#### Notes for Tables 12(A) and 12(B)

1. $V_{max}$, $I_{max}$, and $VA_{max}$ are determined as follows:

$V_{max}$: Maximum output voltage regardless of load with rated input applied.

$I_{max}$: Maximum output current under any noncapacitive load, including short circuit, and with overcurrent protection bypassed if used. Where a transformer limits the output current, $I_{max}$ limits apply after 1 minute of operation. Where a current-limiting impedance, listed for the purpose, is used in combination with a nonpower-limited transformer or a stored energy source, e.g., storage battery, to limit the output current, $I_{max}$ limits apply after 5 seconds.

$VA_{max}$: Maximum volt-ampere output after 1 minute of operation regardless of load and overcurrent protection bypassed if used. Current limiting impedance shall not be bypassed when determining $I_{max}$ and $VA_{max}$.

2. If the power source is a transformer, $VA_{max}$ is 350 or less when $V_{max}$ is 15 or less.

# Annex A Product Safety Standards

*Annex A is not a part of the requirements of this NFPA document but is included for informational purposes only.*

This informational annex provides a list of product safety standards used for product listing where that listing is required by this *Code*. It is recognized that this list is current at the time of publication but that new standards or modifications to existing standards can occur at any time while this edition of the *Code* is in effect.

This annex does not form a mandatory part of the requirements of this *Code* but is intended only to provide *Code* users with informational guidance about the product characteristics about which *Code* requirements have been based.

| Product Standard Name | Product Standard Number |
|---|---|
| Antenna-Discharge Units | UL 452 |
| Armored Cable | UL 4 |
| Attachment Plugs and Receptacles | UL 498 |
| Audio/Video and Musical Instrument Apparatus for Household, Commercial, and Similar General Use | UL 6500 |
| Audio-Video Products and Accessories | UL 1492 |
| Busways and Associated Fittings | UL 857 |
| Cables — Thermoplastic-Insulated Underground Feeder and Branch-Circuit Cables | UL 493 |
| Cables — Thermoplastic-Insulated Wires and Cables | UL 83 |
| Cables — Thermoset-Insulated Wires and Cables | UL 44 |
| Cables for Non–Power-Limited Fire-Alarm Circuits | UL 1425 |
| Cables for Power-Limited Fire-Alarm Circuits | UL 1424 |
| Cellular Metal Floor Raceways and Fittings | UL 209 |
| Class 2 Power Units | UL 1310 |
| Commercial Audio Equipment | UL 813 |
| Communication Circuit Accessories | UL 1863 |
| Communications Cables | UL 444 |
| Community-Antenna Television Cables | UL 1655 |
| Conduit — Type EB and A Rigid PVC Conduit and HDPE Conduit | UL 651A |
| Continuous Length High Density Polyethylene Conduit | UL 651B |
| Control Centers for Changing Message Type Electric Signs | UL 1433 |
| Cord Sets and Power-Supply Cords | UL 817 |
| Data-Processing Cable | UL 1690 |
| Dead-Front Switchboards | UL 891 |
| Electric Signs | UL 48 |
| Electric Spas, Equipment Assemblies, and Associated Equipment | UL 1563 |

| Product Standard Name | Product Standard Number |
|---|---|
| Electric Water Heaters for Pools and Tubs | UL 1261 |
| Electrical Apparatus for Use in Class I, Zone 1 Hazardous (Classified) Locations Type of Protection — Encapsulation "m" | ISA S12.23.01 |
| Electrical Apparatus for Use in Class I, Zones O & 1 Hazardous (Classified) Locations: General Requirements | ISA 12.0.01 |
| Electrical Apparatus for Use in Class I, Zone 1 Hazardous (Classified) Locations: Type of Protection — Increased Safety "e" | ISA S12.16.01 |
| Electrical Apparatus for Use in Class I, Zone 1 Hazardous (Classified) Locations: Type of Protection — Flameproof "d" | ISA S12.22.01 |
| Electrical Apparatus for Use in Class I, Zone 1 Hazardous (Classified) Locations: Type of Protection — Powder Filling "q" | ISA S12.25.01 |
| Electrical Apparatus for Use in Class I, Zone 1 Hazardous (Classified) Locations: Type of Protection — Oil-Immersion "O" | ISA S12.26.01 |
| Electrical Equipment for Use in Class I, Zone 0, 1, and 2 Hazardous (Classified) Locations | UL 2279 |
| Electrical Metallic Tubing | UL 797 |
| Electrical Nonmetallic Tubing | UL 1653 |
| Electric-Battery-Powered Industrial Trucks | UL 583 |
| Electrode Receptacles for Gas-Tube Signs | UL 879 |
| Enclosed and Dead-Front Switches | UL 98 |
| Enclosures for Electrical Equipment | UL 50 |
| Energy Management Equipment | UL 916 |
| Fire Pump Controllers | UL 218 |
| Fittings for Cable and Conduit | UL 514B |
| Flexible Cord and Fixture Wire | UL 62 |
| Flexible Metal Conduit | UL 1 |
| Fluorescent Lighting Fixtures | UL 1570 |
| Fluorescent-Lamp Ballasts | UL 935 |
| Gas-Tube-Sign and Ignition Cable | UL 814 |

| Product Standard Name | Product Standard Number |
|---|---|
| General-Use Snap Switches | UL 20 |
| Ground-Fault Circuit-Interrupters | UL 943 |
| Ground-Fault Sensing and Relaying Equipment | UL 1053 |
| Grounding and Bonding Equipment | UL 467 |
| High Intensity Discharge Lighting Fixtures | UL 1572 |
| High-Intensity-Discharge Lamp Ballasts | UL 1029 |
| Incandescent Lighting Fixtures | UL 1571 |
| Industrial Battery Chargers | UL 1564 |
| Industrial Control Equipment | UL 508 |
| Instrumentation Tray Cable | UL 2250 |
| Insulated Wire Connector Systems for Underground Use or in Damp or Wet Locations | UL 486D |
| Intermediate Metal Conduit | UL 1242 |
| Isolated Power Systems Equipment | UL 1047 |
| Junction Boxes for Swimming Pool Lighting Fixtures | UL 1241 |
| Liquid-Tight Flexible Nonmetallic Conduit | UL 1660 |
| Liquid-Tight Flexible Steel Conduit | UL 360 |
| Low Voltage Landscape Lighting Systems | UL 1838 |
| Low-Voltage Fuses — Part 1: General Requirements | UL 248-1 |
| Low-Voltage Fuses — Part 2: Class C Fuses | UL 248-2 |
| Low-Voltage Fuses — Part 3: Class CA and CB Fuses | UL 248-3 |
| Low-Voltage Fuses — Part 4: Class CC Fuses | UL 248-4 |
| Low-Voltage Fuses — Part 5: Class G Fuses | UL 248-5 |
| Low-Voltage Fuses — Part 6: Class H Non-Renewable Fuses | UL 248-6 |
| Low-Voltage Fuses — Part 7: Class H Renewable Fuses | UL 248-7 |
| Low-Voltage Fuses — Part 8: Class J Fuses | UL 248-8 |
| Low-Voltage Fuses — Part 9: Class K Fuses | UL 248-9 |
| Low-Voltage Fuses — Part 10: Class L Fuses | UL 249-10 |
| Low-Voltage Fuses — Part 11: Plug Fuses | UL 248-11 |
| Low-Voltage Fuses — Part 12: Class R Fuses | UL 248-12 |
| Low-Voltage Fuses — Part 15: Class T Fuses | UL 248-15 |
| Machine-Tool Wires and Cables | UL 1063 |
| Manufactured Wiring Systems | UL 183 |
| Medical and Dental Equipment | UL 544 |
| Medium-Voltage Power Cables | UL 1072 |
| Metal-Clad Cables | UL 1569 |
| Metal-Clad Cables and Cable-Sealing Fittings for Use in Hazardous (Classified) Locations | UL 2225 |
| Metallic Outlet Boxes | UL 514A |
| Mobile Home Pipe Heating Cable | UL 1462 |

| Product Standard Name | Product Standard Number |
|---|---|
| Molded-Case Circuit Breakers, Molded-Case Switches, and Circuit-Breaker Enclosures | UL 489 |
| Molded-Case Switches | UL 1087 |
| Neon Transformers and Power Supplies | UL 2161 |
| Nonincendive Electrical Equipment for Use in Class I and II, Division 2 and Class III, Divisions 1 and 2 Hazardous (Classified) Locations | ISA S12.12 |
| Nonmetallic Outlet Boxes, Flush-Device Boxes, and Covers | UL 514C |
| Nonmetallic Surface Raceways and Fittings | UL 5A |
| Nonmetallic Underground Conduit with Conductors | UL 1990 |
| Office Furnishings | UL 1286 |
| Optical Fiber Cable | UL 1651 |
| Optical Fiber Cable Raceway | UL 2024 |
| Panelboards | UL 67 |
| Personal Protection Systems for Electric Vehicle Supply Circuits: General Requirements | UL 2231-1 |
| Personal Protection Systems for Electric Vehicle Supply Circuits: Particular Requirements for Protection Devices for Use in Charging Systems | UL 2231-2 |
| Portable Electric Lamps | UL 153 |
| Potting Compounds for Swimming Pool, Fountain, and Spa Equipment | UL 676A |
| Power Outlets | UL 231 |
| Power Units Other Than Class 2 | UL 1012 |
| Power-Limited Circuit Cables | UL 13 |
| Professional Video and Audio Equipment | UL 1419 |
| Protectors for Coaxial Communications Circuits | UL 497C |
| Protectors for Data Communication and Fire Alarm Circuits | UL 497B |
| Protectors for Paired Conductor Communications Circuits | UL 497 |
| Radio Receivers, Audio Systems, and Accessories | UL 1270 |
| Reference Standard for Electrical Wires, Cables, and Flexible Cords | UL 1581 |
| Reinforced Thermosetting Resin Conduit (RTRC) and Fittings | UL 1684 |
| Residential Pipe Heating Cable | UL 2049 |
| Rigid Metal Conduit | UL 6 |
| Roof and Gutter De-Icing Cable Units | UL 1588 |
| Safety of Information Technology Equipment, Including Electrical Business Equipment | UL 1950 |

| Product Standard Name | Product Standard Number |
|---|---|
| Schedule 40 and 80 Rigid PVC Conduit | UL 651 |
| Secondary Protectors for Communications Circuits | UL 497A |
| Service-Entrance Cables | UL 854 |
| Smoke Detectors for Fire Protective Signaling Systems | UL 268 |
| Specialty Transformers | UL 506 |
| Splicing Wire Connectors | UL 486C |
| Static Inverters and Charge Controllers for Use in Photovoltaic Power Systems | UL 1741 |
| Strut-Type Channel Raceways and Fittings | UL 5B |
| Surface Metal Raceways and Fittings | UL 5 |
| Surface Raceways and Fittings for Use with Data, Signal and Control Circuits | UL 5C |
| Surge Arresters — Gapped Silicon-Carbide Surge Arresters for AC Power Circuits | IEEE C62.1 |

| Product Standard Name | Product Standard Number |
|---|---|
| Surge Arresters — Metal-Oxide Surge Arresters for AC Power Circuits | IEEE C62.11 |
| Swimming Pool Pumps, Filters, and Chlorinators | UL 1081 |
| Telephone Equipment | UL 1459 |
| Transfer Switch Equipment | UL 1008 |
| Transient Voltage Surge Suppressors | UL 1449 |
| Underfloor Raceways and Fittings | UL 884 |
| Underwater Lighting Fixtures | UL 676 |
| Vacuum Cleaners, Blower Cleaners, and Household Floor Finishing Machines | UL 1017 |
| Wire Connectors and Soldering Lugs for Use with Copper Conductors | UL 486A |
| Wire Connectors for Use with Aluminum Conductors | UL 486B |
| Wireways, Auxiliary Gutters, and Associated Fittings | UL 870 |

# Annex B   Application Information for Ampacity Calculation

*This annex is not a part of the requirements of this NFPA document but is included for informational purposes only.*

**B.310.15(B)(1) Formula Application Information.** This annex provides application information for ampacities calculated under engineering supervision.

**B.310.15(B)(2) Typical Applications Covered by Tables.** Typical ampacities for conductors rated 0 through 2000 volts are shown in Table B.310.1 through Table B.310.10. Underground electrical duct bank configurations, as detailed in Figure B.310.3, Figure B.310.4, and Figure B.310.5, are utilized for conductors rated 0 through 5000 volts. In Figure B.310.2 through Figure B.310.5, where adjacent duct banks are used, a separation of 1.5 m (5 ft) between the centerlines of the closest ducts in each bank or 1.2 m (4 ft) between the extremities of the concrete envelopes is sufficient to prevent derating of the conductors due to mutual heating. These ampacities were calculated as detailed in the basic ampacity paper, AIEE Paper 57-660, *The Calculation of the Temperature Rise and Load Capability of Cable Systems,* by J. H. Neher and M. H. McGrath. For additional information concerning the application of these ampacities, see IEEE/ICEA Standard S-135/P-46-426, *Power Cable Ampacities,* and IEEE Standard 835-1994, *Standard Power Cable Ampacity Tables.*

Typical values of thermal resistivity (Rho) are as follows:
Average soil (90 percent of USA) = 90
Concrete = 55
Damp soil (coastal areas, high water table) = 60
Paper insulation = 550
Polyethylene (PE) = 450
Polyvinyl chloride (PVC) = 650
Rubber and rubber-like = 500
Very dry soil (rocky or sandy) = 120

*Thermal resistivity,* as used in this annex, refers to the heat transfer capability through a substance by conduction. It is the reciprocal of thermal conductivity and is normally expressed in the units °C-cm/watt. For additional information on determining soil thermal resistivity (Rho), see ANSI/IEEE Standard 442-1996, *Guide for Soil Thermal Resistivity Measurements.*

**B.310.15(B)(3) Criteria Modifications.** Where values of load factor and Rho are known for a particular electrical duct bank installation and they are different from those shown in a specific table or figure, the ampacities shown in the table or figure can be modified by the application of factors derived from the use of Figure B.310.1.

Where two different ampacities apply to adjacent portions of a circuit, the higher ampacity can be used beyond the point of transition, a distance equal to 3 m (10 ft) or 10 percent of the circuit length figured at the higher ampacity, whichever is less.

Where the burial depth of direct burial or electrical duct bank circuits are modified from the values shown in a figure or table, ampacities can be modified as shown in (a) and (b) as follows:

(a)   Where burial depths are increased in part(s) of an electrical duct run to avoid underground obstructions, no decrease in ampacity of the conductors is needed, provided the total length of parts of the duct run increased in depth to avoid obstructions is less than 25 percent of the total run length.

(b)   Where burial depths are deeper than shown in a specific underground ampacity table or figure, an ampacity derating factor of 6 percent per increased 300 mm (foot) of depth for all values of Rho can be utilized. No rating change is needed where the burial depth is decreased.

**B.310.15(B)(4) Electrical Ducts.** The term *electrical duct(s)* is defined in 310.60.

**B.310.15(B)(5) Tables B.310.6 and B.310.7.**

(a)   To obtain the ampacity of cables installed in two electrical ducts in one horizontal row with 190-mm (7.5-in.) center-to-center spacing between electrical ducts, similar to Figure B.310.2, Detail 1, multiply the ampacity shown for one duct in Table B.310.6 and Table B.310.7 by 0.88.

(b)   To obtain the ampacity of cables installed in four electrical ducts in one horizontal row with 190-mm (7.5-in.) center-to-center spacing between electrical ducts, similar to Figure B.310.2, Detail 2, multiply the ampacity shown for three electrical ducts in Table B.310.6 and Table B.310.7 by 0.94.

**B.310.15(B)(6) Electrical Ducts Used in Figure B.310.2.** If spacing between electrical ducts, as shown in Figure B.310.2, is less than specified in Figure B.310.2, where electrical ducts enter equipment enclosures from underground, the ampacity of conductors contained within such electrical ducts need not be reduced.

**B.310.15(B)(7) Examples Showing Use of Figure B.310.1 for Electrical Duct Bank Ampacity Modifications.** Figure B.310.1 is used for interpolation or extrapolation for values of Rho and load factor for cables installed in electrical ducts. The upper family of curves shows the variation in ampacity and Rho at unity load factor in terms of $I_1$, the ampacity for Rho = 60, and 50 percent load factor. Each curve is designated for a particular ratio $I_2/I_1$, where $I_2$ is the ampacity at Rho = 120 and 100 percent load factor.

The lower family of curves shows the relationship between Rho and load factor that will give substantially the same ampacity as the indicated value of Rho at 100 percent load factor.

As an example, to find the ampacity of a 500 kcmil copper cable circuit for six electrical ducts as shown in Table B.310.5: At the Rho = 60, LF = 50, $I_1$ = 583; for Rho = 120 and LF = 100, $I_2$ = 400. The ratio $I_2/I_1$ = 0.686. Locate Rho = 90 at the bottom of the chart and follow the 90 Rho line to the intersection with 100 percent load factor where the equivalent Rho = 90. Then follow the 90 Rho line to $I_2/I_1$ ratio of 0.686 where $F$ = 0.74. The desired ampacity = 0.74 × 583 = 431, which agrees with the table for Rho = 90, LF = 100.

To determine the ampacity for the same circuit where Rho = 80 and LF = 75, using Figure B.310.1, the equivalent Rho = 43, F = 0.855, and the desired ampacity = 0.855 × 583 = 498 amperes.

## Table B.310.1 Ampacities of Two or Three Insulated Conductors, Rated 0 Through 2000 Volts, Within an Overall Covering (Multiconductor Cable), in Raceway in Free Air Based on Ambient Air Temperature of 30°C (86°F)

| Size (AWG or kcmil) | Temperature Rating of Conductor. (See Table 310.13.) | | | | | | Size (AWG or kcmil) |
|---|---|---|---|---|---|---|---|
| | 60°C (140°F) | 75°C (167°F) | 90°C (194°F) | 60°C (140°F) | 75°C (167°F) | 90°C (194°F) | |
| | Types TW, UF | Types RHW, THHW, THW, THWN, XHHW, ZW | Types THHN, THHW, THW-2, THWN-2, RHH, RWH-2, USE-2, XHHW, XHHW-2, ZW-2 | Type TW | Types RHW, THHW, THW, THWN, XHHW | Types THHN, THHW, THW-2, THWN-2, RHH, RWH-2, USE-2, XHHW, XHHW-2, ZW-2 | |
| | COPPER | | | ALUMINUM OR COPPER-CLAD ALUMINUM | | | |
| 14 | 16* | 18* | 21* | | | | 14 |
| 12 | 20* | 24* | 27* | 16* | 18* | 21* | 12 |
| 10 | 27* | 33* | 36* | 21* | 25* | 28* | 10 |
| 8 | 36 | 43 | 48 | 28 | 33 | 37 | 8 |
| 6 | 48 | 58 | 65 | 38 | 45 | 51 | 6 |
| 4 | 66 | 79 | 89 | 51 | 61 | 69 | 4 |
| 3 | 76 | 90 | 102 | 59 | 70 | 79 | 3 |
| 2 | 88 | 105 | 119 | 69 | 83 | 93 | 2 |
| 1 | 102 | 121 | 137 | 80 | 95 | 106 | 1 |
| 1/0 | 121 | 145 | 163 | 94 | 113 | 127 | 1/0 |
| 2/0 | 138 | 166 | 186 | 108 | 129 | 146 | 2/0 |
| 3/0 | 158 | 189 | 214 | 124 | 147 | 167 | 3/0 |
| 4/0 | 187 | 223 | 253 | 147 | 176 | 197 | 4/0 |
| 250 | 205 | 245 | 276 | 160 | 192 | 217 | 250 |
| 300 | 234 | 281 | 317 | 185 | 221 | 250 | 300 |
| 350 | 255 | 305 | 345 | 202 | 242 | 273 | 350 |
| 400 | 274 | 328 | 371 | 218 | 261 | 295 | 400 |
| 500 | 315 | 378 | 427 | 254 | 303 | 342 | 500 |
| 600 | 343 | 413 | 468 | 279 | 335 | 378 | 600 |
| 700 | 376 | 452 | 514 | 310 | 371 | 420 | 700 |
| 750 | 387 | 466 | 529 | 321 | 384 | 435 | 750 |
| 800 | 397 | 479 | 543 | 331 | 397 | 450 | 800 |
| 900 | 415 | 500 | 570 | 350 | 421 | 477 | 900 |
| 1000 | 448 | 542 | 617 | 382 | 460 | 521 | 1000 |

### Correction Factors

| Ambient Temp. (°C) | For ambient temperatures other than 30°C (86°F), multiply the ampacities shown above by the appropriate factor shown below. | | | | | | Ambient Temp. (°F) |
|---|---|---|---|---|---|---|---|
| 21–25 | 1.08 | 1.05 | 1.04 | 1.08 | 1.05 | 1.04 | 70–77 |
| 26–30 | 1.00 | 1.00 | 1.00 | 1.00 | 1.00 | 1.00 | 79–86 |
| 31–35 | 0.91 | 0.94 | 0.96 | 0.91 | 0.94 | 0.96 | 88–95 |
| 36–40 | 0.82 | 0.88 | 0.91 | 0.82 | 0.88 | 0.91 | 97–104 |
| 41–45 | 0.71 | 0.82 | 0.87 | 0.71 | 0.82 | 0.87 | 106–113 |
| 46–50 | 0.58 | 0.75 | 0.82 | 0.58 | 0.75 | 0.82 | 115–122 |
| 51–55 | 0.41 | 0.67 | 0.76 | 0.41 | 0.67 | 0.76 | 124–131 |
| 56–60 | — | 0.58 | 0.71 | — | 0.58 | 0.71 | 133–140 |
| 61–70 | — | 0.33 | 0.58 | — | 0.33 | 0.58 | 142–158 |
| 71–80 | — | — | 0.41 | — | — | 0.41 | 160–176 |

*Unless otherwise specifically permitted elsewhere in this Code, the overcurrent protection for these conductor types shall not exceed 15 amperes for 14 AWG, 20 amperes for 12 AWG, and 30 amperes for 10 AWG copper; or 15 amperes for 12 AWG and 25 amperes for 10 AWG aluminum and copper-clad aluminum.

## Table B.310.3 Ampacities of Multiconductor Cables with Not More Than Three Insulated Conductors, Rated 0 Through 2000 Volts, in Free Air Based on Ambient Air Temperature of 40°C (104°F) (For Types TC, MC, MI, UF, and USE Cables)

| Size (AWG or kcmil) | Temperature Rating of Conductor. (See Table 310.13.) | | | | | | | | Size (AWG or kcmil) |
|---|---|---|---|---|---|---|---|---|---|
| | 60°C (140°F) | 75°C (167°F) | 85°C (185°F) | 90°C (194°F) | 60°C (140°F) | 75°C (167°F) | 85°C (185°F) | 90°C (194°F) | |
| | COPPER | | | | ALUMINUM OR COPPER-CLAD ALUMINUM | | | | |
| 18 | — | — | — | 11* | — | — | — | — | 18 |
| 16 | — | — | — | 16* | — | — | — | — | 16 |
| 14 | 18* | 21* | 24* | 25* | — | — | — | — | 14 |
| 12 | 21* | 28* | 30* | 32* | 18* | 21* | 24* | 25* | 12 |
| 10 | 28* | 36* | 41* | 43* | 21* | 28* | 30* | 32* | 10 |
| 8 | 39 | 50 | 56 | 59 | 30 | 39 | 44 | 46 | 8 |
| 6 | 52 | 68 | 75 | 79 | 41 | 53 | 59 | 61 | 6 |
| 4 | 69 | 89 | 100 | 104 | 54 | 70 | 78 | 81 | 4 |
| 3 | 81 | 104 | 116 | 121 | 63 | 81 | 91 | 95 | 3 |
| 2 | 92 | 118 | 132 | 138 | 72 | 92 | 103 | 108 | 2 |
| 1 | 107 | 138 | 154 | 161 | 84 | 108 | 120 | 126 | 1 |
| 1/0 | 124 | 160 | 178 | 186 | 97 | 125 | 139 | 145 | 1/0 |
| 2/0 | 143 | 184 | 206 | 215 | 111 | 144 | 160 | 168 | 2/0 |
| 3/0 | 165 | 213 | 238 | 249 | 129 | 166 | 185 | 194 | 3/0 |
| 4/0 | 190 | 245 | 274 | 287 | 149 | 192 | 214 | 224 | 4/0 |
| 250 | 212 | 274 | 305 | 320 | 166 | 214 | 239 | 250 | 250 |
| 300 | 237 | 306 | 341 | 357 | 186 | 240 | 268 | 280 | 300 |
| 350 | 261 | 337 | 377 | 394 | 205 | 265 | 296 | 309 | 350 |
| 400 | 281 | 363 | 406 | 425 | 222 | 287 | 317 | 334 | 400 |
| 500 | 321 | 416 | 465 | 487 | 255 | 330 | 368 | 385 | 500 |
| 600 | 354 | 459 | 513 | 538 | 284 | 368 | 410 | 429 | 600 |
| 700 | 387 | 502 | 562 | 589 | 306 | 405 | 462 | 473 | 700 |
| 750 | 404 | 523 | 586 | 615 | 328 | 424 | 473 | 495 | 750 |
| 800 | 415 | 539 | 604 | 633 | 339 | 439 | 490 | 513 | 800 |
| 900 | 438 | 570 | 639 | 670 | 362 | 469 | 514 | 548 | 900 |
| 1000 | 461 | 601 | 674 | 707 | 385 | 499 | 558 | 584 | 1000 |

### Correction Factors

| Ambient Temp. (°C) | For ambient temperatures other than 40°C (104°F), multiply the ampacities shown above by the appropriate factor shown below. | | | | | | | | Ambient Temp. (°F) |
|---|---|---|---|---|---|---|---|---|---|
| 21–25 | 1.32 | 1.20 | 1.15 | 1.14 | 1.32 | 1.20 | 1.15 | 1.14 | 70–77 |
| 26–30 | 1.22 | 1.13 | 1.11 | 1.10 | 1.22 | 1.13 | 1.11 | 1.10 | 79–86 |
| 31–35 | 1.12 | 1.07 | 1.05 | 1.05 | 1.12 | 1.07 | 1.05 | 1.05 | 88–95 |
| 36–40 | 1.00 | 1.00 | 1.00 | 1.00 | 1.00 | 1.00 | 1.00 | 1.00 | 97–104 |
| 41–45 | 0.87 | 0.93 | 0.94 | 0.95 | 0.87 | 0.93 | 0.94 | 0.95 | 106–113 |
| 46–50 | 0.71 | 0.85 | 0.88 | 0.89 | 0.71 | 0.85 | 0.88 | 0.89 | 115–122 |
| 51–55 | 0.50 | 0.76 | 0.82 | 0.84 | 0.50 | 0.76 | 0.82 | 0.84 | 124–131 |
| 56–60 | — | 0.65 | 0.75 | 0.77 | — | 0.65 | 0.75 | 0.77 | 133–140 |
| 61–70 | — | 0.38 | 0.58 | 0.63 | — | 0.38 | 0.58 | 0.63 | 142–158 |
| 71–80 | — | — | 0.33 | 0.44 | — | — | 0.33 | 0.44 | 160–176 |

*Unless otherwise specifically permitted elsewhere in this Code, the overcurrent protection for these conductor types shall not exceed 15 amperes for 14 AWG, 20 amperes for 12 AWG, and 30 amperes for 10 AWG copper; or 15 amperes for 12 AWG and 25 amperes for 10 AWG aluminum and copper-clad aluminum.

**Table B.310.5 Ampacities of Single Insulated Conductors, Rated 0 through 2000 Volts, in Nonmagnetic Underground Electrical Ducts (One Conductor per Electrical Duct), Based on Ambient Earth Temperature of 20°C (68°F), Electrical Duct Arrangement per Figure B.310.2, Conductor Temperature 75°C (167°F)**

| Size (kcmil) | 3 Electrical Ducts (Fig. B.310.2, Detail 2) Types RHW, THHW, THW, THWN, XHHW, USE — COPPER | | | 6 Electrical Ducts (Fig. B.310.2, Detail 3) Types RHW, THHW, THW, THWN, XHHW, USE — COPPER | | | 9 Electrical Ducts (Fig. B.310.2, Detail 4) Types RHW, THHW, THW, THWN, XHHW, USE — COPPER | | | 3 Electrical Ducts (Fig. B.310.2, Detail 2) Types RHW, THHW, THW, THWN, XHHW, USE — ALUMINUM OR COPPER-CLAD ALUMINUM | | | 6 Electrical Ducts (Fig. B.310.2, Detail 3) Types RHW, THHW, THW, THWN, XHHW, USE — ALUMINUM | | | 9 Electrical Ducts (Fig. B.310.2, Detail 4) Types RHW, THHW, THW, THWN, XHHW, USE — ALUMINUM | | | Size (kcmil) |
|---|---|---|---|---|---|---|---|---|---|---|---|---|---|---|---|---|---|---|---|
| | RHO 60 LF 50 | RHO 90 LF 100 | RHO 120 LF 100 | RHO 60 LF 50 | RHO 90 LF 100 | RHO 120 LF 100 | RHO 60 LF 50 | RHO 90 LF 100 | RHO 120 LF 100 | RHO 60 LF 50 | RHO 90 LF 100 | RHO 120 LF 100 | RHO 60 LF 50 | RHO 90 LF 100 | RHO 120 LF 100 | RHO 60 LF 50 | RHO 90 LF 100 | RHO 120 LF 100 | |
| 250 | 410 | 344 | 327 | 386 | 295 | 275 | 369 | 270 | 252 | 320 | 269 | 256 | 302 | 230 | 214 | 288 | 211 | 197 | 250 |
| 350 | 503 | 418 | 396 | 472 | 355 | 330 | 446 | 322 | 299 | 393 | 327 | 310 | 369 | 277 | 258 | 350 | 252 | 235 | 350 |
| 500 | 624 | 511 | 484 | 583 | 431 | 400 | 545 | 387 | 360 | 489 | 401 | 379 | 457 | 337 | 313 | 430 | 305 | 284 | 500 |
| 750 | 794 | 640 | 603 | 736 | 534 | 494 | 674 | 469 | 434 | 626 | 505 | 475 | 581 | 421 | 389 | 538 | 375 | 347 | 750 |
| 1000 | 936 | 745 | 700 | 864 | 617 | 570 | 776 | 533 | 493 | 744 | 593 | 557 | 687 | 491 | 453 | 629 | 432 | 399 | 1000 |
| 1250 | 1055 | 832 | 781 | 970 | 686 | 632 | 854 | 581 | 536 | 848 | 668 | 627 | 779 | 551 | 508 | 703 | 478 | 441 | 1250 |
| 1500 | 1160 | 907 | 849 | 1063 | 744 | 685 | 918 | 619 | 571 | 941 | 736 | 689 | 863 | 604 | 556 | 767 | 517 | 477 | 1500 |
| 1750 | 1250 | 970 | 907 | 1142 | 793 | 729 | 975 | 651 | 599 | 1026 | 796 | 745 | 937 | 651 | 598 | 823 | 550 | 507 | 1750 |
| 2000 | 1332 | 1027 | 959 | 1213 | 836 | 768 | 1030 | 683 | 628 | 1103 | 850 | 794 | 1005 | 693 | 636 | 877 | 581 | 535 | 2000 |

| Ambient Temp. (°C) | Correction Factors | | | | | | Ambient Temp. (°F) |
|---|---|---|---|---|---|---|---|
| 6–10 | 1.09 | 1.09 | 1.09 | 1.09 | 1.09 | 1.09 | 43–50 |
| 11–15 | 1.04 | 1.04 | 1.04 | 1.04 | 1.04 | 1.04 | 52–59 |
| 16–20 | 1.00 | 1.00 | 1.00 | 1.00 | 1.00 | 1.00 | 61–68 |
| 21–25 | 0.95 | 0.95 | 0.95 | 0.95 | 0.95 | 0.95 | 70–77 |
| 26–30 | 0.90 | 0.90 | 0.90 | 0.90 | 0.90 | 0.90 | 79–86 |

**Table B.310.6 Ampacities of Three Insulated Conductors, Rated 0 through 2000 Volts, Within an Overall Covering (Three-Conductor Cable) in Underground Electrical Ducts (One Cable per Electrical Duct) Based on Ambient Earth Temperature of 20°C (68°F), Electrical Duct Arrangement per Figure B.310.2, Conductor Temperature 75°C (167°F)**

| Size (AWG or kcmil) | 1 Electrical Duct (Fig. B.310.2, Detail 1) Types RHW, THHW, THW, THWN, XHHW, USE — COPPER | | | 3 Electrical Ducts (Fig. B.310.2, Detail 2) Types RHW, THHW, THW, THWN, XHHW, USE — COPPER | | | 6 Electrical Ducts (Fig. B.310.2, Detail 3) Types RHW, THHW, THW, THWN, XHHW, USE — COPPER | | | 1 Electrical Duct (Fig. B.310.2, Detail 1) Types RHW, THHW, THW, THWN, XHHW, USE — ALUMINUM OR COPPER-CLAD ALUMINUM | | | 3 Electrical Ducts (Fig. B.310.2, Detail 2) Types RHW, THHW, THW, THWN, XHHW, USE — ALUMINUM | | | 6 Electrical Ducts (Fig. B.310.2, Detail 3) Types RHW, THHW, THW, THWN, XHHW, USE — ALUMINUM | | | Size (AWG or kcmil) |
|---|---|---|---|---|---|---|---|---|---|---|---|---|---|---|---|---|---|---|---|
| | RHO 60 LF 50 | RHO 90 LF 100 | RHO 120 LF 100 | RHO 60 LF 50 | RHO 90 LF 100 | RHO 120 LF 100 | RHO 60 LF 50 | RHO 90 LF 100 | RHO 120 LF 100 | RHO 60 LF 50 | RHO 90 LF 100 | RHO 120 LF 100 | RHO 60 LF 50 | RHO 90 LF 100 | RHO 120 LF 100 | RHO 60 LF 50 | RHO 90 LF 100 | RHO 120 LF 100 | |
| 8 | 58 | 54 | 53 | 56 | 48 | 46 | 53 | 42 | 39 | 45 | 42 | 41 | 43 | 37 | 36 | 41 | 32 | 30 | 8 |
| 6 | 77 | 71 | 69 | 74 | 63 | 60 | 70 | 54 | 51 | 60 | 55 | 54 | 57 | 49 | 47 | 54 | 42 | 39 | 6 |
| 4 | 101 | 93 | 91 | 96 | 81 | 77 | 91 | 69 | 65 | 78 | 72 | 71 | 75 | 63 | 60 | 71 | 54 | 51 | 4 |
| 2 | 132 | 121 | 118 | 126 | 105 | 100 | 119 | 89 | 83 | 103 | 94 | 92 | 98 | 82 | 78 | 92 | 70 | 65 | 2 |
| 1 | 154 | 140 | 136 | 146 | 121 | 114 | 137 | 102 | 95 | 120 | 109 | 106 | 114 | 94 | 89 | 107 | 79 | 74 | 1 |
| 1/0 | 177 | 160 | 156 | 168 | 137 | 130 | 157 | 116 | 107 | 138 | 125 | 122 | 131 | 107 | 101 | 122 | 90 | 84 | 1/0 |
| 2/0 | 203 | 183 | 178 | 192 | 156 | 147 | 179 | 131 | 121 | 158 | 143 | 139 | 150 | 122 | 115 | 140 | 102 | 95 | 2/0 |
| 3/0 | 233 | 210 | 204 | 221 | 178 | 158 | 205 | 148 | 137 | 182 | 164 | 159 | 172 | 139 | 131 | 160 | 116 | 107 | 3/0 |
| 4/0 | 268 | 240 | 232 | 253 | 202 | 190 | 234 | 168 | 155 | 209 | 187 | 182 | 198 | 158 | 149 | 183 | 131 | 121 | 4/0 |
| 250 | 297 | 265 | 256 | 280 | 222 | 209 | 258 | 184 | 169 | 233 | 207 | 201 | 219 | 174 | 163 | 202 | 144 | 132 | 250 |
| 350 | 363 | 321 | 310 | 340 | 267 | 250 | 312 | 219 | 202 | 285 | 252 | 244 | 267 | 209 | 196 | 245 | 172 | 158 | 350 |
| 500 | 444 | 389 | 375 | 414 | 320 | 299 | 377 | 261 | 240 | 352 | 308 | 297 | 328 | 254 | 237 | 299 | 207 | 190 | 500 |
| 750 | 552 | 478 | 459 | 511 | 388 | 362 | 462 | 314 | 288 | 446 | 386 | 372 | 413 | 314 | 293 | 374 | 254 | 233 | 750 |
| 1000 | 628 | 539 | 518 | 579 | 435 | 405 | 522 | 351 | 321 | 521 | 447 | 430 | 480 | 361 | 336 | 433 | 291 | 266 | 1000 |

| Ambient Temp. (°C) | Correction Factors | | | | | | Ambient Temp (°F) |
|---|---|---|---|---|---|---|---|
| 6–10 | 1.09 | 1.09 | 1.09 | 1.09 | 1.09 | 1.09 | 43–50 |
| 11–15 | 1.04 | 1.04 | 1.04 | 1.04 | 1.04 | 1.04 | 52–59 |
| 16–20 | 1.00 | 1.00 | 1.00 | 1.00 | 1.00 | 1.00 | 61–68 |
| 21–25 | 0.95 | 0.95 | 0.95 | 0.95 | 0.95 | 0.95 | 70–77 |
| 26–30 | 0.90 | 0.90 | 0.90 | 0.90 | 0.90 | 0.90 | 79–86 |

**Table B.310.7 Ampacities of Three Single Insulated Conductors, Rated 0 Through 2000 Volts, in Underground Electrical Ducts (Three Conductors per Electrical Duct) Based on Ambient Earth Temperature of 20°C (68°F), Electrical Duct Arrangement per Figure B.310.2, Conductor Temperature 75°C (167°F)**

| | 1 Electrical Duct (Fig. B.310.2, Detail 1) Types RHW, THHW, THW, THWN, XHHW, USE | | | 3 Electrical Ducts (Fig. B.310.2, Detail 2) Types RHW, THHW, THW, THWN, XHHW, USE | | | 6 Electrical Ducts (Fig. B.310.2, Detail 3) Types RHW, THHW, THW, THWN, XHHW, USE | | | 1 Electrical Duct (Fig. B.310.2, Detail 1) Types RHW, THHW, THW, THWN, XHHW, USE | | | 3 Electrical Ducts (Fig. B.310.2, Detail 2) Types RHW, THHW, THW, THWN, XHHW, USE | | | 6 Electrical Ducts (Fig. B.310.2, Detail 3) Types RHW, THHW, THW, THWN, XHHW, USE | | | |
|---|---|---|---|---|---|---|---|---|---|---|---|---|---|---|---|---|---|---|---|
| | COPPER | | | | | | | | | ALUMINUM OR COPPER-CLAD ALUMINUM | | | | | | | | | |
| Size (AWG or kcmil) | RHO 60 LF 50 | RHO 90 LF 100 | RHO 120 LF 100 | RHO 60 LF 50 | RHO 90 LF 100 | RHO 120 LF 100 | RHO 60 LF 50 | RHO 90 LF 100 | RHO 120 LF 100 | RHO 60 LF 50 | RHO 90 LF 100 | RHO 120 LF 100 | RHO 60 LF 50 | RHO 90 LF 100 | RHO 120 LF 100 | RHO 60 LF 50 | RHO 90 LF 100 | RHO 120 LF 100 | Size (AWG or kcmil) |
| 8 | 63 | 58 | 57 | 61 | 51 | 49 | 57 | 44 | 41 | 49 | 45 | 44 | 47 | 40 | 38 | 45 | 34 | 32 | 8 |
| 6 | 84 | 77 | 75 | 80 | 67 | 63 | 75 | 56 | 53 | 66 | 60 | 58 | 63 | 52 | 49 | 59 | 44 | 41 | 6 |
| 4 | 111 | 100 | 98 | 105 | 86 | 81 | 98 | 73 | 67 | 86 | 78 | 76 | 79 | 67 | 63 | 77 | 57 | 52 | 4 |
| 3 | 129 | 116 | 113 | 122 | 99 | 94 | 113 | 83 | 77 | 101 | 91 | 89 | 83 | 77 | 73 | 84 | 65 | 60 | 3 |
| 2 | 147 | 132 | 128 | 139 | 112 | 106 | 129 | 93 | 86 | 115 | 103 | 100 | 108 | 87 | 82 | 101 | 73 | 67 | 2 |
| 1 | 171 | 153 | 148 | 161 | 128 | 121 | 149 | 106 | 98 | 133 | 119 | 115 | 126 | 100 | 94 | 116 | 83 | 77 | 1 |
| 1/0 | 197 | 175 | 169 | 185 | 146 | 137 | 170 | 121 | 111 | 153 | 136 | 132 | 144 | 114 | 107 | 133 | 94 | 87 | 1/0 |
| 2/0 | 226 | 200 | 193 | 212 | 166 | 156 | 194 | 136 | 126 | 176 | 156 | 151 | 165 | 130 | 121 | 151 | 106 | 98 | 2/0 |
| 3/0 | 260 | 228 | 220 | 243 | 189 | 177 | 222 | 154 | 142 | 203 | 178 | 172 | 189 | 147 | 138 | 173 | 121 | 111 | 3/0 |
| 4/0 | 301 | 263 | 253 | 280 | 215 | 201 | 255 | 175 | 161 | 235 | 205 | 198 | 219 | 168 | 157 | 199 | 137 | 126 | 4/0 |
| 250 | 334 | 290 | 279 | 310 | 236 | 220 | 281 | 192 | 176 | 261 | 227 | 218 | 242 | 185 | 172 | 220 | 150 | 137 | 250 |
| 300 | 373 | 321 | 308 | 344 | 260 | 242 | 310 | 210 | 192 | 293 | 252 | 242 | 272 | 204 | 190 | 245 | 165 | 151 | 300 |
| 350 | 409 | 351 | 337 | 377 | 283 | 264 | 340 | 228 | 209 | 321 | 276 | 265 | 296 | 222 | 207 | 266 | 179 | 164 | 350 |
| 400 | 442 | 376 | 361 | 394 | 302 | 280 | 368 | 243 | 223 | 349 | 297 | 284 | 321 | 238 | 220 | 288 | 191 | 174 | 400 |
| 500 | 503 | 427 | 409 | 460 | 341 | 316 | 412 | 273 | 249 | 397 | 338 | 323 | 364 | 270 | 250 | 326 | 216 | 197 | 500 |
| 600 | 552 | 468 | 447 | 511 | 371 | 343 | 457 | 296 | 270 | 446 | 373 | 356 | 408 | 296 | 274 | 365 | 236 | 215 | 600 |
| 700 | 602 | 509 | 486 | 553 | 402 | 371 | 492 | 319 | 291 | 488 | 408 | 389 | 443 | 321 | 297 | 394 | 255 | 232 | 700 |
| 750 | 632 | 529 | 505 | 574 | 417 | 385 | 509 | 330 | 301 | 508 | 425 | 405 | 461 | 334 | 309 | 409 | 265 | 241 | 750 |
| 800 | 654 | 544 | 520 | 597 | 428 | 395 | 527 | 338 | 308 | 530 | 439 | 418 | 481 | 344 | 318 | 427 | 273 | 247 | 800 |
| 900 | 692 | 575 | 549 | 628 | 450 | 415 | 554 | 355 | 323 | 563 | 466 | 444 | 510 | 365 | 337 | 450 | 288 | 261 | 900 |
| 1000 | 730 | 605 | 576 | 659 | 472 | 435 | 581 | 372 | 338 | 597 | 494 | 471 | 538 | 385 | 355 | 475 | 304 | 276 | 1000 |

| Ambient Temp. (°C) | Correction Factors | | | | | | | | | | | | | | | | | | Ambient Temp. (°F) |
|---|---|---|---|---|---|---|---|---|---|---|---|---|---|---|---|---|---|---|---|
| 6–10 | 1.09 | | | 1.09 | | | 1.09 | | | 1.09 | | | 1.09 | | | 1.09 | | | 43–50 |
| 11–15 | 1.04 | | | 1.04 | | | 1.04 | | | 1.04 | | | 1.04 | | | 1.04 | | | 52–59 |
| 16–20 | 1.00 | | | 1.00 | | | 1.00 | | | 1.00 | | | 1.00 | | | 1.00 | | | 61–68 |
| 21–25 | 0.95 | | | 0.95 | | | 0.95 | | | 0.95 | | | 0.95 | | | 0.95 | | | 70–77 |
| 26–30 | 0.90 | | | 0.90 | | | 0.90 | | | 0.90 | | | 0.90 | | | 0.90 | | | 79–86 |

**Table B.310.8 Ampacities of Two or Three Insulated Conductors, Rated 0 Through 2000 Volts, Cabled Within an Overall (Two- or Three-Conductor) Covering, Directly Buried in Earth, Based on Ambient Earth Temperature of 20°C (68°F), Arrangement per Figure B.310.2, 100 Percent Load Factor, Thermal Resistance (Rho) of 90**

| Size (AWG or kcmil) | 1 Cable (Fig. B.310.2, Detail 5) | | 2 Cables (Fig. B.310.2, Detail 6) | | 1 Cable (Fig. B.310.2, Detail 5) | | 2 Cables (Fig. B.310.2, Detail 6) | | Size (AWG or kcmil) |
|---|---|---|---|---|---|---|---|---|---|
| | 60°C (140°F) | 75°C (167°F) | 60°C (140°F) | 75°C (167°F) | 60°C (140°F) | 75°C (167°F) | 60°C (140°F) | 75°C (167°F) | |
| | TYPES | | | | TYPES | | | | |
| | UF | RHW, THHW, THW, THWN, XHHW, USE | UF | RHW, THHW, THW, THWN, XHHW, USE | UF | RHW, THHW, THW, THWN, XHHW, USE | UF | RHW, THHW, THW, THWN, XHHW, USE | |
| | COPPER | | | | ALUMINUM OR COPPER-CLAD ALUMINUM | | | | |
| 8 | 64 | 75 | 60 | 70 | 51 | 59 | 47 | 55 | 8 |
| 6 | 85 | 100 | 81 | 95 | 68 | 75 | 60 | 70 | 6 |
| 4 | 107 | 125 | 100 | 117 | 83 | 97 | 78 | 91 | 4 |
| 2 | 137 | 161 | 128 | 150 | 107 | 126 | 110 | 117 | 2 |
| 1 | 155 | 182 | 145 | 170 | 121 | 142 | 113 | 132 | 1 |
| 1/0 | 177 | 208 | 165 | 193 | 138 | 162 | 129 | 151 | 1/0 |
| 2/0 | 201 | 236 | 188 | 220 | 157 | 184 | 146 | 171 | 2/0 |
| 3/0 | 229 | 269 | 213 | 250 | 179 | 210 | 166 | 195 | 3/0 |
| 4/0 | 259 | 304 | 241 | 282 | 203 | 238 | 188 | 220 | 4/0 |
| 250 | — | 333 | — | 308 | — | 261 | — | 241 | 250 |
| 350 | — | 401 | — | 370 | — | 315 | — | 290 | 350 |
| 500 | — | 481 | — | 442 | — | 381 | — | 350 | 500 |
| 750 | — | 585 | — | 535 | — | 473 | — | 433 | 750 |
| 1000 | — | 657 | — | 600 | — | 545 | — | 497 | 1000 |
| Ambient Temp. (°C) | Correction Factors | | | | | | | | Ambient Temp. (°F) |
| 6–10 | 1.12 | 1.09 | 1.12 | 1.09 | 1.12 | 1.09 | 1.12 | 1.09 | 43–50 |
| 11–15 | 1.06 | 1.04 | 1.06 | 1.04 | 1.06 | 1.04 | 1.06 | 1.04 | 52–59 |
| 16–20 | 1.00 | 1.00 | 1.00 | 1.00 | 1.00 | 1.00 | 1.00 | 1.00 | 61–68 |
| 21–25 | 0.94 | 0.95 | 0.94 | 0.95 | 0.94 | 0.95 | 0.94 | 0.95 | 70–77 |
| 26–30 | 0.87 | 0.90 | 0.87 | 0.90 | 0.87 | 0.90 | 0.87 | 0.90 | 79–86 |

**Table B.310.9 Ampacities of Three Triplexed Single Insulated Conductors, Rated 0 Through 2000 Volts, Directly Buried in Earth Based on Ambient Earth Temperature of 20°C (68°F), Arrangement per Figure B.310.2, 100 Percent Load Factor, Thermal Resistance (Rho) of 90**

| Size (AWG or kcmil) | See Fig. B.310.2, Detail 7 | | See Fig. B.310.2, Detail 8 | | See Fig. B.310.2, Detail 7 | | See Fig. B.310.2, Detail 8 | | Size (AWG or kcmil) |
|---|---|---|---|---|---|---|---|---|---|
| | 60°C (140°F) | 75°C (167°F) | 60°C (140°F) | 75°C (167°F) | 60°C (140°F) | 75°C (167°F) | 60°C (140°F) | 75°C (167°F) | |
| | TYPES | | | | TYPES | | | | |
| | UF | USE | UF | USE | UF | USE | UF | USE | |
| | COPPER | | | | ALUMINUM OR COPPER-CLAD ALUMINUM | | | | |
| 8 | 72 | 84 | 66 | 77 | 55 | 65 | 51 | 60 | 8 |
| 6 | 91 | 107 | 84 | 99 | 72 | 84 | 66 | 77 | 6 |
| 4 | 119 | 139 | 109 | 128 | 92 | 108 | 85 | 100 | 4 |
| 2 | 153 | 179 | 140 | 164 | 119 | 139 | 109 | 128 | 2 |
| 1 | 173 | 203 | 159 | 186 | 135 | 158 | 124 | 145 | 1 |
| 1/0 | 197 | 231 | 181 | 212 | 154 | 180 | 141 | 165 | 1/0 |
| 2/0 | 223 | 262 | 205 | 240 | 175 | 205 | 159 | 187 | 2/0 |
| 3/0 | 254 | 298 | 232 | 272 | 199 | 233 | 181 | 212 | 3/0 |
| 4/0 | 289 | 339 | 263 | 308 | 226 | 265 | 206 | 241 | 4/0 |
| 250 | — | 370 | — | 336 | — | 289 | — | 263 | 250 |
| 350 | — | 445 | — | 403 | — | 349 | — | 316 | 350 |
| 500 | — | 536 | — | 483 | — | 424 | — | 382 | 500 |
| 750 | — | 654 | — | 587 | — | 525 | — | 471 | 750 |
| 1000 | — | 744 | — | 665 | — | 608 | — | 544 | 1000 |
| Ambient Temp. (°C) | Correction Factors | | | | | | | | Ambient Temp. (°F) |
| 6–10 | 1.12 | 1.09 | 1.12 | 1.09 | 1.12 | 1.09 | 1.12 | 1.09 | 43–50 |
| 11–15 | 1.06 | 1.04 | 1.06 | 1.04 | 1.06 | 1.04 | 1.06 | 1.04 | 52–59 |
| 16–20 | 1.00 | 1.00 | 1.00 | 1.00 | 1.00 | 1.00 | 1.00 | 1.00 | 61–68 |
| 21–25 | 0.94 | 0.95 | 0.94 | 0.95 | 0.94 | 0.95 | 0.94 | 0.95 | 70–77 |
| 26–30 | 0.87 | 0.90 | 0.87 | 0.90 | 0.87 | 0.90 | 0.87 | 0.90 | 79–86 |

Note: For ampacities of Type UF cable in underground electrical ducts, multiply the ampacities shown in the table by 0.74.

**Table B.310.10 Ampacities of Three Single Insulated Conductors, Rated 0 Through 2000 Volts, Directly Buried in Earth Based on Ambient Earth Temperature of 20°C (68°F), Arrangement per Figure B.310.2, 100 Percent Load Factor, Thermal Resistance (Rho) of 90**

| Size (AWG or kcmil) | See Fig. B.310.2, Detail 9 | | See Fig. B.310.2, Detail 10 | | See Fig. B.310.2, Detail 9 | | See Fig. B.310.2, Detail 10 | | Size (AWG or kcmil) |
|---|---|---|---|---|---|---|---|---|---|
| | 60°C (140°F) | 75°C (167°F) | 60°C (140°F) | 75°C (167°F) | 60°C (140°F) | 75°C (167°F) | 60°C (140°F) | 75°C (167°F) | |
| | **TYPES** | | | | **TYPES** | | | | |
| | UF | USE | UF | USE | UF | USE | UF | USE | |
| | **COPPER** | | | | **ALUMINUM OR COPPER-CLAD ALUMINUM** | | | | |
| 8 | 84 | 98 | 78 | 92 | 66 | 77 | 61 | 72 | 8 |
| 6 | 107 | 126 | 101 | 118 | 84 | 98 | 78 | 92 | 6 |
| 4 | 139 | 163 | 130 | 152 | 108 | 127 | 101 | 118 | 4 |
| 2 | 178 | 209 | 165 | 194 | 139 | 163 | 129 | 151 | 2 |
| 1 | 201 | 236 | 187 | 219 | 157 | 184 | 146 | 171 | 1 |
| 1/0 | 230 | 270 | 212 | 249 | 179 | 210 | 165 | 194 | 1/0 |
| 2/0 | 261 | 306 | 241 | 283 | 204 | 239 | 188 | 220 | 2/0 |
| 3/0 | 297 | 348 | 274 | 321 | 232 | 272 | 213 | 250 | 3/0 |
| 4/0 | 336 | 394 | 309 | 362 | 262 | 307 | 241 | 283 | 4/0 |
| 250 | — | 429 | — | 394 | — | 335 | — | 308 | 250 |
| 350 | — | 516 | — | 474 | — | 403 | — | 370 | 350 |
| 500 | — | 626 | — | 572 | — | 490 | — | 448 | 500 |
| 750 | — | 767 | — | 700 | — | 605 | — | 552 | 750 |
| 1000 | — | 887 | — | 808 | — | 706 | — | 642 | 1000 |
| 1250 | — | 979 | — | 891 | — | 787 | — | 716 | 1250 |
| 1500 | — | 1063 | — | 965 | — | 862 | — | 783 | 1500 |
| 1750 | — | 1133 | — | 1027 | — | 930 | — | 843 | 1750 |
| 2000 | — | 1195 | — | 1082 | — | 990 | — | 897 | 2000 |
| Ambient Temp. (°C) | **Correction Factors** | | | | | | | | Ambient Temp. (°F) |
| 6–10 | 1.12 | 1.09 | 1.12 | 1.09 | 1.12 | 1.09 | 1.12 | 1.09 | 43–50 |
| 11–15 | 1.06 | 1.04 | 1.06 | 1.04 | 1.06 | 1.04 | 1.06 | 1.04 | 52–59 |
| 16–20 | 1.00 | 1.00 | 1.00 | 1.00 | 1.00 | 1.00 | 1.00 | 1.00 | 61–68 |
| 21–25 | 0.94 | 0.95 | 0.94 | 0.95 | 0.94 | 0.95 | 0.94 | 0.95 | 70–77 |
| 26–30 | 0.87 | 0.90 | 0.87 | 0.90 | 0.87 | 0.90 | 0.87 | 0.90 | 79–86 |

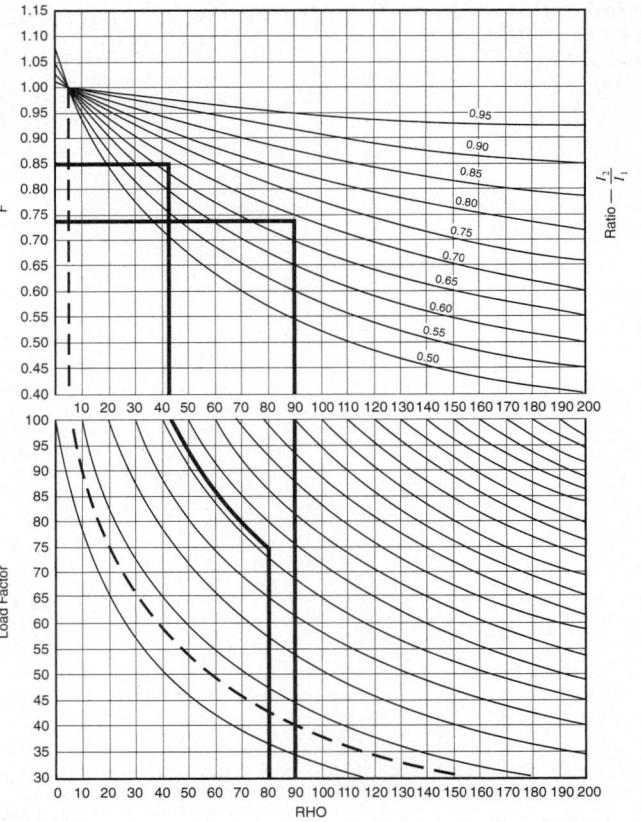

**Figure B.310.1 Interpolation chart for cables in a duct bank**
$I_1$ = ampacity for Rho = 60, 50 LF; $I_2$ = ampacity for Rho = 120, 100 LF ( load factor); desired ampacity = F × $I_1$.

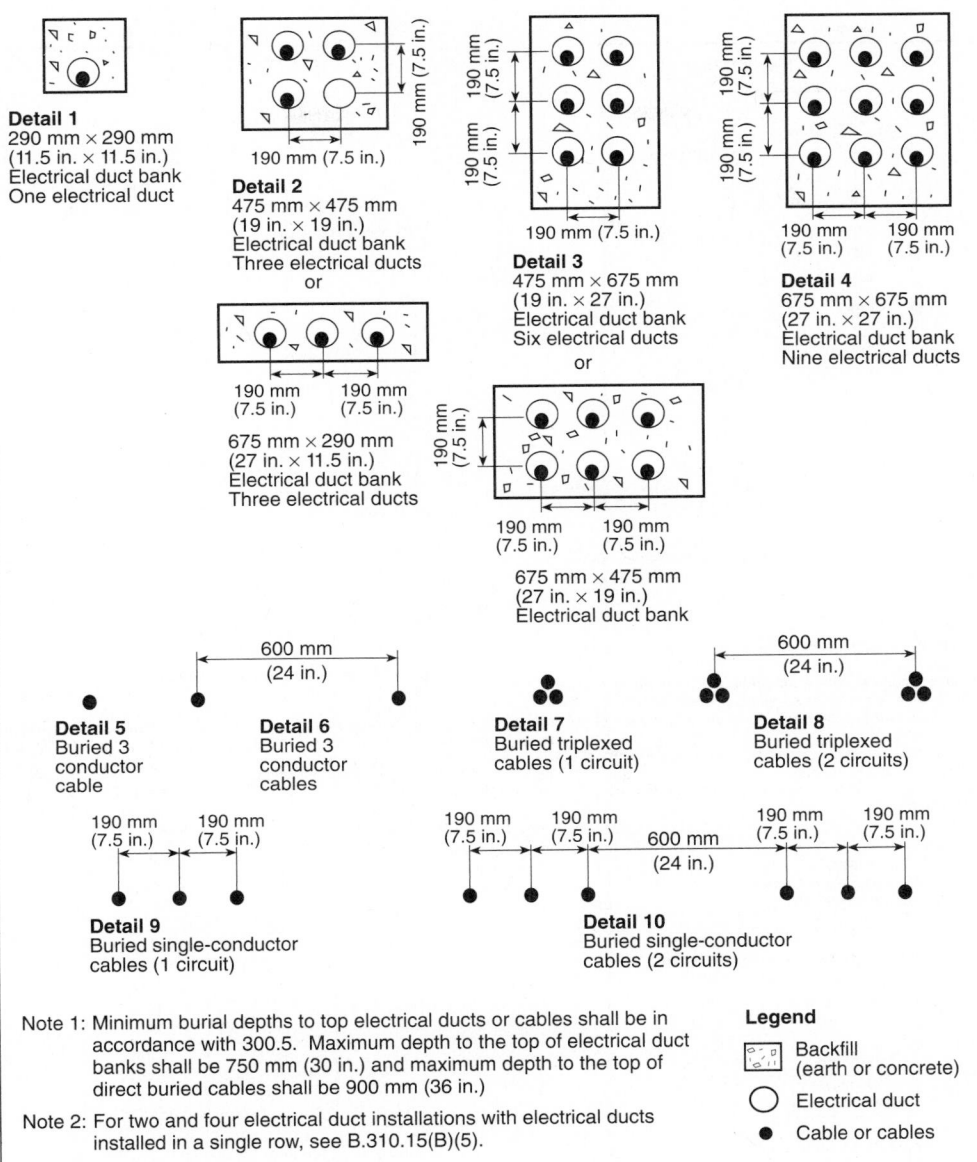

**Detail 1**
290 mm × 290 mm
(11.5 in. × 11.5 in.)
Electrical duct bank
One electrical duct

**Detail 2**
475 mm × 475 mm
(19 in. × 19 in.)
Electrical duct bank
Three electrical ducts
or

675 mm × 290 mm
(27 in. × 11.5 in.)
Electrical duct bank
Three electrical ducts

**Detail 3**
475 mm × 675 mm
(19 in. × 27 in.)
Electrical duct bank
Six electrical ducts
or

675 mm × 475 mm
(27 in. × 19 in.)
Electrical duct bank

**Detail 4**
675 mm × 675 mm
(27 in. × 27 in.)
Electrical duct bank
Nine electrical ducts

**Detail 5**
Buried 3
conductor
cable

**Detail 6**
Buried 3
conductor
cables

**Detail 7**
Buried triplexed
cables (1 circuit)

**Detail 8**
Buried triplexed
cables (2 circuits)

**Detail 9**
Buried single-conductor
cables (1 circuit)

**Detail 10**
Buried single-conductor
cables (2 circuits)

Note 1: Minimum burial depths to top electrical ducts or cables shall be in accordance with 300.5. Maximum depth to the top of electrical duct banks shall be 750 mm (30 in.) and maximum depth to the top of direct buried cables shall be 900 mm (36 in.)

Note 2: For two and four electrical duct installations with electrical ducts installed in a single row, see B.310.15(B)(5).

**Legend**

Backfill (earth or concrete)

Electrical duct

Cable or cables

**Figure B.310.2  Cable installation dimensions for use with Table B.310.5 through Table B.310.10.**

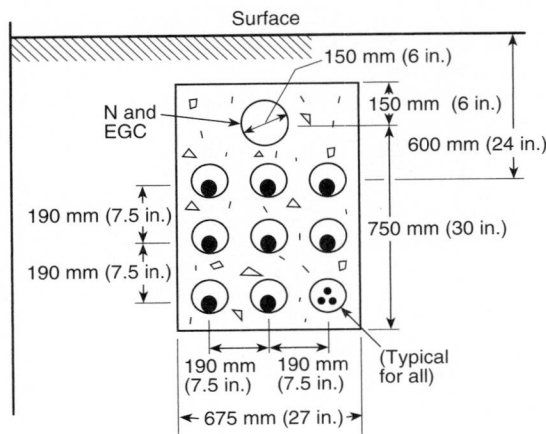

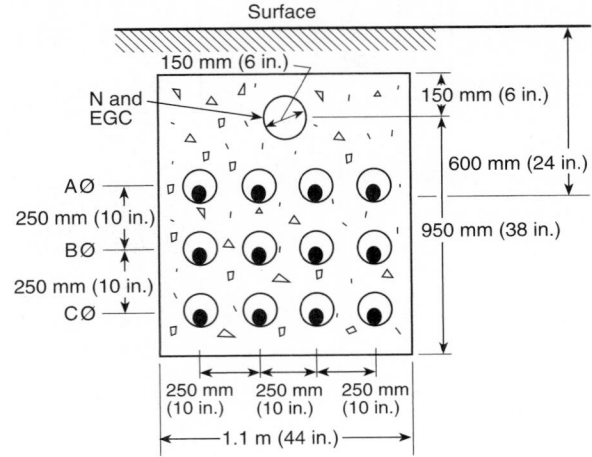

Design Criteria
Neutral and Equipment
  Grounding conductor (EGC)
    Duct = 150 mm (6 in.)
Phase Ducts = 75 to 125 mm (3 to 5 in.)
Conductor Material = Copper
Number of Cables per Duct = 3

Number of Cables per Phase = 9
Rho concrete = Rho Earth – 5

Rho PVC Duct = 650
Rho Cable Insulation = 500
Rho Cable Jacket = 650

Notes:
1. Neutral configuration per 300.5(I), Exception No. 2, for isolated phase installations in nonmagnetic ducts.
2. Phasing is A, B, C in rows or columns. Where magnetic electrical ducts are used, conductors are installed A, B, C per electrical duct with the neutral and all equipment grounding conductors in the same electrical duct. In this case, the 6-in. trade size neutral duct is eliminated.
3. Maximum harmonic loading on the neutral conductor cannot exceed 50 percent of the phase current for the ampacities shown in the table.
4. Metallic shields of Type MV-90 cable shall be grounded at one point only where using A, B, C phasing in rows or columns.

| Size kcmil | TYPES RHW, THHW, THW, THWN, XHHW, USE, OR MV-90* | | | Size kcmil |
|---|---|---|---|---|
| | Total per Phase Ampere Rating | | | |
| | RHO EARTH 60 LF 50 | RHO EARTH 90 LF 100 | RHO EARTH 120 LF 100 | |
| 250 | 2340 (260A/Cable) | 1530 (170A/Cable) | 1395 (155A/Cable) | 250 |
| 350 | 2790 (310A/Cable) | 1800 (200A/Cable) | 1665 (185A/Cable) | 350 |
| 500 | 3375 (375A/Cable) | 2160 (240A/Cable) | 1980 (220A/Cable) | 350 |

| Ambient Temp. (°C) | For ambient temperatures other than 20°C (68°F), multiply the ampacities shown above by the appropriate factor shown below. | | | | | Ambient Temp. (°F) |
|---|---|---|---|---|---|---|
| 6–10 | 1.09 | 1.09 | 1.09 | 1.09 | 1.09 | 43–50 |
| 11–15 | 1.04 | 1.04 | 1.04 | 1.04 | 1.04 | 52–59 |
| 16–20 | 1.00 | 1.00 | 1.00 | 1.00 | 1.00 | 61–68 |
| 21–25 | 0.95 | 0.95 | 0.95 | 0.95 | 0.95 | 70–77 |
| 26–30 | 0.90 | 0.90 | 0.90 | 0.90 | 0.90 | 79–86 |

*Limited to 75°C conductor temperature.

**FPN Figure B.310.3 Ampacities of single insulated conductors rated 0 through 5000 volts in underground electrical ducts (three conductors per electrical duct), nine single-conductor cables per phase based on ambient earth temperature of 20°C (68°F), conductor temperature 75°C (167°F).**

Design Criteria
Neutral and Equipment
  Grounding Conductor (EGC)
    Duct = 150 mm (6 in.)
Phase Ducts = 75 mm (3 in.)
Conductor Material = Copper
Number of Cables per Duct = 1

Number of Cables per Phase = 4
Rho Concrete = Rho Earth – 5
Rho PVC Duct = 650

Rho Cable Insulation = 500
Rho Cable Jacket = 650

Notes:
1. Neutral configuration per 300.5(I), Exception No. 2.
2. Maximum harmonic loading on the neutral conductor cannot exceed 50 percent of the phase current for the ampacities shown in the table.
3. Metallic shields of Type MV-90 cable shall be grounded at one point only.

| Size kcmil | TYPES RHW, THHW, THW, THWN, XHHW, USE, OR MV-90* | | | Size kcmil |
|---|---|---|---|---|
| | Total per Phase Ampere Rating | | | |
| | RHO EARTH 60 LF 50 | RHO EARTH 90 LF 100 | RHO EARTH 120 LF 100 | |
| 750 | 2820 (705A/Cable) | 1860 (465A/Cable) | 1680 (420A/Cable) | 750 |
| 1000 | 3300 (825A/Cable) | 2140 (535A/Cable) | 1920 (480A/Cable) | 1000 |
| 1250 | 3700 (925A/Cable) | 2380 (595A/Cable) | 2120 (530A/Cable) | 1250 |
| 1500 | 4060 (1015A/Cable) | 2580 (645A/Cable) | 2300 (575A/Cable) | 1500 |
| 1750 | 4360 (1090A/Cable) | 2740 (685A/Cable) | 2460 (615A/Cable) | 1750 |

| Ambient Temp. (°C) | For ambient temperatures other than 20°C (68°F), multiply the ampacities shown above by the appropriate factor shown below. | | | | | Ambient Temp. (°F) |
|---|---|---|---|---|---|---|
| 6–10 | 1.09 | 1.09 | 1.09 | 1.09 | 1.09 | 43–50 |
| 11–15 | 1.04 | 1.04 | 1.04 | 1.04 | 1.04 | 52–59 |
| 16–20 | 1.00 | 1.00 | 1.00 | 1.00 | 1.00 | 61–68 |
| 21–25 | 0.95 | 0.95 | 0.95 | 0.95 | 0.95 | 70–77 |
| 26–30 | 0.90 | 0.90 | 0.90 | 0.90 | 0.90 | 79–86 |

*Limited to 75°C conductor temperature.

**FPN Figure B.310.4 Ampacities of single insulated conductors rated 0 through 5000 volts in nonmagnetic underground electrical ducts (one conductor per electrical duct), four single-conductor cables per phase based on ambient earth temperature of 20°C (68°F), conductor temperature 75°C (167°F).**

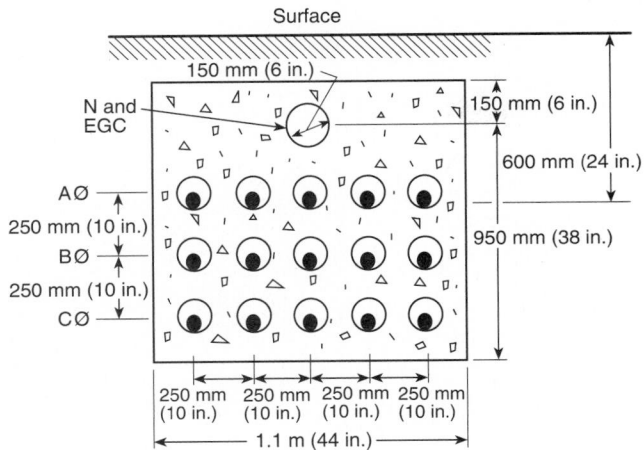

Surface

Design Criteria
Neutral and Equipment
  Grounding Conductor (EGC)
  Duct = 150 mm (6 in.)
Phase Ducts = 75 mm (3 in.)
Conductor Material = Copper
Number of Cables per Duct = 1

Number of Cables per Phase = 5
Rho Concrete = Rho Earth − 5
Rho PVC Duct = 650

Rho Cable Insulation = 500
Rho Cable Jacket = 650

Notes:
1. Neutral configuration per 300.5(I), Exception No. 2.
2. Maximum harmonic loading on the neutral conductor cannot exceed 50 percent of the
   phase current for the ampacities shown in the table.
3. Metallic shields of Type MV-90 cable shall be grounded at one point only.

| Size kcmil | TYPES RHW, THHW, THW, THWN, XHHW, USE, OR MV-90* | | | Size kcmil |
|---|---|---|---|---|
| | Total per Phase Ampere Rating | | | |
| | RHO EARTH 60 LF 50 | RHO EARTH 90 LF 100 | RHO EARTH 120 LF 100 | |
| 2000 | 5575 (1115A/Cable) | 3375 (675A/Cable) | 3000 (600A/Cable) | 2000 |

| Ambient Temp. (°C) | For ambient temperatures other than 20°C (68°F), multiply the ampacities shown above by the appropriate factor shown below. | | | | | Ambient Temp. (°F) |
|---|---|---|---|---|---|---|
| 6–10 | 1.09 | 1.09 | 1.09 | 1.09 | 1.09 | 43–50 |
| 11–15 | 1.04 | 1.04 | 1.04 | 1.04 | 1.04 | 52–59 |
| 16–20 | 1.00 | 1.00 | 1.00 | 1.00 | 1.00 | 61–68 |
| 21–25 | 0.95 | 0.95 | 0.95 | 0.95 | 0.95 | 70–77 |
| 26–30 | 0.90 | 0.90 | 0.90 | 0.90 | 0.90 | 79–86 |

*Limited to 75°C conductor temperature.

**FPN Figure B.310.5 Ampacities of single insulated conductors rated 0 through 5000 volts in nonmagnetic underground electrical ducts (one conductor per electrical duct), five single-conductor cables per phase based on ambient earth temperature of 20°C (68°F), conductor temperature 75°C (167°F).**

Values for using Figure B.310.1 are found in the electrical duct bank ampacity tables of this annex.

Where the load factor is less than 100 percent and can be verified by measurement or calculation, the ampacity of electrical duct bank installations can be modified as shown. Different values of Rho can be accommodated in the same manner.

**Table B.310.11 Adjustment Factors for More Than Three Current-Carrying Conductors in a Raceway or Cable with Load Diversity**

| Number of Current-Carrying Conductors | Percent of Values in Tables as Adjusted for Ambient Temperature if Necessary |
|---|---|
| 4 – 6 | 80 |
| 7 – 9 | 70 |
| 10 – 24 | 70* |
| 25 – 42 | 60* |
| 43 – 85 | 50* |

*These factors include the effects of a load diversity of 50 percent.

FPN: The ampacity limit for the number of current-carrying conductors in 10 through 85 is based on the following formula. For greater than 85 conductors, special calculations are required that are beyond the scope of this table.

$$A_2 = \sqrt{\frac{0.5N}{E}} \times (A_1) \text{ or } A_1, \text{whichever is less}$$

where:

$A_1$ = ampacity from Tables 310.16; 310.18; B.310.1; B.310.6; and B.310.7 multiplied by the appropriate factor from Table B.310.11.

$N$ = total number of conductors used to obtain multiplying factor from Table B.310.11

$E$ = desired number of current-carrying conductors in the raceway or cable

$A_2$ = ampacity limit for the current-carrying conductors in the raceway or cable

**Example 1**
Calculate the ampacity limit for twelve 14 AWG THWN current-carrying conductors (75°C) in a raceway that contains 24 conductors.

$$A_2 = \sqrt{\frac{(0.5)(24)}{12}} \times 20(0.7)$$
$$= 14 \text{ amperes } (\text{i. e., } 50 \text{ percent diversity})$$

**Example 2**
Calculate the ampacity limit for eighteen 14 AWG THWN current-carrying conductors (75°C) in a raceway that contains 24 conductors.

$$A_2 = \sqrt{\frac{(0.5)(24)}{18}} \times 20(0.7) = 11.5 \text{ amperes}$$

# Annex C
## Conduit and Tubing Fill Tables for Conductors and Fixture Wires of the Same Size

*This annex is not a part of the requirements of this NFPA document but is included for informational purposes only.*

| Table | Page |
|---|---|
| C1 — Electrical Metallic Tubing (EMT) | 645 |
| C1(A)* — Electrical Metallic Tubing (EMT) | 646 |
| C2 — Electrical Nonmetallic Tubing (ENT) | 647 |
| C2(A)* — Electrical Nonmetallic Tubing (ENT) | 648 |
| C3 — Flexible Metal Conduit (FMC) | 648 |
| C3(A)* — Flexible Metal Conduit (FMC) | 650 |
| C4 — Intermediate Metal Conduit (IMC) | 650 |
| C4(A)* — Intermediate Metal Conduit (IMC) | 651 |
| C5 — Liquidtight Flexible Nonmetallic Conduit (Type LFNC-B) | 652 |
| C5(A)* — Liquidtight Flexible Nonmetallic Conduit (Type LFNC-B) | 653 |
| C6 — Liquidtight Flexible Nonmetallic Conduit (Type LFNC-A) | 654 |
| C6(A)* — Liquidtight Flexible Nonmetallic Conduit (Type LFNC-A) | 655 |
| C7 — Liquidtight Flexible Metal Conduit (LFML) | 655 |
| C7(A)* — Liquidtight Flexible Metal Conduit (LFML) | 657 |
| C8 — Rigid Metal Conduit (RMC) | 658 |
| C8(A)* — Rigid Metal Conduit (RMC) | 659 |
| C9 — Rigid PVC Conduit, Schedule 80 | 659 |
| C9(A)* — Rigid PVC Conduit, Schedule 80 | 661 |
| C10 — Rigid PVC Conduit, Schedule 40 and HDPE Conduit | 662 |
| C10(A)* — Rigid PVC Conduit, Schedule 40 and HDPE Conduit | 662 |
| C11 — Type A, Rigid PVC Conduit | 663 |
| C11(A)* — Type A, Rigid PVC Conduit | 664 |
| C12 — Type EB, PVC Conduit | 665 |
| C12(A)* — Type EB, PVC Conduit | 666 |

*Where this table is used in conjunction with Tables C1 through C12, the conductors installed must be of the compact type.

**Table C1 Maximum Number of Conductors or Fixture Wires in Electrical Metallic Tubing (EMT)** (*Based on Table 1, Chapter 9*)

| | | CONDUCTORS | | | | | | | | | |
|---|---|---|---|---|---|---|---|---|---|---|---|
| | | Metric Designator (Trade Size) | | | | | | | | | |
| Type | Conductor Size (AWG/kcmil) | 16 (½) | 21 (¾) | 27 (1) | 35 (1¼) | 41 (1½) | 53 (2) | 63 (2½) | 78 (3) | 91 (3½) | 103 (4) |
| RHH, RHW, RHW-2 | 14 | 4 | 7 | 11 | 20 | 27 | 46 | 80 | 120 | 157 | 201 |
| | 12 | 3 | 6 | 9 | 17 | 23 | 38 | 66 | 100 | 131 | 167 |
| | 10 | 2 | 5 | 8 | 13 | 18 | 30 | 53 | 81 | 105 | 135 |
| | 8 | 1 | 2 | 4 | 7 | 9 | 16 | 28 | 42 | 55 | 70 |
| | 6 | 1 | 1 | 3 | 5 | 8 | 13 | 22 | 34 | 44 | 56 |
| | 4 | 1 | 1 | 2 | 4 | 6 | 10 | 17 | 26 | 34 | 44 |
| | 3 | 1 | 1 | 1 | 4 | 5 | 9 | 15 | 23 | 30 | 38 |
| | 2 | 1 | 1 | 1 | 3 | 4 | 7 | 13 | 20 | 26 | 33 |
| | 1 | 0 | 1 | 1 | 1 | 3 | 5 | 9 | 13 | 17 | 22 |
| | 1/0 | 0 | 1 | 1 | 1 | 2 | 4 | 7 | 11 | 15 | 19 |
| | 2/0 | 0 | 1 | 1 | 1 | 2 | 4 | 6 | 10 | 13 | 17 |
| | 3/0 | 0 | 0 | 1 | 1 | 1 | 3 | 5 | 8 | 11 | 14 |
| | 4/0 | 0 | 0 | 1 | 1 | 1 | 3 | 5 | 7 | 9 | 12 |
| | 250 | 0 | 0 | 0 | 1 | 1 | 1 | 3 | 5 | 7 | 9 |
| | 300 | 0 | 0 | 0 | 1 | 1 | 1 | 3 | 5 | 6 | 8 |
| | 350 | 0 | 0 | 0 | 1 | 1 | 1 | 3 | 4 | 6 | 7 |
| | 400 | 0 | 0 | 0 | 1 | 1 | 1 | 2 | 4 | 5 | 7 |
| | 500 | 0 | 0 | 0 | 0 | 1 | 1 | 2 | 3 | 4 | 6 |
| | 600 | 0 | 0 | 0 | 0 | 1 | 1 | 1 | 3 | 4 | 5 |
| | 700 | 0 | 0 | 0 | 0 | 0 | 1 | 1 | 2 | 3 | 4 |
| | 750 | 0 | 0 | 0 | 0 | 0 | 1 | 1 | 2 | 3 | 4 |
| | 800 | 0 | 0 | 0 | 0 | 0 | 1 | 1 | 2 | 3 | 4 |
| | 900 | 0 | 0 | 0 | 0 | 0 | 1 | 1 | 1 | 3 | 3 |
| | 1000 | 0 | 0 | 0 | 0 | 0 | 1 | 1 | 1 | 2 | 3 |
| | 1250 | 0 | 0 | 0 | 0 | 0 | 0 | 1 | 1 | 1 | 2 |
| | 1500 | 0 | 0 | 0 | 0 | 0 | 0 | 1 | 1 | 1 | 1 |
| | 1750 | 0 | 0 | 0 | 0 | 0 | 0 | 1 | 1 | 1 | 1 |
| | 2000 | 0 | 0 | 0 | 0 | 0 | 0 | 1 | 1 | 1 | 1 |
| TW, THHW, THW, THW-2 | 14 | 8 | 15 | 25 | 43 | 58 | 96 | 168 | 254 | 332 | 424 |
| | 12 | 6 | 11 | 19 | 33 | 45 | 74 | 129 | 195 | 255 | 326 |
| | 10 | 5 | 8 | 14 | 24 | 33 | 55 | 96 | 145 | 190 | 243 |
| | 8 | 2 | 5 | 8 | 13 | 18 | 30 | 53 | 81 | 105 | 135 |

**Table C1** *Continued*

| | | CONDUCTORS | | | | | | | | | |
|---|---|---|---|---|---|---|---|---|---|---|---|
| | | Metric Designator (Trade Size) | | | | | | | | | |
| Type | Conductor Size (AWG/kcmil) | 16 (½) | 21 (¾) | 27 (1) | 35 (1¼) | 41 (1½) | 53 (2) | 63 (2½) | 78 (3) | 91 (3½) | 103 (4) |
| RHH*, RHW*, RHW-2* | 14 | 6 | 10 | 16 | 28 | 39 | 64 | 112 | 169 | 221 | 282 |
| | 12 | 4 | 8 | 13 | 23 | 31 | 51 | 90 | 136 | 177 | 227 |
| | 10 | 3 | 6 | 10 | 18 | 24 | 40 | 70 | 106 | 138 | 177 |
| | 8 | 1 | 4 | 6 | 10 | 14 | 24 | 42 | 63 | 83 | 106 |
| RHH*, RHW*, RHW-2*, TW, THW, THHW, THW-2 | 6 | 1 | 3 | 4 | 8 | 11 | 18 | 32 | 48 | 63 | 81 |
| | 4 | 1 | 1 | 3 | 6 | 8 | 13 | 24 | 36 | 47 | 60 |
| | 3 | 1 | 1 | 3 | 5 | 7 | 12 | 20 | 31 | 40 | 52 |
| | 2 | 1 | 1 | 2 | 4 | 6 | 10 | 17 | 26 | 34 | 44 |
| | 1 | 1 | 1 | 1 | 3 | 4 | 7 | 12 | 18 | 24 | 31 |
| | 1/0 | 0 | 1 | 1 | 2 | 3 | 6 | 10 | 16 | 20 | 26 |
| | 2/0 | 0 | 1 | 1 | 1 | 3 | 5 | 9 | 13 | 17 | 22 |
| | 3/0 | 0 | 1 | 1 | 1 | 2 | 4 | 7 | 11 | 15 | 19 |
| | 4/0 | 0 | 0 | 1 | 1 | 1 | 3 | 6 | 9 | 12 | 16 |
| | 250 | 0 | 0 | 1 | 1 | 1 | 3 | 5 | 7 | 10 | 13 |
| | 300 | 0 | 0 | 1 | 1 | 1 | 2 | 4 | 6 | 8 | 11 |
| | 350 | 0 | 0 | 0 | 1 | 1 | 1 | 4 | 6 | 7 | 10 |
| | 400 | 0 | 0 | 0 | 1 | 1 | 1 | 3 | 5 | 7 | 9 |
| | 500 | 0 | 0 | 0 | 1 | 1 | 1 | 3 | 4 | 6 | 7 |
| | 600 | 0 | 0 | 0 | 1 | 1 | 1 | 2 | 3 | 4 | 6 |
| | 700 | 0 | 0 | 0 | 0 | 1 | 1 | 1 | 3 | 4 | 5 |
| | 750 | 0 | 0 | 0 | 0 | 1 | 1 | 1 | 3 | 4 | 5 |
| | 800 | 0 | 0 | 0 | 0 | 1 | 1 | 1 | 3 | 3 | 5 |
| | 900 | 0 | 0 | 0 | 0 | 0 | 1 | 1 | 2 | 3 | 4 |
| | 1000 | 0 | 0 | 0 | 0 | 0 | 1 | 1 | 2 | 3 | 4 |
| | 1250 | 0 | 0 | 0 | 0 | 0 | 1 | 1 | 1 | 2 | 3 |
| | 1500 | 0 | 0 | 0 | 0 | 0 | 1 | 1 | 1 | 1 | 2 |
| | 1750 | 0 | 0 | 0 | 0 | 0 | 0 | 1 | 1 | 1 | 2 |
| | 2000 | 0 | 0 | 0 | 0 | 0 | 1 | 1 | 1 | 1 | 1 |
| THHN, THWN, THWN-2 | 14 | 12 | 22 | 35 | 61 | 84 | 138 | 241 | 364 | 476 | 608 |
| | 12 | 9 | 16 | 26 | 45 | 61 | 101 | 176 | 266 | 347 | 443 |
| | 10 | 5 | 10 | 16 | 28 | 38 | 63 | 111 | 167 | 219 | 279 |
| | 8 | 3 | 6 | 9 | 16 | 22 | 36 | 64 | 96 | 126 | 161 |
| | 6 | 2 | 4 | 7 | 12 | 16 | 26 | 46 | 69 | 91 | 116 |
| | 4 | 1 | 2 | 4 | 7 | 10 | 16 | 28 | 43 | 56 | 71 |
| | 3 | 1 | 1 | 3 | 6 | 8 | 13 | 24 | 36 | 47 | 60 |
| | 2 | 1 | 1 | 3 | 5 | 7 | 11 | 20 | 30 | 40 | 51 |
| | 1 | 1 | 1 | 1 | 4 | 5 | 8 | 15 | 22 | 29 | 37 |
| | 1/0 | 1 | 1 | 1 | 3 | 4 | 7 | 12 | 19 | 25 | 32 |
| | 2/0 | 0 | 1 | 1 | 2 | 3 | 6 | 10 | 16 | 20 | 26 |
| | 3/0 | 0 | 1 | 1 | 1 | 3 | 5 | 8 | 13 | 17 | 22 |
| | 4/0 | 0 | 1 | 1 | 1 | 2 | 4 | 7 | 11 | 14 | 18 |
| | 250 | 0 | 0 | 1 | 1 | 1 | 3 | 6 | 9 | 11 | 15 |
| | 300 | 0 | 0 | 1 | 1 | 1 | 3 | 5 | 7 | 10 | 13 |
| | 350 | 0 | 0 | 1 | 1 | 1 | 2 | 4 | 6 | 9 | 11 |
| | 400 | 0 | 0 | 0 | 1 | 1 | 1 | 4 | 6 | 8 | 10 |
| | 500 | 0 | 0 | 0 | 1 | 1 | 1 | 3 | 5 | 6 | 8 |
| | 600 | 0 | 0 | 0 | 1 | 1 | 1 | 2 | 4 | 5 | 7 |
| | 700 | 0 | 0 | 0 | 1 | 1 | 1 | 2 | 3 | 4 | 6 |
| | 750 | 0 | 0 | 0 | 0 | 1 | 1 | 1 | 3 | 4 | 5 |
| | 800 | 0 | 0 | 0 | 0 | 1 | 1 | 1 | 3 | 4 | 5 |
| | 900 | 0 | 0 | 0 | 0 | 1 | 1 | 1 | 3 | 3 | 4 |
| | 1000 | 0 | 0 | 0 | 0 | 1 | 1 | 1 | 2 | 3 | 4 |
| FEP, FEPB, PFA, PFAH, TFE | 14 | 12 | 21 | 34 | 60 | 81 | 134 | 234 | 354 | 462 | 590 |
| | 12 | 9 | 15 | 25 | 43 | 59 | 98 | 171 | 258 | 337 | 430 |
| | 10 | 6 | 11 | 18 | 31 | 42 | 70 | 122 | 185 | 241 | 309 |
| | 8 | 3 | 6 | 10 | 18 | 24 | 40 | 70 | 106 | 138 | 177 |
| | 6 | 2 | 4 | 7 | 12 | 17 | 28 | 50 | 75 | 98 | 126 |
| | 4 | 1 | 3 | 5 | 9 | 12 | 20 | 35 | 53 | 69 | 88 |
| | 3 | 1 | 2 | 4 | 7 | 10 | 16 | 29 | 44 | 57 | 73 |
| | 2 | 1 | 1 | 3 | 6 | 8 | 13 | 24 | 36 | 47 | 60 |
| PFA, PFAH, TFE | 1 | 1 | 1 | 2 | 4 | 6 | 9 | 16 | 25 | 33 | 42 |
| PFAH, TFE PFA, PFAH, TFE, Z | 1/0 | 1 | 1 | 1 | 3 | 5 | 8 | 14 | 21 | 27 | 35 |
| | 2/0 | 0 | 1 | 1 | 3 | 4 | 6 | 11 | 17 | 22 | 29 |
| | 3/0 | 0 | 1 | 1 | 2 | 3 | 5 | 9 | 14 | 18 | 24 |
| | 4/0 | 0 | 1 | 1 | 1 | 2 | 4 | 8 | 11 | 15 | 19 |

**Table C1** *Continued*

### CONDUCTORS

| Type | Conductor Size (AWG/kcmil) | 16 (½) | 21 (¾) | 27 (1) | 35 (1¼) | 41 (1½) | 53 (2) | 63 (2½) | 78 (3) | 91 (3½) | 103 (4) |
|------|------|------|------|------|------|------|------|------|------|------|------|
| Z | 14 | 14 | 25 | 41 | 72 | 98 | 161 | 282 | 426 | 556 | 711 |
| | 12 | 10 | 18 | 29 | 51 | 69 | 114 | 200 | 302 | 394 | 504 |
| | 10 | 6 | 11 | 18 | 31 | 42 | 70 | 122 | 185 | 241 | 309 |
| | 8 | 4 | 7 | 11 | 20 | 27 | 44 | 77 | 117 | 153 | 195 |
| | 6 | 3 | 5 | 8 | 14 | 19 | 31 | 54 | 82 | 107 | 137 |
| | 4 | 1 | 3 | 5 | 9 | 13 | 21 | 37 | 56 | 74 | 94 |
| | 3 | 1 | 2 | 4 | 7 | 9 | 15 | 27 | 41 | 54 | 69 |
| | 2 | 1 | 1 | 3 | 6 | 8 | 13 | 22 | 34 | 45 | 57 |
| | 1 | 1 | 1 | 2 | 4 | 6 | 10 | 18 | 28 | 36 | 46 |
| XHH, XHHW, XHHW-2, ZW | 14 | 8 | 15 | 25 | 43 | 58 | 96 | 168 | 254 | 332 | 424 |
| | 12 | 6 | 11 | 19 | 33 | 45 | 74 | 129 | 195 | 255 | 326 |
| | 10 | 5 | 8 | 14 | 24 | 33 | 55 | 96 | 145 | 190 | 243 |
| | 8 | 2 | 5 | 8 | 13 | 18 | 30 | 53 | 81 | 105 | 135 |
| | 6 | 1 | 3 | 6 | 10 | 14 | 22 | 39 | 60 | 78 | 100 |
| | 4 | 1 | 2 | 4 | 7 | 10 | 16 | 28 | 43 | 56 | 72 |
| | 3 | 1 | 1 | 3 | 6 | 8 | 14 | 24 | 36 | 48 | 61 |
| | 2 | 1 | 1 | 3 | 5 | 7 | 11 | 20 | 31 | 40 | 51 |
| XHH, XHHW, XHHW-2 | 1 | 1 | 1 | 1 | 4 | 5 | 8 | 15 | 23 | 30 | 38 |
| | 1/0 | 1 | 1 | 1 | 3 | 4 | 7 | 13 | 19 | 25 | 32 |
| | 2/0 | 0 | 1 | 1 | 2 | 3 | 6 | 10 | 16 | 21 | 27 |
| | 3/0 | 0 | 1 | 1 | 1 | 3 | 5 | 9 | 13 | 17 | 22 |
| | 4/0 | 0 | 1 | 1 | 1 | 2 | 4 | 7 | 11 | 14 | 18 |
| | 250 | 0 | 0 | 1 | 1 | 1 | 3 | 6 | 9 | 12 | 15 |
| | 300 | 0 | 0 | 1 | 1 | 1 | 3 | 5 | 8 | 10 | 13 |
| | 350 | 0 | 0 | 1 | 1 | 1 | 2 | 4 | 7 | 9 | 11 |
| | 400 | 0 | 0 | 0 | 1 | 1 | 1 | 4 | 6 | 8 | 10 |
| | 500 | 0 | 0 | 0 | 1 | 1 | 1 | 3 | 5 | 6 | 8 |
| | 600 | 0 | 0 | 0 | 1 | 1 | 1 | 2 | 4 | 5 | 6 |
| | 700 | 0 | 0 | 0 | 0 | 1 | 1 | 2 | 3 | 4 | 6 |
| | 750 | 0 | 0 | 0 | 0 | 1 | 1 | 1 | 3 | 4 | 5 |
| | 800 | 0 | 0 | 0 | 0 | 1 | 1 | 1 | 3 | 4 | 5 |
| | 900 | 0 | 0 | 0 | 0 | 1 | 1 | 1 | 3 | 3 | 4 |
| | 1000 | 0 | 0 | 0 | 0 | 0 | 1 | 1 | 2 | 3 | 4 |
| | 1250 | 0 | 0 | 0 | 0 | 0 | 1 | 1 | 1 | 2 | 3 |
| | 1500 | 0 | 0 | 0 | 0 | 0 | 1 | 1 | 1 | 1 | 3 |
| | 1750 | 0 | 0 | 0 | 0 | 0 | 0 | 1 | 1 | 1 | 2 |
| | 2000 | 0 | 0 | 0 | 0 | 0 | 0 | 1 | 1 | 1 | 1 |

### FIXTURE WIRES

| Type | Conductor Size (AWG/kcmil) | 16 (½) | 21 (¾) | 27 (1) | 35 (1¼) | 41 (1½) | 53 (2) |
|------|------|------|------|------|------|------|------|
| FFH-2, RFH-2, RFHH-3 | 18 | 8 | 14 | 24 | 41 | 56 | 92 |
| | 16 | 7 | 12 | 20 | 34 | 47 | 78 |
| SF-2, SFF-2 | 18 | 10 | 18 | 30 | 52 | 71 | 116 |
| | 16 | 8 | 15 | 25 | 43 | 58 | 96 |
| | 14 | 7 | 12 | 20 | 34 | 47 | 78 |
| SF-1, SFF-1 | 18 | 18 | 33 | 53 | 92 | 125 | 206 |
| RFH-1, RFHH-2, TF, TFF, XF, XFF | 18 | 14 | 24 | 39 | 68 | 92 | 152 |
| RFHH-2, TF, TFF, XF, XFF | 16 | 11 | 19 | 31 | 55 | 74 | 123 |
| XF, XFF | 14 | 8 | 15 | 25 | 43 | 58 | 96 |
| TFN, TFFN | 18 | 22 | 38 | 63 | 108 | 148 | 244 |
| | 16 | 17 | 29 | 48 | 83 | 113 | 186 |
| PF, PFF, PGF, PGFF, PAF, PTF, PTFF, PAFF | 18 | 21 | 36 | 59 | 103 | 140 | 231 |
| | 16 | 16 | 28 | 46 | 79 | 108 | 179 |
| | 14 | 12 | 21 | 34 | 60 | 81 | 134 |
| ZF, ZFF, ZHF, HF, HFF | 18 | 27 | 47 | 77 | 133 | 181 | 298 |
| | 16 | 20 | 35 | 56 | 98 | 133 | 220 |
| | 14 | 14 | 25 | 41 | 72 | 98 | 161 |
| KF-2, KFF-2 | 18 | 39 | 69 | 111 | 193 | 262 | 433 |
| | 16 | 27 | 48 | 78 | 136 | 185 | 305 |
| | 14 | 19 | 33 | 54 | 93 | 127 | 209 |
| | 12 | 13 | 23 | 37 | 64 | 87 | 144 |
| | 10 | 8 | 15 | 25 | 43 | 58 | 96 |
| KF-1, KFF-1 | 18 | 46 | 82 | 133 | 230 | 313 | 516 |
| | 16 | 33 | 57 | 93 | 161 | 220 | 362 |
| | 14 | 22 | 38 | 63 | 108 | 148 | 244 |
| | 12 | 14 | 25 | 41 | 72 | 98 | 161 |
| | 10 | 9 | 16 | 27 | 47 | 64 | 105 |
| XF, XFF | 12 | 4 | 8 | 13 | 23 | 31 | 51 |
| | 10 | 3 | 6 | 10 | 18 | 24 | 40 |

Note: This table is for concentric stranded conductors only. For compact stranded conductors, Table C1(A) should be used.

*Types RHH, RHW, and RHW-2 without outer covering.

**Table C1(A) Maximum Number of Compact Conductors in Electrical Metallic Tubing (EMT)** *(Based on Table 1, Chapter 9)*

### COMPACT CONDUCTORS

| Type | Conductor Size (AWG/kcmil) | 16 (½) | 21 (¾) | 27 (1) | 35 (1¼) | 41 (1½) | 53 (2) | 63 (2½) | 78 (3) | 91 (3½) | 103 (4) |
|------|------|------|------|------|------|------|------|------|------|------|------|
| THW, THW-2, THHW | 8 | 2 | 4 | 6 | 11 | 16 | 26 | 46 | 69 | 90 | 115 |
| | 6 | 1 | 3 | 5 | 9 | 12 | 20 | 35 | 53 | 70 | 89 |
| | 4 | 1 | 2 | 4 | 6 | 9 | 15 | 26 | 40 | 52 | 67 |
| | 2 | 1 | 1 | 3 | 5 | 7 | 11 | 19 | 29 | 38 | 49 |
| | 1 | 1 | 1 | 1 | 3 | 4 | 8 | 13 | 21 | 27 | 34 |
| | 1/0 | 1 | 1 | 1 | 3 | 4 | 7 | 12 | 18 | 23 | 30 |
| | 2/0 | 0 | 1 | 1 | 2 | 3 | 5 | 10 | 15 | 20 | 25 |
| | 3/0 | 0 | 1 | 1 | 1 | 3 | 5 | 8 | 13 | 17 | 21 |
| | 4/0 | 0 | 1 | 1 | 1 | 2 | 4 | 7 | 11 | 14 | 18 |
| | 250 | 0 | 0 | 1 | 1 | 1 | 3 | 5 | 8 | 11 | 14 |
| | 300 | 0 | 0 | 1 | 1 | 1 | 3 | 5 | 7 | 9 | 12 |
| | 350 | 0 | 0 | 1 | 1 | 1 | 2 | 4 | 6 | 8 | 11 |
| | 400 | 0 | 0 | 1 | 1 | 1 | 1 | 4 | 6 | 8 | 10 |
| | 500 | 0 | 0 | 0 | 1 | 1 | 1 | 3 | 5 | 6 | 8 |
| | 600 | 0 | 0 | 0 | 1 | 1 | 1 | 2 | 4 | 5 | 7 |
| | 700 | 0 | 0 | 0 | 1 | 1 | 1 | 2 | 3 | 4 | 6 |
| | 750 | 0 | 0 | 0 | 0 | 1 | 1 | 1 | 3 | 4 | 5 |
| | 1000 | 0 | 0 | 0 | 0 | 1 | 1 | 1 | 2 | 3 | 4 |
| THHN, THWN, THWN-2 | 8 | — | — | — | — | — | — | — | — | — | — |
| | 6 | 2 | 4 | 7 | 13 | 18 | 29 | 52 | 78 | 102 | 130 |
| | 4 | 1 | 3 | 4 | 8 | 11 | 18 | 32 | 48 | 63 | 81 |
| | 2 | 1 | 1 | 3 | 6 | 8 | 13 | 23 | 34 | 45 | 58 |
| | 1 | 1 | 1 | 2 | 4 | 6 | 10 | 17 | 26 | 34 | 43 |
| | 1/0 | 1 | 1 | 1 | 3 | 5 | 8 | 14 | 22 | 29 | 37 |
| | 2/0 | 1 | 1 | 1 | 3 | 4 | 7 | 12 | 18 | 24 | 30 |
| | 3/0 | 0 | 1 | 1 | 2 | 3 | 6 | 10 | 15 | 20 | 25 |
| | 4/0 | 0 | 1 | 1 | 1 | 3 | 5 | 8 | 12 | 16 | 21 |
| | 250 | 0 | 1 | 1 | 1 | 1 | 4 | 6 | 10 | 13 | 16 |
| | 300 | 0 | 0 | 1 | 1 | 1 | 3 | 5 | 8 | 11 | 14 |
| | 350 | 0 | 0 | 1 | 1 | 1 | 3 | 5 | 7 | 10 | 12 |
| | 400 | 0 | 0 | 1 | 1 | 1 | 2 | 4 | 6 | 9 | 11 |
| | 500 | 0 | 0 | 0 | 1 | 1 | 1 | 4 | 5 | 7 | 9 |
| | 600 | 0 | 0 | 0 | 1 | 1 | 1 | 3 | 4 | 6 | 7 |
| | 700 | 0 | 0 | 0 | 1 | 1 | 1 | 2 | 4 | 5 | 7 |
| | 750 | 0 | 0 | 0 | 1 | 1 | 1 | 2 | 4 | 5 | 6 |
| | 1000 | 0 | 0 | 0 | 0 | 1 | 1 | 1 | 3 | 3 | 4 |
| XHHW, XHHW-2 | 8 | 3 | 5 | 8 | 15 | 20 | 34 | 59 | 90 | 117 | 149 |
| | 6 | 1 | 4 | 6 | 11 | 15 | 25 | 44 | 66 | 87 | 111 |
| | 4 | 1 | 3 | 4 | 8 | 11 | 18 | 32 | 48 | 63 | 81 |
| | 2 | 1 | 1 | 3 | 6 | 8 | 13 | 23 | 34 | 45 | 58 |
| | 1 | 1 | 1 | 2 | 4 | 6 | 10 | 17 | 26 | 34 | 43 |
| | 1/0 | 1 | 1 | 1 | 3 | 5 | 8 | 14 | 22 | 29 | 37 |
| | 2/0 | 1 | 1 | 1 | 3 | 4 | 7 | 12 | 18 | 24 | 31 |
| | 3/0 | 0 | 1 | 1 | 2 | 3 | 6 | 10 | 15 | 20 | 25 |
| | 4/0 | 0 | 1 | 1 | 1 | 3 | 5 | 8 | 13 | 17 | 21 |
| | 250 | 0 | 1 | 1 | 1 | 2 | 4 | 7 | 10 | 13 | 17 |
| | 300 | 0 | 0 | 1 | 1 | 1 | 3 | 6 | 9 | 11 | 14 |
| | 350 | 0 | 0 | 1 | 1 | 1 | 3 | 5 | 8 | 10 | 13 |
| | 400 | 0 | 0 | 1 | 1 | 1 | 2 | 4 | 7 | 9 | 11 |
| | 500 | 0 | 0 | 0 | 1 | 1 | 1 | 4 | 6 | 7 | 9 |
| | 600 | 0 | 0 | 0 | 1 | 1 | 1 | 3 | 4 | 6 | 8 |
| | 700 | 0 | 0 | 0 | 1 | 1 | 1 | 2 | 4 | 5 | 7 |
| | 750 | 0 | 0 | 0 | 1 | 1 | 1 | 2 | 3 | 5 | 6 |
| | 1000 | 0 | 0 | 0 | 0 | 1 | 1 | 1 | 3 | 4 | 5 |

Definition: *Compact stranding* is the result of a manufacturing process where the standard conductor is compressed to the extent that the interstices (voids between strand wires) are virtually eliminated.

**Table C2 Maximum Number of Conductors or Fixture Wires in Electrical Nonmetallic Tubing (ENT)** (*Based on Table 1, Chapter 9*)

| Type | Conductor Size (AWG/kcmil) | 16 (½) | 21 (¾) | 27 (1) | 35 (1¼) | 41 (1½) | 53 (2) |
|---|---|---|---|---|---|---|---|
| RHH, RHW, RHW-2 | 14 | 3 | 6 | 10 | 19 | 26 | 43 |
| | 12 | 2 | 5 | 9 | 16 | 22 | 36 |
| RHH, RHW, RHW-2 | 10 | 1 | 4 | 7 | 13 | 17 | 29 |
| | 8 | 1 | 1 | 3 | 6 | 9 | 15 |
| | 6 | 1 | 1 | 3 | 5 | 7 | 12 |
| | 4 | 1 | 1 | 2 | 4 | 6 | 9 |
| | 3 | 1 | 1 | 1 | 3 | 5 | 8 |
| | 2 | 0 | 1 | 1 | 3 | 4 | 7 |
| | 1 | 0 | 1 | 1 | 1 | 3 | 5 |
| | 1/0 | 0 | 0 | 1 | 1 | 2 | 4 |
| | 2/0 | 0 | 0 | 1 | 1 | 1 | 3 |
| | 3/0 | 0 | 0 | 1 | 1 | 1 | 3 |
| | 4/0 | 0 | 0 | 1 | 1 | 1 | 2 |
| | 250 | 0 | 0 | 0 | 1 | 1 | 1 |
| | 300 | 0 | 0 | 0 | 1 | 1 | 1 |
| | 350 | 0 | 0 | 0 | 1 | 1 | 1 |
| | 400 | 0 | 0 | 0 | 1 | 1 | 1 |
| | 500 | 0 | 0 | 0 | 0 | 1 | 1 |
| | 600 | 0 | 0 | 0 | 0 | 1 | 1 |
| | 700 | 0 | 0 | 0 | 0 | 0 | 1 |
| | 750 | 0 | 0 | 0 | 0 | 0 | 1 |
| | 800 | 0 | 0 | 0 | 0 | 0 | 1 |
| | 900 | 0 | 0 | 0 | 0 | 0 | 1 |
| | 1000 | 0 | 0 | 0 | 0 | 0 | 1 |
| | 1250 | 0 | 0 | 0 | 0 | 0 | 0 |
| | 1500 | 0 | 0 | 0 | 0 | 0 | 0 |
| | 1750 | 0 | 0 | 0 | 0 | 0 | 0 |
| | 2000 | 0 | 0 | 0 | 0 | 0 | 0 |
| TW, THHW, THW, THW-2 | 14 | 7 | 13 | 22 | 40 | 55 | 92 |
| | 12 | 5 | 10 | 17 | 31 | 42 | 71 |
| | 10 | 4 | 7 | 13 | 23 | 32 | 52 |
| | 8 | 1 | 4 | 7 | 13 | 17 | 29 |
| RHH*, RHW*, RHW-2* | 14 | 4 | 8 | 15 | 27 | 37 | 61 |
| RHH*, RHW*, RHW-2* | 12 | 3 | 7 | 12 | 21 | 29 | 49 |
| | 10 | 3 | 5 | 9 | 17 | 23 | 38 |
| RHH*, RHW*, RHW-2* | 8 | 1 | 3 | 5 | 10 | 14 | 23 |
| RHH*, RHW*, RHW-2*, TW, THW, THHW, THW-2 | 6 | 1 | 2 | 4 | 7 | 10 | 17 |
| | 4 | 1 | 1 | 3 | 5 | 8 | 13 |
| | 3 | 1 | 1 | 2 | 5 | 7 | 11 |
| | 2 | 1 | 1 | 2 | 4 | 6 | 9 |
| | 1 | 0 | 1 | 1 | 3 | 4 | 6 |
| | 1/0 | 0 | 1 | 1 | 2 | 3 | 5 |
| | 2/0 | 0 | 1 | 1 | 1 | 3 | 5 |
| | 3/0 | 0 | 0 | 1 | 1 | 2 | 4 |
| | 4/0 | 0 | 0 | 1 | 1 | 1 | 3 |
| | 250 | 0 | 0 | 1 | 1 | 1 | 2 |
| | 300 | 0 | 0 | 0 | 1 | 1 | 2 |
| | 350 | 0 | 0 | 0 | 1 | 1 | 1 |
| | 400 | 0 | 0 | 0 | 1 | 1 | 1 |
| | 500 | 0 | 0 | 0 | 1 | 1 | 1 |
| | 600 | 0 | 0 | 0 | 0 | 1 | 1 |
| | 700 | 0 | 0 | 0 | 0 | 1 | 1 |
| | 750 | 0 | 0 | 0 | 0 | 1 | 1 |
| | 800 | 0 | 0 | 0 | 0 | 1 | 1 |
| | 900 | 0 | 0 | 0 | 0 | 0 | 1 |
| | 1000 | 0 | 0 | 0 | 0 | 0 | 1 |
| | 1250 | 0 | 0 | 0 | 0 | 0 | 1 |
| | 1500 | 0 | 0 | 0 | 0 | 0 | 0 |
| | 1750 | 0 | 0 | 0 | 0 | 0 | 0 |
| | 2000 | 0 | 0 | 0 | 0 | 0 | 0 |

**Table C2** *Continued*

| Type | Conductor Size (AWG/kcmil) | 16 (½) | 21 (¾) | 27 (1) | 35 (1¼) | 41 (1½) | 53 (2) |
|---|---|---|---|---|---|---|---|
| THHN, THWN, THWN-2 | 14 | 10 | 18 | 32 | 58 | 80 | 132 |
| | 12 | 7 | 13 | 23 | 42 | 58 | 96 |
| | 10 | 4 | 8 | 15 | 26 | 36 | 60 |
| | 8 | 2 | 5 | 8 | 15 | 21 | 35 |
| | 6 | 1 | 3 | 6 | 11 | 15 | 25 |
| | 4 | 1 | 1 | 4 | 7 | 9 | 15 |
| | 3 | 1 | 1 | 3 | 5 | 8 | 13 |
| | 2 | 1 | 1 | 2 | 5 | 6 | 11 |
| | 1 | 1 | 1 | 1 | 3 | 5 | 8 |
| | 1/0 | 0 | 1 | 1 | 3 | 4 | 7 |
| | 2/0 | 0 | 1 | 1 | 2 | 3 | 5 |
| | 3/0 | 0 | 1 | 1 | 1 | 3 | 4 |
| | 4/0 | 0 | 0 | 1 | 1 | 2 | 4 |
| | 250 | 0 | 0 | 1 | 1 | 1 | 3 |
| | 300 | 0 | 0 | 1 | 1 | 1 | 2 |
| | 350 | 0 | 0 | 0 | 1 | 1 | 2 |
| | 400 | 0 | 0 | 0 | 1 | 1 | 1 |
| | 500 | 0 | 0 | 0 | 1 | 1 | 1 |
| | 600 | 0 | 0 | 0 | 1 | 1 | 1 |
| | 700 | 0 | 0 | 0 | 0 | 1 | 1 |
| | 750 | 0 | 0 | 0 | 0 | 1 | 1 |
| | 800 | 0 | 0 | 0 | 0 | 1 | 1 |
| | 900 | 0 | 0 | 0 | 0 | 1 | 1 |
| | 1000 | 0 | 0 | 0 | 0 | 0 | 1 |
| FEP, FEPB, PFA, PFAH, TFE | 14 | 10 | 18 | 31 | 56 | 77 | 128 |
| | 12 | 7 | 13 | 23 | 41 | 56 | 93 |
| | 10 | 5 | 9 | 16 | 29 | 40 | 67 |
| | 8 | 3 | 5 | 9 | 17 | 23 | 38 |
| | 6 | 1 | 4 | 6 | 12 | 16 | 27 |
| | 4 | 1 | 2 | 4 | 8 | 11 | 19 |
| | 3 | 1 | 1 | 4 | 7 | 9 | 16 |
| | 2 | 1 | 1 | 3 | 5 | 8 | 13 |
| PFA, PFAH, TFE | 1 | 1 | 1 | 1 | 4 | 5 | 9 |
| PFA, PFAH, TFE, Z | 1/0 | 0 | 1 | 1 | 3 | 4 | 7 |
| | 2/0 | 0 | 1 | 1 | 2 | 4 | 6 |
| | 3/0 | 0 | 1 | 1 | 1 | 3 | 5 |
| | 4/0 | 0 | 1 | 1 | 1 | 2 | 4 |
| Z | 14 | 12 | 22 | 38 | 68 | 93 | 154 |
| | 12 | 8 | 15 | 27 | 48 | 66 | 109 |
| | 10 | 5 | 9 | 16 | 29 | 40 | 67 |
| | 8 | 3 | 6 | 10 | 18 | 25 | 42 |
| | 6 | 1 | 4 | 7 | 13 | 18 | 30 |
| | 4 | 1 | 3 | 5 | 9 | 12 | 20 |
| | 3 | 1 | 1 | 3 | 6 | 9 | 15 |
| | 2 | 1 | 1 | 3 | 5 | 7 | 12 |
| | 1 | 1 | 1 | 2 | 4 | 6 | 10 |
| XHH, XHHW, XHHW-2, ZW | 14 | 7 | 13 | 22 | 40 | 55 | 92 |
| | 12 | 5 | 10 | 17 | 31 | 42 | 71 |
| | 10 | 4 | 7 | 13 | 23 | 32 | 52 |
| | 8 | 1 | 4 | 7 | 13 | 17 | 29 |
| | 6 | 1 | 3 | 5 | 9 | 13 | 21 |
| | 4 | 1 | 1 | 4 | 7 | 9 | 15 |
| | 3 | 1 | 1 | 3 | 6 | 8 | 13 |
| | 2 | 1 | 1 | 2 | 5 | 6 | 11 |
| XHH, XHHW, XHHW-2 | 1 | 1 | 1 | 1 | 3 | 5 | 8 |
| | 1/0 | 0 | 1 | 1 | 3 | 4 | 7 |
| | 2/0 | 0 | 1 | 1 | 2 | 3 | 6 |
| | 3/0 | 0 | 1 | 1 | 1 | 3 | 5 |
| | 4/0 | 0 | 0 | 1 | 1 | 2 | 4 |
| | 250 | 0 | 0 | 1 | 1 | 1 | 3 |
| | 300 | 0 | 0 | 1 | 1 | 1 | 3 |
| | 350 | 0 | 0 | 1 | 1 | 1 | 2 |
| | 400 | 0 | 0 | 0 | 1 | 1 | 1 |
| | 500 | 0 | 0 | 0 | 1 | 1 | 1 |
| | 600 | 0 | 0 | 0 | 1 | 1 | 1 |
| | 700 | 0 | 0 | 0 | 0 | 1 | 1 |
| | 750 | 0 | 0 | 0 | 0 | 1 | 1 |
| | 800 | 0 | 0 | 0 | 0 | 1 | 1 |
| | 900 | 0 | 0 | 0 | 0 | 1 | 1 |
| | 1000 | 0 | 0 | 0 | 0 | 0 | 1 |
| | 1250 | 0 | 0 | 0 | 0 | 0 | 1 |
| | 1500 | 0 | 0 | 0 | 0 | 0 | 1 |
| | 1750 | 0 | 0 | 0 | 0 | 0 | 0 |
| | 2000 | 0 | 0 | 0 | 0 | 0 | 0 |

## FIXTURE WIRES

| Type | Conductor Size (AWG/kcmil) | 16 (½) | 21 (¾) | 27 (1) | 35 (1¼) | 41 (1½) | 53 (2) |
|---|---|---|---|---|---|---|---|
| FFH-2, RFH-2, RFHH-3 | 18 | 6 | 12 | 21 | 39 | 53 | 88 |
| | 16 | 5 | 10 | 18 | 32 | 45 | 74 |
| SF-2, SFF-2 | 18 | 8 | 15 | 27 | 49 | 67 | 111 |
| | 16 | 7 | 13 | 22 | 40 | 55 | 92 |
| | 14 | 5 | 10 | 18 | 32 | 45 | 74 |
| SF-1, SFF-1 | 18 | 15 | 28 | 48 | 86 | 119 | 197 |
| RFH-1, RFHH-2, TF, TFF, XF, XFF | 18 | 11 | 20 | 35 | 64 | 88 | 145 |
| RFHH-2, TF, TFF, XF, XFF | 16 | 9 | 16 | 29 | 51 | 71 | 117 |
| XF, XFF | 14 | 7 | 13 | 22 | 40 | 55 | 92 |
| TFN, TFFN | 18 | 18 | 33 | 57 | 102 | 141 | 233 |
| | 16 | 13 | 25 | 43 | 78 | 107 | 178 |
| PF, PFF, PGF, PGFF, PAF, PTF, PTFF, PAFF | 18 | 17 | 31 | 54 | 97 | 133 | 221 |
| | 16 | 13 | 24 | 42 | 75 | 103 | 171 |
| | 14 | 10 | 18 | 31 | 56 | 77 | 128 |
| ZF, ZFF, ZHF, HF, HFF | 18 | 22 | 40 | 70 | 125 | 172 | 285 |
| | 16 | 16 | 29 | 51 | 92 | 127 | 210 |
| | 14 | 12 | 22 | 38 | 68 | 93 | 154 |
| KF-2, KFF-2 | 18 | 31 | 58 | 101 | 182 | 250 | 413 |
| | 16 | 22 | 41 | 71 | 128 | 176 | 291 |
| | 14 | 15 | 28 | 49 | 88 | 121 | 200 |
| | 12 | 10 | 19 | 33 | 60 | 83 | 138 |
| | 10 | 7 | 13 | 22 | 40 | 55 | 92 |
| KF-1, KFF-1 | 18 | 38 | 69 | 121 | 217 | 298 | 493 |
| | 16 | 26 | 49 | 85 | 152 | 209 | 346 |
| | 14 | 18 | 33 | 57 | 102 | 141 | 233 |
| | 12 | 12 | 22 | 38 | 68 | 93 | 154 |
| | 10 | 7 | 14 | 24 | 44 | 61 | 101 |
| XF, XFF | 12 | 3 | 7 | 12 | 21 | 29 | 49 |
| | 10 | 3 | 5 | 9 | 17 | 23 | 38 |

Note: This table is for concentric stranded conductors only. For compact stranded conductors, Table C2(A) should be used.

*Types RHH, RHW, and RHW-2 without outer covering.

**Table C2(A) Maximum Number of Compact Conductors in Electrical Nonmetallic Tubing (ENT)** (*Based on Table 1, Chapter 9*)

### COMPACT CONDUCTORS

| Type | Conductor Size (AWG/kcmil) | 16 (½) | 21 (¾) | 27 (1) | 35 (1¼) | 41 (1½) | 53 (2) |
|---|---|---|---|---|---|---|---|
| THW, THW-2, THHW | 8 | 1 | 3 | 6 | 11 | 15 | 25 |
| | 6 | 1 | 2 | 4 | 8 | 11 | 19 |
| | 4 | 1 | 1 | 3 | 6 | 8 | 14 |
| | 2 | 1 | 1 | 2 | 4 | 6 | 10 |
| | 1 | 0 | 1 | 1 | 3 | 4 | 7 |
| | 1/0 | 0 | 1 | 1 | 3 | 4 | 6 |
| | 2/0 | 0 | 1 | 1 | 2 | 3 | 5 |
| | 3/0 | 0 | 1 | 1 | 1 | 3 | 4 |
| | 4/0 | 0 | 0 | 1 | 1 | 2 | 4 |
| | 250 | 0 | 0 | 1 | 1 | 1 | 3 |
| | 300 | 0 | 0 | 1 | 1 | 1 | 2 |
| | 350 | 0 | 0 | 0 | 1 | 1 | 2 |
| | 400 | 0 | 0 | 0 | 1 | 1 | 1 |
| | 500 | 0 | 0 | 0 | 1 | 1 | 1 |
| | 600 | 0 | 0 | 0 | 1 | 1 | 1 |
| | 700 | 0 | 0 | 0 | 0 | 1 | 1 |
| | 750 | 0 | 0 | 0 | 0 | 1 | 1 |
| | 1000 | 0 | 0 | 0 | 0 | 0 | 1 |
| THHN, THWN, THWN-2 | 8 | — | — | — | — | — | — |
| | 6 | 1 | 4 | 7 | 12 | 17 | 28 |
| | 4 | 1 | 2 | 4 | 7 | 10 | 17 |
| | 2 | 1 | 1 | 3 | 5 | 7 | 12 |
| | 1 | 1 | 1 | 2 | 4 | 5 | 9 |
| | 1/0 | 1 | 1 | 1 | 3 | 5 | 8 |
| | 2/0 | 0 | 1 | 1 | 3 | 4 | 6 |
| | 3/0 | 0 | 1 | 1 | 2 | 3 | 5 |
| | 4/0 | 0 | 1 | 1 | 1 | 2 | 4 |
| | 250 | 0 | 0 | 1 | 1 | 1 | 3 |
| | 300 | 0 | 0 | 1 | 1 | 1 | 3 |
| | 350 | 0 | 0 | 1 | 1 | 1 | 2 |
| | 400 | 0 | 0 | 0 | 1 | 1 | 2 |
| | 500 | 0 | 0 | 0 | 1 | 1 | 1 |
| | 600 | 0 | 0 | 0 | 1 | 1 | 1 |
| | 700 | 0 | 0 | 0 | 1 | 1 | 1 |

**Table C2(A)** *Continued*

### COMPACT CONDUCTORS

| Type | Conductor Size (AWG/kcmil) | 16 (½) | 21 (¾) | 27 (1) | 35 (1¼) | 41 (1½) | 53 (2) |
|---|---|---|---|---|---|---|---|
| THHN, THWN, THWN-2 | 750 | 0 | 0 | 0 | 1 | 1 | 1 |
| | 1000 | 0 | 0 | 0 | 0 | 1 | 1 |
| XHHW, XHHW-2 | 8 | 2 | 4 | 8 | 14 | 19 | 32 |
| | 6 | 1 | 3 | 6 | 10 | 14 | 24 |
| | 4 | 1 | 2 | 4 | 7 | 10 | 17 |
| | 2 | 1 | 1 | 3 | 5 | 7 | 12 |
| | 1 | 1 | 1 | 2 | 4 | 5 | 9 |
| | 1/0 | 1 | 1 | 1 | 3 | 5 | 8 |
| | 2/0 | 0 | 1 | 1 | 3 | 4 | 7 |
| | 3/0 | 0 | 1 | 1 | 2 | 3 | 5 |
| | 4/0 | 0 | 1 | 1 | 1 | 3 | 4 |
| | 250 | 0 | 0 | 1 | 1 | 1 | 3 |
| | 300 | 0 | 0 | 1 | 1 | 1 | 3 |
| | 350 | 0 | 0 | 1 | 1 | 1 | 3 |
| | 400 | 0 | 0 | 1 | 1 | 1 | 2 |
| | 500 | 0 | 0 | 0 | 1 | 1 | 1 |
| | 600 | 0 | 0 | 0 | 1 | 1 | 1 |
| | 700 | 0 | 0 | 0 | 1 | 1 | 1 |
| | 750 | 0 | 0 | 0 | 1 | 1 | 1 |
| | 1000 | 0 | 0 | 0 | 0 | 1 | 1 |

Definition: *Compact stranding* is the result of a manufacturing process where the standard conductor is compressed to the extent that the interstices (voids between strand wires) are virtually eliminated.

**Table C3 Maximum Number of Conductors or Fixture Wires in Flexible Metal Conduit (FMC)** (*Based on Table 1, Chapter 9*)

### CONDUCTORS

| Type | Conductor Size (AWG/kcmil) | 16 (½) | 21 (¾) | 27 (1) | 35 (1¼) | 41 (1½) | 53 (2) | 63 (2½) | 78 (3) | 91 (3½) | 103 (4) |
|---|---|---|---|---|---|---|---|---|---|---|---|
| RHH, RHW, RHW-2 | 14 | 4 | 7 | 11 | 17 | 25 | 44 | 67 | 96 | 131 | 171 |
| | 12 | 3 | 6 | 9 | 14 | 21 | 37 | 55 | 80 | 109 | 142 |
| RHH, RHW, RHW-2 | 10 | 3 | 5 | 7 | 11 | 17 | 30 | 45 | 64 | 88 | 115 |
| | 8 | 1 | 2 | 4 | 6 | 9 | 15 | 23 | 34 | 46 | 60 |
| | 6 | 1 | 1 | 3 | 5 | 7 | 12 | 19 | 27 | 37 | 48 |
| | 4 | 1 | 1 | 2 | 4 | 5 | 10 | 14 | 21 | 29 | 37 |
| | 3 | 1 | 1 | 1 | 3 | 5 | 8 | 13 | 18 | 25 | 33 |
| | 2 | 1 | 1 | 1 | 3 | 4 | 7 | 11 | 16 | 22 | 28 |
| | 1 | 0 | 1 | 1 | 1 | 2 | 5 | 7 | 10 | 14 | 19 |
| | 1/0 | 0 | 1 | 1 | 1 | 2 | 4 | 6 | 9 | 12 | 16 |
| | 2/0 | 0 | 1 | 1 | 1 | 1 | 3 | 5 | 8 | 11 | 14 |
| | 3/0 | 0 | 0 | 1 | 1 | 1 | 3 | 5 | 7 | 9 | 12 |
| | 4/0 | 0 | 0 | 1 | 1 | 1 | 2 | 4 | 6 | 8 | 10 |
| | 250 | 0 | 0 | 0 | 1 | 1 | 1 | 3 | 4 | 6 | 8 |
| | 300 | 0 | 0 | 0 | 1 | 1 | 1 | 2 | 4 | 5 | 7 |
| | 350 | 0 | 0 | 0 | 1 | 1 | 1 | 2 | 3 | 5 | 6 |
| | 400 | 0 | 0 | 0 | 0 | 1 | 1 | 1 | 3 | 4 | 6 |
| | 500 | 0 | 0 | 0 | 0 | 1 | 1 | 1 | 3 | 4 | 5 |
| | 600 | 0 | 0 | 0 | 0 | 1 | 1 | 1 | 2 | 3 | 4 |
| | 700 | 0 | 0 | 0 | 0 | 0 | 1 | 1 | 1 | 3 | 3 |
| | 750 | 0 | 0 | 0 | 0 | 0 | 1 | 1 | 1 | 2 | 3 |
| | 800 | 0 | 0 | 0 | 0 | 0 | 1 | 1 | 1 | 2 | 3 |
| | 900 | 0 | 0 | 0 | 0 | 0 | 1 | 1 | 1 | 2 | 3 |
| | 1000 | 0 | 0 | 0 | 0 | 0 | 1 | 1 | 1 | 1 | 3 |
| | 1250 | 0 | 0 | 0 | 0 | 0 | 0 | 1 | 1 | 1 | 1 |
| | 1500 | 0 | 0 | 0 | 0 | 0 | 0 | 1 | 1 | 1 | 1 |
| | 1750 | 0 | 0 | 0 | 0 | 0 | 0 | 1 | 1 | 1 | 1 |
| | 2000 | 0 | 0 | 0 | 0 | 0 | 0 | 0 | 1 | 1 | 1 |
| TW, THHW, THW, THW-2 | 14 | 9 | 15 | 23 | 36 | 53 | 94 | 141 | 203 | 277 | 361 |
| | 12 | 7 | 11 | 18 | 28 | 41 | 72 | 108 | 156 | 212 | 277 |
| | 10 | 5 | 8 | 13 | 21 | 30 | 54 | 81 | 116 | 158 | 207 |
| | 8 | 3 | 5 | 7 | 11 | 17 | 30 | 45 | 64 | 88 | 115 |
| RHH*, RHW*, RHW-2* | 14 | 6 | 10 | 15 | 24 | 35 | 62 | 94 | 135 | 184 | 240 |
| RHH*, RHW*, RHW-2* | 12 | 5 | 8 | 12 | 19 | 28 | 50 | 75 | 108 | 148 | 193 |
| | 10 | 4 | 6 | 10 | 15 | 22 | 39 | 59 | 85 | 115 | 151 |
| RHH*, RHW*, RHW-2* | 8 | 1 | 4 | 6 | 9 | 13 | 23 | 35 | 51 | 69 | 90 |

**Table C3** *Continued*

| | | CONDUCTORS | | | | | | | | | |
|---|---|---|---|---|---|---|---|---|---|---|---|
| | | Metric Designator (Trade Size) | | | | | | | | | |
| Type | Conductor Size (AWG/kcmil) | 16 (½) | 21 (¾) | 27 (1) | 35 (1¼) | 41 (1½) | 53 (2) | 63 (2½) | 78 (3) | 91 (3½) | 103 (4) |
| RHH*, RHW*, RHW-2*, TW, THW, THHW, THW-2 | 6 | 1 | 3 | 4 | 7 | 10 | 18 | 27 | 39 | 53 | 69 |
| | 4 | 1 | 1 | 3 | 5 | 7 | 13 | 20 | 29 | 39 | 51 |
| | 3 | 1 | 1 | 3 | 4 | 6 | 11 | 17 | 25 | 34 | 44 |
| | 2 | 1 | 1 | 2 | 4 | 5 | 10 | 14 | 21 | 29 | 37 |
| | 1 | 1 | 1 | 1 | 2 | 4 | 7 | 10 | 15 | 20 | 26 |
| | 1/0 | 0 | 1 | 1 | 1 | 3 | 6 | 9 | 12 | 17 | 22 |
| | 2/0 | 0 | 1 | 1 | 1 | 3 | 5 | 7 | 10 | 14 | 19 |
| | 3/0 | 0 | 1 | 1 | 1 | 2 | 4 | 6 | 9 | 12 | 16 |
| | 4/0 | 0 | 0 | 1 | 1 | 1 | 3 | 5 | 7 | 10 | 13 |
| | 250 | 0 | 0 | 1 | 1 | 1 | 3 | 4 | 6 | 8 | 11 |
| | 300 | 0 | 0 | 1 | 1 | 1 | 2 | 3 | 5 | 7 | 9 |
| | 350 | 0 | 0 | 0 | 1 | 1 | 1 | 3 | 4 | 6 | 8 |
| | 400 | 0 | 0 | 0 | 1 | 1 | 1 | 3 | 4 | 6 | 7 |
| | 500 | 0 | 0 | 0 | 1 | 1 | 1 | 2 | 3 | 5 | 6 |
| | 600 | 0 | 0 | 0 | 0 | 1 | 1 | 1 | 3 | 4 | 5 |
| | 700 | 0 | 0 | 0 | 0 | 1 | 1 | 1 | 2 | 3 | 4 |
| | 750 | 0 | 0 | 0 | 0 | 1 | 1 | 1 | 2 | 3 | 4 |
| | 800 | 0 | 0 | 0 | 0 | 1 | 1 | 1 | 1 | 3 | 4 |
| | 900 | 0 | 0 | 0 | 0 | 0 | 1 | 1 | 1 | 3 | 3 |
| | 1000 | 0 | 0 | 0 | 0 | 0 | 1 | 1 | 1 | 2 | 3 |
| | 1250 | 0 | 0 | 0 | 0 | 0 | 1 | 1 | 1 | 1 | 2 |
| | 1500 | 0 | 0 | 0 | 0 | 0 | 0 | 1 | 1 | 1 | 1 |
| | 1750 | 0 | 0 | 0 | 0 | 0 | 0 | 1 | 1 | 1 | 1 |
| | 2000 | 0 | 0 | 0 | 0 | 0 | 0 | 1 | 1 | 1 | 1 |
| THHN, THWN, THWN-2 | 14 | 13 | 22 | 33 | 52 | 76 | 134 | 202 | 291 | 396 | 518 |
| | 12 | 9 | 16 | 24 | 38 | 56 | 98 | 147 | 212 | 289 | 378 |
| | 10 | 6 | 10 | 15 | 24 | 35 | 62 | 93 | 134 | 182 | 238 |
| | 8 | 3 | 6 | 9 | 14 | 20 | 35 | 53 | 77 | 105 | 137 |
| | 6 | 2 | 4 | 6 | 10 | 14 | 25 | 38 | 55 | 76 | 99 |
| | 4 | 1 | 2 | 4 | 6 | 9 | 16 | 24 | 34 | 46 | 61 |
| | 3 | 1 | 1 | 3 | 5 | 7 | 13 | 20 | 29 | 39 | 51 |
| | 2 | 1 | 1 | 3 | 4 | 6 | 11 | 17 | 24 | 33 | 43 |
| | 1 | 1 | 1 | 1 | 3 | 4 | 8 | 12 | 18 | 24 | 32 |
| | 1/0 | 1 | 1 | 1 | 2 | 4 | 7 | 10 | 15 | 20 | 27 |
| | 2/0 | 0 | 1 | 1 | 1 | 3 | 6 | 9 | 12 | 17 | 22 |
| | 3/0 | 0 | 1 | 1 | 1 | 2 | 5 | 7 | 10 | 14 | 18 |
| | 4/0 | 0 | 1 | 1 | 1 | 1 | 4 | 6 | 8 | 12 | 15 |
| | 250 | 0 | 0 | 1 | 1 | 1 | 3 | 5 | 7 | 9 | 12 |
| | 300 | 0 | 0 | 1 | 1 | 1 | 3 | 4 | 6 | 8 | 11 |
| | 350 | 0 | 0 | 1 | 1 | 1 | 2 | 3 | 5 | 7 | 9 |
| | 400 | 0 | 0 | 0 | 1 | 1 | 1 | 3 | 5 | 6 | 8 |
| | 500 | 0 | 0 | 0 | 1 | 1 | 1 | 2 | 4 | 5 | 7 |
| | 600 | 0 | 0 | 0 | 0 | 1 | 1 | 1 | 3 | 4 | 5 |
| | 700 | 0 | 0 | 0 | 0 | 1 | 1 | 1 | 3 | 4 | 5 |
| | 750 | 0 | 0 | 0 | 0 | 1 | 1 | 1 | 2 | 3 | 4 |
| | 800 | 0 | 0 | 0 | 0 | 1 | 1 | 1 | 2 | 3 | 4 |
| | 900 | 0 | 0 | 0 | 0 | 0 | 1 | 1 | 1 | 3 | 4 |
| | 1000 | 0 | 0 | 0 | 0 | 0 | 1 | 1 | 1 | 3 | 3 |
| FEP, FEPB, PFA, PFAH, TFE | 14 | 12 | 21 | 32 | 51 | 74 | 130 | 196 | 282 | 385 | 502 |
| | 12 | 9 | 15 | 24 | 37 | 54 | 95 | 143 | 206 | 281 | 367 |
| | 10 | 6 | 11 | 17 | 26 | 39 | 68 | 103 | 148 | 201 | 263 |
| | 8 | 4 | 6 | 10 | 15 | 22 | 39 | 59 | 85 | 115 | 151 |
| | 6 | 2 | 4 | 7 | 11 | 16 | 28 | 42 | 60 | 82 | 107 |
| | 4 | 1 | 3 | 5 | 7 | 11 | 19 | 29 | 42 | 57 | 75 |
| | 3 | 1 | 2 | 4 | 6 | 9 | 16 | 24 | 35 | 48 | 62 |
| | 2 | 1 | 1 | 3 | 5 | 7 | 13 | 20 | 29 | 39 | 51 |
| PFA, PFAH, TFE | 1 | 1 | 1 | 2 | 3 | 5 | 9 | 14 | 20 | 27 | 36 |
| PFA, PFAH, TFE, Z | 1/0 | 1 | 1 | 1 | 3 | 4 | 8 | 11 | 17 | 23 | 30 |
| | 2/0 | 1 | 1 | 1 | 2 | 3 | 6 | 9 | 14 | 19 | 24 |
| | 3/0 | 0 | 1 | 1 | 1 | 3 | 5 | 8 | 11 | 15 | 20 |
| | 4/0 | 0 | 1 | 1 | 1 | 2 | 4 | 6 | 9 | 13 | 16 |
| Z | 14 | 15 | 25 | 39 | 61 | 89 | 157 | 236 | 340 | 463 | 605 |
| | 12 | 11 | 18 | 28 | 43 | 63 | 111 | 168 | 241 | 329 | 429 |
| | 10 | 6 | 11 | 17 | 26 | 39 | 68 | 103 | 148 | 201 | 263 |
| | 8 | 4 | 7 | 11 | 17 | 24 | 43 | 65 | 93 | 127 | 166 |
| | 6 | 3 | 5 | 7 | 12 | 17 | 30 | 45 | 65 | 89 | 117 |
| | 4 | 1 | 3 | 5 | 8 | 12 | 21 | 31 | 45 | 61 | 80 |
| | 3 | 1 | 2 | 4 | 6 | 8 | 15 | 23 | 33 | 45 | 58 |
| | 2 | 1 | 1 | 3 | 5 | 7 | 12 | 19 | 27 | 37 | 49 |
| | 1 | 1 | 1 | 2 | 4 | 6 | 10 | 15 | 22 | 30 | 39 |
| XHH, XHHW, XHHW-2, ZW | 14 | 9 | 15 | 23 | 36 | 53 | 94 | 141 | 203 | 277 | 361 |
| | 12 | 7 | 11 | 18 | 28 | 41 | 72 | 108 | 156 | 212 | 277 |
| | 10 | 5 | 8 | 13 | 21 | 30 | 54 | 81 | 116 | 158 | 207 |
| | 8 | 3 | 5 | 7 | 11 | 17 | 30 | 45 | 64 | 88 | 115 |
| | 6 | 1 | 3 | 5 | 8 | 12 | 22 | 33 | 48 | 65 | 85 |
| | 4 | 1 | 2 | 4 | 6 | 9 | 16 | 24 | 34 | 47 | 61 |
| | 3 | 1 | 1 | 3 | 5 | 7 | 13 | 20 | 29 | 40 | 52 |
| | 2 | 1 | 1 | 3 | 4 | 6 | 11 | 17 | 24 | 33 | 44 |

**Table C3** *Continued*

| | | CONDUCTORS | | | | | | | | | |
|---|---|---|---|---|---|---|---|---|---|---|---|
| | | Metric Designator (Trade Size) | | | | | | | | | |
| Type | Conductor Size (AWG/kcmil) | 16 (½) | 21 (¾) | 27 (1) | 35 (1¼) | 41 (1½) | 53 (2) | 63 (2½) | 78 (3) | 91 (3½) | 103 (4) |
| XHH, XHHW, XHHW-2 | 1 | 1 | 1 | 1 | 3 | 5 | 8 | 13 | 18 | 25 | 32 |
| | 1/0 | 1 | 1 | 1 | 2 | 4 | 7 | 10 | 15 | 21 | 27 |
| | 2/0 | 0 | 1 | 1 | 2 | 3 | 6 | 9 | 13 | 17 | 23 |
| | 3/0 | 0 | 1 | 1 | 1 | 3 | 5 | 7 | 10 | 14 | 19 |
| | 4/0 | 0 | 1 | 1 | 1 | 2 | 4 | 6 | 9 | 12 | 15 |
| | 250 | 0 | 0 | 1 | 1 | 1 | 3 | 5 | 7 | 10 | 13 |
| | 300 | 0 | 0 | 1 | 1 | 1 | 3 | 4 | 6 | 8 | 11 |
| | 350 | 0 | 0 | 1 | 1 | 1 | 2 | 4 | 5 | 7 | 9 |
| | 400 | 0 | 0 | 0 | 1 | 1 | 1 | 3 | 5 | 6 | 8 |
| | 500 | 0 | 0 | 0 | 1 | 1 | 1 | 3 | 4 | 5 | 7 |
| | 600 | 0 | 0 | 0 | 0 | 1 | 1 | 1 | 3 | 4 | 5 |
| | 700 | 0 | 0 | 0 | 0 | 1 | 1 | 1 | 3 | 4 | 5 |
| | 750 | 0 | 0 | 0 | 0 | 1 | 1 | 1 | 2 | 3 | 4 |
| | 800 | 0 | 0 | 0 | 0 | 1 | 1 | 1 | 2 | 3 | 4 |
| | 900 | 0 | 0 | 0 | 0 | 0 | 1 | 1 | 1 | 3 | 4 |
| | 1000 | 0 | 0 | 0 | 0 | 0 | 1 | 1 | 1 | 3 | 3 |
| | 1250 | 0 | 0 | 0 | 0 | 0 | 1 | 1 | 1 | 1 | 3 |
| | 1500 | 0 | 0 | 0 | 0 | 0 | 1 | 1 | 1 | 1 | 2 |
| | 1750 | 0 | 0 | 0 | 0 | 0 | 0 | 1 | 1 | 1 | 1 |
| | 2000 | 0 | 0 | 0 | 0 | 0 | 0 | 1 | 1 | 1 | 1 |

*Types RHH, RHW, and RHW-2 without outer covering.

| | | FIXTURE WIRES | | | | | |
|---|---|---|---|---|---|---|---|
| | | Metric Designator (Trade Size) | | | | | |
| Type | Conductor Size (AWG/kcmil) | 16 (½) | 21 (¾) | 27 (1) | 35 (1¼) | 41 (1½) | 53 (2) |
| FFH-2, RFH-2, RFHH-3 | 18 | 8 | 14 | 22 | 35 | 51 | 90 |
| | 16 | 7 | 12 | 19 | 29 | 43 | 76 |
| SF-2, SFF-2 | 18 | 11 | 18 | 28 | 44 | 64 | 113 |
| | 16 | 9 | 15 | 23 | 36 | 53 | 94 |
| | 14 | 7 | 12 | 19 | 29 | 43 | 76 |
| SF-1, SFF-1 | 18 | 19 | 32 | 50 | 78 | 114 | 201 |
| RFH-1, RFHH-2, TF, TFF, XF, XFF | 18 | 14 | 24 | 37 | 58 | 84 | 148 |
| RFHH-2, TF, TFF, XF, XFF | 16 | 11 | 19 | 30 | 47 | 68 | 120 |
| XF, XFF | 14 | 9 | 15 | 23 | 36 | 53 | 94 |
| TFN, TFFN | 18 | 23 | 38 | 59 | 93 | 135 | 237 |
| | 16 | 17 | 29 | 45 | 71 | 103 | 181 |
| PF, PFF, PGF, PGFF, PAF, PTF, PTFF, PAFF | 18 | 22 | 36 | 56 | 88 | 128 | 225 |
| | 16 | 17 | 28 | 43 | 68 | 99 | 174 |
| | 14 | 12 | 21 | 32 | 51 | 74 | 130 |
| ZF, ZFF, ZHF, HF, HFF | 18 | 28 | 47 | 72 | 113 | 165 | 290 |
| | 16 | 20 | 35 | 53 | 83 | 121 | 214 |
| | 14 | 15 | 25 | 39 | 61 | 89 | 157 |
| KF-2, KFF-2 | 18 | 41 | 68 | 105 | 164 | 239 | 421 |
| | 16 | 28 | 48 | 74 | 116 | 168 | 297 |
| | 14 | 19 | 33 | 51 | 80 | 116 | 204 |
| | 12 | 13 | 23 | 35 | 55 | 80 | 140 |
| | 10 | 9 | 15 | 23 | 36 | 53 | 94 |
| KF-1, KFF-1 | 18 | 48 | 82 | 125 | 196 | 285 | 503 |
| | 16 | 34 | 57 | 88 | 138 | 200 | 353 |
| | 14 | 23 | 38 | 59 | 93 | 135 | 237 |
| | 12 | 15 | 25 | 39 | 61 | 89 | 157 |
| | 10 | 10 | 16 | 25 | 40 | 58 | 103 |
| XF, XFF | 12 | 5 | 8 | 12 | 19 | 28 | 50 |
| | 10 | 4 | 6 | 10 | 15 | 22 | 39 |

Note: This table is for concentric stranded conductors only. For compact stranded conductors, Table C3(A) should be used.

**Table C3(A) Maximum Number of Compact Conductors in Flexible Metal Conduit (FMC)** (*Based on Table 1, Chapter 9*)

| | | COMPACT CONDUCTORS | | | | | | | | | |
|---|---|---|---|---|---|---|---|---|---|---|---|
| | Conductor | Metric Designator (Trade Size) | | | | | | | | | |
| Type | Size (AWG/kcmil) | 16 (½) | 21 (¾) | 27 (1) | 35 (1¼) | 41 (1½) | 53 (2) | 63 (2½) | 78 (3) | 91 (3½) | 103 (4) |
| THW, | 8 | 2 | 4 | 6 | 10 | 14 | 25 | 38 | 55 | 75 | 98 |
| THHW, | 6 | 1 | 3 | 5 | 7 | 11 | 20 | 29 | 43 | 58 | 76 |
| THW-2 | 4 | 1 | 2 | 3 | 5 | 8 | 15 | 22 | 32 | 43 | 57 |
| | 2 | 1 | 1 | 2 | 4 | 6 | 11 | 16 | 23 | 32 | 42 |
| | 1 | 1 | 1 | 1 | 3 | 4 | 7 | 11 | 16 | 22 | 29 |
| | 1/0 | 1 | 1 | 1 | 2 | 3 | 6 | 10 | 14 | 19 | 25 |
| | 2/0 | 0 | 1 | 1 | 1 | 3 | 5 | 8 | 12 | 16 | 21 |
| | 3/0 | 0 | 1 | 1 | 1 | 2 | 4 | 7 | 10 | 14 | 18 |
| | 4/0 | 0 | 1 | 1 | 1 | 1 | 4 | 6 | 8 | 11 | 15 |
| | 250 | 0 | 0 | 1 | 1 | 1 | 3 | 4 | 7 | 9 | 12 |
| | 300 | 0 | 0 | 1 | 1 | 1 | 2 | 4 | 6 | 8 | 10 |
| | 350 | 0 | 0 | 1 | 1 | 1 | 2 | 3 | 5 | 7 | 9 |
| | 400 | 0 | 0 | 0 | 1 | 1 | 1 | 3 | 5 | 6 | 8 |
| | 500 | 0 | 0 | 0 | 1 | 1 | 1 | 3 | 4 | 5 | 7 |
| | 600 | 0 | 0 | 0 | 0 | 1 | 1 | 1 | 3 | 4 | 6 |
| | 700 | 0 | 0 | 0 | 0 | 1 | 1 | 1 | 3 | 4 | 5 |
| | 750 | 0 | 0 | 0 | 0 | 1 | 1 | 1 | 2 | 3 | 5 |
| | 1000 | 0 | 0 | 0 | 0 | 0 | 1 | 1 | 1 | 3 | 4 |
| THHN, | 8 | — | — | — | — | — | — | — | — | — | — |
| THWN, | 6 | 3 | 4 | 7 | 11 | 16 | 29 | 43 | 62 | 85 | 111 |
| THWN-2 | 4 | 1 | 3 | 4 | 7 | 10 | 18 | 27 | 38 | 52 | 69 |
| | 2 | 1 | 1 | 3 | 5 | 7 | 13 | 19 | 28 | 38 | 49 |
| | 1 | 1 | 1 | 2 | 3 | 5 | 9 | 14 | 21 | 28 | 37 |
| | 1/0 | 1 | 1 | 1 | 3 | 4 | 8 | 12 | 17 | 24 | 31 |
| | 2/0 | 1 | 1 | 1 | 2 | 4 | 6 | 10 | 14 | 20 | 26 |
| | 3/0 | 0 | 1 | 1 | 1 | 3 | 5 | 8 | 12 | 17 | 22 |
| | 4/0 | 0 | 1 | 1 | 1 | 2 | 4 | 7 | 10 | 14 | 18 |
| | 250 | 0 | 1 | 1 | 1 | 1 | 3 | 5 | 8 | 11 | 14 |
| | 300 | 0 | 0 | 1 | 1 | 1 | 3 | 5 | 7 | 9 | 12 |
| | 350 | 0 | 0 | 1 | 1 | 1 | 3 | 4 | 6 | 8 | 10 |
| | 400 | 0 | 0 | 1 | 1 | 1 | 2 | 3 | 5 | 7 | 9 |
| | 500 | 0 | 0 | 0 | 1 | 1 | 1 | 3 | 4 | 6 | 8 |
| | 600 | 0 | 0 | 0 | 1 | 1 | 1 | 2 | 3 | 5 | 6 |
| | 700 | 0 | 0 | 0 | 0 | 1 | 1 | 1 | 3 | 4 | 6 |
| | 750 | 0 | 0 | 0 | 0 | 1 | 1 | 1 | 3 | 4 | 5 |
| | 1000 | 0 | 0 | 0 | 0 | 0 | 1 | 1 | 1 | 3 | 4 |
| XHHW, | 8 | 3 | 5 | 8 | 13 | 19 | 33 | 50 | 71 | 97 | 127 |
| XHHW-2 | 6 | 2 | 4 | 6 | 9 | 14 | 24 | 37 | 53 | 72 | 95 |
| | 4 | 1 | 3 | 4 | 7 | 10 | 18 | 27 | 38 | 52 | 69 |
| | 2 | 1 | 1 | 3 | 5 | 7 | 13 | 19 | 28 | 38 | 49 |
| | 1 | 1 | 1 | 2 | 3 | 5 | 9 | 14 | 21 | 28 | 37 |
| | 1/0 | 1 | 1 | 1 | 3 | 4 | 8 | 12 | 17 | 24 | 31 |
| | 2/0 | 1 | 1 | 1 | 2 | 4 | 7 | 10 | 15 | 20 | 26 |
| | 3/0 | 0 | 1 | 1 | 1 | 3 | 5 | 8 | 12 | 17 | 22 |
| | 4/0 | 0 | 1 | 1 | 1 | 2 | 4 | 7 | 10 | 14 | 18 |
| | 250 | 0 | 1 | 1 | 1 | 1 | 4 | 5 | 8 | 11 | 14 |
| | 300 | 0 | 0 | 1 | 1 | 1 | 3 | 5 | 7 | 9 | 12 |
| | 350 | 0 | 0 | 1 | 1 | 1 | 3 | 4 | 6 | 8 | 11 |
| | 400 | 0 | 0 | 1 | 1 | 1 | 2 | 4 | 5 | 7 | 10 |
| | 500 | 0 | 0 | 0 | 1 | 1 | 1 | 3 | 4 | 6 | 8 |
| | 600 | 0 | 0 | 0 | 1 | 1 | 1 | 2 | 3 | 5 | 6 |
| | 700 | 0 | 0 | 0 | 0 | 1 | 1 | 1 | 3 | 4 | 6 |
| | 750 | 0 | 0 | 0 | 0 | 1 | 1 | 1 | 3 | 4 | 5 |
| | 1000 | 0 | 0 | 0 | 0 | 1 | 1 | 1 | 2 | 3 | 4 |

Definition: *Compact stranding* is the result of a manufacturing process where the standard conductor is compressed to the extent that the interstices (voids between strand wires) are virtually eliminated.

**Table C4 Maximum Number of Conductors or Fixture Wires in Intermediate Metal Conduit (IMC)** (*Based on Table 1, Chapter 9*)

| | | CONDUCTORS | | | | | | | | | |
|---|---|---|---|---|---|---|---|---|---|---|---|
| | Conductor | Metric Designator (Trade Size) | | | | | | | | | |
| Type | Size (AWG/kcmil) | 16 (½) | 21 (¾) | 27 (1) | 35 (1¼) | 41 (1½) | 53 (2) | 63 (2½) | 78 (3) | 91 (3½) | 103 (4) |
| RHH, | 14 | 4 | 8 | 13 | 22 | 30 | 49 | 70 | 108 | 144 | 186 |
| RHW, RHW-2 | 12 | 4 | 6 | 11 | 18 | 25 | 41 | 58 | 89 | 120 | 154 |
| RHH, | 10 | 3 | 5 | 8 | 15 | 20 | 33 | 47 | 72 | 97 | 124 |
| RHW, | 8 | 1 | 3 | 4 | 8 | 10 | 17 | 24 | 38 | 50 | 65 |
| RHW-2 | 6 | 1 | 1 | 3 | 6 | 8 | 14 | 19 | 30 | 40 | 52 |
| | 4 | 1 | 1 | 3 | 5 | 6 | 11 | 15 | 23 | 31 | 41 |
| | 3 | 1 | 1 | 2 | 4 | 6 | 9 | 13 | 21 | 28 | 36 |
| | 2 | 1 | 1 | 1 | 3 | 5 | 8 | 11 | 18 | 24 | 31 |
| | 1 | 0 | 1 | 1 | 2 | 3 | 5 | 7 | 12 | 16 | 20 |
| | 1/0 | 0 | 1 | 1 | 1 | 3 | 4 | 6 | 10 | 14 | 18 |
| | 2/0 | 0 | 1 | 1 | 1 | 2 | 4 | 6 | 9 | 12 | 15 |
| | 3/0 | 0 | 0 | 1 | 1 | 1 | 3 | 5 | 7 | 10 | 13 |
| | 4/0 | 0 | 0 | 1 | 1 | 1 | 3 | 4 | 6 | 9 | 11 |
| | 250 | 0 | 0 | 1 | 1 | 1 | 1 | 3 | 5 | 6 | 8 |
| | 300 | 0 | 0 | 0 | 1 | 1 | 1 | 3 | 4 | 6 | 7 |
| | 350 | 0 | 0 | 0 | 1 | 1 | 1 | 2 | 4 | 5 | 7 |
| | 400 | 0 | 0 | 0 | 1 | 1 | 1 | 2 | 3 | 5 | 6 |
| | 500 | 0 | 0 | 0 | 1 | 1 | 1 | 1 | 3 | 4 | 5 |
| | 600 | 0 | 0 | 0 | 0 | 1 | 1 | 1 | 2 | 3 | 4 |
| | 700 | 0 | 0 | 0 | 0 | 1 | 1 | 1 | 2 | 3 | 4 |
| | 750 | 0 | 0 | 0 | 0 | 1 | 1 | 1 | 1 | 3 | 4 |
| | 800 | 0 | 0 | 0 | 0 | 0 | 1 | 1 | 1 | 3 | 3 |
| | 900 | 0 | 0 | 0 | 0 | 0 | 1 | 1 | 1 | 2 | 3 |
| | 1000 | 0 | 0 | 0 | 0 | 0 | 1 | 1 | 1 | 2 | 3 |
| | 1250 | 0 | 0 | 0 | 0 | 0 | 1 | 1 | 1 | 1 | 2 |
| | 1500 | 0 | 0 | 0 | 0 | 0 | 0 | 1 | 1 | 1 | 1 |
| | 1750 | 0 | 0 | 0 | 0 | 0 | 0 | 1 | 1 | 1 | 1 |
| | 2000 | 0 | 0 | 0 | 0 | 0 | 0 | 0 | 1 | 1 | 1 |
| TW, | 14 | 10 | 17 | 27 | 47 | 64 | 104 | 147 | 228 | 304 | 392 |
| THHW, | 12 | 7 | 13 | 21 | 36 | 49 | 80 | 113 | 175 | 234 | 301 |
| THW, | 10 | 5 | 9 | 15 | 27 | 36 | 59 | 84 | 130 | 174 | 224 |
| THW-2 | 8 | 3 | 5 | 8 | 15 | 20 | 33 | 47 | 72 | 97 | 124 |
| RHH*, RHW*, RHW-2 | 14 | 6 | 11 | 18 | 31 | 42 | 69 | 98 | 151 | 202 | 261 |
| RHH*, RHW*, | 12 | 5 | 9 | 14 | 25 | 34 | 56 | 79 | 122 | 163 | 209 |
| RHW-2* | 10 | 4 | 7 | 11 | 19 | 26 | 43 | 61 | 95 | 127 | 163 |
| RHH*, RHW*, RHW-2* | 8 | 2 | 4 | 7 | 12 | 16 | 26 | 37 | 57 | 76 | 98 |
| RHH*, RHW*, | 6 | 1 | 3 | 5 | 9 | 12 | 20 | 28 | 43 | 58 | 75 |
| RHW-2*, | 4 | 1 | 2 | 4 | 6 | 9 | 15 | 21 | 32 | 43 | 56 |
| TW, THW, | 3 | 1 | 1 | 3 | 6 | 8 | 13 | 18 | 28 | 37 | 48 |
| THHW, | 2 | 1 | 1 | 3 | 5 | 6 | 11 | 15 | 23 | 31 | 41 |
| THW-2 | 1 | 1 | 1 | 1 | 3 | 4 | 7 | 11 | 16 | 22 | 28 |
| | 1/0 | 1 | 1 | 1 | 3 | 4 | 6 | 9 | 14 | 19 | 24 |
| | 2/0 | 0 | 1 | 1 | 2 | 3 | 5 | 8 | 12 | 16 | 20 |
| | 3/0 | 0 | 1 | 1 | 1 | 3 | 4 | 6 | 10 | 13 | 17 |
| | 4/0 | 0 | 1 | 1 | 1 | 2 | 4 | 5 | 8 | 11 | 14 |
| | 250 | 0 | 0 | 1 | 1 | 1 | 3 | 4 | 7 | 9 | 12 |
| | 300 | 0 | 0 | 1 | 1 | 1 | 2 | 4 | 6 | 8 | 10 |
| | 350 | 0 | 0 | 1 | 1 | 1 | 2 | 3 | 5 | 7 | 9 |
| | 400 | 0 | 0 | 0 | 1 | 1 | 1 | 3 | 4 | 6 | 8 |
| | 500 | 0 | 0 | 0 | 1 | 1 | 1 | 2 | 4 | 5 | 7 |
| | 600 | 0 | 0 | 0 | 1 | 1 | 1 | 1 | 3 | 4 | 5 |
| | 700 | 0 | 0 | 0 | 0 | 1 | 1 | 1 | 3 | 4 | 5 |
| | 750 | 0 | 0 | 0 | 0 | 1 | 1 | 1 | 2 | 3 | 4 |
| | 800 | 0 | 0 | 0 | 0 | 1 | 1 | 1 | 2 | 3 | 4 |
| | 900 | 0 | 0 | 0 | 0 | 1 | 1 | 1 | 2 | 3 | 4 |
| | 1000 | 0 | 0 | 0 | 0 | 0 | 1 | 1 | 1 | 3 | 3 |
| | 1250 | 0 | 0 | 0 | 0 | 0 | 1 | 1 | 1 | 1 | 3 |
| | 1500 | 0 | 0 | 0 | 0 | 0 | 1 | 1 | 1 | 1 | 2 |
| | 1750 | 0 | 0 | 0 | 0 | 0 | 0 | 1 | 1 | 1 | 1 |
| | 2000 | 0 | 0 | 0 | 0 | 0 | 0 | 1 | 1 | 1 | 1 |

**Table C4** *Continued*

## CONDUCTORS

| Type | Conductor Size (AWG/kcmil) | Metric Designator (Trade Size) | | | | | | | | | |
|------|------|------|------|------|------|------|------|------|------|------|------|
| | | 16 (½) | 21 (¾) | 27 (1) | 35 (1¼) | 41 (1½) | 53 (2) | 63 (2½) | 78 (3) | 91 (3½) | 103 (4) |
| THHN, THWN, THWN-2 | 14 | 14 | 24 | 39 | 68 | 91 | 149 | 211 | 326 | 436 | 562 |
| | 12 | 10 | 17 | 29 | 49 | 67 | 109 | 154 | 238 | 318 | 410 |
| | 10 | 6 | 11 | 18 | 31 | 42 | 68 | 97 | 150 | 200 | 258 |
| | 8 | 3 | 6 | 10 | 18 | 24 | 39 | 56 | 86 | 115 | 149 |
| | 6 | 2 | 4 | 7 | 13 | 17 | 28 | 40 | 62 | 83 | 107 |
| | 4 | 1 | 3 | 4 | 8 | 10 | 17 | 25 | 38 | 51 | 66 |
| | 3 | 1 | 2 | 4 | 6 | 9 | 15 | 21 | 32 | 43 | 56 |
| | 2 | 1 | 1 | 3 | 5 | 7 | 12 | 17 | 27 | 36 | 47 |
| | 1 | 1 | 1 | 2 | 4 | 5 | 9 | 13 | 20 | 27 | 35 |
| | 1/0 | 1 | 1 | 1 | 3 | 4 | 8 | 11 | 17 | 23 | 29 |
| | 2/0 | 1 | 1 | 1 | 3 | 4 | 6 | 9 | 14 | 19 | 24 |
| | 3/0 | 0 | 1 | 1 | 2 | 3 | 5 | 7 | 12 | 16 | 20 |
| | 4/0 | 0 | 1 | 1 | 1 | 2 | 4 | 6 | 9 | 13 | 17 |
| | 250 | 0 | 0 | 1 | 1 | 1 | 3 | 5 | 8 | 10 | 13 |
| | 300 | 0 | 0 | 1 | 1 | 1 | 3 | 4 | 7 | 9 | 12 |
| | 350 | 0 | 0 | 1 | 1 | 1 | 2 | 4 | 6 | 8 | 10 |
| | 400 | 0 | 0 | 1 | 1 | 1 | 2 | 3 | 5 | 7 | 9 |
| | 500 | 0 | 0 | 0 | 1 | 1 | 1 | 3 | 4 | 6 | 7 |
| | 600 | 0 | 0 | 0 | 1 | 1 | 1 | 2 | 3 | 5 | 6 |
| | 700 | 0 | 0 | 0 | 1 | 1 | 1 | 1 | 3 | 4 | 5 |
| | 750 | 0 | 0 | 0 | 1 | 1 | 1 | 1 | 3 | 4 | 5 |
| | 800 | 0 | 0 | 0 | 0 | 1 | 1 | 1 | 3 | 4 | 5 |
| | 900 | 0 | 0 | 0 | 0 | 1 | 1 | 1 | 2 | 3 | 4 |
| | 1000 | 0 | 0 | 0 | 0 | 1 | 1 | 1 | 2 | 3 | 4 |
| FEP, FEPB, PFA, PFAH, TFE | 14 | 13 | 23 | 38 | 66 | 89 | 145 | 205 | 317 | 423 | 545 |
| | 12 | 10 | 17 | 28 | 48 | 65 | 106 | 150 | 231 | 309 | 398 |
| | 10 | 7 | 12 | 20 | 34 | 46 | 76 | 107 | 166 | 221 | 285 |
| | 8 | 4 | 7 | 11 | 19 | 26 | 43 | 61 | 95 | 127 | 163 |
| | 6 | 3 | 5 | 8 | 14 | 19 | 31 | 44 | 67 | 90 | 116 |
| | 4 | 1 | 3 | 5 | 10 | 13 | 21 | 30 | 47 | 63 | 81 |
| | 3 | 1 | 3 | 4 | 8 | 11 | 18 | 25 | 39 | 52 | 68 |
| | 2 | 1 | 2 | 4 | 6 | 9 | 15 | 21 | 32 | 43 | 56 |
| PFA, PFAH, TFE | 1 | 1 | 1 | 2 | 4 | 6 | 10 | 14 | 22 | 30 | 39 |
| PFA, PFAH, TFE, Z | 1/0 | 1 | 1 | 1 | 4 | 5 | 8 | 12 | 19 | 25 | 32 |
| | 2/0 | 1 | 1 | 1 | 3 | 4 | 7 | 10 | 15 | 21 | 27 |
| | 3/0 | 0 | 1 | 1 | 2 | 3 | 6 | 8 | 13 | 17 | 22 |
| | 4/0 | 0 | 1 | 1 | 1 | 3 | 5 | 7 | 10 | 14 | 18 |
| Z | 14 | 16 | 28 | 46 | 79 | 107 | 175 | 247 | 381 | 510 | 657 |
| | 12 | 11 | 20 | 32 | 56 | 76 | 124 | 175 | 271 | 362 | 466 |
| | 10 | 7 | 12 | 20 | 34 | 46 | 76 | 107 | 166 | 221 | 285 |
| | 8 | 4 | 7 | 12 | 21 | 29 | 48 | 68 | 105 | 140 | 180 |
| | 6 | 3 | 5 | 9 | 15 | 20 | 33 | 47 | 73 | 98 | 127 |
| | 4 | 1 | 3 | 6 | 10 | 14 | 23 | 33 | 50 | 67 | 87 |
| | 3 | 1 | 2 | 4 | 7 | 10 | 17 | 24 | 37 | 49 | 63 |
| | 2 | 1 | 1 | 3 | 6 | 8 | 14 | 20 | 30 | 41 | 53 |
| | 1 | 1 | 1 | 3 | 5 | 7 | 11 | 16 | 25 | 33 | 43 |
| XHH, XHHW, XHHW-2, ZW | 14 | 10 | 17 | 27 | 47 | 64 | 104 | 147 | 228 | 304 | 392 |
| | 12 | 7 | 13 | 21 | 36 | 49 | 80 | 113 | 175 | 234 | 301 |
| | 10 | 5 | 9 | 15 | 27 | 36 | 59 | 84 | 130 | 174 | 224 |
| | 8 | 3 | 5 | 8 | 15 | 20 | 33 | 47 | 72 | 97 | 124 |
| | 6 | 1 | 4 | 6 | 11 | 15 | 24 | 35 | 53 | 71 | 92 |
| | 4 | 1 | 3 | 4 | 8 | 11 | 18 | 25 | 39 | 52 | 67 |
| | 3 | 1 | 2 | 4 | 7 | 9 | 15 | 21 | 33 | 44 | 56 |
| | 2 | 1 | 1 | 3 | 5 | 7 | 12 | 18 | 27 | 37 | 47 |
| XHH, XHHW, XHHW-2 | 1 | 1 | 1 | 2 | 4 | 5 | 9 | 13 | 20 | 27 | 35 |
| | 1/0 | 1 | 1 | 1 | 3 | 5 | 8 | 11 | 17 | 23 | 30 |
| | 2/0 | 1 | 1 | 1 | 3 | 4 | 6 | 9 | 14 | 19 | 25 |
| | 3/0 | 0 | 1 | 1 | 2 | 3 | 5 | 7 | 12 | 16 | 20 |
| | 4/0 | 0 | 1 | 1 | 1 | 2 | 4 | 6 | 10 | 13 | 17 |
| | 250 | 0 | 0 | 1 | 1 | 1 | 3 | 5 | 8 | 11 | 14 |
| | 300 | 0 | 0 | 1 | 1 | 1 | 3 | 4 | 7 | 9 | 12 |
| | 350 | 0 | 0 | 1 | 1 | 1 | 3 | 4 | 6 | 8 | 10 |
| | 400 | 0 | 0 | 1 | 1 | 1 | 2 | 3 | 5 | 7 | 9 |
| | 500 | 0 | 0 | 0 | 1 | 1 | 1 | 3 | 4 | 6 | 8 |
| | 600 | 0 | 0 | 0 | 1 | 1 | 1 | 2 | 3 | 5 | 6 |
| | 700 | 0 | 0 | 0 | 1 | 1 | 1 | 1 | 3 | 4 | 5 |
| | 750 | 0 | 0 | 0 | 1 | 1 | 1 | 1 | 3 | 4 | 5 |
| | 800 | 0 | 0 | 0 | 0 | 1 | 1 | 1 | 3 | 4 | 5 |
| | 900 | 0 | 0 | 0 | 0 | 1 | 1 | 1 | 2 | 3 | 4 |
| | 1000 | 0 | 0 | 0 | 0 | 1 | 1 | 1 | 2 | 3 | 4 |
| | 1250 | 0 | 0 | 0 | 0 | 0 | 1 | 1 | 1 | 2 | 3 |
| | 1500 | 0 | 0 | 0 | 0 | 0 | 1 | 1 | 1 | 1 | 2 |
| | 1750 | 0 | 0 | 0 | 0 | 0 | 1 | 1 | 1 | 1 | 2 |
| | 2000 | 0 | 0 | 0 | 0 | 0 | 0 | 1 | 1 | 1 | 1 |

## FIXTURE WIRES

| Type | Conductor Size (AWG/kcmil) | Metric Designator (Trade Size) | | | | | |
|------|------|------|------|------|------|------|------|
| | | 16 (½) | 21 (¾) | 27 (1) | 35 (1¼) | 41 (1½) | 53 (2) |
| FHH-2, RFH-2, RFHH-3 | 18 | 9 | 16 | 26 | 45 | 61 | 100 |
| | 16 | 8 | 13 | 22 | 38 | 51 | 84 |
| SF-2, SFF-2 | 18 | 12 | 20 | 33 | 57 | 77 | 126 |
| | 16 | 10 | 17 | 27 | 47 | 64 | 104 |
| | 14 | 8 | 13 | 22 | 38 | 51 | 84 |
| SF-1, SFF-1 | 18 | 21 | 36 | 59 | 101 | 137 | 223 |
| RFH-1, RFHH-2, TF, TFF, XF, XFF | 18 | 15 | 26 | 43 | 75 | 101 | 165 |
| RFH-2, TF, TFF, XF, XFF | 16 | 12 | 21 | 35 | 60 | 81 | 133 |
| XF, XFF | 14 | 10 | 17 | 27 | 47 | 64 | 104 |
| TFN, TFFN | 18 | 25 | 42 | 69 | 119 | 161 | 264 |
| | 16 | 19 | 32 | 53 | 91 | 123 | 201 |
| PF, PFF, PGF, PGFF, PAF, PTF, PTFF, PAFF | 18 | 23 | 40 | 66 | 113 | 153 | 250 |
| | 16 | 18 | 31 | 51 | 87 | 118 | 193 |
| | 14 | 13 | 23 | 38 | 66 | 89 | 145 |
| ZF, ZFF, ZHF, HF, HFF | 18 | 30 | 52 | 85 | 146 | 197 | 322 |
| | 16 | 22 | 38 | 63 | 108 | 145 | 238 |
| | 14 | 16 | 28 | 46 | 79 | 107 | 175 |
| KF-2, KFF-2 | 18 | 44 | 75 | 123 | 212 | 287 | 468 |
| | 16 | 31 | 53 | 87 | 149 | 202 | 330 |
| | 14 | 21 | 36 | 60 | 103 | 139 | 227 |
| | 12 | 14 | 25 | 41 | 70 | 95 | 156 |
| | 10 | 10 | 17 | 27 | 47 | 64 | 104 |
| KF-1, KFF-1 | 18 | 52 | 90 | 147 | 253 | 342 | 558 |
| | 16 | 37 | 63 | 103 | 178 | 240 | 392 |
| | 14 | 25 | 42 | 69 | 119 | 161 | 264 |
| | 12 | 16 | 28 | 46 | 79 | 107 | 175 |
| | 10 | 10 | 18 | 30 | 52 | 70 | 114 |
| XF, XFF | 12 | 5 | 9 | 14 | 25 | 34 | 56 |
| | 10 | 4 | 7 | 11 | 19 | 26 | 43 |

Note: This table is for concentric stranded conductors only. For compact stranded conductors, Table C4(A) should be used.

*Types RHH, RHW, and RHW-2 without outer covering.

**Table C4(A)  Maximum Number of Compact Conductors in Intermediate Metal Conduit (IMC)** (*Based on Table 1, Chapter 9*)

### COMPACT CONDUCTORS

| Type | Conductor Size (AWG/kcmil) | Metric Designator (Trade Size) | | | | | | | | | |
|------|------|------|------|------|------|------|------|------|------|------|------|
| | | 16 (½) | 21 (¾) | 27 (1) | 35 (1¼) | 41 (1½) | 53 (2) | 63 (2½) | 78 (3) | 91 (3½) | 103 (4) |
| THW, THW-2, THHW | 8 | 2 | 4 | 7 | 13 | 17 | 28 | 40 | 62 | 83 | 107 |
| | 6 | 1 | 3 | 6 | 10 | 13 | 22 | 31 | 48 | 64 | 82 |
| | 4 | 1 | 2 | 4 | 7 | 10 | 16 | 23 | 36 | 48 | 62 |
| | 2 | 1 | 1 | 3 | 5 | 7 | 12 | 17 | 26 | 35 | 45 |
| | 1 | 1 | 1 | 1 | 4 | 5 | 8 | 12 | 18 | 25 | 32 |
| | 1/0 | 1 | 1 | 1 | 3 | 4 | 7 | 10 | 16 | 21 | 27 |
| | 2/0 | 0 | 1 | 1 | 3 | 4 | 6 | 9 | 13 | 18 | 23 |
| | 3/0 | 0 | 1 | 1 | 2 | 3 | 5 | 7 | 11 | 15 | 20 |
| | 4/0 | 0 | 1 | 1 | 1 | 2 | 4 | 6 | 9 | 13 | 16 |
| | 250 | 0 | 0 | 1 | 1 | 1 | 3 | 5 | 7 | 10 | 13 |
| | 300 | 0 | 0 | 1 | 1 | 1 | 3 | 4 | 6 | 9 | 11 |
| | 350 | 0 | 0 | 1 | 1 | 1 | 2 | 4 | 6 | 8 | 10 |
| | 400 | 0 | 0 | 1 | 1 | 1 | 2 | 3 | 5 | 7 | 9 |
| | 500 | 0 | 0 | 0 | 1 | 1 | 1 | 3 | 4 | 6 | 8 |
| | 600 | 0 | 0 | 0 | 1 | 1 | 1 | 2 | 3 | 5 | 6 |
| | 700 | 0 | 0 | 0 | 1 | 1 | 1 | 1 | 3 | 4 | 5 |
| | 750 | 0 | 0 | 0 | 1 | 1 | 1 | 1 | 3 | 4 | 5 |
| | 1000 | 0 | 0 | 0 | 0 | 1 | 1 | 1 | 2 | 3 | 4 |

Table C4(A)   *Continued*

| | | COMPACT CONDUCTORS | | | | | | | | | |
|---|---|---|---|---|---|---|---|---|---|---|---|
| | **Conductor Size** | **Metric Designator (Trade Size)** | | | | | | | | | |
| **Type** | **(AWG/kcmil)** | **16** (½) | **21** (¾) | **27** (1) | **35** (1¼) | **41** (1½) | **53** (2) | **63** (2½) | **78** (3) | **91** (3½) | **103** (4) |
| THHN, | 8 | — | — | — | — | — | — | — | — | — | — |
| THWN, | 6 | 3 | 5 | 8 | 14 | 19 | 32 | 45 | 70 | 93 | 120 |
| THWN-2 | 4 | 1 | 3 | 5 | 9 | 12 | 20 | 28 | 43 | 58 | 74 |
| | 2 | 1 | 1 | 3 | 6 | 8 | 14 | 20 | 31 | 41 | 53 |
| | 1 | 1 | 1 | 3 | 5 | 6 | 10 | 15 | 23 | 31 | 40 |
| | 1/0 | 1 | 1 | 2 | 4 | 5 | 9 | 13 | 20 | 26 | 34 |
| | 2/0 | 1 | 1 | 1 | 3 | 4 | 7 | 10 | 16 | 22 | 28 |
| | 3/0 | 0 | 1 | 1 | 3 | 4 | 6 | 9 | 14 | 18 | 24 |
| | 4/0 | 0 | 1 | 1 | 2 | 3 | 5 | 7 | 11 | 15 | 19 |
| | 250 | 0 | 1 | 1 | 1 | 2 | 4 | 6 | 9 | 12 | 15 |
| | 300 | 0 | 0 | 1 | 1 | 1 | 3 | 5 | 7 | 10 | 13 |
| | 350 | 0 | 0 | 1 | 1 | 1 | 3 | 4 | 7 | 9 | 11 |
| | 400 | 0 | 0 | 1 | 1 | 1 | 2 | 4 | 6 | 8 | 10 |
| | 500 | 0 | 0 | 1 | 1 | 1 | 2 | 3 | 5 | 7 | 9 |
| | 600 | 0 | 0 | 0 | 1 | 1 | 1 | 2 | 4 | 5 | 7 |
| | 700 | 0 | 0 | 0 | 1 | 1 | 1 | 2 | 3 | 5 | 6 |
| | 750 | 0 | 0 | 0 | 1 | 1 | 1 | 1 | 3 | 4 | 6 |
| | 1000 | 0 | 0 | 0 | 0 | 1 | 1 | 1 | 2 | 3 | 4 |
| XHHW, | 8 | 3 | 6 | 9 | 16 | 22 | 37 | 52 | 80 | 107 | 138 |
| XHHW-2 | 6 | 2 | 4 | 7 | 12 | 16 | 27 | 38 | 59 | 80 | 103 |
| | 4 | 1 | 3 | 5 | 9 | 12 | 20 | 28 | 43 | 58 | 74 |
| | 2 | 1 | 1 | 3 | 6 | 8 | 14 | 20 | 31 | 41 | 53 |
| | 1 | 1 | 1 | 3 | 5 | 6 | 10 | 15 | 23 | 31 | 40 |
| | 1/0 | 1 | 1 | 2 | 4 | 5 | 9 | 13 | 20 | 26 | 34 |
| | 2/0 | 1 | 1 | 1 | 3 | 4 | 7 | 11 | 17 | 22 | 29 |
| | 3/0 | 0 | 1 | 1 | 3 | 4 | 6 | 9 | 14 | 18 | 24 |
| | 4/0 | 0 | 1 | 1 | 2 | 3 | 5 | 7 | 11 | 15 | 20 |
| | 250 | 0 | 1 | 1 | 1 | 2 | 4 | 6 | 9 | 12 | 16 |
| | 300 | 0 | 0 | 1 | 1 | 1 | 3 | 5 | 8 | 10 | 13 |
| | 350 | 0 | 0 | 1 | 1 | 1 | 3 | 4 | 7 | 9 | 12 |
| | 400 | 0 | 0 | 1 | 1 | 1 | 3 | 4 | 6 | 8 | 11 |
| | 500 | 0 | 0 | 1 | 1 | 1 | 2 | 3 | 5 | 7 | 9 |
| | 600 | 0 | 0 | 0 | 1 | 1 | 1 | 2 | 4 | 5 | 7 |
| | 700 | 0 | 0 | 0 | 1 | 1 | 1 | 2 | 3 | 5 | 6 |
| | 750 | 0 | 0 | 0 | 1 | 1 | 1 | 1 | 3 | 4 | 6 |
| | 1000 | 0 | 0 | 0 | 0 | 1 | 1 | 1 | 2 | 3 | 4 |

*Definition: Compact stranding* is the result of a manufacturing process where the standard conductor is compressed to the extent that interstices (voids between strand wires) are virtually eliminated.

**Table C5 Maximum Number of Conductors or Fixture Wires in Liquidtight Flexible Nonmetallic Conduit (Type LFNC-B\*)** *(Based on Table 1, Chapter 9)*

| | | CONDUCTORS | | | | | | |
|---|---|---|---|---|---|---|---|---|
| | **Conductor Size** | **Metric Designator (Trade Size)** | | | | | | |
| **Type** | **(AWG/kcmil)** | **12** (⅜) | **16** (½) | **21** (¾) | **27 (1)** | **35** (1¼) | **41** (1½) | **53 (2)** |
| RHH, RHW, RHW-2 | 14 | 2 | 4 | 7 | 12 | 21 | 27 | 44 |
| | 12 | 1 | 3 | 6 | 10 | 17 | 22 | 36 |
| RHH, RHW, RHW-2 | 10 | 1 | 3 | 5 | 8 | 14 | 18 | 29 |
| | 8 | 1 | 1 | 2 | 4 | 7 | 9 | 15 |
| | 6 | 1 | 1 | 1 | 3 | 6 | 7 | 12 |
| | 4 | 0 | 1 | 1 | 2 | 4 | 6 | 9 |
| | 3 | 0 | 1 | 1 | 1 | 4 | 5 | 8 |
| | 2 | 0 | 1 | 1 | 1 | 3 | 4 | 7 |
| | 1 | 0 | 0 | 1 | 1 | 1 | 3 | 5 |
| | 1/0 | 0 | 0 | 1 | 1 | 1 | 2 | 4 |
| | 2/0 | 0 | 0 | 1 | 1 | 1 | 1 | 3 |
| | 3/0 | 0 | 0 | 0 | 1 | 1 | 1 | 3 |
| | 4/0 | 0 | 0 | 0 | 1 | 1 | 1 | 2 |
| | 250 | 0 | 0 | 0 | 0 | 1 | 1 | 1 |
| | 300 | 0 | 0 | 0 | 0 | 1 | 1 | 1 |
| | 350 | 0 | 0 | 0 | 0 | 1 | 1 | 1 |
| | 400 | 0 | 0 | 0 | 0 | 1 | 1 | 1 |
| | 500 | 0 | 0 | 0 | 0 | 1 | 1 | 1 |
| | 600 | 0 | 0 | 0 | 0 | 0 | 1 | 1 |
| | 700 | 0 | 0 | 0 | 0 | 0 | 0 | 1 |
| | 750 | 0 | 0 | 0 | 0 | 0 | 0 | 1 |
| | 800 | 0 | 0 | 0 | 0 | 0 | 0 | 1 |
| | 900 | 0 | 0 | 0 | 0 | 0 | 0 | 1 |
| | 1000 | 0 | 0 | 0 | 0 | 0 | 0 | 1 |
| | 1250 | 0 | 0 | 0 | 0 | 0 | 0 | 0 |
| | 1500 | 0 | 0 | 0 | 0 | 0 | 0 | 0 |
| | 1750 | 0 | 0 | 0 | 0 | 0 | 0 | 0 |
| | 2000 | 0 | 0 | 0 | 0 | 0 | 0 | 0 |
| TW, THHW, THW, THW-2 | 14 | 5 | 9 | 15 | 25 | 44 | 57 | 93 |
| | 12 | 4 | 7 | 12 | 19 | 33 | 43 | 71 |
| | 10 | 3 | 5 | 9 | 14 | 25 | 32 | 53 |
| | 8 | 1 | 3 | 5 | 8 | 14 | 18 | 29 |

Table C5   *Continued*

| | | CONDUCTORS | | | | | | |
|---|---|---|---|---|---|---|---|---|
| | **Conductor Size** | **Metric Designator (Trade Size)** | | | | | | |
| **Type** | **(AWG/kcmil)** | **12** (⅜) | **16** (½) | **21** (¾) | **27 (1)** | **35** (1¼) | **41** (1½) | **53 (2)** |
| RHH†, RHW†, RHW-2† | 14 | 3 | 6 | 10 | 16 | 29 | 38 | 62 |
| RHH†, RHW†, RHW-2† | 12 | 3 | 5 | 8 | 13 | 23 | 30 | 50 |
| | 10 | 1 | 3 | 6 | 10 | 18 | 23 | 39 |
| RHH†, RHW†, RHW-2† | 8 | 1 | 1 | 4 | 6 | 11 | 14 | 23 |
| RHH†, RHW†, RHW-2†, | 6 | 1 | 1 | 3 | 5 | 8 | 11 | 18 |
| TW, THW, | 4 | 1 | 1 | 1 | 3 | 6 | 8 | 13 |
| THHW, THW-2 | 3 | 1 | 1 | 1 | 3 | 5 | 7 | 11 |
| | 2 | 0 | 1 | 1 | 2 | 4 | 6 | 9 |
| | 1 | 0 | 1 | 1 | 1 | 3 | 4 | 7 |
| | 1/0 | 0 | 0 | 1 | 1 | 2 | 3 | 6 |
| | 2/0 | 0 | 0 | 1 | 1 | 2 | 3 | 5 |
| | 3/0 | 0 | 0 | 1 | 1 | 1 | 2 | 4 |
| | 4/0 | 0 | 0 | 0 | 1 | 1 | 1 | 3 |
| | 250 | 0 | 0 | 0 | 1 | 1 | 1 | 3 |
| | 300 | 0 | 0 | 0 | 1 | 1 | 1 | 2 |
| | 350 | 0 | 0 | 0 | 0 | 1 | 1 | 1 |
| | 400 | 0 | 0 | 0 | 0 | 1 | 1 | 1 |
| | 500 | 0 | 0 | 0 | 0 | 1 | 1 | 1 |
| | 600 | 0 | 0 | 0 | 0 | 1 | 1 | 1 |
| | 700 | 0 | 0 | 0 | 0 | 0 | 1 | 1 |
| | 750 | 0 | 0 | 0 | 0 | 0 | 1 | 1 |
| | 800 | 0 | 0 | 0 | 0 | 0 | 1 | 1 |
| | 900 | 0 | 0 | 0 | 0 | 0 | 0 | 1 |
| | 1000 | 0 | 0 | 0 | 0 | 0 | 0 | 1 |
| | 1250 | 0 | 0 | 0 | 0 | 0 | 0 | 1 |
| | 1500 | 0 | 0 | 0 | 0 | 0 | 0 | 0 |
| | 1750 | 0 | 0 | 0 | 0 | 0 | 0 | 0 |
| | 2000 | 0 | 0 | 0 | 0 | 0 | 0 | 0 |
| THHN, | 14 | 8 | 13 | 22 | 36 | 63 | 81 | 133 |
| THWN, | 12 | 5 | 9 | 16 | 26 | 46 | 59 | 97 |
| THWN-2 | 10 | 3 | 6 | 10 | 16 | 29 | 37 | 61 |
| | 8 | 1 | 3 | 6 | 9 | 16 | 21 | 35 |
| | 6 | 1 | 2 | 4 | 7 | 12 | 15 | 25 |
| | 4 | 1 | 1 | 2 | 4 | 7 | 9 | 15 |
| | 3 | 1 | 1 | 1 | 3 | 6 | 8 | 13 |
| | 2 | 1 | 1 | 1 | 3 | 5 | 7 | 11 |
| | 1 | 0 | 1 | 1 | 1 | 4 | 5 | 8 |
| | 1/0 | 0 | 1 | 1 | 1 | 3 | 4 | 7 |
| | 2/0 | 0 | 0 | 1 | 1 | 2 | 3 | 6 |
| | 3/0 | 0 | 0 | 1 | 1 | 1 | 3 | 5 |
| | 4/0 | 0 | 0 | 1 | 1 | 1 | 2 | 4 |
| | 250 | 0 | 0 | 0 | 1 | 1 | 1 | 3 |
| | 300 | 0 | 0 | 0 | 1 | 1 | 1 | 3 |
| | 350 | 0 | 0 | 0 | 1 | 1 | 1 | 2 |
| | 400 | 0 | 0 | 0 | 0 | 1 | 1 | 1 |
| | 500 | 0 | 0 | 0 | 0 | 1 | 1 | 1 |
| | 600 | 0 | 0 | 0 | 0 | 1 | 1 | 1 |
| | 700 | 0 | 0 | 0 | 0 | 1 | 1 | 1 |
| | 750 | 0 | 0 | 0 | 0 | 0 | 1 | 1 |
| | 800 | 0 | 0 | 0 | 0 | 0 | 1 | 1 |
| | 900 | 0 | 0 | 0 | 0 | 0 | 1 | 1 |
| | 1000 | 0 | 0 | 0 | 0 | 0 | 0 | 1 |
| FEP, FEPB, | 14 | 7 | 12 | 21 | 35 | 61 | 79 | 129 |
| PFA, PFAH, | 12 | 5 | 9 | 15 | 25 | 44 | 57 | 94 |
| TFE | 10 | 4 | 6 | 11 | 18 | 32 | 41 | 68 |
| | 8 | 1 | 3 | 6 | 10 | 18 | 23 | 39 |
| | 6 | 1 | 2 | 4 | 7 | 13 | 17 | 27 |
| | 4 | 1 | 1 | 3 | 5 | 9 | 12 | 19 |
| | 3 | 1 | 1 | 2 | 4 | 7 | 10 | 16 |
| | 2 | 1 | 1 | 1 | 3 | 6 | 8 | 13 |
| PFA, PFAH, TFE | 1 | 0 | 1 | 1 | 2 | 4 | 5 | 9 |
| PFA, PFAH TFE, Z | 1/0 | 0 | 1 | 1 | 1 | 3 | 4 | 7 |
| | 2/0 | 0 | 1 | 1 | 1 | 3 | 4 | 6 |
| | 3/0 | 0 | 0 | 1 | 1 | 2 | 3 | 5 |
| | 4/0 | 0 | 0 | 1 | 1 | 1 | 2 | 4 |
| Z | 14 | 9 | 15 | 26 | 42 | 73 | 95 | 156 |
| | 12 | 6 | 10 | 18 | 30 | 52 | 67 | 111 |
| | 10 | 4 | 6 | 11 | 18 | 32 | 41 | 68 |
| | 8 | 2 | 4 | 7 | 11 | 20 | 26 | 43 |
| | 6 | 1 | 3 | 5 | 8 | 14 | 18 | 30 |
| | 4 | 1 | 1 | 3 | 5 | 9 | 12 | 20 |
| | 3 | 1 | 1 | 2 | 4 | 7 | 9 | 15 |
| | 2 | 0 | 1 | 1 | 3 | 6 | 7 | 12 |
| | 1 | 0 | 1 | 1 | 2 | 5 | 6 | 10 |

**Table C5** *Continued*

| | | CONDUCTORS | | | | | | |
|---|---|---|---|---|---|---|---|---|
| | | Metric Designator (Trade Size) | | | | | | |
| Type | Conductor Size (AWG/kcmil) | 12 (3/8) | 16 (1/2) | 21 (3/4) | 27 (1) | 35 (1 1/4) | 41 (1 1/2) | 53 (2) |
| XHH, XHHW, XHHW-2, ZW | 14 | 5 | 9 | 15 | 25 | 44 | 57 | 93 |
| | 12 | 4 | 7 | 12 | 19 | 33 | 43 | 71 |
| | 10 | 3 | 5 | 9 | 14 | 25 | 32 | 53 |
| | 8 | 1 | 3 | 5 | 8 | 14 | 18 | 29 |
| | 6 | 1 | 1 | 3 | 6 | 10 | 13 | 22 |
| | 4 | 1 | 1 | 2 | 4 | 7 | 9 | 16 |
| | 3 | 1 | 1 | 1 | 3 | 6 | 8 | 13 |
| | 2 | 1 | 1 | 1 | 3 | 5 | 7 | 11 |
| XHH, XHHW, XHHW-2 | 1 | 0 | 1 | 1 | 1 | 4 | 5 | 8 |
| | 1/0 | 0 | 1 | 1 | 1 | 3 | 4 | 7 |
| | 2/0 | 0 | 0 | 1 | 1 | 2 | 3 | 6 |
| | 3/0 | 0 | 0 | 1 | 1 | 1 | 3 | 5 |
| | 4/0 | 0 | 0 | 1 | 1 | 1 | 2 | 4 |
| | 250 | 0 | 0 | 0 | 1 | 1 | 1 | 3 |
| | 300 | 0 | 0 | 0 | 1 | 1 | 1 | 3 |
| | 350 | 0 | 0 | 0 | 1 | 1 | 1 | 2 |
| | 400 | 0 | 0 | 0 | 0 | 1 | 1 | 1 |
| | 500 | 0 | 0 | 0 | 0 | 1 | 1 | 1 |
| | 600 | 0 | 0 | 0 | 0 | 1 | 1 | 1 |
| | 700 | 0 | 0 | 0 | 0 | 1 | 1 | 1 |
| | 750 | 0 | 0 | 0 | 0 | 0 | 1 | 1 |
| | 800 | 0 | 0 | 0 | 0 | 0 | 1 | 1 |
| | 900 | 0 | 0 | 0 | 0 | 0 | 1 | 1 |
| | 1000 | 0 | 0 | 0 | 0 | 0 | 0 | 1 |
| | 1250 | 0 | 0 | 0 | 0 | 0 | 0 | 1 |
| | 1500 | 0 | 0 | 0 | 0 | 0 | 0 | 1 |
| | 1750 | 0 | 0 | 0 | 0 | 0 | 0 | 0 |
| | 2000 | 0 | 0 | 0 | 0 | 0 | 0 | 0 |
| **FIXTURE WIRES** | | | | | | | | |
| FFH-2, RFH-2 | 18 | 5 | 8 | 15 | 24 | 42 | 54 | 89 |
| | 16 | 4 | 7 | 12 | 20 | 35 | 46 | 75 |
| SF-2, SFF-2 | 18 | 6 | 11 | 19 | 30 | 53 | 69 | 113 |
| | 16 | 5 | 9 | 15 | 25 | 44 | 57 | 93 |
| | 14 | 4 | 7 | 12 | 20 | 35 | 46 | 75 |
| SF-1, SFF-1 | 18 | 11 | 19 | 33 | 53 | 94 | 122 | 199 |
| RFH-1, RFHH-2, TF, TFF, XF, XFF | 18 | 8 | 14 | 24 | 39 | 69 | 90 | 147 |
| RFHH-2, TF, TFF, XF, XFF | 16 | 7 | 11 | 20 | 32 | 56 | 72 | 119 |
| XF, XFF | 14 | 5 | 9 | 15 | 25 | 44 | 57 | 93 |
| TFN, TFFN | 18 | 14 | 23 | 39 | 63 | 111 | 144 | 236 |
| | 16 | 10 | 17 | 30 | 48 | 85 | 110 | 180 |
| PF, PFF, PGF, PGFF, PAF, PTF, PTFF, PAFF | 18 | 13 | 21 | 37 | 60 | 105 | 136 | 223 |
| | 16 | 10 | 16 | 29 | 46 | 81 | 105 | 173 |
| | 14 | 7 | 12 | 21 | 35 | 61 | 79 | 129 |
| HF, HFF, ZF, ZFF, ZHF | 18 | 17 | 28 | 48 | 77 | 136 | 176 | 288 |
| | 16 | 12 | 20 | 35 | 57 | 100 | 129 | 212 |
| | 14 | 9 | 15 | 26 | 42 | 73 | 95 | 156 |
| KF-2, KFF-2 | 18 | 24 | 40 | 70 | 112 | 197 | 255 | 418 |
| | 16 | 17 | 28 | 49 | 79 | 139 | 180 | 295 |
| | 14 | 12 | 19 | 34 | 54 | 95 | 123 | 202 |
| | 12 | 8 | 13 | 23 | 37 | 65 | 85 | 139 |
| | 10 | 5 | 9 | 15 | 25 | 44 | 57 | 93 |
| KF-1, KFF-1 | 18 | 29 | 48 | 83 | 134 | 235 | 304 | 499 |
| | 16 | 20 | 34 | 58 | 94 | 165 | 214 | 350 |
| | 14 | 14 | 23 | 39 | 63 | 111 | 144 | 236 |
| | 12 | 9 | 15 | 26 | 42 | 73 | 95 | 156 |
| | 10 | 6 | 10 | 17 | 27 | 48 | 62 | 102 |
| XF, XFF | 12 | 3 | 5 | 8 | 13 | 23 | 30 | 50 |
| | 10 | 1 | 3 | 6 | 10 | 18 | 23 | 39 |

Note: This table is for concentric stranded conductors only. For compact stranded conductors, Table C5(A). should be used.

*Corresponds to 356.2(2).

†Types RHH, RHW, and RHW-2 without outer covering.

**Table C5(A). Maximum Number of Compact Conductors in Liquidtight Flexible Nonmetallic Conduit (Type LFNC-B\*)** *(Based on Table 1, Chapter 9)*

| | | COMPACT CONDUCTORS | | | | | | |
|---|---|---|---|---|---|---|---|---|
| | Conductor Size (AWG/kcmil) | Metric Designator (Trade Size) | | | | | | |
| Type | | 12 (3/8) | 16 (1/2) | 21 (3/4) | 27 (1) | 35 (1 1/4) | 41 (1 1/2) | 53 (2) |
| THW, THW-2, THHW | 8 | 1 | 2 | 4 | 7 | 12 | 15 | 25 |
| | 6 | 1 | 1 | 3 | 5 | 9 | 12 | 19 |
| | 4 | 1 | 1 | 2 | 4 | 7 | 9 | 14 |
| | 2 | 1 | 1 | 1 | 3 | 5 | 6 | 11 |
| | 1 | 0 | 1 | 1 | 1 | 3 | 4 | 7 |
| | 1/0 | 0 | 1 | 1 | 1 | 3 | 4 | 6 |
| | 2/0 | 0 | 0 | 1 | 1 | 2 | 3 | 5 |
| | 3/0 | 0 | 0 | 1 | 1 | 1 | 3 | 4 |
| | 4/0 | 0 | 0 | 1 | 1 | 1 | 2 | 4 |
| | 250 | 0 | 0 | 0 | 1 | 1 | 1 | 3 |
| | 300 | 0 | 0 | 0 | 1 | 1 | 1 | 2 |
| | 350 | 0 | 0 | 0 | 1 | 1 | 1 | 2 |
| | 400 | 0 | 0 | 0 | 0 | 1 | 1 | 1 |
| | 500 | 0 | 0 | 0 | 0 | 1 | 1 | 1 |
| | 600 | 0 | 0 | 0 | 0 | 1 | 1 | 1 |
| | 700 | 0 | 0 | 0 | 0 | 1 | 1 | 1 |
| | 750 | 0 | 0 | 0 | 0 | 0 | 1 | 1 |
| | 1000 | 0 | 0 | 0 | 0 | 0 | 1 | 1 |
| THHN, THWN, THWN-2 | 8 | — | — | — | — | — | — | — |
| | 6 | 1 | 2 | 4 | 7 | 13 | 17 | 28 |
| | 4 | 1 | 1 | 3 | 4 | 8 | 11 | 17 |
| | 2 | 1 | 1 | 1 | 3 | 6 | 7 | 12 |
| | 1 | 0 | 1 | 1 | 2 | 4 | 6 | 9 |
| | 1/0 | 0 | 1 | 1 | 1 | 4 | 5 | 8 |
| | 2/0 | 0 | 1 | 1 | 1 | 3 | 4 | 6 |
| | 3/0 | 0 | 0 | 1 | 1 | 2 | 3 | 5 |
| | 4/0 | 0 | 0 | 1 | 1 | 1 | 3 | 4 |
| | 250 | 0 | 0 | 1 | 1 | 1 | 1 | 3 |
| | 300 | 0 | 0 | 0 | 1 | 1 | 1 | 3 |
| | 350 | 0 | 0 | 0 | 1 | 1 | 1 | 2 |
| | 400 | 0 | 0 | 0 | 1 | 1 | 1 | 2 |
| | 500 | 0 | 0 | 0 | 0 | 1 | 1 | 1 |
| | 600 | 0 | 0 | 0 | 0 | 1 | 1 | 1 |
| | 700 | 0 | 0 | 0 | 0 | 1 | 1 | 1 |
| | 750 | 0 | 0 | 0 | 0 | 1 | 1 | 1 |
| | 1000 | 0 | 0 | 0 | 0 | 0 | 1 | 1 |
| XHHW, XHHW-2 | 8 | 1 | 3 | 5 | 9 | 15 | 20 | 33 |
| | 6 | 1 | 2 | 4 | 6 | 11 | 15 | 24 |
| | 4 | 1 | 1 | 3 | 4 | 8 | 11 | 17 |
| | 2 | 1 | 1 | 1 | 3 | 6 | 7 | 12 |
| | 1 | 0 | 1 | 1 | 2 | 4 | 6 | 9 |
| | 1/0 | 0 | 1 | 1 | 1 | 4 | 5 | 8 |
| | 2/0 | 0 | 1 | 1 | 1 | 3 | 4 | 7 |
| | 3/0 | 0 | 0 | 1 | 1 | 2 | 3 | 5 |
| | 4/0 | 0 | 0 | 1 | 1 | 1 | 3 | 4 |
| | 250 | 0 | 0 | 1 | 1 | 1 | 1 | 3 |
| | 300 | 0 | 0 | 0 | 1 | 1 | 1 | 3 |
| | 350 | 0 | 0 | 0 | 1 | 1 | 1 | 3 |
| | 400 | 0 | 0 | 0 | 1 | 1 | 1 | 2 |
| | 500 | 0 | 0 | 0 | 0 | 1 | 1 | 1 |
| | 600 | 0 | 0 | 0 | 0 | 1 | 1 | 1 |
| | 700 | 0 | 0 | 0 | 0 | 1 | 1 | 1 |
| | 750 | 0 | 0 | 0 | 0 | 1 | 1 | 1 |
| | 1000 | 0 | 0 | 0 | 0 | 0 | 1 | 1 |

*Corresponds to 356.2(2).

Definition: *Compact stranding* is the result of a manufacturing process where the standard conductor is compressed to the extent that the interstices (voids between strand wires) are virtually eliminated.

**Table C6 Maximum Number of Conductors or Fixture Wires in Liquidtight Flexible Nonmetallic Conduit (Type LFNC-A*)** (*Based on Table 1, Chapter 9*)

| Type | Conductor Size (AWG/kcmil) | 12 (⅜) | 16 (½) | 21 (¾) | 27 (1) | 35 (1¼) | 41 (1½) | 53 (2) |
|---|---|---|---|---|---|---|---|---|
| RHH, RHW, RHW-2 | 14 | 2 | 4 | 7 | 11 | 20 | 27 | 45 |
| | 12 | 1 | 3 | 6 | 9 | 17 | 23 | 38 |
| | 10 | 1 | 3 | 5 | 8 | 13 | 18 | 30 |
| | 8 | 1 | 1 | 2 | 4 | 7 | 9 | 16 |
| | 6 | 1 | 1 | 1 | 3 | 5 | 7 | 13 |
| | 4 | 0 | 1 | 1 | 2 | 4 | 6 | 10 |
| | 3 | 0 | 1 | 1 | 1 | 4 | 5 | 8 |
| | 2 | 0 | 1 | 1 | 1 | 3 | 4 | 7 |
| | 1 | 0 | 0 | 1 | 1 | 1 | 3 | 5 |
| | 1/0 | 0 | 0 | 1 | 1 | 1 | 2 | 4 |
| | 2/0 | 0 | 0 | 1 | 1 | 1 | 1 | 4 |
| | 3/0 | 0 | 0 | 0 | 1 | 1 | 1 | 3 |
| | 4/0 | 0 | 0 | 0 | 1 | 1 | 1 | 3 |
| | 250 | 0 | 0 | 0 | 0 | 1 | 1 | 1 |
| | 300 | 0 | 0 | 0 | 0 | 1 | 1 | 1 |
| | 350 | 0 | 0 | 0 | 0 | 1 | 1 | 1 |
| | 400 | 0 | 0 | 0 | 0 | 1 | 1 | 1 |
| | 500 | 0 | 0 | 0 | 0 | 0 | 1 | 1 |
| | 600 | 0 | 0 | 0 | 0 | 0 | 1 | 1 |
| | 700 | 0 | 0 | 0 | 0 | 0 | 0 | 1 |
| | 750 | 0 | 0 | 0 | 0 | 0 | 0 | 1 |
| | 800 | 0 | 0 | 0 | 0 | 0 | 0 | 1 |
| | 900 | 0 | 0 | 0 | 0 | 0 | 0 | 1 |
| | 1000 | 0 | 0 | 0 | 0 | 0 | 0 | 1 |
| | 1250 | 0 | 0 | 0 | 0 | 0 | 0 | 0 |
| | 1500 | 0 | 0 | 0 | 0 | 0 | 0 | 0 |
| | 1750 | 0 | 0 | 0 | 0 | 0 | 0 | 0 |
| | 2000 | 0 | 0 | 0 | 0 | 0 | 0 | 0 |
| TW, THHW, THW, THW-2 | 14 | 5 | 9 | 15 | 24 | 43 | 58 | 96 |
| | 12 | 4 | 7 | 12 | 19 | 33 | 44 | 74 |
| | 10 | 3 | 5 | 9 | 14 | 24 | 33 | 55 |
| | 8 | 1 | 3 | 5 | 8 | 13 | 18 | 30 |
| RHH†, RHW†, RHW-2† | 14 | 3 | 6 | 10 | 16 | 28 | 38 | 64 |
| | 12 | 3 | 4 | 8 | 13 | 23 | 31 | 51 |
| | 10 | 1 | 3 | 6 | 10 | 18 | 24 | 40 |
| | 8 | 1 | 1 | 4 | 6 | 10 | 14 | 24 |
| RHH†, RHW†, RHW-2†, TW, THW, THHW, THW-2 | 6 | 1 | 1 | 3 | 4 | 8 | 11 | 18 |
| | 4 | 1 | 1 | 1 | 3 | 6 | 8 | 13 |
| | 3 | 1 | 1 | 1 | 3 | 5 | 7 | 11 |
| | 2 | 0 | 1 | 1 | 2 | 4 | 6 | 10 |
| | 1 | 0 | 1 | 1 | 1 | 3 | 4 | 7 |
| | 1/0 | 0 | 0 | 1 | 1 | 2 | 3 | 6 |
| | 2/0 | 0 | 0 | 1 | 1 | 1 | 3 | 5 |
| | 3/0 | 0 | 0 | 1 | 1 | 1 | 2 | 4 |
| | 4/0 | 0 | 0 | 0 | 1 | 1 | 1 | 3 |
| | 250 | 0 | 0 | 0 | 1 | 1 | 1 | 3 |
| | 300 | 0 | 0 | 0 | 1 | 1 | 1 | 2 |
| | 350 | 0 | 0 | 0 | 0 | 1 | 1 | 1 |
| | 400 | 0 | 0 | 0 | 0 | 1 | 1 | 1 |
| | 500 | 0 | 0 | 0 | 0 | 1 | 1 | 1 |
| | 600 | 0 | 0 | 0 | 0 | 1 | 1 | 1 |
| | 700 | 0 | 0 | 0 | 0 | 0 | 1 | 1 |
| | 750 | 0 | 0 | 0 | 0 | 0 | 1 | 1 |
| | 800 | 0 | 0 | 0 | 0 | 0 | 1 | 1 |
| | 900 | 0 | 0 | 0 | 0 | 0 | 0 | 1 |
| | 1000 | 0 | 0 | 0 | 0 | 0 | 0 | 1 |
| | 1250 | 0 | 0 | 0 | 0 | 0 | 0 | 1 |
| | 1500 | 0 | 0 | 0 | 0 | 0 | 0 | 1 |
| | 1750 | 0 | 0 | 0 | 0 | 0 | 0 | 0 |
| | 2000 | 0 | 0 | 0 | 0 | 0 | 0 | 0 |

**Table C6  *Continued***

| Type | Conductor Size (AWG/kcmil) | 12 (⅜) | 16 (½) | 21 (¾) | 27 (1) | 35 (1¼) | 41 (1½) | 53 (2) |
|---|---|---|---|---|---|---|---|---|
| THHN, THWN, THWN-2 | 14 | 8 | 13 | 22 | 35 | 62 | 83 | 137 |
| | 12 | 5 | 9 | 16 | 25 | 45 | 60 | 100 |
| | 10 | 3 | 6 | 10 | 16 | 28 | 38 | 63 |
| | 8 | 1 | 3 | 6 | 9 | 16 | 22 | 36 |
| | 6 | 1 | 2 | 4 | 6 | 12 | 16 | 26 |
| | 4 | 1 | 1 | 2 | 4 | 7 | 9 | 16 |
| | 3 | 1 | 1 | 1 | 3 | 6 | 8 | 13 |
| | 2 | 1 | 1 | 1 | 3 | 5 | 7 | 11 |
| | 1 | 0 | 1 | 1 | 1 | 4 | 5 | 8 |
| | 1/0 | 0 | 1 | 1 | 1 | 3 | 4 | 7 |
| | 2/0 | 0 | 0 | 1 | 1 | 2 | 3 | 6 |
| | 3/0 | 0 | 0 | 1 | 1 | 1 | 3 | 5 |
| | 4/0 | 0 | 0 | 1 | 1 | 1 | 2 | 4 |
| | 250 | 0 | 0 | 0 | 1 | 1 | 1 | 3 |
| | 300 | 0 | 0 | 0 | 1 | 1 | 1 | 3 |
| | 350 | 0 | 0 | 0 | 1 | 1 | 1 | 2 |
| | 400 | 0 | 0 | 0 | 0 | 1 | 1 | 1 |
| | 500 | 0 | 0 | 0 | 0 | 1 | 1 | 1 |
| | 600 | 0 | 0 | 0 | 0 | 1 | 1 | 1 |
| | 700 | 0 | 0 | 0 | 0 | 1 | 1 | 1 |
| | 750 | 0 | 0 | 0 | 0 | 0 | 1 | 1 |
| | 800 | 0 | 0 | 0 | 0 | 0 | 1 | 1 |
| | 900 | 0 | 0 | 0 | 0 | 0 | 1 | 1 |
| | 1000 | 0 | 0 | 0 | 0 | 0 | 0 | 1 |
| FEP, FEPB, PFA, PFAH, TFE | 14 | 7 | 12 | 21 | 34 | 60 | 80 | 133 |
| | 12 | 5 | 9 | 15 | 25 | 44 | 59 | 97 |
| | 10 | 4 | 6 | 11 | 18 | 31 | 42 | 70 |
| | 8 | 1 | 3 | 6 | 10 | 18 | 24 | 40 |
| | 6 | 1 | 2 | 4 | 7 | 13 | 17 | 28 |
| | 4 | 1 | 1 | 3 | 5 | 9 | 12 | 20 |
| | 3 | 1 | 1 | 2 | 4 | 7 | 10 | 16 |
| | 2 | 1 | 1 | 1 | 3 | 6 | 8 | 13 |
| PFA, PFAH, TFE | 1 | 0 | 1 | 1 | 2 | 4 | 5 | 9 |
| PFA, PFAH, TFE, Z | 1/0 | 0 | 1 | 1 | 1 | 3 | 5 | 8 |
| | 2/0 | 0 | 1 | 1 | 1 | 3 | 4 | 6 |
| | 3/0 | 0 | 0 | 1 | 1 | 2 | 3 | 5 |
| | 4/0 | 0 | 0 | 1 | 1 | 1 | 2 | 4 |
| Z | 14 | 9 | 15 | 25 | 41 | 72 | 97 | 161 |
| | 12 | 6 | 10 | 18 | 29 | 51 | 69 | 114 |
| | 10 | 4 | 6 | 11 | 18 | 31 | 42 | 70 |
| | 8 | 2 | 4 | 7 | 11 | 20 | 26 | 44 |
| | 6 | 1 | 3 | 5 | 8 | 14 | 18 | 31 |
| | 4 | 1 | 1 | 3 | 5 | 9 | 13 | 21 |
| | 3 | 1 | 1 | 2 | 4 | 7 | 9 | 15 |
| | 2 | 1 | 1 | 1 | 3 | 6 | 8 | 13 |
| | 1 | 1 | 1 | 1 | 2 | 4 | 6 | 10 |
| XHH, XHHW, XHHW-2, ZW | 14 | 5 | 9 | 15 | 24 | 43 | 58 | 96 |
| | 12 | 4 | 7 | 12 | 19 | 33 | 44 | 74 |
| | 10 | 3 | 5 | 9 | 14 | 24 | 33 | 55 |
| | 8 | 1 | 3 | 5 | 8 | 13 | 18 | 30 |
| | 6 | 1 | 1 | 3 | 5 | 10 | 13 | 22 |
| | 4 | 1 | 1 | 2 | 4 | 7 | 10 | 16 |
| | 3 | 1 | 1 | 1 | 3 | 6 | 8 | 14 |
| | 2 | 1 | 1 | 1 | 3 | 5 | 7 | 11 |
| XHH, XHHW, XHHW-2 | 1 | 0 | 1 | 1 | 1 | 4 | 5 | 8 |
| | 1/0 | 0 | 1 | 1 | 1 | 3 | 4 | 7 |
| | 2/0 | 0 | 0 | 1 | 1 | 2 | 3 | 6 |
| | 3/0 | 0 | 0 | 1 | 1 | 1 | 3 | 5 |
| | 4/0 | 0 | 0 | 1 | 1 | 1 | 2 | 4 |
| | 250 | 0 | 0 | 0 | 1 | 1 | 1 | 3 |
| | 300 | 0 | 0 | 0 | 1 | 1 | 1 | 3 |
| | 350 | 0 | 0 | 0 | 1 | 1 | 1 | 2 |
| | 400 | 0 | 0 | 0 | 0 | 1 | 1 | 1 |
| | 500 | 0 | 0 | 0 | 0 | 1 | 1 | 1 |
| | 600 | 0 | 0 | 0 | 0 | 1 | 1 | 1 |
| | 700 | 0 | 0 | 0 | 0 | 1 | 1 | 1 |
| | 750 | 0 | 0 | 0 | 0 | 0 | 1 | 1 |
| | 800 | 0 | 0 | 0 | 0 | 0 | 1 | 1 |
| | 900 | 0 | 0 | 0 | 0 | 0 | 1 | 1 |
| | 1000 | 0 | 0 | 0 | 0 | 0 | 0 | 1 |
| | 1250 | 0 | 0 | 0 | 0 | 0 | 0 | 1 |
| | 1500 | 0 | 0 | 0 | 0 | 0 | 0 | 1 |
| | 1750 | 0 | 0 | 0 | 0 | 0 | 0 | 0 |
| | 2000 | 0 | 0 | 0 | 0 | 0 | 0 | 0 |
| **FIXTURE WIRES** | | | | | | | | |
| FFH-2, RFH-2, RFHH-3 | 18 | 5 | 8 | 14 | 23 | 41 | 55 | 92 |
| | 16 | 4 | 7 | 12 | 20 | 35 | 47 | 77 |
| SF-2, SFF-2 | 18 | 6 | 11 | 18 | 29 | 52 | 70 | 116 |
| | 16 | 5 | 9 | 15 | 24 | 43 | 58 | 96 |
| | 14 | 4 | 7 | 12 | 20 | 35 | 47 | 77 |

**Table C6** *Continued*

| Type | Conductor Size (AWG/kcmil) | 12 (⅜) | 16 (½) | 21 (¾) | 27 (1) | 35 (1¼) | 41 (1½) | 53 (2) |
|---|---|---|---|---|---|---|---|---|
| | | **CONDUCTORS** | | | | | | |
| | | **Metric Designator (Trade Size)** | | | | | | |
| SF-1, SFF-1 | 18 | 12 | 19 | 33 | 52 | 92 | 124 | 205 |
| RFH-1, RFHH-2, TF, TFF, XF, XFF | 18 | 8 | 14 | 24 | 39 | 68 | 91 | 152 |
| RFHH-2, TF, TFF, XF, XFF | 16 | 7 | 11 | 19 | 31 | 55 | 74 | 122 |
| XF, XFF | 14 | 5 | 9 | 15 | 24 | 43 | 58 | 96 |
| TFN, TFFN | 18 | 14 | 22 | 39 | 62 | 109 | 146 | 243 |
| | 16 | 10 | 17 | 29 | 47 | 83 | 112 | 185 |
| PF, PFF, PGF, PGFF, PAF, PTF, PTFF, PAFF | 18 | 13 | 21 | 37 | 59 | 103 | 139 | 230 |
| | 16 | 10 | 16 | 28 | 45 | 80 | 107 | 178 |
| | 14 | 7 | 12 | 21 | 34 | 60 | 80 | 133 |
| HF, HFF, ZF, ZFF, ZHF | 18 | 17 | 27 | 47 | 76 | 133 | 179 | 297 |
| | 16 | 12 | 20 | 35 | 56 | 98 | 132 | 219 |
| | 14 | 9 | 15 | 25 | 41 | 72 | 97 | 161 |
| KF-2, KFF-2 | 18 | 25 | 40 | 69 | 110 | 193 | 260 | 431 |
| | 16 | 17 | 28 | 48 | 77 | 136 | 183 | 303 |
| | 14 | 12 | 19 | 33 | 53 | 94 | 126 | 209 |
| | 12 | 8 | 13 | 23 | 36 | 64 | 86 | 143 |
| | 10 | 5 | 9 | 15 | 24 | 43 | 58 | 96 |
| KF-1, KFF-1 | 18 | 29 | 48 | 82 | 131 | 231 | 310 | 514 |
| | 16 | 21 | 33 | 57 | 92 | 162 | 218 | 361 |
| | 14 | 14 | 22 | 39 | 62 | 109 | 146 | 243 |
| | 12 | 9 | 15 | 25 | 41 | 72 | 97 | 161 |
| | 10 | 6 | 10 | 17 | 27 | 47 | 63 | 105 |
| XF, XFF | 12 | 3 | 4 | 8 | 13 | 23 | 31 | 51 |
| | 10 | 1 | 3 | 6 | 10 | 18 | 24 | 40 |

Note: This table is for concentric stranded conductors only. For compact stranded conductors, Table C6(A) should be used.

*Corresponds to 356.2(1).

†Types RHH, RHW, and RHW-2 without outer covering.

**Table C6(A) Maximum Number of Compact Conductors in Liquidtight Flexible Nonmetallic Conduit (Type LFNC-A*)** (*Based on Table 1, Chapter 9*)

| Type | Conductor Size (AWG/kcmil) | 12 (⅜) | 16 (½) | 21 (¾) | 27 (1) | 35 (1¼) | 41 (1½) | 53 (2) |
|---|---|---|---|---|---|---|---|---|
| | | **COMPACT CONDUCTORS** | | | | | | |
| | | **Metric Designator (Trade Size)** | | | | | | |
| THW, THW-2, THHW | 8 | 1 | 2 | 4 | 6 | 11 | 16 | 26 |
| | 6 | 1 | 1 | 3 | 5 | 9 | 12 | 20 |
| | 4 | 1 | 1 | 2 | 4 | 7 | 9 | 15 |
| | 2 | 1 | 1 | 1 | 3 | 5 | 6 | 11 |
| | 1 | 0 | 1 | 1 | 1 | 3 | 4 | 8 |
| | 1/0 | 0 | 1 | 1 | 1 | 3 | 4 | 7 |
| | 2/0 | 0 | 0 | 1 | 1 | 2 | 3 | 5 |
| | 3/0 | 0 | 0 | 1 | 1 | 1 | 3 | 5 |
| | 4/0 | 0 | 0 | 1 | 1 | 1 | 2 | 4 |
| | 250 | 0 | 0 | 0 | 1 | 1 | 1 | 3 |
| | 300 | 0 | 0 | 0 | 1 | 1 | 1 | 3 |
| | 350 | 0 | 0 | 0 | 1 | 1 | 1 | 2 |
| | 400 | 0 | 0 | 0 | 0 | 1 | 1 | 1 |
| | 500 | 0 | 0 | 0 | 0 | 1 | 1 | 1 |
| | 600 | 0 | 0 | 0 | 0 | 1 | 1 | 1 |
| | 700 | 0 | 0 | 0 | 0 | 0 | 1 | 1 |
| | 750 | 0 | 0 | 0 | 0 | 0 | 1 | 1 |
| | 1000 | 0 | 0 | 0 | 0 | 0 | 1 | 1 |
| THHN, THWN, THWN-2 | 8 | — | — | — | — | — | — | — |
| | 6 | 1 | 2 | 4 | 7 | 13 | 18 | 29 |
| | 4 | 1 | 1 | 3 | 4 | 8 | 11 | 18 |
| | 2 | 1 | 1 | 1 | 3 | 6 | 8 | 13 |
| | 1 | 0 | 1 | 1 | 2 | 4 | 6 | 10 |
| | 1/0 | 0 | 1 | 1 | 1 | 3 | 5 | 8 |
| | 2/0 | 0 | 1 | 1 | 1 | 3 | 4 | 7 |
| | 3/0 | 0 | 0 | 1 | 1 | 2 | 3 | 6 |
| | 4/0 | 0 | 0 | 1 | 1 | 1 | 3 | 5 |
| | 250 | 0 | 0 | 1 | 1 | 1 | 1 | 3 |
| | 300 | 0 | 0 | 0 | 1 | 1 | 1 | 3 |
| | 350 | 0 | 0 | 0 | 1 | 1 | 1 | 3 |
| | 400 | 0 | 0 | 0 | 1 | 1 | 1 | 2 |
| | 500 | 0 | 0 | 0 | 0 | 1 | 1 | 1 |
| | 600 | 0 | 0 | 0 | 0 | 1 | 1 | 1 |
| | 700 | 0 | 0 | 0 | 0 | 1 | 1 | 1 |
| | 750 | 0 | 0 | 0 | 0 | 1 | 1 | 1 |
| | 1000 | 0 | 0 | 0 | 0 | 0 | 1 | 1 |

**Table C6(A)** *Continued*

| Type | Conductor Size (AWG/kcmil) | 12 (⅜) | 16 (½) | 21 (¾) | 27 (1) | 35 (1¼) | 41 (1½) | 53 (2) |
|---|---|---|---|---|---|---|---|---|
| | | **COMPACT CONDUCTORS** | | | | | | |
| | | **Metric Designator (Trade Size)** | | | | | | |
| XHHW, XHHW-2 | 8 | 1 | 3 | 5 | 8 | 15 | 20 | 34 |
| | 6 | 1 | 2 | 4 | 6 | 11 | 15 | 25 |
| | 4 | 1 | 1 | 3 | 4 | 8 | 11 | 18 |
| | 2 | 1 | 1 | 1 | 3 | 6 | 8 | 13 |
| | 1 | 0 | 1 | 1 | 2 | 4 | 6 | 10 |
| | 1/0 | 0 | 1 | 1 | 1 | 3 | 5 | 8 |
| | 2/0 | 0 | 1 | 1 | 1 | 3 | 4 | 7 |
| | 3/0 | 0 | 0 | 1 | 1 | 2 | 3 | 6 |
| | 4/0 | 0 | 0 | 1 | 1 | 1 | 3 | 5 |
| | 250 | 0 | 0 | 1 | 1 | 1 | 2 | 4 |
| | 300 | 0 | 0 | 0 | 1 | 1 | 1 | 3 |
| | 350 | 0 | 0 | 0 | 1 | 1 | 1 | 3 |
| | 400 | 0 | 0 | 0 | 1 | 1 | 1 | 2 |
| | 500 | 0 | 0 | 0 | 0 | 1 | 1 | 1 |
| | 600 | 0 | 0 | 0 | 0 | 1 | 1 | 1 |
| | 700 | 0 | 0 | 0 | 0 | 1 | 1 | 1 |
| | 750 | 0 | 0 | 0 | 0 | 1 | 1 | 1 |
| | 1000 | 0 | 0 | 0 | 0 | 0 | 1 | 1 |

*Corresponds to 356.2(1).

Definition: *Compact stranding* is the result of a manufacturing process where the standard conductor is compressed to the extent that the interstices (voids between strand wires) are virtually eliminated.

**Table C7 Maximum Number of Conductors or Fixture Wires in Liquidtight Flexible Metal Conduit (LFMC)** (*Based on Table 1, Chapter 9*)

| Type | Conductor Size (AWG/kcmil) | 16 (½) | 21(¾) | 27 (1) | 35 (1¼) | 41 (1½) | 53 (2) | 63 (2½) | 78 (3) | 91 (3½) | 103 (4) |
|---|---|---|---|---|---|---|---|---|---|---|---|
| | | **CONDUCTORS** | | | | | | | | | |
| | | **Metric Designator (Trade Size)** | | | | | | | | | |
| RHH, RHW, RHW-2 | 14 | 4 | 7 | 12 | 21 | 27 | 44 | 66 | 102 | 133 | 173 |
| | 12 | 3 | 6 | 10 | 17 | 22 | 36 | 55 | 84 | 110 | 144 |
| | 10 | 3 | 5 | 8 | 14 | 18 | 29 | 44 | 68 | 89 | 116 |
| | 8 | 1 | 2 | 4 | 7 | 9 | 15 | 23 | 36 | 46 | 61 |
| | 6 | 1 | 1 | 3 | 6 | 7 | 12 | 18 | 28 | 37 | 48 |
| | 4 | 1 | 1 | 2 | 4 | 6 | 9 | 14 | 22 | 29 | 38 |
| | 3 | 1 | 1 | 1 | 4 | 5 | 8 | 13 | 19 | 25 | 33 |
| | 2 | 1 | 1 | 1 | 3 | 4 | 7 | 11 | 17 | 22 | 29 |
| | 1 | 1 | 1 | 1 | 1 | 3 | 5 | 7 | 11 | 14 | 19 |
| | 1/0 | 0 | 1 | 1 | 1 | 2 | 4 | 6 | 10 | 13 | 16 |
| | 2/0 | 0 | 1 | 1 | 1 | 1 | 3 | 5 | 8 | 11 | 14 |
| | 3/0 | 0 | 0 | 1 | 1 | 1 | 3 | 4 | 7 | 9 | 12 |
| | 4/0 | 0 | 0 | 1 | 1 | 1 | 2 | 4 | 6 | 8 | 10 |
| | 250 | 0 | 0 | 0 | 1 | 1 | 1 | 3 | 4 | 6 | 8 |
| | 300 | 0 | 0 | 0 | 1 | 1 | 1 | 2 | 4 | 5 | 7 |
| | 350 | 0 | 0 | 0 | 1 | 1 | 1 | 2 | 3 | 5 | 6 |
| | 400 | 0 | 0 | 0 | 1 | 1 | 1 | 1 | 3 | 4 | 6 |
| | 500 | 0 | 0 | 0 | 1 | 1 | 1 | 1 | 3 | 4 | 5 |
| | 600 | 0 | 0 | 0 | 0 | 1 | 1 | 1 | 2 | 3 | 4 |
| | 700 | 0 | 0 | 0 | 0 | 1 | 1 | 1 | 1 | 3 | 3 |
| | 750 | 0 | 0 | 0 | 0 | 1 | 1 | 1 | 1 | 2 | 3 |
| | 800 | 0 | 0 | 0 | 0 | 1 | 1 | 1 | 1 | 2 | 3 |
| | 900 | 0 | 0 | 0 | 0 | 1 | 1 | 1 | 1 | 2 | 3 |
| | 1000 | 0 | 0 | 0 | 0 | 0 | 1 | 1 | 1 | 1 | 3 |
| | 1250 | 0 | 0 | 0 | 0 | 0 | 0 | 1 | 1 | 1 | 1 |
| | 1500 | 0 | 0 | 0 | 0 | 0 | 0 | 1 | 1 | 1 | 1 |
| | 1750 | 0 | 0 | 0 | 0 | 0 | 0 | 1 | 1 | 1 | 1 |
| | 2000 | 0 | 0 | 0 | 0 | 0 | 0 | 1 | 1 | 1 | 1 |
| TW, THHW, THW, THW-2 | 14 | 9 | 15 | 25 | 44 | 57 | 93 | 140 | 215 | 280 | 365 |
| | 12 | 7 | 12 | 19 | 33 | 43 | 71 | 108 | 165 | 215 | 280 |
| | 10 | 5 | 9 | 14 | 25 | 32 | 53 | 80 | 123 | 160 | 209 |
| | 8 | 3 | 5 | 8 | 14 | 18 | 29 | 44 | 68 | 89 | 116 |
| RHH*, RHW*, RHW-2* | 14 | 6 | 10 | 16 | 29 | 38 | 62 | 93 | 143 | 186 | 243 |

**Table C7** *Continued*

### CONDUCTORS

| Type | Conductor Size (AWG/kcmil) | 16 (½) | 21(¾) | 27 (1) | 35 (1¼) | 41 (1½) | 53 (2) | 63 (2½) | 78 (3) | 91 (3½) | 103 (4) |
|---|---|---|---|---|---|---|---|---|---|---|---|
| RHH*, | 12 | 5 | 8 | 13 | 23 | 30 | 50 | 75 | 115 | 149 | 195 |
| RHW*, | 10 | 3 | 6 | 10 | 18 | 23 | 39 | 58 | 89 | 117 | 152 |
| RHW-2* | 8 | 1 | 4 | 6 | 11 | 14 | 23 | 35 | 53 | 70 | 91 |
| RHH*, | 6 | 1 | 3 | 5 | 8 | 11 | 18 | 27 | 41 | 53 | 70 |
| RHW*, | 4 | 1 | 1 | 3 | 6 | 8 | 13 | 20 | 30 | 40 | 52 |
| RHW-2*, | 3 | 1 | 1 | 3 | 5 | 7 | 11 | 17 | 26 | 34 | 44 |
| TW, THW, | 2 | 1 | 1 | 2 | 4 | 6 | 9 | 14 | 22 | 29 | 38 |
| THHW, | 1 | 1 | 1 | 1 | 3 | 4 | 7 | 10 | 15 | 20 | 26 |
| THW-2 | 1/0 | 0 | 1 | 1 | 2 | 3 | 6 | 8 | 13 | 17 | 23 |
| | 2/0 | 0 | 1 | 1 | 2 | 3 | 5 | 7 | 11 | 15 | 19 |
| | 3/0 | 0 | 1 | 1 | 1 | 2 | 4 | 6 | 9 | 12 | 16 |
| | 4/0 | 0 | 0 | 1 | 1 | 1 | 3 | 5 | 8 | 10 | 13 |
| | 250 | 0 | 0 | 1 | 1 | 1 | 3 | 4 | 6 | 8 | 11 |
| | 300 | 0 | 0 | 1 | 1 | 1 | 2 | 3 | 5 | 7 | 9 |
| | 350 | 0 | 0 | 0 | 1 | 1 | 1 | 3 | 5 | 6 | 8 |
| | 400 | 0 | 0 | 0 | 1 | 1 | 1 | 3 | 4 | 6 | 7 |
| | 500 | 0 | 0 | 0 | 1 | 1 | 1 | 2 | 3 | 5 | 6 |
| | 600 | 0 | 0 | 0 | 1 | 1 | 1 | 1 | 3 | 4 | 5 |
| | 700 | 0 | 0 | 0 | 0 | 1 | 1 | 1 | 2 | 3 | 4 |
| | 750 | 0 | 0 | 0 | 0 | 1 | 1 | 1 | 2 | 3 | 4 |
| | 800 | 0 | 0 | 0 | 0 | 1 | 1 | 1 | 2 | 3 | 4 |
| | 900 | 0 | 0 | 0 | 0 | 0 | 1 | 1 | 1 | 3 | 3 |
| | 1000 | 0 | 0 | 0 | 0 | 0 | 1 | 1 | 1 | 2 | 3 |
| | 1250 | 0 | 0 | 0 | 0 | 0 | 1 | 1 | 1 | 1 | 2 |
| | 1500 | 0 | 0 | 0 | 0 | 0 | 0 | 1 | 1 | 1 | 2 |
| | 1750 | 0 | 0 | 0 | 0 | 0 | 0 | 1 | 1 | 1 | 1 |
| | 2000 | 0 | 0 | 0 | 0 | 0 | 0 | 1 | 1 | 1 | 1 |
| THHN, | 14 | 13 | 22 | 36 | 63 | 81 | 133 | 201 | 308 | 401 | 523 |
| THWN, | 12 | 9 | 16 | 26 | 46 | 59 | 97 | 146 | 225 | 292 | 381 |
| THWN-2 | 10 | 6 | 10 | 16 | 29 | 37 | 61 | 92 | 141 | 184 | 240 |
| | 8 | 3 | 6 | 9 | 16 | 21 | 35 | 53 | 81 | 106 | 138 |
| | 6 | 2 | 4 | 7 | 12 | 15 | 25 | 38 | 59 | 76 | 100 |
| | 4 | 1 | 2 | 4 | 7 | 9 | 15 | 23 | 36 | 47 | 61 |
| | 3 | 1 | 1 | 3 | 6 | 8 | 13 | 20 | 30 | 40 | 52 |
| | 2 | 1 | 1 | 3 | 5 | 7 | 11 | 17 | 26 | 33 | 44 |
| | 1 | 1 | 1 | 1 | 4 | 5 | 8 | 12 | 19 | 25 | 32 |
| | 1/0 | 1 | 1 | 1 | 3 | 4 | 7 | 10 | 16 | 21 | 27 |
| | 2/0 | 0 | 1 | 1 | 2 | 3 | 6 | 8 | 13 | 17 | 23 |
| | 3/0 | 0 | 1 | 1 | 1 | 3 | 5 | 7 | 11 | 14 | 19 |
| | 4/0 | 0 | 1 | 1 | 1 | 2 | 4 | 6 | 9 | 12 | 15 |
| | 250 | 0 | 0 | 1 | 1 | 1 | 3 | 5 | 7 | 10 | 12 |
| | 300 | 0 | 0 | 1 | 1 | 1 | 3 | 4 | 6 | 8 | 11 |
| | 350 | 0 | 0 | 1 | 1 | 1 | 2 | 3 | 5 | 7 | 9 |
| | 400 | 0 | 0 | 0 | 1 | 1 | 1 | 3 | 5 | 6 | 8 |
| | 500 | 0 | 0 | 0 | 1 | 1 | 1 | 2 | 4 | 5 | 7 |
| | 600 | 0 | 0 | 0 | 1 | 1 | 1 | 1 | 3 | 4 | 6 |
| | 700 | 0 | 0 | 0 | 1 | 1 | 1 | 1 | 3 | 4 | 5 |
| | 750 | 0 | 0 | 0 | 0 | 1 | 1 | 1 | 3 | 3 | 5 |
| | 800 | 0 | 0 | 0 | 0 | 1 | 1 | 1 | 2 | 3 | 4 |
| | 900 | 0 | 0 | 0 | 0 | 0 | 1 | 1 | 1 | 2 | 4 |
| | 1000 | 0 | 0 | 0 | 0 | 0 | 1 | 1 | 1 | 2 | 3 |
| FEP, | 14 | 12 | 21 | 35 | 61 | 79 | 129 | 195 | 299 | 389 | 507 |
| FEPB, | 12 | 9 | 15 | 25 | 44 | 57 | 94 | 142 | 218 | 284 | 370 |
| PFA, | 10 | 6 | 11 | 18 | 32 | 41 | 68 | 102 | 156 | 203 | 266 |
| PFAH, | 8 | 3 | 6 | 10 | 18 | 23 | 39 | 58 | 89 | 117 | 152 |
| TFE | 6 | 2 | 4 | 7 | 13 | 17 | 27 | 41 | 64 | 83 | 108 |
| | 4 | 1 | 3 | 5 | 9 | 12 | 19 | 29 | 44 | 58 | 75 |
| | 3 | 1 | 2 | 4 | 7 | 10 | 16 | 24 | 37 | 48 | 63 |
| | 2 | 1 | 1 | 3 | 6 | 8 | 13 | 20 | 30 | 40 | 52 |
| PFA, PFAH, TFE | 1 | 1 | 1 | 2 | 4 | 5 | 9 | 14 | 21 | 28 | 36 |
| PFA, | 1/0 | 1 | 1 | 1 | 3 | 4 | 7 | 11 | 18 | 23 | 30 |
| PFAH, | 2/0 | 1 | 1 | 1 | 3 | 4 | 6 | 9 | 14 | 19 | 25 |
| TFE, Z | 3/0 | 0 | 1 | 1 | 2 | 3 | 5 | 8 | 12 | 16 | 20 |
| | 4/0 | 0 | 1 | 1 | 1 | 2 | 4 | 6 | 10 | 13 | 17 |
| Z | 14 | 20 | 26 | 42 | 73 | 95 | 156 | 235 | 360 | 469 | 611 |
| | 12 | 14 | 18 | 30 | 52 | 67 | 111 | 167 | 255 | 332 | 434 |
| | 10 | 8 | 11 | 18 | 32 | 41 | 68 | 102 | 156 | 203 | 266 |
| | 8 | 5 | 7 | 11 | 20 | 26 | 43 | 64 | 99 | 129 | 168 |
| | 6 | 4 | 5 | 8 | 14 | 18 | 30 | 45 | 69 | 90 | 118 |
| | 4 | 2 | 3 | 5 | 9 | 12 | 20 | 31 | 48 | 62 | 81 |
| | 3 | 2 | 2 | 4 | 7 | 9 | 15 | 23 | 35 | 45 | 59 |
| | 2 | 1 | 1 | 3 | 6 | 7 | 12 | 19 | 29 | 38 | 49 |
| | 1 | 1 | 1 | 2 | 5 | 6 | 10 | 15 | 23 | 30 | 40 |

**Table C7** *Continued*

### CONDUCTORS

| Type | Conductor Size (AWG/kcmil) | 16 (½) | 21(¾) | 27 (1) | 35 (1¼) | 41 (1½) | 53 (2) | 63 (2½) | 78 (3) | 91 (3½) | 103 (4) |
|---|---|---|---|---|---|---|---|---|---|---|---|
| XHH, | 14 | 9 | 15 | 25 | 44 | 57 | 93 | 140 | 215 | 280 | 365 |
| XHHW, | 12 | 7 | 12 | 19 | 33 | 43 | 71 | 108 | 165 | 215 | 280 |
| XHHW-2, | 10 | 5 | 9 | 14 | 25 | 32 | 53 | 80 | 123 | 160 | 209 |
| ZW | 8 | 3 | 5 | 8 | 14 | 18 | 29 | 44 | 68 | 89 | 116 |
| | 6 | 1 | 3 | 6 | 10 | 13 | 22 | 33 | 50 | 66 | 86 |
| | 4 | 1 | 2 | 4 | 7 | 9 | 16 | 24 | 36 | 48 | 62 |
| | 3 | 1 | 1 | 3 | 6 | 8 | 13 | 20 | 31 | 40 | 52 |
| | 2 | 1 | 1 | 3 | 5 | 7 | 11 | 17 | 26 | 34 | 44 |
| XHH, | 1 | 1 | 1 | 1 | 4 | 5 | 8 | 12 | 19 | 25 | 33 |
| XHHW, | 1/0 | 1 | 1 | 1 | 3 | 4 | 7 | 10 | 16 | 21 | 28 |
| XHHW-2 | 2/0 | 0 | 1 | 1 | 2 | 3 | 6 | 9 | 13 | 17 | 23 |
| | 3/0 | 0 | 1 | 1 | 1 | 3 | 5 | 7 | 11 | 14 | 19 |
| | 4/0 | 0 | 1 | 1 | 1 | 2 | 4 | 6 | 9 | 12 | 16 |
| | 250 | 0 | 0 | 1 | 1 | 1 | 3 | 5 | 7 | 10 | 13 |
| | 300 | 0 | 0 | 1 | 1 | 1 | 3 | 4 | 6 | 8 | 11 |
| | 350 | 0 | 0 | 1 | 1 | 1 | 2 | 3 | 5 | 7 | 10 |
| | 400 | 0 | 0 | 0 | 1 | 1 | 1 | 3 | 5 | 6 | 8 |
| | 500 | 0 | 0 | 0 | 1 | 1 | 1 | 2 | 4 | 5 | 7 |
| | 600 | 0 | 0 | 0 | 1 | 1 | 1 | 1 | 3 | 4 | 6 |
| | 700 | 0 | 0 | 0 | 1 | 1 | 1 | 1 | 3 | 4 | 5 |
| | 750 | 0 | 0 | 0 | 0 | 1 | 1 | 1 | 3 | 3 | 5 |
| | 800 | 0 | 0 | 0 | 0 | 1 | 1 | 1 | 2 | 3 | 4 |
| | 900 | 0 | 0 | 0 | 0 | 1 | 1 | 1 | 2 | 3 | 4 |
| | 1000 | 0 | 0 | 0 | 0 | 0 | 1 | 1 | 1 | 3 | 3 |
| | 1250 | 0 | 0 | 0 | 0 | 0 | 1 | 1 | 1 | 1 | 3 |
| | 1500 | 0 | 0 | 0 | 0 | 0 | 1 | 1 | 1 | 1 | 2 |
| | 1750 | 0 | 0 | 0 | 0 | 0 | 0 | 1 | 1 | 1 | 2 |
| | 2000 | 0 | 0 | 0 | 0 | 0 | 0 | 1 | 1 | 1 | 2 |

*Types RHH, RHW, and RHW-2 without outer covering.

### FIXTURE WIRES

| Type | Conductor Size (AWG/kcmil) | 16 (½) | 21 (¾) | 27 (1) | 35 (1¼) | 41 (1½) | 53 (2) |
|---|---|---|---|---|---|---|---|
| FFH-2, | 18 | 8 | 15 | 24 | 42 | 54 | 89 |
| RFH-2, RFHH-3 | 16 | 7 | 12 | 20 | 35 | 46 | 75 |
| SF-2, SFF-2 | 18 | 11 | 19 | 30 | 53 | 69 | 113 |
| | 16 | 9 | 15 | 25 | 44 | 57 | 93 |
| | 14 | 7 | 12 | 20 | 35 | 46 | 75 |
| SF-1, SFF-1 | 18 | 19 | 33 | 53 | 94 | 122 | 199 |
| RFH-1, RFHH-2, TF, TFF, XF, XFF | 18 | 14 | 24 | 39 | 69 | 90 | 147 |
| RFHH-2, TF, TFF, XF, XFF | 16 | 11 | 20 | 32 | 56 | 72 | 119 |
| XF, XFF | 14 | 9 | 15 | 25 | 44 | 57 | 93 |
| TFN, TFFN | 18 | 23 | 39 | 63 | 111 | 144 | 236 |
| | 16 | 17 | 30 | 48 | 85 | 110 | 180 |
| PF, PFF, PGF, PGFF, PAF, PTF, PTFF, PAFF | 18 | 21 | 37 | 60 | 105 | 136 | 223 |
| | 16 | 16 | 29 | 46 | 81 | 105 | 173 |
| | 14 | 12 | 21 | 35 | 61 | 79 | 129 |
| HF, HFF, ZF, ZFF, ZHF | 18 | 28 | 48 | 77 | 136 | 176 | 288 |
| | 16 | 20 | 35 | 57 | 100 | 129 | 212 |
| | 14 | 15 | 26 | 42 | 73 | 95 | 156 |
| KF-2, KFF-2 | 18 | 40 | 70 | 112 | 197 | 255 | 418 |
| | 16 | 28 | 49 | 79 | 139 | 180 | 295 |
| | 14 | 19 | 34 | 54 | 95 | 123 | 202 |
| | 12 | 13 | 23 | 37 | 65 | 85 | 139 |
| | 10 | 9 | 15 | 25 | 44 | 57 | 93 |
| KF-1, KFF-1 | 18 | 48 | 83 | 134 | 235 | 304 | 499 |
| | 16 | 34 | 58 | 94 | 165 | 214 | 350 |
| | 14 | 23 | 39 | 63 | 111 | 144 | 236 |
| | 12 | 15 | 26 | 42 | 73 | 95 | 156 |
| | 10 | 10 | 17 | 27 | 48 | 62 | 102 |
| XF, XFF | 12 | 5 | 8 | 13 | 23 | 30 | 50 |
| | 10 | 3 | 6 | 10 | 18 | 23 | 39 |

Note: This table is for concentric stranded conductors only. For compact stranded conductors, Table C7(A) should be used.

**Table C7(A) Maximum Number of Compact Conductors in Liquidtight Flexible Metal Conduit (LFMC)** (*Based on Table 1, Chapter 9*)

| | | COMPACT CONDUCTORS | | | | | | | | | | |
|---|---|---|---|---|---|---|---|---|---|---|---|---|
| | Conductor Size | Metric Designator (Trade Size) | | | | | | | | | | |
| Type | (AWG/kcmil) | 12 (⅜) | 16 (½) | 21 (¾) | 27 (1) | 35 (1¼) | 41 (1½) | 53 (2) | 63 (2½) | 78 (3) | 91 (3½) | 103 (4) |
| THW, | 8 | 1 | 2 | 4 | 7 | 12 | 15 | 25 | 38 | 58 | 76 | 99 |
| THW-2, | 6 | 1 | 1 | 3 | 5 | 9 | 12 | 19 | 29 | 45 | 59 | 77 |
| THHW | 4 | 1 | 1 | 2 | 4 | 7 | 9 | 14 | 22 | 34 | 44 | 57 |
| | 2 | 1 | 1 | 1 | 3 | 5 | 6 | 11 | 16 | 25 | 32 | 42 |
| | 1 | 0 | 1 | 1 | 1 | 3 | 4 | 7 | 11 | 17 | 23 | 30 |
| | 1/0 | 0 | 1 | 1 | 1 | 3 | 4 | 6 | 10 | 15 | 20 | 26 |
| | 2/0 | 0 | 0 | 1 | 1 | 2 | 3 | 5 | 8 | 13 | 16 | 21 |
| | 3/0 | 0 | 0 | 1 | 1 | 1 | 3 | 4 | 7 | 11 | 14 | 18 |
| | 4/0 | 0 | 0 | 1 | 1 | 1 | 2 | 4 | 6 | 9 | 12 | 15 |
| | 250 | 0 | 0 | 0 | 1 | 1 | 1 | 3 | 4 | 7 | 9 | 12 |
| | 300 | 0 | 0 | 0 | 1 | 1 | 1 | 2 | 4 | 6 | 8 | 10 |
| | 350 | 0 | 0 | 0 | 1 | 1 | 1 | 2 | 3 | 5 | 7 | 9 |
| | 400 | 0 | 0 | 0 | 0 | 1 | 1 | 1 | 3 | 5 | 6 | 8 |
| | 500 | 0 | 0 | 0 | 0 | 1 | 1 | 1 | 3 | 4 | 5 | 7 |
| | 600 | 0 | 0 | 0 | 0 | 1 | 1 | 1 | 1 | 3 | 4 | 6 |
| | 700 | 0 | 0 | 0 | 0 | 1 | 1 | 1 | 1 | 3 | 4 | 5 |
| | 750 | 0 | 0 | 0 | 0 | 0 | 1 | 1 | 1 | 3 | 3 | 5 |
| | 1000 | 0 | 0 | 0 | 0 | 0 | 1 | 1 | 1 | 1 | 3 | 4 |
| THHN, | 8 | — | — | — | — | — | — | — | — | — | — | — |
| THWN, | 6 | 1 | 2 | 4 | 7 | 13 | 17 | 28 | 43 | 66 | 86 | 112 |
| THWN-2 | 4 | 1 | 1 | 3 | 4 | 8 | 11 | 17 | 26 | 41 | 53 | 69 |
| | 2 | 1 | 1 | 1 | 3 | 6 | 7 | 12 | 19 | 29 | 38 | 50 |
| | 1 | 0 | 1 | 1 | 2 | 4 | 6 | 9 | 14 | 22 | 28 | 37 |
| | 1/0 | 0 | 1 | 1 | 1 | 4 | 5 | 8 | 12 | 19 | 24 | 32 |
| | 2/0 | 0 | 1 | 1 | 1 | 3 | 4 | 6 | 10 | 15 | 20 | 26 |
| | 3/0 | 0 | 0 | 1 | 1 | 2 | 3 | 5 | 8 | 13 | 17 | 22 |
| | 4/0 | 0 | 0 | 1 | 1 | 1 | 3 | 4 | 7 | 10 | 14 | 18 |
| | 250 | 0 | 0 | 1 | 1 | 1 | 1 | 3 | 5 | 8 | 11 | 14 |
| | 300 | 0 | 0 | 0 | 1 | 1 | 1 | 3 | 4 | 7 | 9 | 12 |
| | 350 | 0 | 0 | 0 | 1 | 1 | 1 | 2 | 4 | 6 | 8 | 11 |
| | 400 | 0 | 0 | 0 | 1 | 1 | 1 | 2 | 3 | 5 | 7 | 9 |
| | 500 | 0 | 0 | 0 | 0 | 1 | 1 | 1 | 3 | 5 | 6 | 8 |
| | 600 | 0 | 0 | 0 | 0 | 1 | 1 | 1 | 2 | 4 | 5 | 6 |
| | 700 | 0 | 0 | 0 | 0 | 1 | 1 | 1 | 1 | 3 | 4 | 6 |
| | 750 | 0 | 0 | 0 | 0 | 1 | 1 | 1 | 1 | 3 | 4 | 5 |
| | 1000 | 0 | 0 | 0 | 0 | 0 | 1 | 1 | 1 | 2 | 3 | 4 |
| XHHW, | 8 | 1 | 3 | 5 | 9 | 15 | 20 | 33 | 49 | 76 | 98 | 129 |
| XHHW-2 | 6 | 1 | 2 | 4 | 6 | 11 | 15 | 24 | 37 | 56 | 73 | 95 |
| | 4 | 1 | 1 | 3 | 4 | 8 | 11 | 17 | 26 | 41 | 53 | 69 |
| | 2 | 1 | 1 | 1 | 3 | 6 | 7 | 12 | 19 | 29 | 38 | 50 |
| | 1 | 0 | 1 | 1 | 2 | 4 | 6 | 9 | 14 | 22 | 28 | 37 |
| | 1/0 | 0 | 1 | 1 | 1 | 4 | 5 | 8 | 12 | 19 | 24 | 32 |
| | 2/0 | 0 | 1 | 1 | 1 | 3 | 4 | 7 | 10 | 16 | 20 | 27 |
| | 3/0 | 0 | 0 | 1 | 1 | 2 | 3 | 5 | 8 | 13 | 17 | 22 |
| | 4/0 | 0 | 0 | 1 | 1 | 1 | 3 | 4 | 7 | 11 | 14 | 18 |
| | 250 | 0 | 0 | 1 | 1 | 1 | 1 | 3 | 5 | 8 | 11 | 15 |
| | 300 | 0 | 0 | 0 | 1 | 1 | 1 | 3 | 5 | 7 | 9 | 12 |
| | 350 | 0 | 0 | 0 | 1 | 1 | 1 | 3 | 4 | 6 | 8 | 11 |
| | 400 | 0 | 0 | 0 | 1 | 1 | 1 | 2 | 4 | 6 | 7 | 10 |
| | 500 | 0 | 0 | 0 | 0 | 1 | 1 | 1 | 3 | 5 | 6 | 8 |
| | 600 | 0 | 0 | 0 | 0 | 1 | 1 | 1 | 2 | 4 | 5 | 6 |
| | 700 | 0 | 0 | 0 | 0 | 1 | 1 | 1 | 1 | 3 | 4 | 6 |
| | 750 | 0 | 0 | 0 | 0 | 1 | 1 | 1 | 1 | 3 | 4 | 5 |
| | 1000 | 0 | 0 | 0 | 0 | 0 | 1 | 1 | 1 | 2 | 3 | 4 |

Definition: *Compact stranding* is the result of a manufacturing process where the standard conductor is compressed to the extent that the interstices (voids between strand wires) are virtually eliminated.

**Table C8 Maximum Number of Conductors or Fixture Wires in Rigid Metal Conduit (RMC)** (*Based on Table 1, Chapter 9*)

| | | CONDUCTORS | | | | | | | | | | | |
|---|---|---|---|---|---|---|---|---|---|---|---|---|---|
| | Conductor Size | Metric Designator (Trade Size) | | | | | | | | | | | |
| Type | (AWG/kcmil) | 16 (½) | 21 (¾) | 27 (1) | 35 (1¼) | 41 (1½) | 53 (2) | 63 (2½) | 78 (3) | 91 (3½) | 103 (4) | 129 (5) | 155 (6) |
| RHH, | 14 | 4 | 7 | 12 | 21 | 28 | 46 | 66 | 102 | 136 | 176 | 276 | 398 |
| RHW, | 12 | 3 | 6 | 10 | 17 | 23 | 38 | 55 | 85 | 113 | 146 | 229 | 330 |
| RHW-2 | 10 | 3 | 5 | 8 | 14 | 19 | 31 | 44 | 68 | 91 | 118 | 185 | 267 |
| | 8 | 1 | 2 | 4 | 7 | 10 | 16 | 23 | 36 | 48 | 61 | 97 | 139 |
| | 6 | 1 | 1 | 3 | 6 | 8 | 13 | 18 | 29 | 38 | 49 | 77 | 112 |
| | 4 | 1 | 1 | 2 | 4 | 6 | 10 | 14 | 22 | 30 | 38 | 60 | 87 |
| | 3 | 1 | 1 | 2 | 4 | 5 | 9 | 12 | 19 | 26 | 34 | 53 | 76 |
| | 2 | 1 | 1 | 1 | 3 | 4 | 7 | 11 | 17 | 23 | 29 | 46 | 66 |
| | 1 | 0 | 1 | 1 | 1 | 3 | 5 | 7 | 11 | 15 | 19 | 30 | 44 |
| | 1/0 | 0 | 1 | 1 | 1 | 2 | 4 | 6 | 10 | 13 | 17 | 26 | 38 |
| | 2/0 | 0 | 1 | 1 | 1 | 2 | 4 | 5 | 8 | 11 | 14 | 23 | 33 |
| | 3/0 | 0 | 0 | 1 | 1 | 1 | 3 | 4 | 7 | 10 | 12 | 20 | 28 |
| | 4/0 | 0 | 0 | 1 | 1 | 1 | 3 | 4 | 6 | 8 | 11 | 17 | 24 |
| | 250 | 0 | 0 | 0 | 1 | 1 | 1 | 3 | 4 | 6 | 8 | 13 | 18 |
| | 300 | 0 | 0 | 0 | 1 | 1 | 1 | 2 | 4 | 5 | 7 | 11 | 16 |
| | 350 | 0 | 0 | 0 | 1 | 1 | 1 | 2 | 4 | 5 | 6 | 10 | 15 |
| | 400 | 0 | 0 | 0 | 1 | 1 | 1 | 1 | 3 | 4 | 6 | 9 | 13 |
| | 500 | 0 | 0 | 0 | 1 | 1 | 1 | 1 | 3 | 4 | 5 | 8 | 11 |
| | 600 | 0 | 0 | 0 | 0 | 1 | 1 | 1 | 2 | 3 | 4 | 6 | 9 |
| | 700 | 0 | 0 | 0 | 0 | 1 | 1 | 1 | 1 | 3 | 4 | 6 | 8 |
| | 750 | 0 | 0 | 0 | 0 | 0 | 1 | 1 | 1 | 3 | 3 | 5 | 8 |
| | 800 | 0 | 0 | 0 | 0 | 0 | 1 | 1 | 1 | 2 | 3 | 5 | 7 |
| | 900 | 0 | 0 | 0 | 0 | 0 | 1 | 1 | 1 | 2 | 3 | 5 | 7 |
| | 1000 | 0 | 0 | 0 | 0 | 0 | 1 | 1 | 1 | 1 | 3 | 4 | 6 |
| | 1250 | 0 | 0 | 0 | 0 | 0 | 0 | 1 | 1 | 1 | 1 | 3 | 5 |
| | 1500 | 0 | 0 | 0 | 0 | 0 | 0 | 1 | 1 | 1 | 1 | 3 | 4 |
| | 1750 | 0 | 0 | 0 | 0 | 0 | 0 | 1 | 1 | 1 | 1 | 2 | 4 |
| | 2000 | 0 | 0 | 0 | 0 | 0 | 0 | 0 | 1 | 1 | 1 | 2 | 3 |
| TW, | 14 | 9 | 15 | 25 | 44 | 59 | 98 | 140 | 216 | 288 | 370 | 581 | 839 |
| THHW, | 12 | 7 | 12 | 19 | 33 | 45 | 75 | 107 | 165 | 221 | 284 | 446 | 644 |
| THW, | | | | | | | | | | | | | |
| THW-2 | 10 | 5 | 9 | 14 | 25 | 34 | 56 | 80 | 123 | 164 | 212 | 332 | 480 |
| | 8 | 3 | 5 | 8 | 14 | 19 | 31 | 44 | 68 | 91 | 118 | 185 | 267 |
| RHH*, RHW*, RHW-2* | 14 | 6 | 10 | 17 | 29 | 39 | 65 | 93 | 143 | 191 | 246 | 387 | 558 |
| RHH*, RHW*, RHW-2* | 12 | 5 | 8 | 13 | 23 | 32 | 52 | 75 | 115 | 154 | 198 | 311 | 448 |
| | 10 | 3 | 6 | 10 | 18 | 25 | 41 | 58 | 90 | 120 | 154 | 242 | 350 |
| RHH*, RHW*, RHW-2* | 8 | 1 | 4 | 6 | 11 | 15 | 24 | 35 | 54 | 72 | 92 | 145 | 209 |
| RHH*, RHW*, RHW-2*, TW, THW, THHW, THW-2 | 6 | 1 | 3 | 5 | 8 | 11 | 18 | 27 | 41 | 55 | 71 | 111 | 160 |
| | 4 | 1 | 1 | 3 | 6 | 8 | 14 | 20 | 31 | 41 | 53 | 83 | 120 |
| | 3 | 1 | 1 | 3 | 5 | 7 | 12 | 17 | 26 | 35 | 45 | 71 | 103 |
| | 2 | 1 | 1 | 2 | 4 | 6 | 10 | 14 | 22 | 30 | 38 | 60 | 87 |
| | 1 | 1 | 1 | 1 | 3 | 4 | 7 | 10 | 15 | 21 | 27 | 42 | 61 |
| | 1/0 | 0 | 1 | 1 | 2 | 3 | 6 | 8 | 13 | 18 | 23 | 36 | 52 |
| | 2/0 | 0 | 1 | 1 | 2 | 3 | 5 | 7 | 11 | 15 | 19 | 31 | 44 |
| | 3/0 | 0 | 1 | 1 | 1 | 2 | 4 | 6 | 9 | 13 | 16 | 26 | 37 |
| | 4/0 | 0 | 0 | 1 | 1 | 1 | 3 | 5 | 8 | 10 | 14 | 21 | 31 |
| | 250 | 0 | 0 | 1 | 1 | 1 | 3 | 4 | 6 | 8 | 11 | 17 | 25 |
| | 300 | 0 | 0 | 1 | 1 | 1 | 2 | 3 | 5 | 7 | 9 | 15 | 22 |
| | 350 | 0 | 0 | 0 | 1 | 1 | 1 | 3 | 5 | 6 | 8 | 13 | 19 |
| | 400 | 0 | 0 | 0 | 1 | 1 | 1 | 3 | 4 | 6 | 7 | 12 | 17 |
| | 500 | 0 | 0 | 0 | 1 | 1 | 1 | 2 | 3 | 5 | 6 | 10 | 14 |
| | 600 | 0 | 0 | 0 | 1 | 1 | 1 | 1 | 3 | 4 | 5 | 8 | 12 |
| | 700 | 0 | 0 | 0 | 0 | 1 | 1 | 1 | 2 | 3 | 4 | 7 | 10 |
| | 750 | 0 | 0 | 0 | 0 | 1 | 1 | 1 | 2 | 3 | 4 | 7 | 10 |
| | 800 | 0 | 0 | 0 | 0 | 1 | 1 | 1 | 2 | 3 | 4 | 6 | 9 |
| | 900 | 0 | 0 | 0 | 0 | 1 | 1 | 1 | 1 | 3 | 4 | 6 | 8 |
| | 1000 | 0 | 0 | 0 | 0 | 0 | 1 | 1 | 1 | 2 | 3 | 5 | 8 |
| | 1250 | 0 | 0 | 0 | 0 | 0 | 1 | 1 | 1 | 1 | 2 | 4 | 6 |
| | 1500 | 0 | 0 | 0 | 0 | 0 | 1 | 1 | 1 | 1 | 2 | 3 | 5 |
| | 1750 | 0 | 0 | 0 | 0 | 0 | 0 | 1 | 1 | 1 | 1 | 3 | 4 |
| | 2000 | 0 | 0 | 0 | 0 | 0 | 0 | 1 | 1 | 1 | 1 | 3 | 4 |

**Table C8** *Continued*

| | | CONDUCTORS | | | | | | | | | | | |
|---|---|---|---|---|---|---|---|---|---|---|---|---|---|
| | Conductor Size | Metric Designator (Trade Size) | | | | | | | | | | | |
| Type | (AWG/kcmil) | 16 (½) | 21 (¾) | 27 (1) | 35 (1¼) | 41 (1½) | 53 (2) | 63 (2½) | 78 (3) | 91 (3½) | 103 (4) | 129 (5) | 155 (6) |
| THHN, THWN, THWN-2 | 14 | 13 | 22 | 36 | 63 | 85 | 140 | 200 | 309 | 412 | 531 | 833 | 1202 |
| | 12 | 9 | 16 | 26 | 46 | 62 | 102 | 146 | 225 | 301 | 387 | 608 | 877 |
| | 10 | 6 | 10 | 17 | 29 | 39 | 64 | 92 | 142 | 189 | 244 | 383 | 552 |
| | 8 | 3 | 6 | 9 | 16 | 22 | 37 | 53 | 82 | 109 | 140 | 221 | 318 |
| | 6 | 2 | 4 | 7 | 12 | 16 | 27 | 38 | 59 | 79 | 101 | 159 | 230 |
| | 4 | 1 | 2 | 4 | 7 | 10 | 16 | 23 | 36 | 48 | 62 | 98 | 141 |
| | 3 | 1 | 1 | 3 | 6 | 8 | 14 | 20 | 31 | 41 | 53 | 83 | 120 |
| | 2 | 1 | 1 | 3 | 5 | 7 | 11 | 17 | 26 | 34 | 44 | 70 | 100 |
| | 1 | 1 | 1 | 1 | 4 | 5 | 8 | 12 | 19 | 25 | 33 | 51 | 74 |
| | 1/0 | 1 | 1 | 1 | 3 | 4 | 7 | 10 | 16 | 21 | 27 | 43 | 63 |
| | 2/0 | 0 | 1 | 1 | 2 | 3 | 6 | 8 | 13 | 18 | 23 | 36 | 52 |
| | 3/0 | 0 | 1 | 1 | 1 | 3 | 5 | 7 | 11 | 15 | 19 | 30 | 43 |
| | 4/0 | 0 | 1 | 1 | 1 | 2 | 4 | 6 | 9 | 12 | 16 | 25 | 36 |
| | 250 | 0 | 0 | 1 | 1 | 1 | 3 | 5 | 7 | 10 | 13 | 20 | 29 |
| | 300 | 0 | 0 | 1 | 1 | 1 | 3 | 4 | 6 | 8 | 11 | 17 | 25 |
| | 350 | 0 | 0 | 1 | 1 | 1 | 2 | 3 | 5 | 7 | 10 | 15 | 22 |
| | 400 | 0 | 0 | 1 | 1 | 1 | 2 | 3 | 5 | 7 | 8 | 13 | 20 |
| | 500 | 0 | 0 | 0 | 1 | 1 | 1 | 2 | 4 | 5 | 7 | 11 | 16 |
| | 600 | 0 | 0 | 0 | 1 | 1 | 1 | 1 | 3 | 4 | 6 | 9 | 13 |
| | 700 | 0 | 0 | 0 | 1 | 1 | 1 | 1 | 3 | 4 | 5 | 8 | 11 |
| | 750 | 0 | 0 | 0 | 0 | 1 | 1 | 1 | 3 | 4 | 5 | 7 | 11 |
| | 800 | 0 | 0 | 0 | 0 | 1 | 1 | 1 | 2 | 3 | 4 | 7 | 10 |
| | 900 | 0 | 0 | 0 | 0 | 1 | 1 | 1 | 2 | 3 | 4 | 6 | 9 |
| | 1000 | 0 | 0 | 0 | 0 | 1 | 1 | 1 | 1 | 3 | 4 | 6 | 8 |
| FEP, FEPB, PFA, PFAH, TFE | 14 | 12 | 22 | 35 | 61 | 83 | 136 | 194 | 300 | 400 | 515 | 808 | 1166 |
| | 12 | 9 | 16 | 26 | 44 | 60 | 99 | 142 | 219 | 292 | 376 | 590 | 851 |
| | 10 | 6 | 11 | 18 | 32 | 43 | 71 | 102 | 157 | 209 | 269 | 423 | 610 |
| | 8 | 3 | 6 | 10 | 18 | 25 | 41 | 58 | 90 | 120 | 154 | 242 | 350 |
| | 6 | 2 | 4 | 7 | 13 | 17 | 29 | 41 | 64 | 85 | 110 | 172 | 249 |
| | 4 | 1 | 3 | 5 | 9 | 12 | 20 | 29 | 44 | 59 | 77 | 120 | 174 |
| | 3 | 1 | 2 | 4 | 7 | 10 | 17 | 24 | 37 | 50 | 64 | 100 | 145 |
| | 2 | 1 | 1 | 3 | 6 | 8 | 14 | 20 | 31 | 41 | 53 | 83 | 120 |
| PFA, PFAH, TFE | 1 | 1 | 1 | 2 | 4 | 6 | 9 | 14 | 21 | 28 | 37 | 57 | 83 |
| PFA, PFAH, TFE, Z | 1/0 | 1 | 1 | 1 | 3 | 5 | 8 | 11 | 18 | 24 | 30 | 48 | 69 |
| | 2/0 | 1 | 1 | 1 | 3 | 4 | 6 | 9 | 14 | 19 | 25 | 40 | 57 |
| | 3/0 | 0 | 1 | 1 | 2 | 3 | 5 | 8 | 12 | 16 | 21 | 33 | 47 |
| | 4/0 | 0 | 1 | 1 | 1 | 2 | 4 | 6 | 10 | 13 | 17 | 27 | 39 |
| Z | 14 | 15 | 26 | 42 | 73 | 100 | 164 | 234 | 361 | 482 | 621 | 974 | 1405 |
| | 12 | 10 | 18 | 30 | 52 | 71 | 116 | 166 | 256 | 342 | 440 | 691 | 997 |
| | 10 | 6 | 11 | 18 | 32 | 43 | 71 | 102 | 157 | 209 | 269 | 423 | 610 |
| | 8 | 4 | 7 | 11 | 20 | 27 | 45 | 64 | 99 | 132 | 170 | 267 | 386 |
| | 6 | 3 | 5 | 8 | 14 | 19 | 31 | 45 | 69 | 93 | 120 | 188 | 271 |
| | 4 | 1 | 3 | 5 | 9 | 13 | 22 | 31 | 48 | 64 | 82 | 129 | 186 |
| | 3 | 1 | 2 | 4 | 7 | 9 | 16 | 22 | 35 | 47 | 60 | 94 | 136 |
| | 2 | 1 | 1 | 3 | 6 | 8 | 13 | 19 | 29 | 39 | 50 | 78 | 113 |
| | 1 | 1 | 1 | 2 | 5 | 6 | 10 | 15 | 23 | 31 | 40 | 63 | 92 |
| XHH, XHHW, XHHW-2, ZW | 14 | 9 | 15 | 25 | 44 | 59 | 98 | 140 | 216 | 288 | 370 | 581 | 839 |
| | 12 | 7 | 12 | 19 | 33 | 45 | 75 | 107 | 165 | 221 | 284 | 446 | 644 |
| | 10 | 5 | 9 | 14 | 25 | 34 | 56 | 80 | 123 | 164 | 212 | 332 | 480 |
| | 8 | 3 | 5 | 8 | 14 | 19 | 31 | 44 | 68 | 91 | 118 | 185 | 267 |
| | 6 | 1 | 3 | 6 | 10 | 14 | 23 | 33 | 51 | 68 | 87 | 137 | 197 |
| | 4 | 1 | 2 | 4 | 7 | 10 | 16 | 24 | 37 | 49 | 63 | 99 | 143 |
| | 3 | 1 | 1 | 3 | 6 | 8 | 14 | 20 | 31 | 41 | 53 | 84 | 121 |
| | 2 | 1 | 1 | 3 | 5 | 7 | 12 | 17 | 26 | 35 | 45 | 70 | 101 |

**Table C8** *Continued*

| | | CONDUCTORS | | | | | | | | | | | |
|---|---|---|---|---|---|---|---|---|---|---|---|---|---|
| | Conductor Size | Metric Designator (Trade Size) | | | | | | | | | | | |
| Type | (AWG/kcmil) | 16 (½) | 21 (¾) | 27 (1) | 35 (1¼) | 41 (1½) | 53 (2) | 63 (2½) | 78 (3) | 91 (3½) | 103 (4) | 129 (5) | 155 (6) |
| XHH, XHHW, XHHW-2 | 1 | 1 | 1 | 1 | 4 | 5 | 9 | 12 | 19 | 26 | 33 | 52 | 76 |
| | 1/0 | 1 | 1 | 1 | 3 | 4 | 7 | 10 | 16 | 22 | 28 | 44 | 64 |
| | 2/0 | 0 | 1 | 1 | 2 | 3 | 6 | 9 | 13 | 18 | 23 | 37 | 53 |
| | 3/0 | 0 | 1 | 1 | 1 | 3 | 5 | 7 | 11 | 15 | 19 | 30 | 44 |
| | 4/0 | 0 | 1 | 1 | 1 | 2 | 4 | 6 | 9 | 12 | 16 | 25 | 36 |
| | 250 | 0 | 0 | 1 | 1 | 1 | 3 | 5 | 7 | 10 | 13 | 20 | 30 |
| | 300 | 0 | 0 | 1 | 1 | 1 | 3 | 4 | 6 | 9 | 11 | 18 | 25 |
| | 350 | 0 | 0 | 1 | 1 | 1 | 2 | 3 | 6 | 7 | 10 | 15 | 22 |
| | 400 | 0 | 0 | 1 | 1 | 1 | 2 | 3 | 5 | 7 | 9 | 14 | 20 |
| | 500 | 0 | 0 | 0 | 1 | 1 | 1 | 2 | 4 | 5 | 7 | 11 | 16 |
| | 600 | 0 | 0 | 0 | 1 | 1 | 1 | 1 | 3 | 4 | 6 | 9 | 13 |
| | 700 | 0 | 0 | 0 | 1 | 1 | 1 | 1 | 3 | 4 | 5 | 8 | 11 |
| | 750 | 0 | 0 | 0 | 0 | 1 | 1 | 1 | 3 | 4 | 5 | 7 | 11 |
| | 800 | 0 | 0 | 0 | 0 | 1 | 1 | 1 | 2 | 3 | 4 | 7 | 10 |
| | 900 | 0 | 0 | 0 | 0 | 1 | 1 | 1 | 2 | 3 | 4 | 6 | 9 |
| | 1000 | 0 | 0 | 0 | 0 | 1 | 1 | 1 | 1 | 3 | 4 | 6 | 8 |
| | 1250 | 0 | 0 | 0 | 0 | 0 | 1 | 1 | 1 | 2 | 3 | 4 | 6 |
| | 1500 | 0 | 0 | 0 | 0 | 0 | 1 | 1 | 1 | 1 | 2 | 4 | 5 |
| | 1750 | 0 | 0 | 0 | 0 | 0 | 0 | 1 | 1 | 1 | 1 | 3 | 5 |
| | 2000 | 0 | 0 | 0 | 0 | 0 | 0 | 1 | 1 | 1 | 1 | 3 | 4 |

| | | FIXTURE WIRES | | | | |
|---|---|---|---|---|---|---|
| | Conductor Size | Metric Designator (Trade Size) | | | | |
| Type | (AWG/kcmil) | 16 (½) | 21 (¾) | 27 (1) | 35 (1¼) | 41 (1½) | 53 (2) |
| FFH-2, RFH-2, RFHH-3 | 18 | 8 | 15 | 24 | 42 | 57 | 94 |
| | 16 | 7 | 12 | 20 | 35 | 48 | 79 |
| SF-2, SFF-2 | 18 | 11 | 19 | 31 | 53 | 72 | 118 |
| | 16 | 9 | 15 | 25 | 44 | 59 | 98 |
| | 14 | 7 | 12 | 20 | 35 | 48 | 79 |
| SF-1, SFF-1 | 18 | 19 | 33 | 54 | 94 | 127 | 209 |
| RFH-1, RFHH-2, TF, TFF, XF, XFF | 18 | 14 | 25 | 40 | 69 | 94 | 155 |
| RFHH-2, TF, TFF, XF, XFF | 16 | 11 | 20 | 32 | 56 | 76 | 125 |
| XF, XFF | 14 | 9 | 15 | 25 | 44 | 59 | 98 |
| TFN, TFFN | 18 | 23 | 40 | 64 | 111 | 150 | 248 |
| | 16 | 17 | 30 | 49 | 84 | 115 | 189 |
| PF, PFF, PGF, PGFF, PAF, PTF, PTFF, PAFF | 18 | 21 | 38 | 61 | 105 | 143 | 235 |
| | 16 | 16 | 29 | 47 | 81 | 110 | 181 |
| | 14 | 12 | 22 | 35 | 61 | 83 | 136 |
| HF, HFF, ZF, ZFF, ZHF | 18 | 28 | 48 | 79 | 135 | 184 | 303 |
| | 16 | 20 | 36 | 58 | 100 | 136 | 223 |
| | 14 | 15 | 26 | 42 | 73 | 100 | 164 |
| KF-2, KFF-2 | 18 | 40 | 71 | 114 | 197 | 267 | 439 |
| | 16 | 28 | 50 | 80 | 138 | 188 | 310 |
| | 14 | 19 | 34 | 55 | 95 | 129 | 213 |
| | 12 | 13 | 23 | 38 | 65 | 89 | 146 |
| | 10 | 9 | 15 | 25 | 44 | 59 | 98 |
| KF-1, KFF-1 | 18 | 48 | 84 | 136 | 235 | 318 | 524 |
| | 16 | 34 | 59 | 96 | 165 | 224 | 368 |
| | 14 | 23 | 40 | 64 | 111 | 150 | 248 |
| | 12 | 15 | 26 | 42 | 73 | 100 | 164 |
| | 10 | 10 | 17 | 28 | 48 | 65 | 107 |
| XF, XFF | 12 | 5 | 8 | 13 | 23 | 32 | 52 |
| | 10 | 3 | 6 | 10 | 18 | 25 | 41 |

Note: This table is for concentric stranded conductors only. For compact stranded conductors, Table C8(A) should be used.

*Types RHH, RHW, and RHW-2 without outer covering.

**Table C8(A) Maximum Number of Compact Conductors in Rigid Metal Conduit (RMC)**
*(Based on Table 1, Chapter 9)*

### COMPACT CONDUCTORS

| Type | Conductor Size (AWG/kcmil) | Metric Designator (Trade Size) | | | | | | | | | | | |
|---|---|---|---|---|---|---|---|---|---|---|---|---|---|
| | | 16 (½) | 21 (¾) | 27 (1) | 35 (1¼) | 41 (1½) | 53 (2) | 63 (2½) | 78 (3) | 91 (3½) | 103 (4) | 129 (5) | 155 (6) |
| THW, THW-2, THHW | 8 | 2 | 4 | 7 | 12 | 16 | 26 | 38 | 59 | 78 | 101 | 158 | 228 |
| | 6 | 1 | 3 | 5 | 9 | 12 | 20 | 29 | 45 | 60 | 78 | 122 | 176 |
| | 4 | 1 | 2 | 4 | 7 | 9 | 15 | 22 | 34 | 45 | 58 | 91 | 132 |
| | 2 | 1 | 1 | 3 | 5 | 7 | 11 | 16 | 25 | 33 | 43 | 67 | 97 |
| | 1 | 1 | 1 | 1 | 3 | 5 | 8 | 11 | 17 | 23 | 30 | 47 | 68 |
| | 1/0 | 1 | 1 | 1 | 3 | 4 | 7 | 10 | 15 | 20 | 26 | 41 | 59 |
| | 2/0 | 0 | 1 | 1 | 2 | 3 | 6 | 8 | 13 | 17 | 22 | 34 | 50 |
| | 3/0 | 0 | 1 | 1 | 1 | 3 | 5 | 7 | 11 | 14 | 19 | 29 | 42 |
| | 4/0 | 0 | 1 | 1 | 1 | 2 | 4 | 6 | 9 | 12 | 15 | 24 | 35 |
| | 250 | 0 | 0 | 1 | 1 | 1 | 3 | 4 | 7 | 9 | 12 | 19 | 28 |
| | 300 | 0 | 0 | 1 | 1 | 1 | 3 | 4 | 6 | 8 | 11 | 17 | 24 |
| | 350 | 0 | 0 | 1 | 1 | 1 | 2 | 3 | 5 | 7 | 9 | 15 | 22 |
| | 400 | 0 | 0 | 1 | 1 | 1 | 1 | 3 | 5 | 7 | 8 | 13 | 20 |
| | 500 | 0 | 0 | 0 | 1 | 1 | 1 | 3 | 4 | 5 | 7 | 11 | 17 |
| | 600 | 0 | 0 | 0 | 1 | 1 | 1 | 1 | 3 | 4 | 6 | 9 | 13 |
| | 700 | 0 | 0 | 0 | 1 | 1 | 1 | 1 | 3 | 4 | 5 | 8 | 12 |
| | 750 | 0 | 0 | 0 | 0 | 1 | 1 | 1 | 3 | 4 | 5 | 7 | 11 |
| | 1000 | 0 | 0 | 0 | 0 | 1 | 1 | 1 | 1 | 3 | 4 | 6 | 9 |
| THHN, THWN, THWN-2 | 8 | — | — | — | — | — | — | — | — | — | — | — | — |
| | 6 | 2 | 5 | 8 | 13 | 18 | 30 | 43 | 66 | 88 | 114 | 179 | 258 |
| | 4 | 1 | 3 | 5 | 8 | 11 | 18 | 26 | 41 | 55 | 70 | 110 | 159 |
| | 2 | 1 | 1 | 3 | 6 | 8 | 13 | 19 | 29 | 39 | 50 | 79 | 114 |
| | 1 | 1 | 1 | 2 | 4 | 6 | 10 | 14 | 22 | 29 | 38 | 60 | 86 |
| | 1/0 | 1 | 1 | 1 | 4 | 5 | 8 | 12 | 19 | 25 | 32 | 51 | 73 |
| | 2/0 | 1 | 1 | 1 | 3 | 4 | 7 | 10 | 15 | 21 | 26 | 42 | 60 |
| | 3/0 | 0 | 1 | 1 | 2 | 3 | 6 | 8 | 13 | 17 | 22 | 35 | 51 |
| | 4/0 | 0 | 1 | 1 | 1 | 3 | 5 | 7 | 10 | 14 | 18 | 29 | 42 |
| | 250 | 0 | 1 | 1 | 1 | 2 | 4 | 5 | 8 | 11 | 14 | 23 | 33 |
| | 300 | 0 | 0 | 1 | 1 | 1 | 3 | 4 | 7 | 10 | 12 | 20 | 28 |
| | 350 | 0 | 0 | 1 | 1 | 1 | 3 | 4 | 6 | 8 | 11 | 17 | 25 |
| | 400 | 0 | 0 | 1 | 1 | 1 | 2 | 3 | 5 | 7 | 10 | 15 | 22 |
| | 500 | 0 | 0 | 0 | 1 | 1 | 1 | 3 | 5 | 6 | 8 | 13 | 19 |
| | 600 | 0 | 0 | 0 | 1 | 1 | 1 | 2 | 4 | 5 | 6 | 10 | 15 |
| | 700 | 0 | 0 | 0 | 1 | 1 | 1 | 1 | 3 | 4 | 6 | 9 | 13 |
| | 750 | 0 | 0 | 0 | 1 | 1 | 1 | 1 | 3 | 4 | 5 | 9 | 13 |
| | 1000 | 0 | 0 | 0 | 0 | 1 | 1 | 1 | 2 | 3 | 4 | 6 | 9 |
| XHHW, XHHW-2 | 8 | 3 | 5 | 9 | 15 | 21 | 34 | 49 | 76 | 101 | 130 | 205 | 296 |
| | 6 | 2 | 4 | 6 | 11 | 15 | 25 | 36 | 56 | 75 | 97 | 152 | 220 |
| | 4 | 1 | 3 | 5 | 8 | 11 | 18 | 26 | 41 | 55 | 70 | 110 | 159 |
| | 2 | 1 | 1 | 3 | 6 | 8 | 13 | 19 | 29 | 39 | 50 | 79 | 114 |
| | 1 | 1 | 1 | 2 | 4 | 6 | 10 | 14 | 22 | 29 | 38 | 60 | 86 |
| | 1/0 | 1 | 1 | 1 | 4 | 5 | 8 | 12 | 19 | 25 | 32 | 51 | 73 |
| | 2/0 | 1 | 1 | 1 | 3 | 4 | 7 | 10 | 16 | 21 | 27 | 43 | 62 |
| | 3/0 | 0 | 1 | 1 | 2 | 3 | 6 | 8 | 13 | 17 | 22 | 35 | 51 |
| | 4/0 | 0 | 1 | 1 | 1 | 3 | 5 | 7 | 11 | 14 | 19 | 29 | 42 |
| | 250 | 0 | 1 | 1 | 1 | 2 | 4 | 5 | 8 | 11 | 15 | 23 | 34 |
| | 300 | 0 | 0 | 1 | 1 | 1 | 3 | 5 | 7 | 10 | 13 | 20 | 29 |
| | 350 | 0 | 0 | 1 | 1 | 1 | 3 | 4 | 6 | 9 | 11 | 18 | 25 |
| | 400 | 0 | 0 | 1 | 1 | 1 | 2 | 4 | 6 | 8 | 10 | 16 | 23 |
| | 500 | 0 | 0 | 0 | 1 | 1 | 1 | 3 | 5 | 6 | 8 | 13 | 19 |
| | 600 | 0 | 0 | 0 | 1 | 1 | 1 | 2 | 4 | 5 | 7 | 10 | 15 |
| | 700 | 0 | 0 | 0 | 1 | 1 | 1 | 1 | 3 | 4 | 6 | 9 | 13 |
| | 750 | 0 | 0 | 0 | 1 | 1 | 1 | 1 | 3 | 4 | 5 | 8 | 12 |
| | 1000 | 0 | 0 | 0 | 0 | 1 | 1 | 1 | 2 | 3 | 4 | 7 | 10 |

Definition: *Compact stranding* is the result of a manufacturing process where the standard conductor is compressed to the extent that the interstices (voids between strand wires) are virtually eliminated.

**Table C9 Maximum Number of Conductors or Fixture Wires in Rigid PVC Conduit, Schedule 80** *(Based on Table 1, Chapter 9)*

### CONDUCTORS

| Type | Conductor Size (AWG/kcmil) | Metric Designator (Trade Size) | | | | | | | | | | | |
|---|---|---|---|---|---|---|---|---|---|---|---|---|---|
| | | 16 (½) | 21 (¾) | 27 (1) | 35 (1¼) | 41 (1½) | 53 (2) | 63 (2½) | 78 (3) | 91 (3½) | 103 (4) | 129 (5) | 155 (6) |
| RHH, RHW, RHW-2 | 14 | 3 | 5 | 9 | 17 | 23 | 39 | 56 | 88 | 118 | 153 | 243 | 349 |
| | 12 | 2 | 4 | 7 | 14 | 19 | 32 | 46 | 73 | 98 | 127 | 202 | 290 |
| | 10 | 1 | 3 | 6 | 11 | 15 | 26 | 37 | 59 | 79 | 103 | 163 | 234 |
| | 8 | 1 | 1 | 3 | 6 | 8 | 13 | 19 | 31 | 41 | 54 | 85 | 122 |
| | 6 | 1 | 1 | 2 | 4 | 6 | 11 | 16 | 24 | 33 | 43 | 68 | 98 |
| | 4 | 1 | 1 | 1 | 3 | 5 | 8 | 12 | 19 | 26 | 33 | 53 | 77 |
| | 3 | 0 | 1 | 1 | 3 | 4 | 7 | 11 | 17 | 23 | 29 | 47 | 67 |
| | 2 | 0 | 1 | 1 | 3 | 4 | 6 | 9 | 14 | 20 | 25 | 41 | 58 |
| | 1 | 0 | 1 | 1 | 1 | 2 | 4 | 6 | 9 | 13 | 17 | 27 | 38 |
| | 1/0 | 0 | 0 | 1 | 1 | 1 | 3 | 5 | 8 | 11 | 15 | 23 | 33 |
| | 2/0 | 0 | 0 | 1 | 1 | 1 | 3 | 4 | 7 | 10 | 13 | 20 | 29 |
| | 3/0 | 0 | 0 | 1 | 1 | 1 | 3 | 4 | 6 | 8 | 11 | 17 | 25 |
| | 4/0 | 0 | 0 | 0 | 1 | 1 | 2 | 3 | 5 | 7 | 9 | 15 | 21 |
| | 250 | 0 | 0 | 0 | 1 | 1 | 1 | 2 | 4 | 5 | 7 | 11 | 16 |
| | 300 | 0 | 0 | 0 | 1 | 1 | 1 | 2 | 3 | 5 | 6 | 10 | 14 |
| | 350 | 0 | 0 | 0 | 1 | 1 | 1 | 1 | 3 | 4 | 5 | 9 | 13 |
| | 400 | 0 | 0 | 0 | 0 | 1 | 1 | 1 | 3 | 4 | 5 | 8 | 12 |
| | 500 | 0 | 0 | 0 | 0 | 1 | 1 | 1 | 2 | 3 | 4 | 7 | 10 |
| | 600 | 0 | 0 | 0 | 0 | 0 | 1 | 1 | 1 | 3 | 3 | 6 | 8 |
| | 700 | 0 | 0 | 0 | 0 | 0 | 1 | 1 | 1 | 2 | 3 | 5 | 7 |
| | 750 | 0 | 0 | 0 | 0 | 0 | 1 | 1 | 1 | 2 | 3 | 5 | 7 |
| | 800 | 0 | 0 | 0 | 0 | 0 | 1 | 1 | 1 | 2 | 3 | 4 | 7 |
| | 1000 | 0 | 0 | 0 | 0 | 0 | 1 | 1 | 1 | 1 | 2 | 4 | 5 |
| | 1250 | 0 | 0 | 0 | 0 | 0 | 0 | 1 | 1 | 1 | 1 | 3 | 4 |
| | 1500 | 0 | 0 | 0 | 0 | 0 | 0 | 1 | 1 | 1 | 1 | 2 | 4 |
| | 1750 | 0 | 0 | 0 | 0 | 0 | 0 | 0 | 1 | 1 | 1 | 2 | 3 |
| | 2000 | 0 | 0 | 0 | 0 | 0 | 0 | 0 | 1 | 1 | 1 | 1 | 3 |
| TW, THHW, THW, THW-2 | 14 | 6 | 11 | 20 | 35 | 49 | 82 | 118 | 185 | 250 | 324 | 514 | 736 |
| | 12 | 5 | 9 | 15 | 27 | 38 | 63 | 91 | 142 | 192 | 248 | 394 | 565 |
| | 10 | 3 | 6 | 11 | 20 | 28 | 47 | 67 | 106 | 143 | 185 | 294 | 421 |
| | 8 | 1 | 3 | 6 | 11 | 15 | 26 | 37 | 59 | 79 | 103 | 163 | 234 |
| RHH*, RHW*, RHW-2* | 14 | 4 | 8 | 13 | 23 | 32 | 55 | 79 | 123 | 166 | 215 | 341 | 490 |
| | 12 | 3 | 6 | 10 | 19 | 26 | 44 | 63 | 99 | 133 | 173 | 274 | 394 |
| | 10 | 2 | 5 | 8 | 15 | 20 | 34 | 49 | 77 | 104 | 135 | 214 | 307 |
| | 8 | 1 | 3 | 5 | 9 | 12 | 20 | 29 | 46 | 62 | 81 | 128 | 184 |
| RHH*, RHW*, RHW-2*, TW, THW, THHW, THW-2 | 6 | 1 | 1 | 3 | 7 | 9 | 16 | 22 | 35 | 48 | 62 | 98 | 141 |
| | 4 | 1 | 1 | 3 | 5 | 7 | 12 | 17 | 26 | 35 | 46 | 73 | 105 |
| | 3 | 1 | 1 | 2 | 4 | 6 | 10 | 14 | 22 | 30 | 39 | 63 | 90 |
| | 2 | 1 | 1 | 1 | 3 | 5 | 8 | 12 | 19 | 26 | 33 | 53 | 77 |
| | 1 | 0 | 1 | 1 | 2 | 3 | 6 | 8 | 13 | 18 | 23 | 37 | 54 |
| | 1/0 | 0 | 1 | 1 | 1 | 3 | 5 | 7 | 11 | 15 | 20 | 32 | 46 |
| | 2/0 | 0 | 1 | 1 | 1 | 2 | 4 | 6 | 10 | 13 | 17 | 27 | 39 |
| | 3/0 | 0 | 0 | 1 | 1 | 1 | 3 | 5 | 8 | 11 | 14 | 23 | 33 |
| | 4/0 | 0 | 0 | 1 | 1 | 1 | 3 | 4 | 7 | 9 | 12 | 19 | 27 |
| | 250 | 0 | 0 | 0 | 1 | 1 | 2 | 3 | 5 | 7 | 9 | 15 | 22 |
| | 300 | 0 | 0 | 0 | 1 | 1 | 1 | 3 | 5 | 6 | 8 | 13 | 19 |
| | 350 | 0 | 0 | 0 | 1 | 1 | 1 | 2 | 4 | 6 | 7 | 12 | 17 |
| | 400 | 0 | 0 | 0 | 1 | 1 | 1 | 2 | 4 | 5 | 7 | 10 | 15 |
| | 500 | 0 | 0 | 0 | 1 | 1 | 1 | 1 | 3 | 4 | 5 | 9 | 13 |
| | 600 | 0 | 0 | 0 | 0 | 1 | 1 | 1 | 2 | 3 | 4 | 7 | 10 |
| | 700 | 0 | 0 | 0 | 0 | 1 | 1 | 1 | 2 | 3 | 4 | 6 | 9 |
| | 750 | 0 | 0 | 0 | 0 | 0 | 1 | 1 | 1 | 3 | 4 | 6 | 8 |
| | 800 | 0 | 0 | 0 | 0 | 0 | 1 | 1 | 1 | 3 | 3 | 6 | 8 |
| | 900 | 0 | 0 | 0 | 0 | 0 | 1 | 1 | 1 | 2 | 3 | 5 | 7 |
| | 1000 | 0 | 0 | 0 | 0 | 0 | 1 | 1 | 1 | 2 | 3 | 5 | 7 |
| | 1250 | 0 | 0 | 0 | 0 | 0 | 0 | 1 | 1 | 1 | 2 | 4 | 5 |
| | 1500 | 0 | 0 | 0 | 0 | 0 | 0 | 1 | 1 | 1 | 1 | 3 | 4 |
| | 1750 | 0 | 0 | 0 | 0 | 0 | 0 | 1 | 1 | 1 | 1 | 3 | 4 |
| | 2000 | 0 | 0 | 0 | 0 | 0 | 0 | 0 | 1 | 1 | 1 | 2 | 3 |

Table C9  *Continued*

| | | CONDUCTORS | | | | | | | | | | | |
|---|---|---|---|---|---|---|---|---|---|---|---|---|---|
| | | Metric Designator (Trade Size) | | | | | | | | | | | |
| Type | Conductor Size (AWG/kcmil) | 16 (½) | 21 (¾) | 27 (1) | 35 (1¼) | 41 (1½) | 53 (2) | 63 (2½) | 78 (3) | 91 (3½) | 103 (4) | 129 (5) | 155 (6) |
| THHN, THWN, THWN-2 | 14 | 9 | 17 | 28 | 51 | 70 | 118 | 170 | 265 | 358 | 464 | 736 | 1055 |
| | 12 | 6 | 12 | 20 | 37 | 51 | 86 | 124 | 193 | 261 | 338 | 537 | 770 |
| | 10 | 4 | 7 | 13 | 23 | 32 | 54 | 78 | 122 | 164 | 213 | 338 | 485 |
| | 8 | 2 | 4 | 7 | 13 | 18 | 31 | 45 | 70 | 95 | 123 | 195 | 279 |
| | 6 | 1 | 3 | 5 | 9 | 13 | 22 | 32 | 51 | 68 | 89 | 141 | 202 |
| | 4 | 1 | 1 | 3 | 6 | 8 | 14 | 20 | 31 | 42 | 54 | 86 | 124 |
| | 3 | 1 | 1 | 3 | 5 | 7 | 12 | 17 | 26 | 35 | 46 | 73 | 105 |
| | 2 | 1 | 1 | 2 | 4 | 6 | 10 | 14 | 22 | 30 | 39 | 61 | 88 |
| | 1 | 0 | 1 | 1 | 3 | 4 | 7 | 10 | 16 | 22 | 29 | 45 | 65 |
| | 1/0 | 0 | 1 | 1 | 2 | 3 | 6 | 9 | 14 | 18 | 24 | 38 | 55 |
| | 2/0 | 0 | 1 | 1 | 1 | 3 | 5 | 7 | 11 | 15 | 20 | 32 | 46 |
| | 3/0 | 0 | 1 | 1 | 1 | 2 | 4 | 6 | 9 | 13 | 17 | 26 | 38 |
| | 4/0 | 0 | 0 | 1 | 1 | 1 | 3 | 5 | 8 | 10 | 14 | 22 | 31 |
| | 250 | 0 | 0 | 1 | 1 | 1 | 3 | 4 | 6 | 8 | 11 | 18 | 25 |
| | 300 | 0 | 0 | 0 | 1 | 1 | 2 | 3 | 5 | 7 | 9 | 15 | 22 |
| | 350 | 0 | 0 | 0 | 1 | 1 | 1 | 3 | 5 | 6 | 8 | 13 | 19 |
| | 400 | 0 | 0 | 0 | 1 | 1 | 1 | 3 | 4 | 6 | 7 | 12 | 17 |
| | 500 | 0 | 0 | 0 | 1 | 1 | 1 | 2 | 3 | 5 | 6 | 10 | 14 |
| | 600 | 0 | 0 | 0 | 0 | 1 | 1 | 1 | 3 | 4 | 5 | 8 | 12 |
| | 700 | 0 | 0 | 0 | 0 | 1 | 1 | 1 | 2 | 3 | 4 | 7 | 10 |
| | 750 | 0 | 0 | 0 | 0 | 1 | 1 | 1 | 2 | 3 | 4 | 7 | 9 |
| | 800 | 0 | 0 | 0 | 0 | 1 | 1 | 1 | 2 | 3 | 4 | 6 | 9 |
| | 900 | 0 | 0 | 0 | 0 | 0 | 1 | 1 | 1 | 3 | 3 | 6 | 8 |
| | 1000 | 0 | 0 | 0 | 0 | 0 | 1 | 1 | 1 | 2 | 3 | 5 | 7 |
| FEP, FEPB, PFA, PFAH, TFE | 14 | 8 | 16 | 27 | 49 | 68 | 115 | 164 | 257 | 347 | 450 | 714 | 1024 |
| | 12 | 6 | 12 | 20 | 36 | 50 | 84 | 120 | 188 | 253 | 328 | 521 | 747 |
| | 10 | 4 | 8 | 14 | 26 | 36 | 60 | 86 | 135 | 182 | 235 | 374 | 536 |
| | 8 | 2 | 5 | 8 | 15 | 20 | 34 | 49 | 77 | 104 | 135 | 214 | 307 |
| | 6 | 1 | 3 | 6 | 10 | 14 | 24 | 35 | 55 | 74 | 96 | 152 | 218 |
| | 4 | 1 | 2 | 4 | 7 | 10 | 17 | 24 | 38 | 52 | 67 | 106 | 153 |
| | 3 | 1 | 1 | 3 | 6 | 8 | 14 | 20 | 32 | 43 | 56 | 89 | 127 |
| | 2 | 1 | 1 | 3 | 5 | 7 | 12 | 17 | 26 | 35 | 46 | 73 | 105 |
| PFA, PFAH, TFE | 1 | 1 | 1 | 1 | 3 | 5 | 8 | 11 | 18 | 25 | 32 | 51 | 73 |
| PFA, PFAH, TFE, Z | 1/0 | 0 | 1 | 1 | 3 | 4 | 7 | 10 | 15 | 20 | 27 | 42 | 61 |
| | 2/0 | 0 | 1 | 1 | 2 | 3 | 5 | 8 | 12 | 17 | 22 | 35 | 50 |
| | 3/0 | 0 | 1 | 1 | 1 | 2 | 4 | 6 | 10 | 14 | 18 | 29 | 41 |
| | 4/0 | 0 | 0 | 1 | 1 | 1 | 4 | 5 | 8 | 11 | 15 | 24 | 34 |
| Z | 14 | 10 | 19 | 33 | 59 | 82 | 138 | 198 | 310 | 418 | 542 | 860 | 1233 |
| | 12 | 7 | 14 | 23 | 42 | 58 | 98 | 141 | 220 | 297 | 385 | 610 | 875 |
| | 10 | 4 | 8 | 14 | 26 | 36 | 60 | 86 | 135 | 182 | 235 | 374 | 536 |
| | 8 | 3 | 5 | 9 | 16 | 22 | 38 | 54 | 85 | 115 | 149 | 236 | 339 |
| | 6 | 2 | 4 | 6 | 11 | 16 | 26 | 38 | 60 | 81 | 104 | 166 | 238 |
| | 4 | 1 | 2 | 4 | 8 | 11 | 18 | 26 | 41 | 55 | 72 | 114 | 164 |
| | 3 | 1 | 2 | 3 | 5 | 8 | 13 | 19 | 30 | 40 | 52 | 83 | 119 |
| | 2 | 1 | 1 | 2 | 5 | 6 | 11 | 16 | 25 | 33 | 43 | 69 | 99 |
| | 1 | 0 | 1 | 2 | 4 | 5 | 9 | 13 | 20 | 27 | 35 | 56 | 80 |
| XHH, XHHW, XHHW-2, ZW | 14 | 6 | 11 | 20 | 35 | 49 | 82 | 118 | 185 | 250 | 324 | 514 | 736 |
| | 12 | 5 | 9 | 15 | 27 | 38 | 63 | 91 | 142 | 192 | 248 | 394 | 565 |
| | 10 | 3 | 6 | 11 | 20 | 28 | 47 | 67 | 106 | 143 | 185 | 294 | 421 |
| | 8 | 1 | 3 | 6 | 11 | 15 | 26 | 37 | 59 | 79 | 103 | 163 | 234 |
| | 6 | 1 | 2 | 4 | 8 | 11 | 19 | 28 | 43 | 59 | 76 | 121 | 173 |
| | 4 | 1 | 1 | 3 | 6 | 8 | 14 | 20 | 31 | 42 | 55 | 87 | 125 |
| | 3 | 1 | 1 | 3 | 5 | 7 | 12 | 17 | 26 | 36 | 47 | 74 | 106 |
| | 2 | 1 | 1 | 2 | 4 | 6 | 10 | 14 | 22 | 30 | 39 | 62 | 89 |

Table C9  *Continued*

| | | CONDUCTORS | | | | | | | | | | | |
|---|---|---|---|---|---|---|---|---|---|---|---|---|---|
| | | Metric Designator (Trade Size) | | | | | | | | | | | |
| Type | Conductor Size (AWG/kcmil) | 16 (½) | 21 (¾) | 27 (1) | 35 (1¼) | 41 (1½) | 53 (2) | 63 (2½) | 78 (3) | 91 (3½) | 103 (4) | 129 (5) | 155 (6) |
| XHH, XHHW, XHHW-2 | 1 | 0 | 1 | 1 | 3 | 4 | 7 | 10 | 16 | 22 | 29 | 46 | 66 |
| | 1/0 | 0 | 1 | 1 | 2 | 3 | 6 | 9 | 14 | 19 | 24 | 39 | 56 |
| | 2/0 | 0 | 1 | 1 | 1 | 3 | 5 | 7 | 11 | 16 | 20 | 32 | 46 |
| | 3/0 | 0 | 1 | 1 | 1 | 2 | 4 | 6 | 9 | 13 | 17 | 27 | 38 |
| | 4/0 | 0 | 0 | 1 | 1 | 1 | 3 | 5 | 8 | 11 | 14 | 22 | 32 |
| | 250 | 0 | 0 | 1 | 1 | 1 | 3 | 4 | 6 | 9 | 11 | 18 | 26 |
| | 300 | 0 | 0 | 1 | 1 | 1 | 2 | 3 | 5 | 7 | 10 | 15 | 22 |
| | 350 | 0 | 0 | 0 | 1 | 1 | 1 | 3 | 5 | 6 | 8 | 14 | 20 |
| | 400 | 0 | 0 | 0 | 1 | 1 | 1 | 3 | 4 | 6 | 7 | 12 | 17 |
| | 500 | 0 | 0 | 0 | 1 | 1 | 1 | 2 | 3 | 5 | 6 | 10 | 14 |
| | 600 | 0 | 0 | 0 | 0 | 1 | 1 | 1 | 3 | 4 | 5 | 8 | 11 |
| | 700 | 0 | 0 | 0 | 0 | 1 | 1 | 1 | 2 | 3 | 4 | 7 | 10 |
| | 750 | 0 | 0 | 0 | 0 | 1 | 1 | 1 | 2 | 3 | 4 | 6 | 9 |
| | 800 | 0 | 0 | 0 | 0 | 1 | 1 | 1 | 1 | 3 | 4 | 6 | 9 |
| | 900 | 0 | 0 | 0 | 0 | 0 | 1 | 1 | — | 3 | 3 | 5 | 8 |
| | 1000 | 0 | 0 | 0 | 0 | 0 | 1 | 1 | 1 | 2 | 3 | 5 | 7 |
| | 1250 | 0 | 0 | 0 | 0 | 0 | 1 | 1 | 1 | 1 | 2 | 4 | 6 |
| | 1500 | 0 | 0 | 0 | 0 | 0 | 0 | 1 | 1 | 1 | 1 | 3 | 5 |
| | 1750 | 0 | 0 | 0 | 0 | 0 | 0 | 1 | 1 | 1 | 1 | 3 | 4 |
| | 2000 | 0 | 0 | 0 | 0 | 0 | 0 | 1 | 1 | 1 | 1 | 2 | 4 |

| | | FIXTURE WIRES | | | | | |
|---|---|---|---|---|---|---|---|
| | | Metric Designator (Trade Size) | | | | | |
| Type | Conductor Size (AWG/kcmil) | 16 (½) | 21 (¾) | 27 (1) | 35 (1¼) | 41 (1½) | 53 (2) |
| FFH-2, RFH-2, RFHH-3 | 18 | 6 | 11 | 19 | 34 | 47 | 79 |
| | 16 | 5 | 9 | 16 | 28 | 39 | 67 |
| SF-2, SFF-2 | 18 | 7 | 14 | 24 | 43 | 59 | 100 |
| | 16 | 6 | 11 | 20 | 35 | 49 | 82 |
| | 14 | 5 | 9 | 16 | 28 | 39 | 67 |
| SF-1, SFF-1 | 18 | 13 | 25 | 42 | 76 | 105 | 177 |
| RFH-1, RFHH-2, TF, TFF, XF, XFF | 18 | 10 | 18 | 31 | 56 | 77 | 130 |
| RFHH-2, TF, TFF, XF, XFF | 16 | 8 | 15 | 25 | 45 | 62 | 105 |
| XF, XFF | 14 | 6 | 11 | 20 | 35 | 49 | 82 |
| TFN, TFFN | 18 | 16 | 29 | 50 | 90 | 124 | 209 |
| | 16 | 12 | 22 | 38 | 68 | 95 | 159 |
| PF, PFF, PGF, PGFF, PAF, PTF, PTFF, PAFF | 18 | 15 | 28 | 47 | 85 | 118 | 198 |
| | 16 | 11 | 22 | 36 | 66 | 91 | 153 |
| | 14 | 8 | 16 | 27 | 49 | 68 | 115 |
| HF, HFF, ZF, ZFF, ZHF | 18 | 19 | 36 | 61 | 110 | 152 | 255 |
| | 16 | 14 | 27 | 45 | 81 | 112 | 188 |
| | 14 | 10 | 19 | 33 | 59 | 82 | 138 |
| KF-2, KFF-2 | 18 | 28 | 53 | 88 | 159 | 220 | 371 |
| | 16 | 19 | 37 | 62 | 112 | 155 | 261 |
| | 14 | 13 | 25 | 43 | 77 | 107 | 179 |
| | 12 | 9 | 17 | 29 | 53 | 73 | 123 |
| | 10 | 6 | 11 | 20 | 35 | 49 | 82 |
| KF-1, KFF-1 | 18 | 33 | 63 | 106 | 190 | 263 | 442 |
| | 16 | 23 | 44 | 74 | 133 | 185 | 310 |
| | 14 | 16 | 29 | 50 | 90 | 124 | 209 |
| | 12 | 10 | 19 | 33 | 59 | 82 | 138 |
| | 10 | 7 | 13 | 21 | 39 | 54 | 90 |
| XF, XFF | 12 | 3 | 6 | 10 | 19 | 26 | 44 |
| | 10 | 2 | 5 | 8 | 15 | 20 | 34 |

Note: This table is for concentric stranded conductors only. For compact stranded conductors, Table C9(A) should be used.

*Types RHH, RHW, and RHW-2 without outer covering.

**Table C9(A) Maximum Number of Compact Conductors in Rigid PVC Conduit, Schedule 80** (*Based on Table 1, Chapter 9*)

COMPACT CONDUCTORS

| Type | Conductor Size (AWG/kcmil) | 16 (½) | 21 (¾) | 27 (1) | 35 (1¼) | 41 (1½) | 53 (2) | 63 (2½) | 78 (3) | 91 (3½) | 103 (4) | 129 (5) | 155 (6) |
|---|---|---|---|---|---|---|---|---|---|---|---|---|---|
| THW, THW-2, THHW | 8 | 1 | 3 | 5 | 9 | 13 | 22 | 32 | 50 | 68 | 88 | 140 | 200 |
| | 6 | 1 | 2 | 4 | 7 | 10 | 17 | 25 | 39 | 52 | 68 | 108 | 155 |
| | 4 | 1 | 1 | 3 | 5 | 7 | 13 | 18 | 29 | 39 | 51 | 81 | 116 |
| | 2 | 1 | 1 | 1 | 4 | 5 | 9 | 13 | 21 | 29 | 37 | 60 | 85 |
| | 1 | 0 | 1 | 1 | 3 | 4 | 6 | 9 | 15 | 20 | 26 | 42 | 60 |
| | 1/0 | 0 | 1 | 1 | 2 | 3 | 6 | 8 | 13 | 17 | 23 | 36 | 52 |
| | 2/0 | 0 | 1 | 1 | 1 | 3 | 5 | 7 | 11 | 15 | 19 | 30 | 44 |
| | 3/0 | 0 | 0 | 1 | 1 | 2 | 4 | 6 | 9 | 12 | 16 | 26 | 37 |
| | 4/0 | 0 | 0 | 1 | 1 | 1 | 3 | 5 | 8 | 10 | 13 | 22 | 31 |
| | 250 | 0 | 0 | 1 | 1 | 1 | 2 | 4 | 6 | 8 | 11 | 17 | 25 |
| | 300 | 0 | 0 | 0 | 1 | 1 | 2 | 3 | 5 | 7 | 9 | 15 | 21 |
| | 350 | 0 | 0 | 0 | 1 | 1 | 1 | 3 | 5 | 6 | 8 | 13 | 19 |
| | 400 | 0 | 0 | 0 | 1 | 1 | 1 | 3 | 4 | 6 | 7 | 12 | 17 |
| | 500 | 0 | 0 | 0 | 1 | 1 | 1 | 2 | 3 | 5 | 6 | 10 | 14 |
| | 600 | 0 | 0 | 0 | 0 | 1 | 1 | 1 | 3 | 4 | 5 | 8 | 12 |
| | 700 | 0 | 0 | 0 | 0 | 1 | 1 | 1 | 2 | 3 | 4 | 7 | 10 |
| | 750 | 0 | 0 | 0 | 0 | 1 | 1 | 1 | 2 | 3 | 4 | 7 | 10 |
| | 1000 | 0 | 0 | 0 | 0 | 0 | 1 | 1 | 1 | 2 | 3 | 5 | 8 |
| THHN, THWN, THWN-2 | 8 | — | — | — | — | — | — | — | — | — | — | — | — |
| | 6 | 1 | 3 | 6 | 11 | 15 | 25 | 36 | 57 | 77 | 99 | 158 | 226 |
| | 4 | 1 | 1 | 3 | 6 | 9 | 15 | 22 | 35 | 47 | 61 | 98 | 140 |
| | 2 | 1 | 1 | 2 | 5 | 6 | 11 | 16 | 25 | 34 | 44 | 70 | 100 |
| | 1 | 1 | 1 | 1 | 3 | 5 | 8 | 12 | 19 | 25 | 33 | 53 | 75 |
| | 1/0 | 0 | 1 | 1 | 3 | 4 | 7 | 10 | 16 | 22 | 28 | 45 | 64 |
| | 2/0 | 0 | 1 | 1 | 2 | 3 | 6 | 8 | 13 | 18 | 23 | 37 | 53 |
| | 3/0 | 0 | 1 | 1 | 1 | 3 | 5 | 7 | 11 | 15 | 19 | 31 | 44 |
| | 4/0 | 0 | 0 | 1 | 1 | 2 | 4 | 6 | 9 | 12 | 16 | 25 | 37 |
| | 250 | 0 | 0 | 1 | 1 | 1 | 3 | 4 | 7 | 10 | 12 | 20 | 29 |
| | 300 | 0 | 0 | 1 | 1 | 1 | 3 | 4 | 6 | 8 | 11 | 17 | 25 |
| | 350 | 0 | 0 | 0 | 1 | 1 | 2 | 3 | 5 | 7 | 9 | 15 | 22 |
| | 400 | 0 | 0 | 0 | 1 | 1 | 1 | 3 | 5 | 6 | 8 | 13 | 19 |
| | 500 | 0 | 0 | 0 | 1 | 1 | 1 | 2 | 4 | 5 | 7 | 11 | 16 |
| | 600 | 0 | 0 | 0 | 1 | 1 | 1 | 1 | 3 | 4 | 6 | 9 | 13 |
| | 700 | 0 | 0 | 0 | 0 | 1 | 1 | 1 | 3 | 4 | 5 | 8 | 12 |
| | 750 | 0 | 0 | 0 | 0 | 1 | 1 | 1 | 3 | 4 | 5 | 8 | 11 |
| | 1000 | 0 | 0 | 0 | 0 | 0 | 1 | 1 | 1 | 3 | 3 | 5 | 8 |
| XHHW, XHHW-2 | 8 | 1 | 4 | 7 | 12 | 17 | 29 | 42 | 65 | 88 | 114 | 181 | 260 |
| | 6 | 1 | 3 | 5 | 9 | 13 | 21 | 31 | 48 | 65 | 85 | 134 | 193 |
| | 4 | 1 | 1 | 3 | 6 | 9 | 15 | 22 | 35 | 47 | 61 | 98 | 140 |
| | 2 | 1 | 1 | 2 | 5 | 6 | 11 | 16 | 25 | 34 | 44 | 70 | 100 |
| | 1 | 1 | 1 | 1 | 3 | 5 | 8 | 12 | 19 | 25 | 33 | 53 | 75 |
| | 1/0 | 0 | 1 | 1 | 3 | 4 | 7 | 10 | 16 | 22 | 28 | 45 | 64 |
| | 2/0 | 0 | 1 | 1 | 2 | 3 | 6 | 8 | 13 | 18 | 24 | 38 | 54 |
| | 3/0 | 0 | 1 | 1 | 1 | 3 | 5 | 7 | 11 | 15 | 19 | 31 | 44 |
| | 4/0 | 0 | 0 | 1 | 1 | 2 | 4 | 6 | 9 | 12 | 16 | 26 | 37 |
| | 250 | 0 | 0 | 1 | 1 | 1 | 3 | 5 | 7 | 10 | 13 | 21 | 30 |
| | 300 | 0 | 0 | 1 | 1 | 1 | 2 | 3 | 6 | 8 | 11 | 17 | 25 |
| | 350 | 0 | 0 | 0 | 1 | 1 | 1 | 3 | 5 | 7 | 10 | 15 | 22 |
| | 400 | 0 | 0 | 0 | 1 | 1 | 1 | 3 | 5 | 7 | 9 | 14 | 20 |
| | 500 | 0 | 0 | 0 | 1 | 1 | 1 | 2 | 4 | 5 | 7 | 11 | 17 |
| | 600 | 0 | 0 | 0 | 1 | 1 | 1 | 1 | 3 | 4 | 6 | 9 | 13 |
| | 700 | 0 | 0 | 0 | 0 | 1 | 1 | 1 | 3 | 4 | 5 | 8 | 12 |
| | 750 | 0 | 0 | 0 | 0 | 1 | 1 | 1 | 2 | 3 | 5 | 7 | 11 |
| | 1000 | 0 | 0 | 0 | 0 | 0 | 1 | 1 | 1 | 3 | 3 | 6 | 8 |

Definition: *Compact stranding* is the result of a manufacturing process where the standard conductor is compressed to the extent that the interstices (voids between strand wires) are virtually eliminated.

**Table C10 Maximum Number of Conductors or Fixture Wires in Rigid PVC Conduit, Schedule 40 and HDPE Conduit** (*Based on Table 1, Chapter 9*)

CONDUCTORS

| Type | Conductor Size (AWG/kcmil) | 16 (½) | 21 (¾) | 27 (1) | 35 (1¼) | 41 (1½) | 53 (2) | 63 (2½) | 78 (3) | 91 (3½) | 103 (4) | 129 (5) | 155 (6) |
|---|---|---|---|---|---|---|---|---|---|---|---|---|---|
| RHH, RHW, RHW-2 | 14 | 4 | 7 | 11 | 20 | 27 | 45 | 64 | 99 | 133 | 171 | 269 | 390 |
| | 12 | 3 | 5 | 9 | 16 | 22 | 37 | 53 | 82 | 110 | 142 | 224 | 323 |
| | 10 | 2 | 4 | 7 | 13 | 18 | 30 | 43 | 66 | 89 | 115 | 181 | 261 |
| | 8 | 1 | 2 | 4 | 7 | 9 | 15 | 22 | 35 | 46 | 60 | 94 | 137 |
| | 6 | 1 | 1 | 3 | 5 | 7 | 12 | 18 | 28 | 37 | 48 | 76 | 109 |
| | 4 | 1 | 1 | 2 | 4 | 6 | 10 | 14 | 22 | 29 | 37 | 59 | 85 |
| | 3 | 1 | 1 | 1 | 4 | 5 | 8 | 12 | 19 | 25 | 33 | 52 | 75 |
| | 2 | 1 | 1 | 1 | 3 | 4 | 7 | 10 | 16 | 22 | 28 | 45 | 65 |
| | 1 | 0 | 1 | 1 | 1 | 3 | 5 | 7 | 11 | 14 | 19 | 29 | 43 |
| | 1/0 | 0 | 1 | 1 | 1 | 2 | 4 | 6 | 9 | 13 | 16 | 26 | 37 |
| | 2/0 | 0 | 0 | 1 | 1 | 1 | 3 | 5 | 8 | 11 | 14 | 22 | 32 |
| | 3/0 | 0 | 0 | 1 | 1 | 1 | 3 | 4 | 7 | 9 | 12 | 19 | 28 |
| | 4/0 | 0 | 0 | 1 | 1 | 1 | 2 | 4 | 6 | 8 | 10 | 16 | 24 |
| | 250 | 0 | 0 | 0 | 1 | 1 | 1 | 3 | 4 | 6 | 8 | 12 | 18 |
| | 300 | 0 | 0 | 0 | 1 | 1 | 1 | 2 | 4 | 5 | 7 | 11 | 16 |
| | 350 | 0 | 0 | 0 | 1 | 1 | 1 | 2 | 3 | 5 | 6 | 10 | 14 |
| | 400 | 0 | 0 | 0 | 1 | 1 | 1 | 1 | 3 | 4 | 6 | 9 | 13 |
| | 500 | 0 | 0 | 0 | 0 | 1 | 1 | 1 | 3 | 4 | 5 | 8 | 11 |
| | 600 | 0 | 0 | 0 | 0 | 1 | 1 | 1 | 2 | 3 | 4 | 6 | 9 |
| | 700 | 0 | 0 | 0 | 0 | 0 | 1 | 1 | 1 | 3 | 3 | 6 | 8 |
| | 750 | 0 | 0 | 0 | 0 | 0 | 1 | 1 | 1 | 2 | 3 | 5 | 8 |
| | 800 | 0 | 0 | 0 | 0 | 0 | 1 | 1 | 1 | 2 | 3 | 5 | 7 |
| | 900 | 0 | 0 | 0 | 0 | 0 | 1 | 1 | 1 | 2 | 3 | 5 | 7 |
| | 1000 | 0 | 0 | 0 | 0 | 0 | 1 | 1 | 1 | 1 | 3 | 4 | 6 |
| | 1250 | 0 | 0 | 0 | 0 | 0 | 0 | 1 | 1 | 1 | 1 | 3 | 5 |
| | 1500 | 0 | 0 | 0 | 0 | 0 | 0 | 1 | 1 | 1 | 1 | 3 | 4 |
| | 1750 | 0 | 0 | 0 | 0 | 0 | 0 | 1 | 1 | 1 | 1 | 2 | 3 |
| | 2000 | 0 | 0 | 0 | 0 | 0 | 0 | 0 | 1 | 1 | 1 | 2 | 3 |
| TW, THHW, THW, THW-2, RHH*, RHW*, RHW-2* | 14 | 8 | 14 | 24 | 42 | 57 | 94 | 135 | 209 | 280 | 361 | 568 | 822 |
| | 12 | 6 | 11 | 18 | 32 | 44 | 72 | 103 | 160 | 215 | 277 | 436 | 631 |
| | 10 | 4 | 8 | 13 | 24 | 32 | 54 | 77 | 119 | 160 | 206 | 325 | 470 |
| | 8 | 2 | 4 | 7 | 13 | 18 | 30 | 43 | 66 | 89 | 115 | 181 | 261 |
| RHH*, RHW*, RHW-2* | 14 | 5 | 9 | 16 | 28 | 38 | 63 | 90 | 139 | 186 | 240 | 378 | 546 |
| | 12 | 4 | 8 | 12 | 22 | 30 | 50 | 72 | 112 | 150 | 193 | 304 | 439 |
| | 10 | 3 | 6 | 10 | 17 | 24 | 39 | 56 | 87 | 117 | 150 | 237 | 343 |
| | 8 | 1 | 3 | 6 | 10 | 14 | 23 | 33 | 52 | 70 | 90 | 142 | 205 |
| TW, THW, THHW, THW-2 | 6 | 1 | 2 | 4 | 8 | 11 | 18 | 26 | 40 | 53 | 69 | 109 | 157 |
| | 4 | 1 | 1 | 3 | 6 | 8 | 13 | 19 | 30 | 40 | 51 | 81 | 117 |
| | 3 | 1 | 1 | 3 | 5 | 7 | 11 | 16 | 25 | 34 | 44 | 69 | 100 |
| | 2 | 1 | 1 | 2 | 4 | 6 | 10 | 14 | 22 | 29 | 37 | 59 | 85 |
| | 1 | 0 | 1 | 1 | 3 | 4 | 7 | 10 | 15 | 20 | 26 | 41 | 60 |
| | 1/0 | 0 | 1 | 1 | 2 | 3 | 6 | 8 | 13 | 17 | 22 | 35 | 51 |
| | 2/0 | 0 | 1 | 1 | 1 | 3 | 5 | 7 | 11 | 15 | 19 | 30 | 43 |
| | 3/0 | 0 | 1 | 1 | 1 | 2 | 4 | 6 | 9 | 12 | 16 | 25 | 36 |
| | 4/0 | 0 | 0 | 1 | 1 | 1 | 3 | 5 | 8 | 10 | 13 | 21 | 30 |
| | 250 | 0 | 0 | 1 | 1 | 1 | 3 | 4 | 6 | 8 | 11 | 17 | 25 |
| | 300 | 0 | 0 | 1 | 1 | 1 | 2 | 3 | 5 | 7 | 9 | 15 | 21 |
| | 350 | 0 | 0 | 0 | 1 | 1 | 1 | 3 | 5 | 6 | 8 | 13 | 19 |
| | 400 | 0 | 0 | 0 | 1 | 1 | 1 | 3 | 4 | 6 | 7 | 12 | 17 |
| | 500 | 0 | 0 | 0 | 1 | 1 | 1 | 2 | 3 | 5 | 6 | 10 | 14 |
| | 600 | 0 | 0 | 0 | 0 | 1 | 1 | 1 | 3 | 4 | 5 | 8 | 11 |
| | 700 | 0 | 0 | 0 | 0 | 1 | 1 | 1 | 2 | 3 | 4 | 7 | 10 |
| | 750 | 0 | 0 | 0 | 0 | 1 | 1 | 1 | 2 | 3 | 4 | 6 | 10 |
| | 800 | 0 | 0 | 0 | 0 | 1 | 1 | 1 | 2 | 3 | 4 | 6 | 9 |
| | 900 | 0 | 0 | 0 | 0 | 0 | 1 | 1 | 1 | 3 | 3 | 6 | 8 |
| | 1000 | 0 | 0 | 0 | 0 | 0 | 1 | 1 | 1 | 2 | 3 | 5 | 7 |
| | 1250 | 0 | 0 | 0 | 0 | 0 | 1 | 1 | 1 | 1 | 2 | 4 | 6 |
| | 1500 | 0 | 0 | 0 | 0 | 0 | 1 | 1 | 1 | 1 | 1 | 3 | 5 |
| | 1750 | 0 | 0 | 0 | 0 | 0 | 0 | 1 | 1 | 1 | 1 | 3 | 4 |
| | 2000 | 0 | 0 | 0 | 0 | 0 | 0 | 1 | 1 | 1 | 1 | 3 | 4 |

2002 Edition

Table C10   *Continued*

### CONDUCTORS

| Type | Conductor Size (AWG/kcmil) | 16 (½) | 21 (¾) | 27 (1) | 35 (1¼) | 41 (1½) | 53 (2) | 63 (2½) | 78 (3) | 91 (3½) | 103 (4) | 129 (5) | 155 (6) |
|---|---|---|---|---|---|---|---|---|---|---|---|---|---|
| THHN, THWN, THWN-2 | 14 | 11 | 21 | 34 | 60 | 82 | 135 | 193 | 299 | 401 | 517 | 815 | 1178 |
| | 12 | 8 | 15 | 25 | 43 | 59 | 99 | 141 | 218 | 293 | 377 | 594 | 859 |
| | 10 | 5 | 9 | 15 | 27 | 37 | 62 | 89 | 137 | 184 | 238 | 374 | 541 |
| | 8 | 3 | 5 | 9 | 16 | 21 | 36 | 51 | 79 | 106 | 137 | 216 | 312 |
| | 6 | 1 | 4 | 6 | 11 | 15 | 26 | 37 | 57 | 77 | 99 | 156 | 225 |
| | 4 | 1 | 2 | 4 | 7 | 9 | 16 | 22 | 35 | 47 | 61 | 96 | 138 |
| | 3 | 1 | 1 | 3 | 6 | 8 | 13 | 19 | 30 | 40 | 51 | 81 | 117 |
| | 2 | 1 | 1 | 3 | 5 | 7 | 11 | 16 | 25 | 33 | 43 | 68 | 98 |
| | 1 | 1 | 1 | 1 | 3 | 5 | 8 | 12 | 18 | 25 | 32 | 50 | 73 |
| | 1/0 | 1 | 1 | 1 | 3 | 4 | 7 | 10 | 15 | 21 | 27 | 42 | 61 |
| | 2/0 | 0 | 1 | 1 | 2 | 3 | 6 | 8 | 13 | 17 | 22 | 35 | 51 |
| | 3/0 | 0 | 1 | 1 | 1 | 3 | 5 | 7 | 11 | 14 | 18 | 29 | 42 |
| | 4/0 | 0 | 1 | 1 | 1 | 2 | 4 | 6 | 9 | 12 | 15 | 24 | 35 |
| | 250 | 0 | 0 | 1 | 1 | 1 | 3 | 4 | 7 | 10 | 12 | 20 | 28 |
| | 300 | 0 | 0 | 1 | 1 | 1 | 3 | 4 | 6 | 8 | 11 | 17 | 24 |
| | 350 | 0 | 0 | 1 | 1 | 1 | 2 | 3 | 5 | 7 | 9 | 15 | 21 |
| | 400 | 0 | 0 | 0 | 1 | 1 | 1 | 3 | 5 | 6 | 8 | 13 | 19 |
| | 500 | 0 | 0 | 0 | 1 | 1 | 1 | 2 | 4 | 5 | 7 | 11 | 16 |
| | 600 | 0 | 0 | 0 | 1 | 1 | 1 | 1 | 3 | 4 | 5 | 9 | 13 |
| | 700 | 0 | 0 | 0 | 0 | 1 | 1 | 1 | 3 | 4 | 5 | 8 | 11 |
| | 750 | 0 | 0 | 0 | 0 | 1 | 1 | 1 | 2 | 3 | 4 | 7 | 11 |
| | 800 | 0 | 0 | 0 | 0 | 1 | 1 | 1 | 2 | 3 | 4 | 7 | 10 |
| | 900 | 0 | 0 | 0 | 0 | 1 | 1 | 1 | 2 | 3 | 4 | 6 | 9 |
| | 1000 | 0 | 0 | 0 | 0 | 1 | 1 | 1 | 1 | 3 | 3 | 6 | 8 |
| FEP, FEPB, PFA, PFAH, TFE | 14 | 11 | 20 | 33 | 58 | 79 | 131 | 188 | 290 | 389 | 502 | 790 | 1142 |
| | 12 | 8 | 15 | 24 | 42 | 58 | 96 | 137 | 212 | 284 | 366 | 577 | 834 |
| | 10 | 6 | 10 | 17 | 30 | 41 | 69 | 98 | 152 | 204 | 263 | 414 | 598 |
| | 8 | 3 | 6 | 10 | 17 | 24 | 39 | 56 | 87 | 117 | 150 | 237 | 343 |
| | 6 | 2 | 4 | 7 | 12 | 17 | 28 | 40 | 62 | 83 | 107 | 169 | 244 |
| | 4 | 1 | 3 | 5 | 8 | 12 | 19 | 28 | 43 | 58 | 75 | 118 | 170 |
| | 3 | 1 | 2 | 4 | 7 | 10 | 16 | 23 | 36 | 48 | 62 | 98 | 142 |
| | 2 | 1 | 1 | 3 | 6 | 8 | 13 | 19 | 30 | 40 | 51 | 81 | 117 |
| PFA, PFAH, TFE | 1 | 1 | 1 | 2 | 4 | 5 | 9 | 13 | 20 | 28 | 36 | 56 | 81 |
| PFA, PFAH, TFE, Z | 1/0 | 1 | 1 | 1 | 3 | 4 | 8 | 11 | 17 | 23 | 30 | 47 | 68 |
| | 2/0 | 0 | 1 | 1 | 3 | 4 | 6 | 9 | 14 | 19 | 24 | 39 | 56 |
| | 3/0 | 0 | 1 | 1 | 2 | 3 | 5 | 7 | 12 | 16 | 20 | 32 | 46 |
| | 4/0 | 0 | 1 | 1 | 1 | 2 | 4 | 6 | 9 | 13 | 16 | 26 | 38 |
| Z | 14 | 13 | 24 | 40 | 70 | 95 | 158 | 226 | 350 | 469 | 605 | 952 | 1376 |
| | 12 | 9 | 17 | 28 | 49 | 68 | 112 | 160 | 248 | 333 | 429 | 675 | 976 |
| | 10 | 6 | 10 | 17 | 30 | 41 | 69 | 98 | 152 | 204 | 263 | 414 | 598 |
| | 8 | 3 | 6 | 11 | 19 | 26 | 43 | 62 | 96 | 129 | 166 | 261 | 378 |
| | 6 | 2 | 4 | 7 | 13 | 18 | 30 | 43 | 67 | 90 | 116 | 184 | 265 |
| | 4 | 1 | 3 | 5 | 9 | 12 | 21 | 30 | 46 | 62 | 80 | 126 | 183 |
| | 3 | 1 | 2 | 4 | 6 | 9 | 15 | 22 | 34 | 45 | 58 | 92 | 133 |
| | 2 | 1 | 1 | 3 | 5 | 7 | 12 | 18 | 28 | 38 | 49 | 77 | 111 |
| | 1 | 1 | 1 | 2 | 4 | 6 | 10 | 14 | 23 | 30 | 39 | 62 | 90 |
| XHH, XHHW, XHHW-2, ZW | 14 | 8 | 14 | 24 | 42 | 57 | 94 | 135 | 209 | 280 | 361 | 568 | 822 |
| | 12 | 6 | 11 | 18 | 32 | 44 | 72 | 103 | 160 | 215 | 277 | 436 | 631 |
| | 10 | 4 | 8 | 13 | 24 | 32 | 54 | 77 | 119 | 160 | 206 | 325 | 470 |
| | 8 | 2 | 4 | 7 | 13 | 18 | 30 | 43 | 66 | 89 | 115 | 181 | 261 |
| | 6 | 1 | 3 | 5 | 10 | 13 | 22 | 32 | 49 | 66 | 85 | 134 | 193 |
| | 4 | 1 | 2 | 4 | 7 | 9 | 16 | 23 | 35 | 48 | 61 | 97 | 140 |
| | 3 | 1 | 1 | 3 | 6 | 8 | 13 | 19 | 30 | 40 | 52 | 82 | 118 |
| | 2 | 1 | 1 | 3 | 5 | 7 | 11 | 16 | 25 | 34 | 44 | 69 | 99 |
| XHH, XHHW, XHHW-2 | 1 | 1 | 1 | 1 | 3 | 5 | 8 | 12 | 19 | 25 | 32 | 51 | 74 |
| | 1/0 | 1 | 1 | 1 | 3 | 4 | 7 | 10 | 16 | 21 | 27 | 43 | 62 |
| | 2/0 | 0 | 1 | 1 | 2 | 3 | 6 | 8 | 13 | 17 | 23 | 36 | 52 |
| | 3/0 | 0 | 1 | 1 | 1 | 3 | 5 | 7 | 11 | 14 | 19 | 30 | 43 |
| | 4/0 | 0 | 1 | 1 | 1 | 2 | 4 | 6 | 9 | 12 | 15 | 24 | 35 |
| | 250 | 0 | 0 | 1 | 1 | 1 | 3 | 5 | 7 | 10 | 13 | 20 | 29 |
| | 300 | 0 | 0 | 1 | 1 | 1 | 3 | 4 | 6 | 8 | 11 | 17 | 25 |
| | 350 | 0 | 0 | 1 | 1 | 1 | 2 | 3 | 5 | 7 | 9 | 15 | 22 |
| | 400 | 0 | 0 | 1 | 1 | 1 | 1 | 3 | 5 | 6 | 8 | 13 | 19 |
| | 500 | 0 | 0 | 0 | 1 | 1 | 1 | 2 | 4 | 5 | 7 | 11 | 16 |
| | 600 | 0 | 0 | 0 | 1 | 1 | 1 | 1 | 3 | 4 | 5 | 9 | 13 |
| | 700 | 0 | 0 | 0 | 0 | 1 | 1 | 1 | 3 | 4 | 5 | 8 | 11 |
| | 750 | 0 | 0 | 0 | 0 | 1 | 1 | 1 | 2 | 3 | 4 | 7 | 11 |
| | 800 | 0 | 0 | 0 | 0 | 1 | 1 | 1 | 2 | 3 | 4 | 7 | 10 |
| | 900 | 0 | 0 | 0 | 0 | 1 | 1 | 1 | 2 | 3 | 4 | 6 | 9 |
| | 1000 | 0 | 0 | 0 | 0 | 0 | 1 | 1 | 1 | 3 | 3 | 6 | 8 |
| | 1250 | 0 | 0 | 0 | 0 | 0 | 1 | 1 | 1 | 1 | 3 | 4 | 6 |
| | 1500 | 0 | 0 | 0 | 0 | 0 | 1 | 1 | 1 | 1 | 2 | 4 | 5 |
| | 1750 | 0 | 0 | 0 | 0 | 0 | 0 | 1 | 1 | 1 | 1 | 3 | 5 |
| | 2000 | 0 | 0 | 0 | 0 | 0 | 0 | 1 | 1 | 1 | 1 | 3 | 4 |

### FIXTURE WIRES

| Type | Conductor Size (AWG/kcmil) | 16 (½) | 21 (¾) | 27 (1) | 35 (1¼) | 41 (1½) | 53 (2) |
|---|---|---|---|---|---|---|---|
| FFH-2, RFH-2, RFHH-3 | 18 | 8 | 14 | 23 | 40 | 54 | 90 |
| | 16 | 6 | 12 | 19 | 33 | 46 | 76 |
| SF-2, SFF-2 | 18 | 10 | 17 | 29 | 50 | 69 | 114 |
| | 16 | 8 | 14 | 24 | 42 | 57 | 94 |
| | 14 | 6 | 12 | 19 | 33 | 46 | 76 |
| SF-1, SFF-1 | 18 | 17 | 31 | 51 | 89 | 122 | 202 |
| RFHH-2, TF, TFF, XF, XFF RFH-1, | 18 | 13 | 23 | 38 | 66 | 90 | 149 |
| RFHH-2, TF, TFF, XF, XFF | 16 | 10 | 18 | 30 | 53 | 73 | 120 |
| XF, XFF | 14 | 8 | 14 | 24 | 42 | 57 | 94 |
| TFN, TFFN | 18 | 20 | 37 | 60 | 105 | 144 | 239 |
| | 16 | 16 | 28 | 46 | 80 | 110 | 183 |
| PF, PFF, PGF, PGFF, PAF, PTF, PTFF, PAFF | 18 | 19 | 35 | 57 | 100 | 137 | 227 |
| | 16 | 15 | 27 | 44 | 77 | 106 | 175 |
| | 14 | 11 | 20 | 33 | 58 | 79 | 131 |
| HF, HFF, ZF, ZFF, ZHF | 18 | 25 | 45 | 74 | 129 | 176 | 292 |
| | 16 | 18 | 33 | 54 | 95 | 130 | 216 |
| | 14 | 13 | 24 | 40 | 70 | 95 | 158 |
| KF-2, KFF-2 | 18 | 36 | 65 | 107 | 187 | 256 | 424 |
| | 16 | 26 | 46 | 75 | 132 | 180 | 299 |
| | 14 | 17 | 31 | 52 | 90 | 124 | 205 |
| | 12 | 12 | 22 | 35 | 62 | 85 | 141 |
| | 10 | 8 | 14 | 24 | 42 | 57 | 94 |
| KF-1, KFF-1 | 18 | 43 | 78 | 128 | 223 | 305 | 506 |
| | 16 | 30 | 55 | 90 | 157 | 214 | 355 |
| | 14 | 20 | 37 | 60 | 105 | 144 | 239 |
| | 12 | 13 | 24 | 40 | 70 | 95 | 158 |
| | 10 | 9 | 16 | 26 | 45 | 62 | 103 |
| XF, XFF | 12 | 4 | 8 | 12 | 22 | 30 | 50 |
| | 10 | 3 | 6 | 10 | 17 | 24 | 39 |

Note: This table is for concentric stranded conductors only. For compact stranded conductors, Table C10(A) should be used.

*Types RHH, RHW, and RHW-2 without outer covering.

**Table C10(A)  Maximum Number of Compact Conductors in Rigid PVC Conduit, Schedule 40 and HDPE Conduit** (*Based on Table 1, Chapter 9*)

### COMPACT CONDUCTORS

| Type | Conductor Size (AWG/kcmil) | 16 (½) | 21 (¾) | 27 (1) | 35 (1¼) | 41 (1½) | 53 (2) | 63 (2½) | 78 (3) | 91 (3½) | 103 (4) | 129 (5) | 155 (6) |
|---|---|---|---|---|---|---|---|---|---|---|---|---|---|
| THW, THW-2, THHW | 8 | 1 | 4 | 6 | 11 | 15 | 26 | 37 | 57 | 76 | 98 | 155 | 224 |
| | 6 | 1 | 3 | 5 | 9 | 12 | 20 | 28 | 44 | 59 | 76 | 119 | 173 |
| | 4 | 1 | 1 | 3 | 6 | 9 | 15 | 21 | 33 | 44 | 57 | 89 | 129 |
| | 2 | 1 | 1 | 2 | 5 | 6 | 11 | 15 | 24 | 32 | 42 | 66 | 95 |
| | 1 | 1 | 1 | 1 | 3 | 4 | 7 | 11 | 17 | 23 | 29 | 46 | 67 |
| | 1/0 | 0 | 1 | 1 | 3 | 4 | 6 | 9 | 15 | 20 | 25 | 40 | 58 |
| | 2/0 | 0 | 1 | 1 | 2 | 3 | 5 | 8 | 12 | 16 | 21 | 34 | 49 |
| | 3/0 | 0 | 1 | 1 | 1 | 3 | 5 | 7 | 10 | 14 | 18 | 29 | 42 |
| | 4/0 | 0 | 1 | 1 | 1 | 2 | 4 | 5 | 9 | 12 | 15 | 24 | 35 |
| | 250 | 0 | 0 | 1 | 1 | 1 | 3 | 4 | 7 | 9 | 12 | 19 | 27 |
| | 300 | 0 | 0 | 1 | 1 | 1 | 2 | 4 | 6 | 8 | 10 | 16 | 24 |
| | 350 | 0 | 0 | 1 | 1 | 1 | 2 | 3 | 5 | 7 | 9 | 15 | 21 |
| | 400 | 0 | 0 | 0 | 1 | 1 | 1 | 3 | 5 | 6 | 8 | 13 | 19 |
| | 500 | 0 | 0 | 0 | 1 | 1 | 1 | 2 | 4 | 5 | 7 | 11 | 16 |
| | 600 | 0 | 0 | 0 | 1 | 1 | 1 | 1 | 3 | 4 | 5 | 9 | 13 |
| | 700 | 0 | 0 | 0 | 0 | 1 | 1 | 1 | 3 | 4 | 5 | 8 | 12 |
| | 750 | 0 | 0 | 0 | 0 | 1 | 1 | 1 | 2 | 3 | 5 | 7 | 11 |
| | 1000 | 0 | 0 | 0 | 0 | 1 | 1 | 1 | 1 | 3 | 4 | 6 | 9 |

Table C10(A)　*Continued*

### COMPACT CONDUCTORS

| Type | Conductor Size (AWG/kcmil) | 16 (½) | 21 (¾) | 27 (1) | 35 (1¼) | 41 (1½) | 53 (2) | 63 (2½) | 78 (3) | 91 (3½) | 103 (4) | 129 (5) | 155 (6) |
|---|---|---|---|---|---|---|---|---|---|---|---|---|---|
| THHN, | 8 | — | — | — | — | — | — | — | — | — | — | — | — |
| THWN, | 6 | 2 | 4 | 7 | 13 | 17 | 29 | 41 | 64 | 86 | 111 | 175 | 253 |
| THWN-2 | 4 | 1 | 2 | 4 | 8 | 11 | 18 | 25 | 40 | 53 | 68 | 108 | 156 |
|  | 2 | 1 | 1 | 3 | 5 | 8 | 13 | 18 | 28 | 38 | 49 | 77 | 112 |
|  | 1 | 1 | 1 | 2 | 4 | 6 | 9 | 14 | 21 | 29 | 37 | 58 | 84 |
|  | 1/0 | 1 | 1 | 1 | 3 | 5 | 8 | 12 | 18 | 24 | 31 | 49 | 72 |
|  | 2/0 | 0 | 1 | 1 | 3 | 4 | 7 | 9 | 15 | 20 | 26 | 41 | 59 |
|  | 3/0 | 0 | 1 | 1 | 2 | 3 | 5 | 8 | 12 | 17 | 22 | 34 | 50 |
|  | 4/0 | 0 | 1 | 1 | 1 | 3 | 4 | 6 | 10 | 14 | 18 | 28 | 41 |
|  | 250 | 0 | 0 | 1 | 1 | 1 | 3 | 5 | 8 | 11 | 14 | 22 | 32 |
|  | 300 | 0 | 0 | 1 | 1 | 1 | 3 | 4 | 7 | 9 | 12 | 19 | 28 |
|  | 350 | 0 | 0 | 1 | 1 | 1 | 3 | 4 | 6 | 8 | 10 | 17 | 24 |
|  | 400 | 0 | 0 | 1 | 1 | 1 | 2 | 3 | 5 | 7 | 9 | 15 | 22 |
|  | 500 | 0 | 0 | 0 | 1 | 1 | 1 | 3 | 4 | 6 | 8 | 13 | 18 |
|  | 600 | 0 | 0 | 0 | 1 | 1 | 1 | 2 | 4 | 5 | 6 | 10 | 15 |
|  | 700 | 0 | 0 | 0 | 1 | 1 | 1 | 1 | 3 | 4 | 5 | 9 | 13 |
|  | 750 | 0 | 0 | 0 | 1 | 1 | 1 | 1 | 3 | 4 | 5 | 8 | 12 |
|  | 1000 | 0 | 0 | 0 | 0 | 1 | 1 | 1 | 2 | 3 | 4 | 6 | 9 |
| XHHW, | 8 | 3 | 5 | 8 | 14 | 20 | 33 | 47 | 73 | 99 | 127 | 200 | 290 |
| XHHW-2 | 6 | 1 | 4 | 6 | 11 | 15 | 25 | 35 | 55 | 73 | 94 | 149 | 215 |
|  | 4 | 1 | 2 | 4 | 8 | 11 | 18 | 25 | 40 | 53 | 68 | 108 | 156 |
|  | 2 | 1 | 1 | 3 | 5 | 8 | 13 | 18 | 28 | 38 | 49 | 77 | 112 |
|  | 1 | 1 | 1 | 2 | 4 | 6 | 9 | 14 | 21 | 29 | 37 | 58 | 84 |
|  | 1/0 | 1 | 1 | 1 | 3 | 5 | 8 | 12 | 18 | 24 | 31 | 49 | 72 |
|  | 2/0 | 1 | 1 | 1 | 3 | 4 | 7 | 10 | 15 | 20 | 26 | 42 | 60 |
|  | 3/0 | 0 | 1 | 1 | 2 | 3 | 5 | 8 | 12 | 17 | 22 | 34 | 50 |
|  | 4/0 | 0 | 1 | 1 | 1 | 3 | 5 | 7 | 10 | 14 | 18 | 29 | 42 |
|  | 250 | 0 | 0 | 1 | 1 | 1 | 4 | 5 | 8 | 11 | 14 | 23 | 33 |
|  | 300 | 0 | 0 | 1 | 1 | 1 | 3 | 4 | 7 | 9 | 12 | 19 | 28 |
|  | 350 | 0 | 0 | 1 | 1 | 1 | 3 | 4 | 6 | 8 | 11 | 17 | 25 |
|  | 400 | 0 | 0 | 1 | 1 | 1 | 2 | 3 | 5 | 7 | 10 | 15 | 22 |
|  | 500 | 0 | 0 | 1 | 1 | 1 | 1 | 3 | 4 | 6 | 8 | 13 | 18 |
|  | 600 | 0 | 0 | 0 | 1 | 1 | 1 | 2 | 4 | 5 | 6 | 10 | 15 |
|  | 700 | 0 | 0 | 0 | 1 | 1 | 1 | 1 | 3 | 4 | 5 | 9 | 13 |
|  | 750 | 0 | 0 | 0 | 1 | 1 | 1 | 1 | 3 | 4 | 5 | 8 | 12 |
|  | 1000 | 0 | 0 | 0 | 0 | 1 | 1 | 1 | 2 | 3 | 4 | 6 | 9 |

Definition: *Compact stranding* is the result of a manufacturing process where the standard conductor is compressed to the extent that the interstices (voids between strand wires) are virtually eliminated.

**Table C11 Maximum Number of Conductors or Fixture Wires in Type A, Rigid PVC Conduit** (*Based on Table 1, Chapter 9*)

### CONDUCTORS

| Type | Conductor Size (AWG/kcmil) | 16 (½) | 21 (¾) | 27 (1) | 35 (1¼) | 41 (1½) | 53 (2) | 63 (2½) | 78 (3) | 91 (3½) | 103 (4) |
|---|---|---|---|---|---|---|---|---|---|---|---|
| RHH, | 14 | 5 | 9 | 15 | 24 | 31 | 49 | 74 | 112 | 146 | 187 |
| RHW | 12 | 4 | 7 | 12 | 20 | 26 | 41 | 61 | 93 | 121 | 155 |
| RHW-2 | 10 | 3 | 6 | 10 | 16 | 21 | 33 | 50 | 75 | 98 | 125 |
|  | 8 | 1 | 3 | 5 | 8 | 11 | 17 | 26 | 39 | 51 | 65 |
|  | 6 | 1 | 2 | 4 | 6 | 9 | 14 | 21 | 31 | 41 | 52 |
|  | 4 | 1 | 1 | 3 | 5 | 7 | 11 | 16 | 24 | 32 | 41 |
|  | 3 | 1 | 1 | 3 | 4 | 6 | 9 | 14 | 21 | 28 | 36 |
|  | 2 | 1 | 1 | 2 | 4 | 5 | 8 | 12 | 18 | 24 | 31 |
|  | 1 | 0 | 1 | 1 | 2 | 3 | 5 | 8 | 12 | 16 | 20 |
|  | 1/0 | 0 | 1 | 1 | 2 | 3 | 5 | 7 | 10 | 14 | 18 |
|  | 2/0 | 0 | 1 | 1 | 1 | 2 | 4 | 6 | 9 | 12 | 15 |
|  | 3/0 | 0 | 1 | 1 | 1 | 1 | 3 | 5 | 8 | 10 | 13 |
|  | 4/0 | 0 | 0 | 1 | 1 | 1 | 3 | 4 | 7 | 9 | 11 |
|  | 250 | 0 | 0 | 1 | 1 | 1 | 1 | 3 | 5 | 7 | 8 |
|  | 300 | 0 | 0 | 1 | 1 | 1 | 1 | 3 | 4 | 6 | 7 |
|  | 350 | 0 | 0 | 0 | 1 | 1 | 1 | 2 | 4 | 5 | 7 |
|  | 400 | 0 | 0 | 0 | 1 | 1 | 1 | 2 | 4 | 5 | 6 |
|  | 500 | 0 | 0 | 0 | 1 | 1 | 1 | 1 | 3 | 4 | 5 |
|  | 600 | 0 | 0 | 0 | 0 | 1 | 1 | 1 | 2 | 3 | 4 |
|  | 700 | 0 | 0 | 0 | 0 | 1 | 1 | 1 | 2 | 3 | 4 |
|  | 750 | 0 | 0 | 0 | 0 | 1 | 1 | 1 | 1 | 3 | 4 |
|  | 800 | 0 | 0 | 0 | 0 | 1 | 1 | 1 | 1 | 3 | 3 |
|  | 900 | 0 | 0 | 0 | 0 | 0 | 1 | 1 | 1 | 2 | 3 |
|  | 1000 | 0 | 0 | 0 | 0 | 0 | 1 | 1 | 1 | 2 | 3 |
|  | 1250 | 0 | 0 | 0 | 0 | 0 | 1 | 1 | 1 | 1 | 2 |
|  | 1500 | 0 | 0 | 0 | 0 | 0 | 0 | 1 | 1 | 1 | 1 |
|  | 1750 | 0 | 0 | 0 | 0 | 0 | 0 | 1 | 1 | 1 | 1 |
|  | 2000 | 0 | 0 | 0 | 0 | 0 | 0 | 1 | 1 | 1 | 1 |
| TW, | 14 | 11 | 18 | 31 | 51 | 67 | 105 | 157 | 235 | 307 | 395 |
| THHW, | 12 | 8 | 14 | 24 | 39 | 51 | 80 | 120 | 181 | 236 | 303 |
| THW, | 10 | 6 | 10 | 18 | 29 | 38 | 60 | 89 | 135 | 176 | 226 |
| THW-2 | 8 | 3 | 6 | 10 | 16 | 21 | 33 | 50 | 75 | 98 | 125 |

Table C11　*Continued*

### CONDUCTORS

| Type | Conductor Size (AWG/kcmil) | 16 (½) | 21 (¾) | 27 (1) | 35 (1¼) | 41 (1½) | 53 (2) | 63 (2½) | 78 (3) | 91 (3½) | 103 (4) |
|---|---|---|---|---|---|---|---|---|---|---|---|
| RHH*, | 14 | 7 | 12 | 20 | 34 | 44 | 70 | 104 | 157 | 204 | 262 |
| RHW*, | 12 | 6 | 10 | 16 | 27 | 35 | 56 | 84 | 126 | 164 | 211 |
| RHW-2* | 10 | 4 | 8 | 13 | 21 | 28 | 44 | 65 | 98 | 128 | 165 |
|  | 8 | 2 | 4 | 8 | 12 | 16 | 26 | 39 | 59 | 77 | 98 |
| RHH, RHW* | 6 | 1 | 3 | 6 | 9 | 13 | 20 | 30 | 45 | 59 | 75 |
| TW, THW, | 4 | 1 | 2 | 4 | 7 | 9 | 15 | 22 | 33 | 44 | 56 |
| THHW, | 3 | 1 | 1 | 4 | 6 | 8 | 13 | 19 | 29 | 37 | 48 |
| THW-2 | 2 | 1 | 1 | 3 | 5 | 7 | 11 | 16 | 24 | 32 | 41 |
|  | 1 | 1 | 1 | 1 | 3 | 5 | 7 | 11 | 17 | 22 | 29 |
|  | 1/0 | 1 | 1 | 1 | 3 | 4 | 6 | 10 | 14 | 19 | 24 |
|  | 2/0 | 0 | 1 | 1 | 2 | 3 | 5 | 8 | 12 | 16 | 21 |
|  | 3/0 | 0 | 1 | 1 | 1 | 3 | 4 | 7 | 10 | 13 | 17 |
|  | 4/0 | 0 | 1 | 1 | 1 | 2 | 4 | 6 | 9 | 11 | 14 |
|  | 250 | 0 | 0 | 1 | 1 | 1 | 3 | 4 | 7 | 9 | 12 |
|  | 300 | 0 | 0 | 1 | 1 | 1 | 2 | 4 | 6 | 8 | 10 |
|  | 350 | 0 | 0 | 1 | 1 | 1 | 2 | 3 | 5 | 7 | 9 |
|  | 400 | 0 | 0 | 1 | 1 | 1 | 1 | 3 | 5 | 6 | 8 |
|  | 500 | 0 | 0 | 0 | 1 | 1 | 1 | 2 | 4 | 5 | 7 |
|  | 600 | 0 | 0 | 0 | 1 | 1 | 1 | 1 | 3 | 4 | 5 |
|  | 700 | 0 | 0 | 0 | 1 | 1 | 1 | 1 | 3 | 4 | 5 |
|  | 750 | 0 | 0 | 0 | 1 | 1 | 1 | 1 | 3 | 3 | 4 |
|  | 800 | 0 | 0 | 0 | 0 | 1 | 1 | 1 | 2 | 3 | 4 |
|  | 900 | 0 | 0 | 0 | 0 | 1 | 1 | 1 | 2 | 3 | 4 |
|  | 1000 | 0 | 0 | 0 | 0 | 1 | 1 | 1 | 1 | 3 | 3 |
|  | 1250 | 0 | 0 | 0 | 0 | 0 | 1 | 1 | 1 | 1 | 3 |
|  | 1500 | 0 | 0 | 0 | 0 | 0 | 1 | 1 | 1 | 1 | 2 |
|  | 1750 | 0 | 0 | 0 | 0 | 0 | 0 | 1 | 1 | 1 | 1 |
|  | 2000 | 0 | 0 | 0 | 0 | 0 | 0 | 1 | 1 | 1 | 1 |
| THHN, | 14 | 16 | 27 | 44 | 73 | 96 | 150 | 225 | 338 | 441 | 566 |
| THWN, | 12 | 11 | 19 | 32 | 53 | 70 | 109 | 164 | 246 | 321 | 412 |
| THWN-2 | 10 | 7 | 12 | 20 | 33 | 44 | 69 | 103 | 155 | 202 | 260 |
|  | 8 | 4 | 7 | 12 | 19 | 25 | 40 | 59 | 89 | 117 | 150 |
|  | 6 | 3 | 5 | 8 | 14 | 18 | 28 | 43 | 64 | 84 | 108 |
|  | 4 | 1 | 3 | 5 | 8 | 11 | 17 | 26 | 39 | 52 | 66 |
|  | 3 | 1 | 2 | 4 | 7 | 9 | 15 | 22 | 33 | 44 | 56 |
|  | 2 | 1 | 1 | 3 | 6 | 8 | 12 | 19 | 28 | 37 | 47 |
|  | 1 | 1 | 1 | 2 | 4 | 6 | 9 | 14 | 21 | 27 | 35 |
|  | 1/0 | 1 | 1 | 2 | 4 | 5 | 8 | 11 | 17 | 23 | 29 |
|  | 2/0 | 1 | 1 | 1 | 3 | 4 | 6 | 10 | 14 | 19 | 24 |
|  | 3/0 | 0 | 1 | 1 | 2 | 3 | 5 | 8 | 12 | 16 | 20 |
|  | 4/0 | 0 | 1 | 1 | 1 | 3 | 4 | 6 | 10 | 13 | 17 |
|  | 250 | 0 | 1 | 1 | 1 | 2 | 3 | 5 | 8 | 10 | 14 |
|  | 300 | 0 | 0 | 1 | 1 | 1 | 3 | 4 | 7 | 9 | 12 |
|  | 350 | 0 | 0 | 1 | 1 | 1 | 2 | 4 | 6 | 8 | 10 |
|  | 400 | 0 | 0 | 1 | 1 | 1 | 2 | 3 | 5 | 7 | 9 |
|  | 500 | 0 | 0 | 1 | 1 | 1 | 1 | 3 | 4 | 6 | 7 |
|  | 600 | 0 | 0 | 0 | 1 | 1 | 1 | 2 | 3 | 5 | 6 |
|  | 700 | 0 | 0 | 0 | 1 | 1 | 1 | 1 | 3 | 4 | 5 |
|  | 750 | 0 | 0 | 0 | 1 | 1 | 1 | 1 | 3 | 4 | 5 |
|  | 800 | 0 | 0 | 0 | 1 | 1 | 1 | 1 | 3 | 4 | 5 |
|  | 900 | 0 | 0 | 0 | 0 | 1 | 1 | 1 | 2 | 3 | 4 |
|  | 1000 | 0 | 0 | 0 | 0 | 1 | 1 | 1 | 2 | 3 | 4 |
| FEP, | 14 | 15 | 26 | 43 | 70 | 93 | 146 | 218 | 327 | 427 | 549 |
| FEPB, | 12 | 11 | 19 | 31 | 51 | 68 | 106 | 159 | 239 | 312 | 400 |
| PFA, | 10 | 8 | 13 | 22 | 37 | 48 | 76 | 114 | 171 | 224 | 287 |
| PFAH, | 8 | 4 | 8 | 13 | 21 | 28 | 44 | 65 | 98 | 128 | 165 |
| TFE | 6 | 3 | 5 | 9 | 15 | 20 | 31 | 46 | 70 | 91 | 117 |
|  | 4 | 1 | 4 | 6 | 10 | 14 | 21 | 32 | 49 | 64 | 82 |
|  | 3 | 1 | 3 | 5 | 8 | 11 | 18 | 27 | 40 | 53 | 68 |
|  | 2 | 1 | 2 | 4 | 7 | 9 | 15 | 22 | 33 | 44 | 56 |
| PFA, PFAH, TFE | 1 | 1 | 1 | 3 | 5 | 6 | 10 | 15 | 23 | 30 | 39 |
| PFA, | 1/0 | 1 | 1 | 2 | 4 | 5 | 8 | 13 | 19 | 25 | 32 |
| PFAH, | 2/0 | 1 | 1 | 1 | 3 | 4 | 7 | 10 | 16 | 21 | 27 |
| TFE, Z | 3/0 | 1 | 1 | 1 | 3 | 3 | 6 | 9 | 13 | 17 | 22 |
|  | 4/0 | 0 | 1 | 1 | 2 | 3 | 5 | 7 | 11 | 14 | 18 |
| Z | 14 | 18 | 31 | 52 | 85 | 112 | 175 | 263 | 395 | 515 | 661 |
|  | 12 | 13 | 22 | 37 | 60 | 79 | 124 | 186 | 280 | 365 | 469 |
|  | 10 | 8 | 13 | 22 | 37 | 48 | 76 | 114 | 171 | 224 | 287 |
|  | 8 | 5 | 8 | 14 | 23 | 30 | 48 | 72 | 108 | 141 | 181 |
|  | 6 | 3 | 6 | 10 | 16 | 21 | 34 | 50 | 76 | 99 | 127 |
|  | 4 | 2 | 4 | 7 | 11 | 15 | 23 | 35 | 52 | 68 | 88 |
|  | 3 | 1 | 3 | 5 | 8 | 11 | 17 | 25 | 38 | 50 | 64 |
|  | 2 | 1 | 2 | 4 | 7 | 9 | 14 | 21 | 32 | 41 | 53 |
|  | 1 | 1 | 1 | 3 | 5 | 7 | 11 | 17 | 26 | 33 | 43 |
| XHH, | 14 | 11 | 18 | 31 | 51 | 67 | 105 | 157 | 235 | 307 | 395 |
| XHHW, | 12 | 8 | 14 | 24 | 39 | 51 | 80 | 120 | 181 | 236 | 303 |
| XHHW-2, | 10 | 6 | 10 | 18 | 29 | 38 | 60 | 89 | 135 | 176 | 226 |
| ZW | 8 | 3 | 6 | 10 | 16 | 21 | 33 | 50 | 75 | 98 | 125 |
|  | 6 | 2 | 4 | 7 | 12 | 15 | 24 | 37 | 55 | 72 | 93 |
|  | 4 | 1 | 3 | 5 | 8 | 11 | 18 | 26 | 40 | 52 | 67 |
|  | 3 | 1 | 2 | 4 | 7 | 9 | 15 | 22 | 34 | 44 | 57 |
|  | 2 | 1 | 1 | 3 | 6 | 8 | 12 | 19 | 28 | 37 | 48 |

**Table C11** *Continued*

### CONDUCTORS

| Type | Conductor Size (AWG/kcmil) | 16 (½) | 21 (¾) | 27 (1) | 35 (1¼) | 41 (1½) | 53 (2) | 63 (2½) | 78 (3) | 91 (3½) | 103 (4) |
|---|---|---|---|---|---|---|---|---|---|---|---|
| XHH, | 1 | 1 | 1 | 3 | 4 | 6 | 9 | 14 | 21 | 28 | 35 |
| XHHW, | 1/0 | 1 | 1 | 2 | 4 | 5 | 8 | 12 | 18 | 23 | 30 |
| XHHW-2 | 2/0 | 1 | 1 | 1 | 3 | 4 | 6 | 10 | 15 | 19 | 25 |
| | 3/0 | 0 | 1 | 1 | 2 | 3 | 5 | 8 | 12 | 16 | 20 |
| | 4/0 | 0 | 1 | 1 | 1 | 3 | 4 | 7 | 10 | 13 | 17 |
| | 250 | 0 | 1 | 1 | 1 | 2 | 3 | 5 | 8 | 11 | 14 |
| | 300 | 0 | 0 | 1 | 1 | 1 | 3 | 5 | 7 | 9 | 12 |
| | 350 | 0 | 0 | 1 | 1 | 1 | 3 | 4 | 6 | 8 | 10 |
| | 400 | 0 | 0 | 1 | 1 | 1 | 2 | 3 | 5 | 7 | 9 |
| | 500 | 0 | 0 | 1 | 1 | 1 | 1 | 3 | 4 | 6 | 8 |
| | 600 | 0 | 0 | 0 | 1 | 1 | 1 | 2 | 3 | 5 | 6 |
| | 700 | 0 | 0 | 0 | 1 | 1 | 1 | 1 | 3 | 4 | 5 |
| | 750 | 0 | 0 | 0 | 1 | 1 | 1 | 1 | 3 | 4 | 5 |
| | 800 | 0 | 0 | 0 | 1 | 1 | 1 | 1 | 3 | 4 | 5 |
| | 900 | 0 | 0 | 0 | 0 | 1 | 1 | 1 | 2 | 3 | 4 |
| | 1000 | 0 | 0 | 0 | 0 | 1 | 1 | 1 | 2 | 3 | 4 |
| | 1250 | 0 | 0 | 0 | 0 | 0 | 1 | 1 | 1 | 2 | 3 |
| | 1500 | 0 | 0 | 0 | 0 | 0 | 1 | 1 | 1 | 1 | 2 |
| | 1750 | 0 | 0 | 0 | 0 | 0 | 1 | 1 | 1 | 1 | 2 |
| | 2000 | 0 | 0 | 0 | 0 | 0 | 0 | 1 | 1 | 1 | 1 |

### FIXTURE WIRES

| Type | Conductor Size (AWG/kcmil) | 16 (½) | 21 (¾) | 27 (1) | 35 (1¼) | 41 (1½) | 53 (2) |
|---|---|---|---|---|---|---|---|
| FFH-2, RFH-2, RFHH-3 | 18 | 10 | 18 | 30 | 48 | 64 | 100 |
| | 16 | 9 | 15 | 25 | 41 | 54 | 85 |
| SF-2, SFF-2 | 18 | 13 | 22 | 37 | 61 | 81 | 127 |
| | 16 | 11 | 18 | 31 | 51 | 67 | 105 |
| | 14 | 9 | 15 | 25 | 41 | 54 | 85 |
| SF-1, SFF-1 | 18 | 23 | 40 | 66 | 108 | 143 | 224 |
| RFH-1, RFHH-2, TF, TFF, XF, XFF | 18 | 17 | 29 | 49 | 80 | 105 | 165 |
| RFHH-2, TF, TFF, XF, XFF | 16 | 14 | 24 | 39 | 65 | 85 | 134 |
| XF, XFF | 14 | 11 | 18 | 31 | 51 | 67 | 105 |
| TFN, TFFN | 18 | 28 | 47 | 79 | 128 | 169 | 265 |
| | 16 | 21 | 36 | 60 | 98 | 129 | 202 |
| PF, PFF, PGF, PGFF, PAF, PTF, PTFF, PAFF | 18 | 26 | 45 | 74 | 122 | 160 | 251 |
| | 16 | 20 | 34 | 58 | 94 | 124 | 194 |
| | 14 | 15 | 26 | 43 | 70 | 93 | 146 |
| HF, HFF, ZF, ZFF, ZHF | 18 | 34 | 58 | 96 | 157 | 206 | 324 |
| | 16 | 25 | 42 | 71 | 116 | 152 | 239 |
| | 14 | 18 | 31 | 52 | 85 | 112 | 175 |
| KF-2, KFF-2 | 18 | 49 | 84 | 140 | 228 | 300 | 470 |
| | 16 | 35 | 59 | 98 | 160 | 211 | 331 |
| | 14 | 24 | 40 | 67 | 110 | 145 | 228 |
| | 12 | 16 | 28 | 46 | 76 | 100 | 157 |
| | 10 | 11 | 18 | 31 | 51 | 67 | 105 |
| KF-1, KFF-1 | 18 | 59 | 100 | 167 | 272 | 357 | 561 |
| | 16 | 41 | 70 | 117 | 191 | 251 | 394 |
| | 14 | 28 | 47 | 79 | 128 | 169 | 265 |
| | 12 | 18 | 31 | 52 | 85 | 112 | 175 |
| | 10 | 12 | 20 | 34 | 55 | 73 | 115 |
| XF, XFF | 12 | 6 | 10 | 16 | 27 | 35 | 56 |
| | 10 | 4 | 8 | 13 | 21 | 28 | 44 |

Note: This table is for concentric stranded conductors only. For compact stranded conductors, Table C11(A) should be used.

*Types RHH, RHW, and RWH-2 without outer covering.

**Table C11(A) Maximum Number of Compact Conductors in Type A, Rigid PVC Conduit** *(Based on Table 1, Chapter 9)*

### COMPACT CONDUCTORS

| Type | Conductor Size (AWG/kcmil) | 16 (½) | 21 (¾) | 27 (1) | 35 (1¼) | 41 (1½) | 53 (2) | 63 (2½) | 78 (3) | 91 (3½) | 103 (4) |
|---|---|---|---|---|---|---|---|---|---|---|---|
| THW, | 8 | 3 | 5 | 8 | 14 | 18 | 28 | 42 | 64 | 84 | 107 |
| THW-2, | 6 | 2 | 4 | 6 | 10 | 14 | 22 | 33 | 49 | 65 | 83 |
| THHW | 4 | 1 | 3 | 5 | 8 | 10 | 16 | 24 | 37 | 48 | 62 |
| | 2 | 1 | 1 | 3 | 6 | 7 | 12 | 18 | 27 | 36 | 46 |
| | 1 | 1 | 1 | 2 | 4 | 5 | 8 | 13 | 19 | 25 | 32 |
| | 1/0 | 1 | 1 | 1 | 3 | 4 | 7 | 11 | 16 | 21 | 28 |
| | 2/0 | 1 | 1 | 1 | 3 | 4 | 6 | 9 | 14 | 18 | 23 |
| | 3/0 | 0 | 1 | 1 | 2 | 3 | 5 | 8 | 12 | 15 | 20 |
| | 4/0 | 0 | 1 | 1 | 1 | 3 | 4 | 6 | 10 | 13 | 17 |
| | 250 | 0 | 1 | 1 | 1 | 1 | 3 | 5 | 8 | 10 | 13 |
| | 300 | 0 | 0 | 1 | 1 | 1 | 3 | 4 | 7 | 9 | 11 |
| | 350 | 0 | 0 | 1 | 1 | 1 | 2 | 4 | 6 | 8 | 10 |
| | 400 | 0 | 0 | 1 | 1 | 1 | 2 | 3 | 5 | 7 | 9 |
| | 500 | 0 | 0 | 1 | 1 | 1 | 1 | 3 | 4 | 6 | 8 |
| | 600 | 0 | 0 | 0 | 1 | 1 | 1 | 2 | 3 | 5 | 6 |
| | 700 | 0 | 0 | 0 | 1 | 1 | 1 | 1 | 3 | 4 | 5 |
| | 750 | 0 | 0 | 0 | 1 | 1 | 1 | 1 | 3 | 4 | 5 |
| | 1000 | 0 | 0 | 0 | 0 | 1 | 1 | 1 | 2 | 3 | 4 |
| THHN, | 8 | — | — | — | — | — | — | — | — | — | — |
| THWN, | 6 | 3 | 5 | 9 | 15 | 20 | 32 | 48 | 72 | 94 | 121 |
| THWN-2 | 4 | 1 | 3 | 6 | 9 | 12 | 20 | 30 | 45 | 58 | 75 |
| | 2 | 1 | 2 | 4 | 7 | 9 | 14 | 21 | 32 | 42 | 54 |
| | 1 | 1 | 1 | 3 | 5 | 7 | 10 | 16 | 24 | 31 | 40 |
| | 1/0 | 1 | 1 | 2 | 4 | 6 | 9 | 13 | 20 | 27 | 34 |
| | 2/0 | 1 | 1 | 1 | 3 | 5 | 7 | 11 | 17 | 22 | 28 |
| | 3/0 | 1 | 1 | 1 | 3 | 4 | 6 | 9 | 14 | 18 | 24 |
| | 4/0 | 0 | 1 | 1 | 2 | 3 | 5 | 8 | 11 | 15 | 19 |
| | 250 | 0 | 1 | 1 | 1 | 2 | 4 | 6 | 9 | 12 | 15 |
| | 300 | 0 | 1 | 1 | 1 | 1 | 3 | 5 | 8 | 10 | 13 |
| | 350 | 0 | 0 | 1 | 1 | 1 | 3 | 4 | 7 | 9 | 11 |
| | 400 | 0 | 0 | 1 | 1 | 1 | 2 | 4 | 6 | 8 | 10 |
| | 500 | 0 | 0 | 1 | 1 | 1 | 2 | 3 | 5 | 7 | 9 |
| | 600 | 0 | 0 | 0 | 1 | 1 | 1 | 3 | 4 | 5 | 7 |
| | 700 | 0 | 0 | 0 | 1 | 1 | 1 | 2 | 3 | 5 | 6 |
| | 750 | 0 | 0 | 0 | 1 | 1 | 1 | 2 | 3 | 4 | 6 |
| | 1000 | 0 | 0 | 0 | 0 | 1 | 1 | 1 | 2 | 3 | 4 |
| XHHW, | 8 | 4 | 6 | 11 | 18 | 23 | 37 | 55 | 83 | 108 | 139 |
| XHHW-2 | 6 | 3 | 5 | 8 | 13 | 17 | 27 | 41 | 62 | 80 | 103 |
| | 4 | 1 | 3 | 6 | 9 | 12 | 20 | 30 | 45 | 58 | 75 |
| | 2 | 1 | 2 | 4 | 7 | 9 | 14 | 21 | 32 | 42 | 54 |
| | 1 | 1 | 1 | 3 | 5 | 7 | 10 | 16 | 24 | 31 | 40 |
| | 1/0 | 1 | 1 | 2 | 4 | 6 | 9 | 13 | 20 | 27 | 34 |
| | 2/0 | 1 | 1 | 1 | 3 | 5 | 7 | 11 | 17 | 22 | 29 |
| | 3/0 | 1 | 1 | 1 | 3 | 4 | 6 | 9 | 14 | 18 | 24 |
| | 4/0 | 0 | 1 | 1 | 2 | 3 | 5 | 8 | 12 | 15 | 20 |
| | 250 | 0 | 1 | 1 | 1 | 2 | 4 | 6 | 9 | 12 | 16 |
| | 300 | 0 | 1 | 1 | 1 | 1 | 3 | 5 | 8 | 10 | 13 |
| | 350 | 0 | 0 | 1 | 1 | 1 | 3 | 5 | 7 | 9 | 12 |
| | 400 | 0 | 0 | 1 | 1 | 1 | 3 | 4 | 6 | 8 | 11 |
| | 500 | 0 | 0 | 1 | 1 | 1 | 2 | 3 | 5 | 7 | 9 |
| | 600 | 0 | 0 | 0 | 1 | 1 | 1 | 3 | 4 | 5 | 7 |
| | 700 | 0 | 0 | 0 | 1 | 1 | 1 | 2 | 3 | 5 | 6 |
| | 750 | 0 | 0 | 0 | 1 | 1 | 1 | 2 | 3 | 4 | 6 |
| | 1000 | 0 | 0 | 0 | 0 | 1 | 1 | 1 | 2 | 3 | 4 |

Definition: *Compact stranding* is the result of a manufacturing process where the standard conductor is compressed to the extent that the interstices (voids between strand wires) are virtually eliminated.

**Table C12 Maximum Number of Conductors in Type EB, PVC Conduit** (*Based on Table 1, Chapter 9*)

| Type | Conductor Size (AWG/kcmil) | 53 (2) | 78 (3) | 91 (3½) | 103 (4) | 129 (5) | 155 (6) |
|---|---|---|---|---|---|---|---|
| RHH, RHW, RHW-2 | 14 | 53 | 119 | 155 | 197 | 303 | 430 |
| | 12 | 44 | 98 | 128 | 163 | 251 | 357 |
| RHH, RHW, RHW-2 | 10 | 35 | 79 | 104 | 132 | 203 | 288 |
| | 8 | 18 | 41 | 54 | 69 | 106 | 151 |
| | 6 | 15 | 33 | 43 | 55 | 85 | 121 |
| | 4 | 11 | 26 | 34 | 43 | 66 | 94 |
| | 3 | 10 | 23 | 30 | 38 | 58 | 83 |
| | 2 | 9 | 20 | 26 | 33 | 50 | 72 |
| | 1 | 6 | 13 | 17 | 21 | 33 | 47 |
| | 1/0 | 5 | 11 | 15 | 19 | 29 | 41 |
| | 2/0 | 4 | 10 | 13 | 16 | 25 | 36 |
| | 3/0 | 4 | 8 | 11 | 14 | 22 | 31 |
| | 4/0 | 3 | 7 | 9 | 12 | 18 | 26 |
| | 250 | 2 | 5 | 7 | 9 | 14 | 20 |
| | 300 | 1 | 5 | 6 | 8 | 12 | 17 |
| | 350 | 1 | 4 | 5 | 7 | 11 | 16 |
| | 400 | 1 | 4 | 5 | 6 | 10 | 14 |
| | 500 | 1 | 3 | 4 | 5 | 9 | 12 |
| | 600 | 1 | 3 | 3 | 4 | 7 | 10 |
| | 700 | 1 | 2 | 3 | 4 | 6 | 9 |
| | 750 | 1 | 2 | 3 | 4 | 6 | 9 |
| | 800 | 1 | 2 | 3 | 4 | 6 | 8 |
| | 900 | 1 | 1 | 2 | 3 | 5 | 7 |
| | 1000 | 1 | 1 | 2 | 3 | 5 | 7 |
| | 1250 | 1 | 1 | 1 | 2 | 3 | 5 |
| | 1500 | 0 | 1 | 1 | 1 | 3 | 4 |
| | 1750 | 0 | 1 | 1 | 1 | 3 | 4 |
| | 2000 | 0 | 1 | 1 | 1 | 2 | 3 |
| TW, THHW, THW, THW-2 | 14 | 111 | 250 | 327 | 415 | 638 | 907 |
| | 12 | 85 | 192 | 251 | 319 | 490 | 696 |
| | 10 | 63 | 143 | 187 | 238 | 365 | 519 |
| | 8 | 35 | 79 | 104 | 132 | 203 | 288 |
| RHH*,RHW*, RHW-2* | 14 | 74 | 166 | 217 | 276 | 424 | 603 |
| RHH*, RHW*, RHW-2* | 12 | 59 | 134 | 175 | 222 | 341 | 485 |
| RHH*, RHW*, RHW-2* | 10 | 46 | 104 | 136 | 173 | 266 | 378 |
| RHH*, RHW*, RHW-2* | 8 | 28 | 62 | 81 | 104 | 159 | 227 |
| RHH*, RHW*, RHW-2*, TW, THW, THHW, THW-2 | 6 | 21 | 48 | 62 | 79 | 122 | 173 |
| | 4 | 16 | 36 | 46 | 59 | 91 | 129 |
| | 3 | 13 | 30 | 40 | 51 | 78 | 111 |
| | 2 | 11 | 26 | 34 | 43 | 66 | 94 |
| | 1 | 8 | 18 | 24 | 30 | 46 | 66 |
| | 1/0 | 7 | 15 | 20 | 26 | 40 | 56 |
| | 2/0 | 6 | 13 | 17 | 22 | 34 | 48 |
| | 3/0 | 5 | 11 | 14 | 18 | 28 | 40 |
| | 4/0 | 4 | 9 | 12 | 15 | 24 | 34 |
| | 250 | 3 | 7 | 10 | 12 | 19 | 27 |
| | 300 | 3 | 6 | 8 | 11 | 17 | 24 |
| | 350 | 2 | 6 | 7 | 9 | 15 | 21 |
| | 400 | 2 | 5 | 7 | 8 | 13 | 19 |
| | 500 | 1 | 4 | 5 | 7 | 11 | 16 |
| | 600 | 1 | 3 | 4 | 6 | 9 | 13 |
| | 700 | 1 | 3 | 4 | 5 | 8 | 11 |
| | 750 | 1 | 3 | 4 | 5 | 7 | 11 |
| | 800 | 1 | 3 | 3 | 4 | 7 | 10 |
| | 900 | 1 | 2 | 3 | 4 | 6 | 9 |
| | 1000 | 1 | 2 | 3 | 4 | 6 | 8 |
| | 1250 | 1 | 1 | 2 | 3 | 4 | 6 |
| | 1500 | 1 | 1 | 1 | 2 | 4 | 6 |
| | 1750 | 1 | 1 | 1 | 2 | 3 | 5 |
| | 2000 | 0 | 1 | 1 | 1 | 3 | 4 |

**Table C12 Continued**

| Type | Conductor Size (AWG/kcmil) | 53 (2) | 78 (3) | 91 (3½) | 103 (4) | 129 (5) | 155 (6) |
|---|---|---|---|---|---|---|---|
| THHN, THWN, THWN-2 | 14 | 159 | 359 | 468 | 595 | 915 | 1300 |
| | 12 | 116 | 262 | 342 | 434 | 667 | 948 |
| | 10 | 73 | 165 | 215 | 274 | 420 | 597 |
| | 8 | 42 | 95 | 124 | 158 | 242 | 344 |
| | 6 | 30 | 68 | 89 | 114 | 175 | 248 |
| | 4 | 19 | 42 | 55 | 70 | 107 | 153 |
| | 3 | 16 | 36 | 46 | 59 | 91 | 129 |
| | 2 | 13 | 30 | 39 | 50 | 76 | 109 |
| | 1 | 10 | 22 | 29 | 37 | 57 | 80 |
| | 1/0 | 8 | 18 | 24 | 31 | 48 | 68 |
| | 2/0 | 7 | 15 | 20 | 26 | 40 | 56 |
| | 3/0 | 5 | 13 | 17 | 21 | 33 | 47 |
| | 4/0 | 4 | 10 | 14 | 18 | 27 | 39 |
| | 250 | 4 | 8 | 11 | 14 | 22 | 31 |
| | 300 | 3 | 7 | 10 | 12 | 19 | 27 |
| | 350 | 3 | 6 | 8 | 11 | 17 | 24 |
| | 400 | 2 | 6 | 7 | 10 | 15 | 21 |
| | 500 | 1 | 5 | 6 | 8 | 12 | 18 |
| | 600 | 1 | 4 | 5 | 6 | 10 | 14 |
| | 700 | 1 | 3 | 4 | 6 | 9 | 12 |
| | 750 | 1 | 3 | 4 | 5 | 8 | 12 |
| | 800 | 1 | 3 | 4 | 5 | 8 | 11 |
| | 900 | 1 | 3 | 3 | 4 | 7 | 10 |
| | 1000 | 1 | 2 | 3 | 4 | 6 | 9 |
| FEP, FEPB, PFA, PFAH, TFE | 14 | 155 | 348 | 454 | 578 | 888 | 1261 |
| | 12 | 113 | 254 | 332 | 422 | 648 | 920 |
| | 10 | 81 | 182 | 238 | 302 | 465 | 660 |
| | 8 | 46 | 104 | 136 | 173 | 266 | 378 |
| | 6 | 33 | 74 | 97 | 123 | 189 | 269 |
| | 4 | 23 | 52 | 68 | 86 | 132 | 188 |
| | 3 | 19 | 43 | 56 | 72 | 110 | 157 |
| | 2 | 16 | 36 | 46 | 59 | 91 | 129 |
| PFA, PFAH, TFE | 1 | 11 | 25 | 32 | 41 | 63 | 90 |
| PFA, PFAH, TFE, Z | 1/0 | 9 | 20 | 27 | 34 | 53 | 75 |
| | 2/0 | 7 | 17 | 22 | 28 | 43 | 62 |
| | 3/0 | 6 | 14 | 18 | 23 | 36 | 51 |
| | 4/0 | 5 | 11 | 15 | 19 | 29 | 42 |
| Z | 14 | 186 | 419 | 547 | 696 | 1069 | 1519 |
| | 12 | 132 | 297 | 388 | 494 | 759 | 1078 |
| | 10 | 81 | 182 | 238 | 302 | 465 | 660 |
| | 8 | 51 | 115 | 150 | 191 | 294 | 417 |
| | 6 | 36 | 81 | 105 | 134 | 206 | 293 |
| | 4 | 24 | 55 | 72 | 92 | 142 | 201 |
| | 3 | 18 | 40 | 53 | 67 | 104 | 147 |
| | 2 | 15 | 34 | 44 | 56 | 86 | 122 |
| | 1 | 12 | 27 | 36 | 45 | 70 | 99 |
| XHH, XHHW, XHHW-2, ZW | 14 | 111 | 250 | 327 | 415 | 638 | 907 |
| | 12 | 85 | 192 | 251 | 238 | 490 | 696 |
| | 10 | 63 | 143 | 187 | 238 | 365 | 519 |
| | 8 | 35 | 79 | 104 | 132 | 203 | 288 |
| | 6 | 26 | 59 | 77 | 98 | 150 | 213 |
| | 4 | 19 | 42 | 56 | 71 | 109 | 155 |
| | 3 | 16 | 36 | 47 | 60 | 92 | 131 |
| | 2 | 13 | 30 | 39 | 50 | 77 | 110 |
| XHH, XHHW, XHHW-2 | 1 | 10 | 22 | 29 | 37 | 58 | 82 |
| | 1/0 | 8 | 19 | 25 | 31 | 48 | 69 |
| | 2/0 | 7 | 16 | 20 | 26 | 40 | 57 |
| | 3/0 | 6 | 13 | 17 | 22 | 33 | 47 |
| | 4/0 | 5 | 11 | 14 | 18 | 27 | 39 |
| | 250 | 4 | 9 | 11 | 15 | 22 | 32 |
| | 300 | 3 | 7 | 10 | 12 | 19 | 28 |
| | 350 | 3 | 6 | 9 | 11 | 17 | 24 |
| | 400 | 2 | 6 | 8 | 10 | 15 | 22 |
| | 500 | 1 | 5 | 6 | 8 | 12 | 18 |
| | 600 | 1 | 4 | 5 | 6 | 10 | 14 |
| | 700 | 1 | 3 | 4 | 6 | 9 | 12 |
| | 750 | 1 | 3 | 4 | 5 | 8 | 12 |
| | 800 | 1 | 3 | 4 | 5 | 8 | 11 |
| | 900 | 1 | 3 | 3 | 4 | 7 | 10 |
| | 1000 | 1 | 2 | 3 | 4 | 6 | 9 |
| | 1250 | 1 | 1 | 2 | 3 | 5 | 7 |
| | 1500 | 1 | 1 | 1 | 3 | 4 | 6 |
| | 1750 | 1 | 1 | 1 | 2 | 4 | 5 |
| | 2000 | 0 | 1 | 1 | 1 | 3 | 5 |

Note: This table is for concentric stranded conductors only. For compact stranded conductors, Table C12(A) should be used.

*Types RHH, RHW, and RHW-2 without outer covering.

**Table C12(A)  Maximum Number of Compact Conductors in Type EB, PVC Conduit**
(*Based on Table 1, Chapter 9*)

| Type | Conductor Size (AWG/kcmil) | 53 (2) | 78 (3) | 91 (3½) | 103 (4) | 129 (5) | 155 (6) |
|------|------|------|------|------|------|------|------|
| THW, | 8 | 30 | 68 | 89 | 113 | 174 | 247 |
| THW-2, | 6 | 23 | 52 | 69 | 87 | 134 | 191 |
| THHW | 4 | 17 | 39 | 51 | 65 | 100 | 143 |
|  | 2 | 13 | 29 | 38 | 48 | 74 | 105 |
|  | 1 | 9 | 20 | 26 | 34 | 52 | 74 |
|  | 1/0 | 8 | 17 | 23 | 29 | 45 | 64 |
|  | 2/0 | 6 | 15 | 19 | 24 | 38 | 54 |
|  | 3/0 | 5 | 12 | 16 | 21 | 32 | 46 |
|  | 4/0 | 4 | 10 | 14 | 17 | 27 | 38 |
|  | 250 | 3 | 8 | 11 | 14 | 21 | 30 |
|  | 300 | 3 | 7 | 9 | 12 | 19 | 26 |
|  | 350 | 3 | 6 | 8 | 11 | 17 | 24 |
|  | 400 | 2 | 6 | 7 | 10 | 15 | 21 |
|  | 500 | 1 | 5 | 6 | 8 | 12 | 18 |
|  | 600 | 1 | 4 | 5 | 6 | 10 | 14 |
|  | 700 | 1 | 3 | 4 | 6 | 9 | 13 |
|  | 750 | 1 | 3 | 4 | 5 | 8 | 12 |
|  | 1000 | 1 | 2 | 3 | 4 | 7 | 9 |
| THHN, | 8 | — | — | — | — | — | — |
| THWN, | 6 | 34 | 77 | 100 | 128 | 196 | 279 |
| THWN-2 | 4 | 21 | 47 | 62 | 79 | 121 | 172 |
|  | 2 | 15 | 34 | 44 | 57 | 87 | 124 |
|  | 1 | 11 | 25 | 33 | 42 | 65 | 93 |
|  | 1/0 | 9 | 22 | 28 | 36 | 56 | 79 |
|  | 2/0 | 8 | 18 | 23 | 30 | 46 | 65 |
|  | 3/0 | 6 | 15 | 20 | 25 | 38 | 55 |
|  | 4/0 | 5 | 12 | 16 | 20 | 32 | 45 |
|  | 250 | 4 | 10 | 13 | 16 | 25 | 35 |
|  | 300 | 4 | 8 | 11 | 14 | 22 | 31 |
|  | 350 | 3 | 7 | 9 | 12 | 19 | 27 |
|  | 400 | 3 | 6 | 8 | 11 | 17 | 24 |
|  | 500 | 2 | 5 | 7 | 9 | 14 | 20 |
|  | 600 | 1 | 4 | 6 | 7 | 11 | 16 |
|  | 700 | 1 | 4 | 5 | 6 | 10 | 14 |
|  | 750 | 1 | 4 | 5 | 6 | 9 | 14 |
|  | 1000 | 1 | 3 | 3 | 4 | 7 | 10 |

**Table C12(A)**  *Continued*

COMPACT CONDUCTORS

| Type | Conductor Size (AWG/kcmil) | 53 (2) | 78 (3) | 91 (3½) | 103 (4) | 129 (5) | 155 (6) |
|------|------|------|------|------|------|------|------|
| XHHW, | 8 | 39 | 88 | 115 | 146 | 225 | 320 |
| XHHW-2 | 6 | 29 | 65 | 85 | 109 | 167 | 238 |
|  | 4 | 21 | 47 | 62 | 79 | 121 | 172 |
|  | 2 | 15 | 34 | 44 | 57 | 87 | 124 |
|  | 1 | 11 | 25 | 33 | 42 | 65 | 93 |
|  | 1/0 | 9 | 22 | 28 | 36 | 56 | 79 |
|  | 2/0 | 8 | 18 | 24 | 30 | 47 | 67 |
|  | 3/0 | 6 | 15 | 20 | 25 | 38 | 55 |
|  | 4/0 | 5 | 12 | 16 | 21 | 32 | 46 |
|  | 250 | 4 | 10 | 13 | 17 | 26 | 37 |
|  | 300 | 4 | 8 | 11 | 14 | 22 | 31 |
|  | 350 | 3 | 7 | 10 | 12 | 19 | 28 |
|  | 400 | 3 | 7 | 9 | 11 | 17 | 25 |
|  | 500 | 2 | 5 | 7 | 9 | 14 | 20 |
|  | 600 | 1 | 4 | 6 | 7 | 11 | 16 |
|  | 700 | 1 | 4 | 5 | 6 | 10 | 14 |
|  | 750 | 1 | 3 | 5 | 6 | 9 | 13 |
|  | 1000 | 1 | 3 | 4 | 5 | 7 | 10 |

Definition: *Compact stranding* is the result of a manufacturing process where the standard conductor is compressed to the extent that the interstices (voids between strand wires) are virtually eliminated.

# Annex D   Examples

*This annex is not a part of the recommendations of this NFPA document but is included for informational purposes only.*

**Selection of Conductors.** In the following examples, the results are generally expressed in amperes (A). To select conductor sizes, refer to the 0 through 2000 volt (V) ampacity tables of Article 310 and the rules of 310.15 that pertain to these tables.

**Voltage.** For uniform application of Articles 210, 215, and 220, a nominal voltage of 120, 120/240, 240, and 208Y/120 V is used in computing the ampere load on the conductor.

**Fractions of an Ampere.** Except where the computations result in a major fraction of an ampere (0.5 or larger), such fractions are permitted to be dropped.

**Power Factor.** Calculations in the following examples are based, for convenience, on the assumption that all loads have the same power factor (PF).

**Ranges.** For the computation of the range loads in these examples, Column C of Table 220.19 has been used. For optional methods, see Columns A and B of Table 220.19. Except where the computations result in a major fraction of a kilowatt (0.5 or larger), such fractions are permitted to be dropped.

**SI Units.** For metric conversions, $0.093 \text{ m}^2 = 1 \text{ ft}^2$ and $0.3048 \text{ m} = 1 \text{ ft}$.

### Example D1(a) One-Family Dwelling

The dwelling has a floor area of 1500 ft², exclusive of an unfinished cellar not adaptable for future use, unfinished attic, and open porches. Appliances are a 12-kW range and a 5.5-kW, 240-V dryer. Assume range and dryer kW ratings equivalent to kVA ratings in accordance with 220.18 and 220.19.

**Computed Load** *[see 220.10]*

General Lighting Load: 1500 ft² at 3 VA per ft² = 4500 VA

**Minimum Number of Branch Circuits Required** *[see 210.11(A)]*

General Lighting Load: 4500 VA ÷ 120 V = 37.5 A
  This requires three 15-A, 2-wire or two 20-A, 2-wire circuits.
Small Appliance Load: Two 2-wire, 20-A circuits *[see 210.11(C)(1)]*
Laundry Load: One 2-wire, 20-A circuit *[see 210.11(C)(2)]*
Bathroom Branch Circuit: One 2-wire, 20-A circuit (no additional load calculation is required for this circuit) *[see 210.11(C)(3)]*

**Minimum Size Feeder Required** *[see 220.10]*

| | | |
|---|---|---:|
| General Lighting | | 4500 VA |
| Small Appliance | | 3000 VA |
| Laundry | | 1500 VA |
| | Total | 9000 VA |
| 3000 VA at 100% | | 3000 VA |
| 9000 VA – 3000 VA = 6000 VA at 35% | | 2100 VA |
| | Net Load | 5100 VA |
| Range *(see Table 220.19)* | | 8000 VA |
| Dryer Load *(see Table 220.18)* | | 5500 VA |
| | Net Computed Load | 18,600 VA |

**Net Computed Load for 120/240-V, 3-wire, single-phase service or feeder**

$$18,600 \text{ VA} \div 240 \text{ V} = 77.5 \text{ A}$$

Sections 230.42(B) and 230.79 require service conductors and disconnecting means rated not less than 100 amperes.

### Calculation for Neutral for Feeder and Service

| | | |
|---|---|---:|
| Lighting and Small Appliance Load | | 5100 VA |
| Range: 8000 VA at 70% *(see 220.22)* | | 5600 VA |
| Dryer: 5500 VA at 70% *(see 220.22)* | | 3850 VA |
| | Total | 14,550 VA |

**Computed Load for Neutral**

$$14,550 \text{ VA} \div 240 \text{ V} = 60.6 \text{ A}$$

### Example D1(b) One-Family Dwelling

Assume same conditions as Example No. D1(a), plus addition of one 6-A, 230-V, room air-conditioning unit and one 12-A, 115-V, room air-conditioning unit,* one 8-A, 115-V, rated waste disposer, and one 10-A, 120-V, rated dishwasher. See Article 430 for general motors and Article 440, Part VII, for air-conditioning equipment. Motors have nameplate ratings of 115 V and 230 V for use on 120-V and 240-V nominal voltage systems.

*(For feeder neutral, use larger of the two appliances for unbalance.)
  From Example D1(a), feeder current is 78 A (3-wire, 240 V).

| | Line A | Neutral | Line B |
|---|---|---|---|
| Amperes from Example D1(a) | 78 | 61 | 78 |
| One 230-V air conditioner | 6 | — | 6 |
| One 115-V air conditioner and 120-V dishwasher | 12 | 12 | 10 |
| One 115-V disposer | — | 8 | 8 |
| 25% of largest motor *(see 430.24)* | 3 | 3 | 2 |
| Total amperes per line | 99 | 84 | 104 |

Therefore, the service would be rated 110 A.

### Example D2(a) Optional Calculation for One-Family Dwelling, Heating Larger Than Air Conditioning *[see 220.30]*

The dwelling has a floor area of 1500 ft², exclusive of an unfinished cellar not adaptable for future use, unfinished attic, and open porches. It has a 12-kW range, a 2.5-kW water heater, a 1.2-kW dishwasher, 9 kW of electric space heating installed in five rooms, a 5-kW clothes dryer, and a 6-A, 230-V, room air-conditioning unit. Assume range, water heater, dishwasher, space heating, and clothes dryer kW ratings equivalent to kVA.

**Air Conditioner kVA Calculation**

$$6 \text{ A} \times 230 \text{ V} \div 1000 = 1.38 \text{ kVA}$$

This 1.38 kVA [item 1 from 220.30(C)] is less than 40% of 9 kVA of separately controlled electric heat [item 6 from 220.30(C)], so the 1.38 kVA need not be included in the service calculation.

**General Load**

| | | |
|---|---|---:|
| 1500 ft² at 3 VA | | 4500 VA |
| Two 20-A appliance outlet circuits at 1500 VA each | | 3000 VA |
| Laundry circuit | | 1500 VA |
| Range (at nameplate rating) | | 12,000 VA |
| Water heater | | 2500 VA |
| Dishwasher | | 1200 VA |
| Clothes dryer | | 5000 VA |
| | Total | 29,700 VA |

**Application of Demand Factor** *[see 220.30(B)]*

| | | |
|---|---|---:|
| First 10 kVA of general load at 100% | | 10,000 VA |
| Remainder of general load at 40% (19.7 kVA × 0.4) | | 7880 VA |
| | Total of general load | 17,880 VA |
| 9 kVA of heat at 40% (9000 VA × 0.4) = | | 3600 VA |
| | Total | 21,480 VA |

## Calculated Load for Service Size

$$21.48 \text{ kVA} = 21,480 \text{ VA}$$

$$21,480 \text{ VA} \div 240 \text{ V} = 89.5 \text{ A}$$

Therefore, the minimum service rating would be 100 A in accordance with 230.42 and 230.79.

## Feeder Neutral Load, per 220.22

| | | |
|---|---|---|
| 1500 ft² at 3 VA | | 4500 VA |
| Three 20-A circuits at 1500 VA | | 4500 VA |
| | Total | 9000 VA |
| 3000 VA at 100% | | 3000 VA |
| 9000 VA − 3000 VA = 6000 | | |
| VA at 35% | | 2100 VA |
| | Subtotal | 5100 VA |
| Range: 8 kVA at 70% | | 5600 VA |
| Clothes dryer: 5 kVA at 70% | | 3500 VA |
| Dishwasher | | 1200 VA |
| | Total | 15,400 VA |

## Calculated Load for Neutral

$$15,400 \text{ VA} \div 240 \text{ V} = 64.2 \text{ A}$$

### Example D2(b) Optional Calculation for One-Family Dwelling, Air Conditioning Larger Than Heating
*[see 220.30(A) and 220.30(C)]*

The dwelling has a floor area of 1500 ft², exclusive of an unfinished cellar not adaptable for future use, unfinished attic, and open porches. It has two 20-A small appliance circuits, one 20-A laundry circuit, two 4-kW wall-mounted ovens, one 5.1-kW counter-mounted cooking unit, a 4.5-kW water heater, a 1.2-kW dishwasher, a 5-kW combination clothes washer and dryer, six 7-A, 230-V room air-conditioning units, and a 1.5-kW permanently installed bathroom space heater. Assume wall-mounted ovens, counter-mounted cooking unit, water heater, dishwasher, and combination clothes washer and dryer kW ratings equivalent to kVA.

## Air Conditioning kVA Calculation

$$\text{Total amperes} = 6 \text{ units} \times 7 \text{ A} = 42 \text{ A}$$

$$42 \text{ A} \times 240 \text{ V} \div 1000 = 10.08 \text{ kVA (assume PF} = 1.0)$$

## Load Included at 100%

Air Conditioning: Included below *[see item 1 in 220.30(C)]*
Space Heater: Omit *[see item 5 in 220.30(C)]*

## General Load

| | | |
|---|---|---|
| 1500 ft² at 3 VA | | 4500 VA |
| Two 20-A small appliance | | |
| circuits at 1500 VA each | | 3000 VA |
| Laundry circuit | | 1500 VA |
| Two ovens | | 8000 VA |
| One cooking unit | | 5100 VA |
| Water heater | | 4500 VA |
| Dishwasher | | 1200 VA |
| Washer/dryer | | 5000 VA |
| | Total general load | 32,800 VA |
| First 10 kVA at 100% | | 10,000 VA |
| Remainder at 40% | | |
| (22.8 kVA × 0.4 × 1000) | | 9120 VA |
| | Subtotal general load | 19,120 VA |
| Air conditioning | | 10,080 VA |
| | Total | 29,200 VA |

## Calculated Load for Service

$$29,200 \text{ VA} \div 240 \text{ V} = 122 \text{ A (service rating)}$$

## Feeder Neutral Load, per 220.22.

Assume that the two 4-kVA wall-mounted ovens are supplied by one branch circuit, the 5.1-kVA counter-mounted cooking unit by a separate circuit.

| | | |
|---|---|---|
| 1500 ft² at 3 VA | | 4500 VA |
| Three 20-A circuits at 1500 VA | | 4500 VA |
| | Subtotal | 9000 VA |
| 3000 VA at 100% | | 3000 VA |
| 9000 VA − 3000 VA = 6000 VA at 35% | | 2100 VA |
| | Subtotal | 5100 VA |

Two 4-kVA ovens plus one 5.1-kVA cooking unit = 13.1 kVA. Table 220.19 permits 55% demand factor or 13.1 kVA × 0.55 = 7.2 kVA feeder capacity.

| | | |
|---|---|---|
| Subtotal from above | | 5100 VA |
| Ovens and cooking unit: 7200 VA × 70% for neutral load | | 5040 VA |
| Clothes washer/dryer: 5 kVA × 70% for neutral load | | 3500 VA |
| Dishwasher | | 1200 VA |
| | Total | 14,840 VA |

## Calculated Load for Neutral

$$14,840 \text{ VA} \div 240 \text{ V} = 61.83 \text{ A (use 62 A)}$$

### Example D2(c) Optional Calculation for One-Family Dwelling with Heat Pump (Single-Phase, 240/120-Volt Service)
*(see 220.30)*

The dwelling has a floor area of 2000 ft², exclusive of an unfinished cellar not adaptable for future use, unfinished attic, and open porches. It has a 12-kW range, a 4.5-kW water heater, a 1.2-kW dishwasher, a 5-kW clothes dryer, and a 2½-ton (24-A) heat pump with 15 kW of backup heat.

## Heat Pump kVA Calculation

$$24 \text{ A} \times 240 \text{ V} \div 1000 = 5.76 \text{ kVA}$$

This 5.76 kVA is less than 15 kVA of the backup heat; therefore, the heat pump load need not be included in the service calculation *[see 220.30(C)]*.

## General Load

| | | |
|---|---|---|
| 2000 ft² at 3 VA | | 6000 VA |
| Two 20-A appliance outlet circuits at 1500 VA each | | 3000 VA |
| Laundry circuit | | 1500 VA |
| Range (at nameplate rating) | | 12,000 VA |
| Water heater | | 4500 VA |
| Dishwasher | | 1200 VA |
| Clothes dryer | | 5000 VA |
| | Subtotal general load | 33,200 VA |
| First 10 kVA at 100% | | 10,000 VA |
| Remainder of general load at 40% (23,200 VA × 0.4) | | 9280 VA |
| | Total net general load | 19,280 VA |

## Heat Pump and Supplementary Heat*

$$240 \text{ V} \times 24 \text{ A} = 5760 \text{ VA}$$

## 15-kW Electric Heat:

$$5760 \text{ VA} + 15,000 \text{ VA} = 20,760 \text{ VA or} = 20.76 \text{ kVA}$$

$$20.76 \text{ kVA} \times 100\% = 20.76 \text{ kVA}$$

*If supplementary heat is not on at same time as heat pump, heat pump kVA need not be added to total.

## Totals

| | | |
|---|---|---|
| Net general load | | 19,280 VA |
| Heat pump and supplementary heat | | 20,760 VA |
| | Total | 40,040 VA |

## Calculated Load for Service

$$40.04 \text{ kVA} \times 1000 \div 240 \text{ V} = 166.8 \text{ A}$$

Therefore, this dwelling unit would be permitted to be served by a 175-A service.

## Example D3 Store Building

A store 50 ft by 60 ft, or 3000 ft$^2$, has 30 ft of show window. There are a total of 80 duplex receptacles. The service is 120/240 V, single phase 3-wire service. Actual connected lighting load is 8500 VA.

### Computed Load *(see 220.10)*

**Noncontinuous Loads**

| Receptacle Load *(see 220.13)* | |
|---|---|
| 80 receptacles at 180 VA | 14,400 VA |
| 10,000 VA at 100% | 10,000 VA |
| 14,400 VA − 10,000 VA = 4400 at 50% | 2,200 VA |
| Subtotal | 12,200 VA |

**Continuous Loads**

| General Lighting* | |
|---|---|
| 3000 ft$^2$ at 3 VA per ft$^2$ | 9000 VA |
| Show Window Lighting Load | |
| 30 ft at 200 VA per ft | 6000 VA |
| Outside Sign Circuit *[see 220.3(B)(6)]* | 1200 VA |
| Subtotal | 16,200 VA |
| Subtotal from noncontinuous | 12,200 VA |
| Total noncontinuous loads + continuous loads = | 28,400 VA |

*In the example, 125% of the actual connected lighting load (8500 VA × 1.25 = 10,625 VA) is less than 125% of the load from Table 220.3(A), so the minimum lighting load from Table 220.3(A) is used in the calculation. Had the actual lighting load been greater than the value computed from Table 220.3(A), 125% of the actual connected lighting load would have been used.

### Minimum Number of Branch Circuits Required

General Lighting: Branch circuits need only be installed to supply the actual connected load *[see 210.11(B)]*.

$$8500 \text{ VA} \times 1.25 = 10,625 \text{ VA}$$

$$10,625 \text{ VA} \div 240 \text{ V} = 44 \text{ A for 3-wire, 120/240 V}$$

The lighting load would be permitted to be served by 2-wire or 3-wire, 15- or 20-A circuits with combined capacity equal to 44 A or greater for 3-wire circuits or 88 A or greater for 2-wire circuits. The feeder capacity as well as the number of branch-circuit positions available for lighting circuits in the panelboard must reflect the full calculated load of 9000 VA × 1.25 = 11,250 VA.

### Show Window

$$6000 \text{ VA} \times 1.25 = 7500 \text{ VA}$$

$$7500 \text{ VA} \div 240 \text{ V} = 31 \text{ A for 3-wire, 120/240 V}$$

The show window lighting is permitted to be served by 2-wire or 3-wire circuits with a capacity equal to 31 A or greater for 3-wire circuits or 62 A or greater for 2-wire circuits.

Receptacles required by 210.62 are assumed to be included in the receptacle load above if these receptacles do not supply the show window lighting load.

### Receptacles

Receptacle Load: 14,400 VA ÷ 240 V = 60 A for 3-wire, 120/240 V

The receptacle load would be permitted to be served by 2-wire or 3-wire circuits with a capacity equal to 60 A or greater for 3-wire circuits or 120 A or greater for 2-wire circuits.

### Minimum Size Feeder (or Service) Overcurrent Protection
*[see 215.3 or 230.90]*

| Subtotal noncontinuous loads | 12,200 VA |
|---|---|
| Subtotal continuous load at 125% (16,200 VA × 1.25) | 20,250 VA |
| Total | 32,450 VA |

$$32,450 \text{ VA} \div 240 \text{ V} = 135 \text{ A}$$

The next higher standard size is 150 A *(see 240.6)*.

### Minimum Size Feeders (or Service Conductors) Required
*[see 215.2, 230.42(A)]*

For 120/240-V, 3-wire system,

$$32,450 \text{ VA} \div 240 \text{ V} = 135 \text{ A}$$

Service or feeder conductor is 1/0 Cu per 215.3 and Table 310.16 (with 75°C terminations).

## Example D4(a) Multifamily Dwelling

A multifamily dwelling has 40 dwelling units.

Meters are in two banks of 20 each with individual feeders to each dwelling unit.

One-half of the dwelling units are equipped with electric ranges not exceeding 12 kW each. Assume range kW rating equivalent to kVA rating in accordance with 220.19. Other half of ranges are gas ranges.

Area of each dwelling unit is 840 ft$^2$.

Laundry facilities on premises are available to all tenants. Add no circuit to individual dwelling unit.

### Computed Load for Each Dwelling Unit *(see Article 220)*

General Lighting: 840 ft$^2$ at 3 VA per ft$^2$ = 2520 VA

Special Appliance: Electric range *(see 220.19)* = 8000 VA

### Minimum Number of Branch Circuits Required for Each Dwelling Unit *[see 210.11(A)]*

General Lighting Load: 2520 VA ÷ 120 V = 21 A or two 15-A, 2-wire circuits; or two 20-A, 2-wire circuits

Small Appliance Load: Two 2-wire circuits of 12 AWG wire *[see 210.11(C)(1)]*

Range Circuit: 8000 VA ÷ 240 V = 33 A or a circuit of two 8 AWG conductors and one 10 AWG conductor as permitted by 220.22 *[see 210.19(C)]*

### Minimum Size Feeder Required for Each Dwelling Unit *(see 215.2)*

| Computed Load *(see Article 220)*: | |
|---|---|
| General Lighting | 2520 VA |
| Small Appliance (two 20-ampere circuits) | 3000 VA |
| Subtotal Computed Load (without ranges) | 5520 VA |

**Application of Demand Factor** *(see Table 220.11)*

| First 3000 VA at 100% | 3000 VA |
|---|---|
| 5520 VA − 3000 VA = 2520 VA at 35% | 882 VA |
| Net Computed Load (without ranges) | 3882 VA |
| Range Load | 8000 VA |
| Net Computed Load (with ranges) | 11,882 VA |

### Size of Each Feeder *(see 215.2)*

For 120/240-V, 3-wire system (without ranges)

Net computed load of 3882 VA ÷ 240 V = 16.2 A

For 120/240-V, 3-wire system (with ranges)

Net computed load, 11,882 VA ÷ 240 V = 49.5 A

### Feeder Neutral

| Lighting and Small Appliance Load | 3882 VA |
|---|---|
| Range Load: 8000 VA at 70% *(see 220.22)* | 5600 VA |
| (only for apartments with electric range) | 5600 VA |
| Net Computed Load (neutral) | 9482 VA |

### Calculated Load for Neutral

$$9482 \text{ VA} \div 240 \text{ V} = 39.5 \text{ A}$$

### Minimum Size Feeders Required from Service Equipment to Meter Bank (For 20 Dwelling Units — 10 with Ranges)

| Total Computed Load: | |
|---|---|
| Lighting and Small Appliance 20 units × 5520 VA | 110,400 VA |
| Application of Demand Factor First 3000 VA at 100% | 3000 VA |
| 110,400 VA − 3000 VA = 107,400 VA at 35% | 37,590 VA |

| | |
|---|---|
| Net Computed Load | 40,590 VA |

Range Load: 10 ranges (less than 12 kVA)
(*see Col. C, Table 220.19*) — 25,000 VA
Net Computed Load (with ranges) — 65,590 VA

Net computed load for 120/240-V, 3-wire system,

$$65{,}590 \text{ VA} \div 240 \text{ V} = 273 \text{ A}$$

**Feeder Neutral**

| | |
|---|---|
| Lighting and Small Appliance Load | 40,590 VA |

Range Load: 25,000 VA at 70%
(*see 220.22*) — 17,500 VA
Computed Load (neutral) — 58,090 VA

**Calculated Load for Neutral**

$$58{,}090 \text{ VA} \div 240 \text{ V} = 242 \text{ A}$$

**Further Demand Factor** (*220.22*)

| | |
|---|---|
| 200 A at 100% | 200 A |
| 242 A − 200 A = 42 A at 70% | 29 A |
| Net Computed Load (neutral) | 229 A |

**Minimum Size Main Feeders (or Service Conductors) Required (Less House Load)** (For 40 Dwelling Units — 20 with Ranges)

Total Computed Load:
Lighting and Small Appliance Load
40 units × 5520 VA — 220,800 VA

**Application of Demand Factor** (*from Table 220.11*)

| | |
|---|---|
| First 3000 VA at 100% | 3000 VA |
| Next 120,000 VA − 3000 VA = 117,000 VA at 35% | 40,950 VA |
| Remainder 220,800 VA − 120,000 VA = 100,800 VA at 25% | 25,200 VA |
| Net Computed Load | 69,150 VA |

Range Load: 20 ranges (less than 12 kVA)
(*see Col. C, Table 220.19*) — 35,000 VA
Net Computed Load — 104,150 VA

For 120/240-V, 3-wire system

Net computed load of 104,150 VA ÷ 240 V = 434 A

**Feeder Neutral**

| | |
|---|---|
| Lighting and Small Appliance Load | 69,150 VA |
| Range: 35,000 VA at 70% | 24,500 VA |

(*see 220.22*)
Computed Load (neutral) — 93,650 VA

$$93{,}650 \text{ VA} \div 240 \text{ V} = 390 \text{ A}$$

**Further Demand Factor** (*see 220.22*)

| | |
|---|---|
| 200 A at 100% | 200 A |
| 390 A − 200 A = 190 A at 70% | 133 A |
| Net Computed Load (neutral) | 333 A |

[*See Tables 310.16 through 310.21, and 310.15(B)(2) and (B)(4).*]

### Example D4(b) Optional Calculation for Multifamily Dwelling

A multifamily dwelling equipped with electric cooking and space heating or air conditioning and has 40 dwelling units.

Meters are in two banks of 20 each plus house metering and individual feeders to each dwelling unit.

Each dwelling unit is equipped with an electric range of 8-kW nameplate rating, four 1.5-kW separately controlled 240-V electric space heaters, and a 2.5-kW, 240-V electric water heater. Assume range, space heater, and water heater kW ratings equivalent to kVA.

A common laundry facility is available to all tenants [*see 210.52(F), Exception No. 1*].

Area of each dwelling unit is 840 ft$^2$.

**Computed Load for Each Dwelling Unit** (*see Article 220*)

General Lighting Load:
| | |
|---|---|
| 840 ft$^2$ at 3 VA per ft$^2$ | 2520 VA |
| Electric range | 8000 VA |
| Electric heat: 6 kVA (or air conditioning if larger) | 6000 VA |
| Electric water heater | 2500 VA |

**Minimum Number of Branch Circuits Required for Each Dwelling Unit**

General Lighting Load: 2520 VA ÷ 120 V = 21 A or two 15-A, 2-wire circuits, or two 20-A, 2-wire circuits
Small Appliance Load: Two 2-wire circuits of 12 AWG [*see 210.11(C)(1)*]
Range Circuit: 8000 VA × 80% ÷ 240 V = 27 A on a circuit of three 10 AWG conductors as permitted in Column B of Table 220.19
Space Heating: 6000 VA ÷ 240 V = 25 A
Number of circuits (*see 210.11*)

**Minimum Size Feeder Required for Each Dwelling Unit** (*see 215.2*)

Computed Load (*see Article 220*):
| | |
|---|---|
| General Lighting | 2520 VA |
| Small Appliance (two 20-A circuits) | 3000 VA |
| Subtotal Computed Load (without range and space heating) | 5520 VA |

**Application of Demand Factor**

| | |
|---|---|
| First 3000 VA at 100% | 3000 VA |
| 5520 VA − 3000 VA = 2520 VA at 35% | 882 VA |
| Net Computed Load (without range and space heating) | 3882 VA |
| Range | 6400 VA |
| Space Heating (*see 220.15*) | 6000 VA |
| Water Heater | 2500 VA |
| Net Computed Load (for individual dwelling unit) | 18,782 VA |

**Size of Each Feeder**
For 120/240-V, 3-wire system,
Net computed load of 18,782 VA ÷ 240 V = 78 A

**Feeder Neutral** (*see 220.22*)

| | |
|---|---|
| Lighting and Small Appliance | 3882 VA |
| Range Load: 6400 VA at 70% (*see 220.22*) | 4480 VA |
| Space and Water Heating (no neutral): 240 V | 0 VA |
| Net Computed Load (neutral) | 8362 VA |

**Calculated Load for Neutral**

$$8362 \text{ VA} \div 240 \text{ V} = 35 \text{ A}$$

**Minimum Size Feeder Required from Service Equipment to Meter Bank** (For 20 Dwelling Units)

Total Computed Load:
| | |
|---|---|
| Lighting and Small Appliance Load 20 units × 5520 VA | 110,400 VA |
| Water and Space Heating Load 20 units × 8500 VA | 170,000 VA |
| Range Load: 20 × 8000 VA | 160,000 VA |
| Net Computed Load (20 dwelling units) | 440,400 VA |
| Net Computed Load Using Optional Calculation (*see Table 220.32*) 440,400 VA × 0.38 | 167,352 VA |

$$167{,}352 \text{ VA} \div 240 \text{ V} = 697 \text{ A}$$

**Minimum Size Main Feeder Required (Less House Load)** (For 40 Dwelling Units)

Computed Load:
| | |
|---|---|
| Lighting and Small Appliance Load 40 units × 5520 VA | 220,800 VA |

Water and Space Heating Load 340,000 VA
40 units × 8500 VA
Range: 40 ranges × 8000 VA 320,000 VA

Net Computed Load (40 dwelling units) 880,800 VA

Net Computed Load Using Optional Calculation (*see Table 220.32*)

880,800 VA × 0.28 = 246,624 VA

246,624 VA ÷ 240 V = 1028 A

### Feeder Neutral Load for Feeder from Service Equipment to Meter Bank (For 20 Dwelling Units)

Lighting and Small Appliance Load
20 units × 5520 VA 110,400 VA
First 3000 VA at 100% 3000 VA
110,400 VA – 3000 VA = 107,400 VA 37,590 VA
  at 35%
            Net Computed Load 40,590 VA
20 ranges: 35,000 VA at 70% (*see Table* 24,500 VA
*220.19 and 220.22*)
           Total 65,090 VA

65,090 VA ÷ 240 V = 271 A

### Further Demand Factor (*see 220.22*)

First 200 A at 100% 200 A
Balance: 271 A – 200 A = 71 A at 70% 50 A
          Total 250 amperes

### Feeder Neutral Load of Main Feeder (Less House Load)
(For 40 Dwelling Units)

Lighting and Small Appliance Load
40 units × 5520 VA 220,800 VA
First 3000 VA at 100% 3000 VA
Next 120,000 VA – 3000 VA = 117,000 VA 40,950 VA
  at 35%
Remainder 220,800 VA – 120,000 VA =
  100,800 VA at 25% 25,200 VA
          Net Computed Load 69,150 VA
40 ranges: 55,000 VA at 70% 38,500 VA
  (*see Table 220.19 and 220.22*)
          Total 107,650 VA

107,650 VA ÷ 240 V = 449 A

### Further Demand Factor (*see 220.22*)

First 200 A at 100% 200 A
Balance: 449 – 200 A = 249 A at 70% 174 A
          Total 374 A

### Example D5(a) Multifamily Dwelling Served at 208Y/120 Volts, Three Phase

All conditions and calculations are the same as for the multifamily dwelling [Example D4(a)] served at 120/240 V, single phase except as follows:
    Service to each dwelling unit would be two phase legs and neutral.

### Minimum Number of Branch Circuits Required for Each Dwelling Unit (*see 210.11*)

Range Circuit: 8000 VA ÷ 208 V = 38 A or a circuit of two 8 AWG conductors and one 10 AWG conductor as permitted by 210.19, Column B

### Minimum Size Feeder Required for Each Dwelling Unit (*see 215.2*)

For 120/208-V, 3-wire system (without ranges),
Net computed load of 3882 VA ÷ 2 legs ÷ 120 V/leg = 16.2 A
For 120/208-V, 3-wire system (with ranges),
  Net computed load (range) of 8000 VA ÷ 208 V = 38.5 A
  Total load (range + lighting) = 38.5 A + 16.2 A = 54.7 A
Feeder neutral: (range) of 8000 VA × 70% = 5600 VA ÷ 208 V = 26.9 A
Total load: (range + lighting) = 26.9 A + 16.2 A = 43.1 A

### Minimum Size Feeders Required from Service Equipment to Meter Bank (For 20 Dwelling Units — 10 with Ranges)

For 208Y/120-V, 3-phase, 4-wire system,

Ranges: Maximum number between any two phase legs = 4
  2 × 4 = 8.
  Table 220.19 demand = 23,000 VA
  Per phase demand = 23,000 VA ÷ 2 = 11,500 VA
  Equivalent 3-phase load = 34,500 VA
Net Computed Load (total):

40,590 VA + 34,500 VA = 75,090 VA

75,090 VA ÷ (208 V)(1.732) = 208.4 A

### Feeder Neutral Size:
Net Computed Lighting and Appliance Load & Equivalent Range Load:

40,590 VA + (34,500 VA × 70%) = 64,740 VA

Net Computed Neutral Load:

64,700 VA ÷ (208 V)(1.732) = 179.7 A

### Minimum Size Main Feeder (Less House Load)
(For 40 Dwelling Units — 20 with Ranges)

For 208Y/120-V, 3-phase, 4-wire system,

Ranges:
  Maximum number between any two phase legs = 7
  2 × 7 = 14.
  Table 220.19 demand = 29,000 VA
  Per phase demand = 29,000 VA ÷ 2 = 14,500 VA
  Equivalent 3-phase load = 43,500 VA
Net Computed Load (total):

69,150 VA + 43,500 VA = 112,650 VA

112,650 VA ÷ (208 V)(1.732) = 312.7 A

Main Feeder Neutral Size:

69,150 VA + (43,500 VA at 70%) = 99,600 VA

99,600 VA ÷ (208 V)(1.732) = 276.5 A

### Further Demand Factor (*see 220.22*)

200 A at 100% 200.0 A
276.5 A – 200 A = 76.5 A at 70% 53.6 A
    Net Computed Load (neutral) 253.6 A

### Example D5(b) Optional Calculation for Multifamily Dwelling Served at 208Y/120 Volts, Three Phase

All conditions and calculations are the same as for Optional Calculation for the Multifamily Dwelling [Example D4(b)] served at 120/240 V, single phase except as follows:
    Service to each dwelling unit would be two phase legs and neutral.

### Minimum Number of Branch Circuits Required for Each Dwelling Unit (*see 210.11*)

Range Circuit: 8000 VA at 80% ÷ 208 V = 30.7 A or a circuit of two 8 AWG conductors and one 10 AWG conductor as permitted by 210.19(B)
Space Heating: 6000 VA ÷ 208 V = 28.8 A
Two 20-ampere, 2-pole circuits required, 12 AWG conductors

### Minimum Size Feeder Required for Each Dwelling Unit

120/208-V, 3-wire circuit

Net computed load of 18,782 VA ÷ 208 V = 90.3 A
Net computed load (lighting line to neutral):

3882 VA ÷ 2 legs ÷ 120 V per leg = 16.2 amperes

Line to line = 14,900 VA ÷ 208 V = 71.6 A
Total load = 16.2 A + 71.6 A = 87.8 A

### Minimum Size Feeder Required for Service Equipment to Meter Bank (For 20 Dwelling Units)

### Net Computed Load

167,352 VA ÷ (208 V)(1.732) = 464.9 A

### Feeder Neutral Load:

65,080 VA ÷ (208 V)(1.732) = 180.65 A

### Minimum Size Main Feeder Required (Less House Load)
(For 40 Dwelling Units)

**Net computed load:**

$$246,624 \text{ VA} \div (208 \text{ V})(1.732) = 684.6 \text{ A}$$

**Main Feeder Neutral Load:**

$$107,650 \text{ VA} \div (208 \text{ V})(1.732) = 298.8 \text{ A}$$

**Further Demand Factor** (see 220.22)

| | |
|---|---|
| 200 A at 100% | 200.0 A |
| 298.8 A − 200 A = 98.8 A at 70% | 69.2 A |
| Net Computed Load (neutral) | 269.2 A |

### Example D6 Maximum Demand for Range Loads

Table 220.19, Column C applies to ranges not over 12 kW. The application of Note 1 to ranges over 12 kW (and not over 27 kW) and Note 2 to ranges over 8¾ kW (and not over 27 kW) is illustrated in the following two examples.

**A. Ranges All the Same Rating** (see Table 220.19, Note 1)
Assume 24 ranges, each rated 16 kW.

From Table 220.19, Column C, the maximum demand for 24 ranges of 12-kW rating is 39 kW. 16 kW exceeds 12 kW by 4.
5% × 4 = 20% (5% increase for each kW in excess of 12)
39 kW × 20% = 7.8 kW increase
39 + 7.8 = 46.8 kW (value to be used in selection of feeders)

**B. Ranges of Unequal Rating** (see Table 220.19, Note 2)
Assume 5 ranges, each rated 11 kW; 2 ranges, each rated 12 kW; 20 ranges, each rated 13.5 kW; 3 ranges, each rated 18 kW.

| | | |
|---|---|---|
| 5 ranges × 12 kW | = | 60 kW (use 12 kW for range rated less than 12) |
| 2 ranges × 12 kW | = | 24 kW |
| 20 ranges × 13.5 kW | = | 270 kW |
| 3 ranges × 18 kW | = | 54 kW |
| 30 ranges   Total kW | | 408 kW |

408 ÷ 30 ranges = 13.6 kW (average to be used for computation)
From Table 220.19, Column C, the demand for 30 ranges of 12-kW rating is 15 kW + 30 (1 kW × 30 ranges) = 45 kW. 13.6 kW exceeds 12 kW by 1.6 kW (use 2 kW).

$$5\% \times 2 = 10\% \text{ (5\% increase for each kW in excess of 12 kW)}$$

$$45 \text{ kW} \times 10\% = 4.5 \text{ kW increase}$$

$$45 \text{ kW} + 4.5 \text{ kW} = 49.5 \text{ kW (value to be used in selection of feeders)}$$

### Example D8 Motor Circuit Conductors, Overload Protection, and Short-Circuit and Ground-Fault Protection (see 240.6, 430.6, 430.22, 430.23, 430.24, 430.32, 430.52, and 430.62, Tables 430.52 and 430.150)

Determine the minimum required conductor ampacity, the motor overload protection, the branch-circuit short-circuit and ground-fault protection, and the feeder protection, for three induction-type motors on a 480-V, 3-phase feeder, as follows:

(a) One 25-hp, 460-V, 3-phase, squirrel-cage motor, nameplate full-load current 32 A, Design B, Service Factor 1.15
(b) Two 30-hp, 460-V, 3-phase, wound-rotor motors, nameplate primary full-load current 38 A, nameplate secondary full-load current 65 A, 40°C rise.

**Conductor Ampacity**
The full-load current value used to determine the minimum required conductor ampacity is obtained from Table 430.150[see 430.6(A)] for the squirrel-cage motor and the primary of the wound-rotor motors. To obtain the minimum required conductor ampacity, the full-load current is multiplied by 1.25 [see 430.22 and 430.23(A)].
For the 25-hp motor,

$$34 \text{ A} \times 1.25 = 42.5 \text{ A}$$

For the 30-horsepower motors,

$$40 \text{ A} \times 1.25 = 50 \text{ A}$$

$$65 \text{ A} \times 1.25 = 81.25 \text{ A}$$

**Motor Overload Protection**
Where protected by a separate overload device, the motors are required to have overload protection rated or set to trip at not more than 125% of the nameplate full-load current [see 430.6(A) and 430.32(A)(1) ].
For the 25-hp motor,

$$32 \text{ A} \times 1.25 = 40.0 \text{ A}$$

For the 30-hp motors,

$$38 \text{ A} \times 1.25 = 47.5 \text{ A}$$

Where the separate overload device is an overload relay (not a fuse or circuit breaker), and the overload device selected at 125% is not sufficient to start the motor or carry the load, the trip setting is permitted to be increased in accordance with 430.32(C).

**Branch-Circuit Short-Circuit and Ground-Fault Protection**
The selection of the rating of the protective device depends on the type of protective device selected, in accordance with 430.52 and Table 430.52. The following is for the 25-hp motor.

(a) Nontime-Delay Fuse: The fuse rating is 300% × 34 A = 102 A. The next larger standard fuse is 110 A [see 240.6and 430.52(C)(1), Exception No. 1]. If the motor will not start with a 110-A nontime-delay fuse, the fuse rating is permitted to be increased to 125 A because this rating does not exceed 400% [see 430.52(C)(1), Exception No. 2(a)].
(b) Time-Delay Fuse:The fuse rating is 175% × 34 A = 59.5 A. The next larger standard fuse is 60 A [see 240.6 and 430.52(C)(1), Exception No. 1]. If the motor will not start with a 60-A time-delay fuse, the fuse rating is permitted to be increased to 70 A because this rating does not exceed 225% [see 430.52(C)(1), Exception No. 2(b)].

**Feeder Short-Circuit and Ground-Fault Protection**
The rating of the feeder protective device is based on the sum of the largest branch-circuit protective device (example is 110 A) plus the sum of the full-load currents of the other motors, or 110 A + 40 A + 40 A = 190 A. The nearest standard fuse that does not exceed this value is 175 A [see 240.6 and 430.62(A)].

### Example D9 Feeder Ampacity Determination for Generator Field Control [see 215.2, 430.24, 430.24 Exception No. 1, 620.13, 620.14, 620.61, and Tables 430.22(E) and 620.14]

Determine the conductor ampacity for a 460-V 3-phase, 60-Hz ac feeder supplying a group of six elevators. The 460-V ac drive motor nameplate rating of the largest MG set for one elevator is 40 hp and 52 A, and the remaining elevators each have a 30-hp, 40-A, ac drive motor rating for their MG sets. In addition to a motor controller, each elevator has a separate motion/operation controller rated 10 A continuous to operate microprocessors, relays, power supplies, and the elevator car door operator. The MG sets are rated continuous.

**Conductor Ampacity.** Conductor ampacity is determined as follows:

(a) Per 620.13(D) and 620.61(B)(1), use Table 430.22(B), for intermittent duty (elevators). For intermittent duty using a continuous rated motor, the percentage of nameplate current rating to be used is 140%.
(b) For the 30-hp ac drive motor,

$$140\% \times 40 \text{ A} = 56 \text{ A}.$$

(c) For the 40-hp ac drive motors,

$$140\% \times 52 \text{ A} = 73 \text{ A}.$$

(d) The total conductor ampacity is the sum of all the motor currents:

$$(1 \text{ motor} \times 73 \text{ A}) + (5 \text{ motors} \times 56 \text{ A}) = 353 \text{ A}$$

(e) Per 620.14 and Table 620.14, the conductor (feeder) ampacity would be permitted to be reduced by the use of a demand factor. Constant loads are not included (see 620.14, FPN). For six elevators, the demand factor is 0.79. The feeder diverse ampacity is, therefore, 0.79 × 353 A = 279 A.
(f) Per 430.24 and 215.3, the controller continuous current is 125% × 10 A = 12.5 A
(g) The total feeder ampacity is the sum of the diverse current and all the controller continuous current.

$$I_{\text{total}} = 279 \text{ A} + (6 \text{ elevators} \times 12.5 \text{ A}) = 354 \text{ A}$$

(h) This ampacity would be permitted to be used to select the wire size.

See Figure D9.

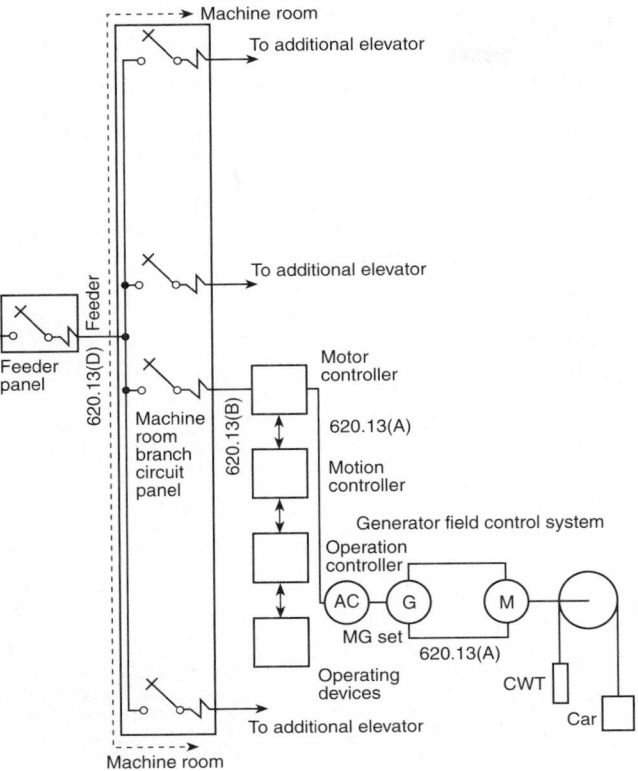

**Figure D9  Generator field control.**

**Example D10 Feeder Ampacity Determination for Adjustable Speed Drive Control** *[see 215.2, 430.24, 430.24, 620.13, 620.14, 620.61, and Tables 430.22(B), and 620.14]*

Determine the conductor ampacity for a 460-V, 3-phase, 60-Hz ac feeder supplying a group of six identical elevators. The system is adjustable-speed SCR dc drive. The power transformers are external to the drive (motor controller) cabinet. Each elevator has a separate motion/operation controller connected to the load side of the main line disconnect switch rated 10 A continuous to operate microprocessors, relays, power supplies, and the elevator car door operator. Each transformer is rated 95 kVA with an efficiency of 90%.

**Conductor Ampacity.**

**Conductor ampacity is determined as follows:**

(a) Calculate the nameplate rating of the transformer:

$$I = \frac{95\,\text{kVA} \times 1000}{\sqrt{3} \times 460\,\text{V} \times 0.90_{\text{eff.}}} = 133\,\text{A}$$

(b) Per 620.13(D), for six elevators, the total conductor ampacity is the sum of all the currents.

$$6\ \text{elevators} \times 133\ \text{A} = 798\ \text{A}$$

(c) Per 620.14 and Table 620.14, the conductor (feeder) ampacity would be permitted to be reduced by the use of a demand factor. Constant loads are not included *(see 620.13, FPN No. 2)*. For six elevators, the

demand factor is 0.79. The feeder diverse ampacity is, therefore, 0.79 × 798 A = 630 A.

(d) Per 430.24 and 215.3, the controller continuous current is 125% × 10 A = 12.5 A.

(e) The total feeder ampacity is the sum of the diverse current and all the controller constant current.

$$I_{\text{total}} = 630\ \text{A} + (6\ \text{elevators} \times 12.5\ \text{A}) = 705\ \text{A}$$

(f) This ampacity would be permitted to be used to select the wire size. See Figure D10.

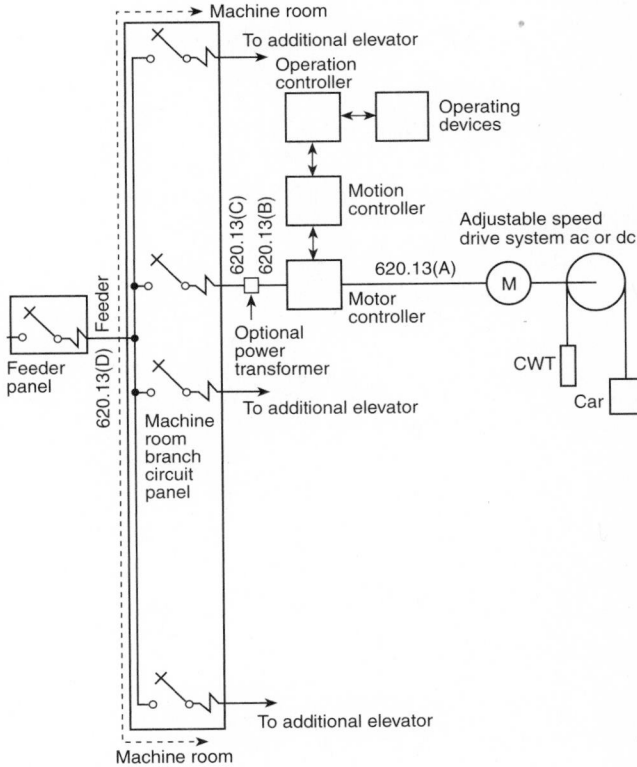

**Figure D10  Adjustable speed drive control.**

**Example D11 Mobile Home**

A mobile home floor is 70 ft by 10 ft and has two small appliance circuits; a 1000-VA, 240-V heater; a 200-VA, 120-V exhaust fan; a 400-VA, 120-V dishwasher; and a 7000-VA electric range.

**Lighting and Small Appliance Load**

| | |
|---|---:|
| Lighting (70 ft × 10 ft × 3 VA per ft²) | 2100 VA |
| Small appliance (1500 VA × 2 circuits) | 3000 VA |
| Laundry (1500 VA × 1 circuit) | 1500 VA |
| Subtotal | 6600 VA |
| First 3000 VA at 100% | 3000 VA |
| Remainder (6600 VA – 3000 VA = 3600 VA ) × 35% | 1260 VA |
| Total | 4260 VA |

4260 VA ÷ 240 V = 17.75 A per leg

## Amperes per Leg

|  | Leg A | Leg B |
|---|---|---|
| Lighting and appliances | 17.75 | 17.75 |
| Heater (1000 VA ÷ 240 V) | 4.20 | 4.20 |
| Fan (200 VA × 125% ÷ 120 V) | 2.08 | — |
| Dishwasher (400 VA ÷ 120 V) | — | 3.30 |
| Range (7000 VA × 0.8 ÷ 240 V) | 23.30 | 23.30 |
| Total amperes per leg | 47.33 | 48.55 |

Based on the higher current calculated for either leg, a minimum 50-A supply cord would be required.

For SI units, 0.093 m$^2$ = 1 ft$^2$ and 0.3048 m = 1 ft.

### Example D12 Park Trailer

A park trailer floor is 40 ft by 10 ft and has two small appliance circuits, a 1000-VA, 240-V heater, a 200-VA, 120-V exhaust fan, a 400-VA, 120-V dishwasher, and a 7000-VA electric range.

## Lighting and Small Appliance Load

| | | |
|---|---|---|
| Lighting (40 ft × 10 ft × 3 VA per ft$^2$) | | 1200 VA |
| Small appliance (1500 VA × 2 circuits) | | 3000 VA |
| Laundry (1500 VA × 1 circuit) | | 1500 VA |
| | Subtotal | 5700 VA |
| First 3000 VA at 100% | | 3000 VA |
| Remainder (5700 VA – 3000 VA = 2700 VA) × 35% | | 945 VA |
| | Total | 3945 VA |

3945 VA ÷ 240 V = 16.44 A per leg

## Amperes per Leg

|  | Leg A | Leg B |
|---|---|---|
| Lighting and appliances | 16.44 | 16.44 |
| Heater (1000 VA ÷ 240 V) | 4.20 | 4.20 |
| Fan (200 VA × 125% ÷ 120 V) | 2.08 | — |
| Dishwasher (400 VA ÷ 120 V) | — | 3.3 |
| Range (7000 VA × 0.8 ÷ 240 V) | 23.30 | 23.30 |
| Totals | 46.02 | 47.24 |

Based on the higher current calculated for either leg, a minimum 50-A supply cord would be required.

For SI units, 0.093 m$^2$ = 1 ft$^2$ and 0.3048 m = 1 ft.

# Annex E   Types of Construction

*This annex is not part of the requirements of this NFPA document but is included for informational purposes only.*

**Table E.1  Fire Resistance Ratings (in hours) for Type I through Type V Construction**

| | Type I | | Type II | | | Type III | | Type IV | Type V | |
|---|---|---|---|---|---|---|---|---|---|---|
| | 443 | 332 | 222 | 111 | 000 | 211 | 200 | 2HH | 111 | 000 |
| **Exterior Bearing Walls –** Supporting more than one floor, columns, or other bearing walls | 4 | 3 | 2 | 1 | $0^1$ | 2 | 2 | 2 | 1 | $0^1$ |
| Supporting one floor only | 4 | 3 | 2 | 1 | $0^1$ | 2 | 2 | 2 | 1 | $0^1$ |
| Supporting a roof only | 4 | 3 | 1 | 1 | $0^1$ | 2 | 2 | 2 | 1 | $0^1$ |
| **Interior Bearing Walls –** Supporting more than one floor, columns, or other bearing walls | 4 | 3 | 2 | 1 | 0 | 1 | 0 | 2 | 1 | 0 |
| Supporting one floor only | 3 | 2 | 2 | 1 | 0 | 1 | 0 | 1 | 1 | 0 |
| Supporting roofs only | 3 | 2 | 1 | 1 | 0 | 1 | 0 | 1 | 1 | 0 |
| **Columns –** Supporting more than one floor, columns, or other bearing walls | 4 | 3 | 2 | 1 | 0 | 1 | 0 | $H^2$ | 1 | 0 |
| Supporting one floor only | 3 | 2 | 2 | 1 | 0 | 1 | 0 | $H^2$ | 1 | 0 |
| Supporting roofs only | 3 | 2 | 1 | 1 | 0 | 1 | 0 | $H^2$ | 1 | 0 |
| **Beams, Girders, Trusses & Arches –** Supporting more than one floor, columns, or other bearing walls | 4 | 3 | 2 | 1 | 0 | 1 | 0 | $H^2$ | 1 | 0 |
| Supporting one floor only | 3 | 2 | 2 | 1 | 0 | 1 | 0 | $H^2$ | 1 | 0 |
| Supporting roofs only | 3 | 2 | 2 | 1 | 0 | 1 | 0 | $H^2$ | 1 | 0 |
| **Floor Construction** | 3 | 2 | 2 | 1 | 0 | 1 | 0 | $H^2$ | 1 | 0 |
| **Roof Construction** | 2 | 1½ | 1 | 1 | 0 | 1 | 0 | $H^2$ | 1 | 0 |
| **Exterior Nonbearing Walls³** | $0^1$ | $0^1$ | $0^1$ | $0^1$ | $0^1$ | $0^1$ | $0^1$ | $0^1$ | $0^1$ | $0^1$ |

▮ Those members that shall be permitted to be of approved combustible material.

Source: Table 3.1 from NFPA 220, *Standard on Building Construction,* 1999.

[1] See A-3-1 in NFPA 220.

[2] "H" indicates heavy timber members; see text for requirements.

[3] Exterior nonbearing walls meeting the conditions of acceptance of NFPA 285, *Standard Method of Test for the Evaluation of Flammability Characteristics of Exterior Non-Load-Bearing Wall Assemblies Containing Combustible Components Using the Intermediate-Scale, Multistory Test Apparatus,* shall be permitted to be used.

# Annex F   Cross-Reference Tables

*This annex is not a part of the requirements of this Code but is included for informational purposes only.*

**Table F.1 Chapter 3 Cross-Reference from the 2002 *NEC* to the 1999 *NEC***

| 2002 *NEC* | 1999 *NEC* | Article Title |
|---|---|---|
| 300 | 300 | Wiring Methods |
| 310 | 310 | Conductors for General Wiring |
| 312 | 373 | Cabinets, Cutout Boxes, and Meter Socket Enclosures |
| 314 | 370 | Outlet, Device, Pull, and Junction Boxes; Conduit Bodies; Fittings; and Manholes |
| 320 | 333 | Armored Cable: Type AC |
| 322 | 363 | Flat Cable Assemblies: Type FC |
| 324 | 328 | Flat Conductor Cable: Type FCC |
| 326 | 325 | Integrated Gas Spacer Cable: Type IGS |
| 328 | 326 | Medium Voltage Cable: Type MV |
| 330 | 334 | Metal-Clad Cable: Type MC |
| 332 | 330 | Mineral-Insulated, Metal-Sheathed Cable: Type MI |
| 334 | 336 | Nonmetallic-Sheathed Cable: Types NM, NMC, and NMS |
| 336 | 340 | Power and Control Tray Cable: Type TC |
| 338 | 338 | Service-Entrance Cable: Types SE and USE |
| 340 | 339 | Underground Feeder and Branch-Circuit Cable: Type UF |
| 342 | 345 | Intermediate Metal Conduit: Type IMC |
| 344 | 346 | Rigid Metal Conduit: Type RMC |
| 348 | 350 | Flexible Metal Conduit: Type FMC |
| 350 | 351 (Part A) | Liquidtight Flexible Metal Conduit: Type LFMC |
| 352 | 347 | Rigid Nonmetallic Conduit: Type RNC |
| 354 | 343 | Nonmetallic Underground Conduit with Conductors: Type NUCC |
| 356 | 351 (Part B) | Liquidtight Flexible Nonmetallic Conduit: Type LFNC |
| 358 | 348 | Electrical Metallic Tubing: Type EMT |
| 360 | 349 | Flexible Metallic Tubing: Type FMT |
| 362 | 331 | Electrical Nonmetallic Tubing: Type ENT |
| 366 | 374 | Auxiliary Gutters |
| 368 | 364 | Busways |
| 370 | 365 | Cablebus |
| 372 | 358 | Cellular Concrete Floor Raceways |
| 374 | 356 | Cellular Metal Floor Raceways |
| 376 | 362 (Part A) | Metal Wireways |
| 378 | 362 (Part B) | Nonmetallic Wireways |
| 380 | 353 | Multioutlet Assembly |
| 382 | 342 | Nonmetallic Extensions |
| 384 | 352 (Part C) | Strut-Type Channel Raceway |
| 386 | 352 (Part A) | Surface Metal Raceways |
| 388 | 352 (Part B) | Surface Nonmetallic Raceways |
| 390 | 354 | Underfloor Raceways |
| 392 | 318 | Cable Trays |
| 394 | 324 | Concealed Knob-and-Tube Wiring |
| 396 | 321 | Messenger Supported Wiring |
| 398 | 320 | Open Wiring on Insulators |
| 404 | 380 | Switches |
| 408 | 384 | Switchboards and Panelboards |
| 527 | 305 | Temporary Installations |

**Table F.2  Chapter 3 Cross-Reference from the 1999 *NEC* to the 2002 *NEC***

| 1999 *NEC* | 2002 *NEC* | Article Title |
|---|---|---|
| 300 | 300 | Wiring Methods |
| 305 | 527 | Temporary Installations |
| 310 | 310 | Conductors for General Wiring |
| 318 | 392 | Cable Trays |
| 320 | 398 | Open Wiring on Insulators |
| 321 | 396 | Messenger Supported Wiring |
| 324 | 394 | Concealed Knob-and-Tube Wiring |
| 325 | 326 | Integrated Gas Spacer Cable: Type IGS |
| 326 | 328 | Medium Voltage Cable: Type MV |
| 328 | 324 | Flat Conductor Cable: Type FCC |
| 330 | 332 | Mineral-Insulated, Metal-Sheathed Cable: Type MI |
| 331 | 362 | Electrical Nonmetallic Tubing: Type ENT |
| 333 | 320 | Armored Cable: Type AC |
| 334 | 330 | Metal-Clad Cable: Type MC |
| 336 | 334 | Nonmetallic-Sheathed Cable: Types NM, NMC, and NMS |
| 338 | 338 | Service-Entrance Cable: Types SE and USE |
| 339 | 340 | Underground Feeder and Branch-Circuit Cable: Type UF |
| 340 | 336 | Power and Control Tray Cable: Type TC |
| 342 | 382 | Nonmetallic Extensions |
| 343 | 354 | Nonmetallic Underground Conduit with Conductors: Type NUCC |
| 345 | 342 | Intermediate Metal Conduit: Type IMC |
| 346 | 344 | Rigid Metal Conduit: Type RMC |
| 347 | 352 | Rigid Nonmetallic Conduit: Type RNC |
| 348 | 358 | Electrical Metallic Tubing: Type EMT |
| 349 | 360 | Flexible Metallic Tubing: Type FMT |
| 350 | 348 | Flexible Metal Conduit: Type FMC |
| 351 (Part A) | 350 | Liquidtight Flexible Metal Conduit: Type LFMC |
| 351 (Part B) | 356 | Liquidtight Flexible Nonmetallic Conduit: Type LFNC |
| 352 (Part C) | 384 | Strut-Type Channel Raceway |
| 352 (Part A) | 386 | Surface Metal Raceways |
| 352 (Part B) | 388 | Surface Nonmetallic Raceways |
| 353 | 380 | Multioutlet Assembly |
| 354 | 390 | Underfloor Raceways |
| 356 | 374 | Cellular Metal Floor Raceways |
| 358 | 372 | Cellular Concrete Floor Raceways |
| 362 (Part A) | 376 | Metal Wireways |
| 362 (Part B) | 378 | Nonmetallic Wireways |
| 363 | 322 | Flat Cable Assemblies: Type FC |
| 364 | 368 | Busways |
| 365 | 370 | Cablebus |
| 370 | 314 | Outlet, Device, Pull, and Junction Boxes; Conduit Bodies; Fittings; and Manholes |
| 373 | 312 | Cabinets, Cutout Boxes, and Meter Socket Enclosures |
| 374 | 366 | Auxiliary Gutters |
| 380 | 404 | Switches |
| 384 | 408 | Switchboards and Panelboards |

**Table F.3 Chapter 3 Alphabetical Cross-Reference, 2002–1999 *NEC***

| Article Title | 2002 *NEC* | 1999 *NEC* |
|---|---|---|
| Armored Cable: Type AC | 320 | 333 |
| Auxiliary Gutters | 366 | 374 |
| Busways | 368 | 364 |
| Cabinets, Cutout Boxes, and Meter Socket Enclosures | 312 | 373 |
| Cable Trays | 392 | 318 |
| Cablebus | 370 | 365 |
| Cellular Concrete Floor Raceways | 372 | 358 |
| Cellular Metal Floor Raceways | 374 | 356 |
| Concealed Knob-and-Tube Wiring | 394 | 324 |
| Conductors for General Wiring | 310 | 310 |
| Electrical Metallic Tubing: Type EMT | 358 | 348 |
| Electrical Nonmetallic Tubing: Type ENT | 362 | 331 |
| Flat Cable Assemblies: Type FC | 322 | 363 |
| Flat Conductor Cable: Type FCC | 324 | 328 |
| Flexible Metal Conduit: Type FMC | 348 | 350 |
| Flexible Metallic Tubing: Type FMT | 360 | 349 |
| Integrated Gas Spacer Cable: Type IGS | 326 | 325 |
| Intermediate Metal Conduit: Type IMC | 342 | 345 |
| Liquidtight Flexible Metal Conduit: Type LFMC | 350 | 351 (Part A) |
| Liquidtight Flexible Nonmetallic Conduit: Type LFNC | 356 | 351 (Part B) |
| Medium Voltage Cable: Type MV | 328 | 326 |
| Messenger Supported Wiring | 396 | 321 |
| Metal Wireways | 376 | 362 (Part A) |
| Metal-Clad Cable: Type MC | 330 | 334 |
| Mineral-Insulated, Metal-Sheathed Cable: Type MI | 332 | 330 |
| Multioutlet Assembly | 380 | 353 |
| Nonmetallic Extensions | 382 | 342 |
| Nonmetallic Underground Conduit with Conductors: Type NUCC | 354 | 343 |
| Nonmetallic Wireways | 378 | 362 (Part B) |
| Nonmetallic-Sheathed Cable: Types NM, NMC, and NMS | 334 | 336 |
| Open Wiring on Insulators | 398 | 320 |
| Outlet, Device, Pull, and Junction Boxes; Conduit Bodies; Fittings; and Manholes | 314 | 370 |
| Power and Control Tray Cable: Type TC | 336 | 340 |
| Rigid Metal Conduit: Type RMC | 344 | 346 |
| Rigid Nonmetallic Conduit: Type RNC | 352 | 347 |
| Service-Entrance Cable: Types SE and USE | 338 | 338 |
| Strut-Type Channel Raceway | 384 | 352 (Part C) |
| Surface Metal Raceways | 386 | 352 (Part A) |
| Surface Nonmetallic Raceways | 388 | 352 (Part B) |
| Switchboards and Panelboards | 408 | 384 |
| Switches | 404 | 380 |
| Temporary Installations | 527 | 305 |
| Underfloor Raceways | 390 | 354 |
| Underground Feeder and Branch-Circuit Cable: Type UF | 340 | 339 |
| Wiring Methods | 300 | 300 |

# Index

## -A-

**AC and DC conductors in same enclosure,** 300.3(C)(1), 725.26

**AC armored cable.** *see* Armored cable (Type AC)

**AC resistance and reactance conversion,** Chap. 9, Table 9

**AC systems**
Conductor to be grounded, 250.26
Grounding connections, 250.24(A)
Grounding electrode conductor, 250.66
Grounding of, 250.20, 250.30
In same metallic enclosures, 215.4(B), 300.20
Sensitive electronic equipment, 647.3, 647.5

**Access and working space.** *see also* Working space
Elevator machine room, 620.71
Manholes, 314.55
Not over 600 volts, 110.26
Over 600 volts, 110-III
Portable substations, 530.62
Switchboards, 408.8
Transformers, electric signs, 600.21(D)
Vaults and tunnels, 314.56

**Accessible**
Air-conditioning and refrigeration disconnects, 440.14
Conduit bodies, junction, pull, and outlet boxes, 314.29
Definition, Art. 100-I
Grounding electrode connection, 250.68
Motor disconnects, 430.107
Overcurrent devices, 240.24
Sealing fittings, 501.5(C)(1), 502.5, 505.16(D)(1). *see also* Hazardous (classified) locations
Services, 230.70
Splices and taps in auxiliary gutters, 366.9(A)
Splices and taps in wireways, 376.56, 378.56
Transformers, signs, outlet lighting, 600.21(A)
Transformers and vaults, 450.13
Unfinished attics and roof spaces, knob-and-tube wiring, 394.23(A)

**AC-DC general-use snap switches**
Marking, 404.15
Motors, 430.83(C)
Panelboards, use in, 408.16(C)
Ratings, type loads, 408.14

**Administration and enforcement,** Art. 80
Adoption of Code, 80.5
Application of Code, 80.9
Authority, 80.13
Connection to electricity supply, 80.25
Definitions, 80.2
Effective date, 80.35
Electrical board, 80.15
Inspector's qualifications, 80.27
Liability for damages, 80.29
Notice of violations, penalties, 80.23
Occupancy of building or structure, 80.11
Permits and approvals, 80.19
Plans review, 80.21
Records and reports, 80.17
Repeal of conflicting acts, 80.33
Title of Code, 80.7
Validity of Art. 80, 80.31

**Aerial cables**
Identification, 200.6(A)
Installation, 820.10
Messenger supported wiring, Art. 396
Network-powered broadband communications systems-aerial cables, 830.11

**Agricultural buildings,** Art. 547
Bonding and equipotential plane, 547.9
Definitions, 547.2
Equipotential planes, bonding, 547.10
Luminaires (lighting fixtures), 547.8
Motors, 547.7
Service equipment, separately derived systems, feeders, disconnecting means, and grounding, 547.9
Surface temperatures, 547.4
Switches, receptacles, circuit breakers, controllers and fuses, 547.6
Wiring methods, 547.5

**Air conditioners, room,** 440-VII
Branch-circuit requirements, 440.62
Definition, 440.60
Disconnecting means, 440.63
Grounding, 440.61
Nameplate marking, 440.4

**Air conditioning and refrigerating equipment,** Art. 440
Branch circuit
Conductors, 440-IV

Ampacity, determination of, 440.6
Combination loads for, 440.34
General, 440.31
Multimotor equipment for, 440.35
Several motor-compressors for, 440.33
Single motor-compressor for, 440.32
Outlets, 210.63
Selection current, 440-I
Definition, 440.2
Marking on nameplate, 440.4(C)
Short-circuit and ground-fault protection, 440-III
Equipment for, 440.22(B): General, 440.21; Individual motor-compressor for, 440.22(A)
Controllers for, 440-V
Marking, 440.5
Rating, 440.41
Definition of hermetic refrigerant motor-compressor, 440.2
Definitions, 440.2, 550.2, 551.2
Disconnecting means, 440-II
Cord-connected as, 440.13
Ratings and interrupting capacity, 440.12
Nameplate requirements, 440.4
Overload protection, 440-VI
Application and selection of, 440.52
Branch-circuit conductors for, 440.52(B)
General, 440.51
Motor-compressors for, 440.52(A)
On 15- or 20-ampere branch circuits, 440.54
Rated-load current
Definition, 440.4
Marking on nameplate, 440.4(A)
Room air conditioners. *see* Air conditioners, room
Single machine, when considered, 440.8

**Air plenums.** *see* Plenums

**Aircraft hangars,** Art. 513
Aircraft batteries, 513.10
Aircraft electrical systems, 513.10(A)
Classification of locations, 513.3
Definition, 513.2
Equipment, 513.4, 513.7, 513.8, 513.10
External power sources, 513.16(C)

Grounding, 513.16
Locations, classification of, 513.3
Mobile servicing equipment, 513.7(F), 513.10(D)
Portable equipment, 513.10(E)
Scope, 513.1
Sealing, 513.9
Stanchions, rostrums, and docks, 513.7(E)
Wiring, 513.4, 513.7, 513.8

**Airport runways, underground conductors,** Table 300.50 Ex. 5

**Alarm systems, health care facilities,** 517.32(C), 517.42(C)

**Alarms**
Burglar. *see* Remote-control, signaling, and power-limited circuits
Fire, 230.82(4), 230.94 Ex. 4. *see also* Fire alarm systems

**Alternate power source**
Definition, 517.2

**Alternators.** *see* Generators

**Aluminum conduit.** *see* Rigid metal conduit

**Aluminum siding, grounding of,** 250.116 FPN

**Ambient temperature,** 310.10 FPN. *see also* Temperature limitations

**Ambulatory health care centers**
Definition, 517.2
Essential electrical systems, 517.45

**Ampacities**
Conductors, 310.15, Tables 310.16 through 310.21, 310.61 through 310.86, 366.7, B.310.1 through B.310.10
Crane and hoist conductors, Table 610.14(A)
Definition, Art. 100-I
Fixture wires, Table 402.3
Flexible cords and cables, Table 402.5
Grounding, 250.122
Tables, 0–2000 volts, 310.16 through 310.21, Annex B

**Anesthetizing locations,** 517-IV. *see also* Health care facilities
Flammable
Definition, 517.2

**Antenna discharge units,** 810.20, 810.97

**Antenna systems, radio and television.** *see* Community antenna television and radio distribution (CATV) systems; Radio and television equipment

**Apparatus**
  Associated, grounding of, 504.50(A)
    Definition, 504.2
  Intrinsically safe, grounding of, 504.50(A)
    Definition, 504.2
  Simple
    Definition, 504.2

**Appliances,** Art. 422. *see also* Motors
  Battery-powered, 517.64(E)
  Branch circuits. *see* Branch circuits, appliances
  Branch circuits for (utilization equipment), 210.23
  Branch-circuit calculations, 220.3
  Cords. *see* Cords, flexible
  Definition, Art. 100-I
  Disconnecting means, 422-III
  Dwelling units, outlets, 210.52
  Feeder calculations for, 220.10
  Fixed
    Definition, 550.2, 551.2
  Grounding, 250-VI, 550.16, 551.54
  Hazardous (classified) locations. *see* Hazardous (classified) locations
  Household cooking, demand loads for, Table 220.19
  Installation, 422-II
  Load calculations, 220.3, 220.10, Table 220.19
  Marking, 422-V
  Mobile homes, in, 550.12(B) and (C), 550.14
  Motor-operated, load, 220.4(A)
  Overcurrent protection, 422.11
    Single, 210.20, 422.11(E)
  Portable
    Definition, 550.2, 551.2
  Recreational vehicles, in, 551.42
  Signal transmission between, 517.82
  Stationary
    Definition, 550.2, 551.2
  Subject to immersion, 422.21
  Terminals, identification, 200.10(E)

**Approval, approved**
  Definition, Art. 100-I
  Equipment and materials, 80.19(F), 90.4, 90.7, 110.2, 500.8(A)

**Arc lamps, portable**
  Motion picture studios, 530.17
  Stage equipment, 410.2, 520.61

**Arc welders.** *see* Electric welders

**Arc-fault circuit interrupters,** 210.12

**Arc-fault circuit-interrupter**
  Definition, 210.12(A)
  Protection, 210.12(B), 550.25

**Arcing parts,** 240.41
  Enclosure of, 110.18, 430.14(B), 511.7(A), 513.7(C), 514.7, 515.7(B), 516.7(B), 517.61(B)(2)

**Arc-welder motors,** 630-II
  Ampacities of conductors, 630.11
  Nameplate marking, 630.14

**Arenas,** Art. 518, Art. 520
  Emergency lighting system, Art. 700

**Armored cable (Type AC),** Art. 320
  Accessible attics, 320.23
  Ampacity, 320.80
  Bends, 320.24
  Boxes and fittings, 320.40
  Conductor type, 320.104
  Construction, 320.100, 320-III
  Definition, 320.2
  Equipment grounding, 320.108
  Exposed work, 320.15
  Marking, 320.120
  Supports, 320.30
  Through or parallel to framing members, 320.17
  Uses not permitted, 320.12
  Uses permitted, 320.10

**Armories,** Art. 518
  Emergency lighting system, Art. 700

**Array, solar photovoltaic systems**
  Definition, 690.2

**Arresters, surge.** *see* Lightning (surge) arresters

**Askarels**
  Definition, Art. 100-I
  Transformers, 450.25

**Assembly, places of,** Art. 518
  Classifications, 518.2
  Emergency lighting system, Art. 700
  Supply, 518.5
  Theaters, audience areas, Art. 520
  Wiring methods, 518.4

**Atmospheric groups.** *see* Hazardous (classified) locations

**Attachment plugs (caps).** *see also* Hazardous (classified) locations
  Construction of, 406.6
  Definition, Art. 100-I
  Flexible cords on, 400.7(B), 400.24
  Grounding type, 406.9
  Polarized
    Terminal identification, 200.10(B)

**Attics**
  Armored cable, 320.23
  Knob and tube, 394.23
  Metal clad cable, 330.23
  Nonmetallic-sheathed cable, 334.23
  Open wiring, 398.23
  Sign transformers in, 600.21(E)

**Audio equipment, underwater,** 680.27(A), 680.43(G)

**Audio signal processing, amplification, and reproduction equipment,** Art. 640
  Access, 640.5, 640.46
  Conduit or tubing, 640.23
  Definitions, 640.2
  Environmental protection, 640.44
  Flexible cords, use of, 640.21, 640.42
  Grounding, 640.7
  Grouping of conductors, 640.8
  Locations and other articles, 640.3
  Loudspeakers in fire resistant construction, 640.25
  Mechanical execution of work, 640.6
  Motion picture projectors, 540.50
  Permanent installations, 640-II
  Portable and temporary installations, 640-III
  Protection of electrical equipment, 640.4
  Theaters, 520.4
  Transformers, 640.9(D)
    Definition, 640.2
  Underwater, 680.27(A), 680.43(G)
  Water, near, 640.10
  Wireways, gutters, auxiliary gutters, 640.24
  Wiring methods, 640.9
  Wiring of equipment racks, 640.22, 640.43

**Auditoriums,** Art. 518, Art. 520
  Emergency lighting system, Art. 700

**Authority having jurisdiction,** 90.4. *see also* Approval

**Automatic**
  Definition, Art. 100-I

**Autotransformers,** 450.4, 450.5
  Audio, 640.9(D)
  Ballast for lighting units, 410.78
  Branch circuits, 210.9
  Feeders, 215.11
  Motor starting, 430.82(B)

**Auxiliary gutters,** Art. 366
  Ampacities, 366.7
  Clearance, bare live parts, 366.8
  Conductors, installation, 366.10
  Construction, installation, 366.10
  Covers, 366.5
  Extension beyond equipment, 366.3
  Number of conductors, 366.6
  Sound recording, similar equipment, 640.24
  Splices, taps, 366.9
  Supports, 366.4
  Use, 366.2

## -B-

**Back fed devices,** 408.16(F)
**Backfill,** 300.5(F), 300.50(D)

**Ballasts, electric discharge lamps,** 410-XIII, 410-XIV
  Protection in fluorescent fixtures, 410.73(E)
  Protection in recessed HID fixtures, 410.73(F)

**Bare conductors**
  Ampacities, 310.15(B)(3)
  Contact conductors, cranes and hoists, 610-III
  Definition, Art. 100-I
  Service entrance, Type SE, USE, 338.100
  Sizing conduits for, Chap. 9, Note 3 to Table 1
  Underground conduits, grounded neutral, 230.30 Ex., 230.41 Ex.

**Barriers,** 408.3(A)(2)

**Basements (cellars)**
  Armored cable, 320.15
  Nonmetallic extensions, 382.15
  Nonmetallic sheathed cable, 334.15
  Receptacles in, 210.8(A), 210.52(G), 406.8(A)
  Unfinished, 210.8(A)(5)

**Bathroom**
  Definition, Art. 100-I
  Receptacles in, 210.8(A)(1), 210.8(B)(1)

**Bathtubs**
  Hydromassage, 680-VI, 680-VII
    Definition, 680.2
  Luminaires (lighting fixtures), 410.4(D)

**Batteries.** *see also* Storage batteries
  Aircraft, 513.10
  Installations, Art. 480, 551.10(D), 690.4, 690.71
  Nominal voltage
    Definition, 480.2
  Sealed
    Definition, 480.2

**Battery charging equipment,** 511.8
  Aircraft hangars, 513.10

**Bedrooms**
  Arc-fault circuit-interrupter protection, 210.12(B), 550.25

**Bell circuits.** *see* Remote-control, signaling, and power-limited circuits

**Bends,** 300.34. *see also* wiring system involved

**Block (city, town, or village)**
  Definition, 800.2

**Blocking diode, solar photovoltaic systems**
  Definition, 690.2

**Boatyards.** *see* Marinas and boatyards, Art. 555

**Bodies, conduit.** *see* Conduit bodies

**Boilers**
  Electrode-type, 424-VIII
  Resistance-type, 424-VII

**Bonding,** 250-V
　CATV and radio distribution
　　systems, 820.42(B)
　Definition, Art. 100-I
　Grounding-type receptacles,
　　250.146
　Hazardous (classified) locations.
　　*see* Hazardous (classi-
　　fied) locations
　Intrinsically safe systems,
　　504.60
　Loosely jointed raceways,
　　250.98
　Network-powered broadband
　　communications sys-
　　tems, 830.40(D),
　　830.42(B)
　Other enclosures, 250.96
　Outside raceway, 250.102(E)
　Over 250 volts, 250.97
　Piping systems and exposed
　　structural steel, 250.104
　Receptacles, 250.146
　Service equipment, 250.92,
　　250.94
　Swimming pools, 680.26
**Bonding jumpers.** *see* Jumpers,
　bonding
**Bored holes through studs, joists,**
　300.4(A)(1)
**Bowling alleys,** Art. 518
　Emergency lighting system, Art.
　　700
**Boxes (outlet, device, pull, and**
　**junction),** Art. 314. *see*
　*also* Hazardous (classi-
　fied) locations
　Accessibility, 314.29
　Concealed work, 314.20
　Conductors, number in box,
　　314.16
　　Entering boxes, conduit bod-
　　　ies or fittings, 314.17
　Construction specifications,
　　314.III
　Covers, 314.25, 314.28(C),
　　314.41, 314.42
　Cutout. *see* Cabinets, cutout
　　boxes, and meter socket
　　enclosures
　Damp locations, 314.15
　Depth, 314.24
　Drop, 520.46
　　Definition, theaters, etc.,
　　　520.2
　Exposed surface extensions,
　　314.22
　Fill calculations, 314.16(B)
　Floor, for receptacles, 314.27(C)
　Grounding, 250.148, 250-VI
　Insulating. *see* Boxes, nonmetal-
　　lic
　Junction, pull. *see* Junction
　　boxes
　Lighting (luminaire) outlets,
　　314.27(A), 410.16(A),
　　410-III
　Luminaire outlets. *see* Lighting
　　outlets
　Metal
　　Construction, 314.40, 314-III
　　Grounding, 314.4

　Installation, 314.II
　Nonmetallic, 314.3, 314.17(C),
　　314.43, 334.40
　Over 600 volts, 314.V
　Plugging
　　Definition, motion picture and
　　　TV studios, 530.2
　Portable, in theaters, 520.62
　Remote control, signal circuits,
　　725.25, 725.51
　Repairing plaster around, 314.21
　Required location, 300.15
　Round, 314.2
　Secured supports, 300.11,
　　314.23
　Snap switches over 300 volts,
　　not ganged, 404.8(B)
　Unused openings, closed,
　　110.12(A)
　Vertical raceway runs, 300.19
　Volume calculations, 314.16(A)
　Wall or ceiling, 314.20
　Wet locations, 314.15
**Branch circuits,** Art. 210, Art. 220
　Air conditioners, 440-IV, 440-
　　VII
　Appliances, 422-II
　　Definition, Art. 100-I
　Arc-fault circuit-interrupter pro-
　　tection, 210.12
　Audio equipment, 640.10(A)
　Busways as branch circuits,
　　368.12, 368.14
　Calculation of loads, 220.3, An-
　　nex D
　Color code, 210.5
　Common area, 210.25
　Conductors, minimum ampacity
　　and size, 210.19
　Critical
　　Definition, 517.2
　Definition, Art. 100-I
　Fixed electric space heating
　　equipment, 424.4
　General, 210-I
　General purpose
　　Definition, Art. 100-I
　Health care facilities, 517.18(A)
　Individual
　　Definition, Art. 100-I
　　Overcurrent protection,
　　　210.20
　　Permissible loads, 210.23
　　Rating or setting, 430.52
　　Required, 490.72,
　　　520.53(F)(2), 600.5,
　　　605.8(B), 620.22
　Infrared lamps, 422.14
　Isolated power systems, 517.160
　Maximum loads, 220.4
　Mobile homes, 550.7
　Motor, on individual branch cir-
　　cuit, 430-II
　Multiwire, 210.4, 501.18,
　　502.18, 505.21
　　Definition, Art. 100-I
　Outside, Art. 225
　Overcurrent protection, 210.20,
　　240.3
　Patient bed location, 517.18(A),
　　517.19(A)
　Permissible loads, 210.23

　Ratings, 210.3, 210-II
　Recreational vehicles, 551.42
　Requirements for, Table 210.24
　Selection current, 440.4(C)
　　Definition, 440.2
　Small appliance, 210.11(C)(1)
　Stage or set, 530.23
　Taps from, 210.19(A)(4),
　　210.24, 240.4(E),
　　240.5(B)(1) and (2)
　Through luminaires (fixtures),
　　410.11, 410.31
　Two or more outlets on, 210.24
　Voltage drop, 210.19(A) FPN
　　No. 4
　Voltage limitations, 210.6
　X-ray equipment, 517-V, 660.4,
　　660.6(A)
**Branch-circuit selection current**
　Definition, 440.2
**Building component**
　Definition, 545.3
**Building system**
　Definition, 545.3
**Building wire.** *see* Conductors
　Definition, Art. 100-I
　First floor of, 334.12(A)(1)
**Bulk storage plants,** Art. 515. *see*
　　*also* Hazardous (classi-
　　fied) locations
　Class I locations, 515.3
　Definition, 515.2
　Gasoline dispensing, 515.10
　Grounding, 515.16
　Sealing, 515.9
　Underground wiring, 515.8
　Wiring and equipment, 515.4,
　　515.7
**Bull switches**
　Definition, 530.2
**Burglar alarm systems.** *see*
　　Remote-control, signal-
　　ing, and power-limited
　　circuits
**Buried grounding conductor con-**
　**nection,** 250.68(A), Ex.
**Busbars**
　Aboveground wiring methods,
　　300.37
　Auxiliary gutters, 366.8
　Switchboards, 408.31
**Bushings**
　Fiber. *see* Fiber bushings
　Generators, 445.16
　Insulated, 300.4(F), 312.6(C)
　　Lampholders attached to flex-
　　　ible cord, 410.30(A)
　　Securing fittings or raceways,
　　　300.4(F)
　Intermediate metal conduit,
　　342.46
　Motors, 430.13
　Outlet boxes, conduit bodies,
　　and outlet fittings,
　　314.17(D), 314.42
　Rigid metal conduit, 344.46
　Rigid nonmetallic conduit,
　　352.46
　Snow-melting and deicing
　　equipment, 426.22(C)
　Underground installations,
　　300.5(H)

　Use in lieu of box or terminal
　　fitting, 300.16(B)
**Busways,** Art. 368
　Branches from, 368.8
　Dead ends closed, 368.7
　Definition, 368.2
　Extension through walls and
　　floors, 368.6
　Feeder or branch circuits,
　　368.12
　Marking, 368.15
　Over 600 volts, 368-II
　Overcurrent protection, 368.9,
　　368.10, 368.13
　Reduction in size, 368.11
　Support, 368.5
　Through walls and floors, 368.6
　Under 600 volts, 368.30
　Use, 368.4
**Bypass isolation switch**
　Definition, Art. 100.I

**-C-**

**Cabinets, cutout boxes, and**
　**meter socket enclo-**
　**sures,** Art. 312
　Construction specifications,
　　373-II
　Damp, wet, or hazardous (classi-
　　fied) locations, 312.2
　Definitions, Art. 100-I
　Deflection of conductors, 312.6
　Installation, 373-I
　Insulation at bushings, 300.4(F),
　　312.6(C)
　Position in walls, 312.3
　Switch enclosures, splices, taps,
　　312.8
　Unused openings, 312.5(A)
　Wire bending space at terminals,
　　312.6(B)
　Wiring space, side or back,
　　312.9
**Cable bending.** *see* type of cable
**Cable end fittings, changing from**
　**cable to knob-and-**
　**tube,** 300.16(A)
**Cable sheath**
　Definition, 800.2
**Cable suspension,** 300.19
**Cable trays,** Art. 392
　Ampacity, 392.11, 392.13
　Cable fill, 392.9, 392.10, 392.12
　Cable installation, 392.8
　Conduits and cables supported
　　from, 392.6(J)
　Construction specifications,
　　392.5
　Definition, 392.2
　Grounding, 392.7
　Installation, 392.6
　Separation, 392.6(F)(2)
　Uses not permitted, 392.4
　Uses permitted, 392.3, 392.6(A)
**Cablebus,** Art. 370
　Conductors, 370.4
　　Overcurrent protection, 370.5
　　Terminations, 370.8
　Definition, 370.2
　Fittings, 370.7

Grounding, 370.9
Marking, 370.10
Support and extension through walls and floors, 370.6
Use, 370.3
**Cables**
Aerial, 820.10, 830.10
Armored (Type AC). see Armored cable (Type AC), Art. 320
Border lights, theater, 520.44(B)
Bundled
Definition, 520.2
CATV, Art. 820
Continuity, 300.12
Definition, 800.2
Flat cable assemblies (Type FC). see Flat cable assemblies (Type FC)
Flat conductor (Type FCC). see Flat conductor cable (Type FCC)
Grouped
Definition, 520.2
Heating. see Heating cables
Installation in cable trays, 392.8
Installed in shallow grooves, 300.4(E)
Instrumentation tray (Type ITC). see Instrumentation tray cable (Type ITC)
Integrated gas spacer cable (Type IGS). see Integrated gas spacer cable (Type IGS)
Medium voltage cable (Type MV). see Medium voltage cable (Type MV)
Metal-clad cable (Type MC). see Metal-clad cable (Type MC)
Mineral-insulated metal-sheathed (Type MI). see Mineral-insulated metal-sheathed cable (Type MI)
Nonmetallic extension. see Nonmetallic extensions
Nonmetallic underground conduit with conductors. see Nonmetallic underground conduit with conductors
Nonmetallic-sheathed (Types NM, NMC, and NMS). see Nonmetallic-sheathed cable (Types NM, NMC, and NMS)
Optical fiber. see Optical fiber cables
Other types of. see names of systems
Point of entrance
Definition, 800.2, 820.2, 830.2
Portable. see Cords, flexible
Power and control tray cable (Type TC). see Power and control tray cable (Type TC)

Preassembled in nonmetallic conduit. see Nonmetallic underground conduit with conductors
Protection against physical damage, 300.4
Sealing, 501.5(D), 501.5(E), 505.16(A)(2), 505.16(B)(5), 505.16(B)(6), 505.16(B)(7), 505.16(C)(2)
Secured, 300.11, 314(B) and (C)
Service. see Service cables
Service-entrance (Types SE and USE). see Service-entrance cable (Types SE and USE)
Splices in boxes, 300.15
Stage, 530.18(A)
Supported from cable trays, 392.6(J)
Through studs, joists, rafters, 300.4
Underground, 230-III, 300.5, 300.50
Underground feeder and branch circuit Type UF. see Underground feeder and branch-circuit cable (Type UF)
**Calculations,** Annex D. see also Loads
**Camping trailer**
Definition, 551.2
**Canopies**
Boxes and fittings, 314.25
Live parts, exposed, 410.3
Luminaires (lighting) fixtures
Conductors, space for, 410.10
Cover
At boxes, 410.12
Combustible finishes, covering required between canopy and box, 410.13
**Capacitors,** Art. 460. see also Hazardous (classified) locations
Enclosing and guarding, 460.2
Induction and dielectric heating, 665.24
600 volts and under, 460-I
Conductors, 460.8
Grounding, 460.10
Marking, 460.12
Rating or setting of motor overload device, 460.9
Over 600 volts, 460-II
Grounding, 460.27
Identification, 460.26
Means for discharge, 460.28
Overcurrent protection, 460.25
Switching, 460.24
X-ray equipment, 660.36
**Capacity, interrupting.** see Interrupting capacity
**Caps.** see Attachment plugs
**Carnivals, circuses, fairs, and similar events,** Art. 525
Conductor overhead clearance, 525.5

Grounding and bonding, 525-V
Equipment bonding, 525.30
Equipment grounding, 525.31
Grounding conductor continuity assurance, 525.32
Power sources, 525-III
Separately derived systems, 525.10
Services, 525.11
Protection of electrical equipment, 525.6
Wiring methods, 525-III, 525.20
Concessions, 525.21
Ground-fault circuit-interrupter protection, 525.23
Portable distribution or terminal boxes, 525.22
Rides, 525.21
Tents, 525.21
**Cartridge fuses,** 240-VI
Disconnection, 240.40
**CATV systems.** see Community antenna television and radio distribution (CATV) systems
**Ceiling fans,** 680.22(B), 680.43(B)
Support of, 314.27(D), 422.18
**Cell**
Cellular concrete floor raceways
Definition, 372.2
Electrolytic
Definition, 668.2
Sealed, storage batteries
Definition, 480.2
Solar
Definition, 690.2
**Cell line, electrolytic cells**
Attachments and auxiliary equipment
Definition, 668.2
Definition, 668.2
**Cellars.** see Basements
**Cellular concrete floor raceways,** Art. 372
Connection to cabinets and other enclosures, 372.6
Definitions, 372.2
Discontinued outlets, 372.13
Header, 372.5
Inserts, 372.9
Junction boxes, 372.7
Markers, 372.8
Number of conductors, 372.11
Other articles, 372.3
Size of conductors, 372.10
Splices and taps, 372.12
Uses not permitted, 372.4
**Cellular metal floor raceways,** Art. 374
Connection to cabinets and extension from cells, 374.11
Construction, 374.12
Definitions, 374.2
Discontinued outlets, 374.7
Inserts, 374.10
Installation, 374-I
Junction boxes, 374.9
Markers, 374.8
Number of conductors, 374.5
Size of conductors, 374.4

Splices and taps, 374.6
Uses not permitted, 374.3
**Chair lifts.** see Elevators, dumbwaiters, escalators, moving walks, wheelchair lifts, and stairway chair lifts
**Churches,** Art. 518
**Cinder fill**
Intermediate or rigid metal conduits and electrical metallic tubing, in or under, 342.10(C), 344.10(C), 348.12(3)
Rigid nonmetallic conduit, 352.10
**Circuit breakers,** Art. 240. see also Hazardous (classified) locations
Accessibility and grouping, 404.8
Circuits over 600 volts, 490.21
Definition, Art. 100-I
Disconnection of grounded circuits, 404.2(B), 514.11(A)
Enclosures, 404.3
General, 110.9, 240-I
Overcurrent protection, 230.208, 240-I, 240-VII
Generators, 445.12
Motors, 430.52(A), 430.58, 430.110, 430.111
Transformers, 450.3
Panelboards, 408-III
Rating, motor branch circuits, 430.58
Rating, nonadjustable trip, 240.6(A), 240.83(C)
Service overcurrent protection, 230.90, 230.91
Services, disconnecting means, 230.70, 230.205
Switches, use as, 240.83(D), 404.11, 410.81
Wet locations, in, 404.4
**Circuit directory, panelboards,** 110.22
**Circuit-interrupters, ground-fault.** see Ground-fault circuit-interrupters
**Circuits**
Anesthetizing locations, 517.63(F)
Bonding jumper
Definition, Art. 100-I
Branch. see Branch circuits
Burglar alarm. see Remote-control, signaling, and power-limited circuits
Central station. see Fire alarm systems
Communication. see Communication circuits
Fire alarm
Definitions, 760.2
Fuel cell systems, 692-II
Grounding, Art. 250
Impedance, 110.10
Information technology equipment, 645.5
Intrinsically safe, 504.30

Definition, 504.2
Inverter input circuit
  Definition, 690.2
Less than 50 volts, Art. 720
  Class 1, 725-II
  Grounding, 250.20(A)
More than 600 volts. *see* Over
  600 volts
  Services, 230-IX
Motor, 430-II
Motor control, 430-VI
  Definition, 430.71
Nonincendive
  Definition, Art. 100-I
  Number of, in enclosures,
    90.8(B)
Photovoltaic output
  Definition, 690.2
Photovoltaic source
  Definition, 690.2
Power-limited. *see* Remote-
  control, signaling, and
  power-limited circuits
Protectors required, 800.12,
  800.30, 800.31, 800.32
Remote-control, Art. 725
  Definition, Art. 100-I
  Motors, controllers, 430-VI
Signal. *see* Remote-control, sig-
  naling, and power-
  limited circuits
Telegraph. *see* Communication
  circuits
Telephone. *see* Communication
  circuits
Underground. *see* Communica-
  tion circuits
Ungrounded, 210.10, 215.7,
  410.48
Circuses. *see* Carnivals, circuses,
  fairs, and similar events
**Clamp fill, boxes,** 314.16(B)(2)
**Clamps, ground,** 250.8, 250.10,
  250.70
**Class 1, 2, and 3 circuits.** *see*
  Remote-control, signal-
  ing, and power-limited
  circuits
**Class I, II, and III locations.** *see*
  Hazardous (classified)
  locations
**Clean surfaces, grounding con-
  ductor connections,**
  250.12
**Clearances.** *see also* Enclosures;
  Space
  Antennas, 810.3, 810.18, 810.54
  Community antenna systems,
    Art. 820
  Conductors
    Open, outside branch circuits
      and feeders, 225.18,
      225.19
    Service drop, 230-II
  Lighting luminaires (fixtures),
    410.66, 410.76(B)
  Live parts
    Auxiliary gutters, 366.8
    Circuits over 600 volts, 110-
      III

Network-powered broadband
  communications sys-
  tems, 830.11
Swimming pools, 680.8
Switchboards, 408.7, 408.8,
  408.10
**Climbing space, line conductors
  on poles,** 225.14(D)
**Closed-loop and programmed
  power distribution,**
  Art. 780
  Cables and conductors, Art. 334,
    780.6
  Control, 780.3
    Hybrid cable, 780.6(A)
    Noninterchangeability, 780.7
    Power limitation, in signaling
      circuits, 780.5
    In same cabinet, panel, or
      box, 780.6(B)
**Clothes closets**
  Heating, 424.38(C)
  Luminaires (lighting fixtures),
    410.8
  Overcurrent devices, 240.24(D)
**Clothes dryers**
  Calculations for, 220.18
  Feeder demand factors, Table
    220.18
  Grounding, 250.140
  Mobile homes, 550.15(E)
**Clothing manufacturing plants,**
  500.5(D), Art. 503. *see
  also* Hazardous (classi-
  fied) locations
**CO/ALR**
  Receptacles, 406.2(C)
  Switches, 404.14(C)
**Collector rings,** 490.54, 675.11
  Definition, 675.2
**Collectors, cranes and hoists,**
  610.22
**Color code**
  Branch circuits, 210.4(D), 210.5
  Conductors, 310.12, 504.80(C),
    647.4(C)
  Grounded conductor, 200.6
  Grounding conductor, 250.119,
    310.12(B), 400.23
  Heating cables, 424.35
  Higher voltage to ground
    Feeders, 215.8
    Panel boards, 408.3(E)
    Sensitive electronic equip-
      ment, 647.4(C)
    Service-entrance conductors,
      230.56
**Combustible dusts,** Art. 502. *see
  also* Hazardous (classi-
  fied) locations
**Commercial garages.** *see* Garages,
  commercial
**Common grounding electrode.**
  *see* Electrodes, ground-
  ing, common
**Common neutral**
  Feeders, 215.4
  Outside wiring, 225.7(B)
**Communications circuits,** Art.
  800
  Access to electrical equipment,
    800.5

Bonding, 800.40(D), 800.41(B)
Cable marking, 800.50
Cable substitution, 800.53(G)
Circuits requiring primary pro-
  tectors, 800.12
Conductors
  Other circuits with, 800.53(E)
  Overhead, 800.10
  Within buildings, 800-V
Definitions, 800.2
Essential electrical systems,
  517.32(D), 517.42(D)
Fire resistance, 800.49
Grounding, 800.33, 800-IV
Hazardous (classified) locations,
  800.8
Health care facilities, 517.32(D),
  517.42(D), 517-VI
Hybrid power and communica-
  tions cables, 800.3
Installation, 800.7, 800.52
Lightning conductors, 800.13
Listing of communications wires
  and cables, 800.50,
  800.51
Listing of equipment, 800.4
Marking, 800.50
Mechanical execution of work,
  800.6
Mobile homes, 800.41
Protection, 800-III
  Devices, 800.30
  Grounding, 800.40, 800.41
  Requirements, 800.31,
    800.32, 800.40, 800.41
Types of cable, 800.51
Underground circuits entering
  buildings, 800.11
**Community antenna television
  and radio distribution
  (CATV) systems,** Art.
  820
  Access to electrical equipment,
    820.5
  Cables
    Aerial entrance, 820.10
    Within buildings, 820-V
    Entering buildings, 820.11
    Fire resistance of, 820.49
    Listing and marking, 820.50,
      820.51
    Outside, 820.10
    Types of, 820.51
    Underground, 820.11(A)
  Definitions, 820.2
  Energy limitations, 820.4
  Grounding, 820-IV
    Cables, 820.33, 820.40,
      820-IV
    Equipment, 820.41
  Installation of cables and equip-
    ment, 820.52
  Installation of systems, 820.7
  Locations, 820.3
  Mechanical execution of work,
    820.6
  Mobile homes, 820.42
  Protection, 820.40(A)(6), 820-III
**Compressors, refrigeration,** Art.
  440
**Computer rooms.** *see* Information
  technology equipment

**Concealed**
  Definition, Art. 100-I
  Knob-and-tube wiring. *see*
    Knob-and-tube wiring
**Concentric knockouts.** *see*
  Knockouts, bonding
  service equipment
**Concrete**
  Electrodes encased in,
    250.52(A)(3)
  Metal raceways and equipment
    in, 300.6(B)
**Conductive surfaces, exposed**
  Definition, 517.2
**Conductor fill**
  Audio systems, 640.23(A)
  Auxiliary gutters, 366.6
  Boxes, 314.16(B)(1) and (5)
  Cable trays, 392.9, 392.10
  Cellular concrete floor raceways,
    372.11
  Cellular metal floor raceways,
    374.5
  Control circuits, 517.74(A)
  Electrical metallic tubing,
    358.22
  Electrical nonmetallic tubing,
    362.22
  Elevators, 620.32, 620.33
  Fixture wire, 402.7
  Flexible metal conduit, 348.22,
    Table 348.22
  Flexible metallic tubing, 358.22
  General installations, 300.17
  Hazardous (classified) locations,
    sealing fittings,
    501.5(C)(6),
    505.16(D)(5)
  Intermediate metal conduit,
    342.22
  Liquidtight flexible metal con-
    duit, 350.22
  Liquidtight flexible nonmetallic
    conduit, 356.22
  Maximum permitted, 300.17
  Outlet boxes, etc., 314.16
  Remote control circuits, 725.28
  Rigid metal conduit, 344.22
  Rigid nonmetallic conduit,
    352.22
  Signs, 600.31(C)
  Surface raceways, 386.22,
    388.22
  Theaters, 520.6
  Underfloor raceways, 390.5
  Wireways, 376.22, 378.22
  X-ray equipment, 660.8
**Conductors.** *see also* Cords, flex-
  ible; Fixture wires
  Aluminum, conductor material,
    310.14
  Aluminum, properties of, Chap.
    9, Table 8
  Motors, 430-II
  Ampacities of, 310.15, Tables
    310.16 through 310.21,
    Tables 310.69 through
    310.86, B.310.1 through
    B.310.10
  Application, 310.13

Armored cable, Type AC. see Armored cable (Type AC)
Bare. see Bare conductors
Bending radius, 300.34
Boxes and fittings, junction, 314.16, 314.17
Branch circuits, Art. 210
Bundled, in theaters, etc. Definition, 520.2
Buried, 310.7
Busways. see Busways
Cabinets and cutout boxes, 312.5 through 312.7
Cablebus. see Cablebus
Capacitors, 460.8
Cellular concrete floor raceways. see Cellular concrete floor raceways
Cellular metal floor raceways. see Cellular metal floor raceways
Circuit
    Communications, Art. 800
    Control, health care facilities, 517.74
    Fire alarm, Art. 760
    Less than 50 volts, Art. 720
    Over 600 volts, Art. 490
    Signal, remote-control, power-limited, Art. 725
Combinations, Chap. 9, Table 1
Communications circuits, 800-II, 800-V
Computations of, examples, Annex D
Concealed knob-and-tube. see Knob-and-tube wiring
Conduit or tubing, number in, 342.22, 344.22, 348.22, 350.22, 352.22, 354.22, 356.22, 358.22, 360.22, 362.22, Annex C, Chap. 9 Tables, Table 348.22
Construction, general, Table 310.13
Cooling of electric equipment, 110.13(B)
Copper, 110.5, 505.18(A), 760.27(C)
    Ampacities, Tables 310.16 through 310.21, Tables 310.69 through 310.86, B.310.1 through B.310.10
    Properties, Chap. 9, Table 8
Copper-clad aluminum Definition, Art. 100-I
Cords, flexible. see Cords, flexible
Corrosive conditions, 300.6, 310.9, 501.13, 509.18(B)
Covered
    Definition, Art. 100-I
Crane and hoists, 610.41, 610-II, 610-III
Definitions, Art. 100-I
Deflection of. see Deflection of conductors
Different systems, 300.3(C), 725.26

Dimensions of, Chap. 9, Table 4
Electrical metallic tubing. see Electrical metallic tubing
Electrical nonmetallic tubing. see Electrical nonmetallic tubing
Elevators, dumbwaiters, escalators, moving walks, 620-II through 620-IV
Enclosure, grounding, 250.1, 250-IV
Equipment grounding. see Equipment grounding conductors
Feeder, Art. 215
Fire alarm systems, 760.26, 760.27, 760.52
Fixture wires, Art. 402
In free air, Tables 310.17 and 310.19, 310.20
Gauges (AWG), general provisions, 110.6
General wiring, Art. 310
Generators, size, 445.13
Grounded
    Alternating current systems, 250.24, 250.26
        Identification, 200.6, 210.5(A)
    Change in size, 240.23
    Definition, Art. 100-I
    Different systems, 200.6(D)
    Fuses in, 430.36
    Overcurrent protection, 240.22
    Services
        Disconnection of, 230.70, 230.75
        Overcurrent protection, 230.90(B)
        Switches, disconnection, 404.2(B)
    Use and identification, Art. 200
Grounded power systems in anesthetizing locations, 517.63
Grounding. see Grounding conductors
Grounding electrode. see Grounding electrode conductors
Grouped
    Definition theaters, etc., 520.2
Identification, 310.12
Induction and dielectric heating equipment, ampacity of supply, 665.10
Installation with other systems, 300.8
Insulated
    Definition, Art. 100-I
Insulating materials, 310.13
Insulation, Art. 310
    Fire alarm systems, 760.27(B)
    Fixtures, 410.24, 410.67
    Hazardous (classified) locations, Class I, 501.13, 505.18(B)

Motion picture projectors, 540.14
    Theaters, 520.42
Insulation at bushings, No. 4 and larger, 300.4(F), 312.6(C)
Intrinsically safe, 504.30, 504.50
Length in boxes, 300.14
Length in transient voltage surge suppressors, 285.12
Lightning rods, spacing from, 250.106, 800.13, 820.10(F)(3), 820.10(I)(3)
Liquidtight flexible metal conduit. see Liquidtight flexible metal conduit
Liquidtight flexible nonmetallic conduit. see Liquidtight flexible nonmetallic conduit
Metal enclosures, spacing from lightning rods, 250.106
Metal-clad cable. see Metal-clad cable
Mineral-insulated metal-sheathed cable. see Mineral-insulated metal-sheathed cable (Type MI)
Minimum size, 230.42, 310.5, 517.4(B), 620.12, 650.5(A), 660.9, 725.27(A), 760.27(A), 760.58, 810.17
Motion picture projectors, sizes, 540.13
Motor circuits, 430-II
Multioutlet assembly. see Multioutlet assembly
Multiple, 250.102(C), 250.122(F), 300.20, 310.4, 392.8(D), 520.27(A)(2), 620.12(A)(1)
Nonmetallic extensions. see Nonmetallic extensions
Nonmetallic sheathed cable, Types NM, NMC, and NMS. see Nonmetallic-sheathed cable (Types NM, NMC, and NMS)
Number of, in. see Conductor fill
Organs, 650.5, 650.6
Outlet boxes, temperature limits, 410.11
Outside wiring, Art. 225
Over 600 volts, Art. 490
Overcurrent protection, 240.3, 240.4
Overhead, 800.10
Paralleled, 250.102(C), 250.122(F), 300.20, 310.4, 392.8(D), 520.27(A)(2), 620.12(A)(1)
Pendant fixtures, 410.27
Properties of, Chap. 9, Table 8
Raceways, number of conductors in. see Conductor fill
Radio and television

Antenna distribution systems, 810.4, Art. 820
Antenna systems, Art. 810
Equipment, Art. 810
Railway, power and light from, 110.19
Recreational vehicle site feeder circuit
    Definition, 551.2
Remote-control circuits, Art. 725
Rigid metal conduit. see Rigid metal conduit
Same circuit, 300.5(I)
Selection of, examples, Annex D
Service. see Service cables; Service-entrance cable (Types SE and USE); Service-entrance conductors
    Definition, Art. 100-I
Service-entrance. see Service-entrance conductors
Signal circuits, Art. 725, Art. 760
Signs and outline lighting, 600.31
Single, 392.8(D), 392.10, 392.11(B), 392.13(B)
Sizes, 110.6, Chap. 9, Tables 5 and 5A. see also Conductors, minimum size
Stranded, 310.3
Support of, in vertical raceways, 300.19
Surface raceway. see Surface metal raceways; Surface nonmetallic raceways; Strut-type channel raceway
Switchboards and panelboards, 408.3(A), 408.9
Temperature, limitations of, 310.10, 338.10(B)(3)
Theaters, portable stage equipment, 520.68
Underfloor raceway. see Underfloor raceways
Underground feeder and branch-circuit cable. see Underground feeder and branch-circuit cable (Type UF)
Ungrounded
    Identification, 210.4(D)
    Services, 230.90(A)
Welders, electric
    Ampacity of supply, 630.11, 630.31
    Overcurrent protection, 630.12(B), 630.32(B)
Wet locations, 225.4, 310.8(C)
Wireways. see Wireways, metal; Wireways, nonmetallic
X-ray equipment, 517-V, 660.6 through 660.8
**Conduit bodies.** see also Boxes, outlet, device, pull, and junction
    Definition, Art. 100-I

Number of conductors, 314.16(C)

Pull and junction box, 314.28

Short radius, 314.5

**Conduit fill.** *see also* Conductor fill

Equipment grounding conductors, Chap. 9, Note 3 to Table 1

Grounding conductor, 310.15(B)(5)

**Conduit nipples,** Chap. 9, Note 4 to Table 1

**Conduits**

Boxes supporting, 314.23(H)(2)

Circuits, physical protection of remote control, 725.8(B)

Conductors, number in, Chap. 9, Table 1 and Annex C, Tables C1 through C12(A)

Dimensions, Chap. 9, Table 4

Electrical metallic tubing. *see* Electrical metallic tubing

Electrical nonmetallic tubing. *see* Electrical nonmetallic tubing

Flexible metal. *see* Flexible metal conduit

Intermediate metal. *see* Intermediate metal conduit

Liquidtight flexible metal. *see* Liquidtight flexible metal conduit

Liquidtight flexible nonmetallic. *see* Liquidtight flexible nonmetallic conduit

Metallic, grounding runs 7.5 m (25 ft), 250.86, Ex.1

Number of fixture wires in, 402.7

Outside wiring, 225.10

Rigid metal. *see* Rigid metal conduit

Rigid nonmetallic. *see* Rigid nonmetallic conduit

**Connections**

Aluminum, copper conductors, 110.14

Cellular metal floor raceways to cabinets and extensions, 374.11

Essential electrical systems, 517.34, 517.42, 517.43, 517.45

Feed-through. *see* Feed-through connections of neutral conductors

Fuel cell systems to other circuits, 692-VII

Grounding conductor, 250-III, 250-VI

High-impedance grounded neutral system, 250.164, 250.186(C)

Integrity of, 110.12(C)

Multiple circuit. *see* Multiple circuit connections, grounding

Point of interconnected power production sources, 705.12

Splices, 110.14(B)

Switches, 404.2

Temperature limitations, 110.14(C), 110.40

Terminals, general provisions, 110.14(A)

X-ray installations, 517.71

**Connectors**

Armored cable, 320.40

Boxes, 314.17

Cabinets and cutout boxes, 312.5(C)

Cable, theater, 520.67

Electric vehicle

Definition, 625.2

Electrical metallic tubing, 358.42

Flexible metal conduit, 348.42

Intermediate metal conduit, 342.42

Liquidtight flexible metal conduit, 350.42

Liquidtight flexible nonmetallic conduit, 356.42

Pressure (solderless)

Definition, Art. 100-I

Rigid metal conduit, 344.42

Single-pole separable

Definition, 530.2

**Constant-voltage generators,** Art. 445

Overcurrent protection for, 445.12(A)

**Construction, closed**

Definition, 545.3

**Construction sites**

Assured equipment grounding conductor program, 527.6(B)(2)

Extension cord sets, 527.6(A) and (B)(2)

Ground-fault circuit-interrupter, protection for, 527.6(A) and (B)(1)

**Continuity**

Electrical, metal raceways and cables, 250.96, 250.97

Service equipment, 250.92, 250.94

Grounding, metal boxes, grounding-type receptacles, 250.146, 250.148

Mechanical, raceways and cables, 300.12

Conductors, 300.13

Grounded conductor of multi-wire circuits, 300.13(B)

**Continuous duty**

Definition, Art. 100-I

**Continuous industrial processes**

Coordination, electric system, 240.12

Ground-fault protection, 230.95 Ex. 1

Orderly shutdown, 430.44

Power loss hazard, 240.3(A)

**Continuous load**

Applications, 210.19(A), 210.20(A), 215.2(A)

Definition, Art. 100-I

**Control.** *see also* Hazardous (classified) locations

Remote

Induction and dielectric heating equipment, 665.7

Overcurrent protection, 240.3(G)

Power-limited and signal circuits, Art. 725

X-ray equipment, 517-V, 660-II

**Control centers**

Guarding live parts, 110.27

Headroom, 110.26(E), 110.32

Illumination at, 110.26(D), 110.34(D)

Motor, 430-VIII

Definition, Art. 100-I

Working space, 110.26, 110.33, 110.34

**Control circuit,** 430.71

**Control drawing,** 504.10(A)

Definition, 504.2

**Control panels**

Working space, 110.26

**Control system**

Definition, 620.2

**Controllers.** *see also* Hazardous (classified) locations

Cases, grounding, 250.112(D)

Definition, Art. 100-I, 430.81(A)

Enclosures, grounding, 250.112(D)

Motion

Definition, 620.2

Motor, 430-VII

Definition, 620.2

Hazardous (classified) locations, 501.6, 502.6, 503.4

Operation

Definition, 620.2

Ratings, 430.83

Resistors and reactors. *see* Resistors and reactors

**Convenience outlets.** *see* Receptacles

**Conversion table, AC conductor resistances and reactances,** Chap. 9, Table 9

**Converters**

Definition, 551.2

Phase. *see* Phase converters

Recreational vehicle

Definition, 551.2

**Cooking unit, counter-mounted.** *see also* Ranges; Ovens, wall-mounted

Branch circuits, 210.19(A)(3), Note 4 to Table 220.19

Definition, Art. 100-I

**Cooling of equipment,** 110.13(B)

**Coordination, electrical systems,** 240.12

Feeders, 240.100(C), 517.17

**Copper conductors.** *see* Conductors, copper

**Copper-clad aluminum conductors.** *see* Conductors, copper-clad aluminum

**Cord sets.** *see* Extension cord sets

**Cords.** *see also* Hazardous (classified) locations

Flexible, Art. 400, 422.16, 422.43

Ampacities, Table 400.5

Audio signal processing, amplification; and reproduction equipment grouping conductors, 640.8

Identified conductors within, 200.6(C)

Lampholders, insulating bushings, 410.30(A)

Motion picture projectors, 540.15

Over 600 volts, 400-III

Overcurrent protection, 240.4, 400.13

Power-limited and signal circuits, remote-control, 725.27

Pull at joints and terminals, 400.10

Repair of, hard service, 400.9

Splices and taps, 400.9, 530.12(A) and (B)

Construction sites, 527.4(G)

Types, Table 400.4

Heater, Tables 400.4 and 422.43(A)

Tinsel

Minimum size, 400.12

Overcurrent protection, 400.13

Uses not permitted, 400.8

Uses permitted, 400.7

**Correction factors, ambient temperature,** Tables 310.16 through 310.20, B.310.1 through B.310.10

**Corrosion protection.** *see* Protection, corrosion

**Corrosive conditions**

Bushing, 430.13

Conductor insulation, 310.9, 501.13, 505.18(B)

Deteriorating agents, 110.11

**Countertops, outlets,** 210.52

**Couplings**

Electrical metallic tubing, 358.42

Intermediate metal conduit, 342.42

Rigid metal conduit, 344.42

Rigid nonmetallic conduit, expansion fittings, 352.44

Running threads at, 342.42(B), 344.42(B)

Threaded and threadless, 250.94

**Cove lighting, space for,** 410.9

**Covers**

Auxiliary gutters, 366.5

Boxes and fittings, 314.25, 314.28(C)

Cable, 650.5(D)

Electrically operated pool, 680.27(B)

Faceplates. *see* Faceplates

Pool

Definition, 680.2

Electrically operated, 680.27(B)
Wireways, 376.2, 378.2
**Cranes.** Art. 610. *see also* Hoists
Conductors, 610-II, 610-III
Control, 610-VI
Disconnecting means, 610-IV
Electrolytic cells, 668.32
Flexible cords, 400.7(A)(5)
Grounding, 250.22(A), 250.112(E), 250.116(1), 610-VII
Hazardous (classified) locations, 503.13, 610.2(A)
Motors and controllers, disconnecting means, 430.112, Ex.
Overcurrent protection, 610-V
**Critical branch**
Definition, 517.2
**Critical care areas,** 517.19
Definition, 517.2
**Cross sectional areas**
Conductors, Chap. 9, Tables 5 through 8
Conduits, Chap. 9, Table 4
**Current-carrying capacities.** *see* Ampacities
**Current-limiting devices.** *see also* Circuit breakers; Fuses
Definition, 240.2
**Curtain machines, theaters,** 520.48
**Cutout bases.** *see* Fuseholders
**Cutout boxes.** *see also* Cabinets, cutout boxes, and meter socket enclosures
Definition, Art. 100-I
**Cutouts, thermal.** *see* Thermal cutouts
**Cutting tables, motion picture,** 530-IV

### -D-

**Damp or wet locations,** 110.11. *see also* Wet locations
Boxes and fittings, 314.15(A)
Cabinets and cutout boxes, 312.2(A)
Definition, Art. 100-I
Lampholders, 410.49
Luminaires (lighting fixtures), 410.4(A)
Open wiring, 398.15(B)
Overcurrent protection, enclosure, 240.32
Panelboards, 408.17
Receptacles, 406.8
**Dampers (flue), control**
Fixed stage equipment, smoke ventilator control, 520.49
Ventilating openings, transformer vaults, 450.45(E)
**Dance halls,** Art. 518
**Data processing systems.** *see* Information technology equipment
**DC systems, grounding,** 250.162, 250.164

**Dead ends**
Busways, 368.7
Cablebus, 370.7(2)
Flat cable assemblies, 322.40(A)
Wireways, 376.58, 378.58
**Dead front**
Definition, Art. 100-I, 551.2
Recreational vehicles, 551.2
Stage switchboards, 520.21
**Definitions,** Art. 100
**Deflection of conductors**
Auxiliary gutters, 366.10(D)
Cabinets and cutout boxes, 312.6
**Deicing installations.** *see* Fixed outdoor electric deicing and snow-melting equipment
**Delta breakers,** 408.16(E)
**Delta-connected**
Identifying high leg, 3-phase supply, 4-wire, 215.8, 230.56, 408.3(E)
**Demand factor.** *see also* Loads
Definition, Art. 100-I
**Detached buildings, oil-insulated transformers in,** 450.26 Ex. 5
**Detearing equipment, electrostatic,** 516.4(E)
**Deteriorating agents,** 110.11
**Device.** *see also* Switches; Receptacles
Definition, Art. 100-I
**Device or equipment fill for boxes,** 314.16(B)(4)
**Diagnostic equipment,** 517.73(A)
**Diagrams**
Adjustable speed drive control, Fig. D.10
Autotransformer overcurrent protection, Fig. 450.4
Cable installation dimensions, underground, Fig. 310.60 for Tables 310.77 through 310.86, Figs. B.310.2 through B.310.5 for use with Tables 310.5 through B.310.10
Cable substitution hierarchy Communications systems, Fig. 800.53
Community antenna TV and radio distribution systems, Fig. 820.53
Fire alarm systems, Fig. 760.61
Optical fiber cables, Fig. 770.53
Remote control, signaling, and power limited circuits, Fig. 725.61
Elevators, dumbwaiters, escalators, etc.
Control system, Fig. 610.2
Single line diagram, Fig. 620.13
Feeders, Fig. 215.5
Generator field control, Fig. D.9

Hazardous (classified) locations, Class I, Zones 0, 1, and 2, marking, 505.9(C)(2)
Hazardous (classified) locations adjacent to dispensers, Fig. 514.3
Health care facility, 517.30, Figs. 517.30(A), 517.30(B)(4), 517.41(A) and (B)
Luminaires, closet storage space, Fig. 410.8
Marine terminal handling flammable liquids, Fig. 515.3
Mobile homes, receptacle and attachment plug, Fig. 550.10(C)
Motor feeder and branch circuits, Fig. 430.1
Park trailers, receptacles and attachment plugs, Fig. 552.44(C)
Recreational vehicles, receptacles and attachment plugs, Fig. 551.46(C)
Remote control, signaling, and power limited circuits, Class 2 and 3 circuits, Fig. 725.41
Services, Fig. 230.1
Solar photovoltaic systems, identification of components, Fig. 690.1(A) and (B)
Spray application, dipping, and coating, Figs. 513.3(B)(1), (B)(2), (B)(4), and (B)(5)
Swimming pools, clearance from pool structures, Fig. 680.8
**Dielectric heat generating equipment.** *see* Induction and dielectric heating equipment
**Dielectric heating.** *see also* Induction and dielectric heating equipment
Definition, 665.2
**Different systems, conductors in same enclosure,** 300.3(C), 392.6(F), 725.26, 725.55(B), 760.26, 760.55, 800.11(A), 800.52(A), 810.18(C), 820.52(A)
**Dimensions**
Conductors, Chap. 9, Tables 5 through 8
Conduits and tubing, Chap. 9, Table 4
**Dimmers, stage switchboard,** 520.25, 520.53(E)
**Dining areas, essential electrical systems,** 517.42(E)
**Dip tanks,** Art. 516
**Direct burial**
Liquidtight flexible metal conduit, 350.10(3)
Liquidtight flexible nonmetallic conduit, 352.10(G)

Over 600 volts, 300.50
Rigid nonmetallic conduit, 300.50, 352.10(G), Table 300.5
Bulk storage plants, 515.8
Service stations, 514.8
Underground feeder and branch-circuit cable, 340.10(1)
Underground service cable, 230.49, 310.7, Table 300.5
**Disconnecting means**
Air-conditioning and refrigerating equipment, 440-II
Appliances, 422-III
Capacitors, 460.8(C)
Cranes, 610-IV
Definitions, Art. 100-I, 551.2
Electric space heating equipment, 424-III
Electrolytic cells, 668.13
Electroplating systems, 669.8
Elevators, 620.51, 620-VI
Fuel cell systems, 692-III
Fuses and thermal cutouts, 240.40
Identification, 110.22
Induction and dielectric heating equipment, 665.12
Information technology equipment, 645.2(1), 645.10
Interconnected electric power production sources, 705.20, 705.21, 705.22
Irrigation machines, 675.8
Mobile homes, 550.11
Motors and controllers, 430.127, 430-IX
Pools, spas, and hot tubs, 680.12
Recreational vehicles, 551.45
Sensitive electronic equipment, lighting equipment, 647.8(A)
Separate building on same premises, 250.32(D)
Services, 230-VI
Connections, ahead of, 230.82
Over 600 volts, 230.205, 230.206
Signs and outline lighting, 600.6
Welders, resistance, 630.33
X-ray equipment, 517.72, 660.5, 660.24
**Discontinued outlets.** *see* Outlets, discontinued
**Dissimilar metals.** *see* Metals, dissimilar
**Doors, transformer vaults,** 450.43
**Double insulated, appliances and tools,** 250.114 Ex., 422.16(B)(1) Ex., 422.16(B)(2) Ex.
**Double locknuts.** *see* Locknuts, double, required
**Drainage**
Equipment, 501.5(F), 505.16(E)
Oil-insulated outdoor transformers, 450.27
Raceways. *see* Raceways, drainage
Transformer vaults, 450.46

**Dressing rooms**
 Motion picture studios, 530-III
 Theaters, 520-VI
**Drip loops**
 Conductors entering buildings,
  225.11, 230.52
 Service heads, 230.54(F)
**Driveways**
 Clearance of conductors, 225.18
 Clearance of service drop,
  230.24(B)
 Protection of service cables,
  230.50(A)
**Drop.** *see* Service drops; Voltage
  and volts, drop
**Dry location.** *see also* Damp or
  wet locations
 Definition, Art. 100-I
**Dryers, clothes.** *see* Clothes dryers
**Dry-type transformers,** 450.1 Ex.
  2, 450.21, 450.22
**Dual-voltage motor, locked-rotor**
  **rating,** 430.7(B)(3)
**Duct heaters, installation of,**
  424-VI
**Ducts**
 Electrical, 310.15,
  B.310.15(B)(4)
 Luminaires (lighting fixtures) in,
  410.4(C)
 Wiring in, 300.21, 300.22,
  725.3(C), 760.3(B),
  770.3(B), 800.51(A),
  800.53(A), 830.54(A),
  830.55(B)
**Dumbwaiters.** *see* Elevators,
  dumbwaiters, escalators,
  moving walks, wheel-
  chair lifts, and stairway
  chair lifts
**Dust-ignitionproof,** 500.2
**Dustproof**
 Definition, 500.2, Art. 100-I
**Dusts.** *see* Hazardous (classified)
  locations
**Dusttight**
 Definition, 500.2, Art. 100-I
**Duty, types**
 Definitions, Art. 100-I
**Duty cycle, welding,** 630.31(B)
  FPN
**Dwellings.** *see also* Appliances;
  Boxes; Branch circuits;
  Fixtures; Grounding,
  and similar general cat-
  egories
 Branch-circuit voltages,
  210.6(A)
 Definitions, Art. 100-I
 Farm, services, 220.40(A),
  220.41
 Feeder load, calculations for,
  220.3(B)(10), 220.30
  through 220.32, Annex
  D
 Lighting loads for, Table
  220.3(A)
 Luminaires (lighting fixtures),
  410.75, 410.80(B)
 Panelboards as services,
  408.3(C), 408.16(A)

Receptacle circuits required,
  210.11
Receptacle outlets required,
  210.52, 680.22(A)
Protection by ground-fault
  circuit-interrupters,
  210.8(A), 680.22(A)(5)

**-E-**

**Eccentric knockouts.** *see* Knockouts, bonding service
  equipment
**Elbows, metal, protection from**
  **corrosion,** 300.6
**Electric discharge lighting**
 1000 volts or less, 410-VIII
 Connection to luminaires (lighting fixtures), 410.14,
  410.30(C)
 Definition, 600.2
 More than 1000 volts, 410-XIV
 Signs and outline lighting, Art.
  600
 Wiring, equipment, 410-VIII,
  410-XIV
**Electric power production**
  **sources.** *see* Generators;
  Interconnected electric
  power production
  sources; Solar photovol-
  taic systems
**Electric signs.** *see* Signs, electric
**Electric space heating equip-**
  **ment.** *see* Fixed electric
  space heating equipment
**Electric vehicle charging system**
  **equipment,** Art. 625
 Automatic de-energization,
  625.19
 Cable, 625.17
 Coupler, 625.9
 Coupling method, 625.16
 Definitions, 625.2
 Disconnecting means, 625.23
 Listed, 625.5
 Marking, 625.15
 Protection and control, 625-IV
 Rating, 625.14
 Supply equipment, 625.13
  Construction, 625-III
  Definition, 625.2
  Interlock, 625.18
  Locations, 625-V
  Markings, 625.15
 Ventilation not required,
  625.29(C)
 Ventilation required, 625.29(D),
  Table 625.29(D)
 Voltages, 625.4
 Wiring methods, 625-II
**Electric vehicles**
 Definition, 625.2
**Electric welders.** *see* Welders,
  electric
**Electrical ducts**
 Definition, 310.60(A)
**Electrical life support equipment**
 Definition, 517.2
**Electrical metallic tubing (Type**
  **EMT),** Art. 358
 Bends, how made, 358.24

Bends, number in one run,
  358.26
Connectors and couplings,
  358.42
Construction, 358.100
Construction specifications, 358-
  III
Definition, 358.2
Grounding, 358.60
Installation, 358-II
Listing, 358.6
Marking, 358.120
Maximum number of conductors
  and fixture wires in,
  Tables C1 and C1A
Number of conductors in,
  358.22
Reaming and threading, 358.28
Securing and supporting, 358.30
Size, 358.20
Splices and taps, 358.56
Uses not permitted, 358.12
Uses permitted, 358.10
Wet locations, in, 358.6(C)
**Electrical noise.** *see also* Sensitive
  electronic equipment,
  647.3
 Grounding, 250.96(B),
  250.146(D)
 Panelboards, 408.20 Ex.
 Receptacles, 406.10(D)
**Electrical nonmetallic tubing**
  **(Type ENT),** Art. 362
 Bends, 362.24, 362.26
 Bushings, 362.46
 Construction specifications, 362-
  III
 Definition, 362.2
 Grounding, 362.60
 Installation, 362-II
 Joints, 362.48
 Listing, 362.6
 Maximum number of conductors
  and fixture wires in,
  Tables C2 and C2A
 Number of conductors in,
  362.22
 Securing and supporting, 362.30
 Size, 362.20
 Splices and taps, 362.56
 Through metal framing mem-
  bers, 300.4(B)
 Trimming, 362.28
 Uses not permitted, 362.12
 Uses permitted, 362.10
**Electrically operated pool covers.**
  *see* Covers, pool, elec-
  trically operated
**Electrodes, grounding**
 Aluminum, 250.52(B)
 Common, 250.58
 Concrete encased, 250.52(A)(3)
 Gas piping as, 250.52(B)
 Made, 250.50, 250.52
 Metal frame of building as,
  250.52(A)(2)
 Metal water piping system,
  250.52(A)(1)
 Resistance to ground of, 250.56
 Separately derived systems,
  250.30(A)(4)

**Electrolytic cells,** Art. 668
 Auxiliary nonelectric connec-
  tions, 668.31
 Cell line conductors, 668.12
 Cell line working zone, 668.10
 Cranes and hoists, 668.32
 DC cell line process power sup-
  ply, 668.11
 Definitions, 668.2
 Disconnecting means, 668.13
 Fixed electrical equipment,
  668.30
 Grounding, 668.15
 Portable electrical equipment,
  668.20, 668.21, 668.30
 Shunting means, 668.14
**Electronic computer/data pro-**
  **cessing equipment.** *see*
  Information technology
  equipment
**Electronically actuated fuse**
 Definition, Art. 100-I
**Electroplating systems,** Art. 669
 Branch-circuit conductors, 669.5
 Disconnecting means, 669.8
 Overcurrent protection, 669.9
 Warning signs, 669.7
 Wiring methods, 669.6
**Electrostatic equipment, spraying**
  **and detearing paint,**
  516.4(E), 516.10
**Elevators, dumbwaiters, escala-**
  **tors, moving walks,**
  **wheelchair lifts, and**
  **stairway chair lifts,**
  Art. 620
 Cables, traveling, 620-V
 Clearances, working, 620.5
 Conductors, 620-II through
  620-IV
 Definitions, 620.2
 Disconnecting means and con-
  trol, 620-VI
 Emergency and standby power
  systems, 620-X
 Essential electrical systems,
  517.32(F), 517.42(G)
 Ground-fault circuit interrupters,
  620.85
 Grounding, 250.112, 620-IX
 Live parts, 620.4
 Machine room, 620-VIII
 Overcurrent protection, 620-VII
 Voltage limitations, 620.3
 Wiring, 620-III
 Working clearances, 620.5
**Emergency systems,** Art. 700. *see*
  *also* Legally required
  standby systems; Op-
  tional standby systems
 Capacity, 700.5
 Circuit wiring, 700-II
 Circuits for lighting and power,
  700-IV, 700-V
 Connections at services, 230.82
 Control, 700-V
 Definitions, 517.2
 Equipment approval, 700.3
 Health care facilities, 517.30,
  517.31
 Overcurrent protection, 700-VI

Service disconnecting means, 230.72(B)
Services, 230.2(A)(2)
Signals, 700.7
Signs, 700.8
Sources of power, 700-III
Tests and maintenance, 700.4
Transfer equipment, 695.4, 700.6
Unit equipment, 700.12(E)
Wiring, 700.9
**Enamel, coating of**
Electrical metallic tubing, 358.10(B)
General equipment, indoor use only, 300.6(A)
Intermediate metal conduit, 342.10(B)
Metal boxes, 314.40(A)
Metal cabinets, 312.10(A)
Removing for grounding connections, 250.96(A)
Rigid metal conduit, 344.10(B)
**Enclosed**
Definition, Art. 100-I
**Enclosures.** *see also* Hazardous (classified) locations
Arcing parts, 110.18
Bonding, 250.96, 250.146
Cabinets and cutout boxes, 312.7 through 312.9, 312.11
Circuits in, number of, 90.8(B)
Cranes, hoists, resistors, 610.2(B)
Definition, Art. 100-I
Elevators, dumbwaiters, escalators, moving walks, wheelchair lifts, and stairway chair lifts, 620.4, 620-VIII
Grounding, 250-IV
Hazardous (classified) locations. *see* Hazardous (classified) locations involved
High-intensity discharge lamp auxiliary equipment, 410.54(A)
Induction and dielectric heating, 665.20
Installations over 600 volts, 110.31
Intrinsically safe conductors in, 504.30(A)(3)
Overcurrent protection, 240-III
Panelboards, 408.18
Radio equipment, 810.71(A)
Signs, 600.8, 600.42(G)
Subsurface, 110.12(B), 314.50
Switches, 404.3
Underground, 110.12(B), 314.50
**Energized**
Definition, Art. 100-I
**Energized parts.** *see also* Live parts
Motors, motor circuits, and controllers, 430-XI
**Energy-limited, Class 2 and Class 3 control and signal circuits,** 725.41
**Enforcement of code,** 90.4, 510.2, Art. 80

**Equipment.** *see also* specific types of
Approval, 80.19, 90.4, 110.2
Cooling of, 110.13(B)
Definition, Art. 100-I
Examination of, 90.7, 110.3
Grounding, 250-VI
Installation, general provisions, Art. 110
Less than 50 volts, Art. 720
More than 600 volts, Art. 490
Mounting. *see* Mounting of equipment
Sealable. *see* Sealable equipment
Service
    Definition, Art. 100-I
Signal
    Definition, 620.2
**Equipment grounding conductor fill and boxes,** 314.16(B)(5)
**Equipment grounding conductors**
Connections at outlets, 250.146, 250.148
Definition, Art. 100-I
Fuel cell systems, 692.44
Installation, 215.6, 250.120, 250.130
Sizing, 250.122
Types recognized, 250.118, 250.120, 250.134
**Equipment system**
Definition, 517.2
**Escalators.** *see* Elevators, dumbwaiters, escalators, moving walks, wheelchair lifts, and stairway chair lifts
**Essential electrical systems.** *see also* Health care facilities
Definition, 517.2
**Examples (computations),** Annex D
**Exciters**
Leads in enclosure with other conductors, 300.3(C)(2)(c)
Overcurrent protection for, 445.12(A)
**Exhaust ducts.** *see* Ducts
**Exit signs.** *see* Signs, exit
**Exits, emergency lighting for,** 700.1 FPN No. 3. *see also* Health care facilities
**Expansion fittings,** 352.44
**Expansion joints (fittings),** 250.98, 300.7(B), 352.44, 366.8, 368.28, 424.44(C), 424.98(C), 426.20(E), 426.22(D), 427.16
**Explanatory material FPN,** 90.5
**Explosionproof apparatus,** 500.2, 500.7(A), 502.1, 505.16(B)(2)
Definition, Art. 100-I

**Explosive atmospheres, gases, liquids, dusts, ignitable fibers, lints and flyings.** *see* Hazardous (classified) locations
**Exposed.** *see also* Enclosures
Conductive surfaces
    Definition, 517.2
Extensions, boxes, and fittings, 314.22
Live parts, 110.26
    Definition, Art. 100-I
Structural steel, grounding, 250.52(A)(2)
Wiring methods
    Definition, Art. 100-I
**Extension cord sets**
On construction sites, 527.6(A) and (B)(2)
Overcurrent protection, 240.4
**Extensions**
Auxiliary gutters, 366.3
Boxes and fittings, exposed, 314.22
Cellular metal floor raceways, 374.11
Flat cable assemblies, 322.40(D)
Nonmetallic. *see* Nonmetallic extensions
Wireways, 376.70, 378.70
**Externally operable.** *see* Operable, externally

**-F-**
**Faceplates**
Grounding, 250.86, 250.110, 404.9(B), 406.5(B), 550.15(D)
Material, 404.9(C), 406.5
Minimum thicknesses for, 404.9(C), 406.5
Mobile homes, 550.15(D)
Mounting surfaces, against, 404.9(A), 404.10(B), 406.4(A), 406.4(B)
**Fairs.** *see* Carnivals, circuses, fairs, and similar events
**Fans, ceiling.** *see* Ceiling fans
**Farm buildings, feeder calculations,** 220.40. *see also* Agricultural buildings
**Feeder assembly, mobile homes**
Definitions, 550.2
**Feeders,** Art. 215, Art. 220
Busways, 368.12
Calculation of loads, 215.2(A), 220.10, 220.30, Annex D
Computing loads optional calculations, 220-III
Definition, Art. 100-I
Farm buildings, for, 220.40
Ground-fault circuit-interrupters, with, 215.9
Grounding means, 215.6
Hoistways and machine rooms, 620.37(C)
Identification, 225.37
Kitchen equipment, commercial, 220.20

Loads. *see* Loads, feeder
Mobile homes and parks, 550.10(A), 550.33, 550-III
Motion picture studios, 530.18(B)
Motors, 430-II
Outside, Art. 225
Overcurrent protection, motor, 430-V
Restaurants, 220.36
Services, 220-II
Stage switchboard, 520.27
Taps, 240.21(B), 430.28
Television studio sets, 530.19
**Feed-through connections of neutral conductors,** 300.13(B)
**Ferrous metals, protection from corrosion,** 300.6
**Festoon lighting**
Conductor size, 225.6(B)
Definition, Art. 100-I
**Festoons, portable stage equipment,** 520.65
**Fiber bushings, AC cable ends,** 320.40
**Fibers, lint, flyings, ignitible,** Art. 503. *see also* Hazardous (classified) locations
**Figures.** *see* Diagrams
**Films, motion picture.** *see also* Projectors, motion picture
Storage vaults, 530-V
Viewing, cutting, patching tables, 530-IV
**Fine Print Notes FPN, mandatory rules, permissive rules, and explanatory material,** 90.5
**Finish rating**
Definition, 362.10 FPN 2
**Fire alarm systems,** Art. 760
Access to electrical equipment, 760.5
Circuits extending beyond one building, 760.11
Classification, 760.15
Connections at services, 230.82(4), 230.94 Ex. 4
Definitions, 760.2
Grounding, 250.112(I), 760.9
Health care facilities, 517-VI
Identification, 760.10
Installation, 760.7
Location, 760.3
Mechanical execution of work, 760.6
Nonpower-limited circuits, 760-II
    Cable marking, 760.31, Table 760.31(G)
    Conductors, 760.27
    Conductors of different circuits in same raceway, etc., 760.26
    Listing, 760.31
    Location of overcurrent devices, 760.24
    Marking, 760.31
    Multiconductor cable, 760.30

Number of conductors in raceway, etc., and derating, 760.28
Overcurrent protection, 760.23, 760.24
Power source requirements, 760.21
Requirements, 760.15
Wiring method, 760.25, 760.30(A)
Power-limited circuits, 760-III
Cable markings, 760.71
Cable substitutions, 760.61(D), Table 760.61
Circuit integrity cable, 760.71(G)
Circuit markings, 760.42
Conductor size, 760.58
Conductor support, 760.57
Installation, 760.54, 760.56
Line-type fire detectors, 760.59, 760.71(K)
Listing, 760.71
Location of overcurrent devices, Chap. 9, Tables 12(A) and 12(B)
Overcurrent protection, Chap. 9, Tables 12(A) and 12(B)
Power sources, 760.41
Separation of conductors, 760.55
Wiring methods and materials
Load side, 760.52
Supply side, 760.51
**Fire alarms.** *see* Alarm systems, health care facilities; Alarms; Fire alarm systems
**Fire detectors, line-type,** 760.71(K)
**Fire pumps,** Art. 695
Connection at services, 230.72(A) Ex., 230.82(4), 230.94 Ex. 4
Continuity of power, 695.4
Control wiring, 695.14
Definitions, 695.2
Emergency power supply, Art. 700
Equipment location, 695.12
Listing, 695.7
Motor, overcurrent protection, 430.31 FPN
Power sources to electric motor-driven, 695.3
Power wiring, 695.6
Remote-control circuits, 430.72
Service equipment overcurrent protection, 230.90 Ex. 4
Services, 230.2(A)(1)
Supervised connection, 695.4(B)
Transformers, 695.5
Voltage drop, 695.7
**Fire spread**
Prevention of, wiring methods, 300.21, 725.3(B), 760.3(A), 770.3(A), 800.52(B), 820.52(B), 830.58(B)
**Fire-stopped partitions,** 300.21. *see also* Firewalls

**Firewalls, wiring through,** 300.21, 800.52(B), 820.52(B), 830.58(B)
**Fittings,** Art. 314. *see also* Wiring Methods, types and materials
Conduit bodies, 314.16
Definition, Art. 100-I
Expansion, 352.44
Insulation, 300.4(F)
**Fixed electric heating equipment for pipelines and vessels,** Art. 427
Application of other controls, 427.3
Branch circuit sizing, 427.4
Control and protection, 427-VII
Controls, 427.56
Disconnecting means, 427.55
Overcurrent protection, 427.57
Definitions, 427.2
Impedance heating, 427-IV
Grounding, 427.29
Induced currents, 427.28
Isolation transformer, 427.26
Personnel protection, 427.25
Secondary conductor sizing, 427.30
Voltage limitations, 427.27
Induction heating, 427-V
Induced current, 427.37
Personnel protection, 427.36
Scope, 427.35
Installation, 427-II
General, 427.10
Identification, 427.13
Thermal protection, 427.12
Use, 427.11
Resistance heating elements, 427-III
Electrical connections, 427.19
Equipment protection, 427.22
Expansion and construction, 427.16
Flexural capability, 427.17
Grounded conductive covering, 427.23
Marking, 427.20
Not in direct contact, 427.15
Power supply leads, 427.18
Secured, 427.14
Skin effect heating, 427-VI
Conductor ampacity, 427.45
Grounding, 427.45
Pull boxes, 427.46
Single conductor in enclosure, 427.47
**Fixed electric space heating equipment,** Art. 424
Boilers, protection of, 424.72(A)
Branch circuits for, 424.3
Cables, 424-V
Area restrictions, 424.38
Clearances, wiring
Ceilings, 424.36
Other objects, openings, 424.39
Walls, exterior, 424.37
Construction, 424.34
Finished ceilings, 424.42
Installation

Cables, 424.44
Nonheating leads, 424.43
Marking, 424.35
Splices, 424.40, 424.41(D)
Tests and inspections, 424.45
Control and protection, 424-III
Controllers, 424.20(A)
Disconnecting means, 424.19 through 424.21
Indicating, 424.21
Overcurrent protection, 424.22
Duct heaters, 424-VI
Air flow, 424.59
Condensation, 424.62
Disconnecting means, location, 424.65
Elevated inlet temperature, 424.60
Fan circuit interlock, 424.63
Identification, 424.58
Installation, 424.61, 424.66
Limit controls, 424.64
Electrode-type boilers, 424-VIII
Installation, 424-II
Location
Exposed to physical damage, 424.12(A)
Wet areas, 424.12(B)
Spacing, combustible materials, 424.13
Special permission, 424.10
Supply conductors, 424.11
Marking, 424-IV
Heating elements, 424.29
Nameplate, 424.28
Overcurrent protection for, 424.22
Radiant heating panels and heating panel sets, 424-IX
Clearances, wiring
Ceilings, 424.94
Walls, 424.95
Connection, 424.96
Definitions, 424.91
Installation, 424.93, 424.98, 424.99
Markings, 424.92
Nonheating leads, 424.97
Resistance-type boilers, 424-VII
**Fixed equipment, grounding,** 250.110, 250.112, 250.116, 250.134, 250.136, 517.13
**Fixed outdoor electric deicing and snow-melting equipment,** Art. 426
Branch circuits for, 210.23, 426.4
Connections, 426.24
Control and protection, 426-VI
Cord- and plug-connected, 426.54
Definitions, 426.2
Disconnecting means, 426.50
Equipment protection, 426.28
General, 426.10
Grounding, 426.27, 426.34, 426.44
Identification of, 426.13
Impedance heating, 426-IV
Induced currents, 426.33

Isolation transformer, 426.31
Personnel protection, 426.30
Voltage limitations, 426.32
Installation, 426-II
Marking, 426.25
Nonheating leads, installation of, 426.22, 426.33
Overcurrent protection of, 426.52
Protection, corrosion, 426.26, 426.43
Protection, thermal, 426.12
Resistance heating elements, 426-III
Embedded, 426.20
Exposed, 426.21
Skin effect heating, 426-V
Conductor ampacity, 426.40
Pull boxes, 426.41
Single conductor in enclosure, 426.42
Special permission, 426.14
Thermostats for, 426.51
Use of, 426.11
**Fixture wires,** Art. 402
Ampacities, 402.5
Grounded conductor, identification, 402.8
Marking, 402.9
Minimum size, 402.6
Number in conduits or tubing, 402.7
Overcurrent protection, 240.4, 402.12
Types, 402.3
Uses not permitted, 402.11
Uses permitted, 402.10
**Fixtures.** *see* Luminaires (lighting fixtures)
**Flame spread.** *see* Fire spread
**Flammable anesthetics.** *see* Anesthetics, flammable
**Flammable gases and vapors.** *see* Hazardous (classified) locations
**Flammable liquids.** *see* Hazardous (classified) locations
**Flashers, time switches, etc.,** 404.5, 600.6(B)
**Flat cable assemblies (Type FC),** Art. 322
Boxes and fittings, 322.40
Branch-circuit rating, 322.10(1)
Conductor insulations for, 322.112
Conductor size, 322.104
Construction, 322-III
Dead ends, 322.40(A)
Definition, 322.2
Extensions from, 322.40(D)
Fittings for, 322.40(C)
Identification grounded conductor, 322.120(B)
Luminaire (fixture) hangers, 322.40(B)
Marking, 322.120
Number of conductors in, 322.100
Size of conductors in, 322.104
Splices, 322.56
Supports for, 322.30

Terminal block identification, 322.120(C)
Uses not permitted, 322.12
Uses permitted, 322.10
**Flat conductor cable (Type FCC),** Art. 324
Branch-circuit rating, 324.10(B)
Cable connections and ends, 324.40(A)
Construction, 324-III
  Conductor identification, 324.120(B)
  Corrosion resistance, 324.10(I)
  Insulation, 324.112
  Markings, 324.120(A)
  Receptacles and housings, 324.42(B)
  Transition assemblies, 324.56(B)
Definitions, 324.2
Installation, 324-II
  Alterations, 324.56(A)
  Boxes and fittings, 324.40
  Cable connections and insulating ends, 324.40
  Connections to other systems, 324.40(D)
  Coverings, 324.10(H)
  Crossings, 324.18
  Enclosure and shield connections, 324.10(I)
  Heated floors, 324.10(F)
  Metal shield connectors, 324.10(J)
  Receptacles, 324.42
  Shields, 324.40(C)
  Supports, 324.30
  System height, 324.10(G)
Polarization, 324.40(B)
Systems alterations, 324.56(A)
Uses not permitted, 324.12
Uses permitted, 324.10
**Flatirons,** 422.46
Signals, 422.42
Stands for, 422.45
Temperature-limiting means, 422.42, 422.46
**Flexible cords.** *see* Cords, flexible
**Flexible metal conduit (Type FML),** Art. 348
Bends, 348.24 and 348.26
Couplings and connectors, 348.42
Definition, 348.2
Grounding and bonding, 348.60
Installation, 348-II
Liquidtight. *see* Liquidtight flexible metal conduit
Listing, 348.6
Maximum number of conductors and fixture wires in, Tables C3 and C3A
Number of conductors, 348.22
Securing and supporting, 348.30
Size, 348.20
Splices and taps, 348.56
Trimming, 348.28
Uses not permitted, 348.12
Uses permitted, 348.10

**Flexible metallic tubing (Type FMT),** Art. 360
Bends, 360.24
Boxes and fittings, 360.40
Construction specifications, 360-II
Definition, 360.2
Grounding, 360.60
Installation, 360-II
Listing, 360.6
Marking, 360.120
Number of conductors, 360.22
Sizes, 360.20
Splices and taps, 360.56
Uses not permitted, 360.12
Uses permitted, 360.10
**Floating buildings,** Art. 553
Bonding of noncurrent carrying metal parts, 553.11
Definition, 553.2
Feeder conductors, 553.6
Grounding, 553.8, 553.10, 553-III
Insulated neutral, 553.9
Service conductors, 553.5
Service equipment location, 553.4
Services and feeders, installation, 553.7
**Floor pockets,** 520.46
**Floors, receptacles,** 210.52(A)(3), 250.146(C), 314.27(C)
**Flue damper control.** *see* Dampers (flue), control
**Fluorescent luminaires (lighting fixtures),** 410-XIII, 410-XIV
Autotransformers in ballasts, 410.78
Auxiliary equipment, remote from, 410.77
Ballast protection required, 410.73(E)
Circuit breakers used to switch, 240.83(D)
Connection of, 410.14, 410.30(C)
Load calculations
  Ampere ratings, ballasts, 220.4(B)
  Branch circuits, 210.23
Raceways, 410.31
Snap switches for, 404.14(A)(1) and (B)(2)
Surface-mounted or recessed, 410.8(B)(2), 410.8(D)(2) and (4)
Thermal protection, 410.73(E)
**Fluoroscopic equipment.** *see* X-ray equipment
**Footlights, theaters,** 520.43
Definition, 520.2
Disappearing, 520.43(C)
**Formal interpretations,** 90.6
**Forming shells, underwater pool lighting fixtures,** 680.23(B)
Definition, 680.2
**Fountains.** *see* Swimming pools, fountains, and similar installations

**FPN.** *see* Fine Print Notes, explanatory material
**Frame**
Definition, 551.2
**Fuel cell systems,** Art. 692
Circuit requirements, 692-II
Circuit sizing, 692.8
Connection to other circuits, 692-VII
  Identified interactive equipment, 692.60
  Loss of interactive system power, 692.62
  Output characteristics, 692.61
  Point of connection, 692.65
  Transfer switch, 692.59
  Unbalanced interconnection, 692.64
Definitions, 692.2
Disconnecting means, 692-III
  All conductors, 692.13
  Provisions of, 225.31, 225.33 through 225.40, and 692.14
  Switch or circuit breaker, 692.17
Grounding, 692-V
  Equipment grounding conductor, 692.44
  Grounding electrode system, 692.47
  Size of equipment grounding conductor, 692.55
  System grounding, 692.41
Installation, 692.4
Marking, 692-VI
  Fuel cell power sources, 692.53
  Fuel shutoff, 692.54
  Stored energy, 692.56
Over 600 volts, 692-VIII
  General, 692.80
Overcurrent protection, 692.9
Stand-alone system, 692.10
Wiring methods, 692-IV
  Wiring systems, 692.31
**Full-load current motors**
Alternating current
  Single-phase, Table 430.148
  Three-phase, Table 430.150
  Two-phase, Table 430.149
Direct current, Table 430.147
**Furnaces.** *see* Heating equipment, central
**Fuseholders**
Cartridge fuses, 240-VI
Over 600 volts, 490.21(B)
Plug fuses, 240-V
Rating, motor controllers, 430.90
Type S, 240.53, 240.54
**Fuses,** Art. 240. *see also* Hazardous (classified) locations
Cartridge. *see* Cartridge fuses
Definition, over 600 volts, 100-II
Disconnecting means, 240.40
Electronically actuated
  Definition, Art. 100-II
Enclosures, 240-III
Generators, constant-voltage, 445.12(A)

Group installation of motors, 430.53(C)
Location in circuit, 240.21
Location on premises, 240.24
Motor branch circuit protection, 430-IV
Motor overload protection, 430-III
Multiple. *see* Multiple fuses
In parallel, not permitted, 240.8, 404.17
Plug, Edison-base type, 240-V
Potential (voltage) transformers, 450.3(C)
Ratings, 240.6
Services over 600 volts, 230.208
Type S, 240.53, 240.54
**Future expansion and convenience,** 90.8(A)


-G-
**Gages (AWG), conductors,** 110.6
**Garages, commercial,** Art. 511
Battery charging equipment, 511.10(A)
Classification of locations, 511.3
Definition, Art. 100-I
Electric vehicle charging, 511.10(B)
Elevators, dumbwaiters, escalators, moving walks, in, 620.38
Equipment, 511.4, 511.7
Ground-fault circuit-interruption protection, 511.12
Grounding, 250.112(F), 511.16
Locations, 511.2, 511.3
Sealing, 511.9
Special equipment, 511.10
Wiring, 511.4, 511.7
**Gas pipe**
As grounding electrode, 250.52(B)(1)
**Gases.** *see* Hazardous (classified) locations
**Gasoline bulk storage plants.** *see* Bulk storage plants; *see also* Hazardous (classified) locations
**Gasoline dispensing and service stations.** *see* Motor fuel dispensing facilities; *see also* Hazardous (classified) locations
**General care areas,** 517.18
Definition, 517.2
**General provisions, installation of wiring and equipment,** Art. 110
**General requirements for wiring methods,** Art. 300. *see also* Wiring methods
Boxes, conduit bodies, or fittings, where required, 300.15
Changing raceway or cable to open wiring, 300.16
Conductors, 300.3
  Different systems, 300.3(C)
  Installation with other systems, 300.8

Insulated fittings, 300.4(F), 300.4(H)
Number and sizes of, in raceways, 300.17
Supporting of conductors in vertical raceways, 300.19
Exhaust and ventilating ducts, wiring, 300.22
Expansion joints, raceways, 300.7(B)
Feed-through neutral connections, 300.13(B)
Free length of wire at outlets, junction and switch points, 300.14
Induced currents in metal enclosures, 300.20
Mechanical and electrical continuity, conductors, raceways, cables, 300.10, 300.12, 300.13
Over 600 volts, 110-III, 300-II, Art. 490
Protection against
  Corrosion, 300.6
  Physical damage, 300.4
Raceway installations, 300.18
Raceways and cables with other systems, 300.8
Raceways and fittings, metric designator and trade size, 300.1(C)
Raceways as a means of support, 300.11(B)
Raceways exposed to different temperatures, 300.7
Sealing, raceways, change in temperature, 300.7(A)
Securing and supporting, 300.11
Spread of fire or products of combustion, 300.21
Temperature limitations. *see* 310.10
Through studs, joists, rafters, 300.4(A)
Underground installations, 300.5, 300.50
Voltage limitations, 300.2
Wiring in ducts, plenums, air-handling spaces, 300.22
**Generators,** Art. 445
Bushings, 445.16
Conductor, ampacity of, 445.13
Emergency systems, 700.12(B)
Essential electrical systems, task illumination, 517.32(E), 517.42(F)
Grounding, 250.34, 250.112
Guards for attendants, 445.15
Health care facilities, 517.30, 517.35, 517.44
Location, 445.10
Marking, 445.11
Overcurrent protection, 445.12
Portable, 250.34
Protection of live parts, 445.14
Recreational vehicles, 551.30
Standby systems, Art. 701, 702
**Goosenecks, service cables,** 230.54

**Grooves, shallow, cables and raceways installed in,** 300.4(E)
**Ground**
Definition, Art. 100-I
Voltage to
  Definition, Art. 100-I
**Ground clamps,** 250.8, 250.10, 250.70
**Ground ring,** 250.50(4), 250.166(E)
**Grounded**
Definition, Art. 100-I
Effectively, 250.50(2)
  CATV system, 820.40, 820-IV
  Communication system, 800-IV
  Definition, 100-I
Solidly
  Definition, 230.95
**Grounded conductor.** *see* Conductors, Grounded; Neutral
**Ground-fault circuit-interrupters**
Accessory buildings, dwelling units, 210.8(A)(2)
Basements, dwelling units, unfinished, 210.8(A)(5)
Bathtubs, hydromassage, 680.71
Definition, Art. 100-I
Deicing and snow-melting equipment, 426.32
Electrically operated pool covers, 680.27(B)(2)
Fountains, 680.51(A)
Garages, commercial, 511.12
Garages, dwelling units, 210.8(A)(2)
Permitted uses, 210.8, 215.9
Personnel, protection for, 426.32
Pipeline heating, 427.27
Pools and tubs for therapeutic use, 680.62(A)
Receptacles, 210.8
  Bathrooms, in dwelling units, 210.8(A)(1)
  Bathrooms in other than dwelling units, 210.8(B)(1)
  Boathouses, 555.19(B)
  Construction sites, at, 527.6
  Existing, 406.3(D)(2)
  Garages, in dwelling units, 210.8(A)(2)
  Health care facilities, 517.2(A)
  Kitchens in dwelling units, 210.8(A)(6)
  Marinas and boatyards, 555.19(B)
  Mobile homes, 550.13(B), 550.32(E)
  Outdoors, dwelling units, 210.8(A)(3)
  Park trailers, 552.41(C)
  Pools, 680.5, 680.6, 680.22, 680.23, 680.32
  Recreational vehicles, 551.40(C), 551.41(C), 551.71
  Required, 210.8

Rooftops in other than dwelling units, 210.8(B)(2)
Sensitive electronic equipment, 647.7(A)
Spas and hot tubs, 680.42, 680.43, 680.44, 680.57
Wet bar sinks, 210.8(A)(7)
**Ground-fault protection**
Deicing and snow melting equipment, 426.28
Emergency systems, not required, 700.26
Equipment, 215.10, 240.13
  Definition, Art.100-I
Health care facilities, 517.17
Personnel. *see* Ground-fault circuit-interrupters
Pipeline heaters, 427.22
Service disconnecting means, 230.95
Solar photovoltaic systems, 690.5
**Grounding,** Art. 250
AC systems, 250.20, 250.24, 250.26
Air-conditioning units, 250.114
Anesthetizing locations, 517.62
Antenna, 810.15, 810.21, 810.58, 810.71(B)
Appliances, 250.114
Audio signal processing, amplification, and reproduction equipment, 640.7
Bonding, 250-V
Busways, over 600 volts, 368.22
Cablebus, 370.9
Capacitors, 460.10
CATV systems, 820-IV
Circuits, 250-I, 250-II
  Less than 50 volts, 250.20(A), 720.10
Clothes dryers, 250.140
Clothes washers, 250.114
Communications systems, 800.33, 800-IV
Continuity, 250.96
Control, radio equipment, 810.71(B)
Control circuits, 725.9
Cranes and hoists, 610.61
DC systems, 250.162, 250.164
Deicing, snow-melting systems, 426.27
Dishwashers, 250.114
Electrode system, 250-III
Elevators, 620-X
Enclosures, 250-IV
Equipment, cord- and plug-connected, 250.114
Fire alarm systems, 250.112(I), 760.9
Fixed equipment, 250.110, 250.112, 250.134, 517.13
Fixtures, lampholders, etc., 410-VI
Fountains, 680.54, 680.55
Freezers, 250.114
Fuel cell systems, 692-V
Generators, 445.1
  Portable and vehicle mounted, 250.34

Hazardous (classified) locations, 501.16, 502.16, 503.16, 505.25
Health care facilities, 517.13, 517.19
Induction and dielectric heating equipment, 665-II
Information technology equipment, 645.15
Instrument transformers, relays, etc., 250-X
Intrinsically safe systems, 504.50
Lightning surge arresters, 280.25
Metal boxes, 314.40(D)
Metal enclosures for conductors, 250.86
Metal faceplates, 404.9(B), 406.5(B)
Metal siding, 250.116 FPN
Methods, 250-VII
Mobile homes, 550.16
More than 600 volts between conductors, 300.40
Motion picture studios, 530.20
Motors and controllers, 430-XII
Nonelectrical equipment, 250.116
Organs, 650.4
Over 600 volts, 300.40, 490.74
Over 1000 volts, 250-X
Panelboards, 408.20, 517.19(D)
Patient care areas, 517.13, 517.19(C)
Portable equipment, 250.114
Radio and television, 810.15, 810.21, 810.58, 810.71
Ranges and similar appliances, 250.140
Receptacles, 210.7, 250.146, 250.148, 406.3, 517.13, 517.19(G)
Recreational vehicles, 551.54, 551.55, 551.75, 551.76
Refrigerators, 250.114
Sensitive electronic equipment, 647.6
Separate buildings, 250.32
Separately derived systems, 250.20(D), 250.30
Signs and outline lighting, 600.7
Spas and tubs, 680.6, 680.7, 680.43(F)
Surge arresters, 280.25
Swimming pools, 680.6, 680.7, 680.24(D), 680.25
Switchboards, 408.12
Switches, 404.12
Systems, 250-I, 250-II
Theaters and similar locations, 520.81
Tools, motor operated, 250.114
Transformers, 450.10
Transient voltage surge suppressors, 285.25
X-ray equipment, 517.78, 660-IV
**Grounding conductors,** 250-III, 250-VI
Definition, Art. 100-I
Earth as, 250.4(A)(5), 250.54
Enclosures, 250-IV

Identification, multiconductor cable, 250.119
Installation, 250.64, 250.120
Material, 250.62
Objectionable current over, 250.6
Sizes, 250.122
**Grounding electrode conductors**
Connection to electrodes, 250-I, 250-III
Definition, Art. 100-I
Installation, 250.64
Material, 250.62, 250.118
Sizing, 250.66, 250.166
**Grounding point**
Patient equipment
Definition, 517.2
Reference
Definition, 517.2
**Grounding-type attachment plugs.** *see* Attachment plugs, grounding-type
**Group installation, motors,** 430.53
**Grouping, switches, circuit breakers,** 404.8. *see also* Accessible
**Grouping of disconnects,** 230.72
**Guarded**
Definition, Art. 100-I
**Guarding, guards.** *see also* Enclosures; Live parts
Circuit breaker handles, 240.41(B)
Construction sites, 527.7
Elevators, dumbwaiters, escalators, moving walks, 620.71
Generators, 445.15
Handlamps, portable, 410.42(B)
Induction and dielectric heating equipment, 665-II
Lamps, theaters, dressing rooms, etc., 520.44(A)(3), 520.47, 520.65, 520.72
Live parts
General, 110.27
On stage switchboards, 520.22
In theaters, 520.7
Motion picture studios, 530.15, 530.62
Motors and motor controllers, 430.143, 430-XI
Over 600 volts, 110.34, 300.50(B), 527.7
Transformers, 450.8
X-ray installations, 517.78
**Guest rooms, outlets,** 210.60
**Gutters, auxiliary.** *see* Auxiliary gutters

**-H-**

**Hallways, outlets,** 210.52(H)
**Handlamps, portable,** 410.42(B)
**Hangars, aircraft.** *see* Aircraft hangars
**Hazard current**
Definition, 517.2
**Hazardous areas.** *see* Hazardous (classified) locations

**Hazardous atmospheres,** Art. 500
Class I locations, 500.5(B)
Class II locations, 500.5(C)
Class III locations, 500.5(D)
Groups A through G, 500.6
Specific occupancies, Art. 510
**Hazardous (classified) locations, Class I,** 500.5(B), 500.6(A), Art. 501. *see also* Class I, Zone 0, 1, and 2 locations
Bonding, 501.16(A), 504.60
Circuit breakers, 501.6
Conductor insulation, 501.13
Control transformers and resistors, 501.7
Cranes and hoists, 610.2
Drainage of equipment, 501.5(F)
Flexible cords, 501.11
Fuses, 501.6
Grounding, 501.16
Hermetically sealed, 500.7(J)
Induction and dielectric heating equipment, 665.4
Live parts, 501.15
Luminaires (lighting fixtures), 501.9
Meters, instruments, and relays, 501.3
Motor controllers, 501.6
Motors and generators, 501.8
Nonincendive circuit, 500.7(F)
Nonincendive component, 500.7(H)
Nonincendive equipment, 500.7(G)
Oil immersion, 500.7(I)
Receptacles and attachment plugs, 501.12
Sealing, 501.5
Signaling, alarm, remote-control, and communications systems, 501.14
Surge protection, 501.17
Switches, 501.6
Transformers and capacitors, 501.2
Utilization equipment, 501.10
Wiring methods, 501.4
Zone 0, 1, and 2 locations. *see* Hazardous (classified) locations, Class I, Zone 0, 1, and 2
**Hazardous (classified) locations, Class I, Zone 0, 1, and 2 locations,** Art. 505
Class I, Zone 0, 1, and 2 group classifications, 505.6(A)
Classification of locations, 505.5, 505.5(B)
Class I, Zone 0, 505.5(B)(1)
Class I, Zone 1, 505.5(B)(2)
Class I, Zone 2, 505.5(B)(3)
Conductors and conductor insulator, 505.18
Definitions, 505.2
Documentation, 505.4(A)
Equipment construction, 505.9
Listing, 505.9(B)
Marking, 505.9(C)
Suitability, 505.9(A)
Equipment for use in, 505.20

General, 505.4
Grounding and bonding, 505.25
Live parts, 505.19
Material groups, 505.6
Protection techniques, 505.8
Combustible gas detection system, 505.8(I)
Encapsulation, 505.8(G)
Flameproof, 505.8(A)
Increased safety, 505.8(F)
Intrinsic safety, 505.8(C)
Oil immersion, 505.8(E)
Powder filling, 505.8(H)
Purged and pressurized, 505.8(B)
Type of protection "n," 505.8(D)
Reference standards, 505.4(B)
Sealing and drainage, 505.16
Special precaution, 505.7
Wiring methods, 505.15
**Hazardous (classified) locations, Class II,** 500.5(C), Art. 502
Bonding, 502.16(A)
Circuit breakers, 502.6
Control transformers and resistors, 502.7
Cranes and hoists, 610.2
Flexible cords, 502.12
Fuses, 502.6
Grounding, 502.16
Live parts, 502.15
Luminaires (lighting fixtures), 502.11
Motor controllers, 502.6
Motors and generators, 502.8
Multiwire branch circuits, 502.18
Receptacles and attachment plugs, 502.13
Sealing, 502.5
Signaling, alarm, remote-control, and communications systems, meters, instruments, and relays, 502.14
Surge protection, 502.17
Switches, 502.6
Transformers and capacitors, 502.2
Utilization equipment, 502.10
Ventilating piping, 502.9
Wiring methods, 502.4
**Hazardous (classified) locations, Class III,** 500.5(D), Art. 503
Bonding, 503.16(A)
Circuit breakers, 503.4
Control transformers and resistors, 503.5
Cranes and hoists, 503.13, 610.2
Flexible cords, 503.10
Fuses, 503.4
Grounding, 503.16
Live parts, 503.15
Luminaires (lighting fixtures), 503.9
Motor controllers, 503.4
Motors and generators, 503.6
Receptacles and attachment plugs, 503.11

Signaling, alarm, remote-control, and local loudspeaker intercommunications, 503.12
Storage battery charging equipment, 503.14
Switches, 503.4
Transformers and capacitors, 503.2
Utilization equipment, 503.8
Ventilation piping, 503.7
Wiring methods, 503.3
**Hazardous (classified) locations, Classes I, II, and III, Divisions 1 and 2,** Art. 500. *see also* Hazardous (classified) locations, Class I, Zones 0, 1, and 2
Aircraft hangars, 513.3
Anesthetizing locations, 517.60, 517.61
Approval for class and properties, 500.8(A)
Bulk storage plants, 515.3
Class I. *see* Hazardous (classified) locations, Class I
Class I, Zone 0, 1, and 2. *see* Hazardous (classified) locations, Class I, Zone 0, 1, and 2
Class II. *see* Hazardous (classified) locations, Class II
Class III. *see* Hazardous (classified) locations, Class III
Combustible gas detection system, 500.7(K)
Communications circuits, 800.8
Definitions, 500.2
Dust ignition proof, 500.7(B)
Dusts, explosive, Art. 502
Dusttight, 500.7(C)
Equipment, 500.8
Explosion proof apparatus, 500.7(A)
Fiber optic cable assembly, 500.8(E)
Fibers, flyings, lint, highly combustible material, Art. 503
Flammable liquids, Arts. 500 and 501
Garages, commercial, 511.2, 511.3
Gases, flammable, Arts. 500 and 501
Gasoline service stations. *see* Gasoline dispensing and service stations
General, 500.4
Group classifications, 500.6
Hoists, 503.13
Inhalation anesthetizing locations. *see* Anesthetizing locations
Intrinsic safety, 500.7(E). *see also* Art. 504
Lighting systems, less than 30 volts, 411.7
Marking, 500.8(B)
Temperature, 500.8 (C)
Material groups, 500.6

Protection techniques, 500.7
Specific occupancies, Art. 510
Spray application, dipping and coating processes, 516.3
Threading, 500.8(D)
Vapors, flammable, Art. 500, 501

**Hazardous (classified locations), specific,** Art. 510

**Headers**
Cellular concrete floor raceways, 372.2, 372.5, 372.9
Cellular metal floor raceways, 374.1, 374.2, 374.6
Definition, 372.2, 374.2

**Health care facilities,** Art. 517
Communications, signaling systems, data systems, less than 120 volts, nominal, 517-VI
Other-than-patient care areas, 517.81
Patient care areas, 517.80
Signal transmission between appliances, 517.82
Definitions, 517.2
Essential electrical systems, 517-III
Ambulatory health care centers, 517.45
Clinics, medical and dental offices, and other health care facilities, 517.50
Connections
Alternate power source, 517.34
Critical branch, 517.43
Life safety branch, 517.42
Critical branch, 517.33
Emergency system, 517.31
Essential electrical systems, 517.41
Hospitals, 517.30
Life safety branch, 517.32
Nursing homes and limited care facilities, 517.40
Power sources, 517.35, 517.44
General, 517.1
Inhalation anesthetizing locations, 517-IV
Classification, 517.60
Grounded power systems in anesthetizing locations, 517.63
Grounding, 517.62
Line isolation monitor, 517.160(B)
Low-voltage equipment and instruments, 517.64
Wiring and equipment, 517.61, 517.160
Isolated power systems, 517.19(F), 517-VII
Wiring and protection, 517-II
Applicability, 517.10
Critical care areas, 517.19
Fixed electrical equipment and grounding of receptacles, 517.13
General care areas, 517.18

General
installation/construction criteria, 517.11
Ground-fault circuit-interrupter protection, 517.21
Ground-fault protection, 517.17
Grounding, 517.13
Panelboard bonding, 517.14
Receptacles, 517.13, 517.16
Wet locations, 517.20
Wiring methods, 517.12
X-ray installations, 517-V
Connection to supply circuit, 517.71
Control circuit conductors, 517.74
Disconnecting means, 517.72
Equipment installations, 517.75
Guarding and grounding, 517.78
High tension x-ray cables, 517.77
Overcurrent protection, 517.73
Rating of supply conductors, 517.73
Transformers and capacitors, 517.76

**Heat generating equipment.** *see* Induction and dielectric heating equipment

**Heat tape outlet, mobile home,** 550.13(E)

**Heater cords.** *see* Cords, heater

**Heating.** *see also* Heating systems
Dielectric
Definition, 665.2
Fixed electric space. *see* Fixed electric space heating equipment
Induction
Definition, 665.2

**Heating appliances,** Art. 422

**Heating cables,** Art. 424

**Heating elements**
Marking, 422.61
Resistance
Definition, 427.2

**Heating equipment**
Central, 422.12
Definition, 665.2
Outside, 550.20(B)

**Heating panels and heating panel sets, radiant,** 424-IX, 427.27(B)

**Heating systems**
Impedance
Definition, 427.2
Induction
Definition, 427.2
Integrated
Definition, 427.2
Skin effect
Definition, 427.2

**Heavy-duty lampholders**
Branch circuits, 210.21(A), 210.23(B), 210.24, Table 210.24
Unit loads, 220.3(B)(5)

**Hermetic refrigerant motor compressors.** *see also* Air-conditioning and refrigerating equipment
Ampacity and rating, 440.12(A)
Definition, 440.2
Marking, 440.4
Rating and interrupting capacity, 440.12(A)

**Hermetically sealed,** 500.2, 500.7(J)

**High-impedance grounded neutral systems,** 250.21(5), 250.36

**Hoists,** Art. 610. *see also* Cranes
Conductors, contact, 610-III
Control, 610-VI
Electrolytic cells, 668.32
Flexible cords, 400.7(A)(5)
Grounding, 250.22(A), 250.112(E), 250.116(1), 610-VII
Hazardous (classified) locations, 503.13
Motors and controllers, disconnecting means, 430.112 Ex., 610-IV
Overcurrent protection, 610-V
Wiring, 610-II

**Hoistways**
Definition, Art. 100-I
Wiring in, 620.37, 725.55(I), 760.52(B)(2), 760.55(F)

**Hood, metal, stage switchboard,** 520.24

**Hoods for commercial cooking, lighting in,** 410.4(C)

**Hospitals.** *see also* Health care facilities
Definition, 517.2
Essential electrical systems, 517.30
Patient care areas
Definition, 517.2
Psychiatric
Definition, 517.2

**Hot tubs.** *see* Spas and hot tubs

**Hotels and motels**
Lighting load, Table 220.3(A)
Receptacle ground-fault circuit interrupter protection, 210.8(B)
Receptacle outlets required, guest rooms, 210.60

**Houseboats.** *see* Floating buildings

**Hydromassage bathtubs,** 680-VII
Definition, 680.2

**-I-**

**Identification**
Disconnecting means, 110.22, 620.51(D)
Flexible cords
Grounded-conductor, 400.22
Grounding-conductor, 400.23
Fuel cell systems, 692.4(B)
Grounded conductors, Art. 200
High leg, 215.8, 230.56, 408.3(E)
Intrinsically safe systems, 504.80

Panelboard circuits, 408.13
Service disconnecting means, 230.70(B)
Ungrounded conductors, 210.4(D)
Wiring device terminals, 250.126, 504.80

**Identified**
Definition, Art. 100-I

**Illumination**
Means of egress, 517.32(A), 517.42(A)
Task, 517.33(A)
Definition, 517.2

**Immersible appliances,** 422.41

**Immersion heaters, cord- and plug-connected,** 422.44

**In sight from**
Air-conditioning or refrigerating equipment, 440.14
Center pivot irrigation machines, 675.8(B)
Definition, Art. 100-I
Duct heaters, 424.65
Electric-discharge lighting, 410.81(B)
Fixed electric space heating equipment, 424.19
Motor driven appliances, 422.32
Motors, motor controllers, 430.102
Room air conditioners, 440.63
Signs, 600.6(A)

**Incandescent lamps,** Art. 410. *see also* Hazardous (classified) locations
Guards
Aircraft hangars, 513.7(C)
Garages, 511.7(A) and (B)
Theater dressing rooms, 520.72
Lamp wattage, marking on luminaire (fixture), 410.70
Medium and mogul bases, 410.53
Snap switches for, 404.14(B)(3)

**Independent**
Circuits for emergency lighting, 700.17
Supports, services, over buildings, 230.29
Wiring, emergency circuits, 700.9
X-ray control, 660.24

**Individual branch circuits.** *see* Branch circuits, individual

**Induced currents, metal enclosures,** 300.20, 330.31

**Induction and dielectric heating equipment,** Art. 665
Ampacity of supply conductors, 665.10
Definitions, 665.2
Disconnecting means, 665.12
Guarding, grounding, and labeling, 665-II
Access to interior equipment, 665.22
Capacitors, 665.24
Component interconnection, 665.19

Control panels, 665.21
Enclosures, 665.20
Grounding and bonding, 665.26
Marking, 665.27
Shielding, 665.25
Warning labels or signs, 665.23
Hazardous (classified) locations, 665.4
Output circuits, 665.5
Overcurrent protection, 665.11
Remote control, 665.7
**Induction heating**
Definition, 665.2
**Inductive loads**
Motors, 430.83, 430.109
Signs, 600.6(B)
Switches, types, ratings, 404.14
**Industrial machinery,** Art. 670
Clearance, 670.5
Definition, 670.2
Nameplate data, 670.3
Supply conductors and overcurrent protection, 670.4
**Industrial manufacturing system**
Definition, 670.2
**Information technology equipment,** Art. 645
Cables not in information technology equipment rooms, 645.6
Disconnecting means, 645.10
Grounding, 645.15
Marking, 645.16
Penetrations, 645.7
Special requirements for information technology equipment rooms, 645.2
Supply circuits and interconnecting cables, 645.5
Uninterruptible power supplies (UPS), 645.11
**Infrared lamp industrial heating appliances,** 422.14
Branch circuits, 210.23(C), 422.11(C), 424.3(A)
Overcurrent protection, 422.11(C)
**Inhalation anesthetizing location.** *see* Anesthetizing locations
**Inhalation anesthetizing locations,** 517-IV. *see also* Hazardous (classified) locations
Definition, 517.2
**Inserts**
Cellular concrete floor raceways, 372.9
Cellular metal floor raceways, 374.10
Underfloor raceways, 390.14
**Institutions, emergency lighting,** Art. 700
**Instructions,** 110.3(B), 230.95(C)
**Instrument transformers.** *see* Transformers, instrument, grounding
**Instrumentation tray cable (Type ITC),** Art. 727
Allowable ampacity, 727.8

Bends, 727.10
Construction, 727.6
Definition, 727.2
Marking, 727.7
Overcurrent protection, 727.9
Uses not permitted, 727.5
Uses permitted, 727.4
**Instruments, meters, relays**
Grounding, 250-IX
Hazardous (classified) location, 501.3, 502.14
Low-voltage, 517.64
**Insulation**
Conductors, Art. 310
Construction and application, 310.13
Hazardous (classified) location, 501.13, 505.18(B)
Identification, 310.12
Line-type fire detectors, 760.71(K)
Marking, 310.11
Motion picture projectors, 540.14
Theaters, 520.42
Double. *see* Double insulated, appliances and tools
Equipment, 110.3(A)(4)
Fixture wire, 402.3, Table 402.3
Flexible cords, 400.4, 400-II, Table 400.4
Heating cables, 424-V
Integrity, 110.7
Service conductors, 230.22, 230.30, 230.41
Splices and joints, 110.14(B), 400.9, 527.4(G), 530.12
**Insulation levels**
Definitions,100 percent,133 percent, Table 310.64
**Insulators, nonabsorbent,** 230.27, 394.30(A)
**Insulators, open wiring.** *see* Open wiring on insulators
**Integrated electrical systems,** Art. 685
Applications of other articles, 685.2
DC system grounding, 685.12
Orderly shutdown, 645.10 Ex., 645.11, 685-II
Overcurrent protection, location of, 685.10
Ungrounded control circuits, 685.14
**Integrated gas spacer cable (Type IGS),** Art. 326
Construction, 326-III
Conductors, 326.104
Conduit, 326.116
Insulation, 326.112
Marking, 326.120
Installation, 326-II
Ampacity, 326.80
Bending radius, 326.24
Bends, 326.26
Uses not permitted, 326.12
Uses permitted, 326.10
**Interactive systems**
Definition, 690.2, 705.2
Fuel cell systems, connection to, 692-VII

**Intercommunications systems.** *see* Communications circuits; Hazardous (classified) locations
**Interconnected electric power production sources,** Art. 705
Connection point, 705.12
Definition, 705.2
Directory, 705.10
Disconnect device, 705.22
Disconnecting means, equipment, 705.21
Disconnecting means, sources, 705.20
Ground-fault protection, 705.32
Grounding, 705.50
Interrupting and short-circuit current rating, 705.16
Loss of primary source, 705.40
Output characteristics, 705.14
Overcurrent protection, 705.30
Point of connection, 705.12
Synchronous generators, 705.43
Unbalanced interconnections, 705.42
**Intermediate metal conduit (Type IMC),** Art. 342
Bends, 342.24, 342.26
Bushings, 342-III
Construction, 342-III
Couplings and connectors, 342.42
Definition, 342.2
Dissimilar metals, 342.14
Installation, 342-II
Listing, 342.6
Maximum number of conductors and fixture wires in, Tables C4 and C4A
Number of conductors, 342.22
Reaming and threading, 342.28
Size, 342.20
Splices and taps, 342.56
Standard lengths, 342.130
Supports and securing, 342.30
Uses permitted, 342.10
Wet locations, 342.10(D)
**Intermittent duty**
Definition, Art. 100-I
Motors, 430.22(E)
**Interpretations, formal.** *see* Formal interpretations
**Interrupter switch.** *see* Switches, interrupter
**Interrupting rating,** 110.9, 240.60(C)(3), 240.83(C)
Definition, Art. 100-I
**Intrinsically safe apparatus**
Definition, 504.2
Grounding, 504.50(A)
**Intrinsically safe circuits**
Definition, 504.2
Different
Definition, 504.2
**Intrinsically safe systems,** Art. 504
Bonding, 504.60
Conductors, separation of, 504.30
Definitions, 504.2
Equipment installation, 504.10

Equipment listing, 504.4
Grounding, 504.50
Identification, 504.80
Sealing, 504.70
Separation of intrinsically safe conductors, 504.30
Wiring methods, 504.20
**Introduction,** Art. 90
**Inverter**
Definition, 690.2
**Irons.** *see* Flatirons
**Irrigation machines,** Art. 675
Bonding, 675.14
Center pivot
Definition, 675.2
Center pivot irrigation machines, 675-II
Collector rings, 675.11
Conductors, 675.9
Derating of, 675.5
Current ratings, equivalent, 675.7
Definitions, 675.2
Disconnecting means, 675.8
Grounding, 675.12, 675.13
Irrigation cable, 675.4
Lightning protection, 675.15
Marking, 675.6
Motors on branch circuit, 675.10
Supply source, more than one, 675.16
**Isolated**
Arcing parts, 110.18
Circuits, low voltage, 517.64(C)
Definition, Art. 100-I
Equipment grounding conductor, 250.146 FPN
Phase installations, 300.5(I) Ex. 2
Power systems, 517.160, 517-VII
Definition, 517.2
Essential electrical systems, 517.30(C)(2)
Grounding, 517.19(F), 647.7(A)
Installation, 517.160(A)
**Isolating means, over 600 volts,** 490.22
**Isolating switch.** *see* Switches, isolating
**Isolation by elevation**
Circuits over 600 volts, 110.34(E)
Circuits under 600 volts, 110.27
**Isolation transformer**
Definition, 517.2

-J-

**Joints.** *see also* Splices and taps
Expansion. *see* Expansion joints
Grounding electrode conductor, 250.64(B)
Insulating, fixtures, 410.16(E)
Insulation of, 110.14(B)
Strain at, 400.10
**Joists**
Air-handling, space, 300.22(C) Ex.
Armored cable, 320.23(A)

Concealed knob-and-tube wiring, 394.23
Electric space heating cables, 424.41(J)
Holes through or notches in, 300.4(A)
Nonmetallic sheathed cable, 334.15(C)
Open wiring, crossing, 398.15(C)
Parallel to framing members, cables and raceways, 300.4(D)
**Jumpers, bonding,** 250.28, 250.102, 250.168
Circuit
    Definition, Art. 100-I
    Equipment, 250.102
    Definition, Art. 100-I
Expansion joints, telescoping sections of raceways, 250.98
Grounding-type receptacles, 250.146
Hazardous (classified) locations, 250.100, 501.16(A), 502.16(A), 503.16(A)
Health care facilities, 517.19(C) and (D)
Main
    Definition, Art. 100-I
    Piping systems, 250.104
    Service equipment, 250.28, 250.94, 250.102(C)
**Junction boxes.** see also Boxes; Pull boxes
Accessibility, 314.29
Cellular concrete floor raceways, 372.7
Cellular metal floor raceways, 374.9
Covers, 314.28(C)
Deicing and snow-melting cables, 426.24(B)
Motor controllers and disconnects, 430.10
Nonheating cable leads, 424.43
Sensitive electronic equipment, 647.4(B)
Separation from motors, 430.145(B)
Size
    Conductors No. 4 and larger, 314.16, 314.28(A)
    Conductors No. 6 and smaller, 314.16(C)(1)
    Construction specifications, 314-III
Supports, 314.23
Swimming pools, 680.29
Switch enclosures, 110.59, 312.8, 404.3
Underfloor raceways, 390.13

-K-

**Kitchen equipment, commercial,** 220.20
**Knife switches**
Butt contacts, 404.6(C)
Connection, 404.6(C)

Construction specifications, 404-II
Enclosures, 404.3
General-use, 404.13(C)
Interrupt current, 404.13(B)
Isolating, 404.13(A)
Motor-circuit, 404.13(D)
Position, enclosed and open types, 404.6
Ratings, 404.14
**Knob-and-tube wiring,** Art. 394
Accessible attics, 394.23
Clearances, 394.19
Conductors, 394.104
Definition, 394.2
Devices, 394.42
Installation, 394-II
Securing and supporting, 394.30
Splices and taps, 394.56
Through or parallel to framing members, 394.17
Uses not permitted, 394.12
Uses permitted, 394.10
**Knockouts**
Bonding service equipment, 250.94
Openings to be closed, 110.12(A), 314.17(A)

-L-

**Labeled**
Definition, Art. 100-I
**Labels required,** 550.20(B), 551.46(D), 551.47(Q)(3), 551.47(R)(4), 552.44(D), 552.48(P)(3), 552.59(B)
**Lacquers and paints**
Application, Art. 516
Atmospheres, 500.5(B), 500.6(A), 505.5(B), Art. 501, Art. 505
**Lampholders**
Branch circuits supplying, 210.23
Circuits less than 50 volts, 720.5
Construction, 410-IX
Damp or wet locations, 410.4(A), 410.49
Double-pole switched, 410.48
Heavy-duty. see Heavy-duty lampholders
Infrared lamps, 422.14
Mogul base. see Mogul base lampholders
Outdoor, 225.24
Pendant
    Bathrooms, 410.4(D)
    Not in clothes closets, 410.8(C)
    Not in theater dressing rooms, 520.71
Screw-shell types for lamps only, 410.47
Unswitched over combustible material, 410.6

**Lamps,** Art. 410. see also Lighting; Lighting fixtures; Hazardous (classified) locations
Arc. see Arc lamps, portable
Backstage (bare bulb), 520.47
Clothes closets, in, 410.8
Electric discharge, 410.30(C), 410.54, 410-XIII, 410-XIV
Electric discharge, enclosure, 410.54(A)
Fluorescent. see Fluorescent lighting fixtures
Guards. see Guarding, guards
Headlamps. see Handlamps, portable
Incandescent. see Incandescent lamps
Infrared. see Infrared lamp heating appliances
Motion picture projectors, 540.14, 540.20
Motion picture studios
    Film storage vaults, 530.51
    Stages, portable, 530.16, 530.17
    Viewing, cutting tables, 530.41
Outdoor, location, 225.25
Portable
    Flexible cords for, 400.7
    Motion picture studios, in, 530.16, 530.17
    Show windows, show cases, 400.11
Stand
    Definition, 520.2, 530.2
Theaters
    Border and proscenium, 520.44
    Dressing rooms, lamp guards, 520.72
    Festoons, 520.65
    Footlights, 520.43
    Stage, arc, portable, 520.61
    Switchboards, pilot lights, 520.53(G)
Wattage marking, flush and recessed luminaires (fixtures), 410.70
**Laundry, outlets, dwellings,** 210.11(C)(2), 210.50(C), 210.52(F)
**Laundry area, mobile home parks**
Definition, 550.2
**Legally required standby systems,** Art. 701
Accessibility, 701.15
Approval, equipment, 701.4
Capacity and rating, 701.6
Circuit wiring, 701.10, 701-II
Definition, 701.2
Ground-fault protection of equipment, 701.17
Overcurrent protection, 701-IV
Signals, 701.8
Signs, 701.9
Sources or power, 701-III
Tests and maintenance, 701.5
Transfer equipment, 701.7

Wiring, 701.10
**Lengths**
Branches from busways, 368.8
Free conductors at outlets and switches, 300.14, 424.43, 426.23
Intermediate metal conduit, 342.130
Open wiring in nonmetallic flexible tubing, 398.15(A)
Pull and junction boxes, 314.28, 314.71
Rigid metal conduit, 346.16(A) and (C)
Rigid nonmetallic conduit, marking, 352.120
Space heating cable, nonheating leads, 424.34
Taps, 210.19(A)(4)Ex. 1, 240.21
Motor branch circuit, 430.53(D)
Motor feeders, 430.28
**Life safety branch**
Definition, 517.2
Essential electrical systems, 517.32, 517.42
**Life support equipment, electrical**
Definition, 517.2
**Lighting**
Branch circuits, calculation of load, 220.3(A)
Cove, 410.9
Electric discharge. see Electric discharge lighting
Emergency, Art. 700
Feeders, calculation of load, 220.11, 220.30
Festoon. see Festoon lighting
Fixtures. see Lighting fixtures
Outlets, 210.70
Outline. see Outline lighting
Sensitive electronic equipment, 647.8
Systems. see Lighting systems, 30 volts or less
Track. see Lighting track
**Lighting assembly, cord- and plug-connected**
Definition, 680.2
**Lighting outlets,** 210.70
**Lighting systems, 30 volts or less,** Art. 411
Branch circuit, 411.6
Definition, 411.2
Hazardous (classified) locations, 411.7
Listing required, 411.3
Locations not permitted, 411.4
Secondary circuits, 411.5
**Lighting track,** 220.3(C), 220.12(B), 410-XV
Construction requirements, 410.105
Definition, 410.100
Fastening, 410.104
Heavy-duty, 410.103
Installation, 410.101
Load calculations, 220.12(B)
**Lightning rods**
As a ground, 250.60
Irrigation machines, 675.15

Spacing from, 250.106
**Lightning (surge) arresters,** Art. 280
Antenna discharge units, 810.20
Connections at services, 230.82(3)
Definition, 280.2
Grounding, 250.60, 250.106, 280.25
Radio and television equipment
Receiving stations (antenna discharge units), 810.20
Transmitting stations, antenna discharge units, 810.57
Services over 600 volts, 230.209
**Lightning (surge) protection,** Art. 280
Communication circuit conductors, 800.13
Hazardous (classified) locations, 501.17, 502.17
Hoistways and machine rooms, 620.37(B)
Network-powered broadband communications systems, 830.10(I)(3)
**Lights.** *see also* Lamps; Lighting
Border
Cables for, theaters, 520.44(B)
Definition, 520.2
From railway conductors, 110.19
Scenery, theaters, halls, 520.63
Strip
Definition, 520.2
**Limited care facility**
Definition, 517.2
**Line isolation monitor,** 517.160(B)
Definition, 517.2
**Lint, flyings,** Art. 503
**Liquidtight flexible metal conduit (Type LFMC),** Art. 350
Bends, 350.24, 350.26
Couplings and connectors, 350.42
Definition, 350.2
Grounding and bonding, 350.60
Installation, 350-II
Listing, 350.6
Maximum number of conductors and fixture wires in, Tables C7 and C7A
Number of conductors and cables in, 350.22
Securing and supporting, 350.30
Size, 350.20
Uses not permitted, 350.12
Uses permitted, 350.10
**Liquidtight flexible nonmetallic conduit (Type LFNC),** Art. 356
Bends, 356.24, 356.26
Construction specifications, 356-III
Definition, 356.2
Grounding and bonding, 350.60
Installation, 356-II
Listing, 356.6
Marking, 356.120

Maximum number of conductors and fixture wires in, Tables C5 through C6A
Number of conductors or cables in, 356.22
Securing and supporting, 350.30
Size, 356.20
Splices and taps, 356.56
Trimming, 356.28
Uses not permitted, 356.12
Uses permitted, 356.10
**Listed**
Definition, Art. 100-I
**Live parts.** *see also* Enclosures; Energized parts; Guarding, guards
Capacitors, 460.2
Definition, Art. 100-I
Exposed, 110.26
Guarding, 110.27
Hazardous (classified) locations, 501.15, 502.15, 503.15, 505.19
Lamp terminals and lampholders, 410.82
Lighting systems, electric discharge, 410-XIII, 410-XIV
Luminaires (lighting fixtures), lampholders, lamps, 410.3, 410.46
In theaters, 520.7
Transformers, guarding, 450.8(C)
**Loads**
Appliances, household cooking, demand table, 220.19
Branch circuits
Calculations, Annex D, Art. 220
Maximum, 210.25, 220.4, Table 210.4
Mobile homes, 550.13
Permissible, 210.23, 210.24, Table 210.24
Continuous. *see* Continuous load
Demand
Clothes dryers, 220.18
Household cooking appliances, 220.19
Farm, 220.40
Feeder, 220-II, Annex D
Calculations, Art. 220
Inductive. *see* Inductive loads
Mobile home parks, 550.31
Mobile homes, 550.18
Motors, conductors, 430-II
Nonlinear
Definition, Art. 100-I
Stage equipment, circuit loads, 520.41
**Location board**
Definition, 530.2
**Locations**
Arc welders, 630-II
Capacitors, 460.2(A)
Crane and hoist disconnecting means, 610-IV
Damp or wet. *see* Damp or wet locations
Definition, Art. 100-I
Dry. *see* Dry location

Electric discharge lighting transformers, over 1000 volts, 410.84
Elevator motor disconnecting means, 620.51(C)
Foreign pipes, accessories, transformer vaults, 450.47
Generators, 445.10
Grounding connections at electrodes, 250.68
Hazardous. *see* Hazardous (classified) locations
Lamps, outdoors, 225.25
Luminaires (lighting fixtures), 410-II
Mobile homes disconnecting means and branch-circuit protective equipment, 550.11
Motion picture projection equipment, 540.11
Motor disconnecting means, 430.102
Motor feeder taps, 430.28 Ex.
Motors, 430.14
Outlet boxes and conduit boxes, 314.29
Overcurrent devices, 240-II
Overhead service, 230.54
Panelboards, damp or wet, 408.17
Protective devices for communications circuits, 800.30, 830.30
Recreational vehicle disconnecting means and distribution panelboard, 551.45(B)
Resistors and reactors, 470.2
Service disconnecting means, 230.70(A), 230.72(A) Ex.
Service overcurrent protection, 230.91, 230.92
Shooting
Definition, 530.2
Sign switches, 600.6(A)
Splices and taps
Auxiliary gutters, 366.9
Wireways, 376.56, 378.56
Surge arresters, 280.11
Swimming pool junction box and transformer enclosures, 680.24
Switchboards, 408.5 through 408.7
Switches, wet, 404.4
System grounding connections, 250-II
Transformers and vaults, 450.13, Art. 450
Transient voltage surge suppressors, 258.11
Ventilation openings for transformer vaults, 450.45(A)
Wet. *see* Wet locations; Damp or wet locations
**Locked rotor motor current**
Code letters, 430.7(B), Table 430.7(B)

Conversion, Tables 430.151(A) and (B)
Hermetic refrigerant motor-compressors, 440.4(A)
**Locknuts, double, required**
Hazardous (classified) locations, 501.16(A)
Mobile homes, 550.15(F)
Over 250 volts to ground, 250.97
Recreational vehicles, 551.47(B)
**Low-voltage circuits.** *see also* Remote-control, signaling, and power-limited circuits
Definition, 551.2
Less than 50 volts, Art. 720
**Low-voltage equipment and instruments**
Definition, 517.2
**Lugs**
Connection of service conductors, 230.81
Connection to terminals, 110.14(A)
Solderless type at electrodes, 250.70
**Luminaire (fixture) stud construction,** 410.16(D)
**Luminaires (lighting fixtures),** Art. 410. *see also* Hazardous (classified) locations
Arc, portable, 520.61, 530.17
Autotransformers
Ballasts supplying fluorescent luminaires, 420.78
Supply circuits, 210.9, 215.11
Auxiliary equipment, 410.77
Bathtubs, near, 410.4(D)
Boxes, canopies, pans, 410-III
Branch circuits
Computation of, 210.19(A), 220.3(A) and (B)
Sizes, 210.23, 220.4
Voltages, 210.6, 410.73
Clothes closets, 410.8
Definition, 410.8(A)
Combustible material, near, 410.5, 410.76
Connection, fluorescent, 410.14, 410.30(C)
Construction, 410.105, 410-V, 410-VI, 410-X
Corrosive, 410.4(B)
Damp, wet, or corrosive locations, 410.4(A) and (B)
Dry-niche, 680.22(B)
Definition, 680.2
Ducts or hoods, in, 410.4(C)
Electric discharge. *see* Electric discharge lighting
Fluorescent. *see* Fluorescent lighting luminaires (fixtures)
Flush, 410-XI, 410-XII
Fountains, 680.51
Grounding, 410-V
Live parts, 410.3
Location, 410-II
Mounting, 410.76, 410.77
No-niche, 680.23(D)

Definition, 680.2
Outlets required, 210.70
Overcurrent protection, wires and cords, 240.4
Polarization, 410.23
Raceways, 410.31
Recessed. see Recessed lighting fixtures
Recreational vehicles, 551.53
Show windows, 410.7
Showers, near, 410.4(D)
Spas and hot tubs, 680.43(B)
Supports, 410-IV
Swimming pools, 680.22, 680.23
Theaters, Art. 520
Wet, 410.4(A)
Wet-niche, 680.23(B)
Definition, 680.2
Wiring, 410-VI

-M-

Machine rooms
Guarding equipment, 620.71
Wiring, 620.37
Machine tools. see Industrial machinery
Made electrodes, 250.52
Mandatory rules, 90.5
Manholes, 314-IV
Access, 314.55
Covers, 314.55(D)
Dimensions, 314.55(A)
Location, 314.55(C)
Marking, 314.55(D)
Obstructions, 314.55(B)
Cabling work space, 314.52
Conductors
Class 1, of different circuits, 725.55(F)
Class 2, 3 installation, 725.54, 725.55(F)
Conductors, over 600 volts, 300.3(C)(2)(e)
Manufactured buildings, Art. 545
Bonding and grounding, 545.11
Boxes, 545.9
Component interconnections, 545.13
Definitions, 545.3
Grounding electrode conductor, 545.12
Protection of conductors and equipment, 545.8
Receptacle or switch with integral enclosure, 545.10
Service equipment, 545.7
Service-entrance conductors, 545.5, 545.6
Supply conductors, 545.5
Wiring methods, 545.4
Manufactured home. see also Mobile homes; Recreational vehicles
Definition, 550.2
Manufactured phase
Definition, 455.2
Manufactured wiring systems, Art. 604
Construction, 604.6

Definition, 604.2
Unused outlets, 604.7
Uses not permitted, 604.5
Uses permitted, 604.4
Marinas and boatyards, Art. 555
Circuit breakers, 555.11
Connections, 555.9
Cranes, 555.23
Definition, 555.2
Disconnecting means, 555.17
Distribution system, 555.4
Enclosures, 555.10
Gasoline dispensing, 555.21
Ground-fault circuit-interrupters, 555.19(B)(1)
Grounding, 555.15
Hazardous (classified) locations, 555.21
Hoists, 555.23
Load calculations, 555.12
Marine power outlets, 555.11
Definition, 555.2
Panelboards, 555.11
Railways, 555.23
Receptacles, 555.19
Service equipment, location, 555.7
Switches, 555.11
Transformers, 555.5
Wiring, installation, 555.13(B)
Wiring, methods, 555.13(A)
Markings. see articles on wiring and equipment involved
Means of egress
Health care facilities, 517.32(A)
Illumination, 517.32(A), 517.42(A)
Mechanical execution of work, 110.12, 720.11, 725.6, 760.6, 770.8, 800.6, 820.6, 830.7
Medium voltage cable (Type MV), Art. 328
Definition, 328.2
Uses permitted, 328.10
Messenger supported wiring, Art. 396
Definition, 396.2
Grounding, 396.60
Installation, 396-II
Messenger support, 396.30
Splices and taps, 396.56
Uses not permitted, 396.12
Uses permitted, 396.10
Metal frame of building
Grounding electrode, 250.52(A)(2)
Not permitted as equipment grounding conductor, 250.136(A)
Metal hood, stage switchboard. see Hood, metal, stage switchboard
Metal outlet boxes. see Boxes, metal
Metal siding, grounding of, 250.116 FPN
Metal wireways. see Wireways, metal
Metal working machine tools and plastic machinery. see Industrial machinery

Metal-clad cable (Type MC), Art. 330
Accessible attics, 330.23
Ampacity, 330.80
Bends, 330.24
Boxes and fittings, 330.40
Conductors, 330.104
Construction specifications, 330-III
Definition, 330.2
Grounding, 330.108
Installation, 330-II
Marking, 310.11(B)(2) Ex. 3
Single conductors, 330.31
Supports, 330.30
Through or parallel to framing members, 300.4(A), 330.17
Uses not permitted, 330.12
Uses permitted, 330.10
Metal-enclosed switchgear. see Switchgear, metal-enclosed
Metals
Dissimilar, 110.14, 250.70, 342.14, 344.14
Ferrous. see Ferrous metals, protection from corrosion
Meter socket enclosures, 312.1. see also Cabinets, cutout boxes, and meter socket enclosures
Meters
Connection and location at services, 230.82(2), 230.94 Ex. 5
Grounding of cases, 250.174, 250.176
Grounding to grounded circuit conductor, 250.142(B) Ex. 2
Hazardous (classified) locations, 501.3, 502.14
Metric equivalents
Raceways, trade sizes, 300.1(C)
Metric units of measurement, 90.9
Mineral-insulated metal-sheathed cable (Type MI), Art. 332
Accessible attics, 320.23
Ampacity, 332.80
Bends, 332.24
Conductors, 332.104
Construction specifications, 330-III
Definition, 332.2
Fittings, 332.40
Installation, 332-II
Insulation, 332.112
Outer sheath, 332.116
Single conductors, 332.31
Supports, 332.30
Terminal seals, 332.40(B)
Through or parallel to framing members, 300.4, 332.17
Uses not permitted, 332.12
Uses permitted, 332.10
Mobile home lot
Definition, 550.2
Mobile home parks, Art. 550
Definitions, 550.2

Distribution system, 550.30
Electrical wiring system
Definition, 550.2
Feeder and service demand factors, 550.18, 550-III, 550.31, Table 550.31
Minimum allowable demand factor, 550.31
Mobile home service equipment, 550.23
Definition, 550.2
Mobile homes, Art. 550. see also Park trailers; Recreational vehicles
Accessory buildings or structures
Definitions, 550.2
Appliances, 550.14
Arc-fault circuit interrupter protection, 550.25
Branch circuit protective equipment, 550.11
Branch circuits, 550.12
Calculations of loads, 550.18
Communications circuits, 800.41
Definitions, 550.2
Disconnecting means, 550.11
Expandable and dual units, wiring, 550.19
Feeder, 550.33
General requirements, 550.4
Ground-fault circuit interrupter, 550.13(B)
Grounding, 550.16
Heat tape outlet, 550.8(E)
Insulated neutral required, 550.16(A)(1)
Luminaires (fixtures), 550.14
Multiple section, wiring, 550.19
Nameplates, 550.11(D)
Outdoor outlets, luminaires, air-cooling equipment, 550.20
Power supply, 550.10
Receptacle outlets, 550.13
Service equipment, 550.32
Definition, 550.2
Testing, 550.17
Wiring methods and materials, 550.15
Mobile X-ray equipment, 660.4(B)
Definition, 660.2
Module, solar voltaic systems
Definition, 690.2
Mogul base lampholders, 210.7(C)(3), 410.53
Monorails, Art. 610
Motion picture and television studios, Art. 530
Definitions, 530.2
Dressing rooms, 530-III
Feeder conductors, sizing, 530.19
Film storage vault, 530-V
Grounding, 530.20
Lamps, portable, 530.16, 530.17
Live parts, 530.15
Overcurrent protection, 530.18
Portable equipment, 530.6
Stage or set, 530-II
Substations, 530-VI

Viewing, cutting, and patching tables, 530-IV
Wiring, permanent, 530.11
Wiring, portable, 530.12
**Motion picture projector rooms.** see Projector rooms, motion picture
**Motion picture theaters.** see Theaters
**Motor control center**
Definition, Art. 100-I
**Motor controller rating**
Definition, Art. 100-I
**Motor fuel dispensing facilities,** Art. 514. see also Hazardous (classified) locations
Circuit disconnects, 514.11
Classification of locations, 514.3
Equipment, 514.4, 514.7
Grounding, 514.16
Maintenance and servicing, 514.13
Sealing, 514.9
Underground wiring, 514.8
Wiring, 514.4, 514.7, 514.8
**Motor home**
Definition, 551.2
**Motor-circuit switches.** see Switches, motor circuit
**Motor-generator arc welders.** see Arc welders, 630-II
**Motors,** Art. 430
Air conditioning units, Art. 440
Appliances, motor driven, 422.10(A)
Branch circuits, 430-II
Combination loads, 430.25, 430.63
Continuous duty, 430.22(A)
Intermittent duty, 430.22(E)
Motor and other loads, 430.24
Several motors, 430.24
Single motor, 430.22
Taps, 430.53
Torque motors, 430.52(D)
Wound rotor secondary, 430.23
Bushing, 430.13
Capacitor, 460.9
Circuit conductors, 430-II
Code letters, Table 430.7(B)
Combined overcurrent protection, 430.55
Conductors, 430-II
Control circuits, 430-VI
Controllers. see Controllers, motor
Current, full load. see Full load current motors
Curtain, theater, 520.48
Disconnecting means, 430-IX
Feeder demand factor, 430.26
Feeders, calculation of load, 220.10, 220.30
Full-load current. see Full-load current motors
Fuseholder, size of, 430.57
General, 430-I
Ground-fault protection, 430-IV, 430-V

Grounding, 250-VI, 430.12(E), 430-XII
Grouped, 430.24, 430.42, 430.53, 430.87, 430.112
Guards for attendants, 430.133
Hazardous (classified) locations, 501.8, 502.8, 503.6, 505.16(E)(2), 505.22
Highest-rated or smallest-rated, 530.17
Industrial machinery, Art. 670
Liquids, protection from, 430.11
Location, 430.14
Maintenance, 430.14(A)
Marking, 430.7
Terminals, 430.9(A)
Motor control centers, 430-VIII
Multispeed. see Multispeed motors
Over 600 volts, 430-X
Overheating, dust accumulations, 430.16
Overload protection, 430-III
Part winding. see Part-winding motors
Protection of live parts, 430-XI
Rating or setting of branch-circuit short-circuit and ground-fault devices, Table 430.152
Restarting, automatic, 430.43
Short circuit protection, 430-IV, 430-V
Starting, shunting, 430.35
Tables, 430-XIII
Taps, 430.28
Terminal housings, size, 430.12
Terminals, 430.9
Three overload units, Table 430.37
Ultimate trip current, 430.32(A)(2)
Ventilation, 430.14(A)
Wiring diagram, Fig. 430.1
Wiring space in enclosures, 430.10
**Mounting of equipment,** 110.13, 314.23, 404.10, 410-IV
**Moving walks.** see Elevators, dumbwaiters, escalators, moving walks, wheelchair lifts, and stairway chair lifts
**Multioutlet assembly,** Art. 380
Calculation of load, 220.3(B)(8)
Definition, Art. 100-I
Metal, through dry partitions, 380.3
Use, 380.2
**Multiple circuit connections, grounding,** 250.144
**Multiple conductors (conductors in parallel).** see Conductors, multiple
**Multiple fuses (fuses in parallel),** 240.8, 404.17
Definition, 100-II
**Multispeed motors**
Branch circuits, 430.22(B)
Locked-rotor code letters, 430.7(B)
Marking, 430.7

Overload protection, 430.32(A) and (C)
**Multiwire branch circuit,** 210.4
Definition, Art. 100-I
Hazardous locations, 501.18, 502.18, 505.21

**-N-**

**Nameplates.** see articles on wiring and equipment involved
**Neat and workmanlike installation,** 110.12, 720.11, 725.6, 760.6, 770.8, 800.6, 820.6, 830.7
**Neon tubing,** 600.41
Definition, 600.2
**Network-powered broadband communications systems,** Art. 830
Buildings, 830-II, 830-V
Cables outside and entering buildings, 830-II
Aerial cables, 830.11: Above roofs 830.11(F); Between buildings 830.11(H); Clearance from ground 830.11(D); Climbing space 830.11(B); Final spans 830.11(G); Lead-in clearance 830.11(C); On buildings 830.11(I); On poles 830.11(A); 830.11; Over pools 830.11(E)
Entrance cables, 830.10: Low power circuits 830.10(B); Medium power circuits 830.10(A)
Underground circuits entering buildings, 830.12
Direct-buried cables and raceways, 830.12(B)
Mechanical protection, 830.12(C)
Pools, 830.12(D)
Underground systems, 830.12(A)
Wiring methods within, 830-V
Bends, 830.57
Installation of, 830.58
Low power wiring, 830.55: Ducts, plenums, other air handling spaces, 830.55(A); Other wiring 830.55(C); Riser, 830.55(B)
Medium power wiring, 830.54: Ducts, plenums, other air handling spaces 830.54(A); Other wiring 830.54(C), Riser 830.54(B)
Protection against physical damage, 830.56
General, 830-I
Access to electrical equipment behind panels, 830.6

Definitions, 830.2
Equipment and cables, 830.5
Listing and marking, 830.5(A)
Hazardous (classified) locations, 830.9
Installation of circuits, 830.8
Locations and other articles, 830.3
Mechanical execution of work, 830.7
Power limitations, 830.4
Scope, 830.1
Grounding methods, 830-IV
Bonding and grounding at mobile homes, 830.42
Cable network interface unit, and primary protection, 830.40
Protection, 830-III
Grounding of metallic members, 830.33
Primary electrical protection, 830.30
**Neutral.** see also Conductors, grounded
Bare, 230.22 Ex., 230.30 Ex., 230.41 Ex.
Bonding to service equipment, 250.94
Common. see Common neutral
Conductor, Art. 310 Notes 3 and 10 to Ampacity Tables 0–2000 volts, Tables B.310.1 through B.310.10
Continuity of, 300.13
Definition, Art. 100-I
Equipment, grounding to, 250.142
Feeder load, 220.22
Grounding of
AC systems, 250.4, 250.24, 250.26, 250.34, 250.36
DC systems, 250.4, 250.34, 250.36, 250.160(B)
Identification, Art. 200
Ranges and dryers, grounding, 250.140, 250.142
Uninsulated, where permitted, 230.22 Ex., 230.30 Ex., 230.41 Ex., 250.140(3), 338.10(B)
**Nightclubs,** Art. 518
**Nipples, conduit.** see Conduit nipples
**Noise.** see Electrical noise
**Nonautomatic**
Definition, Art. 100-I
**Nonelectrical equipment, grounding,** 250.116
**Nongrounding-type receptacles, replacements.** see Receptacles, nongrounding type, replacement
**Nonincendive circuits**
Definition, 100-I
Hazardous (classified) locations, 500.7(F)
**Nonincendive component,** 500.7(H)

**Nonincendive equipment,** 500.7(G)
**Nonlinear load**
Definition, Art. 100-I
**Nonmetallic boxes.** *see* Boxes, nonmetallic
**Nonmetallic conduit.** *see* Liquidtight flexible nonmetallic conduit; Nonmetallic underground conduit with conductors
**Nonmetallic extensions,** Art. 382
Boxes and fittings, 382.40
Definition, 382.2
Exposed, 382.15
Installation, 382-II
Splices and taps, 382.56
Uses not permitted, 382.12
Uses permitted, 382.10
**Nonmetallic rigid conduit.** *see* Rigid nonmetallic conduit
**Nonmetallic underground conduit with conductors (Type NUCC),** Art. 354
Bends, 354.24, 354.26
Bushings, 354.46
Conductor terminations, 354.50
Construction, 354.100
Construction specifications, 354-III
Definition, 354.2
Grounding, 344.60
Installation, 354-II
Joints, 354.48
Listing, 344.6
Marking, 354.120
Number of conductors, 354.22
Size, 354.20
Splices and taps, 344.56
Trimming, 354.28
Uses not permitted, 354.12
Uses permitted, 354.10
**Nonmetallic wireways.** *see* Wireways, nonmetallic
**Nonmetallic-sheathed cable (Types NM, NMC, and NMS),** Art. 334
Accessible attics, 334.23
Bends, 334.24
Boxes and fittings, 334.40
Conductors, 334.104
Construction, 334.100, 336-III
Definition, 334.2
Devices of insulating material, 334.40(B)
Devices with integral enclosures, 334.40(C)
Exposed work, 334.15
Grounding, 334.108
Installation, 334-II
Insulation, 334.112
Listed, 334.6
Marking, 310.11
Nonmetallic outlet boxes, 334.40(A)
Sheath, 334.116
Supports, 334.30
Three floor limitation, 334.12(A)(1)
Through or parallel to framing members, 300.4, 334.17

Unfinished basements, 344.15(C)
Uses not permitted, 334.12
Uses permitted, 334.10
**Nonpower-limited fire alarm circuit (NPLFA),** 760-II
Definition, 760.2
**Nontamperable**
Circuit breakers, 240.82
Type S fuses, 240.54(D)
**Number of services,** 230.2
**Nurses' stations**
Definition, 517.2
**Nursing homes**
Definition, 517.2
Essential electrical systems, 517.40

-O-
**Occupancy, lighting loads,** 220.3, Table 220.3(A)
**Office furnishings,** Art. 605
General, 605.2
Lighting accessories, 605.5
Partitions
Fixed-type, 605.6
Freestanding type, 605.7, 605.8
Interconnections, 605.4
Wireways, 605.3
**Official interpretations.** *see* Formal interpretations
**Oil (filled) cutout**
Definition, 100-II
**Oil immersion,** 500.7(I)
**Open wiring on insulators,** Art. 398
Accessible attics, 398.23
Clearances, 398.19
Conductors, types permitted, 398.104
Construction specifications, 398-III
Definition, 398.2
Exposed work, 398.15
Flexible nonmetallic tubing, 398.15(A)
Installation, 398-II
Securing and supporting, 398.30
Through or parallel to framing members, 398.17
Uses not permitted, 398.12
Uses permitted, 398.10
**Openings in equipment to be closed,** 110.12(A), 312.5(A), 314.17(A)
**Operable, externally**
Definition, Art. 100-I
**Operating device**
Definition, 620.2
**Operating rooms, hospital,** 517-IV
Definition, Art. 100-I
Emergency lighting system, Art. 700
**Optical fiber cables,** Art. 770
Access to electrical equipment, 770.7
Within buildings, 770-III
Cable trays, 770.6
Cables

Application, 770.53
Marking, 770.50
Substitutions, 770.53(F)
Definition, 770.2, 770.4
Fire resistance of, 770.49
Grounding of entrance cables, 770.33
Installation, 770.9, 770.50, 770.52
Listing requirements, 770.50, 770.51
Location, 770.3
Marking, 770.50
Mechanical execution of work, 770.8
Optical fibers and electrical conductors, 770.52
Protection, 770-II
Raceway system, 770.6
Types, 770.5, 770.6, 770.51
**Optional standby systems,** Art. 702
Capacity and rating, 702.5
Circuit wiring, 702-II
Definition, 702.2
Equipment approval, 702.4
Grounding, 702-III
Portable generator grounding, 702.10
Signals, 702.7
Signs, 702.8
Transfer equipment, 702.6
**Organs**
Electronic, 640.1
Pipe, Art. 650
Conductors, 650.5
Grounding, 250.112(B), 650.4
Installation of conductors, 650.6
Overcurrent protection, 650.7
Source of energy, 650.3
**Outdoor receptacles,** 210.8(A)(3), 210.52(E), 406.8, 680.22(A)
**Outlet boxes.** *see* Boxes
**Outlet spacing, dwelling baseboard heaters,** 210.52
**Outlets**
Appliance, 210.50(C)
Definition, Art. 100-I
Devices, branch circuits, 210.7
Discontinued
Cellular concrete floor raceways, 372.13
Cellular metal floor raceways, 374.7
Underfloor raceways, 390.7
Heating, air-conditioning, and refrigeration equipment, 210.63, 550.12(D)
Laundry, 210.11(C)(2), 210.52(F), 220.3(B), 550.13(D)(7)
Lighting, 210.70
Definition, Art. 100-I
Multioutlet assembly. *see* Multioutlet assembly
Power
Definition, Art. 100-I
Receptacle
Definition, Art. 100-I
Required, 210-III

**Outline lighting,** Art. 600. *see also* Signs
Definition, 600.2
Grounding, conductor size, 600.7
**Output circuits**
Amplifiers, 640.9(C)
Fuel cell systems, 692.61, 692-VIII
Heat generating equipment, 665.5
**Outside branch circuits and feeders,** Art. 225
Calculation of load, 225.3
Branch circuits, 220-I, 220.3, 225.3(A)
Feeders, 220-II, 225.3(B)
Circuit entrances and exits, buildings, 225.11
Conductors
Clearance
From buildings, 225.19
From ground, 225.18
Covering, 225.4
Size, 225.5, 225.6
Disconnection, 225.31
Lighting equipment installed outdoors, 225.7
Location of outdoor lamps, 225.25
Mechanical protection, 225.20
More than one building or structure, 225-II
Access to occupants, 225.35
Access to overcurrent protective devices, 225.40
Disconnect
Construction, 225.38
Disconnecting means, 225.31
Grouping of, 225.34
Location of, 225.31
Maximum number, 225.33
Rating of, 225.39
Suitable for service equipment, 225.36
Identification, 225.37
Number of supplies, 225.30
Outdoor lampholders, 225.24, 225.25
Overcurrent protection, 225.9
Point of attachment, 225.16
Spacing, open conductors, 225.14
Supports
Open conductors, 225.12
Over buildings, 225.15
Vegetation, 225.26
Wiring on buildings, 225.10
**Outside of buildings, when services considered,** 230.6
**Ovens, wall-mounted.** *see also* Cooking units, counter-mounted; Ranges
Branch circuits, 210.19(A)(3), 210.23, 220.19, Table 220.19 Note 4
Connections, 422.16(B)(3), 422.31(B)
Definition, Art. 100-I
Demand loads, Table 220.19
Grounding, 250.134, 250.140

**Over 600 volts,** Art. 490
  Ampacity, 310.15, 310.60, 392.11, 392.13
  Bending, 300.34
  Boxes, pull and junction, 314-V
  Busways, 368-II
  Capacitors, 460-II
  Circuit breakers, 490.21(A)
  Circuit conductors, 110.36
  Circuit interrupting devices, 490.21
  Conductors, 110-III, 300.39
  Definition, 490.2, Art. 100-II
  Distribution cutouts, 490.21(C)
  Electrode-type boilers, 490-V
    Branch circuit requirements, 490.72
    Electricity supply system, 490.71
    General, 490.70
    Grounding, 490.74
    Pressure and temperature limit control, 490.73
  Elevation of unguarded live parts, 110.34(E)
  Enclosure for electrical installations, 110.31
  Equipment, 490-II, 490-III, 490-IV
  Fuel cell systems, 692-VIII
  Fuses and fuseholders, 490.21(B)
  Grounding, Art. 250, 490.36, 490.37
  Headroom above working spaces, 110.32
  Illumination of working spaces, 110.34(D)
  Indoor installations, 110.31(A)
  Insulation shielding, 300.40
  Isolating means, 490.22
  Load interrupters, 490.21(E)
  Metal-enclosed equipment, 110.31(C), 490-III
  Metal-enclosed power switchgear and industrial control assemblies, 490-III
    Accessibility of energized parts, 490.35
    Arrangement of devices in assemblies, 490.31
    Circuit breakers — interlocks, 490.45
    Clearance for cable conductors entering enclosure, 490.34
    Door stops and cover plates, 490.38
    Fused interrupter switches, 490.44
    Gas discharge from interrupting devices, 490.39
    General, 490.30
    Grounding, 490.36
    Grounding of devices, 490.37
    Guarding, 490.32, 490.33
    Inspection windows, 490.40
    Interlocks — interrupter switches, 490.42, 490.45
    Location of devices, 490.41
    Stored energy for opening, 490.43

Mobile and portable equipment, 490-IV
  Collector rings, 490.54
  Enclosures, 490.53
  General, 490.51
  High voltage cables for main power supply, 490.56
  Overcurrent protection, 490.52
  Power cable connections to mobile machines, 490.55
Moisture, mechanical protection, metal-sheathed cables, 300.42
Oil filled equipment, 490.3
Oil-filled cutouts, 490.21(D)
Outdoor installations, 110.31(C)
Overcurrent protection, 240.100, 240.101, 240-IX
Protection of equipment, 110.34(F)
Resistors and reactors, 470-II
Separation, 490.24, Table 490.24
Services, 230-VIII
Shielding solid dielectric-insulated conductors, 310.6
Temporary, Art. 527
Tray installations, Art. 392
Tunnel installation, 110-IV
Voltage regulators, 490.23
Wiring methods, 300.37, 300.50, 300-II, Table 300.50
**Overcurrent**
  Definition, Art. 100-I
**Overcurrent devices.** *see also* Art. 240
  Enclosed, 230.208(B)
  Standard, 240.6
**Overcurrent protection,** Art. 240
  Air-conditioning and refrigerating equipment, 440-III
  Appliances, 422.11
  Branch circuits, 210.20
  Busways, 368.9 through 368.13
  Capacitors, 460.8(B)
  Circuit breakers, 240-VII
  Circuits, remote control, signaling, and power-limited
    Class 1 systems, 725.23
    Class 2 and Class 3 systems, Chap. 9, Tables 11(A) and 11(B)
  Circuits less than 50 volts, 720.8
  Communications systems. *see* Protector, communications systems
  Cords, flexible and tinsel, 240.5, 400.13
  Cranes and hoists, 610-V
  Current-limiting
    Definition, 240.2
  Disconnecting and guarding, 240-IV
  Electric space heating equipment, 424.22
  Electroplating systems, 669.9
  Elevators, dumbwaiters, escalators, moving walks, wheelchair lifts, stairway chair lifts, 620.61

Emergency systems, 700-VI
Enclosures, 240-III
Feeder taps, 240.21, 240.92, 430.28
Fire alarm systems, 760.23, Chap. 9, Tables 12(A) and 12(B)
Fixture wires, 240.4, 402.12
Flexible cords. *see* Cords, flexible
Fuel cell systems, 692.8, 692.9
Fuses and fuseholders, 240-V, 240-VI
Generators, 445.12
Grounded conductor, 240.22
Induction and dielectric heating equipment, 665.11
Interconnected electric power production sources, 705.30
Legally required standby systems, 701-IV
Lighting track, 410.103
Location, 230.91, 240.92, 240-II
Mobile home, 550.11
Motion picture and television studios, 530.18, 530.63
Motors, motor circuits, controllers, 430-IV
  Control circuits, 430.72
  Over 600 volts, 430.125
Multiple fuses and circuit breakers (in parallel), 240.8, 404.17
Occupant access to, 240.24(B)
Organs, 650.7
Outside branch circuits and feeders, 225.9
Over 600 volts, 110.52, 240.100, 240.101, 240-IX, 460.25, 490.52
Panelboards, 408.15, 408.16
Paralleled fuses and circuit breakers, 240.8, 404.17
Recreational vehicles, 551.10(E), 551.43
Remote-control circuits, 240.3(G), Art. 725
Sensitive electronic equipment, 647.4(A)
Services
  Equipment, 230-VII
  Over 600 volts, 230.208
Single appliance, 210.20
Solar photovoltaic systems, 690.9
Supervised industrial installations, 240-VIII
Supplementary, 240.10
Switchboards, 408.2
Television studios. *see* Motion picture and television studios
Theaters, stage switchboards, 520.23, 520.25(A), 520.52, 520.62(B)
Transformers, 450.3 through 450.5
Vertical position, enclosures, 240.33
Welders, 630.12, 630.32
X-ray equipment, 517.73, 660.6

**Overhead spans,** 225.6(A)
**Overload**
  Definition, Art. 100-I


-P-
**Panel, solar photovoltaic systems**
  Definition, 690.2
**Panelboards,** Art. 408
  Bonding, 517.14
  Circuit directory, 408.4
  Component parts, 408.33
  Damp or wet locations, 408.17
  Definition, Art. 100-I
  Distribution
    Definition, Art. 100-I, 550.2, 551.2
  Enclosure, 408.18
  General, 408.13
  Grounding, 408.20, 517.19(D)
  Installation, 110.26(F)
  Lighting and appliance branch-circuit
    Definition, 408.14(A)
    Number of overcurrent devices, 408.15
  Overcurrent protection, 408.16
  Relative arrangement of switches and fuses, 408.19
  Sensitive electronic equipment, 647.4(A)
  Service equipment, 230-VI, 230-VIII, 408.3(C), 408.16(A)
  Spacing, minimum, 408.36
  Support for busbars and conductors, 408.3
  Use as enclosure, 312.8
  Wire bending space, 408.35
**Pans, fixture,** 410.13
**Paralleled**
  Alternate sources, Art. 705
  Elevators, dumbwaiters, escalators, moving walks, 620.12(A)(1)
  Power production sources, Art. 705
**Paralleled circuit breakers and fuses.** *see* Fuses, in parallel, not permitted
**Paralleled conductors.** *see* Conductors, paralleled
**Park trailers,** Art. 552. *see also* Mobile homes; Recreational vehicles
  Appliance accessibility and fastening, 552.58
  Bonding, 552.57
  Branch circuits, 552.46
  Calculations, 552.47
  Combination electrical systems, 552.20
  Conductors and boxes, 552.48, 552.49
  Connection of terminals and splices, 552.51
  Cord, 552.44
  Definition, 552.2
  Distribution panelboard, 552.45
  Grounded conductors, 552.50

Ground-fault circuit interrupters, 552.41(C)
Grounding, 552.55, 552.56
Low-voltage systems, 552.10
Luminaires (lighting fixtures), 552.54
Nominal 120- or 120/240-volt systems, 552.40
Outdoor outlets, fixtures, equipment, 552.59
Power supply, 552.43
Receptacle outlets required, 552.41, 552.53
Switches, 552.52
Tests, factory, 552.60
Wiring methods, 552.48
**Parts.** see specific type such as Live parts
**Part-winding motors,** 430.3
Code letter markings, 430.7(B)(5)
**Patching tables, motion picture,** 530-IV
**Path, grounding,** 250.2, 250.4(A)(5), 250.28
**Patient bed location,** 517.18(A), 517.19(A)
Definition, 517.2
**Patient care areas,** 517.13(A), 517.80. see also Health care facilities
Definition, 517.2
**Patient vicinity**
Definition, 517.2
**Pediatric locations,** 517.18(C)
**Pendant conductors, lamps,** 410.27
**Pendants**
Aircraft hangars, 513.7(B)
Anesthetizing locations, 517.61(B)(3) Ex. 2, 517.61(C)(1) Ex.
Bathrooms, 410.4(D)
Clothes closets, 410.8(C)
Connector, cord, 210.50(A)
Dressing rooms, theater, 520.71
Flexible cord, 400.7(A)(1)
Garages, commercial, 511.7(A)(2)
Mobile homes, 550.14(C)
**Periodic duty**
Definition, Art. 100-I
**Permanent plaque or directory,** 225.37, 230.2(E), 705.10
**Permission, special**
Definition, Art. 100-I
**Person, qualified**
Definition, Art. 100-I
**Phase converters, Art.** 455
Capacitors, 455.23
Conductors, 455.6
Connection of single-phase loads, 445.9
Definition, 455.2
Different types of, 455-II
Disconnecting means, 455.8, 455.20
Equipment grounding connection, 455.5
Marking, 455.4
Overcurrent protection, 455.7

Power interruption, 455.22
Rotary
Definition, 455.2
Start-up, 455.21
Static
Definition, 455.2
Terminal housings, 455.10
**Photovoltaic systems.** see Solar photovoltaic systems
**Physical damage.** see Protection, physical damage
**Pipe, gas.** see Gas pipe
**Pipe electrodes,** 250.52(A)(1), (A)(5), and (A)(7)
**Pipe organs.** see Organs, pipe
**Pipeline**
Definition, 427.2
**Piping systems, bonding,** 250.104
**Places of assembly.** see Assembly, places of
**Plants**
Bulk storage, Art. 515
Cleaning and dyeing, 500.5(B)(1), Art. 501, Art. 505
Clothing manufacturing, 500.5(D), Art. 503
**Plate electrodes,** 250.52(A)(6)
**Plenums,** 300.22
Definition, Art. 100-I
Wiring in, 300.22
**Pliable raceways.** see Raceways, pliable
**Plugging boxes, motion picture studios,** 530.14, 530.18(E)
Definition, 530.2
**Plugs, attachment.** see Attachment plugs
**Point of entrance**
Definition, 800.2, 830.2
**Polarization**
Appliances, 422.40
Connections, 200.11
Luminaires (lighting fixtures), 410.23
Plugs, receptacles, and connectors, 200.10(B)
Portable handlamps, 410.42(A)
Receptacle adapters, 406.9(B)(3)
Screw-shell type lampholders, 410.47
**Poles**
Climbing space, conductors, 225.14(D)
Conductors, mechanical protection, 225.20, 230.50
Supporting luminaires (lighting fixtures), 410.15(B)
**Pool cover.** see Covers, pool, electrically operated
**Pools.** see also Swimming pools, fountains, and similar installations
Definitions, 680.2
**Portable appliances.** see Appliances
**Portable equipment**
Definition, 520.2
Disconnecting means, 517.17(C)
Double insulation, 250.114
Generators, 250.34

Grounding, 250.114, 250.138
Grounding, conductor size, 250.122, Table 250.122
Stage and studio, 530.6
X-ray, 660.4(B), 660.21
Definition, 660.2
**Portable handlamps.** see Handlamps, portable
**Portable lamps,** 410.42, 511.4(B)(2), 513.10(E)(1), 515.7(C), 516.4(D), 530.16, 530.17. see also Lamps, portable
**Portable power distribution unit,** 520.62
Definition, 520.2
**Portable stage equipment.** see Stage equipment, theaters, portable
**Portable switchboards.** see Switchboards, portable, theater stages
**Portable wiring, motion picture studios,** 530.12
**Positive-pressure ventilation,** 500.7(D), 501.8(A)(2), 502.8(A)(2), 502.8(B), 505.8(B)
**Power, emergency systems.** see Emergency systems
**Power and control tray cable (Type TC), Art.** 336
Ampacity, 336.80
Bends, 336.24
Conductors, 336.104
Construction, 336.100
Construction specifications, 336-III
Definition, 336.2
Installation, 336-II
Jacket, 336.116
Marking, 336.120
Uses not permitted, 336.12
Uses permitted, 336.10
**Power factor**
Definition, Annex D
**Power outlet.** see Outlets, power
**Power production sources.** see Interconnected electric power production sources
**Power source, alternate**
Definition, 517.2
**Power supply, mobile homes,** 550.5
**Power supply assembly, recreational vehicles,** 551.44
Definition, 551.2
**Power-limited fire alarm circuit (PLFA)**
Definition, 760.2
**Power-limited tray cable (Type PLTC),** 725.61, 725.71(E)
Class I, Division 2 locations, 501.4(B)(4)
Marking, 310.11
**Preassembled cable in nonmetallic conduit.** see Nonmetallic underground conduit with conductors

**Premises wiring (system)**
Definition, Art. 100-I
**Pressure connector.** see Connectors, pressure
**Prevention of fire spread.** see Fire spread
**Programmed power distribution.** see Closed-loop and programmed power distribution
**Projector rooms, motion picture, Art.** 540
Audio signal equipment, 540.50, 540-IV
Definitions, 540.2
Projectors, nonprofessional, 540-III
Listing, 540.32
Projection rooms, 540.31
Projectors, professional type, 540-II
Conductor size, 540.13
Conductors on hot equipment, 540.14
Flexible cords, 540.15
Listing, 540.20
Location of equipment, 540.11
Marking, 540.21
Projector room, 540.10
Work space, 540.12
**Proscenium**
Arc-fault circuit-interrupter. see Arc-fault circuit-interrupter
Definition, 500.2
**Protection**
Combustible material, appliances, 422.17
Corrosion
Boxes, metal, 314.40(A), 314.72(A)
Conductors, 310.9
Electrical metallic tubing, 358.10(B)
General equipment, 300.6
Intermediate metal conduit, 342.10(B)
Metal-clad cable, 330.16
MI cable, 332.12
Nonmetallic sheathed cable, 334.15(B)
Rigid metal conduit, 344.10(B)
Underfloor raceways, 390.2(B)
Ground fault. see Ground fault protection
Ground fault circuit-interrupter. see Ground fault circuit-interrupter
Hazardous (classified) locations, 500.7, 505.8
Liquids, motors, 430.11
Live parts, 110.27, 445.14, 450.8(C)
Motor overload, 430-III
Overcurrent. see Overcurrent protection
Physical damage
Armored cable, 320.10
Busways, 368.4(B)

.4, 300.50(B)
ic tubing,

allic tubing,

discharge
lighting, 410.85
Lighting track, 410.101(C)(1)
Liquidtight flexible metal
conduit, 350.12(1)
Liquidtight flexible nonmetal-
lic conduit, 356.12(1)
Metal-clad cable, 330.10(A)
Mineral-insulated metal-
sheathed cable,
332.10(10)
Multioutlet assembly,
380.2(B)(2)
Nonmetallic-sheathed cable,
334.15(B)
Open conductors and cables,
230.50
Open wiring, 398.15(A) and
(C)
Overcurrent devices,
240.24(C)
Raceways, 300.5(D),
300.50(B)
Recreational vehicle park
underground branch
circuits and feeders,
551.80(B)
Resistors and reactors,
470.18(A)
Rigid nonmetallic conduit,
352.12(C)
Space heating systems,
424.12(A)
Surface raceways, 386.12(1),
388.12(2)
Transformers, 450.8(A)
UF cable, 340.12(10)
Underground installations,
230.49, 300.5(D) and
(J)
Wireways, 376.12(1),
378.12(1)
**Protective devices.** see Arc-fault
circuit-interrupters; Cir-
cuit breakers; Fuses;
Ground-fault circuit-
interrupters; Overcurrent
protection; Thermal
cutouts
**Protective equipment.** see Guard-
ing, guards
**Protector, communications sys-
tems,** 800.12, 800.30,
800.32, 830.30
**Public address systems,** Art. 640
Emergency power systems,
700.1 FPN 3
**Public assembly places,** Art. 518
Emergency lighting system, Art.
700
**Pull boxes.** see also Boxes; Junc-
tion boxes
Accessibility, 314.29
Construction specifications, 314-
III, 314.72
Sizes

No. 4 and larger conductors,
314.28(A)
No. 6 and smaller conductors,
314.16
Over 600 volts, 314.71
**Pumps.** see also Fire pumps
Canned, Class I hazardous (clas-
sified) locations,
501.5(F)(3)
Pool, double insulated, 680.31

-Q-

**Qualified person.** see Person,
qualified

-R-

**Raceways**
Bonding, 250-V, 501.16(A),
502.16(A), 503.16(A),
505.25(A)
Busways. see Busways
Cellular concrete floor. see Cel-
lular concrete floor race-
ways
Cellular metal floor. see Cellular
metal floor raceways
Conductors in service, 230.7
Continuity, 300.10, 300.12
Definition, Art. 100-I
Drainage, 225.22, 230.53
Electrical metallic tubing (Type
EMT). see Electrical
metallic tubing (Type
EMT)
Electrical nonmetallic tubing
(Type ENT). see Electri-
cal nonmetallic tubing
(Type ENT)
Emergency circuits, indepen-
dent, 700.9(B)
Expansion joints, 250.98,
300.7(B), 352.44
Exposed to different tempera-
tures, 300.7
Flexible metal conduit (Type
FMC). see Flexible
metal conduit (Type
FMC)
Flexible metallic tubing (Type
FMT). see Flexible me-
tallic tubing (Type
FMT)
Grounding, 250.132, 250-IV
Short sections, 250.86 Ex. 2,
250.132
Induced currents, 300.20
Installed in shallow grooves,
300.4(E)
Insulating bushings, 300.4(F),
300.16(B)
Intermediate metal conduit
(Type IMC). see Inter-
mediate metal conduit
(Type IMC)
Liquidtight flexible metal con-
duit (Type LFMC). see
Liquidtight flexible
metal conduit (Type
LFMC)

Liquidtight flexible nonmetallic
conduit (Type LFNC).
see Liquidtight flexible
nonmetallic conduit
(Type LFNC)
Luminaires (fixtures) as, 410.31
Metal-trough, audio signal pro-
cessing, amplification,
and reproduction equip-
ment, 640.7
Nonmetallic underground con-
duit with conductors
(Type NUCC)
Number of conductors, 300.17
Pliable, 362.2
Rigid metal conduit (Type
RMC). see Rigid metal
conduit (Type RMC)
Rigid nonmetallic conduit (Type
RNC). see Rigid non-
metallic conduit (Type
RNC)
Secured, 300.11(A)
Service. see Service raceways
Signaling Class I circuits,
725.26, 725.28
Strut-type channel. see Strut-
type channel raceway
Support for nonelectrical equip-
ment, 300.11(B)
Supporting conductors, vertical,
300.19
Surface metal. see Surface metal
raceways
Surface nonmetallic. see Surface
nonmetallic raceways
Underfloor. see Underfloor race-
ways
Underground, 300.5(C)
Wireways. see Wireways, metal;
Wireways, nonmetallic
Wiring, exterior surfaces of
buildings, 225.22
**Radio and television equipment,**
Art. 810. see also Com-
munity antenna televi-
sion and radio distribu-
tion (CATV) system
Amateur transmitting and re-
ceiving stations, 810-III
Community television antenna,
810.4
Definitions, 810.2
Interior installation — transmit-
ting stations, 810-IV
Noise suppressors, 810.5
Receiving equipment — an-
tenna systems, 810-II
**Radiographic equipment,**
660.23(A). see also
517-V, Art. 660
**Railway conductors, power and
light,** 110.19
**Rainproof**
Definition, Art. 100-I
**Raintight**
Definition, Art. 100-I
**Ranges,** 422.16(B)(3), 422.33(B).
see also Cooking units,
counter-mounted; Ov-
ens, wall-mounted
Branch circuits

Calculation of load, 220.19,
Table 220.19
Conductors, 210.19
Maximum load, 220.4
Feeders, calculation of load,
220.10, 220.30
Grounding, 250.140
Loads, demand, Table 220.19
Receptacles, 210.7, 250.140(4)
**Rated load current,** 440.4(A)
Definition, 440.2
**Reactors.** see Resistors and reac-
tors
**Readily accessible.** see Accessible,
readily
**Reaming, ends of metal conduits,**
342.28, 344.28, 358.28
**Re-bar electrodes,** 250.52(A)(3)
**Receptacles, cord connectors and
attachment plugs
(caps),** Art. 406
Anesthetizing locations,
517.64(F)
Branch circuits, 210.7, 210.52
Configurations, 550.10(C),
551.46(C), 552.44(C)
Critical branch, 517.33(A)
Definition, Art. 100-I
Disconnecting means, 422.33,
440.63
Faceplates, 406.5
Grounding type, 210.7,
250.130(C), 250.146,
406.9, 517.13,
517.19(G)
Hazardous (classified) locations,
501.12, 502.13, 503.11
Health care facilities, 517.13,
517.18, 517.19
Insulated grounded terminals,
250.146(D), 517.16
Less than 50 volts, 720.6, 720.7
Marinas and boatyards, 555.19
Maximum cord- and plug-
connected load to,
210.21(B)(2), 210.23,
Table 210.21(B)(2)
Minimum ratings, 406.2(B)
Mobile homes, 550.13
Nongrounding-type, replace-
ment, 210.7(D),
250.130(C)
Outdoor. see Outdoor recep-
tacles
Outlet
Definition, Art. 100-I
Outlets, where required, dwell-
ings, 210.52
Patient bed location, 517.18(B),
517.19(B)
Ratings for various size circuits,
210.21(B)(3), Table
210.21(B)(3)
Recreational vehicles, 551.52
Replacement, 210.7(D)
Selected, health care facilities,
517.33(A)
Definition, 517.2
Sensitive electronic equipment,
647.7
Show windows, in, 210.62
Stages and sets, 530.21

Swimming pools, 680.22
Temporary installations,
    527.4(D), 527.6
Terminals, identification,
    200.10(B)
Theaters, 520.45
**Recessed luminaires (lighting
    fixtures), 410-XI, 410-
    XII**
Clearances, installation, 410.66
Construction, 410-XII
As raceways, 410.11, 410.31
Temperatures, 410.65
Wiring, 410.67
**Recording systems, Art. 640**
**Recreational areas, and dining
    essential electrical sys-
    tems, health care facil-
    ity, 517.42(E)**
**Recreational vehicle parks, 551-
    VII**
Calculated load, 551.73
    Demand factors, Table 551.73
Definitions, 551.2
Disconnecting means, 551.77(B)
Distribution system, 551.72
Ground-fault circuit-interrupter
    protection, 551.71
Grounding, 551.75, 551.76
Outdoor equipment, protection
    of, 551.78
Overcurrent protection, 551.74
Overhead conductors, clearance
    for, 551.79
Receptacles, 551.71, 551.81
Underground wiring, 551.80
**Recreational vehicle site**
Definition, 551.2
Supply equipment, 551.71
    Grounding, 551.76
**Recreational vehicle stand**
Definition, 551.2
**Recreational vehicles (camping
    trailers, motor homes,
    park trailers, travel
    trailers, truck camp-
    ers), Art. 551**
Alternate power source, restric-
    tion, 551.33
Appliance accessibility and fas-
    tening, 551.57
Attachment plugs, 551.20(F),
    551.46(C)
Battery installations, 551.10(D)
Bonding, 551.56
Branch circuits required, 551.42
Calculation for loads, 551.42(D)
Combination electrical systems,
    551.20, 551-III
Conductors, 551.10(B), 551.48
Connections
    Grounding, 551.54, 551.55
    Power supply, 551.46
    Splices and terminals, 551.50
Definitions, 551.2
Distribution panelboard, 551.45
Expandable units, connection,
    551.47(P)
Generator installations, 551.30
Ground-fault circuit-interrupter,
    551.40(C), 551.41(C)
Grounding, 551.54, 551.55

Grounding conductor splices,
    551.50
Identification of grounded con-
    ductor, 551.49
Low voltage systems, 551-II
Luminaires (lighting fixtures),
    551.53
Multiple supply source, 551.31
Other power sources, 551.32
Outlet boxes, 551.48
Overcurrent protection
    Branch circuit, 551.43
    Distribution panelboard,
        551.45
    Low voltage wiring, 551.10
    Power sources, other, 551-IV
Power supply assembly, 551.44,
    551.46
Receptacles, 551.20(F), 551.41,
    551.52
Supply source 120-volt or
    120/240-volt system,
    551-V
Switches, 551.51
System voltages, 551-V
Tags, labels, and marking,
    551.46(D)
Tests, factory, 551-VI
Wiring methods, 551.10(C),
    551.47
**Refrigeration compressor motors
    and controls, Art. 440**
**Refrigeration equipment.** *see* Air-
    conditioning and refrig-
    eration equipment
**Refrigerators, grounding, 250.114**
**Regulator bypass switch**
Definition, 100-II
**Relays**
Hazardous (classified) locations,
    501.3, 502.14
Overload, motor overcurrent
    protection, 430.40
Reverse-current, transformers,
    450.6(B)
**Remote-control, signaling, and
    power-limited circuits,
    250.112, Art. 725**
Access to electrical equipment,
    725.5
Class 1 circuits, 725-II
    Circuits extending beyond
        one building, 725.29
    Conductors, 725.27
        Different circuits, 725.26
        Extending beyond one
            building, 725.29
        Insulation, 725.27(B)
        Number in raceways,
            725.28
        Overcurrent protection,
            725.23
        Size and use, 725.27(A)
    Grounding, 725.9
    Locations, 725.3, 725.24
    Overcurrent protection,
        725.23, 725.24
    Physical protection, 725.8(B)
    Power limitations, 725.21
    Wiring methods, 725.25
Class 2 and Class 3 circuits,
    725-III

Applications of PLTC cables,
    725.61
    Circuits extending beyond
        one building, 725.57
    Conductors, 725.52, 725.58
    Installation, 725.54
    Interconnection of power sup-
        plies, 725.41(B)
    Listing, 725.71
    Locations, 725.3
    Marking, 725.42, 725.71
    Overcurrent protection, Chap.
        9, Tables 11(A) and
        11(B)
    Power sources, 725.41(A),
        Chap. 9 Tables 11(A)
        and (B)
    Separation, 725.55
    Wiring methods
        Load side, 725.52
        Supply side, 725.51
Classifications, definitions, 725.2
    Class 1, 725-II
    Class 2 and Class 3, 725-III
Grounding, 725.9
Identification, 725.10
Installation of circuits, 725.7
Location and other articles,
    725.3
Mechanical execution of work,
    725.6
Motors, 430-VI
Safety-control equipment, 725.8
**Remote-control circuit.** *see* Cir-
    cuits, remote-control
**Remote-control switches, 517.160**
**Requirements for electrical in-
    stallations, Art. 110**
**Residential occupancies.** *see*
    Dwellings
**Resistance**
AC resistance and reactance,
    cables, Chap. 9, Table 9
Conductor properties, Chap. 9,
    Table 8
Insulation, 110.7
Welders, 630-III
**Resistance to ground, made elec-
    trodes, 250.56**
**Resistors and reactors, Art. 470**
Combustible material, on, near,
    470.3
Conductor insulation, 470.4
Location, 470.2
Over 600 volts, 470-II
    General, 470.18
    Grounding, 470.19
    Oil-filled reactors, 470.20
    Space separation, 470.3
**Rheostats, construction specifica-
    tions, 430.82(C)**
**Rigid metal conduit (Type
    RMC), Art. 344**
Aluminum, in concrete and in
    earth, 300.6(B)
Bends, 344.24, 344.26
Bushings, 344.46
Cinder fill, 344.10(C)
Construction specifications, 344-
    III
Couplings and connectors,
    344.42

Definition, 344.2
Dissimilar metals, 344.14
Expansion joints, 300.7(B)
Grounding, 344.60
Installation, 344-II
Listing, 344.6
Marking, 344.120
Maximum number of conductors
    and fixture wires in,
    Tables C8 and C8(A)
Number of conductors, 344.22,
    Chap. 9, Table 1
Reaming and threading, 344.28
Size, 344.20
Splices and taps, 344.56
Standard lengths, 344.130
Supporting and securing,
    314.23(E) and (F),
    344.30
Uses permitted, 344.10
Wet locations, 344.10(D)
**Rigid nonmetallic conduit (Type
    RNC), Art. 352**
Bends, 352.24, 352.26
Bushings, 352.46
Construction specifications,
    352-II
Expansion fittings, 300.7(B),
    352.44, Tables
    352.44(A) and
    352.44(B)
Joints, 352.48
Maximum number of conductors
    and fixture wires in,
    Tables C9 through
    C12(A)
Number of conductors, 352.22
PVC Schedule 80, 300.5(D),
    300.50, 551.80(B)
Size, 352.20
Splices and taps, 352.56
Supporting and securing,
    352.30, Table 352.30(B)
Trimming ends, 352.28
Uses not permitted, 352.12
Uses permitted, 352.10
**Road show connection panel,
    520.50**
**Rod electrodes, 250.52(A)(5)**
**Room air conditioners.** *see* Air
    conditioners, room
**Rooms, motion picture projector.**
    *see* Projector rooms,
    motion picture
**Rotary phase converter**
Definition, 455.2
**Running threads, 342.42(B),
    344.42(B)**

**-S-**
**Safety, examination of equipment
    for, 90.7**
**Screw shells**
Identification
    Polarity, 200.10(C), 410.23
    Terminals, 200.10
Lampholders, 410-IX
**Sealable equipment**
Definition, Art. 100-I

**Sealing.** *see also* Hazardous (classified) locations
  Conduit systems, 501.5, 502.5, 505.16
    Intrinsically safe systems, 504.70
    Raceway seal, underground service, 230.8
    Temperature changes, 300.7(A)
    Hermetically, 500.2
**Secondary ties, transformers,** 450.6
**Sensitive electronic equipment,** Art. 647
  Grounding, 647.6
  Lighting equipment, 647.8
  Receptacles, 647.7
  Single-phase supply system, 647.3
  Three-phase supply system, 647.5
  Wiring methods, 647.4
**Separately derived systems,** 250.20(D), 250.30
  Definition, Art. 100-I
**Service conductors.** *see* Conductors, service
**Service drops**
  Clearances, 230.24
  Connections, service head, 230.54
  Definition, Art. 100-I
  Means of attachment, 230.27
  Minimum size, 230.23
  Point of attachment, 230.26, 230.28
  Supports over buildings, 230.29
**Service equipment**
  Definition, Art. 100-I
  Overcurrent protection, 230-VII
**Service lateral**
  Definition, Art. 100-I
**Service loads, calculations,** Annex D, Art. 220
**Service raceways**
  Conductors, others permitted in, 230.7
  Drainage, raintight, 230.53
  Service head, 230.54
  Underground, 230-III
**Service stations, gasoline.** *see* Gasoline dispensing and service stations
**Service-entrance cable (Types SE and USE),** Art. 338
  Bends, 338.24
  Branch circuits or feeders, 338.10(B)
  Construction, 338.100, 338-III
  Definition, 338.2
  Grounding frames of ranges and clothes dryers, 250.140(3)
  Installation, 338-II
  Installation methods, for branch circuits and feeders, 338.10(B)(4)
  Marking, 338.120
  Service-entrance conductors, Art. 230-IV, 338.10(A)
  Uses permitted, 338.10

**Service-entrance conductors,** 230-IV, 338.10(A)
  Conductor sets, number of, 230.40
  Considered outside of building, 230.6
  Definitions, Art. 100-I
  Disconnecting means, 230-VI
  Drip loops, 230.52
  Insulation, 230.41
  Over 600 volts, 230-VIII
  Overcurrent protection, 230.90, 230.91, 230.208
  Physical damage, 230.49, 230.50
    Underground, 230.49
  Service head, 230.54
  Size, 230.42
  Splices, 230.46
  Underground, 230.49
  Wiring methods, 230.43
**Service-entrance equipment**
  Disconnecting means, 230-VI
    Connections, supply side, 230.82
    Connections to terminals, 230.81
    Disconnection of grounded conductor, 230.75
    Electrically operated, 230.94 Ex. 6
    Ground-fault, protection at, 230.95, 705.32
    Indicating, 230.77
    Location, 230.70(A)
    Marking, 230.66, 230.70(B)
    Maximum number of disconnects, six switch rule, 230.71
    Multiple occupancy buildings, 230.72(C)
    Over 600 volts, 230.205, 230.206
    Rating, 230.79
    Simultaneous openings, 230.74
    Six switch rule, 230.71
    Suitable for use, 230.70(C)
  Guarding, 230.62
  Overcurrent protection, 230-VI
    Location, 230.91
    Over 600 volts, 230.208
    Relative location, 230.94
    Specific circuits, 230.93
    Ungrounded conductors, 230.90(A)
  Panelboards as, 408-III
**Services,** Art. 230
  Definition, Art. 100-I
  Emergency systems separate service, 700.12(D)
  Farm, 220.40
  Ground-fault protection, 230.95
  Insulation, 230.22
  Number, 230.2
  Over 600 volts, 230-VIII
  Overhead supply, 230-II
  Supply to one building not through another, 230.3
  Two or more buildings, 250.32
  Underground, 230-III
**Setting (of circuit breaker)**
  Definition, Art. 100-I

**Shielding.** *see* Guarding, guards
**Short-circuit current rating**
  Definition, 110.10
  Transient voltage surge suppressors, 285.6
**Short-time duty**
  Definition, Art. 100-I
**Show cases, wall cases,** 410.29
**Show windows**
  Definition, Art. 100-I
  Flexible cords, 400.11
  Luminaires (lighting fixtures), 410.7
    Branch circuits, 220.3(B)(7)
    Feeders, 220.10, 220.12(A)
  Receptacles, 210.62, 314.27(C) Ex.
**Sidelights, borders and proscenium,** 520.44
**Sign body**
  Definition, 600.2
**Signaling circuits.** *see also* Fire alarm systems; Remote-control, signaling, and power-limited circuits
  Definition, Art. 100-I
  Health care facilities, 517-VI
  Installation requirements, Art. 725, Art. 760
**Signals for heated appliances,** 422.42
**Signs**
  Discharge, lighting, electric, 410-XIII, 410-XIV
  Electric, Art. 600
    Ballasts, transformers and electronic power supplies, 600.21 through 600.23
    Branch circuits, 600.5
    Definition, Art. 100-I
    Definitions, 600.2
    Disconnects, 600.6
    Enclosures, 600.8
    Field installed skeleton tubing, 600-II
      Applicability, 600.30; Electrode connections 600.42; Neon secondary circuit conductors, 1000 volts or less, 600.31; Neon secondary circuit conductors over 1000 volts, 600.32; Neon tubing, 600.41
    Grounding, 600.7
    Listing, 600.3
    Location, 600.9
    Markings, 600.4
    Portable or mobile, 600.10
  Exit, health care facilities, 517.32(B), 517.42(B)
  Grounding, 250.112(G)
  Mandated standby, 701.9(A)
  Outline lighting, Art. 600
  Standby, 702.8
  Warning. *see* Warning signs
**Skeleton tubing,** 600-II
  Definition, 600.2
**Smoke ventilator control, stage,** 520.49

**Snap switches**
  Accessibility, grouping, 404.8
  Definition, Art. 100-I
  Motors, 430.82(C), 430.109(C)
  Panelboards, 404.16(C), 404.19
  Ratings, 404.14
**Snow melting.** *see* Fixed outdoor electric deicing and snow-melting equipment
**Solar cell**
  Definition, 690.2
**Solar photovoltaic systems,** Art. 690
  A-C modules, 690.6
  Circuit requirements, 690-II
  Circuit sizing and current, 690.8
  Connection to other sources, 690-VII
  Definitions, Art. 100-I, 690.2
  Disconnecting means, 690-III
  Ground-fault protection, 690.5
  Grounding, 690-V
  Installation, 690.4
  Marking, 690-VI
  Maximum voltage, 690.7
  Over 600 volts, 690-IX
  Overcurrent protection, 690.9
  Stand-alone systems, 690.10
  Storage batteries, 690-VIII
  Wiring methods, 690-IV
**Solderless (pressure) connector.** *see* Connectors, pressure (solderless)
**Solidly grounded**
  Definition, 230.95
**Sound recording equipment,** Art. 640
  Audio signal processing, amplification, and reproduction equipment, 540.50
  Theaters, 520.4
**Space**
  Cabinets and cutout boxes, 312.7, 312.9, 312.11
  Climbing. *see* Climbing space, line conductors on poles
  Lightning rods, conductor enclosures, equipment, 250.60, 250.106
  Outside branch circuits and feeders, 225.14
  Over 600 volts, separation, 110.33, 110.34
  Working. *see* Working space
**Space heating, fixed.** *see* Fixed electric space heating equipment
**Spacing between bare metal parts,** 408.36, Table 408.36
**Spas and hot tubs,** 680-IV
  Definitions, 680.2
  Indoor installations, 680.43
  Outdoor installations, 680.42
  Packaged equipment assembly Definition, 680.2
  Protection, 680.42(A)(2), 680.43
**Special permission.** *see* Permission, special
**Spider (cable splicing block)**
  Definition, 530.2

**Splices and taps**
Antennas, 810.14
Auxiliary gutters, 366.9
Cabinets and cutout boxes, 312.8
Cellular concrete floor raceways, 372.12
Cellular metal floor raceways, 374.6
Concealed knob-and-tube, 394.56
Conduit bodies, 314.16(C)(2)
Construction sites, 527.4(G)
Deicing and snow-melting, 426.24(B)
Electrical metallic tubing, 358.56
Flat cable assemblies, 322.56
Flexible cords, 400.9
General provisions, 110.14
Messenger supported wiring, 396.56
Nonmetallic extensions, 382.56
Space heating cables, 424.40
Surface raceways, 386.56, 388.56
Underfloor raceways, 390.6
Underground, 300.5(E)
Wireways, 376.56, 378.56
**Spray application, dipping, and coating processes,** Art. 516
Classification of locations, 516.3
Definitions, 516.2
Equipment, 516.4, 516.7, 516.8
Grounding, 516.16
Wiring and equipment, 516.4, 516.7
**Spread of fire or products of combustion.** see Fire spread
**Stage effect (special effect)**
Definition, 530.2
**Stage equipment, theaters**
Fixed, 520-III
Portable, 520-V
**Stage property**
Definition, 530.2
**Stage set**
Definition, 530.2
**Stages, motion picture and television,** 530-II
Definition, 530.2
**Stairway chair lifts,** Art. 620. see also Elevators, and stairway chair lifts, dumbwaiters, escalators, moving walks, wheelchair lifts
**Stand-alone system**
Definition, 690.2
**Standby systems.** see Emergency systems; Legally required standby systems; Optional standby systems
**Static phase converter**
Definition, 455.2
**Steel, exposed structural, grounding,** 250.104(C)
**Steel siding,** 250.116 FPN

**Storage batteries,** Art. 480
Aircraft hangars, 513.10
Charging equipment, 503.14
Definition, 480.2
Electric vehicle nonvented
Definition, 625.2
Emergency systems, 700.12(A) and (B)(4)
Garages, 511.10
Grounding, 480.5
Installation, 690.71
Insulation, 480.6, 480.7
Locations, 480.9
Racks and trays, 480.8
Recreational vehicles, 551.4(B)
Solar photovoltaic systems, 690-VIII
Sound recording equipment, 640.9(B)
Vents, 480.10
**Strut-type channel raceway,** Art. 354
Construction, 384.100
Construction specifications, 384-III
Definition, 384.2
Grounding, 384.60
Marking, 384.120
Number of conductors, 384.22
Securing and supporting, 384.30
Size of conductors, 384.21
Uses not permitted, 384.12
Uses permitted, 384.10
**Submersible equipment,** 680.51
**Substations**
Motion picture and television studios, 530-VI
Over 600 volts, 490-III
**Subsurface**
Enclosures, 110.12(B)
**Support fittings fill, boxes,** 314.16(B)(3)
**Supports.** see articles on wiring and equipment
**Suppressors, radio noise,** 810.5
**Surface metal raceways,** Art. 386
Combination raceways, 386.70
Construction, 386.100
Construction specifications, 386-III
Definition, 386.2
Grounding, 386.60
Installation, 386-II
Listing, 386.6
Number of conductors or cables, 386.22
Size of conductors, 386.21
Splices and taps, 386.56
Uses not permitted, 386.12
Uses permitted, 386.10
**Surface nonmetallic raceways,** Art. 388
Combination raceways, 388.70
Construction, 388.100
Construction specifications, 388-III
Definition, 388.2
Grounding, 388.60
Listing, 388.6
Marking, 388.120
Number of conductors or cables in, 388.22

Size of conductors, 388.21
Splices and taps, 388.56
Uses not permitted, 388.12
Uses permitted, 388.10
**Surfaces, exposed conductive**
Definition, 517.2
**Surge arresters.** see Lightning (surge) arresters
**Surge protection.** see Lightning (surge) protection
**Surge suppressors.** see Transient voltage surge suppressors (TVSSs)
**Swimming pools, fountains, and similar installations,** Art. 680
Approval of equipment, 680.4
Bonding, 680.26
Ceiling fans, 680.22
Cord- and plug-connected equipment, 680.7
Deck area heating, 680.27(C)
Definitions, 680.2
Fountains, 680-V
Ground-fault circuit-interrupters, 680.5
Fountains, 680.51(A)
Hydromassage bathtubs, 680.71
Junction boxes for, 680.24
Luminaires (lighting fixtures), 680.23(A)(3), 680.24(B)(4)
Pool covers, 680.27(B)(2)
Receptacles, 680.22(A)(5), 680.62(E)
Signs, 680.57(B)
Spas, hot tubs, 680.42, 680.43, 680.44
Storable pool equipment ground-fault circuit-interrupter required for, 680.32, 680.33(B)(3)
Therapeutic pools, 680.62(A), 680.62(E)
Wiring to, 680.24(B)
Grounding, 680.6, 680.24(F), 680.25(B), 680.54, 680.55
Hydromassage bathtubs, 680-VI, 680-VII
Junction boxes and enclosures, 680.24
Lighting, 680.22, 680.23
Overhead conductor clearances, 680.8
Permanently installed, 680-II
Receptacles, location and protection, 680.22(A)
Spas and hot tubs, 680-IV
Storable, 680-III
Definition, 680.2
Switching devices, 680.22(C)
Therapeutic pools and tubs, 680-VI
Transformers, 680.4, 680.23(A)(2), 680.24(B)
Underwater audio equipment, 680.23
Underwater luminaires (lighting fixtures), 680.23

**Switchboards,** Art. 408
Clearances, 110.26, 408.8
Combustible material, location relative to, 408.7
Conductor insulation, 408.9
Construction specifications, 384-IV
Definition, Art. 100-I
Easily ignitable materials, near, 408.7
Grounding instruments, 408.12
Guarding live parts, 110.27
Illumination, 110.26(D)
Installation, indoor and outdoor, 110.26
Location, 408.5 through 408.7
Portable, theater stages, 520-IV
Stage, 520-II
Support, busbars, conductors, 408.3
Wet locations, 408.6
Working spaces about, 110.26
**Switches,** Art. 404. see also Hazardous (classified) locations; specific types of switches
AC general use snap switch, 404.14(A)
Accessibility and grouping, 404.8
AC-DC general use snap switches. see AC-DC general use snap switches
Air-conditioning and refrigerating equipment, 440-II
Appliances, 422.34
Bypass isolation
Definition, Art. 100-I
Circuit breakers used as, 240.83(D)
Definitions, Art. 100-I
Devices over 600 volts
Definition, 100-II
Disconnecting means
Appliances, 422-III
Motors, controller, 430-IX
Services, 230-VI
Emergency systems, 700-V
X-ray equipment, 517.72, 660-II
Enclosures, installation in, 450.8(C)
Essential electrical systems, transfer switches, 517.41(B)
General use, Art. 404
Definition, Art. 100-I
Identification, 110.22
Interrupter
Definition, 100-II
Isolating, 501.6(B)(2)
Definition, Art. 100-I
Motors over 100 HP, 430.109(E)
Services over 600 volts, 230.204
Knife. see Knife switches
Limit, cranes and hoists, 610.55
Manually operable, Art. 404
Motor circuit
Definition, Art. 100-I

Motor controllers, 430-VII
Panelboards, 408.16(C), 408.19
Regulator bypass. *see* Regulator bypass switch
Remote-control, 517.160(A)(1)
Service, 230-VI
Signs, outline lighting, 600.6
Snap. *see* Snap switches
Theater dressing rooms, 520.73
Transfer
    Definition, Art. 100-I
    Unit, appliances, 422.34
**Switchgear, metal-enclosed,** 110.34(F), 230.211, 490-III

## -T-

**Tables,** Chap. 9, Annex B and Annex C
AC resistance and reactance cables, Chap. 9 Table 9
Ampacities
    Cable insulated, over 2000 volts
        Three conductor aluminum in isolated conduit in air, Table 310.76
        Three conductor aluminum in underground electrical ducts, Table 310.80
        Three conductor aluminum isolated in air, Table 310.72
        Three conductor copper cable in isolated conduit in air, Table 310.75
        Three conductor copper in underground electrical ducts, Table 310.79
        Three conductor copper isolated in air, Table 310.71
        Triplexed or three single conductor aluminum in isolated conduit in air, Table 310.74
        Triplexed or three single conductor aluminum in underground electrical ducts, Table 310.77
        Triplexed or three single conductor copper in isolated conduit in air, Table 310.73
    Conductor, single insulated, isolated in air, over 2000 volts
        Aluminum, Table 310.70
        Copper, Table 310.69
    Conductors, three single insulated in underground electrical ducts, over 2000 volts, Table 310.78
    Crane and hoist motor conductors, Table 610.14(A)
    Fixture wire, Table 402.3
    Flexible cord, Table 400.4
    General conductors, 0 through 2000 volts

Aluminum, copper, or copper-clad aluminum, single conductor in free air, Tables 310.17, 310.19
Aluminum, copper, or copper-clad aluminum, two or three single-insulated conductors supported on messenger, Table 310.20
Aluminum, copper, or copper-clad aluminum in raceways or cables types AC, NM, NMC SE, Tables 310.16, 310.18
Bare or covered conductors, Table 310.21
Multiconductor cables, types TC, MC, and MI in free air, Table B.310.3
Three conductor cable in raceway in free air, Table B.310.1
Three insulated conductors in cable in underground electrical ducts, Table B.310.6
Three single insulated conductors directly buried in earth, types UF, USE, Table B.310.10
Three single insulated conductors in nonmagnetic underground electrical ducts, Table B.310.5
Three single insulated conductors in underground electrical ducts, Table B.310.7
Three triplexed single insulated conductors directly buried in earth (UF and USE cables), Table B.310.9
Two or three insulated conductors cabled within an overall covering directly buried in earth, Table B.310.8
Wound-rotor secondaries, Table 430.23(C)
Bare metal parts, spacings between switchboard and panelboard, Table 408.36
Branch-circuit requirements, Table 210.24
Cable markings, Tables 725.71, 760.31(G), 760.71(J), 770.50, 800.50, 820.50
Cable substitutions, 725.61, 760.61, 770.93, 800.53, 820.53, 830.58
Cable trays, grounding, Table 392.7(B)
Cable fill, Tables 392.9, 392.9(E), 392.9(F), 392.10

Calculation of feeder loads by occupancies, Table 220.11
Conductors
    Application, Tables 310.13, 310.61, 402.3
    Clearances, conductors entering bus enclosures, 408.10
    Clearances, services, Table 230.51(C)
    Conduit and tubing fill for, Tables Annex C
    Deflection, minimum bending space in cabinets, cutout boxes, Tables 312.6(A), 312.6(B)
    Dimensions
        Compact aluminum building wiring, Chap. 9 Table 5A
        Insulated conductors, and fixture wires, Chap. 9 Table 5
        Rubber, thermoplastic-covered, Chap. 9 Table 5
    Fixture wires, Chap. 9, Tables 5 and 402.3
    Flexible cords and cables, types, Table 400.4
    Grounding, size
        For AC equipment, Table 250.122
        For grounded systems, Table 250.66
    Hazardous (classified) locations, Class I, Zones 0, 1, and 2
        Classification of maximum surface temperature of Group II equipment, Table 505.9(D)(1)
        Gas classification groups, Table 505.9(C)
        Minimum distance of obstructions from flameproof flange openings, Table 505.7(D)
        Types of protection designation, Table 505.9(C)(2)
    Hazardous (classified) locations, Classes I, II, and III, Divisions 1 and 2
        Class III temperatures, Table 500.8(C)(2)
        Classification of maximum surface temperature, Table 500.8(B)
        Insulations, Tables 310.13, 310.61 through 310.63
    Maximum number, Tables Annex C
    Maximum number in
        Electrical metallic tubing, Tables C1, C1(A)
        Electrical nonmetallic tubing, Tables C2, C2(A)
        Flexible metal conduit, Tables C3, C3(A)

Intermediate metal conduit, Tables C4, C4(A)
Liquidtight flexible metal conduit, Tables C7, C7(A)
Liquidtight flexible nonmetallic conduit, Tables C5, C5(A), C6, C6(A)
Rigid metal conduit, Tables C8, C8(A)
Rigid nonmetallic conduit, Tables C9 through C12(A)
Number in metal boxes, Table 314.16(A)
Over 2000 to 35,000 volts
    Ampacities, Tables 310.81 through 310.86
    Shielding, solid dielectric-insulated conductors, Table 310.64
Properties, Chap. 9 Table 8
Support, vertical raceways, 300.19(A)
    Volume required per conductor, Table 314.16(B)
Conduit and tubing fill, for conductors and fixture wires, Tables Annex C
Conduit or tubing
    Combination of conductors, percent area fill, Chap. 9 Table 1
    Dimensions, Chap. 9 Table 4
    Expansion characteristics, Tables 352.44(A), 352.44(B)
    Flexible metal (⅜ in.), Table 348.12
    Number of conductors in, Annex C Tables
    PVC rigid nonmetallic, expansion characteristics, 352.44(A)
    Radius of bends, rigid metal conduit, Table 344
    Supports, Tables 344.30(B)(2), 352.30(B)
Cooking appliances, demand factors and loads, Tables 220.19, 220.20
Farm load, method for computing, Tables 220.40, 220.41
Fixture wires
    Conduit and tubing fill for, Annex C Tables
    Maximum number in
        Electrical metallic tubing, Tables C1, C1(A)
        Electrical nonmetallic tubing, Tables C2, C2(A)
        Flexible metal conduit, Tables C3, C3(A)
        Intermediate metal conduit, Tables C4, C4(A)
        Liquidtight flexible metal conduit, Tables C7, C7(A)
        Liquidtight flexible nonmetallic conduit, Tables C6, C6(A)

Rigid metal conduit, Tables C8, C8(A)

Rigid nonmetallic conduit, Tables C9 through C12(A)

General lighting unit loads by occupancies, Table 220.3(A)

Household clothes dryers, demand loads, Table 220.18

Household ranges and similar cooking appliances, demand loads, Table 220.19

Live parts, separation
Over 600 volts
Air separation, Table 490.24
Elevation, Table 110.34(E)
Working space, Table 110.34(A)
Working clearances, Table 110.26(A)(1)

Minimum cover, underground wiring, Table 300.5

Minimum distance from fence to live parts, Table 110.31

Minimum size of conductors, Table 310.5

Minimum wire bending space, Tables 312.6(A), 312.6(B)

Mobile home park demand factors, Table 550.31

Motors
Conductor rating factors for power resistors, Table 430.29
Controller enclosure selection, Table 430.91
Duty cycle service, Table 430.22(E)
Full-load currents, Tables 430.147 through 430.150
Locked-rotor, code letters, Table 430.7(B)
Locked-rotor current conversion, Tables 430.151(A) and (B)
Maximum rating or setting, branch circuit protective devices, Table 430.52
Maximum rating or setting, control circuit overcurrent protective device, Table 430.72(B)
Minimum spacings between bare live parts, motor control centers, Table 430.97
Number and location, overload units, Table 430.37
Other articles, Table 430.5
Secondary ampacity, Table 430.23(C)
Terminal, spacing and housing, Tables 430.12(B), 430.12(C)(1), (C)(2)

Multifamily dwellings, optional calculation demand factors, Table 220.32

Network-powered broadband communications systems
Cable substitution, Table 830.58
Cover requirements, Table 830.12
Limitations, Table 830.4

Optional calculations, three or more multifamily units, Table 220.32

Radio and TV equipment, antenna sizes
Amateur stations, Table 810.52
Receiving stations, Table 810.16(A)

Rating factors for power resistors, Table 430.29

Receptacle loads, nondwelling units, Table 220.13

Recreational vehicle park demand factors, Table 551.73

Reinforced thermosetting resin conduit expansion characteristics, 354.44(B)

Restaurants, optional method load calculation, Table 220.36

Schools, optional method load calculation, Table 220.34

Support services, Table 230.51(C)

Transformers, medium and high voltage, Tables 450.3(A), 450.3(B)

Underground wiring, minimum cover, Tables 300.5, 300.50

**Tamperability**
Circuit breakers, nontamperable, 240.82
Type S fuses, nontamperable, 240.54(D)

**Tamperproof receptacles.** see Receptacles

**Taps.** see also Splices and taps
Branch circuit, 210.19
Busways, 368.12 Ex. 1
Feeders. see Feeders, taps
Overcurrent protection, 240.21
Service-entrance conductors, 230.46

**Task illumination,** 517.33(A)
Definition, 517.2

**Telegraph systems.** see Communication circuits

**Telephone equipment,** 800.4. see also Communication circuits

**Telephone exchanges, circuit load,** 220.3(B), Ex.

**Telephone systems.** see Communication circuits

**Television and radio distribution systems.** see Community antenna television and radio distribution (CATV) systems

**Television equipment.** see Radio and television equipment

**Television studios,** Art. 520, Art. 530

**Temperature limitations**
Conductors, 310.10
Nonmetallic raceways and tubing. see raceway or tubing type
In outlet boxes for luminaires (fixtures), 410.11
Service-entrance cable, 338.10(B)(3)

**Temporary installations,** Art. 527
All wiring installations, 527.2
Branch circuits, 527.4(C)
Disconnecting means, 527.4(E)
Feeders, 527.4(B)
Ground-fault protection, 527.6
Guarding over 600 volts, 527.7
Lamp protection, 527.4(F)
Protection from accidental damage, 527.4(H)
Receptacles, 527.4(D), 527.6(A), 527.6(B)
Services, 527.4(A)
Splices, 527.4(G)
Terminations at devices, 527.4(I)
Time constraints, 527.3

**Terminal housing, grounding through motor,** 430.145(A)

**Terminals**
Connections to, 110.14, 250.8, 250.68
Electric discharge tubing, signs, etc., 600.42
Identification
Flat conductor cable, 322.120
Motors, controllers, 430.9(A)
Polarity, 200.9 through 200.11
Wiring device, 250.126

**Tests**
Emergency systems, 700.4
Ground-fault protection, 230.95(C)
Insulation resistance, space heating cables, 424.45
Legally required standby systems, 701.5
Luminaires (lighting fixtures), 410.45
Mobile homes, 550.17
Park trailers, 552.60
Recreational vehicles, 551.60

**Theaters,** Art. 520
Audio signal processing, 520.4
Branch circuits, 520.9
Conductors, number in raceway, 520.6
Definitions, 520.2
Dressing room, 520-VI
Emergency systems, Art. 700
Grounding, 520-VII
Fixed electric equipment, 250.112(F)
Live parts, 520.7
Portable equipment, 520.10
Stage equipment
Fixed, 520-II

Portable, 520-V
Switchboard
Fixed, 520-II
Portable, 520-IV
Wiring methods, 520.5

**Therapeutic equipment,** 517.73(B)

**Therapeutic high-frequency diathermy equipment**
Definition, 517.2

**Therapeutic pools and tubs,** 680.2, 680-VI

**Thermal cutouts**
Definition, Art. 100-I
Disconnecting means, 240.40

**Thermal devices**
Overcurrent protection, 240.9

**Thermal protector**
Definition, Art. 100-I

**Thermal resistivity,** 310.30, B.310.15(B)(2)

**Thermally protected (thermal protection)**
Definition, Art. 100-I
Fixtures, recessed, 410.65(C)
Fluorescent lamp ballasts in luminaires (fixtures), 410.73(E)
Luminaires (fixtures), recessed, 410.65(C)

**Three overload units, motors,** Table 430.37

**Tools**
Double insulated, 250.114 Ex.
Metal working machine. see Industrial machinery
Motor-operated, hand-held, grounding, 250.114(3)(C), 250.114(4)(C)

**Track lighting.** see Lighting track

**Trailers, types of**
Definition, 551.2

**Transfer switch**
Definition, Art. 100-I
Fuel cell systems, 692.59

**Transformer vaults,** 450-III
Doorways, 450.43
Drainage, 450.46
Location, 450.41
Storage, 450.48
Ventilation openings, 450.45
Walls, roofs, and floors, 450.42
Water pipes and accessories, 450.47

**Transformers,** Art. 450. see also Hazardous (classified) locations
Arc welders, 630-II
Askarel-insulated, 450.25
Autotransformers, 210.9, 215.11, 410.78, 430.82(B), 450.4, 450.5
Capacitors, Art. 460
Installation, Art. 450
X-ray equipment, 517.76, 660-III
Definitions, 450.2, 551.2
Dry-type, 450.21, 450.22
Electric discharge lighting systems
1000 volts or less, 410-XIII

More than 1000 volts, 410-XIV
Fire pumps, 695.5
Grounding, 450.10
Guarding, 450.8
Installations, indoor and outdoor, 450.21 through 450.27
Instrument, grounding, 250-IX
Connections at services, 230.82(3)
Isolation
Anesthetizing locations and patient care areas, 517.19(F), 517.20, 517.63(E) and (F)
Definition, 517.2
Less-flammable liquid-insulated, 450.23
Location, accessibility, 450.13
Marking, 450.11
Modification of, 450.28
Motor control circuit, 430.72(C)
Nonflammable fluid-insulated, 450.24
Oil-insulated
Indoors, 450.26
Outdoors, 450.27
Overcurrent protection, 450.3
Power-limited and signaling circuits, remote control, 725.21(A)(1), 725.41
Remote control circuits for, 430.74(B), 725.21, 725.41
Research and development, 450.1 Ex.8
Secondary ties, 450.6
Signs and outline lighting, 600.23
Specific provisions, 450-II
Terminal wiring space, 450.12
Two-winding, underwater lighting, 680.23(A)(2)
Vaults, 450-III
Ventilation, 450.9
X-ray equipment, 517.76, 660-III
Transient voltage surge suppressors: (TVSSs), Art. 285
Conductor routing, 285.12
Connection, 285.21, 285-III
Definition, 285.2
Installation, 285-II
Listing, 285.5
Location, 285.11
Number required, 285.4
Short circuit current rating, 285.6
Uses not permitted, 285.3
Transmitting stations, radio and television, 810-III
Travel trailer
Definition, 551.2
Trays, storage batteries, 480.7(B)
Trees, luminaires (lighting fixtures) supported by, 410.16(H)
Truck camper
Definition, 551.2
Tubing. see also Conduits
Definitions, 600.2

Electric discharge, signs, etc., 600.41
Electrical metallic. see Electrical metallic tubing
Electrical nonmetallic. see Electrical nonmetallic tubing
Flexible metallic, Art. 360. see Flexible metallic tubing
TV. see Television equipment
Two-fer
Definition, 520.2

-U-
Under 600 volts, 368.30
Underfloor raceways, Art. 390
Conductors
Number in raceway, 390.5
Size of, 390.4
Connections to cabinets, wall outlets, 390.15
Covering, 390.3
Dead ends, 390.10
Discontinued outlets, 390.7
Inserts, 390.14
Junction boxes, 390.13
Laid in straight lines, 390.8
Markers, 390.9
Splices and taps, junction boxes, 390.6
Uses not permitted, 390.2(B)
Uses permitted, 390.2(A)
Underground circuits, communication, 800.11
Underground enclosures, 110.12(B)
Underground feeder and branch-circuit cable (Type UF), Art. 340
Ampacity, 340.80
Bending radius, 340.20
Conductors, 340.104
Construction specifications, 340-III
Definition, 340.2
Equipment grounding conductor, 340.108
Installation, 340-II
Insulation, 340.112
Sheath, 340.116
Uses not permitted, 340.12
Uses permitted, 340.10
Underground wiring. see also Hazardous (classified) locations
Buried conductors, Types USE, UF, 338.10(B)(2), 340.10
Conductor types in raceways, 310.8
Ground movement and, 300.5(J)
Intermediate metal conduit, 342.10(B) and (D)
Liquidtight flexible metal conduit, 350.10(3)
Minimum cover requirements, 300.5(A)
Over 600 volts, 300.50
Protection of, 300.5(D), 300.5(J)
Rigid metal conduit, 344.10
Rigid nonmetallic conduit, 352.10(B) and (D)

"S" loops, 300.5(J) FPN
Services, 230.30, 230.32
Splices and taps, 300.5(E)
Swimming pools, 680.10
Wet locations, 310.8(C) and (D)
Uninterruptible power supplies (UPS), 645.11
Unit equipment, emergency and standby systems, 700.12(E), 701.11(F)
Unused openings
Boxes and fittings, 110.12
Utilization equipment
Definition, Art. 100-I

-V-
Vapors, flammable. see Hazardous (classified) locations
Varying duty
Definition, Art. 100-I
Vaults
Capacitors, 460.2
Film storage, 530-V
Service 600 volts or less, 230.6(3)
Service over 600 volts, 110.31, 230.212
Transformers, 450-III
Vehicles. see Electric vehicles; Recreational vehicles
Ventilated
Definition, Art. 100-I
Ventilating ducts, wiring, 300.21, 300.22
Ventilating piping for motors, etc., 502.9, 503.7
Ventilation
Aircraft hangars, 513.3(D)
Battery locations, 480.9(A)
Equipment, general, 110.13(B)
Motor fuel dispensing facilities, lubrication and service rooms — without dispensing, Table 514.3(B)(1)
Motors, 430.14(A), 430.16
Transformers, 450.9, 450.45
Vessel, fixed electric heating equipment for
Definition, 427.2
Viewing tables, motion picture, 530-IV
Volatile flammable liquid
Definition, Art. 100-I
Voltage and volts
Branch circuits, limits, 210.6
Circuit
Definition, Art. 100-I
Drop
Branch circuits, 210.19(A) FPN No. 4
Conductors, 310.15(A)(1) FPN No. 1
Feeders, 215.2(A)(4) FPN No. 2
Sensitive electronic equipment, 647.4(D)
Electric discharge lighting, 410-XIII, 410-XIV
General provisions, 110.4

Ground to
Definition, Art. 100-I
High
Definition, 490.2
Less than 50, Art. 720
Limitations, elevators, dumbwaiters, escalators, moving walks, 620.3
Low
Definition, 551.2
Marking, 240.83(E)
Nominal
Definition, Art. 100-I
Nominal battery
Definition, 480.2
Over 600 volts, Art. 490
Swimming pool underwater lighting fixtures, 680.23(A)(4)
Wiring methods, 300.2

-W-
Wading pools
Definition, 680.2
Wall or show cases. see Show cases, wall cases
Wall-mounted ovens. see Ovens, wall-mounted
Warning signs (labels), at equipment
Aircraft hangars, 513.10
Electroplating, 669.7
Electrostatic hand spraying, 516.10(A)(8)
Guarding live parts 600 volts or less, 110.27(C)
Induction and dielectric heating, 665.23
Locked room or enclosure with live parts over 600 volts, 110.34(C), 490.21(B)(7)Ex., 490.21(C)(2), 490.21(E), 490.44(B), 490.53, 490.55
Transformers, 450.8(D)
Water heaters, 422.11(F)(3), 422.13
Controls, 422.47
Nameplate load, 220.32(C)(3)
Protection, 422.11(E) and (F)(3)
Water pipe
Bonding (metal), 250.104(A)
Connections, 250.8, 250.68
As grounding electrode, 250.52(A)
Watertight
Definition, Art. 100-I
Weatherproof
Definition, Art. 100-I
Welders, electric, Art. 630
Arc, 630-II
Definition, Art. 100-I
Resistance, 630-III
Welding cable, 630-IV
Wet locations. see also Damp or wet locations
Conductors, types, 310.8(C), 310.13, Table 310.13
Definition, Art. 100-I

Electrical metallic tubing, 358.10(C)
Enameled equipment, 300.6(A)
Health care facilities, 517.20
Intermediate metal conduit, 342.10(D)
Luminaires (lighting fixtures) in, 410.4(A)
Mounting of equipment, 300.6(C)
Rigid metal conduit, 344.10(D)
Rigid nonmetallic conduit, 352.10(D)
Switchboards, 408.6
Switches, 404.4
**Wheelchair lifts.** *see* Elevators, dumbwaiters, escalators, moving walks, wheelchair lifts, and stairway chair lifts
**Windows, show.** *see* Show windows
**Wired luminaire (fixture) sections,** 410.77(C)
**Wires.** *see also* Conductors; Cords; Fixture wires
In concrete footings, electrodes, 250.52(A)(3)

Definition, 800.2
**Wireways, metal,** Art. 376
Construction specifications, 376-III
Dead ends, 376.58
Definition, 376.2
Deflected insulated, 376.23(A)
Extensions, 376.70
Grounding, 362.13
Insulated conductors, 376.23
Marking, 376.120
Number of conductors, 376.22
Securing and supporting, 376.30
Size of conductors, 376.21
Splices and taps, 376.56
Uses not permitted, 376.12
Uses permitted, 376.10
**Wireways, nonmetallic,** Art. 378
Construction specifications, 378-III
Dead ends, 378.58
Definition, 378.2
Expansion fittings, 378.44
Extensions, 378.70
Grounding, 378.60
Installation, 378-II
Insulated conductors, 378.23
Marking, 378.120

Number of conductors, 378.22
Securing and supporting, 378.30
Size of conductors, 378.21
Splices and taps, 378.56
Uses not permitted, 378.12
Uses permitted, 378.10
**Wiring methods,** Art. 300
Ducts, 300.21, 300.22
Exposed
Definition, Art. 100-I
General requirements for. *see* General requirements for wiring methods
Hazardous (classified) locations. *see* article on hazardous location involved
Health care facilities, 517-II
Intrinsically safe systems, 504.20
Manufactured homes, 550.10
Mobile home parks, 550.10
Mobile homes, 550.10
Planning, 90.8
Temporary. *see* Temporary installations
Theaters, 520.5
Types and materials, Chap. 3

**Within sight from.** *see* In sight from
**Working space**
About electrical equipment, 110.26, 110.32 through 110.34
Adjacent to live parts (circuits over 600 volts), 110.33, 110.34
Motion picture projectors, 540.12
Switchboards, 110.26, 408.8
**Workmanlike installation,** 110.12, 720.11, 725.6, 760.6, 800.6, 820.6, 830.7

-X-
**X-ray equipment,** 517-V, Art. 660
Control, 517.74, 660-II
Fixed and stationary
Definition, 660.2
Guarding and grounding, 517.78, 660-IV
Installations, 517-V
Definition, 517.3
Transformers and capacitors, 517.76, 660-III

## 2005 *National Electrical Code*® Schedule
### (2004 World Fire Safety Congress and Exposition™)

| Date | Event |
|---|---|
| November 1, 2002 (5:00 PM EST) | Receipt of Proposals |
| January 13-25, 2003 | Code-Making Panel Meetings (ROP) |
| April 28-May 2, 2003 | Correlating Committee Meeting |
| July 11, 2003 | ROP to Mailing House |
| October 31, 2003 (5:00 EST) | Receipt of Comments |
| December 1-13, 2003 | Code-Making Panel Meetings (ROC) |
| February 23-27, 2004 | Correlating Committee Meeting |
| April 8, 2004 | ROC to Mailing House |
| May 23-27, 2004 | NFPA World Fire Safety Congress and Exposition™ (Salt Lake City, Utah) |
| July 2004 | Standards Council Issuance |

Anyone may submit proposals to amend the 2002 *Code*. A sample form for this purpose may be obtained from the Secretary of the Standards Council at NFPA headquarters, and a copy is included in this *Code*.

### Method of Submitting a Proposal to Revise the *National Electrical Code*®

The following is based on the NFPA Regulations Governing Committee Projects, adopted by the Board of Directors on December 3, 1977 (last amended on November 16, 1997).

A proposal to revise the 2002 edition of the *National Electrical Code* must be submitted so that the proposal is received at NFPA headquarters by November 1, 2002, as indicated in the time schedule for the 2005 *National Electrical Code*. A proposal received after this date will be returned to the submitter. The proposal is to be sent to the Secretary of the Standards Council at NFPA Headquarters, 1 Batterymarch Park, P.O. Box 9101, Quincy, MA 02269-9101.

Each proposal must include the following:
1. Identification of the submitter (the person's name) and his or her affiliation (i.e., committee, organization, company), where appropriate
2. An indication that the proposal is for revision of the 1999 *National Electrical Code* and identification of the specific section number, table number (or equivalent identification) of the section, etc., to be revised
3. A statement of the problem and substantiation for proposal
4. The proposed text of the proposal, including the wording to be added, revised (and how revised), or deleted

Proposals that do not include all of the above information may not be acted on by the National Electrical Code Committee.

It is preferred that the forms available from NFPA for submittal of proposals be used. A separate proposal form should be used for revision of each section of the *Code*.

A proposal form appears on the following page.

# FORM FOR PROPOSALS FOR 2005 *NATIONAL ELECTRICAL CODE*®

Mail to:  **Secretary, Standards Council**
**National Fire Protection Association**
**1 Batterymarch Park, P.O. Box 9101**
**Quincy, Massachusetts 02269-9101**
**Fax to: 617-770-3500**

```
┌─────────────────────────────────┐
│ FOR OFFICE USE ONLY             │
│ Log # _____ │
│ Date Rec'd_____ │
└─────────────────────────────────┘
```

Notes:  1. All proposals must be received by 5:00 p.m. EST on Friday, November 1, 2002
Proposals received after 5:00 p.m. EST, Friday, November 1, 2002 will be returned to the submitter
2. Type or print legibly in black ink. Limit each proposal to a SINGLE section. Use a separate copy for each proposal.
3. If supplementary material (photographs, diagrams, reports, etc.) is included, you may be required to submit sufficient copies for all members and alternates of the technical committee.

Please indicate in which format you wish to receive your ROP/ROC: ❑ electronic or ❑ paper ❑ download

Date _____ Name _____ Tel. No. _____

Company _____

Street Address _____

Please Indicate Organization Represented (if any) _____

1. Section/Paragraph _____

2. Proposal Recommends: (Check one)   ❑ new text   ❑ revised text   ❑ deleted text

3. Proposal (include proposed new or revised wording, or identification of wording to be deleted): (Note: Proposed text should be in legislative format: i.e., use underscore to denote wording to be inserted (inserted wording) and strike-through to denote wording to be deleted (deleted wording)

4. Statement of Problem and Substantiation for Proposal: (Note· State the problem that will be resolved by your recommendation; give the specific reason for your proposal including copies of tests, research papers, fire experience, etc  If more than 200 words, it may be abstracted for publication )

5. ❑ This Proposal is original material. (Note  Original material is considered to be the submitter's own idea based on or as a result of his/her own experience, thought, or research and, to the best of his/her knowledge, is not copied from another source )

❑ This Proposal is not original material; its source (if known) is as follows: _____

```
┌──────────────────────────────────────────────────────────────────────────┐
│     If you need further information on the standards-making process, please contact the    │
│          Standards Administration Department at 617-984-7249.               │
│        For technical assistance, please call NFPA at 617-770-3000.          │
└──────────────────────────────────────────────────────────────────────────┘
```

*I hereby grant the NFPA the non-exclusive, royalty-free rights, including non-exclusive, royalty-free rights in copyright, in this proposal and I understand that I acquire no rights in any publication of NFPA in which this proposal, in this or another similar or analogous form, is used.*

_____
Signature (Required)

**PLEASE USE SEPARATE FORM FOR EACH PROPOSAL**

## Sequence of Events Leading to Publication of an NFPA Committee Document

Call goes out for proposals to amend existing document or for recommendations on new document.

Committee meets to act on proposals, to develop its own proposals, and to prepare its report.

Committee votes on proposals by letter ballot. If two-thirds approve, report goes forward. Lacking two-thirds approval, report returns to committee.

Report — *Report on Proposals* (ROP) — is published for public review and comment.

Committee meets to act on each public comment received.

Committee votes on comments by letter ballot. If two-thirds approve, supplementary report goes forward. Lacking two-thirds approval, supplementary report returns to committee.

Supplementary report — *Report on Comments* (ROC) — is published for public review.

NFPA membership meets (Annual or Fall Meeting) and acts on committee report (ROP or ROC).

Committee votes on any amendments to report approved at NFPA Annual or Fall Meeting.

Appeals to Standards Council on Association action must be filed within 20 days of the NFPA Annual or Fall Meeting.

Standards Council decides, based on all evidence, whether or not to issue standard or to take other action, including upholding any appeals.

## Committee Membership Classifications

The following classifications apply to Technical Committee members and represent their principal interest in the activity of a committee.

M    *Manufacturer:* A representative of a maker or marketer of a product, assembly, or system, or portion thereof, that is affected by the standard.

U    *User:* A representative of an entity that is subject to the provisions of the standard or that voluntarily uses the standard.

I/M    *Installer/Maintainer:* A representative of an entity that is in the business of installing or maintaining a product, assembly, or system affected by the standard.

L    *Labor:* A labor representative or employee concerned with safety in the workplace.

R/T    *Applied Research/Testing Laboratory:* A representative of an independent testing laboratory or independent applied research organization that promulgates and/or enforces standards.

E    *Enforcing Authority:* A representative of an agency or an organization that promulgates and/or enforces standards.

I    *Insurance:* A representative of an insurance company, broker, agent, bureau, or inspection agency.

C    *Consumer:* A person who is, or represents, the ultimate purchaser of a product, system, or service affected by the standard, but who is not included in the *User* classification.

SE    *Special Expert:* A person not representing any of the previous classifications, but who has special expertise in the scope of the standard or portion thereof.

NOTE 1: "Standard" connotes code, standard, recommended practice, or guide.

NOTE 2: A representative includes an employee.

NOTE 3: While these classifications will be used by the Standards Council to achieve a balance for Technical Committees, the Standards Council may determine that new classifications of members or unique interests need representation in order to foster the best possible committee deliberations on any project. In this connection, the Standards Council may make such appointments as it deems appropriate in the public interest, such as the classification of "Utilities" in the National Electrical Code Committee.

NOTE 4: Representatives of subsidiaries of any group are generally considered to have the same classification as the parent organization.